Books From California
51 W. Easy St
Simi Valley, CA 93065
UNITED STATES
customerservice@booksfromca.com

Books From California
51 W. Easy St
Simi Valley, CA 93065
UNITED STATES

To: Alibris APEX DC 76926405-99 - APEX
800 Avondale Ave.

Grandview Heights, OH 43212-3473
UNITED STATES

If this order is shipping from the US to an international buyer, remember to attach proper documentation. Complete and attach USPS Form CN-22 for all Priority Mail International Envelopes or First-Class Mail International shipments (available at your post office).

Order Number:	34032847
Ship Method:	Standard
Customer Name:	Alibris APEX DC 76926405-99 - APEX
Order Date:	9/15/2026
Alibris Order #:	76926405-99
Email:	

Items:

Qty	Item	Locator	Item ID	Condition
1	Solid-State Materials in Pharmaceutical Chemistry: Propertie...erization, and Applications Byrn, Stephen R.,Zografi, George,Chen, Xiaoming (Sean) SKU: mon0004189289 ISBN: 139425444X - Books	C -2-01-256-001-2123	76926405-99	Very Good

Notes:

Very good Very Clean Copy-Over 500, 000 Internet Orders Filled. hardcover

alibris

Alibris guarantees the condition of every item as it is described on our Web site. If you are dissatisfied for any reason, return your purchase within 60 days of receipt for a refund of the item price. We'll also refund shipping costs if the return was a result of our error. All return requests must be submitted via the Alibris Web site. Any item returned without accompanying paperwork and/or more than 60 days after its shipment date will be discarded and you will not be eligible to receive a refund.

If you have any questions or concerns regarding this order, please contact us at customerservice@booksfromca.com
Thanks for your order!

Solid-State Materials in Pharmaceutical Chemistry

Solid-State Materials in Pharmaceutical Chemistry

Properties, Characterization, and Applications

Stephen R. Byrn, George Zografi, Xiaoming (Sean) Chen

Purdue University
University of Wisconsin
Improved Pharma
USA

Second Edition

WILEY

This edition first published 2017

Edition History
John Wiley & Sons, Inc. (1e, 2017)

Published by John Wiley & Sons, Inc., Hoboken, New Jersey.

For general information on our other products and services or for technical support, please contact our Customer Care Department within the United States at (800) 762-2974, outside the United States at (317) 572-3993 or fax (317) 572-4002.

Wiley also publishes its books in a variety of electronic formats. Some content that appears in print may not be available in electronic formats. For more information about Wiley products, visit our web site at www.wiley.com.

Library of Congress Cataloging-in-Publication Data:

Names: Byrn, Stephen R. author | Zografi, George author | Chen, Xiaoming (Sean) author
Title: Solid-state materials in pharmaceutical chemistry : properties, characterization, and applications / Stephen R. Byrn, George Zografi, Xiaoming (Sean) Chen, Purdue University, University of Wisconsin, Improved Pharma, USA.
Other titles: Solid-state properties of pharmaceutical materials
Description: Second edition. | Hoboken, New Jersey : Wiley, [2026] | Previous edition: 2017. | Includes bibliographical references and index.
Identifiers: LCCN 2025019692 | ISBN 9781394254446 (hardback) | ISBN 9781394254477 (epdf) | ISBN 9781394254460 (epub) | ISBN 9781394254484 (obook)
Subjects: LCSH: Solid state chemistry | Solid dosage forms—Properties
Classification: LCC QD478 .B96 2026
LC record available at https://lccn.loc.gov/2025019692

Cover Design: Wiley
Cover Image: © Argonne National Laboratory, managed and operated by UChicago Argonne, LLC, for the U.S. Department of Energy under Contract No. DE-AC02-06CH11357

Table of Contents

Biography of Authors

Stephen R. Byrn, PhD, is Charles B. Jordan Professor of Medicinal Chemistry in the Department of Industrial and Physical Pharmacy, Purdue University, West Lafayette, Indiana, USA. He received his BA degree in Chemistry from DePauw University and his PhD degree in Chemistry from the University of Illinois, Urbana, USA. He did postdoctoral research at UCLA. His research focuses on Solid-State Chemistry, Polymorphism, Stability, Manufacturing Science, Quality by Design, and Medicinal Chemistry. Dr. Byrn set in motion the development of the field of Solid-State Chemistry of Drugs with his books, short courses, and papers on the subject, the first of which were first published in the mid-1970s. He has also educated over 50 PhD students and postdoctoral fellows and taught a wide range of courses at Purdue. Dr. Byrn has had numerous grants, including one of the first 13 NIH Centers for AIDS Research. Dr. Byrn is a cofounder of Purdue's graduate programs in regulatory and quality compliance. These programs now constitute the Biotechnology Innovation and Regulatory Science (BIRS) MS program. He is also a cofounder of the Purdue-Kilimanjaro School of Pharmacy Sustainable Medicines in Africa project in Moshi, Tanzania. Dr. Byrn has served as the chair of the Pharmaceutical Sciences Advisory Committee to the FDA and chaired several USP committees. Dr. Byrn is also a cofounder of SSCI, Inc. (Solid-State Chemical Information), a cGMP research and information company. SSCI, Inc. is now owned by AMRI. Dr. Byrn has taught a range of courses and short courses involving Medicinal Chemistry, Industrial Pharmacy, Physical Pharmacy, and Solid-State Chemistry. Dr. Byrn is an elected Fellow of the AAPS and has received several awards for his research and entrepreneurial activities, including the AAPS David Grant Award for Research Achievement and the AAPS Wurster Award in Pharmaceutics and Formulation. Dr. Byrn also received the Purdue University Morrill Award.

George Zografi, PhD, Edward Kremers Professor Emeritus of Pharmaceutical Sciences, School of Pharmacy, University of Wisconsin-Madison, received his BS degree (1956) from Columbia University and MS (1958) and PhD (1960) degrees in Pharmaceutics from the University of Michigan, USA. After serving on the faculties of Columbia University and the University of Michigan and a sabbatical year at the University of Utrecht, Dr. Zografi joined the Wisconsin faculty in 1972. From 1975 to 1980, he served as the dean of the School of Pharmacy. His research interests are in interfacial phenomena, including the properties of lipid, protein, and polymer monolayers and the hydration of crystalline and amorphous solids. In 1989, Dr. Zografi was elected to the Institute of Medicine of the National Academy of Sciences. He received the APhA Ebert Prize in 1984 and 2001, the AAPS Dale E. Wurster Award for Pharmaceutics in 1990, the Distinguished Pharmaceutical Scientist Award from the AAPS in 1995, and the Volwiler Research Achievement Award from the American Association of Colleges of Pharmacy in 1996.

Xiaoming (Sean) Chen, PhD, works in the field of pharmaceutical product development and held various positions at Schering-Plough, OSI Pharmaceuticals, Astellas Pharma, Shionogi Inc, Antares Pharma, Arvinas, and Improved Pharma. Dr. Chen received his BS degree in Biochemistry from Fudan University, Shanghai, China, and his PhD in Industrial and Physical Pharmacy from Purdue University. During his tenure in the Pharma Industry, Dr. Chen has participated in numerous projects involving preformulation, formulation development, process development, technology transfer for solid dosage forms, injectables, and combinational products and contributed to the regulatory approvals and the launching of multiple products, including Osphena and Xyosted. He has won numerous awards, including a Schering-Plough President Award and a Shionogi Corporation President Award. He has published over 16 papers in peer-reviewed journals and is a coinventor of five U.S. patents.

Preface

Since the publication of the 2017 edition of *Solid-State Properties of Pharmaceutical Materials*, the landscape of pharmaceutical research, development, and manufacturing has undergone significant transformations, presenting both novel challenges and unprecedented opportunities.

The global impact of the COVID-19 pandemic, while mitigated by the rapid development of nucleotide-based vaccines, underscored the critical importance of understanding and controlling the solid-state properties of biopharmaceuticals and advanced delivery systems like lipid nanoparticles, particularly in relation to stability and storage. Moreover, the trend toward larger, poorly water-soluble, small-molecule drugs has positioned amorphous dispersions as a default formulation strategy, demanding innovative approaches to leverage solid-state material science for the development of high-quality active pharmaceutical ingredients (APIs) and stable drug products. The increasing demand for drug-device combination products for self-administration and diagnostics has further underlined the importance of compatibility studies and the interplay between material properties and device performance. Furthermore, the pharmaceutical industry has grappled with the withdrawal of life-saving drugs due to nitrosamine contamination, highlighting the urgent need for advanced analytical techniques and mitigation strategies.

To address these evolving challenges, this second edition, building upon the successful framework of the 2017 edition, is substantially expanded and retitled *Solid-State Materials in Pharmaceutical Chemistry: Properties, Characterization, and Applications* to better reflect its contents and structure. For instance, Chapter 1 now includes discussions on bulk and surface properties of solids and lipid nanoparticles, while Chapter 3 offers an in-depth review of pharmaceutical solvates, underscoring their applications in impurity purging and final form preparation during API crystallization. A new Chapter 21 explores the physical and chemical properties of therapeutic nucleotides, reflecting the growing importance of these agents in life-saving medicines and vaccines. Additionally, a new Chapter 23 addresses challenges in API process development, including impurity purging, chiral separation, final form preparation, particle size reduction, and nitrosamine control. Chapter 24 has been expanded to include the analysis of particulate contaminants, crucial for drug product investigations and compliance. Finally, the drug product development chapter (Chapter 25) now features expanded sections on oral delivery peptides such as semaglutide, injectable drug-device combination products, and *N*-nitrosamine control in drug products.

Through these enhancements, this edition aims to provide pharmaceutical scientists with a comprehensive and up-to-date resource for understanding and controlling the solid-state properties of pharmaceutical materials, ultimately contributing to the development of safe and effective medicines, including small-molecule compounds, peptides, proteins, and nucleotides.

Preface to the First Edition

The aim of this book is to illustrate the importance of understanding the fundamental solid-state properties of pharmaceutical materials during the development of solid pharmaceutical products and to lay out general strategies for the physical characterization of solids using various analytical tools. Generally, great emphasis is understandably placed on the discovery of new APIs for the cure, treatment, and prevention of various acute and chronic diseases. However, it has been firmly established that the ability to obtain successful drug products in an efficient and timely manner strongly depends on the formulation and manufacture of stable and bioavailable drugs into useful products, where various physical and chemical characteristics play an essential role. In essence, it can be said, therefore, that a "drug" is more than a molecule, rather being part of a complex mixture of materials with physical and chemical characteristics that can determine therapeutic success or failure.

The book is divided into four parts. The first part focuses on the various phases or forms that solids can assume, including polymorphs, solvates/hydrates, salts, cocrystals, amorphous forms, crystal mesophases, and nanocrystals, and various issues related to the relative stability and tendencies to undergo transformations. The second part focuses on the key methods of solid-state analysis, such as X-ray crystallography, X-ray powder diffraction, thermal analysis, microscopy, vibrational spectroscopy, and solid-state NMR. The third part reviews critical physical attributes of pharmaceutical materials, mainly related to drug substances, including particle size/surface area, hygroscopicity, mechanical properties, solubility, and physical and chemical stability. The fourth part of the book builds on the first three parts to illustrate how an understanding of the various properties of pharmaceutical materials may be used for (i) the rational selection of drug solid form, (ii) the analysis of mixtures of various solid forms within the drug substance and the drug product, (iii) establishing rational protocols and strategies for carrying out efficient and successful product development, and (iv) applications of appropriate manufacturing and control procedures, using quality by design, and other strategies that lead to safe and effective products with a minimum of resources and time. Furthermore, we have attempted to design this book in such a way that it can be used by preformulation and formulation scientists, process engineers, analytical chemists, quality assurance and quality control managers, regulators, and other researchers, who all contribute to the drug development process. We hope that by presenting a mixture of fundamental solid-state science and its practical applications to the drug development process we will have helped all involved to gain a greater perspective of the importance of both aspects.

Acknowledgment

Stephen Byrn credits his wife, Sally, and his family, without whom this would not have been possible.

George Zografi expresses his gratitude to his wife, Dorothy, and his family for their continuous support throughout his professional career.

Xiaoming (Sean) Chen is deeply grateful for the love and support from his wife, Feifei, and his sons, Albert and Austin, during the preparation and revision of this book.

The authors extend their sincere appreciation to Bob Esposito, Kshitija Iyer, Purvi Patel, Michael Leventhal, and Melissa Yanuzzi at Wiley and Suresh Srinivasan at Aptara for their support and patience during the preparation and publication of the 2017 edition.

The authors are also very thankful to Summers Scholl and Elizabeth Rose Amaldoss at Wiley for their encouragement and support in preparing the new edition.

1

Solid-State Properties and Pharmaceutical Development

1.1 Introduction

Solid-state chemistry and the solid-state properties of pharmaceutical materials play an ever-increasing and important role in pharmaceutical development. There is much more emphasis on physical characterization since the ICH Q6A guidance on specifications was released. This guidance directs the scientist to determine what solid form is present in the drug substance (API) and drug product. It directs the manufacturer to "know what they have." Additionally, the ICH Q8 guidance on development and the ICH Q9 guidance on risk require a firm understanding of how the medicine was developed and any risks involved.

There are many more poorly soluble drugs under development. In many cases, the solid form of the API and the solid form and formulation in the drug product determine apparent solubility, which in turn determines blood levels. That is, the formulation determines bioavailability and therapeutic response. In these cases, it is even more important to physically characterize the API form and the formulations. Furthermore, most medicines (drug products) are solids, and those drug products that are not solids often start with solid APIs. In addition to solubility and bioavailability, the solid form may affect stability, flow, compression, hygroscopicity, and several other properties.

This book focuses on solid-state properties of pharmaceutical materials and methods of determining these properties. The authors have made every effort to include examples and case studies to illustrate the importance of knowing what you have. This book will focus on solid-state properties and general strategies for physical characterization. Case studies and practical examples will be emphasized. This book will illustrate that a medicine is more than a molecule in many aspects. Additional goals include providing a full physical/analytical/operational definition of the different solid forms, as well as other terms frequently used in pharmaceutical materials science, including polymorph, solvate, amorphous form, habit, nucleation, transformation, dissolution, solubility, and stability.

1.2 Solid-state Forms

Pharmaceutical materials can exist in a crystalline or amorphous state. Figure 1.1 illustrates the crystalline state as a perfectly ordered solid with molecules (circles) packed in an orderly array. Figure 1.1 also illustrates an amorphous material as a disordered material with only short-range order. Crystalline materials give an X-ray diffraction pattern because Bragg planes exist in the material (see Figure 1.2). Amorphous materials do not give a diffraction pattern (see Figure 1.2). Of course, there are many interesting cases where a pharmaceutical material shows an intermediate degree of order falling somewhere between the highly ordered crystalline state and the completely disordered amorphous state. From a thermodynamic point of view, crystalline materials are more stable, but the rate of transformation of amorphous materials to crystalline materials can be highly variable [1].

Crystals of a pharmaceutical material from different sources can vary greatly in their size and shape. Typical particles in different samples may resemble, for example, needles, rods, plates, prisms. Such differences in shape are collectively referred to as differences in morphology. This term merely acknowledges the fact of different shapes. It does not distinguish among the many possible reasons for the different shapes. Naturally, when different compounds are involved, different crystal shapes would be expected as a matter of course. However, when batches of the same substance display crystals with different morphology, further work is needed to determine whether the different shapes are indicative of polymorphs,

Solid-State Materials in Pharmaceutical Chemistry: Properties, Characterization, and Applications, Second Edition. Stephen R. Byrn, George Zografi and Xiaoming (Sean) Chen.

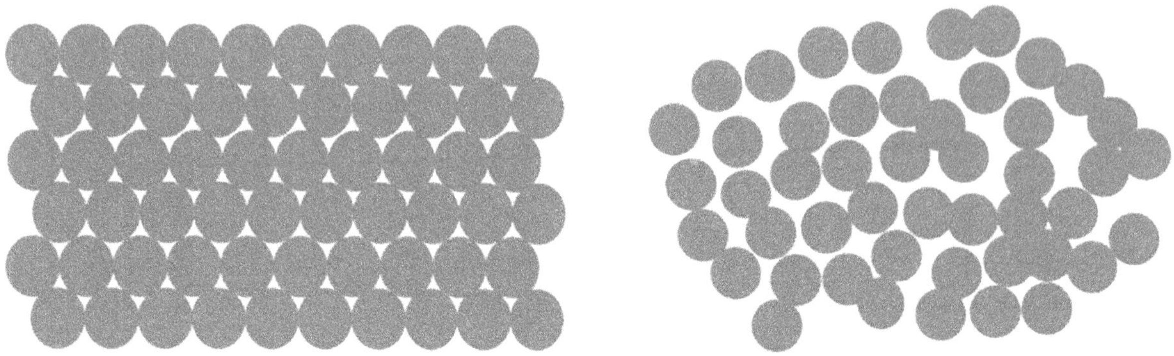

Figure 1.1 Idealized view of crystalline (left panel) and amorphous (right panel) material. In this two-dimensional figure, the molecules are viewed as circles.

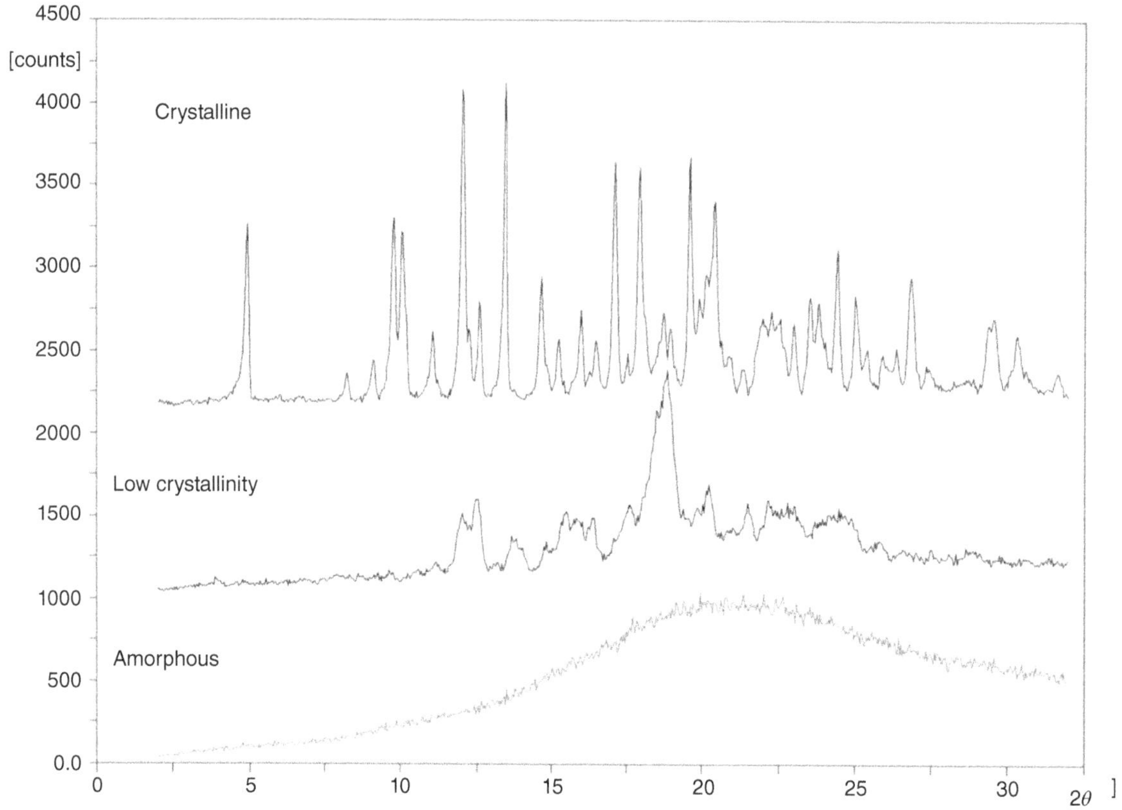

Figure 1.2 X-ray diffraction pattern of three samples: crystalline, low crystallinity, and amorphous.

solvates, or just habits. Because these distinctions can have a profound impact on drug performance, their careful definition is very important to our discourse. At this time, only brief definitions are presented:

- Polymorphs – When two crystals have the *same chemical composition* but *different internal structure* (molecular packing), they are polymorphic modifications or polymorphs (think of the three forms of carbon: diamond, graphite, and fullerenes). Polymorphs can result from different molecular packing, different molecular conformation, tautomeric structure, or combinations of these.
- Solvates – In addition to containing molecules of the same given substance, these crystal forms contain *molecules of solvent* regularly incorporated into a unique structure (think of wet, setting plaster: $CaSO_4 + 2H_2O \rightarrow CaSO_4 \cdot 2H_2O$).
- Habits – Crystals are said to have different habits when samples have the *same* chemical composition and the *same* crystal structure (i.e. the same polymorph and unit cell) but display different shapes (think of snowflakes).

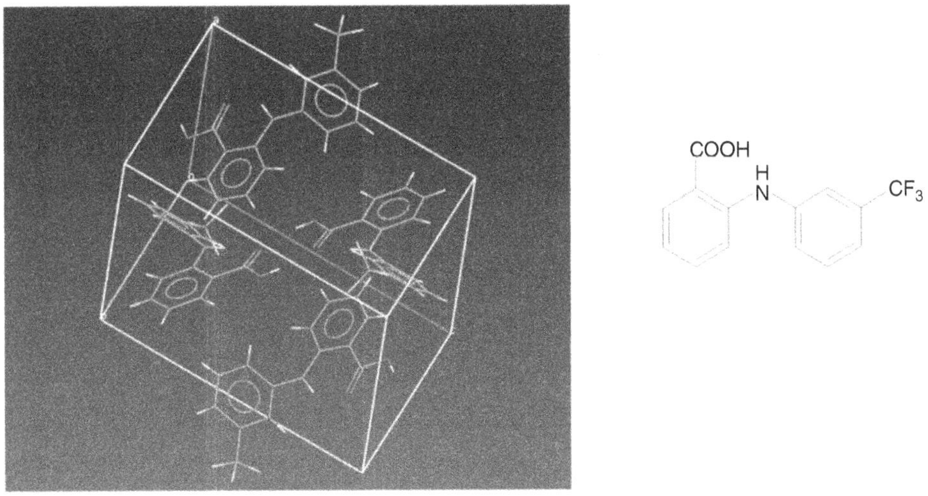

Figure 1.3 Single crystal structure of the Form I polymorph of flufenamic acid (structure shown on right panel).

Together, these solid-state physical modifications of a compound are referred to as crystalline forms. When differences between early batches of a substance are found by microscopic examination, for example, a reference to "form" is particularly useful in the absence of information that allows the more accurate description of a given variant batch (i.e. polymorph, solvate, habit, or amorphous material). The term "pseudopolymorphism" is applied frequently to designate solvates. These solid-state modifications have different physical properties.

To put these important definitions into a practical context, we consider two cases (aspirin and flufenamic acid) in which a drug was crystallized from several different solvents, and different-shaped crystals resulted in each experiment. Although dramatically different shapes were sometimes obtained on changing solvents for the various crystallizations, the final interpretations in the two cases are different. For aspirin, X-ray powder diffraction showed that all crystals regardless of shape had the same diffraction pattern. Thus, the different-shaped crystals are termed crystal habits. For flufenamic acid, the different-shaped crystals had different X-ray powder diffraction patterns. Subsequent analysis showed that the crystals did not contain solvent. Thus, these different crystals are polymorphs.

Further analysis of the crystals from this case provides the single crystal structure. The single crystal structure gives the locations of the atoms relative to a hypothetical unit cell. The unit cell is the smallest building block of a crystal. Figure 1.3 shows the unit cell of Form I of flufenamic acid. This unit cell contains four flufenamic acid molecules. Figure 1.4 shows a space-filling model of the contents of the flufenamic acid Form I unit cell. This figure illustrates Kitaigorodski's close packing theory, which requires that the molecules pack to minimize free volume [2].

Amorphous materials will be discussed in Chapter 6. In this introductory chapter, as previously mentioned briefly, amorphous materials have no long-range order and are thermodynamically metastable. An amorphous solid is characterized by a unique glass transition temperature T_g, the temperature at which it changes from a glass to a super-cooled liquid or rubbery state. When T rises above T_g, the rigid solid can flow, and the corresponding increase in molecular mobility can result in crystallization or increased chemical reactivity of the solid. Several historic papers describe some additional details of amorphous materials. Pikal and coworkers [3] at Eli Lilly showed that amorphous materials can have lower chemical stability, and Fukoka et al. [4] showed that amorphous materials had a tendency to crystallize. Nevertheless, in some cases, amorphous forms have been historically used as products. An excellent example is novobiocin [5], which exists in a crystalline and an amorphous form. The crystalline form, when administered orally, is poorly absorbed and does not provide therapeutic blood levels; in contrast, the amorphous form is readily absorbed and is therapeutically active. Further studies show that the solubility rate of the amorphous form is 70 times greater than the crystalline form in 0.1 N HCl at 25°C when particles <10 microns are used.

Making a "top 10" list of the differences between crystalline and amorphous materials is possible. Crystalline materials have the following characteristics:

1. higher purity,
2. more physically and chemically stable, *crystalline hydrate > anhydrous crystal > amorphous*,
3. lower solubility,

Figure 1.4 Space-filling drawing of the unit cell of flufenamic acid form I.

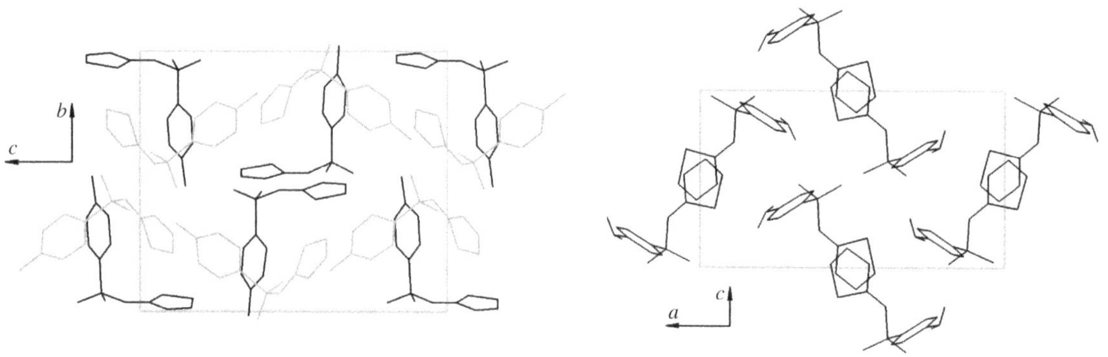

Figure 1.5 Crystal packing and unit cells (gray) of Forms I (left panel) and II (right panel) of sulfathiazole. The gray and black molecules in Form I indicate two unrelated molecules in the asymmetric unit. *Source*: Kruger and Gafner [7, 8]. Redrawn from data published.

4. narrow and (usually) higher melting point range,
5. harder,
6. brittle – slip and cleavage,
7. directionally dependent properties – anisotropy,
8. less compressible,
9. better flow and handling characteristics, and
10. less hygroscopic.

From this list, crystalline materials are generally more desirable unless they are so insoluble that they cannot be used as medicines.

Not only do polymorphs show different X-ray powder diffraction patterns, but they also have different unit cells and different properties, including thermal properties [6]. Figure 1.5 shows the different crystal packing of the Forms I and II of sulfathiazole [7, 8].

Additionally, polymorphs are characterized as monotropic or enantiotropic depending on their thermal properties [9, 10].

- Monotropic polymorphs exist if the transition temperature between forms occurs after the melt. In monotropic polymorphs, one form is most stable throughout the temperature range.
- Enantiotropic polymorphs exist if the transition temperature between forms occurs before melting. In this case, one form is more stable at one temperature. At a different temperature, the other form is most stable. For flufenamic acid, Form I is most stable above the transition temperature of 42 °C, and Form III is most stable below the transition temperature. Practically, this means that slurrying at room temperature will convert Form I to Form III.

Crystalline solvates contain solvents regularly incorporated into the crystal lattice. When the solvent is water, the solid form is called a hydrate. Solvates and hydrates do not have the same composition as unsolvated materials. Solvates and hydrates are sometimes referred to as pseudopolymorphs or solvatomorphs. Interestingly, solvates and hydrates can be polymorphic. In such a case, polymorphic solvates exist. Kuhnert Brandstatter in her 1971 book showed photomicrographs of 16 solvates of estradiol [11]. Figure 1.6 shows the crystal structure of caffeine monohydrate. The crystal of caffeine is built up by stacking the layers shown in Figure 1.6 on top of each other. Thus, the hydrate molecules are in tunnels in this solid form.

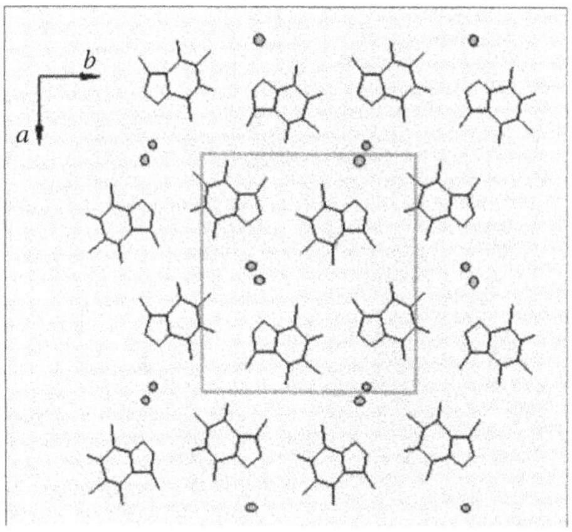

Figure 1.6 Projection of the crystal structure of caffeine hydrate on the *ab* plane. *Source*: Burger and Ramberger [9, 10]. Reproduced with the permission of Springer.

It is important to note that the Food and Drug Administration (FDA) has defined polymorphs as "different crystalline forms of the same drug substance. This may include solvation or hydration products (also known as pseudopolymorphs) and amorphous forms. Per the current regulatory scheme, different polymorphic forms are considered the same active ingredients." Thus, for purposes of registration, scientists are directed to define polymorphs more broadly to include amorphous forms, solvates, and hydrates.

Cocrystals, that is, two-component crystals, are another solid material of interest. Like solvates, the new crystalline structure imparts different properties to the material, including solubility, stability, and mechanical properties. Of special interest are cocrystals with altered solubility or stability. Figure 1.7 shows the crystal structure of a cocrystal of 2-methoxy-4-nitrophenol–4-(dimethylamino)pyridine and phenol (2 : 1) [12]. The FDA has recently released a guidance defining cocrystals as "solids that are crystalline materials composed of two or more molecules in the same crystal lattice."

Pharmaceutical salts are substances formed by the reaction of an acid and a base. The FDA has suggested the following definition of salts: "Any of numerous compounds that result from replacement of part or all of the acid hydrogens of an acid by a metal or a radical acting like a metal: an ionic or electrovalent crystalline compound. Per the current regulatory scheme, different salt forms of the same active moiety are considered different active ingredients." When a carboxylic acid reacts with an amine, a salt is typically formed (Scheme 1.1). However, the degree of proton transfer can vary depending on the acidity and basicity of the reacting groups. The FDA definition seems to encompass all these materials.

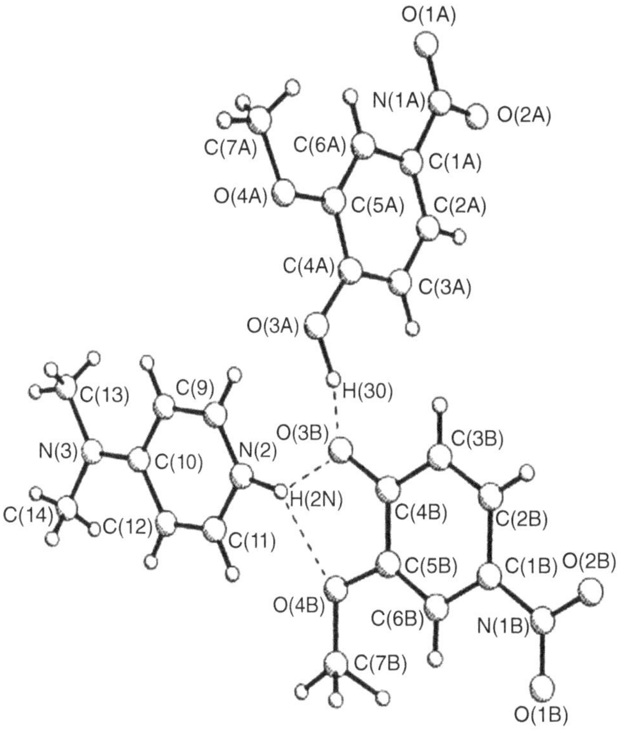

Figure 1.7 Crystal structure of a cocrystal (2-methoxy-4-nitrophenol–4-(dimethylamino) pyridine and phenol (2 : 1)). The unit cell parameters are $a = 6.880$, $b = 38.40$, $c = 8.454$, and the space group is $Pna2_1$. *Source*: Reproduced from [12]/with permission of Royal Society of Chemistry/CC BY 3.0.

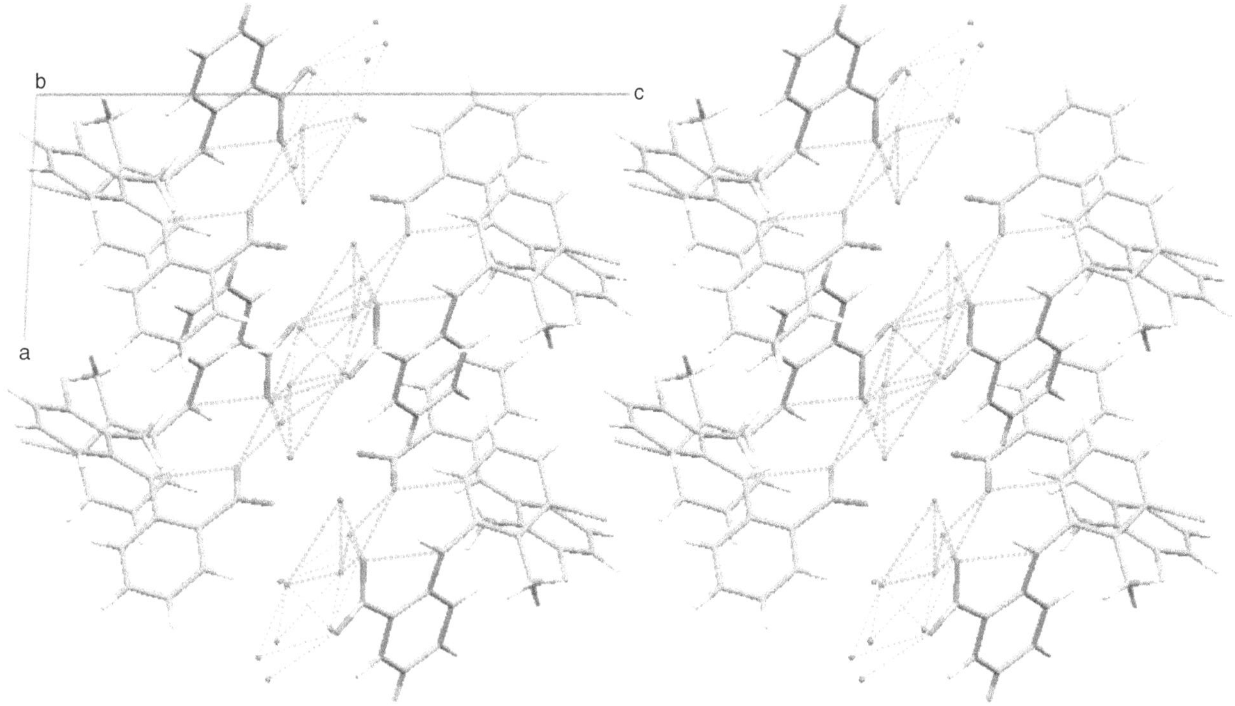

Figure 1.8 Crystal packing of calcium tolfenamate trihydrate showing hydrogen bonding network. The directions of the unit cell axis are (a) vertical, (c) across, and (b) out of the plane of the paper. *Source*: Adapted from [13].

Scheme 1.1

$$\mathrm{RCOOH + H_2N - R' \rightarrow RCOO^{-}\cdots H_3N^{+} - R}$$

Figure 1.8 shows the crystal structure of calcium tolfenamate trihydrate [13]. The unit cell is composed of regions containing mostly hydrocarbon functional groups and regions containing polar functionalities. This type of crystal packing is typical for salts.

1.3 Bulk and Surface Properties of Solids

As we can anticipate from this previous discussion of solid forms, molecules in the solid state, whether in a crystalline or amorphous state, will pack together in various geometric arrangements with different interaction energies, which ultimately determine their various physical and chemical properties. Such arrangements and interactions will primarily depend on the chemical structure, molecular size and shape, and modes of bringing about solidification, that is, modes of crystallization or vitrification. Most importantly, such structural features will determine physical and chemical properties, described in this book, such as rates of crystallization, polymorphic transformations, chemical instability, solubility and dissolution, mechanical responses to various stresses, liquid crystal formation, protein denaturation, and modes of hygroscopicity. It is also important to recognize that such properties will be quite different for molecules in the bulk and surface phases. In the bulk phase, a molecule is generally locally surrounded in all dimensions by other molecules, as schematically illustrated in Figure 1.1, and forces on any molecule will be evenly balanced. On the other hand, interactions across an immiscible phase boundary will generally be unbalanced and greater in the direction of the bulk phase, giving rise to lower densities and greater molecular mobility at the surface. This gives rise to the surface tension not only observed with liquids but also demonstrated in solids [14]. Usually, the mass of molecules in the bulk phase is much greater than that at the surface; that is, there is a very small surface-to-volume ratio, and surface properties will not necessarily be important in determining overall thermodynamic and dynamic properties. However, for extremely small particles with very high surface-to-volume ratios, such as nanocrystals, porous solids, or crystal nuclei, surface properties can be the determining factor in influencing both surface and bulk physical and chemical properties. Such surface properties include vapor adsorption by solids, the rate of crystal nucleation, and the ability of liquids to spontaneously spread over solid surfaces or penetrate highly porous structures, that is, wetting [15, 16]. Throughout this book various properties of solids will be discussed in the context of structural arrangements and intermolecular interactions in the bulk phase, with additional surface effects under certain conditions.

1.4 ICH Q6A Decision Trees

In 1995, Byrn, Pfeiffer, Ganey, Poochikian, and Hoiberg from Purdue University and the FDA published a paper using decision trees to describe a strategy to identify the best solid form early in development. In this way, it is possible to ensure uniformity of a solid form in clinical trials and resolve solid-state issues before critical stages of development. The decision trees also suggested appropriate analytical methods for control. Four decision trees were presented: polymorphs, hydrates/solvates, desolvated solvates, and amorphous forms [17].

In the late 1990s, the International Committee on Harmonization (ICH) used a similar decision tree approach to describe how specifications for the solid form in drug substances (API) and drug products should be determined. The ICH Q6A document presented several decision trees, including decision trees on particle size and polymorphs [18]. The ICH utilized the broadened definition of polymorphs that includes hydrates, solvates, and amorphous forms. The ICH decision trees are divided into three questions, as shown in Figures 1.9–1.11.

These three decision trees outline a strategy widely used during drug development. Most firms conduct an early polymorph screen to address question number 1. Once new forms have been identified, they are physically characterized (solubility, stability, melting point), and an effort is made to understand whether these differences in properties will affect drug product safety, performance, or efficacy. If the different solid forms can affect safety, efficacy, or performance, then question 3 is addressed by determining whether drug product testing can detect changes in ratios of these forms. Additionally, the ratios of forms are monitored during stability studies to make sure changes that affect performance, safety, or efficacy do not occur. Using this strategy, it is possible to find the best solid form for rapid development.

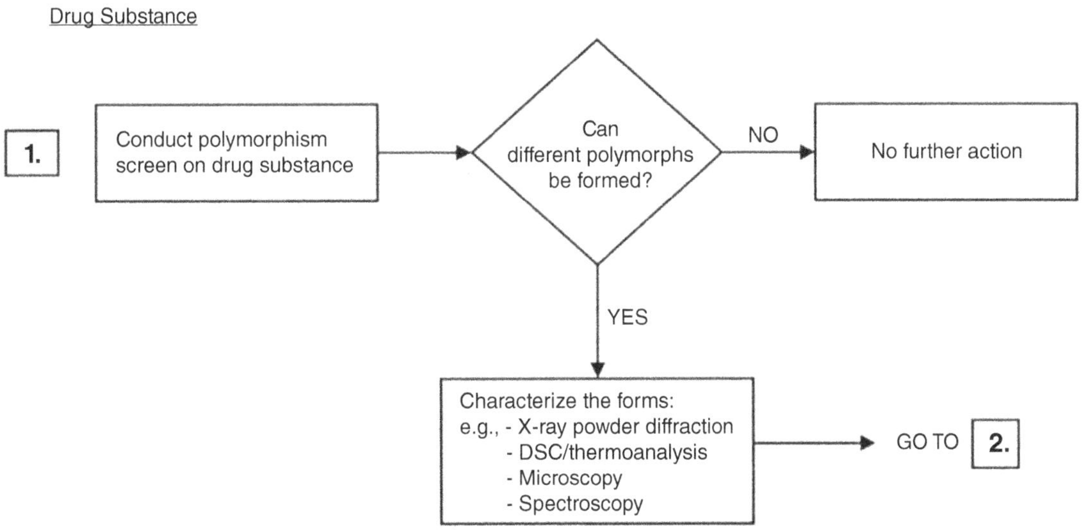

Figure 1.9 ICH Q6A question 1 on polymorphs: Can different polymorphs be formed? *Source*: ICH Harmonized Tripartite Guideline [18]. Reproduced with the permission of ICH.

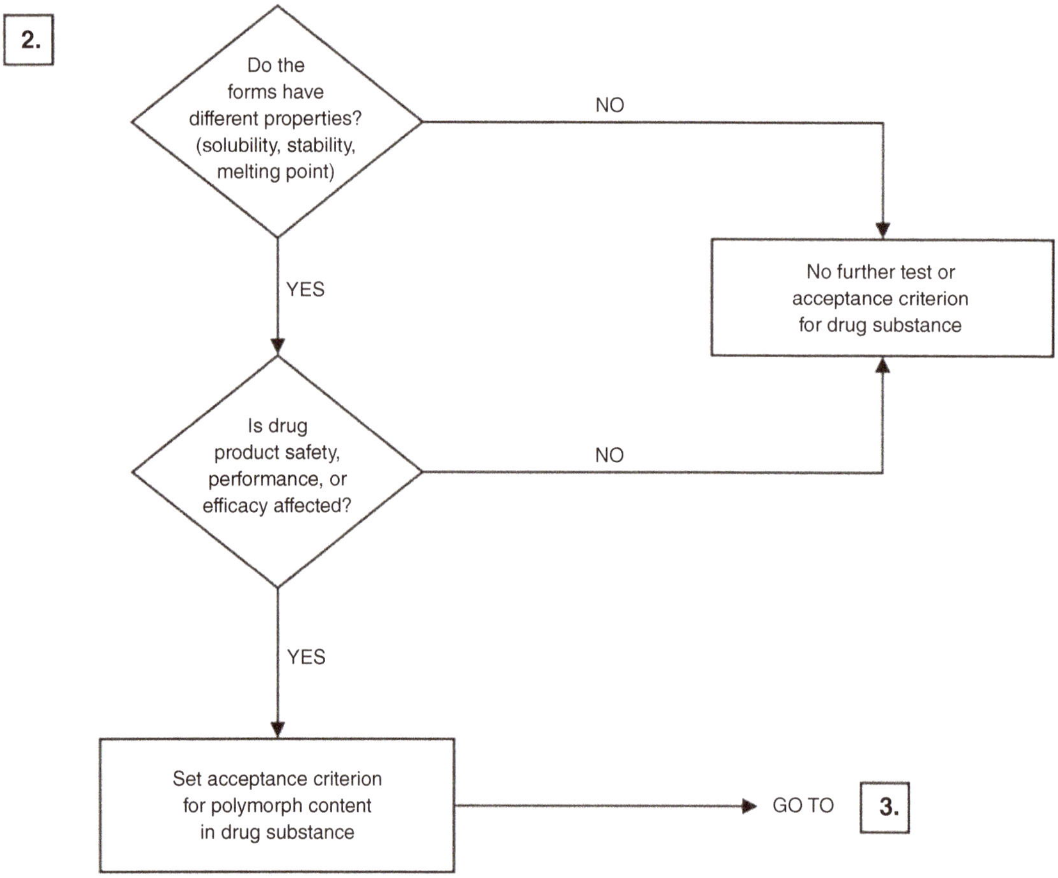

Figure 1.10 ICH Q6A question 2 on polymorphs: Do the forms have different properties (solubility, stability, and melting point)? *Source*: ICH Harmonized Tripartite Guideline [18]. Reproduced with permission of ICH.

1.5 "Big Questions" for Drug Development

In addition to selecting the solid form, Table 1.1 lists other critical issues/measurements required for drug development.

Another way to think about drug development is to think of this process in terms of answering a series of questions we call the "Big Questions." These must be answered to be able to develop a drug for clinical trials and beyond. These questions are:

- What is the structure of the compound?
- What is the likely dose?
- What is the route of administration and desired dosage form?
- What is the indication?
- How difficult is it to synthesize?
- How soluble is the compound/formulation?
- How well is the compound absorbed? What is its Biopharmaceutical Classification System (BCS) class?
- What is the toxicology of the compound? What is its no-observed-adverse-effect-level? What is its maximum tolerated dose (MTD)?
- What biomarkers are available to monitor clinical trials?
- What doses should be used for the Phase 2 clinical trial?
- What are its solid-state properties and physical characterization?
- How chemically stable is the compound?
- How physically stable is the compound?
- How well will the powder flow?
- Is moisture an issue?
- What is the design, composition, and manufacturing procedure of the formulation/product?

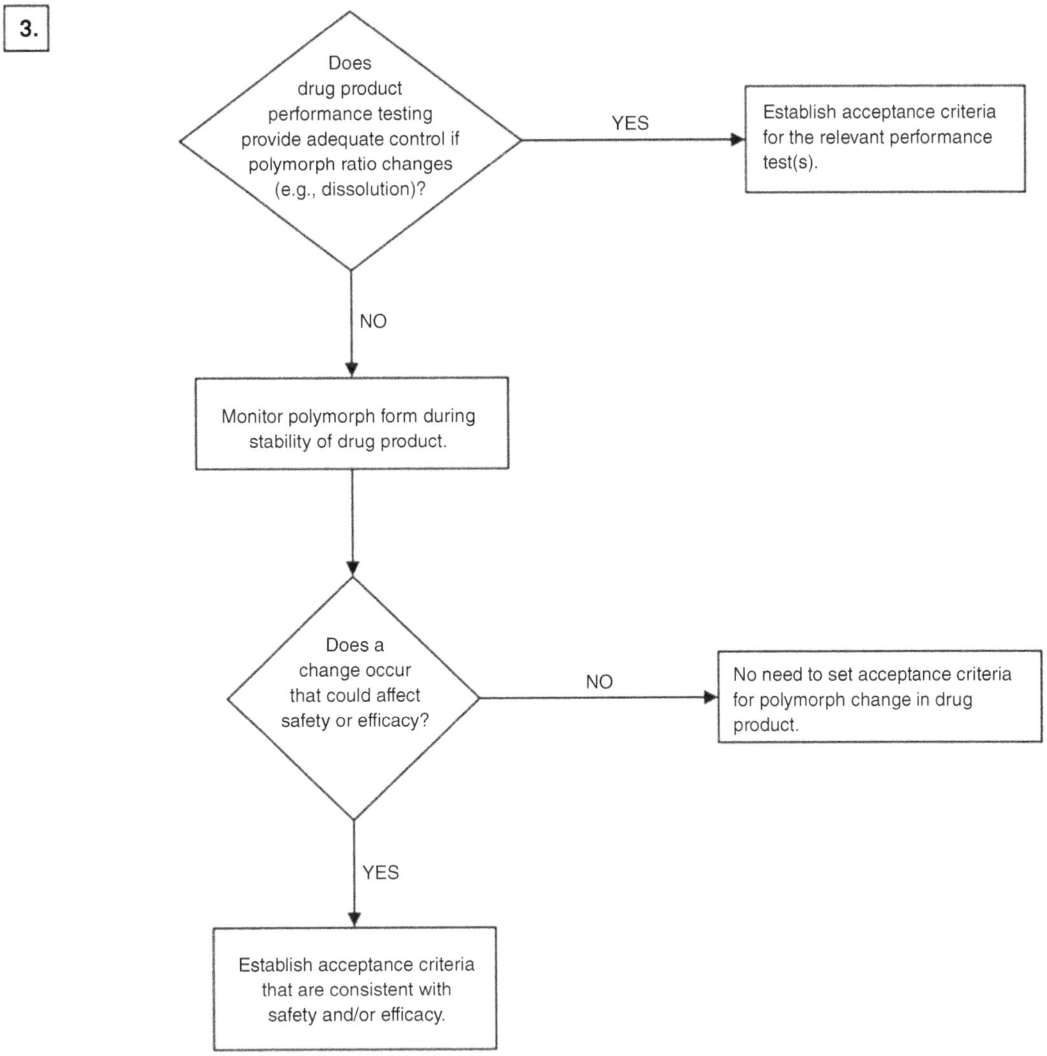

Figure 1.11 ICH Q6A question 3 on polymorphs: Does product performance testing provide adequate control in polymorph ratio changes (e.g. dissolution)? *Source*: ICH Harmonized Tripartite Guideline (1999) [18]. Reproduced with permission of ICH.

Table 1.1 Critical issues for accelerated drug development.

Polymorph selection
Chemical synthesis
Salt selection (optional)
Assay/impurities
Particle size
Physical characterization and properties
Dissolution/solubility
Consistency
Stability
Validated methods/processes
Regulatory issues
Intellectual property

Table 1.2 Biopharmaceutical classification system (BCS).

BCS classification	Solubility	Permeability
BCS class I	High	High
BCS class II	Low	High
BCS class III	High	Low
BCS class IV	Low	Low

As has already been discussed, the solubility of the compound is a critical quality attribute important for specifications and development. The solubility of a solid substance is the concentration at which the solution phase is in equilibrium with a given solid phase at a stated temperature and pressure. Under these conditions, the solid neither dissolves nor continues to crystallize. Note that the definition implies the presence of a specific solid phase. Once determined under the stated conditions, however, we can talk about the "solubility" of a given phase (e.g. a specific polymorph or pseudopolymorph) as a quantity, even in the absence of that solid phase. The use of the term "equilibrium" in connection with crystallizing systems requires clarification. When a substance exists in more than one crystal form, that is, when other polymorphs are possible, only the least soluble of these at a given temperature is considered the most physically stable form at that temperature; all others are metastable forms. In given cases, a solution of a substance may be in apparent equilibrium with one of these metastable phases for a long time, in which case, the system is in metastable equilibrium and is expressing the thermodynamic solubility of that solid form.

It is important to stress the difference between polymorphs and solvates (pseudopolymorphs) at this point. If a solvate/pseudopolymorph exists, it is always (with few exceptions) the most stable form in the solvent that produces the pseudopolymorph.

Undersaturation pertains to solutions at a lower concentration than the saturation value (i.e. diluted solutions). Crystals will dissolve in undersaturated solutions. Saturation is the state of a system where the solid is in equilibrium with the solution, or, in other words, the solution will neither dissolve crystals nor let them grow (i.e. the concentration of the solution represents the solubility value for that crystalline phase). Supersaturation pertains to solutions that, for one reason or another (e.g. rapid cooling of a saturated solution without forming crystals), are at a higher concentration than the saturation value. Supersaturation is required for crystals to grow.

The solubility and permeability are combined to determine the BCS class (Table 1.2). BCS class I drugs dissolve easily and are easily transported into the bloodstream because they are highly permeable with respect to the membranes in the gastrointestinal (GI) tract. BCS class III and IV drugs have poor permeability and are generally difficult to develop. BCS class II drugs are of the greatest interest to pharmaceutical scientists because the structure of the solid, the formulation, its physical character, and many other factors are likely to have a significant effect on bioavailability and ultimately safety, performance, and efficacy. Several important drugs that are widely prescribed are BCS class II, including atorvastatin calcium, celecoxib, efavirenz, irbesartan, lopinavir, medroxyprogesterone acetate, raloxifene hydrochloride, simvastatin, and warfarin sodium. Of the marketed drugs nearly 70% are in BCS class I or II with 31% being in BCS class II. It has been estimated that as high as 80% of the drugs under development are BCS class II.

1.6 Accelerating Drug Development

Accelerating drug development has been a goal of pharmaceutical scientists for many years. One of the earliest discussions of accelerated development can be traced back to 1958 when the National Cancer Chemotherapy Program reported that the pharmaceutical industry was launching major anticancer drug development efforts to accelerate the search for effective agents [19]. This early program was, of course, responsible for the major advances in the treatment and eradication of cancer. In 1995, Colin Gardner of Merck introduced a flow chart showing the synthesis of the API and the development of clinical supplies for first-in-human trials in 1 year. In this early flow chart, drug substance synthesis and process development were carried out in parallel with preformulation/formulation design/development and safety studies. Despite the early introduction of the concept that an investigational new drug (IND) can be submitted in 1 year, it has been difficult to achieve this goal except in very favorable cases. One of the difficulties is the availability of the API, and another one of the difficulties is accelerating toxicology studies. In 1997, Pisano wrote a book for Harvard University Press entitled *The Development Factory* that addressed rapid drug development. Many of his examples came from Bristol Myers Squibb

(BMS) [20]. He described drug development as a learning process and concluded that learning before doing was the most reliable approach to rapid drug development.

In 2007, Aptuit/Solid State Chemical Information (SSCI) introduced their IND-I-GO (INDIGO) program, offering fast development in a Contract Research Organization (CRO) environment. INDIGO was tailored to work with BCS class II compounds and poorly soluble compounds. This INDIGO offering was supported by an example case study on the poorly soluble drug itraconazole and was summarized in a recent publication [21]. Additionally, Byrn et al. and Byrn and Henck outlined strategies based on solid-state chemistry for reducing development time [22, 23]. These publications contained much more detail on how to carry out screens and are discussed in more detail in the following text. In this same timeframe, Chorus, a Lilly-based firm, focused on fast development, introducing their strategy for accelerated development to Phase II [24]. In this early publication, they suggested that it was possible to develop a compound to Phase II in 30 months for $3 million in contrast to the industry average of 42 months and $30 million. The Chorus approach involves a virtual company that heavily uses preferred CROs. Chorus has been quite successful, and Lilly has now established Chorus as an independent entity. In 2015 the success of Chorus over the past 10 years was summarized [25]. Chorus has developed 41 molecules over the years and maintains a portfolio of 15–17 active projects at any one time.

In this same timeframe, PricewaterhouseCoopers introduced a concept of limited launch with what they have termed a "live license" that permits a company to market a drug under very restricted conditions. In one manifestation of this concept, they suggested the launch of a drug after 1.5 years. The details of this strategy are not clear, but it appears that the proposal involves introducing the drug for a limited population and then expanding use as clinical data allows. This, of course, would require the FDA to license drugs differently. The main point of this strategy for our context is that the introduction of a medicine after 1.5 years would require very rapid development and very rapid FDA review. Regardless of the details, the solid form, the formulation, the synthesis of the drug, and all of the other critical steps outlined in Table 1.1 would have to be accomplished quite rapidly.

Recently, the National Institute for Pharmaceutical Technology & Education held a conference aimed at addressing the question: What does Industry Need to Accelerate Drug Product and Process Development? [26] The workshop identified five fundamental barriers to overcome: (1) an intermediate designation by the FDA for accelerated development (This designation would not be an emergency use authorization but also not be a standard NDA. It would be something in between); (2) a mechanism to allow faster stability testing based on artificial intelligence (AI) and structural analysis of the drug; (3) flexible development and manufacturing including continuous and semi-continuous manufacturing; (4) advanced enabling technologies including process analytical technology (PAT) and on-line methods; and (5) integrated product and process development similar to the INDIGO methods outlined earlier and including formulation by design. They also suggested the possible use of AI methods and other technologies to advance animal testing.

Regardless of the model, it is clear that the development process must be accelerated in the future. Accelerated development is especially dependent on the first year of activities. During the first year, three critical steps must be accomplished. The drug must be synthesized. The toxicology must be determined to figure out the initial dose for first-in-human trials, and the solid form and formulation must be developed so that a medicine is available for first-in-human trials. This requires three groups to work together: API synthesis, toxicology, and pharmaceutical sciences. Figure 1.12 shows a detailed 52-week strategy.

The top three bars show the synthesis of the API. Depending on the dose, several kilograms of API will be required for toxicology studies, which are shown in the fourth bar. A key aspect of the toxicology study is the determination of the toxicology formulation. As pointed out by the Merck group, the toxicology formulation can be critical for poorly soluble drugs. If not enough of the drug can be dissolved, then it is impossible to advance that lead [27]. The four bars on the bottom outline the formulation and manufacture of clinical supplies and the regulatory activities needed to file an IND. Engers, Byrn, Newman, and coworkers were able to meet the timelines in Figure 1.12 and develop a mock IND for itraconazole [21].

Even more acceleration of the development timeframe can be achieved if, during the first year, adaptive clinical trials are designed, biomarkers for the clinical trial endpoints are defined, and assays for these biomarkers are developed. Adaptive clinical trials allow merging of phase I and phase II studies and can sometimes accelerate proof-of-concept studies [28]. Adaptive clinical trials are often considered in two phases, exploratory and confirmatory, and use biomarkers. Biomarkers are quantifiable substances that can be correlated to a clinical response, for example, glucose levels for diabetes. Biomarkers can include genes, gene products, enzymes, cytokines, or even cells. Finding a biomarker can greatly advance early decision making and accelerate clinical trials.

One of the goals of accelerated development is to eliminate compounds that are clinically unacceptable or cannot be developed early, so that more resources are available to focus on compounds that are truly acceptable. For example, if a material is so insoluble that its maximum absorbable dose (MAD) or MTD cannot be determined in animal studies, then this must be determined as early as possible. Likewise, if a compound has unacceptable neurotoxicity, this needs to be determined as early as possible. Thus, a strategy that focuses on the most soluble formulation early in the process will provide important

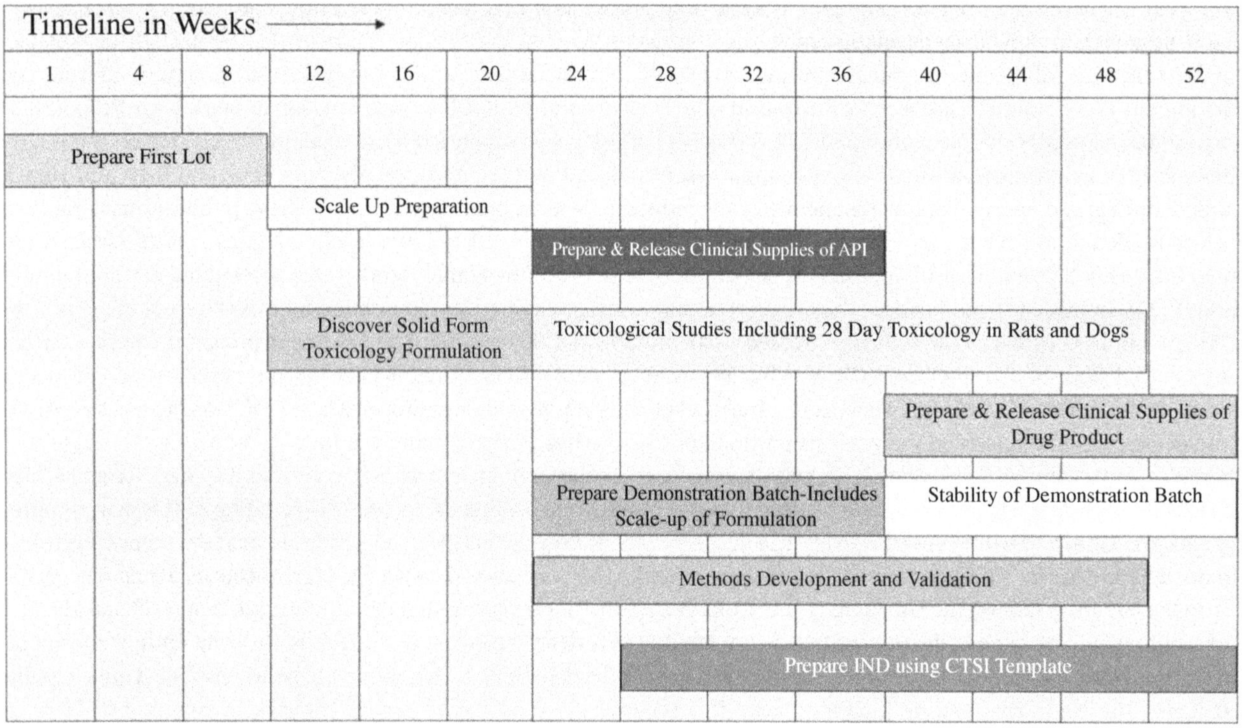

Figure 1.12 Detailed 52-week strategy for accelerated drug development.

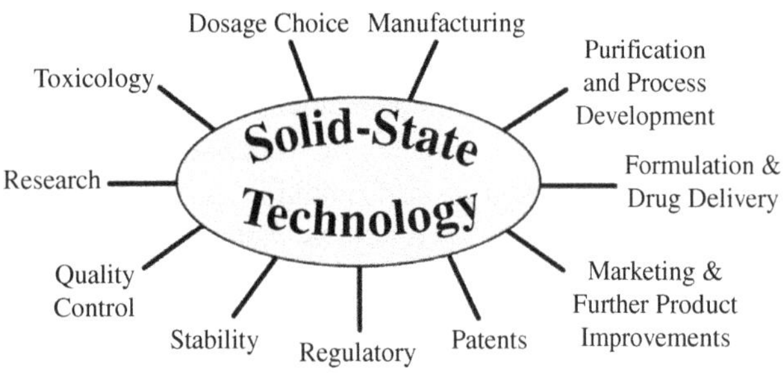

Figure 1.13 Role of solid-state technology in the pharmaceutical industry.

information in both cases. For this reason, some experts suggest that initial formulations should use amorphous forms to identify poorly soluble and highly toxic materials.

1.7 Solid-state Chemistry in Preformulation and Formulation

Figure 1.13 illustrates the important role that solid-state technology plays in the entire development process, including research. In fact, a group from Merck (see above) in a *Journal of Medicinal Chemistry* review has advocated moving solid-state chemistry earlier in drug discovery to ensure that a developable solid form is discovered [27]. As mentioned earlier, the solid form is important in toxicology since an insoluble solid-state form will appear to be nontoxic, perhaps falsely. The solid form is important in dosage choice, manufacturing, purification, development, and formulation. In some cases, product improvements and marketing can be facilitated by finding the best solid form. For example, the Kaletra product (containing lopinavir and ritonavir) is now marketed in an amorphous form that is stable and does not require refrigeration, as did the previous soft gel capsule. Solid forms are quite important for patents and regulatory filings, as discussed earlier. Finally, the stability of the product can be critically related to the solid form.

Particle size, like polymorphism, is one of the most critical aspects of solid-state chemistry. The incorrect particle size can cause a change in the rate of dissolution and affect safety, efficacy, and performance. Figure 1.14 shows a classic study of the effect of particle size of a suspension on blood levels of phenobarbital [29]. This data, as well as the ICH Q6A document on specifications, make it clear that the particle size of APIs (drug substances) must be controlled, especially if they are poorly soluble.

Figure 1.12 illustrates one possibility for when solid-state chemistry is done during preformulation and formulation design. It is important to find a form with acceptable aqueous solubility during this study. In many cases, a form with fast kinetic solubility must be found. If possible, the high solubility should be maintained until the drug is absorbed

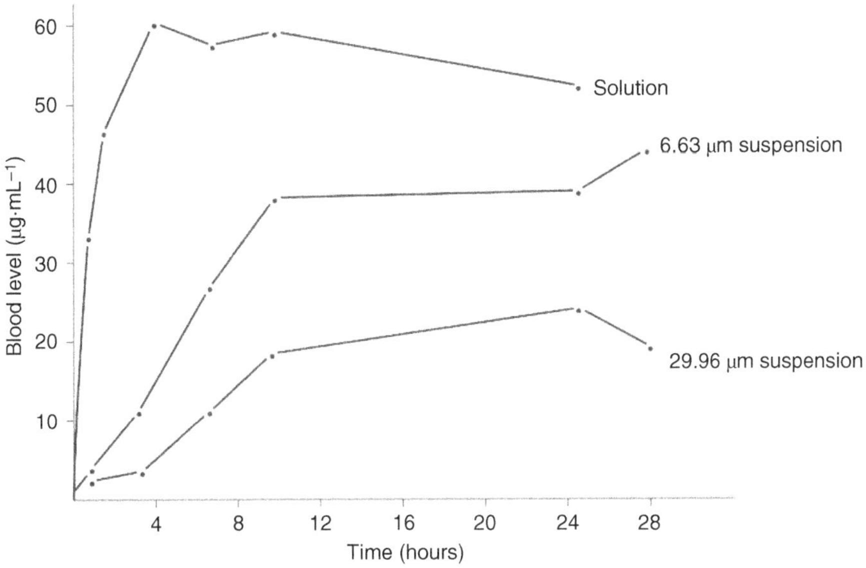

Figure 1.14 Effect of particle size on blood levels. *Source*: Adapted from [29].

from the GI tract. It may be necessary to use crystallization inhibitors to prevent premature precipitation of the solid form. Additionally, the solid form must be chemically stable enough not to decompose during an IND trial. An important goal is to find the form before the final step of the API synthesis. In this way the final form can be prepared in the last step. Additionally, if this can be achieved, the same form can be used for both toxicology and first-in-human studies.

After the first year, additional solid-state and particle size studies will be needed. First, it will be important to confirm the results of the early solid-state studies. It is especially important to do some additional screening to confirm that the best solid form has been selected. There is time for additional salt screening studies and perhaps even cocrystal or nanocrystal screening studies. Furthermore, more complex formulations can be developed. For example, if possible, a simple granulation process should be developed. Alternatively, a roller compaction procedure can be investigated.

In many cases, when the dose of the drug is dissolved, it will be in a supersaturated solution with respect to the most stable crystal form. Thermodynamically, there will be a tendency for this most stable form to crystallize. This system has been termed a supersaturating drug delivery system [30].

Screening is the approach to use to find the best solid form, a manufacturing process, and a crystallization inhibitor, if needed. Polymorph screening was suggested in the mid-1990s for regulatory purposes and to find the form that resulted in the most desirable bioavailability. During the intervening years it has become clear that screening is critical. In the early studies, during the first year, abbreviated screens should be used.

A typical abbreviated screen was described in the early review by Byrn et al. [17]. Solvents should include those used in the final crystallization steps and those used during formulation and processing, such as water, methanol, ethanol, propanol, isopropanol, acetone, acetonitrile, ethyl acetate, hexane, and mixtures if appropriate. New crystal forms can often be obtained by cooling hot, saturated solutions or partly evaporating clear, saturated solutions. The solids produced are analyzed using X-ray diffraction and at least one of the other methods. In these analyses, care must be taken to show that the method of sample preparation (i.e. drying, grinding) has not affected the solid form.

Later, more complete screening studies are recommended. If a supersaturated solution is created it is important to screen for a crystallization inhibitor [30]. As indicated, screening is typically done for polymorphs, including amorphous forms, salts, cocrystals, and nanoparticulate formulations.

In addition to screening and selection of the best solid form and optimization/control of particle size, preformulation experiments are carried out. These experiments include the determination of the partition coefficient (log *P*). This reflects the hydrophobicity of the drug and can be useful in determining the BCS class. The solubility of all available forms and the degree of precipitation of any solid forms are determined. Since solubility can depend on solid form, the solubility is typically determined in aqueous buffers, organic solvents, surfactant solutions, and perhaps cyclodextrins and lipids. Solubility can be difficult to determine and should be determined under equilibrium conditions. The solution and solid-state stability

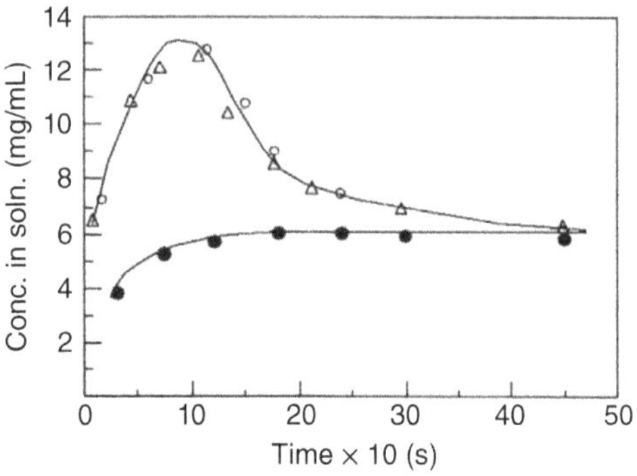

Figure 1.15 Dissolution study of two different forms of theophylline. *Source*: Shefter and Higuchi [31]. Reproduced with the permission of Elsevier.

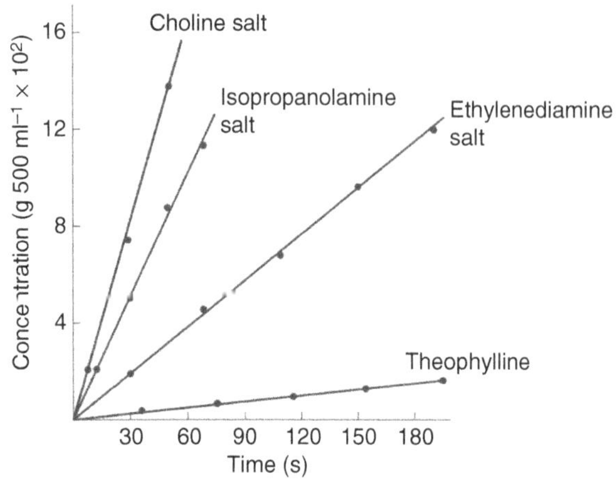

Figure 1.16 Enhancement of solubility and dissolution rate by salt formation. *Source*: Nelson [32]. Reproduced with the permission of Elsevier.

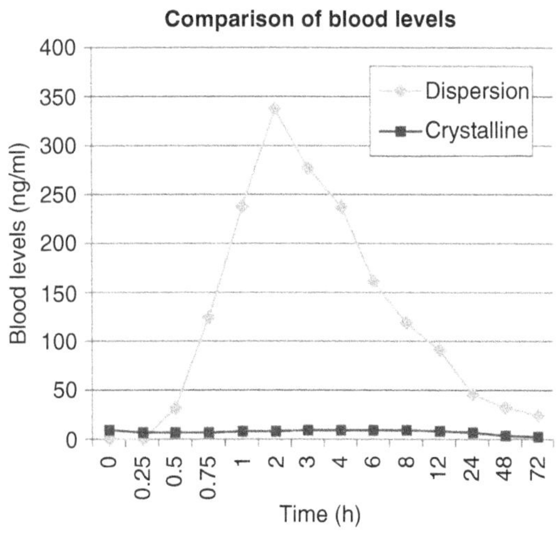

Figure 1.17 Comparison of the bioavailability of a 2:1 HPMC-P:itraconazole dispersion in dogs. *Source*: Adapted from [21]. Redrawn from data published.

of the API is determined under stress conditions, including extreme pH, temperature, light, and humidity. This provides information on the system's intrinsic chemical stability; this knowledge is critical in formulation development. The pK_a is also determined or calculated. This provides important information on the acidity or basicity of the material.

Once the initial solid form has been selected based on the previous screening experiments, the stability of that form is determined under stress and accelerated conditions. This provides important information on how to handle that particular form. The dissolution properties of this form are also monitored. This provides important information on what might happen in the GI tract.

Figure 1.15 illustrates the importance of studying the dissolution rate of different forms [31]. In this study, the hydrate showed normal dissolution behavior, reaching a solubility of about 6 mg/mL. In contrast, the anhydrate showed rapid kinetic dissolution to form a supersaturated solution. This solution then crystallized after about 100 seconds into the hydrate. At the end of the experiment (500 seconds), equilibrium had been reached, and the solubilities of both forms were the same and equal to the hydrate form, which was the solid form present.

Figure 1.16 shows a similar experiment by Nelson on salts of theophylline [32]. This experiment's timeframe was short enough that theophylline hydrate did not crystallize. This figure shows the rapid dissolution rate achieved by some salts. In fact, for some pharmaceutical salts, such as sodium phenytoin, a solubility enhancement of about 1 000 000 is achieved. This clearly shows the desirability of finding salt forms and explains why many drugs are developed as salts. Of course, salt forms are also known to increase the bioavailability of solid forms greatly.

A viable alternative to salt formation, especially in cases where a salt cannot be formed, is to develop an amorphous form/formulation. There are several products containing amorphous forms on the market, including Kaletra and Sporanox. A review reports that amorphous formulations can result in as much as an 82× increase in bioavailability [33]. Law and coworkers reported greatly enhanced plasma concentrations of amorphous ritonavir over crystalline material. Ritonavir is one of the components in Kaletra [34]. Figure 1.17 shows the results of studies of a 2:1 hydroxypropyl methylcellulose (HPMC)-P:itraconazole dispersion in dogs. Clearly, the dispersion results in a large increase in bioavailability.

An alternative to salt formation is cocrystal formation. Figure 1.18 shows that a cocrystal enhanced the bioavailability of an amide-containing API by about 4× [35].

In another interesting study of cocrystals, Childs and coworkers at SSCI showed that cocrystals of fluoxetine hydrochloride could be formed that had faster and slower dissolution rates than the parent [36]. This remains one of the only examples where cocrystals resulted in both enhanced and reduced dissolution rates.

As discussed earlier, amorphous and cocrystalline formulations are classified as supersaturated drug delivery systems. Crystallization inhibitors are sometimes used to prevent premature crystallization for these formulations.

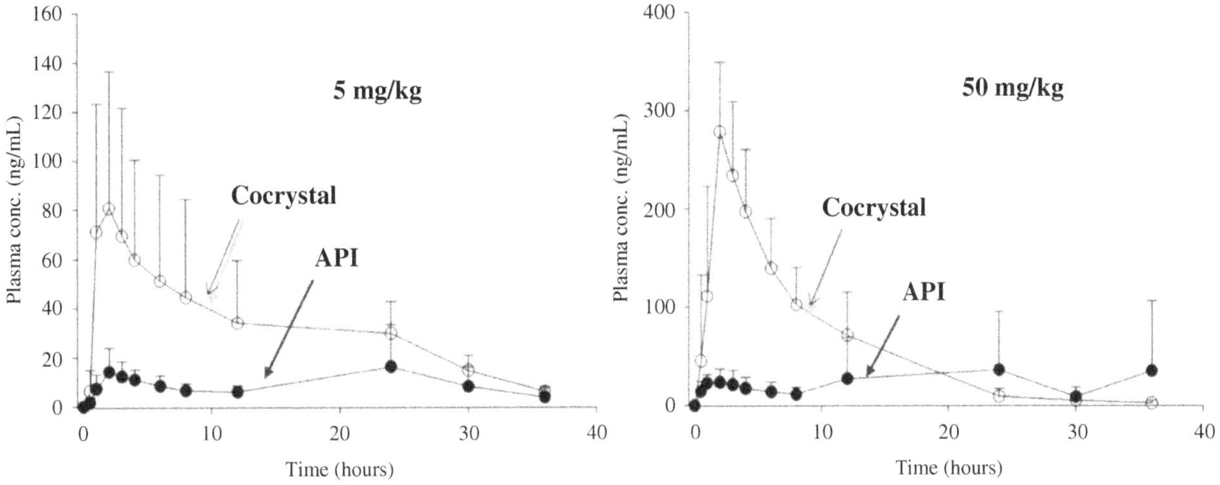

Figure 1.18 Enhanced bioavailability of a cocrystal over the parent drug. *Source*: McNamara et al. [35]. Reproduced with the permission of Springer.

In addition to increasing solubility/dissolution rate, the solid form can influence several other properties important for formulation, including milling, blending, tableting, dry filling, suspension formulation, and lyophilization. Transformations to other forms can also occur during these processes.

1.8 Solid–Lipid Nanoparticles

Solid–lipid nanoparticles (SLNs) provide another strategy for the formulation of drugs and nucleic acids. As the name implies, SLNs are lipidic compositions that are in a solid form. The COVID vaccines, SpikeVax and Comirnaty, contain SLN composed of the cationic lipid, distearoylphosphatidylcholine (DSPC), cholesterol, and polyethylene glycol (PEG) lipid. PEG lipid is a class of polyethylene glycol derivatives attached to a lipid moiety. C&E News stated that, without these lipid shells, there would have been no COVID vaccine. These vaccines have been administered to millions of people with minimal side effects. Cationic lipid-based SLN have been introduced into clinical medicine because of the rapid development of SARS-CoV-2 vaccines. These vaccines utilize second-generation cationic lipids that are easily metabolized and do not exhibit many of the side effects of earlier cationic lipids [37]. The cationic lipids in the SARS-CoV-2 vaccines encapsulate the anionic mRNA by forming a hydrophobic ion pair between the lipid's amine group and the phosphate groups on the nucleic acid backbone [38].

There has been a significant interest in SLN materials for many years before the COVID pandemic. Mehnert and Mader provided an excellent review of the development of SLN up to 2012 [39]. In this review they mention many SLN incorporating drugs, but, curiously, they do not describe the incorporation of nucleic acid and RNA in SLN. A more recent review by Duan and coworkers describes the use of SLN as a dominant formulation strategy with the advantages of liposomes and polymeric nanoparticles, partly because they are non-biotoxic and stable [40]. Montoto describes the current state of this field as the "Golden Age of Pharmaceutical Nanocarriers" [41]. He also points out that SLNs are easy to produce and non-toxic and describes a wide range of diseases that SLNs can treat, including infectious diseases, CNS diseases, and a variety of site-specific treatments. Additionally, nucleic acid SLNs can be used to develop gene therapies, cancer vaccines, malaria vaccines, and a wide range of other therapies.

SLNs have a range of advantages in their use, including improved active agent stability, the potential for controlled release, the potential to increase the solubility of poorly soluble drugs, reduced drug leakage, and more specific drug targeting. SLNs can also often be easily manufactured, for example, freeze-dried and, in favorable cases, reconstituted for injection or utilized directly. For example, celecoxib SLN were freeze-dried with cryoprotectants to yield an acceptable non-collapsed, freeze-dried cake with good redispersal properties, minimal changes in particle size, and long-term stability at refrigerator temperatures [42]. Santonocito and coworkers also reported successful freeze drying of SLN [43]. Freeze drying has also been used to prepare dry powder SLN containing siRNA for inhalation to treat pulmonary disorders. One study successfully used thin film freeze drying to produce SLNs with good inhalation properties [44]. Spray drying is another well-known method for preparing dry powder SLN. In one study lactose and sodium dodecyl sulphate (SDS)

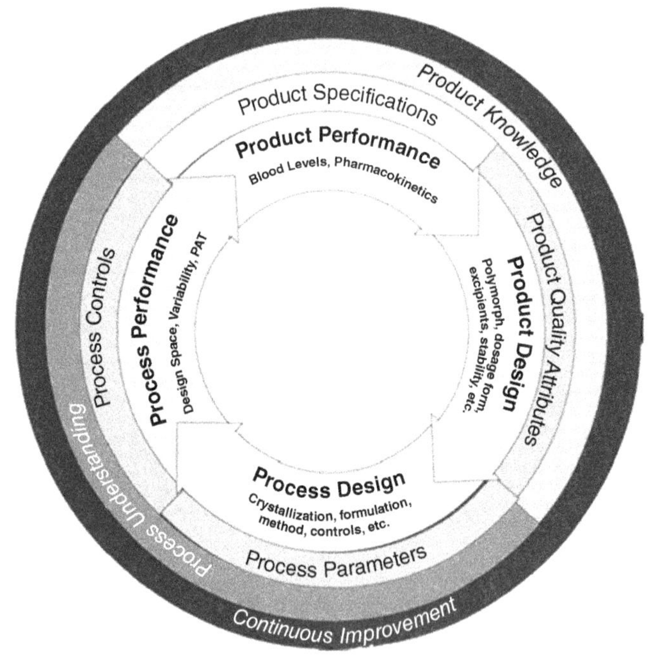

Figure 1.19 Quality-by-design strategies. *Source*: Byrn and Henck [23]. Reproduced with the permission of Elsevier.

were used as a matrix material in the spray drying of tristearin containing fenofibrate [45]. The spray-dried product had good re-dispersibility and produced particle sizes less than 300 nm. Spray drying has also been used to prepare siRNA SLN for pulmonary use. The nanoparticles contained an ionizable cationic lipid, which is a sulfur-containing analog of DLin-MC3-DMA, a helper lipid, cholesterol, and PEG-DMG encapsulating siRNA [46].

1.9 Learning Before Doing and Quality by Design

It is possible to achieve accelerated development using quality by design. Details are presented in Chapter 26. Figure 1.19 shows the quality-by-design wheel used by both the FDA and Pfizer. This wheel illustrates the importance of product design and process design. In effect, we have been discussing product design in this chapter. Using the optimum solid form, it is possible to design a product with the desired solubility. Once this has been achieved, the goal is to develop a formulation that optimizes other properties (e.g. flow) to make a viable product available. Additionally, a process that can reproducibly make the formulation must be achieved. As outlined earlier, often, with proper attention to the solid form, a quality product can be rapidly designed. Thus, in many respects, quality by design is based on preformulation studies.

A simple crystallization/amorphization process followed by filling a solid form in a capsule is often sufficient for early studies. This type of product can be manufactured from solid produced in a variety of ways including spray drying or crystallization followed by capsule filling using one of the commercially available automated capsule-filling machines.

First, good analytical methods for the physical characterization of the designed product must be developed. Then, critical sources of variability in the designed product need to be addressed: Is the solid form stable during its shelf life? What about extremes in temperature or humidity? What factors affect stability?

Stability issues, both chemical and physical, are a major impediment to quality by design. Often it is possible to identify stability problems during preformulation. Exposing the solid form to stress conditions and accelerated stability conditions can provide invaluable information on stability. Armed with this information, adjustments in the designed product can be made.

As development progresses beyond the initial product design and process design stage, multivariate experiments using sensors are typically used to develop a robust process for producing the medicine. If possible, PAT methods are used. A control strategy based on risk is developed to reduce the probability of catastrophic failures. Specifications based on desired product performance are established.

All this information is filed according to the ICH Common Technical Document (CTD). Table 1.3 shows where important quality-by-design data are submitted in an IND. Material submitted in the summary section for drug substance (S.2.x) and drug product (P.2.x) includes a description of critical quality attributes like dissolution rate, process description, process controls, and development history.

Table 1.3 shows the location of quality-by-design elements in the quality overall summary of the CTD, a prescribed format for submissions developed by the ICH.

Figure 1.20 shows the areas of study (learning) and then the formulation, product, and process design in a quality-by-design regime. The formulation is designed by utilizing information about the solid-state chemistry (polymorph), biopharmaceutics, stability, and mechanical properties. This is the product design. Then, a process is designed to make the dosage form.

Table 1.3 QbD elements in the quality overall summary, Section 2.

Question	Drug substance (DS)	Drug product (DP)	Comment
How do the manufacturing processes and controls ensure consistent production of the drug substance/product?	2.3.S.2	2.3.P.3	
How were potential impurities identified and characterized?	2.3.S.3		Impurity analysis is not always included in the drug product
What is the drug substance/product specification?	2.3.S.4	2.3.P.5	Includes explanation of how specifications ensure quality
What drug substance stability studies support the retest or expiration date and storage conditions for the drug substance?	2.3.S.7	2.3.P.8	
Pharmaceutical development		2.3.P.2	Includes discussion of polymorphic form and particle size
What attributes should the drug product possess? How was the product designed to have these attributes? Were alternative formulations or mechanisms investigated?		2.3.P.2.2	Includes quality-by-design elements. Includes contour plots for QbD

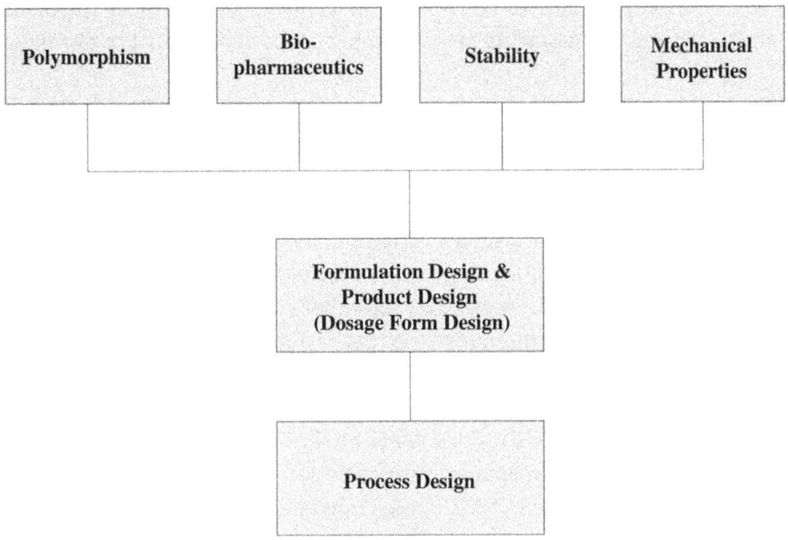

Figure 1.20 Strategy for product and process design.

As outlined in Figure 1.20, the product is designed to meet the target profile based on the system's biopharmaceutics, stability, and mechanical properties. Typically, the target profile is defined by blood levels or biomarkers. The biopharmaceutics includes the BCS class, the solubility, and the permeability. The stability target is a product that is stable throughout clinical trials. Ultimately, a product with 2-year stability is most desirable. The mechanical properties of the system must be addressed. A formulation that does not flow or cannot be compressed will be rejected and must be redesigned to meet the target mechanical profile. In many cases, the initial product design is a powder in a capsule or a powder in a bottle. Typically, modern capsule-filling equipment can fill more than 90% of the available powders into capsules.

For these initial studies the maxim "simple is best" applies. It is important to examine the mechanical properties of the drug substance and/or simple formulation. If the drug substance cannot be filled into capsules, then it is necessary to prepare trial formulations. In this case excipient compatibility must be determined. Also, the formulation should be designed with the manufacturing method in mind. In some cases, a wet or dry granulation will be needed. Alternatively, soft gels containing dissolved API are a viable alternative to improve solubility and bioavailability.

For the initial studies other factors also influence development considerations. In all cases the polymorph in the formulation must be known and controlled. In many instances, it should be possible to prepare a drug product on a small scale and assess its stability and dissolution rate in laboratory equipment. Dissolution studies on formulations filled in capsules can be facilitated very early in development. In some cases, it is helpful to determine the pharmacokinetics of the early formulation in animals. Also, preliminary stability studies are always helpful in understanding the drug substance and drug product. All the initial studies must be done with an eye to the final manufacturing process.

Other challenges faced during initial development are also apparent. Of paramount importance is the short timeline. Another potential problem is that a broad dose range is required. This is because the MTD and MAD are not known. These dose levels will be determined in animal studies and in the first-in-human clinical trials. Until that time, solid-state chemistry and polymorphism studies, as well as formulation design, must be able to accommodate a broad dose range. Additionally, usually minimal amounts of API are available, and sometimes APIs with different attributes are obtained. The early formulation design and solid-state chemistry work must be able to accommodate these variations.

Once the dosage form and formulation have been identified, a process is designed. The process is designed to prepare a drug product that consistently meets the critical quality attributes (design space). Early on, the goal is to have a process that works with the currently available lots of API. As development progresses, the impact of starting materials on the process is understood, and the critical sources of process variability are identified and controlled. The mechanical properties of the formulation must also be understood at this stage. Flow is particularly important as the formulation must flow well to be handled in modern high-speed tablet presses or capsule-filling equipment. Appropriate control strategies are applied to make sure the process is reproducible. Additionally, methods for assessing process performance and product performance are developed, validated, and applied.

As the medicine nears launch and commercial sale, continuous learning occurs. In this process all the elements of Figure 1.20 are reinvestigated in a feed-forward feed-backward process. This allows the development of the final design space. Data are developed to support the design space and control strategy, and all this information is documented.

For polymorphism, the solid-state chemistry of the drug substance (API) and drug product is thoroughly identified. With respect to screening, as discussed earlier, first an accelerated screen is done to develop a dosage form for IND first-in-human trials. Later an intermediate screen should be carried out to make sure the phase II supplies are well controlled. Finally, a comprehensive screen that includes all crystallization solvents and other conditions should be done to try to ensure that all the important solid forms are known. This comprehensive screen is also important for intellectual property purposes.

All the work outlined in Figure 1.20 must be carried out from a risk management perspective. The largest risk is that a successful formulation cannot be reproduced. Thus, no matter how short the timeline the formulation developed must be reproducible. Other risks include unstable formulations with unacceptable purity levels. In addition, it is important to carry out a formal risk assessment at each stage of development (Phase I/IND), Phase II, and Phase III. This risk assessment should be documented and communicated to all affected parties.

One strategy that can be quite helpful during the development process outlined in Figure 1.20 is to utilize a sameness/equivalence protocol. A sameness/equivalence protocol is useful in comparing APIs and formulations made during the development process. A sameness/equivalence protocol includes the parameters outlined in Figure 1.20. If an API or a formulation can be shown to be the same according to measurements of polymorphism, biopharmaceutics, stability, and mechanical properties, then switching between these materials is clearly supported.

In conclusion, it is important to realize that the work outlined in Figure 1.20 is a learning process. Scientists are learning about the compound and its properties. Like any learning process, understanding occurs in stages. During early development, the level of understanding is just enough to answer early questions of clinical efficacy and toxicity. Later learning must also be able to address manufacturability, stability, and polymorphism. In all stages, it is important to maximize learning before doing. Learning before doing emphasizes learning on a laboratory scale and using computer models, if possible. Learning before doing enhances performance and can greatly accelerate drug development programs.

1.10 Performance and Stability in Pharmaceutical Development

Physical characterization is a critical aspect of evaluating the performance and stability of materials under development. The performance of a pharmaceutical material is intimately related to its dissolution in a GI tract full of variability (e.g. variable pH). For this reason, it is important to determine solubility and whether a supersaturated solution is formed. If a supersaturated solution is formed, then it is important to determine if precipitation occurs. Other studies important to evaluating performance include dissolution tests and tests in animals.

As outlined in Figure 1.15, theophylline hydrate dissolves to reach a constant concentration of about 6 mg/mL. In contrast, theophylline anhydrate dissolves to give a significantly higher concentration, but then the theophylline hydrate precipitates. The theophylline anhydrate solution ultimately reaches a final concentration equivalent to that of the hydrate. This physical transformation and precipitation is a slurry conversion of one physical form to another. If such a precipitation occurs in the body, it could affect blood levels. Clearly, an understanding of slurry conversions and precipitation in the GI tract is important.

As with the precipitation (solution-mediated transformation) discussed earlier, solid forms can interconvert in the solid state on storage, in dosage forms, and during processing. Thus, it is important to understand physical stability and these physical interconversions. Of particular interest is conversion from a less stable form to a more stable form, which would have lower solubility and perhaps lower bioavailability. Of particular interest are crystallization of amorphous forms, Oswald ripening of nanocrystalline products, and conversion of polymorphs to more stable forms or hydrates.

In addition to physical stability, solids can transform chemically. Although the historic rule of thumb is that solid-state reactions are 10–100 times slower than solution reactions, solid-state reactions can occur at a rate that can affect the stability of the compound. Overall, solid-state reactions can be broadly classified into three groups:

- Thermal reactions and degradations
- Photochemical reactions
- Oxidation

Each of these reactions can be important for a given compound. Solid-state chemical reactions can be characterized using X-ray diffraction since the solid structure changes, solid-state NMR since the chemical connectivity changes, and by dissolution of the solid and HPLC analysis. A key question with respect to solid-state chemical reactions and solid-state stability is, "Does your product have an acceptable shelf life?"

A detailed examination of physical transformations shows that the system will always tend to produce only the less soluble of the two forms. To be sure, the time it takes to express this tendency depends on kinetic factors and may be quite variable, but, in any event, a less soluble form never converts to the more soluble form under rigorously defined conditions.

A few illustrations of the dissolution behavior of some polymorphic drugs may help to review these relationships as they apply to solutions at constant temperature. When temperature is introduced as a variable, however, further distinctions concerning the relative stability of alternative forms need to be made. The thermodynamic activity (usually observed as solubility) of each form may change quite differently as a function of temperature. Monotropic systems are defined as systems where a single form is always more stable regardless of the temperature. Enantiotropic systems are defined as systems where the relative stability of the two forms inverts at some transition temperature.

In actual practice, it is customary to plot log solubility versus $1/T$ for each solid phase (i.e. as a so-called van't Hoff plot). These plots give, in most cases, the data in a linear form that lends itself to extrapolation, so that transition points can be determined even when complete data for a given solid phase are unreliable or unavailable. Figure 1.21 shows a van't Hoff plot of solubility versus $1/T$ [47]. In this case, there is a transition point where the lines cross and the relative stabilities of the two forms are the same ($\Delta G = 0$). Extrapolation of data 10 K beyond the experimental range is prone to producing large errors and is not reliable.

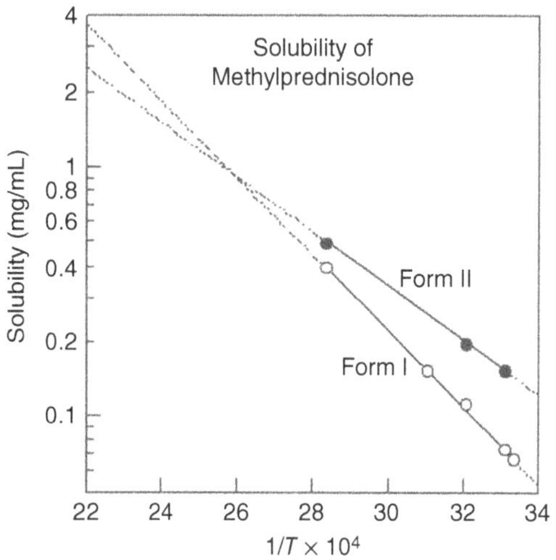

Figure 1.21 A van't Hoff plot of the water solubility of two methylprednisolone crystal forms (log of the solubility as an inverse function of temperature). *Source*: Higuchi et al. [47]. Reproduced with the permission of Elsevier.

Transitions from one solid phase to another can occur in the absence of solvent. The mechanisms and kinetics of such solid-state transitions can be very complex. For example, Kitaigorodskii et al. showed that a pin-prick can initiate the solid-state transformation of *a-p*-dichlorobenzene to *b-p*-dichlorobenzene within a single crystal [48]. The transformation of *a*- to *b-p*-dichlorobenzene is delineated by the spread of the reaction front from the nucleation site through the crystal. A related process is the thermally induced rearrangement of the *a*- to *b*-form of *p*-nitrophenol [49]. In this reaction, needle-shaped single crystals rearrange with the phase boundary moving approximately perpendicular to the needle axis. Grinding or other input of mechanical energy induces the polymorphic transformation of chlorpropamide [50], fostedil [51], chloramphenicol palmitate [52], and several other drugs [53].

1.11 Moisture Uptake

Some crystalline solids take up water from the atmosphere and are termed hygroscopic solids in the literature. Chapter 15 provides a detailed discussion of these materials. Here, the topic is briefly introduced.

Unfortunately, there can be no clear definition of hygroscopic solids because hygroscopicity is a relative term. Hygroscopicity is determined by both a kinetic and a thermodynamic term and is a function of the atmospheric relative humidity. At high relative humidities, many solids are hygroscopic. In atmospheres of low humidity, only a few solids will be hygroscopic. Another factor influencing hygroscopicity is surface area and porosity. The larger the surface area of the solid, the more rapid the uptake of moisture. This is because solids with larger surface areas have more sites for the sorption of water molecules.

When solids that are not solvates contain large amounts of water, it has been hypothesized that water must be taken up into the solid by disordered or high-energy regions, such as defects and amorphous sites. Such effects might be exaggerated by manufacturing processes that reduce particle size, such as micronization, milling, or related processes known to increase the number of high-energy sites. Of course, some solids can take up so much water during these processes that they become damp or even liquefy at RH_0 [54]. This tendency is usually easily detected by microscopic observation.

The formation of crystal hydrates, of course, is another way for water to be incorporated into a solid. In these cases, the water molecules generally occupy a specific crystallographic site in the solid. This site can be determined by X-ray crystallography, which thus unequivocally proves the existence and composition of the hydrate. However, many hydrates exist in which the water is in tunnels within the crystal. The water can be located accurately only by the determination of the crystal structure at low temperatures (if even then). In these cases, the water content may change rather easily with changes in relative humidity.

Plots of vapor pressure versus relative humidity are an excellent way to determine the nature of a solid with respect to water sorption. The different kinds of behavior that these plots may be expected to show include the following:

1. virtually no water uptake,
2. gradual water uptake, characteristic of an amorphous material or a nonstoichiometric hydrate (a hydrate without a simple ratio of water to host molecule), and
3. "stair-step" water uptake, characteristic of a stoichiometric hydrate.

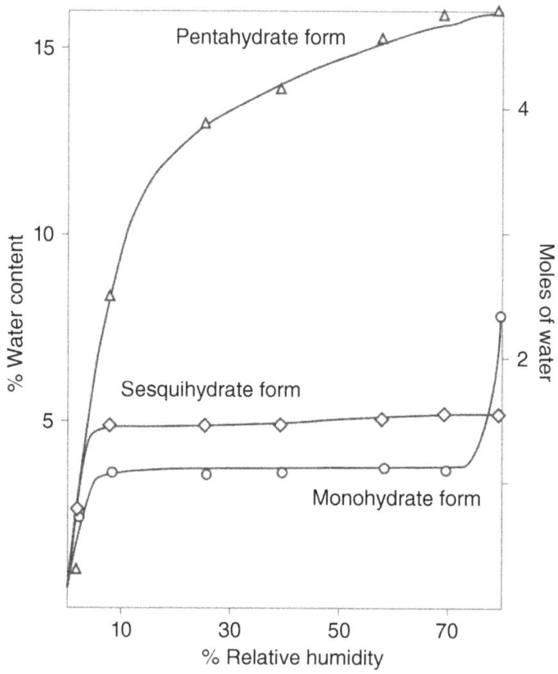

Figure 1.22 Vapor pressure versus relative humidity diagrams for three hydrates of sodium cefazolin. The sesquihydrate and monohydrate behave normally, and the "pentahydrate" is actually a nonstoichiometric hydrate. *Source*: Adapted from [55].

Figure 1.22 shows the behavior of two stoichiometric hydrates (a monohydrate and a sesquihydrate) as well as a nonstoichiometric hydrate of sodium cefazolin [55]. In addition, amorphous sodium cefazolin exists and takes up and gradually loses water as described in the following text.

Sorption of water into amorphous solids or regions of a solid involves dispersion or dissolution of the water molecules within a solid. The more polar a solid is, the greater the amount of water it takes up. Obviously, in such systems, the water content depends on relative humidity.

In summary, Zografi et al. made the following recommendations with regard to water specifications [54]:

1. A complete profile of relative humidity versus water content (weight) should be reported for all reference standards.
2. For amorphous solids, both T_g and W_g should be reported.
3. For deliquescent solids, the RH_0 and an appropriate warning on the label should be provided.
4. For stoichiometric hydrates, the water specifications should reflect the stoichiometry.
5. Attention should be paid to materials that do not form well-defined hydrates and can take up or lose water as the humidity is varied. Any structural changes that accompany changes in water content should be noted. (Some hydrates can lose water without changing their crystal structure. This is due to the formation of an extremely stable crystal packing network by the host molecule.)

1.12 Solid-state Reactions

The scientific discipline of solid-state chemistry of drugs emphasizes the studies of the chemical and physical properties of the various solid forms just discussed. These studies include solid-state phase transformations (polymorphic transformations), reactions in which solvent of crystallization is lost or gained, and a broad range of solid-state chemical reactions.

It is necessary to establish criteria for solid-state reactions to focus on true solid-state reactions. This will avoid a liquid-state reaction being identified as a solid-state reaction. Morawetz suggested four criteria for determining whether a reaction is a true solid-state reaction [56] and a fifth very important criterion can be added from Paul and Curtin [57]:

1. A reaction occurs in the solid when the liquid reaction does not occur or is much slower.
2. A reaction occurs in the solid when pronounced differences are found in the reactivity of closely related compounds.
3. A reaction occurs in the solid when different reaction products are formed in the liquid state.
4. A reaction occurs in the solid if the same reagent in different crystalline modifications has different reactivity or leads to different reaction products.
5. A reaction occurs in the solid phase if it occurs at a temperature below the eutectic point of a mixture of the starting material and products.

Once it has been established that the reaction is occurring in the solid state, the reaction can be understood in terms of a four-step process [57]:

1. Loosening of the molecules at the reaction site
2. Molecular change
3. Solid solution formation
4. Separation of the product phase

It is important to realize that many solid-state reactions of drugs involve drug degradations that have been studied mostly on the macroscopic level. In fact, few studies aimed at determining the molecular aspects of the solid-state chemistry of drugs have been published. These reactions are of interest because of a desire to prevent such degradation. Even for such common drugs as vitamin D_2 and vitamin A, the structures of only a few of the solid-state degradation products have been published. Therefore, in many respects, the solid-state chemistry of drugs is synonymous with drug degradation.

1.13 Noninteracting Formulations – Physical Characterizations

Noninteracting formulations are those in which the components do not interact or alter in any meaningful way the particles of the API. Thus, the term "noninteracting formulations" applies to solid dosage forms. Physical characterization of noninteracting formulations is important, in part, because it shows there is indeed no meaningful interaction between the API and the formulation components.

Hypothetically, it would be impossible to have no interaction between the formulation components. Figure 1.23 shows the functional groups on the surface of R,S-alanine [58]. It is clear that the 001 plane has carboxyl groups exposed, whereas

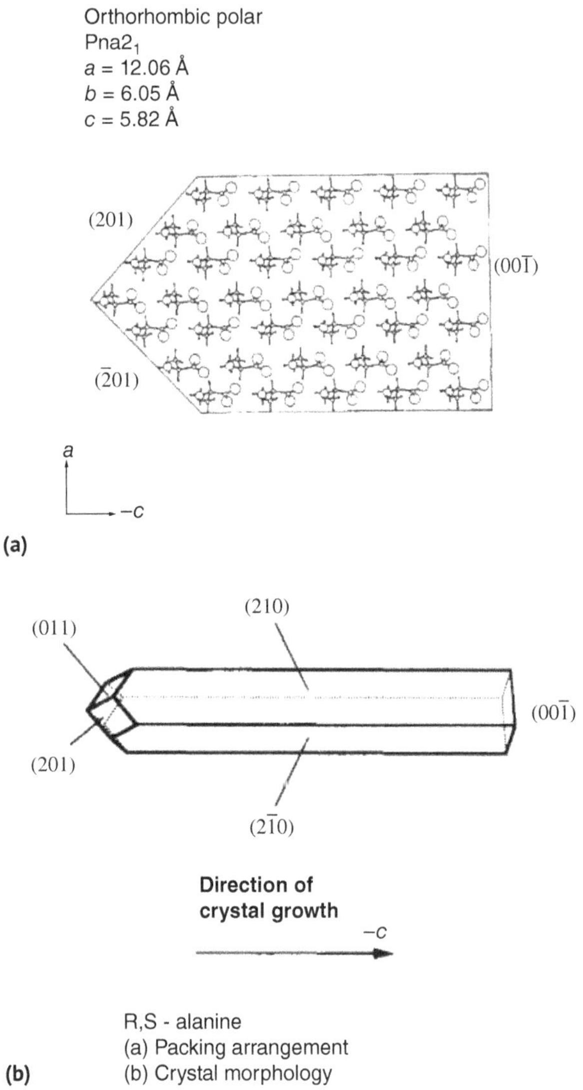

Figure 1.23 Functional groups on the surface of crystals of R,S-alanine. *Source*: Weissbuch et al. [58]. Reproduced with the permission of the American Chemical Society.

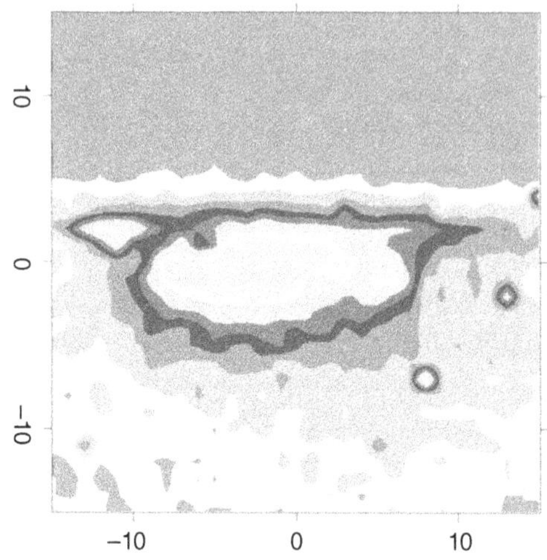

Figure 1.24 A Raman map of cyclobenzaprine showing an oblong particle of API in the center of the figure near the surface of the tablet. *Source*: Adapted from SSCI, West Lafayette.

the 201 plane has amino groups exposed. Curtin and coworkers confirmed that carboxyl groups were on the 001 plane by reacting these crystals with ammonia gas. In such reactions the ammonia reacted preferentially with the 001 face [59].

Consider the situation where R,S-alanine is formulated with a cellulose derivative. Cellulose has many hydroxyl groups. It is reasonable to hypothesize that the OH groups on the cellulose will hydrogen bond to the C=O group of the alanine carboxylic acid (which, in fact, is likely to be ionized due to the zwitterionic nature of the crystals). Likewise, the NH_3^+ group is also likely to hydrogen bond to the OH group of a cellulose molecule. This suggests that there might be surface interactions between solid particles in a formulation. Liversidge and coworkers took advantage of this type of interaction to stabilize nanocrystals using surface modifiers [60].

Raman mapping is a good method for analyzing noninteracting formulation or any formulations for that matter. Figure 1.24 shows a Raman map of cyclobenzaprine drug product. In this map, an oblong drug particle of about 20 microns by 5 microns is visible. This figure clearly shows an intact particle. This figure suggests that Raman mapping can be used to estimate particle size.

Solid-state NMR (SSNMR) is another good method. As shown in the solid-state NMR chapter, different solid forms can be distinguished relatively easily using SSNMR. X-ray diffraction is another powerful method for analyzing formulations. Both SSNMR and X-ray diffraction lack the spatial information available from Raman mapping.

References

1 Van Eerdenbrugh B, Baird JA, Taylor LS. Crystallization tendency of active pharmaceutical ingredients following rapid solvent evaporation—classification and comparison with crystallization tendency from undercooled melts. *J Pharm Sci* 2010; 99: 3826–3838.

2 Kitaigorodskii AI. *Organic Chemical Crystallography*. New York: Consultants Bureau; 1961.

3 Pikal MJ, Lukes AL, Lang JE. Thermal decomposition of amorphous beta lactam antibacterials. *J Pham Sci* 1977; 66: 1312–1316.
4 Fukoka E, Midori M, Yamamura S. Glassy state of pharmaceuticals. V. Relaxation during cooling and heating of glass by differential scanning calorimetry. *Chem Pharm Bull* 1991; 39: 2087–2090.
5 Mullins JD, Macek TJ. Some pharmaceutical properties of novobiocin. *J Am Pharm Assoc Sci Ed* 1960; 49: 245–248.
6 Haleblian JC. Characterization of habits and crystalline modifications of solids and their pharmaceutical applications. *J Pharm Sci* 1975; 64: 1269–1288.
7 Kruger GJ, Gafner G. Crystal structures of polymorphs I and III of sulfathiazole. *Acta Crystallogr Sect B Struct Sci* 1971; B27: 272–283.
8 Kruger GJ, Gafner G. Crystal structures of sulfathiazole II. *Acta Crystallogr Sect B Struct Sci* 1971; B27: 326–333.
9 Burger A, Ramberger R. On the polymorphism of pharmaceuticals and other molecular crystals. I. Theory and thermodynamic rules. *Mikrochim Acta* 1979; II: 259–271.
10 Burger A, Ramberger R. On the polymorphism of pharmaceuticals and other molecular crystals. II. Applicability of thermodynamic rules. *Mikrochim Acta* 1979; 72: 273–316.
11 Kuhnert-Branbdstatter M. *Thermomicroscopy in the Analysis of Pharmaceuticals*. New York: Pergamon; 1971.
12 Huang KS, Britton D, Byrn SR. A novel class of phenol–pyridine co-crystals for second harmonic generation. *J Mater Chem* 1997; 7: 713–720.
13 Atassi F. *General Trends in the Desolvation Behavior of Calcium salts and the Characterization of Mesomorphous Calcium Salts*. Ph.D. thesis. West Lafayette, IN: Department of Industrial and Physical Pharmacy. Purdue University; 2007.
14 Shuttleworth R. The surface tension of solids. *Proc Phys Soc Sec A* 1950; 63: 444–456.
15 Zisman WA. Relation of the equilibrium contact angle to liquid and solid constitution. In: Fowkes FM, ed. *Contact Angle, Wettability, and Adhesion*. Washington, DC: Advances in Chemistry, American Chemical Society, 1964: 1–51.
16 Newman A, Zografi G. Perspectives on the wetting of solids in pharmaceutical systems. *Pharm Res* 2023; 40(12): 3099–3118. https://doi.org/10.1007/s11095-023-03491-3
17 Byrn S, Pfeiffer R, Ganey M, Hoiberg C, Poochikian G. Pharmaceutical solids: a strategic approach to regulatory considerations. *Pharm Res* 1995; 12: 945–954.
18 ICH Q6A. Test procedure and acceptance criteria for new drug substances and new drug products: chemical substances (1999). Retrieved from https://database.ich.org/sites/default/files/Q6A%20Guideline.pdf
19 Endicott KM. The national cancer chemotherapy program. *J Chronic Dis* 1958; 8(1): 171–177.
20 Pisano GP. *The Development Factory: Unlocking the Potential of Process Innovation*. Brighton: Harvard Business Press; 1997.
21 Engers D, Teng J, Jimenez-Novoa J, Gent P, Hossack S, Campbell C, Thomson J, Ivanisevic I, Templeton A, Byrn S, Newman A. A solid state approach to enable early development of compounds: Selection and animal bioavailability studies of an itraconazole solid dispersion. *J Pharm Sci* 2010; 99: 3901–3922.
22 Byrn S, Zografi G, Chen X. Accelerating proof of concept for small molecule drugs using solid-state chemistry. *J Pharm Sci* 2011; 99: 3665–3675.
23 Byrn S, Henck JO. Molecular delivery system in preclinical timeframe. *Drug Discov Today* 2007; 12: 189–196.
24 Longman R. Lilly's chorus experiment. *In-Vivo* 2007; 25: 1–5.
25 Owens PK, Raddad E, Miller JW, Stille JR, Olovich KG, Smith NV, Jones RS, Scherer JC. A decade of innovation in pharmaceutical R&D: the Chorus model. *Nat Rev Drug Discov* 2015; 14(1): 17–28.
26 Chatterjee B, Steiner R, Kaul G. Industry perspective–what does industry need to accelerate drug product and process development? *Pharm Res* 2023; 11: 1–5.
27 Palucki M, Higgins JD, Kwong E, Templeton AC. Strategies at the interface of drug discovery and development: early optimization of the solid state phase and preclinical toxicology formulation for potential drug candidates. *J Med Chem* 2010; 53: 5897–5905.
28 Woodbury-Harris KM, Coull BM. Adaptive clinical trials. *Clin Trials Neruosci* 2009; 25: 71–75.
29 Miller LG, Fincher JH. Influence of drug particle size after intramuscular dosage of phenobarbital to dogs. *J Pham Sci* 1971; 60: 1733–1736.
30 Brouwers J, Brewster ME, Ausustijns P. Supersaturating drug delivery systems: the answer to solubility-limited oral bioavailability? *J Pharm Sci* 2009; 98: 2549–2572.
31 Shefter E, Higuchi T. Dissolution behavior of crystalline solvated and nonsolvated forms of pharmaceuticals. *J Pham Sci* 1963; 52: 781–791.
32 Nelson E. Solution rate of theophylline salts and effects from oral administration. *J Am Pharm Assoc Sci Ed* 1957; 46: 607–614.

33 Newman A, Knipp G, Zografi G. Assessing the performance of amorphous solid dispersions. *J Pharm Sci* 2011; 101: 1355–1377.

34 Law D, Schmitt EA, Marsh KC, Everitt EA, Wang W, Fort JJ, Krill SL, Qiu Y. Ritonavir-PEG 8000 amorphous solid dispersions: In-vitro and in-vivo evaluations. *J Pharm Sci* 2004; 93: 563–570.

35 McNamara DP, Childs SL, Giordano J, Iarriccio A, Cassidy J, Shet MS, Mannion R, O'Donnell E, Park A. Use of a glutaric acid cocrystal to improve oral bioavailability of a low solubility API. *Pharm Res* 2006; 23: 1888–1897.

36 Childs SL1, Chyall LJ, Dunlap JT, Smolenskaya VN, Stahly BC, Stahly GP. Crystal engineering approach to forming cocrystals of amine hydrochlorides with organic acids. Molecular complexes of fluoxetine hydrochloride with benzoic, succinic, and fumaric acids. *J Am Chem Soc* 2004; 126: 13335–133342.

37 Jörgensen AM, Wibel R, Bernkop-Schnürch A. Biodegradable cationic and ionizable cationic lipids: a roadmap for safer pharmaceutical excipients. *Small* 2023; 19(17): 2206968.

38 Brito LA, Kommareddy S, Maione D, Uematsu Y, Giovani C, Scorza FB, Otten GR, Yu D, Mandl CW, Mason PW, Dormitzer PR. Self-amplifying mRNA vaccines. *Adv Genet* 2015; 89: 179–233.

39 Mehnert W, Mäder K. Solid lipid nanoparticles: production, characterization and applications. *Adv Drug Deliv Rev* 2012; 64: 83–101.

40 Duan Y, Dhar A, Patel C, Khimani M, Neogi S, Sharma P, Kumar NS, Vekariya RL. A brief review on solid lipid nanoparticles: part and parcel of contemporary drug delivery systems. *RSC Adv* 2020; 10(45): 26777–26791.

41 Scioli Montoto S, Muraca G, Ruiz ME. Solid lipid nanoparticles for drug delivery: pharmacological and biopharmaceutical aspects. *Front Mol Biosci* 2020; 7: 319.

42 Elbrink K, Van Hees S, Holm R, Kiekens F. Optimization of the different phases of the freeze-drying process of solid lipid nanoparticles using experimental designs. *Int J Pharm* 2023; 635: 122717.

43 Santonocito D, Sarpietro MG, Castelli F, Lauro MR, Torrisi C, Russo S, Puglia C. Development of solid lipid nanoparticles as dry powder: characterization and formulation considerations. *Molecules* 2023; 28(4): 1545.

44 Wang JL, Hanafy MS, Xu H, Leal J, Zhai Y, Ghosh D, Williams III RO, Smyth HD, Cui Z. Aerosolizable siRNA-encapsulated solid lipid nanoparticles prepared by thin-film freeze-drying for potential pulmonary delivery. *Int J Pharm* 2021; 596: 120215.

45 Steiner D, Schumann LV, Bunjes H. Processing of lipid nanodispersions into solid powders by spray drying. *Pharmaceutics* 2022; 14(11): 2464.

46 Zimmermann CM, Baldassi D, Chan K, Adams NB, Neumann A, Porras-Gonzalez DL, Wei X, Kneidinger N, Stoleriu MG, Burgstaller G, Witzigmann D. Spray drying siRNA-lipid nanoparticles for dry powder pulmonary delivery. *J Control Release* 2022; 351: 137–150.

47 Higuchi WI, Lau PK, Higuchi T. Polymorphism and drug availability. Solubility relations in the methylprednisolone system. *J Pharm Sci* 1963; 52: 150–153.

48 Kitaigorodskii AI, Mnyukh YV, Asadov YG. Relations for single-crystal growth during polymorphic transformation. *J Phys Chem Solids* 1965; 26: 463–472.

49 Coppens P, Schmidt GMJ. The crystal structure of the metastable (beta) modification of p-nitrophenol. *Acta Cryst* 1965; 18: 654–663.

50 Otsuka M, Matsumoto T, Kaneniwa N. Effects of the mechanical energy of multi-tableting compression on the polymorphic transformations of chlorpropamide. *J Pharm Pharmacol* 1989; 41(10): 665–669.

51 Takahashi Y, Nakashima K, Ishihara T, Nakagawa H, Sugimoto I. Polymorphism of fostedil: characterization and polymorphic change by mechanical treatments. *Drug Dev Ind Pharm* 1985; 11: 1543–1563.

52 Kaneniwa N, Otsuka M. Effects of grinding on the transformation of polymorphs of chloramphencol palmitate. *Chem Pharm Bull* 1985; 33: 1660–1668.

53 Chan HK, Doelker E. Polymorphic transformation of some drugs under compression. *Drug Dev Ind Pharm* 1985; 11: 315–322.

54 Zografi GR, Hollenbeck G, Laughlin SM, Pikal MJ, Schwartz JP, Campen LV. Report of the advisory panel on moisture specifications. *Pharmacopieal Forum* 1991; 17: 1459–1474.

55 Osawa T, Kamaty MS, DeLuca PP. Hygroscopicity of cefazolin sodium: application to evaluate the crystallinity of freezer-dried products. *Pharm Res* 1988; 7: 421–425.

56 Morawetz H. Polymerization in the solid state. *J Polymer Sci* 1966; 12: 79–88.

57 Paul I, Curtin DY. Thermally induced organic reactions in the solid state. *Acc Chem Res* 1973; 6: 217–225.

58 Weissbuch I, Kuzmenko I, Vaida M, Zait S, Leiserowitz L, Lahav M. Twinned crystals of enantiomorphous morphology of racemic alanine induced by optically resolved. α-amino acids; a stereochemical probe for the early stages of crystal nucleation. *Chem Mater* 1994; 6(8): 1258–1268.

59 Miller RS, Curtin DY, Paul IC. Reaction of ammonia gas withy substituted benzoic anhydrides. *J Am Chem Soc* 1972; 94: 5117–5119.

60 Liversidge GO, Cundy KC. Particle size reduction for improvement of oral bioavailability of hydrophobic drugs: I. Absolute oral bioavailability of nanocrystalline danazol in beagle dogs. *Int J Pharm* 1995; 125: 91–97.

2

Polymorphs

2.1 Introduction

A drug molecule may form multiple crystalline phases by itself, called polymorphs. Polymorphs have the same chemical composition but different internal structures. The internal structures could differ in unit cell dimensions, crystal packing, and/or molecular conformation. Polymorphs also exist for solvates, hydrates, salts, and cocrystals. The ability of a chemical entity to form polymorphs is denoted as polymorphism. It is well recognized that polymorphism is widely present in pharmaceutical materials, especially among drug substances, that is, active pharmaceutical ingredients (APIs). Henck et al. reported that about 30% of pharmaceutical active ingredients listed in the European Pharmacopoeia are polymorphic [1]. Stahly claimed that, based on the results of a 245-polymorph screening, about 50% of the organic compounds studied exhibit polymorphism [2]. Grunenberg disclosed that, from a survey of 62 drug compounds in Bayer AG, 80% of the drugs were polymorphic [3]. Thus, the well-known statement from Ostwald in 1899 [4], "Almost every substance can exist in two or more solid phases provided the experimental conditions are suitable," indeed, still stays somewhat representative for modern pharmaceutics, as to the prevalence of polymorphs.

It is important to note that differences in the internal structures of polymorphs would result in their distinct physical and chemical properties, such as solubility, dissolution, morphology, density, mechanical attributes, and stability. Consequently, different physical and chemical properties could lead to a significant impact on the performance and safety of an API and its formulated product(s). Thus, polymorph screening and polymorph selection are of paramount importance in drug discovery and development. Furthermore, for any new drug entity, it is important for us to understand the polymorphic landscape, to establish scientific rationales for selecting a proper form and to develop a quality-focused strategy in polymorph control, as required by worldwide regulatory agencies and as manifested by ICH Q6A decision trees discussed in Chapter 1.

However, polymorph control is a challenging task to fulfill. Formation of polymorphs cannot be effectively predicted so far and, therefore, must be physically screened. Even when extensively screened, a new polymorph may show up unexpectedly in late-stage development or commercial products. For example, in the 1990s, Abbott Laboratories had to withdraw ritonavir soft gel capsules from the market and reformulate the product due to the unexpected appearance of a more stable polymorph [5]. It is obvious that the incidence of such transformations poses a significant problem for public health and has brought a greater awareness of the need for polymorph control to the pharmaceutical industry. Some polymorphs also may "disappear." Some polymorphs can concomitantly grow together, posing additional challenges for control [6].

In this chapter, we will begin with various structural and thermodynamic principles underlying pharmaceutical polymorphism. Then we will move on to how various physical and chemical properties of an API are affected by polymorphs. Finally, we will focus on some practical aspects, including polymorph conversion, polymorph control, polymorph screening, and polymorph prediction.

2.2 How Are Polymorphs Formed?

The formation of polymorphs is primarily governed by the underlying thermodynamics of the system through the interplay of enthalpy and entropy along with environmental conditions. Polymorphs are different solid phases with distinct Gibbs free energy, enthalpy, and entropy. As illustrated in Figure 2.1 for a hypothetical compound, the polymorphs as crystalline

Solid-State Materials in Pharmaceutical Chemistry: Properties, Characterization, and Applications, Second Edition. Stephen R. Byrn, George Zografi and Xiaoming (Sean) Chen.

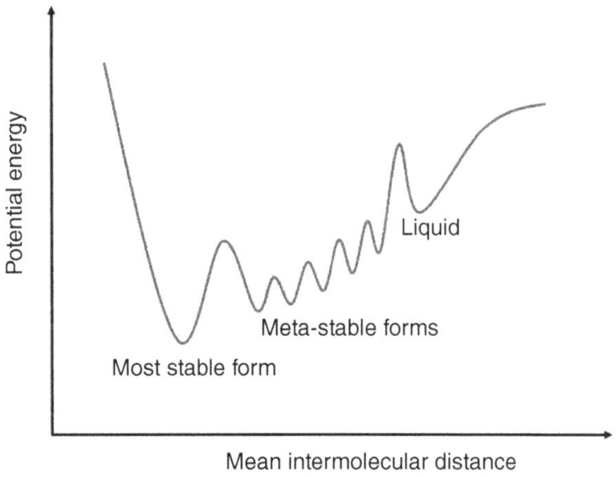

Figure 2.1 Illustration of the energy landscape of various polymorphs related to the liquid state.

materials have lower potential energy than the corresponding amorphous and liquid phases, owing to the ordered molecular structures, more efficient packing, and more robust intramolecular interactions through hydrogen bonding networks and van der Waals interactions. Among the polymorphs, under a specific condition, that is, temperature and pressure, there is a crystalline phase at the ground state, as the most thermodynamically stable form. Between the ground state and the liquid state, there are often multiple possible forms depending on the compound's intermediate energy states. Those energy states, indeed, act more like "valleys," instead of "steps," with "hills" to trap metastable forms kinetically at a certain condition with a certain time span.

The potential existence of a rich polymorphic landscape for many molecules is illustrated by notable cases such as the identification of ice with 17 crystalline phases [7] and the discovery of ROY (5-methyl-2-[(2-nitrophenyl) amino]-3-thiophenecabonitrile) with 13 polymorphs [8]. This phenomenon in terms of the prevalence of polymorphs was much appreciated by McCrone through his pioneering work in this field using hot stage microscopy [9], as reflected by his famous statement:

> It is at least this author's opinion that every compound has different polymorphic forms, and that, in general, the number of forms known for that compound is proportional to the time and money spent in research on that compound.

His insights on this matter, based on his real personal experience, regarded by many as being more intuitive than scientifically rigid, interestingly, are being revalidated by significantly advanced work in molecular simulation as to crystal structure prediction, that is, polymorph prediction. For example, Mortazavi and coworkers, based on the molecular structure of rotigotine, correctly predicted the crystal structure of two experimentally isolated polymorphs, Forms I and II, and their thermodynamic stability energy difference and order, by employing modern first-principles electronic-structure methods [10]. The predicted crystal energy landscape for rotigotine in Figure 2.2 shows that, within a relative lattice energy of 20 kJ/mol, there are about another 20 potential structures that can exist, in addition to the isolated Forms I and II. Utilizing similar first-principle electronic-structure methods for the case of ROY, Beran et al. correctly predicted the crystal structures of nine polymorphs out of 13 known structures, as well as their relative thermodynamic order, and revealed a potential additional 100 structures being present in the energy landscape within relative lattice energy of 10 kJ/mol [11]. In summary, the prevalence of polymorphs is not only thermodynamically sound but also kinetically feasible, as discussed further in the following section.

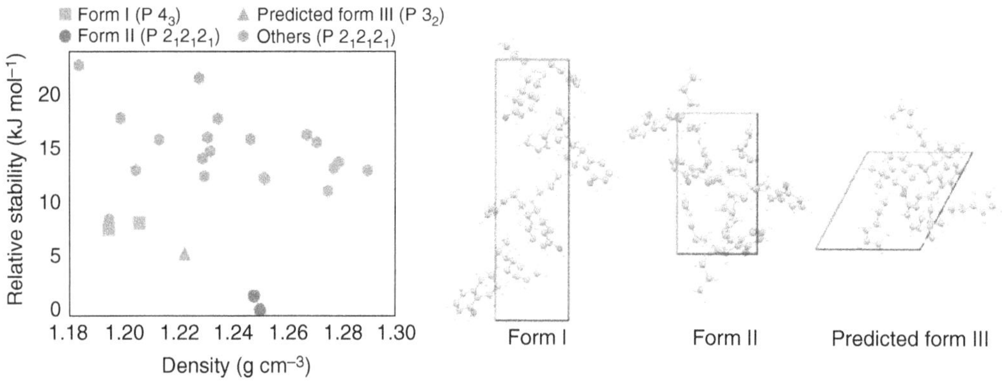

Figure 2.2 Energy-density diagram for the top 25 computer-generated polymorphs for rotigotine and unit cells of the two known forms, I and II, and the predicted form III. *Source*: Mortazavi et al. [10]/Springer Nature/CC BY 4.

In some circumstances, metastable phases or forms are kinetically favored over the most thermodynamically stable form, as described by Ostwald from many observations on the crystallization of minerals in the well-known Ostwald's rule of stages [12]. Ostwald stated that, under a given condition and when a system tends to leave any state driven by thermodynamics, the transition leans to the nearest stable state that is metastable and not the most stable one. The scientific rationales put forth by Ostwald to explain his observations that metastable phases typically appear first and eventually are replaced by increasingly stable phases are based on the hypothesis that metastable phases are closer to the solution thermodynamically and easier to form a crystal that requires less activation energy for nucleation, which aligns with the classical nucleation theory [13, 14].

In the theory, the rate of homogeneous nucleation, J, is expressed as an Arrhenius-type equation:

$$J = Ae^{-\frac{\Delta G}{KT}} \tag{2.1}$$

where A is the kinetic parameter reflecting the ease/difficulty of molecules assembling into the crystal lattice, ΔG the activation energy for nucleation, k the Boltzmann constant, and T the temperature.

Taking into consideration other factors, the equation becomes:

$$J = Ae^{-\frac{16\pi\gamma^3\upsilon^2}{3k^3T^3(\ln S)^2}} \tag{2.2}$$

where γ is the interfacial tension, S the degree of supersaturation, and υ the molecular volume. Generally, different polymorphs have distinct values of A and γ. Especially, γ, as the extra energy required to create the crystal/solution interface, can play a dominating role. It is hypothesized that the most stable form has a greater energy mismatch with the surrounding solution and therefore has a higher interfacial tension, while the least stable form exhibits the lowest interfacial tension and more favorable kinetics to nucleate, explaining Ostwald's rule of stages. Furthermore, solvent–solute interactions, the presence of impurities, temperature, and pressure have significant effects on each form and likely different effects on the interfacial tension, which further complicate the kinetic differences between different forms. In summary, the phenomenon of polymorphism, that is, the complexity and diversity of crystalline solid states for an organic molecule, is the interplay of thermodynamics and kinetics, which involves the various competing kinetics of crystal nucleation, crystal growth, and solution-mediated transformation, as influenced by environmental factors. The diversity of solid-state phases is obviously attributed to the complexity of various conformations an organic molecule can take, different variety of molecular interactions may exist, and various ways of a molecule can pack into an ordered crystal lattice as an ordered supramolecular structure, as further discussed from the structural aspects how polymorphs can easily form and exist for many organic molecules.

2.3 Structural Aspect of Polymorphs

An understanding of the structural aspects of polymorphs starts with some type of classification system. Even though considered somewhat arbitrary, such classification offers a unique perspective on how polymorphs are derived. Vippagunta et al. have classified polymorphs as configurational or conformational, based on two distinct structural aspects [15].

2.3.1 Configurational Polymorphs

Configurational polymorphs usually occur with molecules that have a relatively rigid conformation. Such polymorphs take on a similar or identical molecular conformation but have different three-dimensional structures. Each structure possesses a unique packing motif and a distinct manner of molecular interaction. This mechanism is termed "packing polymorphism" or "configurational polymorphism."

As an example, the drug carbamazepine shows configurational polymorphism. It has a rigid chemical structure and forms at least four polymorphs [16–21]. Form I packs into a triclinic crystal system [21]. Form II is in a trigonal unit cell [19]. Form III, the stable polymorph at room temperature, has a primitive monoclinic unit cell [16–18]. Form IV has a c-centered monoclinic structure [20]. Readers are referred to Chapter 8 for more details on the classification of crystal systems with respect to unit cells.

For all four polymorphs, carbamazepine molecules have a nearly identical molecular conformation and form strong hydrogen bonding motifs, with all polymorphs taking on the motif of anti-carboxamide dimers. However, the packing

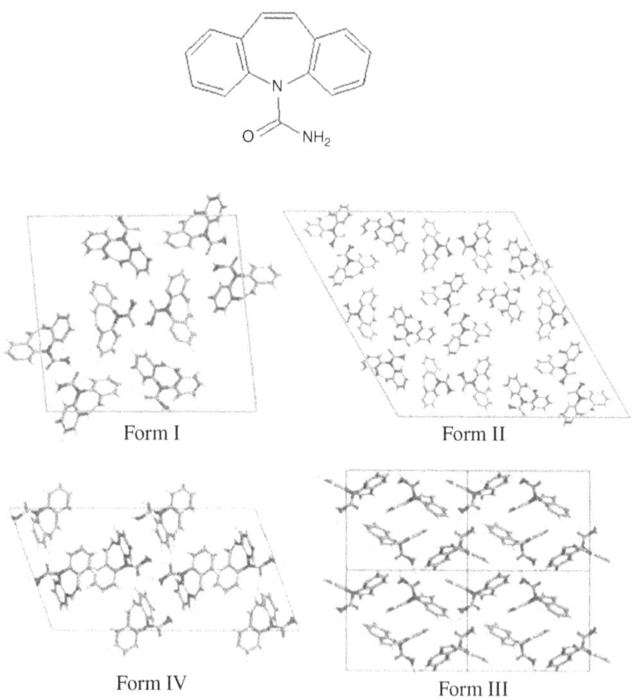

Figure 2.3 Chemical structure of carbamazepine and crystal-packing diagram of carbamazepine Forms I, II, III, and IV. *Source*: Rodríguez-Spong et al. [35]. Reproduced with the permission of Elsevier.

of the dimer units is distinct among polymorphs as shown in Figure 2.3. Form III has an interlocked crystal packing and shows the highest density (1.34 g/mL). Forms III and IV have infinitely long hydrogen bond chains through a vinylic hydrogen of the azepine ring and an oxygen acceptor. The C—H···O interactions that link the dimer units in Forms III and IV have different lengths, with 2.48 Å for Form III and 2.28 Å for Form IV. The hydrogen bond networks in Forms I and II are similar but distinct from Forms III and IV. In Forms I and II, each oxygen atom forms hydrogen bonds with two nearby carbon atoms. One carbon is ortho to the urea from an identical residue of an adjacent unit cell, and the other is para to the urea from the nearest donor molecule. A hydrogen bond chain is formed and spirals along the needle axis of the crystals. In summary, distinctly different packing and hydrogen bond networks are present in carbamazepine polymorphs, despite an unchanged conformation. The different packing causes these polymorphs to have different internal energies and stabilities. At room temperature, the stability order is Form III > Form I > Form IV > Form II [21].

2.3.2 Conformational Polymorphs

Conformational polymorphs commonly occur with drug molecules that have flexible structures. The molecule can take on various conformations and is subsequently packed into alternative crystal phases.

Ritonavir, an antiviral drug, is an example of a molecule exhibiting conformational polymorphism [5]. Only one polymorph, Form I, was discovered during clinical and commercial development, whereas 2 years after the product launch, a new crystal form, Form II, was identified inside the semisolid capsules of various lots of the drug product. Those lots of the drug product failed the dissolution test because the new crystal form, Form II, had much lower aqueous solubility than that of Form I. As shown in Figure 2.4, Form I and Form II display distinct differences in morphology. Form I has a plate-like shape, and Form II has a needle-like morphology. Later, Yao et al. and Parent et al. independently discovered a new polymorph [22, 23], Form III, via melt/cool crystallization, as fine needles (Figure 2.4).

The three polymorphs of ritonavir are conformational polymorphs because they exhibit distinctly different conformations, as shown in Figure 2.4. The "cis" conformation around the carbamate linkage is present in Form I and Form III, while the "trans" conformation exists for Form II. On the other hand, in terms of the *N*-methyl urea group, the "cis" conformation occurs in Form I, whereas the "trans" conformation occurs in Form II and Form III. The conformational differences

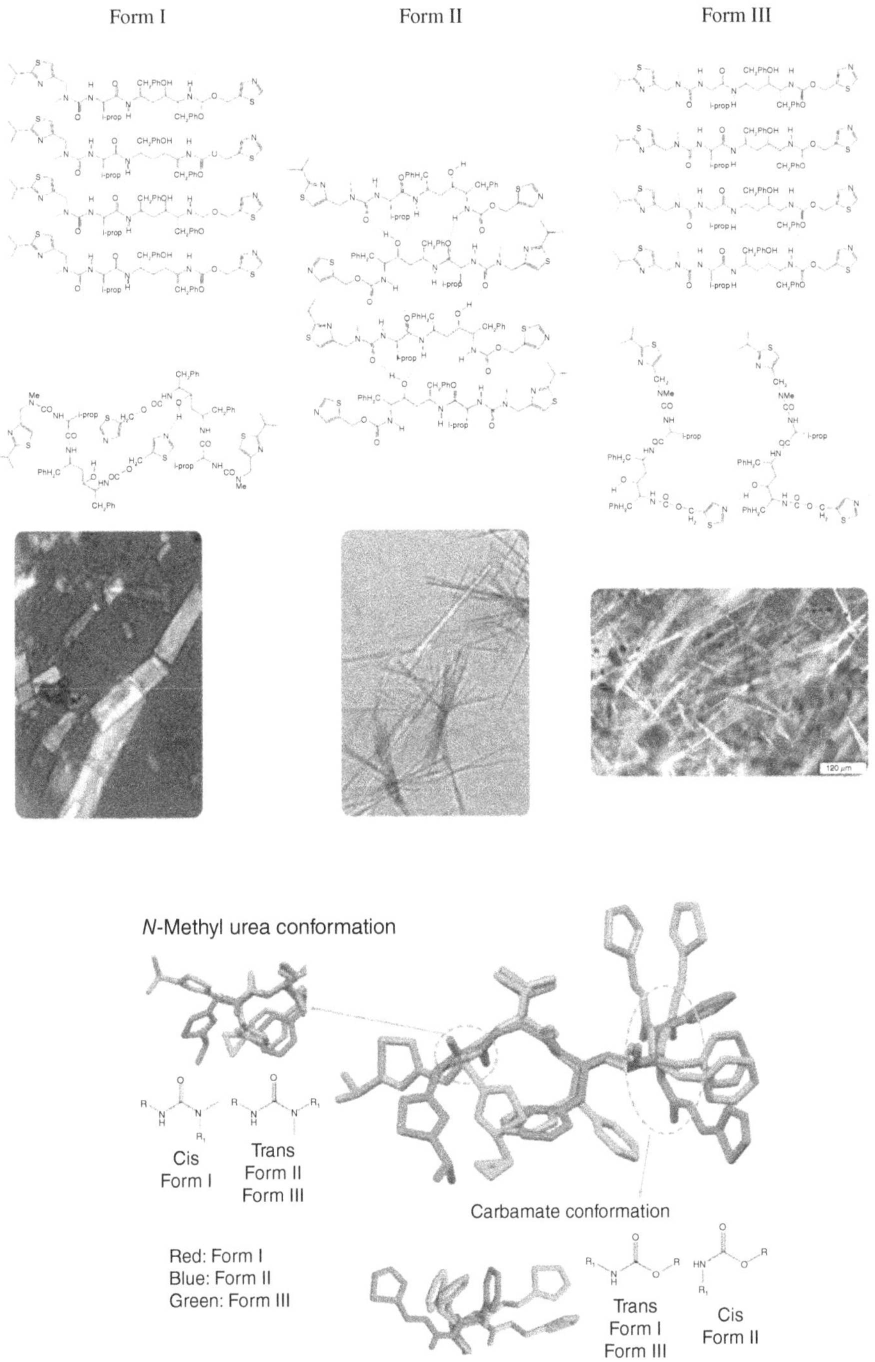

Figure 2.4 Hydrogen bonding networks, microscopic images, and molecular conformations of ritonavir polymorphs. *Source*: Bauer et al. [5] and Yao et al. [22]. Reproduced with permission of Springer and Elsevier.

are also accompanied by very different crystal packing. As shown in Table 2.1, Form I exhibits a $P2_1$ space group; Form II shows a $P2_12_12_1$ space group; and Form III exists in the C2 space group. (Chapter 8 gives an introduction to these different space groups.) Forms I, II, and III also differ in their hydrogen bonding networks, as shown in Figure 2.4. In Form I, the amide linkages of one molecule form hydrogen bonds with the same amide in the next molecule, producing a continuous beta-like layer. These layers are linked by hydrogen bonds between the hydroxyl group of one molecule and the thiazole ring of another molecule in the next layer. In Form II, the hydroxyl group acts as both a hydrogen bond donor and acceptor, by forming bonds with the amide group and keto group of a neighboring molecule. There is also another hydrogen bonding

Table 2.1 Crystallographic information of ritonavir polymorphs.

Parameter	Form I[a]	Form II[a]	Form III[b]
Crystal system	Monoclinic	Orthorhombic	Monoclinic
Space group	P21	P21P21P21	C2
Z value	2	4	4
Density, calculated	1.28 g/cm^3	1.25 g/cm^3	1.33 g/cm^3
Lattice parameters	$a = 13.433$ (1) Å	$a = 10.0236$ (3) Å	$a = 23.3307$ (19) Å
	$b = 5.293$ (2) Å	$b = 18.6744$ (4) Å	$b = 4.9511$ (5) Å
	$c = 27.092$ (4) Å	$c = 20.4692$ (7) Å	$c = 33.638$ (3) Å
	$\beta = 103.102$ (9) Å		$\beta = 91.2745$ (6) Å
	$V = 1876.0$ (8) Å^3	$V = 3831.5$ (2) Å^3	$V = 3884.6$ (6) Å^3

a. Single-crystal diffraction at ambient temperature.
b. Single-crystal diffraction at −173 °C.

Table 2.2 Physical properties of ritonavir polymorphs.

Polymorph	Thermodynamic stability	T_m (°C)	ΔH_m (J/g)	Solubility in ethanol/water (99/1) at 5 °C (mg/mL)
Form II	Stable form	121.5 ± 0.0	91.2 ± 1.4	19
Form I	Less stable	120.8 ± 0.0	74.8 ± 1.2	90
Form III	Least stable	114.6 ± 0.2	53.1 ± 1.0	695–744

dimer between neighboring molecules in Form II, as shown in Figure 2.4. Form III has a similar continuous beta-like layer through hydrogen bonding between neighboring amide bonds, however, with a different way to link layers.

In summary, we can classify polymorphs as configurational polymorphs or conformational polymorphs based on structural aspects. However, the distinction between configuration polymorphism and conformational polymorphism is not clear-cut. Certain packing motifs could impose different conformations on the molecules. On the other hand, different conformations may cause distinctly different packing. Packing and conformation are also closely interrelated with molecular interactions, especially hydrogen bonds. The variation in conformation, packing, and hydrogen bonds, therefore, contributes importantly to different crystal structures and properties. As in the case of ritonavir, three polymorphs display dramatic differences in solubility (Table 2.2) and other physical attributes, therefore, affecting the physical stability as a solid form and its drug product performance as an API [5, 22, 23].

2.4 Physical, Chemical, and Mechanical Properties

The structural differences of polymorphs lead to distinct physical, chemical, and mechanical properties. Some properties that can be impacted by polymorphs are listed in Table 2.3. For example, ritonavir Forms I, II, and III have a density of 1.28, 1.25, and 1.35 g/cm^3, respectively [5, 22]. The density of carbamazepine Forms I, II, III, and IV is 1.31, 1.24, 1.34, and 1.27 g/cm^3, respectively [21].

A polymorph may display a different refractive index due to unique interactions of electronic vibrations with light quanta, distinct thermal conductivity due to unique interactions of molecular motion with heat quanta, and distinct electrical

Table 2.3 Physical properties that may vary for polymorphs.

Packing properties	Molar volume and density
	Refractive index
	Conductivity, electrical, and thermal
	Hygroscopicity
Thermodynamic properties	Melting and sublimation temperatures
	Internal energy (i.e. structural energy)
	Enthalpy (i.e. heat content)
	Heat capacity
	Entropy
	Free energy and chemical potential
	Thermodynamic activity
	Vapor pressure
	Solubility
Spectroscopic properties	Electronic transition (i.e. ultra-visible absorption spectra)
	Vibrational transitions (i.e. infrared absorption spectra and Raman spectra)
	Rotational transitions (i.e. far infrared or microwave absorption spectra)
	Nuclear spin transitions (i.e. nuclear magnetic resonance spectra)
Kinetic properties	Dissolution rate
	Rates of solid-state reactions
	Stability
Surface properties	Surface free energy
	Interfacial tensions
	Habit (i.e. shape)
Mechanical properties	Hardness
	Tensile strength
	Compressibility and tableting
	Handling, flow, and blending

conductivity due to unique movement of the electrons in an electric field. For example, the α-form of sulfanilamide has a lower refractive index than the β-form [24]. Differences in the thermal conductivity of cimetidine forms A and B were observed using scanning thermal microscopy [25]. Changes in electrical conductivity were observed to accompany the polymorphic transitions of potassium nitrate at 1 atmosphere [26].

The most important difference between polymorphs of a compound lies in their varying thermodynamic activities. Distinct molecular interactions in crystal lattices result in unequal internal energy, E, and enthalpy, H. Diverse thermal motion and various spatial locations of molecules contribute to different entropy (S). This means that the free energy $(G = H - TS)$ and the chemical potential, as the partial molar free energy, will be different. Solubility is directly related to such differences in thermodynamic activity.

2.4.1 Solubility

The solubility differences arising between polymorphs will vary with the chemical structure of a particular compound. Pudipeddi and Serajuddin have reviewed a large number of literature reports on solubility or dissolution of polymorphs [27]. The solubility ratios of the polymorphs of 55 compounds were studied, and 81 solubility ratios were reported due to the existence of multiple forms for some compounds. For a compound having several polymorphs, the solubility ratio of each form was presented relative to the least soluble form. The general trend observed was that the ratio of polymorph solubilities is typically less than 2, although, for some cases, there can be a three to five times difference. Only for one case reported, premafloxacin, the solubility ratio in ethyl acetate was shown to be more than 20-fold. Overall, the average solubility ratio for the polymorphs surveyed is 1.7, excluding premafloxacin, and 2.0 with its inclusion.

2.4.2 Chemical Stability

The structural differences between polymorphs of the same drug may be expected to lead to different chemical stability. It has been reported by Chen et al., for example, that two polymorphs of indomethacin, α- and γ-forms, react with ammonia gas at dramatically different rates (Figure 2.5) [28]. The metastable crystal form of indomethacin, the α-form, reacts readily with ammonia gas to produce the corresponding microcrystalline ammonium salt. This reaction is anisotropic and spreads along the *a*-axis of the crystals. The stable crystal form, the γ-form, however, is not reactive to ammonia gas. The reactivity differences between the α- and γ-forms are proposed to be dictated by the arrangement of the molecules within the respective crystal lattices. The carboxylic acid groups of the α-form are open on the {100} faces and are accessible to attack by ammonia gas. The reaction process proceeded along the *a*-axis until the ammonia gas reached the entire crystal. In contrast, the γ-form has a centrosymmetric crystal structure in which the hydrogen-bonded carboxylic acid dimers are not accessible to ammonia gas because they are caged inside a hydrophobic shield comprising the remainder of the indomethacin molecule.

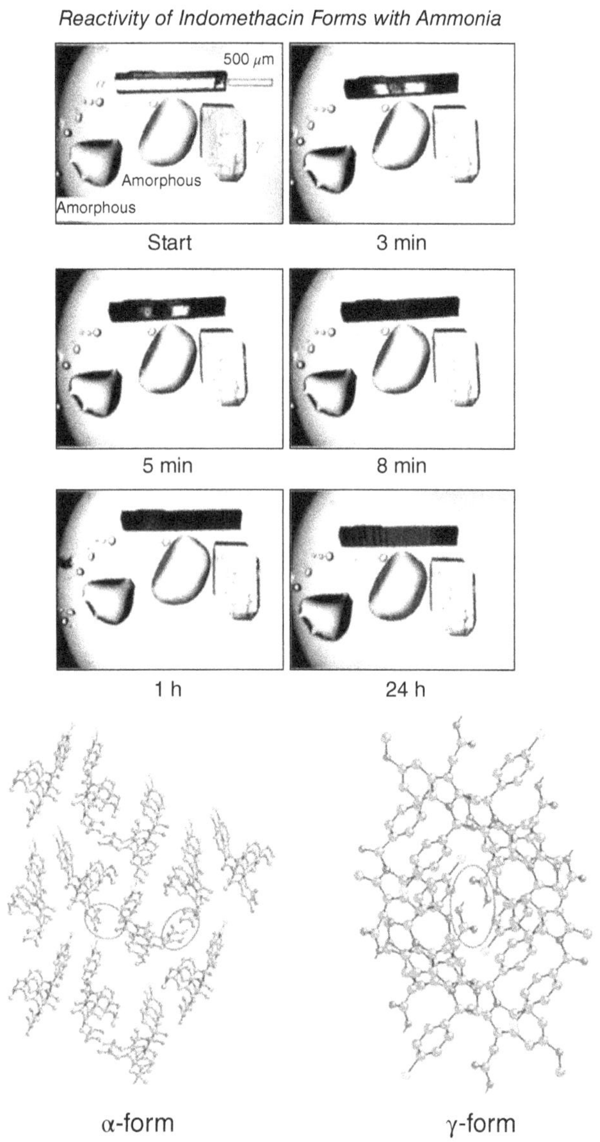

Figure 2.5 Reaction of amorphous form, α-form and γ-form of indomethacin with ammonia. The crystal structures of α-form and γ-form are shown at the bottom. *Source*: Chen et al. [28]. Reproduced with the permission of the American Chemical Society.

2.4.3 Mechanical Properties

Polymorphs might be expected to show differences in mechanical properties such as compressibility, tensile strength, and Young's modulus.

For example, two polymorphs of acetaminophen, Forms I and II, display differences in compressibility related to their distinct crystal structure.

Form I crystallizes in the monoclinic system [29, 30], while Form II has a structure in the space group Pcab [31, 32]. It appears that Forms I and II take different crystal packing (see Figure 2.6).

The molecules in Form I assemble as pleated sheets parallel to the (101) plane [29, 30]. The sheets are stacked along the *b*-axis, and the molecules within a sheet are held together by two types of hydrogen bonds, OH donating to O=C and HO accepting HN. The molecules pack in a head-to-tail configuration in the same direction within the sheet. Van der Waals forces appear to provide the interaction of the molecules between sheets. Interestingly, the molecules on a neighboring sheet orient in the opposite direction and achieve a tighter packing.

A similar hydrogen bond network is present in Form II despite very different packing [31, 32]. The crystal structure of Form II has relatively flat hydrogen-bonded sheets stacked along the *c*-axis, whereas molecules within the sheet pack inversely along the *b*-axis. On the other hand, neighboring sheet stacking occurring down the *c*-axis has nearly identical two-dimensional packing with a shift of one unit cell relative to each other.

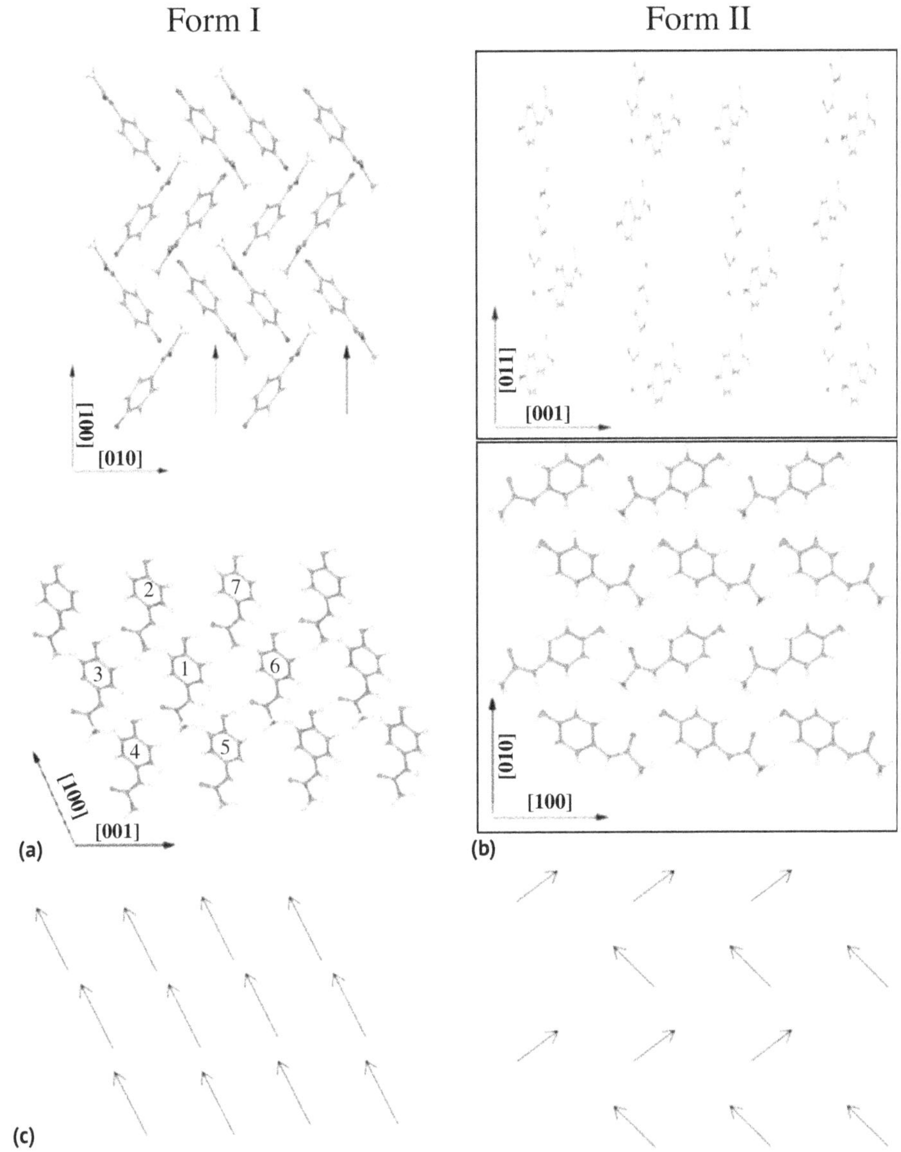

Figure 2.6 Crystal-packing diagram of the two polymorphs of acetaminophen, Forms I and II. *Source*: Espeau et al. [30]. Reproduced with the permission of Elsevier. Haisa et al. [31]. Reproduced with the permission of Wiley-Blackwell.

Form I has poor compressibility because it has a tighter and interlocked packing and lacks a slip plane [32], whereas Form II has a layered packing that provides slip planes, which allow plastic deformation to occur [32]. Unlike Form I, Form II is directly compressible.

2.5 Thermodynamic Stability of Polymorphs

The relative stability of polymorphs is determined by their relative Gibbs free energy, with the more stable form having the lower free energy. With the exception of the transition point, only one polymorph has the lowest free energy at a defined environmental pressure and temperature, and this polymorph is regarded as the thermodynamically stable polymorph. All other polymorphs are metastable forms. A metastable form is thermodynamically unstable but has a finite existence due to the slow kinetics of transformation. Some metastable polymorphs can be kinetically stable for years, whereas other metastable forms exist transiently and are difficult to prepare and store.

2.5.1 Monotropy and Enantiotropy

For any pair of polymorphs, there are two types of stability relationships possible: monotropy and enantiotropy. For a monotropic pair, the relative stability will stay the same below the melting point. As illustrated in Figure 2.7, the G versus T diagrams for solid and liquid phases have negative derivatives based on the following equation:

$$\frac{dG}{dT} = -S \tag{2.3}$$

S is the entropy, always positive above absolute zero. The hypothetical Forms A and B have intersection points with the liquid phase. The temperature of the interaction point is the melting point of Forms A and B. At the melting temperature,

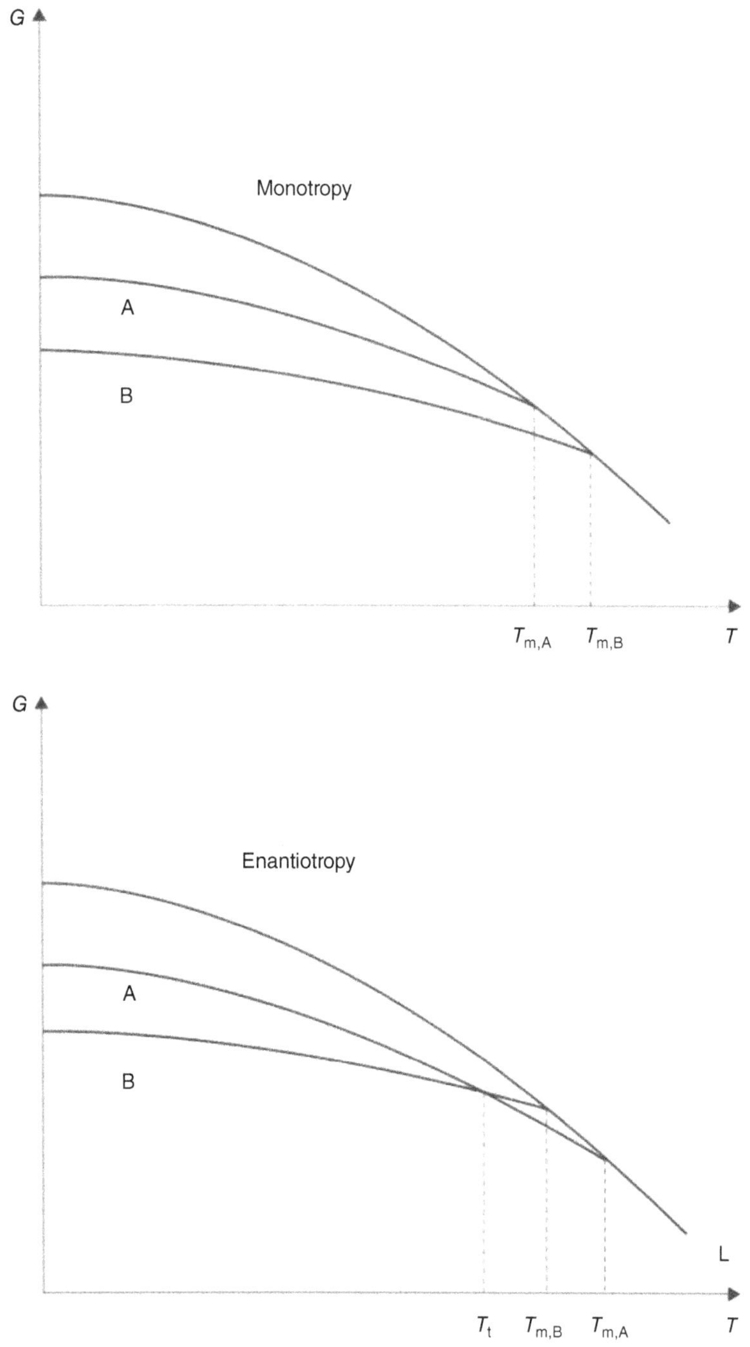

Figure 2.7 Gibbs free energy and temperature diagram for monotropic and enantiotropic pairs of polymorphs.

a specific crystal phase is in equilibrium with the liquid phase, and ΔG is zero. If the pair of polymorphs is monotropic, Form B is always the stable form because it has lower Gibbs free energy than Form A below the melting point; there is a thermodynamic tendency for Form A to change to Form B since the ΔG is less than zero. The transformation is often difficult to observe in the solid state due to a significant kinetic barrier, while it could occur relatively easily in a suspension by means of a solution-mediated mechanism.

An enantiotropic system has a transition point below the melting temperature so that the order of the stability changes above and below the transition temperature. In the bottom diagram of Figure 2.7, Forms A and B represent an enantiotropic pair with a transition temperature at T_t. Below the transition temperature, Form B is more stable. Above the transition temperature, Form A is more stable than Form B.

For a compound with three or more polymorphs, the relative stability becomes complicated. It demands significant resources to establish the full relationship.

We can utilize many qualitative or quantitative approaches to investigate the relative stability of polymorphs. Approaches include thermal analysis, solubility, and interconversion studies, which will be discussed in more detail in the following sections.

2.5.2 Burger and Ramberger's Rules

Burger and Ramberger have proposed several rules for assigning a given polymorphic pair as being either enantiotropic or monotropic [33]. These rules include the heat of transition rule, the heat of fusion rule, and the density rule. Based on the heat of transition rule, two polymorphs are related enantiotropically if an endothermic transition, usually weak, is observed at a certain temperature below the melting points of the two forms. Two polymorphs are related monotropically if an exothermic transition is present at a certain temperature, and no transition occurs at a higher temperature.

The heat of fusion rule is useful if a thermal transition cannot be observed due to the slowness of the transformation. This rule states that a polymorphic pair is monotropic if the higher melting polymorph has the higher heat of fusion. The two polymorphs are enantiotropically related if the higher melting form has the lower heat of fusion.

The density rule asserts that the relative stability of polymorphs is related to the efficiency of crystal packing and the true density. The most stable polymorph will have the highest density. For example, Form I of nabumetone, the stable form, has a density of 1.26 g/cm^3, whereas Form II, the metastable form, has a lower density of 1.21 g/cm^3 [34].

Burger–Ramberger rules have a statistical basis and are useful qualitatively. However, exceptions do exist for these rules. This is so because the density rule is usually applicable only if crystal packing is dominated by van der Waals interactions. However, the most stable form may have a lower density when hydrogen bonding is the prominent molecular interaction. For ritonavir, the stable form, Form II, has a lower density (1.25 g/cm^3) than Form I and Form III, the metastable forms (1.28 and 1.33 g/cm^3) [5, 22]. For another example, the γ-form of indomethacin, the stable form, displays a lower density (1.38 g/cm^3) than the α-form (1.40 g/cm^3), the metastable form [28].

2.5.3 van't Hoff Plot

The relative stability of polymorphs over a narrow temperature range can also be established through a van't Hoff plot, ln(s) versus $1/T$, where s is the equilibrium solubility.

The free energy difference between two polymorphs, A and B, is related to solubility:

$$\Delta G_{A,B} = \mathrm{RT}\ln\left(\frac{\mathrm{S_B}}{\mathrm{S_A}}\right) \tag{2.4}$$

where $\Delta G_{A,B}$ is the free energy difference between A and B; S_A and S_B are the solubilities of forms A and B, respectively. The rank order of solubility at a certain temperature is the same as the thermodynamic stability, and, therefore, the most stable form is the least soluble form.

The van't Hoff plot of ln(s) versus $1/T$ also provides a quantitative way to model the order of stability and possible transitions. The plot is based on the equation:

$$\ln(s) = -\frac{\Delta H}{RT} + c \tag{2.5}$$

where s is the solubility of a given polymorph at an absolute temperature T, R is the gas constant, and c is the constant of integration. The slope of such a plot can be used to calculate the heat of solution, ΔH, for the polymorph. If there is an

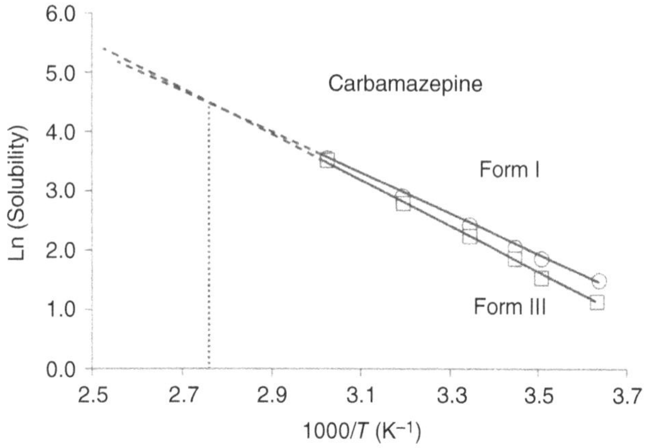

Figure 2.8 van't Hoff plot of carbamazepine Forms I and III. *Source*: Rodríguez-Spong et al. [35]. Reproduced with the permission of Elsevier.

intersection in the plot for two polymorphs and the temperature for the intersection is below the melting points, the two polymorphs are presumably enantiotropic. In contrast, a pair of polymorphs is monotropic if there is no intersection in such a plot or the intersection temperature is above the melting point.

The van't Hoff plots of carbamazepine Forms I and III are shown in Figure 2.8 using solubility data obtained in 2-propanol [35]. The plots of Forms I and III cross at 73 °C (346 K), which is below the melting temperature. The observation, therefore, indicates that Forms I and III are enantiotropic with a transition temperature of 73 °C. Form III has lower solubility than Form I below 73 °C and is more stable than Form I. The heats of solution in 2-propanol derived from the slopes of the lines are 31.54 kJ/mol for Form III and 28.01 kJ/mol for Form I, with the transition from Form III to Form I as endothermic. The heat of transition can be calculated from the difference between the heats of solution for the two forms and is 3.53 kJ/mol.

2.5.4 ΔG/temperature diagram

Yu has demonstrated that the ΔG/temperature diagram could also offer quantitative information about the relative stability of polymorphs [36]. The diagram can be constructed from calorimetric data and/or solubility dependence on temperature.

The calorimetric data used for ΔG/temperature diagram includes melting temperature, heat of fusion, and heat of transition. The first step of this approach is to calculate the free energy difference (ΔG) for two polymorphs at the melting point of the lower melting polymorph and the temperature slope of ΔG at the melting point. ΔG at other temperatures is estimated by extrapolation. For a hypothetical polymorph pair, A and B, A is proposed to be the form of lower melting with a melting point of $T_{m,A}$, and B is the higher melting form with a melting point of $T_{m,B}$. Forms A and B are either monotropic or enantiotropic. The transition of Form A to Form B at $T_{m,A}$ (Process 1) is thermodynamically equivalent to the melting of Form A at $T_{m,A}$, the temperature increase of the melt (liquid) from $T_{m,A}$ to $T_{m,B}$, the crystallization of B at $T_{m,B}$, and the temperature decrease of B from $T_{m,B}$ to $T_{m,A}$ (Process 2) (Figure 2.9).

Process 1: Form $A(T_{m,A}) \rightarrow$Form $B(T_{m,A})$
Process 2: Form $A(T_{m,A}) \rightarrow$Liquid $(T_{m,A}) \rightarrow$Liquid $(T_{m,B}) \rightarrow$Form $B(T_{m,B}) \rightarrow$Form $B(T_{m,A})$

Because the initial and final stages for processes 1 and 2 are the same, the enthalpy and entropy changes for process 1, ΔH_0 and ΔS_0, are equal to the combination of enthalpy and entropy at all stages of process 2, according to the following equations:

$$\Delta H_0 = \Delta H_{m,A} - \Delta H_{m,B} + (C_{p,L} - C_{p,B})(T_{m,B} - T_{m,A}) \tag{2.6}$$

$$\Delta S_0 = \frac{\Delta H_{m,A}}{T_{m,A}} - \frac{\Delta H_{m,B}}{T_{m,B}} + (C_{p,L} - C_{p,B})\ln\frac{T_{m,B}}{T_{m,A}} \tag{2.7}$$

In these equations, $\Delta H_{m,A}$ and $\Delta H_{m,B}$ are the heat of fusion of Forms A and B, respectively. $C_{p,B}$ and $C_{p,L}$ are the respective heat capacities of Form B and the super-cooled liquid at temperatures between $T_{m,A}$ and $T_{m,B}$. ΔG_0 is the free energy difference between Forms A and B at $T_{m,A}$ and is expressed by the following equation:

$$\Delta G_0 = \Delta H_0 - T_{m,A}\Delta S_0 \tag{2.8}$$

If the ΔG dependence on temperature is approximately linear, ΔG for the two polymorphs A and B at other temperatures can be calculated by:

$$\Delta G(T) = \Delta G_0 - \Delta S_0(T - T_{m,A}) \tag{2.9}$$

Thus, the ΔG–temperature diagram can be obtained for a polymorphic form using calorimetric data. Alternatively, ΔG-temperature diagrams can be constructed using solubility data as shown in equation (2.5).

Constructing ΔG and temperature plots is most useful in understanding complicated polymorphic systems such as ROY [37]. The chemical name for ROY is 5-methyl-2-[(2-nitrophenyl)amino]-3-thiophenecarbonitrile. Through decades of research, 13 polymorphs have been discovered experimentally. The name of ROY depicts the nature of color polymorphism demonstrated by this compound. The first six polymorphs can be prepared by crystallization from solution [37]. The six forms are Y (yellow prisms), R (red prisms), ON (orange needles), OP (orange plates), YN (yellow needles), and ORP (orange-red plates). The seventh polymorph, RPL (red plates), was prepared by crystallization from the vapor on succinic acid [38]. Two new polymorphs were identified later through non-solution approaches. Y04 (yellow, discovered in 2004) was obtained from melt crystallization, and YT04 was a product of solid-state transformation from Y04 [39].

Construction of a ΔG–temperature diagram provides a quantitative mapping for relative stabilities of various polymorphs [36, 39]. From the melting data, the free energy and entropy differences relative to the Y polymorph were calculated by using equations (2.6)–(2.9). On the other hand, eutectic melting of the polymorphs against common reference compounds was utilized to reduce the melting points. The data of eutectic melting give rise to the free energy difference ($G_A - G_B$) at eutectic melting temperatures. Figure 2.9 shows the ΔG–temperature diagrams of ROY polymorphs obtained from melting and eutectic melting data [36]. Each line represents the free energy of a polymorph relative to Y. The diagram offers a complete picture of the stability relationship of various polymorphs. For example, R is monotropic to Y, OP, and ON. The diagram shows the enantiotropic relationship for Y/ON and Y/OP. It also provides quantitative information such as the transition point, heat of transition, and entropy of transition. For example, the transition point for Y and ON is calculated at 70 °C with a heat of transition of 2.6 kJ/mol and an entropy of transition of 7.7 J/K/mol. It is clearly elucidated from the ΔG–temperature diagrams that Y is the stable form below 70 °C and ON is the stable form above 70 °C.

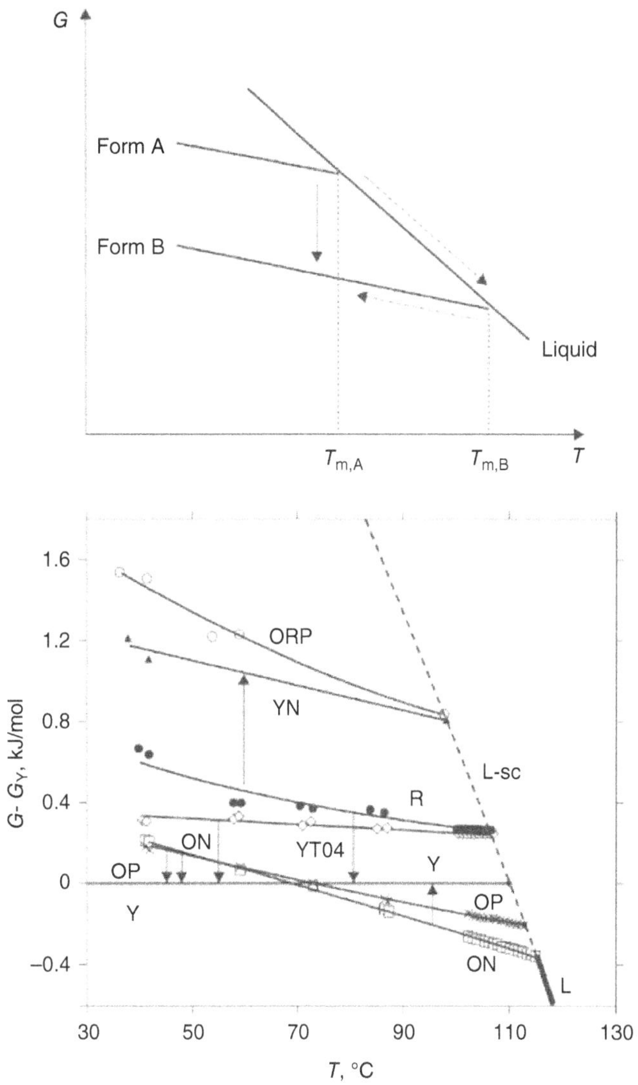

Figure 2.9 Energy temperature diagram for polymorphs. Top: Gibbs free energy and temperature diagram around the melting points of two polymorphs. Bottom: ΔG and temperature diagram for ROY polymorphs. *Source*: Adapted from [39].

2.6 Polymorph Conversion

The thermodynamic approach described earlier provides a map of the relative stability of polymorphs. It tells us if there is a driving force for a certain type of transformation. However, how the transformation occurs is determined by the kinetics of the process. Some metastable forms can be stable for years, whereas others are only present transiently. Therefore, understanding the kinetics pathway and the rate of interconversion is crucial for establishing polymorph control. Such kinetics would be expected to be very different in the solid state versus in solution.

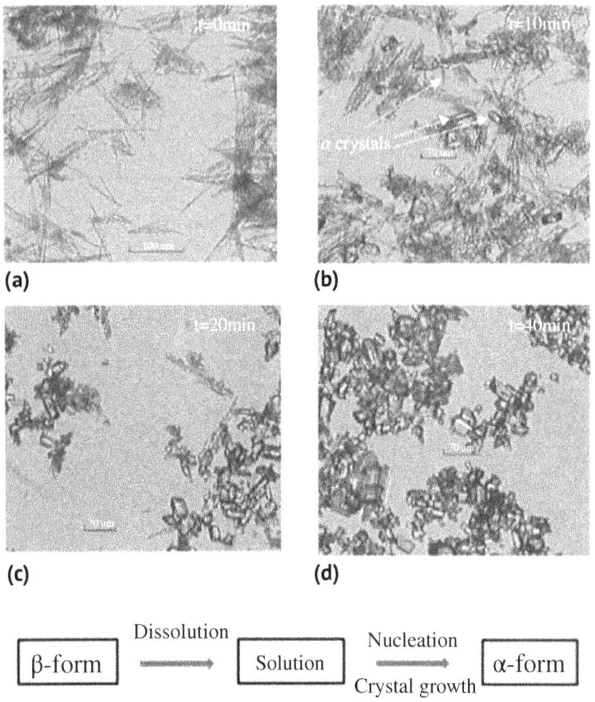

Figure 2.10 Microscopic images recorded during the polymorphic transformation of glycine polymorphs in 20:80 (v/v) water/ethanol mixture. *Source*: Ferrari et al. [40]. Reproduced with the permission of the American Chemical Society.

2.6.1 Solution-mediated Transformation

Solution-mediated transformation from one polymorph to another involves the dissolution of the metastable form and the crystallization of the stable form. In such a system the metastable form generally has a higher solubility than the corresponding stable form, and, therefore, the stable form will nucleate and grow at the expense of the metastable form.

Solution-mediated transformations can be divided into three steps: (i) dissolution of the metastable solid, (ii) nucleation of a more stable solid phase, and (iii) growth of the stable phase. Each step could be rate limiting, while crystallization is generally slower than dissolution. During crystallization, the nucleation rate is usually slower than the rate of crystal growth.

As an example, the transformation of the β form of glycine to the α form in a water/ethanol mixture (20/80 v/v) was studied by Ferrari et al. [40]. As shown in Figure 2.10, the two polymorphs have distinctly different morphologies and are easily differentiated under an optical microscope. The α form is chunk-like, and the β form exists as thin needles. After 10 minutes in the water/ethanol mixture, the crystals of the α form were observed among the thin needles of the β form. The majority of the β form disappeared after 20 minutes, and the transformation was completed after 40 minutes. It was observed further that the conversion rate of glycine polymorphs was related to solvent composition. When the concentration of ethanol in the solvent mixture was increased to 91%, the formation of the first few α crystals was delayed by 30 minutes, whereas the overall time to complete the transformation increased to 3 hours. This reduced conversion rate appears to be related to the decrease in the solubility of glycine in the solvent mixture with 91% ethanol. The solubility of the α polymorph in water/ethanol mixtures falls from about 1.8 g/L in an 80% ethanol solution to 0.43 g/L in a 91% solution.

The effect of solubility on the kinetics of such transformations was also observed for sulfamerazine polymorphs [41]. Here, solubility has an impact on both nucleation and growth of Form II from Form I in various solvents. Generally, the nucleation rate is faster in solvents that give higher solubility. For example, nucleation is extremely slow in solvents in which the solubility of Form II is less than 8 mM. It was hypothesized by Gu et al. that the metastable zone for this system is wider in the solvent, giving a lower solubility [41]. This means that a higher energy barrier for nucleation to occur, so that the metastable form is kinetically more stable in those solvents with lower solubility.

Besides solubility, solvent–solute interaction has a strong influence on the rate of nucleation and crystal growth [41]. Mixtures of methanol and water, containing 10 or 25% of water, give higher nucleation rates than pure methanol for sulfamerazine, though the mixtures have lower solubility. This is probably due to the strong interaction between methanol and sulfamerazine, since stronger solvent–solute interactions would be expected to retard nucleation and crystal growth. Usually, a solvent with moderate strength of solute–solvent interaction but giving a high solubility is desirable to promote solution-mediated transformation. On the other hand, a solvent with low solubility and strong solute–solvent interaction is ideal for stabilizing metastable forms.

2.6.2 Solid-state Conversion

Solid-state conversion is more complicated than solution-mediated transformation because the kinetics and mechanisms involved are not well understood. The biggest challenge to investigate solid-state conversions is to understand the intrinsic heterogeneity of the solid phase. For example, the concepts of concentration and order of reaction associated with solution-mediated transformations are usually not applicable in the solid state.

Like a chemical reaction, solid-state conversions need to overcome the activation energy required to form a more stable phase. Factors that can affect the activation energy include crystal packing, defects, particle size, impurities, temperature, and humidity.

The transformation of a metastable form to a stable form in the solid state is usually quite slow at room temperature but can be accelerated at elevated temperatures. For example, some metastable polymorphs are stable after years of storage, whereas many transitions are often observed during thermal analysis at elevated temperatures. For example, the R form, one of the metastable forms of ROY, is relatively stable at room temperature; that is, conversion to the OP or Y forms has been observed only after years of storage at room temperature. However, R was converted to Y, OP, or ON in hours to days after heating between 70 and 100 °C [37]. Higher temperature increases molecular mobility, and this increase in molecular mobility promotes polymorph conversion, like any other type of physical transformation and chemical reaction.

How temperature and other factors such as pressure and defects affect transformation kinetics is also dependent on the mechanism, which is well recognized as being determined by the degree of similarity between the crystal structures of the pair of polymorphs. Such structural factors include conformation, hydrogen bonding, packing, and the nature of the unit cell. A transformation will have high activation energy if it involves the rotation of certain functional groups, breaking hydrogen bonds, and packing into a new motif, as described as a reconstructive transformation by Buerger [42]. This type of transformation, expected for the majority of polymorph conversion and considered to follow the steps of crystal nucleation and growth, requires significant reorganization and restructuring, which is characterized by a high-energy barrier and, hence, is expected to be slow.

On the other hand, a pair of polymorphs tends to convert easily if the molecular conformation and packing are similar. The new lattice forms by simple differential dilation of the parent lattice or minor displacements or rotations of the molecular species, as characterized as displacive transformation by Buerger [42]. The physical and energy barrier is relatively low considering that there is no making or breaking of strong interactions. This type of phase transition appears cooperative and diffusionless, with a collective and rapid propagation of molecules from one lattice to another, retaining its nearest neighbors for each molecule. Concerted or martensitic transformations have been observed for some organic compounds, including DL-norleucine and DL-methionine [43, 44]. Martensitic transformations are first-order crystal–crystal phase transitions that develop swiftly without diffusion of the atomic or molecular species. Molecules in the crystal lattice move in a concerted way to form the new phase. For example, two polymorphs of a substituted dihydroanthracene can be reversibly interconvertible to the other through a single-crystal to single-crystal phase transformation [45].

Reconstructive transformation is considered the dominant mechanism for drug-related solid-state polymorph conversion, with most polymorphs having distinct packing, conformations, networks of hydrogen bonding, and supramolecular assembly [42].

According to Mnyukh and coworkers [46–48], through their work on molecular single crystals, a reconstructive transformation follows the nucleation and growth mechanism and involves layer-by-layer edgewise growth through molecule-by-molecule structural rearrangement. An appreciable phase boundary between the old crystal phase and the new one is considered to exist, with molecules in the parent phase crossing the interface and assembling into the new crystal phase. Hartshorne et al. and Bradley hypothesized that there is only a small gap, on average, of approximately half a molecular layer in the phase boundary, likely lacking an intermediate amorphous layer to exist [49, 50]. Like polymorph conversion in solution, nucleation is presumably the rate-limiting step and is widely regarded to begin at various crystal defect sites with high energy. Indeed, it was observed that introducing defects by mechanical means can facilitate the phase transition of high-quality single crystals [46]. This can be well understood in the context of Paul and Curtin's theory concerning the steps that must occur for a solid-state physical or chemical transformation to occur [51]. Based on this theory, the process of conversion can be divided into three steps:

1. Molecular loosening (molecular mobility). The transformation begins at one or more defect sites and spreads through the crystal. Molecules at the defect sites are disordered and have high energy.
2. Nucleation of the new solid phase.
3. Growth of the new phase.

The principles developed by Paul and Curtin can serve as a guideline on how to mitigate solid-state polymorph conversion. First and furthermost, the quality of the crystalline phase regarding amorphous content or level of defects needs to be fully understood, since those high-energy portions or regions could be unintentionally or unavoidably introduced by manufacturing processes such as crystallization, milling, micronization, drying, roller compaction, or tablet compression. Second, the kinetics as influenced by temperature, humidity, and pressure need to be thoroughly evaluated to develop a control strategy when dealing with any other instability issues. Another important environmental factor that should be taken into consideration is certain excipients in a drug formulation that can be used to stabilize metastable forms and inhibit the phase transformation. For example, Fan and coworkers reported that the use of 1 mol% citric acid and tricarballylic acid can significantly inhibit the phase transformation of piracetam Form I, the metastable form, to Form II, the stable form, at 40 °C [52]. It is hypothesized that, based on molecular simulation, organic acid molecules like citric acid are absorbed on the crystal surface of Form I, to provide an inhibitory effect on the nucleation of Form II. Baaklini et al. showed that co-spraying of pyrazinamide with 1,3-dimethylurea led to the stabilization of the compound's metastable γ form for up to 12 months at room temperature, in comparison to a 14-day shelf life without the excipient, presumably due to surface interaction [53].

2.7 Control of Polymorphs

Producing polymorphs consistently and reproducibly is desired to maintain optimal quality and stability of drug substances and drug products. Lack of polymorph control can be due to several reasons:

- Very fast and undetected polymorph conversion to stable forms
- Change in types and levels of impurities
- Change in solvents
- Ostwald's rule of stages
- Concomitant crystallization

Uncontrolled polymorph conversion in solution or in the solid state presents a challenge in the preparation of metastable polymorphs. The worst-case scenario can occur when a suddenly appearing or disappearing polymorph is encountered. Suddenly appearing polymorphs sometimes occur in drug products even though comprehensive screening has been implemented during development phases, whereas disappearing polymorphs describe the opposite situation, where a polymorph can no longer be obtained under normal crystallization conditions [54]. In mid-1998, several lots of ritonavir capsules manufactured using Form I failed the dissolution test due to the unanticipated emergence of the more stable Form II [5]. Within weeks, the new polymorph began to appear throughout both the bulk drug and formulation areas.

The sudden appearance of Form II for ritonavir after many years of development and manufacture is still not well understood. The appearance of such new polymorphs and the disappearance of the originally used form could be due to a change of impurities, such as residual reactants, by-products, or degradants, which prevent the nucleation or growth of one polymorph or act as a template to grow another one. Changes in synthetic pathways, purification procedures, or crystallization solvents may result in different impurity profiles and levels, which in turn may promote the crystallization of one polymorph versus another. One hypothesis regarding the formation of ritonavir Form II is that a cyclic carbamate degradant acted as a template or seed to facilitate nucleation of Form II. The degradant was present in significant amounts in late batches, which possibly triggered the appearance of Form II [5].

In another example of such behavior, Blagden et al. have reported that the metastable form of sulphathiazole was stabilized by the presence of ethamidosulphathiazole, a reaction by-product in the synthesis of sulfathiazole [55].

Polymorph formation is also affected by the solvents used. The effect of solvents could be due to differences in solubility, solvent–solute interaction, solvent polarity, hydrogen bonding, and hydrophobic interactions. Trifkovic et al., for example, have reported that crystallization of ranitidine hydrochloride using a non-polar solvent led to Form I, the polymorph containing the enamine tautomer and having strong hydrogen bonding interactions [56], whereas crystallization using a polar solvent such as water or methanol promoted the formation of Form II, the nitronic acid tautomer with weak hydrogen bonding.

Polymorph control could also be complicated by different nucleation mechanisms since such mechanisms are independent and competing processes of homogeneous nucleation for different polymorphs. Nucleation may also follow Ostwald's rule of stages in that an unstable system might transform into another transient state before finally reaching a stable state [12]. The formation of such transient states is preferred due to the smaller loss of free energy with each step. The Ostwald rule indicates that crystallization from solution often starts in such a way that thermodynamically unstable phases appear first, followed by recrystallization to thermodynamically stable phases. Thus, in the process of preparing the stable form, a metastable form may appear first, with sufficient time needed for the metastable form to fully convert to the stable form.

Concomitant crystallization refers to the occurrence of two or more polymorphs simultaneously or in overlapping stages due to competing kinetic and thermodynamic factors. Concomitant crystallization could also be due to cross-nucleation between polymorphs. Chen et al. observed, for example, that certain polymorphs of ROY could not nucleate without the aid of others [39]. Concomitant crystallization makes it difficult to produce metastable forms, even when seeding is used.

In view of such behavior, different strategies to control stable polymorphs versus metastable forms must be instituted. Fast crystallization is one approach that favors the formation of metastable forms. Approaches that produce fast crystallization include high supersaturation, precipitation with anti-solvent, quench-cooling, and pH change. On the other hand, slurring, anti-solvent diffusion, and slow recrystallization are more desirable for obtaining more stable forms. Seeding can be used to prepare both metastable and stable forms; however, cross-nucleation between polymorphs must be fully evaluated. On the other hand, the effect of impurities and solvents should be investigated and could be utilized to control polymorphs.

2.8 Polymorph Screening

The goal of polymorph screening is to find all possible crystal forms by small-scale crystallization and to select an appropriate form that can be manufactured and formulated into a dosage form with acceptable quality, stability, and performance.

Polymorph screening consists of several key steps. The first step is to discover all possible polymorphs. The second step is to evaluate the stability order for those forms. The third step is to study the physical and chemical attributes of one or two leading candidates and select one for development. We shall try to answer the following questions:

1. How many polymorphs can be identified?
2. Which form is the thermodynamically stable polymorph at room temperature?
3. If enantiotropic pairs exist, what are the transition temperatures?

Crystallization, evaporation, precipitation, slurring, and melt crystallization are widely used to discover polymorphic forms. Regarding crystallization, we can choose various solvents, solvent mixtures, cooling rates, or supersaturation levels. Quick cooling and precipitation usually result in metastable forms, as stated by Ostwald's rule [12]. Crystallization in solvents sometimes produces solvates instead of polymorphs (Chapter 3), so various solvent systems or combined solvents shall be explored to increase the hits of unsolvated forms. Slurrying in different solvents is a popular approach used to isolate stable polymorphs. It is based on the occurrence of solution-mediated transformation, as previously discussed in Section 2.6.1. In such cases, a metastable form is dissolved, and, hopefully, a more stable form will grow at the expense of the metastable form.

For solution-based screening, it is well recognized that impurities, especially those with structural similarity to the parent compound, could significantly influence the nucleation and growth of one polymorph versus another. For example, paracetamol Form II, as a metastable form, is hard to prepare from crystallization, while the presence of metacetamol as an additive was found to promote the growth of Form II by inhibiting the crystallization of Form I, the stable form [57]. However, at early development phases, to spike with impurities is not only a bit of over-killing but also challenging to design. Rather, the focus should be on the analysis of impurity types and the levels present in the API batches used in the screening, as well as on the deconvolution of screening results along with solvents, solvent composition, temperature, and saturation level.

Melt crystallization may help discover polymorphs not stable enough to be observed in solution, as exemplified by the generation of ROY Y04 form and ritonavir Form III from melts [23, 39].

More importantly, we also highly recommend screening polymorphs in amorphous solid dispersion systems with commonly used polymers like povidone, co-povidone, hypromellose, and hypromellose succinate acetate, considering the reality that more and more new drugs are being formulated as solid dispersions to enhance bioavailability and to enter clinical studies. Nucleation mechanisms and kinetics in a solvent-free and polymer matrix are expected to differ significantly, which may lead to the crashing out of a totally different crystal phase from the one that exists in the active ingredient to begin with. It is a lesson learned from the very unfortunate case of rotigotine [58, 59]. The late appearance of a new polymorph, Form II, in the marketed amorphous transdermal patch formulation led to a massive global recall of a key therapeutic agent for Parkinson's disease in 2008, and caused a 6-year product off the market in the US until the patch was reformulated and reapproved. The late-appearing Form II was found to be more than eight times less soluble than Form I, the form existing in the original API.

A number of techniques are available for identifying solid forms of a compound, which will be discussed in later chapters. These techniques include X-ray powder diffraction (Chapter 9), thermal analysis (Chapter 10), microscopy (Chapter 11), spectroscopy (Chapter 12), and solid-state NMR (Chapter 13). Usually, a combined use of the above methods is necessary to provide a full understanding of each polymorph.

2.9 Polymorph Prediction

Any attempt to computationally predict the various possible crystal structures, that is, polymorphs, for any compound, based on the molecular structure of a compound, is deemed a significant task that continues to attract much attention. Although there has been steady progress, the task is still considered challenging. To develop any a new drug, an early understanding of the polymorphism design space and the thermodynamical landscape of various crystal structures is much desired to assess the developability of the molecule, to design a development strategy for drug substance and formulation, and to answer some fundamental questions that potentially become roadblocks and significant risks in downstream processes:

- How many polymorphs?
- Which one is the stable form, and what is the stability window against temperature and pressure?
- How many metastable forms have lattice energies close to the stable form, posing risks to polymorph control?
- What are the predicted key physical properties, for example, solubility and compressibility?
- Any potential to form salts or cocrystals to enhance solubility?
- For molecules difficult to crystallize, for example, with flexible conformations or large molecular weights (close to 1000 Daltons), what is the energy landscape for various potential crystalline forms to arise?

Obviously, those questions can be experimentally explored; however, such experimental work is generally deemed too resource driven. For most drug candidates at an early stage, there is usually not sufficient material to cover broad experimental spaces, whereas it is much more economical and efficient to do in silico simulation. Thus, there is substantial interest in the pharmaceutical industry to advance the field of polymorph prediction by developing new algorithms, regardless of the numerous challenges ahead.

Drug molecules generally have flexible structures. To predict polymorphs, we not only need to search through a huge number of possible crystal-packing patterns but must also consider many energetically plausible conformers. Another challenge for polymorph prediction is to determine accurately the relative energy difference between polymorphs, which is usually found to be within 10 kJ/mol of one another, often on the order of 1–2 kJ/mol. A final issue is the possibility of over-prediction. Here, it appears that polymorph predictions often lead to far more thermodynamically viable structures than are ever experimentally observed. It is because prediction algorithms focus on the thermodynamic stability and overall energy landscape, whereas the formation of a polymorph is also heavily influenced by the kinetics of nucleation, growth, and phase transition, which are all difficult to model and predict. Despite many challenges, the accuracy of the prediction of polymorphs is increasing significantly, especially for rigid molecules.

Beran and Bowskill et al. summarized current common crystal structure prediction approaches as the following hierarchical stages [60, 61]:

- Generate ~10^5–10^7 crystal structures with various possible space groups, packing motifs, and intramolecular conformations.
- Rank the candidate structures with an inexpensive force field regarding computational cost, such as random search approaches, or global search algorithms like simulated annealing or evolutionary optimization.
- Refine the ~10^3 lowest-energy structures within 20 kJ/mol of the global minimum, with an intermediate-quality model, as to additional conformational flexibility, intramolecular energetics, multipolar electrostatics, and repulsion/dispersion parameters.
- Use dispersion-corrected density functional theory (DFT), a quantum-mechanical model, to refine a few hundred structures, likely within 10 kJ/mol of the global energy minimum, to provide a rank order.
- Perform final free energy corrections for a small number of the most stable structures to predict their stabilities at finite temperatures and pressures.

The advancement of DFT in late-stage refinement and ranking is considered to be the catalyst in recent successes of polymorph predictions, as evidenced by the outcomes of the Cambridge Crystallographic Data Centre's many blind tests. For example, a DFT-based approach correctly predicted the crystal structures of all blind-tested compounds by the fourth blind test, including three single components and one cocrystal system [62].

The final stage is to validate the structure predication by comparing with known crystal structures of various polymorphs, if they exist, to determine the accuracy of the predicted energy landscape against the order of the known thermodynamic stability established experimentally, and to continuously improve the model, which then can be applied to new compounds with increased confidence.

For example, the ROY molecule, a classic case of conformational polymorphism, having 13 polymorphs discovered so far, was found challenging in many early attempts to predict crystal energy landscapes consistent with the order of experimental thermodynamic stability because of conformational flexibility and the resulting difficulty in simulating the allocation of energy levels [63, 64]. After refining the ROY intramolecular conformation energy contribution to the lattice energy via correlated wave function methods, the predicted crystal energy rankings did align well with experimental ones. The predicted energy landscape for ROY, illustrated in Figure 2.11, clearly shows that all discovered polymorphs mapped out in the landscape aligned well with the respective positions observed experimentally.

The significant improvement in polymorph prediction by in silico means makes it a valuable tool in the search for potential polymorphs, and their relative stabilities, complementary to experimental approaches. For example, Beyer et al. found computational tools beneficial in understanding the polymorphism of acetaminophen [65]. Acetaminophen has two reported metastable polymorphs. One of the metastable forms, the orthorhombic form, has better tableting properties than the stable monoclinic form. A systematic crystal structure prediction was pursued to search for minima in the lattice energy of acetaminophen. The stable monoclinic form was found as the global lattice energy minimum. There are a dozen energetically feasible structures identified, including the orthorhombic form.

The advancement of polymorph prediction will require a collaborative effort between various in silico approaches and experimental work, not to replace experiments but, rather, to make experimental design more efficient. For example, crystal structure predication studies on 4-aminoquinaldine surprisingly revealed that an unknown high-density monohydrate form exists at a global energy minimum, while the existing isolated hydrate, named Hy1A, is actually the metastable form [66]. This discovery drove additional experimental work, leading to the discovery of the thermodynamically most stable hydrate, Hy1B form. It is interesting to note that only the presence of an impurity in the crystallization experiment, likely inhibiting the nucleation of Hy1A, leads to the production of Hy1B.

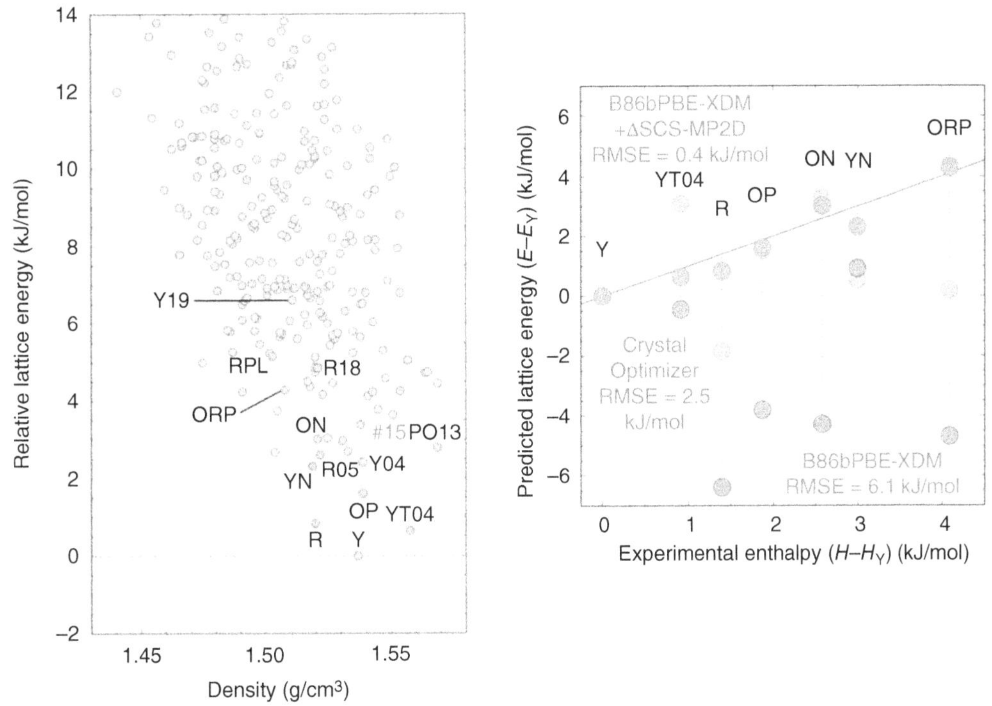

Figure 2.11 Predicted crystal energy landscape for ROY polymorphs. Left: Relative lattice energy versus density; experimentally observed structures as indicated by dark shapes. Right: relative lattice energies for seven polymorphs predicted versus experimentally measured. *Source*: Beran et al. [69]/Royal Society of Chemistry/CC BY 3.0.

Using ROY as a case again, Vasileiadis et al. developed ab initio crystal structure prediction techniques for ROY and predicted successfully seven experimentally resolved structures at that time [64]. More amazingly, the group also predicted three polymorphs whose existence and crystal structures were unknown at that time but were later discovered and solved, namely, PO13, Y19, and Y04. Especially, Y04 was discovered in 2004 by melt crystallization but was found to be physically unstable, converting readily to YT04 or R in solution, thus making it obviously challenging to obtain quality single crystals for any structure elucidation. Later, in 2020, Li et al. crafted a unique melt microdroplet technique and grew high-quality single crystals with Y04 crystal structure being discovered [8], validating previous crystal predictions by Vasileiadis et al. in 2012 [64]. It is interesting to note that the experimental resolved structure of Y04 matches very well with the rank 26th polymorph in the prediction in 2012, regarding conformation and lattice parameters. The developed energy landscape from a computational approach reveals that Y04, as a high-energy form, needs specialized crystallization techniques capable of trapping metastable forms, to enable its existence, which agrees with the unique approach created by Li and coworkers. Thus, another prediction work for ROY by Beran et al., indicating a new crystal form at pressures above 10 GPa becoming enthalpically more stable than all known polymorphs, is worth exploring [60].

Nevertheless, as shown in Figure 2.11 for ROY, there are more predicted crystal structures than experimentally discovered ones, which is also common for many other compounds under such evaluation. It is still not clear whether those extra structures can naturally occur or whether the right experimental conditions simply have not been found. Clearly, more fundamental work in modeling crystallization kinetic landscapes is required to advance this field.

References

1 Henck JO, Griesser UJ, Burger A. Polymorphism of drug substances: an economic challenge. *Pharm Ind* 1997; 59: 165–169.
2 Stahly GP. Diversity in single-and multiple-component crystals. The search for and prevalence of polymorphs and cocrystals. *Cryst Growth Des* 2007; 7(6): 1007–1026.
3 Grunenberg A. Polymorphie und thermische Analyse pharmazeutischer Wirkstoffe. *Pharmazie* 1997; 26(5): 224–231.
4 Ostwald W. *Grundriss der Allgemeinen Chemie*. Leipzig: W. Engelmann; 1899.
5 Bauer J, Spanton S, Henry R, Quick J, Dziki W, Porter W, Morris J. Ritonavir: an extraordinary example of conformational polymorphism. *Pharm Res* 2001; 18(6): 859–866.
6 Yu L. Nucleation of one polymorph by another. *J Am Chem Soc* 2003; 125 (21): 6380–6381.
7 Salzmann CG. Advances in the experimental exploration of water's phase diagram. *J Chem Phys* 2019; 150(6): 060901.
8 Li X, Ou X, Rong H, Huang S, Nyman J, Yu L, Lu M. The twelfth solved structure of ROY: single crystals of Y04 grown from melt microdroplets. *Cryst Growth Des* 2020; 20(11): 7093–7097.
9 McCrone WC. Polymorphism. In: Fox D, Labes MM, Weissberger M, eds. *Physics and Chemistry of the Organic Solid State*, vol. 2. New York: Wiley Interscience; 1965: 725.
10 Mortazavi M, Hoja J, Aerts L, Quéré L, van de Streek J, Neumann MA, Tkatchenko A. Computational polymorph screening reveals late-appearing and poorly-soluble form of rotigotine. *Commun Chem* 2019; 2(1): 70.
11 Beran GJ, Sugden IJ, Greenwell C, Bowskill DH, Pantelides CC, Adjiman CS. How many more polymorphs of ROY remain undiscovered. *Chem Sci* 2022; 13(5): 1288–1297.
12 Ostwald W. Studien uber die Bildung und Umwandlung fester Korper. *Z Phys Chem* 1897; 22: 289–330.
13 Becker D, Döring W. The kinetic treatment of nuclear formation in supersaturated vapors. *Ann Phys* 1935; 24: 719–752.
14 Volmer M, Weber A. Keimbildung in übersättigten Gebilden. *Z Phys Chem* 1925; 119: 277.
15 Vippagunta SR, Brittain HG, Grant DJW. Crystalline solids. *Adv Drug Deliv Rev* 2003; 48(1): 3–26.
16 Kala H, Haack U, Pollandt P, Brezesinski G. Zur Polymorphie des Carbamazepins. *Acta Pharm Technol* 1986; 32: 72–77.
17 Reboul JP, Cristau B, Soyfer JC, Astier JP. 5H-5-Dibenz[b,f]azepinecarboxamide-5 (carbamazepine). *Acta Crystallogr Sect B* 1981; 37(10): 1844–1848.
18 Himes VL, Mighell AD, De Camp WH. Structure of carbamazepine: 5H-Dibenz[b,f]azepine-5-carboxamide. *Acta Crystallogr Sect B: Struct* 1981; 37(12): 2242–2245.
19 Lowes MM, Caira MR, Lotter AP, Watt V, Der JG. Physicochemical properties and X-Ray structural studies of the trigonal polymorph of carbamazepine. *J Pharm Sci* 1987; 76(9): 744–752.
20 Lang M, Kampf JW, Matzger AJ. Form IV of carbamazepine. *J Pharm Sci* 2002; 91(4): 1186–1190.
21 Grzesiak AL, Lang M, Kim K, Matzger AJ. Comparison of the four anhydrous polymorphs of carbamazepine and the crystal structure of form I. *J Pharm Sci* 2003; 92(11): 2260–2271.

22 Yao X, Henry RF, Zhang GG. Ritonavir form III: a new polymorph after 24 years. *J Pharm Sci* 2023; 112(1): 237–242.

23 Parent SD, Smith PA, Purcell DK, Smith DT, Bogdanowich-Knipp SJ, Bhavsar AS, Chan LR, Croom JM, Bauser HC, McCalip A, Byrn SR. Ritonavir form III: a coincidental concurrent discovery. *Crys Growth Des* 2023; 23(1): 320–325.

24 Lin HO, Baenziger NC, Guillory JK. Physical properties of four polymorphic forms of sulfanilamide I: densities, refractive indexes, and X-ray diffraction measurements. *J Pharm Sci* 1974; 63(1): 145–146.

25 Sanders GH, Roberts CJ, Danesh A, Murray AJ, Price DM, Davies MC, Tendler SJ, Wilkins MJ. Discrimination of polymorphic forms of a drug product by localized thermal analysis. *J Microsc* 2000; 198(2); 77–81.

26 Weidenthaler P. Changes in electrical conductivity accompanying the polymorphic transitions in KNO3 at 1 atmosphere. *J Phys Chem Solids* 1964; 25(12): 1491–1493.

27 Pudipeddi M, Serajuddin A. Trends in solubility of polymorphs. *J Pharm Sci* 2005; 94(5): 929–939.

28 Chen X, Morris KR, Griesser UJ, Byrn SR, Stowell JG. Reactivity differences of indomethacin solid forms with ammonia gas. *J Am Chem Soc* 2002; 124(50): 15012–15019.

29 Haisa MA, Kashino S, Kawai R, Maeda H. The monoclinic form of p-hydroxyacetanilide. *Acta Crystallogr B: Struct* 1976; 32(4): 1283–1285.

30 Espeau P, Céolin R, Tamarit JL, Perrin MA, Gauchi JP, Leveiller F. Polymorphism of paracetamol: relative stabilities of the monoclinic and orthorhombic phases inferred from topological pressure-temperature and temperature-volume phase diagrams. *J Pharm Sci* 2004; 94(3): 524–539.

31 Haisa M, Kashino S, Maeda H. The orthorhombic form of p-hydroxyacetanilide. *Acta Crystallogr B: Struct* 1974; 30(10): 2510–2512.

32 Di Martino P, Guyot-Hermann AM, Conflant P, Drache M, Guyot JC. A new pure paracetamol for direct compression: the orthorhombic form. *Int J Pharm* 1996; 128(1): 1–8.

33 Burger A, Ramberger R. On the polymorphism of pharmaceuticals and other molecular crystals. I. *Microchimica Acta* 1979; 72 (3–4): 259–271.

34 Price CP, Grzesiak AL, Lang M, Matzger AJ. Polymorphism of nabumetone. *Cryst Growth Des* 2002; 2(6): 501–503.

35 Rodríguez-Spong B, Price CP, Jayasankar A, Matzger AJ, Rodríguez-Hornedo N. General principles of pharmaceutical solid polymorphism: a supramolecular perspective. *Ad Drug Del Rev* 2004; 56(3): 241–274.

36 Yu L. Inferring thermodynamic stability relationship of polymorphs from melting data. *J Pharm Sci* 1995; 84(8): 966–974.

37 Yu L, Stephenson GA, Mitchell CA, Bunnell CA, Snorek SV, Bowyer JJ, Borchardt TB, Stowell JG, Byrn SR. Thermochemistry and conformational polymorphism of a hexamorphic crystal system. *J Am Chem Soc* 2000; 122(4): 585–591.

38 Mitchell CA, Yu L, Ward MD. Selective nucleation and discovery of organic polymorphs through epitaxy with single crystal substrates. *J Am Chem Soc* 2001; 123(44): 10830–10839.

39 Chen S, Guzei IA, Yu L. New polymorphs of ROY and new record for coexisting polymorphs of solved structures. *J Am Chem Soc* 2005; 127(27): 9881–9885.

40 Ferrari ES, Davey RJ, Cross WI, Gillon AL, Towler CS. Crystallization in polymorphic systems: the solution-mediated transformation of β to α glycine. *Cryst Growth Des* 2003; 3(1): 53–60.

41 Gu CH, Young V, Grant DJ. Polymorph screening: influence of solvents on the rate of solvent-mediated polymorphic transformation. *J Pharm Sci* 2001; 90(11): 1878–1890.

42 Buerger MJ. Polymorphism and phase transformations. *Fortschr Mineral* 1961; 39: 9–24.

43 Anwar J, Tuble SC, Kendrick J. Concerted molecular displacements in a thermally-induced solid-state transformation in crystals of DL-norleucine. *J Am Chem Soc* 2007; 129(9): 2542–2547.

44 Smets MMH, Kalkman E, Krieger A, Tinnemans P, Meekes H, Vlieg E, Cuppen HM. On the mechanism of solid-state phase transitions in molecular crystals - the role of cooperative motion in (quasi)racemic linear amino acids. *IUCrJ* 2020; 7(2): 331–341.

45 Das D, Engel E, Barbour LJ. Reversible single-crystal to single-crystal polymorphic phase transformation of an organic crystal. *Chem Commun* 2010; 46(10): 1676–1678.

46 Mnyukh YV, Petropavlov NN. Polymorphic transitions in molecular crystals-I. Orientations of lattices and interfaces. *J Phys Chem Solids* 1972; 33(11): 2079–IN3.

47 Mnyukh YV, Panfilova NA. Polymorphic transitions in molecular crystals—II. Mechanism of molecular rearrangement at 'contact' interface. *J Phys Chem Solids* 1973; 34(2): 159–170.

48 Mnyukh YV, Panfilova NA, Petropavlov NN, Uchvatova NS. Polymorphic transitions in molecular crystals—III.: Transitions exhibiting unusual behavior. *J Phys Chem Solids* 1975; 36(3): 127–144.

49 Hartshorne NH, Thackray M. 407. Studies in polymorphism. Part VIII. The linear rate of transformation of β-into α-sulphur at low temperatures and at temperatures just below the transition point. *J Chem Soc* 1957; 2122–2131.

50 Bradley RS. The energetics and statistical mechanics of the kinetics of solid → solid reactions. *J Phys Chem* 1956; 60: 1347–1354.

51 Paul IC, Curtin DY. Thermally induced organic reactions in the solid state. *Acc Chem Res* 1973; 6(7): 217–225.

52 Fan F, Xu S, Guo M, Cai T. Effect of organic acids on the solid-state polymorphic phase transformation of piracetam. *Int J Pharm* 2023; 647: 123532.

53 Baaklini G, Dupray V, Coquerel G. Inhibition of the spontaneous polymorphic transition of pyrazinamide γ form at room temperature by co-spray drying with 1, 3-dimethylurea. *Int J Pharm* 2015; 479(1): 163–170.

54 Henck JO, Bernstein J, Ellern A, Boese R. Disappearing and reappearing polymorphs. The benzocaine:picric acid system. *J Am Chem Soc* 2001; 123(9): 1834–1841.

55 Blagden N, Davey RJ, Rowe R, Roberts R. Disappearing polymorphs and the role of reaction by-products: the case of sulphathiazole. *Int J Pharm* 1998; 172(1): 169–177.

56 Trifkovic M, Rohani S, Mirmehrabi M. Polymorphic generation through solvent selection: ranitidine hydrochloride. *Org Proc Res Dev* 2007; 11(1): 138–143.

57 Agnew LR, McGlone T, Wheatcroft HP, Robertson A, Parsons AR, Wilson CC. Continuous crystallization of paracetamol (acetaminophen) form II: selective access to a metastable solid form. *Cryst Growth Des* 2017; 17(5): 2418–2427.

58 Wolff HM, Quere L, Riedner J. Polymorphic form of rotigotine and process for production. *US Patent* 8,232,414; 2012.

59 Rietveld IB, CeÃÅolin R. Rotigotine: unexpected polymorphism with predictable overall monotropic behavior. *J Pharm Sci* 2015; 104: 4117–4122.

60 Beran GJO. Frontiers of molecular crystal structure prediction for pharmaceuticals and functional organic materials. *Chem Sci* 2023; 14(46): 13290–13312.

61 Bowskill DH, Sugden IJ, Konstantinopoulos S, Adjiman CS, Pantelides CC. Crystal structure prediction methods for organic molecules: state of the art. *Annu Rev Chem Biomol Eng* 2021; 12(1): 593–623.

62 Chan HS, Kendrick J, Leusen FJ. Predictability of the polymorphs of small organic compounds: crystal structure predictions of four benchmark blind test molecules. *Phys Chem Phys* 2011; 13(45): 20361–20370.

63 Nyman J, Yu L, Reutzel-Edens SM. Accuracy and reproducibility in crystal structure prediction: the curious case of ROY. *CrystEngComm* 2019; 21: 2080–2088.

64 Vasileiadis M, Kazantsev AV, Karamertzanis PG, Adjiman CS, Pantelides CC. The polymorphs of ROY: application of a systematic crystal structure prediction technique. *Acta Crystallogr Sect B: Struct Sci* 2012; 68: 677–685.

65 Beyer T, Day GM, Price SL. The prediction, morphology, and mechanical properties of the polymorphs of paracetamol. *J Am Chem Soc* 2001; 123(21): 5086–5094.

66 Braun DE, Oberacher H, Arnhard K, Orlova M, Griesser UJ. 4-Aminoquinaldine monohydrate polymorphism: prediction and impurity aided discovery of a difficult to access stable form. *CrystEngComm* 2016; 18(22): 4053–4067.

3

Pseudopolymorphs: Hydrates and Solvates

3.1 Introduction

Pseudopolymorphs, including hydrates and organic solvates, occur commonly for many pharmaceutical materials. Solvates are formed when solvent molecules are incorporated into the crystal lattice of a compound and are considered as molecular complexes between host and solvent molecules. A solvate displays unique unit cell content, dimension, molecular packing, and molecular interactions. This phenomenon is analogous to polymorphism. Consequently, the formation of crystal solid adducts with solvents is often termed "pseudopolymorphism," whereas a group of solvates, with different stoichiometries of the same solvent and compounds, is often called pseudopolymorphs.

Solvates can be formed when a pharmaceutical solid is processed or stored in a solvent during crystallization, slurrying, refluxing, wet granulation, storage, and dissolution. Exposure to solvent vapor can also lead to the formation of solvates. Overall, only a portion of drug molecules form solvates, since not all solvents can incorporate into a crystal lattice to form solvates. The propensity of a drug molecule and a solvent molecule to form a solvate is related to their molecular structures, hydrogen bonding ability, and crystal packing. Hydrates are a specific type of solvates where the guest component is water, while organic solvates refer to the remaining types associated with various organic solvents.

Hydrate formation for organic compounds occurs more frequently than solvate formation with organic solvents, based on a survey of the Cambridge Structural Database [1]. About 50% of entries for solvates of organic and metal organic compounds are hydrates. The significant ability of solids to form hydrates is related to the properties of the water molecule, including its small size and its ability to act as both a hydrogen bond donor and acceptor. Its small size also enables water molecules to fill structural voids. Additionally, its multidirectional hydrogen bonding capability allows it to link with many types of drug molecules and to assume stable crystal structures. Furthermore, water is widely present during pharmaceutical manufacture and product storage. It is not surprising to note that about 30% of drug compounds can form hydrates [2].

Thus, in this chapter, considering the prevalence of pharmaceutical hydrates and their wide use as APIs, we begin with a discussion of crystal hydrates. Then, we extend the discussion to organic solvates to illustrate several unique characteristics of pseudopolymorphs, including isostructural solvates and desolvation behavior. On the one hand, organic solvates can act as useful intermediates for better recovery, purification, and filtration, while, on the other hand, they can serve as a bridge to obtain more desirable crystalline forms, including anhydrous forms and hydrates through desolvation and solvent exchange, respectively.

3.2 Pharmaceutical Importance of Hydrates

The incorporation of water into a crystal lattice leads to changes in unit cell dimension, content, and intermolecular interactions. A hydrate of a pharmaceutical material, therefore, displays different physical, chemical, and mechanical properties not only from its anhydrous polymorph but also from the other hydrates of varying stoichiometry. Consequently, a change in the hydration state of a pharmaceutical material can significantly impact its stability, solubility, performance, and appearance.

Solid-State Materials in Pharmaceutical Chemistry: Properties, Characterization, and Applications, Second Edition. Stephen R. Byrn, George Zografi and Xiaoming (Sean) Chen

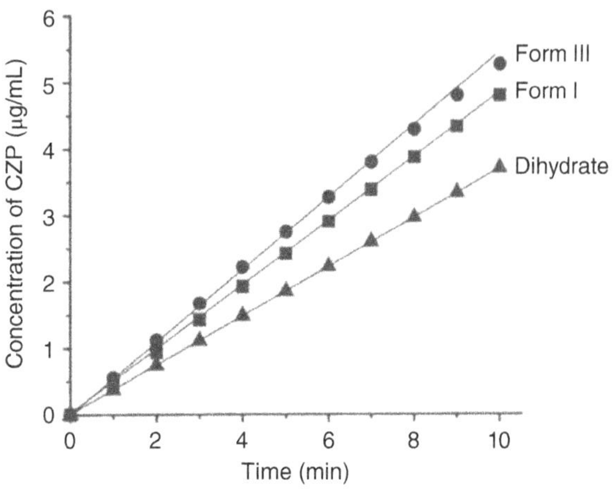

Figure 3.1 Intrinsic dissolution rates of carbamazepine polymorphs and dihydrate in pH 1.2 buffer at 37 °C. *Source:* Kobayashi et al. [6]. Reproduced with the permission of Elsevier.

Pudipeddi and Serajuddin, for example, have studied the solubility ratios of anhydrate/hydrate pairs compiled from literature data [3]. Anhydrous forms are usually two times more soluble than hydrates for pharmaceutical compounds. However, a significantly higher ratio is observed for a few compounds, for example, succinyl sulfathiazole (13 fold) and niclosamide (23 fold). Anhydrous forms are generally more soluble in aqueous media than the corresponding hydrate forms, with a very few exceptions, such as the compound labeled LY334370. The hydrate form of LY334370 HCl is approximately six times more soluble in water than the anhydrous form [4]. It has been confirmed that a monotropic relationship exists between the anhydrous form and the hydrate form, with the anhydrous form being the stable form [4].

Such solubility differences between anhydrous forms and their corresponding hydrates are important because they possibly can lead to distinct differences in dissolution rate according to the Noyes–Whitney equation [5]. Since the more stable polymorph of a compound generally has a lower dissolution rate than its metastable form because of lower solubility, a hydrate will usually have a lower dissolution rate than its corresponding anhydrate. This phenomenon, for example, was observed for the anhydrous forms and the dihydrate of carbamazepine [6]. Dissolution rates of carbamazepine Form I, Form III, and the dihydrate, determined by the static disk method, were shown to correlate with solubility [6] (Figure 3.1). The solubility of Form I, Form III, and the dihydrate in pH 1.2 HCl dissolution media was found to be 501, 460, and 311 μg/mL, respectively. The corresponding intrinsic dissolution rates in the same medium for the three forms are 67.4, 61.8, and 41.8 $\mu g/cm^2/mL$, respectively. Moreover, significant differences in bioavailability in dogs were observed for the anhydrous forms versus the dihydrate of carbamazepine after oral administration at 200 mg/per dog [6]. Compared with a solution formulation, the relative oral bioavailability for Form I, Form III, and dihydrate was estimated to be 48%, 69%, and 33%, respectively. It is interesting to note that Form III has twice as much exposure as the dihydrate and greater bioavailability than Form I, even though Form I has the highest solubility and fastest dissolution rate. It was proposed that Form I may be most rapidly converted to the dihydrate in the gastrointestinal tract of dogs, which results in the bioavailability being lower than expected relative to Form III.

In the fall of 1988, approximately 70 million tablets of carbamazepine produced by a generic manufacturer were withdrawn from the market owing to reports of clinical failure, including the loss of seizure control for some patients taking the generic carbamazepine [7]. The FDA later determined that the manufacturer had switched its source of carbamazepine and inadvertently produced tablets containing the dihydrate instead of Form III.

In another interesting study, it was shown that the state of hydration in a crystal can greatly influence the mechanical properties of the crystal. Dibasic calcium phosphate can exist either as an anhydrate or as a dihydrate. The two forms have a similar flow behavior but display differences in compressibility. The anhydrate compacts were also shown to disintegrate more rapidly than the dihydrate compacts [8]. Lerk et al. discovered that the crushing strength of tablets made of glucose monohydrate was increased with the elevation of the dehydration temperature [9]. Such thermal dehydration appears to change the texture of the particles and to increase the binding capacity.

In another example, the level of hydration of magnesium stearate has been shown to influence the ability of the material to act as a tablet lubricant [10, 11]. Magnesium stearate can exist in four states, from an anhydrous form to the monohydrate, dihydrate, and trihydrate. Higher hydration states exist with weaker interactive forces in the crystal lattice, with the weaker interactive forces leading to easier shearing of the lubricant powder particles. Commercial magnesium stearate often exhibits batch-to-batch variations in lubrication ability, likely because of variations in the state of hydration. The differences in moisture content, therefore, correlate with the effectiveness of lubrication of magnesium stearate, with the lubricating properties decreasing in milled, dried, and stored samples with lower water content. It has also been observed that the hydration state of a crystal hydrate often affects the chemical stability of a compound. For example, the hydrated form of vitamin B_{12} is chemically more stable to light and heat than the anhydrous form [12]. Some hydrated materials can become amorphous upon dehydration, a form that is generally less chemically stable than the various crystalline forms. For example, cephradine dihydrate is stable chemically but undergoes oxidation after dehydration due to the formation of the amorphous form [13].

3.3 Classification of Pharmaceutical Hydrates

Hydrates can be classified into three categories based on different structural aspects, based on how the water molecules are incorporated into the crystal lattice [13].

Class I represents hydrates where the water molecules occupy isolated sites, separated from other water molecules by surrounding drug molecules. In such cases, there is no direct hydrogen bonding between various water molecules. Instead, the water molecules form hydrogen bonds and van der Waals interactions only with drug molecules. This group of hydrates is distinguished by sharp dehydration endotherms in differential scanning calorimetry, narrow temperature ranges of weight loss in thermogravimetric analysis, and distinct hydroxyl bands in infrared spectroscopy [13].

Siramesine hydrochloride monohydrate serves as an example of a system that exists as an isolated site hydrate [14], as shown in the crystal structure illustrated in Figure 3.2. Here, it can be seen that each water molecule is hydrogen bonded to two nearby chloride ions, so that water molecules are relatively strongly bound and physically locked inside the crystal. It was further observed that the loss of water on heating was associated with the disruption of the crystal lattice and the formation of an oily phase [14]. The oily phase was then found to recrystallize into one or more anhydrous crystalline forms.

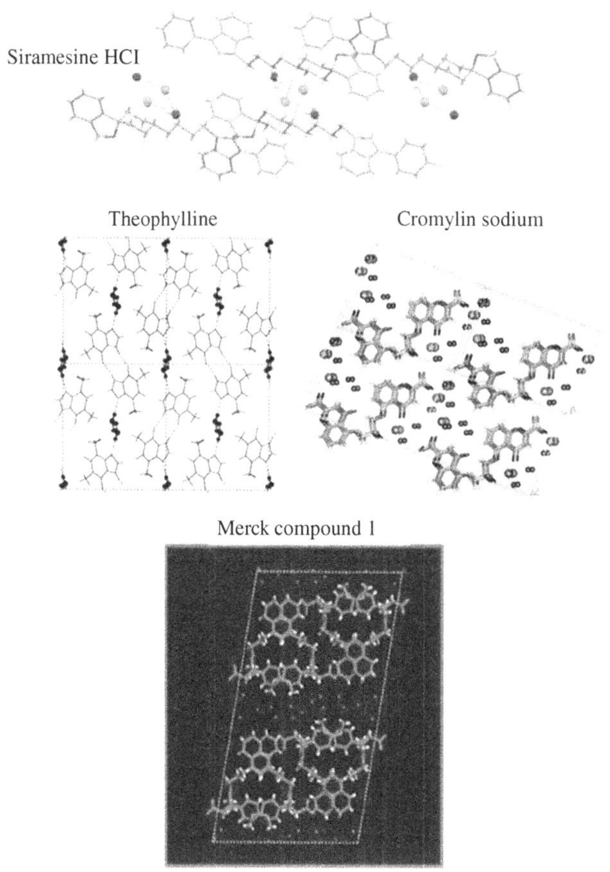

Figure 3.2 Crystal-packing diagrams of various hydrated compounds. *Source*: Zimmermann et al. [14], Sun et al. [15], Chen et al. [16], Kiang et al. [17]. Reproduced with the permission of *Elsevier*, the International Union of Crystallography, and the American Chemical Society.

Crystals classified as Class II hydrates are generally known as channel hydrates. Water molecules in this class lie continuously next to the other water molecules in the crystal lattice, forming "channels" or "layers" along a given crystallographic axis. In such systems, a water molecule forms hydrogen bonds with adjacent water molecules while also interacting with host molecules.

For example, theophylline monohydrate is a channel hydrate with a crystal structure of the monoclinic $P2_1/n$ space group [15]. In the unit cell, two theophylline molecules are centrosymmetrically related and form a dimer through two hydrogen bonds. In the monohydrate, the water molecules form channels along the *a*-axis and organize themselves into infinite hydrogen-bonded chains. In addition, water molecules form hydrogen bonds with neighboring theophylline dimers (Figure 3.2).

Channel hydrates can be further divided into additional subcategories based on the structural changes that take place after hydration and dehydration, including expanded channel hydrates and planar hydrates [13].

For an expanded channel hydrate, additional water molecules can enter the channels when exposed to high humidity, and, consequently, the crystal lattice tends to expand in various possible directions. This lattice expansion, as might be expected, causes changes in unit cell dimensions. Such changes in unit cell dimensions can be observed by X-ray powder diffraction as slight shifts in some or all diffraction peaks. Cromolyn sodium is a case of expanded channel hydrates (Figure 3.2). Cromolyn sodium can be further characterized as a nonstoichiometric hydrate, which sorbs and liberates water in a continuous manner, resulting in continuous changes in crystal lattice parameters as water content increases [16]. The unit cell volume, for example, is increased from 631 to 699 Å^3 when the relative humidity (RH) is changed from 6.4% to 80.9% RH.

In planar hydrates, water molecules form as layers in the crystal lattice. A unique case of a planar hydrate, reported for a Merck compound [17], is a compound that has four hydrates and one anhydrous form. Single-crystal and powder diffraction data have revealed that a two-dimensional layered network is present along the *bc* plane in all five forms (Figure 3.2) [17]. The layered networks stack along the *a*-axis, except for the anhydrate, are separated by a hydrate

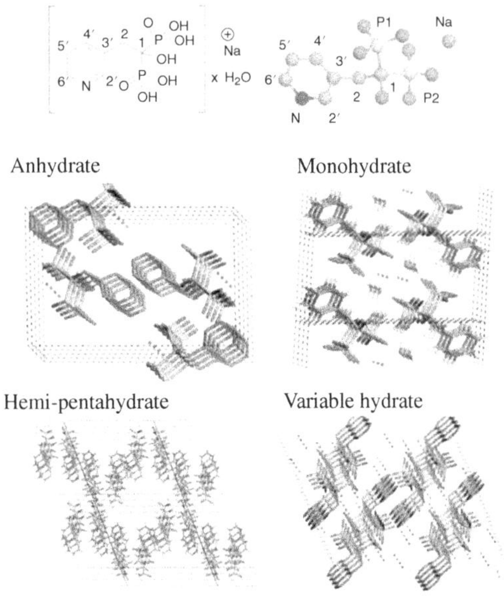

Figure 3.3 Chemical structure of risedronate monosodium salt and comparison of crystal packings of the anhydrate, the monohydrate, the hemipentahydrate, and the variable hydrate of risedronate. *Source*: Redman-Furey et al. [18]. Reproduced with the permission of Elsevier.

water layer. Upon exposure to moisture, water can enter between the drug molecule layers, and the lattice is expanded along the a-axis. The thickness of the resulting water layer is dependent on RH, and, therefore, the stoichiometry of water to drug changes accordingly, from anhydrate to hemihydrates, dihydrate, tetrahydrate, and pentahydrate. The single-crystal structure of the pentahydrate has the unit cell dimensions of a = 36.961(5) Å, b = 7.458(2) Å, c = 20.691(4) Å in the space group of C2. In the expanded crystal structure, water layers stay sandwiched by the compound framework and parallel to the bc plane. Water molecules tend to move in and out of the interlayer space when the RH changes, and therefore this compound has poor physical stability due to the possibility of variable planar hydrates. Some tablets of this compound were shown to crack during storage at high RH because the expansion of interlayer spacing caused a significant increase in lattice volume [17].

Class III hydrates are generally referred to as ion-coordinated site hydrates. In such systems, water molecules form ion–water bonds that are usually much stronger than hydrogen bonds. Risedronate monosodium monohydrate is an example of an ion-coordinated site hydrate [18].

The risedronate monosodium salt, a drug used for the treatment of osteoporosis, has three types of hydrate forms, including a monohydrate, a hemipentahydrate, and a variable hydrate with 4 to 6 moles of water [18]. The crystal packings of the three hydrates and the anhydrous form are compared in Figure 3.3.

The monohydrate is an ion-associated hydrate where each water molecule is tightly bound to a neighboring sodium ion. These structural characteristics enable the monohydrate to have very good thermal stability in that dehydration occurred only above 200 °C, right before the onset of degradation [18].

Interestingly, the hemipentahydrate has two types of lattice locations for water molecules, involving ion-coordinated bonds and present in channels. The water molecules in the channels are characteristically mobile and, hence, can move in and out of the crystal lattice, as determined by RH. It was discovered that the water content of the hemipentahydrate is relatively stable over a range of 20–90% RH [18]. As RH decreases from 9% to 5%, a weight loss of 5% is observed, corresponding to the loss of 1 mole of water per 1 mole of the drug. However, water molecules can reenter the channels as RH increases from 13% to 20% with complete rehydration. After the removal of water molecules from the lattice channels, the lattice seems to remain relatively intact. However, thermal and spectroscopic data revealed that the lattice undergoes an adjustment and appears to be distorted [18].

The variable hydrate of risedronate monosodium is a nonstoichiometric channel hydrate that has noncrystalline character. Because the variable hydrate is prone to dehydration at room temperature, the single-crystal structure of this form was collected at −100 °C under a stream of liquid nitrogen to ensure the hydration [18]. As shown in Figure 3.3, the crystal structure has large channels along the a-axis. However, the water molecules in these channels are disordered, and the crystal was determined to be only partially hydrated. Upon heating, dehydration of the variable hydrate began immediately, and most of the loss occurred before 100 °C.

This study clearly demonstrated in the case of risedronate monosodium that the structural aspects of various hydrate forms determine their physical stability.

3.4 Water Activity

Since a pharmaceutical hydrate can be thought of as a molecular complex of a compound and water, the amount of water in the environment has a significant impact on stability.

Water activity is a thermodynamic measure of the amount of water in the environment that can take part in physical or chemical interactions. Water activity, a_w is defined as

$$a_w = \frac{p}{p_0} \tag{3.1}$$

where p represents the partial vapor pressure of the environment and p_0 is the vapor pressure of pure water at the same temperature. In the atmosphere, water activity is related to RH as

$$RH = a_w \times 100\% \tag{3.2}$$

In a solution, water activity is determined by the vapor pressure of the solution. It is related to the mole fraction of water in the solution according to the equation:

$$a_w = l_w \times x_w \tag{3.3}$$

where l_w is the activity coefficient and x_w is the mole fraction of water in the solution. The activity coefficient of various solutions will be different for different materials and with different concentrations of the same material in solution. Changes in activity coefficient with concentration generally result in nonlinear relationships between RH and concentration. For example, whereas the relationship between water activity and water content is approximately linear for methanol/water mixture [19], an isopropanol/water mixture exhibits strong deviation from a linear relationship (Figure 3.4). This can be attributed to the stronger hydrophobic interaction in the solution, resulting from the larger alkyl group in isopropanol [19]. For example, the water activity of the isopropanol–water mixture is more than 0.8 when the water content is at about 0.5 mole fraction.

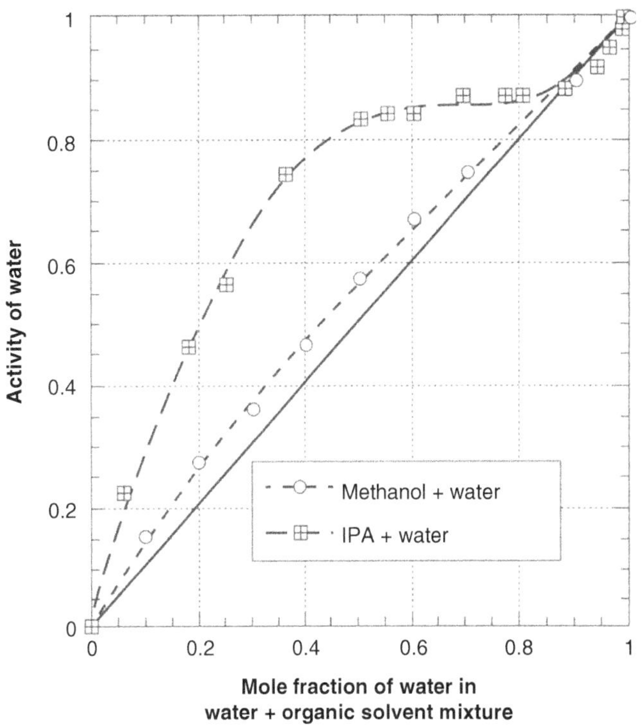

Figure 3.4 Plot of the water activity versus the mole fraction of water in methanol/water and isopropanol/water mixtures at 25 °C. *Source*: Zhu et al. [19]. Reproduced with the permission of Elsevier.

3.5 Stoichiometric Hydrates

A stoichiometric hydrate has a well-defined water content, exhibiting a unique crystal structure when compared to those of the anhydrous forms and other hydrates.

Nonstoichiometric hydrates, on the other hand, have continuously variable water content within a certain range of water activity, and there is no significant corresponding change in the crystal structure when the amount of water varies. Nonstoichiometric hydrates may, in some cases, have substantial noncrystalline character.

Stoichiometric hydrates display stepwise water sorption isotherms with changing RH, with a critical RH for the hydrate/dehydration transition. The transition point is also a function of temperature. Khankari and Grant have treated the thermodynamics of hydrate equilibrium as a reversible chemical reaction [20]. The formation of hydrate crystals from the anhydrous form is described as

$$A_{(s)} + mH_2O \leftrightarrow A \bullet mH_2O_{(s)} \tag{3.4}$$

where $A_{(s)}$ is the anhydrous solid, m is the number of moles of water taken up by 1 mole of the anhydrous form, and $A \bullet m H_2O_{(s)}$ is the hydrate. When the anhydrous and hydrate forms reach equilibrium, the equilibrium constant K_{h0} can be defined according to the following equation:

$$K_{h0} = \frac{a[A \bullet mH_2O_{(s)}]}{a[A_{(s)}] \bullet a_{w0}{}^m} \tag{3.5}$$

where $a[A \bullet mH_2O_{(s)}]$ and $a[A_{(s)}]$ are the thermodynamic activities of the hydrate and the anhydrate, respectively, and a_{w0} is the water activity related to the critical RH at which anhydrous and hydrate forms are in equilibrium. If the solids are treated as states with unit activity, the equilibrium constant can be simplified into

$$K_{h0} = a_{w0}{}^{-m} \tag{3.6}$$

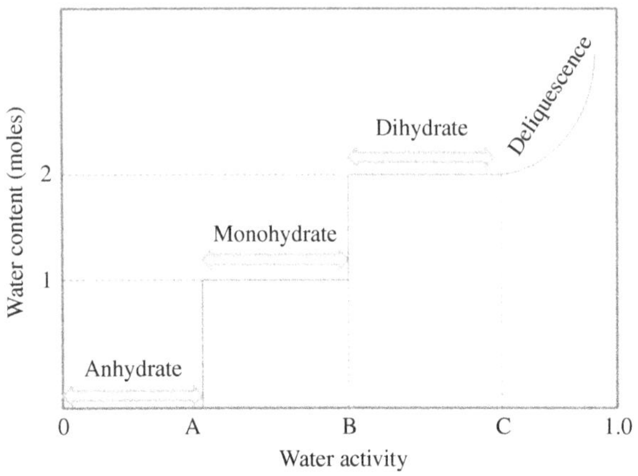

Figure 3.5 Sample isothermal sorption diagram for a compound with stoichiometric hydrates.

As previously discussed, the stability of the hydrate is related to water activity. If $a_w > a_{w0}$, the hydrate is the stable form, and the anhydrate is the metastable form. If $a_w < a_{w0}$, the anhydrous form is the stable form, and the hydrate becomes metastable. This stability relationship is also applicable to multiple hydrates, wherein each hydrate has a stability window within a range of water activities, as depicted in Figure 3.5. Below water activity A, the anhydrate is the stable form. Between A and B, the monohydrate is the most stable, while the dihydrate is stable between B and C. Above C, the compound becomes deliquescent.

The water activity at which an anhydrate/hydrate pair or a lower hydrate/higher hydrate pair reaches equilibrium defines the phase boundary of the two phases. The phase boundary can be determined experimentally using several approaches. A convenient approach is to investigate water sorption/desorption isotherms if compounds show a critical interconversion between different hydration phases. Such water sorption/desorption isotherms can be used to define phase boundaries and stability windows.

If solid-state conversions by water sorption/desorption are more gradual, interconversion experiments of anhydrate and hydrate in mixed solvents with water can be conducted to study their stability relationships. Water and water-miscible organic solvent mixtures of various compositions are used to control water activity. The material attained after slurrying is analyzed to determine which form is present at each selected water activity. Using this approach, the phase diagram for theophylline anhydrate and monohydrate has been determined at room temperature by Zhu et al. [19]. The phase boundary is at a water activity of 0.25. At $a_w < 0.25$, the anhydrate was the only solid phase recovered, no matter which solid form was initially added. At $a_w > 0.25$, the monohydrate was consistently obtained as the most stable form at equilibrium. Zhu et al. concluded that water activity is the major factor determining the nature of the solid phase of theophylline when it crystallizes from a solvent or solvent mixture. The transformation kinetics is faster in solution relative to the solid state, enabling the construction of a phase diagram in a shorter time frame.

Since the equilibrium constant K is a function of temperature, the relative stability between an anhydrate and a hydrate is significantly impacted by temperature, and, therefore, the phase boundary for a pair of anhydrate and hydrate should be shifted by a temperature change [20]. Krzyzaniak et al. have reported that the phase boundaries at different temperatures for caffeine anhydrate and hydrate are approximately 67% RH at 10 °C, 74.5% RH at 25 °C, and 86% RH at 40 °C [21]. As the temperature increases, the phase boundary moves to higher humidity. The importance of the effect of water activity on the transition temperature can be demonstrated with the following example where the transition temperature between carbamazepine anhydrous Form III and dihydrate is shown to be 64.5 °C in water while in an ethanol/water mixture with 31 mol% of water, the transition temperature becomes 14.3 °C [22].

3.6 Nonstoichiometric Hydrates

Nonstoichiometric hydrates display very different thermodynamic properties compared to stoichiometric hydrates. For example, no phase transitions occur for nonstoichiometric hydrates when water molecules move into or out of the crystal lattice, suggesting that these hydrates can be considered as solid solutions. Nonstoichiometric hydrates usually have channel or planar structures, as described earlier. Generally, the crystal form retains the same structure during dehydration or may show some anisotropic changes to accommodate the addition or leaving of water molecules. When the crystal structure is retained on dehydration, the resulting materials are referred to as desolvated solvates, as a unique type of anhydrous form. However, it is common that a nonstoichiometric hydrate may lose crystallinity or collapse to become amorphous when the very last few water molecules desorb [23].

As would be expected, water sorption isotherms of nonstoichiometric hydrates are different from those of stoichiometric hydrates [23]. Schematic shapes of possible isotherms are shown in Figure 3.6. As can be observed, water content changes continuously as water activity varies. The different ways that water interacts within the crystal lattice result in different types of isotherms, which, as shown in the following text, can be given the same classification as that for gas adsorption isotherms.

Type I isotherms are like Langmuir isotherms in gas absorption, where the plateau is interpreted as the state when the water channel is fully occupied. The strong interactions of water molecules with the host molecule cause water to be absorbed significantly as the water activity is slightly increased from the dry state. Hydrogen bonding between water and the host molecule is usually the predominant intermolecular force. For example, the hydration of celiprolol HCl Form III appears to produce a type I isotherm [24]. For RH greater than 40%, the stoichiometric ratio is close to that of a monohydrate. This amounts to 0.14 mole of water per mole of celiprolol HCl, which stays in the crystal lattice even when reducing the RH to about 0% RH.

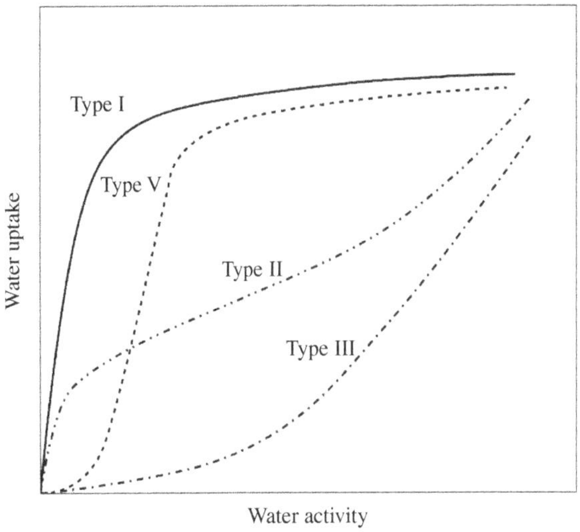

Figure 3.6 Sorption isotherms for nonstoichiometric hydrates. *Source*: Adapted from [23].

Type II isotherms are typically observed for systems that are associated with disordered water molecules in the crystal lattices. Disodium cromolyn is a traditional example of a nonstoichiometric hydrate that exhibits a type II isotherm [25]. It can take up about 9 mole of water per mol of the active compound into channels within the crystal structure. As a result, anisotropic expansion of the unit cell occurs during water sorption. Interestingly, two of the water molecules located in the crystal lattice are ordered enough to be located by X-ray analysis. The others, however, are mainly disordered and located in channels.

Type III isotherms follow the pattern normally observed for solids with disordered structure.

Type V isotherms are generally similar to type I isotherms regarding the limited number of sites in the crystal lattice that water molecules can occupy. In such cases, it seems that the interactions between water molecules are stronger than those with the host compound. Since some water molecules are inside the channel or plane, additional water molecules can more easily enter the lattice until reaching the level of saturation. As an example, the anhydrous form II of RPR102341, an antibacterial agent developed by Rhone Poulenc Rorer, exhibits a type V isotherm of water sorption [23]. It can take up to 1.2 mole water/mole host molecule reversibly if the RH is maintained at lower than 80%, and water incorporation causes the expansion of the unit cell dimensions. It converts to the trihydrate form at an RH greater than 90%.

3.7 Emerging Interests in Organic Solvates

Organic solvents are widely encountered in pharmaceutical development, such as in organic synthesis and purification of drug substances, in polymorph screening, in pharmaceutical analysis, and in some drug product manufacturing processes, including spray drying, wet granulation, and film coating. These operations may lead to the formation of a distinct multicomponent crystalline phase with solvent molecules as part of the crystal structure, as an organic solvate. The prevalence of organic solvates for pharmaceutical compounds is widely recognized. However, they are much less understood than pharmaceutical hydrates owing to their relatively limited use, stemming from safety concerns associated with hazardous solvents. Thus, under conventional thinking, not surprisingly, organic solvates are regarded as "waste land" instead of as resources in the solid form landscape of an API. Regardless, we want to emphasize that organic solvates, in many circumstances during the development process, can offer valuable solutions, for example, including the following:

- Impurity purging during organic synthesis and purification
- Bridging to more developable anhydrous forms or hydrate forms through desolvation or form conversion
- Developing acceptable drug products with class 3 solvents such as ethanol, isopropanol, acetone, and dimethyl sulfoxide (DMSO) when no acceptable anhydrous forms or hydrates are available

It is well recognized that organic solvates can offer a means of impurity purging. For example, Maini et al. disclosed significant challenges in purging *O*-enzalutamide, a major synthetic impurity for *S*-enzalutamide, the active ingredient in the prostate cancer drug Xtandi, through recrystallization as the anhydrous form R1. The difficulty arises due to the formation of a crystalline solid solution of *S*-enzalutamide and *O*-enzalutamide in the anhydrous form [26]. It was interesting to discover that the formation of the isopropanol solvate of *S*-enzalutamide through recrystallization or slurring leads to the effective purging of *O*-enzalutamide, which cannot be incorporated into the crystal lattice of the solvate form. The advantage of organic solvates in breaking the solid-state miscibility of impurities was demonstrated in another literature case on a penultimate intermediate. The compound could crystallize as an anhydrous form or as an acetonitrile solvate. While the anhydrous form was found to incorporate two major impurities into the crystal lattice with minimal utility of impurity purging, the acetonitrile solvate was discovered to have completely rejected the two impurities after recrystallization [27].

It is well recognized that desolvation of certain solvates can be used as a potential pathway to prepare unsolvated polymorphic forms, which could be difficult to obtain by conventional solution crystallization techniques. This is exemplified by a case presented by Desikan and colleagues [28] in the development of a fast-track HIV drug. The drug, given the name Compound A, exhibits prolific pseudopolymorphism, which presents challenges to prepare anhydrous crystal forms during development. The first anhydrous form, designated as Form I, was prepared initially by sublimation, a solvent-free method, but was found challenging to attain through crystallization. Crystallization from toluene/heptane yielded a mixed toluene/heptane solvate, for which desolvation resulted in a material with a significant amount of amorphous content. Further recrystallization in methanol also created a solvate. However, it was of significance to find that the stoichiometric methanol solvate obtained, on desolvation, transformed to Form I with the requisite quality and purity for use as an API, which was further formulated into drug products to support clinical studies for a fast-track project [28].

On the other hand, an organic solvate can serve as a bridging form to yield pharmaceutical hydrates, which are more acceptable for drug products, through solvent exchange. For example, Shiraki reported the successful preparation of a TRK-720 hydrate via solvent exchange from its methanolate. TRK-720 was a compound under development as a potential drug for the treatment of bronchial asthma [29]. To identify a crystal form suitable for development as a dry powder inhaler, a comprehensive polymorph screening was conducted using 42 solvents, yielding only nine organic solvates, all of which were deemed unacceptable as an API. Furthermore, desolvation of the solvates all leads to an amorphous form instead of an anhydrous crystalline form. Interestingly, water vapor substitution of the methanolate resulted in a hydrate, which demonstrated sufficient physical and chemical stability for further formulation work. Likewise, Ramakrishnan et al. reported that, for drug linagliptin, its methanolate or ethanolate as a channel-structured solvate can be converted to an iso-structural hydrate through solvent exchange [30]. Interestingly, the hydrate, which was found to be physically and chemically stable, could not be directly attained through crystallization in water. The solvent exchange process in the solid state was discovered to occur via a cooperative mechanism, preserving the integrity of the crystal lattice [30].

It is generally agreed that an organic solvate with a class 3 solvent is acceptable as a drug substance. However, this option is considered only under circumstances when anhydrous forms or hydrates are difficult to make or exhibit inferior developability, such as poor solubility, stability, or bioavailability, as illustrated in the cases of warfarin and trametinib.

Warfarin, one of the most prescribed oral anticoagulants, is an old drug developed in the 1950s but is still widely used. Owing to the significant solubility advantage, warfarin sodium, instead of the free form, was utilized in the initial oral formulations and intravenous (IV) formulations, as an isopropanol solvate. The discovery of warfarin isopropanol solvate was, indeed, a very fortunate event. During the development of warfarin sodium for IV products, some impurities, such as phenolic products, were very difficult to remove and moreover were detrimental to the product quality owing to color development and precipitation during storage. As can be imagined, without modern sophisticated chromatography tools, crystallization was the sole resort to purify a lifesaving drug. However, despite the small molecule size and relatively simple structure, warfarin sodium does not tend to form crystalline phases easily. As disclosed by Nathan et al., numerous trials with various organic solvents and combinations of solvents failed to yield crystalline materials except the isopropanol solvate [31]. It was found that the use of isopropanol yielded a highly crystalline material as a solvate, effectively separated from non-warfarin impurities that remained in solution. Further work by Sheth et al. revealed that, based on the crystal structure, the isopropanol solvate was shown to be an isolated site solvate with a 1 : 1 ratio between the drug and the solvent (Figure 3.7), which explains the superior physical stability of the solvate, with desolvation only initially occurring at about 150 °C with a peak temperature at 190 °C [32, 33].

Another less notable but interesting case is the trametinib DMSO solvate. Trametinib, a cancer drug developed for the treatment of melanoma, has the DMSO solvate as the active ingredient, because the solvate offers better solubility and more acceptable oral bioavailability than the stable polymorph. It is interesting to note that the initial batches of trametinib presented as a crystalline form with a melting point of 180 °C, fortunately had acceptable oral absorption in animal studies.

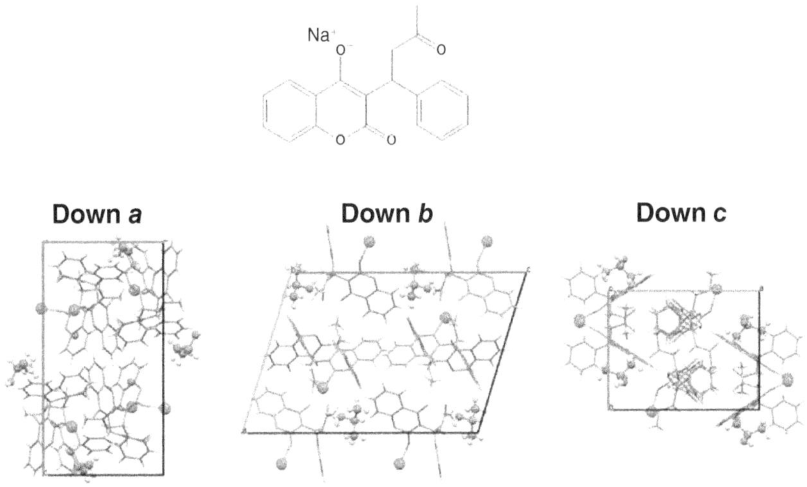

Figure 3.7 Molecular packing of warfarin sodium isopropanol viewed from three axes. *Source*: Shah et al. [33]. Reproduced with the permission of the American Chemical Society.

However, further process development and batch production led to the appearance of a more stable polymorph having a melting point of 300 °C, with poor oral exposure. Due to the lack of ionizable groups, salt formation to enhance solubility is not possible for the compound. Fortunately, several solvates with class 3 solvents, including the DMSO solvate, were identified and found to have comparable bioavailability to the initially discovered metastable form with the melting point at 180 °C. Ultimately, the DMSO solvate was selected due to its best animal exposure and acceptable physicochemical properties [34].

Beyond exploring their utility, the study of organic solvates is also significant for managing the downside of unwanted and unintentional solvate formation. As an increasing number of drug formulations are processed via spray drying using solvents like methanol, dichloromethane, and acetone, it becomes more and more important to screen for and study the potential formation of solvates with these common solvents. Obviously, the formation of a solvate within the spray-drying solution can result in drug precipitation, reduced solubility, and a significantly lower drug load. It is not surprising to note that, based on a report by Werner and Swift [35], the top two organic solvates in the 2020 Cambridge Structural Database are methanol and dichloromethane solvates.

3.8 Isostructural Solvates

One of the interesting properties associated with solvates, which we want to emphasize, is the formation of isostructural solvates. This occurs when multiple solvates of the same compound, sometimes including hydrates, assume a very similar or nearly identical crystal packing. This can potentially result in indistinguishable differences in lattice parameters and practically identical powder X-ray diffraction patterns. Isostructural solvates usually form when the crystal packing assumes a channel structure, where various types of solvent molecules can occupy. Isostructural solvates are commonly observed for organic compounds and are considered one of the main reasons why many compounds form multiple solvates. For example, drug dasatinib, a second-generation tyrosine kinase inhibitor, used for the treatment of a type of chronic myeloid leukemia, has been reported to have more than 60 solvated forms [36]. Among them, Sarceviča et al. isolated 26 organic solvates and found that 21 of them were structurally related as a group of isostructural solvates [36]. As seen from the crystal structures, common channels exist filled with solvent molecules supported by the same hydrogen bonding network, while the degree of disorder for solvent molecules varies with different solvates.

We can further illustrate the isostructural phenomenon using darunavir as an example. Darunavir is a potent HIV protease inhibitor marketed as the brand name Prezista for the treatment of HIV infection, using darunavir ethanolate as the API. It is interesting to note that, along with the discovery of darunavir ethanolate, many other solvates were found to exist, including the hydrate, methanolate, acetonate, dichloromethane, ethyl acetate solvates, and isopropanolate [37].

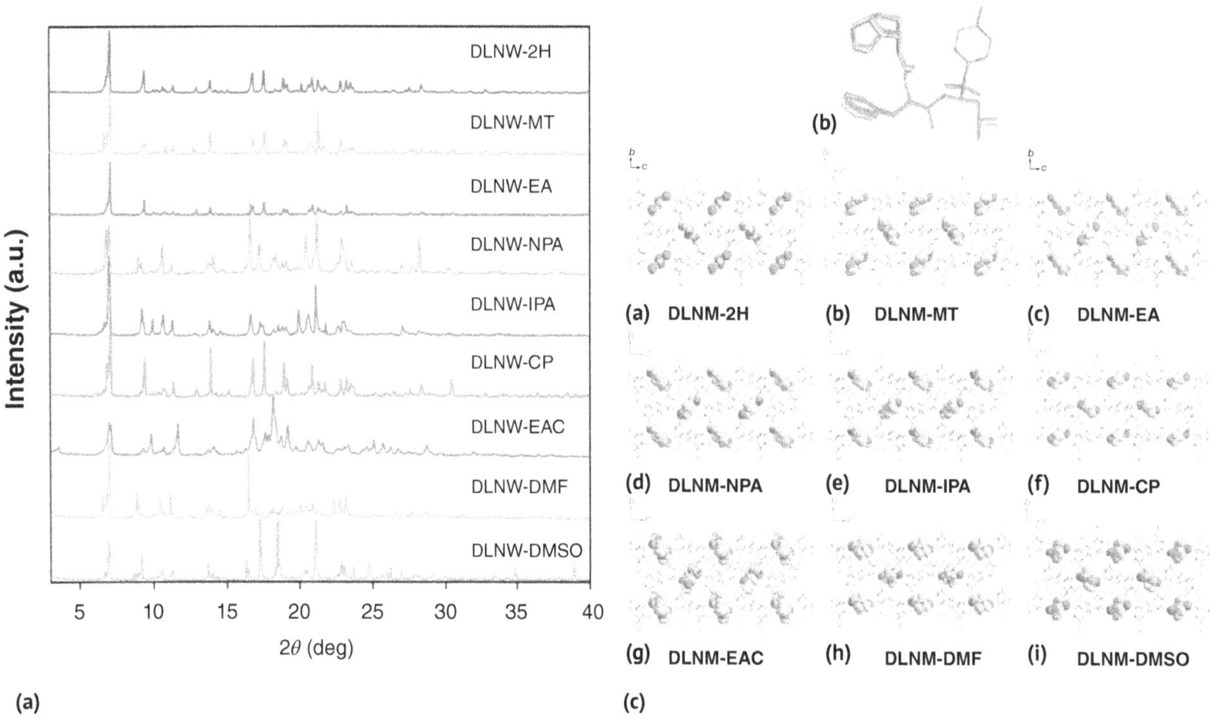

Figure 3.8 Comparison of darunavir isostructural solvates: (a) powder X-ray diffraction patterns; (b) molecular conformation of darunavir in the crystal structures; (c) crystal packing viewed from *a*-axis. *Source*: Yu et al. [38]. Reproduced with the permission of the American Chemical Society.

Further work by Yu et al. revealed that those solvates are isostructural crystal forms, as depicted in Figure 3.8 [38]. It is important to point out that all of the nine solvates investigated crystallize into the same orthorhombic space group $P2_12_12_1$ and exhibit similar unit cell parameters; for example:

- $a = 9.8421(4)–10.2299(6)$ Å,
- $b = 16.5592(8)–17.2427(7)$ Å,
- $c = 18.7465(7)–19.6891(10)$ Å, and
- $V = 3104.8(3)–3335.3(3)$ Å^3.

As can be seen in Figure 3.8, in the structures of the nine solvates, one darunavir molecule is present in an asymmetric unit and exists in a similar conformation with small differences in functional groups, involving hydrogen bonding with various solvents. The most important common structural feature among the solvates, commonly seen for many other isostructural solvates, is a channel structure, along the *a*-axis, with solvent molecules situated inside the channels. The high similarity in crystal structures of the isostructural solvates led to very similar powder X-ray diffraction patterns, as displayed in Figure 3.8, which are seen frequently for many other isostructural solvates.

However, as we can imagine, to fit into channels with similar void volumes, solvents with different molecular sizes, shapes, polarity, and hydrogen bonding potentials would induce a notable diversity in solvent ratio, orientation, and hydrogen bonding, which would obviously lead to different physical and chemical properties of those isostructural solvates. Indeed, for example, smaller solvent molecules like water and methanol form solvates with darunavir at a 1:2 drug:solvent ratio, while larger solvent molecules such as ethanol, isopropanol, and ethyl acetate tend to form a solvate at a 1:1 ratio. In addition, as reported by Yu et al., there are four types of hydrogen bonding patterns in the solvent channels among the nine isostructural solvates. It is obvious that the molecular volume and functional group of a particular solvent molecule dictate different packing modes and hydrogen bonding pattern in the channels, which, in turn, would cause the differentiation of physical and chemical properties as a leading reason why the ethanol solvate was selected as the crystal form to develop instead of the hydrate form.

Among the solvates of darunavir, the ethanolate and the hydrate are both regarded as pharmaceutically acceptable, usually with the hydrate being preferred due to safety advantages. However, the darunavir ethanolate displays much better thermal physical stability. As revealed by thermal analysis data, the darunavir hydrate readily loses water starting from ambient temperatures with a dehydration endotherm at 77 °C, while the darunavir ethanolate exhibits minimal solvent loss below 90 °C, with most desolvation events occurring above 100 °C. This physical stability against desolvation is considered a key quality attribute due to a concern that the amorphous form, potentially generated from dehydration or desolvation, may compromise the quality and stability of the drug product.

3.9 Dehydration and Desolvation

The significance of understanding the process, mechanism, and kinetics of dehydration and desolvation arises from three aspects. First and foremost, such knowledge is essential to establish stability windows for pharmaceutical hydrates or solvates used in drug products as an active ingredient. Second, the resulting solid formed after the loss of solvent can be very diverse, offering a means to enrich the solid form landscape, and sometimes a unique way to prepare certain anhydrous polymorphs. Third, dehydration and desolvation can sometimes completely disrupt the crystalline structure, producing an amorphous material with increased solubility. Such amorphous materials may be challenging to make using conventional methods such as melting, evaporation, spray drying, or freezing drying.

As has been discussed earlier, changes in the hydration state of crystals can be induced by environmental factors such as water activity and temperature. For a nonstoichiometric hydrate, hydrate/dehydration is a continuous process without a well-defined phase boundary. Conversely, hydration and dehydration of a stoichiometric hydrate are regarded as a phase transformation that can occur in the solution or solid states. The hydrate form, generally more stable thermodynamically than the anhydrous form in a particular range of water activity and temperature, nucleates and grows at the expense of the anhydrous form or a lower hydrate.

Dehydration of a hydrate from a crystal lattice can produce an anhydrous form, a lower hydrate with a distinctly different crystal structure, an isomorphic desolvate (desolvated solvate) with a similar crystal lattice to the original hydrate, or a less crystalline/noncrystalline material. The resulting solid form from dehydration is determined by the crystal structure, environmental conditions, and dehydration kinetics. For example, niclosamide, an anthelmintic drug to treat tapeworm infections, can exist as two monohydrates, HA and HB, with distinctive crystal packing and different environments for water molecules. In the monohydrate HA, there are channels along a-axis, whereas in the monohydrate HB, water molecules are located in isolated cavities. It is interesting to see that the water loss from the monohydrate HA triggered by a temperature increase proceeds by diffusion out of the lattice channels and yields an isomorphic desolvate. In contrast, the dehydration of the monohydrate HB results in transformation to a stable anhydrous polymorph F1 through a surface nucleation mechanism, where water loss and phase transformation occur simultaneously [39].

The mass transfer mechanisms and kinetics of water molecules leaving the lattice determine the kinetics of the dehydration of a hydrate, which in turn is significantly impacted by the crystal structure of the hydrate. It is expected that the dehydration rate would be very slow for isolated site hydrates before reaching the onset temperature, which is usually close to the melting point, where the lattice collapses. The rate of dehydration suddenly increases rapidly after the onset because the water molecules are all in similar chemical environments in the crystal lattice. For channel hydrates, dehydration typically begins at a relatively low temperature and may proceed either as a continuous process or in several steps over a broad temperature range. This is because the water molecules in different environments of the lattice have different energies. At a given temperature, therefore, the energy available for dehydration is only enough for some fraction of the water molecules to leave. For example, eprosartan mesylate dihydrate undergoes dehydration in two distinct steps [40]. The first step occurs at 25–70 °C with the loss of 1 mole of water, whereas the loss of the other mole of water in the second step happens at 70–120 °C.

Some hydrates have very loosely bound water and are easily dehydrated. For example, a trihydrate form of acetaminophen, the active ingredient of Tylenol, has been discovered by crystallizing the material from ice-cold water [41]. Though the trihydrate form is physically stable at 0 °C, it was shown to be quickly dehydrated at 22 °C, as shown in Figure 3.9, wherein a transparent and highly crystalline trihydrate crystal becomes opaque within 12 seconds at this elevated temperature. It was shown further that the microcrystalline Form I of acetaminophen was produced after such treatment. A clear phase boundary was observed to move through channels along the long axis. Similar dehydration behavior was also observed for caffeine hydrate [42], where the dehydration occurred along the c-axis where the water molecules pack as channels. The dehydration of the caffeine hydrate took 72 hours to complete at room temperature.

Thus, determining the temperature-dependent dehydration behavior could offer a reasonable assessment of the physical stability of a pharmaceutical hydrate. It is generally agreed that the onset temperature for dehydration from thermal analysis can serve as a good indicator; the higher the onset temperature, the better the stability. After studying various

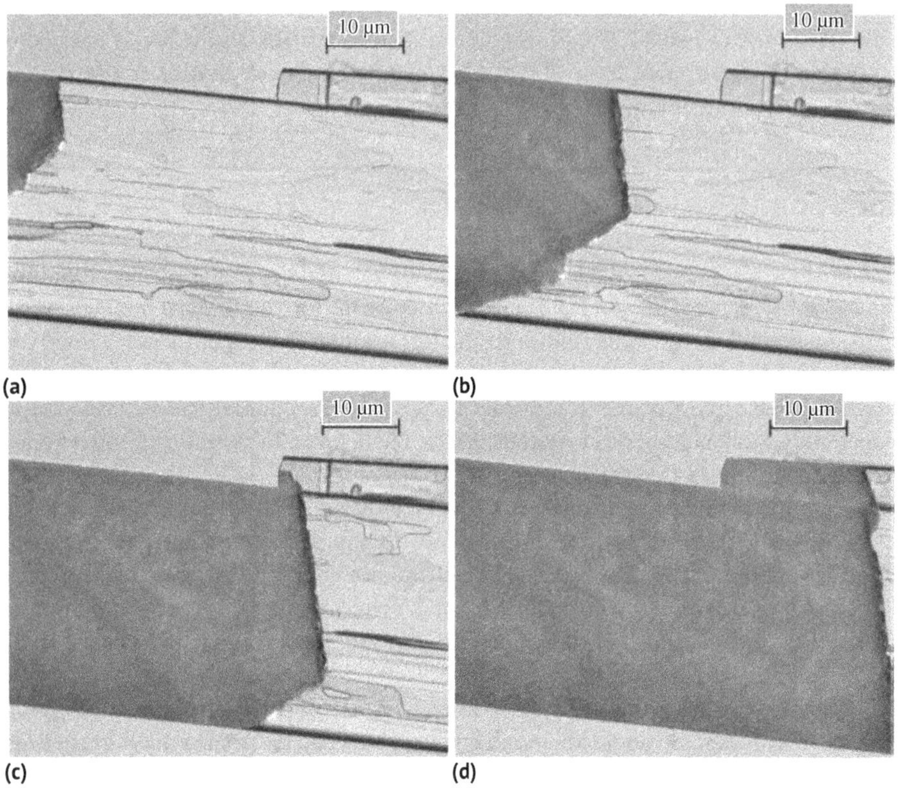

Figure 3.9 Dehydration of acetaminophen trihydrate in air at 22 °C: (a) 0 s, (b) 4 s, (c) 8 s, and (d) 12 s. *Source*: Peterson et al. [41]. Reproduced with the permission of the American Chemical Society.

pharmaceutical hydrates using thermal analysis, Shimanovich et al. concluded that the dehydration activation energy should be taken into consideration along with the onset temperature to assess a hydrate's physical stability against a reference set of pharmaceutical hydrates with known stability profiles as a benchmark [43]. To further improve the validity of using dehydration data to rank order stability of various hydrates, Takahashi and Uekusa developed a unique approach to use both dehydration activation energy and rehydration behavior to classify pharmaceutical hydrates as follows [44]:

- Class I: low dehydration energy and easy and almost reversible rehydration
- Class II: low dehydration energy and difficult rehydration
- Class III: high dehydration energy and difficult rehydration

It appears that the preferred stability order is Class III > Class I > Class II. It is interesting to note that the classification is strongly correlated with the structural characteristics of the hydrates. For example, among 11 pharmaceutical hydrates studied, Class I hydrates have a channel-like structure and likely remain as isomorphous dehydrates after removing the water, with rehydration almost reversible, whereas Class II hydrates tend to become amorphous with the crystal lattice collapsing after water removal owing to layer-like packing, as, for example, amoxicillin trihydrate. It is not surprising to note that Class III hydrates have water molecules residing at isolated lattice sites, using nitrofurantoin monohydrate and tazobactam hemihydrate as examples.

As would be expected, the structural category of solvates, either as an isolated site solvate or a channel solvate, dictates the stability and desolvation behavior for an organic solvate. For example, warfarin sodium isopropanolate, as an isolated site solvate, exhibits a desolvation event above 150 °C [33]. On the other hand, spontaneous desolvation does occur for many unstable organic solvates with channel structures. For example, some solvates desolvate readily right after removal from the mother liquor, as seen by Braun et al. for four solvates of R-cinacalcet hydrochloride, a drug used to treat calcimimetics. The four monosolvates with acetic acid, chloroform, 1,4-dioxane, and tetrachloromethane, were isolated initially as transparent, thin, needle-like crystals, losing transparency within hours after exposure to ambient air. Interestingly, it was discovered that the desolvation of the four solvates generates the metastable form II, which was not identified from solution-based crystallization and screening [45]. Desolvation, therefore, not only provides a unique way to generate novel polymorphs but also serves as an alternative means for producing known polymorphs. As reported by Nowak, fluconazole can form two types of isostructural channel solvates with various solvents. Type I solvates desolvate to become polymorph form I, presumably through a coordinated hydrogen-bonded chain rearrangement, whereas type II solvates, after desolvation, transform to an isomorphic desolvate first and then proceed to yield a polymorph form II [46].

From the previous case, it appears that the structural characteristics of a solvate also determine the primary physical form that occurs after desolvation. This observation is further supported by the work on methyl cholate solvates by Bērziņš et al. [47]. They studied the desolvation mechanism of 25 methyl cholate solvates, which are associated with six isostructural solvate groups, and found out that the primary desolvation product, either as a polymorph or as an isomorphous desolvate, has a similar structure to that of the respective solvate [47]. For example, after desolvation, a type 2 solvate, such as the acetonitrile solvate, yields polymorph IV having similar crystal packing. Likewise, a type 1 solvate, like the methanol solvate, transforms to polymorph III with a comparable crystal structure. It is interesting to note that several thermodynamically favorable polymorphs, including I, II, and VI, which can be readily attained from solution crystallization, were never the primary desolvation products. Furthermore, the amorphous form was obtained for several type 6 solvates after desolvation, revealing that the thermodynamic stability of the desolvated product does not dictate the desolvation mechanism [47].

3.10 Preparation and Characterization of Hydrates and Solvates

Discovering hydrates and organic solvates is usually part of the polymorph screening process. Approaches that can be used include crystallization, slurrying, and evaporation from water, solvents, and solvent mixtures containing water. Some hydrates can be identified during moisture sorption experiments or by exposure to high humidity. A vapor deposition method is also applicable for some organic solvates with solvents having low boiling points. Table 3.1 gives an example showing that different means are available to prepare various solvates as described by Sarceviča and colleagues on the preparation of 26 dasatinib solvates [36]. Among them, six solvates were attained through vapor deposition, ten from slurrying, seven by evaporation, one by desolvation using another solvate followed by vapor deposition, and one by solvent-drop grinding. Zvonicek et al. performed a more systematic screening using ibrutinib as a model compound in 11 solvents and tested several common solution-based methods, including quick cooling, slow cooling, quick evaporation, slow evaporation, slurrying with the metastable form, and slurrying with the stable form. The slurry method with the metastable form turned out to be the most successful one to yield 100% hits [48].

During the design of screening techniques and following characterization experiments for a drug, it is important to be aware of the potential occurrence of solvate/hydrate forms, with solvent and water molecules present in the same crystal structures. For example, Othman et al. showed that finasteride, a steroid for the treatment of benign prostatic hypertrophy, exists in four solvates formed with isopropanol, acetic acid, dioxane, and ethyl acetate, respectively, all found to be hemihydrates as well [49]. All the solvate hydrate forms were found to be isostructural and to have a finasteride to solvent molar ratio of 2:1, with solvent molecules being highly disordered within channels.

Knowledge about a hydrate or a solvate is collected through a comprehensive characterization protocol using various techniques. Commonly used techniques include powder X-ray diffractometry, single-crystal X-ray crystallography, thermogravimetric analysis, differential scanning calorimetry, microscopy, infrared spectroscopy, FT-Raman, solution NMR, gas chromatography, and Karl Fisher titration. The combined use of several of these techniques is necessary to provide a definitive identification of a particular hydrate or solvate. In summary,

- X-ray powder diffraction confirms crystallinity,
- thermogravimetric analysis detects weight loss due to desolvation or dehydration,
- differential scanning calorimetry reveals thermal events related to desolvation or dehydration and phase transformation,
- Karl Fisher titration monitors the amount of water present and hydrate stoichiometry,
- gas chromatography and solution NMR confirm the type and amount of solvents incorporated into solid materials,
- hot stage microscopy allows direct observation of the solid during dehydration or desolvation,
- single-crystal X-ray crystallography confirms the presence and location of water or solvent in the crystal lattice,
- variable temperature X-ray powder diffraction helps to directly monitor the various possible phase transformations that occur after desolvation/dehydration, and
- spectroscopic techniques like FT-IR, FT-Raman, and solid-state NMR provide a means of differentiating various possible hydrates and anhydrates.

Table 3.1 Preparation procedures for dasatinib organic solvates and hydrates.

Solvate	Procedure
Acetonitrile solvate	Vapor deposition, to expose the anhydrous form N6 with the vapor of the solvent of choice in desiccators for 1 week at ambient temperature
Tetrahydrofuran solvate	
Chloroform solvate	
Dichloromethane solvate	
Toluene solvate	
Ethyl acetate solvate	
Methyl ethyl ketone solvate	Slurrying, to slurry the anhydrous form N6 in the corresponding solvent for 1 week at ambient temperature
Methyl acetate solvate	
Cyclohexanone solvate	
2-Methoxylethanol solvate	
n-Heptanol solvate	
Hemi-ethanolate	
Isobutanol solvate	
Methyl tert-butyl ether	Slurrying, to slurry the anhydrous form N6 in the corresponding solvent for 4 hours at 50 °C
Monohydrate	Slurrying, to slurry the anhydrous form N6 in water for 48 hours at ambient temperature
Isopropanol ethyl acetate solvate	Evaporation, to dissolve the anhydrous form N6 in the solvent of interest at 70–80 °C and evaporate at ambient temperature
Butanol solvate	
Isopropanol solvate	
n-Propanol solvate	
1,3-Dixolane solvate	
DMSO solvate	
Hemi-*n*-butyl acetate solvate	
Diethanolate	Desolvation, then vapor deposition, to desolvate the hemi-ethanolate for 24 hours at 80 °C and then to expose the sample to dry ethanol vapor for 48 hours at ambient temperature
Nitromethane solvate	Mechanochemical method, to mill the anhydrous form N6 with a few drops of nitromethane in a ball mill for 15 minutes

Various case studies about solvate characterization are discussed in the following chapters about each of the specific characterization techniques, as well as in Chapter 24 dealing with mixture analysis.

References

1 Rodríguez-Spong B, Price CP, Jayasankar A, Matzger AJ, Rodríguez-Hornedo N. General principles of pharmaceutical solid polymorphism: a supramolecular perspective. *Ad Drug Del Rev* 2004; 56(3): 241–274.

2 Griesser UJ. The importance of solvates. In: Hilfiker R, ed. *Polymorphism in the Pharmaceutical Industry*. Weinheim: Wiley-VCH Verlag GmbH & Co., 2006: 211–257.

3 Pudipeddi M, Serajuddin A. Trends in solubility of polymorphs. *J Pharm Sci* 2005; 94(5): 929–939.

4 Reutzel-Edens SM, Kleemann RL, Lewellen PL, Borghese AL, Antoine LJ. Crystal forms of LY334370 HCl: Isolation, solid-state characterization, and physicochemical properties. *J Pharm Sci* 2003; 92(6): 1196–1205.

5 Noyes AA, Whitney WR. The rate of solution of solid substances in their own solutions. *J Am Chem Soc* 1897; 19(12): 930–934.

6 Kobayashi Y, Ito S, Itai S, Yamamoto K. Physicochemical properties and bioavailability of carbamazepine polymorphs and dihydrate. *Int J Pharm* 2000; 193(2): 137–146.

7 Anderson GD. Understanding the ramifications of switching among AEDs: What are the data. *Adv Stud Pharm* 2008; 5(5): 146–151.

8 Miyazaki T, Sivaprakasam K, Tantry J, Suryanarayanan R. Physical characterization of dibasic calcium phosphate dihydrate and anhydrate. *J Pharm Sci* 2009; 98(3): 905–916.

9 Lerk CF, Zuurman K, Kussendrager K. Effect of dehydration on the binding capacity of particulate hydrates. *J Pharm Pharmcol* 1984; 36(6): 399.

10 Ertel KD, Carstensen JT. Chemical, physical, and lubricant properties of magnesium stearate. *J Pharm Sci* 1988; 77(7): 625–629.

11 Wada Y, Matsuhara T. Pseudopolymorphism and lubricating properties of magnesium stearate. *Powder Technol* 1994; 78(2): 109–114.

12 Haleblian JK. Characterization of habits and crystalline modification of solids and their pharmaceutical applications. *J Pharm Sci* 1975; 64(8): 1269–1288.

13 Morris KR. Structural aspects of hydrates and solvates. In: Britain HG, ed. *Polymorphism in Pharmaceutical Solids*. New York: Marcel Dekker, 1999: 125–181.

14 Zimmermann A, Tian F, De Diego HL, Frydenvang K, Rantanen J, Elema MR, Hovgaard L. Structural characterisation and dehydration behaviour of siramesine hydrochloride. *J Pharm Sci* 2009; 98(10): 3596–3607.

15 Sun C, Zhou D, Grant DJW, Young Jr VG. Theophylline monohydrate. *Acta Cryst E Struct Rep Online* 2002; 58(4): o368–o370.

16 Chen LR, Young Jr VG, Lechuga-Ballesteros D, Grant DJ. Solid-state behavior of cromolyn sodium hydrates. *J Pharm Sci* 1999; 88(11): 1191–1200.

17 Kiang YH, Xu W, Stephens PW, Ball RG, Yasuda N. Layered structure and swelling behavior of a multiple hydrate-forming pharmaceutical compound. *Cryst Growth Des* 2009; 9(4): 1833–1843.

18 Redman-Furey N, Dicks M, Bigalow-Kern A, Cambron RT, Lubey G, Lester C, Vaughn D. Structural and analytical characterization of three hydrates and an anhydrate form of risedronate. *J Pharm Sci* 2005; 94(4): 893–911.

19 Zhu H, Yuen C, Grant DJ. Influence of water activity in organic solvent + water mixtures on the nature of the crystallizing drug phase. 1. Theophylline. *Int J Pharm* 1996; 135(1): 151–160.

20 Khankari RK, Grant DJ. Pharmaceutical hydrates. *Thermochim Acta* 1995; 248: 61–79.

21 Krzyzaniak J, Williams GR, Ni N. Identification of phase boundaries in anydrate/hydrate systems. *J Pharm Sci* 2007; 96(5): 1270–1281.

22 Tian F, Qu H, Zimmermann A, Munk T, Jørgensen AC, Rantanen J. Factors affecting crystallization of hydrates. *J Pharm Pharmcol* 2010; 62(11): 1534–1546.

23 Authelin JR. Thermodynamics of non-stoichiometric pharmaceutical hydrates. *Int J Pharm* 2005; 303(1): 37–53.

24 Burger A, Ratz AW, Zölß G. Polymorphie und pseudopolymorphie von celiprololhydrochlorid. *Acta Pharm Technol* 1988; 34(3): 147–151.

25 Stephenson GA, Diseroad BA. Structural relationship and desolvation behavior of cromolyn, cefazolin and fenoprofen sodium hydrates. *Int J Pharm* 2000; 198(2): 167–177.

26 Maini L, Braga D, Farinella F, Melotto E, Verzini M, Brescello R, Michieletto I, Munari I. Crystal forms of enzalutamide and a crystal engineering route to drug purification. *Cryst Growth Des* 2018; 18(7): 3774–3780.

27 Nordstrom FL, Sirota E, Hartmanshenn C, Kwok TT, Paolello M, Li H, Abeyta V, Bramante T, Madrigal E, Behre T, Capellades G. Prevalence of impurity retention mechanisms in pharmaceutical crystallizations. *Org Process Res Dev* 2023; 27(4): 723–741.

28 Desikan S, Parsons RL, Davis WP, Ward JE, Marshall WJ, Toma PH. Process development challenges to accommodate a late-appearing stable polymorph: a case study on the polymorphism and crystallization of a fast-track drug development compound. *Org Process Res Dev* 2005; 9(6): 933–942.

29 Shiraki M. Physical characterization of TRK-720 hydrate, the very late antigen-4 (VLA-4) inhibitor, as a solid form for inhalation: Preparation of the hydrate by solvent exchange among its solvates and mechanistical considerations. *J Pharm Sci* 2010; 99(9): 3986–4004.

30 Ramakrishnan S, Yarraguntla SR, Peddireddy SR, Kanniah SL, Mudapaka VK, Shekhawat LK, Mahapatra S, Mohammad AB, Vishweshwar P, Stephens PW. Development of pharmaceutically acceptable crystalline forms of drug substances via solid-state solvent exchange. *Org Process Res Dev* 2017; 21(10): 1478–1487.

31 Nathan W, Moe J, Schroeder CH, Paul LK, inventors; Wisconsin Alumni Research Foundation, assignee. *Preparation of Crystalline Warfarin Sodium-Isopropyl Alcohol Complex*. Alexandria: United States patent US; 1966. 3,246,013. Apr 12.

32 Sheth AR, Brennessel WW, Young VG, Muller FX, Grant DJ. Solid-state properties of warfarin sodium 2-propanol solvate. *J Pharm Sci* 2004; 93(11): 2669–2680.

33 Shah HS, Chaturvedi K, Dave RH, Morris KR. Molecular insights into warfarin sodium 2-propanol solvate solid form changes and disproportionation using a low volume two-stage dissolution approach. *Mol Pharm* 2021; 18(4): 1779–1791.

34 Abe H, Kikuchi S, Hayakawa K, Iida T, Nagahashi N, Maeda K, Sakamoto J, Matsumoto N, Miura T, Matsumura K, Seki N. Discovery of a highly potent and selective MEK inhibitor: GSK1120212 (JTP-74057 DMSO solvate). *ACS Med Chem Lett* 2011; 2(4): 320–324.

35 Werner JE, Swift JA. Organic solvates in the Cambridge structural database. *CrystEngComm* 2021; 23(7): 1555–1565.

36 Sarceviča I, Grante I, Belyakov S, Rekis T, Bērziņš K, Actiņš A, Orola L. Solvates of dasatinib: diversity and isostructurality. *J Pharm Sci* 2016; 105(4): 1489–1495.

37 Vermeersch HW, Thoné DJ, Janssens LD, inventors; Tibotec Pharmaceuticals Ltd, assignee. *Pseudopolymorphic Forms of a HIV Protease Inhibitor*. Alexandria: United States patent US; 2010. 7,700,645. Apr 20.

38 Yu K, Shen J, Liu J, Tang G, Hu X. Stability and mechanical properties of darunavir isostructural solvates: an experimental and computational study. *Cryst Growth Des* 2023; 23(4): 2905–2915.

39 Mann JE, Gao R, Swift JA. Dehydration of niclosamide monohydrate polymorphs: different mechanistic pathways to the same product. *Cryst Growth Des* 2023; 7: 5102–5111.

40 Sheng J, Venkatesh GM, Duddu SP, Grant DJ. Dehydration behavior of eprosartan mesylate dihydrate. *J Pharm Sci* 1999; 88(10): 1021–1029.

41 Peterson ML, McIlroy D, Shaw P, Mustonen JP, Oliveira M, Almarsson Ö. Crystallization and transformation of acetaminophen trihydrate. *Cryst Growth Des* 2003; 3(5): 761–765.

42 Lin CT, Byrn SR. Desolvation of solvated organic crystals. *Mol Cryst Liq Cryst* 1979; 50(1): 99–104.

43 Shimanovich R, Cooke M, Peterson ML. A rapid approach to the preliminary assessment of the physical stability of pharmaceutical hydrates. *J Pharm Sci* 2012; 101(10): 4013–4017.

44 Takahashi M, Uekusa H. Dehydration and rehydration mechanisms of pharmaceutical crystals: classification of hydrates by activation energy. *J Pharm Sci* 2022; 111(3): 618–627.

45 Braun DE, Kahlenberg V, Gelbrich T, Ludescher J, Griesser UJ. Solid state characterisation of four solvates of R-cinacalcet hydrochloride. *CrystEngComm* 2008; 10(11): 1617–1625.

46 Nowak M, Dyba AJ, Janczak J, Morritt A, Fabian L, Karolewicz B, Khimyak YZ, Braun DE, Nartowski KP. Directing crystallization outcomes of conformationally flexible molecules: polymorphs, solvates, and desolvation pathways of fluconazole. *Mol Pharmaceutics* 2022; 19(2): 456–471.

47 Bērziņš A, Trimdale A, Kons A, Zvaniņa D. On the formation and desolvation mechanism of organic molecule solvates: A structural study of methyl cholate solvates. *Cryst Growth Des* 2017; 17(11): 5712–5724.

48 Zvonicek V, Skorepova E, Dusek M, Babor M, Zvatora P, Šoóš M. First crystal structures of pharmaceutical ibrutinib: systematic solvate screening and characterization. *Cryst Growth Des* 2017; 17(6): 3116–3127.

49 Othman A, Evans JS, Evans IR, Harris RK, Hodgkinson P. Structural study of polymorphs and solvates of finasteride. *J Pharm Sci* 2007; 96(5): 1380–1397.

4

Pharmaceutical Salts

4.1 Introduction

It is estimated that 50% of all drug molecules present in marketed products are administered as salts [1–3]. A survey of the top 50 prescription drugs in 2018, based on the number of prescriptions, has demonstrated that salt forms are of use in about 54% of widely used drugs [4]. For example, atorvastatin, the active ingredient of Lipitor, is used as a calcium salt, and the hydrogen sulfate salt of clopidogrel is present in the drug product Plavix. Furthermore, among the 129 new drugs approved by the FDA from 2015 to 2019, pharmaceutical salts account for 61 drugs, about 48% [3]. It is interesting to note that the HCl salt is still the most predominant salt for weak bases, whereas the sodium salt is for weak acids, in agreement with the assessment published by Berge, Brighley, and Monkhouse in 1974 [5].

Salt formation is popular because it offers a convenient means of modifying and optimizing the physical and chemical properties of an ionizable compound. The crystal structure of a salt form is completely different from that of the parent compound and is also distinct from the other salts of the same compound. Therefore, a salt form often displays different physical and chemical attributes compared to the parent drug and its other salts. By using a variety of pharmaceutically acceptable counterions, properties such as solubility, dissolution, hygroscopicity, stability, crystallinity, impurity profiles, and crystal habit can be influenced and improved after salt formation.

4.2 Importance of Pharmaceutical Salts

Pharmaceutical salts are of paramount significance for pharmaceutical development, with wide applications to:

- increase solubility, dissolution rates, and oral bioavailability for poorly soluble compounds,
- enable sufficient solubility to develop as injectables,
- decrease solubility to achieve extended release, retard Ostwald ripening, or accomplish taste masking for very soluble compounds,
- improve physical properties such as melting temperature, crystallinity, hygroscopicity, and mechanical properties,
- change chemical stability and overcome incompatibility with excipients,
- improve purity, achieve chiral resolution, and filterability, and
- enhance intellectual property due to improved properties.

We begin with clopidogrel as an example to illustrate the versatility of pharmaceutical salts to facilitate various aspects of pharmaceutical development. Clopidogrel is a potent antiplatelet drug, marketed with the brand name Plavix, which has been widely prescribed for the prevention of vascular thrombotic events in patients at risk. Clopidogrel is a chiral compound, and its S enantiomer is used in its drug products because the corresponding R enantiomer lacks biological activity and shows poor tolerability [6]. Salt formation with a chiral acid, 10-L-camphorsulfonic acid, was utilized to resolve the S enantiomer from the racemic mixture of clopidogrel [7]. Camphor sulfate is not an acceptable counterion for pharmaceutical use; therefore, the camphor sulfate salt was treated with a base to form the S enantiomer of the clopidogrel free base. The free base is an oily semisolid and unstable chemically, in that it has a labile proton in the

Solid-State Materials in Pharmaceutical Chemistry: Properties, Characterization, and Applications, Second Edition. Stephen R. Byrn, George Zografi and Xiaoming (Sean) Chen.

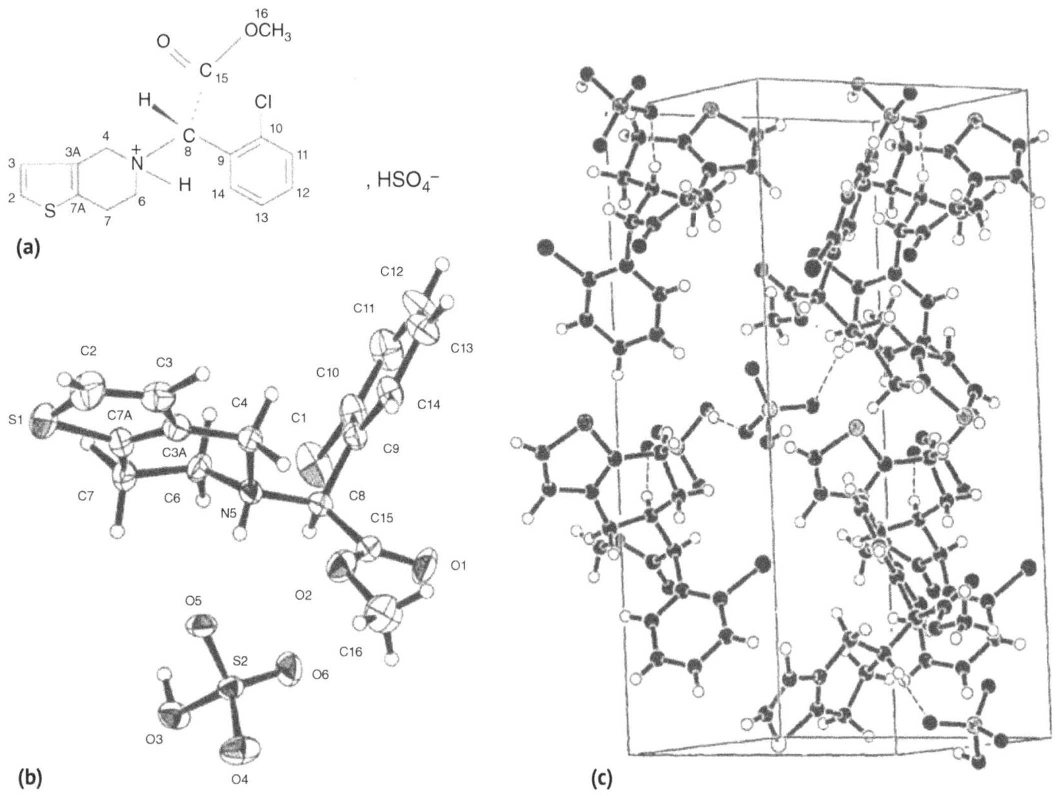

Figure 4.1 Structure (a), Ortep drawing (b), and crystal packing (c) of clopidogrel hydrogen sulfate. *Source*: [8]/John Wiley & Sons.

chiral center and is susceptible to racemization. Consequently, salt formation with a pharmaceutically acceptable counterion is necessary for the improvement of chemical stability and the development of solid dosage forms for clopidogrel [7]. Clopidogrel base is a weak base with a pK_a of 4.55. The low pK_a value dictates that only strong acids like HCl can form a stable salt with it. Though clopidogrel can form a stable salt with HCl, the HCl salt is hygroscopic and unstable. Through comprehensive salt screening using more than 20 counterions, the hydrogen sulfate salt was the only salt with the desirable properties: high melting point, good long-term stability, non-hygroscopicity, and good solubility. The clopidogrel hydrogen sulfate salt has two polymorphs, Forms 1 and 2. Form 2 is the thermodynamically stable polymorph at room temperature and is used in clopidogrel commercial products [7]. It has a melting point of 176 °C and exhibits a crystal structure as shown in Figure 4.1 [8], with an orthorhombic space group $P2_12_12_1$ and unit cell dimensions of $a = 10.321(6)$ Å, $b = 20.118(9)$ Å, $c = 9.187(7)$ Å, and $\alpha = \beta = \gamma = 90°$. As revealed in the crystal structure, the nitrogen atom in the pyridine ring is protonated and has a strong ionic interaction with the hydrogen sulfate anion. In summary, salt formation allowed for the resolution of the S enantiomer of clopidogrel from its racemic mixture and resulted in a physically stable hydrogen sulfate, which can be formulated as a solid dosage form. Interestingly, the calculated pK_a of the hydrogen sulfate in the clopidogrel salt indicates it is a strong enough acid to protonate clopidogrel, yet the crystal structure shows the hydrogen sulfate ion is present in the crystal.

4.3 Weak Acid, Weak Base, and Salt

The chemistry of salt formation essentially involves an acid–base reaction. Proton transfer occurs during any acid–base reaction, and the driving force for the reaction is determined by the ionization equilibrium constants of the acid and the base. The ionization equilibrium of a monoprotic acid HA is given by

$$HA \leftrightarrow H^+ + A^- \tag{4.1}$$

The ionization equilibrium constant K_a is defined as

$$K_a = \frac{[H^+][A^-]}{[HA]} \tag{4.2}$$

For convenience, the ionization of a monobasic compound B is expressed as the dissociation of its conjugated acid BH^+ as follows:

$$BH^+ \leftrightarrow H^+ + B \tag{4.3}$$

The K_a becomes

$$K_a = \frac{[H^+][B]}{[BH^+]} \tag{4.4}$$

K_a is usually expressed as pK_a, the negative logarithm of K_a, as follows:

$$pK_a = -\log K_a \tag{4.5}$$

For example, acetic acid has an K_a of 1.738×10^{-5}, and the calculated pK_a for acetic acid is 4.76. The K_a and pK_a of ammonia, a base, are 4.4×10^{-10} and 9.36, respectively.

The pK_a of a substance is a key parameter in classifying the strength of an acid and a base, as shown in Table 4.1 [9]. Most drugs with ionizable groups fall into the categories of weak/very weak acids or weak/very weak bases.

The ionization equilibrium of weak acids or bases can be conveniently described by the Henderson–Hasselbalch equation, which can be derived by rearranging equation (4.2) to give

$$\frac{1}{[H^+]} = \frac{1}{K_a}\frac{[A^-]}{[HA]} \tag{4.6}$$

Taking the logarithm of both sides of equation (4.6) provides

$$-\log[H^+] = -\log K_a + \log\frac{[A^-]}{[HA]} \tag{4.7}$$

Substituting pH for $-\log[H^+]$ and pK_a for $-\log K_a$ yields the Henderson–Hasselbalch equation

$$pH = pK_a + \log\frac{[A^-]}{[HA]} \tag{4.8}$$

The pH of a weak acid solution can be calculated from equation (4.8) if the molar ratio of ionized to non-ionized species and the pK_a are known. The pH is equal to pK_a when the ratio is one to one.

The Henderson–Hasselbalch equation for a base is given as follows:

$$pH = pK_a + \log\frac{[B]}{[BH^+]} \tag{4.9}$$

Table 4.1 Classification of acids and bases based on pK_a.

	pK_a	
Category	**Acids**	**Bases**
Very strong	<0	>14
Strong	0–4.5	9.5–14
Weak	4.5–9.5	4.5–9.5
Very weak	9.5–14	0–4.5
Extremely weak	>14	< 0

The probability of successful salt formation is related to the difference between the pK_a value of the drug and its counterion. For the formation of a stable salt, it is generally accepted that there should be a minimum difference of about 2–3 units between the pK_a value of a basic component and that of an acidic component [10]. The salt formation process between an acid–base pair HA and B can be described by the following chemical equations:

$$\mathrm{HA} + \mathrm{B} \leftrightarrow \mathrm{A^-} + \mathrm{BH^+} \leftrightarrow \mathrm{A^-}\ \mathrm{BH^+(solid)} \tag{4.10}$$

The first step is a proton transfer from acid HA to base B, and the second step is the crystallization of the salt, $A^-\ BH^+$. The equilibrium constant for the proton transfer is given as follows:

$$K_1 = \frac{[\mathrm{A^-}][\mathrm{BH^+}]}{[\mathrm{HA}][\mathrm{B}]} = \frac{K_a}{[\mathrm{H^+}]} \times \frac{[\mathrm{H^+}]}{K_a'} = \frac{K_a}{K_a'} \tag{4.11}$$

where K_a and K_a' are the ionization constant of HA and BH^+, respectively. If the pK_a' of the base exceeds the pK_a of the acid by more than 2 units, the equilibrium constant for the proton transfer is greater than 100, resulting in significant ionization and favoring salt formation.

In the first step of salt screening for a drug, the following questions should be addressed:

- Are any ionizable functional groups present in the drug molecule?
- Is it a weak acid or weak base?
- What are the pK_a values for those functional groups?
- Can pK_a values be measured by experimental methods such as potentiometric titration?
- What are some potentially useful counterions?

If the pK_a value of a very weak base is less than 3 or that of a very weak acid is greater than 10, salt formation will be very challenging since only a few counterions are available, and the salts formed tend to have poor physical stability in the solid state. Such salts can easily disproportionate in water and even in the solid state under certain conditions.

Various counterions used for pharmaceutical salt formation are listed in Table 4.2. For weakly basic drugs, counter acids for salt preparation include inorganic acids like hydrochloride and sulfate, sulfonic acids like mesylate and isethionate, carboxylic acids like fumarate and maleate, hydroxy acids like citrate and tartrate, and amino acids like aspartate and

Table 4.2 Commonly used counterions for pharmaceutical salts.

Basic drugs		**Acidic drugs**	
Anion	***pK$_a$***	**Cation**	***pK$_a$***
Chloride	−6.10	Potassium	16.00
Sulfate	−3.00, 1.96	Sodium	14.77
Tosylate	−1.34	Lithium	13.82
Mesylate	−1.20	Calcium	12.90
Napsylate	0.17	Magnesium	11.42
Besylate	0.70	Diethanolamine	9.65
Maleate	1.92, 6.23	Zinc	8.96
Phosphate	2.15, 7.20, 12.38	Choline	8.90
Salicylate	3.00	Aluminum	5.00
Tartrate	3.00		
Lactate	3.10		
Citrate	3.13, 4.76, 6.40		
Benzoate	4.20		
Succinate	4.21, 5.64		
Acetate	4.76		

glutamate. Hydrochloride acid salts are typically the most preferred choice for weakly basic drugs because they are easy to form and can be readily recrystallized from organic solvents. However, as pointed out earlier, sometimes hydrochloride salts have unacceptable properties. It is impossible to know ahead of time whether any salt (including the hydrochloride) will have acceptable properties or not. For weakly acidic drugs, salts are typically best formed using inorganic bases such as sodium, magnesium, aluminum, or calcium. Interestingly, salts formed with magnesium, aluminum, or calcium often exhibit disordered, liquid crystalline or amorphous forms. Furthermore, they tend to retain stoichiometric or nearly stoichiometric amounts of water due to the tendency for these ions to bind water. Other bases like triethylamine, ethanolamine, arginine, and lysine have also been used to form salts.

4.4 pH Solubility Profiles of Ionizable Compounds

An ionizable drug, as either an acid or a base, has a pH-dependent solubility profile, with the solubility profile being dictated by the solubility of the free form, the pK_a of the compound, the pH, and the counterion.

The pH solubility profile of a basic drug can be expressed by two independent curves, as shown in Figure 4.2a. One curve depicts the change in ionization of a free form with pH. The other curve shows the variation of salt solubility versus pH. For the top figure in Figure 4.2, the two solubility curves for a weakly basic drug intersect at pH_{max} where the maximum solubility of the drug is reached; pH_{max} is a critical transition point between the salt and the free form.

Before reaching pH_{max}, the solubility of a basic drug will increase as pH decreases because more drug molecules are ionized. The concentration of ionized species at a given pH can be calculated by rearranging equation (4.4) as follows:

$$[BH^+] = \frac{[H^+][B]_s}{K_a} = [B]_s 10^{(pK_a - pH)} \qquad (4.12)$$

where $[B]_s$ is the solubility of the free base. Therefore, the total solubility at that pH is given as

$$S_T = [B]_s + [BH^+] = [B]_s\left(1 + 10^{(pK_a - pH)}\right) \qquad (4.13)$$

For a weakly basic drug A, with a pK_a of 6 and an intrinsic solubility of 0.01 mg/mL, the solubilities at pH 8, 7, 6, 5, and 4 are calculated as 0.01, 0.01, 0.02, 0.11, and 1.01 mg/mL, respectively. The solubility of drug A reaches 10 mg/mL at pH 3, the pH_{max} for its HCl salt.

After reaching pH_{max}, the solubility exceeds the limit of the salt form, leading to the crystallization of the salt form. The equilibrium between salt formation and solubilization can be described by the following chemical equation:

$$(BH^+X^-)_{solid} \leftrightarrow [BH^+]_s + [X^-] \qquad (4.14)$$

where $(BH^+X^-)_{solid}$ represents the activity of the solid salt form, $[BH^+]_s$ is the salt solubility, and $[X^-]$ is the counterion concentration. The equilibrium is commonly described as a solubility product constant K_{sp} based on the following equation:

$$K_{sp} = [BH^+]_s[X^-] \qquad (4.15)$$

K_{sp} is an equilibrium constant that will not change with pH, ionic strength, and the concentration of a counterion. For example, K_{sp} for the HCl salt of drug A (above) is 100 (mg/mL)^2.

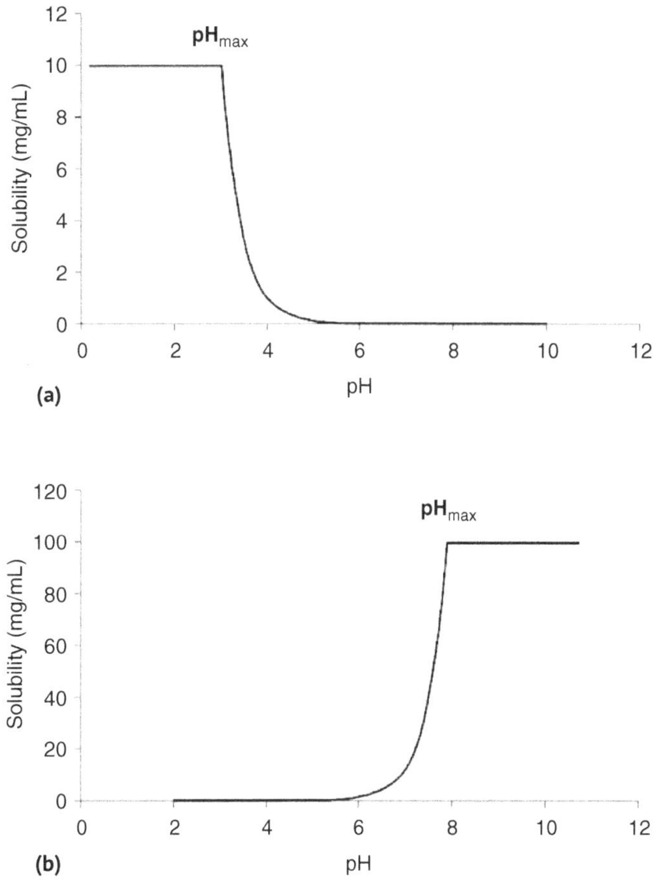

Figure 4.2 pH Solubility profiles of a weak base (a) with a pK_a of 6 and intrinsic solubility of 0.01 mg/mL and a weak acid (b) with a pK_a of 5 and intrinsic solubility of 0.1 mg/mL.

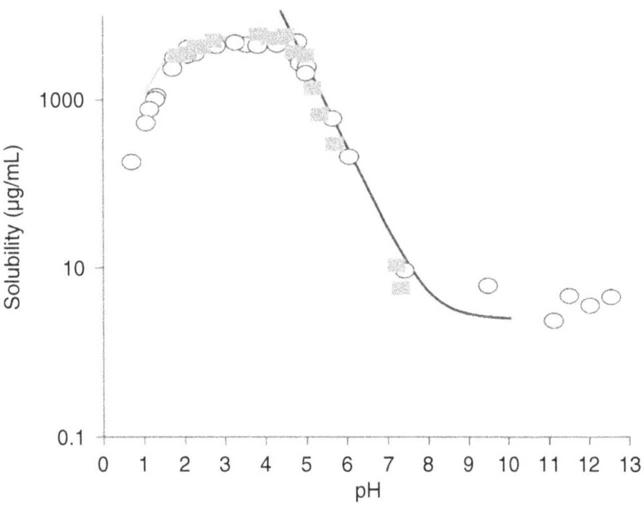

Figure 4.3 pH Solubility profiles of haloperidol free base (solid square) and its HCl salt (open circle). *Source:* [10]/with permission of Springer Nature.

In the absence of excess counterion, the concentration of a counterion is equal to the salt solubility for a 1:1 salt. So, the relationship of K_{sp} and salt solubility becomes

$$\sqrt{K_{sp}} = [BH^+]_s \tag{4.16}$$

When the pH is lower than pH_{max}, the total solubility at a given pH may be expressed as follows:

$$S_T = [BH^+]_s + [B] = [BH^+]_s + \frac{[BH^+]_s K_a}{[H^+]} = [BH^+]_s \left(1 + 10^{(pH - pK_a)}\right) \tag{4.17}$$

Usually, the concentration of a free form is much less than that of the salt solubility, whereas the total solubility is approximately the same as that of the salt solubility. Therefore, the solubility usually remains unchanged when the pH is lower than pH_{max}. However, the salt solubility may decrease when a significant amount of excess counterions is present, due to a common ion effect according to

$$[BH^+]_s = \frac{K_{sp}}{[X^-]} \tag{4.18}$$

The pH solubility profiles of haloperidol given in Figure 4.3 agree with equations (4.13) and (4.17) [11]. Haloperidol is a weak base with a pK_a of 8.0 and an intrinsic solubility of 2.5 μg/mL. Note in Figure 4.3 that solubility profiles generated from the free base and the hydrochloride salt match each other. Based on the pH solubility profile, the pH_{max} of haloperidol is about 5. When the HCl salt was used as the starting material, it was converted to the free base when the pH of the sample was above 5. Similarly, the free base became the HCl salt when the pH of the sample was lowered below 5. Similar species, both in the solid phase and in solution, are in equilibrium at any pH, irrespective of the use of hydrochloride salt or free base as the starting material, which explains the similarity in the two profiles. The solubility profiles reach a plateau between pH 2 and 5, which agrees with equation (4.17), because the solubility value is limited by the solubility of the HCl salt. Below pH 2, the solubility decreases as pH is lowered. This is due to the common ion effect of the chloride ion, which occurs when more HCl is added to lower the pH.

It should also be noted that, at pH_{max}, the two curves intersect with each other. Equation (4.13) matches equation (4.18) at this point. So pH_{max} can be calculated as follows:

$$pH_{max} = pK_a + \log\frac{[B]_s}{[BH^+]_s} \tag{4.19}$$

The pH_{max} value of a salt is determined by the pK_a, the intrinsic solubility of the corresponding base and the salt solubility. It should be noted that the tendency to crystallize can also have an effect unless enough time is given for the system to equilibrate. For example, a pH_{max} will go up by one unit when one of the following factors changes: an increase in the pK_a by one unit, a 10-fold increase in the intrinsic solubility, or a 10-fold decrease in the solubility of the salt.

Theoretically, both the free base and the salt of a drug can coexist as solids at pH_{max}. When the pH is greater than pH_{max}, the free base is the stable solid phase. Conversely, the salt is the stable form when the pH is less than pH_{max}.

Similar equations can be derived for a weakly acidic drug with a division point at pH_{max} (Figure 4.2b). The free acid is the equilibrium species at a pH below pH_{max}, whereas the salt form is the stable phase above pH_{max}.

Below pH_{max}

$$S_T = [AH]_s + [A^-] = [AH]_s\left(1 + \frac{K_a}{[H^+]}\right) = [AH]_s\left(1 + 10^{(pH - pK_a)}\right) \tag{4.20}$$

where S_T is the total solubility and $[AH]_s$ is the equilibrium solubility of the free acid.

Above pH_{max}

$$S_T = [AH] + [A^-]_s = [A^-]_s\left(1 + \frac{[H^+]}{K_a}\right) = [A^-]_s\left(1 + 10^{(pK_a - pH)}\right) \tag{4.21}$$

where $[A^-]_s$ is the equilibrium solubility of the salt form.

For an acid, therefore, pH_{max} is given by

$$pH_{max} = pK_a + \log \frac{[A^-]s}{[AH]_s} \tag{4.22}$$

Different salt forms of the same compound usually display different solubilities and different tendencies to crystallize. As a result, different salt forms have different values of pH_{max}. The salts with higher solubility will have a lower pH_{max} value for a weak base according to equation (4.19). As shown in Figure 4.4, the pH solubility profile of terfenadine, an antihistamine, was affected by the counterion used [9]. The solubility order for the various salts of terfenadine from high to low is: the lactate salt, mesylate salt, HCl salt, and phosphate salt, which have K_{sp} values of, 8.9×10^{-5}, 2.5×10^{-5}, 1.6×10^{-5}, and 2.5×10^{-6} M^2, respectively. The pH_{max} values for the lactate, mesylate salt, HCl salt, and phosphate salt are about 4.4, 4.7, 5.0, and 5.2, respectively. Owing to the common ion effect, solubility was decreased for all four salts as pH was lowered to less than pH 4. The terfenadine salts were precipitated from solution due to the presence of extra anion from the pH adjustment. The salts with different solubilities, therefore, offer opportunities for selecting a salt with enhanced bioavailability of a poorly soluble compound.

4.5 Solubility, Dissolution, and Bioavailability of Pharmaceutical Salts

The strategy to use salt formation to achieve higher solubility becomes more critical for new drug discovery and development, because in recent years the solubilities of most new drug molecules have decreased sharply along with an increase in the average molecular weight and hydrophobicity. Serajuddin estimated that one-third of new compounds obtained from the discovery process have aqueous solubilities less than 10 μg/mL [12]. Lipinski et al. concluded that about 30% of a group of thousands of compounds reported in the literature between 1987 and 1994 have solubilities equal to or less than 20 μg/mL [13].

As discussed previously, a salt form usually demonstrates significant improvement in aqueous solubility compared with its free form. The increase can range from several folds to hundreds and thousands of folds. The solubility increase usually results in an enhancement in dissolution and an improvement in bioavailability.

For example, the drug naproxen has a solubility of 0.016 mg/mL in water. The solubility of its sodium salt is 178 mg/mL, which is 11 000-fold

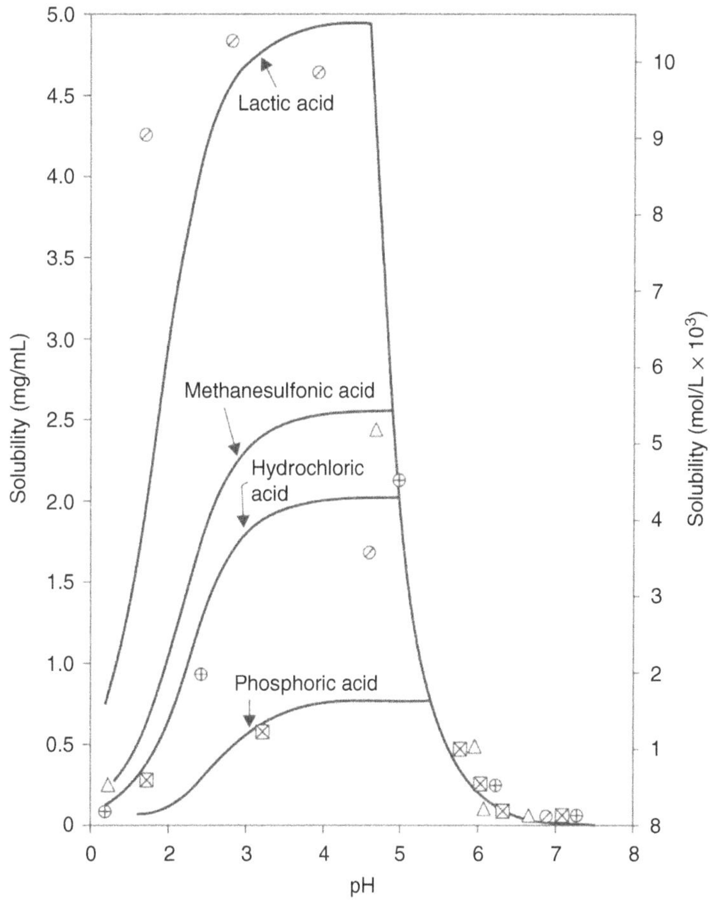

Figure 4.4 pH solubility profile of terfenadine salts. *Source:* [11]/with permission of Elsevier.

more soluble than naproxen free acid [14]. Its tromethamine salt has a solubility of 11 mg/mL, which is 680-fold more soluble than the free acid.

Significant differences in solubilities for different salt forms were also observed for the drug terfenadine [9]. Terfenadine is a weak base with a pK_a of 8.6. The solubility of a weak base is about 60 μg/mL. The solubilities of terfenadine phosphate, HCl, mesylate, and lactate range from 0.5 to 5 mg/mL.

There is no well-established model to predict the rank order of salt solubilities for a specific compound since the differences in solubilities of various salt forms are related to the lattice energy of the salts, the polymorphic form, the acidity/basicity of the counterions, and the hydrophilicity of the corresponding counterions. For a weak base, decreasing melting point, increasing acidity of counterion, or using hydrophilic counterions generally leads to the enhancement of solubility. Consequently, empirical screening is still essential in discovering salt forms with desirable solubility and other physical attributes.

The enhancement of solubility usually results in a significant improvement in dissolution rate. Based on the Noyes–Whitney equation [15], dissolution rate is directly determined by solubility as follows:

$$J = KA(C_s - C) \tag{4.23}$$

where J is the dissolution rate, K is the constant, A is the surface area of the dissolving solid, C_s is the solubility, and C is the concentration in the dissolution medium. The equation can be rewritten by incorporating the Nernst–Brunner diffusion layer theory [16, 17], which hypothesizes that a thin film of saturated solution surrounds a drug particle during the dissolution process, and the rate of transfer of the dissolved drug is limited by the rate of diffusion of drug molecules through this layer. Equation (4.23) can then be rewritten as:

$$J = \frac{DA}{h}(C_s - C) \tag{4.24}$$

where D is the diffusion coefficient of the drug in this layer and h is the diffusion layer thickness. If the surface area is constant and dissolution is under sink conditions, the dissolution rate will be linearly related to solubility.

The dissolution rate of a salt form is usually much greater than that of the parent drug. For naproxen, the dissolution rates of two salt forms and the free form at pH 3 follow the same trend as their solubility [12], wherein the dissolution rates of the sodium and THAM salts are 4200 and 200 times faster than for naproxen, respectively.

In some cases, the rates of dissolution of a salt form can be influenced by factors other than those already discussed. For example, sodium salicylate has a dissolution rate that is 100 times faster than salicylic acid at pH 1.5 and pH 6.8, although the equilibrium solubilities for the salt and the free form at these two pH values are almost identical [18]. After equilibrating at pH 1.5, sodium salicylate tends to form salicylic acid, while, at pH 6.8, salicylic acid is ionized and forms the salicylate ion. The phenomenon of faster solubility of the salt form under these conditions is related to a self-buffering action, which causes different pH values in the diffusion layer than those in the bulk media. At bulk pH 1.5, sodium salicylate has a diffusion layer pH of 8.8. The pH of the diffusion layer governs the dissolution rate rather than the bulk pH, thus promoting the faster dissolution of sodium salicylate. Conversely, salicylic acid has a diffusion layer pH of 2.4 at bulk pH 6.8, and the pH in the diffusion layer depresses the dissolution rate. Thus, the dissolution rate of salicylic acid at pH 6.8 is still substantially slower than that of sodium salicylate. Serajuddin and Jarowski have presented a graphical representation to depict the buffering effect of salicylic acid and sodium salicylate in the diffusion layers of various media (Figure 4.5) [18].

Similar observations have been reported for haloperidol, a weakly basic drug with a pK_a value of 8.0 [10]. The intrinsic solubility for the free base is 2.5 μg/mL, and the solubility for its mesylate is 29 mg/mL. The dissolution rates of haloperidol mesylate at pH 3, 5, and 7 are similar despite the fact that the solubility at pH 3 and 5 is 1000 times higher than the solubility at pH 7. It was hypothesized that the pH at the surface of haloperidol mesylate at bulk pH 7 is about 3 and that the amount of drug solubilized at the surface is close to the solubility of the mesylate, 29 mg/mL [10]. In comparison, the dissolution of haloperidol free base at pH 7 is undetectable, whereas the dissolution rate at pH 5 for the free base is 1000 times less than the mesylate. The difference in the dissolution rate between the mesylate and free base forms at pH 3 is reduced by 33-fold. This difference occurs because the surface pH for the free base is decreased, and the solubility at the surface is increased. The difference in dissolution rate is diminished at pH 1 and 2. Thus, it appears that the surface pH in such a system dominates the dissolution rate instead of the bulk pH. This phenomenon can perhaps explain in part why the salt form in such cases presents a significant enhancement of dissolution rate in comparison to that of the parent compound. Another factor is the rate of nucleation and crystal growth of the neutral drug. If either of these steps is slow, then the drug can be absorbed before it precipitates.

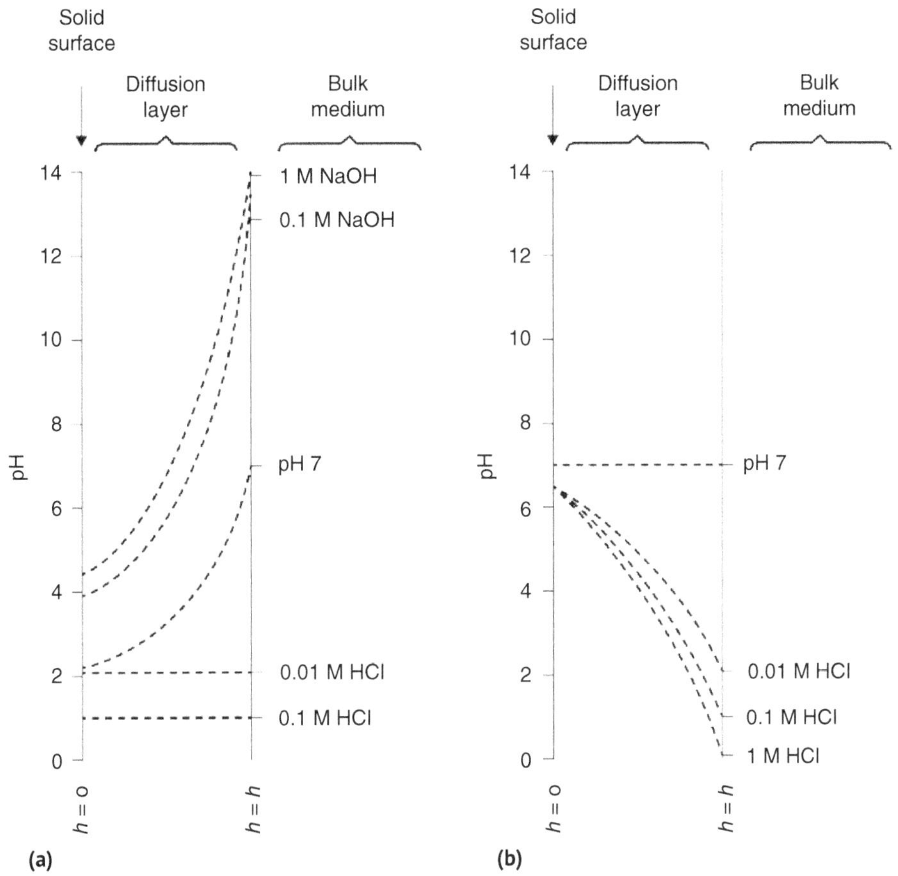

Figure 4.5 Graphical descriptions of the self-buffering effect of (a) salicylic acid and (b) sodium salicylate in the diffusion layers of various media. *Source:* [18]/with permission of Elsevier.

The enhancement of dissolution rates using a salt form may lead to the improvement of oral bioavailability, but the significant enhancement in solubility and dissolution by salt formation may not be linearly translated into the same extent of increase in oral absorption. During exposure to the gastrointestinal tract (GI), precipitation and redissolution may occur, depending on the nature of the compound, the GI contents, and the pH solubility profile. In addition, the permeability of the drug and the effects of any related transporters present in the GI can play significant roles in oral absorption. Nevertheless, the benefits of using a salt form to improve oral absorption have been demonstrated in many case studies. Enhanced bioavailability of piroxicam, for example, was observed for its salts with ethanolamine [19]. Piroxicam is a potent non-steroidal anti-inflammatory drug and belongs to the Biopharmaceutics Classification System (BCS) II because of its poor aqueous solubility. It tends to form a zwitterion with two pK_a values, 1.86 and 5.46, and can form salts with ethanolamine as the mono, di, or triethanolamine salts. All three salts are more bioavailable than piroxicam at the dose of 10 mg/kg in rats, with the most significant increase of oral absorption occurring for the monoethanolamine salt. The C_{max} and AUC_{0-24} of the monoethanolamine salt are about twice as great as those of piroxicam.

Superior performance of pharmaceutical salts in overcoming the bioavailability challenges of poorly soluble drugs is further illustrated in the following case concerning PNU-140690. As a HIV-1 protease inhibitor, PNU-140690 is a poorly soluble weak acid with two pK_a values of 6.2 and 7.6 [20]. The solubility of the weak acid is about 3 μg/mL, and the intrinsic dissolution rate is less than 0.2 μg/cm^2/s. The monosodium salt and disodium salt have dissolution rates of 115 and 450 μg/cm^2/s, respectively, which are hundreds of times faster than that of the free acid. Though the salt forms tend to precipitate in the stomach, both salt forms still lead to a significant improvement in bioavailability; the bioavailability of PNU-140690 free acid in dogs is only 3%, while the bioavailability for the monosodium salt and disodium salt is increased to 15% and 50%, respectively. The disodium salt was used in clinical studies and was successfully developed as a 250 mg capsule and a 100 mg/mL oral solution with the brand name of Tipranavir [21].

Another important antiviral drug, dasabuvir, also relies on salt formation to enhance its oral bioavailability. Dasabuvir, a drug used for the treatment of hepatitis C virus infection, is an extremely weak acid (pK_a > 8) and exhibits very low

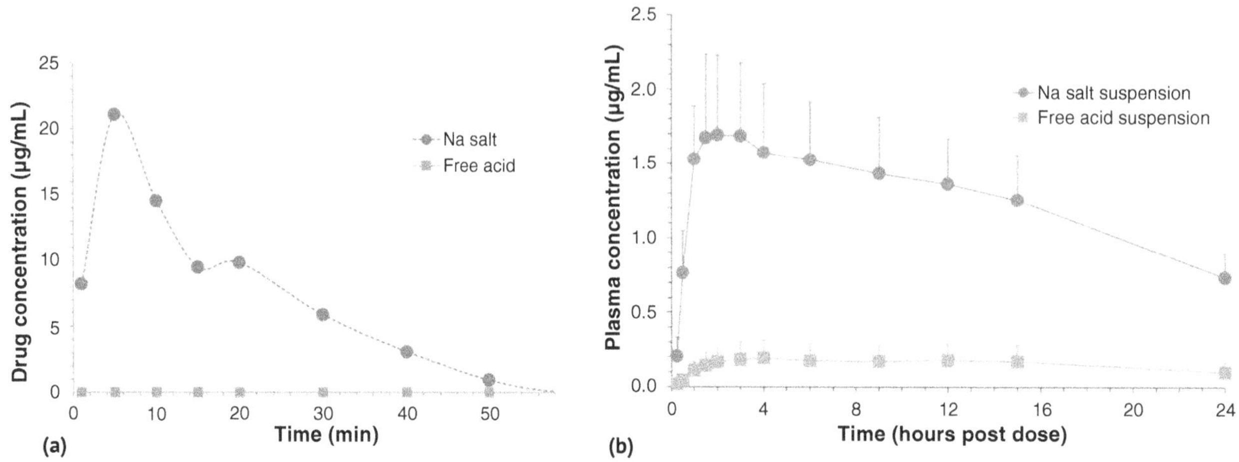

Figure 4.6 Comparison of dasabuvir monosodium salt and free acid. (a) In vitro dissolution in water at 37 °C; (b) pharmacokinetic profile in dogs, dose: 25 mg/dog, $n = 6$ dogs. *Source:* [22]/with permission of American Chemical Society.

aqueous solubility (< 0.2 μg/mL at pH 1–6.8). Crystalline dasabuvir free acid was found, not surprisingly, to have exceedingly poor oral bioavailability in animal pharmacokinetic studies. Through salt screening, Chen and colleagues identified a stable dasabuvir monosodium salt, which, as expected, dissolved rapidly in water and reached a peak solubility of over 20 μg/mL within 5 minutes [22]. After that, the salt form started to disproportionate to form the free acid, leading to a continued decrease of drug concentration until a full loss of supersaturation in 60 min, as illustrated in Figure 4.6. In contrast, the free acid did not dissolve at all with concentrations measured below the detection limit. To further leverage the solubility potential of the sodium salt, an attempt to screen crystallization inhibitors led to finding several polymers, including povidone, co-povidone, and hydroxypropyl methylcellulose (HPMC), to be effective in delaying precipitation and maintaining supersaturation. With the aid of crystallization inhibitors, as mentioned earlier, the superior solubility advantage of the sodium salt can be generally translated into a significant improvement in the performance of drug products. As reported in Figure 4.6, the sodium salt elicited a seven-fold bioavailability increase in dogs dosed as a suspension in 0.2% HPMC, in comparison to the free acid. It is interesting to note that co-povidone, an effective crystallization inhibitor, exists as a key ingredient in the commercial drug product for dasabuvir [23].

The variation in the solubility of different salt forms can lead to different bioavailabilities, as demonstrated in the case of the drug LY333531 [24]. LY333531 exists as a competitive, reversible inhibitor of protein kinase C_β and has been evaluated for the treatment of complications from diabetes. It is a weak base with poor aqueous solubility (< 1 μg/mL). Salt selection resulted in the discovery that the HCl salt and the mesylate salt exhibited acceptable physical and chemical properties. The mesylate salt is five times more soluble than the HCl salt in water, 0.5 mg/mL versus 0.1 mg/mL, while the mesylate salt showed 2.6 times more exposure than the HCl salt based on a pharmacokinetic study in dogs with a 20 mg/kg dose. Therefore, the mesylate was selected as the active pharmaceutical ingredient for clinical development.

Sometimes, salts exhibiting less solubility than a highly soluble form are chosen to mitigate "Ostwald ripening" in suspension formulations, to promote taste masking, or to produce an extended-release form. For example, erythromycin estolate has about one-twelfth the solubility of the free base and was used to formulate a tasteless suspension [25]. The HCl salt of chlorpromazine, an antipsychotic agent, is water soluble but has a very bitter taste. The embonate salt of chlorpromazine is less soluble and has been used in suspension products to increase patient acceptance and to extend the duration of action [26]. Jashnani and Byron reported that albuterol stearate is much less soluble in water than the free base and other salt forms [27]. However, when the stearate salt is aerosolized from an inhaler, it shows the least sensitivity to temperature and humidity, with the potential to extend the duration of the drug in the lung as an inhalation product. For another example, the salt formation of olanzapine, an antipsychotic drug for the treatment of schizophrenia, with pamoic acid, leads to an insoluble pamoate salt, which forms the main component of a long-acting injectable suspension with a half-life of about 30 days [28, 29]. After being injected intramuscularly, the microcrystalline olanzapine pamoate monohydrate slowly dissolves and offers a sustained release.

In summary, the strategy to select a salt form varies with compound properties and the target dosage form. For a poorly soluble compound, salt forms with high solubility and faster dissolution can help improve bioavailability. For a highly soluble compound, a less soluble salt may be chosen to enhance various properties such as taste masking or sustained release.

4.6 Physical Stability of Pharmaceutical Salts

The physical stability of pharmaceutical salts is essential for their performance, processibility, and chemical stability. A salt may disproportionate to the free form due to a change in pH in solution, due to an interaction with moisture in the solid state, or when in contact with certain excipients in a drug product. The disproportionation of a salt often results in a change in solubility because the free form is usually much less soluble than the corresponding salt form. The conversion to the free form may also compromise chemical stability if the free form has poor crystallinity or possesses a significant amount of amorphous content.

Disproportionation of a salt is the reverse reaction of salt formation. For the salt of a weak base, B, for example, disproportionation occurring in the solution phase can be described as follows:

$$BH^+A^-(\text{solid}) \leftrightarrow BH^+ + A^- \leftrightarrow HA + B \leftrightarrow B(\text{solid}) \tag{4.25}$$

A key parameter controlling the disproportionation of a salt is the pH. If the pH is less than pH_{max}, the salt form of a basic drug is the stable form, and no disproportionation should occur. Conversely, disproportionation tends to occur if the pH is greater than pH_{max}. The greater the pH difference from pH_{max}, the greater will be the driving force for the free form to precipitate because the concentration of free form will increase. As shown in Figure 4.3, the pH profile of haloperidol HCl matches that of the free base. Above pH 5, the pH_{max}, the HCl salt starts to convert to the free base, whereas above pH 9, the HCl salt completely converts to the free base.

Disproportionation is a common problem for the salts of very weak bases or very weak acids. For a very weak base, it is only possible to form stable salts with strong acids. Thus, if a stable salt can be made, the pH_{max} will usually be below the physiological pH. Stahl and Nakamo, for example, have reported a study using a sparingly soluble weak base with a pK_a of 4.7 to form a mesylate salt as a free-flowing crystalline material [20]. However, the mesylate salt is readily dissociated in water, and wet granulation during the tableting process could not be used. Interestingly, dissociation even occurred in 0.1 N HCl, indicating that the pH_{max} for this mesylate is less than pH 1. Despite the instability of the mesylate salt in solution, the mesylate salt had three times more bioavailability than the micronized free form, probably because it precipitated to a very fine particle of the free form with a greater dissolution rate than that of the micronized free form or perhaps because its precipitation was inhibited to some extent in the GI tract where crystallization inhibitors may be present.

Pharmaceutical salts are usually quite stable in the solid state; however, disproportionation of pharmaceutical salts in the solid state can occur if the solid is exposed to an unfavorable local pH. Serajuddin and Jarowski have suggested that the local hydrogen ion activity "pH" at a solid surface may be very relevant in determining the so-called microenvironmental pH [18]. This microenvironment is believed, most likely, to be an aqueous layer located at the surface of a drug or excipient particle, or water absorbed into amorphous regions or perhaps porous capillaries.

Serajuddin and Jarowski also established that the microenvironmental pH of a solid pharmaceutical could be estimated by measuring the pH of its saturated solution in water [18]. Table 4.3 provides a list of commonly used excipients and their estimated microenvironmental pH values based on the saturated solution method. Most excipients exhibit a certain degree of acidity or basicity as influenced by their chemical structure or residual impurities. For example, calcium carbonate has a microenvironmental pH of about 8, which is attributed to its intrinsic basic nature. Povidone is essentially a neutral polymer, but, as a solid, it exhibits a microenvironmental pH of about 4, due most likely to impurities carried over from its synthesis.

Guerrieri and Taylor studied the effects of excipients on the disproportionation of salts in the solid state [30]. The mesylate salts of two weakly basic compounds, miconazole and benzocaine, tend to disproportionate to the corresponding free base after being mixed with certain basic excipients and exposed to moderate relative humidities. Interestingly, the napsylate salts of those two compounds showed much better stability against disproportionation with undetectable free base. The distinct stability of napsylate salts versus mesylate salts appeared to be related to their relative solubilities. The solubility of the napsylate salts was found to be three orders of magnitude lower than the corresponding mesylate salts. The lower solubility enables the napsylate salts to have a higher pH_{max} than the mesylate salts; the pH_{max} for the napsylate salt of miconazole is 6.8, whereas that of the mesylate is 1.4. The higher pH_{max} for the napsylate leads to a wider stability window because the change of microenvironmental pH below pH_{max} would not lead to disproportionation.

Table 4.3 Excipient and measured microenvironmental pH.

Excipient	pH
Dibasic calcium phosphate anhydrous USP	2.2
Microcrystalline cellulose (Avicel PH102)	4.0
Lactose monohydrate NF (Fast Flo 316)	4.2
Mannitol	4.7
Sodium starch glycolate NF (Explotab)	4.8
Crospovidone	4.9
Colloidal silica	5.2
Sodium croscarmellose	5–7
Maize starch	6.3
Calcium carbonate USP (Vicron 75-17-FG)	6.6
Magnesium stearate	7.5
Hydrogenated castor oil	7.5
Calcium carbonate USP (Precarb 150)	7.7

Controlling the pH of a microenvironment in a solid formulation is critical to mitigating the disproportionation of a salt form. Zannou et al. reported that the adjustment of the microenvironmental pH in a solid dosage form resulted in the stabilization of the maleate salt of the basic drug, dibenzo [b,f]oxepen-10-ylmethyl-methyl-prop-2-ynyl-amine (compound I) [31]. A major loss in potency and a lack of mass balance were observed for a tablet formulation of the maleate salt of compound I on storage under accelerated stability testing conditions. The loss in potency was observed to be due to disproportionation of the maleate salt to the free form and to subsequent sublimation of the free form during storage. The conversion of the salt to the free form during storage was shown to be directly related to the microenvironmental pH of the tablet formulation, which was found to be about 4.3. Further study of the pH solubility profile of compound I revealed that the maleate salt has a pH_{max} between 3.3 and 3.6, and, therefore, above the pH_{max}, the maleate salt converted to the free base while no such disproportionation occurred below this pH. It was concluded that the microenvironmental pH of 4.3 favored the disproportionation of the maleate salt and the formation of the free form. Adding citric acid to the formulation lowered the microenvironmental pH of tablets from 4.3 to less than 3.0 and prevented the disproportionation. The new formulation was found to be very stable, with a shelf-life of more than 3 years. In another study, Koranne et al. compared five organic acids, including oxalic acid, maleic acid, tartaric acid, fumaric acid, and glutaric acid, on the disproportionation of pioglitazone hydrochloride in the presence of basic excipients and found the effectiveness to stabilize the salt against disproportionation correlated with the strength of acidity [32]. For example, oxalic and maleic acids, as stronger acids with the lower pK_as (1.2 and 1.9, respectively), completely inhibited disproportionation, while glutaric acid, as the weakest acid with a pK_a value of 4.3, exhibited the most amount of free form [32].

In summary, maintaining the pH below pH_{max} for salts of weak bases is essential to ensure salt physical stability in solution, suspension, and the solid state. Conversely, the salts of weak acids are more stable at a pH above pH_{max}.

4.7 Strategies for Salt Selection

Many physical and chemical properties of salt forms should be considered during the salt selection process. These include hygroscopicity, solubility, dissolution, physical stability, chemical stability, and processibility. Several key questions should be addressed during the process:

- What is the intended use of the salt form? Tablets, capsules, suspensions, injectables, process intermediates?
- What are the critical issues associated with the inadequacy of the free form? Solubility, stability, hygroscopicity, crystallinity, mechanical properties, or impurity levels?
- Is salt formation feasible? What is the pK_a value? What are the available counterions to form a stable salt?

The way a salt is selected depends on the intended use. If a salt form is to be used in solid dosage forms, such as tablets and capsules, low hygroscopicity and high crystallinity are desired to ensure the integrity of the salt form after various unit operations and long-term storage. In a solid dosage form, a drug is formulated with multiple excipients. Therefore, a hygroscopic salt may take up moisture from other ingredients or the environment, often resulting in unfavorable physical changes. In a capsule formulation, moisture migration from capsule shells to a hygroscopic salt has been shown to cause physical changes in the capsule shells, thus leading to product failure. High crystallinity of a salt would ensure the physical integrity after exposure to mechanical and thermal stresses arising from various pharmaceutical processes such as milling, micronization, drying, and compaction.

For parental use, hygroscopicity obviously is of less concern since the drug substance is either directly formulated as a solution or processed as a lyophilized product that is processed and stored in containers protected from moisture absorption. In such cases, it is apparent that a high solubility is desirable for the ease of reconstitution of lyophilized products. In addition, the pH of the salt solution should be within an acceptable range of 3 to 8.

For a suspension formulation, in addition to ensuring optimal therapeutic use, factors such as taste masking and the reduction of Ostwald ripening must be considered. For example, tetracycline calcium is used in suspension formulations because of its relatively low solubility and, therefore, minimal taste response in the mouth. Conversely, parental formulations of tetracycline are produced as the HCl salt to take advantage of the superior solubility of this salt form.

The intended use of a salt form also determines the possible counterions that can be used to optimize therapeutic use. In this regard, the toxicity of counterions should be fully assessed. Otherwise, extra safety studies will be required to obtain regulatory approval to initiate clinical studies or to pursue marketing. Some counterions have been reported to have safety concerns. For example, lithium cations in small quantities have barely any toxic effect, but the ingestion of a large amount can cause irreversible damage to the kidneys [33]. Tartrate anions have renal toxicity if they reach the circulation in high concentrations [33]. It was also reported that pravadoline maleate caused renal tubular lesions in a dog, attributed to the maleate anions [34]. Indeed, regulatory agencies like the FDA and EMEA have expressed some concerns about the safety of mesylate salts, in general [35, 36]. For example, mesylate salts have the potential to form genotoxic alkyl mesylates, when methane sulfonic acid is added to the free base dissolved in an alcoholic solvent. Nowadays, regulatory agencies require applicants to prove that the synthesis and purification employed do not lead to any detectable levels of alkyl mesylates.

Safety and physical properties are of lower priority if salt formation is used only as an intermediate step for chemical processing. Here, the focus is generally on the ease of crystallization and efficacy of removing impurities.

Different salt forms usually display a broad range of properties that can benefit or detract from producing an optimized formulation. For example, a salt having good solubility may exhibit a high level of hygroscopicity, or a salt form with excellent chemical stability may have multiple polymorphs. A balanced approach, therefore, is recommended when selecting a salt form, based on a careful risk/benefit analysis.

Serajuddin and Pudipeddi have established a multiple-tier strategy for salt selection targeted for oral formulation [37]:

- Prepare and crystallize all possible salt forms, evaluate crystallinity and aqueous solubility, and select salts with acceptable crystallinity and solubility.
- Evaluate thermal properties and hygroscopicity and choose several for tier 3 evaluation.
- Study physical stability regarding formation of polymorphs, pseudopolymorphs, or disproportionation, and choose 2–3 for tier 4.
- Evaluate bioavailability, chemical stability, compatibility, and processability, and choose one for development and select one as a backup.

A similar tier-based approach was employed to select the optimal salt form for BMS-180431, a novel HMG-CoA reductase inhibitor and a candidate for oral dosage form development [38]. Seven salts of BMS-180431, including those with sodium, potassium, calcium, zinc, magnesium, arginine, and lysine, were evaluated. At tier 1, the hygroscopicity of all salts was assessed, and four salts, including those with sodium, potassium, calcium, and zinc, were eliminated due to substantial moisture uptake between 30% and 50% RH at ambient conditions (Figure 4.7). The remaining three salts, magnesium, arginine, and lysine, were subjected to tier 2 studies such as aqueous solubility and physical stability against moisture. The magnesium salt was removed due to its relatively low solubility (3.7 mg/mL) and humidity-dependent changes in its crystal structure. The arginine and lysine salts were moved to tier 3 evaluation due to their high aqueous solubility (> 200 mg/mL) and good physical stability, including resistance to humidity. The arginine salt was selected for further development after tier 3 evaluation concerning solid-state stability, ease of manufacture, ease of analysis, and potential impurities. The tier-based approach has the advantages of saving time and resources. The least time-consuming

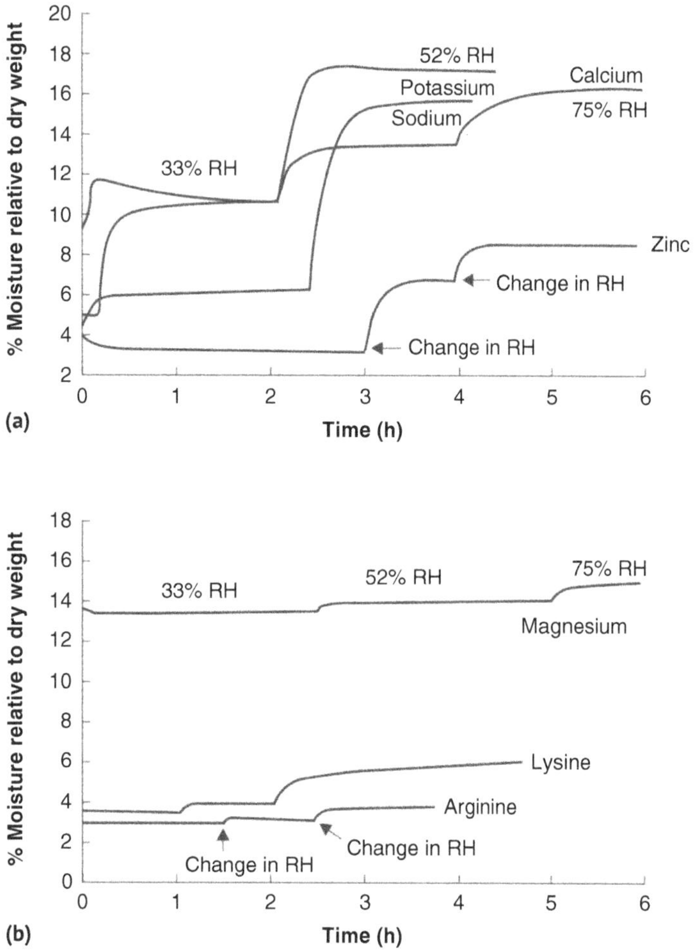

Figure 4.7 Moisture sorption by (a) potassium, sodium, calcium, and zinc salts; (b) magnesium, lysine, and arginine salts of BMS-18043. *Source:* [38]/with permission of Elsevier.

experiments are conducted in earlier tiers. A "go/ no-go" decision is made after each tier to avoid generating extensive data on all available salt forms. The number of tiers necessary to reach a decision will depend on the physicochemical properties desired and the number of salts available.

In summary, selection of a salt form should be tailored to its intended use, and a balanced approach must be taken with consideration of various properties such as solubility, hygroscopicity, crystallinity, physical stability, chemical stability, and processability.

References

1 Griesser UJ, Stowell JG. Solid-state analysis and polymorphism. In: Lee DC, Webb ML, eds. *Pharmaceutical Analysis*. Oxford: Blackwell Publishing Ltd, 2003: 240–294.

2 Wermuth CG, Stahl PH. Introduction. In: Wermuth CG, Stahl PH, eds. *Handbook of Pharmaceutical Salts: Properties, Selection and Use*. Germany: Wiley-VCH, 2002: 1–7.

3 Bharate SS. Recent developments in pharmaceutical salts: FDA approvals from 2015 to 2019. *Drug Discov Today* 2021; 26(2): 384–398.

4 Fuentes AV, Pineda MD, Venkata KC. Comprehension of top 200 prescribed drugs in the US as a resource for pharmacy teaching, training and practice. *Pharmacy* 2018; 6(2): 43.

5 Berge SM, Bighley LD, Monkhouse DC. Pharmaceutical salts. *J Pharm Sci* 1977; 66(1): 1–9.

6 Herbert JM, Frehel D, Vallee E, Kieffer G, Gouy D, Berger Y, Necciari J, Defreyn G, Maffrand JP. Clopidogrel, a novel antiplatelet and antithrombotic agent. *Cardiovascular Drug Rev* 1993; 11(2): 180–198.

7 Alain B, Daniel F, inventors; Sanofi, assignee. Dextrorotatory enantiomer of alpha-(4,5,6,7-tetrahydrothieno[3,2-c]pyrid-5-yl)(2-chlorophenyl)methyl acetate, process for its preparation and pharmaceutical compositions containing it. EP 0281459 (A1). 1988 Sep 7.

8 Bousquet A, Castro B, Saint-Germain J, inventors; Sanofi, assignee. Polymorphic clopidogrel hydrogen sulphate. US Patent 6429210B1. 10 June 1999.

9 Maurin MB, Grant DJW, Stahl PH. The physiochemical background: fundamental of ionic equilibria. In: Stahl PH, Wermuth CG, eds. *Handbook of Pharmaceutical Salts: Properties, Selection and Use*. Weinheim, Germany: Wiley–VCH, 2002: 9–18.

10 Li S, Wong S, Sethia S, Almoazen H, Joshi YM, Serajuddin ATM. Investigation of solubility and dissolution of a free base and two different salt forms as a function of pH. *Pharm Res* 2005; 22(4): 628–635.

11 Streng WH, His SK, Helms PE, Tan HGH. General treatment of pH-solubility profiles of weak acids and bases and the effects of different acids on the solubility of a weak base. *J Pharm Sci* 1984; 73(12): 1679–1684.

12 Serajuddin ATM. Salt formation to improve drug solubility. *Ad Drug Del Rev* 2007; 59: 603–616.

13 Lipinski CA, Lombardo F, Dominy BW, Feeney PJ. Experimental and computational approaches to estimate solubility and permeability in drug discovery and development settings. *Adv Drug Deliv Rev* 1997; 23 (1–3): 3–25.

14 Gu L, Strickley RG. Preformulation salt selection. Physical property comparisons of the Tris(hydroxymethyl) aminomethane (THAM) salts of four analgesic/anti-inflammatory agents with the sodium salts and the free acids. *Pharm Res* 1987; 4(3): 255–257.

15 Noyes AA, Whitney WR. The rate of solution of solid substances in their own solutions. *J Am Chem Soc* 1897; 19: 930–934.

16 Nernst W. Theorie der Reaktionsgeschwindigkeit in heterogenen Systemen. *Z Phys Chem* 1904; 47: 52–55.

17 Brunner E. Reaktionsgeschwindigkeit in heterogenen Systemen. *Z Phys Chem* 1904; 43: 56–102.

18 Serajuddin AT, Jarowski CI. Effect of diffusion layer pH and solubility on the dissolution rate of pharmaceutical acids and their sodium salts. II: Salicylic acid, theophylline, and benzoic acid. *J Pharm Sci* 1985; 74(2): 148–154.

19 Gwaka HS, Choib JS, Choi HK. Enhanced bioavailability of piroxicam via salt formation with ethanolamines. *Int J Pharm* 2005; 297 (1–2): 156–161.

20 Stahl PH, Nakamo M. Pharmaceutical aspects of the drug salt form. In: Stahl PH, Wermuth CG, eds. *Handbook of Pharmaceutical Salts: Properties, Selection and Use*. Weinheim, Germany: Wiley–VCH, 2002: 83–134.

21 Doyon L, Tremblay S, Bourgon L, Wardrop E, Cordingley M. Selection and characterization of HIV-1 showing reduced susceptibility to the non-peptidic protease inhibitor tipranavir. *Antiviral Res* 2005; 68 (1): 27–35.

22 Chen S, Gao Y, Lou X, Henry RF, Stolarik DF, Lipert MP, Sheikh AY, Zhang GG. Overcoming bioavailability challenges of dasabuvir and enabling a triple-combination direct-acting antiviral HCV regimen through a salt of very weak acid for oral delivery. *Mol Pharmaceutics* 2022; 19(7): 2367–2379.

23 European Medicine Agency, Product Information, Exviera, INN-Dasabuvir, Annex I Summary of Product Characteristics. www.ema.europa.eu/en/documents/product-information/exviera-epar-product-information_en.pdf (accessed 14 January 2025).

24 Engel GL, Farid NA, Faul MM, Richardson LA, Winneroski LL. Salt form selection and characterization of LY333531 mesylate monohydrate. *Int J Pharm* 2000; 2(5): 239–247.

25 Stephens VC, Conine JW, Murphy HW. Esters of erythromycin IV. Alkyl sulfate salts. *J Am Pharm Assoc* 1959; 48(11): 620–622.

26 Thomas G. *Medicinal Chemistry: An Introduction*, 2e. New York: Wiley, 2007: 51.

27 Jashnani RN, Byron PR. Dry powder aerosol generation in different environments: performance comparisons of albuterol, albuterol sulfate, albuterol adipate and albuterol stearate. *Int J Pharm* 1996; 130(1): 13–24.

28 Lindenmayer JP. Long-acting injectable antipsychotics: focus on olanzapine pamoate. *Neuropsychiatr Dis Treat* 2010; June 24: 261–267.

29 Park EJ, Amatya S, Kim MS, Park JH, Seol E, Lee H, Shin YH, Na DH. Long-acting injectable formulations of antipsychotic drugs for the treatment of schizophrenia. *Arch Pharmacal Res* 2013; 36: 651–659.

30 Guerrieri P, Taylor LS. Role of salt and excipient properties on disproportionation in the solid-state. *Pharm Res* 2009; 26(8): 2015–2026.

31 Zannou EA, Ji Q, Joshi YM, Serajuddin AT. Stabilization of the maleate salt of a basic drug by adjustment of microenvironmental pH in solid dosage form. *Int J Pharm* 2007; 337 (1–2): 210–218.

32 Koranne S, Lalge R, Suryanarayanan R. Modulation of microenvironmental acidity: a strategy to mitigate salt disproportionation in drug product environment. *Mol Pharmaceutics* 2020; 17(4): 1324–1334.

33 Berge SM, Bighley LD, Monkhouse DC. Pharmaceutical salts. *J Pharm Sci* 1977; 66(1): 1–19.

34 Everett RM, Descotes G, Rollin M, Greener Y, Bradford JC, Benziger DP. Nephrotoxicity of pravadoline maleate in dogs: evidence of maleic acid-induced acute tubular necrosis. *Fundam Appl Toxicol* 1993; 21: 59–65.

35 Robinson DI. Control of genotoxic impurities in active pharmaceutical ingredients: a review and perspective. *Org Process Res Dev* 2010; 14(4): 946–959.

36 Snodin DJ. Residues of genotoxic alkyl mesylates in mesylate salt drug substances: real or imaginary problems? *Regulatory Tox Pharm* 2006; 45(1): 79–90.

37 Serajuddin ATM, Pudipeddi M. Salt selection strategies. In: P.H. Stahl and C.G. Wermuth, eds. *Handbook of Pharmaceutical Salts: Properties, Selection and Use*. Weinheim, Germany: Wiley–VCH, 2002: 135–160.

38 Morris KR, Fakes MG, Thakur AB, Newman AW, Singh AK, Venit JJ, Spagnuolo CJ, Serajuddin ATM. An integrated approach to the selection of optimal salt form for a new drug candidate. *Int J Pharm* 1994; 105: 209–217.

5

Pharmaceutical Cocrystals

5.1 Introduction

A pharmaceutical cocrystal is a crystalline material that contains more than one component within the crystal lattice, one of which is the active pharmaceutical ingredient (API) and the other generally a non-therapeutic coformer. Usually, they are in a defined stoichiometric ratio within the lattice, associated by nonionic and noncovalent interactions. Cocrystals are different from polymorphs and salt forms (Figure 5.1) in that polymorphs contain only the drug within different possible crystal lattices, and salt forms are ionic crystalline phases. Cocrystals consist of components in a neutral state, typically held together by noncovalent interactions such as hydrogen bonding, halogen bonding, van der Waals forces, and π-π interactions. Among these interactions, hydrogen bonding acts as a key driving force to promote the molecular recognition necessary for cocrystal formation. As would be expected, cocrystals usually display significantly different physical and chemical properties than other solid forms of the parent drug, for example, solubility, dissolution rate, melting point, hygroscopicity, chemical stability, and mechanical properties. As a result, cocrystal formation offers an approach to engineer physical and chemical attributes for a drug, especially for a non-ionizable compound, a very weak acid, or a very weak base when salt formation is not possible.

5.2 Cocrystals and Crystal Engineering

The use of cocrystals is an attractive option to design a crystalline form through crystal engineering. Most pharmaceutical compounds can form hydrogen bonding and, therefore, tend to develop supramolecular hydrogen-bonded structures with a guest molecule. Such multicomponent hydrogen-bonded assemblies are also often called supramolecular synthons [1], defined as the smallest structural units by which molecules assemble into supramolecular structures. The synthons for a supramolecular structure contain all the information necessary for molecular recognition events during cocrystal formation. They serve as the algorithms for determining the specific intermolecular interactions required to produce desired supramolecular structures. Hydrogen bonding is the primary force in driving the self-assembly of a cocrystal, although other intermolecular interactions, such as van der Waals and π—π interactions, also contribute to the process. A carboxylic acid dimer arising from the interaction of two carboxylic acid groups represents an example of a simple supramolecular synthon, which is present in many organic crystals. Besides the carboxylic acid dimer, synthons critical for pharmaceutical cocrystals include carboxylic acid chains, carboxylic acid–amide complexes, amide dimers, carboxylic acid–amide chains, and carboxylic acid–pyridine complexes (Figure 5.2).

The development and geometric analysis of synthons are an integral part of the rational design of supramolecular structures in the organic material science field. Rational design is relatively limited in pharmaceutical cocrystals because coformers can only be chosen from a small set of organic acids, organic bases, amino acids, nucleotides, or vitamins, because of safety concerns and regulatory acceptance. However, the principle of synthon analysis is very useful in facilitating the screening of potential coformers.

For example, the molecular structures of itraconazole, theophylline, and caffeine all have a portion of a pyridine-type structure and tend to form carboxylic acid–pyridine heterosynthons. Carboxylic acids like succinic acid, maleic acid, tartaric acid, and malic acid, therefore, are good possible coformers for those three compounds. Interestingly, 6 cocrystals for itraconazole, 11 cocrystals for theophylline, and 13 cocrystals for caffeine have been discovered using various organic acids.

Solid-State Materials in Pharmaceutical Chemistry: Properties, Characterization, and Applications, Second Edition. Stephen R. Byrn, George Zografi and Xiaoming (Sean) Chen.

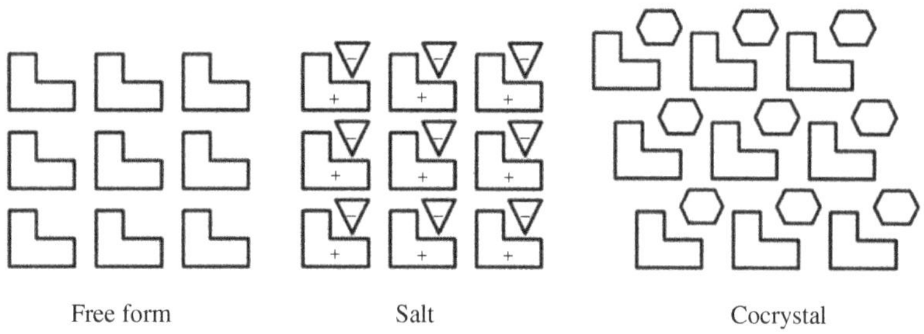

Figure 5.1 Comparison of a free form, a salt form, and a cocrystal for an API.

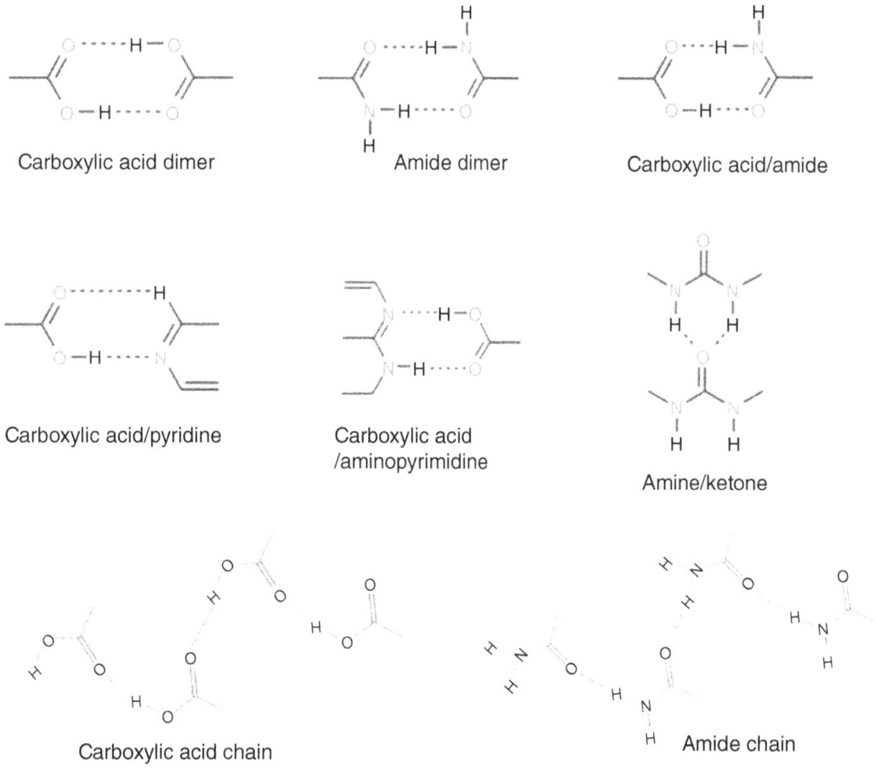

Figure 5.2 Common hydrogen synthons for pharmaceutical cocrystals.

A portion of the crystal structure of the caffeine and citric acid cocrystal is illustrated in Figure 5.3 [2], where we can see that a hydrogen bond with a length of 2.68 Å links citric acid and caffeine together.

Other structural aspects of a drug molecule also affect the formation of cocrystals, including the number and arrangement of hydrogen bond donors and acceptors, conformational flexibility, and molecular weight. Usually, drug molecules that display rigid structures, possess strong hydrogen bonds, and have low molecular weights have more potential to cocrystallize. If a drug has the potential to form hydrates or solvates, there is a good chance that other guest molecules could get into the crystal lattice to form cocrystals. For example, carbamazepine (CBZ) is a rigid molecule and can form a dihydrate and the acetone solvate. More than 20 cocrystals of CBZ have been discovered with a wide range of coformers, including saccharin (SAC), nicotinamide (NCT), formic acid, and succinic acid [3].

As would be expected, computational tools could potentially offer a more efficient way to systematically evaluate the strength of hydrogen bonding potential and other forms of molecular interaction compared to empirical synthon analysis, serving as an effective first step in cocrystal design and engineering. Several computational platforms are gaining widely recognition, including the following:

- Hydrogen bond propensity [4, 5], to quantify the probability of hydrogen bond formation for heteromeric interactions (API-conformer) and homomeric interactions (API–API, conformer–conformer), and to compare the differences between heteromeric and homomeric interactions to rank the likelihood of cocrystallization.

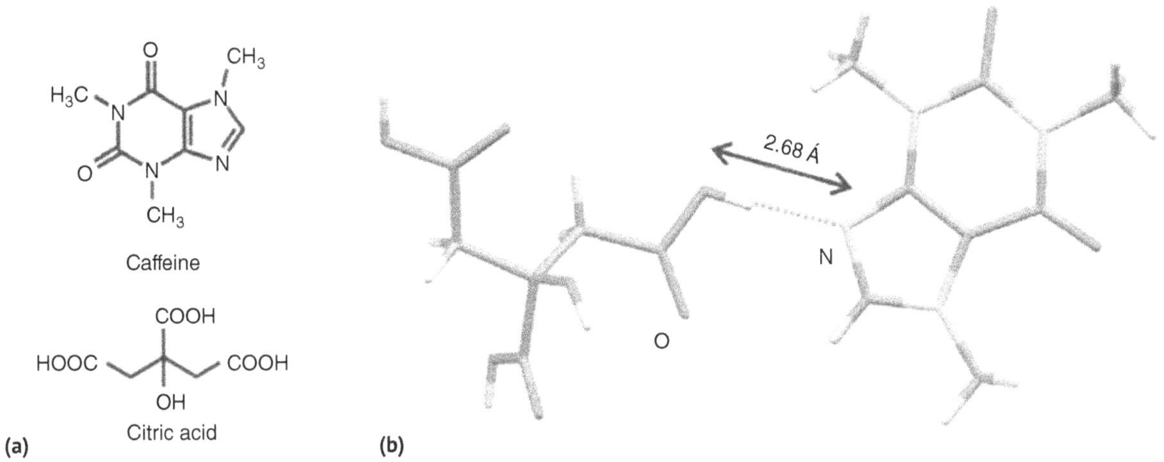

Figure 5.3 Hydrogen bond in caffeine–citric acid cocrystal. *Source*: [2]/with permission of John Wiley & Sons.

- Molecular complementarity [6], to assess the similarity in molecular shape and polarity of an API and a conformer using five calculated numerical descriptors, with three for the shape and two for the polarity, and to predict the propensity of a cocrystal to form, the more similar of the properties, the more likely for the cocrystal to form.
- Molecular electrostatic potential maps [7], to provide an estimate of the energy gain after potential interactions between two molecules, that is, the API and the conformer, the greater the gain, the more likely the cocrystal to form.
- Crystal structure predictions [8], to execute a virtual conformer screening by computing the energy of the cocrystallization interaction via a sublimation thermodynamic cycle as the difference between the calculated lattice energies of the cocrystal and its individual components.

For example, Racher et al. demonstrated that the above-mentioned computational methodologies provided a strategic direction in the discovery of novel cocrystals of dapsone, by significantly narrowing down the number of coformers to study and subsequently reducing the burden of tedious and costly experimental screening [7]. On the other hand, by applying crystal structure predictions, Abramov and coworkers ranked 10 potential cocrystal coformers for MK-8876, an antiviral drug candidate, and successfully predicted the maleic acid cocrystal as the most likely cocrystal to prepare, in agreement with experimental findings [9].

In summary, successful cocrystal formation primarily relies on the ability to form hydrogen-bonded assemblies between a drug and a guest molecule. It also requires achieving an appropriate balance between competing molecular associations, such as those between similar molecules (homomers) and different molecules (heteromers). The application of computational tools apparently offers a more systemic approach to screening coformers that are suitable for cocrystal formation.

5.3 Solubility Phase Diagrams for Cocrystals

The formation and dissolution of a cocrystal, A_aB_b, can be described by the following reversible equation if solution complexation is taken to be negligible, as suggested by Nehm et al. [10]:

$$A_aB_b(\text{solid}) \leftrightarrow a.A + b.B \tag{5.1}$$

In equation (5.1), A represents a drug and B denotes a coformer. Subscripts, a and b, refer to the stoichiometric number of molecules of A or B in the cocrystal. The equilibrium constant for this equation can be defined by the solubility product, K_{sp}, as follows:

$$K_{sp} = [A]^a[B]^b \tag{5.2}$$

where $[A]$ and $[B]$ are the concentrations of drug A and coformer B at equilibrium. K_{sp} is a constant independent of the concentrations. As the concentration of A and/or B is increased until $[A]^a[B]^b$ is greater than K_{sp}, there is a thermodynamic driving force for the cocrystal, A_aB_b, to form. Conversely, if $[A]^a[B]^b$ is lower than K_{sp}, the dissolution of the cocrystal is favored. A solubility phase diagram can be constructed based on equation (5.2). Figure 5.4 shows the solubility phase diagram of CBZ–NCT (1:1) cocrystal as a function of cocrystal component concentration in ethanol at 25°C [10]. The solid line represents the equilibrium

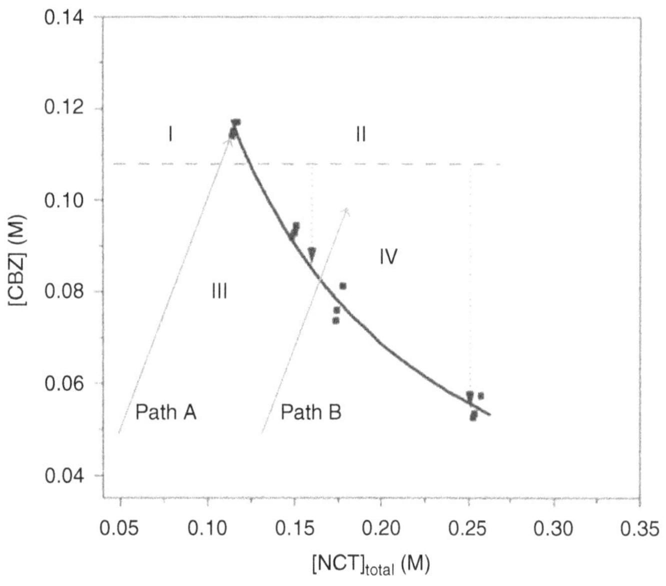

Figure 5.4 Solubility phase diagram of CBZ–NCT (1:1) cocrystal in ethanol at 25 °C. *Source*: [10]/with permission of American Chemical Society.

concentrations of CBZ at different concentrations of NZT based on equation (5.2) using a solubility product K_{sp} of 0.0129 M^2. Because a coformer like NZT is usually very soluble, the concentrations of CBZ in the K_{sp} curve represent the cocrystal solubilities at different levels of NZT. When the concentration of NZT increases, cocrystal solubility decreases as predicted by the solubility phase diagram. This phenomenon is like the common ion effect on salt solubility.

The dashed line in Figure 5.4 represents the solubility of CBZ (Form III) in ethanol, as 0.108 M. The intersection point of the cocrystal solubility curve and the CBZ solubility curve represents the equilibrium point of CBZ and the cocrystal, that is, when the cocrystal solubility is the same as the CBZ solubility. The cocrystal solubility curve and the CBZ solubility line divide the solubility phase diagram into four areas. In area I, the CBZ concentration is supersaturated for CBZ but undersaturated for the CBZ–NZT cocrystal, so CBZ tends to crystallize. In area II, the CBZ concentration is supersaturated for both CBZ and the cocrystal. Both CBZ and the cocrystal, therefore, can form. In area III, the CBZ concentration is undersaturated for both CBZ and the cocrystal. While in area IV, the CBZ concentration is supersaturated for the cocrystal but undersaturated for CBZ. The cocrystal is the thermodynamically stable form in area IV. CBZ tends to transform to the cocrystal by a solution-mediated mechanism because the cocrystal is less soluble than CBZ in that space. Area IV is an ideal place to prepare the cocrystal without the complication of causing the precipitation of CBZ. The phase solubility diagram can be established for any cocrystals based on the cocrystal K_{sp} and the solubility of the free form.

A phase solubility relationship, as illustrated in the previous case, can also be represented as a ternary phase diagram. This diagram uses the API, conformer, and solvent as the vertices of a triangle, offering another worthwhile visualization of the process. As shown in Figure 5.5, assuming no polymorphic forms for any solid phases to simplify the illustration, the phase diagrams would contain three solid phases, that is, S_{API} for the API, $S_{co\text{-}former}$ for the conformer, and S_{cc} for the cocrystal, and the liquid phase to represent the solution in the diagrams [11]. As would be expected, there are six regions in a typical phase diagram, including the following:

- A single-phase region for the solution, as L
- A two-phase region for the solution and the API solid, as $L + S_{API}$
- A two-phase region for the solution and the conformer solid, as $L + S_{coformer}$
- A two-phase region for the solution and the cocrystal, as $L + S_{CC}$
- A three-phase region for the solution, API solid, and the cocrystal, as $L + S_{API} + S_{CC}$
- A three-phase region for the solution, conformer solid, and the cocrystal, as $L + S_{coformer} + S_{CC}$

It is obvious that the two-phase region with the solution and the cocrystal is the most important design space to search for during initial cocrystal screening or later for the optimization of preparation, since the cocrystal is the only thermodynamically stable solid form in that space. It appears that the solubility of the API, the conformer, and the cocrystal in a specific solvent or solvent mixture defines the phase boundaries of each of the six regions, subsequently determining the overall mapping of the phase diagram and, most importantly, the position of the $L + S_{CC}$ region. For example, as illustrated in Figure 5.5, the phase diagram on the left represents the case where the API is significantly more soluble than the conformer, while the one on the right exists for the opposite reason. As expected, the case on the right diagram requires an excess of the conformer to operate in the thermodynamically favored space to generate the cocrystal. The middle case presented in the figure represents an ideal situation where the API and the conformer have enough solubility to have the cocrystal as the thermodynamically stable phase in the solution and is expected to show congruent dissolution behavior.

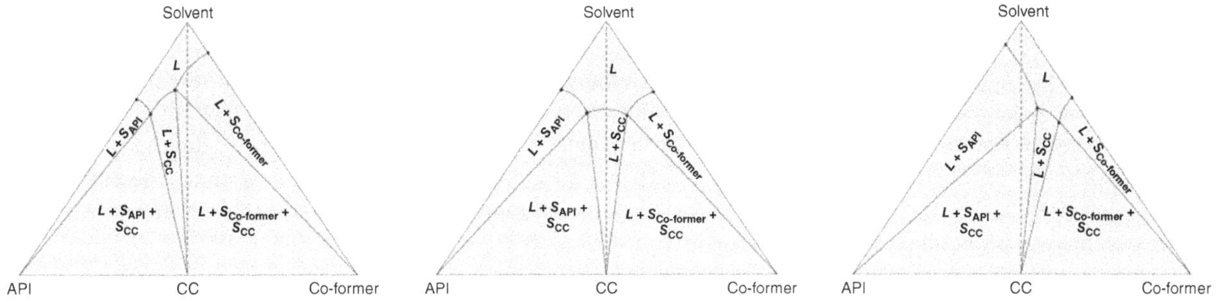

Figure 5.5 Phase diagrams to form cocrystals. L: liquid or solution; S_{API}: crystalline API solid; $S_{co\text{-}former}$: crystalline conformer solid; S_{CC}: crystalline cocrystal solid. *Source*: [11]/ with permission of American Chemical Society.

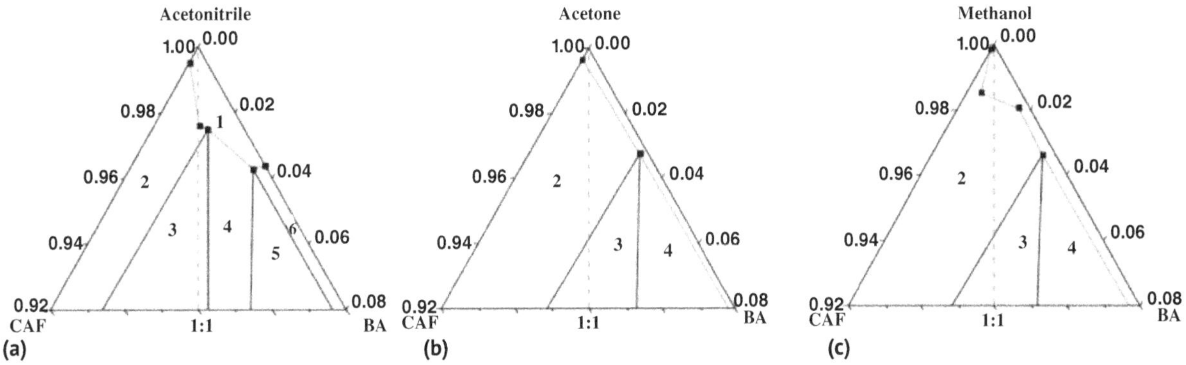

Figure 5.6 Phase diagrams of caffeine–benzoic acid cocrystal in acetonitrile (a), acetone (b), and methanol (c), at 25 °C. Area 1: solution; Area 2: solution + caffeine crystalline solid; Area 3: solution + caffeine crystalline solid + cocrystal solid; Area 4: solution + cocrystal solid; Area 5: solution + benzoic acid crystalline solid + cocrystal solid; Area 6: solution + benzoic acid crystalline solid. *Source*: [12]/ with permission of American Chemical Society.

The dissolution behavior would affect the manufacturing process of cocrystals, especially the cake washing step, since incongruent dissolution caused by a skewed phase diagram can potentially lead to the dissociation of the cocrystal. As reported by Codan et al., solubility ratios dramatically differing from one, that is, below 0.1 and above 10, consequently lead to an asymmetric phase diagram in which the cocrystals dissolved incongruently and the API or the conformer precipitated [11].

As would be expected, for a specific pair of API and conformer, the solvent would have a significant impact on the phase diagram, as shown in the case of a caffeine–benzoic acid cocrystal. As illustrated in Figure 5.6, each solvent, such as acetonitrile, acetone, or methanol, gives a distinct phase diagram [12]. The cocrystal stable zone, region 4 in the diagrams, can be found in the ternary phase diagram of all three solvents, however, located in different areas. It appears that the phase diagram in acetonitrile is more symmetric, indicating that the cocrystal can be attained by the approximate equimolar ratio of caffeine and benzoic acid, in agreement with experimental findings. However, the cocrystal regions in the phase diagrams with acetone and methanol, respectively, are asymmetrically skewed to the benzoic acid side, pinpointing the direction of a large amount of benzoic acid being required for the cocrystal to form. In summary, solvent selection plays an important role in designing conditions to screen for cocrystals or to prepare cocrystals consistently in a controlled manner.

5.4 Preparation of Cocrystals

Initially, establishing a phase solubility diagram is helpful for the preparation of cocrystals. Cocrystals can be prepared by evaporation, reaction crystallization, solution-mediated transformation, solvent-drop grinding, and co-grinding.

Evaporation from solutions of a stoichiometric amount of the drug and coformer is a popular way to prepare cocrystals. For example, fluoxetine hydrochloride cocrystals with succinic acid and fumaric acid were prepared by evaporation [13].

Fluoxetine hydrochloride and these acids in 2 to 1 molar ratio were mixed in acetonitrile and subjected to slow evaporation until crystal formation.

Sometimes, this approach fails to isolate cocrystals. For example, as shown in Figure 5.4, a stoichiometric solution of CBZ and NZT initially resides in area III in the phase solubility diagram. Following Path A in the diagram, as the solvent evaporates, the solution becomes concentrated, and the process path moves toward area I, where the drug may crash out before the cocrystal can form. However, increasing the coformer concentration in the solution can alter the path of the solution in the phase solubility diagram. As illustrated in Path B, the solution will move from area III to IV during the evaporation process, thereby promoting cocrystal formation.

The presence of a significant amount of coformer can also drive cocrystal formation through reaction crystallization and solution-mediated transformation. Rodríguez-Hornedo et al. studied the effect of NCT concentrations on the crystallization of CBZ–NCT cocrystals in suspensions and solutions [14]. The presence of NCT promotes the solution-mediated transformation from CBZ to CBZ–NCT crystals in ethanol. At 0.16 M NCT, the transformation was complete within 3 hours.

A faster conversion of pure CBZ to the cocrystal was observed when the NCT concentration was increased. For example, the synthesis of the cocrystal was complete within 2–3 minutes when the NCT concentration was above 0.25 M. For a high-throughput cocrystal screen for MK-8876 in a 96-well plate, Abramov et al. applied about 1.5 equivalent of conformer relative to MK-8876 and took a material sparing approach using only 10 mg of MK-8876 per well. With less than 1 gram of the API being used, 16 conformers in combination with six solvents were screened in a slurry with mixing further facilitated by magnetic stirring and acoustic mixing, resulting in the cocrystal with maleic acid in three solvents, as shown in Table 5.1 [9].

Table 5.1 Results from the high-throughput cocrystal screening for MK-8876.

	Solvent					
Coformer	**Isopropyl acetate**	**Toluene**	**Water**	**Acetonitrile**	**Methyl ethyl ketone**	**Isopropanol**
Fumaric acid	FF + CF	FF + CF	FF + CF	FF + CF	FF + CF	FF + CF
L-Tartaric acid	FF + CF	FF + CF	FF	FF + CF	FF + CF	FF + CF
Succinic acid	FF + CF	FF + CF	FF	FF	FF + CF	FF + CF
Glutaric acid	FF + CF	FF + CF	FF + CF	FF + CF	FF + CF	FF + CF
Malic acid	FF	FF	FF	FF	FF	FF
Ascorbic acid	FF + CF	FF + CF	FF + CF	FF + CF	FF + CF	FF + CF
Citric acid	FF + CF	FF + CF	FF	FF + CF	FF + CF	FF + CF
α-Ketoglutaric acid	FF	FF + CF	FF	FF	FF	FF
Glutamic acid	FF + CF	FF + CF	FF + CF	FF + CF	FF + CF	FF + CF
Mucic acid	FF	FF	FF + CF	FF + CF	FF + CF	FF + CF
Gentisic acid	FF	FF + CF	FF	FF + CF	FF	FF + CF
Tagatose	FF + CF	FF + CF	FF	FF + CF	FF + CF	FF + CF
Tris	FF + CF	FF + CF	FF + CF	FF + CF	FF + CF	FF + CF
Meglumine	FF	FF + CF	FF	FF + CF	FF + CF	FF + CF
Maleic acid	CC	FF + CF	FF	CC	CC	FF + CF
Benzoic acid	FF	FF	FF	FF	FF	FF
Label	CC		FF		FF + CF	
	Cocrystal		Free form		Free form + coformer	

Source: [9]/with permission of American Chemical Society.

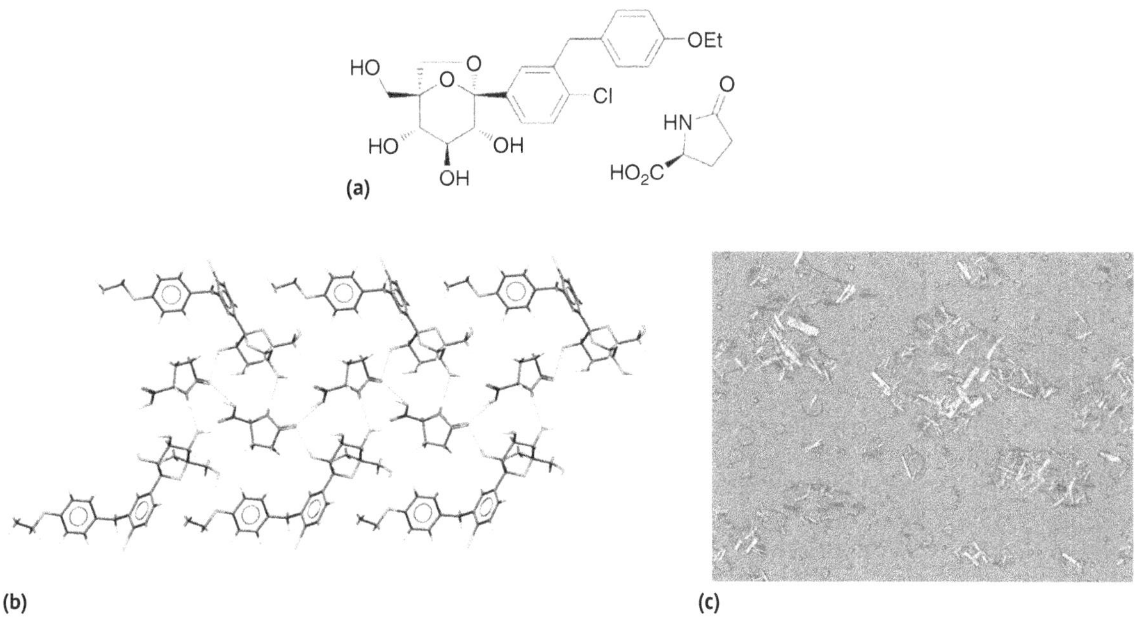

Figure 5.7 Ertugliflozin-l-pyroglutamic acid cocrystal. (a) chemical structure; (b) crystal structure to illustrate hydrogen bond interactions; (c) microscopic image to show biphasic cocrystallization of ertugliflozin and l-pyroglutamic acid in isopropanol and water. *Source*: Adapted from [16]; (c) Reproduced from [16] / with permission of American Chemical Society.

Reaction cocrystallization can be accomplished by mixing solutions of cocrystal components with a significant overage of coformer. It has been reported that CBZ–NCT cocrystals were formed within 10–25 minutes after mixing an ethanol solution of NCT (2.25 mL of 0.8 M) and an ethanol solution of CBZ (3.75 mL of 0.1 M) [14]. Maximizing the coformer concentration by using a saturated or nearly saturated coformer solution is a popular approach for cocrystal screening using reaction crystallization. Thirteen cocrystals of gabapentin with various carboxylic acid coformers, for example, were discovered by this method [15]. In the screening experiments, 1 mL of presaturated acetonitrile or ethanol solution of a coformer was mixed with 50–75 mg of gabapentin. In most cases, cocrystallization occurred within minutes. Bowles et al. reported that a stoichiometric excess L-pyroglutamic acid is critical in the commercial route of preparation of its cocrystal with ertugliflozin (Figure 5.7) in a mixture of isopropanol and water [16]. It is interesting to note that, despite the miscibility of the two solvents, the presence of ertugliflozin and L-pyroglutamic acid caused a phase separation. The more hydrophobic drug resided in isopropanol droplets, while the hydrophilic conformer predominantly occupied the aqueous phase. Consequently, the cocrystallization occurred effectively at the interface, as shown in Figure 5.7.

The saturated solution of a coformer also plays a significant role in solvent-drop grinding, another popular way of screening cocrystals. Solvent-drop grinding involves the grinding of a mixture of a drug and a coformer together with a specific stoichiometry. A minor quantity of a solvent is then added to promote the formation of the cocrystal. The solvent acts like a catalyst and is usually not a component of the final product. Some of the drug and the coformer are dissolved in the solvent and form a saturated solution with both components. A saturated solution of the coformer lowers the cocrystal solubility and provides a significant driving force for the crystallization of the cocrystal form, similarly to reaction cocrystallization. At the same time, grinding reduces the particle size and enhances the dissolution process. Grinding also provides intimate mixing and promotes both the dissolution of components and the crystallization of the cocrystal. Furthermore, high-energy grinding may generate amorphous content and, in this way, increase the solubility of both components, delivering even more driving force to form the cocrystal. Shan et al. reported that the grinding of cyclohexane-1,3-*cis*, 5-*cis*-tricarboxylic acid and 4,40-bipyridine with a few drops of methanol yielded a cocrystal within minutes, while the grinding without solvent failed to produce the cocrystal [17].

The solvent-drop grinding method may also provide better polymorph control of a cocrystal. The polarity of a solvent can affect the type of polymorphs produced, as demonstrated in the case of caffeine and glutaric acid cocrystal [18]. The 1:1 cocrystal of caffeine and glutaric acid has two polymorphs, Forms I and II. Form I has a rod-like shape and belongs to the monoclinic space group for its crystal structure. Form II shows block-type morphology and has a triclinic space group.

It was further reported that solvent-drop grinding in a nonpolar solvent such as *n*-hexane, cyclohexane, or heptane yielded Form I. Conversely, on the addition of a polar solvent, including chloroform, dichloromethane, acetonitrile, and water, the grinding resulted primarily in Form II.

Solvent-drop grinding was found as one of the effective ways to search for a cocrystal in the case of tramadol and celecoxib drug–drug cocrystallization. It further offered valuable seeding materials in the scale-up preparation through reaction crystallization in solvents [19]. Almansa et al. reported that various techniques, such as solvent-drop grinding, slurrying, evaporation, and vapor diffusion, were undertaken, but only solvent-drop grinding in methyl isobutyl ketone or isopropanol yielded the desired drug–drug cocrystal [19]. Using this material as seeds, the cocrystal of tramadol HCl and celecoxib in a 1:1 molar ratio was successfully prepared by crystallization in isopropanol/heptane.

Co-grinding of a drug and a coformer without the addition of a solvent sometimes generates a cocrystal, though it is less efficient and versatile than solvent-drop grinding. When equimolar quantities of anhydrous caffeine and glutaric acid, for example, were ground in a stainless-steel grinding jar with two grinding balls, the cocrystal Form I was predominantly produced [18]. A study of the co-grinding of CBZ and SAC revealed that cocrystal formation during co-grinding and storage was mediated by an amorphous phase. Co-grinding at room temperature and under cryogenic conditions yielded a cocrystal of CBZ–SAC, like the one prepared from solvents [20]. However, the material from co-grinding had very poor crystallinity. The crystallinity of the cocrystal was significantly improved during storage for 1 day at both 0% and 75% relative humidity (RH), but more cocrystal was formed at 75% RH. This observation suggests that an amorphous phase-mediated transformation had occurred because moisture sorption increases the molecular mobility and promotes the crystallization of the amorphous phase.

Cocrystallization from an amorphous phase is also utilized in the melt-crystallization method of obtaining cocrystals. In this procedure, two compounds are melted adjacent to one another on a glass slide and mixed by applying a coverslip over the sample. A cocrystal of the two compounds will form at the interface where the compounds mix. The formation of the cocrystal can be visually determined by a polarized light microscope, since, if there is no interaction between the two compounds, the mixing of the two components would create an eutectic area where no crystallization occurs. On the other hand, crystalline material would form at the interface if cocrystallization had taken place. A cocrystal of 2-[4-(4-chloro-2 fluorophenoxy)phenyl]pyrimidine-4-carboxamide (compound 1), a Purdue Pharma compound, and glutaric acid was discovered using melt-crystallization and then was scaled up by using crystallization from solution with seeding [21].

In summary, techniques widely employed for cocrystal screening fall into two categories: solution-based crystallization methods and solid-based techniques. Solution-based methods include evaporation, slurry conversion, and reaction crystallization. Solubility product, K_{sp}, and phase solubility diagrams are useful to accomplish the control of cocrystallization. Solid-based approaches have the advantage of being solvent-sparing or solvent-free. The solvent-drop method offers high efficiency and control of polymorphs production by using various solvents.

5.5 Dissolution and Bioavailability of Cocrystals

The main motivation to screen for cocrystals for a pharmaceutical purpose is to improve their physical properties and biopharmaceutical performance, primarily dissolution rate and bioavailability. With the advancement of combinatorial chemistry, new chemical entities with limited aqueous solubility are more prevalent in the pipelines of pharmaceutical companies, and the low aqueous solubility of those drug molecules causes their slow dissolution in biological fluids, which in turn may result in undesirable bioavailability and suboptimal clinical efficacy. For neutral molecules or very weak acids/bases, salt formation may not be applicable for enhancing dissolution. For those compounds, cocrystals could offer an alternative to other formulation approaches for enhancing dissolutions, such as amorphous dispersions and nanoparticles.

The enhancement of dissolution rate using a cocrystal occurs when its solubility is greater than that of the parent drug. The solubility mentioned here refers to the solubility of the drug without an overage of the conformer and is the maximum solubility a cocrystal can have. However, the maximum solubility is challenging to assess using the traditional shake flask method when a cocrystal is more soluble than the parent drug, since a solution-mediated transformation to the parent compound tends to occur during the measurement. The transformation process of a cocrystal is like the disproportionation of a salt form. The transformation, for example, of a 1:1 cocrystal can be described by the following equation:

$$\text{AB (solid)} \rightarrow \text{A} + \text{B} \rightarrow \text{A(solid)} + \text{B} \tag{5.3}$$

where AB is the cocrystal, A is the drug, and B is the conformer. To minimize the complications of such transformations, the solubility of a cocrystal can be evaluated by measuring the intrinsic dissolution or by powder dissolution studies. Both techniques measure the dissolution of a drug substance within a relatively short period of time and monitor any possible transformations that might have occurred during the studies.

An intrinsic dissolution rate is measured under the condition of a constant surface area. Based on the Noyes–Nernst equation (equation [5.4]), where the dissolution rate of a drug substance (*dc*/*dt*) is linearly related to its solubility under a constant surface area (*A*) and at a sink condition.

$$\frac{dc}{dt} = \frac{DA}{h}(C_0 - C_h) \tag{5.4}$$

where *D* is the diffusion coefficient of the drug, *h* is the diffusion layer thickness, C_0 is the drug solubility, and C_h is the drug concentration in the bulk solution. At the sink condition, C_h is close to zero.

The intrinsic dissolution of a cocrystal can be compared with that of the drug itself in water, in a buffered solution, or in a biorelevant medium.

Another way to assess the solubility of a cocrystal is to measure the dissolution rate of the powder form. The amount of drug dissolved is captured at different time points when the cocrystal powder is added to a dissolution medium. The dissolution media could be water, 0.1 N HCl, phosphate buffer, simulated gastric fluid, and/or simulated intestinal fluid. Depending on the kinetic stability of the cocrystal during dissolution, measurement of the dissolution profile could be increased over a longer time until reaching a plateau, when no physical transformation occurs during the time scale investigated. If there is a physical transformation during dissolution, the concentration will reach a maximum and then start to decrease. The maximum amount dissolved is also called the apparent solubility of the drug substance. The apparent solubility reflects the real solubility potential of a cocrystal.

As discussed previously, the concentration of a coformer has a significant impact on the solubility of the cocrystal formed, as do the molecular properties of the coformer, such as solubility and ionization. For example, it was reported that the cocrystal CBZ–NCT is 10 times more soluble than CBZ dihydrate [22]. However, the cocrystals with other ligands were only slightly more soluble than CBZ dihydrate. The good solubility of the CBZ–NCT cocrystal can be attributed to the high solubility of NCT.

The ionization of a coformer can further enhance the solubility of a cocrystal. The pH solubility profile for the cocrystal, for example, was predicted by knowing the cocrystal dissociation equilibrium constant and the ionization constant of the coformer [23]. For a 1 : 1 cocrystal between a neutral drug A and an acidic coformer HB, the cocrystal solubility dependence is given as follows:

$$S_{\text{cocrystal}} = \sqrt{K_{\text{sp}}\left(1 + \frac{K_{\text{a}}}{[H^{+}]}\right)} \tag{5.5}$$

where K_{sp} is the solubility product of the cocrystal and K_a is the dissociation constant of the coformer HB.

Glutaric acid, an ionizable coformer, was utilized to form a cocrystal with a poorly soluble development candidate, 2-[4-(4-chloro-2 fluorophenoxy)-phenyl]pyrimidine-4-carboxamide (compound 1) from Purdue Pharma [21]. This compound exhibited extremely low solubility, less than 0.1 μg/m, in water, simulated gastric fluid, and simulated intestinal fluid. It is a very weak base with the estimated p*Ka* of 0.7, which makes it a poor candidate for salt formation. Moreover, the amorphous form of compound 1 was found to be difficult to prepare, while solubilization using cosolvents and surfactants was insufficient for adequately enhancing solubility. Interestingly, a cocrystal containing compound 1 and glutaric acid in a 1 : 1 molecular ratio was discovered (Figure 5.8) and shown to significantly improve the dissolution rate. For instance, the intrinsic dissolution rate of the cocrystal is approximately 18 times higher than compound 1 itself in water at 37 °C. The increase of dissolution rate by the cocrystal resulted in a significant improvement in oral bioavailability in that a single dose pharmacokinetic study in dogs demonstrated that the cocrystal increased the area under the curve (AUC) values by three times at dose levels of 5 and 50 mg/kg, compared with the original compound 1. In addition, the cocrystal at 50 mg/kg exhibited a shorter T_{max} and a higher C_{max}, attributed to its desirable oral absorption (Figure 5.9). In contrast, the oral absorption of the original compound 1 is limited by solubility and displays a prolonged period over which absorption occurred.

Improved bioavailability has also been observed for a 2 : 1 cocrystal of L-883555 and L-tartaric acid (Table 5.2). L-883555 is a phosphodiesterase IV inhibitor with indications for the treatment of asthma and chronic obstructive pulmonary disease. In a pharmacokinetic study with rhesus monkeys, a dose of 3 mg/kg of the cocrystal produced 15 times higher C_{max} and 23 times greater AUC than the parent compound at the same dose [24].

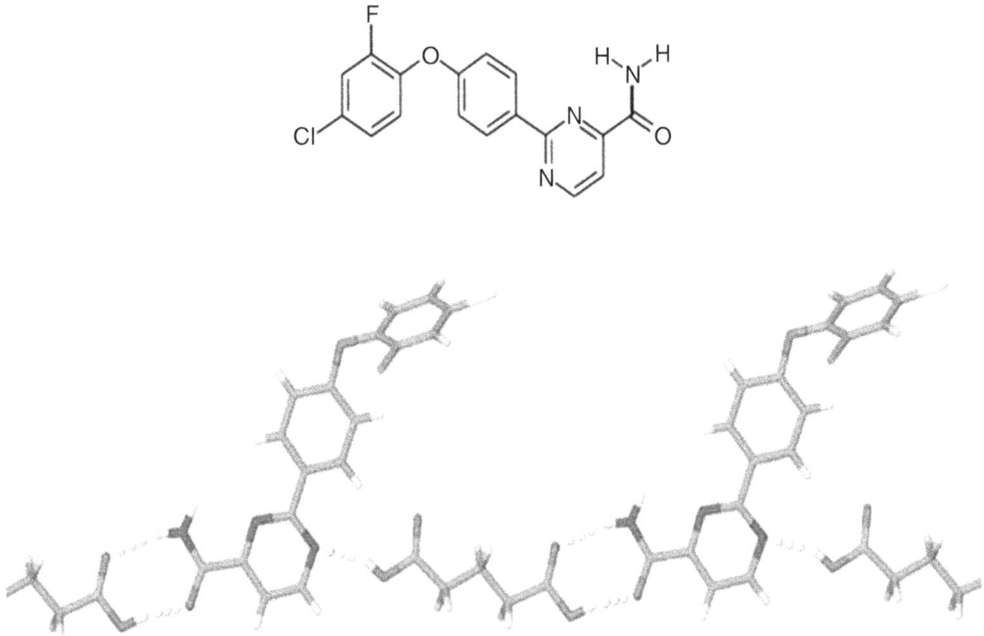

Figure 5.8 Crystal structure of the cocrystal of compound 1 and glutaric acid. *Source*: [21]/with permission of Springer Nature.

Table 5.2 Comparison of pharmacokinetics of L-883555 and L-883555/tartaric acid cocrystal in rhesus monkeys.

Drug substance	AUC (mg/mL×h)	C_{max} (mg/mL)
L-883555	0.24 ± 0.9	0.03 ± 0.009
L-883555/tartaric acid	5.5 ± 4.3	0.44 ± 0.21

Source: [24]/with permission of American Chemical Society.

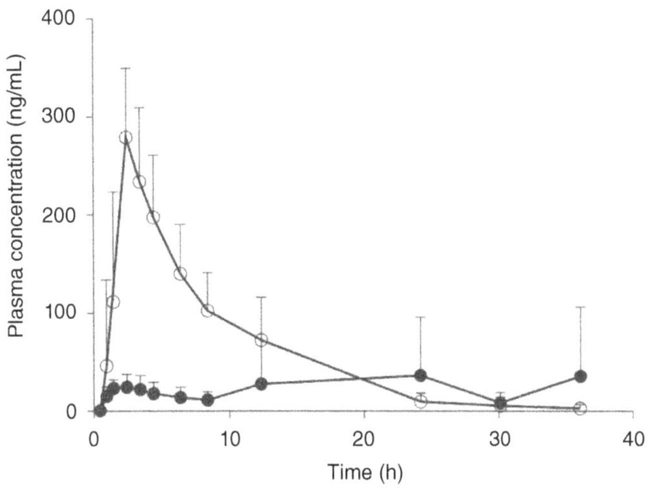

Figure 5.9 Dog plasma concentration versus time for 50 mg/kg dosing of compound 1 (solid circles) and cocrystal of compound 1 and glutaric acid (open circles). Error bars show +1 standard deviation. *Source*: [21]/with permission of Springer Nature.

AMG 517–sorbic acid cocrystal is another case of using a cocrystal to improve bioavailability [25]. AMG 517 is a vanilloid receptor 1 antagonist with poor aqueous solubility. AMG 517 can form a 1:1 cocrystal with sorbic acid, which shows a higher aqueous solubility initially as compared to AMG 517 free base in simulated fasted intestine fluid. The cocrystal, however, was converted back to the AMG 517 hydrate during a prolonged slurry, but a 30 mg/kg dose of the cocrystal had comparable exposure to a 500 mg/kg dose of the AMG 517 free form in a rat bioavailability study.

5.6 Pharmaceutical Applications of Cocrystals

As discussed earlier, the improvement of aqueous solubility and oral bioavailability of many potential drugs is one of the main reasons that cocrystals are gaining more and more attention as a preferred solid form. However, as we shall see, cocrystals have also provided opportunities for engineering processes, such as purification and chiral resolution. Notably, they can enhance physical properties, addressing issues such as poor crystallinity or high hygroscopicity. Furthermore, cocrystallization enables the development of unique combinational products, particularly through drug–drug cocrystals.

Cocrystals can serve as effective intermediates in API purification when other means, such as chromatography and recrystallization, are found not capable of accomplishing the goal. For example, Billot and coworkers encountered significant difficulty in the purification of SAR1, a clinical candidate for the treatment of acute myeloid leukemia, via various conventional ways, including chromatography, impurity adsorption, and repeated recrystallization in multiple solvents [26]. The API was attained at best in 90% purity with a dark-brown color, obviously not acceptable for downstream drug product

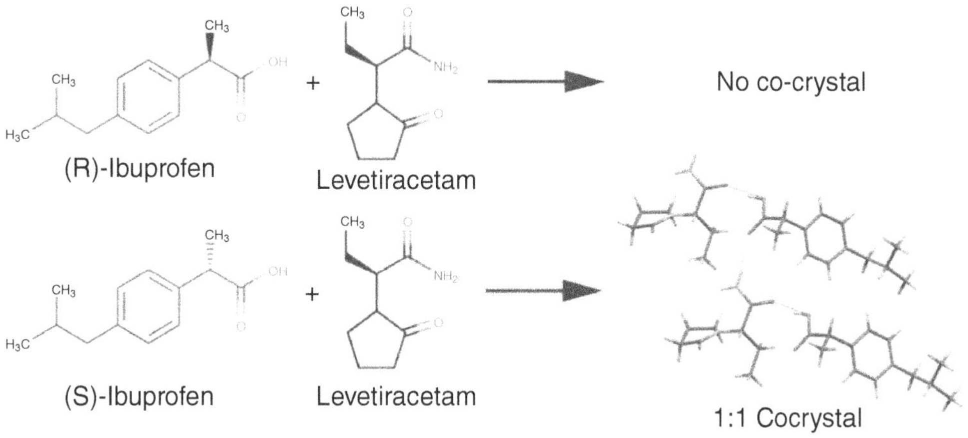

Figure 5.10 Schemes of enantiospecific cocrystal formation between ibuprofen enantiomers and levetiracetam. *Source*: [28]/with permission of American Chemical Society.

formulation. Owing to the existence of various solvates for the compound, an attempt to use solvates as process intermediates to purify the substance and subsequently to desolvate these to form the anhydrous form only led to moderate success. While purity improved to approximately 95%, residual solvents remained challenging to remove without risking chemical degradation. Cocrystallization was explored as an alternative approach, resulting in the isolation of multiple cocrystals with organic acids, including oxalic, malonic, fumaric, succinic, and benzoic acids. Given that the crude API was already dissolved in chlorobenzene, benzoic acid, being the only acid soluble in the solvent, was further studied for cocrystallization. It is interesting to note that a single cocrystallization step with benzoic acid improved the purity from the starting API purity of 65% to 99% at a pilot plant scale, demonstrating its efficacy as a purification strategy [26].

Like salt formation, cocrystallization can be a useful tool for chiral resolution, which often serves as an important step in chemical processing to make the API meet the requirement to have high enantiopurity in drug products. Nulek and coworkers used cocrystallization to obtain the desired enantiomer from the racemic mixture for etiracetam, an antiepileptic drug [27]. Using 2-chloromandelic acid as a conformer, [S]-etiracetam can form a cocrystal with [S]-2-chloromandelic acid but not with [R]-etiracetam. Exploiting this enantiospecific cocrystallization resulted in acceptable chiral resolution with 97% chiral purity and 69% yield [27]. Furthermore, Harmsen and Leyssens explored the feasibility of dual-drug chiral resolution using enantiospecific cocrystallization of ibuprofen with levetiracetam [28]. Interestingly, racemic ibuprofen can be resolved using levetiracetam through the formation of an enantiospecific cocrystal between (S)-ibuprofen and levetiracetam, whereas no cocrystallization occurs for (R)-ibuprofen. This dual-drug cocrystallization not only led to successful chiral resolution but also offered a pathway to a combination product through drug–drug cocrystals, as illustrated in Figure 5.10.

In the case of ipragliflozin, an important drug for the treatment of type 2 diabetes, the application of cocrystallization enabled the improvement of physical attributes for the API, as disclosed by Imamura and coworkers [29]. Ipragliflozin was found to form a clathrate hydrate, which changes reversibly from an anhydrous form to a non-stoichiometric hydrate depending on hygrothermal conditions, posing a significant challenge to maintaining consistent quality as a drug substance. However, since ipragliflozin exists as a neutral compound without ionizable groups at physiologically applicable pH, cocrystal formation became the key approach to improve its physical properties. Through extensive cocrystal screening, fortunately, a cocrystal with L-proline was discovered at a 1 : 1 molar ratio, having acceptable quality and good reproducibility in its preparation. The cocrystal displayed excellent crystallinity and a high melting point exceeding 200 °C. In addition, it provided superior stability against moisture sorption, maintaining moisture contents of 0.7% or less and exhibiting no significant moisture sorption across a wide RH, ranging from 5% to 95%. Indeed, the cocrystal with L-proline led to the development of an important therapeutic agent [29].

Likewise, cocrystal formation also led to acceptable solid dosage forms for valproic acid and valproate sodium for the treatment of epileptic seizures. Valproic acid exists as a liquid at room temperature obviously not feasible for solid dosage formulation, while sodium valproate is a crystalline solid, being highly hygroscopic and thus leading to poor stability. Cocrystallization of valproic acid and sodium valproate from an organic solvent resulted in a one-to-one cocrystal known as divalproex sodium. This cocrystal has a melting point of approximately 100 °C and displays desirable attributes, including nonhygroscopicity, making it suitable for pharmaceutical use [30, 31].

In a case reported by Bowles et al. [16], cocrystallization was found as an indispensable tool for crystallizing ertugliflozin, a drug for the treatment of type 2 diabetes. This method was instrumental not only for purifying the compound but also for yielding a drug substance with acceptable physical and chemical properties to support downstream formulation and manufacturing of the drug product. Ertugliflozin, in its free form, was difficult to crystallize, often leading to an amorphous form characterized by pronounced hygroscopicity and a low glass transition temperature. Furthermore, the amorphous material tended to form an oil after exposure to RH levels exceeding 43% at room temperature, posing significant challenges for handling and stability. Thus, it is important in such a case to isolate a crystalline form for further pharmaceutical development. However, the lack of ionizable groups at physiological pH diminishes the opportunity of salt formation to crystallize the compound. As we can see, solvate and cocrystal screening offers a valuable last resort to seek a crystalline phase. After extensive efforts, two crystalline forms were attained, including L-proline and L-pyroglutamic acid cocrystals. The L-pyroglutamic acid cocrystal, with a 1:1 stoichiometry and as an anhydrous form, demonstrated to have desirable physical properties and became the final crystal form for further development [16, 32], with rationales listed as follows:

- Excellent physicochemical stability at elevated temperatures and RH
- An acceptable melting point, ~140 °C, feasible to endure the various unit operations of drug product manufacturing
- Nonhygroscopic nature, less than 2% water sorption at a RH lower than 90%

Indeed, several marketed cocrystal products consist of drug–drug cocrystals, including Depakote as a valproic acid–valproate sodium 1:1 cocrystal, Entresto as a sacubitril–valsartan 1:1 cocrystal, and Seglentis as a tramadol–celecoxib 1:1 cocrystal. Drug–drug cocrystals not only offer a unique approach for combination therapy but also could lead to improved physicochemical properties and performance matrices over traditional formulations for fixed-dose combinations. For example, Port et al. reported that the dissolution of the cocrystal of tramadol–celecoxib in hydroxypropyl methyl cellulose and buffered solutions led to a significant supersaturation level for celecoxib, a poorly soluble drug [33]. It is interesting to note that tramadol, a soluble drug, was released much more slowly from the cocrystal than the free combination formulation. This observation is consistent with the solubility data. In terms of celecoxib, the cocrystal showed approximately 170 times more solubility than neat celecoxib crystals at the physiological pH range from 1 to 8. More significantly, for tramadol, the cocrystal is 3400 times less soluble than neat tramadol crystals at pH 7, with a more substantial solubility difference at lower pH since tramadol has a pH-dependent solubility, which increases as the pH is lowered. The unique solubility characteristics of the cocrystal produce a different pharmacokinetics profile than the open combination of the two drugs; that is, the rate and extent of absorption for celecoxib are higher, while the rate of tramadol is more delayed and gradual [33, 34].

The cocrystal of valsartan and sacubitril, brand named as Entresto, a first-in-class angiotensin receptor neprilysin inhibitor approved for patients with heart failure, represents another successful use of drug–drug cocrystals. It was surprisingly found that the valsartan component when delivered in the cocrystal form exhibited a higher oral bioavailability than valsartan on its own, that is, about a 1.6-fold increase. More significantly, the cocrystal yielded a much faster onset for valsartan than the valsartan single formulation, with T_{max} 1.8 hours versus 2.6 hours, which is more desirable for the treatment of such an acute condition [35].

5.7 Comparison of Pharmaceutical Salts and Cocrystals

Pharmaceutical salts or cocrystals exist in totally different crystal forms from their parent drug. The inclusion of a counterion or a coformer into the crystal lattice produces a novel material with distinct physical and chemical attributes, such as melting point, solubility, hygroscopicity, and chemical stability. However, salts and cocrystals are very different regarding preparation, stability, and potential application.

5.7.1 Formation

Salt formation is only feasible for ionizable compounds since it relies on the acid–base reaction that occurs between the compound and the counterion; the difference in p*K*a between the compound and counterion should be greater than 2. Theoretically, the formation of cocrystals is more versatile since they can be formed by neutral compounds, ionizable compounds, and salts.

5.7.2 Preparation

The preparation of a salt is essentially a simple acid–base reaction where the salt can be made by mixing a drug with its counterion in a solvent with a defined stoichiometric ratio. The solid is obtained by increasing the supersaturation through evaporation and cooling or by adding an antisolvent.

The preparation of cocrystals in a solvent generally requires a substantial overage of coformers. Otherwise, the parental compounds may crystallize instead of the cocrystals. The solvent-drop method and co-grinding have proved to be most successful in screening for cocrystals.

5.7.3 Polymorphism and Pseudopolymorphism

Pharmaceutical salts and cocrystals may have polymorphs or hydrates. Polymorph screening, therefore, should be part of salt or cocrystal screening. For example, ranitidine HCl, the active ingredient of Zantac, has two polymorphs, Forms I and II [36]. Form II is the stable polymorph and shows more favorable filtration and drying characteristics. Form II was discovered many years after the development of Zantac, the marketed product. From a commercial perspective, the polymorph patent for Form II helped to provide an extension of market exclusivity for 3 years. Lipitor, a very popular lipid-lowering agent, contains as the active ingredient atorvastatin calcium trihydrate, a stable hydrate form.

The cocrystals of CBZ–NCT and CBZ–SAC have displayed a significant amount of polymorphism. Two polymorphs of CBZ–NCT cocrystals and two polymorphs of CBZ–SAC have been identified by Porter et al. through solution crystallization [37]. It has also been reported that the theophylline–citric acid cocrystal can exist as an anhydrous form and a hydrate form. The presence of polymorphism and pseudopolymorphism of a cocrystal makes the identification and characterization process more challenging.

5.7.4 Characterization

X-ray crystallography and X-ray powder diffraction are effective ways to differentiate between a salt form, a cocrystal, and the corresponding free forms since each salt or cocrystal is a unique crystalline phase with its own characteristic crystal structure. X-ray crystallography can determine the stoichiometry and provide a basis for understanding critical molecular interactions. Thermal analysis, such as differential scanning calorimetry (DSC) and thermogravimetric analysis, also provides useful information about melting point, heat of fusion, transition temperature, and formation of solvates or hydrates. Spectroscopic methods such as FT-IR, FT-Raman, and solid-state NMR are also applicable for characterizing salt forms and cocrystals.

5.7.5 Stability

Pharmaceutical salts usually exhibit good physical stability in the solid state. Interactions with excipients in the presence of moisture, however, may cause disproportionation [38]. This can be mitigated through appropriate control of the pH microenvironment or by controlling the moisture content. Pharmaceutical cocrystals may display various degrees of physical instability after exposure to moisture. Some cocrystals are stable after being exposed to moisture. For example, the 1:1 cocrystal of AMG 517 and sorbic acid had a minor uptake of 0.7% water at 90% RH, and no transformations were detected by X-ray powder diffraction [25]. The cocrystal of Purdue Pharma, compound 1 and glutaric acid, discussed previously, also showed excellent stability to moisture [21] since less than 0.08% of moisture was adsorbed under 95% RH after repeated sorption/desorption cycles. Studies by X-ray powder diffraction and DSC detected no physical change for the cocrystal after 2 months of storage at 40 °C/75% RH. In contrast, the physical stability of the cocrystal of CBZ and NCT is very dependent on the starting materials, the RH, and the temperature of the storage conditions. CBZ and NCT cocrystal can be prepared from Form I, Form III, or the dihydrate of the parent crystal by co-milling [39]. The cocrystal made from the dihydrate was stable under all conditions tested, but the cocrystal prepared from Forms I and III reverted to CBZ under the storage conditions of 20 °C/33% RH. At 20 °C/75% RH and 40 °C/33% RH, the transformation from the cocrystal to CBZ occurred initially, but increasing storage time resulted in the formation of the cocrystal. Interestingly, storage at 40 °C/75% RH did not trigger the transformation but, rather, promoted the formation of the cocrystal. It is obvious that such complex stability behavior is not desirable for a drug product.

It is well recognized that the stability of pharmaceutical salts in aqueous media is dependent on the pH. Salts of a weak base are stable when pH is less than pH_{max}, and salts of a weak acid are stable when the pH is greater than pH_{max}. Proper control of pH, therefore, is required to ensure the stability of pharmaceutical salts as a solution or a suspension.

In contrast, pharmaceutical cocrystals are intrinsically unstable in aqueous media if their solubility is higher than that of the parent compound. Solution-mediated transformation to the free form in many cases can occur rapidly for some cocrystals. The kinetics of such transformations vary with the specific cocrystals and are challenging to mitigate. For example, the succinic acid cocrystal of fluoxetine hydrochloride was converted to fluoxetine hydrochloride during dissolution studies. The succinic acid cocrystal reached a peak solubility of 20.2 mg/mL after about 1 minute, but then the solubility started to decrease until reaching the solubility of fluoxetine HCl, 11.4 mg/mL.

5.7.6 Formulation

Pharmaceutical salts have been formulated in various types of dosage forms, such as capsules, tablets, suspensions, and solutions. Proper selection of excipients is required to ensure the stability of a salt form after being subjected, for example, to wet granulation, roller compaction, and long-term storage. However, a cocrystal may undergo a solution-mediated conversion to the less soluble parent compound after wet granulation or storage in the presence of moisture. As would be expected, direct compression or, if granulation is needed, roller compaction is preferred rather than exposure to water during wet granulation. For example, Al-Fayoumi et al. reported that solid dosage forms for the cocrystal of valsartan and sacubitril, brand named as Entresto, were made by a process involving roller compaction, under environmental RH preferably less than at 55%, to mitigate the dissociation of the cocrystal or amorphization of the therapeutic agent [35]. Duggirala et al. reported that the cocrystal of ertugliflozin and L-pyroglutamic acid, tradename Steglatro, though stable at high temperature and humidity in the tablet formulation, dissociated into the amorphous free form of ertugliflozin in the presence of excipients under the accelerated storage condition of 40 °C, 75% RH [32]. Further studies revealed that high surface pH and hygroscopicity associated with certain excipients caused cocrystal dissociation due to the interaction with the acidic conformer in the cocrystal. To mitigate this dissociation, the cocrystal was coated with hydrophobic silica and a pH-lowering agent such as citric acid, resulting in no cocrystal dissociation for 4 weeks [32].

5.7.7 Regulatory

Salt forms have a long history of use in pharmaceutical products. Therefore, there is generally minimal regulatory concern if the counterions have already been used in marketed products. Regarding cocrystals, the FDA has issued a regulatory guidance on the appropriate regulatory classification of pharmaceutical cocrystal solid-state forms [40]. Based on FDA's current thinking, a cocrystal with a pharmaceutically acceptable coformer is considered a polymorphic form and not a new API. A cocrystal comprises two or more APIs, with or without additional inactive coformers, and would be treated as a fixed-dose combination product and not a new single API. Please note that the definition of polymorphs by the FDA is broader than that in this book (Chapter 1), including anhydrous forms, solvation or hydration products, and amorphous forms.

References

1 Rodríguez-Spong B, Price CP, Jayasankar A, Matzgerb AJ, Rodríguez-Hornedo N. General principles of pharmaceutical solid polymorphism: a supramolecular perspective. *Adv Drug Delivery Rev* 2004; 56(3): 241–274.

2 Friščić T, Jones W. Benefits of cocrystallisation in pharmaceutical materials science: an update. *J Pharm Pharmcol* 2010; 62(11): 1547–1559.

3 Schultheiss N, Newman A. Pharmaceutical cocrystals and their physicochemical properties. *Cryst Growth Des* 2009; 9(6): 2950–2967.

4 Galek PT, Allen FH, Fábián L, Feeder N. Knowledge-based H-bond prediction to aid experimental polymorph screening. *CrystEngComm* 2009; 11(12): 2634–2649.

5 Macrae CF, Sovago I, Cottrell SJ, Galek PT, McCabe P, Pidcock E, Platings M, Shields GP, Stevens JS, Towler M, Wood PA. Mercury 4.0: From visualization to analysis, design and prediction. *J Appl Crystallogr* 2020; 53(1): 226–235.

6 Fábián L. Cambridge structural database analysis of molecular complementarity in cocrystals. *Cryst Growth Des* 2009; 9(3): 1436–1443.

7 Racher F, Petrick TL, Braun DE. Exploring the supramolecular interactions and thermal stability of dapsone: Bipyridine cocrystals by combining computational chemistry with experimentation. *Cryst Growth Des* 2023; 23(6): 4638–4654.

8 Sun G, Jin Y, Li S, Yang Z, Shi B, Chang C, Abramov YA. Virtual coformer screening by crystal structure predictions: Crucial role of crystallinity in pharmaceutical cocrystallization. *J Phys Chem Lett* 2020; 11(20): 8832–8838.

9 Abramov YA, Iuzzolino L, Jin Y, York G, Chen CH, Shultz CS, Yang Z, Chang C, Shi B, Zhou T, Greenwell C. Cocrystal synthesis through crystal structure prediction. *Mol Pharm* 2023; 20(7): 3380–3392.

10 Nehm SJ, Rodríguez-Spong B, Rodríguez-Hornedo N. Phase solubility diagrams of cocrystals are explained by solubility product and solution complexation. *Cryst Growth Des* 2006; 6 (2): 592–600.

11 Codan L, Daza L, Sirota E. Method to rapidly identify potential solvent systems for crystallization of cocrystals. *Org Process Res Dev* 2023; 27(3): 513–522.

12 Shen Y, Cruz-Cabeza AJ, Azzouz O, Edkins K. Using prenucleation aggregation of caffeine-benzoic acid as a rapid indication of co-crystallization from solutions. *Mol Pharm* 2023; 20(4): 1942–1950.

13 Childs SL, Chyall LJ, Dunlap JT, Smolenskaya VN, Stahly BC, Stahly GP. Crystal engineering approach to forming cocrystals of amine hydrochlorides with organic acids. Molecular complexes of fluoxetine hydrochloride with benzoic, succinic, and fumaric acids. *J Am Chem Soc* 2004; 126(41): 13335–13342.

14 Rodríguez-Hornedo N, Nehm SJ, Seefeldt KF, Pagán-Torres Y, Falkiewicz CJ. Reaction crystallization of pharmaceutical molecular complexes. *Mol Pharm* 2006; 3(3): 362–367.

15 Reddy LS, Bethune SJ, Kampf JW, Rodríguez-Hornedo N. Cocrystals and salts of gabapentin: pH dependent cocrystal stability and solubility. *Cryst Growth Des* 2009; 9(1): 378–385.

16 Bowles P, Brenek SJ, Caron S, Do NM, Drexler MT, Duan S, Dubé P, Hansen EC, Jones BP, Jones KN, Ljubicic TA. Commercial route research and development for SGLT2 inhibitor candidate ertugliflozin. *Org Process Res Dev* 2014; 18(1): 66–81.

17 Shan N, Toda F, Jones W. Mechanochemistry and cocrystal formation: Effect of solvent on reaction kinetics. *Chem Commun* 2002; 20: 2372–2373.

18 Trask AV, Motherwell WS, Jones W. Solvent-drop grinding: green polymorph control of cocrystallisation. *Chem Commun* 2004; 7: 890–891.

19 Almansa C, Frampton CS, Vela JM, Whitelock S, Plata-Salaman CR. Cocrystals as a new approach to multimodal analgesia and the treatment of pain. *J Pain Res* 2019; 12: 2679–2689.

20 Jayasankar A, Somwangthanaroj A, Shao ZJ, Rodríguez-Hornedo N. Cocrystal formation during cogrinding and storage is mediated by amorphous phase. *Pharm Res* 2006; 23(10): 2381–2392.

21 McNamara DP, Childs SL, Giordano J, Iarriccio A, Cassidy J, Shet MS, Mannion R, O'Donnell R, Park A. Use of a glutaric acid cocrystal to improve oral bioavailability of a low solubility API. *Pharm Res* 2006; 23(8): 1888–1897.

22 Good DJ, Rodríguez-Hornedo N. Solubility advantage of pharmaceutical cocrystals. *Cryst Growth Des* 2009; 9(5): 2252–2264.

23 Bethune SJ, Huang N, Jayasankar A, Rodríguez-Hornedo N. Understanding and predicting the effect of cocrystal components and pH on cocrystal solubility. *Cryst Growth Des* 2009; 9(9): 3976–3988.

24 Variankaval N, Wenslow R, Murry J, Hartman R, Helmy R, Kwong E, Clas SD, Dalton C, Santos I. Preparation and solid-state characterization of nonstoichiometric cocrystals of a phosphodiesterase-IV inhibitor and L-tartaric acid. *Cryst Growth Des* 2006; 6(3): 690–700.

25 Bak A, Gore A, Yanez E, Stanton M, Tufekcic S, Syed R, Akrami A, Rose M, Surapaneni S, Bostick T, King A. The cocrystal approach to improve the exposure of a water-insoluble compound: AMG 517 sorbic acid cocrystal characterization and pharmacokinetics. *J Pharm Sci* 2008; 97(9): 3942–3956.

26 Billot P, Hosek P, Perrin MA. Efficient purification of an active pharmaceutical ingredient via cocrystallization: From thermodynamics to scale-up. *Org Process Res Dev* 2013; 17(3): 505–511.

27 Nulek T, Klaysri R, Cedeno R, Nalaoh P, Bureekaew S, Promarak V, Flood AE. Separation of etiracetam enantiomers using enantiospecific cocrystallization with 2-chloromandelic acid. *ACS Omega* 2022; 7(23): 19465–19473.

28 Harmsen B, Leyssens T. Dual-drug chiral resolution: enantiospecific cocrystallization of (s)-ibuprofen using levetiracetam. *Cryst Growth Des* 2018; 18(1): 441–448.
29 Imamura M, Nakanishi K, Shiraki R, Onda K, Sasuga D, Yuda M, inventors; Kotobuki Seiyaku Co Ltd, Astellas Pharma Inc, assignee. *Cocrystal of C-Glycoside Derivative and L-Proline*. Alexandria: United States patent US; 2012. 8,097,592. Jan 17.
30 Meade EM, inventor; Abbott Laboratories, assignee. *Sodium Hydrogen Divalproate Oligomer*. Alexandria: United States patent US; 1993. 5,212,326. May 18.
31 Sherman BC, inventor. *Solid Substances Comprising Valproic Acid and Sodium Valproate*. Alexandria: United States patent US; 2000. 6,077,542. Jun 20.
32 Duggirala NK, LaCasse SM, Zaworotko MJ, Krzyzaniak JF, Arora KK. Pharmaceutical cocrystals: formulation approaches to develop robust drug products. *Cryst Growth Des* 2019; 20(2): 617–626.
33 Port A, Almansa C, Enrech R, Bordas M, Plata-Salamán CR. Differential solution behavior of the new API–API cocrystal of tramadol–celecoxib (CTC) versus its constituents and their combination. *Cryst Growth Des* 2019; 19(6): 3172–3182.
34 Cebrecos J, Carlson JD, Encina G, Lahjou M, Sans A, Sust M, Vaqué A, Morte A, Gascón N, Plata-Salamán C. Celecoxib-tramadol cocrystal: a randomized 4-way crossover comparative bioavailability study. *Clin Ther* 2021; 43(6): 1051–1065.
35 Al-Fayoumi S, Hu J, Kumaraperumal N, Royce AE, Ruegger C, Zannou EA, inventors; Novartis AG, assignee. *Dual-Acting Pharmaceutical Compositions based on Superstructures of Angiotensin Receptor Antagonist/Blocker (arb) and Neutral Endopeptidase (nep) Inhibitor*. Alexandria: United States patent application US; 2021. 17/109,716. Mar 25.
36 Pratiwi D, Fawcett JP, Gordon KC, Rades T. Quantitative analysis of polymorphic mixtures of ranitidine hydrochloride by Raman spectroscopy and principal components analysis. *Eu J Pharm Biopharm* 2002; 54(3): 337–341.
37 Porter III WW, Elie SC, Matzger AJ. Polymorphism in Carbamazepine Cocrystals. *Cryst Growth Des* 2008; 8 (1): 14–16.
38 Hsieh YL, Taylor LS. Salt stability-effect of particle size, relative humidity, temperature and composition on salt to free base conversion. *Pharm Res* 2015; 32(2): 549–561.
39 Chieng N, Hubert M, Saville D, Rades T, Aaltonen J. Formation kinetics and stability of carbamazepine-nicotinamide cocrystals prepared by mechanical activation. *Cryst Growth Des* 2009; 9(5): 2377–2386.
40 Guidance for Industry. Regulatory Classification of Pharmaceutical Cocrystals. http://www.fda.gov/downloads/Drugs/GuidanceComplianceRegulatoryInformation/Guidances/UCM281764.pdf (accessed 14 January 2025).

6

Amorphous Solids

6.1 Introduction

As we have seen in Chapters 2 through 5, crystalline solids can exist as anhydrates, hydrates, solvates, and cocrystals, as well as in various crystalline polymorphic forms and salt forms. All of these forms are characterized by a well-defined arrangement of atoms and molecules in a crystal lattice with long-range symmetrical repeating unit cells extending in three dimensions. The ability to form crystals in a well-defined lattice is desirable for any molecule, including active pharmaceutical ingredients (APIs), because crystallization generally leads to high levels of purity and solid-state stability. Molecules in the amorphous state, lacking any long-range order, while having the "macroscopic" mechanical properties of a solid, have the "microscopic" or local properties of a liquid, with higher molecular energy and greater molecular mobility than those existing in the crystalline state. Obviously, to maintain a condensed state through intermolecular interaction, amorphous solids must retain a certain level of short-range or "local" order that is limited to a few molecular dimensions, such as at nearest neighbor (NN) and next nearest neighbor (NNN) distances, typically less than 2.0–2.5 nm for small organic molecules. Figure 6.1 schematically depicts the general differences in long-range order for crystalline and amorphous solids, showing the loss of this order in the amorphous state, but with the retention of local order. One way to experimentally observe the difference between the long-range order in crystals and the loss of long-range order in amorphous solids is to compare the powder X-ray diffraction (PXRD) patterns of the crystalline and amorphous forms of the same molecule, as illustrated in Figure 6.2. Here we see for the crystal the characteristic diffraction peaks that represent the various planes of symmetry within the crystal lattice and the complete loss of such diffraction peaks for amorphous solids due to the loss of lattice long-range order. Typically, an amorphous form will exhibit a broad "halo" pattern with the peak intensity characteristically occurring at an angle of about 20° 2θ.

This difference between crystalline and amorphous forms leads to very different intrinsic molecular properties of great importance pharmaceutically. For example, because of a lack of a crystal lattice and the resulting strong lattice energies, API molecules in the amorphous state generally exhibit greater apparent "solubility" (supersaturation relative to that of the crystal) in aqueous solutions and hence the potential to produce greater dissolution rates and oral bioavailability of poorly water-soluble crystals [1]. Thus, the purposeful preparation of the amorphous solid form of molecules having poor aqueous solubility in the crystalline state offers a practical manner in which to influence the biopharmaceutical properties of such poorly soluble crystalline forms of API. However, API molecules in the amorphous state, by having such higher energy and greater molecular mobility, can also be expected to exhibit enhanced chemical and physical instability, a distinct disadvantage when compared to the more stable crystalline state. Thus, any strategy used to take advantage of the enhanced dissolution of molecules in the amorphous state must include approaches that will anticipate and preferably inhibit any instability that might occur over the time course of formulation, manufacture, and use of the product. It is also important to recognize that a wide range of other types of molecules used in pharmaceutical formulations can exist in the amorphous state, including excipients and biopharmaceutical therapeutic agents such as proteins.

An equally important reason for pharmaceutical scientists to have a deep understanding of the amorphous state is the possibility that crystalline solids undergoing various processing conditions, for example, milling, compaction, or drying, can be inadvertently converted to partially amorphous solids at levels that may not be easily detected. Molecules in such amorphous regions, being in a higher energy state, can undergo a variety of unanticipated physical and chemical transformations, not observed with the highly crystalline form. Indeed, it is very likely that such low levels of disorder in "crystalline" solids are the major factors that determine most types of solid-state instability.

Solid-State Materials in Pharmaceutical Chemistry: Properties, Characterization, and Applications, Second Edition. Stephen R. Byrn, George Zografi and Xiaoming (Sean) Chen.

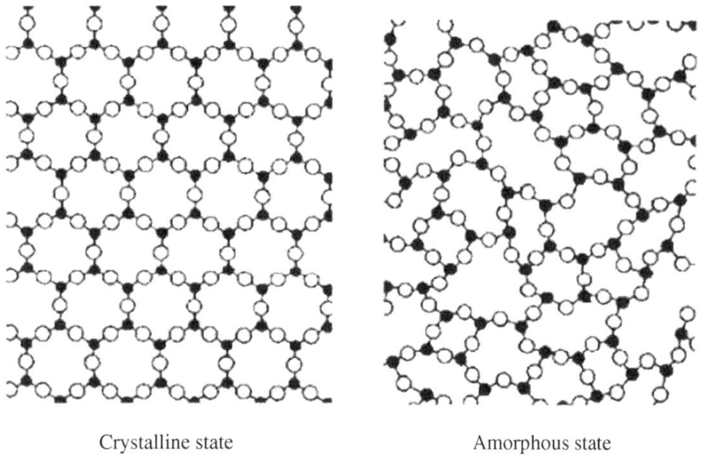

Figure 6.1 Schematic representations of the arrangements of atoms and molecules in the crystalline and amorphous states.

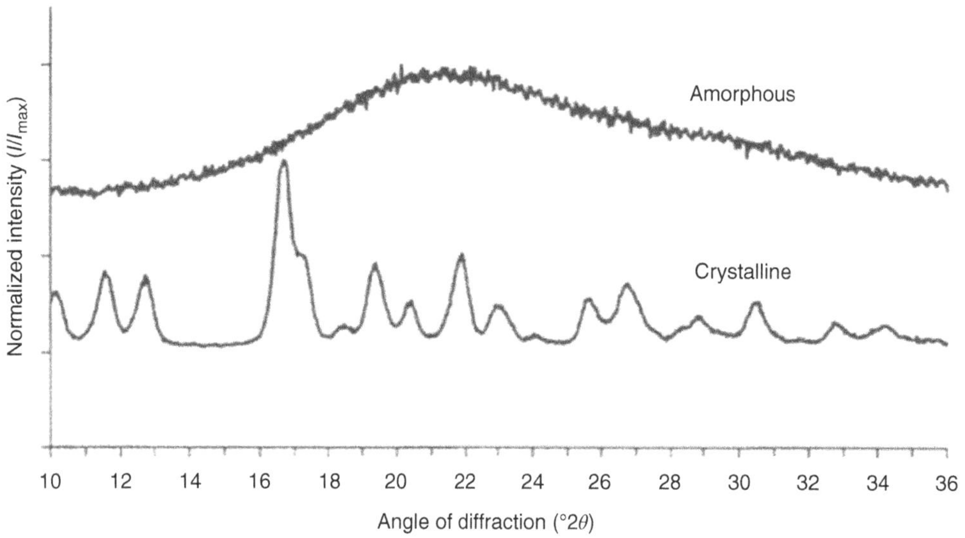

Figure 6.2 Comparison of the PXRD patterns of a crystalline solid and its corresponding amorphous state.

6.2 The Formation of Amorphous Solids

To begin to understand the properties of amorphous solids, let us consider the classical relationship of free energy and temperature between a crystalline form of a molecule and its corresponding liquid form, as schematically shown in Figure 6.3. Here we see that, at temperatures below the melting temperature, T_m, the solid exists at a lower free energy than the liquid, and, therefore, the solid is the thermodynamically stable form. As expected, as the sample is heated and T_m is exceeded, the liquid is now in the lower state of free energy, and the solid spontaneously melts to form the liquid. If we now slowly cool this liquid below T_m, we expect crystallization to occur, thus restoring the crystalline state. However, as represented in Figure 6.3, if we cool the liquid rapidly enough through T_m, avoiding the formation of crystal nuclei, it is possible to form a supercooled liquid, before any crystallization can take place. Thus, in this process, we have formed a supercooled liquid state that is metastable relative to the crystalline form. We also notice in Figure 6.3 that the transformation from the liquid to the supercooled liquid upon cooling occurs with no discontinuity, indicating that the supercooled liquid retains the equilibrium properties of the liquid. Indeed, the supercooled liquid is often referred to as an "equilibrium liquid," which can be treated as such if there are no crystal nuclei present to induce crystallization. As we might expect, the viscosity of the liquid, which is a macroscopic measure of its resistance to flow, will increase as the liquid is cooled, leading to a decrease in molecular mobility. Indeed, the viscosity generally increases from roughly 10^{-2} Pas or 10^{-1} poise for a typical organic small molecule at T_m to about

10^{12} Pas or 10^{13} poise at a temperature where equilibrium can no longer be maintained on the timescale of cooling, and the unstable glassy state forms. The temperature at which the glass forms, uniquely characteristic of the chemical structure of the molecule, is called the glass transition temperature, T_g. In view of the importance of the viscosity increase (or decrease if heating) in determining where T_g occurs, we can conclude that the glass transition represents a kinetically induced phenomenon, rather than a true thermodynamic transition, that is, the value of T_g depends on the rate of temperature and viscosity change and their effect on the ability of the supercooled liquid to remain in equilibrium. This suggests that it should be possible to obtain glasses upon cooling to below T_g that exist in different states of free energy depending on the way the glassy state was formed; the slower the rate of cooling, for example, the lower will be T_g. The slower rate of cooling presumably allows more time for equilibration at the higher viscosities, allowing a lower enthalpy, entropy, and molar volume to be reached before the glass forms. As seen in Figure 6.3, molecules in the glassy state are energetically unstable relative to the supercooled liquid. Thus, it is thermodynamically possible, under certain conditions, for the structure of the glass to "relax" with time, becoming denser and more stable as the thermodynamic properties approach those of the supercooled liquid. Such a slow process of structural change is known as "physical aging" or "annealing," and it can generally be observed by maintaining the glass at a temperature below T_g, where there is still sufficient molecular mobility to observe decreases in molar volume or enthalpy over a measurable period.

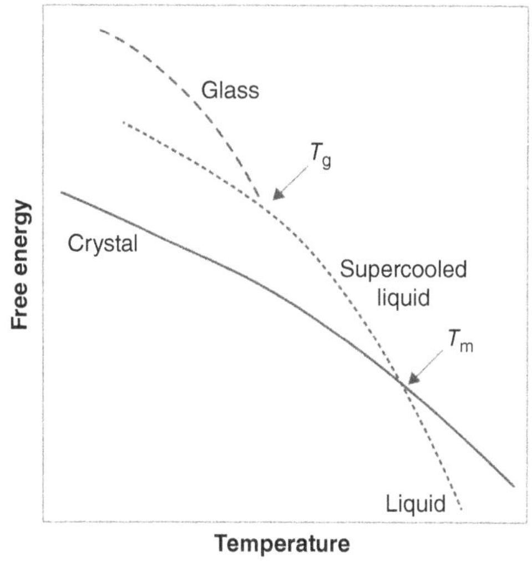

Figure 6.3 Free energy versus temperature of a molecule reflecting different equilibrium and nonequilibrium states.

6.3 Methods of Preparing Amorphous Solids

To this point, we have focused on the process of forming amorphous solids by melting the crystal and then rapidly cooling the sample to bypass recrystallization. However, there are other methods of producing amorphous solids, which either allow molecules in a highly disordered state to avoid crystallization or introduce sufficient energy to cause the crystal lattice to break down into the disordered amorphous form. Figure 6.4 shows a diagram that summarizes the various ways in which this can occur, along with some of the specific processes that form amorphous solids in pharmaceutical situations. First, as in the case of the "supercooling" of liquids below T_m, we can see that it is possible to start with highly disordered molecules in the vapor state and form an amorphous solid by cooling below the melting temperature of the crystalline form. Again, as with the cooling of liquids, the rate of condensation must be such that crystallization is avoided. Also, as in the case of supercooling of a liquid, the rate at which vapor is condensed will give rise to unstable amorphous glasses having different densities and other thermodynamic properties. Indeed, some amorphous solids prepared from the vapor state have been shown to produce as much as 1.5–5% greater density and much more stable glasses than those formed

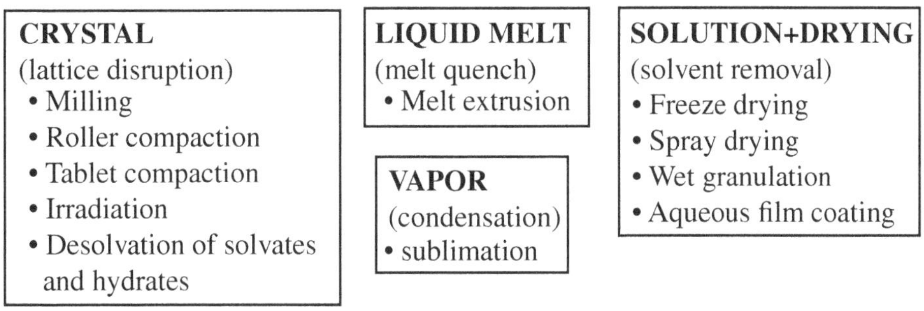

Figure 6.4 Various methods of producing the amorphous state.

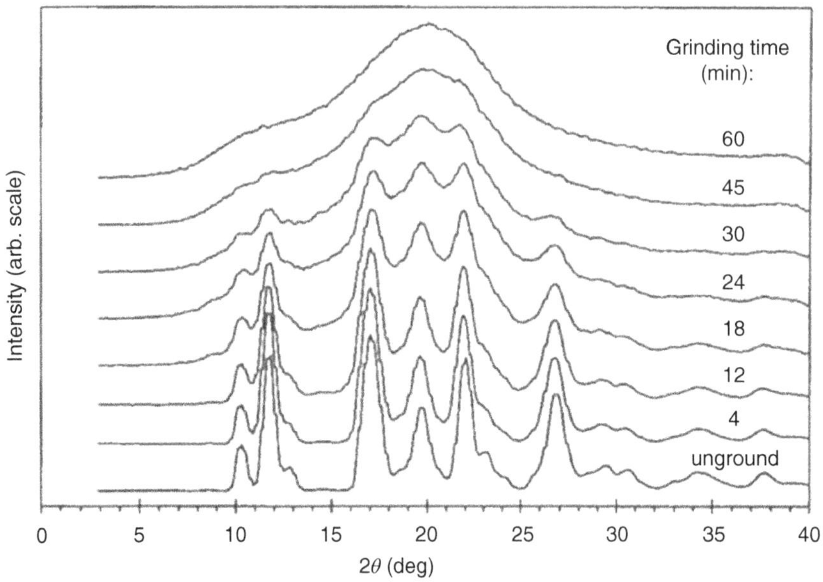

Figure 6.5 PXRD patterns of indomethacin crystals ground for various periods of time. *Source*: [13]/with permission of Springer Nature.

from the melt [2]. As also described in Figure 6.4, one can start with molecules in solution and, therefore, in a highly disordered state, and cause them to precipitate by changing the temperature, pH, or solvent polarity. Ordinarily, this is the way most crystalline solids are formed during the process of solvent-mediated crystallization. However, depending on the manner and conditions used for producing the precipitate, for example, by very rapid precipitation, it is possible to avoid crystallization and form amorphous solids. Indeed, this is the basis of purposefully forming amorphous solids by techniques such as spray-drying or freeze drying (lyophilization). As shown in Figure 6.5, it is also possible to start with a crystal and introduce sufficient mechanical energy to cause a breakdown of the crystal lattice into a completely disordered amorphous form. Removal of water from crystal hydrates, or organic solvents from crystal solvates, has also been shown to produce the amorphous state through collapse of the remaining less dense crystal lattice [3, 4].

6.4 The Glass Transition Temperature

It is clear from the previous discussion that the glass transition temperature is an important parameter, the knowledge of which provides a basis for anticipating major changes in the properties of amorphous solids with changing conditions. We can recall that the change in molar enthalpy with temperature is the molar heat capacity at any temperature, whereas the change in molar volume with temperature is the molar thermal expansion coefficient. From Figure 6.3, we would expect that the T_g observed under a given set of conditions would represent the temperature at which a significant change will occur in heat capacity or thermal expansion coefficient upon cooling from the supercooled liquid to the glass or the reverse if the glass is heated to the supercooled liquid. Therefore, it should be possible to measure T_g by either heating the glass or cooling the supercooled liquid and observing the temperature where a significant change in heat capacity or thermal expansion coefficient occurs. For this purpose, we can use thermal analysis to gain a measure of the heat capacity change or use various types of volumetric or mechanical measurements to gain a measure of changes in thermal expansion. By far the most popular approach in the pharmaceutical field for determining T_g is the use of differential scanning calorimetry (DSC), which allows one to measure differential heat flow in a sample as the sample is scanned at various temperatures and scanning rates; the change in heat flow with temperature is a direct measure of the heat capacity at any temperature. Figure 6.6 illustrates an example of the heat capacity change that takes place at T_g, where the value of T_g is generally taken as either the temperature at the onset of the transition, $T_{g\text{-onset}}$, or the temperature at the midpoint of the transition, $T_{g\text{-mid}}$. Values of T_g are generally reported either way, so it is important in reporting such values in the literature that the point chosen is indicated. It is also important to recognize that T_g is dependent on the rates of cooling and heating so that the rates of heating or cooling must also be specified [5]. To provide for some consistency from study to study, therefore, it is generally recommended that initial attempts to obtain T_g should be carried out at a "standard" scanning rate, for example, 10 K/min.

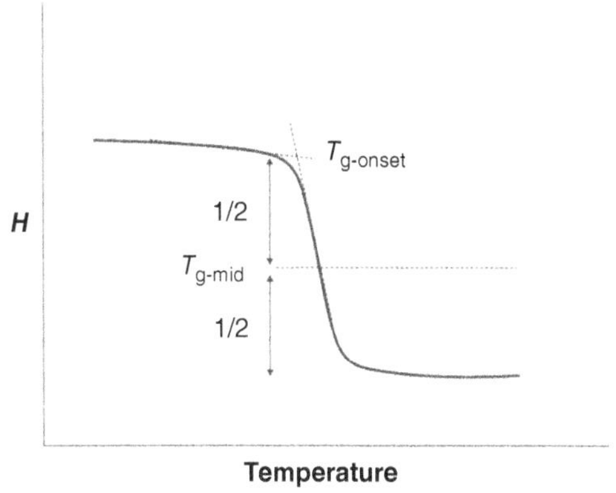

Figure 6.6 Heat capacity change at the glass transition indicating the onset glass transition temperature, $T_{g\text{-onset}}$ and the midpoint glass transition temperature, $T_{g\text{-mid}}$.

It is clear from the discussion so far that the value of T_g relative to any operating temperature, T, is important for characterizing the properties of molecules in the amorphous state. Consequently, at this point, it will be useful to briefly discuss any relationships that might exist between the value of T_g and the structure of the molecule under consideration. There are two major theoretical approaches that can be followed to better understand the nature of the glass transition. One involves the changes that take place in the molar volume with temperature as one passes through the glass transition, and the other involves changes in configurational entropy per mole during this process. A detailed discussion of these theories for the glass transition temperature is beyond the scope of this chapter; however, it is useful to see how these theories can provide a conceptual understanding of the relationship between molecular structure and the glass transition temperature. As a starting point, we can address this question from the perspective of the volume changes that take place at T_g. We can first define the free volume, V_f, of a solid as the difference between the total molar volume, V_{sp}, and the occupied molar volume, V_o, as

$$V_f = V_{sp} - V_o \tag{6.1}$$

Therefore, free volume represents the interstitial space between molecules that exists under a given set of conditions and that acts as a determinant of the ability of molecules to exhibit various types of motions as a function of temperature. The simplest model of such a situation would be that of a fluid made up of molecules acting like hard spheres with occupied volume V_o, packed together to produce a free volume of V_f. We can assume that each sphere occupies a "cage" formed by neighboring spheres. In this way, the "caged" sphere is entrapped by its neighbors and is unable to undergo motion by translational or rotational diffusion unless the free volume of the "cage" exceeds a critical volume, which is related to the occupied volume of the molecule. If we have an amorphous solid well above its T_g in the supercooled liquid state, we expect the value of V_f to be quite large and much greater than the critical volume. As the supercooled liquid is cooled toward T_g, we would expect the value of V_f to decrease significantly closer to V_o and to cause the system to go out of equilibrium, giving rise to the glass transition and the glassy state. Now, if the temperature of the glass is reversibly increased and the free volume of the system increases, a point will be reached where $V_f > V_o$, and the molecule will have sufficient diffusive molecular motion to assume the equilibrium supercooled liquid state at T_g. Based on this conceptual model, we can conclude that various factors associated with molecular structure that can contribute to V_f will play an important role in determining the value of T_g. For example, molecules with high molecular weight that structurally exhibit a strong tendency to interact intermolecularly, as by hydrogen bonding, or pack more densely because of a less bulky molecular structure, would be expected to have a lower V_f in relation to V_o and hence exhibit higher values of T_g. On the other hand, smaller molecular size with weaker intermolecular interactions and less of a tendency to pack closely, by virtue of having a bulky molecular shape and more open molecular packing, should lead to higher values of V_f relative to V_o and lower values of T_g.

Another theoretical approach for understanding the nature of the glass transition temperature considers the changes in entropy that occur during the cooling of the supercooled liquid and in particular the changes in configurational entropy associated with the molecules making up the system. Configurational entropy, S_c, arises primarily from the number of positions in space that can be occupied by a molecule; the greater the number of possible positions or configurations, the greater will be the entropy. The original entropic model for amorphous solids was developed for polymer systems, using fundamental theories based on lattice models introduced by Flory regarding the general thermodynamic properties of polymers [6]. The model indicates that at higher temperatures, for example, near the melting temperature, T_m, because polymers and other molecules would be expected to have more molecular flexibility (high free volume), there will be many molecular arrangements that can be assumed by the "freely moving" molecule, and hence the system would contain a high configurational entropy. As the temperature is reduced and free volume is reduced, the number of allowed molecular configurations will decrease, and hence the configurational entropy will decrease. As cooling of the system is continued, the molecules will have fewer and fewer possible equilibrium configurations until the configurational entropy approaches zero, causing the unstable glass to form before the entropy reaches that of the crystalline form. The temperature at which configurational entropy might closely approach zero, and where the glass would have to form to avoid reaching a value less than that of the crystal, is called the Kauzmann temperature, T_K [7]. In the original theoretical entropic model for the glass transition developed by Gibbs and DiMarzio [8], the temperature at which this transition occurs was termed the ideal glass transition temperature, T_2. This temperature is expected to occur with infinitely slow cooling, and T_2, therefore, is essentially the same as T_K. As described earlier, the experimental values of T_g depend on the rate of cooling (or heating) relative to viscosity changes, and, therefore, they generally occur at temperatures well above the predicted values of T_2. The concept of configurational entropy, however, still provides an excellent basis for a practical understanding of several important phenomena associated with amorphous solids, and it will particularly be useful in later discussions of crystallization in the amorphous state.

Based on the conceptual models involving free volume and configurational entropy, discussed earlier, it is possible to obtain some understanding of the relationship between T_g and molecular weight for various types of molecules of pharmaceutical interest. In the case of polymers, for example, the free volume approach assumes that free polymer chain ends contribute greater excess free volume relative to other parts of the molecule; the lower the molecular weight, the greater the number of free end groups per unit mass and, hence, the greater the free volume and lower T_g. Likewise, the greater the number of free end groups per unit mass, the greater will be the configurational entropy. Therefore, as the molecular weight increases and the number of free end groups per unit mass decreases, there will be a smaller contribution to free volume and configurational entropy, leading to higher values of T_g. This general trend with changing molecular weight can be seen in Table 6.1 for various molecular weight grades of the polymer, poly(vinylpyrrolidone) (PVP), where the values of T_g initially increase significantly with increasing molecular weight, but then tend to change less at the higher molecular weights [9]. This occurs, presumably, because the number of free end groups decreases to a lesser extent with increasing molecular weight at higher molecular weights. Also included in Table 6.1 is the T_g for the copolymer of PVP with poly (vinyl acetate), PVP/VA, in a 60:40 molar ratio of vinyl pyrrolidone (VP) to vinyl acetate (VA) monomer segments. If, for example, we compare the values of T_g for PVP and PVP/VA at approximately the same molecular weight (~50 000 Da), we can see that the substitution of VA segments for VP segments leads to a very significant reduction in T_g. This difference appears to illustrate the additional importance of the ability of molecules to interact with one another through noncovalent interactions. It is known that the carbonyl carbon of VP is generally more acidic than that of VA, since in VP it has greater electron-donating properties than those of the carbonyl group in the VA monomer [9]. VP, therefore, would be expected to intermolecularly interact more strongly than VA and contribute to a higher value of T_g.

Although there is no exact theoretical basis for predicting the value of T_g for a given molecule from its chemical structure, it would be useful to be able to estimate an approximate value before carrying out experiments, thus making it easier, for example, to experimentally locate the T_g. The reason predictions are not possible on theoretical grounds arises primarily from the fact that the glass transition is not a thermodynamic transition but rather a kinetic term dependent on the manner of glass formation, as previously discussed. Also, it is not always clear how molecular size, shape, and ability to

Table 6.1 The glass transition temperature of various molecular weight grades of poly(vinylpyrrolidone) (PVP), and poly(vinylpyrrolidone)-co-vinyl acetate, (PVP/VA).

Sample	Molecular weight (K)	T_g (°C)
PVP K90	1500	177
PVP K30	50	156
PVP K17	10	136
PVP K12	2	101
PVP/VA (60:40)	50	102

Source: Adapted from [9].

Table 6.2 Ratio of glass transition temperature to melting temperature for various small organic molecules.

Solid	T_g (K)	T_m (K)	T_g/T_m
Ketoconazole	319	412	0.77
Indomethacin	315	434	0.73
Sodium indomethacin	393	543	0.72
Nifedipine	323	447	0.72
Felodipine	317	416	0.76
o-Terphenyl	246	328	0.75

intermolecularly interact are interrelated. One useful empirical approach to an a priori estimate of T_g, originally reported with polymers that form both crystalline and amorphous solids, is based on the observation that the values of T_g of the amorphous form and T_m, the melting temperature of the crystal (when these temperatures are expressed in Kelvin), very often produce a ratio of around 2:3 or 0.67 [10]. Table 6.2 illustrates that, indeed, many small organic molecules of varying chemical structure, including salts, do appear to exhibit constant ratios that tend to cluster around a value of 0.70 [11]. A possible explanation for the agreement of such a constancy in this value among many molecules can be found with the recognition that the ratio of T_g to T_m depends on the differences between T_m (fixed for any molecule) and T_g, the temperature at which, upon cooling, the viscosity of the supercooled liquid increases sufficiently to cause the formation of the glass. Hence, we might expect that the relative constancy of this ratio for many polymers and small molecules arises because many organic molecules have very similar dynamic changes of viscosity with temperature as the supercooled liquid is cooled from T_m to a point where the liquid goes out of equilibrium. Consequently, when there is a deviation from the so-called 2/3 rule, it likely means that the system is showing different dynamic behaviors. The significance of this observation will be discussed in more detail when the topics of molecular mobility and the concept of fragility in amorphous solids are discussed in the following text. Despite some inconsistencies in the "2/3 rule," it can be useful as a rapid and convenient way to determine in what temperature ranges T_g might be experimentally found by only knowing the T_m of the crystalline form.

6.5 Structural Features of Amorphous Solids

Having discussed various aspects of the glass transition, we now want to look more closely at molecular events taking place in the amorphous state and to discuss various types of molecular motions that exist in such systems, motions that determine many of their important properties. Before specifically discussing molecular mobility, it will be useful to examine more closely some of the structural features of molecules in the supercooled liquid and glassy states and how these features change as the sample temperature is altered. Of particular interest is the spatial arrangement of atoms and molecules that are linked by the intermolecular interactions that take place in the formation of the supercooled liquid and the glassy states. As mentioned earlier, amorphous solids represent a solid state that lacks the long-range molecular order exhibited by the crystalline state, while maintaining short-range NN, and NNN, order normally associated with a liquid (see Figure 6.1). Of initial interest, therefore, would be to determine how NN and NNN are geometrically arranged relative to any reference molecule, as well as the distance that exists between pairs of NN and NNN molecules. For organic molecules, a critical factor in determining this local order is the covalent structure of the molecule, the nature of the atoms making up the molecule, and the strength and extent of intermolecular hydrogen bonding. Most often, the local arrangement of molecules in the amorphous state retains something close to the basic molecular coordination symmetry of the crystalline unit cell structure. To gain a better understanding of such local structure in amorphous solids, as with liquids, it is possible to use PXRD measurements to determine the pair distribution function (PDF), a parameter that describes the probability, $G(r)$, of finding two atoms contained within a given volume, when separated by a certain radial distance, r, as described in equation (6.2) [12, 13].

$$G(r) = 4\pi r(\rho(r) - \rho_o) \tag{6.2}$$

where $\rho(r)$ and ρ_o are the local and average atomic densities, respectively.

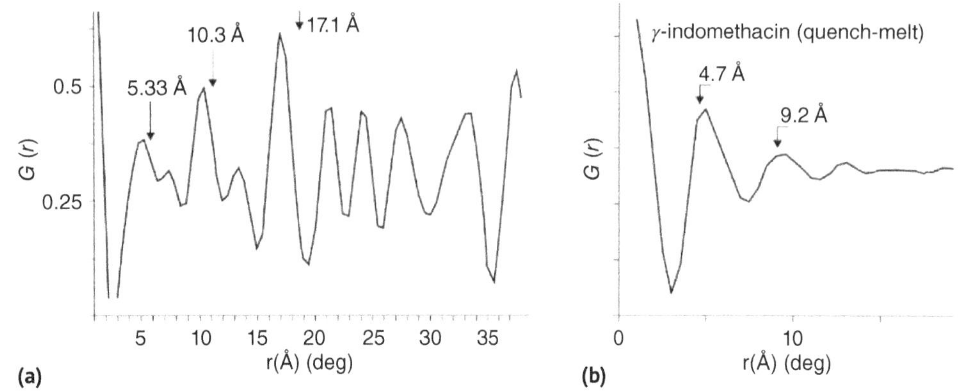

Figure 6.7 Pairwise distribution function for (a) crystalline indomethacin and (b) amorphous indomethacin. *Source*: [13]/with permission of Springer Nature.

Since the PDF for solids is obtained from PXRD data, taking both Bragg and diffuse scattering into account, it can provide information about the short-range atomic ordering in both the crystalline and corresponding amorphous forms. For example, as shown in Figure 6.7, we can observe the PDF for amorphous indomethacin prepared from the melt of the thermodynamically stable γ-crystalline form and the PDF of the γ-crystalline form, where it can be shown that in both cases the appearance of peaks at distinct distances from a reference atom occurs at distances that correspond closely to the NN and NNN molecular dimensions of the indomethacin molecule [13]. Note, however, that, whereas the peaks continue to occur over long distances for the crystal, they are essentially eliminated after the NNN distance for the amorphous form. A more detailed discussion of the PDF and its determination from X-ray diffraction measurements is presented in Chapter 9.

Given that very similar local NN and NNN order exists in an equilibrium liquid and its supercooled liquid and glassy states, we may now ask how these units might tend to organize themselves long-range relative to the corresponding crystalline form and, particularly, how this long-range structure might change as the liquid is cooled below T_m to form the supercooled liquid and then to below T_g when the glassy state forms. There are several theoretical models that have been used to predict molecular properties from the structure of liquids and amorphous forms, a detailed discussion of which is beyond the scope of this chapter. The two primary models that provide some conceptual understandings of structure in noncrystalline materials are the random close-packed (RCP) and continuous random network (CRN) models [14, 15]. In the RCP model, molecules are said to pack together in a manner that minimizes the local free energy but in such a way that makes it possible to avoid more ordered packing associated with crystallization. We can begin by recognizing that the long-range structure of crystals represents a close packing of molecules into a given symmetrical lattice arrangement, which, with intermolecular interactions, leads to what can be called "highly-ordered close packing." In the case of spheres of equal size, for example, this highly ordered closest packing can result in a face-centered cubic structure, as seen in some inorganic crystals, with a maximum occupied volume fraction of 0.74. It is also possible to think of a system where the spheres are allowed to arrange themselves more randomly in what is called "random packing." Such packing, although it can have local short-range order, has no long-range periodic order in the spatial arrangement of particle positions, just as we would expect for a liquid or amorphous solid. The randomly packed network of molecules found in a liquid is generally thought to be made up of a variety of polyhedral domains that pack together loosely with significant free volume and molecular mobility. Upon supercooling below T_m, the structures that are quite homogeneously distributed can maintain the equilibrium properties of the liquid and avoid rapid crystal nucleation and growth as they become more closely packed. Compared with highly ordered close packing volume fractions of 0.74, in RCP, we can expect the closest packing to be on the order of < 0.64 [14]. The RCP model has been particularly successful in describing and predicting the supercooled liquid properties of various monatomic materials, such as liquid metals, assuming that these materials exist as various polyhedral-shaped clusters of atoms randomly packed with significant molecular mobility. Studies of organic molecules in the supercooled state indicate that these more homogeneous liquid-like structures persist upon cooling down to within about 1.2 T_g, at which point a certain level of distorted packing occurs so that the packing arrangement causes the polyhedral to take on more "solid-like behavior." In this region of temperature, the structure of the various intermolecular domains also becomes quite spatially heterogeneous compared to that observed at higher temperatures due to further clustering on the order of a few nanometers in size, with a broadening of cluster size distribution [16]. Experimental studies indicate that molecules in these regions can have molecular mobility that ranges over roughly 10 orders of magnitude at any one temperature. Thus, in this spatially heterogeneous region, the system takes on more solid-like properties and can be considered to be in a "pre-glassy state." Since this spatial heterogeneity leads to systems with more complex dynamic behavior, the temperature where this occurs is often referred to as the "crossover temperature," T_c.

As we have discussed earlier, at T_g, the supercooled liquid is in such a closely packed state that it can no longer maintain the equilibrium properties of the liquid. Under these conditions, with a greatly reduced free volume and molecular mobility, the system is "frozen" into the heterogeneous "solid-like" glassy state. The CRN model for the structure of glassy amorphous solids was originally developed to describe the overall structural properties of covalently bonded inorganic glasses such as SiO_2 and other tetrahedrally structured metal oxides [15]. Essentially, the CRN model recognizes that such glassy materials and their corresponding crystals have very similar free volumes and heat capacities. Since both systems also have very similar NN and NNN local structures, it might be expected that glasses and crystals are structurally similar to the extent that both tend to form three-dimensional networks extending out from this local structure. However, in the crystal, there is long-range periodicity and symmetry in the network, while in the glass, we see that the basic local order extends out in a much more random manner.

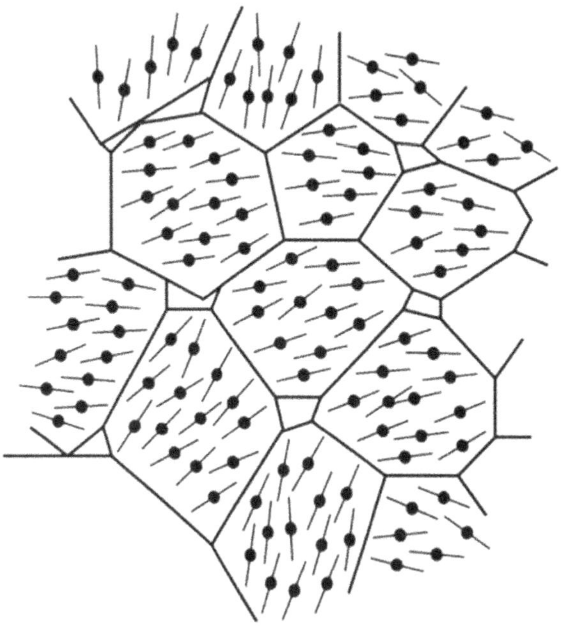

Figure 6.8 Schematic representation of the structure of an amorphous solid in the glassy state. *Source*: [13]/with permission of Springer Nature.

Using X-ray diffraction data for studying the organic compound, indomethacin in the glassy state, it was found that applying PDF data led to the conclusion, as shown in Figure 6.8, that the structure of glassy indomethacin was best described as a system, where more densely packed local domains, on the order of 2.0–2.5 nm size, exist, surrounded by a less densely packed higher energy interfacial region called the "microstructure" [13]. This microstructure serves as the point of interaction between the domains, tending to inhibit the ability of the domains to grow and eventually crystallize. An analogy to such a situation is a polycrystalline system consisting of many small crystallites surrounded by less ordered and higher energy grain boundaries. We will periodically refer to the structure given in Figure 6.8 when we discuss several important properties exhibited in general by amorphous glasses, including molecular mobility and water vapor absorption.

6.6 Molecular Mobility

6.6.1 Overview of Molecular Mobility

During the discussion of amorphous solids so far, we have briefly mentioned molecular mobility in several contexts. We are now in a position to examine this topic in more detail to better understand the important effects that molecular mobility has on the properties of amorphous solids at a fundamental level. The basic premise here is that changes in the properties of amorphous solids require molecular motion on timescales normally encountered during the handling, processing, storage, and use of amorphous materials. Such timescales of interest can amount to as little as fractions of seconds to as much as a few years, a typical expiration date for many solid dosage forms. Generally speaking, atoms and molecules in any solid system can exhibit various types of motion caused not only by thermal fluctuations but also by external electrical, magnetic, and mechanical stresses. The various modes of molecular motions found in solids can be divided into two major categories: primary or α motions, which are diffusive and cooperative, and secondary, non-diffusive and noncooperative motions. The secondary motions can be further subdivided into intermolecular β motions that involve highly restricted whole molecule motions and intramolecular motions involving side-chain rotations and covalent bond vibrations (generally labeled γ, δ, etc.).

It is generally agreed that only the α and β motions play a critical role in determining how molecular mobility influences amorphous solid properties [17].

6.6.2 Viscosity and Molecular Mobility

Two types of primary α motions, translational and rotational self-diffusion, can occur in liquids and in amorphous solids. One way to quantify the dynamics of self-diffusion of molecules is to determine their translational and rotational diffusion constants, D_{trans} and D_{rot}, respectively. The translational diffusion constant of a sphere of radius, r, in a liquid of viscosity, η, and at temperature, T, for example, can be expressed by the Stokes–Einstein equation as,

$$D_{\text{trans}} = \frac{K_B T}{6\pi r\eta} \tag{6.3}$$

where k_B is the Boltzmann constant and $6\pi r$ considers the shape of the sphere of radius r. The rotational diffusion constant for a sphere in a liquid can be expressed by the Debye equation as,

$$D_{\text{rot}} = \frac{K_B T}{8\pi r^3\eta} \tag{6.4}$$

where $8\pi r^3$ considers the volume of a sphere having a radius r. From these equations, we can observe that the three parameters that are most critical for diffusive motion in liquids, and as we shall see in amorphous solids, are the temperature, the molecular size and shape, and the viscosity. In Figure 6.9, the viscosity of a small organic molecule, *tris*-α-naphthylbenzene is reported over an 18 order-of-magnitude change and plotted as an Arrhenius-like plot of $\log \eta$ versus $1/T$ [18]. Here, we can observe that at temperatures above T_m, the melting temperature, there is a gradual increase in viscosity with a reduction of temperature into the supercooled liquid. However, upon further cooling closer to T_g, a significant nonexponential increase in viscosity of many orders of magnitude occurs. Clearly, as the system is cooled more deeply into the supercooled liquid, the factors that influence diffusional motion of the molecule become more complex, and, indeed, the temperature dependence of the diffusional process uncouples to some extent from that of viscosity [19]. It is interesting to note that this region of temperature occurs below the "crossover temperature," T_c, where the supercooled liquid appears to take on a more spatially heterogeneous, solid-like, "pre-glassy" structure. We can also observe in Figure 6.9 that, at temperatures below T_g in the glassy state, the system becomes quite viscous, but

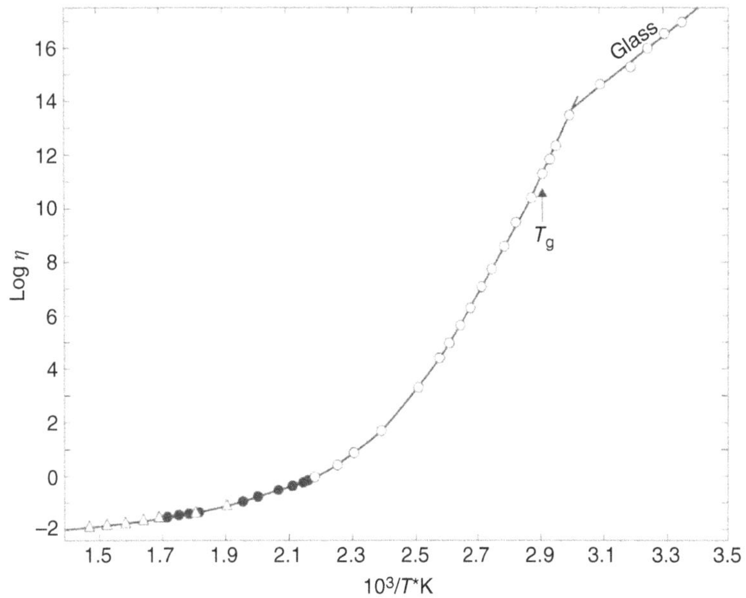

Figure 6.9 Viscosity of amorphous *tris*-α-naphthylbenzene as a function of the reciprocal of temperature. *Source*: [18]/with permission of AIP Publishing.

at some temperature below T_g the change in viscosity appears to exhibit a distinct reduction in the rate of reduction of viscosity with decreasing temperature. This indicates that there is more excess free volume or configurational entropy in the glass than would be expected by extrapolating from the "pre-glassy state."

6.6.3 Relaxation Time

Before further discussing the possible meaning of these observations in the supercooled liquid and before considering molecular mobility in the glassy state, it would be helpful to define a parameter other than viscosity that more generally and more directly indicates the timescales over which molecular mobility occurs. For example, what is the time that it takes for a molecule to complete one molecular rotation or to diffuse translationally across a given reference distance? The general approach for determining such a parameter involves perturbing a molecule, or a specific functional group on the molecule, in such a manner as to cause it to go out of its equilibrium state. When removing the stress, it is then possible to follow the rate at which the system returns or "relaxes" back toward its equilibrium position. If the fraction of relaxation that has occurred as a function of time, $\phi(t)$, follows simple first-order kinetics, it can be expressed as

$$\phi(t) = e^{[-kt]} = e^{[-\frac{t}{\tau}]} \tag{6.5}$$

where t is time, k is a first-order rate constant, and τ is $1/k$, the relaxation time. The relaxation time, τ, is a constant characteristic of the system and analogous, for example, to the half-life, $t_{0.5}$, often used to describe the timescale of chemical reactions, radioactive decay, and pharmacokinetic processes. Several experimental techniques can be used to probe the relaxation times associated with translational and rotational molecular mobility, as well as that of the various intermolecular and intramolecular secondary modes of relaxation mentioned earlier. Table 6.3 lists various techniques that have been used to measure relaxation time, along with the range of relaxation timescales that can be probed with each technique. The methods most widely used with amorphous solids of pharmaceutical interest are dynamic mechanical analysis (DMA) and dielectric relaxation spectroscopy (DES). With DMA, a sinusoidal shear stress (σ) is applied to the system at various frequencies (ω), and the corresponding shear strain (ε), a measure of deformation, is measured [20]. With some equipment, the strain is varied, and the stress is measured. The relationship between stress and strain at any frequency may be expressed as:

$$\sigma(\omega) = G^*(\omega)\varepsilon(\omega) \tag{6.6}$$

where $G^*(\omega)$ is the complex shear modulus that for viscoelastic bodies has a real and imaginary part. The complex shear modulus can be expressed in terms of a storage modulus, $G'(\omega)$, a measure of elastic energy stored, and the loss modulus $G''(\omega)$, a measure of the viscoelastic energy dissipated,

$$G^*(\omega) = G'(\omega) + iG''(\omega) \tag{6.7}$$

From the values of G'' at various frequencies and various temperatures, which provides a measure of molecular motion, it is possible to estimate the viscosity and relaxation time as a function of temperature [20]. Dielectric spectroscopy, DES, measures the dielectric properties of a medium as a function of frequency through the interaction of an applied external electric field with the electric dipole moment of the material [21]. Such dielectric relaxation occurs because of the motion of dipoles in the molecule in response to the applied electric field of a given frequency. In a manner analogous to DMA, for

Table 6.3 Various methods of measuring molecular mobility and the timescales covered.

Method of measuring molecular mobility	Timescale (s)
Dynamic mechanical analysis, DMA	10^{-10}–10^{4}
Dielectric spectroscopy, DES	10^{-11}–10^{4}
Nuclear magnetic resonance, NMR	10^{-11}–10^{3}
Neutron scattering	10^{-12}–10^{-7}
Optical probes	10^{-12}–10^{5}

viscoelastic materials, one can measure the complex dielectric constant $\varepsilon^*(\omega)$ as a function of frequency, which then can be broken down into real and imaginary components, $\varepsilon'(\omega)$ and $\varepsilon''(\omega)$, respectively,

$$\varepsilon^*(\omega) = \varepsilon'(\omega) + i\varepsilon''(\omega) \tag{6.8}$$

Again, as with DMA, $\varepsilon'(\omega)$ represents the storage of energy while $\varepsilon''(\omega)$ represents the loss term that directly reflects molecular mobility. DES is useful for probing the primary relaxation time, τ_α, and secondary relaxation time, τ_β, for amorphous solids, particularly for very fast secondary relaxations and for α slow motions that occur well below the glass transition temperature [21]. As mentioned earlier with DMA, by using DES, it is also possible to follow the primary relaxation time for amorphous solids as a function of time, $\phi(t)$, and at various temperatures. When such data obtained at higher temperatures in the supercooled liquid are analyzed in terms of $\phi(t)$ versus time, they generally follow the simple exponential given in equation (6.5). However, at lower temperatures, below the crossover temperature, T_c, the data appears to follow a "stretch-exponential" relationship, reflected, for example, in the Kohlrausch, Williams Watts (KWW) equation,

$$\phi(t) = e^{\left[-\frac{t}{\tau_{KWW}}\right]^\beta} \tag{6.9}$$

Such a "stretch-exponential" fit indicates that there are multiple relaxation times, where τ_{KWW} is an average relaxation time and β represents the distribution of different relaxation times; the values of β fall between 0 and 1, with a value of β equal to 1 indicating a single exponential relationship and, hence, a single mode of relaxation. Thus, the smaller the value of β the wider will be the distribution of relaxation times. Most amorphous systems involving organic molecules exhibit values of β in the range of 0.3–0.8 [22]. We can conclude at this point, therefore, that, in the supercooled liquid below T_c, τ_{KWW} is an average relaxation time with a distribution of different relaxation times that occur in a more spatially and temporally heterogeneous system. Given these observations, we would expect that, whereas the temperature dependence of τ for liquids and supercooled liquids at higher temperatures would be expected to follow Arrhenius-type kinetics,

$$\tau(T) = \tau_0 e^{\frac{E_a}{RT}} \tag{6.10}$$

where τ_0 is the relaxation time at a high temperature limit, on the order of vibrational timescales of 10^{-14} seconds, and E_a is the activation energy associated with the process; such changes at lower temperatures, closer to T_g, would be best fit with a nonexponential modification of the Arrhenius equation. One such equation, which has been found to be widely applicable, is the empirical Vogel, Tammen, and Fulcher (VTF) equation written as,

$$\tau(T) = \tau_0 e^{\frac{DT_0}{(T-T_0)}} \tag{6.11}$$

where D and T_0 are constants. D has been termed the "strength parameter," a measure of the activation energy of the relaxation process, and T_0 is taken to be equal to the temperature at which $\tau(T)$ would be expected to eventually diverge to "infinity" and produce no molecular motion. Indeed, it has been shown that T_0 corresponds quite closely to the Kauzmann temperature, T_K, obtained from the Gibbs–DiMarzio theory [23]. It is well established that this equation is applicable for average relaxation times obtained between approximately T_c and T_g but not at higher or lower temperatures, where Arrhenius-type behavior is observed. It also should be mentioned that the VTF equation is often expressed in terms of viscosity, η, since τ and η are directly proportional to one another. Thus,

$$\eta(T) = \eta_0 e^{\frac{DT_0}{(T-T_0)}} \tag{6.12}$$

6.6.4 Fragility in Supercooled Liquids

It would be of great interest to those working with amorphous solids to develop a mechanistic understanding of how molecules with different chemical structures and sizes undergo changes in their local structure when subjected to a change in temperature. For example, how might changes in relaxation time or viscosity with increasing temperature above T_g be different for amorphous silicon dioxide, SiO_2, with a T_g of 1600 K, a small organic molecule such as *o*-terphenyl with a T_g of 246 K and another small polar molecule such as glycerol with a T_g of 190 K? With this goal in mind, Angell [24] originally coined the term "fragility" to express the relative change in local structure that occurs as the temperature of a glass is increased across T_g into the supercooled liquid state. For "fragile liquids," the local order is significantly reduced with temperature, producing very large changes in τ or η with temperature. On the other hand, for "strong liquids," the local order is maintained up to higher temperatures, producing smaller decreases in τ or η. At a molecular level, it can be said that strong liquids generally have relatively strong intermolecular interactions and high densities, which cause greater resistance to α motions, thus leading to smaller increases in molecular mobility as the temperature is increased. Fragile liquids with weaker intermolecular interactions are more easily activated by changes in temperature and show more significant changes with temperature. For example, organic molecules, like *o*-terphenyl and glycerol, tend to be more "fragile" than a highly polar inorganic polymeric solid such as SiO_2. To obtain a quantitative measure of fragility, one can construct an "Angell plot" [24] of $\log \tau$ or $\log \eta$ against the normalized reciprocal of T, T_g/T, as shown schematically for SiO_2, glycerol, and *o*-terphenyl in Figure 6.10; using the normalized temperature in this manner allows direct comparison of the dynamics of molecules with very different values of T_g. In Figure 6.10 we can observe an Arrhenius-like pattern with SiO_2 expected for a very strong liquid and a strongly nonexponential pattern for *o*-terphenyl expected for a very fragile liquid, with glycerol exhibiting intermediate behavior.

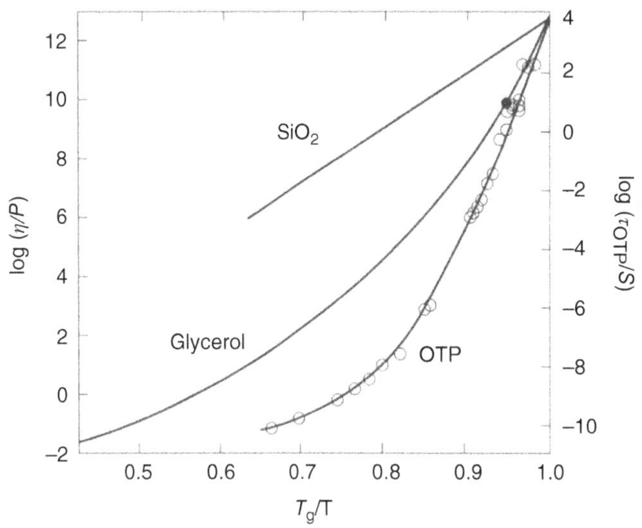

Figure 6.10 Change in relaxation time, α, and viscosity, η, versus the reduced reciprocal of temperature, T_g/T for supercooled liquids exhibiting different fragilities. *Source*: [24]/with permission of Elsevier.

One quantitative measure of fragility introduced by Angell is the fragility index, m,

$$m = \frac{d \log \tau}{d\left(\frac{T_g}{T}\right)}\Big| T = T_g \tag{6.13}$$

where m is the initial slope taken at the limit of $T = T_g$ from an Angell plot, the greater the initial slope, the greater the fragility. Another manner of comparing the fragilities of different compounds is to fit the Angell plot to the VTF equation and obtain values of D and T_0. Equation (6.14) shows the relationship between m, D, and T_0,

$$m = \frac{D\left(\frac{T_0}{T_g}\right)}{\left[1 - \left(\frac{T_0}{T_g}\right)\right]^2 ln(10)} \tag{6.14}$$

Strong liquids are those that exhibit D values that are in the range of 30–100, whereas fragile materials have D values that fall in the range of less than 30. Table 6.4 lists values of D for a variety of small organic molecules of pharmaceutical interest, along with values for the inorganic solid boron trioxide, B_2O_3, expected to be a strong glass-former [25]. Here,

Table 6.4 Glass transition temperature and constants obtained from the VTF equation.

Solid	T_g (K)	T_0 (K)	$T_g - T_0$ (K)	D
B_2O_3	557	299	255	33
Sorbitol	267	215	52	8
o-Terphenyl	249	200	49	10
Indomethacin	315	256	62	9
Sodium indomethacin	389	327	62	7
Ursodeoxycholic acid	379	311	68	8
Felodipine	317	247	70	10

Source: Adapted from [25].

it can be seen that B_2O_3 is distinctly less fragile than the organic compounds, which all generally fall into the range of D equal to 7–15. In Table 6.4, we can also note the corresponding values of T_0, T_g, and $T_g - T_0$, where values of T_0 for these molecules appear to fall into a range of 40–70 below T_g, consistent with the value of roughly 50 K below T_g suggested earlier as a storage temperature at which molecular mobility would be very low, that is, over a timescale of years and, therefore, suitable for optimal long-term storage conditions for stability assessment [26].

6.6.5 Diffusive Relaxation Time in the Glassy State

To more closely examine the possibility of establishing accurate storage temperatures in the glassy state, it will be useful to examine the rates of diffusive molecular mobility more closely in the glassy state relative to those in the supercooled liquid. Up to this point, we have observed that diffusive motions in the supercooled liquid state slow down considerably as the temperature approaches T_g, particularly for fragile liquids as the temperature goes below the crossover temperature, T_c. Typically, however, the measured average relaxation time at T_g is about 10^2 s, and the viscosity is on the order of 10^{12} Pas, indicating that a significant degree of diffusive mobility in the form of α relaxations still exists at T_g relative to the timescales of years over which, for example, solid drug products are most often stored. It is also apparent from Figure 6.9 that, although there is no discontinuity in log η versus $1/T$ right at T_g, a distinct discontinuity generally does occur at temperatures on the order of 10–15 K below T_g, giving values of η that are significantly lower than would be expected if the data were to follow the VTF equation (equation 6.11) deep into the glassy state. Since the VTF equation reflects the assumption that an equilibrium supercooled liquid state exists at all temperatures, this is not surprising. Despite this observation, measurements of molecular mobility in the glassy state at such large relaxation times and viscosities become quite difficult as temperatures are reduced more deeply into the glassy state. However, it is possible to extend viscosity measurements into this region with very sensitive mechanical measurements. It is also possible to use DES at lower frequencies to probe relaxation on very long timescales. Both techniques require specialized equipment and expertise, so only limited studies with pharmaceutical materials have been carried out in this manner [21]. However, one can obtain a measure of molecular mobility in the glassy state more conveniently with the use of thermal analysis [27]. Earlier in this chapter, it was pointed out that when a glass is formed under certain conditions and is stored at various temperatures below T_g, it will undergo relaxation to more ordered states, approaching a degree of order that might be expected to be closer to that of the equilibrium supercooled liquid at that temperature. The rate at which this relaxation process, called physical aging or annealing, occurs is primarily determined by the diffusion of molecules in the glassy state under a particular set of conditions. In Figure 6.11, we can observe a DSC scan for samples of amorphous indomethacin held at a particular temperature, T, below T_g for various time periods and then reheated through T_g. Here, we note that upon reheating the sample an endotherm appears in the scan reflective of the enthalpy associated with the recovery of the annealed sample to the equilibrium supercooled state upon passing through T_g at a given scanning rate. The area under the curve for the endotherm, which directly reflects the change in enthalpy, ΔH_t, associated with the relaxation or recovery processes, appears to increase with time and, thus, reflects the rate at which the solid is relaxing toward the expected equilibrium structure. Figure 6.12 illustrates a plot of ΔH_t versus time for indomethacin at various values of $T_g - T$, and we can see that the rate of physical aging is reduced significantly as the temperature is reduced, exhibiting very slow kinetics at values of $T_g - T$ equal to about 40 K. In equation (6.15), it can be shown further that the extent of "enthalpy relaxation" as a function of time, $\phi(t)$, can be estimated from the values of ΔH_t at any time, t, relative to the maximum enthalpy change, ΔH_{max}, that would occur when complete relaxation to the equilibrium supercooled liquid had occurred,

$$\phi(t) = 1 - \left(\frac{\Delta H_t}{\Delta H_{max}}\right) \tag{6.15}$$

The value of ΔH_{max} at any temperature, T, can be estimated from the following equation [27]:

$$\Delta H_{max} = \Delta C_p(T_g - T) \tag{6.16}$$

where ΔC_p is the change in heat capacity at T_g. Having determined $\phi(t)$ from equations (6.15) and (6.16), one can apply the KWW equation (see equation [6.8]) to obtain the average relaxation time, τ_{KWW}. Thus, in this manner, we can measure the ability of a glass initially "frozen" into a particular glassy state to adjust itself into any number of more ordered states via diffusive motion and obtain some measure of the average α relaxation time associated with this process. It is important to

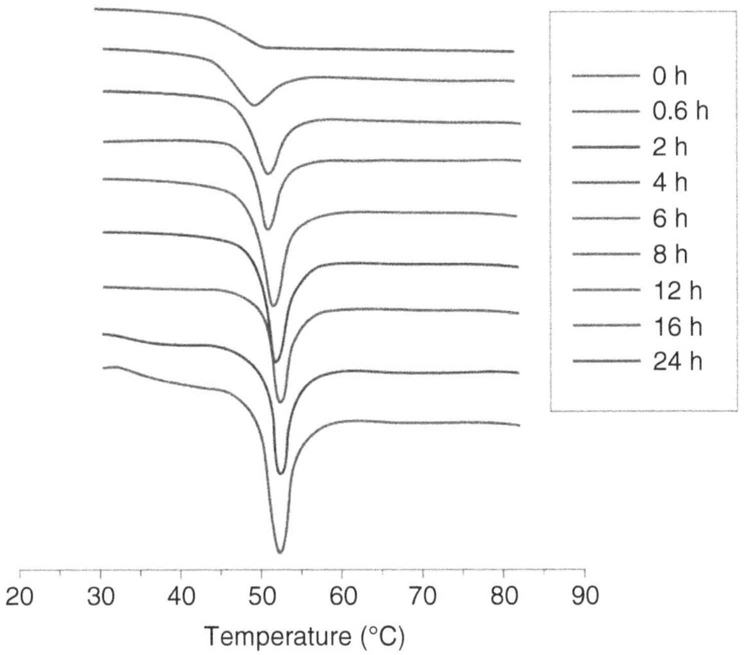

Figure 6.11 DSC scans of amorphous indomethacin previously held at $T_g - 16K$ for various time periods.

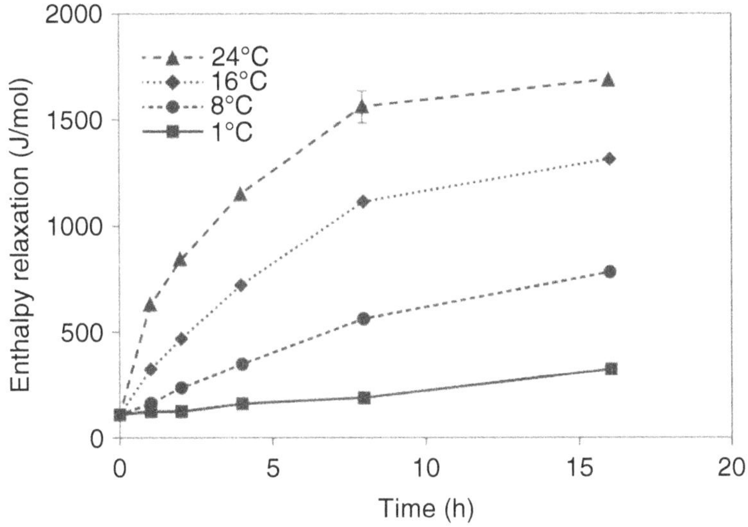

Figure 6.12 Plot of ΔH_t, the enthalpy change for amorphous indomethacin during physical aging versus time for samples held at various temperatures.

recognize at this point that such an approach to measuring relaxation times in the glassy state is complicated by the fact that the process of relaxation can continue to occur over the time required for the measurement of the enthalpy relaxation endotherms. Thus, the values of τ and β obtained from equation (6.9) at various times will have a certain level of uncertainty, particularly at temperatures close to T_g. In this regard, it has been observed that although τ and β in equation (6.9) may change with time, the term τ^β appears to be relatively time-independent [27]. Therefore, it has been suggested, empirically, that for comparison purposes average relaxation times in the glassy state measured by following enthalpy relaxation should be expressed as τ^β. The physical meaning of such a term, however, has not been ascertained. Recognizing this uncertainty, it is still of considerable interest to observe the way τ, obtained from enthalpy relaxation measurements, changes with temperature in the glassy state relative to that for the supercooled liquid. In Figure 6.13, values of log τ versus

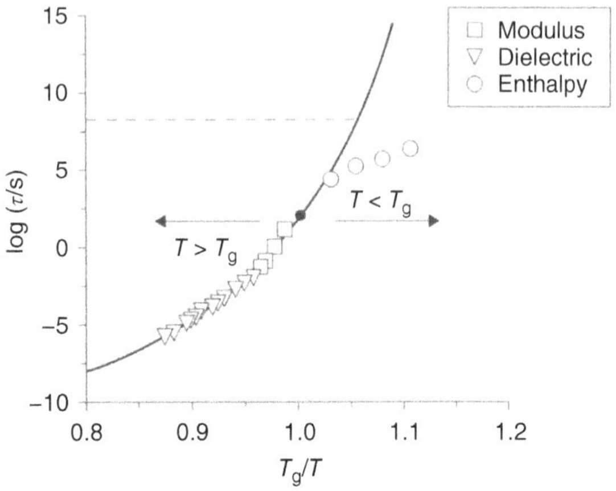

Figure 6.13 Plot of log of relaxation time, τ, for amorphous indomethacin as a function of the reduced reciprocal of temperature, T_g/T. *Source*: Adapted from [28].

T_g/T for amorphous indomethacin are plotted at temperatures above and below T_g [28]. Values above T_g were determined from DMA and DES measurements, while those below T_g were obtained from measurements of enthalpy relaxation. As was observed earlier with measurement of log η versus $1/T$ for *tris*-α-naphthylbenzene (Figure 6.9), the data appear to fit the VTF equation to temperatures that are about 15 K below T_g, followed by a linear Arrhenius-type change and values of τ lower than expected from the VTF plot. Thus, for two systems, at least, it has been shown that at some temperature below T_g molecular mobility in the glass takes on a different pattern of change from that predicted for the supercooled liquid, with greater molecular mobility than expected and Arrhenius-type temperature dependence. It is also interesting to note in Figure 6.13 that extrapolation of the linear portion of the plot to relaxation times of roughly 3 years $(10^8\ \mathrm{s})$ occurs at values of $T_g - T$ roughly equal to 40 K. This is consistent with a "rule of thumb" that suggests that samples should be stored at least 50 K below T_g to provide the optimal storage conditions that would essentially eliminate the α motions that lead to instability on the timescale of years [26].

6.6.6 Secondary Relaxations in Amorphous Solids

Up to this point in our discussions of molecular mobility, we have focused on the diffusive translational and rotational motions that reflect the primary or α relaxations associated with amorphous solids in the supercooled liquid and glassy states. This is most important since it is these molecular motions that primarily determine the properties of amorphous solids that underlie their physical and chemical stability. As discussed earlier briefly, molecules also undergo secondary motions that occur on much shorter timescales than those due to diffusional processes. Such motions are most often probed by high-frequency DES, as well as by solid-state NMR (SSNMR), which also can detect motions that occur intermolecularly on relatively short timescales. Although important in determining the thermodynamic states of the molecule, the intramolecular motions, including bond vibrations and rotations, generally do not appear to be important determinants of physical and chemical instability in the amorphous state [17]. However, it appears that most amorphous materials can undergo a type of secondary intermolecular motion that plays an important role in determining the dynamic properties of amorphous solids. Such secondary intermolecular motions are known as Johari–Goldstein β relaxations, β_{JG}, named after the two scientists, most associated with the early recognition that such molecular motions exist in organic amorphous solids [29]. Critical for the discovery of the β_{JG} relaxations were studies of amorphous solids having molecules that were structurally rigid and thus devoid of any side-chain intramolecular motions and where distinct secondary relaxations were detected by the appearance of an excess high-frequency contribution to the dielectric loss spectra associated with the α relaxation. It was concluded that this excess peak in $\varepsilon''(\omega)$ (see equation (6.8)) arises from hindered noncooperative single-molecule motions within the amorphous structure. It has also been shown that values of the relaxation time associated with the β_{JK} relaxation, τ_β, measured directly from DES data as a function of temperature, follow Arrhenius-type kinetics, with lower activation energies than those observed for the α relaxation [21]. Indeed, as seen in Figure 6.14, at temperatures just below the crossover temperature, T_c, where spatial heterogeneity begins, the α and β_{JG} relaxation times, τ_α and τ_β, respectively, begin to diverge with decreasing temperature, whereas above T_c the values of τ_α and τ_β become indistinguishable with increasing temperature. From Figure 6.14, we can also see that well below T_g, more deeply in the glassy state where α relaxations slow down considerably, the timescales for the β_{JG} relaxations are still significant and only become exceedingly slow at temperatures typically 100 K below T_g. This temperature has been designated $T_{g\beta}$, and it is believed to represent the temperature at which β_{JG} relaxations are no longer detectable, and hence practical effects of molecular mobility on amorphous solid stability should no longer occur [30].

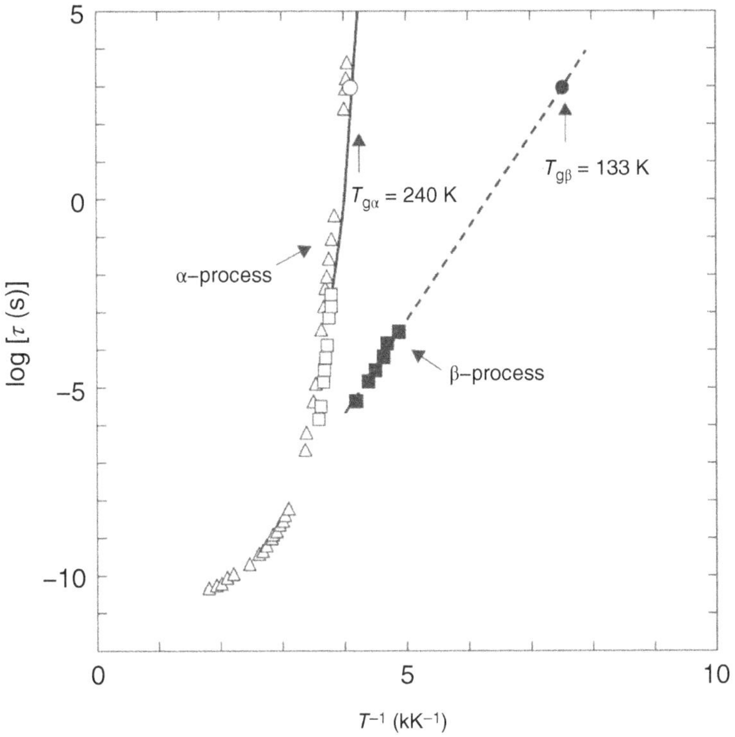

Figure 6.14 Plot of relaxation time, τ, for α and β relaxations of amorphous *o*-terphenyl versus the reciprocal of temperature. *Source*: [30]/with permission of Elsevier.

To better understand the significance of the β_{JG} relaxation process in amorphous solids, it would be helpful to have a clearer picture of the structural features of the glassy state that allow such relaxations to occur. Johari and Goldstein [29] concluded from their studies that the β_{JG} relaxation below T_g most likely takes place in limited regions of the glass where the structure is relatively "loose" and molecules are able to exhibit hindered motions of some significance. They called such regions "islands of mobility." Using the model presented earlier in Figure 6.9 to describe the local structure of molecules in the glassy state, it would appear likely that hindered rotational motion of whole molecules takes place in the microstructure of the glass found between the various domains, primarily making up the glassy state. Combined studies using DES and SSNMR have shown that β_{JG} relaxations involve stepwise hindered rotational motions throughout the solid. However, 80–90% of the molecules appear to have very low amplitude motions that are relatively independent of temperature, while 10–20% of the molecules appear to have higher amplitudes and significant temperature dependence [31]. Considering that the higher energy microstructure regions represent about 10–20% molecules in the glass and that molecular mobility in the microstructure is greater than that in the domains, it would appear likely that, indeed, the most important β_{JG} relaxations are occurring in "islands of mobility" represented by the microstructure. It has also been shown that during the physical aging of a glass, the amplitude of the molecular motions associated with these β_{JG} relaxations are significantly reduced, further suggesting that physical aging leads to a reduction in the higher energy microstructure component of glasses, which in turn leads to a more ordered glassy state as discussed earlier. This is also supported by studies with vapor-deposited glasses that have shown greater densities with vapor-deposited glasses than with physically aged glasses obtained by quenching the supercooled liquid [32]. Here, the vapor-deposited glasses exhibit greater suppression of the secondary β_{JG} relaxations with higher density. Having established that there are intermolecular secondary motions governed by β_{JG} relaxation processes that arise from the hindered rotational motion of molecules in "islands of mobility" and that they are likely located in the microstructure between amorphous domains, it would be appropriate to now consider the possible relationship, if any, that might exist between such secondary motions and the primary diffusive motions that are associated with α relaxations, shown earlier to be fundamental as determinants of various properties of molecules in the amorphous state. Theoretical and experimental studies have led to the conclusion that, indeed, noncooperative motions of individual molecules that lead to the cooperative motions associated with the α relaxations begin with the hindered motions associated with β_{JG} relaxations in "islands of mobility" [33]. Above the critical temperature, $T_{g\beta}$, required for the initiation of β_{JG} relaxations, the hindered motions increase in frequency with temperature until the critical volume needed to initiate diffusive motion is exceeded. Hence, we can conclude that β_{JG} relaxations play a critical role as precursors for the initiation of α relaxations and, therefore, play an important role in initiating the various physical changes that affect the physical and chemical stability of molecules in the amorphous state.

6.7 Mixtures of Amorphous Solids

6.7.1 Overview

There are many situations of pharmaceutical interest where a mixture of substances in a formulation may contain more than one component in the amorphous state. This might occur with a mixture of amorphous APIs in a combination product or, more often, with various amorphous API–excipient combinations. It is possible that the various amorphous components can simply be physically mixed in the formulation and as such exist as separate phases that retain the distinctive properties of each of the individual amorphous components, including any inherent instability. Such a completely phase-separated system would also be expected to exhibit more than one T_g value, each equal to that of the individual components. It is also possible to observe phase separation, and hence multiple T_g values, after subjecting two or more materials to various types of processing ordinarily used to produce amorphous solids. For example, in Figure 6.15, we can observe two values of T_g for a mixture of PVP and dextran, prepared by freeze drying from an aqueous solution of both components [34]. Note that the values are equal to the T_g of each pure component at all concentrations, indicating a lack of miscibility and complete phase separation. On the other hand, since molecules in the amorphous state can be considered to have the properties of liquids, it is very possible that components processed together from solution or from the melt can form a single phase, as would be expected for a solution of two or more miscible liquids. When such miscibility occurs in the amorphous state at any concentration, we would expect there to be only one T_g value intermediate to those of the individual components. For example, Figure 6.16 indicates the value of T_g for pure indomethacin and PVP, as well as two mixtures of indomethacin and PVP prepared by rapidly drying a solution of indomethacin and PVP [35]. Here, we can see that at each composition there is only one T_g that tends to increase as the component having the greater T_g is increased in concentration. At this point, it is important to recognize that molecules in the amorphous state are generally metastable relative to any crystal forms that might be associated with the molecule. Therefore, in most situations where a crystallizable molecule is part of the mixture and the equilibrium solubility of the crystal in the remaining amorphous matrix is exceeded, crystallization of such components may eventually occur upon long-term storage [36]. It is also possible with such metastable systems that processing the mixture under one set of conditions, for example, at a high temperature, might produce a miscible system that upon cooling eventually separates as immiscible amorphous phases, followed by possible crystallization. Thus, when we speak of the miscibility of amorphous components, we must discriminate between those where the crystallizable components are at concentrations less than their equilibrium solubility in the remaining amorphous matrix and thus thermodynamically stable and those situations where the crystallizable component is maintained in a metastable single-phase subject to phase separation into multiple amorphous phases, followed by possible crystallization over time.

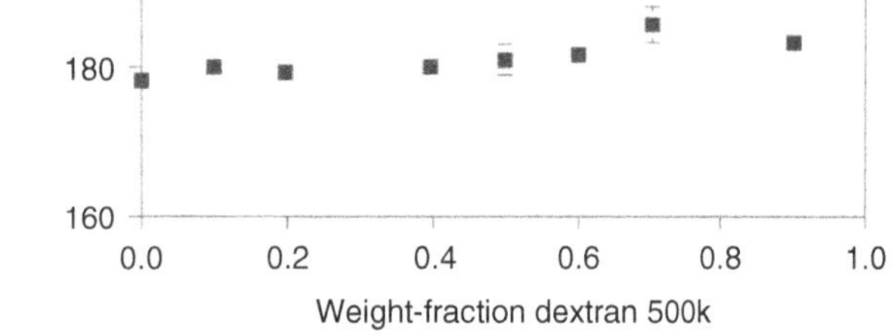

Figure 6.15 Glass transition temperatures for a colyophilized amorphous mixture of dextran and PVP. *Source:* Adapted from [34].

Up to this point, we have discussed the miscibility and immiscibility of amorphous mixtures solely in terms of the presence or absence of more than one glass transition temperature. Indeed, in most cases, this can be a very valid criterion for establishing miscibility. However, if the T_g values of the individual components do not differ by more than roughly

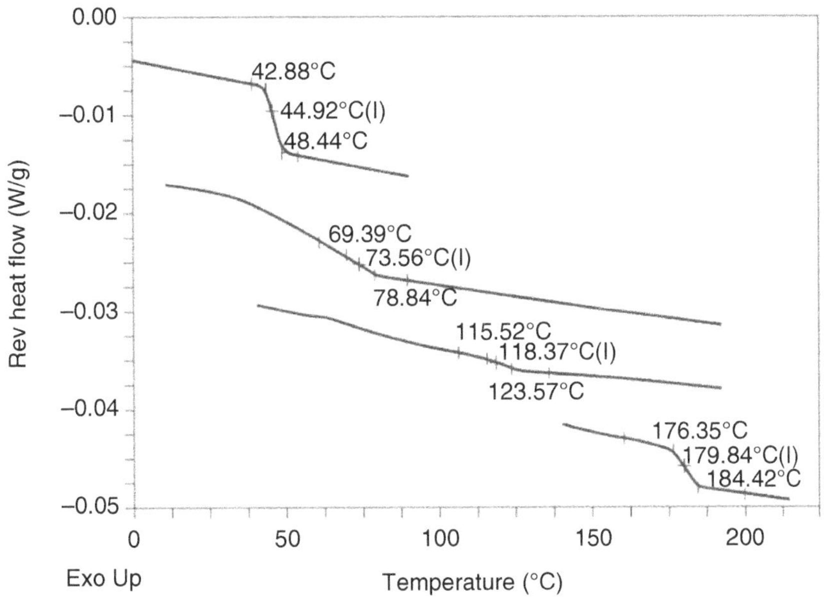

Figure 6.16 Appearance of a single-glass transition temperature for miscible mixtures of 70% and 30% w/w indomethacin and PVP compared to pure components. *Source*: [35]/with permission of Elsevier.

10 K, DSC measurements will not be able to discriminate between the individual T_g values. More importantly, it has been shown that DSC is not able to discriminate phase separation when the dimensions of the separated phases are below the size range of 30–100 nm [35]. In such cases, despite the distinct phase separation, only one value of T_g is observed and miscibility may be incorrectly assumed. To further establish whether amorphous mixtures truly exist in a single phase, it is possible to use PXRD measurements, as well as SSNMR spectroscopy, since both techniques probe molecular interactions at a very local level. For example, in the case of PXRD, one can compare diffraction patterns of the amorphous mixture to those of the individual components, and if the sum of the weighted average of the patterns of the individual components is the same as that of the mixture, the system is completely phase separated [35]. If the calculated values for the mixture do not coincide with those experimentally determined, the system can be considered to exist as a single phase. One can also determine the PDF of the individual components and the mixture and similarly compare the sum of the weighted average of the PDFs of the components to that of the mixture. Again, if the weighted average of the PDFs of the individual components is the same as that of the mixture the system is immiscible, whereas differences indicate miscibility. It is also possible to use SSNMR to detect phase separation down to domain sizes much less than 30 nm using two-dimensional NMR [37, 38] and by measuring T_1 and $T_{1\rho}$ relaxation times, which respond to NN and NNN distance interactions between components [38].

6.7.2 Thermodynamics of Molecular Mixing in Amorphous Solids

We may now ask what it is that primarily determines whether two or more components will be miscible in the amorphous state. First, we can assume that for molecular mixing to occur there must be an overall loss in the Gibbs free energy of mixing, ΔG_{mix}, which can be expressed at temperature, T, by the Gibbs–Helmholtz equation as,

$$\Delta G_{mix} = \Delta H_{mix} - T\Delta S_{mix} \tag{6.17}$$

The mixing of molecules tends to produce a positive entropy change (greater disorder relative to the individual components), which, as shown in equation (6.17), leads to a decrease in free energy; the greater the temperature, T, the greater the decrease in free energy and tendency for mixing. From equation (6.17), we can also see that the sign and magnitude of the enthalpy change, ΔH_{mix}, can either augment or oppose the loss in free energy brought about by the entropy increase. In a binary system, for example, the enthalpy of mixing, ΔH_{mix}, is determined by the relationship between the noncovalent intermolecular interaction between the components, A and B, to form A:B and those between each of the components,

A:A and B:B. For molecules of very similar molecular structure and where the interaction energies of A:A and B:B are very close to that of A:B, the net ΔH_{mix} will be essentially zero and

$$\Delta G_{mix} = -T\Delta S_{mix} \tag{6.18}$$

that is, the negative free energy change for mixing will be primarily determined by the entropy change and temperature. When A:A and B:B interactions are either greater (endothermic) or less (exothermic) than the A:B interactions and there are differences in molecular volume, the mixing tends to be nonideal. An exothermic process ($-\Delta H_{mix}$) will contribute to a free energy loss and enhanced mixing, whereas an endothermic process ($+\Delta H_{mix}$) will contribute to an increase in ΔG_{mix} or a less negative change, which in turn leads to a reduced tendency for mixing or even to complete demixing if ΔH_{mix} is positive and greater than $T\Delta S_{mix}$. Consequently, we can generalize by saying that chemically similar molecules in the liquid and solid states that might strongly interact intermolecularly, for example, by hydrogen bonding, will tend to exhibit a loss in free energy and have a strong tendency to molecularly mix. On the other hand, the more chemically dissimilar the molecules are, and hence the more positive or less negative ΔH_{mix} is, the less will be the tendency to mix, and the more likely that there will be complete phase separation.

To gain a better conceptual and quantitative understanding of molecular mixing in the amorphous state, it would be desirable to have a theoretical framework that describes the free energy of mixing and the miscibility of liquids. For the practical pharmaceutical situation where polymeric excipients may be mixed or mixed with a small molecule API, a useful theory is the Flory–Huggins lattice theory [6], described by the Flory–Huggins equation and originally developed for polymer solutions. In its final form this equation describes the Gibbs free energy of mixing per mole for a polymer and a "solvent," which can include a small molecular weight molecule, in terms of the entropy and enthalpy contributions as

$$\Delta G_{mix} = RT(n_A \ln \Phi_A + n_B \ln \Phi_B + n_A \Phi_B \chi_{AB}) \tag{6.19}$$

Here, R is the gas constant, Φ is the volume fraction of each component, χ_{AB} is the interaction parameter that describes the relative tendency for interaction between components, and n is equal to the number of moles of each component. The first two terms on the left in the parentheses represent the entropic contributions, whereas the third term represents the enthalpic contributions. In this discussion, component A refers to the "solvent" and component B to the polymer. The smaller or more negative the interaction parameter, the stronger the intermolecular interaction. From a conceptual perspective, the Flory–Huggins equation is also very useful in demonstrating the importance of the molecular size in affecting the entropy of mixing. For example, in equation (6.20), the entropy of mixing per mole can be described as,

$$\Delta S_{mix} = -R(n_A \ln \Phi_A + n_B \ln \Phi_B) \tag{6.20}$$

If we now express the number of moles of each component in terms of the mass, m, and molecular weight, mw, we can rewrite equation (6.20) as,

$$\Delta S_{mix} = -R\left[\left(\frac{m_A}{mw_A}\right)\ln \Phi_A + \left(\frac{m_B}{mw_B}\right)\ln \Phi_B\right] \tag{6.21}$$

The important concept that arises from this equation is that the greater the molecular weight of any component for a given mass, the less will be the contribution of entropy to the free energy of mixing. Essentially, by mixing two large molecules that individually have many degrees of conformational freedom, and hence significant configurational entropy, such configurational entropy will be reduced upon mixing of the two components. Consequently, we would expect that tendencies for miscibility of amorphous binary systems involving at least one polymer would be reduced relative to those for two small molecules and that a mixture of two polymers would have a strong tendency to phase separate. Indeed, many amorphous polymer blends, such as in the case of PVP and dextran given above (see Figure 6.15), can be expected to exhibit significant immiscibility primarily for this reason.

6.7.3 The Glass Transition Temperature and Molecular Mobility of Miscible Amorphous Mixtures

From the thermodynamic analysis described earlier, it is shown that miscible mixtures of amorphous solids will exhibit thermodynamic properties that are different than those of the individual components. This difference will depend on the extent of intermolecular interaction and the relative molecular size and shape. For an ideal solution, we would expect, for example, that the molar volume of the mixture, V_{AB}, would be the sum of the weighted average of the molar volume of each component,

$$V_{AB} = n_A V_A + n_B V_B \tag{6.22}$$

where V_A and V_B are the molar volumes of components A and B, respectively. However, in the case of nonideal mixing, we would expect there to be an excess molar volume, V_{excess}, that can be either positive or negative depending on the nature of the intermolecular interactions. This can be written as,

$$V_{AB} = n_A V_A + n_B V_B + V_{excess} \tag{6.23}$$

Thus, very strong interactions between A and B lead to a negative excess molar volume, that is, a reduction in overall volume, as in the well-known case of the mixing of ethanol and water, while stronger interactions between A and A and B and B than between A and B lead to a positive excess molar volume or an increase in overall molar volume. In our earlier discussion of the glass transition temperature, we showed that the molar volume is the sum of the occupied volume plus the free volume (see equation (6.1)) of a system and that free volume plays a very important role in determining the value of T_g ; the greater the free volume, the greater the molecular mobility and the lower T_g. We would expect, therefore, that the T_g of a mixture (T_{gmix}) would be directly related to the molar volume of the mixture relative to the individual components. Indeed, for an ideal mixture, as expressed in equation (6.24), it can be shown that T_{gmix} is the weighted average of the individual T_g values,

$$T_{gmix} = \Phi_A T_{gA} + \Phi_B T_{gB} \tag{6.24}$$

where Φ_A and Φ_B are the volume fractions of component A and component B, respectively, and $\Phi_A + \Phi_B = 1$.

Generally, when working with mixtures of amorphous solids, the concentration of the various ingredients is expressed as weight fraction, w, rather than volume fraction. Consequently, as shown by Gordon and Taylor [39], equation (6.24) can be expressed in terms of weight fraction as,

$$T_{gmix} = \frac{(w_A T_{gA} + w_B T_{gB})}{(w_A + K w_B)} \tag{6.25}$$

where, for ideal mixing,

$$K = \frac{\rho_A T_{gA}}{\rho_B T_{gB}} \tag{6.26}$$

and ρ is the density of each component at its respective T_g. It can be further shown that when the densities of the two components are very similar, as in the case of many polymers, equation (6.25) leads to the Fox equation [40], where

$$\frac{1}{T_{gmix}} = \frac{w_A}{T_{gA}} + \frac{w_B}{T_{gB}} \tag{6.27}$$

The same general equation for ideal mixing, as described in equation (6.25), has been derived on a thermodynamic basis by Couchman and Karasz [41], with K now defined in terms of the heat capacity changes taking place at T_g,

$$K = \frac{\Delta C_{pA}}{\Delta C_{pB}} \tag{6.28}$$

Despite the strong possibility of nonideal mixing in most practical systems encountered, these equations for ideal mixing can be useful for estimating the glass transition temperature of a miscible mixture of amorphous components for several reasons. Initially, one may want to get a rough estimate of what composition of polymer and API might be required to provide a certain value of T_{gmix}, given only the individual T_g values. This can be done very simply by using equation (6.27) or by estimating K in equation (6.26) or equation (6.28) and using equation (6.25). To test for ideality or nonideality in the amorphous mixture, one can use these estimates of T_{gmix} and experimentally determine T_{gmix}. Deviations of experimental results from the expected plot indicate that the mixing of components is nonideal; the deviations can be either negative or positive, meaning that the experimentally determined values of T_{gmix} are either greater or less than predicted. Values greater than those predicted for ideality indicate that the net excess free volume is less than that expected for the weighted sum of individual component free volumes, or that A:B interactions are greater than the A:A and B:B interactions. If the deviations are negative, this means that there is greater excess free volume than expected or stronger A:A and B:B interactions than those between A and B.

From the previous discussion we can see that a miscible mixture of amorphous components having different T_g values should have properties quite different than those of the individual components, as reflected, for example, in the intermediate value of T_{gmix}. As also discussed earlier, a very important property in many pharmaceutical situations involving amorphous solids is molecular mobility and in particular the α and β_{JG} relaxation times. In the case of a binary miscible amorphous mixture, it has been shown that the relaxation times of the component with the lower T_g, and hence with higher molecular mobility, generally will be increased by the presence of a higher T_g component, while the molecular mobility of the higher T_g component will decrease. When molecular mobility is increased in this manner, we say that the molecule has been "plasticized," whereas when molecular mobility is decreased by the presence of another component, we say that the molecule has been "antiplasticized." An important pharmaceutical situation where miscible amorphous dispersions are of great interest involves highly water-insoluble API's that are rendered amorphous to improve their apparent water solubility relative to the crystalline form and thus to enhance dissolution and oral bioavailability but also are subject to recrystallization upon handling and storage. In such cases, amorphous polymers having T_g values significantly greater than the API are used to form "amorphous solid dispersions" and to prevent crystallization from the solid state during storage by primarily decreasing the molecular mobility of the API. The chemical structure and molecular size of the polymer and its composition will determine the extent of miscibility, the T_{gmix}, the molecular mobility and the ability to inhibit crystallization. For example, when an API with a T_g of 300 K forms an amorphous miscible mixture with a polymer having a T_g of 450 K, the presence of the polymer will raise the T_g of the system relative to that of the API, producing an antiplasticizing effect on the other component, with a corresponding decrease in molecular mobility. At the same time, the polymer will be plasticized by the API. To see how significant this effect can be, we can use equation (6.27) for a miscible mixture of 25% w/w API and 75% w/w polymer to obtain a value of T_{gmix} equal to 400 K. Consequently, we would expect that the API would now be in an amorphous environment with an overall T_g that is 50 K higher than that of the API alone, leading to an environment with a lower molecular mobility and hence greater stability.

References

1 Hancock BC, Parks M. What is the true solubility advantage for amorphous solids? *Pharm Res* 2000; 17(4): 397–404.
2 Swallen SF, Kearns KL, Satija S, Traynor K, McMahon RJ, Ediger MD. Molecular view of the isothermal transformation of a stable glass. *J Chem Phys* 2008; 128: 214514.
3 Bates S, Kelly RC, Ivanisevic I, Schields P, Zografi G, Newman AW. Assessment of defects and amorphous structure produced in raffinose pentahydrate upon dehydration. *J Pharm Sci* 2007; 9(5): 1418–1433.
4 Guo Y, Byrn SR, Zografi G. Physical characteristics and chemical degradation of amorphous quinapril hydrochloride. *J Pharm Sci* 2000; 89(1): 128–143.
5 Ediger MD, Angell CA, Nagel SR. Supercooled liquids and glasses. *J Phys Chem* 1996; 100: 13200–13212.
6 Flory, PJ. *Principles of Polymer Chemistry*. Ithaca: Cornell University Press; 1953.
7 Kauzmann W. The nature of the glassy state and behavior of liquids at low temperatures. *Chem Rev* 1948; 43(2): 219–256.
8 Gibbs JH, DiMarzio EA. Nature of the glass transition temperature. *J Chem Phys* 1958; 28(3): 373–383.
9 Matsumoto T, Zografi G. Physical properties of solid molecular dispersions of indomethacin with poly(vinylpyrrolidone) and poly(vinylpyrrolidone-co-acetate) in relation to indomethacin crystallization. *Pharm Res* 1999; 16(11): 1722–1728.
10 Lee WA, Knight GJ. Ratio of the glass transition temperature to the melting point in polymers. *Br Polym J* 1970; 2: 73–80.
11 Fukuoka E, Makita M, Yamamura S. Glassy state of pharmaceuticals. III. Thermal properties and stability of glassy pharmaceuticals. *Chem Pharm Bull* 1989; 37(4): 1047–1050.
12 Billinge JL, Thorpe MF. *Local Structure from Diffraction*. New York: Plenum; 1998.
13 Bates S, Zografi G, Engers D, Morris K, Crowley K, Newman AW. Analysis of amorphous and nanocrystalline solids from their x-ray diffraction patterns. *Pharm Res* 2006; 23(10): 2333–2348.
14 Jalali P, Li M. An estimate of random close packing density in monodisperse hard spheres. *J Chem Phys* 2004; 120(2): 1138–1139.
15 Zachariasen WH. The atomic arrangement in glass. *J Am Chem Soc* 1932; 54: 3841–3851.

16 Silescu H. Heterogeniety at the glass transition: a review. *J Non-Cryst Solids* 1999; 243: 81–108.

17 Ngai KL, Paluch M. Classification of secondary relaxation in glass-formers based on dynamic properties. *J Chem Phys* 2004; 857–873.

18 Plazek DJ, Magill JH. Physical properties of aromatic hydrocarbons. IV. Analysis of the temperature dependence of the viscosity and the compliance of 1,3,5,-tri-α-naphthylbenzene. *J Chem Phys* 1968; 49: 3679–3883.

19 Mapes MK, Swallen SF, Ediger MD. Self-diffusion of super-cooled o-terphenyl near the glass transition temperature. *J Phys Chem B* 2006; 110: 507–511.

20 Andronis V, Zografi G. Molecular mobility of supercooled amorphous indomethacin determined by dynamic mechanical analysis. *Pharm Res* 1997; 14: 410–414.

21 Grzybowaska K, Paluch M, Grzybowski A, Wojnarowska Z, Hawelek I, Kolodziejczyk K, Ngai KL. Molecular dynamics and physical stability of amorphous anti-inflammatory drug: celecoxobid. *J Phys Chem B* 2010; 114: 12792–12801.

22 Shamblin SL, Tang X, Chang L, Hancock BC, Pikal MJ. Characterization of the time scales of molecular motion in pharmaceutically important glasses. *J Phys Chem B* 1999; 103: 4113–4121.

23 Richert R, Angell CA. Dynamics of glass-forming liquids. V. on the link between molecular dynamics and configurational entropy. *J Chem Phys* 1998; 108(21): 9016–9026.

24 Angell CA. Perspectives on the glass transition. *J Phys Chem Solids* 1988; 49(8): 863–871.

25 Crowley K, Zografi G. The use of thermal methods for predicting glass-former fragility. *Thermochim Acta* 2001; 380: 79–83.

26 Hancock BC, Shamblin SL, Zografi G. Molecular mobility of pharmaceutical solids below the glass transition temperature. *Pharma Res* 1995; 12: 799–806.

27 Liu J, Rigsbee DR, Stotz C, Pikal MJ. Dynamics of pharmaceutical solids: study of enthalpy relaxation by isothermal microcalorimetry. *J Pharm Sci* 2002; 91(8): 1853–1862.

28 Andronis V, Zografi G. The molecular mobility of supercooled indomethacin as a function of temperature and relative humidity. *Pharm Res* 1998; 15(6): 835–842.

29 Johari GP, Goldstein M. Viscous liquids and the glass transition. Secondary relaxations in glasses of rigid molecules. *J Chem Phys* 1970; 53(6): 2372–2388.

30 Hikma T, Hanya M, Oguni M. Numerical and morphological approach to the mechanism of homogeneous-nucleation-based crystallization in o-terphenyl. *J Non-cryst Solids* 1998; 235–237: 539–547.

31 Vogel M, Tschirwitz C, Schneider G, Koplin C, Medick P, Rossler E. A 2H NMR and dielectric spectroscopy study of the slow β-process in organic glass formers. *J Non-Cryst Solids* 2002; 307–310: 326–335.

32 Kasting BJ, Beasley MS, Guiseppi-Elie A, Richert R, Ediger MD. Relationship between aged and vapor-deposited organic glasses: secondary relaxation in methyl-m-toluate. *J Chem Phys* 2019; 151: 144502.

33 Hikima T, Hanaya M, Oguni M. Microscopic observation of peculiar crystallization in the glass transition region and the β process as potentially controlling the growth rate in triphenylethylene. *J Mol Struct* 1999; 479: 245–250.

34 Shamblin SL, Taylor LS, Zografi G. Mixing behavior of colyophilized binary systems. *J Pharm Sci* 1998; 87(6): 694–701.

35 Newman A, Engers D, Bates S, Ivanisevic I, Kelly RC, Zografi G. Characterization of amorphous API:polymer mixtures using x-ray diffraction. *J Pharm Sci* 2008; 97(11): 4840–4856.

36 Tao J, Sun Y, Zhang GGZ, Yu L. Solubility of small molecule crystals in polymers: D-mannitol in PVP, indomethacin in PVP/VA and nifedipine in PVP/VA. *Pharm Res* 2009; 26(4): 855–864.

37 Cheung MK, Wang J, Zheng S, Mi Y. Miscibility of poly(epichlorohydrin)/ poly(vinyl acetate) blends investigated with high resolution solid-state 13C nmr. *Polymer* 2000; 41: 1469–1474.

38 Pham TN, Watson SA, Edwards AJ, Chavda M, Clawson JS, Stroheim M, Vogt FG. Analysis of amorphous solid dispersions using 2D solid-state NMR and 1H T1 relaxation measurements. *Mol Pharm* 2010; 7: 2194–2206.

39 Gordon M, Taylor JS. Ideal copolymers and the second-order transition of synthetic rubbers. I., Noncrystalline co-polymers. *J Appl Chem* 1952; 2: 493–500.

40 Fox TG. Influence of diluent and copolymer composition on the glass transition temperature of a polymer system. *Bull Am Phys Soc* 1956; 1: 123.

41 Couchman PR, Karasz FE. A classical thermodynamic discussion on the effect of composition on glass transition temperature. *Macromolecules* 1978; 11: 117–119.

7

Crystal Mesophases and Nanocrystals

7.1 Introduction

In the preceding chapters, we have discussed two distinctly different solid forms: crystalline and amorphous solids. As we have learned, molecules in the crystalline state are characterized by having long-range order and symmetry in three dimensions, whereas, in amorphous solids, as with liquids, molecules lack any long-range order. In ordinary crystals, we can define three types of long-range structural order and symmetry: (i) translational or positional order; (ii) rotational or orientational order; and (iii) internal bond rotational or conformational order [1]. In crystals, the relative position, orientation, and conformation of molecules remain unchanged over very long distances within the crystal, and such order is generally accompanied by a reduction in large-amplitude molecular motions, that is, reduced molecular mobility. There are, however, other solid forms, known as crystal mesophases, that have either fully or partially lost certain types of long-range molecular order while retaining others, thus placing them somewhere between crystals and amorphous solids regarding the structural features that can affect various physical and chemical properties. Because of the loss of some forms of order, crystal mesophases would be expected to have different thermodynamic properties and greater molecular mobility than those that occur in the crystal but still less than those that occur in the amorphous state. There are three types of crystal mesophases: (i) liquid crystals (LCs), (ii) plastic crystals (PCs), and (iii) conformationally disordered crystals – the latter are also known as condis crystals. Although the occurrence of condis crystals and PCs is relatively rare in systems of pharmaceutical interest, many active pharmaceutical ingredients (APIs) and excipients can exist as liquid crystalline forms under various conditions [2]. Consequently, although all types of crystal mesophases will be discussed in this chapter, we will primarily focus on LCs and their behavior relative to crystalline and amorphous solids.

In this chapter, we will also discuss nanocrystals, which exhibit unique structural and thermodynamic features relative to ordinary crystals. Our focus will be on crystalline systems that can exist with particle diameters that are on the order of 100 nm or less. Such particles are in the colloidal size range where surface-to-volume ratios are high enough to allow surface effects to alter bulk thermodynamic properties, producing particles with higher free energy than those of the corresponding crystal having larger particle diameters [3]. In this regard, it is important to recognize that the terms "nanoparticle" and "nanocrystal" are often used in the pharmaceutical sciences to describe particles having diameters less than 1000 nm (1 μm), particularly at particle diameters somewhere between 100 nm and a few hundred nanometers [4]. Although such "submicron" size particles are very useful for enhancing rates of dissolution through their greatly increased specific surface area, molecules in such crystalline particles do not have the higher energy and extra-thermodynamic properties expected of true nanocrystals and, therefore, are not structurally unique forms of a crystal.

7.2 Overview of Crystal Mesophases

Crystal mesophases are thermodynamically stable states of matter that have properties that are intermediate to those of crystals and liquids. As mentioned earlier, such properties are brought about by the absence of different forms of order, relative to that of the complete order found in crystals. Being thermodynamically stable forms, crystal mesophases can undergo distinct first-order transitions, for example, from the crystal to a particular mesophase or by a transition from a mesophase to the liquid melt. At such first-order transitions, we generally expect a discontinuity in the first derivative of the Gibbs free energy, with such a change being equal to the entropy associated with the phase change. In transitions

Solid-State Materials in Pharmaceutical Chemistry: Properties, Characterization, and Applications, Second Edition. Stephen R. Byrn, George Zografi and Xiaoming (Sean) Chen.

involving mesophases, we would expect less entropy change than that observed for transitions of a crystal to its melt, because of the absence of some type of order. This can be illustrated by expressing the total entropy of fusion or the entropy associated with complete disordering of the crystal, ΔS_f, to the melt as

$$\Delta S_f = \Delta S_{pos} + \Delta S_{orient} + \Delta S_{conf} \tag{7.1}$$

where ΔS_{pos}, ΔS_{orient}, and ΔS_{conf} are the contributions from positional, orientational, and conformational disordering, respectively [5]. It has been shown that each of these contributions to the entropy of fusion can be assigned an approximate entropy change, where ΔS_{pos} is generally in the range of 7–14 J/(K · mol), ΔS_{orient} is in the range of 20–50 J/(K · mol), and ΔS_{conf} is in the range of 7–12 J/(K · mol). Consequently, in carrying out thermal analysis with such crystal mesophase systems, we can roughly estimate the entropy change expected with various crystal mesophase transitions that differ in certain types of order and disorder and possibly identify the form of the mesophase. It should also be pointed out that upon rapid supercooling from the isotropic melt crystal mesophases, like liquids, can be "trapped" into a thermodynamically unstable "glassy" state with a distinct glass transition temperature, T_g. Thus, it is possible to speak of glassy condis crystals, glassy PCs, and glassy LCs [1]. In this regard, it is quite possible to mistakenly consider such glass transition temperatures as being due to the glass transition normally associated with an amorphous solid. Therefore, methods of characterizing such systems, in addition to thermal analysis, are required to detect and characterize mesophase systems. Birefringence and powder X-ray diffraction measurements are particularly useful in this regard.

7.3 Liquid Crystals

LCs are, by far, the most frequently observed and most technologically useful crystal mesophase systems. They play an important role in modern LC electronic display technology, as LC thermometers and in optical imaging tools, and occur in many biological systems [6]. Many proteins and lipids naturally exist in various liquid crystalline forms, as do some important drugs and pharmaceutical excipients [2]. As shown schematically in Figure 7.1, LCs are characterized by the common structural feature of exhibiting long-range orientational order, as evidenced by the alignment of all molecules along a common axis, known as the "director," while losing some or all positional order. As such they tend to be more "liquid-like" than the other crystal mesophases that retain the complete positional order of a crystal lattice. Consequently, molecules in the LC state have diffusive mobility close to that of liquids, but in retaining orientational order and some positional order, they have mechanical and electromagnetic properties more like a crystal. The first evidence for such a system appears to have been reported in 1888 when heating of crystalline cholesteryl benzoate led to two phase transitions rather than the one usually associated with melting of the crystal to the liquid form. As the temperature was initially increased, the first transition yielded a cloudy liquid that turned out to be the liquid crystalline form. Further heating produced a transparent liquid that was shown to be the normal isotropic liquid form of cholesteryl benzoate.

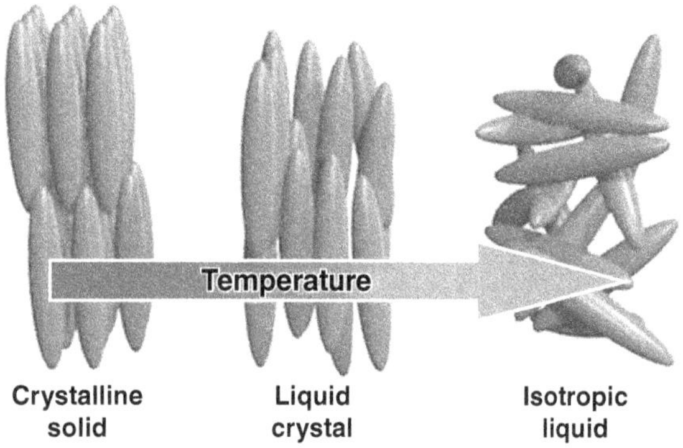

Figure 7.1 A schematic representation of the loss of positional order and retention of orientational order in an LC relative to a crystal and amorphous form. *Source*: [2]/ with permission of Elsevier.

LCs can be broadly divided into two major groups: thermotropic LCs and lyotropic LCs. These are differentiated by the underlying mechanisms that thermodynamically lead to certain types of molecular self-association. Thermotropic systems are those that are induced to form through changes in thermal energy, just as would occur upon heating and cooling crystals and liquids. Indeed, we can speak of enantiotropic LCs that form by both heating the crystal and cooling the liquid and monotropic LCs that can only be formed by either heating the crystal or cooling the liquid but not both. In contrast to thermotropic systems, lyotropic systems require the presence of a solvent to induce self-association. Such systems, for example, are formed in aqueous systems through the hydrophobic interactions responsible for micelle formation and other types of self-associated lipids and surfactants [7]. Lyotropic systems play an important role in the use of lipid formulations as drug delivery systems and occur in certain types of solids that take up water into their crystal structure to form non-stoichiometric hydrates. Some classifications of LCs separate out polymeric LCs as a third group in addition to the thermotropic and lyotropic systems. These are macromolecules that contain monomeric repeating units having the chemical structure characteristic of small molecules that exhibit thermotropic and lyotropic properties. Such systems include aromatic polyesters; cellulosic derivatives, such as hydroxypropyl cellulose; and poly-amino acid peptides such as poly-γ-benzyl-L-glutamate, all capable of exhibiting internal α-helices and other forms of internal secondary structure. The major portion of this discussion of LCs will focus on small molecules that exhibit thermotropic behavior since these systems are the ones that most likely will arise during the evaluation of the solid-state properties of APIs. Lyotropic systems, as they relate to the hygroscopic properties of solids, will be discussed in Chapter 15.

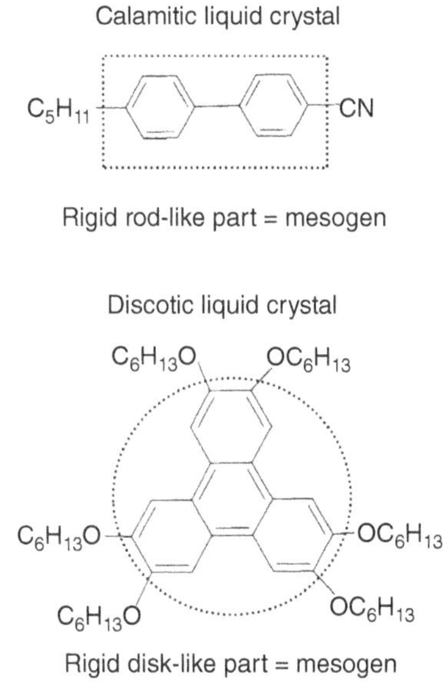

Figure 7.2 Examples of chemical structures that lead to calamitic and discotic liquid crystals.

Molecules that form thermotropic LC systems can be divided into those that form either "calamitic" or "discotic" LCs. Calamitic LCs are formed by rod-shaped molecules that have an elongated, anisotropic shape and high levels of planarity and rigidity brought about by aromatic rings and double and triple bonds within the chemical structure. The portion of the molecule that provides the rigidity is called the "mesogen," which is generally connected to one or more flexible functional groups near the outside of the molecule as depicted in Figure 7.2. The rigid groups produce significant restrictions on bond rotations, allowing the orientational alignment characteristic of LCs, while the flexible groups provide the liquid-like properties characteristic of LCs. As also shown in Figure 7.2, discotic LCs have flat disk-like chemical structures with a central mesogen core of rigid aromatic rings that allow the molecules to form LCs as a two-dimensional column. LCs, both calamitic and discotic, can be structurally categorized further by their ability to possess different combinations of positional and orientational order. The three major categories are (i) nematic LCs, (ii) smectic LCs, and (iii) cholesteric LCs. As shown schematically in Figure 7.3 for a calamitic and a discotic system, nematic LCs have no positional order, but maintain orientational order along a particular director. Lacking any positional order, they are the most liquid-like of the thermotropic LCs. Indeed, the degree of orientational alignment can vary, whereby the less extensive this orientational alignment is, the closer the system will be to an isotropic liquid. Smectic LCs, on the other hand, tend to be more "solid-like" in that they retain some positional order by aligning themselves in layered planes. The variation in positional order can be quite great, and indeed at least 12 different smectic phases have been observed. Two types, the smectic-A and smectic-C phases, occur most commonly and are depicted in Figure 7.3 as calamitic systems. In smectic-A, the orientation of the director is always perpendicular to the plane of molecules, with a loss of positional order in the various planes. In smectic-C, the orientational alignment occurs at various tilt angles relative to the perpendicular direction. Such variations in tilt angle would be expected to give different thermodynamic and dynamic properties that, in turn, would be reflected in differences in physical chemical behavior such as solubility and mechanical properties.

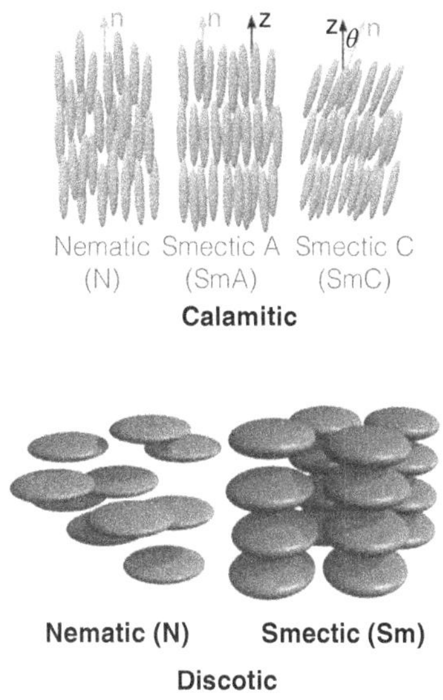

Figure 7.3 Illustration of calamitic and discotic nematic and smectic liquid crystal forms. *Source*: Adapted from [2].

Cholesteric LCs represent the third major group of thermotropic LCs. They are called cholesteric because they were first observed with several cholesterol esters, for example, cholesteryl benzoate. However, other molecules such as hydroxypropyl cellulose and related cellulosic derivatives have also been shown to exhibit such behavior [8]. Cholesteric phases occur with molecules containing chiral groups that can form nematic LCs, generally with a director that tilts in a manner like that of the smectic-C phase. The tilt angle with such molecules allows the LC to form a continuous helix structure with each layer being a thin nematic-like layer, as shown in Figure 7.4. An important characteristic of such cholesteric LCs is the distance associated with each full 360° turn of the director as the helix forms, known as the pitch. The chiral nematic repeats itself every half-pitch or 180°, as shown in Figure 7.4, and it is this pitch and any changes in it that might be changed by external factors that give rise to interesting responses of these systems to electromagnetic radiation and temperature. Indeed, cholesteric LCs are widely used as temperature sensors because of their changing optical properties with temperature.

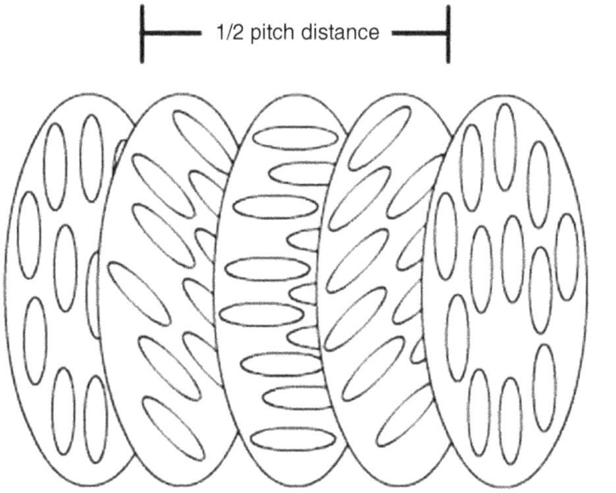

Figure 7.4 Illustration of a cholesteric liquid crystal through half of its pitch.

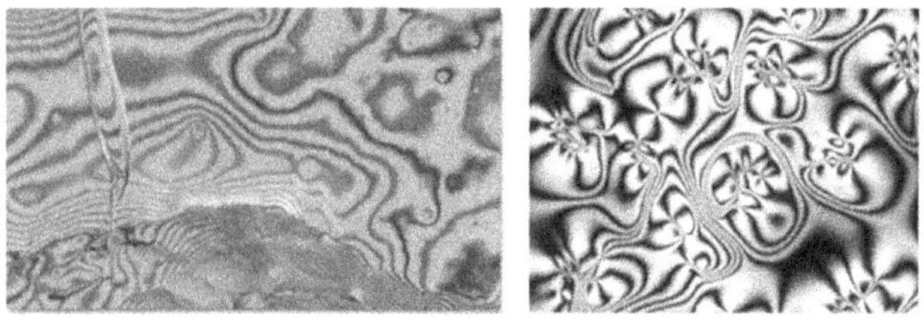

Figure 7.5 Typical birefringence patterns observed for liquid crystals.

We may now ask how LCs are identified and differentiated from crystalline and amorphous solids and liquids. One approach involves detecting the presence or absence of birefringence. Ordinarily, when polarized light is transmitted through a material, its velocity will decrease in proportion to its index of refraction. Birefringence (doubly refracting) arises in anisotropic systems, including most crystals and LCs, because of differences in the index of refraction that occurs in the perpendicular and parallel directions. An isotropic material, like a liquid or an amorphous solid, will have one index of refraction no matter the direction of the transmitted light and, therefore, will not exhibit birefringence. Essentially, the light transmitted through an anisotropic sample splits into two beams of light that are polarized at right angles to each other and are out of phase. The presence of birefringence is typically detected using a polarized light microscope, which is equipped with a polarizer positioned in the light path before the specimen and a second polarizer, called the analyzer and placed between the specimen and the detector. Image contrast arises from the interaction of the plane-polarized light with the birefringent specimen to produce typical birefringent patterns not observed with isotropic materials, as seen in Figure 7.5. The dark and light regions of varying coloration are determined by the nature of the anisotropy, for example, the alignment of the director and the thickness of the sample. Birefringence measurements, therefore, are useful in detecting crystallinity in amorphous samples and differentiating amorphous solids from LC.

As discussed in previous chapters, powder X-ray diffraction techniques are very useful for characterizing the order associated with crystalline materials. As shown earlier in Chapter 6, Figure 6.2, various diffraction peaks occur in the range of roughly 5–40° 2θ because of the different plane orders that exist in a particular crystal structure. This range of 2θ angles is what the X-ray equipment generally used in pharmaceutical research can detect. Amorphous solids, having lost all long-range order, are missing these diffraction peaks and exhibit only a halo appearance with a broad peak at about 20° 2θ (see Figure 6.2). Thermotropic LCs, lacking a certain degree of positional order, most often do not produce diffraction peaks in the range of 5–40° 2θ, thus giving a diffraction pattern that looks very much like an amorphous solid. This is particularly true of nematic LCs, which lack the largest proportion of positional order. Since any order associated with the molecular layering of smectic and nematic phases is generally in the range of 1–100 nm, we would expect that such long-distance order only would be detected at less than 5° 2θ. Indeed, as shown in Figure 7.6, LC systems do produce a diffraction peak at these low angles. To analyze these systems more thoroughly at such small angles, one can use the more specialized small-angle X-ray scattering technique that can detect order in the range of 0.1–5° 2θ [9]. Minimally, therefore, it would be important to carry out both small-angle X-ray and birefringence measurements to distinguish between LCs and amorphous solids.

An important aspect of LCs is their ability to undergo distinct first-order phase transitions and those associated with the glass transition. Critical in this regard is the ability to know what phases might exist at a particular temperature and whether a liquid crystalline system can form more than one type of liquid crystalline phase under a given set of conditions. For example, when might a smectic phase become a nematic phase, or under what conditions can an isotropic liquid form from the heating of a particular LC form? Differential scanning calorimetry is very well suited and convenient for carrying out such studies [1]. Since transitions between phases represent first-order transitions, we would expect to observe a distinct endotherm in going from a more ordered form to one that is less ordered, for example, smectic to nematic, and an exotherm going from the less ordered state to the more ordered state, for example, isotropic liquid to nematic phase. Also, the area under the curve for such endotherms or exotherms reflects the enthalpy change for the phase transition and

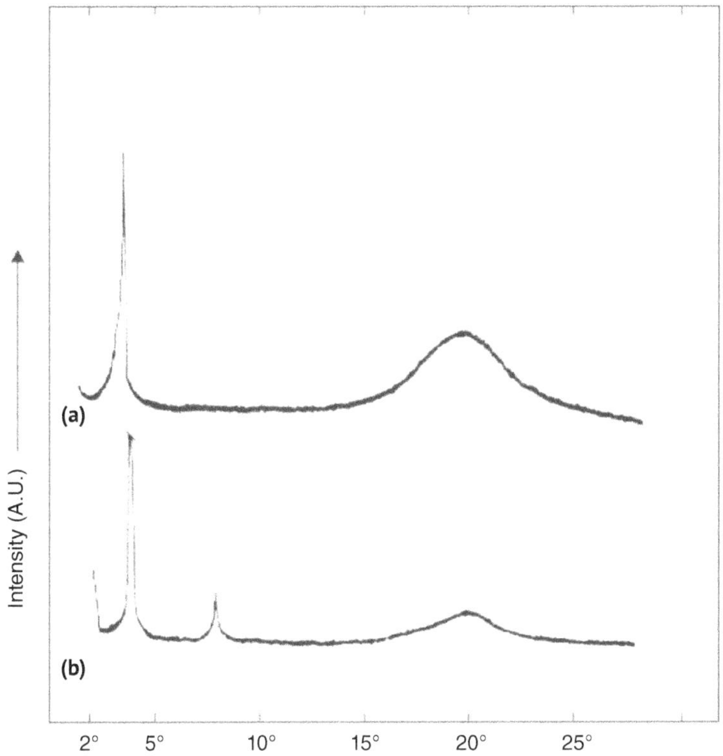

Figure 7.6 Typical X-ray diffraction pattern for a liquid crystal showing small-angle diffraction. *Source*: [11]/with permission of Elsevier.

hence allows a comparison of the intermolecular interactions involved with different phases. Figure 7.7 (top) illustrates an interesting series of thermal events that occur with the sequential heating of crystalline octyloxy cyanobiphenyl that first forms a smectic-A phase at 55 °C, followed by a nematic phase to 67 °C. The barely perceptible endotherm for the smectic-A phase to the nematic phase indicates a very small energetic difference between these phases. This is followed by a small but slightly larger endotherm at about 80 °C for the transition from the nematic phase to the isotropic liquid. This relatively small endotherm is consistent with the earlier observation that the loss of all positional order in a nematic LC brings the structure quite close energetically to that of the isotropic liquid. It is also interesting to note that an exotherm of about the same size as the nematic-to-isotropic liquid transition occurs at slightly below 80 °C upon cooling of the liquid but that the nematic phase upon further cooling appears to go directly to the crystal because of significant supercooling of the smectic-A phase. Such supercooling, presumably due to kinetic effects, could be examined more definitively by carrying out these studies with a wider range of scanning rates and other more subtle experiments. Indeed, as shown in Figure 7.7 (bottom), the author was able to detect the smectic-to-crystal transition exotherm by first heating the sample to just above 55 °C and then cooling to give an exotherm of similar size and temperature to the endotherm. Under such conditions, it is also possible that we would observe the formation of the glassy LC with a distinct glass transition temperature, particularly when rapidly cooling the system [1]. For example, cyanobiphenyl CB5 with the chemical structure shown in Figure 7.2 exhibits a nematic phase, which can form a glassy nematic phase with a glass transition temperature of 305K [10].

Finally, let us conclude this discussion of LCs by considering how what we have learned so far might relate to the task of evaluating solid-state properties of new drug entities. During the preformulation process of characterizing potential new drug entities, it is now common to screen for various polymorphic crystal forms as well as for the conditions that might lead to an amorphous solid. In the case of characterizing a possible amorphous solid, investigators will minimally use powder X-ray diffraction, polarized light microscopy, and thermal analysis, looking for the possible absence of diffraction peaks, a lack of birefringence, and the presence of a glass transition temperature. Additional studies to examine the structural features of the solid can be carried out using vibrational spectroscopy and NMR spectroscopy. Having established that the sample is devoid of any crystallinity, as discussed earlier, there remains the possibility that the sample is an LC and that such a structure may have unexpected physical and chemical characteristics that are determined by the partial order that exists

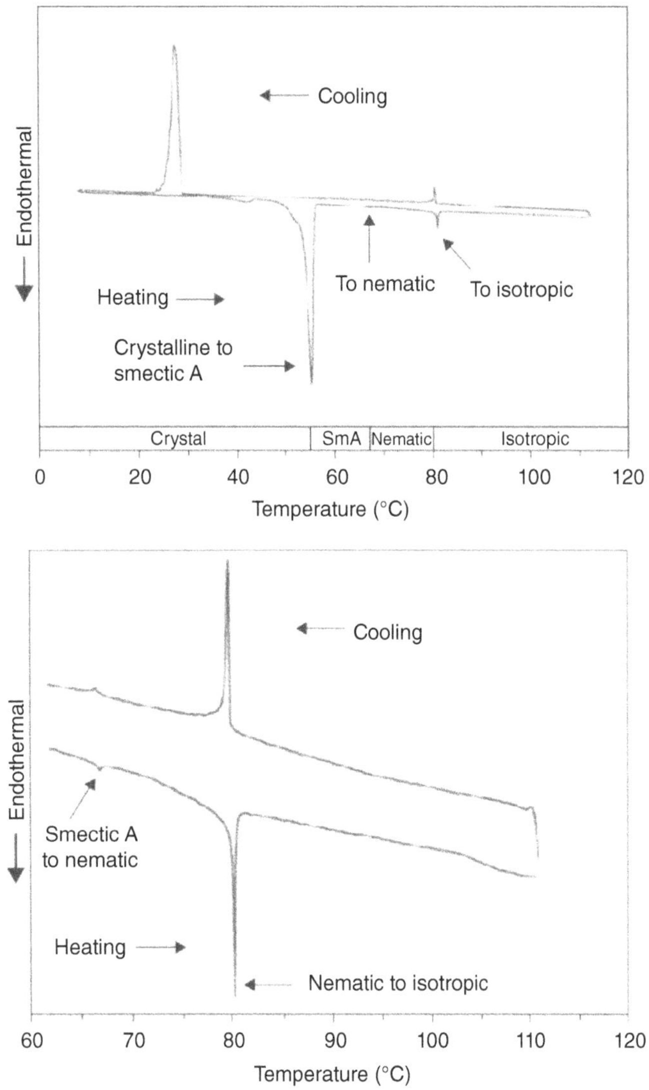

Figure 7.7 Differential scanning calorimetric measurements of various phase changes upon heating and cooling octyloxy cyanobiphenyl; see text for details concerning both plots. Charles Rosenblatt, Case Western Reserve University.

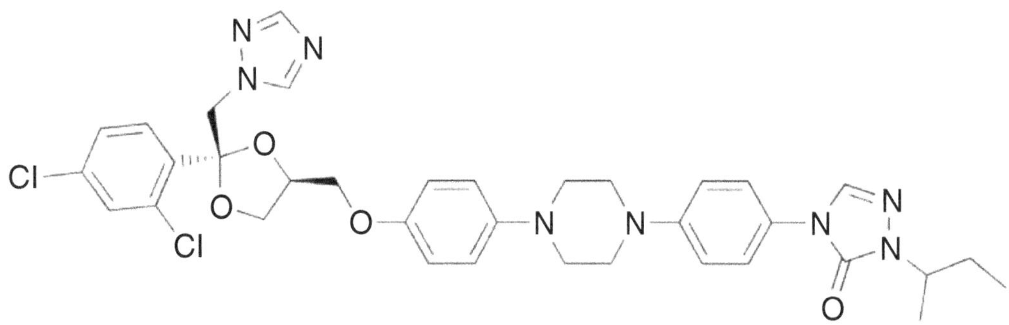

Figure 7.8 Chemical structure of itraconazole.

in the solid. Evidence that many APIs can exist as thermotropic and lyotropic liquid crystalline phases has been compiled in a comprehensive review [2], and indeed many drug molecules do appear to have structural mesogenic characteristics that might lead to LCs under certain conditions. We can illustrate this by examining one study [11] involving the antifungal agent, itraconazole, having the structure given in Figure 7.8. Here, indeed, we can observe a rod-like polyaromatic structure with a relatively large molecular weight and the capability of maintaining significant rigidity in the central portion of the molecule along with greater flexibility at each end. As shown in Figure 7.9, thermal analysis studies revealed that heating the crystal above 90 °C and cooling to below 50 °C yielded a solid that, upon reheating, produced a glass transition at 59 °C and two distinct endotherms at 74 and 90 °C. X-ray diffraction studies revealed that at room temperature the original sample was crystalline but that at 82 °C the X-ray patterns revealed the typical halo at around 20° 2θ and a small peak at 7° 2θ, suggesting a nematic liquid crystalline phase. Further analysis using polarized light microscopy suggested that this molecule was forming a chiral nematic phase. It was concluded that the transition at 90 °C upon cooling represented a transition from the isotropic liquid to the nematic phase, while that at 74 °C was probably caused by further restrictions on the rotational motions of the molecules upon cooling.

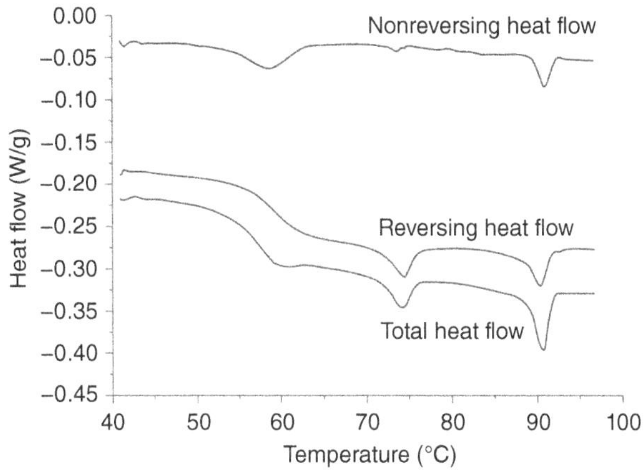

Figure 7.9 Differential scanning calorimetry for the heating of itraconazole crystals above 90 °C, followed by cooling and reheating; see text for details. *Source*: [11]/with permission of Elsevier.

7.4 Conformationally Disordered (Condis) Crystals

As indicated in the Introduction, although the occurrence of conformationally disordered crystals and PCs in pharmaceutical systems appears to be quite rare, some mention should be made concerning the properties of these systems compared to crystals, LCs, and amorphous solids since there is always the possibility that such systems might be encountered during the drug development process. Of the three possible crystal mesophases, conformationally disordered crystals exhibit properties that are closest to those of regular crystals in that they maintain positional and orientational order in the crystal lattice while losing some or all conformational order. Since the disorder associated with these systems arises from the variety of differences that can exist in the orientation of bonds within molecules, the possibility of this occurring is generally greatest for long-chain compounds, particularly linear chain polymers. Such behavior has been reported most often for polymers containing unsaturated groups like, for example, trans-1,4-poly(butadienes) [12] and poly(4-methyl-pentene-1) [13]. The disorder in these systems can arise because of certain orientation-dependent intermolecular interactions or because certain orientations give rise to lower-density packing arrangements not involving specific interactions. Such packing arrangements are often hexagonal. As would be expected, condis crystals exhibit high crystallinity and strong birefringence and, therefore, are best studied by thermal analysis [5].

7.5 Plastic Crystals

PCs generally consist of small molecules that have a compact sphere-like molecular structure. Such molecules generally maintain very good positional and conformational order in the crystal lattice, typically within a cubic crystal structure. They are characterized by the ability of the entire molecule to undergo relatively free rotational motion within the crystal lattice at moderate temperatures. The rotational molecular motion is diffuse and takes place in all directions, unlike the process in LCs, which is principally about one major axis. Because of this, PCs are optically isotropic and do not exhibit birefringence. Furthermore, if the isotropic liquid is rapidly cooled to lower temperatures, such rotational motions can be frozen in to produce a metastable glassy PC with a distinct glass transition temperature, as in the case of adamantane [14]. Because of this structure, PCs are also characterized by a tendency to undergo significant deformation under relatively low levels of mechanical stress and hence the name "plastic crystal." Two well-known examples of molecules that exhibit PC properties are camphor and adamantane, both having the compact sphere-like chemical structures shown in Figure 7.10.

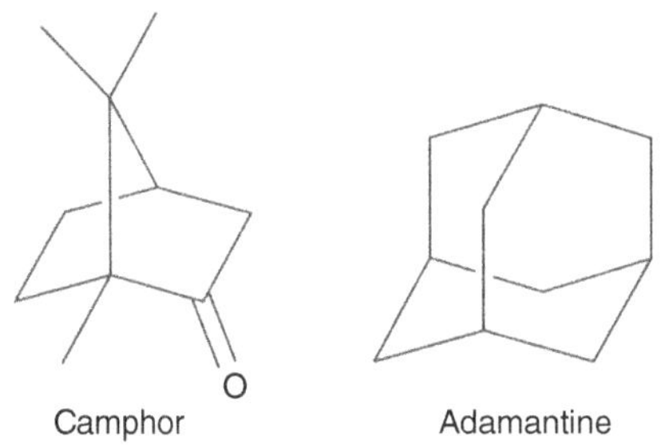

Figure 7.10 Chemical structures of camphor and adamantane.

7.6 Nanocrystals

It is apparent from the discussion of various solid forms in previous chapters that below their melting temperature crystalline solids represent the most thermodynamically stable states of matter. Within this group of crystalline solids, of course, there can be variations in the thermodynamic stability of various polymorphic and solvated forms of the same molecule. Furthermore, the loss of certain structural order within a crystal lattice can give rise to various higher energy crystal mesophases and ultimately to the amorphous state. Within each of these systems, it is also possible for higher energy unstable states, that is, glasses, to form at distinct glass transition temperatures. As indicated in the Introduction section of this chapter, we now would like to consider the properties of purely crystalline solids where the dimensions of the single crystal have been reduced to levels below 100 nm or 0.1 μm. Such systems, generally referred to as nanocrystals [3], exist in a higher state of energy than that of crystals with larger dimensions. They, therefore, can be thought of as being unstable crystalline states. Indeed, if nanocrystals are not properly protected in some manner, they will generally spontaneously undergo crystal growth. This higher energy in nanocrystals is primarily the result of changes in the bulk free energy of the solid due to significant contributions from the surface energy of the system brought about by the very small particle dimensions and the high curvature of the crystal surface, relative to that of the ordinary crystal. An example of the impact of such surface effects on bulk properties of solids can be seen by applying the well-known Gibbs–Thomson equation for conditions that describe the difference between the melting temperature of nanocrystals and that of larger crystals in terms of particle dimensions and interfacial energy. According to this equation, for an isolated spherical particle of diameter, d, at constant pressure, the melting temperature $T_m(d)$ will be depressed from the ordinary melting temperature, $T_m(b)$ by ΔT_m such that

$$\Delta T_m = T_m(b) - T_m(d) = T_m(b)\left[\frac{4\sigma_{sl}}{H_f \rho d}\right] \tag{7.2}$$

where σ_{sl} is the solid–liquid interfacial energy per unit area, H_f is the bulk enthalpy of fusion per gram, ρ is the density of the solid, and d is the particle diameter. Thus, the greater the interfacial energy and the smaller the particle diameter, the greater ΔT_m. Figure 7.11 indicates the effects of particle diameter on the melting temperature of extremely small cadmium sulfide nanocrystals having diameters in the range of 1–5 nm [15]. It shows that to have significant changes in the thermodynamics of solids due to particle size generally requires extremely small particles, well below 100 nm. Indeed, the significant bulk structural change takes place at such small particle diameters can be shown by observing the powder X-ray diffraction patterns of a crystalline material as a function of particle diameter. Figure 7.12 illustrates the X-ray diffraction patterns determined for an inorganic crystalline material, cadmium selenide, as the crystallite size is reduced from 11.5 to 1.2 nm [16]. As can be seen, the individual diffraction peaks undergo continuous broadening with decreasing particle diameter, which is an indication of increasing disorder.

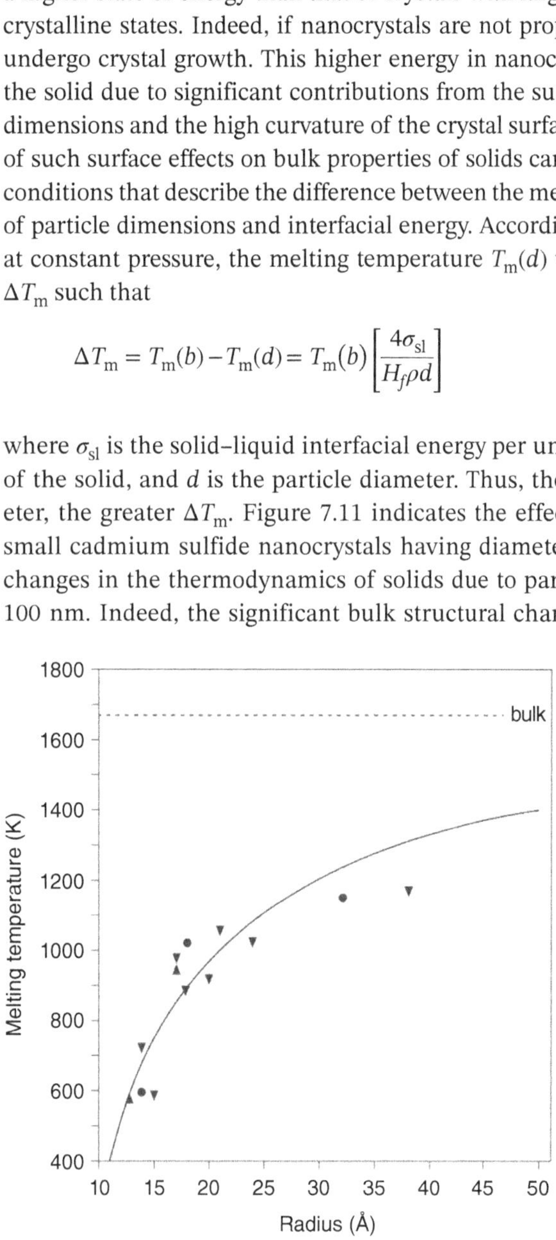

Figure 7.11 Melting temperature of cadmium sulfide nanocrystals as a function of crystallite size. *Source*: [15]/ with permission of American Chemical Society.

From a pharmaceutical perspective, there has been much interest in using nanocrystalline forms of water-insoluble drugs to enhance rates of dissolution beyond that which can be obtained by simply increasing the specific surface area and rather by increasing equilibrium solubility brought about by the extremely small size of nanocrystals [4]. The Gibbs–Thomson equation can be used to determine differences in the equilibrium solubility of nanocrystals and their corresponding larger-size crystal forms at constant temperature in a form that is essentially the same as the Kelvin equation that describes the vapor pressure change for very highly curved liquid surfaces relative to a plane surface at constant temperature [17]. Here, the ratio of nanocrystal solubility, $S(d)$, to bulk solubility, $S(b)$, for spheres is expressed at constant temperature in terms of particle diameter and interfacial energy as

$$\frac{S(d)}{S(b)} = e^{\frac{4\sigma_{sl}M}{d\rho RT}} \tag{7.3}$$

where σ_{sl} is the interfacial energy per unit area, M is the molecular weight, d is the diameter, ρ is the density, R is the gas constant, and T is the temperature. Despite many uncertainties associated with applying the Kelvin equation to solid systems [18, 19], the use of this equation can provide an important general conceptual picture of the degree to which reducing particle sizes into the nanocrystalline range can cause an increase in the equilibrium solubility and hence a certain degree of supersaturation relative to the solubility of the bulk crystal. For example, theoretical considerations indicate that for a monodisperse spherical particle, significant increases in equilibrium solubility should not be expected until the diameter is in the range of 10 – 20 nm or less. For solid systems with particles in the nanometer range, there is likely to be a distribution of particle diameters that covers sizes from over 100 nm to below 10 nm. In such cases, particles on the lower end may likely dissolve as predicted in equation (7.3), while larger particles remain undissolved. If this occurs, those molecules in solution will be supersaturated relative to the solubility of the larger particles, and crystal growth should occur on the larger particle surface. Indeed, this is the basis for the phenomenon known as Ostwald ripening, whereby small particles shrink due to enhanced solubility and larger particles grow. Thus, using nanometer-size particles to enhance equilibrium solubility, as when attempting to increase bioavailability, generally fails to maintain the higher levels of solubility because of Ostwald ripening effects unless a second component can be added to affect the thermodynamics [20]. In recent years, attempts also have been made with some success to kinetically protect nanocrystals from crystal growth and to promote drug delivery by embedding such particles into mesoporous structures such as zeolites, carbon, silicon, silica, and thin polymer films [21–23].

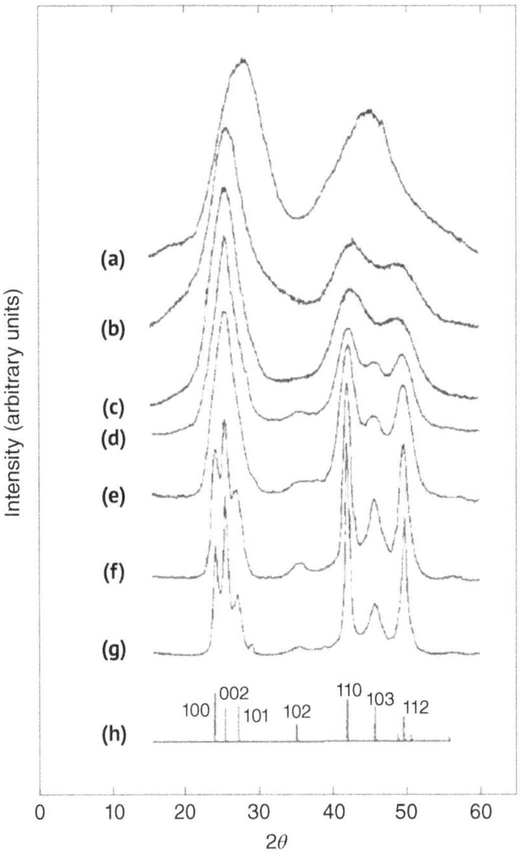

Figure 7.12 Powder X-ray diffraction spectra of CdSe nanocrystals as a function of crystallite particle diameter: (a) 1.2 nm; (b) 1.8 nm; (c) 2.0 nm; (d) 3.7 nm; (e) 4.2 nm; (f) 8.3 nm; (g) 11.5 nm; (h) bulk crystal. *Source*: [16]/with permission of American Chemical Society.

References

1 Wunderlich B. A classification of molecules, phases and transitions as recognized by thermal analysis. *Thermochim Acta* 1999; 340–341: 37–52.

2 Stevenson CL, Bennett DB, Lechuga-Ballesteros D. Pharmaceutical liquid crystals: the relevance of partially ordered systems. *J Pharm Sci* 2005; 94(9): 1861–1880.

3 Pileni MP. Supra-and nanocrystallinities: a new scientific adventure. *J Phys: Condens Matter* 2011; 23(50): 503102.

4 Maravajhala V, Papishetty S, Bandlapali S. Nanotechnology in the development of drug delivery systems. *Int J Pharm Sci Res* 2012; 3(1): 84–96.

5 Wunderlich B. The detection of conformational disorder by thermal analysis. *Pure Appl Chem* 1989; 61(8): 1347–1351.

6 Hoogboom J, Elemans AAW, Rowan AE, Rasing TH, Nolte RJM. The development of self-assembled liquid crystal display alignment layers. *Philos T Roy Soc A* 2007; 365: 1553–1576.

7 Rosevear FR, Liquid crystals: the mesomorphic phases of surfactant concentrations. *J Soc Cosmet Chem* 1968; 19: 581–594.

8 Werbowyj RS, Gray DG. Optical properties of hydroxypropyl cellulose liquid crystals. 1. Cholesteric pitch and polymer concentration. *Macromolecules* 1984; 17(8): 1512–1520.

9 Bradley AE, Hardacre C, Holbrey JD, Johnston S, McMath EJ, Nieuwenhuyzen M. Small-angle X-ray scattering studies of liquid crystalline 1-alkyl-3 methyl imidazolium salts. *Chem Mater* 2002; 14(2): 629–635.

10 Zeller HR. Dielectric relaxation and the glass transition in nematic liquid crystals. *Phys Rev Lett* 1982; 48(5): 334–337.

11 Six K, Verreck G, Peeters J, Binnemans K, Berghmans H, Augustijns P, Kinget R, Van Mooter G. Investigation of thermal properties of glassy itraconazole: identification of a monotropic mesophases. *Thermochim Acta* 2001; 376: 175–181.

12 Van Aerle NAJM, Lemstra PJ, Kanamoto T, Bastiaansen CWM. Solid-state transformations of trans-1,4-polybutadiene. *Polymer* 1991; 32: 34–38.

13 Hohne GWH, Rastogi S, Wunderlich B. High pressure differential scanning calorimetry of poly(4-methyl-pentene-1). *Polymer* 2000; 41: 8869–8879.

14 Affouard F, Willart JF, Descamps M. Orientationally disordered crystals: from the onset of slow dynamics to the glass transition. *J Non-Cryst Solids* 2002; 307–310: 9–15.

15 Alivisatos AP. Perspectives on the physical chemistry of semiconductor nanocrystals. *J Phys Chem* 1996; 100: 13226–13239.

16 Murray CB, Norris DJ, Bawendi MG. Synthesis and characterization of nearly monodisperse CdE (E = S, Se, Te) semiconductor nanocrystallites. *J Am Chem Soc* 1993; 115: 8706–8715.

17 Hammond RB, Pencheva K, Roberts KJ, Auffert T. Quantifying solubility enhancement due to particle size reduction and crystal habit modification: case study of acetyl salicylic acid. *J Pharm Sci* 2007; 96: 1967–1973.

18 Wu W, Nancollas GH. A new understanding of the relationship between solubility and particle size. *J Solution Chem* 1998; 27(5): 521–531.

19 Perez M, Gibbs-Thomson effects in phase transformations. *Scripta Materialia* 2005; 52: 709–712.

20 Liu Y, Kathan K, Saad W, Prud'homme RK. Ostwald ripening of β-carotene nanoparticles. *Phys Rev Lett* 2007; 98: 036102.

21 Salonen J, Latinen L, Kaukonen AM, Tuura J, Bjorkqvist M, Heikkila T, Vaha-Heikkila K, Hirrvonen J, Lehto V-P. Mesoporous silicon microparticles for oral delivery: loading and release of five model drugs. *J Control Release* 2005; 108: 362–374.

22 Mellaerts R, Mols R, Jammaer JA, Aerto CA, Annaert P, Van Humbeek J, Vanden Mooter G, Augustijns P, Martens JA. Increasing bioavailability of the poorly water soluble drug, itraconazole with ordered mesoporous silica. *Eur J Pharm Biopharm* 2008; 69: 223–240.

23 Yan W, Hsiao VKS, Zheng YB, Shariff YM, Gao T, Huang TJ. Toward nanoporous polymer thin film-based drug delivery systems. *Thin Solid Films* 2009; 517: 1794–1798.

8

X-Ray Crystallography and Crystal-Packing Analysis

8.1 Introduction

X-ray crystallography is a powerful tool for the investigation of crystalline solids. In the most favorable cases, it can lead to a complete determination of the structure of the solid and the determination of the crystal-packing relationship among individual molecules in the solid. This knowledge is often crucial to understanding the solid-state chemistry of drugs.

In some respects, X-ray crystallography is analogous to light microscopy. While light microscopy provides an image of objects visible or nearly visible to the naked eye, X-ray crystallography provides an image of objects with atomic dimensions. In light microscopy the image is focused with lenses, whereas in X-ray crystallography the atomic image is determined using Fourier analysis (normally calculated on a computer) of the diffracted radiation whose wavelength is in the atomic size range. This analogy may be helpful to the reader to whom X-ray crystallography is unfamiliar.

In this section, we provide only an overview of crystals and X-ray crystallography. The interested reader should consult one of the books in this area [1, 2] for more details.

8.2 Crystals

As discussed in Chapter 1, crystals contain highly ordered arrays of molecules (or atoms for inorganic crystals). This internal ordering is a fundamental characteristic of crystalline solids and is manifested in the external shapes of crystals. Crystals have sharp edges and flat faces with well-defined angles between them. However, because an object contains sharp edges does not mean it is crystalline (i.e. a window pane is not a crystalline solid but rather a glass, since the atoms are not ordered). The fact that crystals are highly ordered is experimentally demonstrated by showing that a crystal can act as a three-dimensional diffraction grating for X-rays that have a wavelength comparable to the interatomic distances.

Crystals are built up by the repetition of a given unit cell in three dimensions. Sometimes it is convenient to view the unit cell as a box and the crystal as a three-dimensional array of these boxes. Such an array is called the crystal lattice.

For all crystals there are seven classes of unit cells: triclinic, monoclinic, orthorhombic, tetragonal, hexagonal, rhombohedral, and cubic. However, for drugs, only three of these are commonly found: triclinic, monoclinic, and orthorhombic (see Figure 8.1). The triclinic unit cell has $a \neq b \neq c$ with $\alpha \neq \beta \neq \gamma \neq 90°$. For the monoclinic unit cell: $a \neq b \neq c$ with $\alpha \neq \gamma \neq 90°, \beta = 90°$. For the orthorhombic unit cell: $a \neq b \neq c$ with $\alpha = \beta = \gamma = 90°$.

8.3 Miller Indices and Crystal Faces

Figure 8.2 shows a crystal, in two dimensions, formed by orthorhombic unit cells. By inspection, the angles between the faces of the crystal are obviously related to the dimensions of the unit cell, *a* and *b*. The faces labeled (100) and $(\bar{1}00)$ are parallel to the *b*-axis and intersect the *a* axis. (A bar over a number indicates a negative Miller index.) The faces labeled (110), $(1\bar{1}0)$, $(\bar{1}\bar{1}0)$, and $(\bar{1}10)$ intersect both the *a* and *b* axes at 1 unit vector from the origin. The angle between the faces labeled (100) and (110) is dependent on the lengths of *a* and *b*. If $a = b$, then this angle is 45°.

The numbers used to label the faces and planes are termed "Miller indices." The plane with Miller indices *h*, *k*, *l* makes intercepts *a*/*h*, *b*/*k*, and *c*/*l* with the unit-cell axes *a*, *b*, and *c*. Thus, the (100) face has Miller indices 1, 0, 0 and intersects the

Solid-State Materials in Pharmaceutical Chemistry: Properties, Characterization, and Applications, Second Edition. Stephen R. Byrn, George Zografi and Xiaoming (Sean) Chen.

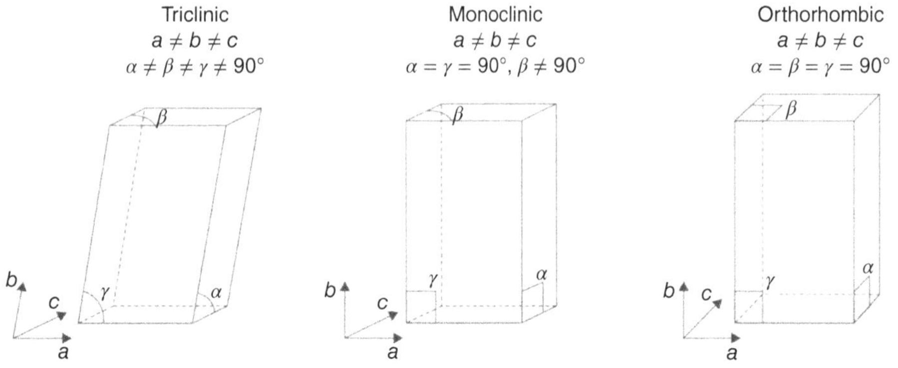

Figure 8.1 The most common unit cells for crystals of drugs.

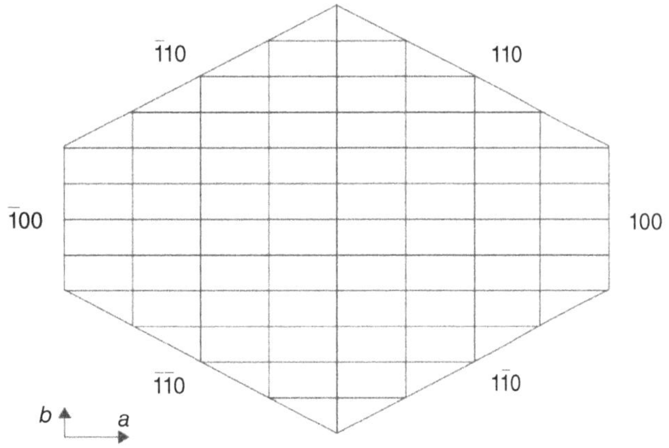

Figure 8.2 A schematic view of a crystal built up from two-dimensional unit cells with axial lengths *a* and *b*. The numbers designate the Miller indices of the crystal faces that are normal to the page. Note that the cells on the (110), $(1\bar{1}0)$, $(\bar{1}\bar{1}0)$, and $(\bar{1}10)$ faces are halved diagonally to produce the diagonal faces. *Source*: Adapted from [3].

a, *b*, and *c* axes at 1, ∞, ∞. Similarly, the (110) plane intersects the *a*, *b*, and *c* axes at 1, 1, ∞. A crystal with a (130) plane will intersect the *a* axis at 1, the *b*-axis at 1/3, and the *c* axis at ∞; a (022) plane will intersect the *a* axis at ∞, the *b*-axis at 1/2, and the *c* axis at 1/2. As indicated earlier, non-intersection with the axis is taken as infinity, and the reciprocal of infinity is, of course, zero. If the plane intersects the axis in a negative direction, then the Miller index is negative. It is important to realize that there are many planes running through a crystal (in theory, an infinite number) and that the Miller indices are used to describe these planes. For example, in Figure 8.2, there are many (100) planes in the crystal, and the (100) plane happens to describe one of the faces of the crystal as well; the same is true for the (110) plane. Although the (010) plane also exists in the crystal, it does not describe one of the faces. Diffraction of X-rays from all the planes in a crystal is responsible for the data that are used to determine a crystal structure and contains information about the atom locations relative to these planes.

It is important to point out a difference typically found between organic and inorganic unit cells in crystals. In a typical inorganic unit cell, such as sodium chloride, atoms lie on the crystallographic planes and the unit-cell faces. For example, in simple cubic structures, atoms usually lie on the (100), (010), and (001) planes. Pictures of these structures are available in many general textbooks. However, in the organic unit cells shown throughout this book, not all of the atoms necessarily lie on the planes. Each set of planes (*hkl*) should instead be thought of as constructs representing the *periodicity* that occurs in the crystal at the corresponding *d*-spacing (the minimum distance between parallel planes) and orientation represented by that set of planes.

The law of rational indices states that the crystal faces have Miller indices that are small whole numbers. This law is extremely helpful in the determination of the Miller indices of the crystal faces and the orientation of the unit cell relative to the external shape of the crystal. By definition, the crystal structure locates the position of atoms within the unit cell. Similarly, the Miller indices define the crystal faces relative to the unit cell. Thus, a combined knowledge of the crystal structure and the Miller indices of the crystal faces allows one to visualize not only the atoms and molecules in the crystal but also the functional groups exposed on the crystal faces. This approach will be used later in this book to describe the orientation of water tunnels relative to the crystal faces. In addition, this approach has been used extensively by Weissbuch et al. [4] to determine the rate of growth and dissolution of different crystal faces. For example, Figure 8.3 shows the crystal packing of *R*,*S*-alanine delineated by the crystal faces as viewed down the *b*-axis and illustrates how the functional groups exposed on the crystal faces can be visualized. The identity of exposed functional groups was used to explain why different faces grow or dissolve faster in different solvents. It was observed that the $(00\bar{1})$ face grew and dissolved faster in aqueous solution, whereas the (101) and $(\bar{1}01)$ faces dissolved and grew faster in methanol. This difference in crystal growth and dissolution was ascribed to the differential interaction of the solvent with the functional groups at the different faces.

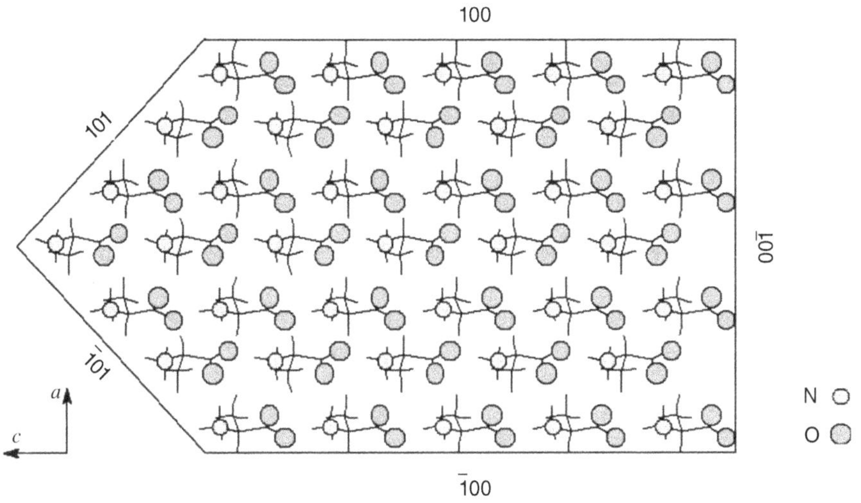

Figure 8.3 Crystal packing of *R,S*-alanine delineated by crystal faces viewed down the *b*-axis. At the $+c$ end, the (101) and ($\bar{1}$01) faces have $-NH_3^+$ and CH_3 groups exposed, whereas, at the $-c$ end, the ($00\bar{1}$) face has $-CO_2^-$ groups exposed. *Source*: [3]/with permission of American Chemical Society.

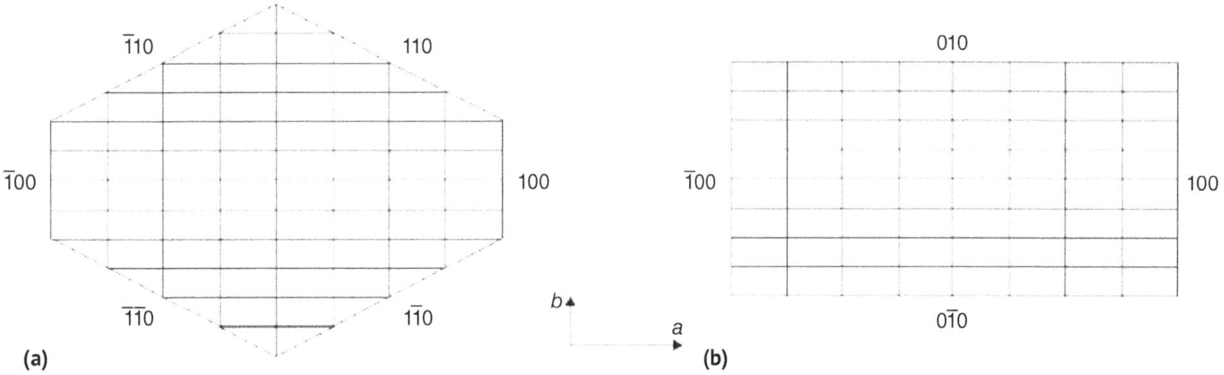

Figure 8.4 Two crystal habits composed of the same two-dimensional unit cell with lengths *a* and *b* but with different external shapes. Habit (a) yields polyhedral crystals, and habit (b) yields rectangular crystals. *Source*: [5]/with permission of Elsevier.

At this point it should be noted that a given unit cell can be propagated in different ways to produce crystals with different external shapes but retain the same internal structure, as shown in Figure 8.4. Such changes are due to differential crystallization conditions and are the consequence of different growth rates at the different faces of the crystal.

8.4 Determination of the Miller Indices of the Faces of a Crystal

The crystal structure describes the position of the atoms of the molecule relative to the unit cell. However, as already discussed, it is often important to know the relationship of the orientation of the unit cell relative to the faces of the crystal. Figure 8.4 describes two habits of the same crystal and shows how the unit cell is related to the faces of the crystal. For these two habits, it is possible to determine the indices of the faces by inspection. For an unknown crystal, the Miller indices of the faces may be unequivocally determined from the interfacial angles (determined from reflective goniometry), the law of rational indices, and X-ray photographs perpendicular to one of the faces of the crystal (or analysis of the faces of the crystal on a diffractometer).

The typical reflective goniometer (see Figures 8.5 and 8.6) [1, 2] consists of four elements: a light source, a crystal support system for the goniometer head (see Figure 8.7), a microscope/telescope for observing the crystal by reflected light, and two circles (graduated in degrees) perpendicular to each other that allow observation of all faces of the crystal (except

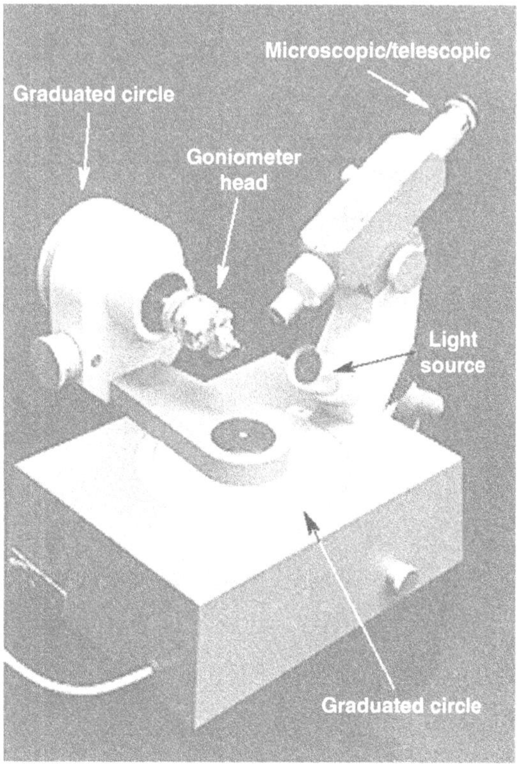

Figure 8.5 A Huber reflective goniometer. *Source*: Reproduced from [1]/with permission of John Wiley & Sons.

those near the point of attachment). These two circles, thus, allow all crystal faces to be positioned such that their normals bisect the angle between the light source and the telescope. In such a position, the crystal face reflects the light.

These reflections are then observed, and the angle is recorded. Usually, a crystal 0.2–0.5 mm on a side is used for these measurements, although slightly smaller crystals can also be studied. The best approach to determining the interfacial angles is to use the axes of the goniometer head to adjust the crystal so that a zone of faces is parallel to the spindle axis. For example, the crystal should be adjusted so that all of the faces on the sides of the crystal are in reflecting position as the crystal is rotated about the spindle axis. Then, the angles between the faces in this zone are measured. Next, the angle between this zone and another zone is measured, and the process is repeated until all of the interfacial angles have been measured.

Once all the interfacial angles have been determined, the crystal is transferred to the precession camera or an X-ray diffractometer, and an X-ray photograph is taken with the X-rays normal to one of the faces. The photograph is examined to see whether it contains reflections that give unit-cell parameters and symmetries corresponding to two of the unit-cell parameters. If this is so, then the face is parallel to these axes and, thus, perpendicular to the third axis (in the orthorhombic crystal system). Thus, if the precession photograph contains the a and c axes, then the face is parallel to a and c and perpendicular to b and has Miller indices of (010). Once the indices of one of the faces have been determined by this procedure, a trial-and-error procedure can be used to determine the indices of the other faces.

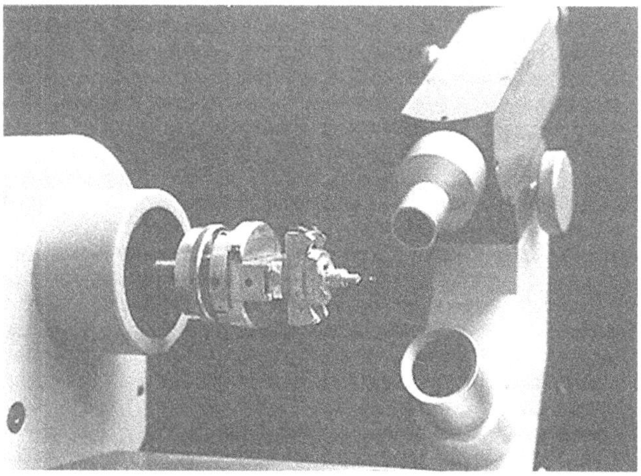

Figure 8.6 A Huber reflective goniometer with a crystal mounted on a two-arc goniometer head. *Source*: Reproduced from [2]/with permission of Oxford University Press.

An alternative procedure involves the use of modern "morphology" computer programs that calculate a theoretical morphology for a given crystal based on the crystal structure and unit-cell parameters. The calculated morphology can be compared to the observed morphology, and the Miller index of one of the faces is checked to ensure that the calculation is correct. A major advantage of these programs is the perspective drawings that can be produced.

8.5 Determination of Crystal Structure

Approaches to the determination of crystal structures are outlined in this section. Crystal structure determination provides important and extremely useful information. However, it is not always possible to grow suitable single crystals of a substance, and the structure information remains unavailable in these cases.

8.5.1 Diffraction of X-Rays

X-rays are scattered by objects in the same way that visible light is scattered. However, there are no known lenses that focus X-rays, so that the determination of the image of the object is not possible directly.

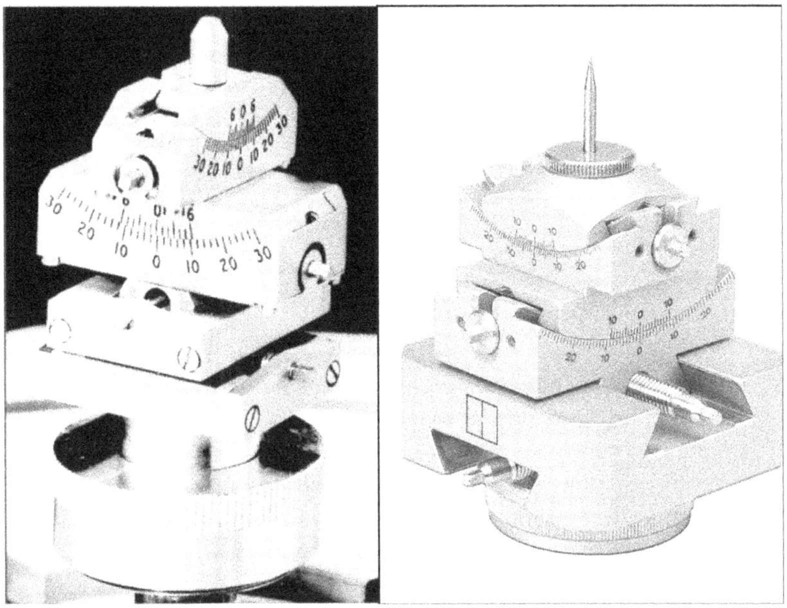

Figure 8.7 Goniometer heads (left: Charles Supper, Co., 1997; right: Huber Diffraction Technology, Ltd, 1998).

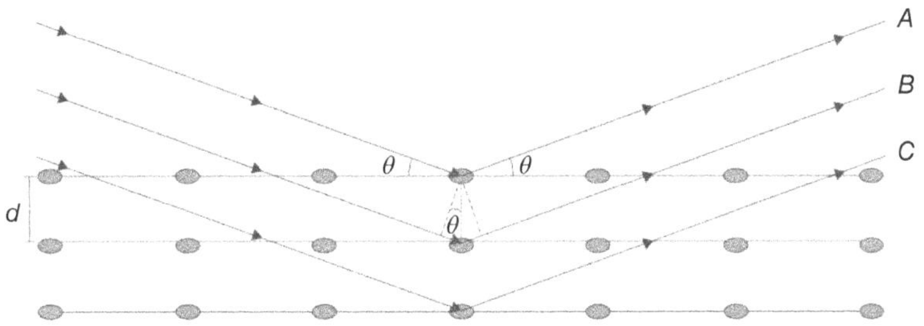

Figure 8.8 X-rays diffracted by planes with spacing d at a diffraction angle of θ.

The pattern of scattered radiation unique to each crystal structure is called the diffraction pattern and was first interpreted by Bragg [6]. He showed that the distribution of scattered radiation could be understood by considering that the diffracted X-ray beams behave as if they were reflected from planes passing through the crystal, as shown schematically in Figure 8.8. If the radiation is viewed as a wave, then the three rays depicted as A, B, and C will reinforce each other only if their path lengths differ by exactly one wavelength (λ).

$$n\lambda = 2d\sin\theta \tag{8.1}$$

This relationship is described by the Bragg equation (equation [8.1]), which states that a multiple (n) of the wavelength (λ) must equal twice the distance (d) between planes multiplied by the sine of the angle of incidence. Thus, for a single crystal and monochromatic X-rays, diffraction for each of the various planes hkl will only be observed for certain values of the angle of incidence (θ). If the crystal is rotated in the X-ray beam, diffraction will occur only at certain rotation angles. The Bragg equation does not predict the intensity of the diffracted radiation. This is determined by the electron density of the atoms in the plane causing the diffraction.

8.5.2 Experimental Measurements

The elucidation of a crystal structure requires the determination of the unit-cell dimensions and the intensities of the diffracted beams from the numerous different diffraction planes of the crystal. The first step is to select a single crystal under a microscope. The crystal is then mounted on a goniometer head with a suitable adhesive.

The intensity of the diffracted radiation, diffracted from the various planes of the crystal, is most conveniently measured using a computer-controlled diffractometer or a diffractometer equipped with a charge-coupled device (CCD) area detector. In a diffractometer using a conventional detector, the intensity of each diffracted beam is measured by an electronic detector as the crystal is positioned in every experimentally accessible diffraction angle with respect to the X-ray beam. If a CCD area detector is used, many reflections are measured simultaneously.

8.5.3 Determination of Space Group Symmetry

The arrangement of molecules in the unit cell is described by one of 230 space groups. All of the 230 possible space groups, their symmetries, and the symmetries of their diffraction patterns are compiled in the *International Tables for Crystallography* [7].

It is possible to determine the space group and the unit-cell type from the intensities of the diffracted X-rays of a crystal by comparing the observed pattern of missing diffraction intensities (called systematic absences) to that predicted for the space group. A knowledge of the space group can simplify the analysis of the diffraction pattern and may reveal an important symmetry element or elements within the unit cell. For example, a determination of a space group in the laboratory of one of the authors (Byrn SR) revealed that the synthetic procedure used to prepare the compound involved a racemization step since the crystals belonged to a space group that required the presence of both *R* and *S* isomers.

8.5.4 Calculation of the Density of the Crystal

The calculated density (D_c) of the crystal can be evaluated from the unit-cell dimensions and the molecular weight of the molecule and any solvent atoms present in the crystal, as described by equation (8.2). This is the true density where D_c is in g/cm^3, M is the molecular weight (in g/mol) of the molecule plus any solvent molecules incorporated into the crystal, Z is the number of crystallographically unique molecular components in the unit cell, and V is the unit-cell volume (in $\AA^3$) calculated from the unit-cell parameters. The number 0.60221 is a conversion factor incorporating Avogadro's number to obtain the correct units of g/cm^3.

$$D_c = \frac{M\,Z}{0.60221\,V} \tag{8.2}$$

The calculated density can be compared to the measured density determined by helium pycnometry.

8.5.5 Structure Determination

The crystal structure provides the atomic locations of all atoms held in regular, repeating positions in the crystal. The problem that must be solved to convert the experimentally determined intensities of the diffracted rays into a structure of the crystal is termed "the phase problem." Solution to this problem involves determining the direction and magnitude of the phase of each recorded intensity [1, 2].

The phase problem can best be understood in terms of equations. To convert the experimentally determined intensities into an image of the atoms in the unit cell, one performs a three-dimensional Fourier summation.

$$\rho(xyz) = \frac{1}{V}\sum_h \sum_k \sum_l F_{hkl} e^{-2\pi(hx+ky+lz)} \tag{8.3}$$

where ρ is the electron density at point (x, y, z), V is the volume of the unit cell, and F_{hkl} is the structure factor for each plane of the corresponding Miller index *hkl*. The structure factor depends on both the intensity and the phase of the diffracted radiation. While the intensities are easily determined, there is no experimental method to determine the phases.

With relatively simple structures, trial-and-error methods can be applied successfully. However, as might be expected, rather complex and elaborate methods have evolved for determining crystal structures. These include the heavy-atom method, the isomorphous replacement method, and direct methods. The heavy-atom method is used to determine the structures of inorganic and organic crystals containing heavy atoms (atoms with atomic weight >17). The isomorphous replacement method is used for the determination of the crystal structures of proteins. Because direct methods are normally used to determine the crystal structures of most drugs, we will confine further discussion to that method.

The direct method is based on the fact that it is possible to derive relationships among the phases of different reflections. These relationships predict the phases of the reflections and also provide a probability that the phase is correct.

The phases are used to calculate the electron density using the Fourier summation shown earlier. In some cases, several electron-density maps must be calculated and inspected until a chemically reasonable structure can be recognized.

Once rough atomic positions are determined from the electron-density maps, the structure is refined by least-squares minimization of the difference between the calculated intensities and the observed intensities. When this refinement process converges, accurate atomic positions and atomic displacement parameters (called temperature factors in the older literature) are obtained.

The atomic positions obtained can then be used to construct pictures of the molecule and/or the contents of the unit cell or several unit cells. These pictures illustrate the structural features of the molecule. In addition, drawings of the contents of contiguous unit cells allow a visual examination of the intermolecular interactions among molecules in the crystal. These intermolecular interactions are often responsible for the solid-state chemistry of drugs, as we shall see throughout this book.

8.5.6 Crystal-packing Drawings

Crystal-packing drawings are a useful way to visualize the arrangement of the molecules in the crystal. These drawings are made by using the atomic positions obtained from the structure determination to show the contents of the unit cell or contiguous unit cells, as shown in Figure 8.9. Display of these drawings on a graphics workstation is a particularly useful way of determining which chemical groups are in close contact within the unit cell. In addition, there are several graphics programs that permit rotation of crystal-packing drawings on the screen, affording the visualization of the details of the molecular packing.

Photomicrographic studies of crystals show that sometimes the crystal behaves anisotropically during a solid-state reaction. For example, during desolvation, some crystals show the anisotropic advancement of an opaque reaction zone through the crystal. When anisotropic behavior is observed, it is often useful to examine crystal-packing drawings viewed from a direction perpendicular to the faces of the crystal. Such drawings require a knowledge of the crystal structure and the Miller indices of the crystal faces determined by reflective goniometry. Since the crystal structure gives the location of the atoms relative to the unit cell and the Miller indices describe the orientation of the unit cell relative to the crystal faces, it is only a routine geometric problem to view the molecules from a direction perpendicular to any crystal face. For example, Figure 8.9 shows a crystal-packing drawing of 5-nitrouracil viewed perpendicular to the *bc* crystal plane.

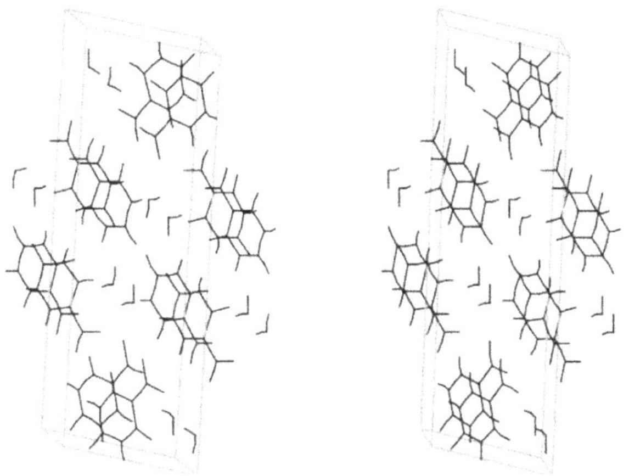

Figure 8.9 Stereoscopic packing drawing of 5-nitrouracil viewed down the *a* crystallographic axis [8]. Viewed through a stereoscope, the drawing appears three-dimensional. *Source*: Adapted from [8].

8.5.7 Atomic Displacement Parameters and Molecular Mobility

In addition to crystal-packing drawings, the refined X-ray structure provides atomic displacement parameters for each atom. These parameters represent the time-averaged displacement of each atom from its equilibrium position. Figure 8.10 shows a typical *Ortep* drawing [9] of the steroid 11β, 17α-dihydroxy-21-oxobutoxypregn-4-ene-3,20-dione (Form II), showing its atomic displacement parameters as thermal ellipsoids drawn at the 50% probability level [5]. From this drawing, it is clear that the ring atoms in this steroid have significantly less mobility than the side chain atoms, an observation that is somewhat

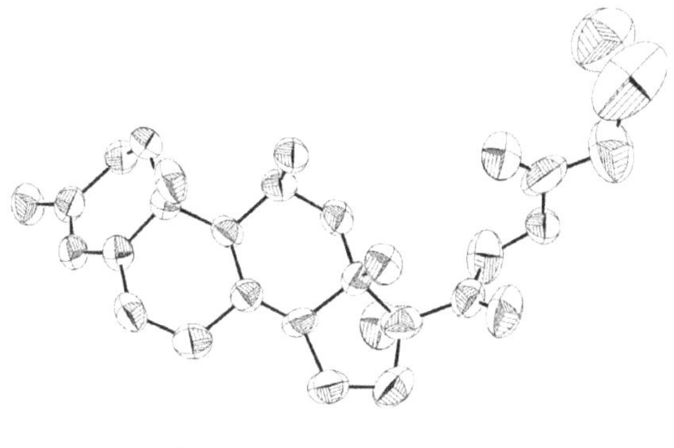

Figure 8.10 *Ortep* drawing of the steroid 11β, 17α-dihydroxy-21-oxobutoxypregn-4-ene-3,20-dione (Form II). *Source*: Adapted from [5].

general for ring compounds that contain side chains. Based on the steps generally involved in a solid-state reaction, it is clear that a knowledge of thermal motion will provide information on the molecular "looseness" of the structure and might provide additional insight into solid-state reactivity. Much more research needs to be done to explore this concept in detail.

8.6 Variable-Temperature X-ray Studies

Variable-temperature (VT) X-ray crystallographic studies have been of interest for many years due to the pioneering work of Trueblood and co-workers. Much of the pioneering work was done in the 1970s and 1980s and is reviewed here. A great deal of information has been presented in two reviews [10, 11]. In biological areas, VT-X-ray crystallography has even been used to "freeze enzyme reactions as they are occurring" [12]. In the pharmaceutical solids area, VT-X-ray crystallography has been used to gain more information about the molecular motion of the functional groups in the molecule and, ultimately, to shed light on the motion of the molecules in the crystal. In this regard, Trueblood and Dunitz [13] have shown that the numerical values of atomic displacement parameters can be treated by simple harmonic oscillator theory and can provide force constants for vibration and motion of molecules. These force constants, as well as the atomic displacement parameters, can provide important information on the motion of molecules. For example, the crystal structure of ibuprofen, indomethacin, and a 4-aminobenzoate ester has been determined, each at several temperatures [14]. The data indicate that the temperature factors vary linearly with temperature until very low temperatures, as expected based on Trueblood and Dunitz's work. Both indomethacin and ibuprofen contain multiple methyl groups. The force constants for vibration of these methyl groups, as determined from variable-temperature X-ray studies, are in the same order as the activation energies for the motion of these groups as determined by solid-state NMR spectroscopy.

References

1 Stout GH, Jensen LH. *X-ray Structure Determination, A Practical Guide*. New York: John Wiley and Sons; 1989.
2 Glusker JP, Trueblook KN. *Crystal Structure Analysis, A Primer*, 2e. New York: Oxford University; 1985.
3 Shimon LJW, Vaida M, Addadi L, Lahav M, Leiserowitz L. Molecular recognition at the solid-solution interface: a 'relay' mechanism for the effect of solvent on crystal growth and dissolution. *J Am Chem Soc* 1990; 112: 6215–6220.
4 Weissbuch I, Addadi L, Leiserowitz L. Molecular recognitions at crystal interfaces. *Science* 1991; 253: 637–645.
5 Byrn SR, Kessler DW. The solid state reactivity of the crystal forms of hydrocortisone esters. *Tetrahedron* 1987; 43: 1335–1343.
6 Bragg WL. Diffraction of short electromagnetic waves by a crystal. *Proc Cambridge Philos Soc* 1913; 17: 43–57.
7 Hahn T. *International Tables for Crystallography*. Boston, MA: International Union of Crystallography, D. Reidel; 1987.
8 Craven BM. The crystal structure of 5-nitrouracil monohydrate. *Acta Cryst* 1967; 23: 376–383.
9 Johnson CK. *ORTEP II, A FORTRAN thermal-ellipsoid plot program for crystal structure illustrations*. Oak Ridge, TN: Oak Ridge National laboratory; 1976.
10 Dunitz JD, Maverick EF, Trueblood KN. Atomic motions in molecular crystals from diffraction measurements. *Angew Chem Intl Ed* 1988; 27: 880–895.
11 Dunitz JD, Schomaker V, Trueblood KN. Interpretation of atomic displacement parameters from diffraction studies of crystals. *J Phys Chem* 1988; 92: 856–867.
12 Petsko GA. Progress and problems in the study of protein dynamics by X-ray diffraction. In: Ehrenberg RRA, Gräslund A, Nilsson L, eds. *Structure, Dynamics and Function of Biomolecules. The First EBSA Workshop, A Marcus Wallenberg Symposium*. Berlin: Springer-Verlag; 1987.
13 Trueblook KN, Dunitz JD. Internal molecular motions in crystals. The estimation of force constants, frequencies and barriers from diffraction data. A feasibility study. *Acta Cryst, Sect B, Struct Sci* 1983; B39: 120–133.
14 Toma PH. Ph.D. Thesis, Purdue University, Atomic Mobility of Selected Pharmaceutical Compounds, 1993.

9

X-Ray Powder Diffraction

9.1 Introduction

X-ray powder diffraction (XRPD) is a technique using X-ray diffraction that allows one to characterize the phase properties of pharmaceutical materials in powder form. As already discussed, pharmaceutical materials can be amorphous, crystalline, mesophases, or perhaps mixtures. Typically, these different materials can be easily distinguished by XRPD, since an amorphous form lacks long-range order and displays diffuse X-ray diffraction, whereas, conversely, a crystalline solid has an ordered structure, showing a distinct diffraction pattern with peaks at various diffraction angles. It can be shown that an XRPD pattern is directly related to the crystal structure of a material, including unit cell dimensions, symmetry, and molecular packing. Thus, XRPD is a key technique for identifying crystalline forms of an active pharmaceutical ingredient (API), providing a distinct "fingerprint" for every crystalline phase, such as polymorph, solvate, hydrate, salt, and cocrystal. XRPD has many applications in characterizing pharmaceutical materials, including the following:

- discriminating crystalline and amorphous materials,
- evaluating the degree of crystallinity,
- differentiating different crystal forms of a compound,
- confirming whether samples are in the same crystal form,
- investigating phase transformation such as dehydration, hydration, polymorph conversion, and the crystallization of amorphous materials,
- quantitatively analyzing the content of polymorphs, hydrates, and amorphous forms in a crystalline sample,
- determining crystal structure from a powder sample, and
- analyzing amorphous materials using pair distribution function (PDF) analysis.

In this chapter, we will discuss the underlying physics of powder diffraction related to the molecular structure of pharmaceutical materials and showcase qualitative/quantitative applications of XRPD in characterizing pharmaceutical solids. We will present some advanced applications of this technique, such as Rietveld analysis, solving crystal structures using powder patterns and studying the structure of amorphous materials by determining their PDFs. Finally, we will briefly describe various types of instrumentation involving XRPD, including synchrotron X-ray diffraction and variable temperature XRPD (VT-XRPD).

Experimentally, XRPD patterns exhibit diffraction intensity as a function of the diffraction angle θ or 2θ (see the previous chapter for some background). Intensity is typically expressed in counts or counts per second.

9.2 XRPD of Crystalline Materials

The characteristics of the X-ray diffraction pattern of a crystalline phase are directly related to its periodic atomic or molecular packing, as described in Chapter 8. Three-dimensional periodic packing of the unit cell can result in many groups of lattice planes. Lattice planes cause incident beams of X-rays to interfere with one another as they leave the crystal, leading to X-ray diffraction.

As discussed in Chapter 8, X-ray diffraction of a crystalline phase follows Bragg's law [1]. As depicted in Figure 9.1, the X-ray waves are scattered from a group of lattice planes separated by an interplanar distance of d, where two scattered waves

Solid-State Materials in Pharmaceutical Chemistry: Properties, Characterization, and Applications, Second Edition. Stephen R. Byrn, George Zografi and Xiaoming (Sean) Chen.

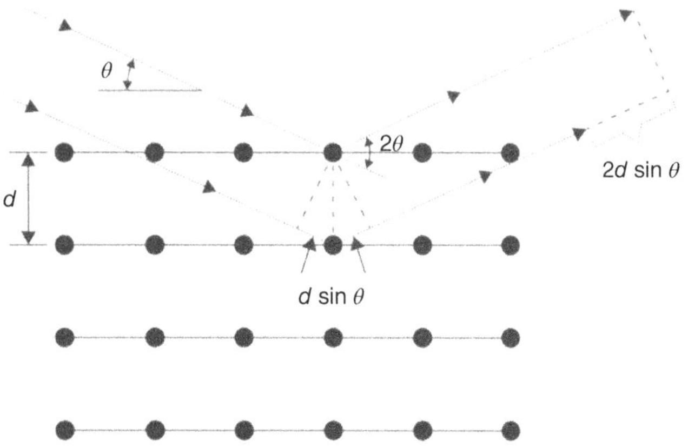

Figure 9.1 X-rays diffracted by planes with inter-plane spacing d at a diffraction angle of θ according to Bragg's law.

will interfere constructively when the path difference is an integer multiple of the wavelength of the X-ray. The path difference between two waves is equal to $2d \sin\theta$, where θ is the angle of incidence of the X-ray beam. Bragg's law is given as follows:

$$n\lambda = 2d \sin\theta \quad (9.1)$$

where n is an integer determined by the order given and λ is the wavelength of the incident X-ray beam. The diffraction angle for a crystalline plane *hkl* is related to its d-spacing and the wavelength of X-rays. A shorter d-spacing will result in a high angle of diffraction, whereas a longer d-spacing will produce a low angle of diffraction. Because the wavelength directly affects the diffraction angle, a monochromatic X-ray is routinely used to ensure specificity.

The most widely used X-ray monochromatic radiation for powder diffraction is copper Kα, which has a wavelength of 1.5418 Å. The copper Kα radiation is the weighted average of Kα1(1.5406 Å) and Kα2 (1.5444 Å). Several other useful sources of Kα radiations are molybdenum at 0.7107 Å, cobalt at 1.7902 Å, and iron at 1.9373 Å.

For the same lattice plane of a crystalline phase, different sources of X-ray radiation will result in different diffraction angles. For example, the 220-lattice plane of sodium chloride has a d-spacing of 1.9707 Å. The diffraction angle is 23° ($2\theta = 46°$) with copper Kα radiation and 10° ($2\theta = 20°$) with molybdenum Kα radiation.

There are two principal applications of X-ray diffraction for characterizing pharmaceutical materials: X-ray crystallography and XRPD. In X-ray crystallography, diffraction signals collected on a single crystal are used to determine the structure of a crystalline phase (Chapter 8). In XRPD, powder samples are analyzed to collect diffraction patterns that can be used to identify distinct crystal forms. The pattern gives the appearance of many discrete diffraction peaks, where useful information such as the position of the diffraction peaks, the peak intensities, the shape of peaks, and the peak intensity distribution can be determined.

As an example, the diffraction pattern of sodium chloride, collected in a 0.2 mm capillary on a STOE STADI P diffractometer, is shown in Figure 9.2 [2]. It is interesting to observe that the reflections with *hkl* even are much stronger than those

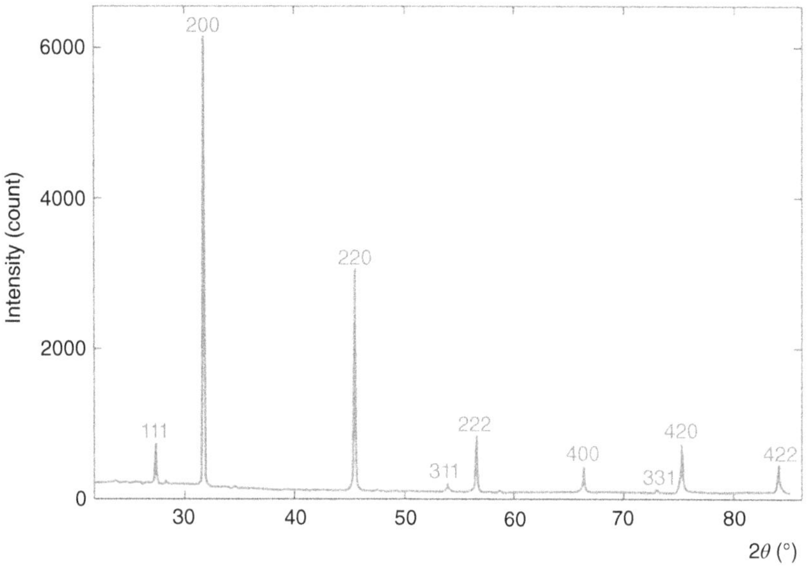

Figure 9.2 XRPD pattern of sodium chloride. The diffraction peaks given by the diffraction lattice planes are labeled. *Source*: Adapted from [2].

with *hkl* odd because the scattering factors of Na and Cl are additive for *hkl* even, and the scattering factors of Na and Cl are opposite for *hkl* odd. This means that the diffraction intensities of the Bragg peaks are related to the atomic positions in the unit cell.

The diffraction intensity of a lattice plane, *hkl*, is linearly related to the structure factor F_{hkl} based on the following equation:

$$I_{hkl} \propto F_{hkl} \tag{9.2}$$

The structure factor is a mathematical function relating the amplitude and phase of radiation diffracted from the lattice plane *h,k,l*.

$$F_{hkl} = \sum_j f_j \exp\left[2\pi i\left(hx_j + ky_j + lz_j\right)\right] \tag{9.3}$$

The structure factor F_{hkl} sums the results of scattering from all the atoms in the (*hkl*) plane of atoms in the crystal lattice to form a diffraction peak. The amplitude of the diffraction peak is determined by the position of atoms on the lattice plane as defined by the fractional coordinate's x_j, y_j, z_j and which atoms are on the lattice plane.

Diffraction intensity is also affected by sample presentation, such as sample size, particle size, and orientation. A random orientation of particles offers a relatively equal opportunity for various lattice planes to provide a diffraction signal. For each lattice plane, there is a small percentage of particles that are properly oriented to diffract. Insufficient sample size, large particles, or needle or plate morphology often challenge obtaining an equal statistical distribution of different lattice planes and cause certain lattice planes to obtain favored diffraction opportunities, resulting in a relatively higher intensity. In reality, only a small portion of crystallites contribute to the observed diffraction pattern for a specific diffraction peak, since many crystals not oriented properly will not produce any diffraction signals. Spinning the sample can help increase the number of crystallites that contribute to the measured diffraction pattern.

The shapes of diffraction peaks are usually bell-like; however, the peak shapes are affected by sample quality and instrumental parameters. Poor crystallinity and small particle size usually result in a broad peak shape. If the radiation is not monochromatic, such as using a combination of Kα1 and Kα2 components, the resulting peaks will include the contributions from both components and exhibit asymmetrical shapes. The shapes of diffraction peaks are also affected by various instrumental and sample variables and are rarely described well by simple Gaussian or Lorentzian functions. Such asymmetric shapes are located between the Gaussian and Lorentzian distributions and can be well modeled using a pseudo-Voigt function or a Pearson VII function [3, 4].

9.3 Qualitative Analysis of Crystalline Materials

XRPD patterns are generally used as the "fingerprints" for crystalline materials because each diffraction pattern is unique and directly related to crystal structure.

XRPD has been used as a primary method to characterize different crystal forms of an API, including polymorphs, salt forms, solvates, hydrates, and cocrystals. In combination with other characterization techniques, such as thermal analysis, microscopy, and spectroscopy, XRPD provides key information for a complete understanding of the nature of a solid form.

The polymorphs of ritonavir, for example, are easily differentiated by XRPD. As shown in Figure 9.3, Form I and Form II of ritonavir have distinctly different diffraction patterns [5], with many unique diffraction peaks below a 30° 2θ angle from each form. Form I has characteristic peaks at 3.3, 6.8, 8.4, and 24.5° 2θ, whereas the unique peaks for Form II are at 9.5, 9.8, 16.1, and 22.2° 2θ. Peaks above 30° 2θ are less intense and have significant overlapping for the two forms. As described earlier, each diffraction peak is related to a specific lattice plane within the crystal lattice. For example, the peak of 6.8° 2θ of Form I is the diffraction peak of the lattice plane (100) with a *d*-spacing of 13.0 Å. The peak of 9.5° 2θ of Form II is the diffraction peak of the lattice plane (020) with a *d*-spacing of 9.3 Å.

XRPD is widely used to distinguish between anhydrous and hydrate forms. For example, mometasone furoate anhydrous Form I and its monohydrate form show distinct differences in XRPD profiles (Figure 9.4) [6]. Form I has characteristic diffraction peaks at 9.3, 9.7, 11.3, 13.8, 14.8, 16.0, 18.9, 19.4° 2θ. Unique peaks for the monohydrate exist

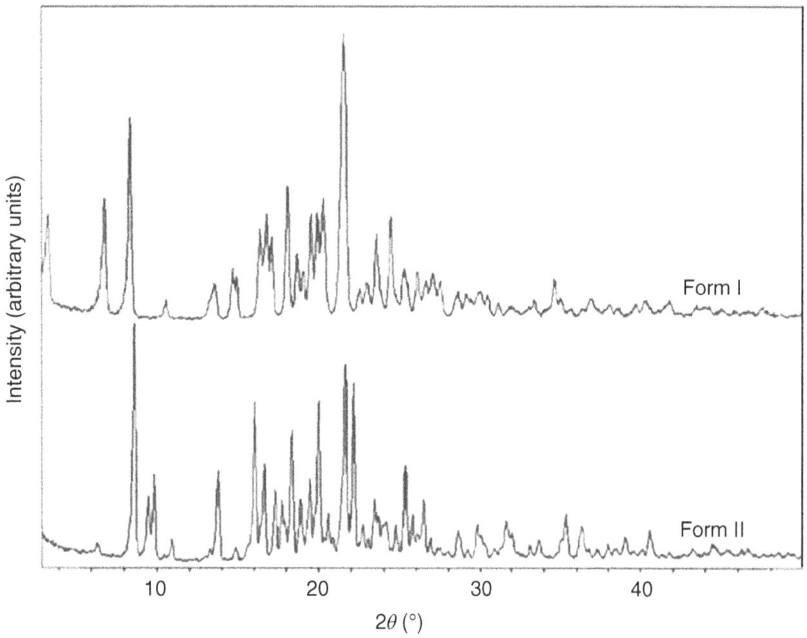

Figure 9.3 Comparison of XRPD patterns of ritonavir Forms I and II. *Source*: [5]/with permission of Taylor & Francis.

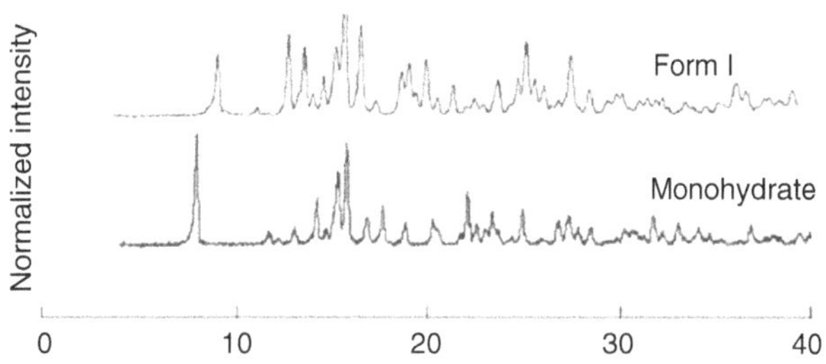

Figure 9.4 Comparison of XRPD patterns of mometasone furoate Form I and monohydrate. *Source*: [6]/ with permission of Elsevier.

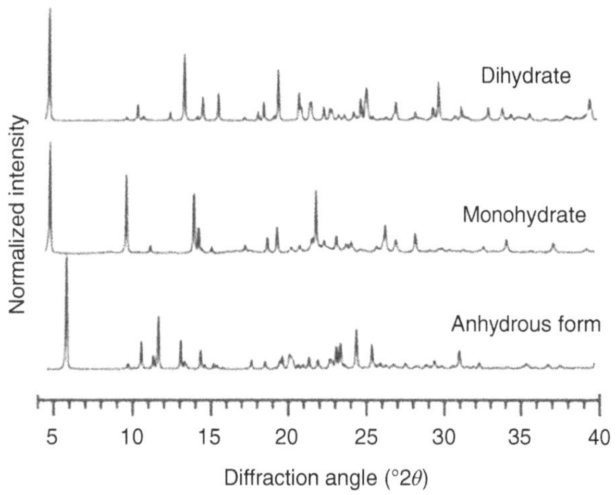

Figure 9.5 Comparison of XRPD patterns of amlodipine besylate dihydrate, monohydrate, and the anhydrous form. *Source*: Adapted from [7].

at 7.8, 14.0, 15.2, and 18.7° 2θ. The diffraction peak at 9.3° 2θ of Form I is associated with the Miller plane (101) in its crystal structure. For the monohydrate, diffraction peaks at 7.8, 15.6, and 31.2° 2θ are related to Miller planes (001), (002), and (004), respectively.

Different hydrate forms for a particular compound have different crystal structures and, therefore, show distinct XRPD patterns as demonstrated in the case of amlodipine besylate [7]. As shown in Figure 9.5, the monohydrate form has characteristic peaks at 4.9, 13.4, 19.4, and 24.9° 2θ, whereas the characteristic peaks for the dihydrate are at 4.8, 9.6, and 13.9° 2θ and at 5.8, 11.6, and 13.0° 2θ for the anhydrous form.

XRPD is the primary method used to guide salt screening since it helps to confirm the formation of certain salt forms and reveal their extent of crystallinity. Figure 9.6 shows the XRPD patterns of several salt forms of trazodone and its free form [8]. The salt forms, including tosylate, pamoate, HCl, and sulfates, have distinctly different XPRD patterns and are also different from the free form. The sulfate salt can be shown to exhibit polymorphism in that the XRPD patterns of the sulfate salt crystallized from ethyl acetate (sulfate (E)) and the one from tetrahydrofuran (sulfate (T)) are different.

The use of XRPD to "fingerprint" various crystal forms relies on detecting non-overlapping or partially overlapping peaks to differentiate crystal forms qualitatively. This approach is also useful in monitoring phase transformations and carrying out some quantitative analysis.

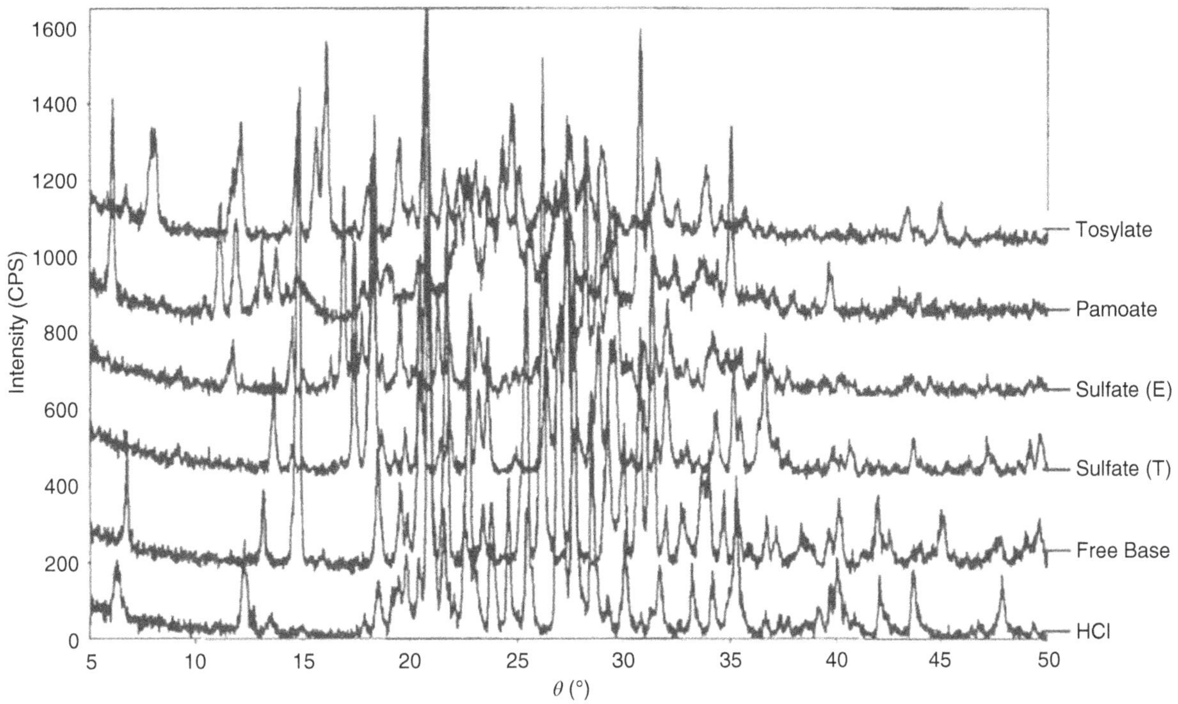

Figure 9.6 XRPD patterns of salt forms of trazodone compared to the free base. *Source*: [8]/with permission of Springer Nature.

9.4 Phase Transformations

Phase transformations related to pharmaceutical materials include polymorph conversion, hydration, dehydration, and crystallization of amorphous materials. The appearance or disappearance of diffraction peaks of particular forms can be used to monitor phase transformations.

XRPD is particularly useful for following the conversion of a metastable polymorph to a stable one, something that can occur spontaneously in the solid state or solution. Such a conversion will result in the disappearance of the metastable form and the appearance of the stable form through changes in the diffraction patterns. Miller et al. applied XRPD to monitor the solution-mediated conversion of ritonavir Forms I to II in different organic solvents [5]. Figure 9.7 shows the XRPD patterns of Form I suspended in 2-propanol over a period of days. After 2 days, the characteristic peaks of Form I at 3.3, 6.8, 8.4° 2θ disappeared completely. The patterns of the material suspended in 2-propanol for 2 days and 2 weeks match the pattern of Form II, indicating that Form I completely transformed to Form II within 2 days.

XRPD is also applicable to study the physical stability of a pharmaceutical material after it has been exposed to moisture. Using XRPD, Kiang et al., for example, discovered four hydrates for Merck compound 1 after storing the anhydrous form at different humidities [9]. X-ray diffraction of compound 1, collected at five controlled relative humidities, 5%, 30%, 60%, 75%, and 95% relative humidity (RH), produced five distinct patterns (Figure 9.8), indicating that each humidity level is associated with a different hydrated state. Moisture sorption data confirmed the presence of anhydrous form at 5% RH, hemihydrate at 30% RH, dihydrate at 60% RH, tetrahydrate at 75% RH, and pentahydrate at 95% RH.

9.5 Quantitative Phase Analysis Using XRPD

XRPD can be utilized for quantitative analysis of pharmaceutical crystalline materials. Quantitative applications include monitoring the content of a crystal form in a mixture and investigating a phase transformation. There are two approaches for carrying out phase analysis using XRPD. One is to use several selected characteristic peaks of a crystalline form, while the other is to analyze the whole diffraction pattern. For a two-component crystalline system, one or more peaks of one phase, with baseline resolution comparable to the other phase, are picked, and the integrated peak intensities are collected.

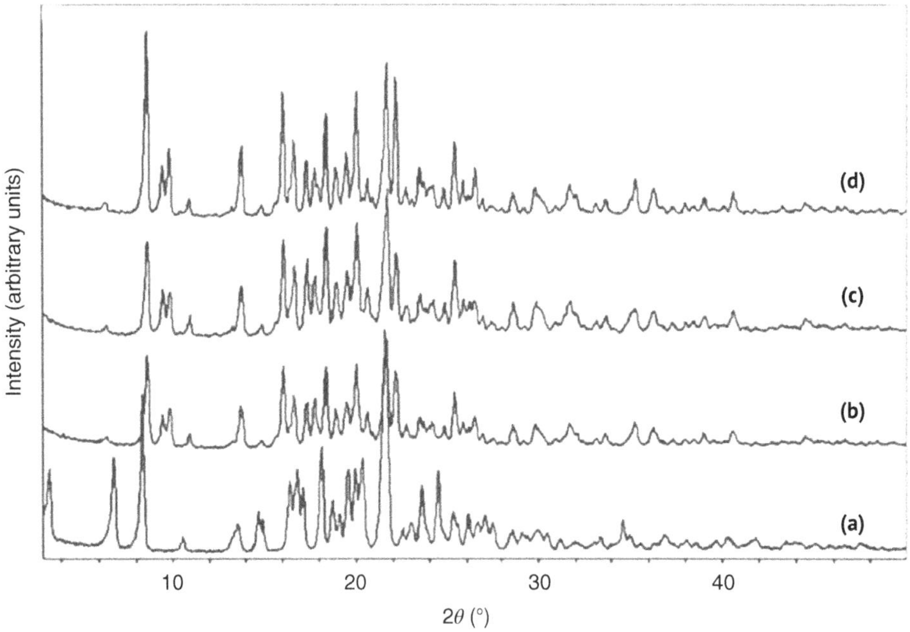

Figure 9.7 Polymorph transformation of ritonavir monitored by XRPD: (a) Form I starting material, (b) Form I suspended in 2-propanol for 2 days, (c) Form I suspended in 2-propanol for 2 weeks, and (d) Form II. *Source*: [5]/with permission of Taylor & Francis.

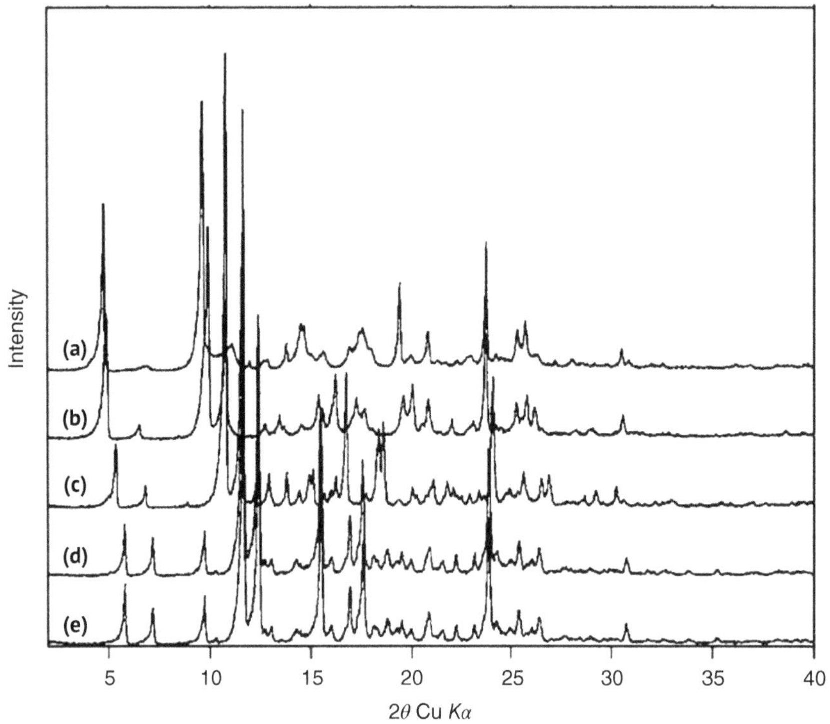

Figure 9.8 XRPD patterns of Merck compound 1 at (a) 95%, (b) 75%, (c) 60%, (d) 30%, and (e) 5% relative humidity. *Source*: [9]/with permission of American Chemical Society.

Subsequently, a calibration method is established with different ratios of the mixture. The goal is to obtain the relative amount of the two components.

Alexander and Klug developed the theoretical basis of using quantitative XRPD to analyze multiple components of crystalline powder mixtures [10]. For a mixture with two components, A and B, the relationship between the integrated intensity of a unique peak i (I_{iA}) of A and its weight fraction (x_A) in the mixture is given by the equation:

$$I_{iA} = \frac{Kx_A}{\rho_A[x_A(\mu_A - \mu_B) + \mu_B]} \tag{9.4}$$

where K is a constant, ρ_A is the density of component A, and μ_A and μ_B are the mass attenuation coefficients (MAPs) of components A and B, respectively. The MAP of a material defines how easily it can be penetrated by a beam of X-ray. The MAP of a compound is related to its chemical composition and equivalent to the weighted average of the MAP of its constituent elements. The integrated intensity of peak i as the pure component A, $(I_{iA})_0$, can be derived from equation (9.4), as follows:

$$(I_{iA})_0 = \frac{K}{\rho_A\mu_A} \tag{9.5}$$

For mixtures with polymorphs or hydrates, MAPs of A and B are the same or approximately the same. The relative intensity for component A in a mixture versus a neat sample can be simplified by the following equation:

$$\frac{I_{iA}}{(I_{iA})_0} = x_A \tag{9.6}$$

Consequently, it follows that a plot of the relative intensity as a function of the sample weight fraction would be expected to be linear. Using this approach, Suryanarayanan developed a quantitative method to monitor the level of anhydrous carbamazepine (CBZI) in a binary mixture with carbamazepine dihydrate (CBZII) [11]. The diffraction peak at d-spacing 6.78 Å for CBZI was chosen since it is unique. The ratio of the integrated intensity of the 6.78 Å peak in a mixture of CBZI and CBZII to the intensity of the 6.78 Å line in a sample consisting of only CBZI was calculated as a function of the weight fraction of CBZI in the mixture. The calculated ratios were in good agreement with theoretical ratios. The samples were compressed into tablets to achieve a similar preferred orientation since the powder samples yielded unsatisfactory results due to inconsistent powder orientation.

Uvarov and Popov carried out a phase analysis for a mixture of clopidogrel bisulfate (CLP) polymorphs using quantitative X-ray diffraction [12]. As shown in Figure 9.9, CLP Forms I and II have a few intense diffraction peaks that do not overlap and, therefore, can be used for quantitation based on Alexander and Klug's method. The non-overlapped peaks for CLP Form I are those with d-spacings 8.13, 5.98, and 4.32 Å, and the peaks for CLP Form II are those with d-spacings 7.24 and 6.88 Å. The calibration curve of relative content of CLP Form I in the binary mixture versus the integrated intensity of the peak with d-spacing 8.13 Å gave a correlation coefficient (R^2) of 0.99.

In some cases, an internal standard was added to the sample mixtures at a constant weight fraction. The absolute peak intensity obtained for such a sample is referenced to that of the internal standard to cancel out instrument variability from run to run and day to day. Generally, inorganic compounds such as corundum, silicon, lithium fluoride, and zinc oxide are used as internal standards for pharmaceutical analysis.

Quantitative analysis using individual peaks is straightforward in method development, but it demands consistent sample preparation, since erratic preferred orientation could bring about significant measurement errors, thus reducing the accuracy. The effect of preferred orientation can be mitigated by reducing particle size, diluting the sample with an amorphous matrix, or compressing the powder into a compact. Roberts et al. showed that a combination of reducing particle size and spinning samples helped to minimize preferred orientation and improve accuracy for the quantitative analysis of mannitol polymorphs [13]. The spinning of samples during analysis enables diffraction signals to be collected from various orientations of particles relative to the X-ray source, resulting in more particles satisfying the Bragg condition and providing diffraction signals. The detection limit of the δ form of mannitol in mixtures with the β form was found to be as low as 1%.

The negative effects of preferred orientation, as described earlier, appear to be reduced considerably when using the whole pattern approach, in that the whole pattern approach takes into consideration all the peaks in the diffraction pattern. Rietveld analysis is the most widely used whole pattern method to quantify crystal phase composition in mixtures. It is a least-squares refinement technique based on the known crystal structures of individual components in the mixture [14].

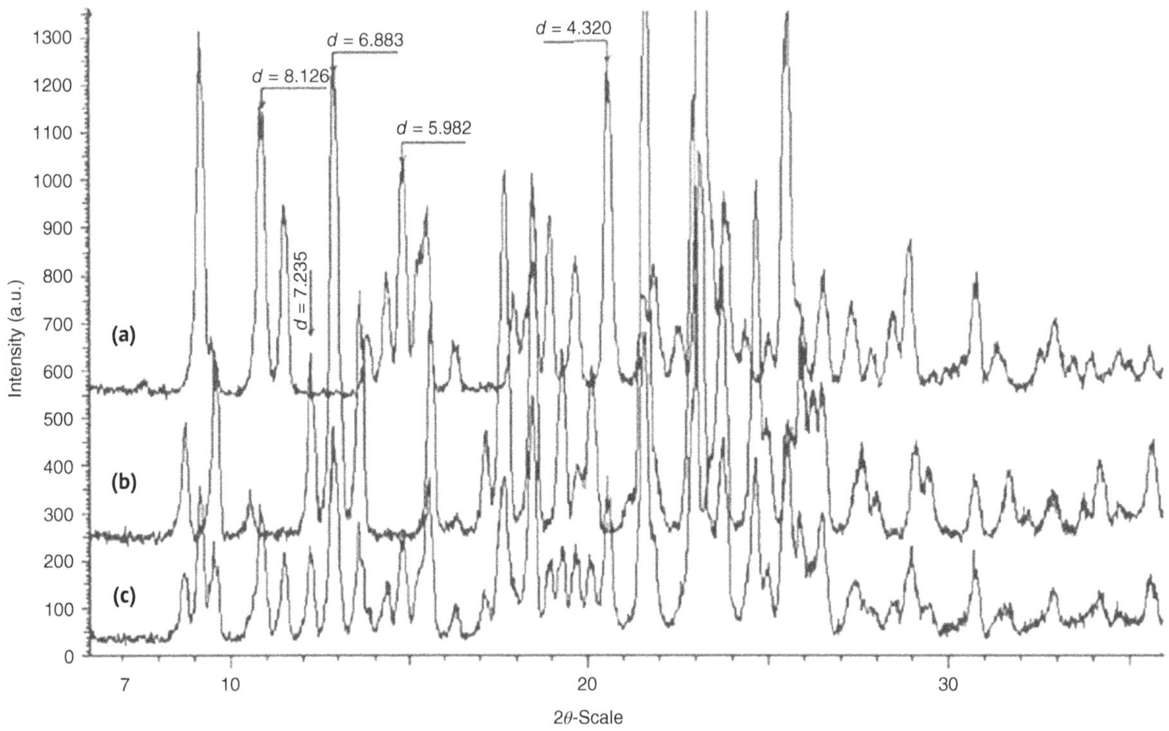

Figure 9.9 XRPD patterns acquired from pure CLP Form I (a), pure Form II (b), and their mixture (c). *Source*: [12]/with permission of Elsevier.

With this approach, a calculated diffraction pattern is produced and refined to fit the experimental data. The crystallographic parameters, the mixture composition, and the data collection parameters are refined to generate the best fit. The goal of the Rietveld method is to minimize the residual function, the weighted sum of squared residuals, using a non-linear least-squares algorithm as follows:

$$\mathrm{WSS} = \sum_i \frac{1}{I_i^{\mathrm{exp}}} (I_i^{\mathrm{exp}} - I_i^{\mathrm{calc}})^2 \tag{9.7}$$

where I_i^{exp} is the experimental diffraction intensity at 2θ point i and I_i^{calc} is the calculated diffraction intensity at the same position. The calculated intensity is given by:

$$I_i^{\mathrm{calc}} = S_F \sum_{j=1}^{Nphases} \frac{f_j}{V_j^2} \sum_{k=1}^{Npeaks} L_k |F_{k,j}|^2 S_j(2\theta_i - 2\theta_{k,j}) P_{k,j} A_j + bkg_i \tag{9.8}$$

where S_F is related to beam intensity, f_j is the phase volume fraction, V_j is the phase cell volume, and L_k is the Lorentz–Polarization factor. Note that L_k depends on instrument settings such as geometry, monochromator, detector, beam size, sample volume, and sample angular position. $|F_{k,j}|^2$ is the structural factor. S_j is the scale factor. $P_{k,J}$ is a term related to preferred orientation. A_j is the absorption factor. bkg_i is related to background contributions.

Many sophisticated software packages have been developed to carry out Rietveld analyses, including TOPAS and GSAS [15, 16]. For example, Antonia et al. employed the TOPAS Academic software v4.1 to quantitatively analyze a mixture of CBZ Forms I and III based on XRPD data [17]. Forms I and III of CBZ are an enantiotropic pair with a transition temperature of 185 °C. Form III is the stable form at room temperature and would be converted to Form I after heating above 185 °C. A Form III sample heated at 190 °C for 30 min was judged as pure Form I based on the diffraction data collected in a laboratory X-ray diffractometer, whereas the analysis using synchrotron radiation revealed a minor contamination of Form III in the heated sample. Rietveld analysis of the diffraction data from the laboratory equipment showed that 4.3% of Form III was present in the heated sample. Figure 9.10 shows the diffraction data after Rietveld analysis. The solid line is the experimental data, and the dotted line is the Rietveld fitting. Under the diffractogram are given the residual plot and diffraction peak position.

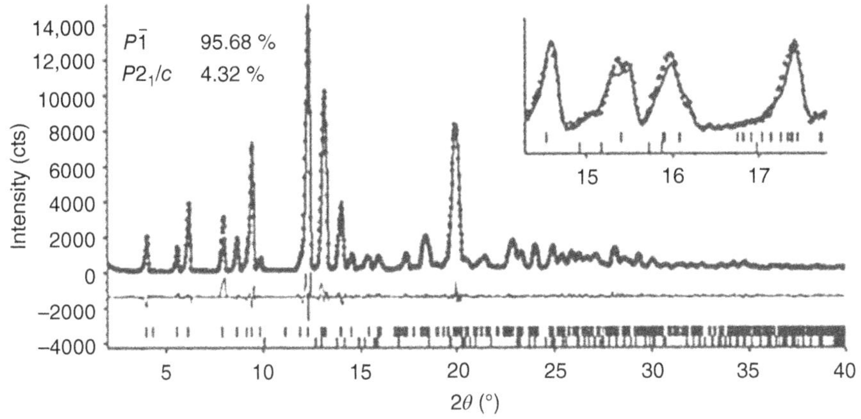

Figure 9.10 Rietveld plot for CBZ Form III heated at 190 °C/30 min. The amount of Form 1 (P-1) and Form III ($P2_1/c$) was calculated by Rietveld analysis. *Source*: [17]/with permission of Elsevier.

9.6 Solving Crystal Structures Using XRPD

Recently, significant progress has been made in using XRPD to solve crystal structures of APIs in the absence of single-crystal data. This is necessary because single crystals with sufficient size and quality are often difficult to prepare for certain crystal forms of an API. For example, some metastable forms are frequently obtained by fast crystallization or precipitation, resulting in microcrystalline aggregates rather than large single crystals.

Very high-quality powder diffraction data are essential for the successful determination of crystal structure using this approach. Although powder diffraction patterns contain the same intrinsic information as single-crystal crystallography, three-dimensional diffraction data are compressed into one dimension in powder patterns, leading to considerable overlap of peaks, which obscures the appropriate measurement of the intensities of individual diffraction maxima. The quality of the powder diffraction pattern can be impacted by the phasic purity of the API form, the accuracy of the peak position, the reliability of peak intensity, peak width, and peak shape. The contamination due to impurities may also cause inaccurate indexing and poor final structure refinement. Obtaining accurate peak positions requires both proper sample presentation, including capillary or transmission modes, and high-resolution methods, which can be achieved with synchrotron radiation. Besides requiring accurate peak positions, successful structure determination from powder diffraction data requires reliable information on the relative intensities of the diffraction maxima. The relative intensities of diffraction peaks are often affected by preferred orientation, a nonrandom distribution of crystallite orientations in the sample. Preferred orientation, however, can be mitigated by reducing particle size or packing the sample in a capillary tube.

Peak width and shape can also be improved by using high-resolution synchrotron radiation. The use of synchrotron radiation provides less ambiguity in extracting the peak positions and peak intensities of individual diffraction peaks. In addition, a shorter total exposure time in the range of tens of seconds reduces radiation potential damage to organic samples. Powder samples are generally packed in capillaries and analyzed with an image plate detector, which reduces preferred orientations. Most of the successful cases of resolving the structure based on powder data for APIs have relied on synchrotron radiation.

Good quality data from laboratory equipment is sometimes also feasible for structure determination if the radiation source is highly monochromatic. Monochromatic radiation can be achieved by placing a primary monochromator between the X-ray tube and the sample. For example, with a copper X-ray tube, CuKα2 and CuKβ are eliminated with only CuKα1 remaining. Appropriate transmission geometry is recommended, with the sample packed in a rotating capillary and as a thin film to minimize the effect of preferred orientation.

Many commercially available software programs are available for solving crystal structures based on powder diffraction data. These include DASH [18], ESPOIR [19], FOX [20], PowderSolve [21], and TOPAS [15].

The general procedure to solve a crystal structure from XRPD is as follows:

- Data collection and data treatment
- Cell indexing and space group determination
- Structure solution
- Structure refinement by the Rietveld method

The first step of crystal structure determination from powder diffraction data involves indexing. Indexing the powder diffraction pattern generates the unit cell dimensions (a, b, c, α, β, γ). Many programs have been developed for indexing powder diffraction data, and in all those programs, the positions for a selected set of diffraction peaks (about 20) are evaluated. Popular programs are Treor [22] and Dicvol [23]. Treor takes a trial-and-error approach to search for the unit cell dimension. The program has separate routines for different symmetrical systems. Dicvol is based on the variation of the lengths in cell edges and interaxial angles in direct space, where the lengths and angles are progressively reduced by a dichotomy procedure. Once the cell parameters and lattice class have been determined, systematic absences are evaluated using Pawley refinement to limit the number of possible space groups. In addition, the information on chirality and density is applied to determine the exact space group.

After indexing, the solution of the structure leads to an approximate crystal structure for the API form. There are two approaches for structure solution: the traditional approach and the direct space method. The traditional approach is similar to single-crystal crystallography, in that the intensities $I(hkl)$ of individual reflections are directly extracted from the powder diffraction pattern and used to solve the structure based on either direct methods or the Patterson method. Owing to peak overlap, it is often challenging to obtain unambiguous values of the intensities $I(hkl)$ of the individual diffraction peaks, which can cause significant difficulties in solving the structure. In contrast, the direct space method is more robust in dealing with powder diffraction data.

For the direct space method, trial crystal structures are generated in direct space by global optimization, which is independent of the experimental data. The merit of each trial structure is assessed by direct comparison of the powder diffraction pattern of the trial structure and the experimental data. The fitness of the two patterns is assessed by the weighted profile R factor Rwp, which is also used in Rietveld refinement. The assessment considers the entire powder diffraction profile instead of the integrated intensities of individual peaks. Peak overlap is already taken into consideration. The goal of the direct space method is to find a trial crystal structure that has the lowest possible R factor and one that is subject to further structure refinement.

The molecular simulation tools for the direct space method include grid search [24], Monte Carlo–simulated annealing [25], and genetic algorithm [26]. Monte Carlo–simulated annealing is the most popular method for global optimization. Both involve a loop of various modifications and evaluations:

- Starting from a trial structure, some random displacement is pursued to generate a new trial structure.
- The powder diffraction pattern for the new trial structure is calculated and compared with the experimental data.
- The trial structure is accepted and rejected based on a defined closeness to the experimental data.
- The accepted structure is taken into the next run.

The aforementioned procedure is repeated until a trial structure is obtained with an acceptable R factor.

The trial structure is further refined against the experimental powder diffraction data using the Rietveld method [25]. The weighted R factor, Rwp, is minimized by optimizing a wide range of refinement parameters such as unit cell, atomic position, diffraction peak profile and asymmetry, crystalline size, strain broadening, preferred orientation, background, zero-point shift, and peak intensities.

Crystal structures of many API forms have been solved based on powder diffraction data. Several examples are presented as follows.

de Armas et al. reported the determination of the crystal structure of carnidazole Form II from synchrotron XRPD [27]. The synchrotron powder diffraction was indexed by the dichotomy method using the program Dicvol. Form II has a monoclinic unit cell with $a = 13.908(2)$ Å, $b = 8.094(2)$ Å, $c = 10.645(2)$ Å, $\beta = 110.82(2)°$. The structure was solved by the program FOX, which utilizes a Monte Carlo–simulated annealing technique. The initial molecular model was taken from the crystal structure of carnidazole Form I. During the simulated annealing, 12 degrees of freedom were defined, including freedom of location and orientation of the whole molecule and freedom of torsional angles. The structure refinement was performed by the Rietveld method implemented in the Fullprof suite of programs. The cell parameters obtained after Rietveld refinement were $a = 13.915(4)$ Å, $b = 8.095(2)$ Å, $c = 10.649(3)$ Å, and $\beta = 110.83(1)°$.

Kiang et al. determined the crystal structure of enalapril maleate Form II from high-resolution XRPD data using the direct space method [28]. The synchrotron diffraction data were indexed with the computer program ITO [29]. The systematic absences indicated $P2_12_12_1$ as a possible space group. The structure determination from the synchrotron XRPD pattern was performed using the Monte Carlo–simulated annealing program PowderSolve [21]. The initial molecular models of enalapril and maleate were taken from the single crystal structures of Form I. By treating the molecules as rigid bodies, the total degrees of freedom of enalapril maleate were reduced from 25 to 12, allowing simulated annealing to be completed within a reasonable computation time. The final *Rwp* of the Rietveld refinement reached 7.61%, indicating the reliability of the final structure. Enalapril maleate Form II has a crystal structure of the space group $P2_12_12_1$, $Z = 4$, with unit cell parameters $a = 33.9898(3)$ Å, $b = 11.2109(1)$ Å, $c = 6.64195(7)$ Å, and $V = 2530.96(5)$ Å^3.

Cheung et al. solved the crystal structure of bupivacaine free base from XRPD data collected using a laboratory diffractometer [30]. The genetic algorithm technique was applied for structure solution, followed by the Rietveld refinement.

9.7 X-ray Diffraction of Amorphous and Crystal Mesophase Forms

One of the main uses of XRPD is its capability of differentiating amorphous, crystal mesophase, and crystalline material for an API. As discussed earlier, a crystalline material will give some strong diffraction peaks, with the peak position determined by the crystal structure. For a crystalline material, diffraction is strong and concentrated in diffraction peaks, whereas diffraction for an amorphous material is weak and spread out throughout a broad range of diffraction angles. Amorphous materials have no long-range order as a super-cooled liquid (Chapter 6). Usually, amorphous phases display diffusive scattering with one or more halos present in the diffraction pattern, related to intermolecular interferences.

The halo position is dependent on the degree of packing of molecules in the amorphous phase, related to the most frequently occurring intermolecular distance, and approximately following an equation derived by Alexander [31]:

$$r \approx 1.22 \frac{\lambda}{2 \sin\theta} \tag{9.9}$$

where r is the intermolecular distance, λ is the wavelength, and θ is the halo position. The lower the density and the higher the intermolecular distance, the smaller the angle for the halo will be.

Mesophases are transitional phases between true liquid and true crystalline phases. Mesophases have different degrees of order and may show few diffraction peaks. As discussed in Chapter 7, mesophases can exist as liquid crystals, plastic crystals, and conformational disordered crystals.

In a nematic liquid crystal phase, molecules have a preferred directional ordering, but the centers of mass are still isotropically distributed. Usually, nematic phases lack any sharp, wide- or low-angle diffraction lines and have patterns similar to amorphous phases. Smectic liquid phases have layered structures and one- or two-dimensional order. The diffraction pattern of a smectic phase usually has a broad, wide-angle diffraction peak and several low-angle diffraction peaks. For example, two weak and relatively broad peaks at approximately 7° and 18° 2θ were observed for dehydrated cefazolin sodium, a smectic phase [32]. The diffraction peaks are related to the interlayer spacing, d. Atassi et al. reported that amorphous, mesomorphous, and crystalline forms of calcium ketoprofen have different XRPD patterns [33] (Figure 9.11). The crystalline form has many sharp diffraction peaks, and the mesophase has one sharp Bragg peak at angle $2\theta = 4.0°$, which suggests the existence of a lamellar structure with a d-spacing of 14.8 Å. The amorphous form has a halo at the same diffraction angle, which means that a similar lamellar structure is present for the amorphous form.

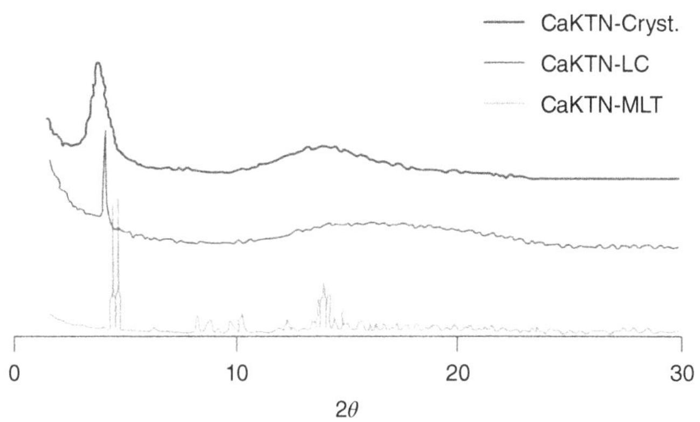

Figure 9.11 XRPD diffraction patterns of (from bottom to top) crystalline, mesomorphous, and amorphous calcium ketoprofen (CaKTn). *Source*: [33]/with permission of Elsevier.

9.8 Pair Distribution Function

Atomic PDF has emerged as a useful technique to characterize pharmaceutical materials based on total diffraction analysis. The PDF offers a fingerprint of the interatomic distances that define the structure of a particular solid form. It is particularly beneficial for characterizing amorphous materials in that it transforms a featureless amorphous halo into a distinct PDF pattern, which can be differentiated from those of crystalline counterparts or other amorphous structures. Also, PDF is useful in probing nanostructured materials and nanocrystals.

The PDF of a material describes the probability of finding two atoms separated by a certain distance. The peaks in the PDF reflect commonly occurring interatomic distances, with the product of the peak area and distance defining the number of atom units with a particular pair separation.

A PDF trace displays a one-dimensional representation of an average three-dimensional structure. Because PDF analysis takes into consideration both the Bragg peaks and diffusive scattering, the information from PDF analysis covers both regions of long-range atomic ordering $(> 100\ Å)$ and short-range ordering in a material. The PDF, $G(r)$ of a powder diffraction pattern, is obtained via the sine Fourier Transformation of the total diffraction pattern $(S(Q))$ as described in the following equation [34–38]:

$$G(r) = \frac{2}{\pi} \int_0^{\infty} Q[S(Q) - 1] \sin (Qr) dQ \tag{9.10}$$

where Q is the magnitude scattering vector, also named as momentum transfer, is described as follows:

$$Q = 4\pi \frac{\sin \theta}{\lambda} \tag{9.11}$$

where θ is the Bragg angle and λ is the incident radiation. The PDF gives the probability of finding an atom at a given distance of "r" from another atom.

PDF is useful as a comparative tool instead of a means to determine any absolute structural information. The nature and level of ordering present in different materials can be compared and related to their physical properties and stability. For example, if a PDF trace generated from the diffuse scattering of an amorphous material has no peak at greater than 5 $Å$, it suggests that there is no atomic order present beyond 5 $Å$.

The procedure to carry out a PDF analysis is as follows:

- X-ray diffraction data is collected with low noise over some range of momentum transfer (Q).
- Experimental data is corrected to eliminate scattering from the air and sample holder. Running a sample under vacuum or a low-density gas is an alternative for reducing the impact of scattering from air. Using a sample holder with a low background is also helpful. The data are then corrected for experimental artifacts and normalized to obtain the reduced structure function, $F(Q) = Q[S(Q) - 1]$.
- $F(Q)$ is then Fourier-transformed to real space.

Figure 9.12 shows a diffraction pattern of amorphous indomethacin prepared by quench-melt and the PDF trace derived from the diffraction pattern [38]. The PDF trace clearly shows the nearest neighbor (NN) and next nearest neighbor (NNN) peaks that correspond to the molecular coordination spheres. The NN PDF peak is related to a molecular packing distance of approximately 4.6–4.7 $Å$, a distance close to the molecular thickness of one indomethacin molecule. The NNN peak has a position of 9.2 $Å$, which is almost equivalent to the thickness of two indomethacin molecules.

The quality of the powder diffraction data obtained is critical for obtaining a reliable PDF profile, since fingerprinting highly disordered amorphous structure becomes challenging if there is too little information in a diffraction data set. For example, it has been demonstrated that there, generally, is insufficient information in the diffraction pattern of an amorphous drug substance in the low-Q (low diffraction angle) region for calculating PDF, whereas in that range crystalline material usually displays strong Bragg's peaks and is easily differentiated from amorphous material. It is recommended, therefore, that diffraction data should be gathered over a broad range of Q values, thus providing more information which is useful in fingerprinting amorphous materials. Besides scanning to high diffraction angles, shorter wavelength X-rays can be utilized to increase the Q value. For example, equipment with copper Kα radiation has a maximum Q range of about 8 $Å^{-1}$, which is equivalent to 160° 2θ. For such high-angle scanning, however, only samples placed in capillary tubes are feasible. The maximum Q range can be increased to 16 $Å^{-1}$ using the molybdenum anode and 22 $Å^{-1}$ for the silver anode. It has been reported, for example, that diffraction data for amorphous CBZ and indomethacin, collected over Q ranges accessible by using molybdenum and silver anodes, provide unique PDFs and reliable fingerprints. In contrast, data from a copper anode was found to be less valuable for PDF analysis of these materials due to its relatively limited Q range [35].

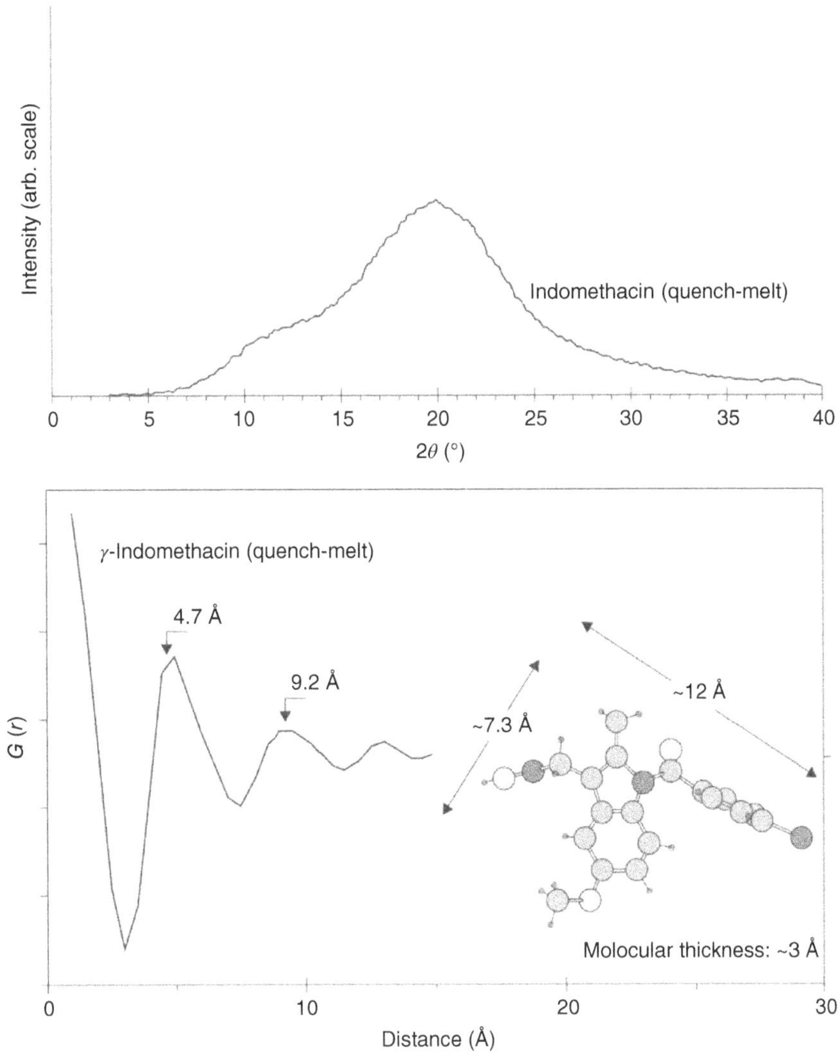

Figure 9.12 PDF and interatomic distances for amorphous indomethacin prepared by quench melting. *Source*: Adapted from [38].

In summary, PDF analysis on data collected from laboratory powder diffractometers is found to have many notable applications:

- Differentiating amorphous materials prepared by different manufacturing processes such as milling and quench cooling [36]
- Characterizing solid dispersions and differentiating the amorphous structure of compounds prepared in different formulations [37] or determining whether a solid dispersion exists of a single phase or in phase-separated mixed phases [38]
- Detecting a disorder/amorphous structure induced into a crystalline drug by processing [39]

Nevertheless, it is widely recognized that high-energy synchrotron radiation is more suitable for the PDF analysis of pharmaceutical materials owing to the following reasons. At first synchrotron can increase the Q range for PDF analysis to as much as 45 $Å^{-1}$ due to the short wavelengths. In addition, the photon flux from synchrotron radiation is usually many magnitudes higher than that from laboratory X-ray equipment, leading to a dramatically shortened run time. For example, the measurement time for a typical X-ray total scattering data from a laboratory powder diffractometer can range from a few hours to a day or so for one data set, whereas the data collection in a synchrotron beamline usually takes seconds or minutes with the use of area detectors [40]. Therefore, the use of synchrotron, owing to its high data quality, superior

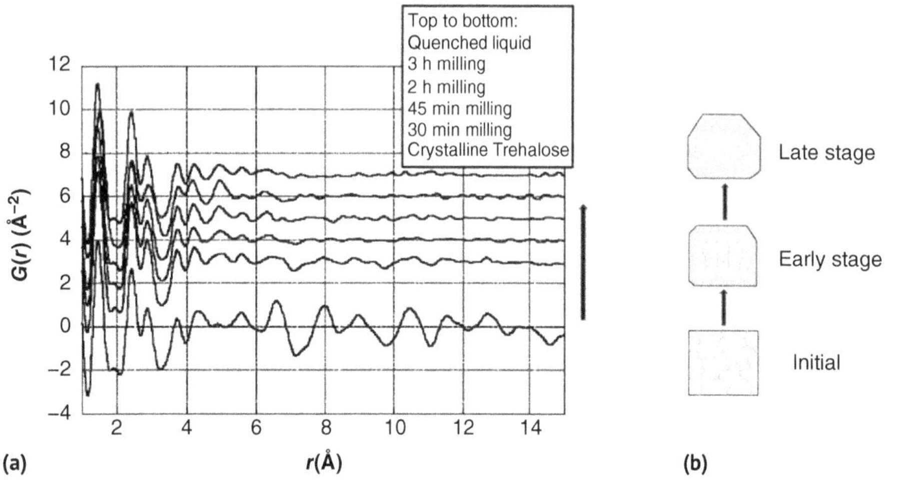

Figure 9.13 PDF analysis of solid-state amorphization of β-trehalose. (a) Evolution of the PDFs of β-trehalose during milling in comparison to the starting crystalline material and the amorphous material made from quench melting. (b) Proposed model of the evolution of a grain substructure upon milling. *Source*: [44]/with permission of American Chemical Society.

resolution, and excellent throughput, significantly facilitates the applications of PDF in many important areas, especially in the understanding of amorphous solid dispersion systems. For example, Chen et al., by applying high-energy X-ray PDF, in the combination of solid-state NMR, elucidated the molecular packing of posaconazole in its amorphous solid dispersions formulated with hypromellose acetate succinate (HPMCAS) and correlated the observed changes in the PDF to the underlying molecular interactions between the drug and the polymer [41]. Bezzon et al. studied the formulation stability of amorphous dispersions of flubendazole in several commonly used polymers, including hypromellose phthalate (HPMCP), HPMCAS, and hypromellose (HPMC), using synchrotron XRPD and PDF analysis. It is of great interest to note that the ion–base interactions between the drug–polymer matrix are attributed to reducing the drug molecule clustering processes. As revealed by PDF analysis, more structural ordering exists for the dispersion with HPMC, and vice versa, less ordering presents for that with HPMCP, correlating with the order of stability, from less to more [42]. In another study intended to understand the local structures of amorphous solid dispersions, using lapatinib as a model compound, De Araujo and coworkers discovered that, after PDF analysis, the intermolecular drug interactions and clustering patterns differ notably between the dispersions with HPMC and those with HPMCP. Particularly, the intermolecular interactions in the range of 3.0–6.0 Å reflecting NN contacts in drug–drug interactions are absent in the dispersions with HPMCP at 1 : 3 drug:polymer ratio, in agreement with the superior physical stability of the formulation [43]. In summary, synchrotron X-ray diffraction and PDF analysis offers a powerful tool for the rational development and structural studies of amorphous solid dispersions, possibly leading to the predication of macroscopic properties and physical stability.

Furthermore, rapid acquisition of PDF with synchrotron radiation opens the way for studying time-resolved events such as physical transformation, including amorphization or phase separation in amorphous systems. For example, Bordet and coworkers studied the amorphization of β-trehalose during high-energy milling via in situ and offline synchrotron powder diffraction and PDF analysis [44]. The PDF analysis not only showed the progression of the phase transformation within 3-hour milling but also, interestingly, pinpointed the mechanism of the amorphization as a core-shell model, for which an amorphous phase forms a shell around a crystalline core (Figure 9.13). It appeared that the proportion and the domain size of the crystalline phase rapidly diminished along with decreased crystallinity, obviously due to an increase in the defect density, during milling until the whole material becomes amorphous after 2 hours.

9.9 X-ray Diffractometers

Currently, XRPD data for pharmaceutical materials are mainly collected by X-ray diffractometers, which consist of an X-ray source, a sample holder, and a detector (Figure 9.14).

The mechanical assembly that holds the X-ray source, the sample holder, the detector arm, and associated gearing is the goniometer. There are two types of goniometers: $\theta - \theta$ and $\theta - 2\theta$ (Figure 9.15). In a $\theta - \theta$ goniometer such as that found in the PANalytical X'Pert Pro diffractometer, the sample holder is stationary, while the X-ray tube and the detector are moved toward each other simultaneously over the angular range θ. For a $\theta - 2\theta$ goniometer, the X-ray source is stationary, while the sample holder is moved by the angle θ and the detector simultaneously moves by the angle 2θ. The $\theta - \theta$ goniometer is generally more useful for loose powders or free-flow materials, since, in the $\theta - 2\theta$ configuration, material may fall off the holder at high diffraction angles.

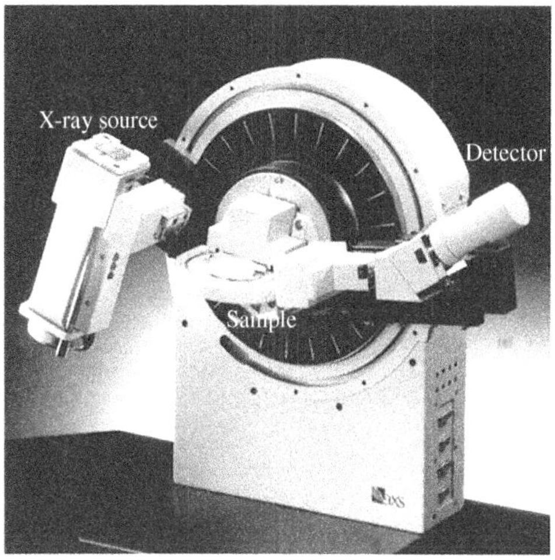

Figure 9.14 A Bruker D8 X-ray diffractometer.

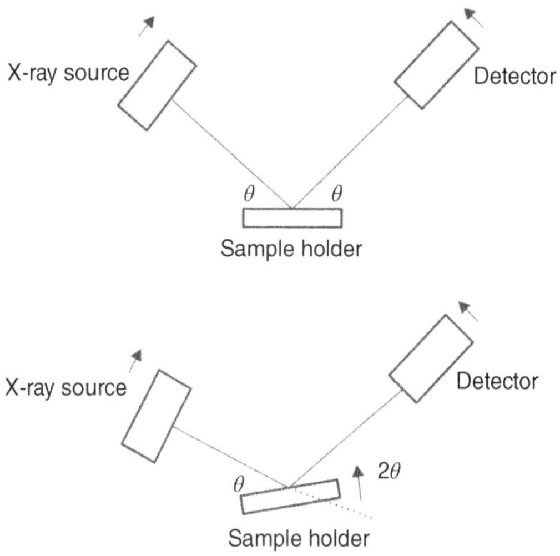

Figure 9.15 Diagrams of $\theta - \theta$ (top) and $\theta - 2\theta$ (bottom) goniometer.

There are three types of geometries that allow the X-ray beam to interact with samples, that is, Bragg-Brentano, parallel beam geometries, and the transmission mode.

Bragg-Brentano provides a parafocusing geometry (Figure 9.16) with good sensitivity, high resolution, and low cost, where the X-ray source provides a divergent monochromatic beam. The divergent beam reaches the sample. The diffraction beams are also divergent and are then focused on the detector. The X-ray tube and the detector sit in the same circle, with the sample holder in the center of the circle. The position of the detector, the focal point, is exactly at the same distance from the sample surface as from the X-ray tube. The sample surface needs to be aligned carefully to avoid defocusing effects. The focusing condition is satisfied only when the angle between the incident beams and the sample surface is equal to the angle between the sample surface and the diffracted rays. The phenomenon is also referred to as the parafocusing effect of a flat sample for Bragg-Brentano optics [45].

Using the parafocusing effect of Bragg-Brentano optics may cause several errors, including axial divergence, flat sample error, sample transparence error, and sample displacement error.

Axial divergence is due to the divergence of the X-ray beam out of the plane of the focusing circle, causing an asymmetric broadening of diffraction peaks toward low diffraction angles and creating negative peak shifts below 90° 2θ. Using soller slits or capillary lenses helps reduce axial divergence. The soller slits are made from a set of closely spaced thin metal plates, whereas the capillary lenses are glass optics made up of thousands or hundreds of capillaries.

Flat sample error is related to the reality that the entire surface of a flat specimen cannot lie on the focusing circle, thus creating an asymmetric broadening toward low diffraction angles. Flat sample error can be reduced by using smaller divergence slits.

Sample transparency error is common for organic samples, since the X-ray beam can penetrate a certain distance below the flat surface due to low absorption. When this happens, the average diffracting surface lies somewhere below the flat surface, with the result that peak positions are shifted to lower angles as much as a tenth of a degree. The error, $\Delta 2\theta$ in radians, is defined by the following equation:

$$\Delta 2\theta = \frac{\sin 2\theta}{2\mu R} \tag{9.12}$$

where 2θ is the diffraction angle, μ is the linear mass absorption coefficient, and R is the radius of the diffractometer circle. Based on this equation, sample transparency error will be more significant at high diffraction angles. Such an error can be mitigated by using thinner films.

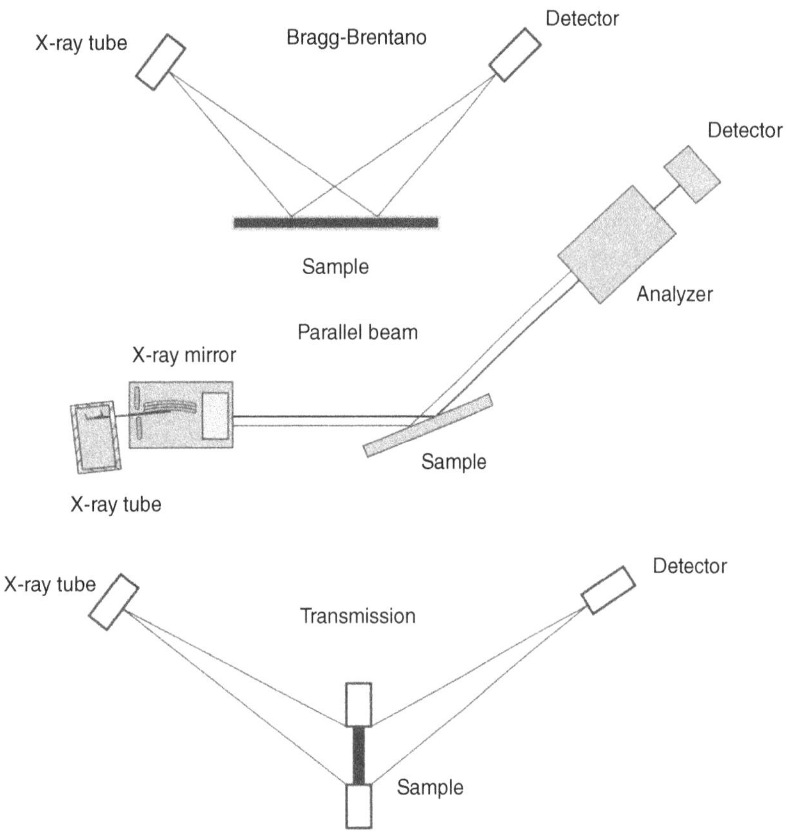

Figure 9.16 Three types of geometry for XRPD.

Sample displacement error is the most significant error for Bragg-Brentano optics. Sample displacement error occurs when the sample surface is not level with the surface of the sample holder, which results in the sample surface not being in the focus circle. The displacement error, $\Delta 2\theta$ in radians, can be defined by the equation as follows:

$$\Delta 2\theta = -\frac{2s\cos\theta}{R} \tag{9.13}$$

where s is the displacement distance from the focus circle. The displacement is positive when the sample surface is lower than the focus circle and vice versa. Due to the cosine function, the displacement error would be expected to be more pronounced at lower diffraction angles. It is estimated that the peak shift in 2θ can be as much as 0.01° for each 15 μm of displacement at low angles. Packing the sample as level as possible to the holder surface is an effective way to mitigate the displacement error. It is also very important that the equipment be properly aligned. Otherwise, the holder surface will not be in the focus circle. The most effective way to eliminate displacement error is to use parallel beam optics.

In parallel beam optics, intense parallel X-ray beams are used instead of divergent beams (Figure 9.16). With such optics, the sample position is more flexible, and the diffractometer is no longer required to maintain the same distance between the X-ray source and sample as between the sample and detector. Parallel beam optics also eliminates errors associated with sample displacement, transparency, and axial divergence since it does not require a flat sample. The measurements, also, are not sensitive to the exact height position of the sample. This geometry is also desirable for irregularly shaped samples since parallel beam optics minimize preferred orientation because samples do not need to be packed and flat. Parallel beam X-ray can be generated by an X-ray mirror or a polycapillary lens.

Transmission X-ray diffraction is another approach that can be used to minimize preferred orientation (Figure 9.16). Samples are packed into a capillary tube, as a thin film, or compressed into a powder compact. X-ray beams are directed at the sample, and the diffraction signal is collected using an area detector or scanned through a range of diffraction angles. When capillaries are used, this approach has the advantage of providing small sample sizes, which is a great advantage for early polymorph screening. The major disadvantage of the transmission mode is that the self-focusing of the diffracted beam is not as precise as in the Bragg-Brentano geometry. Laboratory equipment using diverge beams usually has lower resolution when compared to that using reflection geometry. Transmission X-ray diffraction works well for materials with low absorption and is the standard for powder diffraction using synchrotron radiation because of the strong intensities provided by synchrotron radiation.

Synchrotron beamlines are of great interest to do sophisticated diffraction analysis, including X-ray total scattering data for PDF analysis and powder data for crystal structure determination. Table 9.1 lists the beamlines available for powder diffraction studies [46]. For example, the 1-ID beam line at the advanced photon source located at Argonne National Laboratory has high-energy X-rays operating at a flux of 10^{12} photons/second and in the energy range of 42–130 keV. The X-ray optics are designed for high-energy X-ray usage and quality resolution, with several focusing mechanisms available to enable in-depth analysis of pharmaceutical materials, including PDF measurements, powder diffraction, and small-angle scattering. For data collection, a large area two-dimensional (2D) detector is usually used to attain a broad region of reciprocal space simultaneously, as illustrated in Figure 9.17. For instance, Terban and coworkers used a PerkinElmer 2D flat panel detector located approximately 205 mm from samples in a capillary tube when studying the recrystallization of amorphous lactose from the total scattering PDF [46]. The raw 2D intensity was first adjusted for experimental effects and, therefore, azimuthally integrated to become one-dimensional intensity versus Q plots. The detector-to-sample distance can be adjusted to attain the Q range needed. For example, when testing posaconazole amorphous solid dispersions, Chen et al. placed a Varex CT4343 area detector about 344 mm downstream of the sample in a polyimide capillary tube to attain a Q range from 0.6 to 25.0 $Å^{-1}$ [41].

Sampling geometries including capillary tubes, thin films, which are commonly pursued depending on sample characteristics, whereas acoustic levitation offers a niche approach to extend the capability of synchrotron analysis, for example, in the aspect of screening amorphous solid dispersions. Barbosa and coworkers reported that acoustic levitation and high-resolution synchrotron XRPD lead to a fast-screening approach of niclosamide amorphous solid dispersions. It is

Table 9.1 Synchrotron beamlines open for powder diffraction analysis.

Synchrotron facility	Beamline
Advanced Photon Source, Chicago, Illinois, USA	6-ID-D, 11-ID-B
DESY, Hamburg, Germany	P02.1, P07, P21.1
Diamond, Didcot, United Kingdom	I15-1
ESRF, Grenoble, France	ID11, ID15A, ID31
NSLS-II, Upton, New York, USA	28-ID-1, 28-ID-2
Spring-8, Sayo, Hyogo, Japan	BL04B2, BL08W
LNSL, Campinas, Brazil	XDS-W09A
CHESS, Ithaca, New York, USA	ID1A3
MAX IV, Lund, Sweden	DanMax
ALBA, Barcelona, Spain	BL04-MSPD

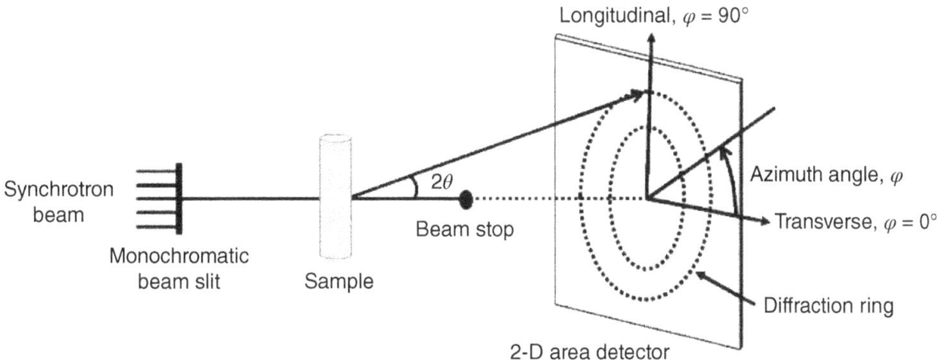

Figure 9.17 A schematic drawing of the optics for synchrotron X-ray diffraction analysis.

interesting to note that only the microlite sample size of niclosamide-polymer solution was being studied by levitating in a multi-emitter single-axis acoustic levitator and, at the same time, analyzed for the formation of solid dispersions when the solvent evaporated during the levitation, by synchrotron X-ray diffraction analysis [47].

9.10 Variable Temperature XRPD

VT-XRPD is a method that allows diffractograms of a single sample to be collected at different temperatures. It is a very useful tool to monitor solid-state transformations that occur during heating or cooling, such as dehydration/desolvation, rehydration, polymorph conversion, melting, and crystallization. It is a complementary means of interpreting the transitions observed with thermal analysis and hot stage microscopy data obtained with changing temperature. Often, the interpretation of certain thermal events observed in differential scanning calorimetry, for example, is not always feasible. By performing VT-XRPD studies, it is generally possible to directly ascertain the nature of the phases before and after the thermal event. For example, after the dehydration of a hydrate form, the dehydrated material could be an amorphous phase, an anhydrous crystalline form, or a dehydrated hydrate.

The dehydration of theophylline monohydrate has been investigated using VT-XRPD [48]. Such studies revealed that dehydration resulted in a metastable anhydrous phase and that this metastable phase was further transformed to the stable anhydrate. Without VT-XRPD, it is unlikely that the metastable phase would have been discovered.

Karjalainen et al. studied the dehydration of nitrofurantoin monohydrate using VT-XRPD [49]. As shown in Figure 9.18, the starting material was very crystalline. As the temperature was elevated above 120 °C, the crystallinity of the monohydrate started to decrease. At 130 °C, the crystallinity was only about 55%, and the monohydrate was barely detectable at 140 °C. The anhydrous form started to appear at 130 °C, and, at 140 °C, the anhydrous form still had poor crystallinity. The crystallinity was increased significantly at 200 °C, and the growth of the anhydrous form was complete at 270 °C. Thus, VT-XRPD provided a very complete description and interpretation of the entire conversion process. Furthermore, the VT-XRPD data in these studies were subjected to a quantitative analysis that established the proportion of nitrofurantoin monohydrate and anhydrate at each temperature, showing that the amount of the monohydrate was decreased, and the anhydrate was increased as a function of temperature.

VT-XRPD is also very useful in following polymorph conversion at elevated temperatures. De Villiers et al., for example, used VT-XRPD to study the polymorphic transformation of mebendazole Form C [50] and showed that Form C was stable below 180 °C and was transformed to polymorph A at higher temperatures (>180 °C); this transformation was complete at 225 °C. Furthermore, this transformation was shown to be a first-order process with an activation energy of 238 ± 16 kJ/mole.

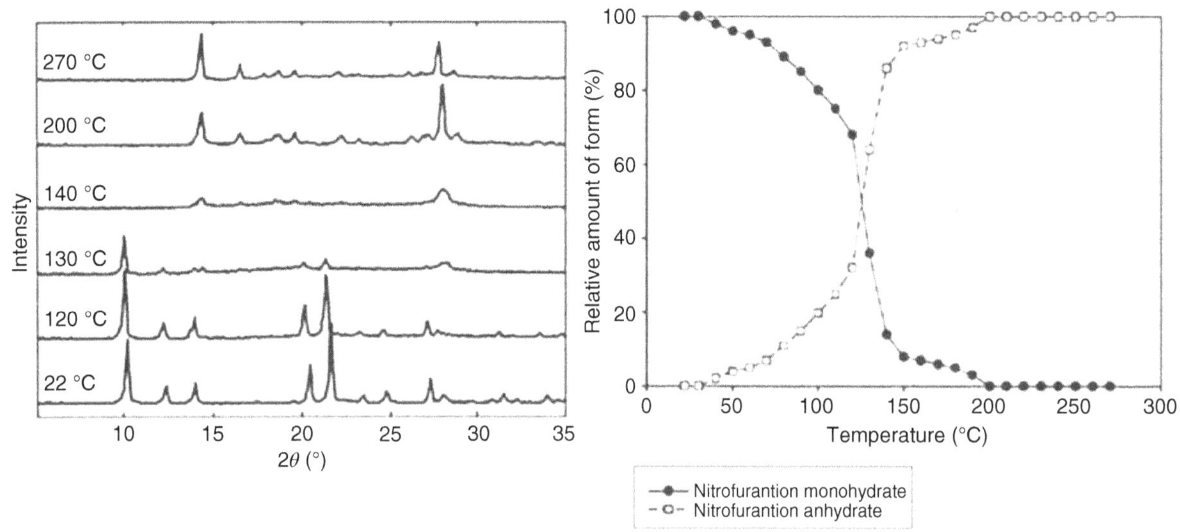

Figure 9.18 VT-XRPD diffraction patterns of nitrofurantoin at different temperatures (left) and calculated relative amount of nitrofurantoin monohydrate and anhydrate in recrystallized nitrofurantoin (right). *Source*: [49]/with permission of Elsevier.

References

1 Bragg WL. The diffraction of short electromagnetic waves by a crystal. *Proc Camb Philos Soc* 1913; 17(43): 4.

2 Attfield M, Barnes P, Cockcroft JK, Driessen H. *Advanced Certificate in Powder Diffraction on the Web.* http://pd.chem.ucl.ac.uk/pdnn/chapter.htm.

3 Young RA, Wiles DB. Application of the Rietveld method for structure refinement with powder diffraction data. In: *Advances in X-ray Analysis.* US: Springer, 1981: 1–23.

4 Young RA, Wiles DB. Profile shape functions in Rietveld refinements. *J Appl Crystallogr* 1982; 15(4): 430–438.

5 Miller JM, Collman BM, Greene LR, Grant DJ, Blackburn AC. Identifying the stable polymorph early in the drug discovery–development process. *Pharm Dev Tech* 2005; 10(2): 291–297.

6 Chen X, Carillo M, Haltiwanger RC, Bradley P. Solid state characterization of mometasone furoate anhydrous and monohydrate forms. *J Pharm Sci* 2005; 94(11): 2496–2509.

7 Rollinger J, Burger A. Physico-chemical characterization of hydrated and anhydrous crystal forms of amlodipine besylate. *J Therm Anal Calorim* 2002; 68(2): 361–372.

8 Ware EC, Lu DR. An automated approach to salt selection for new unique Trazodone salts. *Pharm Res* 2004; 21(1): 177–184.

9 Kiang YH, Xu W, Stephens PW, Ball RG, Yasuda N. Layered structure and swelling behavior of a multiple hydrate-forming pharmaceutical compound. *Cryst Growth Des* 2009; 9(4): 1833–1843.

10 Alexander L, Klug HP. Basic aspects of X-ray absorption in quantitative diffraction analysis of powder mixtures. *Anal Chem* 1948; 20(10): 886–889.

11 Suryanarayanan RR. Determination of the relative amounts of anhydrous carbamazepine ($C_{15}H_{12}N_2O$) and carbamazepine dihydrate ($C_{15}H_{12}N_2O.2H_2O$) in a mixture by powder X-ray diffractometry. *Pharm Res* 1989; 6(12): 1017–1024.

12 Uvarov V, Popov I. Development and metrological characterization of quantitative X-ray diffraction phase analysis for the mixtures of clopidogrel bisulphate polymorphs. *J Pharm Biomed Anal* 2008; 46(4): 676–682.

13 Roberts SN, Williams AC, Grimsey IM, Booth SW. Quantitative analysis of mannitol polymorphs. X-ray powder diffractometry-exploring preferred orientation effects. *J Pharm Biomed Anal* 2002; 28(6): 1149–1159.

14 Bish DL, Howard SA. Quantitative phase analysis using the Rietveld method. *J App Crystallogr* 1988; 21(2): 86–91.

15 Coelho AA, Topas GP. *Structure Analysis Software for Powder Diffraction Data.* V2.0. Karlsruhe, Germany: Bruker AXS GmbH; 2003.

16 Larson AC, Von Dreele R B. *General Structure Analysis System (GSAS).* Report LAUR. New Mexico: Los Alamos National Laboratory, 2000: 86–748.

17 Antonio SG, Benini FR, Ferreira FF, Rosa PC, De Oliveria Paiva-santos CDO. Quantitative phase analyses through the Rietveld method with X-ray powder diffraction data of heat-treated carbamazepine Form III. *J Pharm Sci* 2011; 100(7): 2658–2664.

18 David WI, Shankland K, van de Streek J, Pidcock E, Motherwell WS, Cole JC. DASH: a program for crystal structure determination from powder diffraction data. *J Appl Crystallogr* 2006; 39(6): 910–915.

19 Le Bail AL. ESPOIR: a program for solving structures by Monte Carlo analysis of powder diffraction data. *Mater Sci Forum* 2001; 378–381: 65–70.

20 Favre-Nicolin V, Černý R. FOX: Modular approach to crystal structure determination from powder diffraction. *Mater Sci Forum* 2004; 443: 35–38.

21 Engel GE, Wilke S, König O, Harris KD, Leusen FJ. PowderSolve – a complete package for crystal structure solution from powder diffraction patterns. *J Appl Crystallogr* 1999; 32(6): 1169–1179.

22 Werner PE, Eriksson L, Westdahl M. TREOR, a semi-exhaustive trial-and-error powder indexing program for all symmetries. *J Appl Crystallogr* 1985; 18(5): 367–370.

23 Boultif A, Louër D. Indexing of powder diffraction patterns for low-symmetry lattices by the successive dichotomy method. *J Appl Crystallogr* 1991; 24(6): 987–993.

24 Chernyshev VV. Unknown crystal structure with the use of powder diffraction data. *Z Kristallogr* 1998; 213: 1–3.

25 Hu X, Yuan Z, Lu G. Determination of crystal structures of polymorphic chlorothalonil using Monte Carlo simulated annealing and Rietveld refinement. *Powder Diffr* 2004; 19(4): 325–328.

26 Harris KD, Johnston RL, Kariuki BM. The genetic algorithm: foundations and applications in structure solution from powder diffraction data. *Acta Crystallogr Sect A* 1998; 54(5): 632–645.

27 de Armas HN, Peeters OM, Blaton N, Van den Mooter G, De Ridder DJ, Schenk H. Crystal structure of carnidazole form II from synchrotron X-ray powder diffraction: structural comparison with form I, the hydrated form and the low energy conformations in vacuo. *J Pharm Sci* 2006; 95(10): 2123–2136.

28 Kiang YH, Huq A, Stephens PW, Xu W. Structure determination of enalapril maleate form II from high-resolution X-ray powder diffraction data. *J Pharm Sci* 2003; 92(9): 1844–1853.

29 Visser JW. A fully automatic program for finding the unit cell from powder data. *J Appl Crystallogr* 1969; 2(3): 89–95.

30 Cheung EY, Harris KD, Johnston RL, Kitchin SJ, Hadden KL, Zakrzewski M. Rationalizing the structural properties of bupivacaine base—A local anesthetic—directly from powder X-ray diffraction data. *J Pharm Sci* 2004; 93(3): 667–674.

31 Alexander LE. *X-ray Diffraction Methods in Polymer Science*. New York: Wiley-Interscience, 1969: 381.

32 Kamat MS, Osawa T, DeAngelis RJ, Koyama Y, DeLuca PP. Estimation of the degree of crystallinity of cefazolin sodium by X-ray and infrared methods. *Pharm Res* 1988; 5(7): 426–429.

33 Atassi F, Mao C, Masadeh AS, Byrn SR. Solid-state characterization of amorphous and mesomorphous calcium ketoprofen. *J Pharm Sci* 2010; 99(9): 3684–3697.

34 Egami T; Billinge SJL. *Underneath the Bragg-peaks: Structural Analysis of Complex Materials*. Amsterdam: Elsevier Science B.V.; 2003.

35 Dykhne T, Taylor R, Florence A, Billinge SJ. Data requirements for the reliable use of atomic pair distribution functions in amorphous pharmaceutical fingerprinting. *Pharm Res* 2011; 28(5): 1041–1048.

36 Boetker JP, Koradia V, Rades T, Rantanen J, Savolainen M. Atomic pairwise distribution function analysis of the amorphous phase prepared by different manufacturing routes. *Pharmaceutics* 2012; 4(1): 93–103.

37 Nollenberger K, Gryczke A, Meier C, Dressman J, Schmidt MU, Brühne S. Pair distribution function X-ray analysis explains dissolution characteristics of felodipine melt extrusion products. *J Pharm Sci* 2009; 98(4): 1476–1486.

38 Newman A, Engers D, Bates S, Ivanisevic I, Kelly RC, Zografi G. Characterization of amorphous API: polymer mixtures using X-ray powder diffraction. *J Pharm Sci* 2008; 97(11): 4840–4856.

39 Bøtker JP, Karmwar P, Strachan CJ, Cornett C, Tian F, Zujovic Z, Rantanen J, Rades T. Assessment of crystalline disorder in cryo-milled samples of indomethacin using atomic pair-wise distribution functions. *Int J Pharm* 2011; 417(1): 112–119.

40 Chupas PJ, Qiu X, Hanson JC, Lee PL, Grey CP, Billinge SJ. Rapid-acquisition pair distribution function (RA-PDF) analysis. *J Appl Crystallogr* 2003; 36(6): 1342–1347.

41 Chen Z, Nie H, Benmore CJ, Smith PA, Du Y, Byrn S, Templeton AC, Su Y. Probing molecular packing of amorphous pharmaceutical solids using X-ray atomic pair distribution function and solid-state NMR. *Mol Pharmaceutics* 2023; 20(11): 5763–5777.

42 Bezzon VD, Ferreira FF, Smith P, Benmore CJ, Byrn SR, de Araujo GL. Amorphous dispersions of flubendazole in hydroxypropyl methylcellulose: formulation stability assisted by pair distribution function analysis. *Int J Pharm* 2021; 600: 120500.

43 De Araujo GL, Benmore CJ, Byrn SR. Local structure of ion pair interaction in lapatinib amorphous dispersions characterized by synchrotron X-ray diffraction and pair distribution function analysis. *Sci Rep* 2017; 7(1): 46367.

44 Bordet P, Bytchkov A, Descamps M, Dudognon E, Elkaïm E, Martinetto P, Pagnoux W, Poulain A, Willart JF. Solid state amorphization of β-trehalose: a structural investigation using synchrotron powder diffraction and PDF analysis. *Cryst Growth Des* 2016; 16(8): 4547–4558.

45 Pecharsky VK, Zavalij PY. *Fundamentals of Powder Diffraction and Structural Characterization of Materials*. New York: Springer; 2009.

46 Terban MW, Billinge SJ. Structural analysis of molecular materials using the pair distribution function. *Chem Rev* 2021; 122(1): 1208–1272.

47 Barbosa EJ, Andrade MA, Gubitoso MR, Bezzon VD, Smith PA, Byrn SR, Bou-Chacra NA, Carvalho FM, de Araujo GL. Acoustic levitation and high-resolution synchrotron X-ray powder diffraction: A fast screening approach of niclosamide amorphous solid dispersions. *Int J Pharm* 2021; 602: 120611.

48 Phadnis NV, Suryanarayanan R. Polymorphism in anhydrous theophylline-implications on the dissolution rate of the theophylline tablets. *J Pharm Sci* 1997; 86(11): 1256–1263.

49 Karjalainen M, Airaksinen S, Rantanen J, Aaltonen J, Yliruusi J. Characterization of polymorphic solid-state changes using variable temperature X-ray powder diffraction. *J Pharm Biomed Anal* 2005; 39(1): 27–32.

50 De Villiers MM, Terblanche RJ, Liebenberg W, Swanepoel E, Dekker TG, Song M. Variable-temperature X-ray powder diffraction analysis of the crystal transformation of the pharmaceutically preferred polymorph C of mebendazole. *J Pharm Biomed Anal* 2005; 38(3): 435–441.

10

Differential Scanning Calorimetry and Thermogravimetric Analysis

10.1 Introduction

Thermal analysis is an important analytical tool used to study the properties of materials as the temperature is changed. Differential scanning calorimetry (DSC) and thermogravimetric analysis (TGA) are two of the most widely used thermal analysis techniques in the characterization of crystalline and amorphous pharmaceutical materials. DSC measures various thermal transitions associated with a material when it is heated or cooled in a controlled manner. The thermal transition could be due to a physical transformation or a chemical reaction. Physical transformations, such as melting, crystallization, polymorph conversion, desolvation, and dehydration, are always associated with a change in enthalpy. This change of enthalpy is detected as a heat-absorbed (endothermic) or heat-released (exothermic) event in DSC. The thermal events associated with melting, dehydration, or desolvation are directly related to the crystal structure of a material and, therefore, provide fundamental information related to its crystal structure.

DSC monitors the change of heat flow as the sample is heated or cooled and reveals any change in heat capacity that occurs during any solid-state transition and is, therefore, generally useful for determining the glass transition temperature of amorphous materials. TGA measures changes in sample mass with temperature within a controlled environment. Such measurement provides information concerning the content of volatile materials, such as residual solvent or moisture, in a pharmaceutical material. TGA can also be used to detect dehydration or desolvation for pharmaceutical solvates/hydrates. DSC and TGA are usually combined to understand the thermal behavior of a pharmaceutical material more thoroughly. Applications of DSC/TGA for the study of pharmaceutical materials sciences include:

- characterize polymorphs, solvates, hydrates, salts, and cocrystals by measuring their melting point and other thermal transitions,
- characterize polymers and other macromolecules,
- characterize amorphous forms and amorphous solid dispersions,
- monitor dehydration and desolvation of crystals,
- optimize the freeze-drying cycle,
- measure chemical purity of organic compounds, and
- investigate solid-state reactions.

In this chapter, we examine the instrumentation used to carry out DSC and TGA, and present case studies that have utilized the two techniques in the study of pharmaceutical materials.

10.2 The Basics of DSC

DSC measures the changes that take place in a material with a change of temperature. Such changes are monitored by measuring heat flow associated with the heating of a sample compared with a reference while they are subjected to the same controlled temperature program. The sample and the reference are maintained at nearly the same temperature throughout the experiment. When the sample experiences a thermal transition, the heat flow to the sample relative to the reference changes to maintain the same temperature. For an endothermic event, heat is absorbed by the sample, and heat flow to the sample is higher than that to the reference. Conversely, heat flow to the sample is lower than that to the

Solid-State Materials in Pharmaceutical Chemistry: Properties, Characterization, and Applications, Second Edition. Stephen R. Byrn, George Zografi and Xiaoming (Sean) Chen.

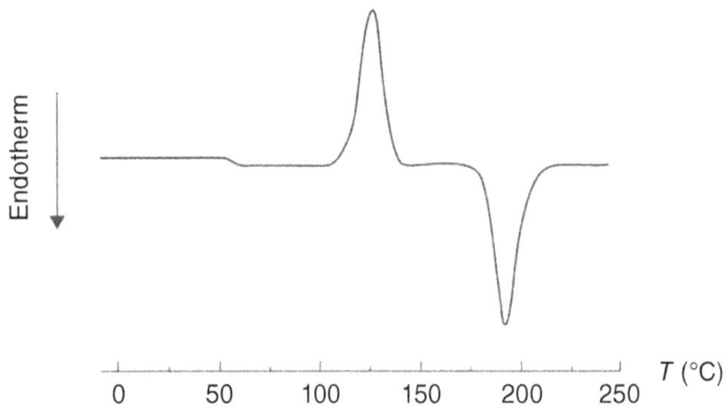

Figure 10.1 DSC thermogram of amorphous carbamazepine. *Source*: [1]/with permission of Elsevier.

reference if the sample is undergoing an exothermic transition. When the sample experiences a change in heat capacity, heat flow to the sample relative to the reference also shifts.

The data obtained by DSC is usually plotted as heat flow versus temperature, producing a thermogram. Figure 10.1 shows the thermogram of amorphous carbamazepine prepared by quench-cooling [1]. A baseline shift exists around 52 °C, followed by a sharp exothermic peak at 125 °C and an endothermic peak at 192 °C.

The shift in baseline around 52 °C is a direct reflection of a change in heat capacity. The heat capacity (C_p) of a material is the quantity of heat needed to elevate the temperature of the material by 1 °C. It is usually expressed in units of Joules/°C and is related to heat flow and heat rate by the following equation:

$$C_p = \frac{q}{\Delta T} = \frac{\frac{q}{t}}{\frac{\Delta T}{t}} = \frac{\text{Heat flow}}{\text{Heat rate}} \tag{10.1}$$

The change in heat capacity displayed in Figure 10.1 is due to the glass transition of amorphous carbamazepine.

An endothermic or exothermic transition is typically displayed as a peak in a thermogram. The area of the peak is related to the enthalpy of the transition as

$$\Delta H = \int_{t_1}^{t_2} \frac{dQ}{dt} dt = \int_{T_2}^{T_1} \frac{\frac{dQ}{dt}}{\frac{dT}{dt}} dT = \frac{\int_{T_1}^{T_2} \frac{dQ}{dt} dT}{\text{Heating rate}} = \frac{\text{Peak area}}{\text{Heating rate}} \tag{10.2}$$

where ΔH is the heat of transition, $\frac{dQ}{dt}$ is the heat flow, $\frac{dT}{dt}$ is the heating rate, t_1 is the time the thermal c transition begins, t_2 is the time at which the transition ends, T_1 is the temperature at which the transition starts, and T_2 is the temperature at which it ends.

The exothermic peak in Figure 10.1 reflects the loss of enthalpy due to crystallization of amorphous carbamazepine to become Form I, whereas the second peak is endothermic and related to the melting of Form I. In Figure 10.1, the endotherm is pointed down and the exotherm is pointed up. However, with some DSC equipment, the endotherm may appear up and the exotherm down.

10.3 Thermal Transitions of Pharmaceutical Materials

Thermal transitions reflect physical and chemical processes that are associated with the release or absorption of heat. Such transitions include melting, the glass transition, crystallization, polymorph conversion, desolvation, dehydration, and chemical reactions.

10.3.1 Melting

The melting of a crystalline solid is a first-order transition that occurs at a distinct melting temperature, which is detected as an endothermic event in DSC. The enthalpy change associated with melting is defined as the heat of fusion. The melting point and heat of fusion are characteristic of a specific solid form. For example, for two polymorphs of indomethacin, the γ form, the stable polymorph, has a melting point of 163 °C and a heat of fusion of 36.5 kJ/mol, whereas the melting point of the α form is 157 °C and the heat of fusion is 35.4 kJ/mol [2]. The melting point for a crystalline small molecule such as indomethacin is usually reported as the onset temperature. Using the onset temperature is more reflective of the true melting point compared to using the peak temperature. In theory, the sample will remain constant at the melting temperature if the solid and liquid are in equilibrium. Consequently, the endotherm would be expected to be infinitely sharp and narrow. In reality, there is a thermal gradient across the sample, so that it takes time for heat to flow into the entire sample.

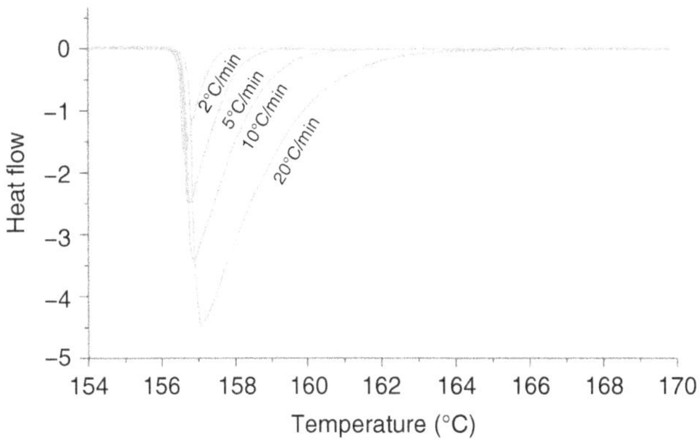

Figure 10.2 Effect of heating rate on the DSC thermogram of indium.

The melting endotherm is usually displayed as an asymmetrical peak with a leading edge. The onset temperature is taken as the intercept point of the leading edge and the baseline. Heating rates significantly affect the peak temperature during melting, as shown for indium in Figure 10.2, whereas onset temperatures remain relatively constant. When the heating rate is higher, peak height and peak area increase, with the increased peak area being proportional to the heating rate as revealed by equation (10.2).

10.3.2 Glass Transition in Amorphous Solids

Just as melting point and heat of fusion are important and distinctive characteristics for a crystalline form, knowing the glass transition temperature of an amorphous material is of equal importance. Amorphous materials lack three-dimensional order, and, as discussed in Chapter 6, they exhibit the "macroscopic" mechanical properties of a solid and have the "microscopic" or local properties of a liquid. As previously discussed in Chapter 6, rapidly cooling a liquid below the melting temperature produces a supercooled liquid with the equilibrium properties of the liquid but much more viscous as the temperature is reduced. When the viscosity reaches roughly 10^{13} poise, upon further cooling, molecular motions are reduced enough that equilibrium cannot be maintained, and the system goes into the unstable glassy state. The temperature at which this occurs is the glass transition temperature, T_g.

The glass transition can be detected by DSC because a change in heat capacity occurs when the material transforms from a supercooled liquid state to the glassy state or vice versa. The glass transition temperature is detected by DSC as a shift in the baseline of the thermogram. Generally, it is taken to be the midpoint temperature for the range of temperatures over which the transition occurs.

Figure 10.3 shows the glass transition temperature of amorphous clopidogrel hydrogen sulfate. The amorphous material, prepared from an ethanol solution, has a T_g of 90 °C [3].

10.3.3 Enthalpy Relaxation for Amorphous Solids

Enthalpy relaxation can be observed during a DSC scan of an amorphous material as an endotherm in the vicinity of the glass transition temperature. The appearance and amplitude of the endotherm depend on the thermal history of the material (Chapter 6). The endotherm arises, since over the period of storage or thermal treatment, the amorphous material "relaxes" into a more ordered and lower energy state approaching the energy expected for a supercooled liquid. This "relaxation" results in a decrease in the molar volume and an increase in density. Higher temperatures, closer to T_g, and longer aging time produces a larger enthalpy relaxation. As shown in Figure 10.4, the enthalpy relaxation of amorphous lactose is increased with temperature after annealing for 24 h [4].

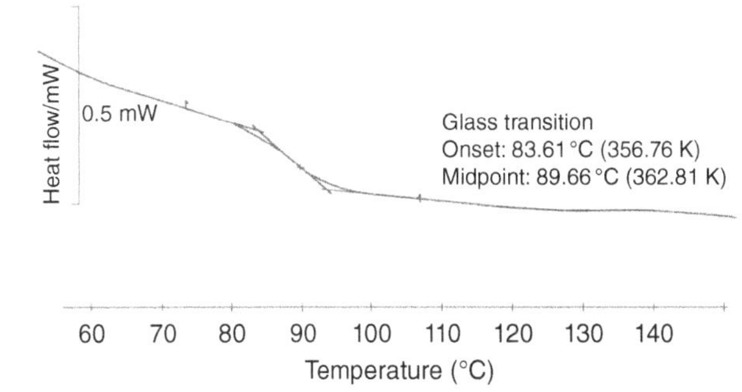

Figure 10.3 Glass transition temperature of amorphous clopidogrel hydrogen sulfate. *Source:* [3]/with permission of Springer Nature.

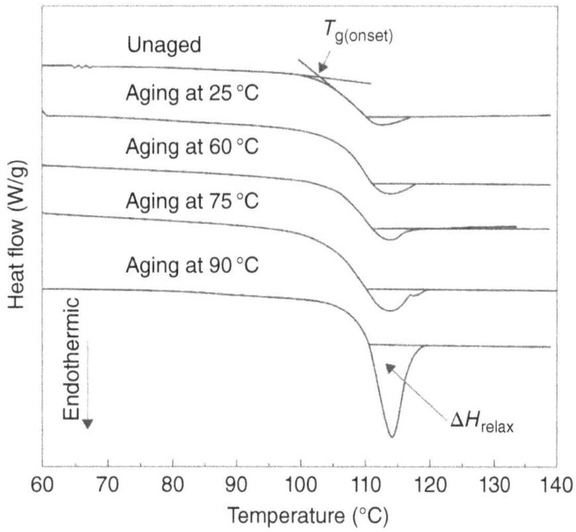

Figure 10.4 DSC thermograms of un-aged and aged (for 24 hours) amorphous lactose showing the glass transition and enthalpy relaxation endotherm at the glass transition temperature. *Source*: [4]/with permission of Elsevier.

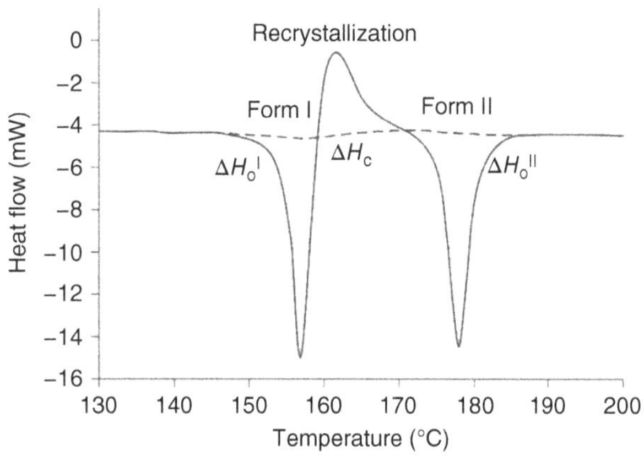

Figure 10.5 DSC scan of a drug substance undergoing physical changes: melting of Form I, recrystallization into Form II, melting of Form II, 10 K/min. *Source*: [5]/with permission of Elsevier.

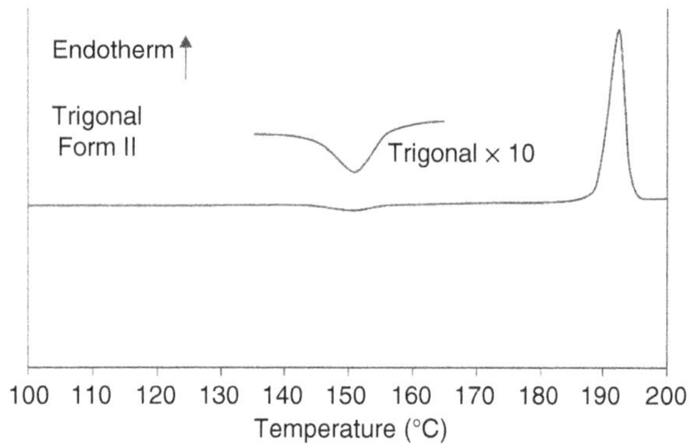

Figure 10.6 DSC scan of carbamazepine Form II. *Source*: [7]/with permission of Elsevier.

10.3.4 Crystallization

Crystallization is a physical transition from a disordered molecular state to an ordered state. It can be detected as an exotherm in a DSC thermogram. Figure 10.1 shows the thermogram for amorphous carbamazepine prepared by quench-cooling, where a glass transition is observed at 52 °C, followed by a sharp crystallization peak at 125 °C. As observed in Figure 10.1, the crystallization resulted in the formation of Form I, which melted at 192 °C [1].

It is possible for an exotherm associated with crystallization to exist in the thermogram of a crystalline form. After the melting of the crystalline form, for example, a higher temperature melting form can crystallize when heated up, as illustrated in Figure 10.5 [5]. Crystallization from the melt is a two-step process: nucleation and crystal growth. Nucleation is the rate-limiting step and kinetically driven, so the presence of the exothermic crystallization peak could be significantly impacted by the heating rate and sample quality. Sometimes the exothermic peak is merged with the melting peak, which makes interpretation challenging.

10.3.5 Crystal Form Transitions

Transitions from one crystal form to another can occur when one polymorph is converted to another one during DSC analysis. The transition can be observed as an endothermic or exothermic event. As discussed in Chapter 2, the relative stability of two polymorphs can be monotropic or enantiotropic. Based on the heat of transition rule [6], an endothermic transition indicates that two polymorphs are possibly related enantiotropically. Conversely, an exothermic transition reveals that the two polymorphs are possibly related monotropically. Compared with those peaks observed with melting and crystallization, the peak area for a polymorphic transition is relatively weak. For example, a small exotherm (−0.35 kcal/mol) is present in the thermogram of carbamazepine trigonal Form II between 140 and 160 °C, as shown in Figure 10.6 [7]; this exotherm represents the transformation from Form II to Form I, whereas the endotherm at 192.1 °C (5.72 kcal/mol) is associated with the melting of Form I. The exothermic transition from Form II to Form I indicates the monotropic relationship between these polymorphs.

In another study, DSC thermograms of pentamidine isethionate Forms A, B, and C revealed a number of thermal transitions [8]. Both Form A and Form B exhibited an endothermic transition to Form C, which melted at 191 °C. The transition temperatures from Form A and Form B to Form C are 124 and 87 °C, respectively. Form A and Form B were shown to be enantiotropically related to Form C.

10.3.6 Desolvation/Dehydration

Desolvation of a crystal solvate (hydrate) at elevated temperatures can be treated as a solid–gas reaction, as follows:

$$A \cdot B\,(\text{solid}) \rightarrow A\,(\text{solid}) + B \uparrow \tag{10.3}$$

where $A \cdot B$ is the solvate, A is the desolvated form, and B is the solvent. Desolvation is an endothermic reaction because energy is required for the solvent to be released from the crystalline lattice and to exist in the vapor phase. Consequently, desolvation and dehydration are displayed as an endotherm in DSC thermograms. Figure 10.7 shows the DSC thermogram of baclofen monohydrate, with an endotherm at 67 °C. In addition, another endotherm at 212 °C occurs at the melting of the anhydrate [9].

10.3.7 Chemical Degradation

During the heating that takes place in DSC analysis, thermal degradation of a material could occur at elevated temperatures. Thermal degradation can be reflected by either an exothermic or an endothermic event. Figure 10.8 shows the DSC thermogram of the palmityl prodrug analog of capecitabine (PAC) [10]. A melting peak occurs at 115 °C, whereas, above the melting peak, a broad endotherm exists at 130 °C due to the degradation of PAC.

Sometimes, thermal degradation events overlap with melting, making the interpretation of such various thermal events more challenging. For a thermally labile compound, thermal degradation also could occur before melting. Since thermal degradation is a kinetic phenomenon, increasing the heating rate in the DSC, to reduce thermal exposure, may reduce the intensity of degradation or mitigate its occurrence.

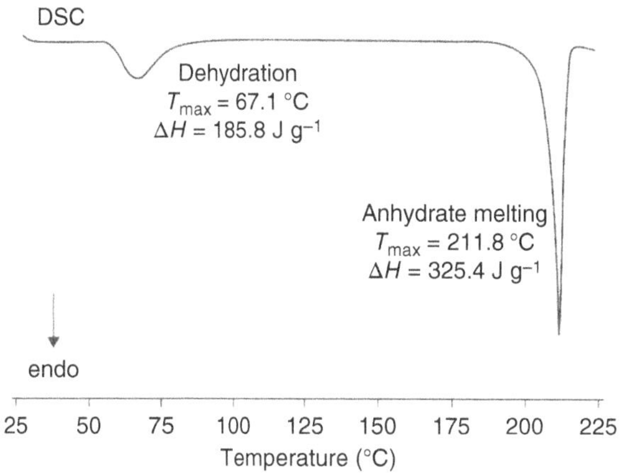

Figure 10.7 DSC thermogram of baclofen monohydrate. *Source*: Adapted from [9].

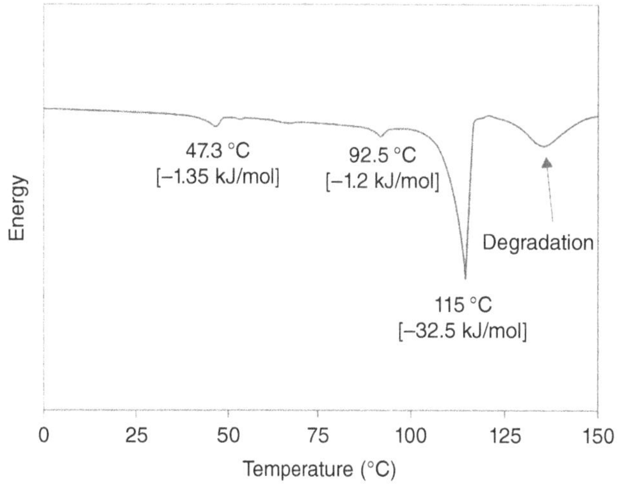

Figure 10.8 DSC thermogram of palmityl prodrug analog of capecitabine. *Source*: [10]/with permission of Elsevier.

10.4 DSC Instrumentation

There are two principal types of DSC instruments with distinct designs: heat-flux DSC and power-compensated DSC.

10.4.1 Heat-flux DSC

For heat-flux DSC, the sample and reference platforms are positioned in the same furnace (Figure 10.9), and a metallic block with high thermal conductivity, the heat-flux plate, links the sample platform and reference platform to ensure a good heat flow. Heat is transferred to the sample S and the reference R through the heat-flux plate in a controlled manner. The enthalpy or heat capacity changes in the sample S cause small temperature differences between the sample and that of reference R. The temperatures of sample S and reference R are monitored through chromel thermocouples, and the temperature difference is recorded and related to differences in heat flow through a mathematical equation that is the thermal equivalent of Ohm's law:

$$q = \frac{\Delta T}{R} \tag{10.4}$$

where q is the sample heat flow, ΔT is the temperature difference between the sample and the reference, and R is the resistance of a thermoelectric disk.

TA DSC 2920, Q2000, and Mettler DSC are examples of heat-flux DSC instruments.

Heat-flux DSC instruments usually give a superior baseline compared to power compensation DSC instruments, because smaller temperature differences exist between the enclosure and the calorimeter in heat-flux DSC.

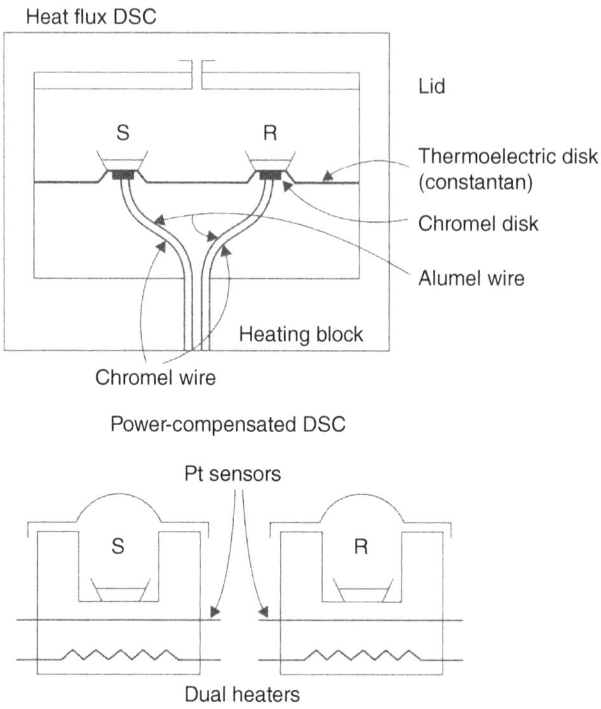

Figure 10.9 Schematic diagrams of (a) heat-flux differential scanning calorimetry and (b) power-compensated differential scanning calorimetry. *Source*: [5]/with permission of Elsevier.

10.4.2 Power-compensated DSC

For power-compensated DSC, the sample and the reference are heated with separate heaters (Figure 10.9). Both the sample and the reference are maintained at the same temperature by supplying needed electric power through a feedback loop. The enthalpy or heat capacity change in the sample causes the change of electric power supplied to the sample furnace, which is directly related to heat flow. In contrast to heat-flux DSC, heat flow is directly measured in power-compensated DSC, also called heat flow DSC. The instrument Perkin Elmer Pyris 1, for example, is a power-compensated DSC.

Both heat-flux DSC and power-compensated DSC are comparable in most applications of pharmaceutical interest. For power-compensated DSC, the mass of the furnace can only be about 1 g compared to 100–200 g for most heat-flux devices. Because of its relatively small heat elements, heating, cooling, and thermal equilibration in power-compensated DSC can occur as quickly as possible. Thus, it can operate at very high heating rates (250 °C/min and above).

10.4.3 Modulated DSC

Traditional DSC is used to analyze thermal transitions for a material using a linear heating rate. Modulated DSC (MDSC), developed by TA Instruments, advances the use of traditional heat-flux DSC by applying a superimposed sinusoidal oscillation to a linear temperature ramp [11]. The oscillations cause a continuously changing small deviation of the real heating rate from the underlying linear ramp, providing some unique capabilities compared to those obtained from traditional DSC, including

- separating reversible and nonreversible thermal events,
- measuring heat capacity directly, and
- increasing sensitivity to detect weak transitions without compromising resolution.

Traditional DSC measures only the sum of all thermal events in the sample. Therefore, if multiple transitions take place in the same temperature range, the results are often difficult to analyze. Owing to its capability to separate reversible and nonreversible thermal events, MDSC offers significant advantages for deconvoluting and interpreting such thermal events. Examples of reversible thermal events are melting and the glass transition, whereas nonreversible events include dehydration, desolvation, chemical degradation, and enthalpy relaxation. For example, enthalpy relaxation for amorphous glasses is an endothermic event that is dependent on the thermal history of the material. In some cases, for example, the method of producing the glass can cause the magnitude of the endotherm to be so great as to make the glass transition appear to

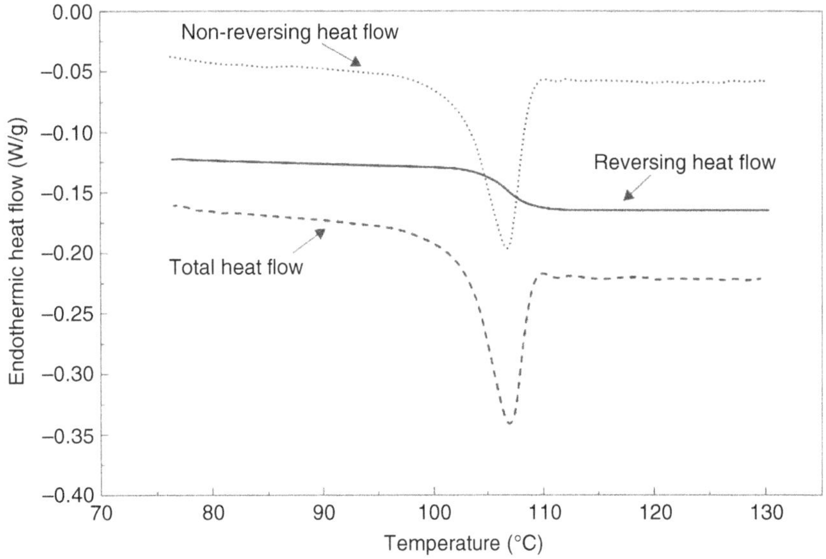

Figure 10.10 The glass transition of saquinavir studied by MDSC. *Source*: [12]/ with permission of Springer Nature.

be a melting transition. The use of MDSC is particularly advantageous when attempting to separate the glass transition from enthalpy relaxation, to give a more precise measurement of the glass transition temperature. Figure 10.10 shows how the glass transition of saquinavir was isolated from the enthalpy relaxation endotherm in MDSC to give a glass transition temperature of 107.0 °C [12].

The Perkin Elmer Co. has developed a technology that is like MDSC, called modulated temperature DSC, where it applies a stepwise scan instead of a sinusoidal oscillation [13]. This approach is also applicable to power-compensated DSC because of its accurate temperature control and the ability to equilibrate rapidly.

10.4.4 Fast-scan DSC

Fast-scan DSC refers to a DSC technique that applies a very high heating or cooling rate to a sample, that is, a range from 100 to 300 °C/min [14]. Fast-scan DSC was originally implemented for use in power-compensated DSC due to the small furnace size and, therefore, a quick thermal response. As shown in equation (10.1), the intensity of heat flow is directly proportional to the heating rate. The advantage of using this technique is that a high heating rate increases the sensitivity of the DSC and mitigates the effects of slower kinetic events. For example, while the response to weak transitions, such as the glass transition, is enhanced by very rapid heating rates, kinetic events such as crystallization and thermal degradation can be mitigated. The mitigation of thermal degradation using fast-scan DSC is advantageous for measuring the melting point and heat fusion of thermally unstable compounds.

One challenge in using very high heating rates in carrying out DSC measurements is thermal lag. While the furnace may be heating at a rapid rate, a finite time delay occurs before the sample has the same heating rate. The extent of the thermal lag is correlated to heating rate, chemical nature of the sample, pan mass, sample- and pan-specific heat capacity, and the thermal resistance between the sensor and the sample. Practically, using a small sample size and ensuring good sample and pan contact helps alleviate thermal lag.

Another technology using infrared radiation as a heating source could enable an even higher heating rate up to 2000 °C/min. Infrared-heated DSC is also called rapid-heating DSC [15].

10.4.5 Operation of DSC Instrumentation

Operating a DSC instrument is relatively straightforward since most DSC equipment comes with computer software to input experimental parameters. Several key experiment parameters need to be considered and controlled to obtain desired analytical results and to ensure consistent data interpretation. These are

- sample pan configuration,
- sample size,

- temperature range,
- scan rate, and
- purge gas.

Pan configurations used in DSC analysis include open pan, pin-hole pan, closed pan, and hermetically sealed pan. Most severe limitations for pan configuration choices exist for samples that contain volatile components or create volatile materials upon heating. Examples of volatile components include residual moisture, residual solvent, water in a hydrate, solvent in a solvate, and volatile degradation products released during DSC analysis. In such cases, therefore, using a hermetically sealed pan would be required to trap the volatile material inside the DSC pan and delay or prevent desolvation. As shown in Figure 10.11, for example, a single melting peak at 68 °C occurs in the thermogram of SAR474832 hydrate Form 1 with a hermetically sealed pan being utilized, whereas, when an open pan or a pan with pin holes is used, events associated with

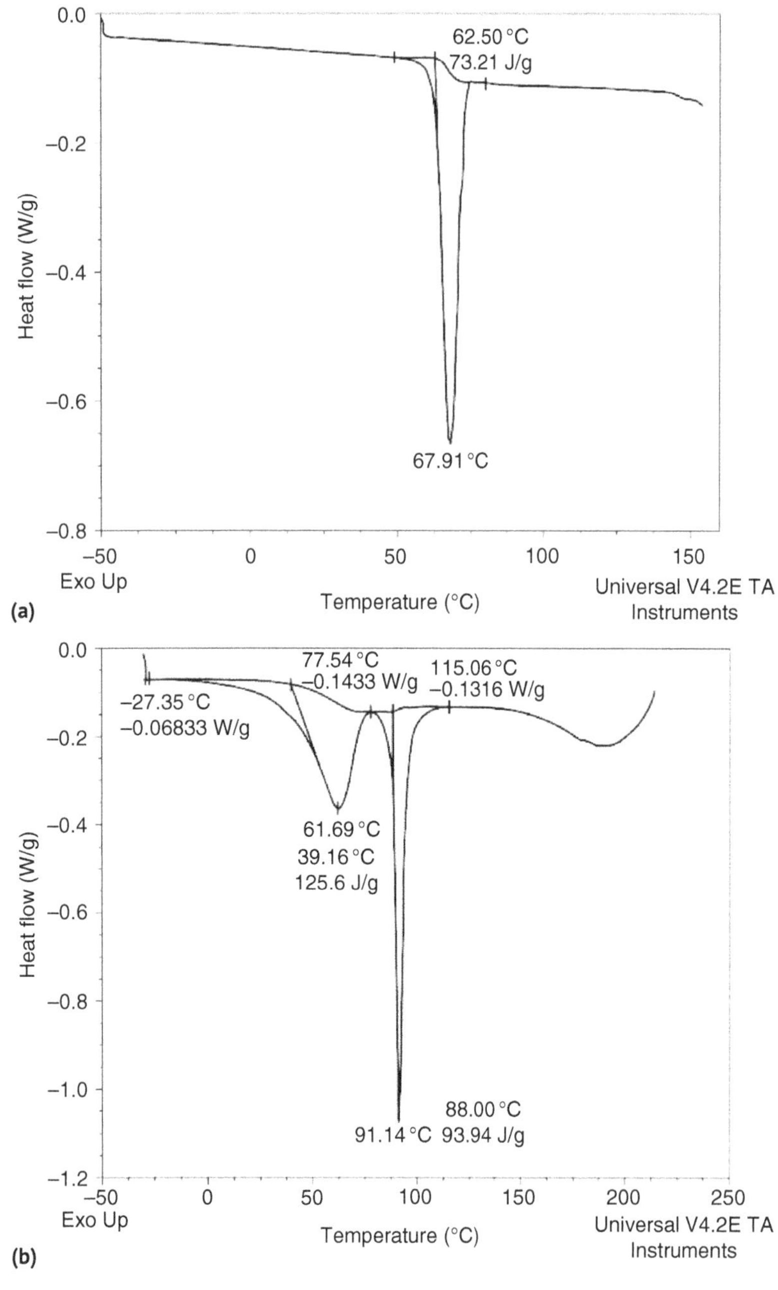

Figure 10.11 DSC plots of SAR474832 crystallographic form 1 in (a) a hermetically sealed steel pan and (b) an aluminum pan with a pin hole. *Source*: Adapted from [16].

desolvation or dehydration takes place in additional to the melting of the desolvated material. As such, the same material gave two endothermic peaks when it was subjected to DSC analysis in a pin-hole pan [16]. The first peak at 62°C reflects dehydration, and the second peak at 91 °C corresponds to the melting of the dehydrated form. This scenario is also applicable to a compound subject to sublimation in that melting can be observed while sublimation is being prevented. This provides good evidence that, when interpreting DSC data, it is critical to take pan configuration into consideration.

DSC is a sensitive analytical technique, so, for conventional DSC analysis, 1–3 mg of sample is generally sufficient to obtain high-quality data. For samples with weak transitions, however, the sample size should be increased to as much as 5–10 mg.

The temperature range for carrying out DSC analysis is chosen by considering the type of analysis to be carried out and the material characteristics of the sample.

Traditionally, the most common scan rate is 10°C/min, but, depending on the instrument design, the heating rate can be varied between 0.001 and 500°C/min. As mentioned earlier, high heating rates would result in greater sensitivity, but, due to thermal gradients that can occur across a sample, the faster the scan rate, potentially the lower the resolution. Consequently, choosing a proper heating rate is a balance between desired sensitivity and resolution. Proper use of heating rates could lead to the separation of two independent thermal events: desolvation and melting. Moreover, different heating rates can be used to detect the occurrence of crystallization during DSC analysis. Figure 10.12, for example, presents the thermograms of nateglinide Form B scanned at various heating rates [17]. At 0.5°C/min, the melting of Form B is merged with an exothermic crystallization peak. As the heating rate is increased to 1 and 2°C/min, the melting peak and the crystallization peak are separated distinctly, where the crystallization occurs at higher temperatures as the heating rate is increased to 3 and 5°C/min. When the heating rate was increased to 10 and 20°C/min, the crystallization peak

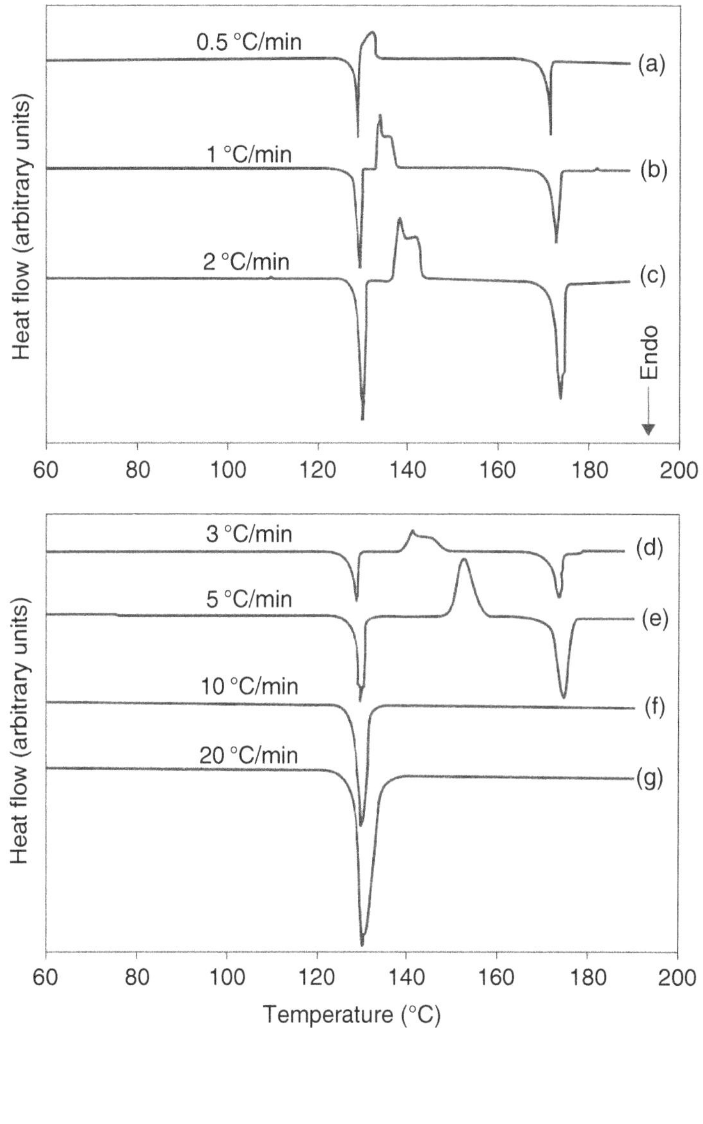

Figure 10.12 DSC scan of nateglinide Form B under various heating rates. *Source*: [17]/with permission of Elsevier.

disappeared. Consequently, the endothermic peak related to the melting of Form S, which is formed during crystallization, was no longer present.

A purge gas is generally used in DSC analysis to control the sample environment, remove volatiles from the system, prevent contamination, reduce signal noise by mitigating internal convection currents, and prevent ice formation at subambient temperatures. Nitrogen is the most used purge gas since it offers an inert environment and prevents oxidative reactions. Helium is an alternative for analysis at very low temperatures or when using fast-scan DSC, due to its low boiling point and high thermal conductivity.

Maintaining cleanness is critical for DSC analysis because contamination will make the interpretation of DSC data difficult and misleading. Also, routine calibration at various heating rates, using calibration standards, should be carried out when the actual runs are made, based on the manufacturer's specifications.

10.5 Thermogravimetric Analysis

TGA is used to measure changes in the mass of a material as a function of time at a selected temperature (i.e. isothermal mode) or over a particular temperature range using a predetermined heating rate under a controlled atmosphere. TGA is carried out with a thermogravimetric analyzer that consists of an electronic microbalance surrounded by a furnace. The microbalance records any mass change as the temperature of the sample is being increased or kept at a constant temperature. The weight of the sample is plotted against time for isothermal studies or as a function of temperature for experiments carried out at a constant heating rate. This technique is widely used to monitor thermal stability and the loss of volatile components for pharmaceutical materials, where a loss of mass occurs because of thermal degradation or desolvation. In desolvation, adsorbed or bound solvents leave the material at elevated temperatures, causing a decrease in sample mass, whereas for thermal degradation the loss in mass occurs when volatile degradation products are formed.

A TGA instrument consists of a sample pan that is linked to a precision balance. The balance can be in either a bottom-loading or hangdown configuration. For example, the Perkin Elmer TGA 400 is bottom loading with a stem support rod to connect the sample pan and the balance. Both the TA instrument TGA 2000 and the Perkin Elmer Pyris 1 TGA use a hangdown design. The change in the mass of the sample is continuously recorded during an experiment, with the sample atmosphere being controlled by purging with a dry gas.

TGA thermal curves obtained are generally displayed with time or temperature as the X-axis and weight or weight percentage as the Y-axis. A typical TGA thermal curve for the dehydration of a sample of SAR474832 hydrate Form 1 is given in Figure 10.13, where there are two steps of weight loss, one due to 3.6% dehydration over the range of 22–26 °C, and another one with 2.8% weight loss from 60 to 120 °C [16].

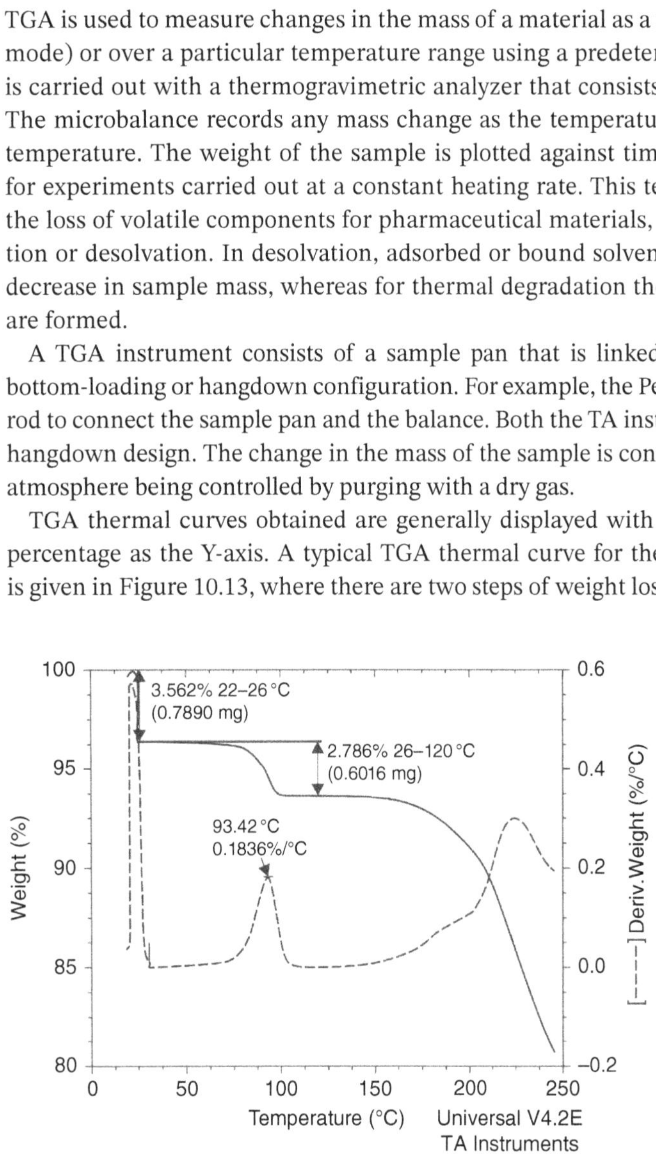

Figure 10.13 TGA plot of SAR474832 crystallographic form 1. Top: weight loss; bottom: first derivative of weight loss. *Source:* Adapted from [16].

From the thermal curve, the exact amount of weight change that occurs in absolute or relative terms can be obtained in a defined time or temperature window. For a stepwise weight change, extrapolation of the baselines before and after the transition to where they cross with the weight loss line, and connection of the points, leads to the onset temperature and end temperature. The weight change between the two points then defines the weight loss for the transition.

Another useful practice for studying stepwise transitions is to plot the first derivative of the weight loss curve (Figure 10.13), so that the stepwise weight loss becomes a peak in the plot using the first derivative. The peak of the first derivative, also known as the inflection point, is directly related to the point of greatest rate of change on the weight loss curve. As shown in Figure 10.13, the second dehydration event of SAR474832 hydrate Form 1 has an inflection point at 93 °C [16].

10.6 Operating a TGA Instrument

Operating a TGA instrument requires initially establishing a few experimental parameters, including sample weight, temperature range, scanning rate, and sample atmosphere. The calibration of the instrument includes calibration of the balance, the operating temperature, and the furnace. Balance calibration is usually carried out by using the calibration weights supplied with the instrument. Curie point reference materials are ferromagnetic materials, such as perkalloy wire and nickel wire, and are utilized to carry out temperature calibration.

Sample weights from 10 to 50 mg are commonly used for TGA analysis. Sample pans, made of either platinum or a ceramic material, are available to accommodate liquids, powders, pellets, compacts, and films. It is also important to recognize that the particle size of the sample can have a significant effect on the kinetics of weight loss, for example, during desolvation and dehydration.

The temperature range is usually selected to cover the transition in the material being tested. If the transition is not defined, a quick survey scan can be carried out at relatively higher heating rates, such as 20 °C/min. After reviewing the survey scan, a proper temperature range and a slower scanning rate for the TGA analysis run, for example, 10 °C/min or 5 °C/min, can be chosen.

When carrying out DSC and TGA analysis on the same sample, it is important to match the heating rates so the data can be compared and interpreted.

In addition, furnace cleanliness is critical for ensuring the successful operation of a TGA instrument. The furnace is generally cleaned by heating the furnace to 900 °C or above while it is exposed to the atmosphere. Under these conditions, we can expect that all organic residues will be burned away.

10.7 Evolved Gas Analysis

Evolved gas analysis is a technique that uses TGA in combination with another methodology to detect and follow the evolution of gas products from solids at increasing temperature. Those techniques include TGA/FT-IR, TGA/MS, and TGA/GC-MS. Evolved gas analysis, using such techniques, provides chemical and structural information for the volatile materials that are released during a TGA operation. The combination of chemical and structural information with TGA is very useful in helping to better understand the mechanistic nature of a thermal transition being followed in TGA.

For TGA/FT-IR, an infrared spectrometer is coupled with a TGA, and the volatile materials generated from a TGA experiment are transferred to the IR cell for measurement and identification. As mentioned earlier, common volatile materials released from pharmaceutical materials during heating include water, carbon dioxide, and various organic solvents. Each of these vapor molecules has a characteristic IR spectrum. Figure 10.14 provides an example of using evolved gas analysis to characterize a pharmaceutical material [18]. The material, exhibiting a weight loss at 73 °C, was shown by FT-IR to be ethanol, indicating that the original solid material was an ethanol solvate.

TGA/MS is a combination of TGA and mass spectrometry, which enables the detection of very low levels of impurities or degradation products that become volatile at elevated temperatures.

TGA/GC-MS is similar to TGA/MS, with the addition of gas chromatography to increase sensitivity. The volatile materials released from TGA analysis are directly transferred to a gas chromatography column for separation and identified by a mass spectrometer. Both techniques provide a powerful tool to identify low levels of impurities or solvents released from pharmaceutical materials during TGA analysis.

10.8 Applications of DSC and TGA

DSC and TGA have diverse applications in the characterization of pharmaceutical materials. Some applications of DSC and TGA are exemplified in the following paragraphs.

10.8.1 The Study of Polymorphs, Solvates, and Hydrates

DSC and TGA are widely used to characterize polymorphs, solvates, and hydrates. At a given heating rate and pan configuration, a polymorph will give a characteristic thermogram reflecting its various physical transitions, such as crystal form

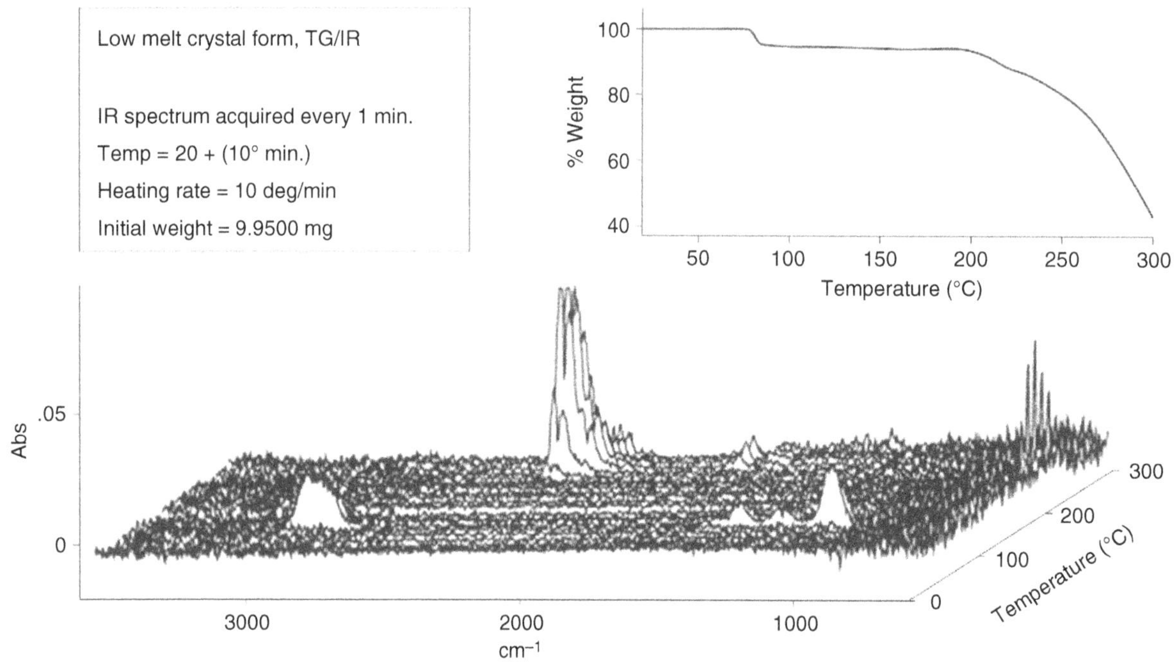

Figure 10.14 TGA weight loss curve and subsequent IR spectra measured as a function of temperature. *Source*: [18]/with permission of Elsevier.

conversion, melting, and recrystallization. Besides recording the temperature associated with the melting point and other phase transitions, DSC data can be used to calculate the heat of fusion and heat of transition. Such physical data can be used to estimate the relative physical stability of polymorphs, based on the heat of fusion rule and heat of transition rule, and to construct temperature and energy diagrams, as discussed in Chapter 2.

DSC analysis is also essential for the study of pseudopolymorphs (solvates and hydrates). Desolvation and dehydration at elevated temperature in DSC measurement appear as endothermic events, usually displayed as a broad peak. Acquiring TGA data from the same system is important to confirm the weight loss in the same temperature range to establish that the thermic event is due to desolvation or dehydration. The position and energy of this endothermic peak are related to the stability of the solvate, while the amount of weight loss can indicate the stoichiometry of the solvate or hydrate, as discussed in Chapter 3.

For example, recrystallization of nevirapine from various solvents resulted in the formation of solvates [19], as identified by using the combination of DSC and TGA (Figure 10.15).

In this system, the presence of broad endotherms in the DSC, accompanied by the weight loss measured in the TGA, over the same temperature range, revealed the formation of a number of solvates, hemi-ethanolate (Form I), hemi-acetonitrilate (Form II), hemichloroformate (Form III), hemi-THF solvate (Form IV), mixed hemi-ethanolate hemi-hydrate (Form V), and hemi-toluenate (Form VI). After heating each of these solvates, in all cases, desolvation produced an anhydrous form with a melting point of 246 °C. Except for the toluenate, the higher desolvation temperatures of all the other solvates relative to their respective boiling point indicated very tight binding of solvent in the crystal lattice. Obtaining a combination of DSC and TGA data is also useful for determining whether a solid is an anhydrous form (polymorphs) or a solvated form (solvate and hydrate). Niederwanger et al., for example, reported the characterization of four crystal polymorphs and a monohydrate of s-bupivacaine hydrochloride using DSC and TGA [20]. The polymorphs and the monohydrate of s-bupivacaine hydrochloride have distinct DSC curves, where the curve of Form A has one endotherm at 84 °C and another endotherm at 261 °C. The first endotherm is related to the phase transition from Form A to Form B. Form B was converted to Form D when it was heated above 180 °C. The second endotherm was due to the melting of Form D. There is also a weak exothermic transition in the DSC curve of Form C between 140 and 180 °C. Form D was formed after this transition; however, pure Form D has no clear thermal events when measuring from room temperature to the melting temperature. The monohydrate has a broad endotherm peak at about 60 °C and a weight loss of 5.8%, observed with TGA measured over the same temperature range.

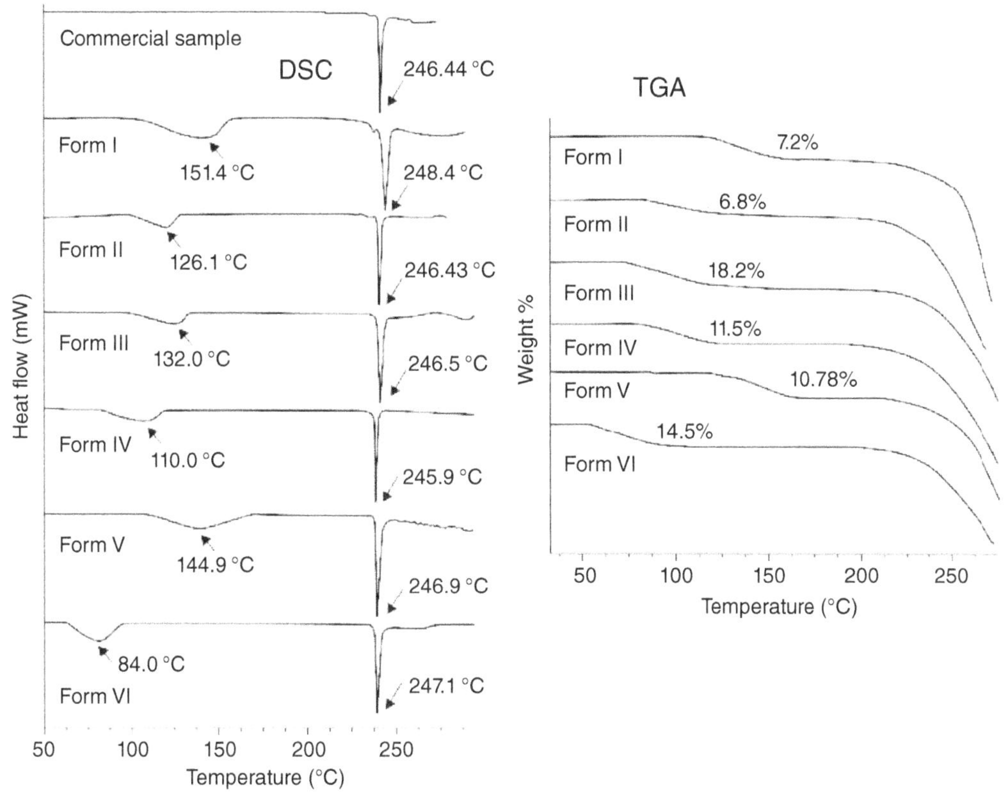

Figure 10.15 DSC thermograms and TGA curves of nevirapine and its solvated forms. *Source*: Adapted from [19].

10.8.2 Polymer Characterization

Many polymers are used pharmaceutically as excipients in solid dosage forms, as packaging materials, and for manufacturing medical devices. DSC is an essential technique to characterize polymers used in pharmaceutical products. Using DSC for the characterization of such polymers provides several important thermal parameters, such as the glass transition temperature (T_g), crystallization temperature (T_c), and melting temperature (T_m), which establish important polymer physical and chemical properties, stability, and processability. For example, plasticizers are often added to a polymer to obtain a proper glass transition temperature for its intended use and processing. The level of plasticizer and type of plasticizer can be selected with the aid of DSC analysis, as illustrated in Figure 10.16, which shows the effect of various plasticizers on the glass transition temperature of ethylcellulose at a 20% and 40% level [21].

DSC analysis is also useful for assessing the physical stability of a polymer after it has undergone a period of storage. Polymers tend to undergo physical aging or curing during storage, possibly leading to a change in mechanical properties. Following any change in the glass transition temperature, which would reflect such behavior, could be carried out by DSC.

DSC is also utilized to assess the miscibility of two polymers when two polymers have distinctly T_g values. If the two polymers are miscible, a single T_g should be observed.

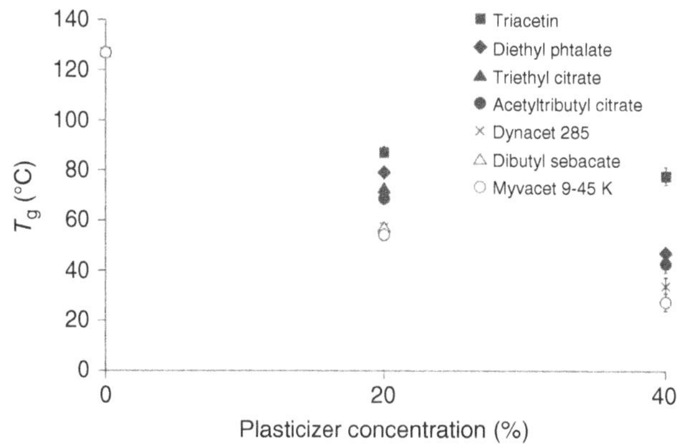

Figure 10.16 Glass transition temperatures of plasticized and pure ethylcellulose films measured with DSC. *Source*: [21]/with permission of Elsevier.

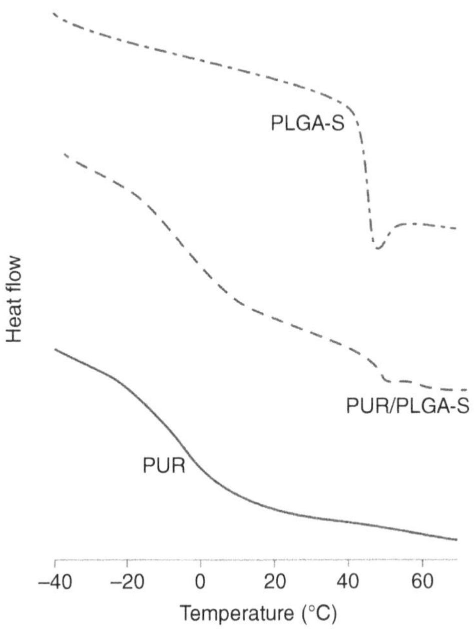

Figure 10.17 DSC thermograms showing glass transitions of poly(lactic-co-glycolic acid) (PLGA-S, top), polyurethane (PUR, bottom), and their mixtures (PUR/PLGA-S, middle). *Source*: Adapted from [22].

Conversely, if the two polymers are not miscible, two independent T_g will be detected in the mixture. Figure 10.17 shows the DSC thermograms of polyurethane, poly(lactic-co-glycolic acid), and their mixtures [22]. Two T_g values were observed in the mixture, indicating that these two biodegradable polymers are not miscible. Consequently, the analysis by DSC of such systems offers guidance in designing appropriate mixtures of polymers.

DSC analysis can also be applied to the measurement of the crystallinity of polymers, since many polymers contain crystalline and amorphous regions, and the relative amounts of amorphous and crystalline regions affect the physical and chemical properties of polymers. For example, L-polylactide (PL) is a biodegradable semicrystalline polymer, widely used as an excipient and matrix for microencapsulation and medical implants. The release rate of the active ingredient has been shown in many cases to be related to the level of amorphous content. For example, Cassel et al. reported a rapid analysis technique to assess the amorphous content of PL using fast MDSC. PL samples were scanned at a relatively high heating rate to mitigate crystallization during heating, producing a melting profile that better reflected the initial proportion of crystalline material [23].

A very important application of TGA for polymers is its ability to provide information about the possible loss of volatile plasticizer as the polymer is heated.

10.8.3 Characterization of Amorphous Forms and Amorphous Solid Dispersions

As has been discussed earlier, the glass transition temperature is a key physical parameter delineating the temperatures at which an amorphous solid exists as a supercooled liquid or a glass with very different physical properties (Chapter 6). The use of DSC to obtain T_g values, as previously discussed, therefore, is central to most studies attempting to understand the physical properties of amorphous materials.

Owing to the increased appearance of poorly water-soluble crystalline active compounds in drug product development, amorphous forms of the drug substances, in the form of amorphous solid dispersions of the drug and the polymer, are widely used to enhance solubility, stability, and bioavailability. Again, as described earlier, DSC can be utilized to determine whether the drug and the polymer are miscible and to evaluate the stability of formulated products.

DiNunzio et al. [24], for example, studied the miscibility of itraconazole with cellulose acetate phthalate (CAP) using modulated DSC (m-DSC). Amorphous solid dispersions of itraconazole with CAP were prepared by ultrarapid freezing in the weight ratio of 1:4, 2:3, 3:2, and 4:1. A single glass transition was observed for all four solid dispersions prepared, indicating that itraconazole is miscible with CAP in the weight ratio investigated. The glass transition temperatures of itraconazole/CAP solid dispersions were found to be 78 °C for 4:1, 88 °C for 3:2, 114 °C for 2:3, and 146 °C for 1:4 ratios, as opposed to the T_g of itraconazole (65 °C) and that of CAP (169 °C). The T_g of each of the solid dispersions increased as the amount of CAP increased in the mixture, with a strong correlation with the Gordon–Taylor equation (Chapter 6).

Rumondor et al. utilized DSC to study the phase behavior of amorphous solid dispersions composed of poly(vinylpyrrolidone) (PVP) and pimozide following exposure to elevated relative humidity [25]. A single T_g was observed for the pimozide–PVP solid dispersion at all compositions, confirming the complete miscibility of the two components. Two separate T_g values were observed for the samples stored under elevated relative humidity (94% RH). It is apparent that the presence of moisture, in this case, induced amorphous–amorphous phase separation.

10.8.4 Dehydration and Desolvation Kinetics

As previously indicated, DSC and TGA have been widely used to quantitatively characterize the dehydration and desolvation of pharmaceutical solvates and hydrates. TGA, for example, is very useful for studying the kinetics of dehydration and desolvation at constant temperature.

Alkhamis et al., for example, studied the dehydration and the desolvation kinetics of fluconazole monohydrate and its ethyl acetate solvate using isothermal TGA [26]. Samples of the two pseudopolymorphs were heated at different temperatures in a TGA instrument, and 18 solid-state reaction mathematical models were applied to test the isothermal TGA data using statistical fitting. Based on the ranking of close fitting, the three-dimensional phase boundary model gave the best fit for the monohydrate, whereas the three-dimensional diffusion model provided the best fit for the ethyl acetate solvate data. The activation energy derived from rate constants was 90 and 153 kJ/mol for fluconazole monohydrate and ethyl acetate solvate, respectively.

Zhu and Grant studied the dehydration of nedocromil magnesium pentahydrate using TGA [27]. Here, nedocromil magnesium pentahydrate gave two distinct steps of weight loss, corresponding to the loss of four water molecules in the first step and one water molecule in the second step. Isothermal TGA was utilized to study the effects of temperature on the kinetics of the two-step dehydration. The Avrami–Erofeev equation gave the best fit for the kinetics of both steps, which suggested a nucleation-controlled mechanism [27]. The activation energy for the second dehydration step was higher than the first step, indicating that the remaining water molecules are tightly bound to magnesium ions and carboxylate oxygen atoms.

10.8.5 Optimization of the Freezing–Drying Cycle in Lyophilization

Many parental products with active ingredients that are chemically unstable, including biological materials, are manufactured as reconstitutable solids using lyophilization. The process of lyophilizing a solution includes, in sequence, freezing, primary drying, and secondary drying. During freezing, most of the liquid water is converted to water ice by cooling the solution to −10 to −45 °C. Throughout the freezing process, the solute is highly concentrated in any remaining solution. After freezing, primary drying is carried out to remove further water to produce a lyophilization “cake.” In secondary drying, this cake is further dried to remove water by slowly increasing the temperature. During primary drying, the control of product temperature is critical to avoid any eutectic melting and other conditions that might cause the “collapse” of the cake.

Many bulking agents, such as mannitol, disaccharides, and polymers, are included in the solution to increase the overall volume of the final cake and to prevent any physical and chemical instability during processing and storage. However, during freezing, some bulking agents, such as mannitol, may crystallize, leading to a collapsed cake and possible instability. For example, bulking agents that crystallize during freezing may form a eutectic mixture with ice that will melt as a single phase at temperatures lower than the freezing point of pure water. Therefore, drying the cake above the eutectic melting temperature is clearly not desirable. The eutectic melting point can be measured by DSC. Using DSC, for example, the eutectic melting point for frozen sodium chloride solution has been determined to be about −20°C [28]. Sundaramurthi et al., using DSC and X-ray powder diffraction (XRPD), studied the crystallization events of succinate buffer components in the frozen solution at different pH and discovered the sequential crystallization of the basic component, disodium succinate, first, then followed by the acidic components, monosodium succinate and succinic acid [29]. As shown in Figure 10.18, the eutectic melting behaviors from DSC display distinctly different pH.

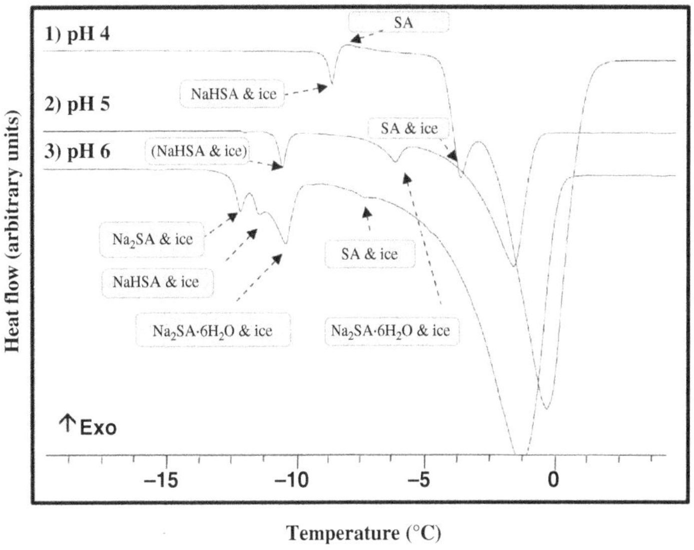

Figure 10.18 DSC thermograms of frozen 200 mM succinate buffers at pH 4, 5, and 6 after heating at 0.5 °C/min. The eutectic melting of various crystalline phases is pointed out. SA: succinic acid; NaHSA: monosodium succinate; Na_2SA: disodium succinate; $Na_2SA{\bullet}6H_2O$: disodium succinate hexahydrate. *Source*: [29]/with permission of American Chemical Society.

- pH 4: a eutectic melting endotherm of monosodium succinate at −8.5 °C, a crystallization exotherm of succinic acid at −8.0 °C, and a eutectic melting endotherm of succinic acid at −3.6 °C.
- pH 5: a eutectic melting endotherm of monosodium succinate at −10.5 °C, and a eutectic melting of disodium succinate hexahydrate at −6.1 °C.
- pH 6: multiple overlapping eutectic melting endotherms, anhydrous disodium succinate at −12.2 °C, monosodium succinate at −11.5 °C, disodium succinate hexahydrate at −10.5 °C, a eutectic melting endotherm of succinic acid at −7.5 °C.

As we shall see, DSC offers a clear elucidation of the phase transition behavior of buffer components in freezing–drying cycles.

Another interesting aspect of the lyophilization process using bulking agents, such as sucrose and trehalose, is that the frozen solutions of these materials exhibit a distinct glass transition event, labeled as T_g' that is lower than that of the completely dried material and is also termed as the glass transition temperature of maximally freeze-concentrated solution. Since the mechanical properties of amorphous sugars change significantly at T_g', if the product temperature during drying is above the glass transition temperature of the product, collapse of the system will occur. Thus, as a commonly pursued good practice, the desirable product temperature during primary drying needs to be maintained 2–3 °C below T_g', which is usually similar to the collapse temperature measured by freeze-dry microscopy.

Using DSC, for example, it has been shown that frozen unannealed trehalose solutions have two glass transition events, with T_{g1}' at −45 °C and T_{g2}' at −31 °C [30]. Annealing resulted in an increase in the lower transition temperature, T_{g1}' while the higher transition temperature, T_{g2}' was unchanged. The two glass transition events were also found in other mono- and disaccharide frozen solutions, including sucrose, maltose, fructose, fucose, glucose, mannose, and lactose [31]. For example, Sacha and Nail reported that a 20% sucrose solution in water exhibited T_{g1}' at −49 °C and T_{g2}' at −35 °C using DSC analysis and hypothesized that the lower temperature transition, that is, T_{g1}', represents a metastable glass mixture containing more water, whereas the higher transition, that is, T_{g2}', signifies the maximally freeze-concentrated solute [31]. Using freeze-dry microscopy, they observed that T_{g2}' was closely related to the collapse temperature and should be paid attention to when designing drying conditions.

Colandene et al. reported using DSC and freeze-dry microscopy to facilitate lyophilization cycle development for a high-concentration monoclonal antibody formulation [32]. They noted that at low protein concentration T_g' was similar to the collapse temperature attained by freeze-dry microscopy, whereas, at higher protein concentrations, the collapse temperature became significantly higher than T_g'. This interesting discovery led to the development of primary drying conditions where the product temperature was above T_g' with drying time being shortened significantly but without any undesirable effect to the product.

10.8.6 Melting-point Depression Method for Purity Analysis and Drug–Polymer Miscibility Assessment

Melting-point depression is a well-known phenomenon where the presence of soluble impurities can decrease the melting point of a crystalline drug in a quantitative manner. The degree of change in the melting point, related to the amount of impurity, can be described by the van't Hoff equation:

$$T_m = T_o - x_2 \frac{RT_o^2}{\Delta H_f} \tag{10.5}$$

where T_m represents the melting temperature of the compound with impurities, T_o is the melting temperature of the pure compound, x_2 is the molar fraction of impurities in the liquid phase, and ΔH_f is the heat of fusion of the compound.

During the melting process, the mole fraction of impurities in the liquid phase constantly decreases as more compound melts into the liquid phase. The molar fraction x_2 can be related to the absolute molar ratio of the impurity in the mixture x_2^* with the following equation:

$$x_2 = x_2^* \frac{1}{F} \tag{10.6}$$

where F equals the melted fraction. F equals 1 when the compound is completely melted. Combination of equations (10.5) and (10.6) gives the following equation:

$$T_m = T_o - \frac{x_2^* RT_o^2}{\Delta H_f} \frac{1}{F} \tag{10.7}$$

This equation predicts a linear correlation between T_m and $1/F$ with an intercept of T_o and a slope of $\frac{x_2^* RT_o^2}{\Delta H_f}$. The slope can be used to derive the original molar ratio of impurity, x_2^*, when the value of other parameters in the equation is known.

The slope is obtained by linear regression of the graph of T_m versus $1/F$ constructed from the melting peak and using the region between 10% and 50% of the peak is for optimal results. The purpose of using this range of peak intensities is to exclude the earliest time points with the highest concentration of impurities in the liquid phase and to avoid the regions where the melting rate is very great. Although data collected from this region usually give good linearity, any nonlinearity observed can be corrected through established mathematical manipulation using computer software provided by instrument vendors.

Figure 10.19, for example, shows the DSC thermogram and purity analysis for a loratadine reference standard [33]. The purity obtained from DSC is found to be 99.69 mol%, which is in good agreement with HPLC data.

The chemical purity of an organic compound can be assessed by DSC if it has a well-defined melting peak. Although the method of impurity analysis by measuring changes in the melting point of a sample relative to that of a highly pure sample is not a popular choice for purity analysis because sophisticated chromatography and detection technology are available, purity analysis by DSC still has its value in certain circumstances, such as when the impurity cannot be separated from the compound by chromatographic techniques or small amounts of impurity cannot be detected.

Other advantages of measuring the extent of impurity using DSC are that it requires only a small sample size and does not require reference standards. There are, however, some limitations in using DSC to measure chemical purity: (1) the method detects only the total amount of impurities and thus lacks the chemical specificity of modern chromatography, NMR and mass spectrometry; and (2) this method is applicable only to cases where the impurities are insoluble in the solid phase but soluble in the melt.

The melting-point depression of a crystalline drug in the presence of polymers is also of significant value in evaluating the miscibility of drug–polymer systems during the development of amorphous solid dispersion formulations. Generally, a DSC study is designed for drug–polymer physical mixtures, and the melting events are studied systematically. As would be expected, qualitatively, a significant melting-point depression can occur if the drug and the polymer are miscible due to the exothermic mixing, whereas the melting-point depression is barely detectable or absent for an immiscible system.

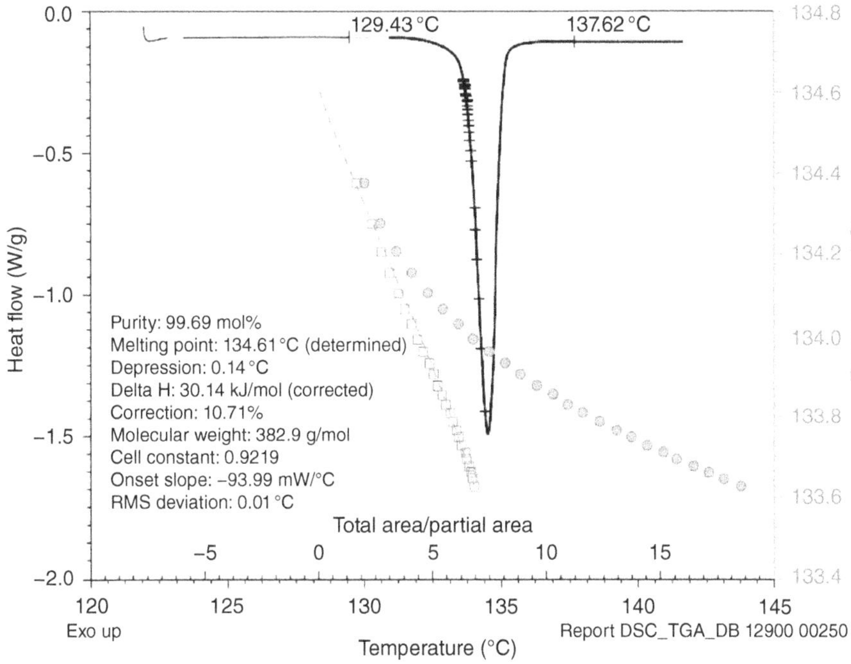

Figure 10.19 DSC scan and purity analysis for loratadine reference standard (heating rate = 1.0 °C/min, sample weight = 1.35 mg). *Source*: [33]/with permission of Elsevier.

To probe the degree of miscibility quantitatively, Marsac et al. pioneered the use of the Flory–Huggins Equation to develop a melting-point depression method [34], as described by equation (10.8),

$$\frac{1}{T_{\text{mix}}} - \frac{1}{T_{\text{pure}}} = -\frac{R}{\Delta H_{\text{fus}}}\left[\ln \phi_{\text{API}} + \left(1 - \frac{1}{m}\right)\phi_{\text{polymer}} + \chi\phi_{\text{polymer}}^2\right] \tag{10.8}$$

where T_{mix} and T_{pure} are the melting points of the physical mixture and the neat drug, respectively; ΔH_{fus} stands for the heat of fusion of the neat drug; ϕ_{API} and $\phi_{polymer}$ represent the volume fraction of drug and polymer, respectively; m is the ratio of the volume of the polymer to that of the drug; and, most importantly, χ symbolizes the Flory–Huggins interaction parameter, commonly used to describe the degree of miscibility. As we shall see, positive χ suggests immiscibility, whereas negative χ implies miscibility of the drug–polymer system. Generally, a greater negative χ indicates a higher degree of miscibility. Using the melting depression method described in equation (10.8), Marsac et al. studied the degree of miscibility of felodipine and nifedipine with poly(vinylpyrrolidone), respectively, and found the predicted miscibility in good agreement with experimental data [34]. Rask and coworkers further evaluated the method in various drug–polymer systems such as celecoxib, indomethacin, carbamazepine, and ritonavir in polyvinylpyrrolidone, hydroxypropyl methylcellulose, and Soluplus, respectively, and, therefore, concluded that a well-defined melting point and the T_g of the polymer being below 20°C + T_m of the drug are generally required to obtain acceptable prediction for miscibility [35]. Tao et al., through the work on D-mannitol in PVP, indomethacin in PVP/VA, and nifedipine in PVP/VA, discovered that cryomilling drug–polymer mixtures before DSC analysis could improve the uniformity of the mixtures and thus enabled the more accurate measurement of the melting-point depression events by facilitating the equilibrium between the drug and the polymer during heating [36].

10.8.7 Study of Solid-state Chemical Stability

It is well established that many pharmaceutical materials undergo chemical change when exposed to elevated temperatures. Often, thermal degradation results in the production of volatile materials, and, since these changes involve weight loss, it is possible to measure the kinetics of these reactions using TGA isothermally or nonisothermally using a linear heating rate.

Leung et al, for example, studied the solid-state degradation of the dipeptides, aspartame and aspartylphenylalanine, using DSC and TGA [37]. The two dipeptides undergo solid-state intramolecular aminolysis with the generation of gas products. Methanol is liberated from aspartame at 167–180°C, whereas water is produced for aspartylphenylalanine at 186–202°C. The rates of degradation of both compounds were investigated using isothermal TGA and temperature-ramp DSC. The kinetics studies concluded that the reaction mechanisms were probably nucleation control with activation energies of about 266 kJ/mol for aspartame and 234 kJ/mol for aspartylphenylalanine.

10.8.8 Characterization of Macromolecules and Their Interactions

Thermal phase transitions and interactions of macromolecules with other molecules involve changes in enthalpy that can be detected by DSC.

One critical physical transition that occurs with proteins is thermal denaturation, whereby the folded three-dimensional structure of the protein is disrupted by heat. Loss of the native three-dimensional structure causes the protein to lose its biological activities.

Protein denaturation involves alteration of several thermodynamic states. The process is highly cooperative with the disruption of many intermolecular forces, including hydrogen bonds, hydrophobic interactions, and other noncovalent interactions.

Owing to its highly cooperative nature, denaturation is displayed as a distinct endotherm in a DSC measurement. The transition midpoint of the transition endotherm is the denaturation temperature, T_m, at approximately 50% of the unfolding of the protein.

For optimal DSC analysis of protein solutions, the concentration of the protein in an aqueous medium should be fairly dilute. Otherwise, the protein molecules generally tend to aggregate, which prevents the detection of denaturation events. A slow heating rate, such as 1–2°C/min, is desired because the denaturation process exhibits relatively slow kinetics. Protein denaturation is also a relatively weak transition with transition energies in the DSC power range of microwatts. Reliable detection of such weak transitions, therefore, requires a very sensitive DSC instrument. Regarding sample preparation, a sealed metal container such as a special stainless-steel capsule, is generally utilized for the analysis of proteins in solution.

References

1 Naima Z, Siro T, Juan-Manuel GD, Chantal C, René C, Jerome D. Interactions between carbamazepine and polyethylene glycol (PEG) 6000: characterisations of the physical, solid dispersed and eutectic mixtures. *Eur J Pharm Sci* 2001; 12(4): 395–404.

2 Urakami K, Shono Y, Higashi A, Umemoto K, Godo M. A novel method for estimation of transition temperature for polymorphic pairs in pharmaceuticals using heat of solution and solubility data. *Chem Pharm Bull* 2002; 50(2): 263–267.

3 Jójárt-Laczkovich O, Szabó-Révész P. Amorphization of a crystalline active pharmaceutical ingredient and thermoanalytical measurements on this glassy form. *J Therm Anal Calorim* 2009; 102(1): 243–247.

4 Haque MK, Kawai K, Suzuki T. Glass transition and enthalpy relaxation of amorphous lactose glass. *Carbohydr Res* 2006; 341(11): 1884–1889.

5 Clas SD, Dalton CR, Hancock BC. Differential scanning calorimetry: applications in drug development. *Pharm Sci Tech Today* 1999; 2(8): 311–320.

6 Burger A, Ramberger R. On the polymorphism of pharmaceuticals and other molecular crystals. I. *Microchim Acta* 1979; 72 (3–4): 259–271.

7 Grzesiak AL, Lang M, Kim K, Matzger AJ. Comparison of the four anhydrous polymorphs of carbamazepine and the crystal structure of form I. *J Pharm Sci* 2003; 92(11): 2260–2271.

8 Chongprasert S, Griesser UJ, Bottorff AT, Williams NA, Byrn SR, Nail SL. Effects of freeze-dry processing conditions on the crystallization of pentamidine isethionate. *J Pharm Sci* 1998; 87(9): 1155–1160.

9 Mirza S, Miroshnyk I, Rantanen J, Aaltonen J, Harjula P, Kiljunen E, Heinämäki J, Yliruusi J. Solid-state properties and relationship between anhydrate and monohydrate of baclofen. *J Pharm Sci* 2007; 96(9): 2399–2408.

10 Gong X, Moghaddam MJ, Sagnella SM, Conn CE, Danon SJ, Waddington LJ, Drummond CJ. Lamellar crystalline self-assembly behaviour and solid lipid nanoparticles of a palmityl prodrug analogue of Capecitabine—a chemotherapy agent. *Colloids Surf B Biointerfaces* 2011; 85(2): 349–359.

11 Danley RL. New modulated DSC measurement technique. *Thermochim Acta* 2003; 402 (1): 91–98.

12 Royall PG, Craig DQ, Doherty C. Characterisation of the glass transition of an amorphous drug using modulated DSC. *Pharm Res* 1998; 15(7): 1117–1121.

13 Schawe JEK. A comparison of different evaluation methods in modulated temperature DSC. *Thermochim Acta* 1995; 260: 1–16.

14 Saunders M, Podluii K, Shergill S, Buckton G. The potential of high speed DSC (Hyper-DSC) for the detection and quantification of small amounts of amorphous content in predominantly crystalline samples. *Int J Phar* 2004; 274(1): 35–40.

15 Yi F, La Van DA. Nanoscale thermal analysis for nanomedicine by nanocalorimetry. *Wiley Interdiscip Rev Nanomed Nanobiotechnol* 2012; 4(1): 31–41.

16 Feth MP, Nagel N, Baumgartner B, Bröckelmann M, Rigal D, Otto B, Spitzenberg M, Schulz M, Becker B, Fischer F, Petzoldt C. Challenges in the development of hydrate phases as active pharmaceutical ingredients—an example. *Eur J Pharm Sci* 2011; 42(1): 116–129.

17 Bruni G, Berbenni V, Milanese C, Girella A, Cardini A, Viganò E, Lanfranconi S, Marini A. Thermodynamic relationships between nateglinide polymorphs. *J Pharm Biomed Anal* 2009; 50(5): 764–770.

18 Bugay DE. Characterization of the solid-state: spectroscopic techniques. *Adv Drug Deliver Rev* 2001; 48(1): 43–65.

19 Chadha R, Arora P, Saini A, Jain DS. Solvated crystalline forms of nevirapine: thermoanalytical and spectroscopic studies. *AAPS PharmSciTech* 2010; 11(3): 1328–1339.

20 Niederwanger V, Gozzo F, Griesser UJ. Characterization of four crystal polymorphs and a monohydrate of s-bupivacaine hydrochloride (levobupivacaine hydrochloride). *J Pharm Sci* 2009; 98(3): 1064–1074.

21 Smikalla M, Mescher A, Walzel P, Urbanetz NA. Impact of excipients on coating efficiency in dry powder coating. *Int J Pharm* 2011; 405(1): 122–131.

22 Li B, Yoshii T, Hafeman AE, Nyman JS, Wenke JC, Guelcher SA. The effects of rhBMP-2 released from biodegradable polyurethane/microsphere composite scaffolds on new bone formation in rat femora. *Biomaterials* 2009; 30(35): 6768–6779.

23 Cassel RB, Riga AT, Biddlecom B. *Methodology for Thermal Analysis of Excipients*. TA317. http://www.tainstruments.co.jp/application/pdf/Thermal_Library/Applications_Briefs/TA317.PDF (accessed 2 January 2016).

24 DiNunzio JC, Miller DA, Yang W, McGinity JW, Williams III RO. Amorphous compositions using concentration enhancing polymers for improved bioavailability of itraconazole. *Mol Pharm* 2008; 5(6): 968–980.
25 Rumondor AC, Marsac PJ, Stanford LA, Taylor LS. Phase behavior of poly(vinylpyrrolidone) containing amorphous solid dispersions in the presence of moisture. *Mol Pharm* 2009; 6(5): 1492–1505.
26 Alkhamis KA, Salem MS, Obaidat RM. Comparison between dehydration and desolvation kinetics of fluconazole monohydrate and fluconazole ethylacetate solvate using three different methods. *J Pharm Sci* 2006; 95(4): 859–870.
27 Zhu H, Grant DJ. Dehydration behavior of nedocromil magnesium pentahydrate. *Int J Pharm* 2001; 215(1): 251–262.
28 Franks F. Solid aqueous solutions. *Pure Appl Chem* 1993; 65(12): 2527–2537.
29 Sundaramurthi P, Shalaev E, Suryanarayanan R. Calorimetric and diffractometric evidence for the sequential crystallization of buffer components and the consequential pH swing in frozen solutions. *J Phys Chem B* 2010; 114(14): 4915–4923.
30 Pyne A, Surana R, Suryanarayanan R. Enthalpic relaxation in frozen aqueous trehalose solutions. *Thermochim Acta* 2003; 405(2): 225–234.
31 Sacha GA, Nail SL. Thermal analysis of frozen solutions: multiple glass transitions in amorphous systems. *J Pharm Sci* 2009; 98(9): 3397–3405.
32 Colandene JD, Maldonado LM, Creagh AT, Vrettos JS, Goad KG, Spitznagel TM. Lyophilization cycle development for a high-concentration monoclonal antibody formulation lacking a crystalline bulking agent. *J Pharm Sci* 2007; 96(6): 1598–1608.
33 Mathkar S, Kumar S, Bystol A, Olawoore K, Min D, Markovich R, Rustum A. The use of differential scanning calorimetry for the purity verification of pharmaceutical reference standards. *J Pharm Biomed Anal* 2009; 49(3): 627–631.
34 Marsac PJ, Shamblin SL, Taylor LS. Theoretical and practical approaches for prediction of drug–polymer miscibility and solubility. *Pharm Res* 2006; 23: 2417–2426.
35 Rask MB, Knopp MM, Olesen NE, Holm R, Rades T. Comparison of two DSC-based methods to predict drug-polymer solubility. *Int J Pharm* 2018; 540 (1–2): 98–105.
36 Tao J, Sun Y, Zhang GG, Yu L. Solubility of small-molecule crystals in polymers: D-mannitol in PVP, indomethacin in PVP/VA, and nifedipine in PVP/VA. *Pharm Res* 2009; 26: 855–864.
37 Leung SS, Grant DJ. Solid state stability studies of model dipeptides: aspartame and aspartylphenylalanine. *J Pharm Sci* 1997; 86(1): 64–71.
38 Ichimura Y. Thermal denaturation of proteins. http://www.hitachi-hightech.com/file/global/pdf/products/science/appli/ana/thermal/application_TA_054e.pdf (accessed 2 January 2016).
39 Cueto M, Dorta MJ, Munguía O. New approach to stability assessment of protein solution formulations by differential scanning calorimetry. *Int J Pharm* 2003; 252(1): 159–166.
40 Privalov PL, Dragan AI. Microcalorimetry of biological macromolecules. *Biophys Chem* 2007; 126(1): 16–24.
41 Barnett GV, Razinkov VI, Kerwin BA, Blake S, Qi W, Curtis RA, Roberts CJ. Osmolyte effects on monoclonal antibody stability and concentration-dependent protein interactions with water and common osmolytes. *J Phys Chem B* 2016; 120(13): 3318–3330.
42 Marbrey S, Sturtevant JM. Investigation of phase transitions of lipids and lipid mixtures by high sensitivity differential scanning calorimetry. *PNAS* 1976; 73(11): 3862–3866.

11

Microscopy

11.1 Introduction

Various types of microscopy have important applications for the characterization of pharmaceutical materials. In general, a microscope is an instrument that provides an enlarged image of an object not visible to the naked eye, with resolution that varies with the type of microscope. The most common microscope is the light or optical microscope, which relies on optical lenses and visible light to magnify and visualize images. For scanning electron microscopy (SEM) and transmission electron microscopy (TEM), an electron beam provides the basis for the formation of images. Atomic force microscopy (AFM) relies on mechanical contact or force to map surface topography. It is obvious that the three types of microscopy are different in resolution, ease of operation, sample preparation, and uses in analyzing pharmaceutical materials. Light microscopy is primarily useful for examining materials at a resolution from a few hundred nanometers (nm) to several millimeters (mm). SEM provides a very broad range of dimensions from 10 to 100 mm, whereas TEM can offer resolution down to 1 nm. AFM is primarily useful for analyzing submicrometer- and nanosize materials. Light microscopes have a low cost, are easy to operate, and need minimal sample preparation, whereas SEM, TEM, and AFM require more sophisticated equipment and demand greater expertise to operate. Extensive sample preparation, such as thorough drying of the sample and coating it with conductive materials, is required for SEM, whereas less attention to such preparation is required when using environmental SEM (ESEM). AFM is only able to image solid materials, although the solid material can be immersed in a liquid medium during the measurement. The resolution range of various microscopes is illustrated in Figure 11.1.

Light microscopy has a broad number of applications for studying pharmaceutical materials. It is utilized to examine particle size, particle morphology, and surface properties. With the help of polarized light, light microscopy is used to observe birefringence for detecting crystallinity, differentiating amorphous and crystalline phases, and monitoring phase transformations. Phase transformation often results in changes in surface characteristics or bulk appearance, which can be detected with light microscopy. With the addition of hot stage capability, physical changes in a material with temperature can be visually observed and correlated with the corresponding results in differential scanning calorimetry (DSC) and thermogravimetric analysis (TGA) measurements. Such changes include melting, recrystallization, polymorph conversion, and dehydration. Light microscopy with a cold stage attachment is useful for studying lyophilization processes. Chemical reactions can also be studied by microscopy if the changes in the physical appearance of the sample occur at the same rate. SEM has a similar capability to light microscopy with better resolution but at a higher cost. Because of its better resolution, SEM is more useful than light microscopy for studying surface topology and nanomaterials.

In this chapter, we will discuss three types of microscopy and their applications in the characterization of pharmaceutical materials, as well as the newly developed microcrystal electron diffraction (ED) using TEM for crystal structure determination.

11.2 Light Microscopy

Light microscopy was invented many centuries ago, but modern light microscopes essentially still have the same principle of design, except for advanced optical lenses, light sources, and digital cameras or recording devices. Microscopes for pharmaceutical use are compound microscopes that consist of a light source, a sample stage, object lens, ocular lens, focus

Solid-State Materials in Pharmaceutical Chemistry: Properties, Characterization, and Applications, Second Edition. Stephen R. Byrn, George Zografi and Xiaoming (Sean) Chen.

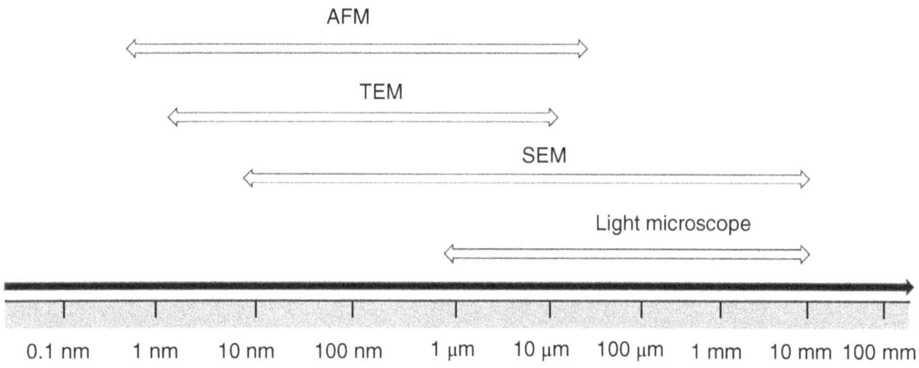

Figure 11.1 Comparison of various types of microscopes regarding resolution and normal operating range.

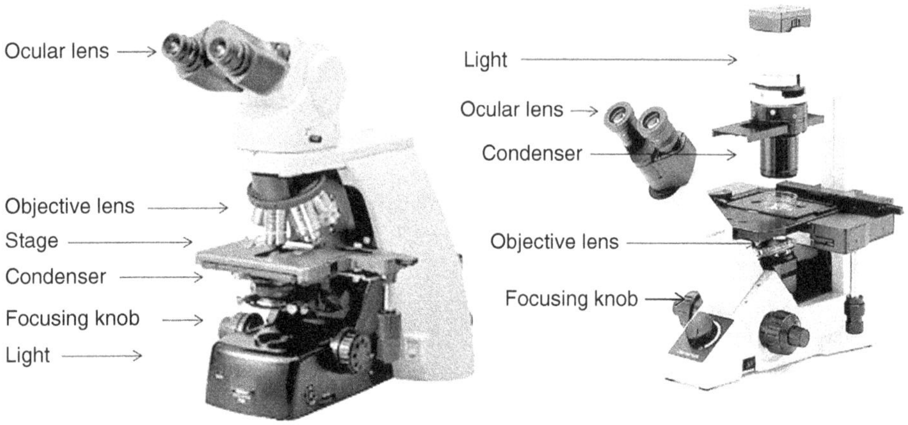

Figure 11.2 Examples of upright and inverted microscopes.

knobs, diaphragms, and condensers (Figure 11.2). For a microscope to observe small objects effectively, it must provide sufficient magnification, effective resolution, and suitable contrast. A compound microscope delivers images by use of convergent light beams. Intermediate images are formed by the objective lens and observed through the eyepiece or camera. To mitigate the optical aberration caused by convergent light beams, a tube lens is included to focus parallel beams. This application is called an infinity-corrected beam path. Usually, an objective lens is constructed from several high-quality lenses, including a tube lens. The ocular lens, or eyepiece, acts as a magnifier for the intermediate image.

There are two types of design for light microscopes: upright and inverted (Figure 11.2). An upright microscope is applicable for a fixed specimen, whereas the inverted microscope offers convenience for directly observing a specimen within a liquid container.

The capability of the microscope is enhanced by manipulating the light to deliver bright field, dark field, or polarized light.

In bright field microscopy, light from a source is directed to a condenser. The condenser, made up of lenses, concentrates the light into a cone and illuminates the object with uniform intensity. Contrast in the sample is caused by different degrees of absorbance of the transmitted light in different areas of the field that may arise from stains, pigmentation, or dense areas. The typical view of bright field microscopy is a dark or highly colored sample on a bright background. Samples observed under bright field microscopy usually tend to have relatively low contrast. Contrast can be increased by staining or enhanced by the intrinsic color of the material.

Dark field microscopy has a different light path and offers higher contrast. An opaque disk is placed under the condenser lens to block the center cone of light and to leave an outer ring of illumination. The light passes through a condenser and is focused on the sample. Light that is transmitted from the sample is eliminated by using a direct illumination block. Only light scattered by the object is collected to observe the object, so that the background is dark and a vivid contrast is obtained. As such, it is very useful for studying biological samples without staining.

11.3 Polarized Light Microscopy

Polarized light microscopy has a wide application in the characterization of pharmaceutical materials. The majority of drug substances encountered are crystalline, and crystalline materials produce birefringence in a polarized light microscope. Birefringence, also called double refraction, is caused by the optical anisotropy of a material. Under a polarized microscope, materials with birefringence are observed as a colorful view contributed by refracted light (see Chapter 7).

In a polarized microscope, light from an incandescent source passes through a polarizer. The polarized light beam passes through a condenser and impinges on an object. The object splits the incoming light beam into an ordinary ray and an extraordinary ray, and the two rays are polarized at right angles to each other. In an optically anisotropic material, the two rays would travel at different velocities, and, by using a second polarizer, the two rays are combined. The retardation in refraction for an anisotropic sample results in the samples being displayed with color images.

Isotropic materials such as liquids, amorphous forms, and cubic crystals give featureless images with minimal contrast from the background. Anisotropic materials, including noncubic crystals, liquid crystals, and semicrystalline samples, display colorful images in a polarized microscope. Polarized light microscopy is very useful to detect material crystallinity in pharmaceutical systems because the majority of organic crystalline materials are anisotropic. Sugano et al. applied polarized light microscopy analysis in a high-throughput solubility measurement [1]. The analysis was able to show that the precipitates arising during solubility assays were either amorphous or crystalline phases. Notably, polarized light microscopy provided the proof of the formation of nanocrystalline celecoxib within a mannitol matrix after spray drying [2].

11.4 Thermal Microscopy

A thermal microscope is a combination of a light microscope and a hot stage apparatus. It is also called a hot stage microscope. The hot stage provides precise temperature control, and the light microscope enables sample observation. Thermal microscopy is widely used to investigate various thermal transitions, including melting, desolvation, dehydration, sublimation, crystallization, and polymorph conversion. It is a complementary tool to thermal analysis and helps to visualize thermal events observed in DSC.

There are two types of equipment widely used in thermal microscopy: the Kofler bench and the Mettler hot stage. The Kofler bench is an offline apparatus for microscopy [3] in that the bench contains a metal strip with a temperature gradient ranging from 50 to 260 °C (Figure 11.3). The temperature is controlled by placing a sample at a certain location of the bench, and, after heating, the sample is examined by a light microscope.

Thermal microscopy of this type was used and extensively promoted by McCrone and coworkers and gained popularity after Mettler introduced its first hot stage apparatus in 1967 [4].

The Mettler hot stage apparatus, in its modern versions, has a precise thermal measuring cell to provide temperature control with a window for visual sample observation (Figure 11.4). Two thin heating plates provide a homogenous temperature field in the temperature range from room temperature to 375 °C. The cooling is accomplished by forced air from an integral fan, with the cooling rate being slower when the temperature of the hot stage is closer to ambient temperature. With the use of liquid nitrogen or another cooling device, the temperature can be decreased much more quickly

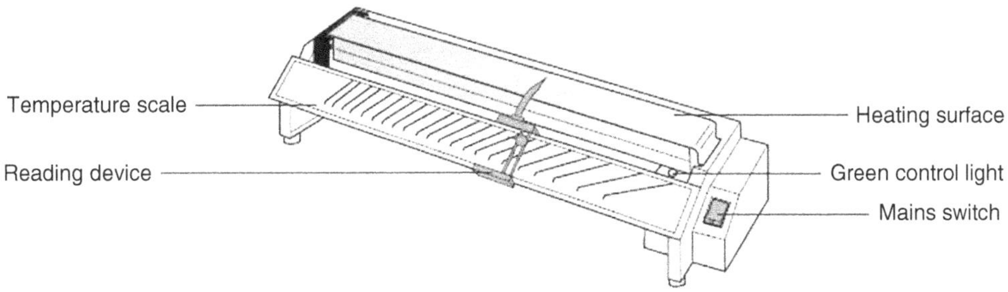

Figure 11.3 Drawing of a Kofler hot stage.

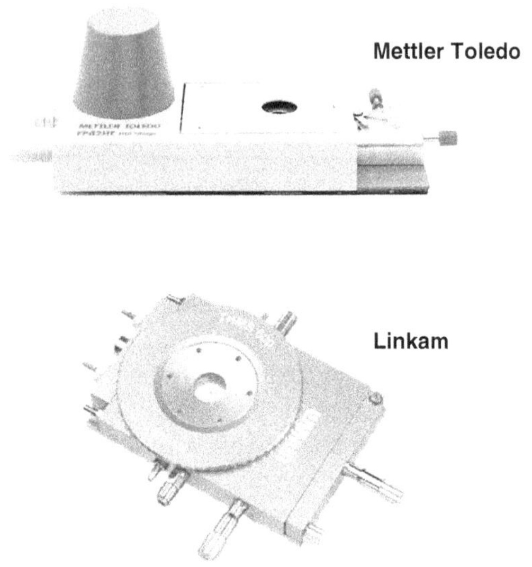

Figure 11.4 Mettler and Linkam hot stage.

and can reach as low as −60 °C. The accuracy of temperature is generally between 0.4 and 0.8 °C, and the material can be investigated by constant heating, cooling, or at a constant temperature. The properties of solids that can be observed include changes in morphology, particle size, birefringence, and the number of particles. The physical changes observed are generally related to phase transitions such as melting, crystallization, sublimation, desolvation, dehydration, and polymorph conversion. As mentioned earlier, the visual observations in a hot stage microscope often facilitate the interpretation of thermal analysis curves obtained from DSC and TGA.

Another popular hot stage with similar capability to the Mettler instrument is made by Linkam [5] (Figure 11.4). The available temperature range is −196 to 600 °C. A cooling pump enables a cooling rate of up to 100 °C/min, and a heating rate up to 150 °C/min. The hot stage microscope with such cooling capability is advantageous for studying the freeze-drying process.

The fast heating and cooling rates, often used with such equipment, require the ability to capture images rapidly, which is made possible by advanced computer and video technology. In general, the application of computer and video technology to microscopy advances the ability to store, evaluate, and share data in digital formats, thus promoting the functionality of light microscopy.

11.5 Functionality of Light Microscopy

The function of a light microscope is to provide quality enlarged images, including magnification, resolution, and depth of field (DOF).

Magnification refers to the degree of image enlargement compared to its original physical size. Total magnification of a microscope is described as the product of the magnification power of the two lenses. For example, using a 40× objective lens and a 10× eyepiece yields a total visual magnification of 400. It is recognized that oil immersion techniques are generally required to obtain the necessary resolution for magnification in the range of 500–1000×.

Resolution is defined as the shortest distance between two separate points in a specimen that can still be distinguished as a distinct entity under a microscope. It is determined by the numerical aperture (NA) of a selected objective lens, the NA of the condenser, and the wavelength of light as expressed in the following equation:

$$r = \frac{1.22\lambda}{(\mathrm{NA_{obj}} + \mathrm{NA_{cond}})} \tag{11.1}$$

where r is the overall resolution, λ is the wavelength of light, $\mathrm{NA_{obj}}$ is the NA of the objective lens, and $\mathrm{NA_{cond}}$ is the NA of the condenser. As revealed by the equation, the resolution is not directly dependent on the magnification, although high magnification lenses usually have a large NA, which can provide better resolution.

According to equation (11.1), resolution can be increased as the wavelength is reduced and the NA is increased. Therefore, in principle, short-wavelength light, such as blue light, can provide better resolution than red light. Under most circumstances, however, white light generated by a tungsten–halogen bulb is used to illuminate the specimen.

The NA of an objective lens or a condenser is a measure of its capability to gather light and emit light. The NA is characteristic of each lens or condenser as determined by its manufacturer. It is a dimensionless value related to the angular aperture of the light cone produced. For a condenser, the light cone comes from the condenser and is focused on the specimen. The light then becomes an inverted cone going from the specimen to the objective lens. The angular aperture of the light cone is related to the distance between the specimen and the lens and the aperture of the lens. The distance between the specimen and the lens is called the focal length. The NA of an objective lens is defined as follows:

$$\mathrm{NA} = n \sin \alpha \tag{11.2}$$

where n represents the refractive index of the medium and α is the one-half angular aperture of the objective lens. The refractive index of air is 1.0, whereas most immersion oils have an n value of about 1.5, which increases the NA of the objective lens. As the focal length decreases, the angular aperture increases. As shown by equation (11.2), the NA increases with the angular aperture and the refractive index of the imaging medium. Practically, the maximum NA that can be achieved is 0.95 in air and 1.4 in oil.

To achieve optimal resolution, the NA of a condenser is adjusted to match the NA of the objective lens. The adjustment is carried out by varying the diameter setting of the condenser aperture. The minimum and maximum NAs vary with different condensers. The technique of matching the NA of a condenser and an objective lens is part of the process required to optimize the light path between the condenser and the objective lens. This is known as Köhler illumination, invented by August Köhler [6]. In addition to providing optimal resolution, Köhler illumination produces specimen illumination that is uniformly bright and free from glare. The method involves aligning the incident light properly and defocusing the image of the light source to produce an evenly dispersed light field.

While resolution concerns lateral resolving power, DOF is associated with vertical resolving ability. By definition, DOF is the distance from the nearest object plane in focus to that of the farthest plane in focus at the same time. It describes the ability to produce a sharp image from a non-flat surface as expressed in the following equation:

$$\text{DOF} \approx \frac{\lambda}{\text{NA}} \tag{11.3}$$

It is obvious that DOF increases as wavelength increases and the NA decreases. As magnification is increased by using a high magnification object lens, resolution usually increases, and DOF decreases. For example, for an objective lens with a magnification of 100× and a NA of 0.95, the DOF is 0.19 μm.

DOF is related to both the NA and magnification of the objective. Under certain circumstances, high NA systems have deeper focus depths than do those systems of low NA, even though the DOF is less.

It is worthwhile to point out that the function of an ocular lens is to magnify intermediate images. The ocular lens has no impact on the resolution and DOF of a microscope.

So, to achieve optimal performance of a microscope there must be a balance between magnification, resolution, and DOF (Table 11.1). The use of new computer technology can enable both resolution and DOF in high magnification. Digital microscopy is one of the successful examples where this is accomplished.

11.6 Digital Microscope

A digital microscope is a high-resolution microscope equipped with a charge-coupled device (CCD) camera and zooming lens. Its design is very different from traditional light microscope, where the optical design focuses on decreasing chromatic and spherical aberration but has limited capability for high magnification above 1000× due to a significant loss in DOF. The use of zoom lens optics enables the digital microscope to gain high resolution under magnifications as high as 5000×. The image is collected using diffraction light instead of transmission light, so it offers better observation of surface characteristics.

Table 11.1 Properties of various optical lenses.

Magnification	Numerical aperture	Depth of field (μm)	Image depth (mm)
4×	0.10	55.5	0.13
10×	0.25	8.5	0.80
20×	0.40	5.8	3.8
40×	0.65	1.0	12.8
60×	0.85	0.40	29.8
100×	0.95	0.19	80.0

A digital microscope has a camera unit and a controller. The camera unit consists of a zooming lens and a high pixel color CCD camera. The zooming lens can provide continuous magnification. The controller is a computer system with built-in software which provides display and recording of the image, as well as image processing and measurement.

A major advantage of the digital microscope is that it offers three-dimensional mapping with a resolution down to a few micrometers. Also, a broad range of magnification from 20× to 2500× can be achieved without switching lens, and the DOF obtained is superior to that of a traditional microscope. For example, KEYENCE's digital microscope can achieve a DOF at least 20 times greater than that of conventional microscopes [7]. It can provide a completely in-focus image at high magnification and has been built to be light-weighted and handheld. Furthermore, it can directly observe material without sampling. A three-dimensional picture can be constructed by combining numerous two-dimensional pictures collected at different depth location through a computer algorithm.

11.7 Application of Light Microscopy to Pharmaceutical Materials

Light microscopy is a very versatile tool and has numerous uses in analyzing pharmaceutical materials. One benefit of light microscopy is that it requires only a small sample size and is generally non-destructive, except when immersion oil or a hot stage is used.

11.7.1 Differentiation of Amorphous and Crystalline Materials

Polarized microscopy is widely used to identify whether a material is amorphous or crystalline. Usually, crystalline materials have well defined shapes while amorphous materials do not. Though such appearance can offer a certain degree of identification, the observation of birefringence provides a more definite differentiation between these two forms. Under a polarized microscope, an amorphous material displays featureless and low contrast images, or even as a dark field, dependent on the relative position of polarization. Conversely, crystalline materials exhibit images with color and various patterns due to birefringence. For identification, a small amount of material is added to a glass slide and examined under a polarized microscope. The use of immersion oil reduces refraction and improves image quality.

Changes in birefringence can also be used to study the physical stability of amorphous materials prepared by melting. A small amount of material is melted on a hot stage and kept at different storage temperatures, allowing crystalline phases to be easily detectable by a polarized microscope. In addition, this technique is used for polymorph screening from the melt. For example, the crystallization of different polymorphs at different temperatures can often be detected because polymorphs generally exhibit different morphologies which can be observed in the microscope. This technique is often referred to as the fusion method or melt crystallization.

Qi et al. studied the crystallization behavior of amorphous paracetamol using the fusion method [8]. Microscopy images in Figure 11.5 show the nucleation and growth of needle-like crystals at 78.4 °C with a clear crystal/amorphous boundary. The amorphous form appears as a featureless background, whereas crystal materials exhibit birefringence. The crystal growth of the needle-like form (Form C) was completed at 87.6 °C, whereas at 102.2 °C another crystal form with a chunk-like appearance was nucleated from Form C. The transformation to the new form was complete at 125.5 °C.

Kestura and Taylor have utilized polarized microscopy to study the role of polymers in influencing crystal growth rates from amorphous felodipine [9]. Amorphous samples of felodipine alone and mixtures with 3% of various polymers were prepared by the fusion method (Figure 11.6). The crystal growth rates of those amorphous samples as a function of temperature (70–110 °C) were studied using polarized microscopy. It was discovered that the various polymers inhibited the crystal growth to different degrees; the order of inhibition effectiveness was poly(vinylpyrrolidone) > poly(vinylpyrrolidone)/vinyl acetate > hydroxypropyl methylcellulose acetate succinate > poly (vinyl acetate). The order of inhibition agreed with the strength/extent of drug–polymer hydrogen bonding interactions.

11.7.2 Characterization of Polymorphs, Hydrates, and Solvates

Microscopy is a very useful qualitative tool for screening, identifying, and selecting active pharmaceutical ingredient (API) forms. The differentiation is accomplished by observing their diverse crystal habits and investigating each unique solid-state transformation. During polymorph screening, a quick analysis using microscopy can provide an assessment of the formation of any new form after comparing the image with that of the starting material. Figure 11.7 shows the distinct morphology of

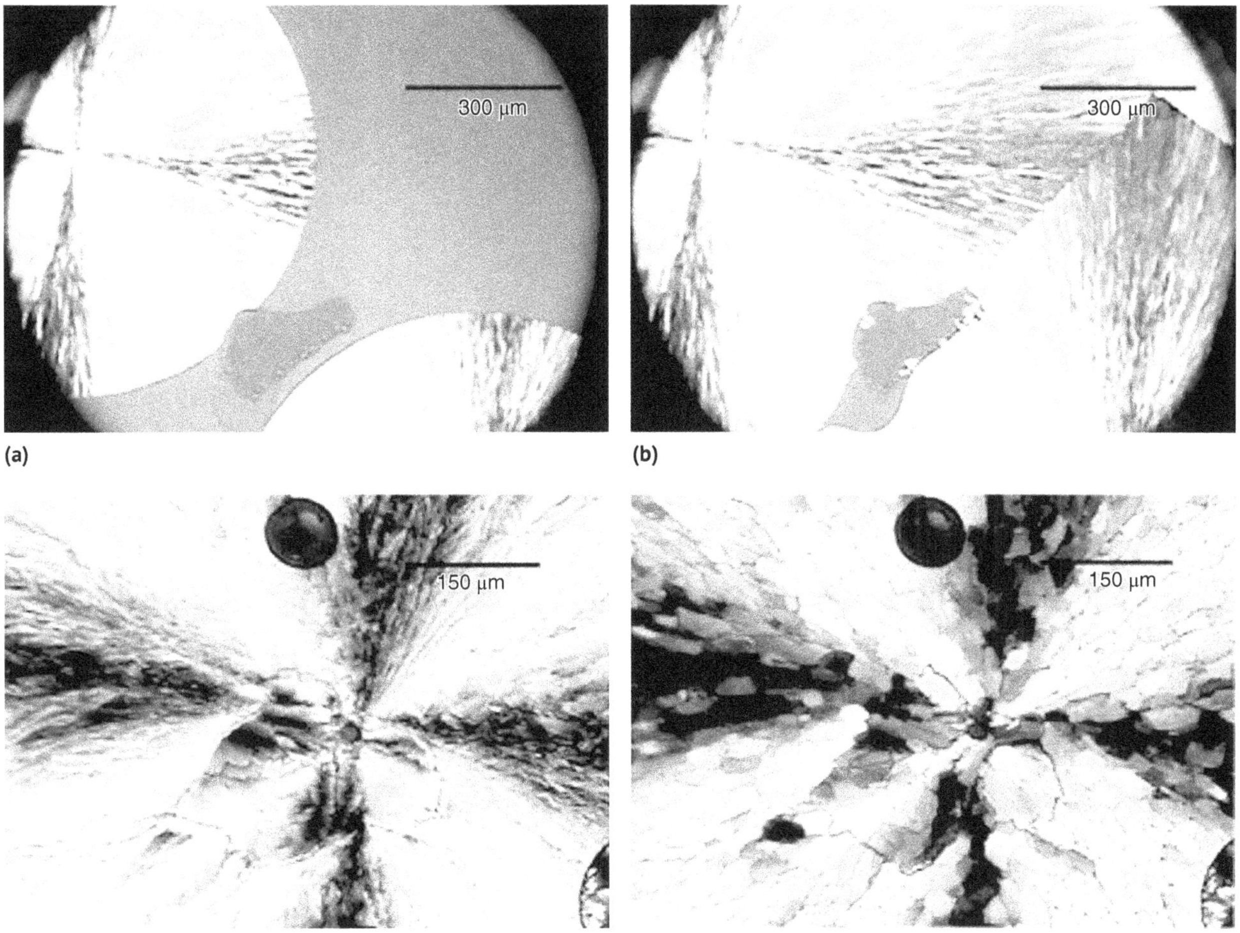

(a) (b) (c) (d)

Figure 11.5 Crystallization of amorphous paracetamol observed by polarized microscope. *Source*: Reproduced from [8]/with permission of Elsevier.

the glutamic acid α-form and β-form. The α-form, the metastable form, has a prismatic shape and the β-form, the stable form, is needle-like [10]. Different shapes, however, are only an indicator of new form, and not definite proof, since in some cases the same form of a compound can have different crystal shapes when produced under different crystallization conditions but retain the same crystal packing.

In such cases the nature of a new form can be identified through additional characterization techniques such as X-ray powder diffraction, thermal analysis, and spectroscopy.

Metformin HCl has two polymorphs, Form A and Form B. Form A is the stable form and Form B is the metastable one. Form B has only been observed under some unusual experimental conditions and transforms quickly to the stable form. Childs et al. isolated Form B using capillary crystallization and studied the form using microscopy [11]. As shown in Figure 11.8, many high-quality single crystals of Form B were obtained. The single-crystal structure of Form B was solved from one of the single crystals.

Long et al. studied four polymorphs of 2-(phenylamino)nicotinic acid using thermal microscopy [12]. The four polymorphs differ in the degree of conjugation of the two aromatic rings and are considered to be conformational polymorphs. Phase transformations observed by thermal microscopy confirmed the thermal transition data observed from DSC and helped to establish the relative physical stability of the four polymorphs.

Under thermal microscopy, the α-form that is the stable form melted at about 158 °C with no other phase transitions observed. The other three polymorphs, the β, γ, and δ forms transformed to the α-form at various temperatures upon heating, as revealed by thermal microscopy. The light yellow crystals of the β form had a phase transformation at around 140 °C, where a new crystal phase nucleated and grew inside a single crystal of the β form. The new phase

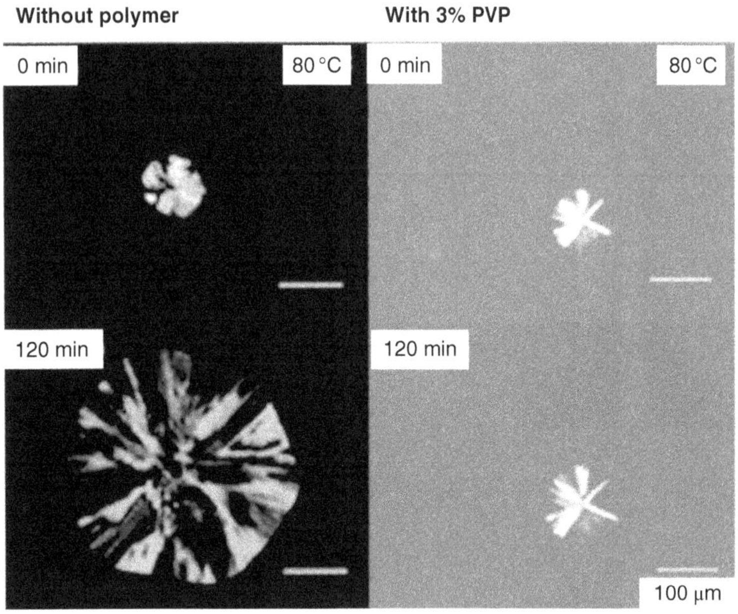

Figure 11.6 Role of polymers in influencing crystal growth rates from amorphous felodipine. *Source*: Reproduced from [9]/with permission of Royal Society of Chemistry.

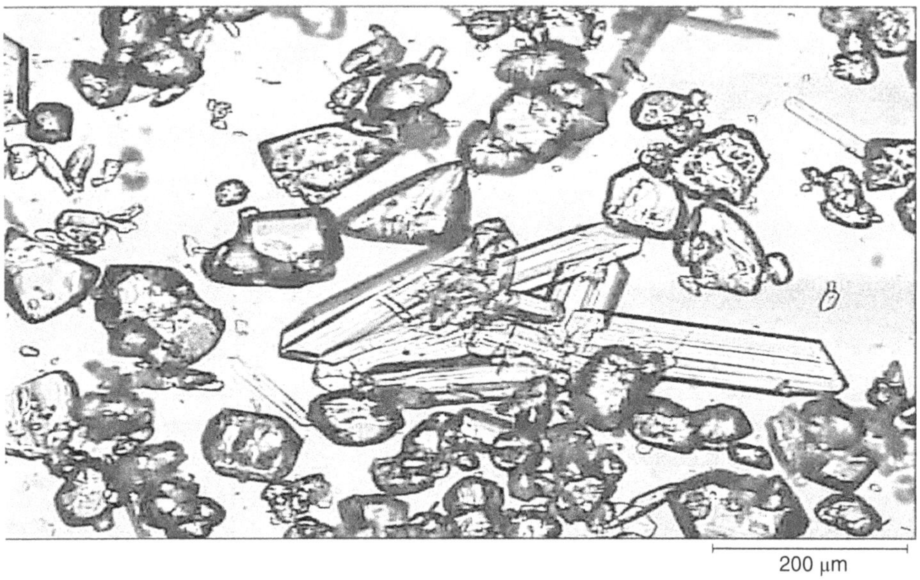

Figure 11.7 Glutamic acid polymorphs. *Source*: Reproduced from [10]/with permission of American Chemical Society.

melted at 158 °C, as the α form. The phase transformation of the γ form to the α form occurred at about 100 °C, which corresponded to an exothermic peak in DSC. The transformation of the colorless δ crystals to the yellow α form took place at about 141 °C.

Thermal microscopy also allows the confirmation of hydrate and solvate forms. Gas evolution of a hydrate and solvate form placed under mineral oil after heating, observed as bubbles, can reveal the dehydration and desolvation temperature. Figure 11.9 illustrates an example of dehydration observed by thermal microscopy, where the evolution of vapor bubbles was observed for norfloxacin dihydrate immersed in mineral oil during heating [13].

The loss of water or solvent from a crystal may occur with melting or with transformation to an anhydrous form, and result in a change in appearance under the microscope. Such physical changes associated with dehydration or desolvation cause decreased light transmittance, wherein the crystals would appear opaque or darker (Figure 11.9).

Figure 11.8 Metformin HCl form B prepared by capillary crystallization. *Source*: Reproduced from [11]/with permission of American Chemical Society.

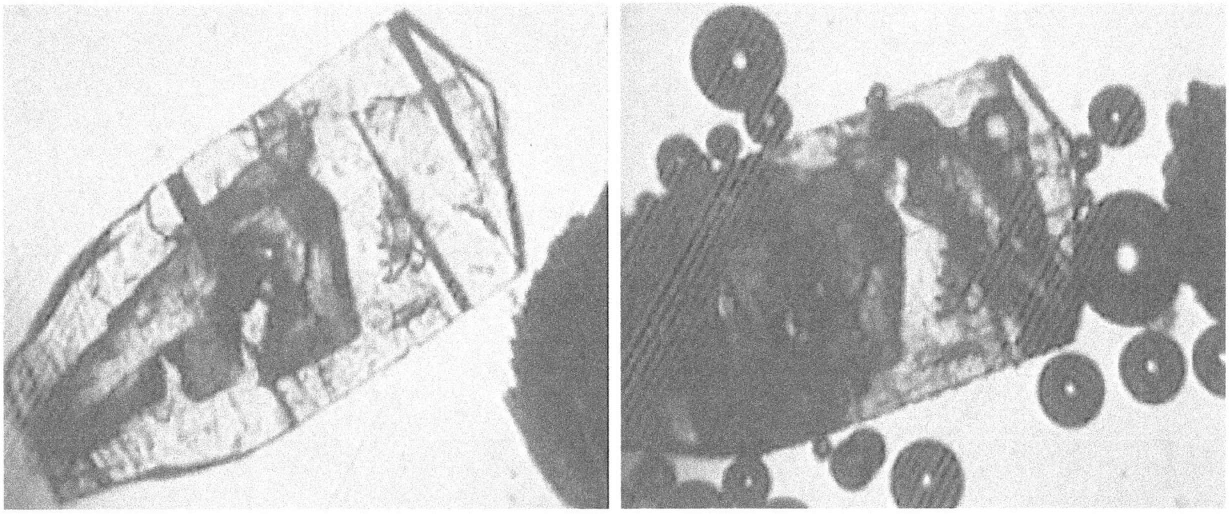

(a) (b)

Figure 11.9 Dehydration of norfloxacin dihydrate. *Source*: Reproduced from [13]/with permission of Elsevier.

Lin and Cheng studied metoclopramide HCl monohydrate using thermal microscopy [14]. The crystals of the monohydrate exhibited no change in appearance until 80 °C, above which the monohydrate lost crystallinity and partially melted. The melting was completed at 97 °C, where the material appeared as a transparent liquid. Several crystals, presumably as an anhydrous form, started to nucleate at around 110 °C, with the crystal growth continuing until 171 °C. The melting of the anhydrous crystals started after 176 °C.

Li and coworkers studied the dehydration of the monohydrate form of an Amgen compound A (Form III) using a hot stage microscope [15]. The Form III crystal changed from being transparent to brown upon dehydration, and the dehydrated form started to melt at 130 °C. A new crystal form was nucleated inside the melted compound after 5 min at 130 °C. The crystal growth continued after 30 min and was completed after 50 min.

11.7.3 Polymorph Conversion

Solution-mediated polymorph conversion is widely used for polymorph screening and the production of stable polymorphs. The metastable form, having higher solubility, dissolves, and the stable form crystallizes out at the expense of the metastable form.

Solution-mediated phase transformation of donepezil hydrochloride hydrate to its anhydrous form in methanol/isopropyl ether mixtures was clearly demonstrated by microscopy [1]. The hydrate was crystallized first from a supersaturated solution.

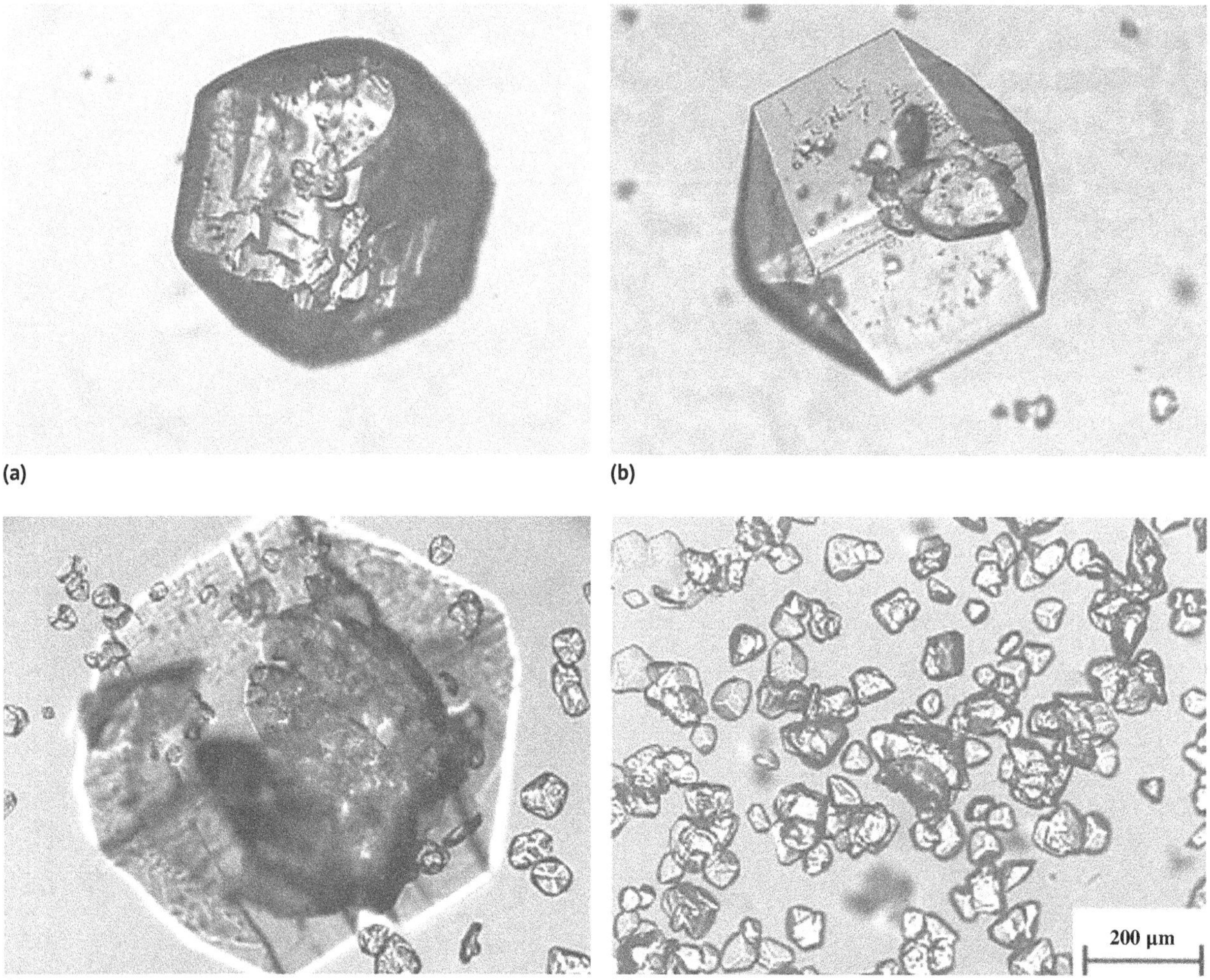

(a) (b) (c) (d)

Figure 11.10 Polymorph conversion of glycine α-form to γ-form in a sodium chloride solution. *Source*: Reproduced from [16]/with permission of Elsevier.

The anhydrous form, Form III, first appeared after 8 hours, and the transformation was completed in 11 hours. Figure 11.10 shows the process of polymorph conversion of glycine α-form to γ-form in a sodium chloride solution at 25°C [16].

11.7.4 Control of Crystallization

Control of crystallization is critical for ensuring the quality of the API obtained upon crystallization, such as knowing which polymorph is present, its morphology, and particle size. The progress of crystallization is easily monitored by microscopy, which provides information in real time. Such information includes the presence of crystals, crystal size, and crystal shape, which are all determined by the rates and conditions of crystal nucleation, crystal growth, and other phase transformations. Direct observation with a microscope provides quick feedback that allows for the control of crystallization parameters such as saturation level, cooling rate, agitation, and seeding. Takiyama reviewed how various supersaturation levels can be used to control crystalline particles during solution crystallization [17]. Examination by microscopy revealed that the size and morphology of taurine crystals were significantly impacted by the supersaturation level, wherein lower supersaturation levels resulted in chunky crystals, and high supersaturation helped to produce fine needles (Figure 11.11).

By the use of an inverted microscope with a flow cell, the crystallization process can be studied in situ, and crystal growth can be monitored quantitatively. In some cases growth rate of each crystal face can be studied independently and correlated with crystal packing.

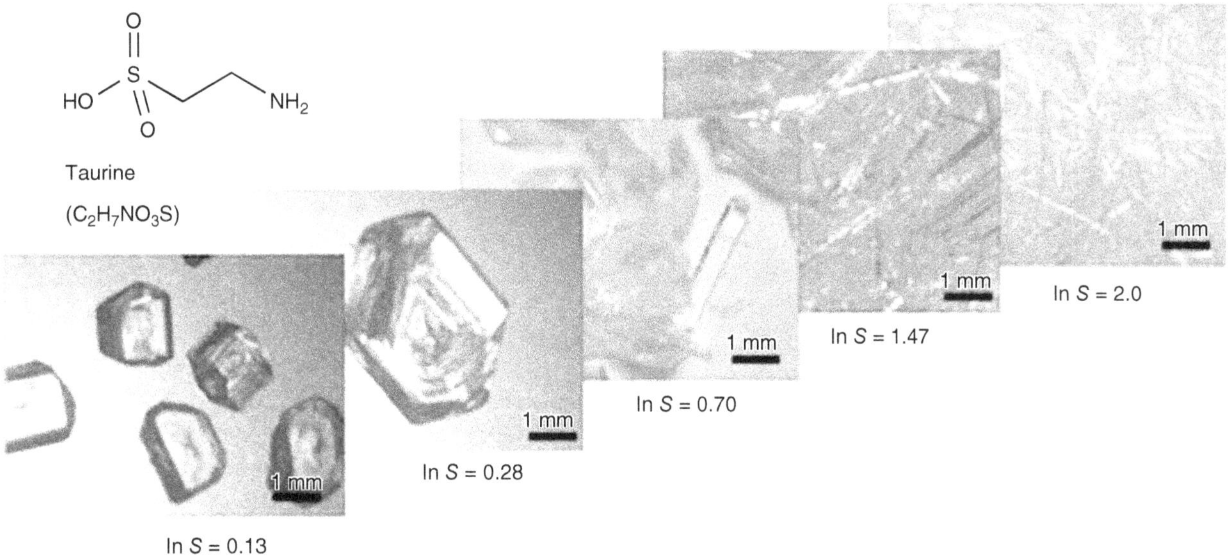

Figure 11.11 Effect of supersaturation on morphology of taurine crystals. *Source*: Reproduced from [17]/with permission of Elsevier.

Microscopy is also useful for identifying nucleation mechanisms. Tao and colleagues, for example, utilized microscopy to study cross-nucleation between mannitol polymorphs during seeded crystallization [18]. Cross-nucleation occurred when the melt of D-mannitol was seeded with the β polymorph, which resulted in the nucleation and growth of the α polymorph at lower temperatures. This is an interesting example of a situation where a metastable polymorph, the α polymorph, was able to nucleate on the stable form, the β polymorph, and had a higher growth rate than the β polymorph.

11.7.5 Screening for Cocrystals

Pharmaceutical cocrystals have attracted significant interest in the past decade or so, as materials with distinct physical and chemical properties from those of the crystalline API alone (Chapter 5).

Thermal microscopy is widely used for cocrystal screening through the mixed fusion method. In this process, the higher melting form of an API crystal is melted, after which a second component, the potential coformer, is added to the microscope slide adjacent to the first component. The second component is then melted and brought into contact with the first component. Some of the first component dissolves in the melt of the second component and forms a mixture of the two components. After the two components solidify, the mixing zone is observed for any physical changes under a hot stage while the temperature is controlled isothermally or at various heating or cooling rates. Multiple molar ratios of the two components are used to see if stoichiometric forms are produced. By using a polarized microscope, crystallization of either individual component or that of a cocrystal can be observed due to the presence of birefringence. From the mixing zone, there are three types of physical changes that can occur, including phase separation to form a conglomerate, formation of cocrystals, and a completely miscible solid solution, all of which can be easily differentiated by microscopy. The solid solution will not show birefringence. The conglomerate will display two crystal phases in the mixing zone, whereas the cocrystal would be expected to be observed as a single-crystal phase, flanked by two eutectic mixtures in the mixing zone. The cocrystal phase tends to show birefringence and is usually clearly distinguishable from the two components by color, appearance, and morphology. The newly formed cocrystal can be characterized in situ for its melting point and used later as seeds for growing cocrystals from solution.

Berry et al. have applied thermal microscopy to the cocrystal screening of nicotinamide as the coformer with seven APIs [19]. The application of thermal microscopy and the mixed fusion method led to the elucidation of the thermodynamic landscape within a binary phase diagram and helped to increase overall screening efficiency. Among the seven APIs used, thermal microscopy showed that five APIs formed cocrystals with nicotinamide. Three new cocrystal systems with nicotinamide were identified (ibuprofen (R/S), ibuprofen (S), and salicylic acid), and their crystal structures were determined.

11.7.6 Analysis of Particle Size

Light microscopy is a convenient and inexpensive tool for determining particle size. USP chapter 776 gives a basic general description for using microscopy for particle characterization [20]. Light microscopy can detect particles ranging from micrometers to millimeters in size. The image of a sample is captured as a two-dimensional digital image, and computer-assisted image analysis of the sample provides a means of calculating a size distribution obtained from the sizes of the individual particles. The limitations of DOF may not allow one to obtain all particles in focus in one image. Unfocused particles can give inaccurate dimensions for image analysis. Also, any diffraction that occurs will become greater as the particles become smaller. The increase in diffraction causes blurring on the edges and inaccurate dimensions, so that the determination of particles of less than 3 μm is not very reliable.

For spherical particles, size is generally defined as the diameter. The evaluation of the effective particle size for irregular shapes can be estimated in a number of ways, including the Feret's diameter, Martin's diameter, and projected area diameter [20]. A Feret's diameter is the distance between two tangents on opposite sides of the particle, parallel to a fixed direction. Martin's diameter is the length of the line that passes through the center of gravity and bisects the particle image. The lines can be in any direction, but the direction should be maintained constantly for all image measurements. The projected area diameter is the diameter of a circle that has the same area as a two-dimensional image of the whole particle.

The use of microscopy for particle size analysis has significant advantages, including low cost, small sample size, and relatively easy method development, while also offering information about particle shape. However, there are some disadvantages existing such as obtaining uniform and repeatable sampling, low throughput, and labor-intensive processes. Some of these disadvantages can be overcome by using automation, where the collection of a large number of images coupled with automatic analysis provides increased statistical significance and thus improved data quality. Particle size data from microscopy is often used as a basis to qualify more advanced methods or help understand particle shape effects.

11.7.7 Contaminant Analysis

Microscopy is widely used in the analysis of contamination in the form of particulates. Contamination can occur in the manufacture of drug substances and drug products, including chemical, particulate, and microbial impurities. Regulatory agencies require that contamination problems be fully investigated in a timely fashion. Most solution and solid drug products must have specifications that limit the level of visible foreign particles so that they are essentially free of such contaminants.

Contaminants can arise from many sources, including contaminated or degraded raw materials, unclean or malfunctioned equipment, and improper operating procedures. Contaminant analysis is essential for identifying the nature and source of these contaminants and for devising effective strategies to eliminate them.

Some of the common particulate contaminants found in drug substances and drug products are listed in Table 11.2.

Light microscopy is very useful in the analysis of particulate contamination because the contamination is usually present in small amounts and the particles exist in minute size, which makes it difficult or impossible to detect using conventional methodologies such as chromatography or spectroscopy. Though some foreign materials can be observed by the naked eye, a more thorough examination generally relies on carrying out microscopic studies. In-depth identification can be accomplished by combining microscopy with analytical tools such as microscopic FT-IR, microscopic FT-Raman, chromatography, or mass spectrometry.

Shearer has published a review about the use of microscopy in the identification of contaminants in pharmaceutical products [21]. Figure 11.12 lists many types of contaminants recovered from pharmaceutical products, including fibers, glass particles, delaminated glass flakes, and corrosive steel particles [21, 22].

Shearer has discussed one case about the identification of silver particles in a tablet product, where at high magnification the particles were reflective and had a plate-like morphology [21]. Further analysis by in situ Raman spectroscopy confirmed that the silver particles are graphite. In another study, this approach was employed to troubleshoot tan discoloration observed on the tablet surface of a drug product. Using microscopy, the discolored spots were extracted

Table 11.2 Common nonmicrobial contaminants in pharmaceuticals.

Glass particles, metal particles, natural fibers, synthetic fibers, silicone, plastics, rubber, vial delamination flakes, skin flakes, hairs, char particles, detergents, lubricant oils, Teflon, graphite.

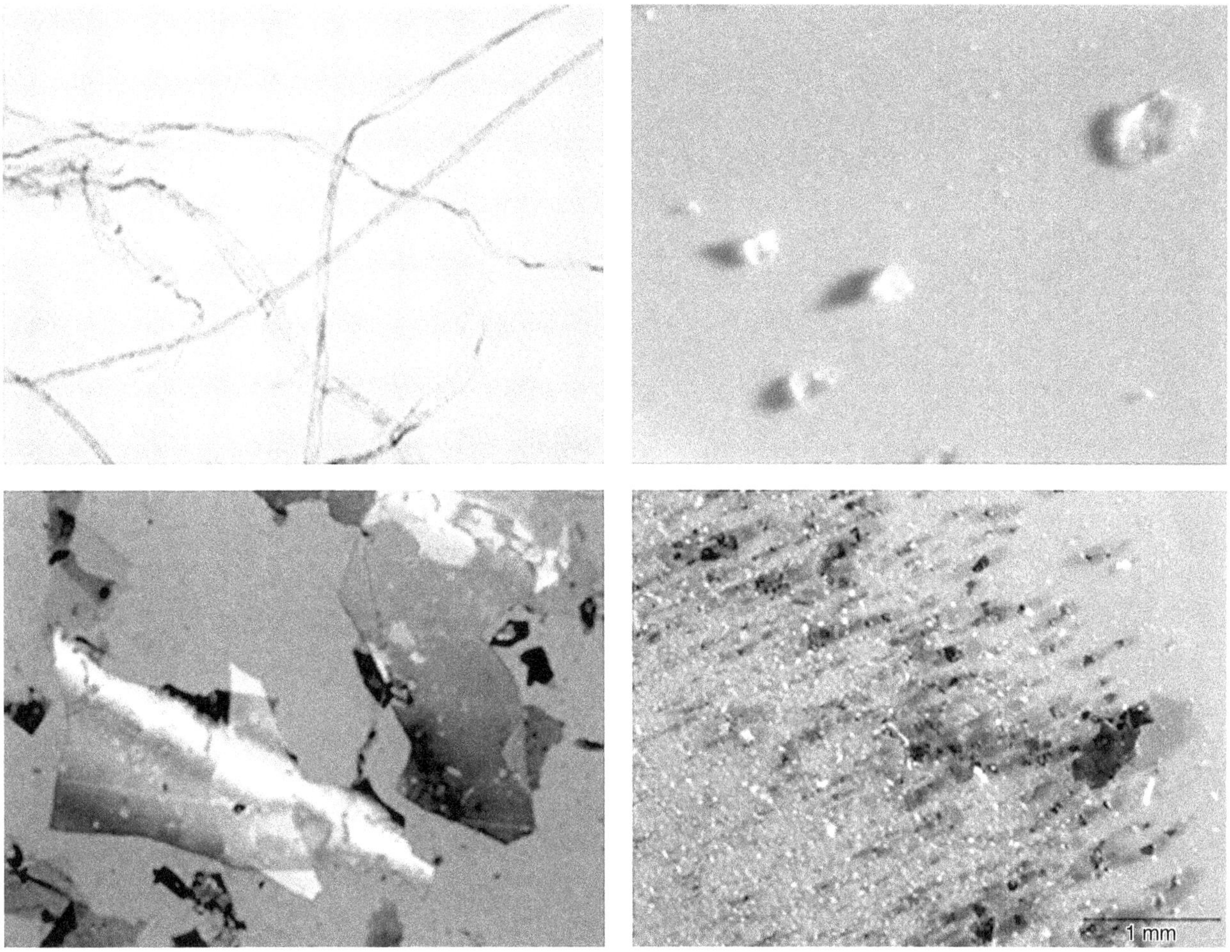

Figure 11.12 Fibers (top right), glass particles (top left), delaminated glass flakes (bottom right), and steel corrosion particles (bottom left) recovered from drug products.

with a glass capillary micropipette, then dissolved in a solvent, and finally identified as a hydrocarbon oil through FT-IR analysis.

Contaminant analysis, especially metallic type, can also be resolved using an SEM to be discussed below.

11.8 Scanning Electron Microscope

An SEM uses electrons instead of visible light to form amplified images of a material. As such, it offers higher resolution, larger DOF, and better control of magnification than a traditional light microscope. The spatial resolution of SEM can reach as low as 50–100 nm, and a larger DOF allows more of a sample to be in focus at one time. SEM also provides more flexibility for the control of magnification since electromagnets are used. In addition, it has unique capabilities to perform chemical analysis at selected locations to determine chemical composition.

The components of a SEM are analogous to those found in a light microscope, as shown in Figure 11.13. Instead of a beam of light, a focused electron beam scans the specimen. The electron beam travels through a "lens," an electromagnetic field, before impacting the specimen. The bombardment of the specimen by high-energy electrons results in the production of X-rays, Auger electrons, and secondary electrons, in addition to primary backscattered electrons. Backscattered electrons and secondary electrons are detected, and the detected signals and the position of the electron beam are combined to produce an image. The use of an electromagnetic field instead of an optical lens offers much better control in achieving magnification, which ranges from 20× to approximately 30 000×.

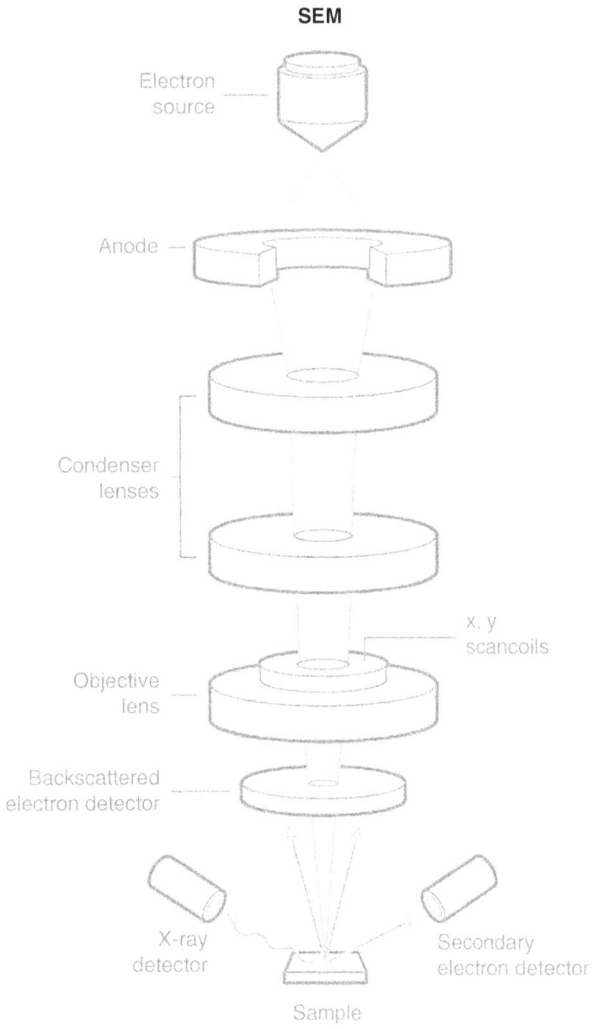

Figure 11.13 Diagram of SEM and TEM.

Because only conductive material can generate a signal in SEM, sample treatment is required for nonconductive materials. This involves coating the surface of the sample with a thin layer of conductive material such as gold. Since the SEM operates under a vacuum, sample treatment to remove water from the specimen is also required.

Signals from SEM provide information about material morphology, chemical composition, and crystalline structure. The impact of secondary electrons on a sample allows the detection of the morphology and topography of samples, whereas backscattered electrons are useful for generating a contrast between multiple components. X-ray signals received during the process are characteristic of the elements in the material and could be used for the determination of chemical composition.

A SEM contains an electron source, electron lenses, a sample stage, detectors for all signals of interest, and a display or data output device. An Everhart–Thornley detector is used in traditional SEM, with a power system and a vacuum system in place to support the functionality of the equipment.

In terms of pharmaceutical applications, for example, SEM was used to study the compatibility improvement of metformin HCl by crystal engineering [23]. Metformin HCl crystallized in the presence of 2% PVP K30 for a duration of four hours gave rise to an impressive improvement in mechanical properties, including flow, compressibility, and compactibility, compared to the untreated metformin HCl. SEM images show that the two materials have distinctly different morphologies. The untreated materials display a needle shape, whereas the engineered particles appear chunky with aggregates. Analysis by DSC and FT-IR confirmed that the two materials are in the same polymorphic form.

SEM is capable of more advanced identification for particulate contaminants observed by light microscopy. Coupled with energy dispersive spectrometry, it can deliver imaging and chemical characterization at the same time.

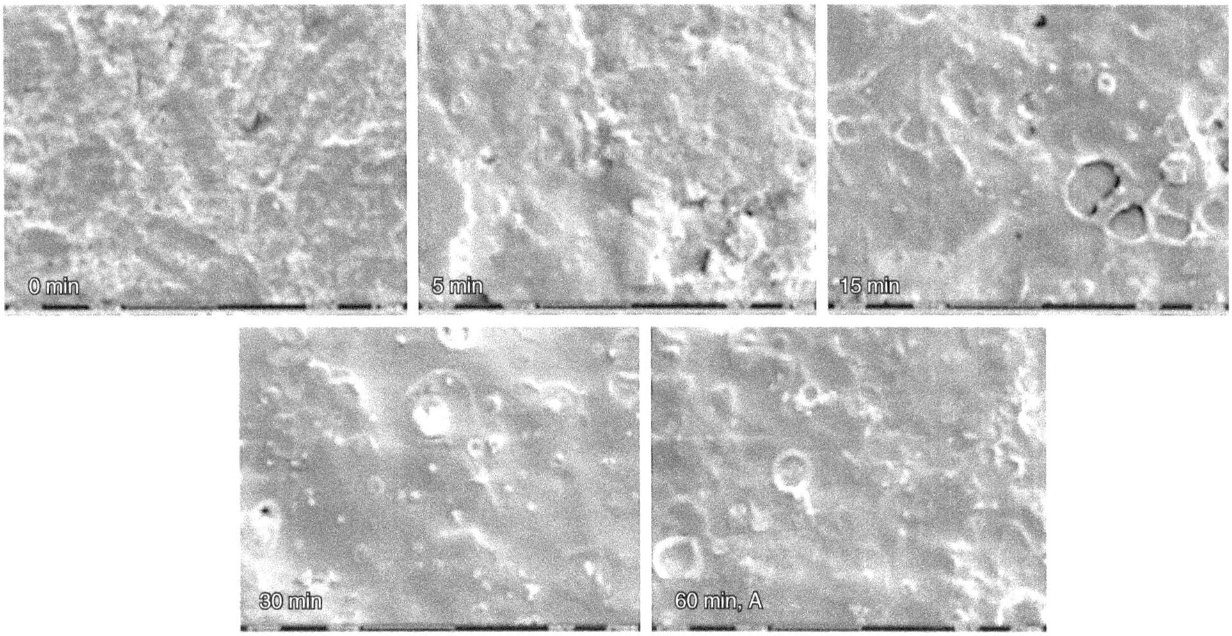

Figure 11.14 The progress of film coating of a tablet product. *Source*: Reproduced from [24]/with permission of Springer Nature.

SEM is also used to aid the development of pharmaceutical coatings. Such coating is widely used for pharmaceutical tablets to improve appearance, taste-masking, moisture protection, or controlled release. SEM can also assist in examining surface properties, measuring coating thickness, and evaluating the quality of coating adhesion to tablet surfaces. Figure 11.14 shows the progress of film coating of a tablet product examined by SEM [24]. As coating progressed, surface roughness decreased, and the tablet cores were completely covered with a thin film after 30 minutes.

SEM has been demonstrated to be a very useful tool for analyzing nanoparticles, generally difficult to study by light microscopy.

In summary, SEM is a useful imaging and analytical technique for material characterization and surface analysis. However, SEM studies must be operated under a high vacuum environment. Samples must be clean, dry, and electrically conductive. For nonconductive specimens, such as the majority of pharmaceutical materials, coating with a conductive film is an essential part of sample preparation. Sample preparation, such as drying and coating, may disrupt sample integrity. Thus, SEM is certainly not suitable for the analysis of wet samples.

11.9 Environmental SEM

An ESEM can remove the constraints of sample preparation but retain the performance advantage of a conventional SEM by altering some aspects of the vacuum system and improving signal detection. For this purpose, a differential pumping system is introduced to vary the sample environment through a range of pressures, temperatures, and gas compositions. The pumping system also provides separation of the electron optics column from the specimen chamber. Consequently, the column still maintains a high vacuum, while the sample chamber may have a pressure as high as 50 Torr. The pumping system eliminates the need for extensive sample preparation, as with SEM, so that ESEM can image wet and oily samples.

The detection system in ESEM also offers high-resolution secondary electron imaging in a gaseous environment with a detector called an environmental secondary detector (ESD), which can function in a non-vacuum environment. The ESD utilizes gas ionization by the secondary electrons emitted by the sample. While a positive potential of a few hundred volts is being applied to the detector, the secondary electrons emitted by the sample are attracted to the detector. As the secondary electrons accelerate in the electric field, the collision with gas molecules produces ionization. The ionization creates additional electrons and positive ions, thereby amplifying the original secondary electron signals. The generated positive ions are attracted to the sample surface, which helps to suppress charging artifacts in nonconductive samples. As such, the samples do not need coating with a conductive film.

ESEM is capable of studying dynamic processes, such as powder compression, heating, cooling, freeze-drying, melting, dehydration, and sublimation.

ESEM has been used successively to analyze a number of pharmaceutical samples in their natural state. For example, Thibert and Hancock reported the direct visualization of super disintegrant hydration using ESEM [25], whereas Stokes et al. directly observed a water–oil emulsion system in the liquid state [26].

11.10 TEM and ED

Like SEM, TEM is commonly used in biological research and material science; however, it manifests very different capabilities in sample analysis due to its transmission mode, as summarized in Table 11.3. First, the transmission mode would incur the strong interaction of the electron beam with the sample and, therefore, require very thin specimens, preferably less than 0.15 μm. Second, as the sample needs to be placed under a high vacuum environment during analysis, TEM is generally not suitable for hydrates or solvates unless they are run under cryogenic conditions. These limitations, not surprisingly, lead to the situation that TEM is less useful in imaging pharmaceutical materials than SEM and ESEM. However, the unique setting of optics enables TEM to be powerful in collecting ED data for the crystal structure elucidation of organic, inorganic compounds, and biological molecules. Indeed, it is often regarded as a complementary tool to traditional X-ray crystallography, and, in certain aspects, even more powerfully, especially as to the capability of working with very small crystals from several microns down to nanosizes. In comparison, single-crystal X-ray diffraction usually requires the typical size of the crystal to be in the tens of microns.

The ability of TEM for crystal structure determination was not fully appreciated until the introduction of rotational methods for sequential data collection, generally referred to as three-dimensional ED (3D ED) [27]. The method involves continuous data collection of ED while a micro- or nanocrystal is continuously rotated by the microscope stage, analogous to the goniometer head in X-ray crystallography (Chapter 8). Interestingly, the data acquisition time is usually much shorter than that of the X-ray diffraction method. More importantly, the data analysis and structure solution are very similar to those of X-ray crystallography, and, indeed, the same software is generally used. For example, the structure solution from ED can be generally accomplished using direct methods such as SHELXT, a widely used platform for integrated space group and crystal structure determination from X-ray diffraction data [28, 29]. Thus, it is not surprising to note that 3D ED, owing to the several unique advantages over the X-ray diffraction method, has emerged as a very promising method in structural biology to determine structures of macromolecules for drug discovery and to visualize drug binding interactions with macromolecular targets. As we shall see, 3D ED, while not yet widely used in the solid-state characterization of pharmaceutical materials, could become another powerful tool for polymorph identification, crystallinity determination, and phase transformation investigation.

Table 11.3 Comparison of SEM and TEM.

Item	SEM	TEM
Optics	Scanning	Transmitting
Electron beam	Fine and focused beam	Broad beam
Electron energy	1–30 kV	60–300 kV
Detection	Scattered electrons	Transmitted electrons, diffracted electrons
Specimen thickness	Any	Ultrathin, typically less than 150 nm
Type of information	Three-dimensional image of surface, topographies	Two-dimensional projection image of inner structure; electron diffraction for crystal structure
Magnification	Up to 1–2 million times	Up to 50 million times
Optimal spatial resolution	~1 nm	~0.1 nm
Operation	Little or no sample preparation	Sophisticated sample preparation needed

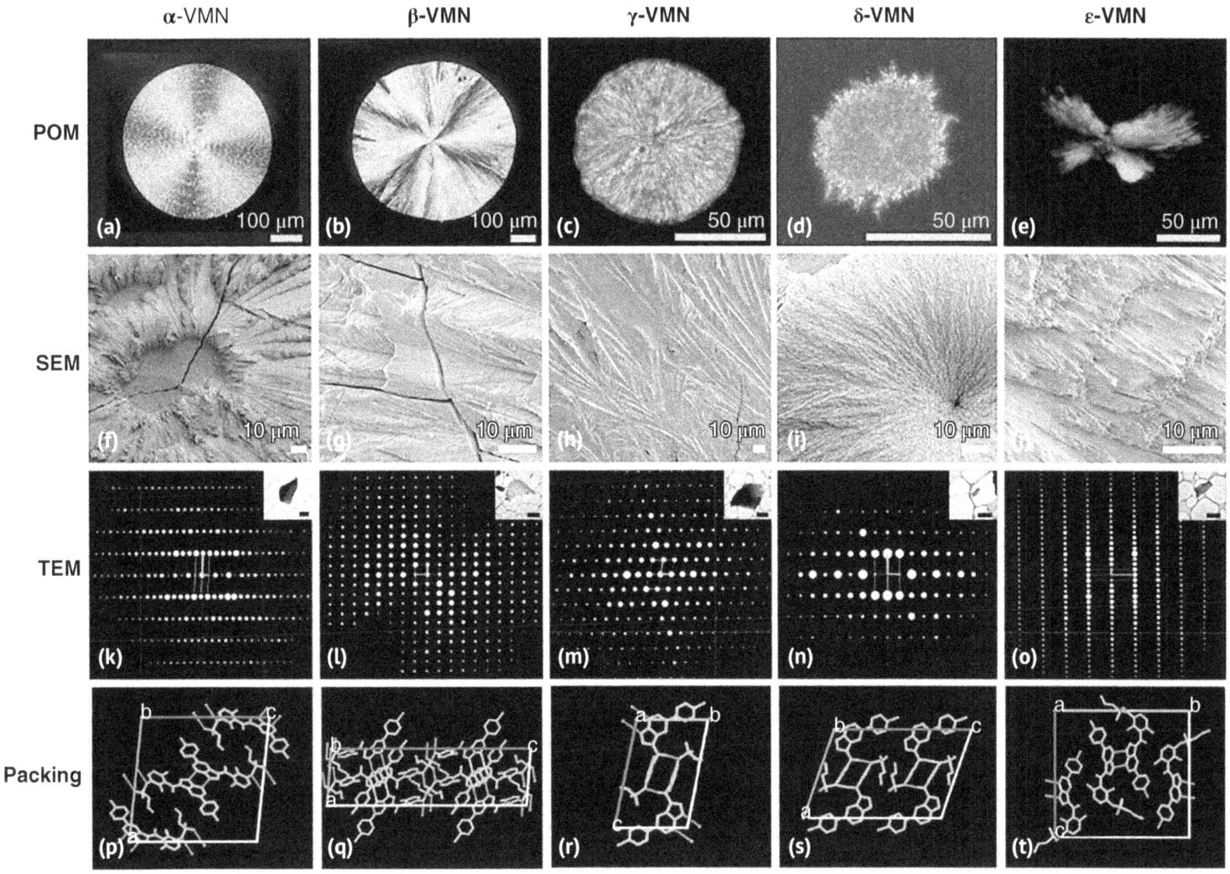

Figure 11.15 Direct structure determination of the five polymorphs of vemurafenib using three-dimensional ED. *Source*: [30] / Springer Nature/CC BY 4.0.

For example, by applying 3D ED, Li et al. have directly solved the crystal structures of several polymorphs of vemurafenib, a cancer drug for the treatment of BRAF-mutated melanomas, from compact spherulites prepared by melt crystallization [30]. As shown in Figure 11.15, those polymorphs grown from the melt produce a spherical aggregate of numerous branching fibrils, which is obviously not applicable for single-crystal X-ray diffraction for structure determination. However, amazingly, by studying a few broken fragments with micrometer sizes obtained from crushed spherulites, Li et al. resolved the crystal structures of the α-, β-, γ-forms in agreement with the structures obtained from single-crystal X-ray crystallography. Furthermore, they were able to solve the unknown crystal structures of the δ- and ε-forms [30]. In another interesting case, Gruene and coworkers successfully determined the crystal structure of paracetamol after directly sampling the crystals from the powder of its capsule dosage form [31]. Likewise, Jones et al. collected high-quality 3D ED data from nanocrystals with particle sizes of about 100 nm and a mass of only 10^{-15} g and subsequently solved crystal structures for progesterone, ibuprofen, carbamazepine, and complex natural products like thiostrepton at an atomic resolution of less than 1 Å [32].

The most striking capability of TEM is the combination of imaging and structure determination. As would be expected, TEM offers superior capability of detecting physical changes to become crystalline forms for amorphous solid dispersions, even in very minor quantities. On the one hand, TEM can deliver nanoscale spatial resolution as an electron microscope. On the other hand, analysis of the ED patterns of drug crystals can ascertain the exact crystal structure information, consequently, the polymorph identity. For example, Ricarte et al. reported that TEM can detect drug crystallites of griseofulvin in its solid dispersions with hypromellose succinate acetate down to 3% crystallinity, with a better sensitivity than wide-angle X-ray diffraction [33]. S'ari and coworkers used a variety of solid-state physical characterization techniques to study an amorphous solid dispersion of felodipine and copovidone made by a hot melt extrusion process [34]. It is interesting to note that the use of X-ray powder diffraction, DSC, or FT-IR could not detect any signs of crystallinity, whereas TEM did indeed pinpoint two areas containing crystals at the edges of milled particles from a total of 55 particles examined. Selected area ED analysis revealed that the crystals are predominantly form II, one of the metastable forms.

11.11 Atomic Force Microscopy

AFM is another useful technique used to visualize submicrometer particles, to study surface characteristics, and to characterize the properties of nanomaterials. It is a high-resolution microscopic technique capable of resolving dimensions as small as an atomic lattice. Besides imaging, AFM offers the possibility of manipulating molecular and atomic-level features. Although the capability of AFM is similar or sometimes superior to the SEM, the mechanism for obtaining an image is distinctly different.

Imaging by AFM is achieved by mapping the surface with a cantilever tip (Figure 11.16). The tip is brought in contact with the surface to be imaged and when the tip reaches a certain distance from the surface, a repulsive force acts on the tip and bends the cantilever in an upward direction. The degree of bending is measured by a laser and collected by a photo detector. As the tip moves across the surface under a constant force, the movement of the tip reflects the surface "roughness" profile. An overall surface topography, therefore, is being established as the tip continues to scan the entire surface of interest.

Several forces can be responsible for the interaction between the tip of the cantilever and the sample surface. These include van der Waals interactions, capillary, adhesion, and electrical double layer forces. Overall, van der Waals forces are the main source of interaction during AFM mapping.

AFM is applicable for imaging any conducting or nonconducting surface, has very high resolution, and offers minimal sample preparation. The resolution in the x–y plane ranges from 0.1 to 1.0 nm, whereas resolution in the vertical z direction can reach 0.001 nm. The utility of AFM can be further improved in situ, for solids and solutions, with the use of available temperature and environmental control.

There are several operating modes of AFM, including contact mode, tapping mode, and noncontact mode. Selecting an operating mode is dependent on the application.

Contact mode is the simplest mode of operation for AFM. It is also often called the repulsive mode. It involves scanning a sample surface with the cantilever tip that makes soft physical contact with the sample as it traces the sample surface. It is applicable to the study of solid surfaces but is less useful for soft materials, since the vertical and shear forces applied by the tip can damage soft materials and cause the samples to lose their integrity.

In view of this deficiency of the contact mode, the tapping mode offers a better alternative for studying soft surfaces. In the tapping mode, the cantilever tip oscillates vertically at a certain frequency while scanning. The tip lightly touches or "taps" the sample surface briefly and then lifts off. The amplitude of the detectable oscillation is usually from 20 to 100 nm. The topographic image is built by recording the force of the oscillating contacts of the tip with the sample surface.

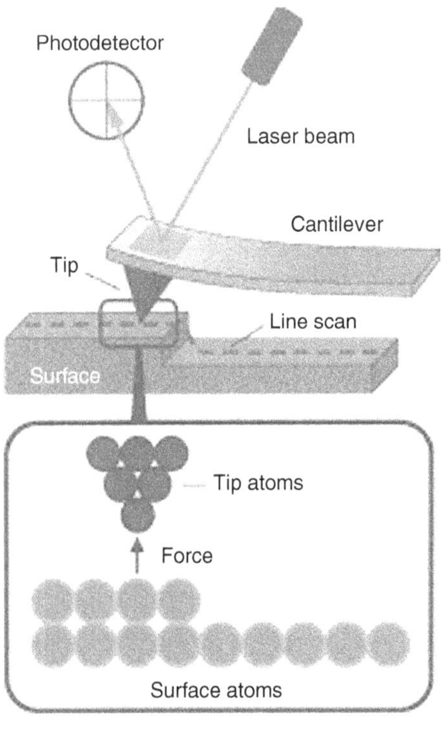

Figure 11.16 A diagram to show AFM.

The noncontact mode does not involve physical contact of the cantilever with the sample by maintaining a wide tip-sample separation, where the cantilever tip moves above the sample surface with a small oscillating amplitude and high frequency. The noncontact mode must be performed under ultra-high vacuum and, therefore, is less extensively used than the contact mode and the tapping mode.

The ability of AFM to study structures on the nanometer scale in ambient and fluid environments makes it a significantly useful tool for pharmaceutical material characterization and research. AFM, for example, has found great success in studying crystal growth, particle interactions, drug dissolution, and phase transformations, helping to reveal mechanisms at the molecular level for those physical changes, as demonstrated by the following case studies.

Lauer et al. were able to detect phase separation induced by exposure to humidity in a solid dispersion of copovidone and a cholesteryl ester transfer protein (CETP) inhibitor using AFM [35]. Figure 11.17 shows the process of phase separation over the course of a week under 40 °C, 75% RH. After 2 hours of exposure to the stress condition, the surface became smooth. At the same time, a limited number of droplets (0.2% of surface area) with dimensions from 50 to 500 nm formed, producing a fluidic

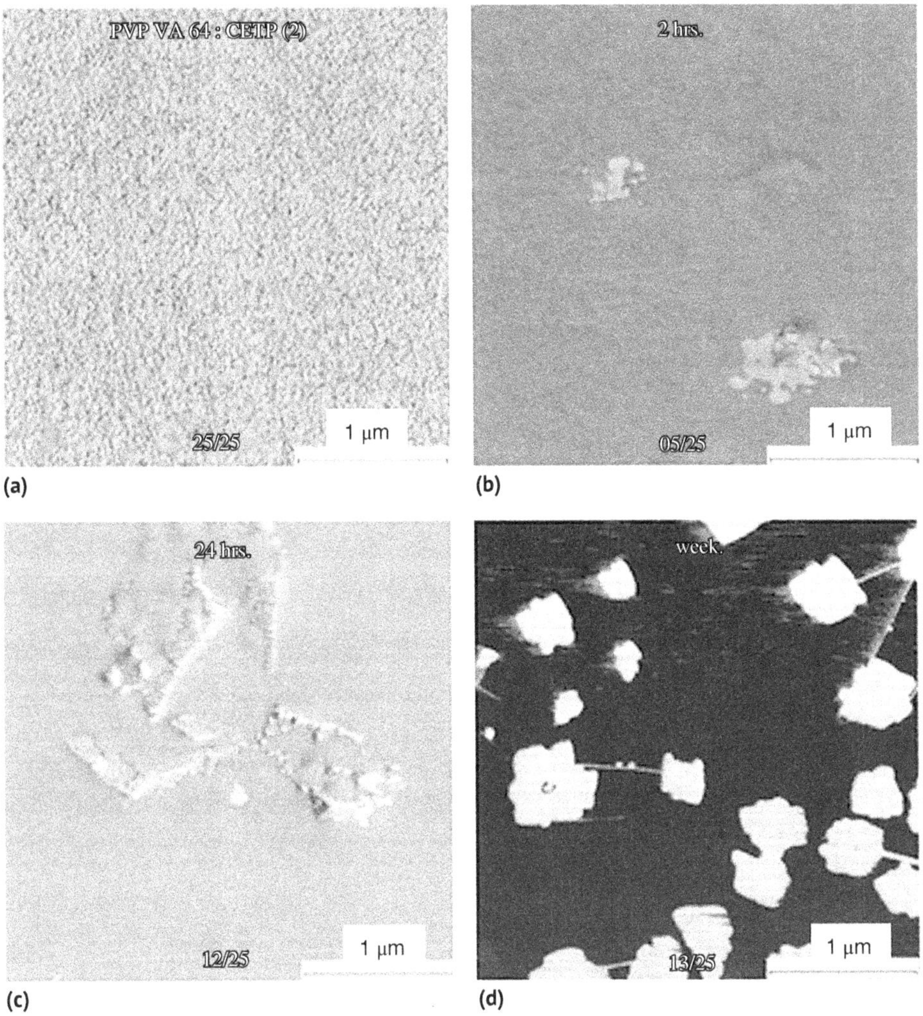

Figure 11.17 AFM phase maps recorded on the fracture surface of CETP(2)/PVP VA 64 film initially (a), after 2 hours (b), after 24 hours (c), and after 1 week (d) exposure to stress conditions (RH = 75%, T = 40 °C). *Source*: [35]/Springer Nature/CC BY-NC 2.0.

morphology. The morphology changed to more regularly sized particles after 24-hour storage, and the surface coverage increased to 5% of the overall imaged area, with particles that appeared to be noncrystalline. Storage up to 1 week gave rise to star-shaped particles with ultrathin needles, indicating the crystallization of the drug. The AFM study concluded that the phase separation process of copovidone and the CETP inhibitor comprises several stages: formation of a separated drug domain, aggregation in the noncrystalline region, and crystallization of the drug.

In addition to its high resolution, AFM has the capability of measuring forces at the nanoscale level. As such, AFM has been used to make measurements of the adhesive forces between drug and excipient in the development of a dry powder inhalation (DPI) formulation [36]. The procedure involved mounting a 1–3 μm drug particle on an AFM cantilever. The adhesion force between the drug particle and a plate of lactose monohydrate, a major excipient used in DPI or stainless steel, representing device wall materials, was investigated. It was interesting to discover that the adhesion force became stronger with increasing surface roughness and atmospheric humidity.

AFM can act as a precise tool for studying physical changes in situ, such as crystal growth and dissolution. A dissolution study of aspirin crystal planes by AFM indicated that the (001) surface has a greater dissolution rate than that of the (100) surface under 0.05 M HCl [37]. In addition, the two planes showed different topographic features during the dissolution process. The dissolution of the (001) plane occurs with receding step edges and has a rate of 0.45 nm/s. In comparison, the rate of dissolution for the (100) plane is approximately six times faster, at 2.93 nm/s, and the (100) plane dissolves by sinking of crystal terraces.

References

1 Sugano K, Kato T, Suzuki K, Keiko K, Sujaku T, Mano T. High throughput solubility measurement with automated polarized light microscopy analysis. *J Pharm Sci* 2006; 95(10):2115–2122.

2 Bhatt V, Shete G, Bansal AK. Mechanism of generation of drug nanocrystals in celecoxib: mannitol nanocrystalline solid dispersion. *Int J Pharm* 2015; 495(1):132–139.

3 Kofler L. Mikroskopische Methoden in der Mikrochemie. *Mikrochemie vereinigt mit Mikrochimica Acta* 1951; 36(1):283–290.

4 McCrone WC, McCrone LB, Delly JG. *Polarized Light Microscopy*. Ann Arbor and Chicago: Ann Arbor Science Publishers Inc. and McCrone Research Institute; 1978.

5 Anonymous. THMS600 heating and freezing stage. Available at http://www.linkam.co.uk/thms600-features. (Accessed 29 December 2015).

6 Köhler A. New method of illumination for photomicrographical purposes. *J Royal Micro Soc* 1894; 14:261–262.

7 Anonymous. VHX-5000 series Digital Microscope. Available at http://www.keyence.ca/products/microscope/digital-microscope/vhx-5000/index.jsp. (Accessed 29 December 2015).

8 Qi S, Avalle P, Saklatvala R, Craig DQ. An investigation into the effects of thermal history on the crystallisation behaviour of amorphous paracetamol. *Eur J Pharm Biopharm* 2008; 69(1):364–371.

9 Kestura US, Taylor LS. Role of polymer chemistry in influencing crystal growth rates from amorphous felodipine. *Cryst Eng Comm* 2010;12(8):2390–2397.

10 Ni XW, Valentine A, Liao A, Sermage SB, Thomson GB, Roberts KJ. On the crystal polymorphic forms of L-glutamic acid following temperature programmed crystallization in a batch oscillatory baffled crystallizer. *Cryst Growth Des* 2004; 4(6):1129–1135.

11 Childs SL, Chyall LJ, Dunlap JT, Coates DA, Stahly BC, Stahly GP. A metastable polymorph of metformin hydrochloride: isolation and characterization using capillary crystallization and thermal microscopy techniques. *Cryst Growth Des* 2004; 4(3):441–449.

12 Long S, Parkin S, Siegler MA, Cammers A, Li T. Polymorphism and phase behaviors of 2-(phenylamino)nicotinic acid. *Cryst Growth Des* 2008; 8:4006–4013.

13 Chongcharoen W, Byrn SR, Sutanthavibul N. Solid state interconversion between anhydrous norfloxacin and its hydrates. *J Pharm Sci* 2008; 97(1):473–489.

14 Lin SY, Cheng WT. The use of hot-stage microscopy and thermal micro-Raman spectroscopy in the study of phase transformation of metoclopramide HCl monohydrate. *J Raman Spectrosc* 2012; 43(8):1166–1170.

15 Li H, Kiang YH, Jona J. Multiple approaches to pharmaceutical polymorphism investigation—a case study. *Eur J Pharm Sci* 2009; 38(5):426–432.

16 Yang X, Lu J, Wang XJ, Ching CB. Effect of sodium chloride on the nucleation and polymorphic transformation of glycine. *J Cryst Growth* 2008; 310(1):604–611.

17 Takiyama H. Supersaturation operation for quality control of crystalline particles in solution crystallization. *Adv Powder Technol* 2012; 23(3):273–278.

18 Tao J, Jones KJ, Yu L. Cross-nucleation between D-mannitol polymorphs in seeded crystallization. *Cryst Growth Des* 2007; 7(12):2410–2414.

19 Berry DJ, Seaton CC, Clegg W, Harrington RW, Coles SJ, Horton PN, Hursthouse MB, Storey R, Jones W, Friscic T, Blagden N. Applying hot-stage microscopy to co-crystal screening: a study of nicotinamide with seven active pharmaceutical ingredients. *Crys Growth Des* 2008; 8(5):1697–1712.

20 USP UP. 34, NF 29. *The United States pharmacopeia and the National formulary*. The United States Pharmacopeial Convention, Rockwille, MD; 2011.

21 Shearer GL. Contaminant identification in pharmaceutical products. *Microscope-London Then Chicago* 2003; 51(1):3–10.

22 Stellmack M, Rhodes K. Metal contamination in biopharmaceutical drugs. *Contract Pharma* 2010; 10:84–87

23 Brake G. *Scanning Electron Microscopy Services for Pharmaceutical Manufacturers*. Available at http://www.semtechsolutions.com/sites/atl.semtechsolutions.com/files/SEM_Services_for_Bio-Pharma_Manufacturers.pdf. (Accessed 29 December 2015).

24 Seitavuopio P, Heinämäki J, Rantanen J, Yliruusi J. Monitoring tablet surface roughness during the film coating process. *AAPS PharmSciTech* 2006; 7(2):E1–E6.

25 Thibert R, Hancock BC. Direct visualization of superdisintegrant hydration using environmental scanning electron microscopy. *J Pharm Sci* 1996; 85(11):1255–1258.

26 Stokes DJ. Characterisation of soft condensed matter and delicate materials using environmental scanning electron microscopy (ESEM). *Adv Eng Mater* 2001; 3(3):126–130.

27 Andrusenko I, Gemmi M. 3D electron diffraction for structure determination of small-molecule nanocrystals: a possible breakthrough for the pharmaceutical industry. *Wiley Interdiscip Rev Nanomed Nanobiotechnol* 2022; 14(5): e1810.

28 Sheldrick GM. SHELXT–Integrated space-group and crystal-structure determination. *Acta Crystallogr A: Found Adv* 2015; 71(1):3–8.

29 Gruene T, Mugnaioli E. 3D electron diffraction for chemical analysis: instrumentation developments and innovative applications. *Chem Rev* 2021; 121(19):11823–11834.

30 Li S, Lightowler M, Ou X, Huang S, Jiang Y, Li X, Zou X, Xu H, Lu M. Direct structure determination of vemurafenib polymorphism from compact spherulites using 3D electron diffraction. *Commun Chem* 2023; 6(1): 18.

31 Gruene T, Wennmacher JT, Zaubitzer C, Holstein JJ, Heidler J, Fecteau-Lefebvre A, De Carlo S, Müller E, Goldie KN, Regeni I, Li T. Rapid structure determination of microcrystalline molecular compounds using electron diffraction. *Angew Chem Int Ed* 2018; 57(50):16313–16317.

32 Jones CG, Martynowycz MW, Hattne J, Fulton TJ, Stoltz BM, Rodriguez JA, Nelson HM, Gonen T. The CryoEM method MicroED as a powerful tool for small molecule structure determination. *ACS Cent Sci* 2018; 4(11):1587–1592.

33 Ricarte RG, Lodge TP, Hillmyer MA. Detection of pharmaceutical drug crystallites in solid dispersions by transmission electron microscopy. *Mol Pharmaceutics* 2015; 12(3):983–990.

34 S'ari M, Blade H, Cosgrove S, Drummond-Brydson R, Hondow N, Hughes LP, Brown A. Characterization of amorphous solid dispersions and identification of low levels of crystallinity by transmission electron microscopy. *Mol Pharm* 2021; 18(5):1905–1919.

35 Lauer ME, Grassmann O, Siam M, Tardio J, Jacob L, Page S, Kindt JH, Engel A, Alsenz J. Atomic force microscopy-based screening of drug-excipient miscibility and stability of solid dispersions. *Pharm Res* 2011; 28(3):572–584.

36 Begat P, Morton DA, Staniforth JN, Price R. The cohesive-adhesive balances in dry powder inhaler formulations I: Direct quantification by atomic force microscopy. *Pharm Res* 2004; 21(9):1591–1597.

37 Danesh A, Connell SD, Davies MC, Roberts CJ, Tendler SJ, Williams PM, Wilkins MJ. An in situ dissolution study of aspirin crystal planes (100) and (001) by atomic force microscopy. *Pharm Res* 2001; 18(3):299–303.

12

Vibrational Spectroscopy

12.1 Introduction

Vibrational spectroscopies widely used in pharmaceutical material analysis include infrared (IR), near-infrared (NIR), Raman, and terahertz spectroscopy. All four spectroscopic techniques share the same underlying principle involving the interaction of electromagnetic waves with a material, leading to various transitions between quantized vibrational energy states of molecules. Each technique is applicable in a different range of vibration energy and, therefore, probes a specific set of molecular vibrations (Figure 12.1). The vibrational energy range used is 14 000–4000 cm^{-1} for NIR spectroscopy (NIR), 4000–400 cm^{-1} for Fourier-transformed infrared (FT-IR) spectroscopy, 133–1 cm^{-1} for terahertz spectroscopy, and 3500–50 cm^{-1} for Raman spectroscopy. NIR spectroscopy detects overtones and combinational vibrations, whereas FT-IR spectroscopy detects fundamental and rotational vibrations. Terahertz spectroscopy is sensitive to lattice vibrations, whereas Raman spectroscopy can detect a broad range of molecular vibrations and rotations and is able to reach the wavenumbers up to those caused by lattice vibrations. NIR, FT-IR, and terahertz spectroscopy are directly related to the absorption of electromagnetic waves by a material, whereas Raman spectroscopy is a scattering technique that detects the scattering of electromagnetic waves after the sample interacts with monochromatic light. Inelastic scattering related to molecular vibrations is collected by a Raman spectrometer and processed as a Raman spectrum.

Vibrational spectroscopy has great utility in pharmaceutical material characterization, including:

- identification and qualification of raw materials,
- differentiating solid forms, such as amorphous forms, polymorphs, hydrates, solvates, salts,
- quantitative analysis of material compositions,
- studying phase transformation at controlled temperature and humidity,
- process analysis for crystallization, blending, granulation, coating, and lyophilization, and
- applications in contamination analysis, forensic studies, and counterfeit detection.

In differentiating solid forms, vibration spectroscopy provides additional confirmation after primary characterization by X-ray powder diffraction and thermal analysis. Spectroscopic techniques are useful quantitation tools in monitoring amorphous content and polymorph composition. Both Raman and NIR spectroscopy are widely used in process analytical technology due to their nondestructive nature and transparency with certain materials such as glass. Terahertz spectroscopy has emerged as a powerful technique in differentiating polymorphs due to its sensitivity to different lattice vibrations.

In this chapter, we will provide some general fundamental understanding of vibrational spectroscopy, and, then, we will discuss the instrumentation associated with FT-IR, NIR, Raman, and terahertz spectroscopy and discuss their applications in pharmaceutical material characterization. Finally, these techniques will be compared regarding sampling techniques and applications to pharmaceutical systems.

Solid-State Materials in Pharmaceutical Chemistry: Properties, Characterization, and Applications, Second Edition. Stephen R. Byrn, George Zografi and Xiaoming (Sean) Chen.

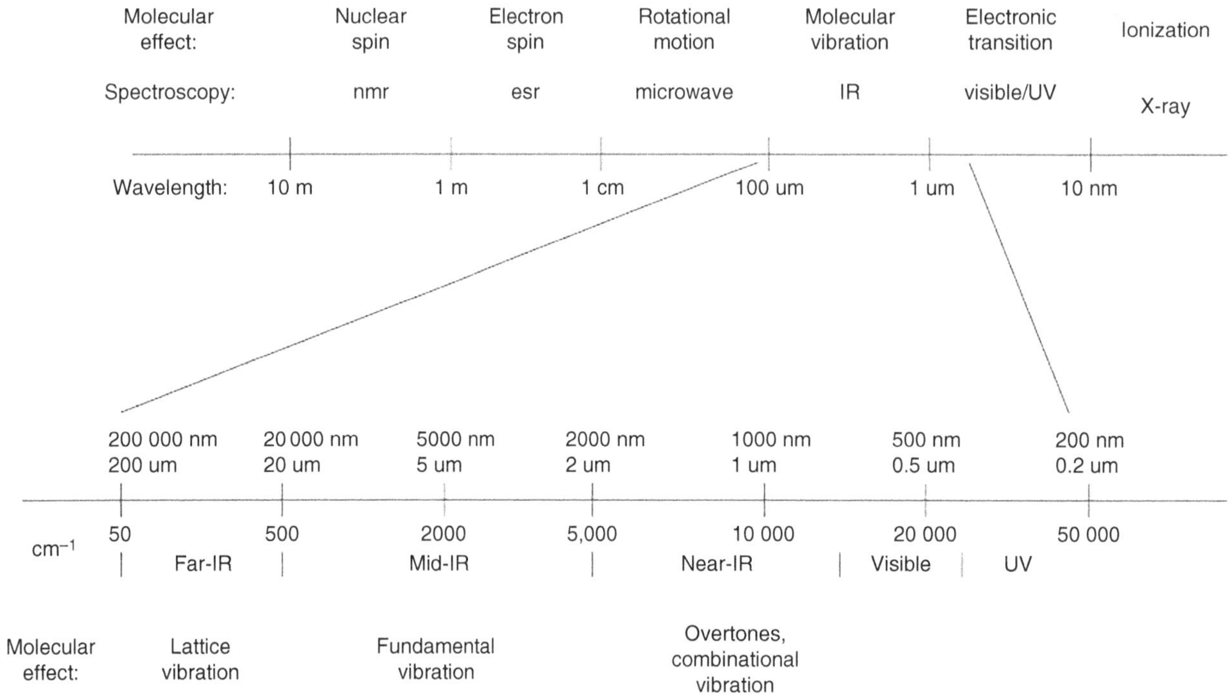

Figure 12.1 Spectroscopic techniques and related electromagnetic wavelengths.

12.2 The Nature of Molecular Vibrations

At temperatures above absolute zero, all the atoms in a molecule are in a constant vibrational state. Molecular vibrations of interest occur when atoms in a molecule are in periodic motion in relation to each other, which is different from the constant translational and rotational motion of the whole molecule. For a given molecule, there are many modes of molecular vibration, determined by composition, structure, and intermolecular interactions.

The types of vibrations include stretching, bending, rocking, wagging, and twisting.

The number of different types of vibrations a molecule can assume is determined by the number of atoms and the molecular structure. It is generally recognized that a molecule with n atoms has $3n$ degrees of freedom, including translational, rational, and vibrational motions. After subtracting three translational degrees of freedom and three rotational degrees of freedom, a nonlinear molecule is expected to have $3n - 6$ fundamental vibration modes, whereas a linear molecule has only two rotational degrees of freedom and, therefore, would have $3n - 5$ fundamental vibration modes. Fundamental vibration modes are also called normal modes of vibration. For example, carbon dioxide, as a linear molecule, has four modes of vibration, whereas propane, with a formula of C_3H_8, has 27 fundamental vibrations. For any molecule, fewer vibrations than the expected fundamental vibrations can be observed with FT-IR or Raman spectroscopy. There are several reasons to account for this:

- The vibration frequency is outside the detected range.
- The vibration is too weak to detect.
- The vibration is too close to other vibrations to resolve it.
- The vibration is not IR active or Raman active.

Among the fundamental vibrations, those that create a net change in the dipole moment of a functional group generally result in IR activity, and those that produce polarizability changes potentially give rise to Raman scattering. Naturally, some vibrations can be both IR- and Raman active.

Fundamental vibrations generally exist in the energy range of 4000–400 cm^{-1}, which can be detected by FT-IR and Raman spectroscopy. Besides fundamental vibrations, overtones and combinational vibrations exist in the middle-IR range and dominate in the NIR region.

Overtones occur when a vibration mode is excited by more than one energy quantum. For example, a first overtone band is related to excitement from $v = 0$ to $v = 2$, and a second overtone is related to that from $v = 0$ to $v = 3$. The first overtone band has approximately twice the frequency of the principal vibration frequency, and the second overtone band has about triple that of the principal one. The intensity of the overtone band is determined by the level of anharmonicity of the principal vibrational band. It is generally recognized that chemical bonds vibrating with large amplitudes, such as with hydrogen bonds, display a high degree of anharmonicity. Combination bands arise when two or more fundamental vibrations are excited simultaneously.

Vibrations detected in the far IR range, less than 400 cm^{-1}, are relatively weak and generally associated with intramolecular distortion, hydrogen bond vibrations, and long-range crystal lattice vibrations. Hydrogen bonds are very important in the formation of molecular crystals. Hydrogen bonds, being much weaker than covalent bonds, give rise to vibrational modes that have relatively low energy and occur in the low-wavenumber region. For example, various vibrational modes of the hydrogen bond dimer in carbamazepine are found between 20 and 110 cm^{-1} [1]. Crystal lattice vibrations, also called phonons, are the collective vibrations of a lattice of molecules that occur in a certain mode. The energetics of phonons are related to crystal packing and therefore are affected by a specific hydrogen bond network. Studying these modes of low-energy vibration is useful for probing long-range order when characterizing pharmaceutical materials regarding crystallinity and polymorphism. Such low-energy vibrations can be detected by low-wavenumber Raman spectroscopy and terahertz spectroscopy but not by FT-IR spectroscopy.

12.3 FT-IR Spectroscopy

FT-IR spectroscopy is an absorption spectroscopy and is generally used to probe the middle-IR range from 400 to 4000 cm^{-1}. A compound can absorb IR radiation in the middle range that corresponds to one of its fundamental vibration modes. Such vibrational modes are generally associated with a permanent dipole moment capable of absorbing IR radiation. Thus, a polar group such as an OH group usually gives strong FT-IR absorption. A nonpolar group, such as a carbon–carbon double bond, in a symmetrical molecule has weak or no FT-IR absorption.

The data obtained from IR spectroscopy is usually presented as an IR spectrum. The spectrum is plotted as an intensity of either transmission or absorption. The intensity of a particular band is proportional to the magnitude of the dipole change. Thus, in the same molecule, OH and NH groups produce more intense bands than does a CH group.

The wavenumber is the inverse of wavelength and is linearly related to frequency as in the following equation:

$$w = \frac{1}{\lambda} = \frac{v}{c} \tag{12.1}$$

where w is the wavenumber, λ is the wavelength, v is the frequency, and c is the speed of light. The unit of wavenumber is cm^{-1}.

The wavenumber is related to the strength of the molecular bond and the mass of the atoms, as described by the following equation:

$$w = \frac{1}{2\pi}\sqrt{\frac{k}{\mu}} \tag{12.2}$$

where k is the force constant and μ is the reduced mass. The reduced mass is calculated from the mass of the two atoms forming the chemical bond.

$$\mu = \frac{m_1 m_2}{m_1 + m_2} \tag{12.3}$$

where m_1 and m_2 are the masses of the two atoms.

It is obvious from the previous equations that a smaller mass and a stronger band energy result in a higher wavenumber.

Consequently, each functional group will give a unique frequency determined by the strength of the bond and the mass of the atoms. The general regions in which various kinds of vibrational bands are found in an IR spectrum are provided in many organic chemistry textbooks. IR spectra found in the 1450–600 cm^{-1} region are usually complicated, and it is difficult to assign all the absorption bands. This is called the fingerprint region. In contrast, absorption bands in the 4000–1450 cm^{-1} region usually have reasonable separation and are related to stretching vibrations of diatomic

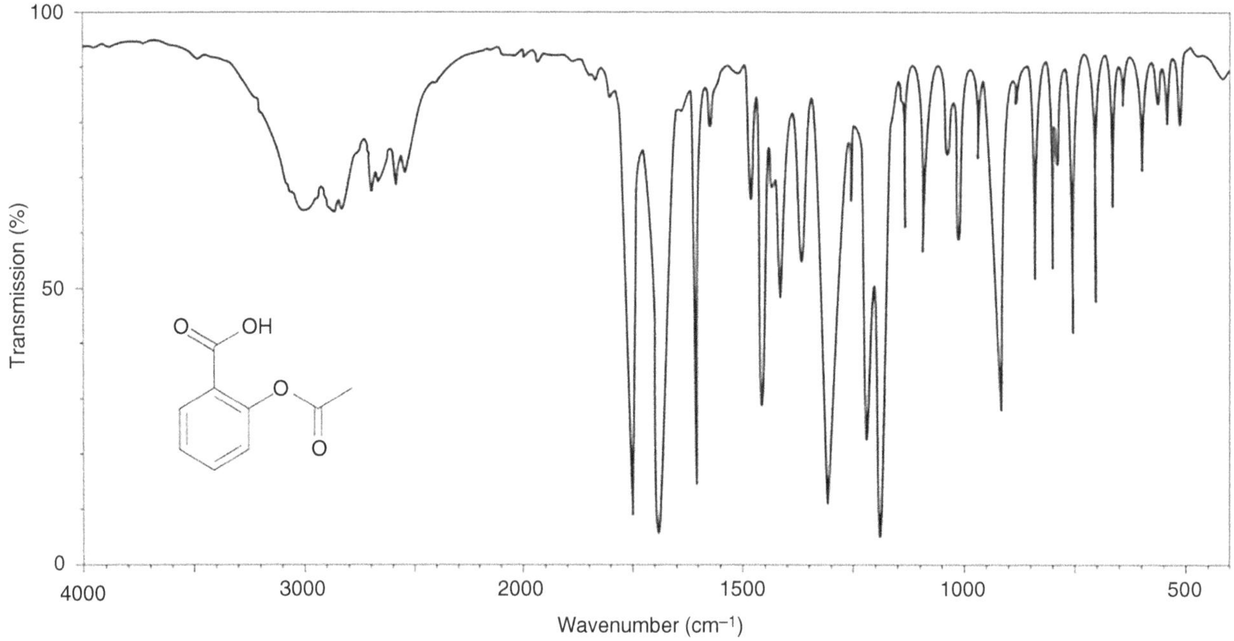

Figure 12.2 FT-IR spectrum of acetylsalicylic acid, the active ingredient of the drug product aspirin. *Source*: Adapted from [2].

units. This region, called the group frequency region, is useful for structure elucidation and the identification of organic compounds.

A pharmaceutical material usually contains various functional groups, so the spectrum of the material reflects a combination of various stretching and vibration bands of those functional groups. Even a simple molecule can give a complicated FT-IR spectrum due to the combination of stretching and vibration bands. Figure 12.2 shows the FT-IR spectrum of aspirin, collected using KBr pellets in preparation of the sample, with the KBr being nonresponsive in the IR region [2]. No two molecular structures give the same FT-IR spectrum, and this is why FT-IR spectroscopy is so useful for compound identification of a known or unknown material.

An FT-IR spectrum is usually plotted in a transmittance format, where the transmittance of a sample is the ratio of the intensity of the IR radiation at a specific wavelength that has passed through the sample to the intensity of the radiation when it entered the sample.

Sometimes, an FT-IR spectrum is plotted in absorbance mode, where absorbance is a measure of the quantity of light that a sample neither transmits nor reflects.

The conversion from transmission mode to absorbance mode can be described by the following equation:

$$A = -\log \frac{T}{100} \tag{12.4}$$

where A is the absorbance and T is the transmittance. The absorbance mode is ideal for quantitative analysis, preferably after some mathematical processing, including subtraction, Kubelka–Munk transformation, or deconvolution with the spectral data.

12.4 Material Characterization by FT-IR Spectroscopy

Since every compound has its unique FT-IR spectrum, the FT-IR method is widely used in the identification of known or unknown compounds. FT-IR also acts as a complementary technique to NMR, elemental analysis, and mass spectrometry for the structure elucidation of unknown organic materials isolated from natural sources. FT-IR is also regarded as a confirmatory test for analyzing a raw material, complementary to assay by HPLC. FT-IR is also widely used for compound identification and qualification. Since substantial spectra of organic compounds are available in the literature, FT-IR is more widely used than Raman and NIR spectroscopy for material identification.

As stated in the US Pharmacopeia (USP), FT-IR offers perhaps the most conclusive evidence of the identity of a substance that can be accomplished from any single test, after comparing the FT-IR spectrum of the substance with a corresponding USP reference standard [3].

In general, material identification has been made very convenient by the ongoing collection of the spectral information using reference standards and comparing unknowns to these referenced spectra. For common excipients or chemicals, spectral library searching has become relatively convenient due to the availability of sophisticated computer software.

As discussed in Chapter 2, polymorphs have identical chemical structure but are molecularly packed differently in the crystal lattice, due to possible different hydrogen bond networks and molecular conformation in the lattice. If this occurs, a chemical moiety of one polymorph will be in a different environment than the other forms, which should give rise to distinctly different vibrational frequencies. Shifts in vibrational frequencies among polymorphs are usually on the order of 1–20 cm^{-1} in an FT-IR spectrum. Figure 12.3 shows the FT-IR spectrum of two polymorphs of indomethacin, the α-form and γ-form, in the range of 1550–1850 cm^{-1}. The FT-IR bands observed in this range can be attributed to carbonyl groups. Consequently, FT-IR bands of the carbonyl moieties in the α- and γ-forms are located at different positions; the spectrum of the α-form has intensive bands at 1733, 1688, 1681, and 1649 cm^{-1}, whereas the spectrum of the γ-form possesses two bands at 1717 and 1698 cm^{-1}. Indomethacin has two carbonyl moieties, one associated with a carboxylic acid and the other in the benzoyl amide group. The two carbonyl moieties exist with different hydrogen bonding in the two polymorphs; in the γ-form, the benzoyl carbonyl is not involved in hydrogen bonding, but the acid carbonyl participates in hydrogen-bonded carboxylic acid dimers. The band at 1717 cm^{-1} is assigned to the acid carbonyl, and the one at 1698 cm^{-1} is assigned to the benzoyl carbonyl [4]. In the α-form, there are two types of hydrogen bonds in the three molecules found in the asymmetric unit of one unit cell. Two of the molecules form mutually hydrogen-bonded carboxylic acid dimers, while the third molecule forms a hydrogen bond between the carboxyl acid and the benzoyl carbonyl in the dimer. The hydrogen bond of benzoyl carbonyl leads to its shift to 1688 cm^{-1} in the α-form. The acid carbonyl of the third molecule does not involve

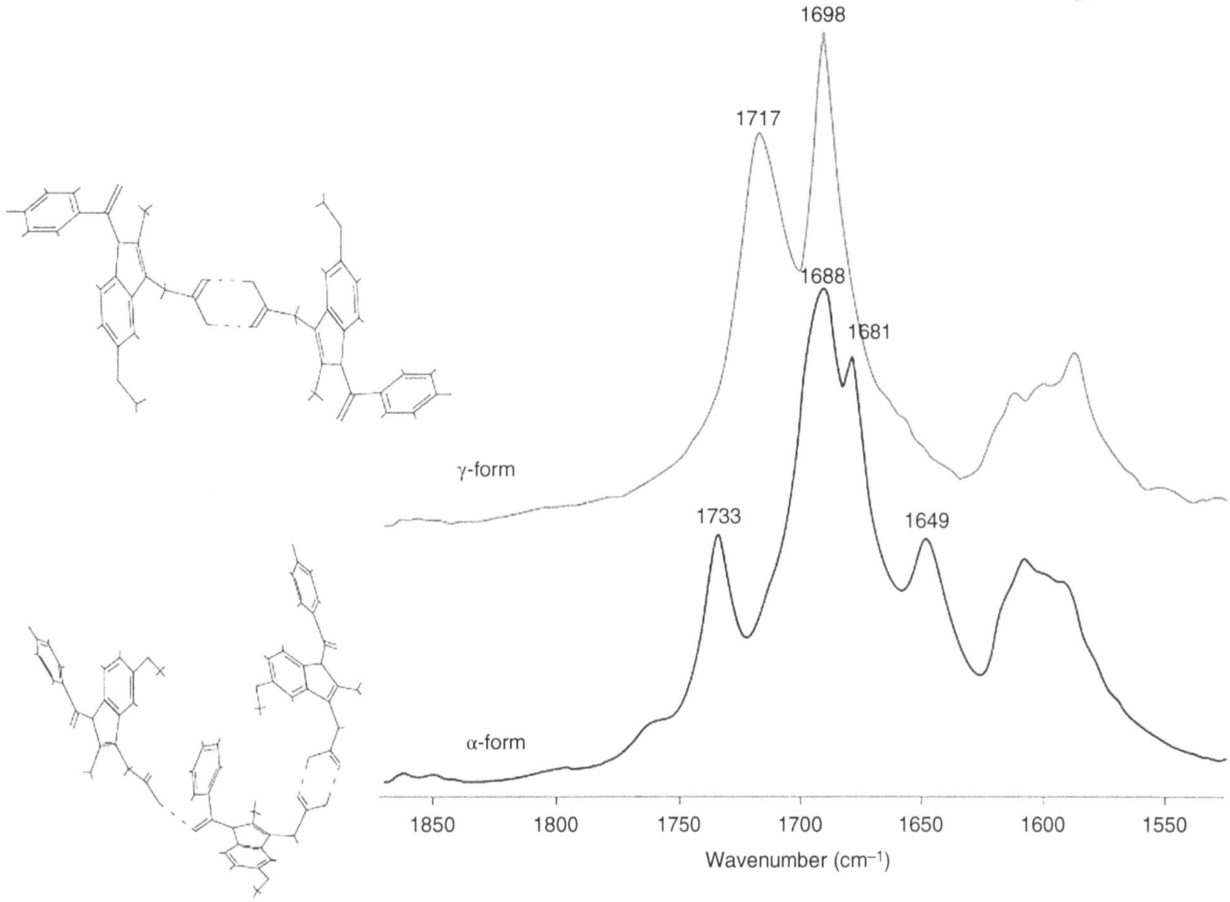

Figure 12.3 FT-IR spectra of indomethacin polymorphs: α-form and γ-form. Asymmetrical units in the crystal structures are provided.

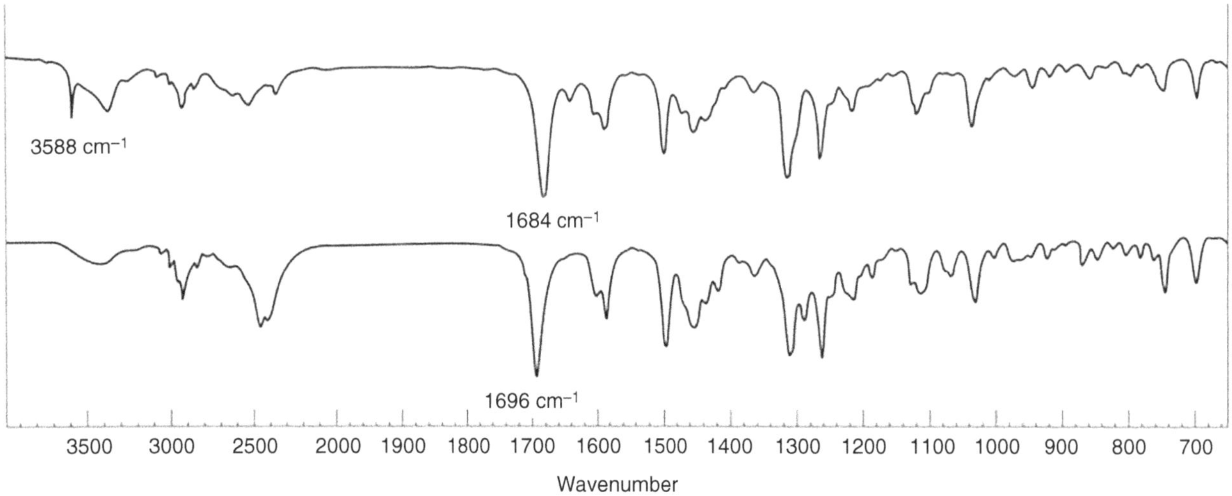

Figure 12.4 FT-IR spectra of donepezil hydrochloride hydrate and anhydrate form. *Source*: Adapted from [6].

hydrogen bonding, which gives rise to its shift to 1733 cm^{-1} [4]. In summary, different hydrogen bond networks of the two indomethacin polymorphs bring about distinctive FT-IR spectra of the two polymorphs. The distinctive FT-IR spectra make it possible to distinguish the two polymorphs and to determine their polymorphic purities. Spychala et al. described a procedure to determine the content of individual polymorphs of indomethacin in their mixtures by FT-IR spectroscopy based on the characteristic bands of the two forms in the range of 1600–1800 cm^{-1} [5].

A hydrate or a solvate can be distinguished from the anhydrous form(s) of a compound in crystal structure, molecular packing, and intermolecular interaction, as described in Chapter 3. In addition, water or solvent molecules are included in crystal lattices. Like the situation of polymorphs, certain chemical moieties within these forms are situated in different chemical environments due to different crystal structures and different intermolecular interactions with water and solvent molecules, which give rise to distinct vibration frequencies of those chemical moieties. Also, there are new vibrational bands possible due to the presence of water or solvent. Consequently, FT-IR spectroscopy can be used to characterize hydrate and solvate forms. The –OH stretching for the water molecules inside a hydrate has a strong characteristic peak at around 3200–3600 cm^{-1} wavenumber. Figure 12.4 shows the FT-IR spectra of donepezil hydrochloride hydrate and anhydrate forms [6]. The hydrate form has a sharp absorption band at around 3588 cm^{-1} that is related to the hydroxyl group of water molecules in the hydrate. In addition, the stretching vibration of C=O group has a lower wavenumber (1684 cm^{-1}) than that of the anhydrous form (1699 cm^{-1}), indicating its involvement in hydrogen bonding.

The capability of FT-IR to distinguish various salt forms of a compound is attributed to the individual crystal structure, molecular interactions, and counterion of each salt form. Also, a cocrystal of an API has a totally different crystal structure and hydrogen bond pattern from the other solid forms of the API, which leads to a new FT-IR spectrum.

Amorphous forms have various chemical moieties in different spatial environments and with different types of hydrogen bonds from those found with the corresponding crystalline form, thus giving a distinctly different FT-IR spectrum than its crystalline counterparts.

In summary, FT-IR spectroscopy is useful in compound identification and is applicable in distinguishing various solid forms.

Currently, IR spectra are collected by IR spectrometers with Fourier transform capability. FT-IR spectrometers are designed to operate in different modes depending on the application, including the transmittance, diffuse reflectance, and attenuated total reflectance (ATR) modes.

12.5 FT-IR Instrumentation

An FT-IR spectrometer can collect FT-IR spectra for solids, liquids, and gases. A modern IR spectrometer is an interferometer, which collects spectral data simultaneously over a wide spectral range. It is distinct from a monochromatically based spectrometer, such as a UV spectrometer, which collects spectral data at an individual wavelength at one time.

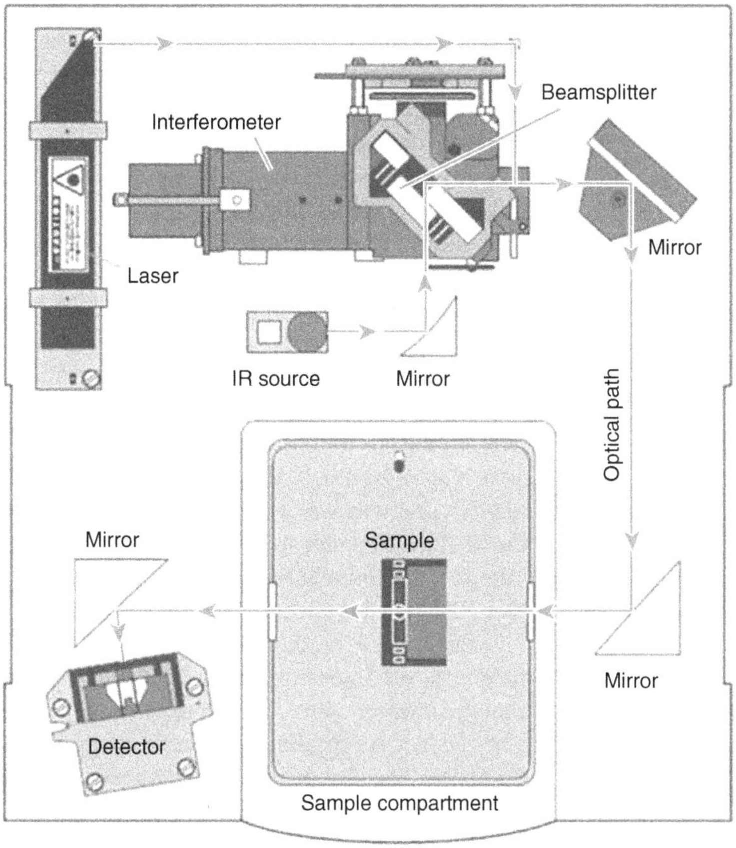

Figure 12.5 Diagram of FT-IR instrumentation. *Source*: Adapted from [7].

Figure 12.5 shows a schematic layout of a typical FT-IR spectrometer, which comprises five basic components: an IR energy source, an interferometer, a sample compartment, a detector, and a computer [7].

The source provides a spectrum of IR radiation emitted from a glowing black-body source that passes through an aperture. The aperture controls the amount of energy used for a particular analysis.

The interferometer is the core component of the spectrometer and consists of a fixed mirror, a moving mirror, and a beam splitter. The interferometer generates an interferogram, which appears as a complex oscillatory signal with decreasing amplitude. The IR beam enters the interferometer and exits as an interferogram.

The sample can be a gas, a liquid, a semisolid, a suspension, or a solid. The interferogram enters the sample compartment and interacts with the sample, and specific frequencies of energy unique for the sample are absorbed, after which the interferogram exits the sample compartment and reaches the detector. The detector, specially designed to acquire the interferogram signal, produces a signal, which is digitized and sent to the computer for Fourier transformation. By means of the Fourier transformation, the computer converts the interferogram to a spectrum.

The diagram of an FT-IR setup, given in Figure 12.5, illustrates how the transmission of the radiation through the sample leads to the acquisition of the FT-IR spectrum. IR radiation passes through a sample, and a fraction of the incident radiation is absorbed at a specific energy. Since organic compounds usually exhibit strong absorption in the middle-IR range, sample dilution and preparation are often required. Gases and liquids can be easily diluted with a carrier gas and a solvent, respectively. A solid material must be pulverized into fine particles and dispersed in an inert matrix.

Sample preparation of solid materials for transmittance analysis commonly uses KBr or KCl pellets as the inert matrix. The mixture of a pharmaceutical material (1–2%) and the inert matrix is generally ground in a mortar/pestle to ensure uniformity and then compressed into a thin pellet for transmission analysis. KBr and KCl exhibit no IR absorption in the middle-IR range and therefore provide a very popular means for carrying out the qualitative analysis and identification of organic materials. The sample size requirement for such studies is generally in the milligram range.

The spectrum obtained from the transmittance method follows traditional Beer's law, which means that the intensity of each band is directly proportional to the concentration of the absorbing species, allowing FT-IR to be used for quantitative analysis.

It is important to recognize that, in some cases, grinding and compression may lead to phase transformations. For example, it has been reported that the pressing of the β-form of indomethacin into a pellet resulted in its conversion to the α- and γ-forms [5]. In addition, the alkali salts tend to be hygroscopic, so that water taken up may reduce the quality of the FT-IR spectrum. In such cases, nondestructive sampling methods, such as diffuse reflectance and ATR IR, can be very useful.

12.6 Diffuse Reflectance FT-IR

Diffuse reflectance FT-IR (DRIFT) functions by focusing IR light onto a sample and detecting scattered IR signals from the sample and transforming the signals to a spectrum. When the IR beam impinges on a sample surface, some of the radiation is transmitted and some is reflected. The relative amount of these two components is related to the reflective index of the

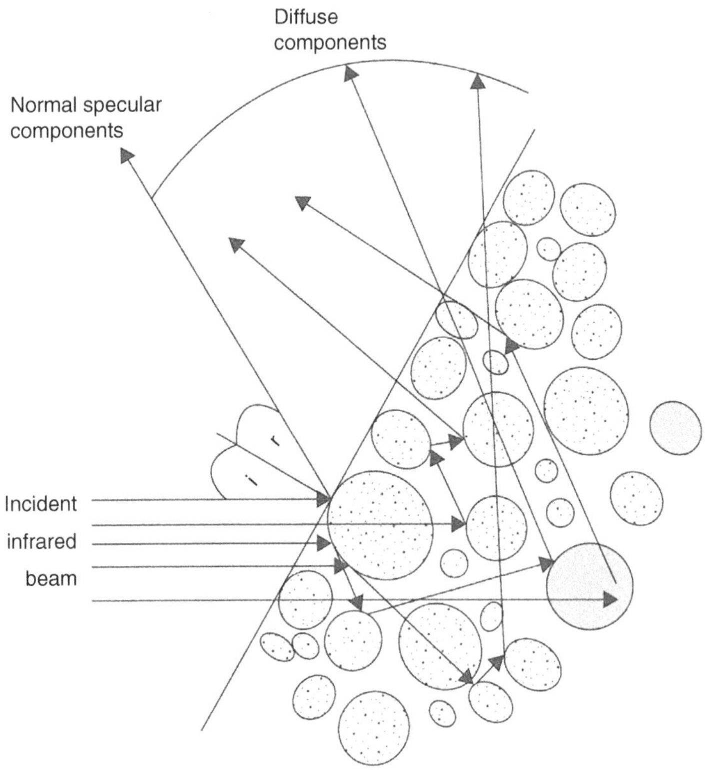

Figure 12.6 Schematic diagram of diffuse reflectance from a powder sample.

two media and the incidence angle. If the surface is smooth, the surface acts as a mirror and reflects the beam in a single direction. This type of reflection is called specular reflection. For a rough surface, the reflection occurs in all directions, producing diffuse reflectance. Diffuse reflectance is collected through an ellipsoid mirror.

In diffuse reflectance, the IR beam has a certain depth of penetration into the material before the diffuse reflection occurs. Some of the beam is reflected multiple times, and the absorption of the radiation by the material is amplified due to the multiple times of penetration (Figure 12.6). The diffusely reflected signals from the sample are collected by a large ellipsoidal mirror and redirected to a detector to produce a reflectance spectrum. The absorption causes the reflectance spectrum to have similar features to a transmission spectrum and is called a diffuse reflectance spectrum.

Owing to the weak reflectance signal that is produced, the intensity of the IR bands does not change linearly with the concentration of the material in a matrix and, therefore, does not follow Beer's law. The introduction of the Kubelka–Munk equation provides the relationship of the intensity of the reflected radiation to the concentration [8]:

$$F(R) = \frac{(1 - R)^2}{2R} = \frac{k}{s} = \frac{Ac}{s} \tag{12.5}$$

where R is the reflectance, k is the absorption coefficient, s is the scattering coefficient, c is the concentration of the absorbing species, and A is the absorbance. Generally, the Kubelka–Munk equation is used to convert the measured reflectance spectrum to a corrected spectrum. The corrected spectrum resembles a transmission spectrum with a linear relationship between band intensity and the sample concentration. For convenience, the diffuse reflection spectrum is usually presented in Kubelka–Munk format rather than in absorbance, where Kubelka–Munk intensity (K–M) is described by the following equation:

$$K\text{–}M = \frac{(1 - R)^2}{2R} \tag{12.6}$$

For diffuse reflectance, sample presentation is relatively simple. A powder sample is placed into a sample cup and subjected to analysis. Some samples can be directly analyzed without dilution. However, for a sample with high absorption, mixing with a nonabsorbent material offers proper dilution. Moreover, dilution allows for a deeper penetration of the IR beam into the sample and increases the portion of the scattered components in the spectrum. DRIFT is sensitive to particle size, so sample presentation must be consistent. A sample with large particles is usually pulverized to reduce the particle size below 10 μm to mitigate any particle size effects. Dilution and particle size reduction both minimize the specular reflection component, which causes changes in band shapes and may produce derivative-shaped bands. DRIFT finds applications in various samples, including nontransparent materials, powders, coatings, and rough surfaces. Kojima et al. used DRIFT to evaluate hydrate formation in crystalline ampicillin and nitrofurantoin powders by monitoring the specific absorption at 3500–3700 cm^{-1} corresponding to crystalline water [9]. Müller et al. developed a DRIFT method for the simultaneous determination of clavulanic acid and amoxicillin in commercial tablets with multivariate calibration [10]. DRIFT was also applied in the simultaneous quantification of carbamazepine crystal forms in ternary mixtures of polymorphs I, III, and IV [11].

DRIFT has the capability of studying samples at elevated temperature and humidity, which allows for the investigation of phase transitions arising from changes related to temperature and moisture content.

12.7 ATR FT-IR

ATR is another desirable FT-IR sampling technique for powder materials that utilizes total internal reflection phenomena. Total internal reflection occurs when light is transmitted from one medium with a higher reflective index to another with a lower reflective index and when the incidence angle is larger than the critical angle. The ratio of the two reflective indices determines the critical angle. ATR FT-IR uses a crystal with a high reflective index and good IR transmission to create total reflectance (Figure 12.7). Diamond, zinc selenide (ZnSe), and germanium (Ge) are used as common ATR crystals, with the sample remaining in close contact with the crystal. Though the radiation is totally reflected from the interface, the radiation tends to have a certain depth of penetration into the sample due to the presence of evanescent waves. An evanescent wave is a transmitted wave associated with total reflectance that exists in the vicinity of the interface and displays exponential decay. The thickness of an evanescent field is determined by the wavelength of the IR radiation, the refractive index of the ATR crystal, the refractive index of the material, and the angle of the incident IR beam. Usually, the field has a depth of several micrometers (0.5–3 μm).

The interaction of the evanescent wave with the material causes absorption in a certain spectral range, and the evanescent wave becomes attenuated. The detection of attenuated IR signals results in an ATR FT-IR spectrum. The ATR FT-IR spectrum is similar to the spectrum obtained in the transmission mode, except that the band intensities at longer wavelengths (shorter wavenumber) are increased. This occurs because longer wavelengths produce deeper penetration of the evanescent wave into the sample. The deeper penetration gives rise to greater absorption. In addition, there are situations where the reflective index of certain materials may vary with wavelength, causing dispersion effects to arise. The penetration depth will change because of variability in the reflective index.

Usually, multiple total reflections occur inside the ATR crystal, which amplify the effect of absorption and enhance the IR signal. This phenomenon is called multiple bounce ATR. The number of reflections generally falls between 5 and 10.

There are several designs of ATR cells, including traditional (vertical) ATR, horizontal ATR, and cylindrical ATR. Among them, horizontal ATR is the most feasible for pharmaceutical materials such as powders, liquids, and pastes. In horizontal ATR, the crystal is aligned as a parallel side plate, with its top surface exposed for sampling.

Since the evanescent wave only penetrates a few micrometers into the sample, close contact of the sample with the ATR crystal is necessary.

ATR analysis requires no or minimal sample preparation. A solid sample is loaded to cover the ATR crystal surface and then is pressed with a clamp to ensure intimate contact. Some materials having large crystals need to be pulverized to ensure intimate contact. Helmy et al. reported the use of ATR FT-IR for the identification and quantitation of two polymorphs of aprepitant. The polymorphic purity results obtained by ATR FT-IR were in good agreement with the value measured by X-ray powder diffraction [12].

Films and liquid samples can be directly loaded for analysis. For example, it has been reported that ATR FT-IR was utilized to measure drug and excipient diffusion in acrylate pressure-sensitive adhesives [13]. Cornel et al. applied ATR FT-IR to monitor in situ solute concentration changes during crystallization upon cooling, and the solvent-mediated polymorph transformation of L-glutamic acid [14].

Components of a liquid should be preliminarily assessed to avoid damaging the ATR crystal surface. For example, ZnSe has low tolerance to strong acids and bases. Diamond has the versatility to handle a wide range of pH values and shows excellent resistance to scratching.

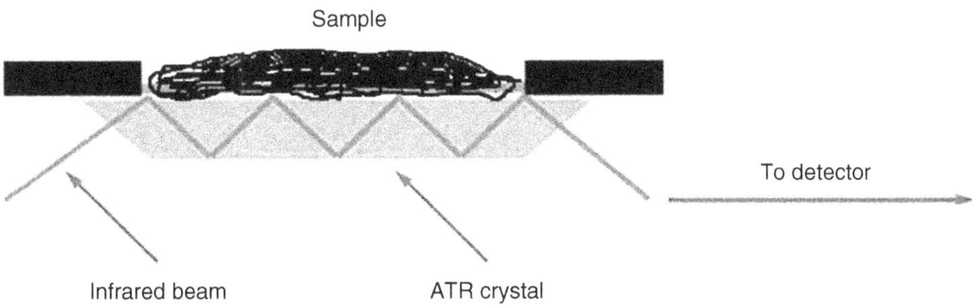

Figure 12.7 Schematic diagram of a multiple reflection ATR system.

12.8 FT-IR Microscopy

FT-IR microscopy is a combination of FT-IR spectroscopy and microscopy. It is also called FT-IR microspectroscopy. FT-IR microscopy can provide both visual and chemical information at the same time, a unique capability that makes it a very useful tool in forensic analysis, identification of contamination, and chemical imaging of pharmaceutical products.

A FT-IR microscope has an optical system for visual observation and an IR spectrometer. The two systems can be easily switched back and forth. The optical system has a visible light source and eyepieces, whereas the IR system has an IR light source and an IR detector. The two systems share the same set of condensers and objective lens. The optical system guides the locating of the spot of interest and collects visual images of the sample area. The IR spectrometer captures the FT-IR spectrum of the sample at the specified spot to create a chemical image. With the combination of an optical microscope and an FT-IR spectrometer, the FT-IR microscope can act as a valuable tool for a rapid, nondestructive analysis of a variety of samples.

The size of the spot on the sample is determined by the diffraction limit of the IR objective lens. The size of the spot provides what is also called spatial resolution. Owing to the long wavelength of IR radiation, the spot can be as large as 10–20 μm.

A FT-IR spectrum can be collected as transmission, reflectance, or ATR, depending on the sample.

Figure 12.8 shows the diagram for an FT-IR microscope in transmittance, reflectance, and ATR mode. In the transmission mode, an IR source emits broadband IR radiation, a portion of which reaches the interferometer and is spectrally encoded. The interferogram passes through the upper aperture and is focused on the sample by the objective lens using Schwarzschild optics. The IR radiation transmitted through the sample is collected by a condensing optic, and passing through the lower aperture, the IR signal reaches the detector. The upper aperture and lower aperture help to improve spatial resolution.

FT-IR microscopy also works with reflected light if the sample is opaque and difficult to transmit. The IR radiation is focused by the objective lens and directed to the sample in transmittance mode. The reflective IR signal is collected by the same objective lens and directed to the detector by a "sluice" mirror, which is removed during transmittance mode.

ATR can also be implemented with an FT-IR microscope. Again, the ATR objective lens uses Schwarzschild optics and polished diamond, germanium, or zinc selenide as its ATR crystal. The ATR crystal is raised during visual viewing of the sample and then lowered to obtain direct contact with the sample for the FT-IR measurement. Very close contact is necessary to minimize the depth of penetration of evanescent waves.

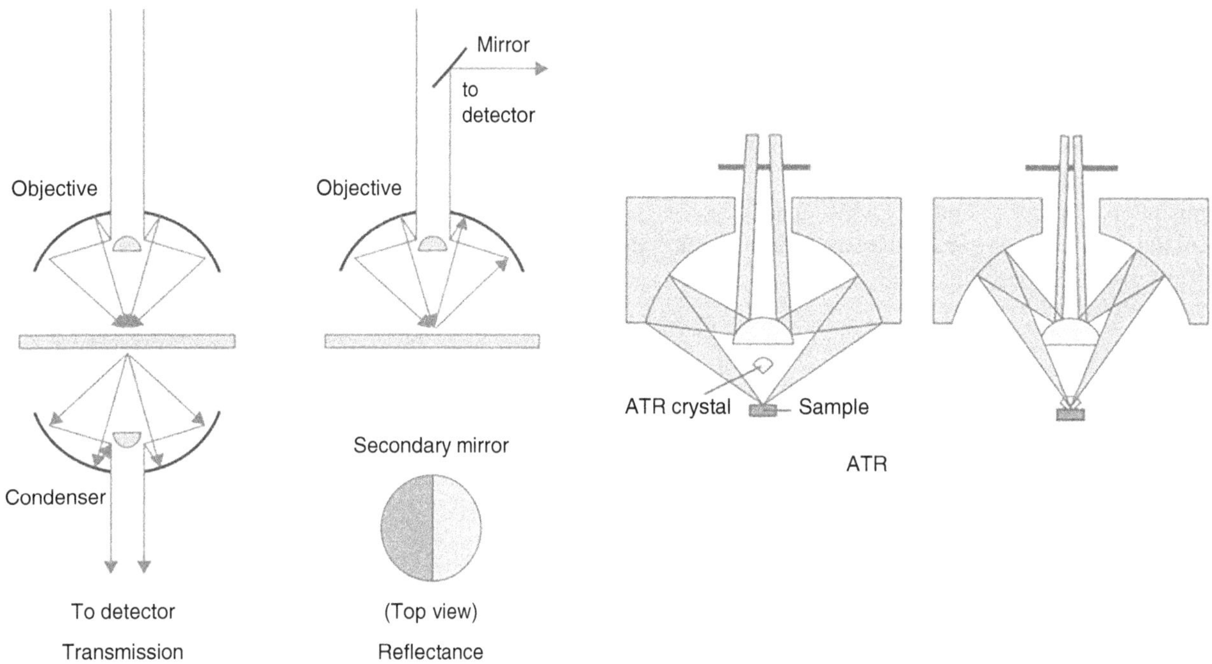

Figure 12.8 Schematic diagram of an FT-IR microscope with different spectroscopic settings.

The FT-IR spectrum collected at every point can be utilized to map a two-dimensional (2D) chemical image. In a mapping experiment, a grid is placed across a sample and is scanned in the *x* and *y* directions sequentially, and these sequential spectra are compiled into a 2D map depicting the distribution of one or more chemicals within a matrix.

The disadvantage of sequential mapping is that it is very time consuming. Each data point requires several minutes to acquire, and a relatively small map can demand hours.

The introduction of a 2D detector significantly improves the throughput of FT-IR imaging. 2D detectors employ focal plane array detectors, usually made from a mercury cadmium telluride (MCT) detector element. The detector collects spectra from all detector pixels simultaneously and exhibits much more efficient data acquisition than ordinary FT-IR mapping. Each pixel collects a full IR spectrum, and an array detector can range from 16 × 16 MCT pixels up to 256 × 256 pixels. Consequently, from 256 to 65 536 individual IR spectra can be collected simultaneously. The number of MCT pixels determines the area of the sample imaged, ranging from 100 μm^2 to 1.2 mm^2. The time for the measurement of each area is just a few seconds.

The use of synchrotron radiation as an IR source is another way to increase the speed of data collection. Synchrotron radiation is hundreds or thousands of times more intense than a standard IR source. In addition, synchrotron also enhances resolution to below 10 μm because of the increase in the signal-to-noise ratio.

FT-IR microscopy is widely used for studying bulk, surface, and composite structures since it is designed for identifying small samples chemically and generating spectra that are specifically related to a visual image of the sample or a part of the sample. Owing to the low energy of IR radiation, no sample damage or photo-induced reaction generally occurs.

Chan et al. used an ATR FT-IR microscope, with a focal plane array detector, to study pharmaceutical formulations [15]. Tablets with caffeine as the active ingredient were chemically imaged to show the spatial distribution of caffeine and excipients, to provide a microscopic level assessment of drug content uniformity in the dosage form. Gotter et al. applied FT-IR microscopy for studying lateral drug diffusion from a semisolid formulation of dithranol [16]. It has also been reported that FT-IR microscopy provided a nondestructive analysis of counterfeit antimalarial tablets [17]. FT-IR imaging delivers chemical specificity and rapid acquisition of data without the need for special sample preparation.

12.9 NIR Spectroscopy

NIR spectroscopy uses electromagnetic radiation in the range of 1000 to 2500 nm (10 000–4000 cm^{-1}), thus detecting radiation that lies between visible light and middle-range IR. NIR spectroscopy detects the absorption of NIR radiation due to molecular vibrations. In contrast to FT-IR spectroscopy, NIR spectroscopy measures absorption bands due to overtones or a combination of fundamental vibrations, predominantly from stretching vibrations involving hydrogen (e.g. C—H, O—H, and N—H chemical bonds).

The intensity of the overtone band is determined by the level of "anharmonicity" of the principal vibrational band. It is generally recognized that chemical bonds vibrating with large amplitudes, such as hydrogen bonds, display a high degree of anharmonicity. Consequently, functional groups including OH and NH with hydrogen-bonding capability dominate a typical NIR spectrum. Combination bands arise when two or more fundamental vibrations are excited simultaneously. They usually have lower energies than overtones and are located at lower frequencies.

Since overtones and combination bands are naturally weaker than principal bands, NIR spectra generally appear very broad, complex, and convoluted. It is extremely difficult, therefore, to interpret peak information for chemical assignments in the traditional manner. Figure 12.9 shows NIR spectra of an API, microcrystalline cellulose, and lactose monohydrate [18]. The API has a sharp absorption band at 5984 cm^{-1}. However, microcrystalline cellulose and lactose monohydrate show broad absorption bands and have no obvious distinct band available for identification. In the second derivative spectrum, however, lactose monohydrate and microcrystalline cellulose give a distinct absorption band at 5168 and 4280 cm^{-1}, respectively.

It is generally recognized that additional processing of the data, such as chemometric analysis and the derivative technique, is needed for further interpretation of a NIR spectrum. For convenience, the NIR spectrum is plotted as intensity versus wavelength instead of wavenumber most of the time.

NIR spectroscopy also provides a rapid and direct measurement of materials with little or no sample preparation. Since the NIR region also exhibits relatively weaker absorption than that in the middle-IR region, sample dilution is not necessary most of the time. In addition, this relatively weak absorption allows for deeper penetration of the IR radiation into the sample, resulting in a larger amount of sample that can interact with the radiation.

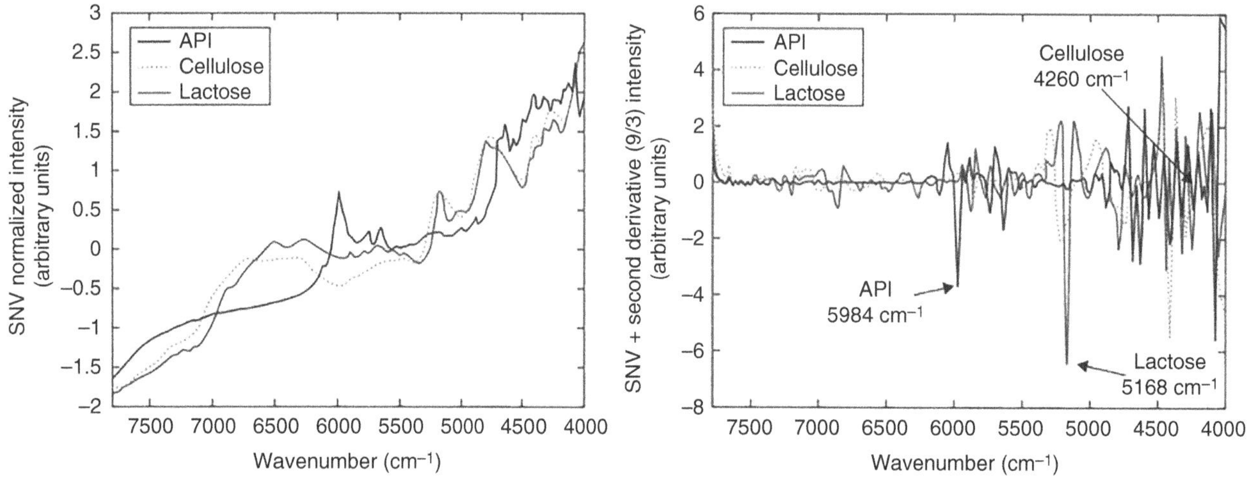

Figure 12.9 NIR spectrum and second derivative spectrum of an API, lactose monohydrate, and microcrystalline cellulose. *Source*: [18]/with permission of Elsevier.

Measurement techniques for NIR spectroscopy include transmittance, diffuse reflectance, and transflection. Diffuse reflectance is preferred for powders, whereas transmission and transflection are applicable for use with liquids and polymer films. Transflection is more advantageous for measuring viscous liquids.

The use of diffuse reflectance in NIR spectroscopy is instrumentally accomplished using an integrating sphere. The integrating sphere collects IR radiation from all angles, for both incoming IR radiation and reflective signals. A detector is located in the sphere, and the sphere is designed as upward looking, with a window located at the top of the sphere, with samples being placed right above the window for analysis.

NIR spectroscopy can analyze a variety of samples, including powder, tablets, films, emulsions, suspensions, and syrups directly and without any special sample preparation. Since glass and quartz barely exhibit any absorption in the NIR region, samples can be directly measured in a glass container or a quartz cuvette. In addition, owing to the minimal absorption of radiation by thin films, such as plastic bag containers, samples can be directly measured in a plastic bag.

NIR spectroscopy can also measure samples using fiber-optic devices, generally called NIR probes. Since NIR radiation has excellent transmission in a fiber-optic system, a fiber-optic cable is positioned near the sample, and the NIR spectrum is collected directly from the sample. The NIR probe can be in the transmission mode or in the ATR mode, as described in Figure 12.10. For transmission mode, a half window is open near the tip of the probe, which allows a liquid sample to be in the optical path when the probe is immersed. The ATR probe has a crystal in the tip and provides ATR detection. An ATR probe is also suitable for use in the analysis of powders.

Clearly, the use of fiber optics advantageously eliminates the need to bring samples into the spectrometer, which offers a significant advantage when measuring hazardous materials and studying substances in process. As such, NIR spectroscopy is very useful for monitoring drug composition during processing and, therefore, is widely used in PAT, which is highly promoted by regulatory agencies and the pharmaceutical industry. Presently, many NIR spectrometers can handle multiplexers, where several NIR probes are installed at different locations of a reactor, with data being collected in real time to monitor a chemical reaction or crystallization.

There are two types of instrumentation that are used with NIR spectroscopy, dispersive and Fourier-transformed. Though the dispersive instrumentation is no longer used with FT-IR spectroscopy, dispersive NIR spectroscopy is still popular for many applications. An example of a dispersive NIR spectrometer is the FOSS NIR system.

A major advantage of NIR spectroscopy is that it allows for rapid analysis, in that modern equipment can complete multiple scans and average them in less than a minute and sometimes in several seconds.

NIR spectroscopy can also provide both qualitative and quantitative analysis. Qualitative analysis confirms the identification of a raw material or a drug product, whereas quantitative assessment allows one to measure the level of specific components within a complex mixture, through appropriate sampling, and in real time.

NIR spectroscopy has been applied to many pharmaceutical systems. The applications include the identification, qualification, and quantification of pharmaceutical starting materials, intermediates, drug substances, and finished products. da Silva et al. demonstrated that NIR spectroscopy was a convenient and fast method for measuring the relative amounts of

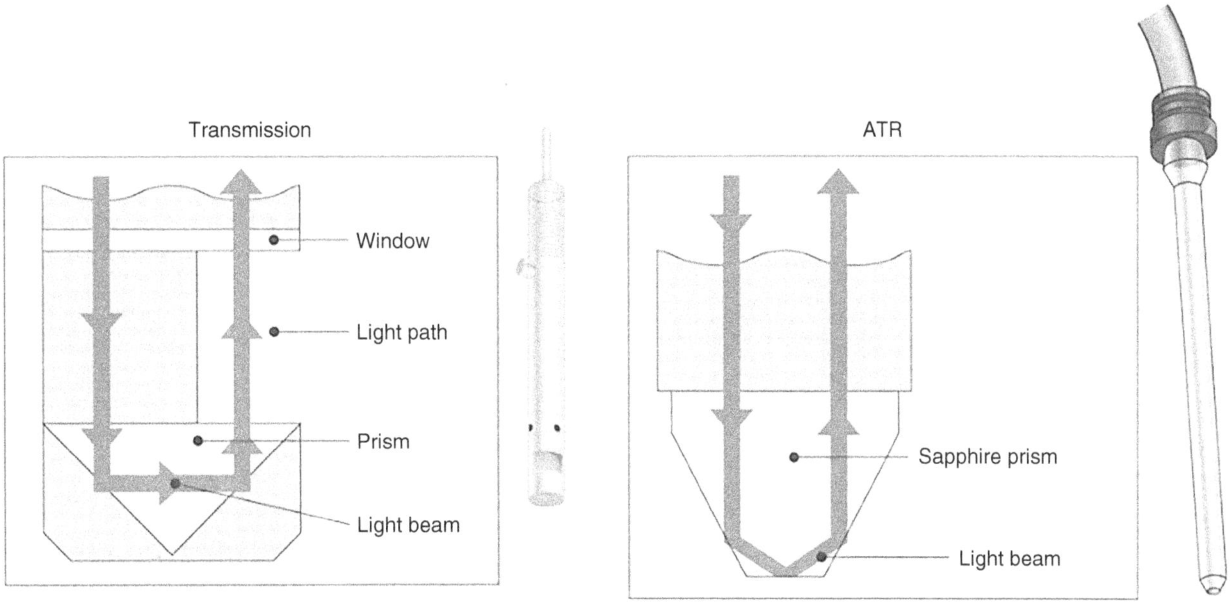

Figure 12.10 Pictures and schematic diagrams of NIR probes.

the three polymorphs of mebendazole in a sample of raw material [19]. NIR spectroscopy was also found to be applicable for detecting very low levels of irbesartan Form B in a tablet formulation when irbesartan Form A is the active ingredient [20]. Form A is more soluble and tends to transform to Form B during the wet granulation process, so it was important to be able to quantitatively detect Form B. Luner and Seyer demonstrated the capability of NIR spectroscopy to determine the degree of crystallinity in sucrose during processing, using calibration standards derived from binary physical mixtures of amorphous and crystalline phases [21]. Brashear et al. reported a NIR spectroscopy method to quantify drugs or excipients within polymeric implants of lomefloxacin HCl [22].

NIR spectroscopy has also been applied to the in-process monitoring of crystallization, polymorph conversion, blending, coating, and lyophilization. Lee et al., for example, reported a study that monitored the solvent-mediated polymorphic transformation of sulfamerazine using inline NIR measurements [23]. A fiber-optic NIR probe was incorporated into the reaction vessel to measure the conversion of Form I to Form II. The NIR spectra in the range of 7500–4900 cm^{-1} were used and computationally examined by multivariate analysis. A study using NIR spectroscopy for indomethacin in an antisolvent crystallization process has demonstrated that inline NIR spectroscopy, combined with principal component analysis, was able to monitor and interpret the crystallization process with respect to the polymorphism and particle size in real time [24]. Shi et al. utilized NIR spectroscopy to characterize powder blending inline for a powder containing acetaminophen, lactose, and microcrystalline cellulose, with one probe located on the top of the blender and the other on the rotation axis. The endpoint for mixing was determined using multivariate models [25]. Hudovornik developed a predictive NIR spectroscopy model for determining coating thicknesses with continuous inline measurements of a pellet coating process [26]. NIR spectroscopy has also been applied for the inline determination of moisture content during the lyophilization process through the monitoring of multiple points. Diffuse reflectance NIR probe heads were located on the freeze-dryer shelf to measure the moisture content in a noninvasive manner [27].

12.10 Raman Spectroscopy

Atoms in a molecule bond essentially consist of a nuclear structure and a cloud of electrons. Those electrons are the basis of chemical bonding. Application of an electromagnetic field causes the electrons to flow toward the positively charged side and away from the negatively charged side of the applied field. The way the electron responds to a given field is a measure of polarizability. Molecular vibrations associated with a chemical bond can cause a change in polarizability, and Raman spectroscopy can measure such a change.

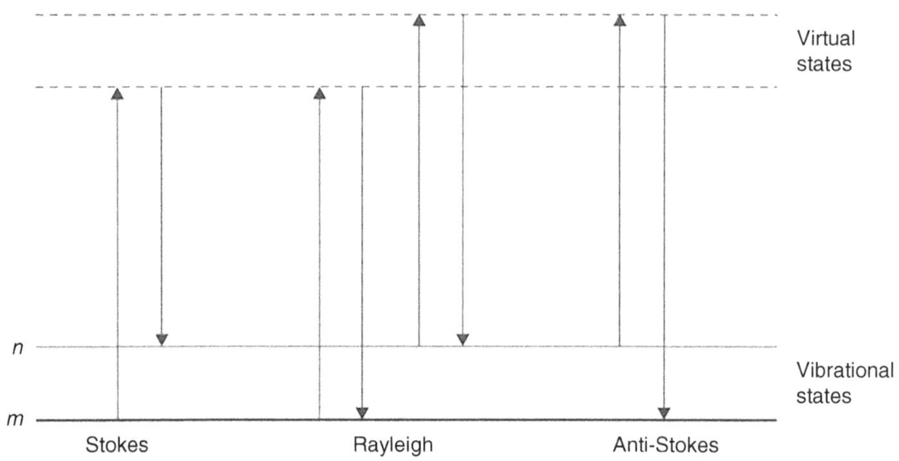

Figure 12.11 Diagram of the Rayleigh and Raman scattering processes.

The principles of Raman spectroscopy are based on its ability to measure Raman scattering, a phenomenon discovered by Sir C. V. Raman [28]. Scattering occurs when monochromatic light shines on a sample and the light interacts with the molecule and polarizes the cloud of electrons, forming a short-lived state called a "virtual state" (Figure 12.11). The virtual state is not stable, and photons are scattered, with most of the scattering occurring as Rayleigh scattering at the same wavelength of the incident light. A very small proportion of photons that scatter are shifted to wavelengths in either direction. This latter scattering is called Raman scattering, which is directly related to the frequency of molecular vibrations. Raman scattering can be designated as either Stokes or anti-Stokes. For Stokes scattering, a molecule is excited from a ground vibrational state to a virtual state by absorbing an incoming photon. When the molecule moves from the virtual state back to a high-energy vibrational state, a photon with less energy and longer wavelength is released. The wavelength difference between the new photon and the incoming photon is called a Raman shift. In anti-Stokes scattering, the molecule moves from the high-energy vibrational state to the ground state after interacting with the photon, and the scattered photon has higher energy and shorter wavelength than the incoming photon. Generally, Stokes scattering exhibits a much higher intensity than anti-Stokes scattering since more molecules are in the ground vibrational state than the high-energy vibrational state. Raman spectroscopy usually measures Stokes shifts, covering an energy level range of 200–4000 cm^{-1}.

Generally, Raman spectral data are presented as Raman intensity versus the Raman shift at a particular wavelength.

Certain functional groups can have both IR and Raman bands. It is generally recognized, however, that centrosymmetric molecules give rise to either IR bands or Raman bands but not to both. In addition, functional groups with the capability for strong IR absorption usually exhibit strong dipole moments and, therefore, display weak Raman scattering. Conversely, nonpolar functional groups, such as carbon double or triple bonds, have weak IR signals but give very strong Raman bands.

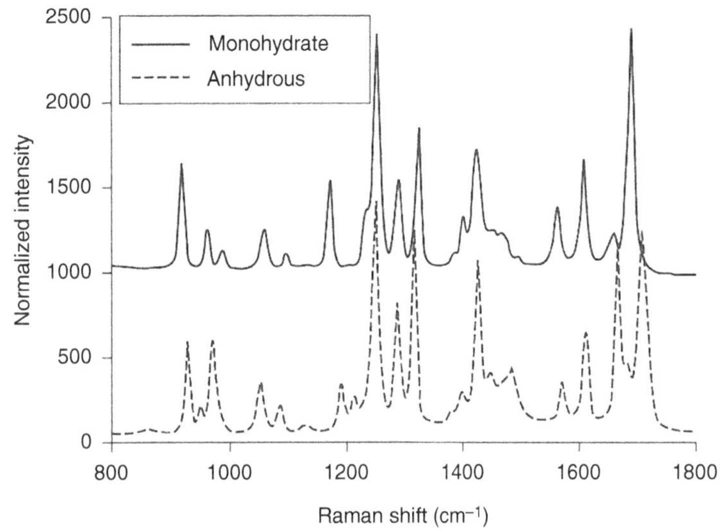

Figure 12.12 Raman spectrum of theophylline anhydrate and hydrate forms. *Source*: [29]/with permission of Elsevier.

As in the case of IR spectra, Raman spectra have bands that arise from fundamental vibrations, and therefore are useful for the detection of functional groups and fingerprint regions, allowing for compound identification. As such, Raman spectroscopy, like FT-IR spectroscopy, can be used to distinguish between solid forms, such as polymorphs, solvates, hydrates, salts, and amorphous materials. Figure 12.12 gives an example where Raman spectroscopy was used to differentiate between hydrate and anhydrate forms of theophylline [29].

12.11 Raman Instrumentation and Sampling

Generally, a Raman spectrometer consists of three primary components: an excitation source, a sampling apparatus, and a detector, with a laser source generally serving as the excitation source. A fiber optics probe or a microscope acts as the sampling apparatus, whereas a spectrometer collects the Raman scattering. In the spectrometer, optical filters are used to remove the Rayleigh scattering, and gratings are applied to spatially separate the different wavelengths. The spatially dispersed beam is directed to a charge-coupled device (CCD) camera. This type of instrument design is referred to as dispersive Raman.

Since Raman spectroscopy measures wavelength shifts, it is imperative that the laser frequency remains stable within a narrow bandwidth. Some equipment is available with a fixed laser, and others offer a selection of several lasers. The laser wavelength affects the strength of Raman scattering and the possible occurrence of fluorescence. It is generally recognized that shorter wavelengths produce stronger Raman signals, whereas they also tend to give rise to fluorescence for many organic compounds. Fluorescence signals can overwhelm the Raman signal and, therefore, compromise the detection of Raman scattering. Shorter wavelength lasers, such as green lasers or UV lasers, are widely used for analyzing inorganic material, whereas blue lasers or NIR lasers are more desirable for organic compounds that tend to fluoresce. Many Raman instruments come with choices of different lasers tailored for specific applications. Table 12.1 lists some common laser sources used in Raman spectroscopy.

Table 12.1 Some common laser sources for Raman spectroscopy.

Laser type	Wavelength (nm)
Argon ion	488.0 or 514.5
Krypton ion	530.9 or 647.1
Helium–neon	632.8
Diode	785 or 840
Nd (neodymium) – Yag (yttrium aluminum garnet)	1064

The sample holder for Raman spectroscopy is relatively simple for use with liquids, suspensions, semisolids, or powders simply placed into a glass container such as a vial or a tube. Tablets and films can be analyzed directly.

To study such systems simply involves placing the sample into the excitation beam and collecting the scattered light (Figure 12.13).

Appropriate detection of generally weak Raman signals requires optical filters and a sensitive CCD. The optical filters remove Rayleigh scattering and prevent the Rayleigh scattering from dominating the detector response. There are four types of optical filters: a long-wave edge filter, a short-wave edge filter, a notch filter, and a laser line filter. Each filter is selected for use as a transmission filter or a laser block filter to allow for the detection of Stokes or anti-Stokes Raman scattering. For example, the combination of a short-wave edge filter in transmission and a long-wave edge filter in block is tailored for Stokes measurements, whereas the use of a laser line filter in transmission and a notch filter in block allows measurement of both Stokes and anti-Stokes scattering simultaneously.

Several ways are generally employed to collect and direct the weak Raman signal to a detector, including back scattering, forward scattering, microscopy, and fiber optics. Fiber optics is particularly very popular as an efficient way to deliver laser light and to collect the Raman signal.

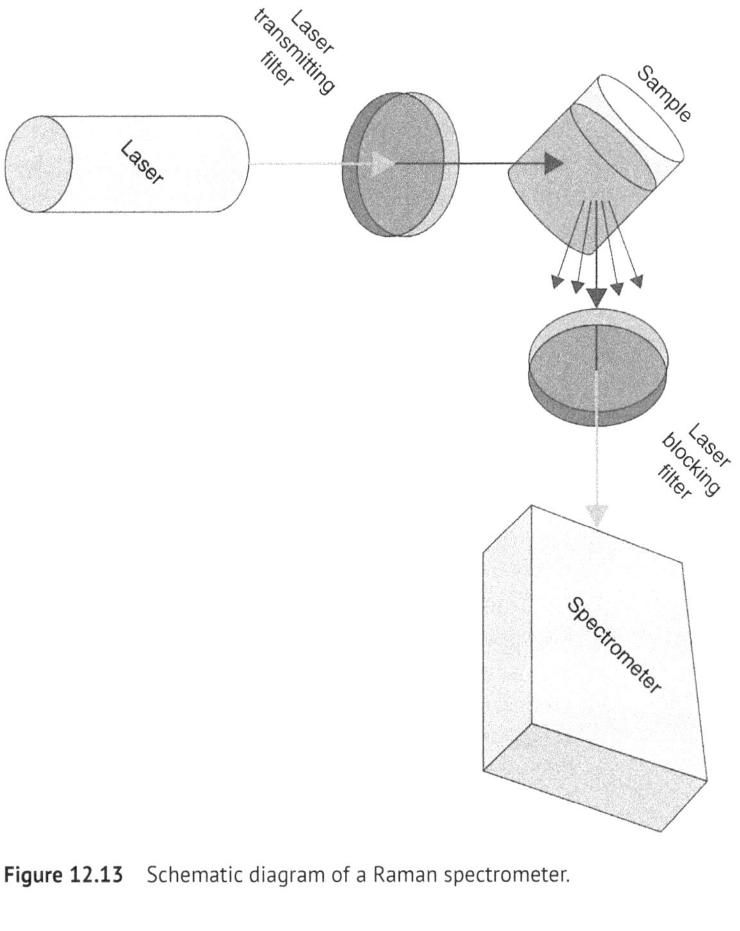

Figure 12.13 Schematic diagram of a Raman spectrometer.

Regarding sampling, a fiber-optic probe offers an effective way to direct the laser light to the sample, collect the Raman scattering, and send the signals to a spectrometer.

A Raman probe consists of an excitation fiber and a collection fiber, where the excitation fiber directs the laser to the sample. The Raman signals then reach the collection fiber through two mirrors. A long-pass filter in the collection fiber removes Rayleigh scattering. After filtering Rayleigh scattering, the Raman signal goes through a grating to reach the detector.

The introduction of high-efficiency multichannel detectors, such as CCD detectors, gives rise to significantly improved sensitivity and performance of the modern Raman spectrometer.

In addition to the improvement in detection, several ways of illumination and sample preparation can be employed to increase the Raman signal intensity. Stimulated Raman scattering uses very strong laser pulses to enhance the Raman signals up to four or five orders of magnitude. Coherent anti-Stokes Raman spectroscopy uses two very strong lasers to irradiate a sample. One of the lasers is kept at constant frequency, whereas the other one is tuned in a way that the frequency difference between the two lasers matches the frequency of a specific Raman scattering. As a result, the signal of the specific Raman signal related to anti-Stokes scattering is strengthened. The use of resonance Raman is associated with certain molecular structures. Resonance Raman occurs when the excitation laser frequency matches the frequencies of electronically excited states and resonates with them. The Raman signals are enhanced three to five orders of magnitude since they originate from an electronic transition. Surface-enhanced Raman spectroscopy can produce five to six orders of magnitude enhancement when materials under study are adsorbed on a certain metal surface [30].

Raman spectroscopy has several advantages over FT-IR in terms of sampling. Since ambient moisture and carbon dioxide barely exhibit any Raman signals, there is no need to control the environment by purging with nitrogen and maintaining a high vacuum. Since glass and plastic packaging, likewise, exhibit weak Raman signals, sample analysis can be carried out inside a glass bottle or plastic bag. In addition, aqueous samples can be directly analyzed since water also has weak Raman signals, in contrast to FT-IR and NIR. Raman spectroscopy is also nondestructive, so there is no need for sample dilution or pellet compression.

The disadvantage of Raman spectroscopy is that the use of a strong laser may cause compound degradation and fluorescence. This can be overcome to some extent using an FT-Raman spectrometer.

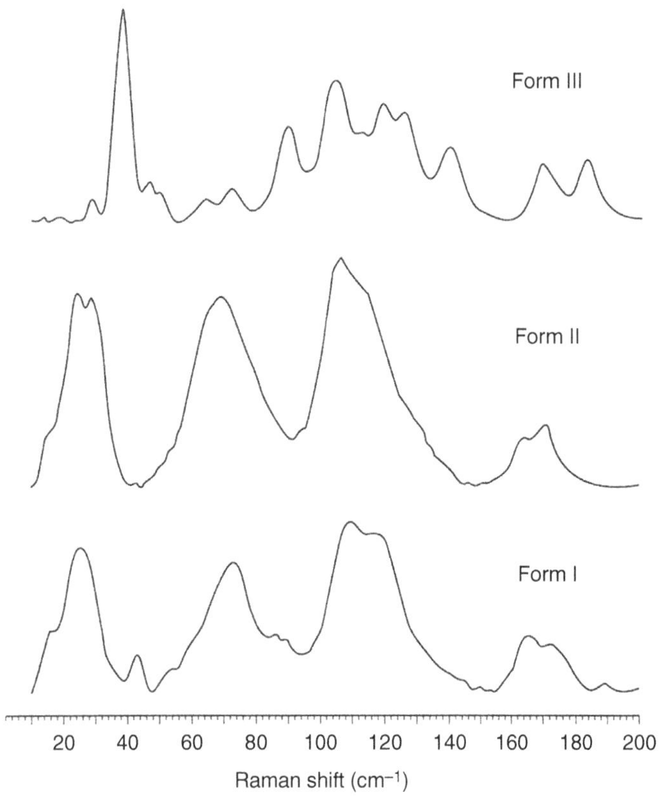

Figure 12.14 The low-wavenumber Raman spectra of carbamazepine polymorphs. *Source*: [32]/with permission of American Chemical Society.

Like FT-IR, FT-Raman spectrometers utilize an interferometer. A laser with a 1 μm wavelength passes through the interferometer and becomes an interferogram. The interferogram interacts with the sample, which then produces a Raman signal in interferogram format. Raman spectra are generated after carrying out a Fourier transformation. FT-Raman offers several advantages over dispersive Raman, including high resolution, measurement of all wavelengths at once, signal averaging to increase the signal-to-noise ratio, and better wavelength accuracy.

In addition, Raman spectroscopy can be used to examine low-wavenumber regions and detect low-energy molecular vibrations. Whereas FT-IR spectroscopy can detect frequencies as low as 400 cm^{-1}, a typical Raman spectrometer can go as low as 150 cm^{-1}. Also, the terahertz region from 10 to 150 cm^{-1} can be detected in Raman spectroscopy by utilizing "ultra-narrowband" filters and wavelength-stabilized lasers [31].

The low-wavenumber region from 10 to 400 cm^{-1} is very useful for providing chemical information related to low-energy molecular vibrations/rotations, lattice vibrations (phonons), and hydrogen-bonding vibrations, which are all directly related to the crystal structure of a material. As such, low-wavenumber Raman spectroscopy has enhanced sensitivity to detect drug polymorphs. Figure 12.14 shows the low-wavenumber

Raman spectra of some carbamazepine polymorphs [32]. The differences are substantial, and, therefore, the spectra are useful in distinguishing between the three polymorphs.

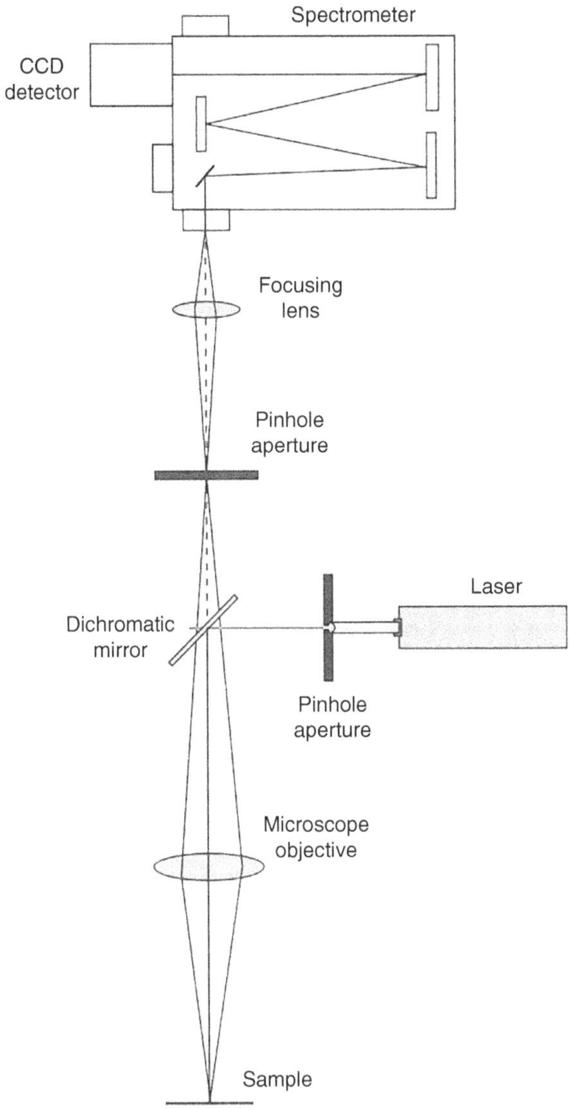

Figure 12.15 Diagram of a Raman microscope.

12.12 Raman Microscopy

Raman microscopy combines Raman spectroscopy and microscopy, typically employed in a dispersive Raman setting. A modern Raman microscope, combining an optical microscope with a Raman spectrometer, functions as a confocal microscope (Figure 12.15). In contrast to wide-field microscopes, the confocal microscope offers point-by-point illumination of a sample, which limits the scattered light. The reduction of scattered light significantly improves the lateral and depth resolution. Point illumination is achieved by setting a pinhole aperture in the front of the light source. After refocusing with the microscope objective lens, a single point is illuminated on the sample. In addition, a second pinhole aperture is placed between the microscope lens and the image and acts as a spatial filter to reject the residual scattered rays. The collection of visual images point by point on a sample enables the reconstruction of a three-dimensional structure. A confocal microscope can utilize visual light or laser light, with laser light offering additional improvement in resolution.

Raman analysis with a confocal microscope uses the same optical path as the light microscope, except that the Raman signals are collected and analyzed instead of visual pixels, since a Raman spectrometer and a CCD detector replace a digital camera. A laser source is controlled by a pinhole aperture, reflected by a dichromatic mirror, and focused on a sample via the objective lens. The Raman scattering from the sample follows the optical path through an objective lens and the second pinhole aperture and is delivered into the spectrometer. The spectrometer has filters and gratings, which separate the spectrum and remove Rayleigh scattering.

A Raman microscope can be used to measure a single point guided by an optical microscope and thus has the capability of obtaining chemical information from a discrete and known location on a sample.

It can also provide a chemical image after scanning a defined area, possibly in three dimensions. Spatial resolution of a Raman microscope is significantly higher than that of an FT-IR microscope, down to 0.2 μm. This capability of a Raman microscope allows for in situ chemical mapping, which can reveal the differences in chemical composition and material structures at a microscale.

Figure 12.16 gives an example of chemical mapping using a Raman microscope in a pharmaceutical system [33]. A low-dose tablet was mapped by assigning the red channel to the API, the green channel to cellulose, and the blue channel to lactose. Characteristic Raman peaks of each component were selected for mapping, with a Raman shift observed at 269 cm^{-1} for the API, 427 cm^{-1} for cellulose, and 347 cm^{-1} for lactose, respectively. The spectra collected from each of the domains specific to each component accurately matched those obtained with the known spectra of each component.

In summary, the coupling of a Raman spectrometer and an optical microscope has many advantages. Confocal light collection greatly improves the quality of a visual image and a chemical image. It can achieve high lateral spatial resolution and excellent depth of resolution.

Because of the confocal setup, it is possible not only to collect information from the sample surface but also to probe deeply into transparent samples and possibly obtain three-dimensional information.

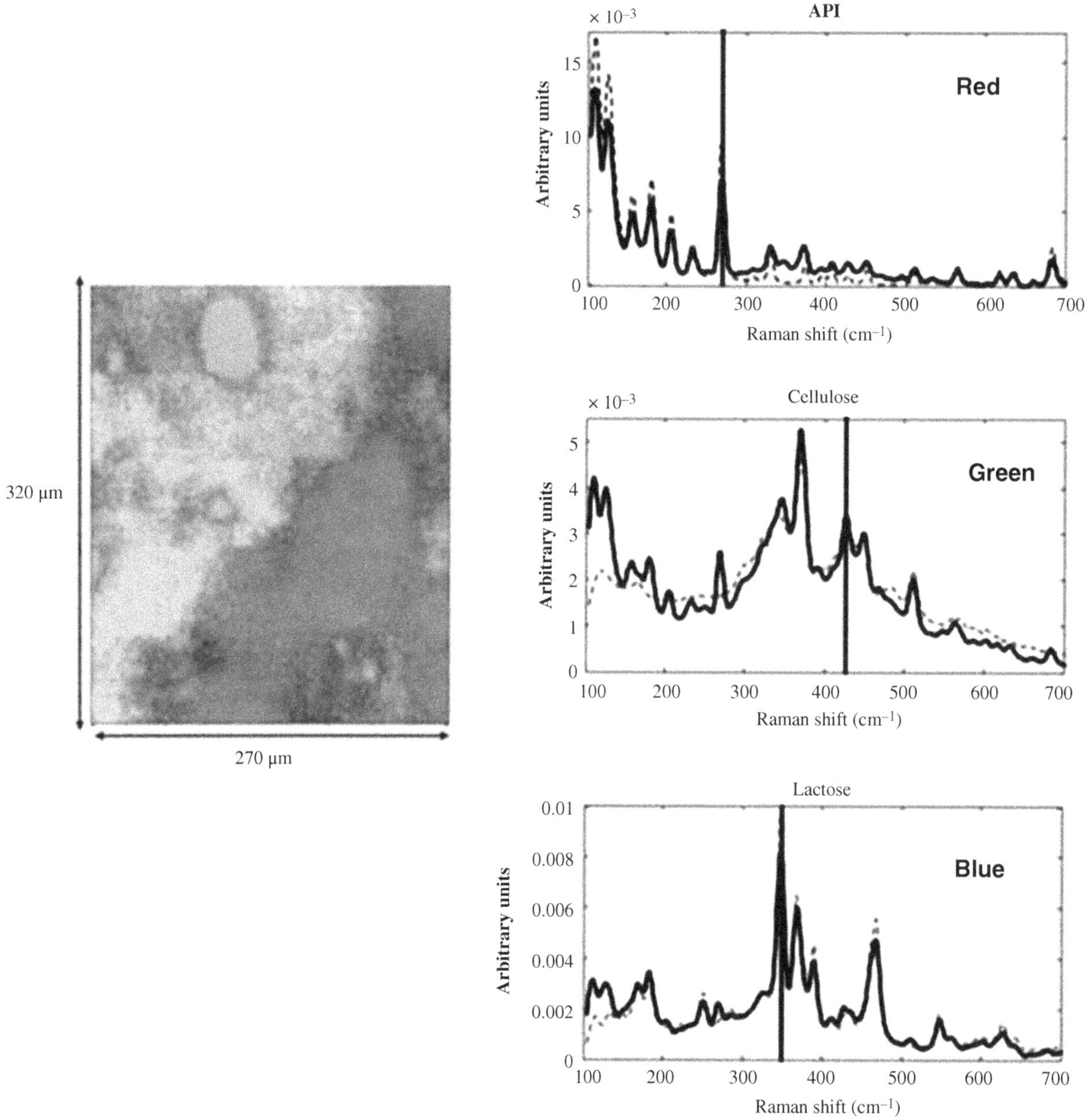

Figure 12.16 Chemical mapping of a low-dose tablet by Raman microscope. *Source*: Adapted from [33].

12.13 Terahertz Spectroscopy

Terahertz spectroscopy utilizes electromagnetic radiation in the terahertz frequency range for material characterization. Terahertz radiation lies between IR and microwave radiation with a frequency from 60 GHz to 4 THz ($1\ \text{THz} = 10^{12}\ \text{Hz}$), corresponding to wavelengths of 2–130 cm^{-1} [34]. Terahertz waves have lower energy than that of most of the intramolecular vibrations exhibited by organic molecules, so that the absorption of terahertz radiation for analytical purposes is restricted to intermolecular vibrations, which are related to translations and vibrations of molecules. Terahertz radiation can induce low-frequency bond distortions, crystalline phonon vibrations, and hydrogen bond vibrations, which are related to intermolecular or collective modes that characterize the crystalline properties of solid materials. Since terahertz spectroscopy directly probes intermolecular interactions and phonon lattice modes, it has a significantly high sensitivity for distinguishing between crystalline forms of an API.

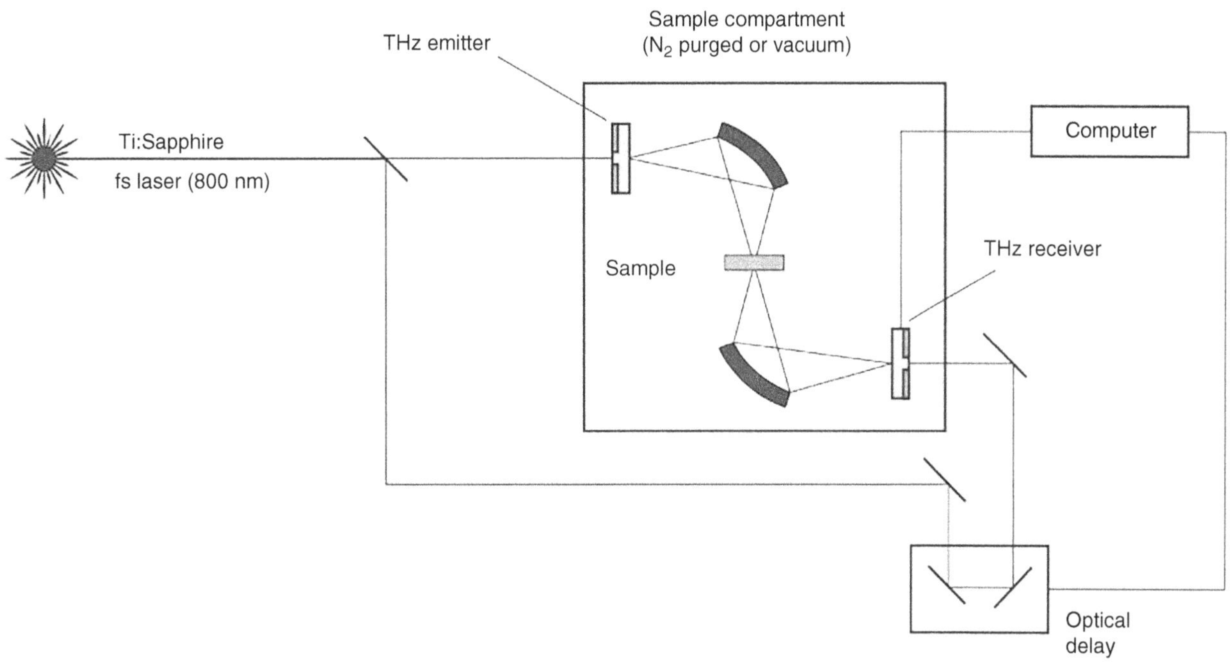

Figure 12.17 Schematic of a typical terahertz pulsed spectroscopy configuration.

In terahertz spectroscopy, short pulses of terahertz waves are applied to a material, and the transmitted terahertz electric field is measured. A typical setup for terahertz spectroscopy is shown in Figure 12.17. Terahertz radiation is generated and detected via a time-gated coherent detection scheme at room temperature, with femtosecond lasers and specially designed photoconductive semiconductor antenna switches. The terahertz wave comes from a terahertz emitter, which produces pulses of terahertz waves when ultrashort laser pulses hit a photoconductive switch. A first parabolic mirror focuses the terahertz waves onto the sample, while another parabolic mirror focuses the transmitted terahertz waves onto a detector. The detector is an optically gated semiconductor receiver with photoconductive capability, like the emitter. Fourier-transforming the resultant time-domain waveform provides frequency information, similar to FT-IR spectroscopy.

Terahertz spectroscopy has the same selection rules as FT-IR spectroscopy and is sensitive to changes in dipole moment. Consequently, terahertz spectra provide a complementary tool to low-wavenumber Raman spectra and allow for a complete picture of the vibrational structure of a material. In addition, the coherent detection used in terahertz spectroscopy allows the absorption coefficient and the refractive index to be measured at the same time. Current terahertz spectrometers can cover a spectral range from 2 to 130 cm^{-1} and reach a spectral resolution of 1 cm^{-1} [34, 35], with terahertz spectra being collected from milliseconds to seconds.

The various sampling techniques include transmission, specular reflection, diffuse reflection, and ATR. As with FT-IR spectroscopy, sample pellets can be used for transmission analysis, with pellets usually having a diameter of 5–30 mm and a thickness of approximately 0.5–3 mm. For materials with good compressibility and weak absorption, sample pellets are generally prepared without diluents. When diluents are required, polyethylene and poly(tetrafluoroethylene), which are transparent to terahertz radiation, are commonly used [34, 35]. As discussed earlier for FT-IR measurements, no sample preparation is required for analysis in the ATR mode. A sample is directly applied to an ATR crystal with a very high refractive index, with silicon being widely used as the crystal for terahertz spectroscopy.

Terahertz spectroscopy is a very useful technique for the characterization of pharmaceutical solid forms because it can probe hydrogen bond networks and phonon modes in molecular crystals with great sensitivity. In this regard, it is superior to FT-IR, NIR, and Raman spectroscopy (middle-range wavenumber) for differentiating solid forms while having comparable sensitivity to X-ray powder diffraction. In contrast to XRPD, it is not affected by preferred orientation.

The terahertz spectra for various solid forms of a compound can be very distinctive. For example, Figure 12.18 shows the spectra of the polymorphs of sulfathiazole, where all five polymorphs of sulfathiazole are clearly distinguished by their terahertz spectra with better specificity than those from FT-IR spectroscopy [36]. Distinct differences in terahertz spectra have also been reported for carbamazepine polymorphs (Forms I and III) and enalapril maleate polymorphs (Forms I and II) [37].

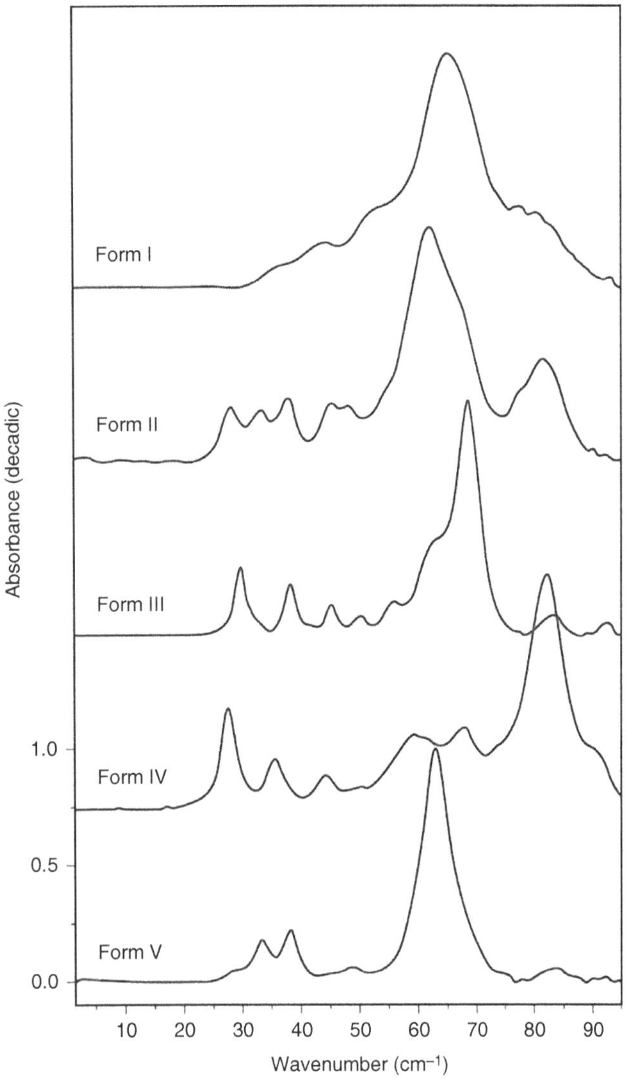

Figure 12.18 Terahertz spectrum of sulfathiazole polymorphs. *Source*: Adapted from Ingle et al. [36].

Owing to the lack of long-range order and the lack of phonon modes, amorphous and liquid crystals do not give useful spectra with terahertz spectroscopy (Figure 12.19). This indicates that crystalline forms of a compound can be readily distinguished from amorphous and liquid crystal forms, as shown in studies of crystalline versus amorphous indomethacin and liquid crystalline versus crystalline fenoprofen calcium [37].

Since high-quality spectra can be acquired in less than 1 second, terahertz spectroscopy has significant advantages for studying solid-state transformations and has been used to characterize the in situ dehydration of theophylline monohydrate [38] and the various temperature-induced phase transitions for the five polymorphic forms of sulfathiazole [39].

12.14 Comparison of FT-IR, NIR, Raman, and Terahertz Spectroscopy

FT-IR, NIR, Raman, and terahertz spectroscopies are grouped together as vibrational spectroscopy. FT-IR, NIR, and terahertz spectroscopies are sensitive to changes in the dipole moment of a particular molecular species, whereas Raman spectroscopy can detect changes in polarizability.

The four spectroscopic techniques require different instrumental designs but give similar capabilities in chemical and physical analysis.

12.14.1 Spectral Information

Although FT-IR and Raman spectroscopies have different selection rules, they usually produce complementary spectral data. Spectral information from FT-IR and Raman spectroscopies is generally simple, is easy to interpret, and

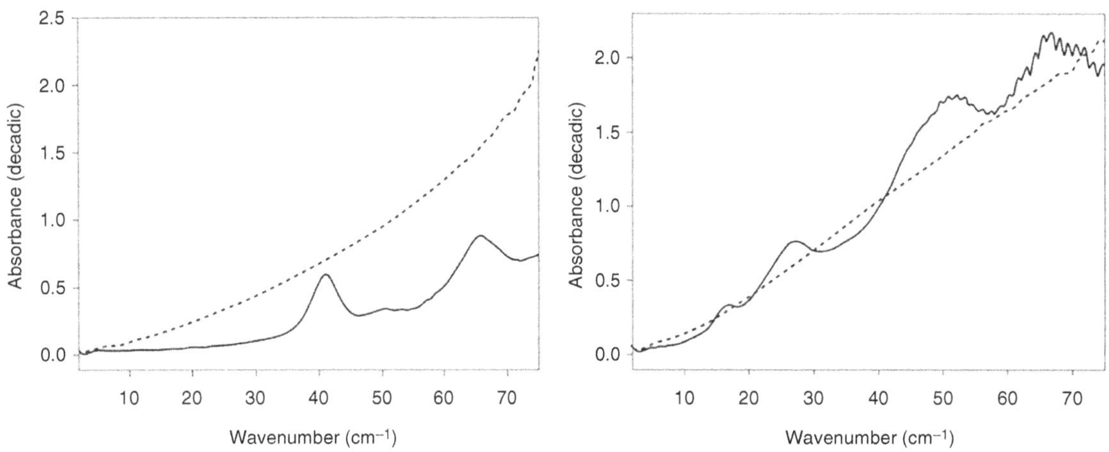

Figure 12.19 Terahertz spectra of indomethacin amorphous and crystalline forms (left) and fenoprofen calcium liquid crystalline and hydrate forms (right). *Source*: [37]/with permission of Elsevier.

can be used to analyze materials directly. For compound identification, FT-IR spectroscopy is more widely used since an extensive historical spectra library is available. Generally, FT-IR spectroscopy covers the wavelength range from 650 to 4000 cm^{-1}, and Raman spectroscopy can collect spectra from 70 to 3600 cm^{-1}. The capability to collect spectral information at lower wavenumbers makes Raman spectroscopy very useful for characterizing inorganic compounds and for differentiating polymorphic forms.

The spectral information from NIR spectroscopy is more complex and not obvious since it collects bands related to overtones and combination bands. Chemometrics software and derivative methods are, therefore, required to deconvolute the chemical information.

Terahertz spectroscopy detects low-energy vibrations such as lattice vibrations and hydrogen bond vibrations with spectral information that is more sensitive to the crystal lattice than to properties of chemical functional groups. In view of this, terahertz spectroscopy has been demonstrated to have superior capability in distinguishing crystal forms of pharmaceutical active ingredients.

12.14.2 Spectral Resolution

FT-IR spectrometers are typically much more accurate in terms of wavelength than Raman spectrometers; the spectral accuracy of FT-IR systems can reach 0.01 cm^{-1}, whereas that of typical dispersive Raman can only reach 0.1 cm^{-1} at best. In dispersive Raman, the stability of the laser significantly affects the wavelength accuracy.

FT-IR spectroscopy usually has spectral resolution on the order of 1–8 cm^{-1}, with high resolution requiring long scanning times. Raman spectroscopy can reach a similar resolution as FT-IR spectroscopy, with a choice of appropriate instrumental settings, such as spectrometer focal length, diffraction grating, and laser wavelength. Normal dispersive NIR spectroscopy has a spectral resolution of 50 cm^{-1}, while FT-NIR spectroscopy can reach 2 cm^{-1}. Terahertz spectroscopy has a spectral resolution of 1 cm^{-1}, whereas high-resolution terahertz spectroscopy can reach below 0.1 cm^{-1}.

12.14.3 Sampling

FT-IR and Raman spectroscopies can be used to analyze solids, liquids, and gases. Raman and NIR spectroscopies are applicable only to the measurement of solids and liquids, whereas terahertz spectroscopy is most useful for analyzing solids.

NIR and Raman spectroscopies are more convenient and flexible in terms of sampling. Since glass vials and plastic bags exhibit minimal NIR and Raman signals, the sample can be directly analyzed in a sample container. NIR and Raman analysis can be transmitted by optical fibers over long distances for remote analysis.

FT-IR spectroscopy allows the use of several sampling techniques, including KBr pellets, Nujol oil, ATR, and diffuse reflectance. Because many organic or inorganic materials have strong IR absorption, a certain degree of sample preparation and dilution is needed. Some techniques are destructive, such as using KBr pellets, and may affect the physical quality of the sample. Owing to the very weak signals associated with water, Raman spectroscopy can handle aqueous solutions. However, FT-IR and NIR spectroscopies are very sensitive to the presence of water due to its significant IR absorbance.

Terahertz spectroscopy uses very similar sampling techniques as FT-IR spectroscopy, including transmission, ATR, and diffuse reflectance.

12.14.4 Environmental Control

Carbon dioxide and water found in the atmosphere significantly absorb radiation in an FT-IR instrument. FT-IR analysis under ambient conditions, therefore, usually involves collecting background spectra and subtracting out such background spectra from the sample spectra to reduce any contributions from carbon dioxide and moisture.

Since water has a strong peak in terahertz spectroscopy, purging the sample with a dry gas is needed before sample analysis.

Conversely, atmospheric carbon dioxide and moisture have minimal effects in NIR and Raman spectroscopies.

12.14.5 Microscopy

FT-IR microscopy and Raman microscopy share similar equipment design and demonstrate similar capability for spot analysis. A Raman microscope has much better spatial resolution due to its use of confocal optics and short wavelengths;

the spatial resolution can be as low as the submicrometer range. Because long-wavelength IR light is used, an FT-IR microscope has spatial resolution that is greater than 10 μm. Raman microscopes, however, are more useful in chemical imaging because of their higher resolution. The high resolution also allows Raman microscopes to be used for depth analysis.

12.14.6 Fluorescence and Photodamage

Since strong lasers are used in Raman analysis, in many cases, this technique may decompose or otherwise damage photo-sensitive materials. In addition, although strong fluorescence often may arise in Raman analysis, fluorescence can be mitigated by using a laser with a longer wavelength or an FT-Raman instrument. On the other hand, using FT-IR and NIR spectroscopies can potentially avoid the problems of fluorescence and photoinstability.

References

1 Rustichelli C, Gamberini G, Ferioli V, Gamberini MC, Ficarra R, Tommasini S. Solid-state study of polymorphic drugs: carbamazepine. *J Pharm Biomed Anal* 2000; 23(1): 41–54.
2 Bunaciu AA, Aboul-Enein HY, Fleschin Ş. FT-IR Spectrophotometric analysis of acetylsalicylic acid and its pharmaceutical formulations. *Can J Analy Sci Spectrosc* 2006; 51: 253–259.
3 USP-NF 2024, General <197> Spectroscopic Identification Tests. US Pharmacopeia (USP) website, www.uspnf.com.
4 Atef E, Chauhan H, Prasad D, Kumari D, Pidgeon C. Quantifying solid-state mixtures of crystalline indomethacin by Raman spectroscopy comparison with thermal analysis. *ISRN Chromatography* 2012; Article ID 892806.
5 Spychala S, Butkiewicz K, Pakula R, Pichnej L. Polymorphism of indomethacin Part II Identification and rapid determination of polymorphic forms of indomethacin by IR spectrometry. *Pol J Pharmacol Pharm* 1976; 29(2): 157–160.
6 Park TJ, Ko DH, Kim YJ, Kim YG. Polymorphic characterization of pharmaceutical solids, donepezil hydrochloride, by 13C CP/MAS solid-state nuclear magnetic resonance spectroscopy. *Bull Korean Chem Soc* 2009; 30(9): 2007–2010.
7 King PL, Ramsey MS, McMillan PF, Swayze G. Laboratory Fourier transform infrared spectroscopy methods for geologic samples. In: King PL, Ramsey MS, McMillan PF, Swayze GA, editors. *Infrared Spectroscopy in Geochemistry, Exploration Geochemistry and Remote Sensing*; 2004. p 57–59.
8 Kubelka P, Munk F. Ein Beitrag zur Optik der Farbanstriche. *Zeit Für Tekn Physik* 1931; 12: 593.
9 Kojima T, Yamauchi Y, Onoue S, Tsuda Y. Evaluation of hydrate formation of a pharmaceutical solid by using diffuse reflectance infrared Fourier-transform spectroscopy. *J Pharm Biomed Anal* 2008; 46(4): 788–791.
10 Müller AL, Picoloto RS, Ferrão MF, da Silva FE, Müller EI, de Moraes Flores ÉM. Simultaneous diffuse reflectance infrared determination of clavulanic acid and amoxicillin using multivariate calibration techniques. *Drug Test Anal* 2012; 4(6): 500–506.
11 Kipouros K, Kachrimanis K, Nikolakakis I, Tserki V, Malamataris S. Simultaneous quantification of carbamazepine crystal forms in ternary mixtures (I, III, and IV) by diffuse reflectance FTIR spectroscopy (DRIFTS) and multivariate calibration. *J Pharm Sci* 2006; 95(11): 2419–2431.
12 Helmy R, Zhou GX, Chen YW, Crocker L, Wang T, Wenslow RM Jr, Vailaya A. Characterization and quantitation of aprepitant drug substance polymorphs by attenuated total reflectance Fourier transform infrared spectroscopy. *Anal Chem* 2003; 75(3): 605–611.
13 Cantor AS. Drug and excipient diffusion and solubility in acrylate adhesives measured by infrared-attenuated total reflectance (IR-ATR) spectroscopy. *J Control Release* 1999; 61(1): 219–231.
14 Cornel J, Lindenberg C, Mazzotti M. Quantitative application of in Situ ATR-FTIR and Raman spectroscopy in crystallization processes. *Ind Eng Chem Res* 2008; 47(14): 4870–4882.
15 Chan KL, Hammond SV, Kazarian SG. Applications of attenuated total reflection infrared spectroscopic imaging to pharmaceutical formulations. *Anal Chem* 2003; 75(9): 2140–2146.
16 Gotter B, Faubel W, Neubert RH. FTIR microscopy and confocal Raman microscopy for studying lateral drug diffusion from a semisolid formulation. *Eur J Pharm Biopharm* 2010; 74(1): 14–20.
17 Ricci C, Nyadong L, Fernandez FM, Newton PN, Kazarian SG. Combined Fourier-transform infrared imaging and desorption electrospray ionization linear ion-trap mass spectrometry for analysis of counterfeit antimalarial tablets. *Anal Bioanal Chem* 2007; 387(2): 551–559.

18 Ravn C, Skibsted E, Bro R. Near-infrared chemical imaging (NIR-CI) on pharmaceutical solid dosage. *J Pharm Biomed Anal* 2008; 48(3): 554–561.

19 da Silva VH, Gonçalves JL, Vasconcelos FV, Pimentel MF, Pereira CF. Quantitative analysis of mebendazole polymorphs in pharmaceutical raw materials using near-infrared spectroscopy. *J Pharm Biomed Anal* 2015; 115: 587–593.

20 Pan D, Crull G, Yin S, Grosso J. Low level drug product API form analysis – Avalide tablet NIR quantitative method development and robustness challenges. *J Pharm Biomed Anal* 2014; 89: 268–275.

21 Luner PE, Seyer JJ. Assessment of crystallinity in processed sucrose by near-infrared spectroscopy and application to lyophiles. *J Pharm Sci* 2014; 103(9): 2884–2895.

22 Brashear RL, Flanagan DR, Luner PE, Seyer JJ, Kemper MS. Diffuse reflectance near-infrared spectroscopy as a nondestructive analytical technique for polymer implants. *J Pharm Sci* 1999; 88(12): 1348–1353.

23 Lee MJ, Seo DY, Wang IC, Chun NH, Lee HE, Jeong MY, Kim WS, Choi GJ. Quantitative in-line monitoring of solvent-mediated polymorphic transformation of sulfamerazine by near-infrared spectroscopy. *J Pharm Sci* 2012; 101(4): 1578–1586.

24 Lee HE, Lee MJ, Kim WS, Jeong MY, Cho YS, Choi GJ. In-line monitoring and interpretation of an indomethacin anti-solvent crystallization process by near-infrared spectroscopy (NIRS). *Int J Pharm* 2011; 420(2): 274–281.

25 Shi Z, Cogdill RP, Short SM, Anderson CA. Process characterization of powder blending by near-infrared spectroscopy: blend end-points and beyond. *J Pharm Biomed Anal* 2008; 47(4): 738–745.

26 Hudovornik G, Korasa K, Vrečer F. A study on the applicability of in-line measurements in the monitoring of the pellet coating process. *Eur J Pharm Sci* 2015; 75: 160–168.

27 Kauppinen A, Toiviainen M, Korhonen O, Aaltonen J, Järvinen K, Paaso J, Juuti M, Ketolainen J. In-line multipoint near-infrared spectroscopy for moisture content quantification during freeze-drying. *Anal Chem* 2013; 85(4): 2377–2384.

28 Raman CV. A new radiation. *Indian J Phy* 1928; 2: 387–398.

29 Wikström H, Kakidas C, Taylor LS. Determination of hydrate transition temperature using transformation kinetics obtained by Raman spectroscopy. *J Pharm Biomed Anal* 2009; 49(2): 247–452.

30 Haynes CL, McFarland AD, Duyne RP. Surface-enhanced Raman spectroscopy. *Anal Chem* 2005; 77(17): 338A–346A.

31 Carriere J, Heyler R. Spectroscopy: Terahertz Raman: Accessing molecular structure with Raman spectroscopy. Available at http://www.laserfocusworld.com/articles/print/volume-49/issue-03/features/spectroscopy--terahertz-raman--accessing-molecular-structure-wit.html. (Accessed 14 January 2025).

32 Roy S, Chamberlin B, Matzger AJ. Polymorph discrimination using low wavenumber Raman spectroscopy. *Org Process Res Dev* 2013; 17(7): 976–980.

33 Gendrin C, Roggo Y, Collet C. Pharmaceutical applications of vibrational chemical imaging and chemometrics: a review. *J Pharm Biomed Anal* 2008; 48(3): 533–553.

34 Zeitler JA, Taday PF, Newnham DA, Pepper M, Gordon KC, Rades T. Terahertz pulsed spectroscopy and imaging in the pharmaceutical setting – a review. *J Pharm Pharmcol* 2007; 59(2): 209–223.

35 Taday PF, Newnham DA. Technological advances in terahertz pulsed systems bring far-infrared spectroscopy into the spotlight. *Spectrosc Eur* 2004; 16(5): 20–24.

36 Ingle LM, Jaiswal AM, Vaidya PV, Kedar AP. Terahertz spectroscopy for pharmaceutical applications. *Int J Pharm Pharm Sci Res* 2013; 3(1): 48–52.

37 Strachan CJ, Rades T, Newnham DA, Gordon KC, Pepper M, Taday PF. Using terahertz pulsed spectroscopy to study crystallinity of pharmaceutical materials. *Chem Phys Lett* 2004; 390(1): 20–24.

38 Zeitler JA, Kogermann K, Rantanen J, Rades T, Taday PF, Pepper M, Aaltonen J, Strachan CJ. Drug hydrate systems and dehydration processes studied by terahertz pulsed spectroscopy. *Int J Pharm* 2007; 334(1): 78–84.

39 Zeitler JA, Newnham DA, Taday PF, Threlfall TL, Lancaster RW, Berg RW, Strachan CJ, Pepper M, Gordon KC, Rades T. Characterization of temperature-induced phase transitions in five polymorphic forms of sulfathiazole by terahertz pulsed spectroscopy and differential scanning calorimetry. *J Pharm Sci* 2006; 95(11): 2486–2498.

13

Solid-State NMR Spectroscopy

13.1 Introduction

Solution proton and carbon nuclear magnetic resonance (^{1}H and ^{13}C NMR) is widely used for the study of drugs, being employed for structure elucidation or identification. In recent years, solid-state NMR has emerged as a powerful tool for the analysis of solids [1–6]. In particular, solid-state ^{13}C NMR, using the techniques of cross polarization and magic-angle spinning (^{13}C CP/MAS NMR), offers access to otherwise elusive information.

Figure 13.1 shows the solid-state ^{13}C NMR spectra of two crystal forms of prednisolone [7]. The differences in the two spectra reflect differences in the environments of some of the C nuclei in the two crystal forms. Some changes in molecular conformation, as well as relative intermolecular environments, are generally expected between different crystal packings and may be quite substantial. As we shall see, this observation has considerable practical value.

Figure 13.2 shows the ^{13}C CP/MAS NMR spectra of prednisolone tablets from three different vendors [8]. The resonances of the prednisolone carbons are readily observed, although the tablets contain only 5 mg of prednisolone in 95 mg of excipients. Furthermore, it is evident that the tablets from vendors A and B contained prednisolone Form I, whereas tablets from vendor C contained Form II. Other examples where these techniques have proven useful include a report by Atalla et al. [5], which shows that different polymorphs of cellulose exhibit different ^{13}C CP/MAS NMR spectra, and a paper by Hill et al. [6], which shows that crystals of L-tartaric acid and crystals of DL-tartaric acid afford different ^{13}C CP/MAS NMR spectra.

Solid-state ^{13}C NMR should be useful to examine conformational and dynamic changes that occur in solid drugs during solid-state transformations and reactions, although no applications of this idea have been published. One advantage of solid-state NMR is that it offers the potential for the analysis of mixed crystals or mixtures because signals from crystallographically distinct but chemically identical sites are often resolvable. Moreover, solid-state ^{13}C CP/MAS NMR spectroscopy offers the possibility for the analysis and characterization of amorphous forms. A particularly important question in this area is: What is the mobility of various functional groups in the molecule in amorphous forms? Solid-state NMR is uniquely suited to the study of molecular dynamics on many time scales. These properties have led to increased use of solid-state NMR spectroscopy in the pharmaceutical industry.

13.2 An Overview of Solid-state 13C CP/MAS NMR Spectroscopy

Solid-state NMR spectroscopy was developed by Schaefer and Stejskal [9], who combined high-powered proton decoupling with CP and MAS, techniques that provided an enormous increase in the resolution and sensitivity of solid-state spectra. Figure 13.3 illustrates this point with the NMR spectra of dideoxyinosine (DDI) [10]. Solid DDI gives no spectrum using typical solution techniques. A solid-state NMR experiment using only high-power decoupling produces broad, featureless spectral signals. The simultaneous application of high-power decoupling and MAS produces an observable spectrum with relatively sharp peaks, but several expected signals are too weak to be seen. Addition of CP to high-power decoupling and MAS increases the per-scan sensitivity and greatly reduces the amount of time needed to achieve a given signal-to-noise (*S/N*) ratio. The resultant spectra exhibit a single line for each chemically (and often crystallographically) distinct carbon in the molecule.

The theory of solid-state NMR has been discussed in several reviews and books [11–14]. One of the major impediments to the measurement of solid-state NMR spectra is a consequence of the restricted motion in the solid state – anisotropic

Solid-State Materials in Pharmaceutical Chemistry: Properties, Characterization, and Applications, Second Edition. Stephen R. Byrn, George Zografi and Xiaoming (Sean) Chen.

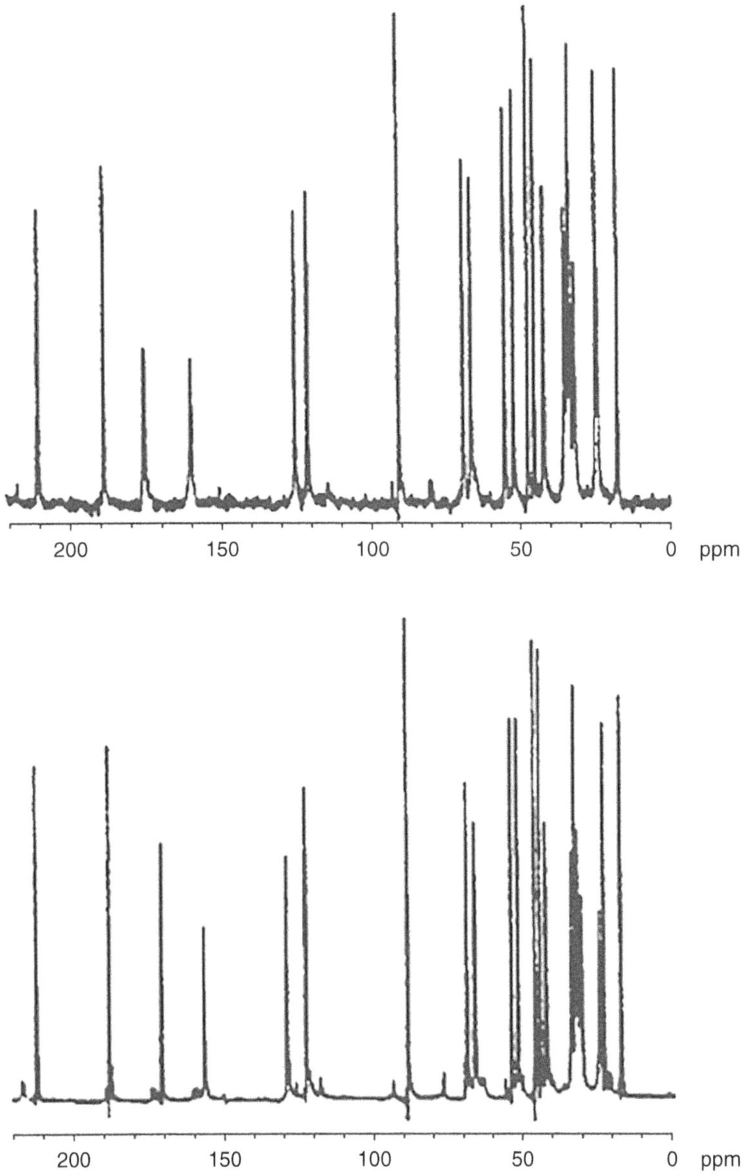

Figure 13.1 Solid-state ^{13}C CP/MAS NMR spectra of prednisolone polymorphs. Forms I (top) and II (bottom).

interactions are not averaged as they are in solution. Table 13.1 shows the types of interactions between carbon nuclei in solids (and solutions) and approaches to removing them. The strength of these interactions, particularly the dipolar and anisotropic effects, underscores the need for special pulse techniques to obtain narrow-line solid-state NMR spectra. Without them, it would not be possible to differentiate between chemically nonequivalent carbons (which cover a range of ~40 ppm), let alone the relatively small changes in chemical shift characteristic differences between polymorphs (~0.04–2 ppm). Finally, quadrupolar coupling also exists for several elements occurring in pharmaceuticals (Cl, Br, and N). This interaction can reduce sensitivity by broadening the signals of adjacent carbons. Fortunately, the effect from ^{14}N (the most common quadrupole in pharmaceuticals) is usually small. To date, no readily available techniques have been developed to eliminate this interaction.

13.2.1 Dipolar Decoupling

The dipolar interaction of ^{13}C nuclei with nearby protons is the primary factor causing line broadening in solid-state NMR. For ^{13}C nuclei in powders, broad featureless resonances are observed because of individual dipolar couplings

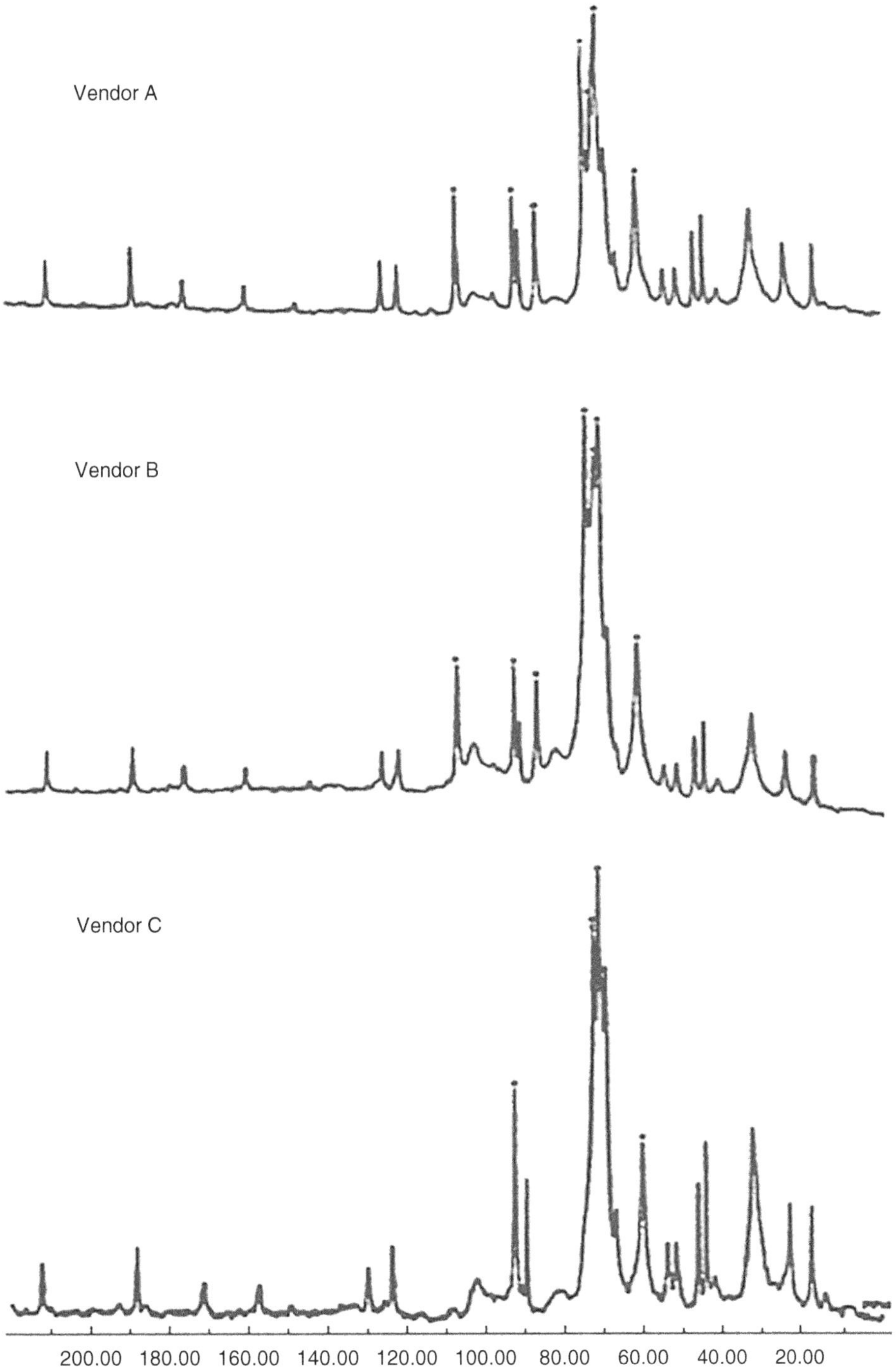

Figure 13.2 Solid-state ^{13}C CP/MAS NMR spectra of prednisolone tablets from three different vendors. The dots denote excipient peaks.

to the many protons and the random orientations of the C—H vectors with respect to the magnetic field. Fortunately, high-power proton decoupling can remove these dipolar effects, as well as the much smaller scalar-coupling effects. This is typically accomplished by irradiating the sample at the proton resonance frequency to remove the observable effects of these interactions from the spectrum. The generation of the decoupling fields needed (on the order of 60 kHz) usually requires power output in excess of 100 W, requiring large amplifiers and sometimes generating considerable amounts of heat in the sample.

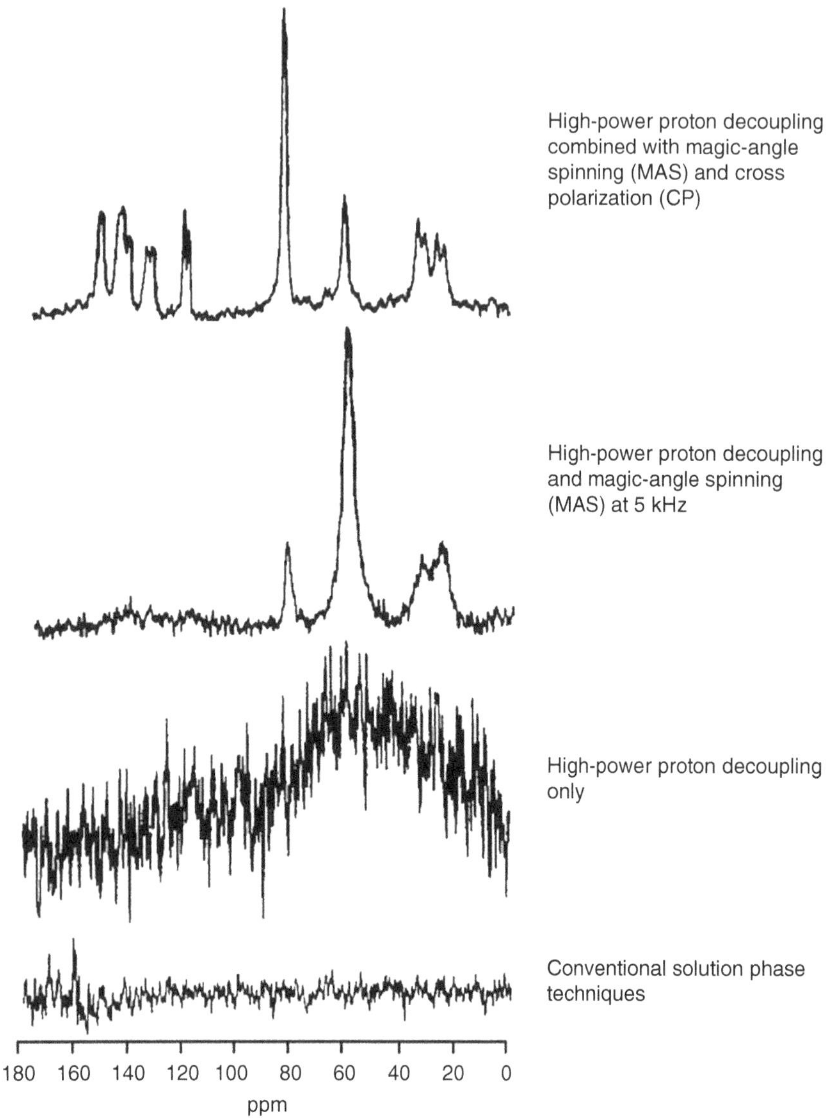

Figure 13.3 Solid-state ^{13}C NMR spectra of DDI acquired by various techniques. In all cases, 512 scans were accumulated. *Source*: [10]/with permission of Springer Nature.

Table 13.1 Carbon-13 nuclear spin interactions in solids.[a]

	Observable magnitude		
Type of interaction	**Solution**	**Solids**	**Approach to removal**
Zeeman interaction with an applied magnetic field	50 MHz	50 MHz	
Dipolar interaction, usually with nearby protons	0 MHz	~50 MHz	Dipolar (high-power) decoupling
Dipolar decoupling of quadrupolar nuclei (quadrupolar broadening by interaction with ^{14}N or other quadrupolar nuclei)	0 MHz	<1 MHz	Reduced but not removed by standard MAS
Chemical shift interaction with all local magnetic fields and dependent on the molecular orientation to the applied magnetic field	$\sigma_{\text{isotropic}}$	>300 ppm (>15 kHz at 4.7 THz)	Reduced to $\sigma_{\text{isotropic}}$ by MAS
Scalar interaction with other diamagnetic nuclei, usually via the molecular electronic scaffold	~200 Hz	~200 Hz	Scalar (low-power) decoupling in liquids High-power decoupling in solids

a. Calculations assume a 4.7 Tesla magnetic field Adapted from [11].

13.2.2 Magic-Angle Spinning

Even after the removal of the dipolar broadening by decoupling, resonance lines for ^{13}C nuclei in solids remain broad (as shown in Figure 13.4) because of the three-dimensional nature of electronic shielding. Thus, the chemical shift in solids is anisotropic (i.e. it depends on the orientation of the nuclei to the static magnetic field). To remove this anisotropy, MAS is employed. This is possible because the chemical shift anisotropy depends on $3\cos^2\theta - 1$, which is equal to zero when θ is 54.7° (the magic angle). In effect, this technique can be thought of as rotating the laboratory frame (z = main field axis) about the body diagonal of a cube fast enough to interconvert the x, y, and z axes. Thus, rapid sample rotation at this angle reduces the observed chemical shift to the isotropic value, concentrating most of the intensity into a strong, narrow band. Figure 13.4 illustrates this effect in carbon monoxide. The residual intensity often appears in one or more pairs of spinning sidebands. These are signals that are equally spaced about the main peak at intervals of the spin rate and whose intensity is inversely proportional to the spin rate, as well as the anisotropic line shape. Since chemical shift anisotropy may range up to 20 kHz, spin rates need to exceed this value to completely remove spinning side bands. Although some instrumentation can attain these speeds, the spectrometers in most laboratories are limited to speeds in the 5–10 kHz range. Thus, spinning sidebands are commonly observed for signals such as aromatic and other unsaturated carbons in the sample. These side bands can be easily identified by varying the spin rate and observing which signals change frequency and can also be attenuated or removed from the spectrum by more sophisticated techniques such as TOSS (total suppression of spinning sidebands) and its derivative.

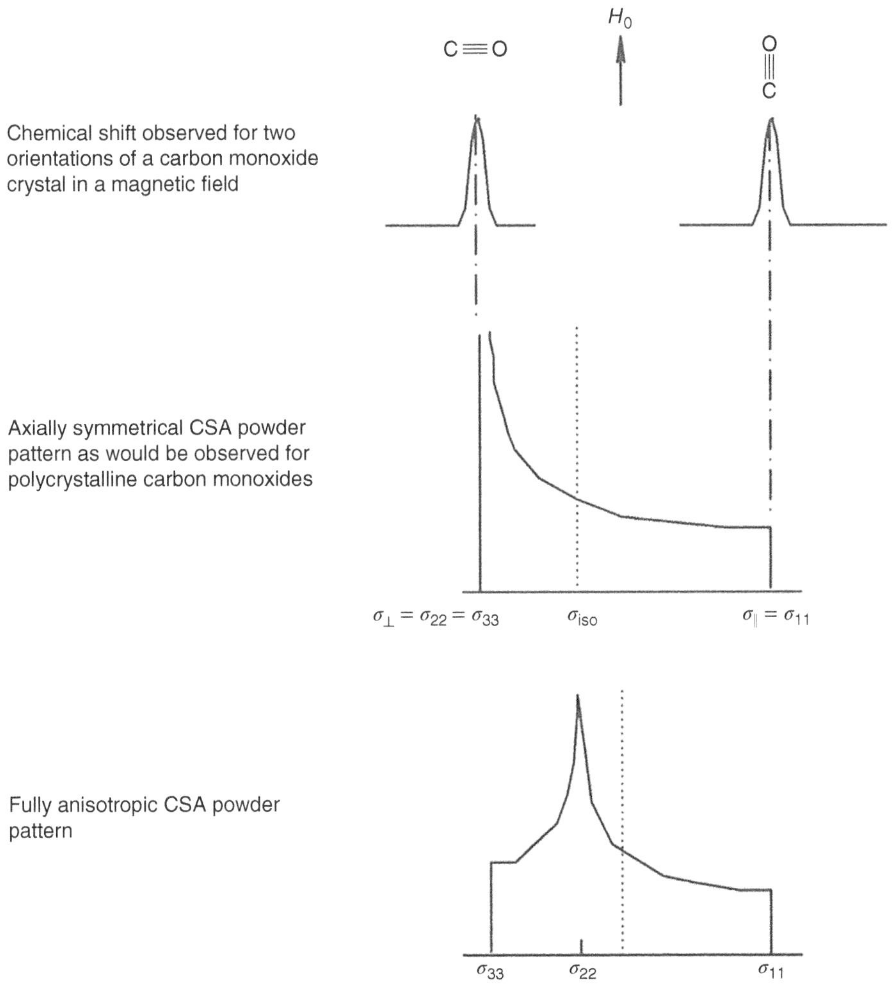

Figure 13.4 Effect of chemical shift anisotropy on the solid-state ^{13}C NMR spectrum of carbon monoxide powder.

13.2.3 Cross Polarization

CP is used to enhance the sensitivity of the solid-state NMR experiment and results in decreased acquisition time. A complete understanding of the CP experiment requires an understanding of the concept of spin temperature and spin diffusion in solids, which has been described in several reviews [11–13]. The complete CP/MAS pulse sequence involves four steps:

1. A 90°*x*-axis radiofrequency (rf) pulse at the proton resonance frequency rotates the net proton magnetization (created by polarization in the main field) from its equilibrium position of alignment along the *z*-axis to the *y*-axis.
2. The phase of the proton rf field is immediately changed by 90°, and the rf field is maintained for a short contact time (usually 1–5 ms). Simultaneously, a second rf field at the carbon resonance frequency is applied for the same interval. During this time, both the proton and carbon magnetization vectors are effectively prevented from precessing in their respective rotating frames by the parallel alignment with the rf fields. Under these conditions, the magnetization vectors are said to be spin-locked. The original equilibrium proton magnetization from the *z*-axis now re-equilibrates along the ~1000-fold weaker rf field with a time constant $T_{1\rho\rho}$ (T_1 in the rotating frame). The initial large nonequilibrium proton magnetization at the beginning of the spin lock may be viewed as a Boltzmann population difference between the two magnetic states corresponding to very little thermal energy; the protons are said to have a low spin temperature in the rotating frame.

   Since Υ_C is about one-fourth Υ_H, the carbon magnetization is initially small, corresponding to a high-temperature Boltzmann distribution (i.e. the ^{13}C reservoir is "hot"). In most common experiments, the recycle rate is fast relative to the ^{13}C T_1, which would normally reduce the steady-state magnetization even further. CP, or proton–carbon magnetization transfer, occurs when the relative amplitudes of the proton and carbon rf fields are carefully adjusted such that the oscillation rates of the two nuclear species along the *z*-axis are synchronized or matched. Under these conditions, the dipolar coupling provides an efficient energy transfer mechanism, and net transfer of polarization or "coolness" from the proton spins to the carbon spins occurs. This is called the Hartmann–Hahn matching condition [14, 15]. When this condition is satisfied, the ^{13}C and ^{1}H spins have identical *z*-axis oscillation frequencies, the smaller carbon reservoir is "cooled" by the larger proton reservoir, and the result is a growth of the ^{13}C magnetization. CP can enhance the signal intensities by up to a factor of four for a single scan. This effect alone represents a potential 16-fold savings in time for signal-averaged experiments (signal-to-noise [S/N] and $\sqrt{N}$ for white noise). In addition, since the carbon signal intensity arises from the much faster relaxing protons, this experiment can be repeated for signal-averaging purposes at a much faster rate than a non-cross-polarized experiment. Practical increases in repetition rate can mean up to a factor of 100 in time savings.
3. After the contact time, the carbon rf field is discontinued, and the carbon magnetization is observed as a free-induction decay. The proton field remains applied during this period for dipolar decoupling.
4. The delay time between pulses needs to be only long enough to allow proton spin relaxation. Another important point is that the contact time used is often a compromise. This is because different carbons cross-polarize at different rates depending on the proximity and the number of "nearby" protons, as well as the amplitude and frequency of any local motions. This has the important consequence that signal intensity comparisons, particularly between carbons in differing environments, must be treated with caution.

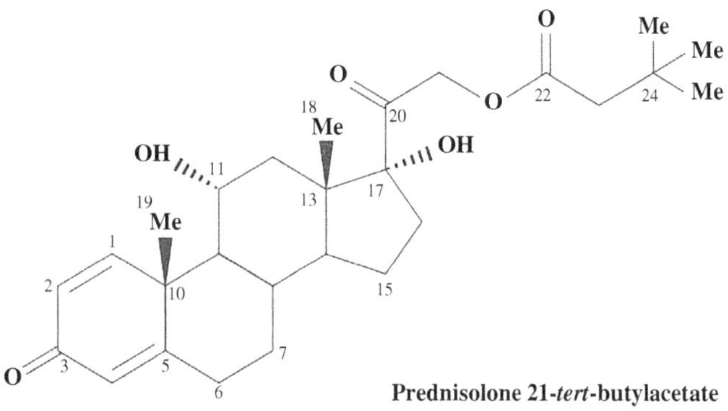

Figure 13.5 Chemical structure of prednisolone 21-*tert*-butylacetate.

13.3 Solid-state NMR Studies of Pharmaceuticals

Byrn et al. have used solid-state NMR to study the different crystal forms of prednisolone *tert*-butylacetate (Figure 13.5) [8, 16]. Table 13.2 shows the unit cell parameters and crystallographic data for these different crystal forms. Figure 13.6 shows the conformation of prednisolone *tert*-butylacetate in the different crystal forms [16].

Analysis of the steroid conformations from single-crystal X-ray data indicates that the steroid rings are generally in

Table 13.2 Cell parameters of the different forms of prednisolone 21-*tert*-butylacetate.

Parameter	Form I	Form II	Form III	Form IV	Form V
Space group	$P2_1$	$P2_1$	$P2_1$	$P2_12_12_1$	$P6_1$
a (Å)	14.901	16.118	11.995	14.664	17.325
b (Å)	7.723	7.974	19.659	9.154	17.331
c (Å)	12.14	12.160	6.292	19.430	15.208
β	89.72	109.82	92.80	90.00	90.00
Z	2	2	2	4	6
ρ_{calc}	1.20	1.18	1.19	1.17	1.16
Solvate	EtOH	IPA	DMF	—	H_2O
Crystal type	B	B	B	C	A

similar conformations but that the *tert*-butyl side chain is *syn* with respect to the steroid ring in Form IV, whereas it is *anti* in the other crystal forms. The crystal packing of the different crystal forms is shown in Figure 13.7. It is clear that the crystal packing is quite different in the different crystal forms, except in Forms I and II, which are similar.

Figure 13.8 shows the solid-state NMR spectrum of the different crystal forms of prednisolone 21-*tert*-butylacetate, and Table 13.3 shows the chemical shifts of the different carbons of prednisolone 21-*tert*-butylacetate [8]. It is clear from these chemical shifts and spectra that different crystal forms have different solid-state NMR spectra. As discussed earlier, this is due to the differences in crystal packing. Although the biggest difference between Forms I and II is the positions for the solvent peaks (in Form I the ethanol solvent shows resonances at 18.7 and 58.2 ppm and in Form II the propanol solvent shows resonances at 26.3 and 64.3 ppm) and these forms have very similar unit cells, the steroid carbons do have slightly different chemical shifts. The solid-state NMR spectra of Forms III, IV, and V are much different from those of Forms I and II, as well as from each other. The biggest difference involves C1 carbon, which shows chemical shifts that cover a range of 8 ppm in the different crystal forms. Several of the other carbons also show fairly large chemical shift differences. Figure 13.9 shows a plot of the difference between the solid-state and solution chemical shift of the C1 carbon versus the hydrogen bond distance to the O3 carbonyl oxygen, which is part of the α, β-unsaturated ketone system that includes the C1 carbon [16]. As can be seen from this plot, there is a more or less gradual shift to larger changes in ppm as the hydrogen bond distance gets shorter. This observation is consistent with other studies of hydrogen bonding in solids. A similar trend is observed in solution studies, where stronger hydrogen bonding produces greater chemical shift changes.

Figure 13.10 shows the studies of the signal intensity from the C18 methyl carbon and the C20 carbonyl carbon as a function of contact time [16]. This study shows that the C18 methyl carbon has a shorter contact time than the C20 carbonyl carbon. This is expected, since there are three protons directly attached to a methyl carbon as opposed to protons that are only near the carbonyl carbon. Thus, a delay time of 500 μs is sufficient for observing the C18 methyl, whereas a 1-ms delay time is needed for the carbonyl carbon.

13.4 Phase Identification in Dosage Forms

Surprisingly little research has been done on the solid-state NMR spectra of pharmaceutical dosage forms. To our knowledge, prior to the study by Saindon et al. [17], solid-state NMR had been used to study only two tablets, both consisting almost entirely of pure drug: aspirin [18, 19] and acetaminophen [20]. Saindon et al. [17] show that solid-state NMR is useful for the study of more complex pharmaceutical products (see Table 13.4). Except for the low-dose enalapril maleate tablet, the spectrum for the drug in each of the tablets listed could be discerned. A search of SciFinder in 2012 revealed only 48 references to the terms "solid-state NMR" and "tablet." In 2015, only 51 references were identified.

For example, Figure 13.11 shows the solid-state NMR spectrum of two ground simvastatin tablets, each of which has labeled amounts of 20 mg [21]. The excipients obscure resonances, which appear at the same chemical shift

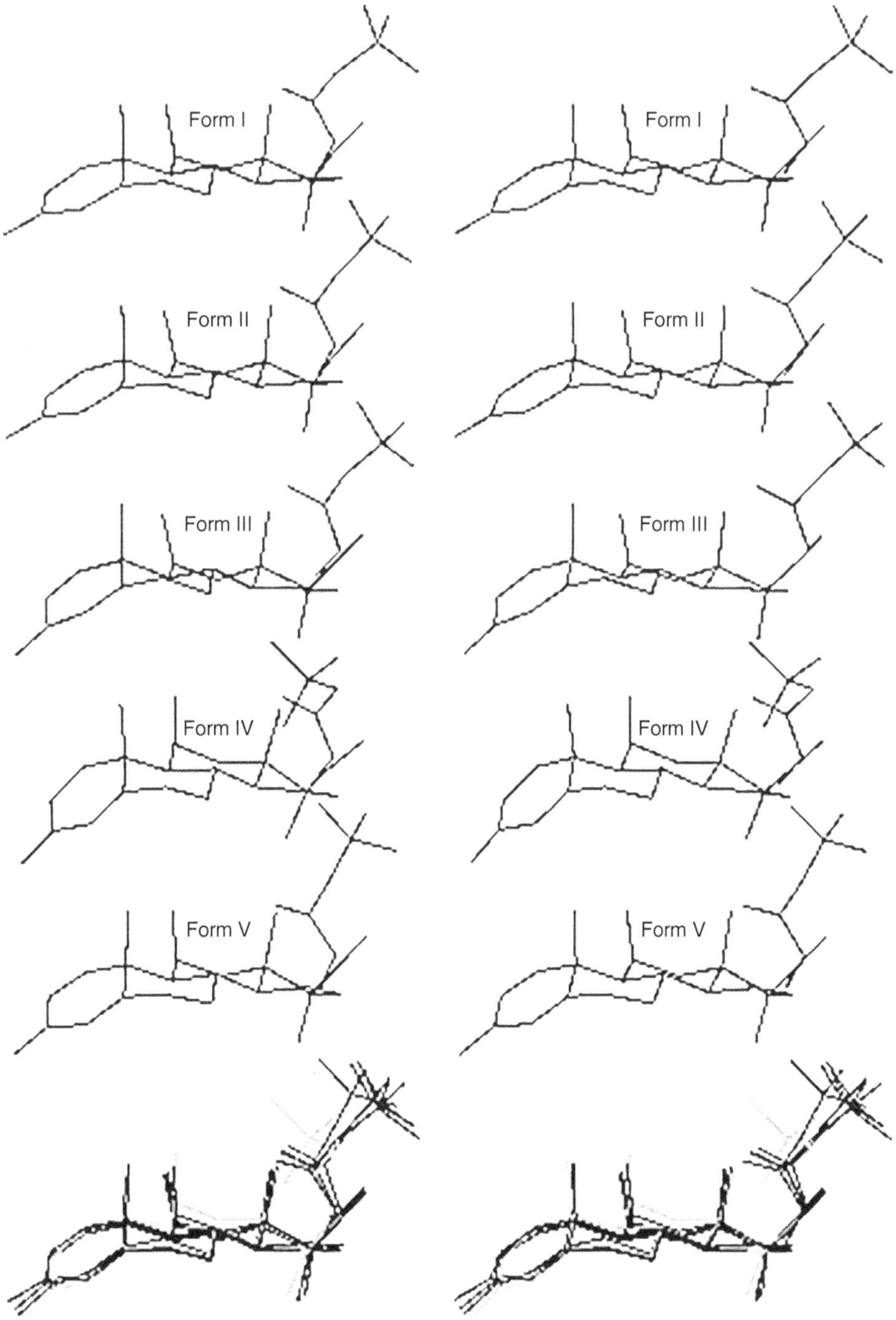

Figure 13.6 Stereoviews of the conformation of prednisolone *tert*-butylacetate Forms I, II, III, IV, and V, respectively. The view is from approximately the same direction perpendicular to the steroid ring nucleus.

(e.g. 35–65 ppm for lactose, 10–25 ppm for microcrystalline cellulose) but have little effect on resonances at other chemical shifts. Similar results were obtained for tablets of another closely related HMG-CoA reductase inhibitor, lovastatin (see Table 13.4).

Enalapril maleate tablets were also studied (Figure 13.12). Spectra of tablets containing different labeled amounts of enalapril maleate were measured under identical conditions. As might be expected, the 20 mg tablets showed significantly

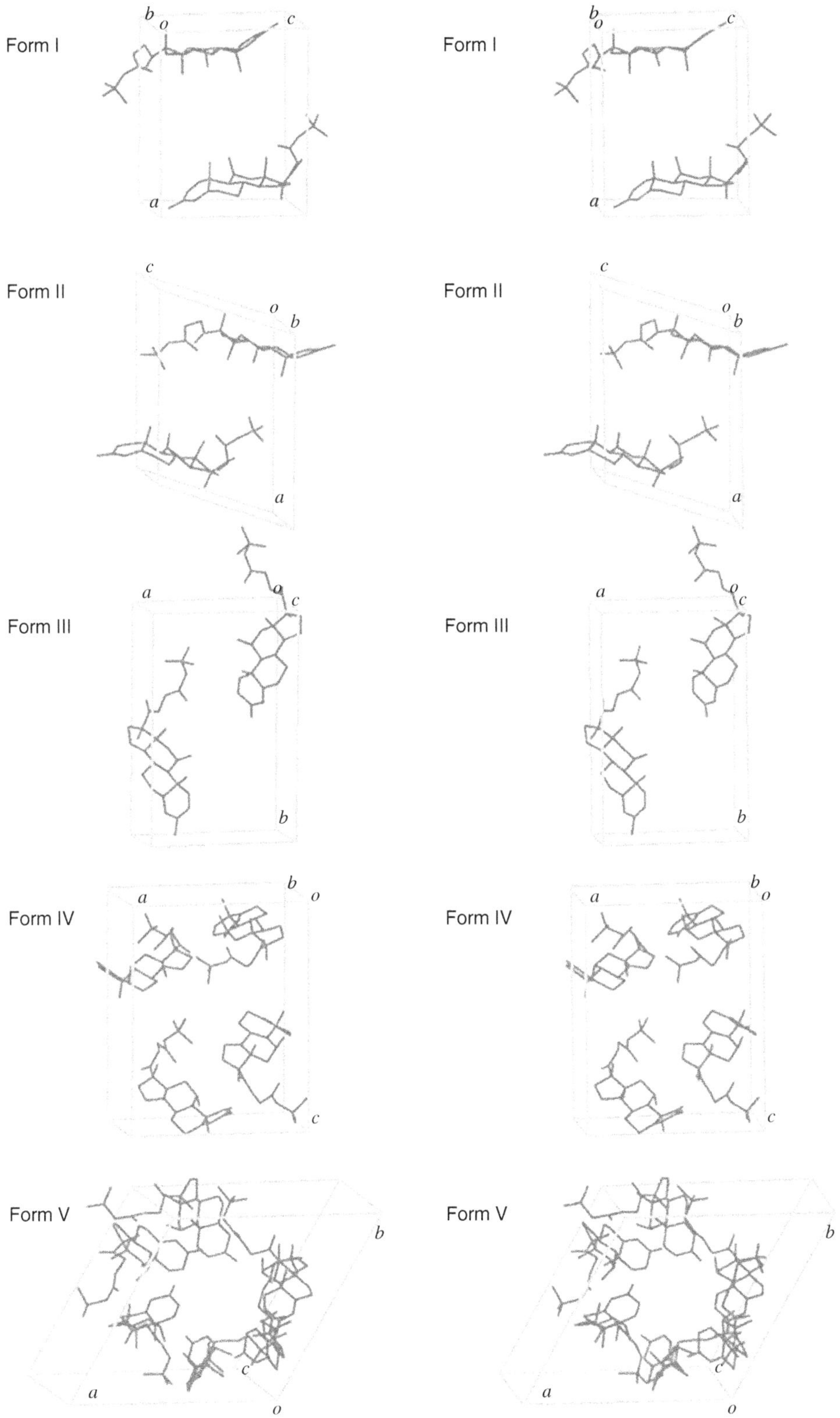

Figure 13.7 Stereoviews of the crystal packing of prednisolone 21-*tert*-butylacetate (solvent omitted for clarity).

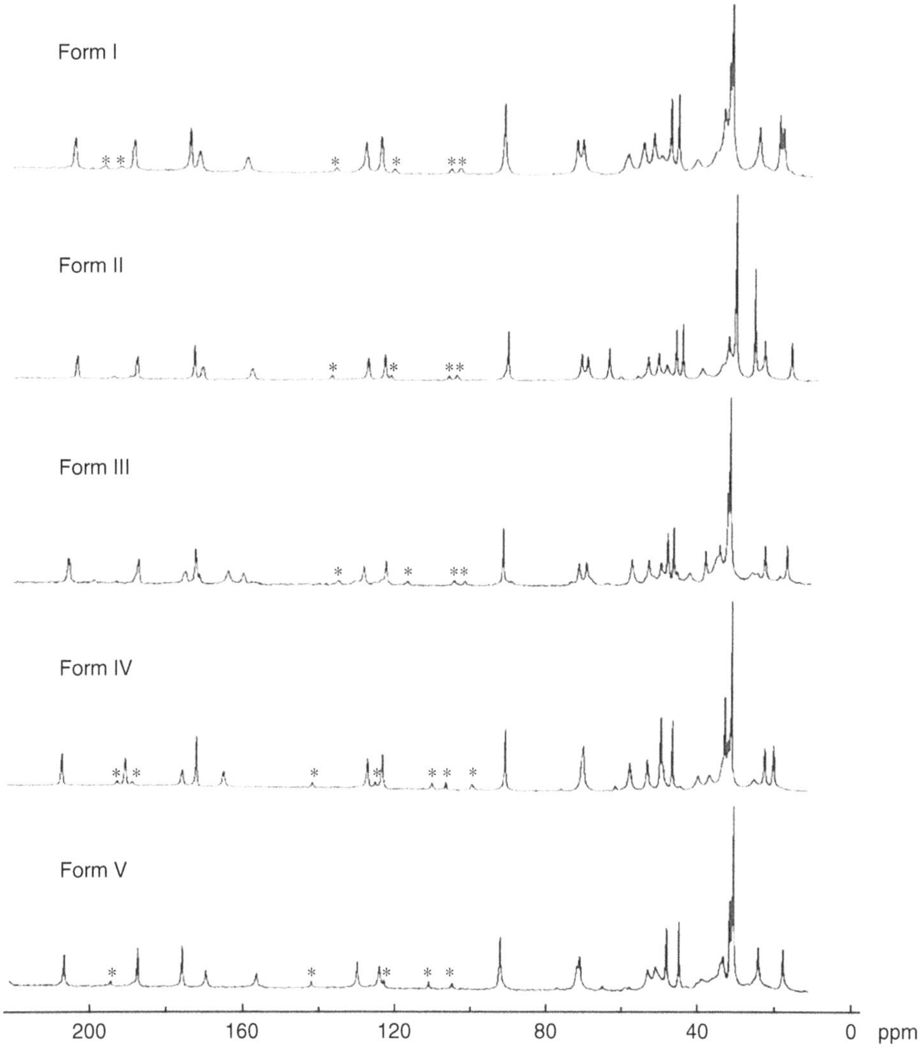

Figure 13.8 Solid-state NMR spectrum of the five crystal forms of prednisolone 21-*tert*-butylacetate (the starred peaks are spinning side bands).

stronger signals than the other tablets (5, 2, and 1.25 mg). The peaks near 177 ppm are carbonyl resonances of the drug and maleate salt. Comparison with the published spectrum [22] and the placebo spectrum shows that the large peaks in the 50–120 ppm range are due to excipients. The small peaks at 14.5, 129.9, and 175.8 ppm are due to the enalapril maleate and are consistent with the spectra shown by Ip et al. [22]. The resonances are not sharp enough to allow a determination of the polymorph present, most likely because of the less crystalline nature of the sample.

Amorphous and less crystalline materials have broad resonances because of the heterogeneity of structural forms found in the sample. This heterogeneity results in a distribution of isotropic shifts for a given carbon. The distribution of shifts is observed as a broadening because of limits on the instrumental resolution imposed by the signal lifetime. Observation of the signals from the active ingredient in a solid dosage form may thus depend not only on the amount of the drug in the formulation but also on its degree of crystallinity.

Figure 13.13 shows the solid-state NMR spectra of prednisolone 21-*tert*-butylacetate: five suspensions that were filtered and dried prior to study (Vendors A, B, C, D, and E) and two bulk drug substances (Vendors Y and Z) [8]. The spectra of the five suspensions include resonances from excipients (60–80 ppm). It is clear that six of the seven spectra contain the same form and that one of the bulk drug substances (vendor Y) contains a different form (see shaded area). It is interesting to note that the spectrum obtained for any one of these products is different from any of the five forms determined by single-crystal X-ray diffraction as seen by their respective solid-state NMR spectra (see Figure 13.8), suggesting that yet another crystal form is involved.

Table 13.3 ^{13}C NMR chemical shifts of prednisolone 21-*tert*-butylacetate in the solid state and in solution for the different crystal forms.

Carbon	Form I	Form II	Form III	Form IV	Form V	Solution
20	203.7	203.7	205.4	205.7	205.6	204.8
3	188.3	188.2	187.0	189.1	186.4	186.6
22	173.8	173.3	172.2	170.7	174.8	172.1
5	171.2	171.1	174.8	174.2	168.6	170.2
1	158.9	158.4	159.6	163.7	155.4	156.4
2	127.5	127.9	127.9	125.7	128.8	127.8
4	123.5	123.5	122.0	121.8	123.0	122.4
17	90.8	91.1	90.9	89.3	91.1	89.8
11[a]	71.8	71.7	70.9	69.0	70.8	70.2
21[a]	70.1	70.0	68.8	68.7	70.1	67.6
9	54.2	54.3	56.9	56.5	52.2	55.4
14	51.5	51.5	52.5	51.8	50.3	51.4
23	49.5	49.4	49.4	47.9		47.8
13	47.2	47.0	47.6	48.5	47.5	47.7
10	45.1	45.2	46.1	45.4	44.1	44.1
12	40.0	39.9	41.6	38.4	38.0	39.6
7[a]	35.0	34.5	37.4	35.5		34.6
16[a]		33.1	34.5		33.0	34.1
6[a]	32.8				32.4	32.0
8[a]						31.3
24	31.6		31.7	31.5	30.7	30.9
25–27	31.1	31.2	31.2	29.9	29.9	29.6
15			25.1	23.9		23.9
19	23.9	23.7	22.0	21.1	23.2	21.1
18	17.7	16.8	16.2	18.9	16.9	17.0
EtOH CH_3	18.7					18.2
EtOH CH_2	58.2					57.7
IPA Me		26.3				25.3
IPA CH		64.3				63.8
DMF Me			31.8			31.1
DMF CH_2			37.6			36.1
DMF C=O			163.8			162.5

a. Assignment of these resonances is tentative since there is extensive overlap with other resonances [8].

13.5 Other Basic Solid-state NMR Experiments Useful for Pharmaceutical Analysis

13.5.1 Interrupted Decoupling for Protonated Carbon Atoms

Application of the interrupted-decoupling technique proved useful in the study of enalapril maleate, both to aid signal assignment and to aid in the identification of resonances of the drug rather than the excipients. A brief delay (typically 40–50 μs) is inserted between the end of the $^{1}H-^{13}C$ contact time and the data acquisition with proton decoupling. During this period, the magnetization of carbons that are directly attached to protons rapidly dephases, with the exception of the methyl groups,

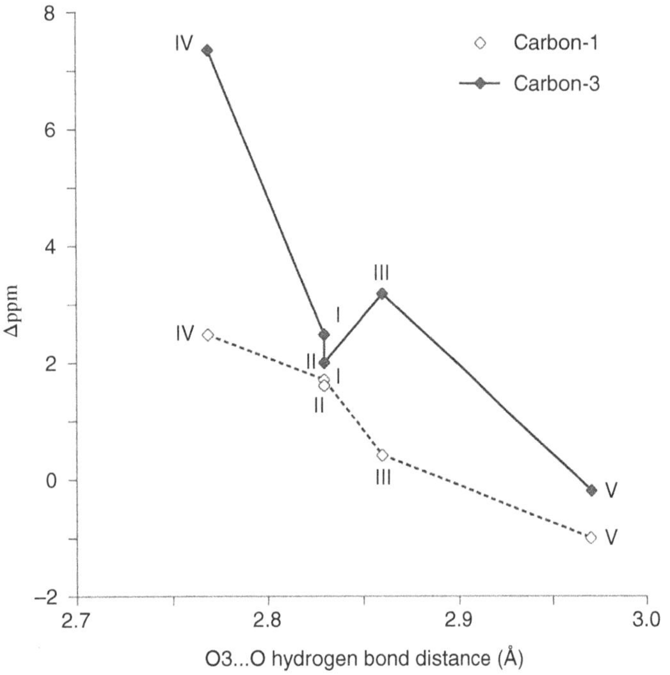

Figure 13.9 Plot of the ^{13}C chemical shift difference between the solid-state and solution spectra (for carbon-1 and carbon-3) versus the C—O···H—O hydrogen bond distance in the five crystal forms of prednisolone 21-*tert*-butylacetate. *Source*: [16]/with permission of American Chemical Society.

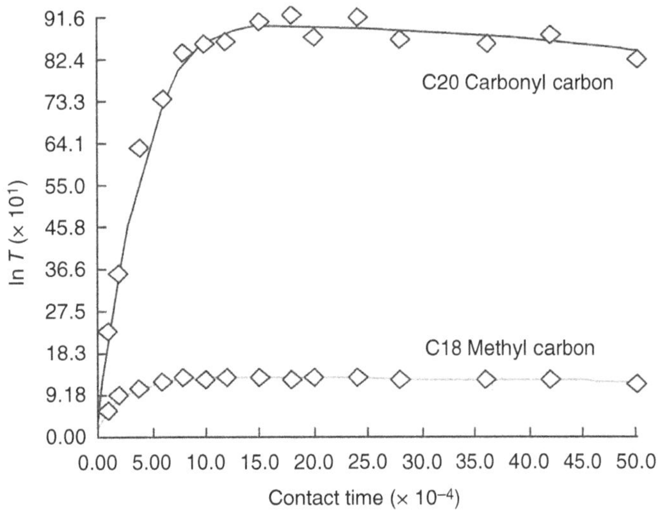

Figure 13.10 Contact time determination for the C18 methyl carbon (lower line) and the C20 carbonyl carbon (upper line) in prednisolone 21-*tert*-butylacetate.

where rapid internal motion partially averages out the ^{1}H—^{13}C dipolar interactions. The resulting spectrum, therefore, shows only significant signals for quaternary and methyl carbons. Since most of the carbon atoms in the common excipients have at least one directly bound proton, this method is also extremely useful in suppressing the excipient background.

The interrupted-decoupling spectrum of the 20 mg enalapril maleate tablets (Figure 13.12) shows two peaks in the methyl region, a peak at 63.1 ppm, and two peaks in the 177–185 ppm range [17]. The interrupted-decoupling spectrum of the placebo tablet (not shown) exhibits only two very small peaks at 15.96 and 63.20 ppm. Therefore, interrupted decoupling provides a convenient and powerful method to suppress the excipient peaks, which frequently have a number of methylene and methine carbons. Hence, the interrupted-decoupling spectra often show only peaks that are due to the methyl and quaternary carbons in the active ingredient, thus facilitating their observation and assignment. This has also been found to be the case in other studies of tablets and capsules.

The CP/MAS spectra of 5 mg enalapril maleate tablets were measured with the same number of scans as the 20 mg tablets. The peaks in this spectrum (not shown) are significantly smaller than those of the 20 mg tablets; this is no doubt because of the smaller amounts of enalapril maleate in these tablets. The interrupted-decoupling spectra of the 5 mg tablets are consistent with this result, in which only the peak at 177.2 ppm is discernible in contrast to the number of peaks observed in the 20 mg tablets (see Figure 13.12). The peak at 16.2 ppm is assumed to be identical to that of the placebo, which appears at 16.0 ppm. The spectra of the 2.5 mg tablets (not shown) are similar to those of the 5 mg tablets. The spectra of the 1.25 mg tablets (not shown) are virtually identical to the spectrum of the placebo; thus, CP/MAS ^{13}C NMR was not sensitive enough to detect the drug in this formulation. In conclusion, it is clear that the intensity of the peaks is consistent with the amount of enalapril maleate present, assuming all other variables are equal.

The solid-state ^{13}C NMR spectra of a number of tablets and capsules of various nonsteroidal anti-inflammatory agents were also obtained (see Table 13.4). As expected from previous results, the spectra of the ibuprofen tablets showed narrow lines with excellent signal-to-noise ratios since the drug itself is crystalline and is present in a high dose in the tablets (see Figure 13.14). In general, most of the resonances from the other nonsteroidal drugs were also clearly distinguishable in the spectra of the dosage forms. The interrupted-decoupling pulse sequence was again found to simplify the appearance of the spectra by suppressing excipient peaks, thereby aiding in signal assignment. It should be noted that this technique is usually capable of determining the principal crystal form directly from a formulated tablet.

Table 13.4 Tablets and capsules of drugs investigated by solid-state ^{13}C NMR.

	Dose	Tablet/capsule weight	Manufacturer	No. of scans[a]	
Drug	**(mg)**	**(mg)**	**(trade name)**	**CP/MAS**	**ID[b]**
Prednisolone tablets	5	100	Vendors X, Y, and Z	20 000	40 000
Enalapril maleate tablets	20	200	Merck Sharp & Dohme (Vasotec)	16 000	33 576
	5	230	Merck Sharp & Dohme (Vasotec)	16 000	33 576
	2.5	—	Merck Sharp & Dohme (Vasotec)	16 000	33 576
	1.25	230	Merck Sharp & Dohme (Vasotec)	16 000	33 576
Lovastatin tablets	20	—	Merck Sharp & Dohme (Mevacor)	—	
Simvastatin tablets	40	400	Merck Sharp & Dohme (Zocor)	4000	4000
Ibuprofen tablets	200	320	Bristol–Meyers Squibb (Nuprin)	2000	4048
	200	330	Upjohn (Haltran)	7264	2692
	400	620	Geneva Generics	6000	9084
Sulindac tablets	200	330	Merck (Clinoril)	3000	8104
Flurbiprofen tablets	100	420	Upjohn (Ansaid)	4000	10 000
Diflunisal tablets	500	840	Merck Sharp & Dohme (Dolobid)	10 000	20 492
Indomethacin tablets	50	350	Merck Sharp & Dohme (Indocin)	12 000	20 000
	75	280	Merck Sharp & Dohme (Indocin-SR)	12 000	28 248
	50	480	United Research Laboratories	10 000	30 000
	50	380	Geneva Generics	10 000	27 772
Mefenamic acid capsules	250	350	Parke–Davis (Ponstel)	7000	6000
Piroxicam capsules	20	300	Pfizer (Feldene)	14 000	19 556

a. The contact time used in each case was 2 ms. The delay times were 3 s for the prednisolone tablets and 2 s for the other tablets and capsules.
b. Interrupted decoupled [17].

This can be an invaluable piece of evidence in deducing whether or not a particular formulation process is altering the desired crystal form.

13.5.2 Block-Decay Experiments for Screening Submolecular Mobility

Another technique, which is sometimes illuminating in solid-state NMR studies of pharmaceuticals, is the Bloch-decay experiment. This experiment is analogous to the most common liquid proton-decoupled ^{13}C survey spectrum, except that decoupling is used only during the acquisition time and at very high power (as above). The experiment is not routinely used because the recycle time depends on the carbon T_1 values (since there are no cross-polarization components), so the recycle time is often 5 to 30 times longer than for a cross-polarized-based experiment. The value of this experiment lies in its discrimination against molecular components that are relatively static on the time scale of 5 s. The more mobile parts of the molecule generate a fluctuating dipolar field, which can provide an important source of relaxation to the atoms involved. Although the recycle time is very slow relative to the proton T_1 values, the recycle time is usually empirically chosen to be on the order of the T_1 values of the methyl carbons, which are much shorter than those of the immobile carbons. Thus, the Bloch-decay experiment discriminates on the basis of carbon T_1 values. This has been found particularly useful in the study of *tert*-butyl compounds and studies of alcohol solvates. The entrained solvent molecules in some of these solids have sufficient mobility that they provide liquid-like spectra.

Harris and coworkers used solid-state NMR to quantify the amount of bambuterol hydrochloride in a formulated product. They used magnesium stearate as the reference material and ratioed the intensity of the bambuterol hydrochloride peaks to the methylene carbon atoms of the stearate [23].

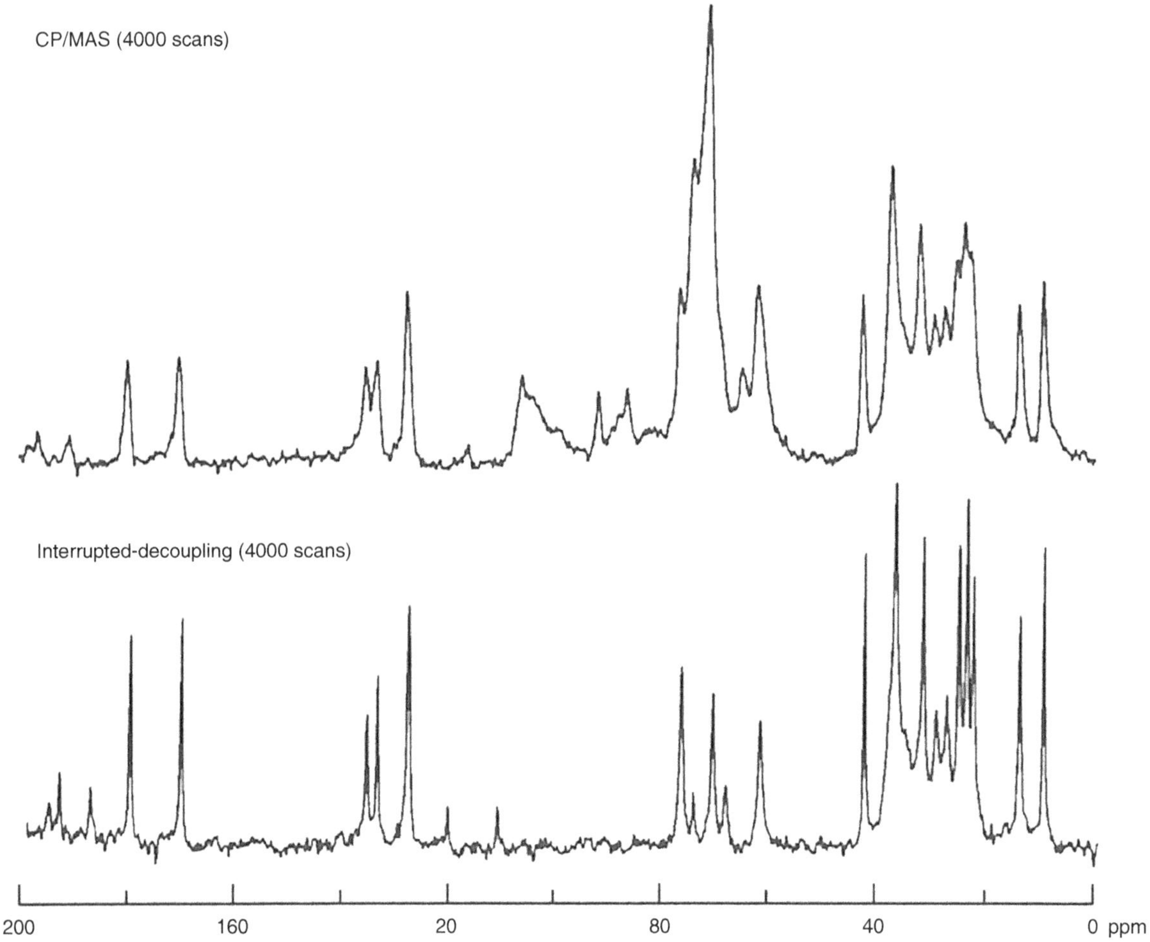

Figure 13.11 Solid-state CP/MAS NMR spectrum of two ground 40 mg simvastatin tablets.

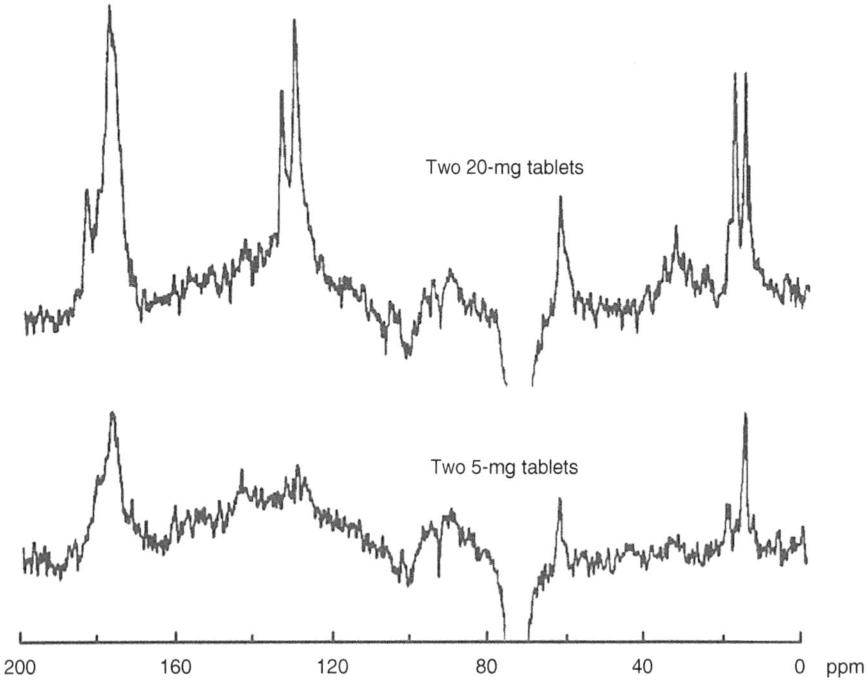

Figure 13.12 Solid-state CP/MAS interrupted-decoupling spectrum of enalapril maleate tablets. (The vertical scale was expanded to show the resonances of the drug more clearly.)

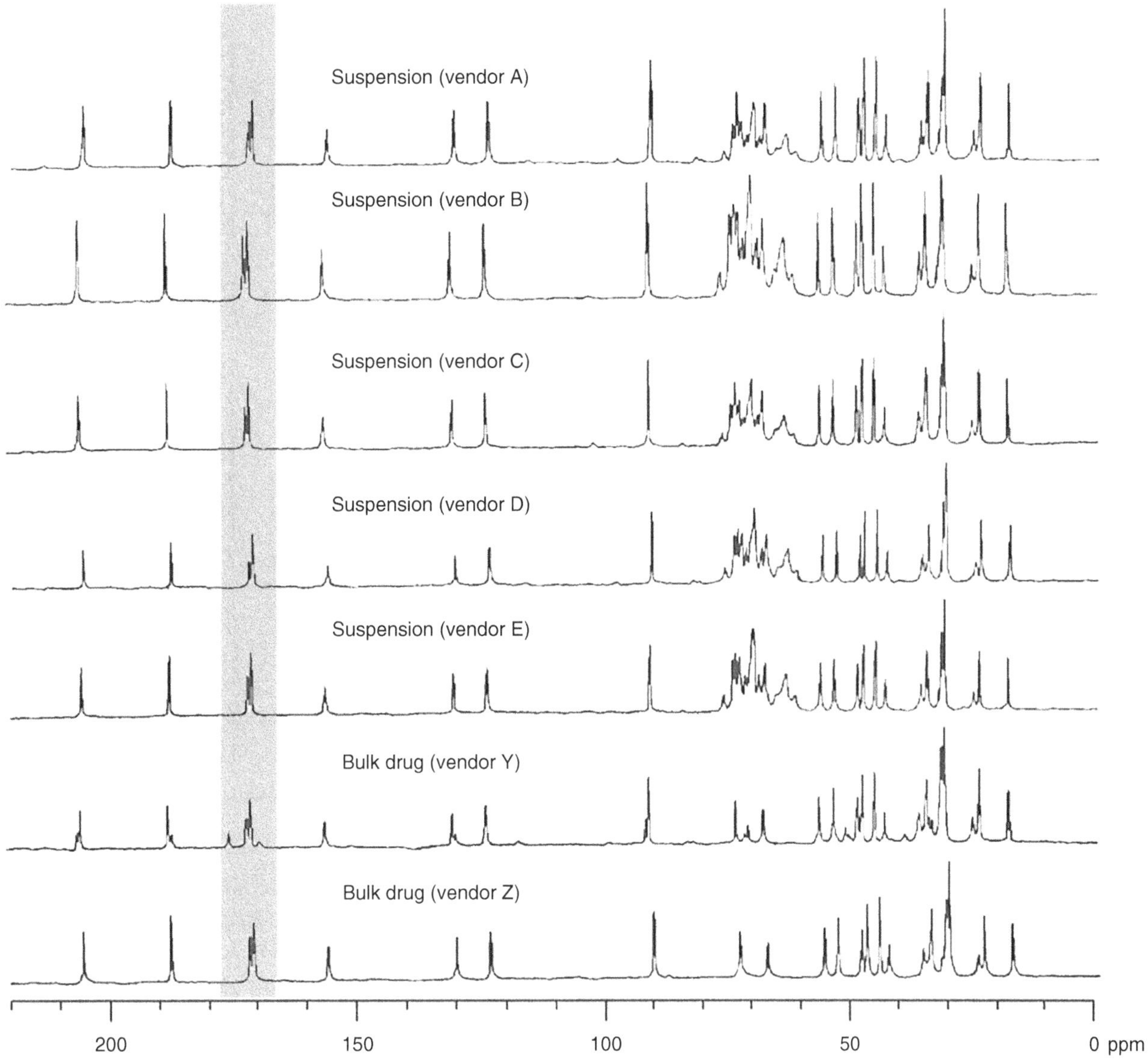

Figure 13.13 Solid-state NMR spectra of prednisolone 21-*tert*-butylacetate filtered suspensions (vendors A–E) and bulk drug substances (vendors Y–Z).

13.6 Determination of the Domain Structure of Amorphous Dispersions Using Solid-state NMR

In 2015, Song et al. studied the nature of amorphous dispersions of lapatinib with several polymers, including hypromellose phthalate (HPMCP), hypromellose acetate succinate (HPMCAS), and copovidone (PVPVA) [24]. These dispersions were prepared by spray drying, and dissolution studies showed that the dissolution rate of lapatinib was greatly accelerated in the HPMCP and HPMCAS dispersions. Preliminary bioequivalence studies of crystalline lapatinib and the lapatinib HPMCP dispersion showed that the HPMCP dispersion had about 10× the bioavailability of the crystalline parent. Solid-state NMR spectra were acquired at Genentech by Dr. Joseph Lubach using a Bruker Advance III HD spectrometer. Figure 13.15 shows the spectra of various dispersions.

It is clear from this figure that, as the drug loading increases, the proportion of the lapatinib N-15 N species with a chemical shift similar to the lapatinib phthalate increases. This is consistent with an increase in the amount of protonated lapatinib as the amount of acidic polymer HPMCP is increased.

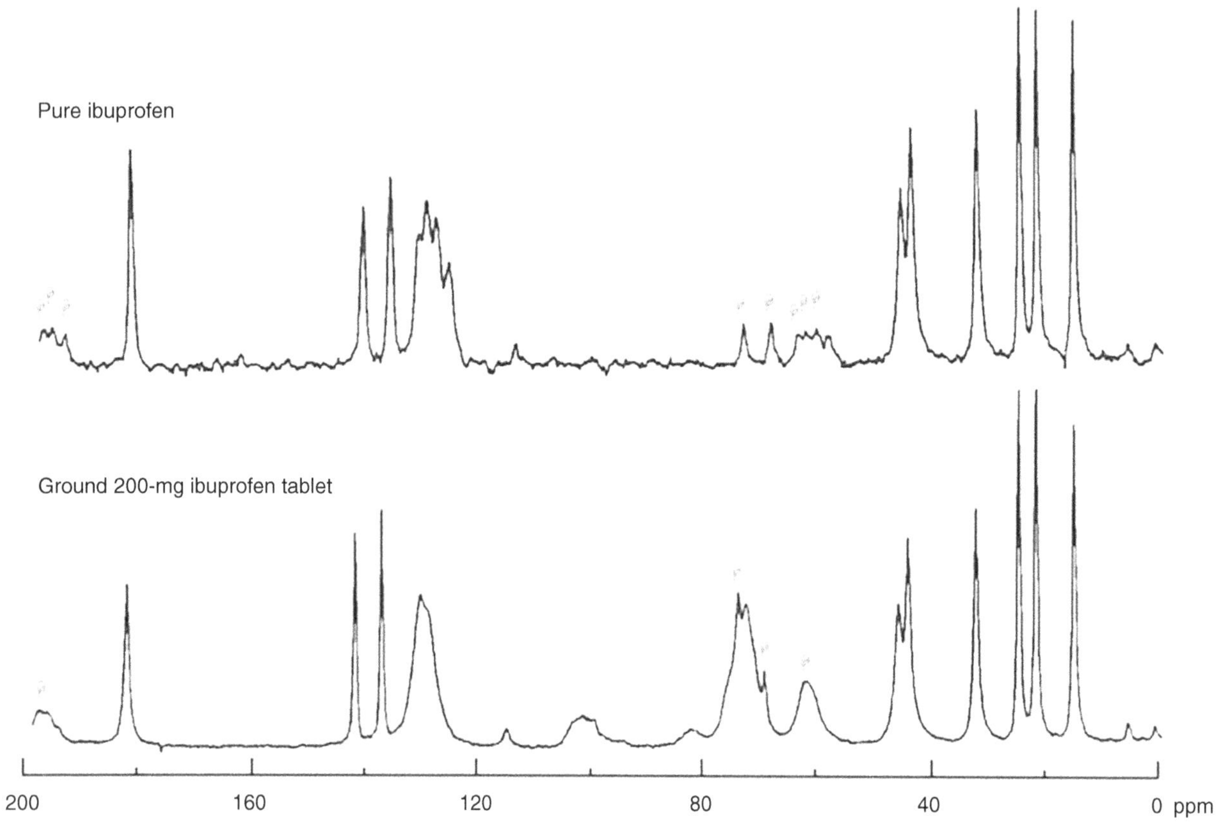

Figure 13.14 Solid-state CP/MAS spectrum of ibuprofen and of 200 mg ibuprofen tablets. (Marked peaks are spinning sidebands.)

In addition, the ^{1}H T_1 and ^{1}H $T_{1'}$ relaxation times of both lapatinib and HPMCP were determined in the amorphous dispersion (ASD). This study will help determine whether there are one or two domains in the ASD. If there are two domains, one would expect to see relaxation times near those of pure amorphous lapatinib and pure amorphous HPMCP. Amorphous lapatinib has ^{1}H T_1 and $T_{1'}$ values of 2.88 s and 11.15 ms, whereas HPMCP has ^{1}H T_1 and $T_{1'}$ values of 2.17 s and 9.54 ms. Figure 13.16 shows the results of these. Similar data have been obtained by Munson in his laboratory study of ASDs [25].

13.7 Solid-state NMR of Amorphous Materials

Solid-state NMR has been used to analyze amorphous drugs and amorphous formulations. Song and coworkers investigated the interaction of lapatinib with Soluplus, polyvinylpyrrolidone vinyl acetate (PVPVA), hydroxypropyl methylcellulose acetate succinate (HPMCAS), and hydroxypropyl methylcellulose phthalate (HPMCP) [24]. The results indicated a strong interaction between the excipients and ^{15}N. Solid-state NMR gave direct evidence for the protonation of lapatinib by the acidic polymer. Relaxation studies were consistent with this conclusion. In addition, HPMCAS and HPMCP, the acidic polymers, inhibited the crystallization of lapatinib. The nonacidic polymers did not stabilize the dispersion against crystallization.

A method for accurate resonance assignments has been developed to provide accurate chemical shift values [25, 26]. This method allows the identification of overlapped resonances and differentiation between the crystalline and amorphous forms.

Solid-state NMR differentiated between two amorphous forms of valsartan [27]. The existence of these two forms is supported by FT-IR and DSC. The one form consists of a mixture of cis-trans conformers, whereas the other form consists of one conformer.

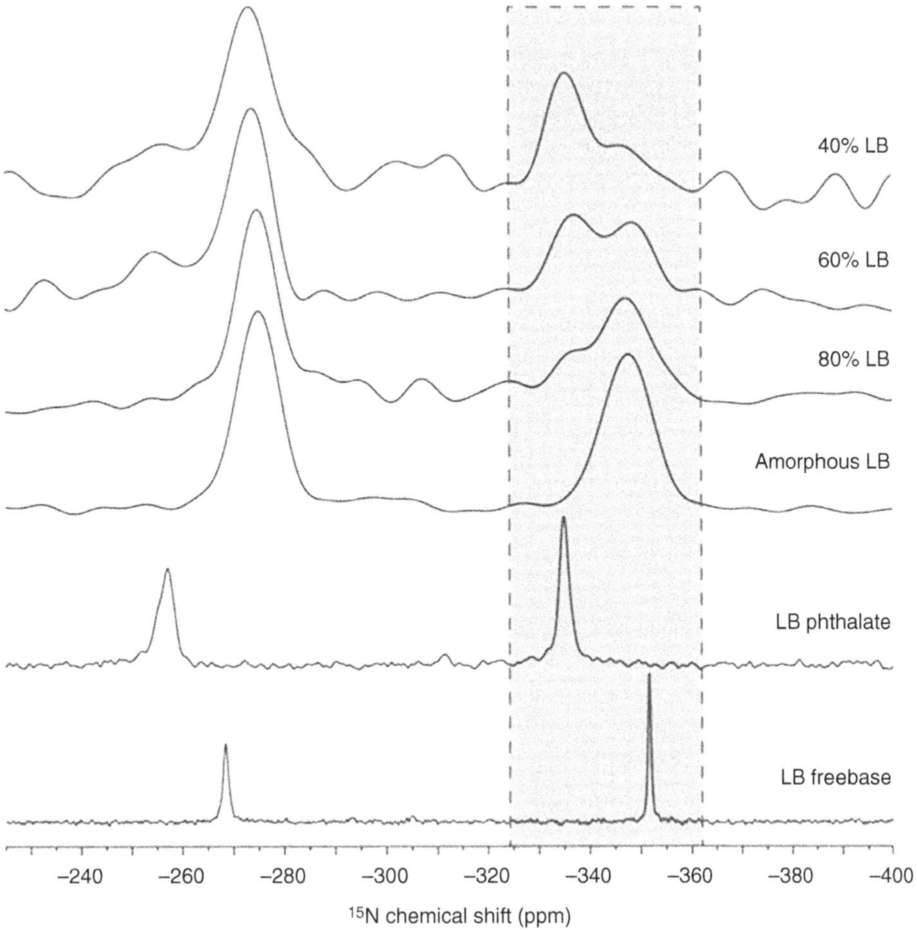

Figure 13.15 ^{15}N CP/MAS NMR spectra of crystalline LB freebase, crystalline LB phthalate, amorphous LB freebase, and LB-HPMCP solid dispersions with drug loadings of 40%, 60%, and 80%, in the spectral range of −225 to −400 ppm. *Source*: [24]/with permission of Springer Nature.

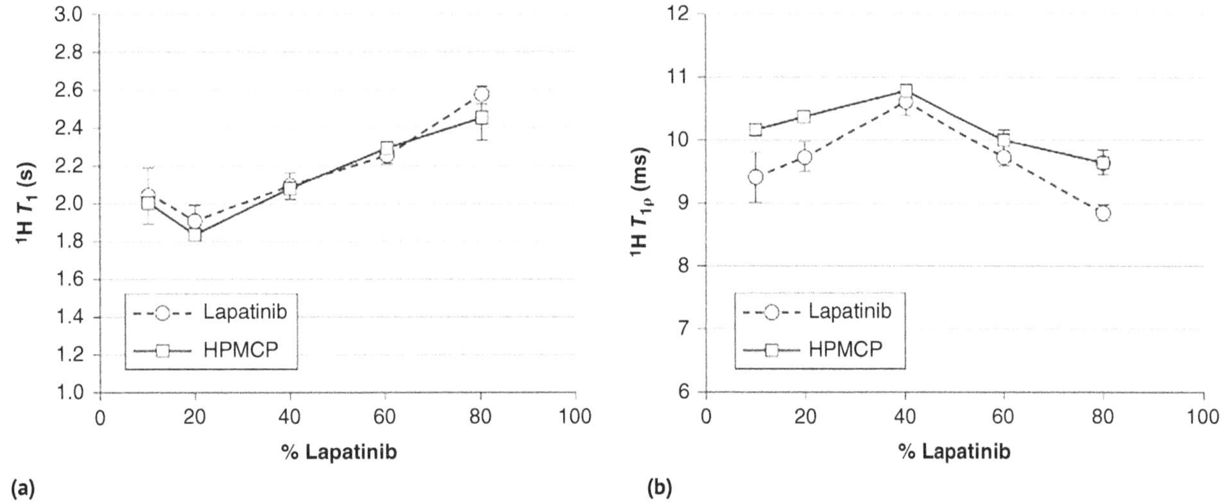

Figure 13.16 Plots of (a) ^1H T_1 and (b) ^1H $T_{1'}$ as a function of drug loading in LB-HPMCP solid dispersions. Error bars represent the error associated with curve fitting the raw T_1 and $T_{1'}$ data. *Source*: [24]/with permission of Springer Nature.

An X-ray pair distribution function (PDF) and ssNMR analysis of amorphous posaconazole showed significant ssNMR ^{19}F—^{13}C in both the PDF and ssNMR data [28]. This type of work promises to assist in the prediction and understanding of amorphous drugs toward crystallization. In a related study, drug–polymer interaction in posaconazole amorphous dispersions was investigated using F-19 NMR [29]. This method identified head-to-head and head-to-tail interactions of posaconazole in the amorphous polymer.

13.8 Summary

This chapter gives a brief overview of solid-state ^{13}C NMR methods as they apply to drug analyses. Examples show that solid-state ^{13}C NMR spectroscopy can be used to determine the crystal form present in low-dose tablets and to identify the active component in a tablet. Although we have not determined the X-ray powder diffraction patterns of the low-dose tablets studied here, we believe that solid-state NMR offers a useful alternative for the direct determination of the crystal form present in the final dosage form.

References

1 Tishmack PE, Bugay DE, Byrn SR. Solid-state nuclear magnetic resonance spectroscopy-pharmaceutical applications. *J Pham Sci* 2003; 92: 441–474.
2 Vogt FG. Evolution of solid state NMR in pharmaceutical analysis. *Future Med Chem* 2010; 2: 915–922.
3 Li M, Xu W, Su Y. Solid-state NMR spectroscopy in pharmaceutical sciences. *Trends Anal Chem* 2021; 135: 116152.
4 Marchetti A, Yin J, Su Y, Kong X. Solid-state NMR in the field of drug delivery: state of the art and new perspectives. *Magn Reson Lett* 2021; 1(1): 28–70.
5 Atalla RH, Gast JC, Sindorf DW, Bartuska VJ, Maciel GE. Carbon-13 NMR spectra of cellulose polymorphs. *J Am Chem Soc* 1980; 102(9): 3249–3251.
6 Hill HD, Zens AP, Jacobus J. Solid-state NMR spectroscopy. Distinction of diastereomers and determination of optical purity. *J Am Chem Soc* 1979; 101(23): 7090–7091.
7 Saindon PJ. Solid pharmaceuticals: analysis and identification by solid-state carbon-13 NMR, in Medicinal Chemistry. Purdue University, West Lafayette, IN; 1992.
8 Byrn SR, Tobias B, Kessler D, Frye J, Sutton P, Saindon P, Kozlowski J. Relationship between solid state NMR spectra and crystal structures of polymorphs and solvates of drugs. *Trans Am Crystallogr Assoc* 1989; 24: 41–54.
9 Schaefer J, Stejskal EO. Carbon-13 nuclear magnetic resonance of polymers spinning at the magic angle. *J Am Chem Soc* 1976; 98(4): 1031–1032.
10 Bugay DE. Solid-state nuclear magnetic resonance spectroscopy: theory and pharmaceutical applications. *Pharm Res* 1993; 10: 317–327.
11 Jelinski LW, Melchior MT. High-resolution NMR of solids. In: *NMR Spectroscopy Techniques*. CRC Press; 1996. p. 431–436.
12 Yannoni CS. High-resolution NMR in solids: the CPMAS experiment. *Acc Chem Res* 1982; 15(7): 201–208.
13 Komoroski RAE. *High Resolution NMR Spectroscopy of Synthetic Polymers in Bulk*. Methods in Stereo-Chemical Analysis Series, Vol 7; Marchard AP, series editor. Deerfield Beach, FL: VCH; 1986.
14 Hartmann SR, Hahn EL. Nuclear double resonance in the rotating frame. *Phys Rev* 1962; 128(5): 2042.
15 Fyfe CA. *Solid State NMR for Chemists*. Guelph, Ontario: CFC; 1983.
16 Byrn SR, Sutton PA, Tobias B, Frye J, Main P. Crystal structure, solid-state NMR spectra, and oxygen reactivity of five crystal forms of prednisolone tert-butylacetate. *J Am Chem Soc* 1988; 110(5): 1609–1614.
17 Saindon PJ, Cauchon NS, Sutton PA, Chang CJ, Peck GE, Byrn SR. Solid-state nuclear magnetic resonance (NMR) spectra of pharmaceutical dosage forms. *Pharm Res* 1993; 10: 197–203.
18 Chang CJ, Díaz LE, Morin F, Grant DM. Solid-state carbon-13 NMR study of drugs: aspirin. *Mag Res Chem* 1986; 24: 768–771.
19 Díaz LE, Frydman L, Olivieri AC, Frydman B. Solid state NMR of drugs: soluble aspirin. *Anal Lett* 1987; 20(10): 1657–1666.
20 Jagannathan NR. High-resolution solid-state carbon-13 nuclear magnetic resonance study of acetaminophen: a common analgesic drug. *Curr Sci* 1987; 56(16): 827–830.

21 Cauchon NS. Solid-state carbon-13 NMR studies of pharmaceutical and chemical modifications of nucleic acids. In: *Medicinal Chemistry*. West Lafayette, IN: Purdue University; 1992.

22 Ip DP, Brenner GS, Stevenson JM, Lindenbaum S, Douglas AW, Klein SD, McCauley JA. High resolution spectroscopic evidence and solution calorimetry studies on the polymorphs of enalapril maleate. *Int J Pharm* 1986; 28 (2–3): 183–191.

23 Harris RK, Hodgkinson P, Larsson T, Muruganantham A. Quantification of bambuterol hydrochloride in a formulated product using solid-state NMR. *J Pharm Biomed Anal* 2005; 38(5): 858–864.

24 Song Y, Yang X, Chen X, Nie H, Byrn S, Lubach JW. Investigation of drug–excipient interactions in lapatinib amorphous solid dispersions using solid-state NMR spectroscopy. *Mol Pharm* 2015; 12(3): 857–866.

25 Yuan X, Sperger D, Munson EJ. Investigating miscibility and molecular mobility of nifedipine-PVP amorphous solid dispersions using solid-state NMR spectroscopy. *Mol Pharm* 2014; 11(1): 329–337.

26 Lu X, Xu W, Hanada M, Jermain SV, Williams III RO, Su Y. Solid-state NMR analysis of crystalline and amorphous Indomethacin: An experimental protocol for full resonance assignments. *J Pharm Biomed Anal* 2019; 165: 47–55.

27 Skotnicki M, Apperley DC, Aguilar JA, Milanowski B, Pyda M, Hodgkinson P. Characterization of two distinct amorphous forms of valsartan by solid-state NMR. *Mol Pharm* 2016; 13(1): 211–222.

28 Chen Z, Nie H, Benmore CJ, Smith PA, Du Y, Byrn S, Templeton AC, Su Y. Probing molecular packing of amorphous pharmaceutical solids using X-ray atomic pair distribution function and solid-state NMR. *Mol Pharm* 2023; 20(11): 5763–5777.

29 Lu X, Li M, Huang C, Lowinger MB, Xu W, Yu L, Byrn SR, Templeton AC, Su Y. Atomic-level drug substance and polymer interaction in posaconazole amorphous solid dispersion from solid-state NMR. *Mol Pharm* 2020; 17(7): 2585–2598.

14

Particle and Powder Analysis

14.1 Introduction

Particle and powder analysis are a critical part of pharmaceutical materials characterization since active pharmaceutical ingredients (APIs) and excipients mainly exist as solids. A powder consists of numerous fine particles, which are loosely associated but not aggregated or agglomerated with each other. Particles can be components of granules, pellets, and beads that are usually fabricated into solid dosage forms such as tablets, capsules, and sachets. Suspensions and nanosuspensions are heterogeneous systems with solid particles suspended in a liquid medium. Lyophilized cakes and dry powder inhalers, likewise, are made up of solid particles of API and excipient formulated together. Particle size and particle size distribution critically affect the quality, performance, processability, stability, and appearance of a product. For example, large differences between the particle sizes of API and excipients may cause segregation and lead to poor content uniformity. For poorly soluble compounds, particle size has a direct impact on dissolution rates and, hence, potentially on bioavailability. Consequently, average particle size and particle size distribution are usually part of any specification required to control product quality.

While the measurement of particle size and size distribution indicates macroscopic particle dimensions, the measurement of specific surface area (area per unit mass) provides a quantitative estimate of the area of the particle exposed to the surrounding environment for a particular particle size. The surface area of particles in a powder can be roughly estimated from the particle size, making a variety of assumptions about the shape of the particles, but this often does not take into consideration the contributions to the surface area by surface roughness and porosity. This is important because the true surface area is an important determinant in a variety of processes involving particles, including dissolution rates, water vapor sorption, chemical reactivity, and a variety of bulk properties. Important bulk properties of powders and granules include density, compressibility, flowability, and electrostatics. These bulk properties, in turn, affect the appearance, processability, and final product quality of solid dosage forms.

In this chapter, various tools used to measure particle size are introduced for particle sizes ranging from nanometers to millimeters. Surface area measurements, based on gas adsorption studies and the use of the Brunauer, Emmett, and Teller (BET) method, will also be discussed. In addition, we will describe equipment that is used to measure size, surface area, and the bulk properties of powders and granules, such as density, compressibility, and flowability.

14.2 Particles in Pharmaceutical Systems

Examples of different types of particles found in pharmaceutical API and drug products are described in Figure 14.1 and can be categorized in order of their dimensions, as discussed in the following text.

14.2.1 Micelles

Micelles are associations of molecules dissolved in aqueous solutions that arise when lipids and other surface-active molecules arrange themselves spontaneously in response to their amphipathic nature. Micelles are generally spherical and have radii that are roughly equal to the length of the hydrocarbon chain of the amphiphile. A sodium lauryl sulfate micelle, for example, has an approximate radius of 1.7 nm, whereas that of Tween 80 is about 10 nm.

Solid-State Materials in Pharmaceutical Chemistry: Properties, Characterization, and Applications, Second Edition. Stephen R. Byrn, George Zografi and Xiaoming (Sean) Chen.

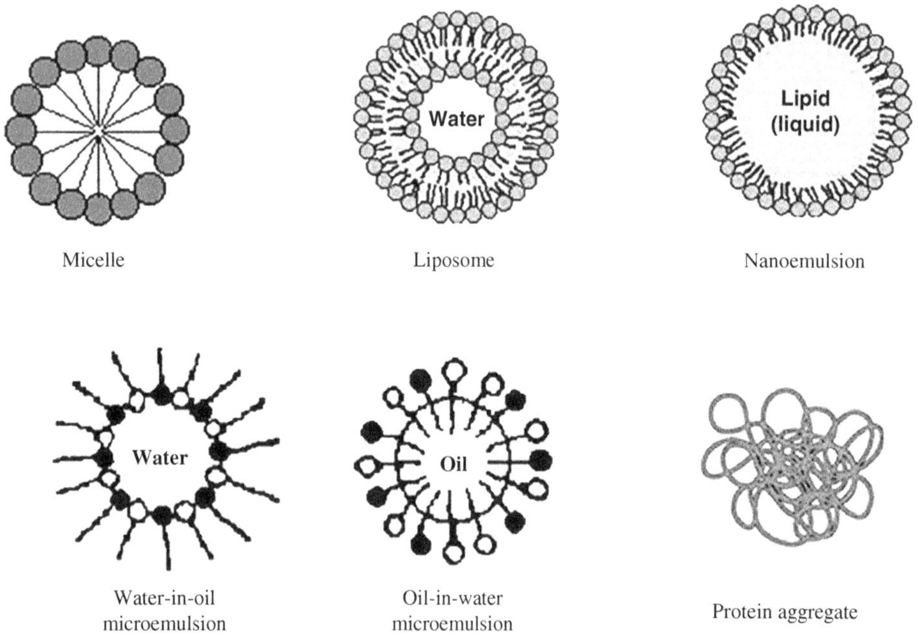

Figure 14.1 Diagrams of pharmaceutical-related particles.

14.2.2 Protein Aggregates

Protein aggregates are assemblies of protein molecules that occur in solution when the protein is denatured, and, therefore, are usually detrimental to the quality of protein products. Protein aggregates can form at various steps in the manufacturing process and during storage, distribution, and handling. Stress from agitation, temperature, pH, or ionic strength often gives rise to protein aggregation. High protein concentration further increases the likelihood of aggregate formation. Protein aggregates come with a broad size range, from nanometer oligomers to micrometer-sized insoluble aggregates.

14.2.3 Liposomes

Liposomes are sphere-shaped vesicles consisting of one or more phospholipid bilayers, typically with a particle size from 0.025 μm (25 nm) to 2.5 μm. Liposomes are extensively studied as possible delivery systems for numerous API molecules in pharmaceutical applications, with the advantages of increasing efficacy, reducing toxicity, and enhancing stability. Liposomes are usually biocompatible, preferably biodegradable, nontoxic, and nonimmunogenic. The particle size of liposomes, which are of interest for delivery, is below the submicron range at around 100–200 nm.

14.2.4 Microemulsions

Microemulsions are clear, thermodynamically stable, isotropic lipid mixtures of oil, water, and surfactants, larger than a micelle. There are three types of microemulsions, including oil-in-water, water-in-oil, and bicontinuous microemulsions. Microemulsions have the advantage of being able to self-emulsify and can solubilize both hydrophilic and lipophilic drugs. Since their particle sizes are usually less than 0.22 μm, microemulsions can be sterilized by filtration. They are widely used as a delivery system in cosmetic and pharmaceutical products.

14.2.5 Nanoemulsions

Nanoemulsions consist of fine oil-in-water dispersions, having particles covering the size range of 100–600 nm. Nanoemulsions are like microemulsions but are thermodynamically unstable. Nanoemulsions use less surfactant than microemulsions and are only kinetically stable. Nanoemulsions generally have larger particle sizes than microemulsions, even though the names appear to suggest the opposite.

Nanoemulsions generally appear translucent and have better stability against flocculation, sedimentation, or creaming than ordinary emulsions. Although nanoemulsions and liposomes have similar particle sizes, nanoemulsions have a lipophilic core, whereas liposomes possess a hydrophilic inner structure.

Owing to their very small particle size, nanoemulsions have attractive properties suitable for transdermal delivery aimed at cosmetic or pharmaceutical applications. The droplet size and size distribution of nanoemulsions depend on the process of emulsification.

14.2.6 Nanosuspensions

Nanosuspensions are submicron colloidal dispersions of drug particles, commonly in the form of crystalline materials made by nanomilling, homogenization, or crystallization. Solid particle sizes in nanosuspensions are usually less than 1 μm and have an average size ranging between 200 and 600 nm. Nanosuspensions can enhance the dissolution of poorly soluble crystalline compounds and, therefore, may enhance their bioavailability.

14.2.7 Nanoparticles

Nanoparticles are generally categorized as solid particles that have submicrometer particle sizes. Generally, this group of particles is used to prepare nanosuspensions or formulations with nanosized drug particles formed by spray drying or melt extrusion.

14.2.8 Aerosols

Aerosols consist of colloidal particles dispersed in a gas and include, as major everyday examples, smoke or fog. Pharmaceutical aerosols are intended for delivery directly into the respiratory system, using nebulizers, metered-dose inhalers, or dry powder inhalers, which deliver a specific quantity of drug. The particle size, size distribution, and shape of aerosols are extremely important for controlling the deposition profile of API in the lung, and it is generally recognized that the optimum aerodynamic particle size distribution for most inhalation aerosols is in the range of 1–5 μm.

14.2.9 Emulsions

Emulsions are a mixture of two or more liquids in which one is present as relatively large droplets and dispersed in the other liquid. They are usually formed by agitating or homogenizing the two liquids. The use of surfactants and polymers can stabilize emulsions but only kinetically, since emulsions are not thermodynamically stable, having stability issues such as flocculation, creaming, coalescence, and demulsification. Emulsions appear milky and have particle sizes from 0.1 μm to hundreds of micrometers. Many topical products for cosmetic and pharmaceutical applications are made of emulsions.

14.2.10 Suspensions

Suspensions are dispersions of finely divided, insoluble solid particles dispersed in liquids, with aqueous media usually being used for pharmaceutical systems that are suitable for oral, parental, and topical administration. API particles in a suspension usually have size ranges of from several micrometers to hundreds of micrometers.

14.2.11 Powders

Powders contain particles having sizes from several micrometers to several hundred micrometers and represent the solid form of a majority of API and excipients used pharmaceutically. The particle size and size distribution of an API or excipient affect bulk properties, such as bulk density, tap density, flow, and compressibility. In addition, the particle size and particle size distribution of an API determine its dissolution rate and consequently the bioavailability of its drug products. Figure 14.2 shows the effect of crystal size on the oral bioavailability of nitrofurantoin, as measured by the rate of urinary recovery. Based on the area under the curve, the fine particles resulted in three times more absorption than the coarse particles [1].

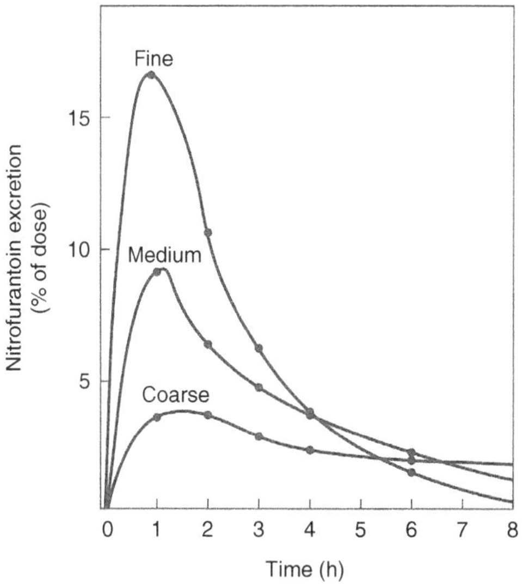

Figure 14.2 Particle size of nitrofurantoin crystals and renal excretion per dose. *Source*: [1]/with permission of Elsevier.

Table 14.1 Techniques to analyze particle size in a pharmaceutical system.

Technique	Particle size	Application
Scanning electron microscopy	1 nm–5 μm	Micelles, microemulsions, nanoemulsions, nanosuspensions, nanoparticles, liposomes
Dynamic light scattering	10 nm–5 μm	Micelles, microemulsions, nanoemulsions, nanosuspensions, nanoparticles, liposomes
Optical microscopy	1 μm–10 mm	Powder, suspensions, emulsions, protein aggregates, granules, pellets
Light diffraction	100 nm–500 μm	Powder, suspensions, emulsions, protein aggregates
Sieve analysis	50 μm–5 mm	Granules, pellets
Dynamic image analysis	30 μm–30 mm	Powder, suspensions, emulsions, protein aggregates, granules, pellets

14.2.12 Granules

Granules are large particles or grains, made by a granulation process. Granulation, which enlarges particle size by agglomeration, is widely used to improve various physical properties of pharmaceutical formulations, including uniformity, wettability, flowability, bulk density, and compressibility. Granules typically have a particle size window from 100 μm to 4.0 mm, depending on their intended use. Granules with a proper particle size range provide advantages for controlling weight uniformity in tablet compression, filling performance of capsules, drug dissolution, and content uniformity of the final dosage.

14.2.13 Pellets

Pellets have similar sizes to granules but are processed to provide particles that are more sphere-like with smooth surfaces. Pellets are most often used as part of multiple particulate systems for controlled-release or delayed-release formulations. They are usually produced by extrusion/spheronization, powder layering, liquid layering, or melt granulation and usually have particle sizes from 0.05 to 2.00 mm.

As will be seen from this chapter, techniques used for any particle size characterization are chosen to fit the size range of the material being investigated, since each technique can cover a certain range of particle size, as shown in Table 14.1.

14.3 Particle Size and Shape

The way a particle size is expressed in a quantitative term is significantly determined by the particle shape. If the particle is approximately spherical, the particle size is simply defined by its radius. Micelles, microemulsions, nanoemulsions, liposomes, and emulsions are considered to have shapes close to a sphere. Conversely, particles in powders, granules, suspensions, and protein aggregates generally have a nonspherical shape. Many techniques do not take particle shape into account, and treatment of the measured data assumes the particles are spherical. In laser diffraction, for example, particle sizes are reported as the radius of a sphere that has a volume equivalent to the tested particles. SEM, microscopy, and dynamic image analysis capture images of particles in two dimensions and can report particle size in several different ways. One simple way is to report the diameter of an equivalent circle with the same projection area as the captured image of the particle. Another way is to use Feret's diameter, which estimates the distance between two parallel tangents of the projected particle area at an arbitrary angle. Usually, a minimum Feret's diameter, maximum Feret's diameter, and diameter at 90° to the direction of the maximum Feret's diameter are reported for particle size comparisons (Figure 14.3).

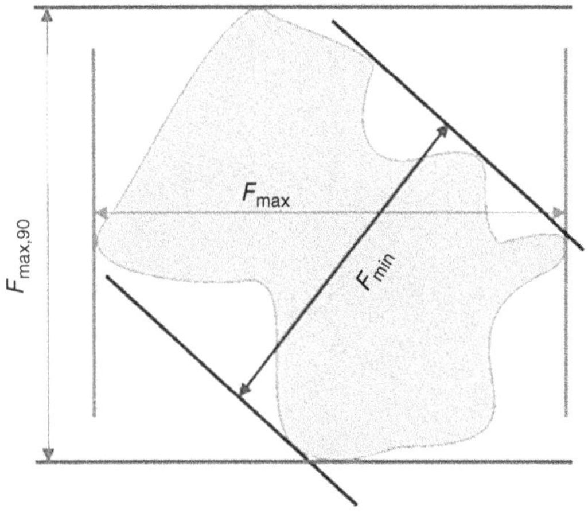

Figure 14.3 Diagram of Feret's diameter of an irregular particle. F_{min}, minimal Feret's diameter; F_{max}, maximal Feret's diameter; $F_{max,90}$, diameter at 90° to F_{max}.

For nonspherical particles, all sizing techniques give different apparent particle sizes, so the specific technique used needs to be specified when reporting particle size data.

14.4 Particle Size Distribution

Since most powders of pharmaceutical interest are heterogeneous with respect to particle size, methods of measurement are generally designed to obtain the statistical distribution of sizes, as well as an average size. Particle size data are commonly reported in plots of size versus frequency as a cumulative distribution curve.

The distribution of each size is a weighted number, volume, or intensity of response of particles. For example, image analysis, such as microscopy, acts as a counting technique and gives a number-weighted distribution. Each particle, no matter how large or small, is treated equally. Laser diffraction provides a volume-weighted distribution since the contribution of each particle to light scattering is related to its volume. Dynamic light scattering delivers an intensity-weighted distribution since each particle contributes to the distribution through the intensity of light scattered by the particle. Based on the Rayleigh approximation, the intensity of light scattering is proportional to $(\text{size})^6$ or $(\text{volume})^2$. Sieve analysis, where the weight of samples retained on each predetermined size sieve, results in a weight-based distribution of sizes. Figure 14.4 describes the different ways to report particle size distributions for the same set of particles with equal quantities of standard monodispersed 5 and 50 μm particles in number, volume, and intensity, respectively.

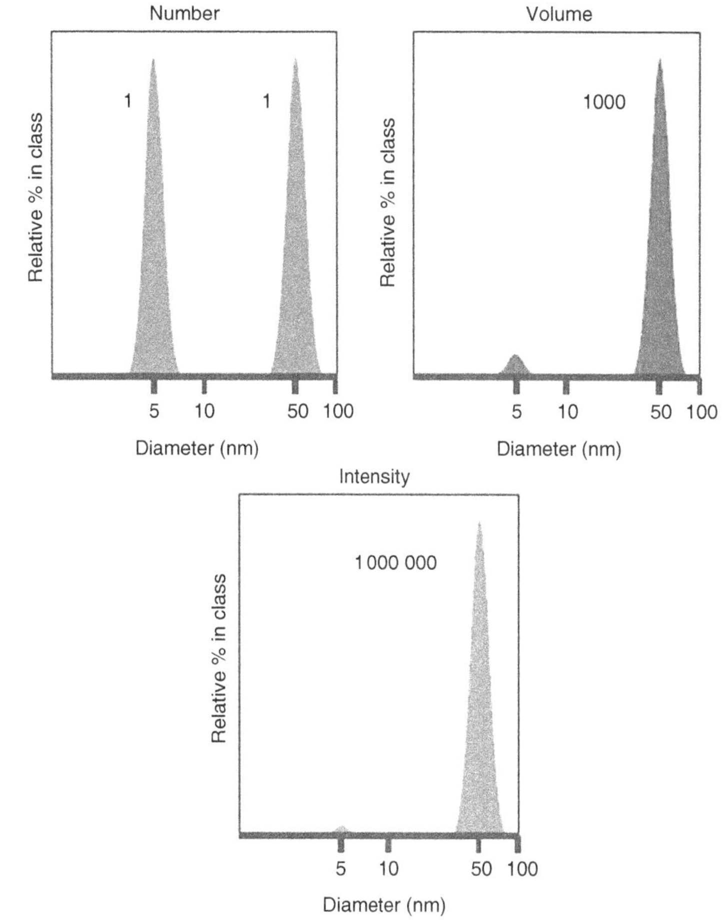

Figure 14.4 Comparison of reported values of particle size as distributions in number, volume, and intensity, respectively.

Various statistical parameters can be calculated from the distribution data and are used to define the particle size distribution. These parameters include the mean, median, mode, d10, d50, and d90, where the mean is the average particle size of a population; the median is the size below or above which 50% of the population exists, and the mode is the particle size with the highest frequency. For a symmetrical "normal" distribution, the mean, median, and mode are equivalent, whereas they are not equivalent for asymmetrical distributions. The term d50 defines the median, where d10 and d90 refer to the sizes where 10% and 90% of the population exist, respectively.

It is important, therefore, to recognize that the exact value of a mean particle size is fully dependent on the measuring technique used and the weighted distribution.

The number length mean, $D[1,0]$, refers to the arithmetic mean obtained in particle-counting applications using imaging techniques where individual particles are measured. In this procedure, the length of each particle is recorded, and the mean is calculated based on the total number of particles counted.

The surface area moment mean, $D[3,2]$, also called the Sauter mean diameter, is defined as the diameter of an equivalent sphere that has the same volume/surface area ratio as the particles of interest based on the following equation:

$$D[3,2] = \frac{\sum dv}{\sum ds} = \frac{\sum d^3 dN}{\sum d^2 dN} \tag{14.1}$$

The surface area moment mean and mean particle size calculated from measurements of specific surface area are correlated with each other, although they are not equal. The surface area moment mean is sensitive to the presence of fine particulates in the size distribution and, therefore, important for analyzing particle size for physical and chemical processes related to the available free surface area, such as chemical reactivity, dissolution, and bioavailability.

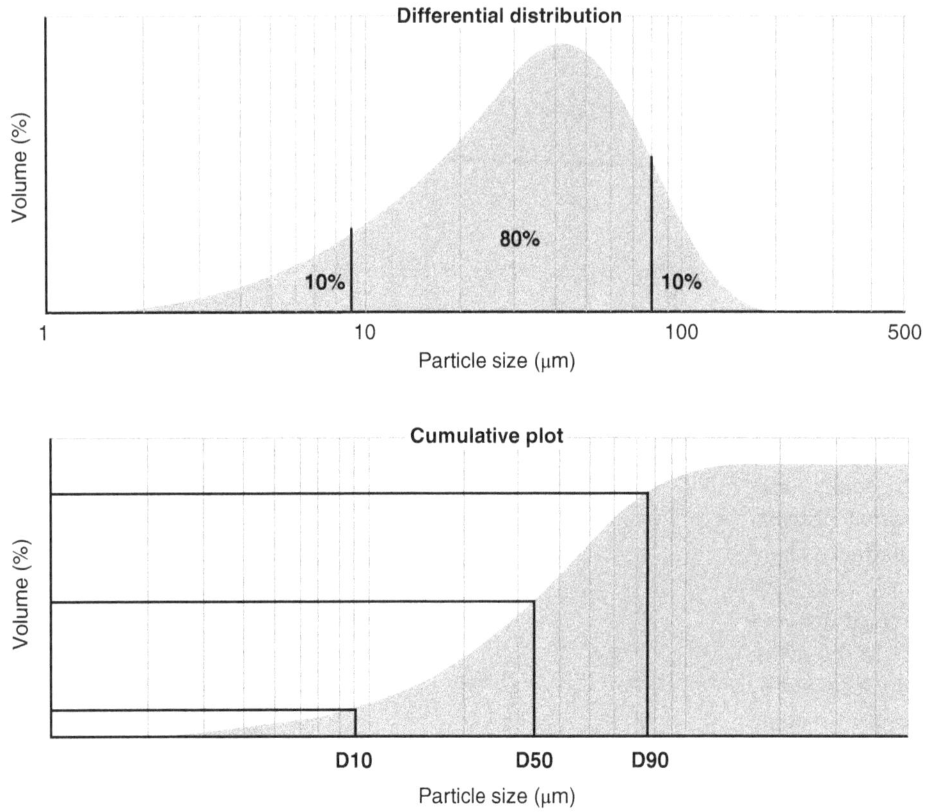

Figure 14.5 Reported particle size distribution as differential distribution or cumulative plot.

The volume moment mean, $D[4,3]$, also known as the de Brouckere mean diameter, can be derived from laser diffraction data and is the weighted average volume diameter, assuming spherical particles of the same volume as the actual particles.

$$D[4,3] = \frac{\sum dm}{\sum ds} = \frac{\sum d^4 dN}{\sum d^3 dN} \tag{14.2}$$

The volume moment mean is more sensitive to larger particles in the distribution, whereas, as mentioned earlier, the surface area mean is dependent on fine particles.

Particle size distributions are generally presented in graphical formats as a differential distribution or as a cumulative plot with increasing particle diameter (Figure 14.5). The differential distribution presents the percentage for each size of a particle range measured. The differential distribution gives a clear picture of the spread of the distribution. The cumulative plot, which reports the cumulative percentage versus particle size, has the advantage of identifying fractional distributions such as d10, d50, and d90.

14.5 Dynamic Light Scattering

Dynamic light scattering is widely used to determine the size of submicrometer particles, colloids, and proteins in a liquid medium. It is also called quasielastic light scattering or photon correlation spectroscopy. Dynamic light scattering determines particle size by measuring the Brownian motion of particles, where bombardment of the particles by the surrounding solvent molecules gives rise to random movements of particles, called Brownian motion.

Smaller particles tend to have higher velocities due to Brownian motion than do larger particles, where the velocity of the Brownian motion is defined by the translational diffusion coefficient.

The Stokes–Einstein equation relates particle size to particle diffusion as follows:

$$d_{\mathrm{h}} = \frac{K_{\mathrm{B}} T}{3\pi\eta D_{\mathrm{t}}} \tag{14.3}$$

where d_h is the hydrodynamic diameter, D_t is the translational diffusion coefficient, K_B is the Boltzmann's constant, T is the temperature, and η is the viscosity. From this equation, we can note that the diffusion coefficient, reflecting particle motion, is inversely related to particle size. The particle size, derived by dynamic light scattering, therefore, is the hydrodynamic size, the size of an equivalent sphere that has the same diffusion coefficient as the measured particle.

The diffusion coefficient, D_t, is extracted from light scattering data based on the following equation:

$$D_t = \frac{\Gamma}{q^2} \tag{14.4}$$

where $\boldsymbol{\Gamma}$ is the decay rate and q is the magnitude of the scattering vector. The magnitude, q, is defined by:

$$q = \frac{4\pi n_0}{\lambda_0} \sin\left(\frac{\theta}{2}\right) \tag{14.5}$$

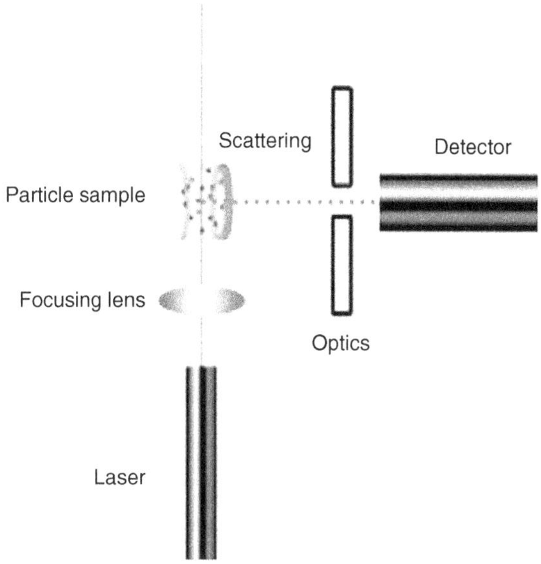

Figure 14.6 Diagram of a dynamic laser diffraction equipment.

where n_0 is the refractive index of the medium, λ_0 is the wavelength of the incident laser light, and θ is the scattering angle.

The decay rate has a linear relationship with the diffusion coefficient and is defined by the time-dependent fluctuation in scattered light caused by the Brownian motion.

When monochromatic laser light is directed to particles, the light scatters from the moving particles (Figure 14.6). The motion of the particles causes randomness in the phase of the scattered light. When the scattered light from two or more particles interacts, a changing destructive or constructive interference occurs, which gives rise to time-dependent fluctuations in the intensity of the scattered light. The time scale for these fluctuations is within a very short time scale, for example, at 1–10 milliseconds. The rate of these fluctuations is determined by how fast the particles move, and it will be slower for larger particles and faster for smaller particles.

The decay rate is derived from the time-dependent fluctuations as in the following equation for a monodisperse sample:

$$g(\tau) = B + \beta e^{-2\Gamma\tau} \tag{14.6}$$

where τ is the decay time, $g(\tau)$ is the correlation function, B is the baseline, and β is the initial amplitude. The decay rate is usually calculated by least squares regression.

Various algorithms are used to calculate the particle size from the correlation function. One simple approach is to fit a single exponential to the correlation function to obtain the mean particle size, which is called the *z*-average diameter. The other common approach is to apply a multiple exponential to the correlation function to obtain the particle size distribution.

The size distribution is plotted as the relative intensity of scattered light versus particle size to give an intensity size distribution. The use of the Rayleigh approximation reveals that the intensity is linearly proportional to d^6, where d is the particle diameter, which means that the intensity data are significantly biased to larger particles. For example, a 50 nm particle will scatter one million times as much light as a 5 nm particle. Consequently, dynamic light scattering can be quite inaccurate in reporting samples with a broad spectrum of particle sizes. This disadvantage can be partially mitigated by converting the intensity-based plot to a volume-based one according to the Mie theory [2, 3].

For any measurement, the mean and standard deviation of the diffusion coefficient D_t are used to calculate the mean hydrodynamic radius and polydispersity index, where the polydispersity index defines the broadness of a size distribution and hence the heterogeneity of a sample. The polydispersity index, PDI, is related to the standard deviation, σ, and the mean radius by the following equation:

$$\text{PDI} = \frac{\sigma^2}{{d_h}^2} \tag{14.7}$$

The sample is generally considered monodisperse when the PDI value is less than 0.04. Polydispersity in % is related to PDI according to equation (14.8).

$$\text{Polydispersity}\ (\%) = 100 \times \sqrt{\text{PDI}} = 100 \times \frac{\sigma}{d_\text{h}} \tag{14.8}$$

To measure colloidal solutions or suspensions, the sample is usually placed into the instrument at a desired concentration, free of large-sized particulate contaminants. The solvent used to dissolve or suspend the sample should also be free of any dust or particulate contamination. Since the intensity of light scattered by particles is proportional to d^6, the presence of dust particles may produce significantly misleading results, especially for sample particles less than 100 nm. For this reason, it is considered a good practice to filter the dispersion media through 0.1 or 0.2 μm filters. If the presence of large contaminating particles is suspected or detected in the prepared sample, an appropriate filter membrane can be used to remove the contaminant but retain the principal sample components. In some cases, a centrifuge can be used to remove large-sized contaminants when measuring low-density particles or colloidal solutions such as those of proteins.

Many samples are measured in aqueous media. However, since pure water is not a desirable dispersion solvent, a small amount of electrolyte, such as 10 mM sodium chloride, is generally used to create a stable electric double layer and mitigate coulombic interaction between charged systems.

A proper concentration of the sample should be selected for consistent results, since a low concentration may lead to a high noise-to-signal ratio and increase the effects of extraneous particles. On the other hand, too high a concentration can cause multiple scattering and produce an inaccurate hydrodynamic size. Hackley and Clogston suggested that the typical sample concentration should be at 1 mg/mL as a starting point for measuring nanoparticles in aqueous media [4]. The choice of concentration should be further evaluated to account for the scattering properties of the sample. To establish a procedure for routine analysis, a validation can be carried out with multiple concentrations to identify a concentration range over which the measured size is minimally concentration dependent.

As a rule of thumb, samples having a clear or slightly hazy appearance should generally be suitable for such analysis.

A noninvasive backscattering technology can be used to mitigate multiple scattering for concentrated suspensions or colloid solutions. In such cases, it is possible to carry out sample analysis without dilution. The scattering angle for such methods is generally chosen to be 173° [5].

If a compound is somewhat soluble in a chosen medium, dissolution of the compound during sample preparation can lead to changes in particle size during the measurement. In such cases, a saturated solution of the compound can be used to maintain the particle stability as demonstrated in a study involving nanosuspension characterization [6].

According to equation (14.3), data obtained to estimate the hydrodynamic diameter are very dependent on temperature, so it is very important that the temperature is controlled and measured accurately. It is also recommended that samples should be equilibrated at the target temperature prior to starting measurements. A minimum of three measurements, 60 seconds or longer in duration and up to 10 measurements per sample, is recommended by ISO/DIS 22412 [7].

In summary, dynamic light scattering has several advantages, including its nondestructive nature, no need for calibration, small sample size, no extensive sample preparation required, and short analysis time. It is widely used for measuring particle size for proteins, polymers, micelles, nanoparticles, colloidal dispersions, emulsions, nanoemulsions, and microemulsions. For example, Yu et al. utilized dynamic light scattering as a process analytical tool to monitor protein folding and aggregation in vaccine manufacturing [8]. Kotta et al. characterized nanoemulsions of efavirenz with improved bioavailability by dynamic light scattering [9]. The optimized nanoemulsion formulations had a mean particle size of less than 30 nm.

14.6 Zeta Potential

In addition to the need to measure the particle size of powders and other systems of pharmaceutical interest, for electrically charged particles, some measure of particle surface charge is essential, particularly for colloidal dispersions, nanosuspensions, suspensions, and emulsions.

Zeta potential measurements can provide an indirect measure of particle surface charge and, therefore, are an essential tool for testing the particle properties of such systems. This is so because maintaining sufficient charge is essential for enhancing suspension and emulsion stability.

The zeta potential is a measure of the surface potential that arises due to the formation of an electrical double layer at the solid–liquid or liquid–liquid interface in any colloidal system, suspension, or emulsion (Figure 14.7). When particles are dispersed in a liquid, most of them carry surface charge for several reasons, including ionization of surface groups, differential ion adsorption from solution, and differential ion dissolution that produce a net charge on the surface. The fixed surface charge tends to attract oppositely charged ions from solution to form an electrical double layer at the surface. The electrical double layer, therefore, has two parts: an inner part, the Stern layer, with more strongly adsorbed oppositely charged ions and an external part, called the diffuse layer, with the opposite charged ions arranged near the surface out in solution under the influence of attractive electrical forces that oppose to some extent the diffusion of these ions away from the surface. These ions are in equilibrium with the solution outside of the double layer.

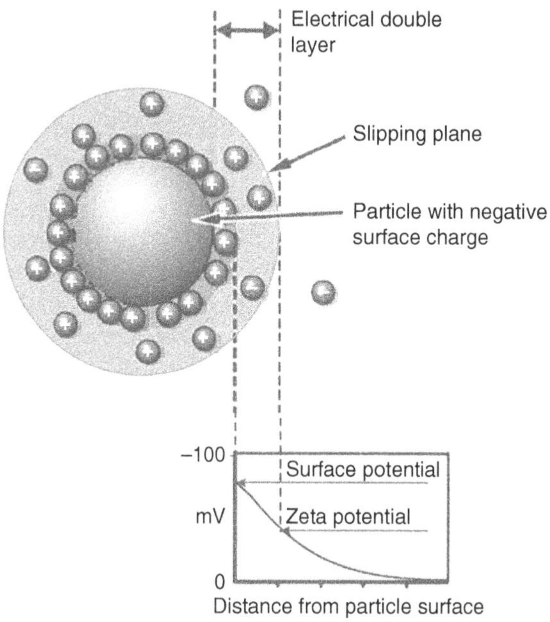

Figure 14.7 Diagram of an electrical double layer and its zeta potential.

The zeta potential represents the electrical potential at the boundary between the electrical double layer and the solution, called the slipping plane, because it is at this location that the charged particle and free oppositely charged ions separate at a certain rate under an applied electric field, as in electrophoresis. Typically, for dispersed particles of pharmaceutical interest, the zeta potential has values ranging from +100 to −100 mV.

The magnitude of the zeta potential, due to the presence of surface charges, correlates very well with the ability of a charged colloid, suspension, or emulsion to be stabilized against natural tendencies for coalescence or aggregation because of the repulsion of like charges. For example, nanoparticles with zeta potential values greater than +25 mV or less than −25 mV typically have a high degree of stability, whereas dispersions exhibiting a low zeta potential will eventually aggregate due to van der Waals interparticle interactions. The magnitude of the zeta potential significantly influences colloidal stability and flocculation processes, as shown in Table 14.2 [10].

The measured value of the zeta potential is significantly affected by the particle physical properties, such as chemical makeup and surface morphology, and solution properties such as pH and ionic strength.

Techniques available for carrying out zeta potential measurements include microscopic, optical, and acoustic methods.

Optical techniques, which are most widely used and most accurate, are based on the application of electrophoretic light scattering. Under an induced electric field, the surface charge on the particle surface causes the particle to move toward either the anode or cathode, depending on the extent of the positive or negative nature of the surface charge. The magnitude of the charge, as reflected by the zeta potential, is determined by the velocity of the particle motion

Table 14.2 Zeta potential and colloid stability.

Zeta potential (mV)	Stability property
0–3	Maximum agglomeration and precipitation
5	Strong agglomeration and precipitation
10–15	Threshold of agglomeration
16–30	Threshold of delicate dispersion
31–40	Moderate stability
41–60	Fairly good stability
61–80	Very good stability
81–100	Extremely good stability

Source: Adapted from [10].

toward the electrode. Other factors affecting the velocity are the temperature, liquid medium viscosity, and dielectric constant.

The particle velocity as a function of the applied electrical field is defined as the electrophoretic mobility. The electrophoretic mobility gives rise to the zeta potential based on the Henry equation [11]:

$$U_E = \frac{2\varepsilon z f(K_a)}{3\eta} \tag{14.9}$$

where U_E is the electrophoretic mobility, ε is the dielectric constant, z is the zeta potential, $f(K_a)$ is Henry's function, and η is the medium viscosity. In aqueous solutions, a value of 1.5 is generally used for Henry's function, whereas a value of 1.0 is applicable for nonpolar solvents based on the Smoluchowski and Hückel approximations [12], respectively.

In an electrophoretic zeta potential measurement, a sample is subject to an electrical field, and the movement of the particles (electrophoretic mobility) is measured by the light scattering of the particles. The measurement of light scattering utilizes the laser Doppler effect since light scattered from a moving particle tends to experience a frequency shift. The frequency shift can be measured by an optical mixing technique where the scattered light from the particles is mixed with a reference beam to generate intensity variations and interferences. The interferences created help to extract the frequency shift in the scattered light sensitively, and the measured frequency shift is then used to determine the particle velocity, and, consequently, the zeta potential.

Thus, laser Doppler electrophoresis, combined with phase analysis light scattering, offers an accurate means to determine zeta potential by tracking the motion of particles dispersed in a liquid medium. It, generally, is applicable for size ranges of 6–6000 nm and concentrations from 0.02 to 0.2 wt%.

14.7 Laser Diffraction

Laser diffraction is the most widely used technique to analyze the particle size of powders, suspensions, protein aggregates, and emulsions. This technique has many advantages in measuring particle size, including

- wide acceptance with an established ISO standard (ISO13320),
- broad size range for measurement from submicrometers to millimeters,
- short measurement time and data generated in minutes,
- excellent repeatability and precision,
- high throughput measurement and desirable productivity,
- convenience for use and feasibility for automation, and
- suitability for both dry samples and wet dispersions.

Laser diffraction is based on a basic physical relationship between particle size and the angle and intensity of scattered light. Smaller particles tend to scatter light at larger angles than larger particles, but larger particles exhibit more intense scattering than smaller particles. Such phenomena are described by the Fraunhofer and Mie Scattering theories [2, 13], where mathematical algorithms based on the two theories are utilized to transform the scattered light data into particle size information. The scattered light data, including both scattered angles and intensities, are collected by a particle size analyzer.

The Fraunhofer theory and the Mie scattering theory both model the situation where a particle scatters light in all directions with intensity patterns and angle information that are physically related to particle size [13]. The two theories have different fundamental assumptions and cover different ranges of particle size.

The Mie theory utilizes the refractive index difference between the particle and the dispersing medium to model the intensity of the scattered light. It also relates the absorption properties of the particle to the different parts of light transmitted, absorbed, or refracted. The Mie theory is derived from Maxwell's electromagnetic field equations [2]. It has several basic assumptions:

- The particles are spherical.
- The dispersion is dilute, so there is no multiple scattering.
- The refractive index of the particles and the medium are known.
- The particle size is homogeneous.

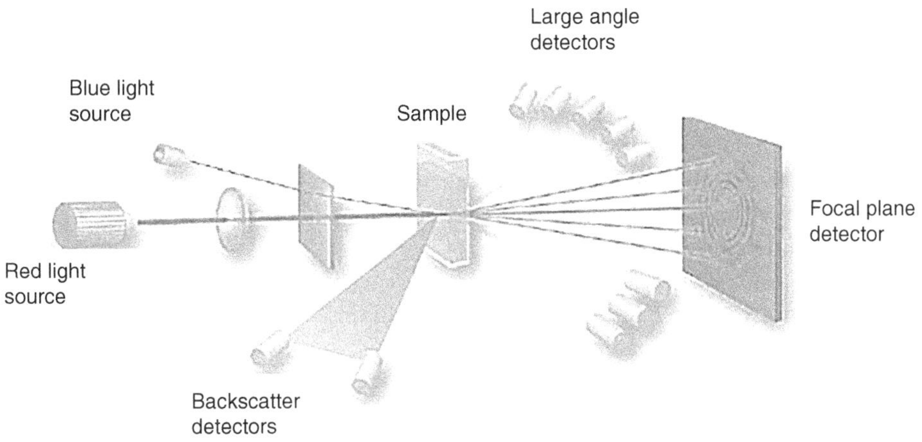

Figure 14.8 Illustration of a laser diffraction equipment. *Source*: [14]/CompareNetworks, Inc.

The Mie theory, which utilizes complex mathematical equations and demands significant calculation power, is optimally applicable for particle sizes less than 50 μm.

The Fraunhofer theory, a simplified version of the Mie theory, performs well when the particle sizes are relatively large (50 μm, at least 10 times the laser wavelength) and the scattering angle is relatively small (less than 30°) [13].

Typically, the Fraunhofer diffraction and Mie scattering theories are used together to ensure that a wide particle size range is covered.

For a laser diffraction measurement, a representative sample is dispersed at a suitable concentration in an acceptable liquid or gas and is passed through the beam of a laser. The signals scattered by the particles at various angles are measured by a multielement detector and subsequently transformed into particle size data through mathematical calculations.

A typical laser diffraction instrument has a laser beam, a particulate dispersion unit, a detector to collect diffraction signals, and a computer to control the instrument, process diffraction signals, and generate particle size results [14] (Figure 14.8).

For particles tending to aggregate, laser diffraction measures the size of the agglomerate rather than the size of the primary particles. Consequently, the complete dispersion of the sample is essential if primary particles are to be measured.

There are two types of sample dispersions that are utilized: dispersion in a liquid and dispersion in air.

For a wet dispersion, a suitable dispersant should be selected to create uniform sampling. Any optically transparent liquid of known refractive index can be used. The solubility of the sample in the liquid should be taken into consideration to avoid sample dissolution during preparation. In such cases, a saturated solution can be used to maintain particle physical stability. Another option is to choose a dispersion medium in which the material exhibits minimal solubility. Sometimes, surfactants are used to increase the wettability of particles and to enhance dispersion. Sonication is often used to create a proper dispersion, and uniformity of the dispersion can be checked with a microscope before sample analysis.

For wet dispersions, the sample concentration must be properly controlled to avoid multiple scattering. Sample concentrations with a 5–35% obscuration are generally acceptable, whereas for samples with small particles a lower concentration is more suitable. Different concentrations should be evaluated to establish an optimum range.

For the measurement of dry powders, the powder is dispersed in a gas or air to create a dry dispersion or aerosol using vacuum or compressed air. Proper control of the vacuum or compressed air is required to create sufficient mechanical forces to break up aggregation but not strong enough to fracture the primary particles. A pressure from 1.5 to 3 bar is usually sufficient for creating an acceptable dry dispersion. The operating parameter should be evaluated for each material in terms of mechanical properties and relative tendency to aggregate. The energy applied for the dispersion of the particles should be varied to observe any change in size distribution. Generally, fines increase as pressure is increased until reaching a plateau. It is important not to exceed pressures that may cause a reduction in primary particle sizes. Therefore, evaluation of pressure is usually included as an important part of method development. Less energy should be used for dispersion if agglomeration is to be monitored at the same time.

An examination by microscopy is strongly encouraged to estimate the size range, particle shape, aggregation, and dispersion uniformity. In addition, a sample should be representative of the whole batch to avoid the limitation of subsampling differences.

Consideration must also be given to limitations due to laser diffraction. First, it must be recognized that the theories behind laser diffraction assume a spherical shape for the particles. Thus, diffraction cannot provide shape information. As such, equipment is usually validated with reference spherical microspheres. Consequently, laser diffraction provides the best accuracy for nontransparent spherical samples, whereas, for acicular particles, such as needles, rods, and plates, poor agreement exists between the results of laser diffraction and image analysis.

Second, laser diffraction produces diffraction signals, and the calculation of particle size is accomplished through a mathematical algorithm. Though all vendors use the same theory, each vendor develops their own algorithm, and, therefore, discrepancies among instruments from different suppliers can occur.

Third, precise measurement requires the proper selection of a lens since each lens covers a specific range of particle size. Therefore, before carrying out a sample analysis, the general range of particle sizes in the sample should be examined by microscopy to ensure that approximately all particles are covered with a particular lens.

Fourth, the algorithms based on the Mie theory require knowing the refractive index of the sample to provide accurate results.

Fifth, the use of laser diffraction assumes random orientation of the sample. For irregularly shaped particles, such as needles, therefore, completely random orientation may not be accomplished during sample analysis.

Laser diffraction has been shown to be very useful for measuring submicrometer particles by using a short-wavelength laser, a converging beam, a larger lens, and a more powerful detector.

In summary, laser diffraction is a powerful technique to measure particle size from the submicrometer to millimeter scale. The measurement includes several simple steps:

- Determine the appropriate particle size range from microscopic data.
- Select wet dispersions or dry dispersions depending on sample properties and analytical needs.
- Align the optical parts properly and select the right lens for the measurement.
- Make a blank measurement.
- Measure the sample.
- Select an appropriate optical model for calculation.
- Report the data.

14.8 Acoustic Spectroscopy

Acoustic spectroscopy, also called ultrasonic extinction or ultrasonic scattering, applies ultrasound attenuation to determine the size distributions of one or more materials dispersed in a liquid, such as colloids, dispersions, slurries, and emulsions. By subjecting a sample to a set of ultrasound frequencies in the megahertz range (MHz), the attenuation spectrum as to the sound propagation velocity and the attenuation coefficient of the sample as a function of frequency are collected and further processed into particle size distributions through mathematical modeling via ultrasonic wave propagation theories, as illustrated in Figure 14.9 [15, 16].

It is interesting to note that acoustic spectroscopy can cover a broad range of particle sizes from 10 nm to 3 mm. Furthermore, acoustic spectroscopy can handle highly concentrated systems without the need for sample dilution and

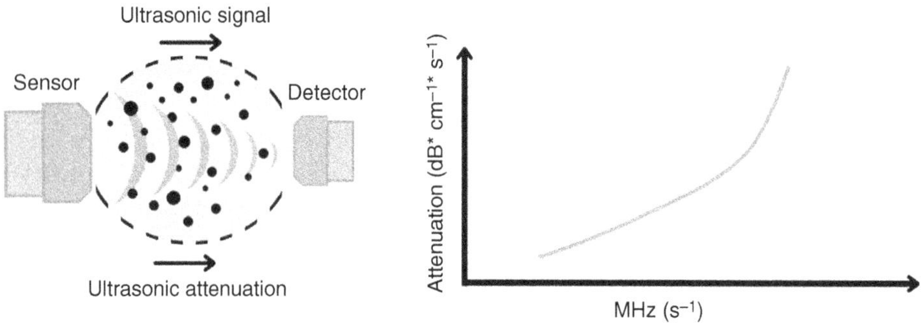

Figure 14.9 Illustrations of acoustic spectroscopy. Right: equipment design; left: attenuation plot. *Source*: [16]/MDPI/CC BY 4.0.

effectively work with optically opaque particle systems as well, which, obviously, offers advantages over the light scattering method discussed earlier. As would be expected, sample dilution, usually needed for optical analyzers, not only requires substantial effort but also potentially changes the sample properties. As we shall see, the method is applicable for concentrations of the dispersed phase from 0.1% to 50% by volume. It is not surprising that the method is useful in monitoring dynamic changes in particle size during various pharmaceutical processes such as crystallization, nanomilling, or homogenization. In addition, it is applicable to studying product stability during storage, for example, particle growth, agglomeration, aggregation, or flocculation. Thus, acoustic spectroscopy can act as a valuable process analytical tool with online capability and a nondestructive nature.

Regarding process analysis, the OPUS system from Symptec can offer real-time particle size and concentration analysis in process environments for measuring particle sizes of suspensions and emulsions, ranging from 0.1 μm to 3 mm, either with an inline or online setting. Kadam et al. used an OPUS sensor as part of PAT tools in skid configuration to enable a rapid process development strategy for the crystallization of pharmaceutical active ingredients [17].

For example, Hou and co-workers applied acoustic spectroscopy to inline or online particle size measurements of various pharmaceutical nanocrystals during nanomilling, with the aid of a computational tool involving neural networks [18].

14.9 Dynamic Image Analysis

Dynamic image analysis allows measurements of particle size in the same range as laser diffraction and is suitable for the analysis of powders, suspensions, granules, and pellets in the range of several micrometers to millimeters. Similarly to light microscopy, dynamic image analysis collects visible images from which particle sizes are directly determined. Though light microscopy can examine smaller particles, dynamic image analysis can provide a much more rapid measurement of a large number of particles. Some advantages of dynamic image analysis can be listed as follows:

- Measures each particle individually
- Obtains shape information at the same time
- Confirms laser diffraction results
- Replaces sieve analysis for obtaining the size distribution of granules and pellets
- Detects agglomeration

Dynamic image analysis is a method that utilizes digital image processing. Particles either in dry form or dispersed as a suspension pass in front of two bright and pulsed LED light sources, while two high-speed digital cameras capture the shadows of the particles and record the images [19] (Figure 14.10). One camera can detect small particles with high resolution, whereas the other camera detects a large field of view and is most suitable for large particles. Each camera is coupled with an LED light source with optimized brightness, pulse length, and field of view. Using a strong LED light source provides shorter exposure time and reduces particle blurriness. The exposure time for the light pulse is generally less than 1 ns, and acquisition rates can be as high as 450 frames per second [19].

At this point, the shadow area of the particle is recorded and used to calculate the maximum diameter or minimum diameter. Notably, the minimum diameter obtained closely aligns with the particle size obtained by sieve analysis [19].

It is important to note that dynamic image analysis employs digital cameras, where the resolution is determined by the number and size of the pixels in the detector. The camera reconstructs the particle images by mapping the square pixels corresponding to the projected shadow.

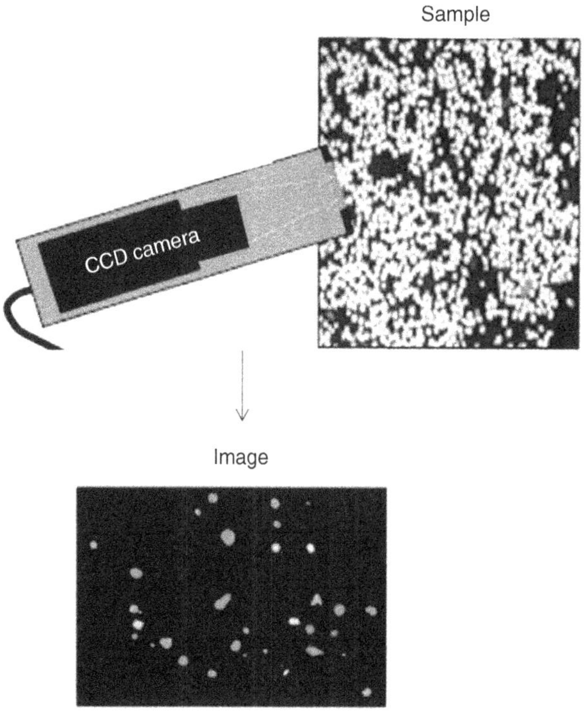

Figure 14.10 Illustration of a dynamic image analyzer.

Sampling techniques for dynamic image analysis include gravity feeding, dry dispersion, and wet dispersion. Gravity feeding is suitable for pourable materials but not agglomerated particles such as granules and pellets. A vibrating feeder promotes material movement into the sample chamber. A dry dispersion is prepared by passing the material through a nozzle under pressure, allowing any aggregates to be separated into primary particles as an aerosol beam. Wet dispersions are prepared by suspending the material in a proper vehicle, often using sonication to enhance dispersion. The sample is then pumped into the measuring cell to collect particle size data.

In summary, dynamic image analysis can provide rapid and high-resolution measurements of various materials with broad particle size distributions. It directly obtains particle size based on the width, length, or projection area derived from the image data. Operation of this technique is relatively convenient since no mathematical algorithms are involved, and the size and shape information are obtained at the same time. To ensure accurate particle size data, special attention should be given to sample size, sample preparation, and image quality.

Another advantage of dynamic image analysis over sieve analysis (to be discussed in the following text) is its applicability as a process analytical tool for inline, online, or offline measurement. Nalluri et al., for example, utilized dynamic image analysis for in-process analysis of pharmaceutical conical milling [20]. A dynamic image analysis system with a xenon flash light and charge-coupled device camera was coupled with a conical mill online, inline, or at-line. The study of various pharmaceutical excipients and granules revealed that the online measurement offered the most robust monitoring of dry milling. Dynamic image analysis has also been found useful in monitoring the process of coating, drug layering, or controlled-release coating. For example, Heinicke and Schwartz reported that dynamic image analysis was capable of measuring differences in diameter equal to a 4 μm coat thickness, which is equivalent to 2% polymer coat weight [21].

Yu and Hancock compared dynamic image analysis to the laser diffraction technique in analyzing different grades of microcrystalline cellulose [22]. The data for spherical particles showed good agreement, but significant discrepancies were observed for rod-shaped particles, which are related to the difference in operating principles of the two techniques. Thus, it is always important to indicate the measuring technique used when reporting particle size data.

14.10 Sieve Analysis

Sieve analysis is a traditional and economical way to measure the particle size distribution for coarse powders, granules, and pellets [23]. A set of selected sieves with different openings is stacked together. After adding a sample, the sieves undergo mechanical shaking to promote particle movement through or retention by various sieve sizes. The amount of material retained on each sieve is measured, and the proportion of each sample associated with a particular screen size is used to form a cumulative distribution plot.

The accuracy of sieve analysis is dependent on the quality and integrity of the sieves since they tend to undergo significant wear and tear during use. It is important to recognize that the time and amplitude of shaking and sample size may affect the measurement. Unlike image analysis, sieve analysis does not provide shape information. Irregular particles as needles or plates, potentially, present problems because they may only pass through the sieves when orienting in a particular position. Sieve analysis with such asymmetric particles usually requires long periods of time since such particles require sufficient time to adapt and to pass through the sieve.

14.11 Bulk Properties of Pharmaceutical Particulates and Powder

Knowledge of bulk properties such as bulk density, tapped density, and flowability is essential for the handling and processing of pharmaceutical particulates and powders. For example, the bulk density of granules affects their filling weight in capsule manufacturing, whereas flowability is important for assuring reproducibility in tablet compression and capsule filling. During tablet compression in a rotary press, for example, the flow of powder or granules from the hopper to the dies impacts the fill weight, hardness, and content uniformity of the product. Powder flow must be optimized for manufacturing processes such as mixing, roller compaction, injection molding, and packaging.

The U.S. Pharmacopoeia (USP) has established acceptable techniques for measuring bulk and tapped density [24]. Bulk density is the mass per unit volume of a loosely packed and untapped powder bed, whereas tapped density is the mass per unit volume of a densely packed powder bed that has been compacted by tapping for a definite period. The bulk density and tapped density are generally expressed in grams per milliliter (g/mL).

According to the USP [24], the bulk density of a powder is determined by pouring a known mass of powder, which may have been passed through a sieve, into a graduated cylinder to record the volume. It is well recognized that preparation, treatment, and storage can affect the bulk properties of a powder. For example, the powder can assume a range of bulk densities with different ways of handling or sampling, since a slight disturbance of the powder bed may lead to a change in bulk density. For this reason, it is very important that the history of the powder be reported when measuring bulk density.

After the bulk density measurement, the tapped density can be obtained by "tapping" the sample with vertical movement in a graduated cylinder until an insignificant volume change is observed [24]. The tapping or vertical movement involves raising the cylinder and allowing it to drop under its own weight over a specific distance. In addition, rotating the cylinder is often usually included in the procedure to minimize any possible separation of the mass during tapping. Tapping is generally carried out in increments of 500, 750, and 1250 taps with 250 drops per minute.

Bulk density and tapped density can be used to calculate the Carr's compressibility index (CI) and the Hausner ratio, which define the propensity of a powder to be compressed [25]. Both parameters can also give a good indication of the ability of the powder to flow.

The Carr's CI is defined as the capacity of the powder to consolidate and is evaluated with the following equation:

$$\mathrm{CI}\ (\%) = \frac{\rho_\mathrm{t} - \rho_\mathrm{b}}{\rho_\mathrm{b}} \times 100 = \frac{V_\mathrm{b} - V_\mathrm{t}}{V_\mathrm{b}} \times 100 \tag{14.10}$$

where ρ_t is the tapped density, ρ_b is the bulk density, V_b is the bulk volume, and V_t is the tapped volume.

The Hausner ratio is the ratio of tapped density to bulk density:

$$\text{Hausner ratio (HR)} = \frac{\rho_\mathrm{t}}{\rho_\mathrm{b}} = \frac{V_\mathrm{b}}{V_\mathrm{t}} \tag{14.11}$$

As mentioned earlier, both the Carr Index and the Hausner ratio provide an indication of flowability. The higher the Hausner ratio, the poorer the flow is, as shown in Table 14.3 [25]. The flowability calculated by the Hausner ratio should be confirmed by another technique since its accuracy can be limited if the powder exhibits significant cohesiveness.

Li et al. compared the flowability of several pharmaceutical ingredients using the Hausner ratio and the direct measurement of flow by bin-flow tests [26]. Based on the Hausner ratio, povidone and crospovidone had very good flow, whereas pregelatinized starch, stearic acid, paracetamol, and magnesium stearate all exhibited very poor flowability. In contrast to these results, flow measured by the bin-flow tests indicated that crospovidone exhibited poor flow, while that of pregelatinized starch was quite good. Such discrepancies in the measurement of flow using different methods are not surprising since flow is dependent on many physical characteristics that might operate differently with different methods, for example, moisture content, particle size, true density, and the presence of flow-enhancing agents.

Additional techniques are available for providing a more quantitative assessment of flow, including flow through an orifice, the angle of repose, and avalanche time.

The flow rate of powder and granules, measured by allowing the sample to freely flow through a specifically sized orifice from a container such as a cylinder, funnel, or hopper, can also be used as a measure of flowability [27].

The angle of repose is a simple method to characterize the flowability of bulk powders. When powder flows through a funnel onto a horizontal plane, it tends to form a cone, where the maximum angle between the slope of the cone and the horizontal plane is defined as the angle of repose. Cohesion between particles reduces flow and creates different angles

Table 14.3 Flowability related to Hausner ratio and CI.

Hausner ratio	Compressibility index (%)	Flow character
1.00–1.11	≤10	Excellent
1.12–1.18	11–15	Good
1.19–1.25	16–20	Fair
1.26–1.34	21–25	Passable
1.35–1.45	26–31	Poor
1.46–1.59	32–37	Very poor
>1.60	>38	Very, very poor

of repose. A free-flowing powder will give rise to a cone with a low angle of repose, while conversely, a poorly flowing material will result in a cone with steeper angles. As recommended by the USP [25], angles less than 30°usually indicate good flow, whereas angles greater than 45° indicate poor flow. The angle of repose is calculated from the cone height, *h*, and the radius, *r*, as follows:

$$\theta = \tan^{-1}\left(\frac{h}{r}\right) \tag{14.12}$$

The angle of response method has the advantage of being very simple to measure. Moreover, a general classification of flowability provided by the USP is available to rank flowability based on the angle of repose. The method is generally best utilized to provide an overall assessment of powder flowability and to give an early warning of potential flow issues.

However, the results of such experiments rely heavily on the method used to form the cone and often lack consistency due to the variability in testing conditions and the operator's experience. In addition, the method may not be able to differentiate between small differences in flowability for some powders as reported by Shan et al. [27].

Dynamic flow measurements are generally considered to be more predictive of flowability in the manufacturing process. One approach is to study powder avalanching in a rotating drum, where a short avalanching time indicates good flow and vice versa [28]. Usually, about 50 grams of sample is placed into a drum with an inner diameter of 13 cm and an inner width of 2.5 cm. A layer of sandpaper is placed around the internal rim of the disc to prevent the powder from slipping. The drum is rotated at a speed of about 2 minutes per rotation. The powder rotates with the drum and builds up on the wall of the drum until an avalanche of powder flow occurs. The avalanche time is measured photoelectrically, where the mean time required for the avalanche to occur is related to flowability. The advantage of the avalanche test is its high throughput and ease of interpretation.

Another dynamic flow measurement that has recently gained popularity is the powder rheometer [29]. A powder rheometer utilizes small amounts of sample to study the rheology of the powder in terms of flowability. During the measurement, a blade rotates through a cylindrical column of powder, and the shear properties are recorded. Shear properties are affected by interparticle interactions and are therefore related to powder flow. Shear stress is measured in a range of normal stresses to generate yield loci, the plots of shear stress versus normal stress. The cohesion of the sample is calculated by extrapolating the yield loci to zero normal stress. In addition, the yield loci are utilized to calculate the unconfined yield strength, the force or stress required to deform or break a material when it is not confined by a container, and the major principal stress, the maximum normal stress that can occur for the material. The flow function is assessed as the ratio of the major principal stress to the unconfined yield strength, as an indicator of powder flow.

Trivedi and Dave studied the physical compatibility between dibasic calcium phosphate and various cohesive APIs using a powder rheometer [29]. They concluded that the powder rheometer provided a better understanding of the flow behavior of binary mixtures with the addition of increasing proportions of API than empirical approaches such as the angle of repose and Carr's CI. Brockbank et al. used a powder rheometer to evaluate the flow properties and caking mechanisms of different powder systems that were affected by chemical-, moisture-, and temperature-based caking [30].

It is generally recognized that a single test is not sufficient to characterize the flow properties of pharmaceutical powders and that rather a combination of different techniques is more reliable to rank flowability and to guide the process development of pharmaceutical formulations.

14.12 Surface Area Measurement

As mentioned earlier, the surface area of solid particulates is an important physical property that affects the quality and performance of many pharmaceutical materials, playing a significant role in various manufacturing processes, such as purification, blending, and tableting. For example, Neusilin® US2 is a fine ultralight granule of magnesium aluminometasilicate that has a very large specific surface area, about 300 m^2/g [31]. Owing to its large surface area, Neusilin US2 can absorb significant amounts of oils or water, thus facilitating the compression of such systems into high-quality tablets. The surface area of particles affects drug dissolution, bioavailability, and stability by affecting the level of particle surface exposure to the environment. In addition, since a powder having a high surface area can adsorb a significant amount of moisture, the uptake of moisture can lead to caking by promoting particle-particle adhesion.

Surface area is generally measured by gas adsorption, which involves exposing solid materials to nitrogen or krypton gas at a variety of vapor pressures and evaluating the amount of gas uptake at a constant temperature. The surface area is calculated from the isothermal data based on the BET equation [32].

When a solid material is exposed to a gas, some of the gas will be absorbed onto the material surface depending on the temperature, pressure, and chemical composition of the gas and solid. Generally, such adsorption is characterized as either chemical adsorption (chemisorption) or physical adsorption (physisorption). Chemisorption involves the interaction of the gas with the solid by means of covalent bonding, generally occurring at high temperatures, whereas physical adsorption involves weak noncovalent intermolecular forces such as van der Waals or hydrogen bonding. Physical adsorption occurs more readily at low temperatures and can lead to multimolecular layers of adsorbed gas. It is easily reversible at ambient temperatures. Due to the nonspecificity of physical adsorption on the chemistry of the solid, it is the most suitable method for measuring surface area [32]. The BET equation was developed to describe the isothermal relationship between the amount of gas adsorbed and the function of gas pressure. The equation is derived based on several assumptions:

- Homogeneous surface.
- No lateral interaction between gas molecules.
- The uppermost layer is in equilibrium with the vapor phase.
- The heat of adsorption represents the interaction energy of gas with the first layer, whereas the energetics of other layers is represented by the heat of condensation.
- An infinite number of layers can form up to a saturation pressure that represents the pressure at which liquid condenses.

The BET equation can be described as

$$v = \frac{v_{\mathrm{m}} C P}{(P_{\mathrm{o}} - P)\left\{1 + (C-1)\frac{P}{P_{\mathrm{o}}}\right\}} \tag{14.13}$$

where v is the volume of gas adsorbed, P is the equilibrium pressure, P_{o} is the saturation pressure, C is a BET constant, and v_{m} is the volume of gas required to form a complete monolayer over the surface, presumably covering the entire solid surface.

Equation (14.13) can be converted to a linear form, which facilitates the graphical estimation of the equation constants:

$$\frac{1}{v\left(\frac{P_{\mathrm{o}}}{P} - 1\right)} = \frac{C-1}{v_{\mathrm{m}} C} \times \frac{P}{P_{\mathrm{o}}} + \frac{1}{v_{\mathrm{m}} C} \tag{14.14}$$

The BET plot, therefore, can be established by plotting $\frac{1}{v\left(\frac{P_{\mathrm{o}}}{P} - 1\right)}$ versus $\frac{P}{P_{\mathrm{o}}}$, where v_{m} and C can be calculated from the slope A and intercept I as follows:

$$v_{\mathrm{m}} = \frac{1}{A + I} \tag{14.15}$$

$$C = 1 + \frac{A}{I} \tag{14.16}$$

The total surface area can be calculated from v_{m} according to the following equation:

$$S = \frac{v_{\mathrm{m}} N s}{V} \tag{14.17}$$

where V is the molar volume of adsorbed gas, N is Avogadro's number, and s is the cross-sectional area of the absorbed gas molecule.

Specific surface area, the area of solid surface per unit mass of material, is derived from the total surface area S and the material mass a:

$$S_{\mathrm{s}} = \frac{S}{a} \tag{14.18}$$

The BET plot of actual samples usually has a limited range of linearity, typically only below $p/p_{\mathrm{o}} \approx 0.3$.

The measurement of surface area using the BET equation is usually carried out at liquid nitrogen temperatures to facilitate physical adsorption. Nitrogen and krypton are the most widely used gases because of their nonpolarity and well-defined cross-sectional areas. Krypton is preferred for powders of low specific surface area $(< 0.2\ \mathrm{m^2/g})$, whereas

Table 14.4 Effect of drying temperature on the surface area of magnesium stearate.

Degas temperature (°C)	Surface area (m^2/g)
35	5.138
50	5.106
65	4.834
80	4.542
100	3.651
125	1.581

Source: Adapted from [33].

nitrogen is favored for samples with relatively large surface areas. Enough material should be used for BET measurement to give a total exposed area of at least 1 m^2 when using nitrogen and at 0.5 m^2 when krypton is used. The sample should be accurately weighed for the calculation of specific surface area.

Before proceeding with the BET measurement, the sample should be sufficiently treated to remove gases or vapors already adsorbed onto the surface. Typically, a sample is put under a vacuum for outgassing, whereas another option is to flush the adsorbent with an inert gas under ambient conditions, typically with helium or nitrogen. Both options often use heat to accelerate outgassing. Special attention should be paid to materials that have thermal instability issues, such as chemical degradation or dehydration/desolvation of crystals, since improper outgassing may lead to significant changes in material characteristics. For example, it has been reported that magnesium stearate should not be outgassed at temperatures above 35°C because melting and sintering tend to occur, leading to decreased surface area as shown in Table 14.4 [33].

The specific surface area of a particle and the particle size can be shown to be directly correlated in the case of nonporous and spherical particles, where the specific surface area is related to the particle diameter (d) or radius (R) in the following equation:

$$S_s = \frac{4\pi R^2}{\frac{4}{3}\pi R^3 \rho} = \frac{3}{\rho R^3} = \frac{6}{\rho d} \tag{14.19}$$

where ρ is the true density and d is the diameter. Based on equation (14.19), specific surface area can be estimated from the average particle size or vice versa.

References

1 Paul HE, Hayes KJ, Paul MF, Borgmann AR. Laboratory studies with nitrofurantoin. Relationship between crystal size, urinary excretion in the rat and man, and emesis in dogs. *J Pharm Sci* 1967; 56(7): 882–885.

2 Mie G. Beiträge zur Optik trüber Medien, speziell kolloidaler Metallösungen. *Annal Phys* 1908; 330(3): 377–445.

3 Wiscombe WJ. *Mie Scattering Calculations: Advances in Technique and Fast, Vector-speed Computer Codes*. Atmospheric Analysis and Prediction Division, National Center for Atmospheric Research; 1979 Jun 29.

4 Hackley V, Clogston J. *Measuring the Size of Nanoparticles in Aqueous Media Using Batch-Mode Dynamic Light Scattering*. NIST Special Publication; 2007.

5 Kaszuba M, Connah MT. Protein and nanoparticle characterisation using light scattering techniques. *Part Part Syst Char* 2006; 23(2): 193–196.

6 Van Eerdenbrugh B, Stuyven B, Froyen L, Van Humbeeck J, Martens JA, Augustijns P, Van den Mooter G. Downscaling drug nanosuspension production: processing aspects and physicochemical characterization. *AAPS PharmSciTech* 2009; 10(1): 44–53.

7 ISO 22412: 2008, Particle size analysis – dynamic light scattering (DLS), International Organization for Standardization, Geneva, Switzerland.

8 Yu Z, Reid J, Yang YP. Utilizing dynamic light scattering as a process analytical technology for protein folding and aggregation monitoring in vaccine manufacturing. *J Pharm Sci* 2013; 102(12): 4284–4290.

9 Kotta S, Khan AW, Ansari SH, Sharma RK, Ali J. Anti HIV nanoemulsion formulation: optimization and in vitro-in vivo evaluation. *Int J Pharm* 2014; 462(1): 129–134.
10 Salopek B, Krasić D, Filipović S. Measurement and application of zeta-potential. *Rudarsko-geološko-naftni fakultet* 1992; 4(1): 147.
11 Henry DC. The cataphoresis of suspended particles. Part I. The equation of cataphoresis. *Proc R Soc A Math Phy Eng Sci* 1931; 133(821): 106–129.
12 Hunter RJ. *Zeta Potential in Colloid Science: Principles and Applications*. London: Academic Press; 2013.
13 de Boer GB, de Weerd C, Thoenes D, Goossens HW. Laser diffraction spectrometry: Fraunhofer diffraction versus Mie scattering. *Part Part Syst Char* 1987; 4(1–4): 14–19.
14 Rawle A, Kippax P. http://www.americanlaboratory.com/913-Technical-Articles/606-Setting-New-Standards-for-Laser-Diffraction-Particle-Size-Analysis (accessed 25 April 2017).
15 Silva CA, Saraiva SV, Bonetti D, Higuti RT, Cunha RL, Pereira LO, Silva FV, Fileti AM. Measurements of bimodal droplet size distribution of emulsions using ultrasonic spectroscopy in the long and intermediate wavelength regimes. *Chem Eng Sci* 2022; 252: 117274.
16 Grumbach C, Czermak P. Process analytical technology for the production of parenteral lipid emulsions according to good manufacturing practices. *Processes* 2022; 10(6): 1174.
17 Kadam SS, Vissers JA, Forgione M, Geertman RM, Daudey PJ, Stankiewicz AI, Kramer HJ. Rapid crystallization process development strategy from lab to industrial scale with PAT tools in skid configuration. *Org Process Res Dev* 2012; 16(5): 769–780.
18 Hou GH, Falola AA, Wang XK, Liang LP, Zou XW, Wu T, Wang XZ. Application of acoustic spectroscopy and neural networks to on-line size measurement of pharmaceutical nanocrystals. *J Acoust Soc Am* 2023; 154(4): 2112–2123.
19 List J, Köhler U, Witt W. *Dynamic Image Analysis Extended to Fine and Coarse Particles. Particle Systems Analysis*. Edinburgh, UK; 2011.
20 Nalluri VR, Schirg P, Gao X, Virdis A, Imanidis G, Kuentz M. Different modes of dynamic image analysis in monitoring of pharmaceutical dry milling process. *Int J Pharm* 2010; 391(1): 107–114.
21 Heinicke G, Schwartz JB. Particle size distributions of inert spheres and pelletized pharmaceutical products by image analysis. *Pharm Dev Tech* 2005; 9(4): 359–367.
22 Yu W, Hancock BC. Evaluation of dynamic image analysis for characterizing pharmaceutical excipient particles. *Int J Pharm* 2008; 361(1): 150–157.
23 Pharmacopeia, US. USP Chapter <786> Particle size distribution estimation by analytical sieving. 2012. Rockville, MD. USP 29/NF 24, ed.
24 Pharmacopeia, US. USP chapter< 616> bulk density and tapped density. 2006. Rockville, MD: USP 29/NF 24, ed.
25 Pharmacopeia, US. USP Chapter <1174> Powder Flow. 2012. Rockville, MD. USP 29/NF 24, ed.
26 Li Q, Rudolph V, Weigl B, Earl A. Interparticle van der Waals force in powder flowability and compactibility. *Int J Pharm* 2004; 280(1): 77–93.
27 Shah RB, Tawakkul MA, Khan MA. Comparative evaluation of flow for pharmaceutical powders and granules. *AAPS PharmSciTech* 2008; 9(1): 250–258.
28 Lavoie F, Cartilier L, Thibert R. New methods characterizing avalanche behavior to determine powder flow. *Pharm Res* 2002; 19(6): 887–893.
29 Trivedi MR, Dave RH. To study physical compatibility between dibasic calcium phosphate and cohesive actives using powder rheometer and thermal methods. *Drug Dev Ind Pharm* 2014; 40(12): 1585–1596.
30 Brockbank K, Armstrong B, Chandorkar Y, Freeman T. Understanding powder caking as a consequence of a range of mechanisms by means of powder rheometry. *Part Sci Technol* 2015; 33(1): 102–108.
31 Fuji C. *Company Literature on Neusilin*. Toyama, Japan: Fuji Chemical Industry, 1997: 3–4.
32 Bruanuer S, Emmett PH, Teller E. Adsorption of gases in multimolecular layers. *J Am Chem Soc* 1938; 60(2): 309–316.
33 Phadke DS, Colliera JL. Effect of degassing temperature on the specific surface area and other physical properties of magnesium stearate. *Drug Dev Ind Pharm* 1994; 20(5): 853–858.

15

Hygroscopic Properties of Solids

15.1 Introduction

The term "hygroscopicity," as it relates to pharmaceutical solids, can be defined as "a phenomenon that occurs when a certain amount of water vapor exposed to a solid at a given temperature and relative humidity is taken up and retained by means of noncovalent interactions, particularly hydrogen bonding, with some effect on the surface and/or bulk properties of the solid" [1]. The significance of hygroscopicity in the context of studying solid-state properties in pharmaceutical systems arises from several factors uniquely associated with water. First, under most climatic conditions, the atmosphere surrounding the earth contains an abundance of water molecules in the vapor state capable of making contact and interacting with an exposed solid. Second, the water molecule has structural characteristics that favor its noncovalent interactions with solids under a wide range of ambient conditions. As seen in Figure 15.1, for instance, water has a small molecular size and a molecular shape that allows it to exhibit strong hydrogen bonding at up to four charge centers, as illustrated by the open hexagonal arrangement of water molecules in the major crystal structure of ice [2]. Also, because of its small molecular size, large numbers of water molecules per unit area can accumulate on solid surfaces and/or penetrate the bulk of the solid.

Pharmaceutical interest in the hygroscopic properties of pharmaceutical solids arises primarily because water associated with solids can act as either an "impurity," capable of having negative effects on the properties of solids, or a "component" of a pharmaceutical formulation required for the functionality of certain excipients. Water acts as an "impurity" in that it can interact with molecules in the solid state to cause chemical and physical instabilities [3, 4]. For example, water is capable of directly interacting with solids to cause hydrolytic chemical degradation. It can also accumulate in solids and act as a polar medium that supports acid–base and free radical reactions brought about by contact with various acidic or basic excipients, electromagnetic radiation, and molecular oxygen. Under circumstances where structural disorder has been introduced into a crystalline solid in the form of defects and partially amorphous regions preferred accumulation of water in these regions can facilitate several phase transformations by enhancing molecular mobility, for example, diffusional motion [3]. These include polymorphic transitions, anhydrate–hydrate conversions, crystallization from amorphous regions, denaturation of proteins, and types of chemical degradation where water is not a reactant. Specific examples of such effects on physical and chemical stability will be given in Chapters 18 and 19.

Situations where water can affect the functionality of excipients include promoting the direct compaction properties of microcrystalline cellulose and spray-dried lactose during tableting; facilitating the lubricating properties of magnesium stearate in tableting; preventing brittleness in gelatin capsules; maintaining mechanical strength in polymeric coatings through plasticizing effects; and promoting the compaction properties of polymeric binders in wet-granulated powders. As we shall see in greater detail in Chapter 16, in each of these cases, the effects of water are primarily on the mechanical properties of the solid system.

It was mentioned earlier that water vapor in the atmosphere exists in high abundance under most relative humidity conditions. Therefore, exposure of solids to the open atmosphere under ambient conditions provides a source of water vapor that can then interact with the solid [5]. It is also possible for "residual water," already associated with various ingredients of a formulation enclosed in various forms of packaging, to re-equilibrate and transfer from one solid to another via the vapor phase [6]. Such "residual water" generally results from incomplete drying during processes that involve the use of water as a solvent. For example, residual water is often retained after (i) crystallization from aqueous solution; (ii) freeze drying (lyophilization) of parenteral products; (iii) aqueous spray drying to form fine powders; (iv) wet granulation

Solid-State Materials in Pharmaceutical Chemistry: Properties, Characterization, and Applications, Second Edition. Stephen R. Byrn, George Zografi and Xiaoming (Sean) Chen.

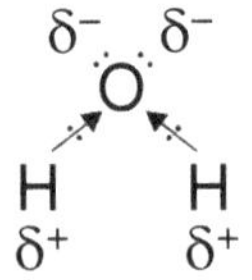

Figure 15.1 Structure of the water molecule indicating the four charge centers capable of interacting through hydrogen bonding.

during the tableting process; and (v) aqueous film coating. The possible impact of such water vapor transfer on physical and chemical instabilities in formulations stored in closed containers will be discussed more fully in Chapters 18 and 19.

15.2 Water Vapor Sorption–Desorption

The process whereby water vapor interacts with solids at a particular temperature and relative humidity is generally referred to as "sorption." Sorption can be further categorized as either adsorption or absorption. Water accumulation by adsorption is restricted to interactions with molecules on the available surface of the solid, whereas absorption occurs by penetration of water molecules into the bulk structure of the solid. Adsorption, therefore, tends to have significant effects on the surface properties of the solid, while absorption also influences many physical-chemical properties associated with the bulk of the solid. The amount of water adsorbed is generally expressed as the mass of water taken up per unit surface area of the solid, for example, kg/m^2. Thus, the total amount of water adsorbed depends on the available surface area. The amount of water absorbed does not depend primarily on surface area since most of the water penetrates the bulk structure, the amount of water absorbed being directly related to the mass of solid available to the water vapor. The most common way that absorption is expressed in the pharmaceutical literature is in terms of mass of water absorbed divided by the combined masses of water and solid. Typically, this is expressed as a fraction, or as a percentage, for example, kilograms of water per 100 kilograms of water and solid. However, some studies express absorption as mass of water per mass of "dry" solid, so it is important to be clear on what units are being used when reporting such data in the literature. Using this latter manner of expressing absorption has the advantage of allowing one to determine the moles of water taken up per mole of molecules in the solid and to detect any possible stoichiometric relationships, as in a crystal hydrate (see Chapter 3).

Desorption occurs when water vapor that has interacted with a solid by sorption leaves the solid and returns to the atmosphere above the solid when temperature or relative humidity is altered; that is, desorption is expected when there is a reduction in relative humidity and/or an increase in temperature. Generally, the equilibrium water content of a solid at a particular relative humidity and temperature should be identical when reached by sorption or desorption processes if the process of water gain or loss does not in some way irreversibly change the properties of the solid. When sorption and desorption values are different, commonly referred to as "sorption/desorption hysteresis," we can assume that either there has been a change in the properties of the solid during the sorption process or that equilibrium has not been attained because of a slower rate of sorption or desorption than the timescale of the measurement [1].

15.3 Water Vapor Sorption Isotherms, Relative Humidity, and Water Activity

The equilibrium relationship between water vapor sorption and relative humidity at constant temperature is generally determined by establishing a water vapor sorption isotherm, where some measure of the amount of water vapor associated with the solid is plotted against the relative humidity. In effect, such isotherms represent the equilibrium relationship between water vapor pressure (relative humidity is related to water vapor pressure, as we shall see in the following text) versus the volume sorbed (related to the mass of water sorbed) and therefore a phase diagram that reflects the thermodynamic state of various water–solid combinations under a given set of conditions. Consequently, any attempt to characterize the hygroscopic properties of solids must begin by establishing the equilibrium sorption/desorption isotherm under various conditions.

So far, in this discussion, relative humidity has been mentioned several times as an important parameter for establishing the hygroscopicity of solids. Therefore, let us explore the meaning of relative humidity in thermodynamic terms, its relevance to hygroscopicity, and how it is generally measured. If we expose a solid to water vapor molecules in a closed container at a given pressure, p, we expect the driving force for any interaction between the solid and water to be determined in large part by the concentration of water vapor in the atmosphere, which can be expressed as n/V, where n is the number of moles and V is the head-space volume of the container. According to the ideal gas law, the pressure, p, can be written as

$$p = \left(\frac{n}{V}\right)RT \tag{15.1}$$

where R is the gas constant and T is the temperature. From this equation, it can be seen that the concentration of water vapor is directly proportional to the pressure, and thus pressure thermodynamically reflects the source of the driving force for the interaction. The basis for this conclusion can also be seen by expressing the chemical potential of liquid water, μ_w, in terms of the vapor pressure of the liquid as

$$\mu_w = \mu_w^o + RT \ln p_o \tag{15.2}$$

where $\mu_w{}^o$ is the standard chemical potential and p_o is the liquid vapor pressure. Since at equilibrium the chemical potential of water in the liquid is equal to that in the vapor state, we can conclude that the vapor pressure is directly linked to the thermodynamic properties of water molecules in the vapor state.

Let us now observe a solution of a solute, 2, dissolved in a solvent, 1, and consider the expected colligative effect of the solute that depresses the vapor pressure of the liquid, p_o. Assuming an ideal solution that follows Raoult's law, we can write

$$p = p_o - X_2 p_o = (1 - X_2)p_o \tag{15.3}$$

where p is the vapor pressure of the liquid from the solution and X_2 is the mole fraction of the solute. As can be seen in equation (15.3), the greater the mole fraction of solute the greater will be the depression of the vapor pressure.

Since

$$(1 - X_2) = X_1 \tag{15.4}$$

where X_1 is the mole fraction of solvent in the solution. By combining equations (15.3) and (15.4), we can write

$$p = X_1 p_o \tag{15.5}$$

and, therefore,

$$X_1 = \frac{p}{p_o} \tag{15.6}$$

The term p/p_0, designated as relative vapor pressure, therefore, is a measure of the concentration of vapor expressed in terms of mole fraction. In the specific case of water, when p/p_0 is multiplied by 100 to express this term as a percent, it becomes the relative humidity (RH%):

$$(RH\%) = \left(\frac{p}{p_o}\right) \cdot 100 \tag{15.7}$$

It can be assumed further from the previous equations that for an ideal aqueous solution the chemical potential of water (w) can be written as:

$$\mu_w = \mu_w^o + RT \ln X_w \tag{15.8}$$

and, therefore,

$$\mu_w = \mu_w^o + RT \ln\left(\frac{p}{p_o}\right)_w \tag{15.9}$$

which directly links the chemical potential of water vapor to the relative pressure and hence to the relative humidity.

It can be shown further that μ_w, expressed for a nonideal solution, can be written in terms of thermodynamic activity, a, where activity coefficients account for differences between concentration and activity:

$$\mu_w = \mu_w{}^o + RT \ln\frac{a}{a_o} \tag{15.10}$$

The term a/a_0 is generally defined as "water activity," a_w, and thus if activity coefficients are ignored one can assume that

$$a_w = \left(\frac{a}{a_o}\right) \approx \left(\frac{p}{p_o}\right) \tag{15.11}$$

From equation (15.7), therefore, we can conclude that,

$$a_w \times 100 \approx (RH\%) \tag{15.12}$$

Thus, the term "water activity" can be used interchangeably with relative humidity when the latter is expressed as a fraction. Furthermore, we can conclude from this discussion that at a relative humidity of 100% (water activity of 1) water

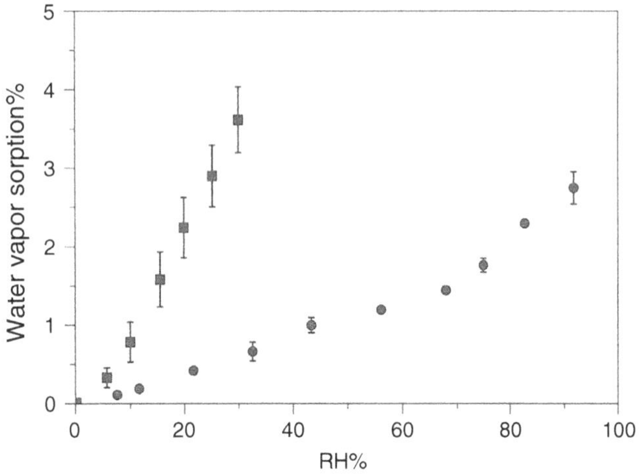

Figure 15.2 Comparison of water vapor absorption of the free acid form (circles) and the sodium salt form (squares) of indomethacin.

vapor will spontaneously and nonspecifically condense on any solid surface as liquid water, whereas at relative humidities of less than 100% (lower water concentrations in the vapor state) condensation of liquid water cannot occur and water vapor must be able to interact with the solid before it can be retained by the solid; the lower the relative humidity required to give a particular level of sorption the lower the thermodynamic activity of the sorbed water will be because of the stronger interactions that have occurred between the solid and water. This is illustrated in Figure 15.2, where the sorption isotherms for amorphous samples of the free acid form of indomethacin and its sodium salt are compared [7]. The ionized carboxylate ion of indomethacin, expected to interact more strongly with water than the unionized free carboxyl group, exhibits a similar amount of sorption at much lower relative humidities and, hence, lower water activities.

Clearly, it is important in studying the hygroscopicity of solids and the relationships between the amount of water sorbed and relative humidity to be able to measure and monitor the relative humidity of the system under consideration. Instruments that are used for such a purpose are called hygrometers, which generally work by being able to respond to the presence of water mechanically or electrically through interaction with components of the hygrometer. For example, less precise hygrometers, normally used in the home, may contain material like a polymer fiber that will change its dimensions when absorbing water vapor. Such dimensional changes may then be calibrated for different relative humidities and indicated on a dial or scale. A more precise hygrometer, widely used in many pharmaceutical applications, consists of a capacitor containing polymeric material enclosed by two metal plates. The absorption of water vapor by the polymer effectively changes the dielectric constant of the polymer in proportion to the amount of water taken up, leading to a capacitance change that can be recorded electrically. Such hygrometers, when properly calibrated, can produce an accuracy of 2% in the range of 5–95 %RH. By far the most accurate and precise hygrometer available for measuring relative humidity and water activity is the chilled mirror dew-point hygrometer. The dew-point, T_d, is taken as the temperature at which water vapor at a given pressure reaches a point of saturation so that further cooling causes liquid condensation to occur. A chilled mirror, connected to an optical and electronic device, is used to precisely detect the temperature at which condensation occurs on the mirror surface, and the relationship between the relative humidity, RH%, and T_d can be expressed by the following equation:

$$RH\% = 100 - 5(T - T_d) \tag{15.13}$$

Using this system, an accuracy of 0.2 °C can be attained, correlating to an accuracy in RH% of about 0.5%.

15.4 Measurement of Water Content and Water Vapor Sorption–Desorption Isotherms

15.4.1 Measurement of Water Content

As a starting point for studies involving the hygroscopicity of solids, it is important to be able to measure the amount of water that has become associated with a solid under a given set of conditions. As mentioned earlier, this "residual water" can arise from previous exposure to certain relative humidity conditions or because a sample previously exposed to liquid water had been incompletely dried. A common method of estimating the amount of residual water is to expose a given weight of solid to a pressure and temperature sufficient to completely remove such water, usually at roughly atmospheric pressure and 105 °C, and then reweigh the sample to measure the mass lost upon drying [8]. This is generally referred to as the "loss on drying" (LOD) method. Since one is not directly measuring the amount of water, it is important to establish that the loss in mass is not due to another volatile material, including material created by chemical decomposition at such temperatures. In the case of possible chemical decomposition upon heating at 105 °C, one can carry out LOD measurements under reduced pressure, where water will be able to completely desorb at reduced temperatures. To facilitate the use of the LOD method, which can be quite time-consuming and tedious if sample weight is monitored manually, automated moisture analyzers have been developed that can reduce the time necessary to just a few minutes. These analyzers combine an electronic balance with a heating element so that the sample can be heated rapidly with concurrent measuring of changes in weight until a constant weight is reached and recorded electronically.

There are other techniques that are able to more directly measure water in solids; the most widely used in the pharmaceutical field are Karl Fischer (KF) titration [9] and near-infrared (NIR) spectroscopy [10]. The KF method uses coulometric or volumetric titration to directly estimate the amounts of residual water in solid samples. In one scenario, water is brought in contact with a protic solvent, such as methyl alcohol, CH_3OH, containing molecular iodine, I_2, sulfur dioxide, SO_2, and an organic base, RN, to give the following reaction that consumes the water and produces a molar equivalent of I^-:

$$I_2 + SO_2 + 3RN + CH_3OH + H_2O \rightarrow [RNH]SO_4CH_3 + 2[RNH]I \tag{15.14}$$

Thus, the amount of I_2 consumed to form I^- is directly used to determine the amount of water present.

In the volumetric technique, a fixed concentration of I_2 in CH_3OH is titrated into the system, producing a redox reaction that consumes a molar amount of I_2 equivalent to the moles of water present. In the coulometric technique, most widely used in pharmaceutical situations, the iodine is generated electrically from iodide ion, I^-, in the electrolytic cell, so that

$$2I^- \rightarrow I_2 + 2e^- \tag{15.15}$$

Here, the amount of water in the sample is estimated potentiometrically from the equivalent moles of electrons used in generating the iodine. The potentiometric electrode detects the slight excess of iodine at the endpoint and stops the titration. Typically, the volumetric method can estimate water content down to 10 ppm, whereas the coulometric method can go down to 1 ppm. As described in more detail in the literature [9], factors that must be considered for optimization of the technique include sample size, sampling techniques, possible side reactions, and the rate of water removal from the solid during the titration. This may be particularly important in determining water held tightly in a crystal hydrate lattice or from solid dosage forms containing polar components.

The NIR technique of measuring water content in liquids and solids uses the fact that water exhibits very strong absorption of electromagnetic radiation in the NIR frequency range, which is roughly between 760 and 2500 nm, with distinct and strong peak intensities at 1430 and 1940 nm [10]. The absorption of radiation by water in the NIR region is primarily related to molecular overtone and combination vibrations associated with the –OH bond. In particular, no other common materials appear to absorb strongly at 1430 and 1940 nm wavelengths, making it possible for the NIR technique to distinguish water from the other components of the sample. Typically, the use of the peak at 1430 nm allows for sensitivity in the range of about 0.1–5% water content, while that at 1940 nm provides a sensitivity of about 0.02–1% water. Another distinct advantage of NIR spectroscopy for this purpose is that it can penetrate very deeply into the bulk of a sample and, therefore, can be used noninvasively to probe the water content throughout a sample. This can be done by probing with either the diffuse reflectance or transmission techniques. This noninvasive characteristic has allowed the NIR technique to be applied to the continuous monitoring of water content during pharmaceutical processing as part of process analytical technology [11].

15.4.2 Measurement of Water Vapor Sorption–Desorption Isotherms

To determine water vapor sorption isotherms, one must be able to control the temperature and relative humidity of the environment in contact with the solid and then measure the mass of water that is taken up by the solid, usually by monitoring the weight gain of the solid after exposure to water vapor. The simplest general approach involves placing the sample in a closed container, like a desiccator, where the sample sits above a saturated salt solution that produces a constant and unique relative humidity at any temperature through its colligative effect on the vapor pressure of water. Table 15.1 indicates some selected values of RH% that are created in the space above various saturated salt solutions at 25 °C due to the colligative effects of the various salts. Typically, this technique is used with relatively large solid samples, which, upon exposure to a particular RH%, are periodically removed and weighed until reaching a constant weight. Desorption experiments are generally conducted by starting with samples that have been equilibrated at a given RH and then exposed to a lower RH%. Although the "desiccator technique" can provide very reproducible results, it is generally very labor-intensive and time-consuming. It is also not useful for small solid samples of API, for example, where the small mass change due to

Table 15.1 Relative humidity at 25 °C for various saturated salt solutions.

Salt	Relative humidity (%)
Cesium fluoride	3.4
Sodium hydroxide	7.6
Lithium chloride	11.3
Potassium acetate	21.5
Magnesium chloride	32.4
Potassium carbonate	43.2
Sodium bromide	56.0
Potassium iodide	67.9
Sodium chloride	75.1
Potassium chloride	83.6
Potassium nitrate	92.3
Potassium dichromate	97.5

sorption or desorption requires a balance with sensitivity in the microgram range. However, once an equilibrium isotherm has been established by any technique, the desiccator method is an excellent tool for exposing large samples of solids at a given RH% to carry out further measurements on the physical and chemical properties of solids exposed to water vapor and knowing the water content of the sample.

The deficiencies in the use of desiccators for measuring sorption/desorption isotherms described earlier led some early workers to develop equipment that allows one to expose small samples to water vapor controlled at different relative humidities, while recording the weight change with time on a continuously recording microelectrobalance until constant weight is reached [12, 13]. The small sample size and resulting small mass of water sorbed generally provide a major advantage in that they lead to very short equilibration times. Classically, gravimetric vapor adsorption experiments, particularly with inorganic materials, have been carried out under vacuum conditions after purging the sample of impurities already adsorbed to the solid surface. However, recognizing that such pretreatments can alter the properties of organic solids, for example, dehydration of a crystal hydrate, it is generally preferable to measure water vapor sorption under ambient pressure and not in a vacuum; moderate temperatures may be used to remove less tightly bound water without altering the solid form [13]. Relative humidity in such electrobalance systems is generally controlled by adding a well-defined mixture of pure nitrogen gas and nitrogen gas saturated with water and continuously measuring the weight change due to sorption over time. The weight change is assumed to be due to sorption of water since the nitrogen only acts as a carrier gas and is not sorbed under typical temperatures of interest. Another major advantage of this technique is that one can use a computer to sequentially ramp up the RH% after a predetermined time, for example, when constant weight has been reached, while continuously recording the RH%, temperature, and weight gain or loss. This allows a complete sorption–desorption isotherm to be measured automatically within a few hours. It is important to stress that, regardless of what technique one uses to measure water vapor sorption–desorption isotherms, it is critical to be sure that one has attained a constant weight change indicative of having reached a state of equilibrium. Generally, too short an equilibration time leads to less apparent sorption and a very significant difference between the sorption and desorption plots [1].

15.5 Modes of Water Vapor Sorption

15.5.1 Introduction

Earlier in this chapter, we presented a broad definition of hygroscopicity containing a number of specific criteria: (i) a certain amount of water vapor, exposed to a solid at a given temperature and relative humidity, is taken up and retained by the solid; (ii) the interactions between water and the solid are noncovalent, particularly involving hydrogen bonding; and (iii) there should be some effect on the surface and/or bulk properties of the solid. In this section, we will show that within

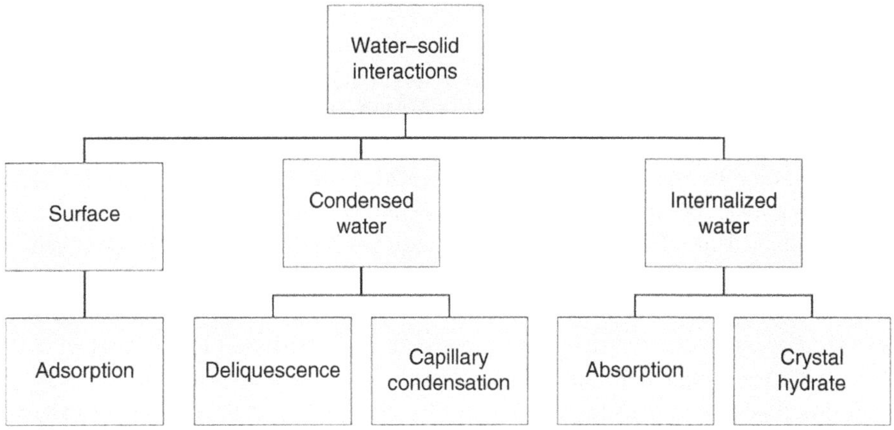

Figure 15.3 Various modes of water vapor–solid interaction.

this definition there can be different modes or mechanisms by which hygroscopicity occurs with very different types of effects on the properties of the solid. As shown in Figure 15.3, these modes of water–solid interaction can be broken down into three major groupings: (i) interaction only at the solid–vapor interface, adsorption; (ii) the appearance of condensed liquid water on the solid at relative humidities less than 100%, as in deliquescence and capillary condensation; and (iii) the absorption of water into the bulk structure of the solid as might occur with amorphous solids and within a crystal lattice with the formation of stoichiometric and nonstoichiometric crystal hydrates. In characterizing the possible hygroscopic properties of solids, after first measuring their water vapor sorption/desorption isotherms, it is important to analyze the results further regarding possible mechanisms [1]. In the following, we will present some of the key features of the major mechanisms of sorption–desorption, other than for crystal hydrate formation and dehydration, which have already been discussed in Chapter 3.

15.5.2 Adsorption

Since water molecules from the vapor state generally cannot penetrate the crystal lattice below the surface, unless there is a tendency to form a new crystal hydrate phase, adsorption is the major means by which crystalline solids directly interact with water vapor. Since the amount of any vapor adsorbed (known as the adsorbate) depends on the available surface area of the solid (known as the adsorbent), adsorption is generally expressed as the amount adsorbed per unit mass divided by the surface area available to the vapor molecules, that is, the mass adsorbed per unit area, for example, kg/m^2. Consequently, the total amount of vapor adsorbed onto a crystalline solid surface will be most significant for solids where specific surface areas are quite great, for example, powders with very small particle sizes (nanoparticles) or particles with a high degree of porosity. Of necessity, therefore, when determining the adsorption of a molecule like water from the vapor state, an independent assessment of solid surface area per unit mass is required.

From a thermodynamic perspective, we can describe the adsorption process for any vapor onto a solid at temperature, T, in terms of the overall free energy change, ΔG_{ads}, that takes place when vapor molecules interact with the molecules at the solid surface, where

$$\Delta G_{ads} = \Delta H_{ads} - T\Delta S_{ads} \tag{15.16}$$

and ΔH_{ads} and ΔS_{ads} are the enthalpy change and entropy change associated with this process, respectively. For vapor adsorption to spontaneously occur, there must be an overall loss in free energy of the system, which is favored by a negative ΔH_{ads} and by a positive ΔS_{ads}. If we assume, as stated earlier, that the adsorption of the vapor does not change the bulk thermodynamic properties of the crystal, that is, that the enthalpy and entropy of the bulk solid do not change significantly upon adsorption, we expect the adsorption of a vapor molecule to a solid surface to be determined primarily by the free energy changes associated with the interaction of the adsorbate molecules with the adsorbent molecules at the surface. It can be shown that, indeed, a major contributor to the overall thermodynamics of adsorption is the fact that molecules at any surface or interface are in a higher state of free energy than those in the bulk phase because of the unsatisfied valence forces that exist for these surface molecules. The extent of this energetic increase relative to the bulk phase is normally expressed as the surface free energy per unit area (specific surface free energy); the greater the polarity of molecules at the surface, the greater the specific surface free energy. For liquids, this term is essentially equal to the surface tension, and it is a constant for any molecule at the surface; the greater the polarity of the molecules in the liquid, the higher the surface tension. For crystalline solids, which exhibit different crystal planes and thus exhibit anisotropic surface properties, a distribution of surface free energies can occur depending on crystal structure and morphology as reflected in the chemical nature of the exposed crystal planes. Also contributing to the distribution of surface free energies for most crystalline samples are numerous higher-energy crystal defects that exist in the lattice and at the surface, including vacancies, impurities, and dislocations. As expected, the greater the energy of the exposed groups on the surface of the solid, the greater will be the overall solid specific surface free energy and, hence, the greater the overall free energy loss when molecules from the vapor state interact with the solid to reduce this free energy.

Since the adsorption of a molecule from the vapor state leads to a loss in the degrees of molecular freedom in the adsorbed species relative to that in the vapor state, there must be a significant loss in the entropy of the vapor upon adsorption, with a resulting reduced ΔG_{ads}. Therefore, for there to be an overall loss in ΔG_{ads}, there must be a large negative ΔH_{ads}. Experimentally, one can estimate ΔH_{ads} in a number of ways. The most direct approach is to calorimetrically measure the heat released upon adsorption of a given amount of vapor at a given temperature and pressure. For example, one can measure the integral heat of adsorption, Q_i, from the total heat released going from zero to n moles of coverage of adsorbate. A more informative parameter is the differential heat of adsorption, q_d, which is the incremental heat change that occurs with the addition of an infinitesimal amount of adsorbate, such that

$$q_d = \left[\frac{dQ_i}{dn}\right]_{T,p,n} \tag{15.17}$$

The value of q_d is obtained experimentally by admitting successive incremental quantities of vapor and measuring the small heat changes upon adsorption after each increment. For an energetically homogeneous solid surface, the value of q_d obtained at lower coverages will remain constant with increasing surface coverage since all sites on the solid surface are energetically equivalent. However, at higher amounts of adsorption, approaching complete monolayer coverage, the adsorbed molecules will begin to interact laterally, releasing additional heat. Further adsorption beyond a monolayer (to be discussed in the following text) will also contribute to the differential heat because of adsorbate–adsorbate interactions. If the surface is energetically heterogeneous because of crystal defects and varying morphologies of crystals, as is usually the case, the differential heats will not be constant at lower coverages but rather will first produce greater heat changes because of adsorption to the higher-energy sites and then less heat released as these high-energy sites are occupied and the vapor interacts with the less energetic sites. This is illustrated for the adsorption of water vapor onto crystalline Ag_2O in Figure 15.4 [14]. Determining the values of q_d as a function of surface coverage is useful for several reasons. It provides (i) an indication of the strength of adsorbate-adsorbent interactions; (ii) an indication of the extent of heterogeneity of the surface; and (iii) a pattern of change that can be used to compare the surface properties of a solid undergoing different processing conditions, as, for example, drying or mechanical stress [3].

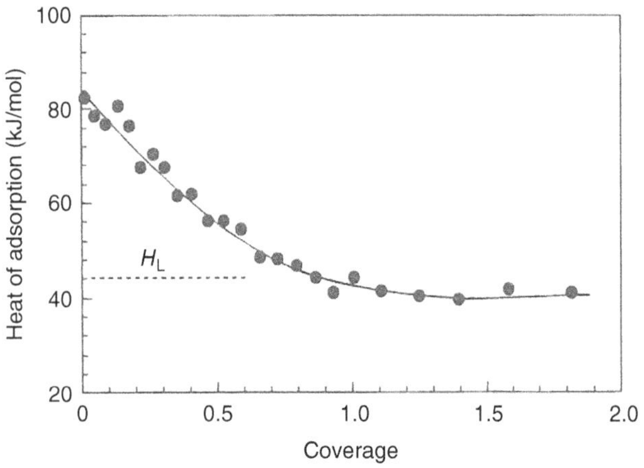

Figure 15.4 Differential heats of adsorption for water vapor on crystalline silver oxide as a function of the extent of surface coverage. *Source*: [14]/with permission of American Chemical Society.

Since vapor adsorption is an exothermic process, we would expect the amount of adsorption at a particular pressure to decrease as the temperature is increased and, therefore, using such data, one can obtain an estimate of the heat changes associated with adsorption by applying the Clausius–Clapeyron equation (equation (15.18)) and calculating the isosteric heat of adsorption, q_{st}:

$$\left\{\frac{d\ \ln p}{dT}\right\}_n = \frac{q_{st}}{R} \tag{15.18}$$

where p is the pressure of vapor causing n moles of vapor adsorption at temperature T, and R is the gas constant; the value of q_{st} is obtained from the slope of ln p versus $1/T$. Data obtained at different temperatures for the adsorption of water on Ag_2O and plotted in this manner are given in Figure 15.5 [14]. Knowing the isosteric heat of adsorption is useful because it can be directly related to the differential heat of adsorption, q_d, by the following expression,

$$q_{st} = q_d + RT \tag{15.19}$$

and, thus, provide a convenient way to analyze the energetic patterns for water vapor adsorption as described in Figure 15.4.

From the previous discussion, adsorption of any vapor onto a solid surface depends very much on the strength of the interaction between exposed chemical groups at the solid and the vapor. Covalent interactions formed between adsorbate and adsorbent molecules, generally described as chemisorption, typically have chemical bond strengths on the order of 100 – 500 kJ/mol and most often occur at high pressures and temperatures. Generally, in such cases, the adsorbate undergoes dissociation of covalent bonds and forms unstable molecular species, which can then undergo specific types of covalent interaction. This process plays a very important role in many chemical reactions where vapor molecules are brought in contact with solids to carry out certain types of catalytic reactions. Because of the specificity of the interaction, it is generally limited to a single molecular layer of adsorbed species, determined by the number of solid surface groups available for interaction. The relationship between the amount adsorbed per unit area and pressure at constant temperature generally can be described by the Langmuir equation [15], as shown in equation (15.20) in terms of the volume adsorbed, v_{ads}, and the maximum volume of vapor that can be adsorbed to form a complete monolayer, V_m, as a function of the pressure, p, within the system:

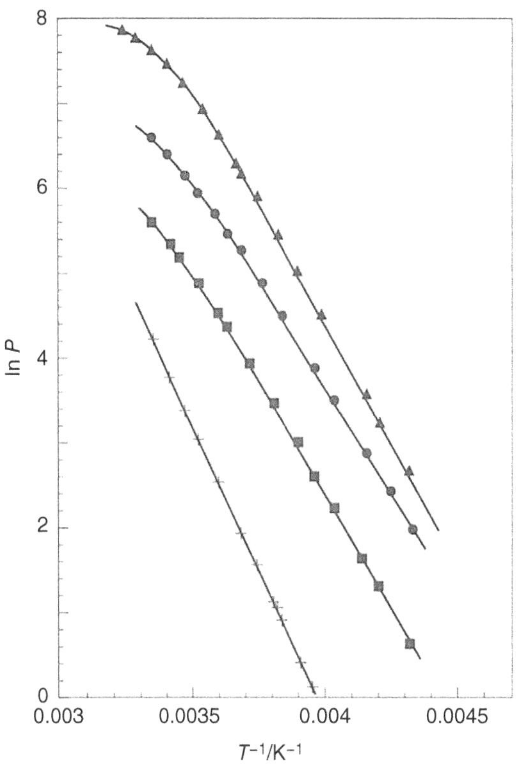

Figure 15.5 Clausius–Clapeyron plots for water vapor adsorption on crystalline silver oxide at various surface coverages. *Source*: [14]/with permission of American Chemical Society.

$$\frac{V_{ads}}{V_m} = \frac{\alpha p}{(1 + \alpha p)} \tag{15.20}$$

Here, α is a constant that reflects the chemical affinity of the vapor for the solid surface groups, related directly to the free energy of adsorption. From equation (15.20), and as represented in the plot shown in Figure 15.6, it can be seen that at very low p, where αp is much less than 1, the change in the fraction of a complete monolayer, V_{ads}/V_m, is expressed as

$$\frac{V_{ads}}{V_m} \approx \alpha p \tag{15.21}$$

and initial plots should be linear. At high values of p, where αp is much greater than 1,

$$\frac{V_{ads}}{V_m} = 1 \tag{15.22}$$

indicating complete coverage of a single molecular layer and no further increase in adsorption with increasing pressure.

Another important and more common category of vapor adsorption is physisorption, which occurs when nonpolar gases or vapors, such as O_2, N_2, CO_2, and CH_3, accumulate at a solid surface by means of intermolecular bonding due to the nonspecific van der Waals forces, with interaction energies on the order of 1 – 2 kJ/m, depending on molecular size. For such physisorption to occur, temperatures must be reduced to values close to the boiling temperature of the adsorbate, since these very low temperatures are required to compensate for the weak enthalpy of adsorption relative to the significant negative entropy changes that occur upon adsorption. Since these interaction energies are very close to those associated with the intermolecular interactions responsible for the condensation of these vapors as liquids

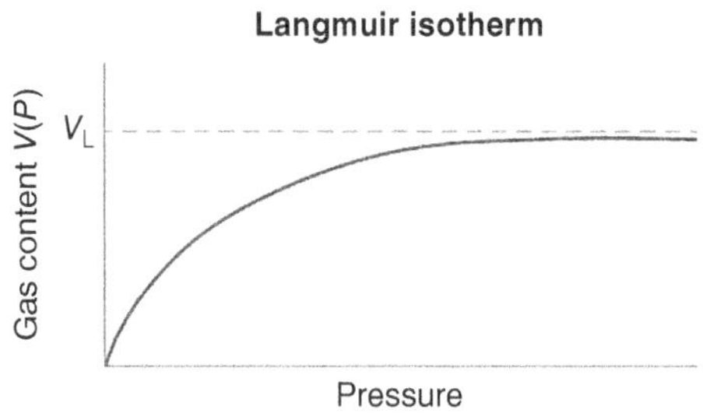

Figure 15.6 Typical gas adsorption isotherm characterized by the Langmuir isotherm equation. *Source*: Reproduced from [15]/with permission of American Physical Society.

Figure 15.7 Schematic representation of multilayer physisorption of gases reflected in the BET equation. *Source*: Adapted from [16].

and because of the small differences between the enthalpy of adsorption and the enthalpy of liquid condensation, physisorption is very nonspecific and easily reversed as temperatures are increased toward ambient conditions. Also, in contrast to chemisorption, at higher pressures and surface coverages, such nonspecificity allows for the possibility of multiple layers of adsorption, as schematically illustrated in Figure 15.7. Many theories and equations have been developed to describe physisorption isotherms, considering the possibility of multilayer adsorption at higher pressures. Conceptually, the simplest and most useful equation is one developed by Brunauer, Emmett, and Teller [16], based in part on the Langmuir equation, and commonly referred to as the BET equation, as shown in equation (15.23),

$$\frac{V_{\text{ads}}}{V_{\text{m}}} = \frac{C\left(\frac{p}{p_{\text{o}}}\right)}{\left[1-\left(\frac{p}{p_{\text{o}}}\right)\right]\cdot\left[1+(C-1)\left(\frac{p}{p_{\text{o}}}\right)\right]} \tag{15.23}$$

Here, as in equation (15.20), V_{ads} is the volume adsorbed, V_{m} is the maximum possible volume adsorbed that can form a closely packed monolayer, p/p_{o} is the relative pressure as defined in equation (15.6), and C is a constant reflecting the affinity of the vapor for the solid surface relative to the affinity of the vapor molecules for each other [16]. Figure 15.8 depicts two types of physisorption isotherms demonstrating different patterns in the amount adsorbed at lower values of p/p_{o}. "Type II" isotherms exhibit values of C that are greater than 1, indicating a stronger interaction between adsorbate and adsorbent than between adsorbate molecules making up the multilayer, while with "Type III" isotherms C values are now less than one due to stronger adsorbate–adsorbate interactions in the multilayer than occurs between the adsorbate and adsorbent in the first layer. Isotherms related to the Langmuir equation (equation [15.20]) are generally referred to as "Type I" isotherms [16].

In general, the more polar the molecules at the solid surface and those of the vapor are, the greater will be the tendency to exhibit higher C values and Type II behavior. An important application of the BET equation is that it is widely used to determine the specific surface area of any solid from physisorption data obtained with nonpolar gases. This is possible because of the nonspecificity of physisorption by very nonpolar gases and the assumption that V_{m} represents the volume of gas that completely covers the available surface of the solid with one molecular layer. Thus, knowing the cross-sectional area per molecule of gas, and assuming a certain packing density, allows one to estimate the total area of the solid surface, S, using equation (15.24).

$$S = \frac{(V_{\text{m}}N_{\text{a}}A_{\text{m}})}{v} \tag{15.24}$$

where N_{a} is Avogadro's number, A_{m} is the cross-sectional area of the nonpolar gas, and v is the molar volume of the absorbent gas. Indeed, physisorption data for small molecules, such as nitrogen and krypton, on most solid surfaces, taken at the boiling temperature and at values of p/p_{o} in the range of 0.05–0.35, follow the BET equation and provide values of specific surface area, in units of area per unit mass that are physically consistent with other methods. In such cases, the values of A_{m} for N_2 and Kr, in m^2/molecule, are generally taken to be 1.62×10^{-17} and 1.95×10^{-17}, respectively.

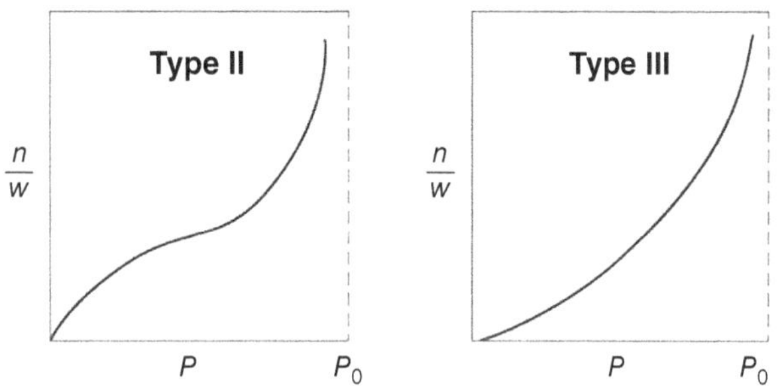

Figure 15.8 Schematic representations of Type II and Type III multilayer gas physisorption isotherms as described by the BET equation. *Source*: Adapted from [16].

Since water is a highly polar molecule, with a strong tendency to undergo dipole–dipole interactions, particularly hydrogen bonding, as well as ion–dipole interactions, we would expect that it would have the ability to adsorb most strongly to those solid surfaces that likewise exhibit significant polarity. Chemisorption of water vapor, involving the dissociation of the water molecule to form the OH and H radicals, has been shown to occur under certain conditions on metal oxide surfaces, such as SiO_2 and ZnO [17]. For example, it was shown that covalent interactions with these radicals occur with ZnO at Zn^+O^- and Zn^+O^{2-} ion pair sites. Recently, evidence has been presented for the first time to show that two organic salts of pharmaceutical interest, exposed to high water vapor pressures under ambient temperature conditions, caused dissociation of the water molecules, leading to chemisorption of the free radical form of oxygen at crystal defect sites [18]. The significance of chemisorption of water in such cases is still uncertain, but since major instabilities in crystals are usually initiated at crystal defects, such chemisorption at elevated relative humidity may have serious implications for solid-state stability.

Since water molecules have a very high potential for hydrogen bonding, we would expect water vapor, in the form of molecular water, to readily undergo physisorption to any solid surface that contains polar groups capable of interacting through hydrogen bonds. Since interaction energies involving hydrogen bonding occur on the order of 5 – 30 kJ/mol, we would expect that such physisorption could occur at higher temperatures than are generally observed with the more nonpolar gases that adsorb only through van der Waals interactions, for example, under ambient temperature and pressure conditions normally encountered in pharmaceutical situations. That the presence of polar groups on the solid surface is essential for water vapor adsorption under ambient conditions is seen by the negligible physisorption of water that occurs on nonpolar solid surfaces, such as pure carbon, solid hydrocarbons, and nonpolar polymers, such as poly(ethylene) and poly(tetrafluoroethylene). On the other hand, it has been shown that if water–solid interactions are strong enough, it is possible for physiosorbed water molecules to form up to three to four multiple layers under ambient conditions. Indeed, an excellent illustration of such behavior can be seen in Figure 15.9, where water vapor adsorption and desorption on the 100 crystal face of crystalline NaCl at 25°C has been reported [19]. Here, we can observe that three to four equivalent monolayers occur upon adsorption and desorption up to pressures just below the relative humidity where water condensation by deliquescence occurs (to be discussed in the following text). It was further established in this study that adsorbed water did not affect the bulk solid properties of the very water-soluble NaCl, as shown by the agreement between adsorption and desorption values at the same pressure. It was shown further that the adsorbed water molecules accumulating at high-energy crystal defect sites solvated the Na^+ and Cl^- ions to the point where the mobility of these ions within the adsorbed layers was increased significantly but not to the point where they were in true aqueous solution [20]. Consequently, we can conclude that water adsorption on crystal surfaces can have an important effect on surface structure and dynamics of the solid but not on bulk properties. In addition, a rough estimate of the number of water molecules that can pack closely to form a complete physiosorbed monolayer, assuming the cross-sectional area of a water molecule to be about $1.2 \times 10^{-7} m^2$, indicates that approximately one monolayer of adsorbed water per m^2 of surface is equivalent to about 0.025% water and that the maximum amount of water adsorbed, four molecular layers, for example, is about 0.1%. Therefore, we can see that ordinarily the total mass of water taken up by a crystalline solid is quite small unless, of course, the solid has an unusually high specific surface area, for example, nanocrystalline particles or nanoporous solids.

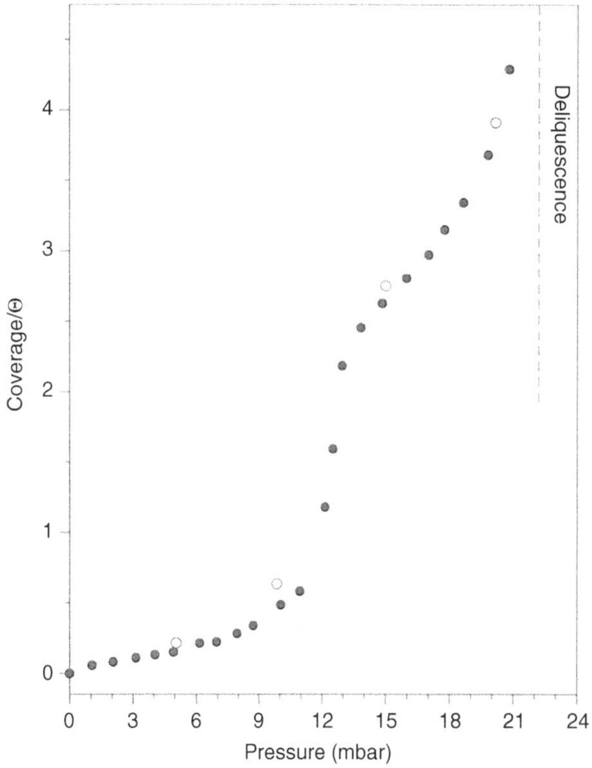

Figure 15.9 Water vapor adsorption on crystalline sodium chloride expressed in terms of the fraction of surface covered as a monolayer or multilayer as a function of water vapor pressure. *Source*: [20]/with permission of AIP Publishing LLC.

15.5.3 Deliquescence

Deliquescence can be generally defined as a process leading to the condensation of water vapor that occurs when the vapor pressure of a saturated aqueous solution of a substance making up the crystalline solid is less than the vapor pressure of water in the ambient air. When water vapor is collected by the pure solid compound a mixture of the solid and liquid or an aqueous solution of the compound, forms until the substance is dissolved and is in equilibrium with its environment; at

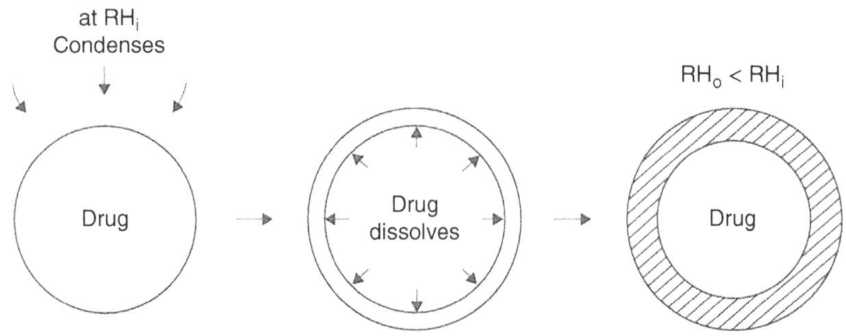

Figure 15.10 Schematic representation of the steps involved in deliquescence of crystalline solids below and above the critical relative humidity, RH_0. *Source*: [21]/with permission of Elsevier.

this point the vapor pressure of water over the aqueous solution will equal the partial pressure of water in the atmosphere in contact with it [21]. In essence, this says that when the relative humidity of the atmosphere, RH_i, exceeds that of a saturated solution of the substance, RH_0, the substance will begin to exhibit deliquescence and start to dissolve. The sequence of events may be observed schematically in Figure 15.10, where at values of RH_i below RH_0, physisorption of water occurs to form adsorbed layers with no effect on the bulk solid. If, however, RH_i just exceeds RH_0 water vapor will spontaneously condense onto the solid surface, causing the solid to dissolve in the condensed water and to form an aqueous film on the surface as a saturated solution. As condensation continues to occur, the concentration of solute in the aqueous solution at the surface of the solid particle is diluted, and the RH above the solid due to the liquid layer is increased. However, more solid dissolves and restores the value of RH_0. As long as RH_i is greater than RH_0, this process of condensation of liquid water, and dissolution will continue until most of the solid has dissolved. Likewise, as shown in Figure 15.10, if RH_i for a stored sample is kept below RH_0, the solid can be maintained free of any tendency to deliquesce. It is also important to recognize that the process of deliquescence, once started, can be reversed, for example, recrystallization of the solid can occur by simply reducing RH_i to values below RH_0. Since deliquescence causes a thermodynamic phase transition, it is most likely initiated at high-energy crystal defect sites. From a pharmaceutical perspective, if this change in RH_i to above and below RH_0 were to occur under uncontrolled storage conditions, it is possible that the recrystallization of the substance could inadvertently lead to a new crystal form with different solid-state properties. It should also be mentioned that the rate of deliquescence will be greater when the system is at a higher RH_i, and it will be further influenced by the specific surface area, rate of dissolution into the aqueous film when the kinetics associated with the dissolution of solids relative to water condensation becomes rate-determining, and by the temperature changes within the aqueous film around the solid due to heat changes associated with condensation and dissolution [21].

From this discussion so far, it is clear that the greater the RH_0 for the solid the greater flexibility the formulator has in maintaining proper storage conditions. Substances having the most ability to deliquesce at lower RH_i values are those with relatively high water solubility and with the ability to produce a significantly nonideal colligative effect, that is, greater activity coefficients, and hence a saturated solution with relatively low values of RH_0. From a pharmaceutical perspective, therefore, we would expect deliquescence to be most prevalent with very water-soluble inorganic and organic salts, as well as low molecular weight sugars. Table 15.2 presents values of RH_0 for various salts and sugars of pharmaceutical interest. There, we can note the significant effect of solubility and extent of polarity by comparing the RH_0 values for three choline salts of varying polarity, choline chloride, bromide, and iodide; the colligative effect is reduced significantly as the polarity of the counterion is decreased. Indeed, this fact can be very helpful as a strategy during the selection of the best salt form of an API salt that seeks to produce a salt form with as high a value of RH_0 as possible to minimize deliquescence. It should be pointed out, however, that the use of less soluble salt forms to raise RH_0values can lead to salts with very poor overall aqueous solubility and bioavailability, so a balanced approach must be used. The implications of deliquescence for drug stability will be discussed in Chapters 18 and 19.

15.5.4 Capillary Condensation

Another means by which vapor, including water vapor, condenses as a liquid below the vapor pressure of the liquid, that is, $p/p_0 < 1$, is capillary condensation. As the name implies, capillary condensation occurs in pores and cracks associated with a solid sample that contain highly curved confined spaces with nanometer dimensions. As shown earlier, ordinarily, any

Table 15.2 Critical relative humidities (RH_o) for deliquescent materials at 25 °C.

Compound	RH_o (%)
Choline chloride	20
Choline bromide	41
Sodium valproate	44
Sorbitol	69
Choline iodide	73
Sodium chloride	75
Ranitidine HCl	76
Citric acid monohydrate	78
Sucrose	85
Mannitol	95

crystalline solid of sufficient polarity will physisorb water molecules to form multilayers when exposed to water vapor. If the solid is relatively water-insoluble with a very high critical relative humidity, RH_o so that deliquescence does not occur at normally encountered RH values, the amount of water associated with the solid surface will be very limited. However, as such multilayers form in these confined spaces, liquid water will spontaneously begin to condense and form a highly curved meniscus at the liquid–vapor interface because of the curvature in the confined space. As discussed in Chapter 7 in the context of nanocrystals, the bulk thermodynamic properties of a material can be significantly altered when the interface is highly curved. For example, it was shown in equation (7.3) that the equilibrium solubility of very fine particles will be greater than the normal bulk values to an extent depending on the specific surface free energy at the solid–liquid interface and the radius or degree of curvature of the particle. Capillary condensation that occurs at $p/p_o < 1$ results from the fact that the vapor pressure of the condensed liquid situated at a highly curved pore surface is generally less than the vapor pressure of an ordinary liquid, and thus condensation can occur at $p < p_o$. The basis of such behavior can be understood by first examining the thermodynamic properties of a liquid when a liquid–vapor surface is created by forming a small spherical droplet in air or a bubble within a liquid. Here, it can be shown that there is a difference between the pressure of the liquid on each side of the interface, ΔP, with pressure always being greater inside the curved interface, that is, on the concave side, so that the pressure inside a bubble or inside a liquid droplet is greater than the pressure on the outside. Such a difference arises because surface forces, due to the surface tension, add to those forces that ordinarily give rise to bulk pressure when the radius of the system becomes quite small, and, therefore, the curvature becomes quite high. This difference for a sphere can be expressed by the Young–Laplace equation [22]:

$$\Delta P = \frac{(2\sigma_{lv})}{r} \tag{15.25}$$

where σ_{lv} is the surface tension at the liquid–vapor interface and r is the radius of a spherical bubble or droplet. From equation (15.25), we can see that the greater the value of σ_{lv} and the smaller the value of r, the greater will be ΔP, referred to as the capillary pressure.

It can be shown further that the presence of a highly curved liquid–vapor surface, and, therefore, the presence of a capillary pressure, can give rise to a change in liquid vapor pressure, p, relative to that of an ordinary "flat" surface, p_o, such that p/p_o can be either greater than or less than one, depending on which side of the curvature the liquid–vapor surface is located, that is, convex or concave. As mentioned in Chapter 7, the effect of surface curvature on the equilibrium solubility of very fine nanocrystalline particles (a measure of free energy) in terms of the radius of the particle and the specific surface free energy at the solid–liquid interface, can be determined from the Gibbs–Thomson equation, equation (7.3), which was derived using the Young–Laplace equation (equation [15.25]); the nature of the curvature always leads to increased solubility. Using the same thermodynamic principles, it is possible to express the relative vapor pressure of a spherical liquid droplet or bubble, p/p_o, with the Kelvin equation [23],

Droplet:

$$\ln\frac{p}{p_o} = \frac{2M\sigma_{lv}}{\rho rRT} \tag{15.26}$$

Bubble:

$$\ln\frac{p}{p_o} = -\frac{2M\sigma_{lv}}{\rho rRT} \quad (15.27)$$

where M is molecular weight, σ_{lv} is the liquid surface tension, ρ is the density of the liquid, r is the radius of the particle, R is the gas constant, and T is the temperature. Since all parameters in the Kelvin equation are positive, equation (15.26) predicts $p > p_o$ above a droplet (higher vapor pressures), while for bubbles equation (15.27) predicts $p > p_o$ inside a bubble (lower vapor pressures), as shown schematically in Figure 15.11. Indeed, this can be seen in Table 15.3 for the vapor pressure of water droplets and bubbles of various sizes. As can also be seen, these capillary effects do not become significant until the dimensions approach values less than 100 nm, as is also true for the solubility of nanocrystals.

We are now in a position to see how capillary condensation might occur by considering a cylindrical solid pore with radius, r, as illustrated in Figure 15.12, that has formed a completely wetted (contact angle of zero) hemispherical meniscus, also of radius, r. Starting with equation (15.27), it can be shown that for a cylindrical pore we can express the Kelvin equation as

$$\ln\frac{p}{p_o} = -\frac{\sigma_{lv}M\cos\Theta}{\rho rRT} \quad (15.28)$$

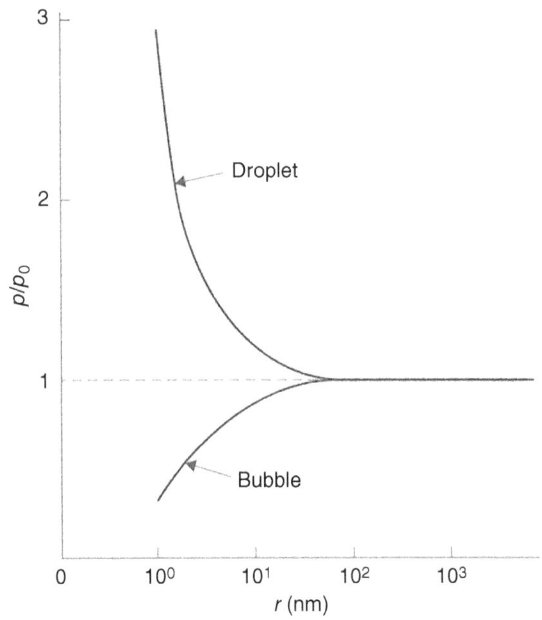

Figure 15.11 Changes in the relative vapor pressure of water (p/p_o) with the radius of a bubble and liquid droplet. *Source*: [22–23].

Table 15.3 Relative vapor pressure of water at curved surfaces with radius, R.

Radius (nm)	p/p_o (drop)	p/p_o (bubble)
1000	1.001	0.999
100	1.011	0.989
10	1.115	0.897
1	2.968	0.337

Source: Adapted from Defay and Prigonine [22] and Fisher et al. [23].

where Θ is the contact angle between the condensed liquid and solid. For a completely wetted system, that is, Θ is 0° and, therefore, $\cos\Theta$ is 1, the value of p/p_o must be less than 1 since all parameters in the equation are positive. Therefore, we can say that water vapor will condense in such a pore at RH values less than 100% if the pore radius is sufficiently small. As expected from what is shown in Table 15.3, and using equation (15.28), we would not expect values of p/p_o to be reduced to any significant degree until the pore diameters are much less than 100 nm. For example, using equation (15.28) and assuming for water that σ_{lv} is 72 mJ/m^2, M is 18 g/mol, ρ is 1×10^{-6}gm/m^3, Θ is 0°, T is 298 K, R is 8.314 JK^{-1}mol^{-1}, and r is 50 nm, we can calculate the value of p/p_o at which condensation will occur to be about 0.99, whereas at a radius of 5 nm the value of p/p_o is about 0.90. Therefore, capillary condensation generally only occurs, with possible major effects, in very highly porous solids containing a high proportion of nanoporous structure that approach molecular dimensions, or in extremely small particle cracks and fissures, and at points of contact between very fine particles. As seen in Figure 15.13, a typical adsorption isotherm for water vapor and a highly porous solid with pore dimensions on the order of nanometers reveals significant amounts of adsorption and a distinct plateau at the point of condensation. Typically, the adsorption–desorption hysteresis shown here is characteristic of capillary condensation, the extent of which is determined by the nature of the porous network. If capillary condensation occurs with solids that are quite water soluble, such as salts or sugars, it is possible to observe dissolution at RH values that are well below the critical relative humidity for deliquescence. For example, using atomic force microscopy (AFM), it was shown that dissolution of NaCl first occurred at the point of contact between the nanosize AFM tip and the NaCl crystal at 40% RH rather than at the deliquescence point of NaCl, that is, 75% RH [24].

15.5.5 Absorption by Amorphous Solids

So far, the modes of water vapor sorption that we have discussed have been shown to occur with crystalline materials either by direct interaction of water molecules with molecules of the solid confined to the crystal surface, followed by possible additional water–water interactions to the extent of three to four molecular layers (adsorption), or by adsorption and then condensation of liquid water below a relative humidity of 100% (deliquescence and capillary condensation). It has been shown that the adsorption of water vapor affects the surface properties of crystalline materials, particularly those at surface defect sites; however, it does not influence the bulk properties of the solid. It has been shown further that only if condensed water, formed during deliquescence or capillary condensation, is capable of significantly dissolving the solid will such sorption affect the bulk properties of the solid. As discussed in detail in Chapter 6, solids can exist in the amorphous state, a state generally categorized as that of a supercooled liquid or glass, depending on the temperature relative to the glass transition temperature, T_g, and as a solid state lacking the three-dimensional order of a crystal having the molecular properties of a liquid. In view of the lack of long-range order in amorphous solids, we would expect an amorphous solid, as with any liquid, to have a greater free volume, lower density, and greater molecular mobility than its corresponding crystalline form. For example, the density of the amorphous form of indomethacin, made by quench-cooling of the melt, is 1.31 g/cm^3, while that of the most stable polymorphic γ crystalline form is 1.38 g/cm^3, [25]. As with miscible liquids, this increase in free volume, plus sufficient polarity of the molecules making up the amorphous solid, leads to a certain degree of "miscibility" and hence penetration or "absorption" of water molecules into the bulk structure of the amorphous solid. In such cases we would expect the total amount of water vapor absorbed at a particular RH to be considerably greater than the amount of water adsorbed to the crystalline form of comparable specific surface area, as is shown in Figure 15.14 for crystalline and amorphous indomethacin made from the quenched melt. Also shown in Figure 15.14 is the water vapor absorption isotherm for a "stable form" of amorphous indomethacin obtained by slow condensation from the vapor state and having a density about 2% higher than the sample prepared from the melt [26]. Note here that the amount of water vapor absorbed by the "stable form" is roughly five times less than that of the sample obtained from the melt. This confirms the importance of the available free volume of the glass in determining levels of absorption. Indeed, water vapor absorption by amorphous solids in the glassy state that have undergone annealing has been shown to absorb less water than an unannealed sample, presumably due to a reduction in free volume brought about by the annealing process [27].

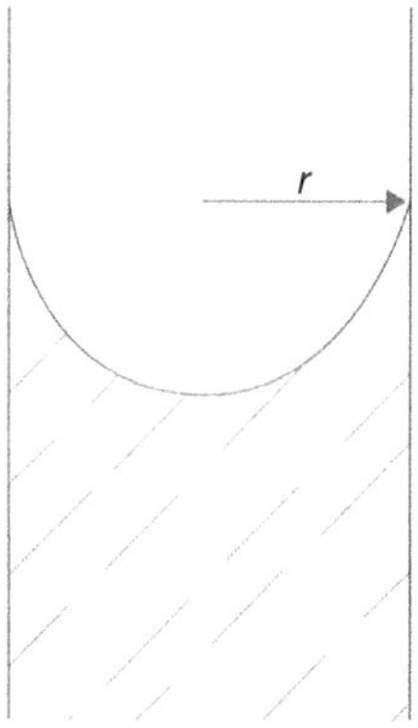

Figure 15.12 Meniscus of condensed water in a completely wetted cylindrical pore of radius, *r*.

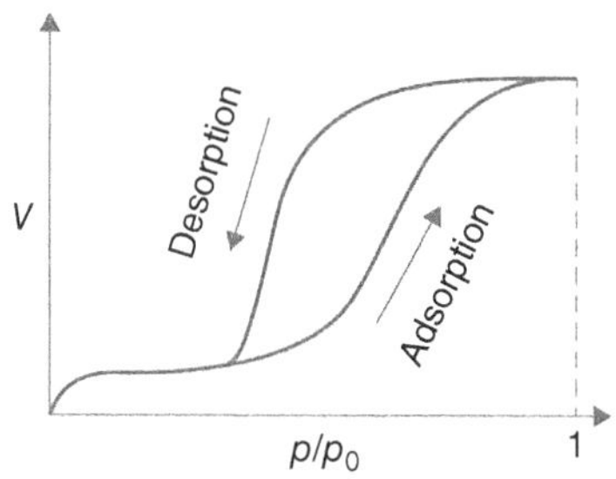

Figure 15.13 Typical adsorption/desorption isotherm for a gas undergoing capillary condensation.

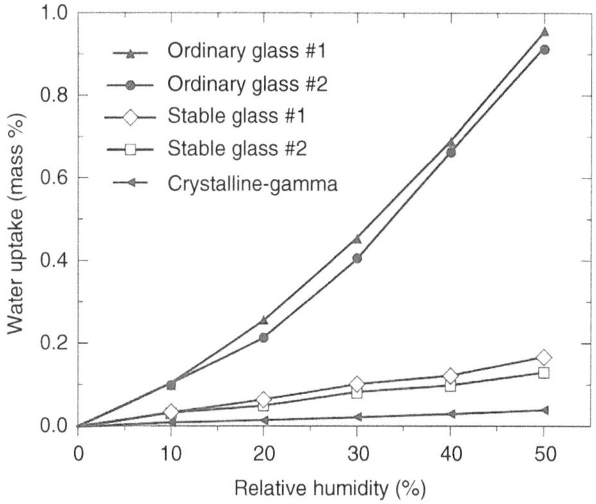

Figure 15.14 Comparison of water vapor sorption by indomethacin in the form of an ordinary glass, stable glass, and crystalline form. *Source*: [26]/with permission of American Chemical Society.

The importance of the polarity of the solid in determining the extent of water "miscibility" in amorphous solids can be seen in Figure 15.2, where the absorption of the more polar sodium salt of indomethacin is shown to be much greater than that of the free acid at comparable values of RH [7]. Another example of the importance of polarity is shown by comparing the water vapor absorption isotherms of poly(vinyl pyrrolidone), PVP, and poly(vinyl acetate), PVAc, where the carbonyl group of PVP is known to exhibit greater hydrogen bonding potential than the carbonyl group of PVAc. As seen in Figure 15.15, and, as expected, PVP exhibits much greater absorption of water molecules than does PVAc [28].

As we learned in Chapter 6, the properties of miscible amorphous mixtures are strongly dependent on the properties of the individual components, particularly in the extent to which the components interact with one another and in the resulting free volume changes of the mixture that occur relative to those of the individual components. This, in turn, is reflected

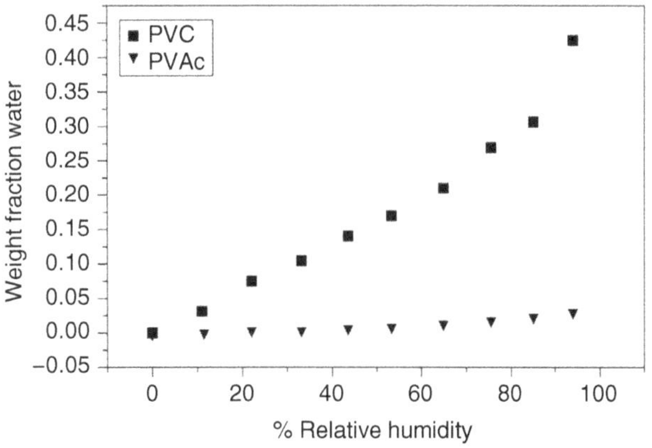

Figure 15.15 Comparison of water vapor absorption of PVP and PVAc. *Source*: [28]/with permission of Elsevier.

in the glass transition temperature of the mixture, with the value of T_g for a miscible mixture generally falling between those of the individual components as predicted in either equations (6.25) or (6.27). For example, given that amorphous water, with a high level of free volume, has a T_g of 136 K, from these equations we would expect the presence of absorbed water to significantly reduce the T_g of any dry amorphous solid and, therefore, to have a pronounced effect on the bulk properties of the solid. This effect of absorbed water on amorphous indomethacin, at relatively low levels of absorption, can be observed in Figure 15.16, where T_g is plotted versus water content [29]. Here, we see that as little as 1% w/w water (56% RH) lowers the T_g of dry indomethacin by roughly 10 K, while 2% w/w absorption (83% RH) reduces the T_g by about 20 K. With more polar amorphous solids, such as PVP (see Figure 15.15), we can expect much more water absorption and greater effects on T_g at comparable RH values. For example, as shown in Figure 15.17, at 75% RH, the roughly 25% w/w water absorbed by PVP produces a T_g value of 300 K, compared to the value of 448 K for dry PVP, or a 118 K decrease that translates into significant plasticization with a greatly increased molecular mobility [30].

From this discussion, two general factors appear to be critical in determining the water vapor absorption isotherm for an amorphous solid: the changes that might take place in free volume relative to the free volume of the solid and the water, and the strength of the interaction between the solid and water through hydrogen bonding. In this context, therefore, the process of absorption by amorphous solids can be thought as the formation of a "solution" of water in a "liquid" governed in large measure by the thermodynamics normally associated with the formation of an aqueous solution, for example, hydration through hydrogen bonding, plus any additional factors that might arise from structural changes that take place in the amorphous solid due to the presence of the water. We can begin to see the importance of these factors by considering the absorption of water vapor into an amorphous polymer above the T_g of the polymer, where it is in the equilibrium liquid or supercooled liquid state. Considering the thermodynamics of this process to be the same as that of a polymer dissolved in water, it has been possible to use the Flory–Huggins equation, as described in Chapter 6 in another context, to express the relationship (sorption isotherm) between the amount of any small vapor molecule absorbed into the polymer at a particular relative pressure, p/p_o :

$$\frac{p}{p_o} = \Phi_1 e^{\{1-(\frac{1}{x})\Phi_2+\chi Q_2^2\}} \tag{15.29}$$

where Φ_1 and Φ_2 are the volume fractions of the absorbed molecule and the polymer, respectively; x is the molar volume ratio of the polymer/to absorbed molecule, and χ is the Flory-Huggins chi interaction parameter, reflecting the small molecule–polymer interaction energy (see equation [6.19] in Chapter 6). This equation has been successfully used to predict the absorption isotherm for a variety of absorbed vapor–polymer combinations with known independently determined values of χ and x, when the polymer is above its T_g. In such situations, both the polymer and vapor molecules have been relatively nonpolar, for example, benzene absorbed by rubber [31]. In principle, equation (15.29) should be applicable to water vapor absorption by amorphous solids, and indeed it is when the temperature of the system is above the T_g of the polymer–water mixture creating the equilibrium liquid state.

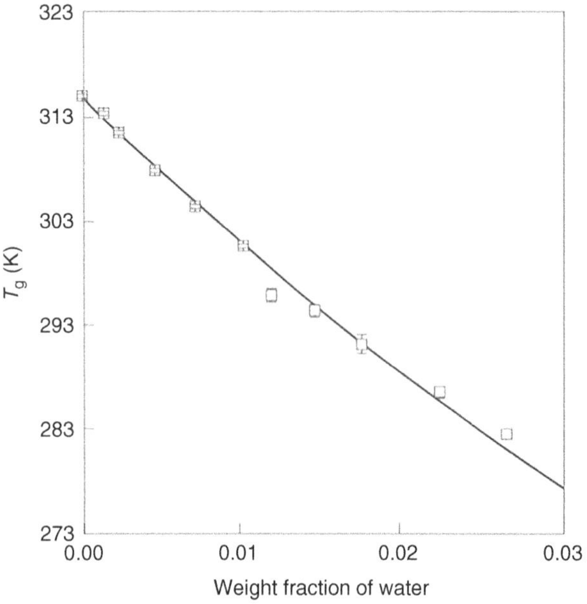

Figure 15.16 The glass transition temperature of amorphous indomethacin as a function of absorbed water content. *Source*: [29]/with permission of Elsevier.

However, polar amorphous solids in which significant amounts of water vapor can be absorbed at ambient temperatures and relative humidities generally have very high values of T_g so that water absorption at lower RH initially occurs in the glassy state until sufficient water has been absorbed to plasticize the system to values of a T_g below that of the operating temperature, T. As shown in Figure 15.18 for the absorption of water by the carbohydrate, maltohexaose, with a "dry" T_g of 448 K [32], equation (15.29) works well at temperatures where the carbohydrate is in the supercooled equilibrium state (above T_g), but greatly underpredicts the absorption in the glassy state [33]. Such behavior above and below T_g also has been reported for water vapor absorption by amorphous PVP with a "dry" T_g of about 450 K [34, 35]. A central question, that we might ask at this point, is, why does the Flory–Huggins solution model for water vapor absorption not work well for more polar amorphous solids at lower relative humidities and temperatures when the solid is in the glassy state but work well at higher relative humidities and when the temperature is above T_g where the solid is in the supercooled liquid state? Stating the question in another way, what factors associated with the glassy state would produce a greater water vapor absorption and, hence, a greater loss in the free energy of absorption than predicted by equations (6.19) and (15.29)?

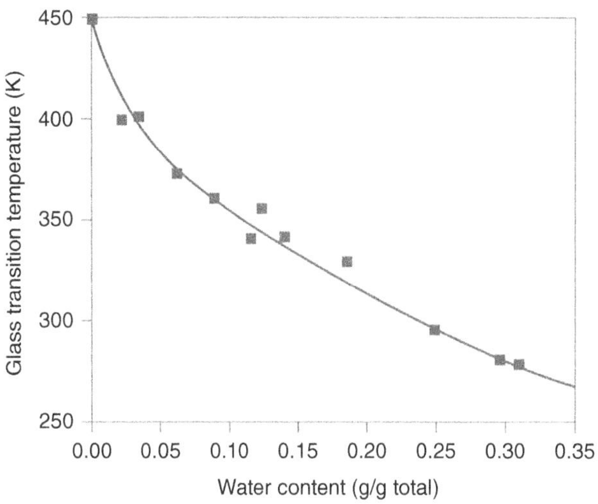

Figure 15.17 The glass transition temperature of amorphous PVP as a function of absorbed water content. *Source*: Adapted from [30].

As described in Chapter 6, when two miscible components are "ideally" mixed, the total free volume of the mixture will be the weighted average of the free volumes of each of the components:

$$V_{f_{\text{idealmix}}} = \Phi_1 V_{f_1} + \Phi_2 V_{f_2} \tag{15.30}$$

where Φ is the volume fraction of each component. With an increase in volume fraction, water, with a T_g of 136 K and thus a high free volume, produces a very significant increased $V_{f_{\text{idealmix}}}$ relative to that of the solid, and this, alone, would explain the significant plasticizing effects of absorbed water described earlier. In most cases, however, such mixing of amorphous solids with water will be thermodynamically nonideal, so that

$$V_{f_{\text{actualmix}}} = V_{f_{\text{idealmix}}} + V_{\text{excess}} \tag{15.31}$$

where V_{excess} is an excess free volume change due to the absorption of water and its interaction with the solid. Here, a negative V_{excess} (loss in volume) would result in a loss of free energy, whereas an increase in V_{excess} (gain in volume) would increase the free energy. Consequently, as expected, any increase in V_{excess} will produce plasticization above that of that expected for ideal mixing, whereas any decrease in V_{excess} will lead to anti-plasticization. Based on these principles, it has been suggested that water vapor absorbed into a glass will distribute itself between high-energy regions of the glass containing excess free volume and lower-energy, more liquid-like, locally ordered domains with less free volume. One possible model for such a glassy state, schematically described in Figure 6.8, has the region of excess free volume represented by a high-energy "microstructure" surrounding the local domains [36]. Here, it is proposed that the water molecules entering the local domains plasticize the

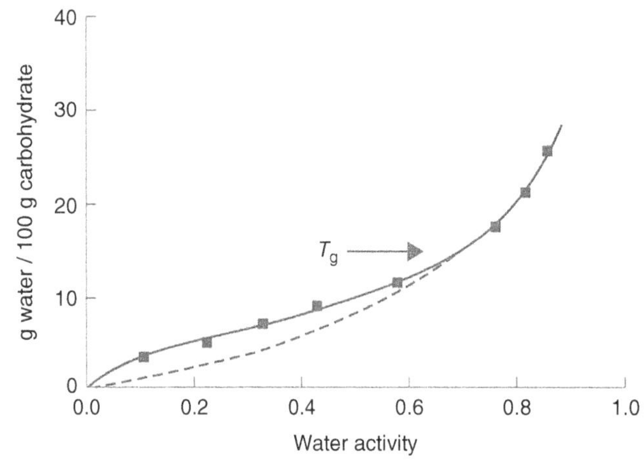

Figure 15.18 Water vapor absorption by maltotetrahexaose as a function of water activity. Dotted lines predicted by the Flory–Huggins equation and compared to actual data. *Source*: Adapted from [32].

domains, which leads to an increase in free volume, as reflected in an increase in diffusive α relaxation times and a decrease in T_g, while water entering the glassy "microstructure" anti-plasticizes that portion of the system by reducing the free volume without any effect on the diffusive molecular mobility due to α relaxations. A plasticizing effect at all concentrations of absorbed water, above and below T_g, would lead to a continuous decrease in T_g, as is generally observed, whereas the anti-plasticizing effects and reduction in free volume in the glass, while not affecting the T_g, causes a free energy loss that is associated with a greater amount of water absorption than that predicted from the Flory–Huggins equation. That such a combination of plasticization and anti-plasticization occurs in the glassy state upon water vapor absorption by various polymeric carbohydrates and the polar polymer, polyamide, has been confirmed experimentally by direct measurement of free volume changes that occur upon water vapor absorption using positron annihilation lifetime spectroscopy [37, 38]. With increasing water contact below T_g the free volume decreases, but at higher water contents the free volume then begins to increase, thus indicating that both processes occur simultaneously with the addition of water, with the anti-plasticization effects dominating in the glass as suggested earlier. Such simultaneous opposing effects of water vapor sorption on a glass appear to be related to opposing effects of absorption on molecular mobility in reducing diffusive α-relaxation rates, while increasing the rates of secondary β- or Johari–Goldstein relaxation rates [39]. How this might translate into a prediction of the absorption isotherm of a polar amorphous solid has been addressed by Vrentas and Vrentas [40, 41], who used free volume theory to estimate the loss in excess free volume below T_g, and the resulting loss of free energy associated with the water vapor absorption in the glass. Combining this phenomenon with the Flory–Huggins equation (equation [15.23]) led to the Vrentas equation:

$$\frac{p}{p_o} = \Phi_1 e^{\left\{1-\left(\frac{1}{x}\right)\Phi_2+\chi Q_2^2\right\}} e^f \tag{15.32}$$

where f is the contribution to absorption from the loss in free volume in the glass,

and

$$f = M_1 w_2^2 \left(C_{pg} - C_p\right)\left(\frac{dT_{gm}}{dw_1}\right)\left(\frac{T}{T_{gm}} - 1\right) \tag{15.33}$$

Here, M_1 is the molecular weight of water; w_1 is the weight fraction of water; w_2 is the weight fraction of the amorphous solid; T is the temperature; T_{gm} is the glass transition temperature at any water content; $C_{pg} - C_P$ is the difference in heat capacity between the glass and equilibrium liquid at T_{gm}; and dT_{gm}/dw_1 is the change in glass transition temperature with water content. Since all of the parameters in equations (15.32) and (15.33) can be independently determined, it is possible to predict the water vapor sorption isotherm *a priori*. From equation (15.33), we can observe that when the temperature, T, is greater than T_{gm}, f is zero, and the antiplasticizing effects of water are no longer dominant in determining the water vapor absorption isotherm. This equation has been tested for the absorption of many organic vapors into glassy polymers [40, 41] and for the water vapor absorption of PVP [34, 35]. As shown in Figure 15.19, the water absorption isotherm is very well predicted at all water contents, supporting the combined Flory–Huggins and Vrentas free volume model for water vapor absorption of amorphous solids. We can conclude, therefore, that water vapor absorption by amorphous solids produces its effects on the properties of the solid through significant solubility in the amorphous matrix coupled with significant effects on free volume, T_g, and molecular mobility.

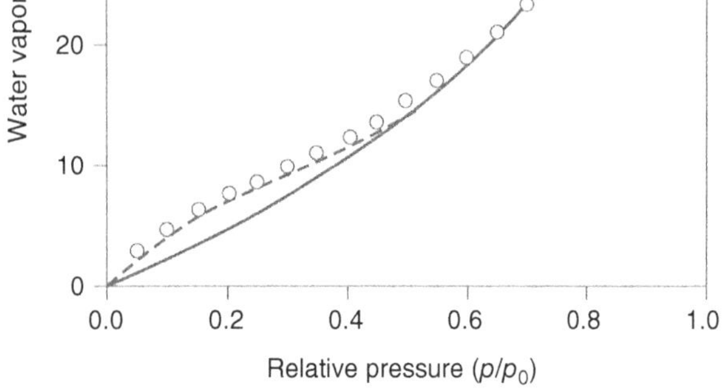

Figure 15.19 Water vapor absorption by PVP as a function of relative pressure (p/p_o), predicted by the Vrentas–Flory–Huggins equation and compared to actual data. *Source*: [35]/ with permission of Elsevier

In summary, we can conclude that pharmaceutical solids of sufficient polarity have the potential to exhibit hygroscopic properties, when exposed to elevated relative humidities [1]. Generally, adsorption of water vapor onto nonporous solids in itself should not be a serious issue since the water is confined to the surface and is present in total amounts of only a fraction of a percent. For many very water-soluble solids, such as salts, deliquescence can occur above a critical relative humidity, leading to dissolution of the solid with significant changes in physical and chemical properties. By far the greatest and most predominant effects of water vapor sorption on the properties of solids occur when the solid is fully or partially in the amorphous state and exhibits sufficient polarity to take up very large amounts of water into the bulk structure with strong plasticizing effects. As will be discussed more fully in Chapters 18 and 19, when crystalline solids are rendered partially amorphous by various pharmaceutical processes, water will preferably concentrate in these amorphous regions and produce effects much greater than would be expected from the overall water content [3].

References

1 Newman AW, Reutzel-Edens SM, Zografi G. Characterization of the hygroscopic properties of active pharmaceutical ingredients. *J Pharm Sci* 2008; 97: 1047–1059.
2 Debenedetti PG. *Metastable Liquids: Concepts and Principles*. Princeton, NJ: Princeton University Press; 1996: 307.
3 Ahlneck C, Zografi G. The molecular basis of moisture effects on the physical and chemical stability of drugs in the solid state. *Int J Pharm* 1990; 62: 87–95.
4 Newman A, Zografi G. An examination of water vapor sorption by multicomponent crystalline and amorphous solids and its effects on their solid-state properties. *J Pharm Sci* 2019; 108: 1061–1080.
5 Kontny MJ, Koppenol S, Graham ET. Use of the sorption-desorption moisture transfer model to assess the utility of a desiccant in a solid product. *Int J Pharm* 1992; 84: 261–271.
6 Zografi G, Grandolfi GP, Kontny MJ, Mendenhall DW. Prediction of moisture transfer in mixtures of solids. *Int J Pharm* 1988; 42: 77–88.
7 Tong P. *Solid-state Amorphous Properties of Sodium Indomethacin in Relation to its Free-acid Form*. PhD Thesis 2000; University of Wisconsin-Madison.
8 United States Pharmacoepia 29 and National Formulary 24 2006; Water Determination <921>.
9 MacLeod SK. Moisture determination using Karl Fischer titrations. *Anal Chem* 1991; 63: 557A–565A.
10 Shenk JS, Workman, Jr JJ, Westerhaus MO. Applications of Near-Infrared spectroscopy to agricultural products. In: Third Edition, Burns DA, Ciurczak EW, eds. *Handbook of Near Infrared Analysis*. London: CRC Press, 2008: Chapter 17: 347–386.
11 Ciurczak EW. Process analytical technologies (PAT) in the pharmaceutical industry. In: Third Edition, Burns DA, Ciurczak EW, eds. London: CRC Press, 2008: Chapter 29: 581–584.
12 Van Campen L, Zografi G, Carstensen JT. An approach to the evaluation of hygroscopicity for pharmaceutical solids. *Int J Pharm* 1980; 5: 1–18.
13 Bergren MS. An automated controlled atmosphere microbalance for the measurement of moisture sorption. *Int J Pharm* 1994; 103: 103–114.
14 Kuroda Y, Watanabe T, Yoshikawa Y, Kumashiro R, Hamano H, Nagao M. Specific features of dielectric behavior of water sorbed on Ag_2O surface. *Langmuir* 1997; 13(14): 3823–3826.
15 Langmuir I. Theory of adsorption. *Phys Rev* 1915; 6: 79–80.
16 Brunauer S, Emmett PH, Teller E. Adsorption of gases in multimolecular layers. *J Am Chem Soc* 1938; 60: 309–319.
17 Sengupta G, Ahluwalia H, Banerjee S, Sen SP. Chemisorption of water vapor on zinc oxide. *J Colloid Interf Sci* 1979; 69: 217–224.
18 Guerrieri P, Zembyanov D, Taylor LS. Dissociation of water on the surface of organic salts studied by X-ray photoelectron spectroscopy. *Langmuir* 2010; 26: 11998–12002.
19 Peters SJ, Ewing GE. Thin film water on NaCl(100)under ambient conditions: An infrared study. *Langmuir* 1997; 13: 6345–6348.
20 Foster MC, Ewing GE. Adsorption of water on NaCl(001) surface. II. An infrared study at ambient temperatures. *J Chem Phys* 2000; 112: 6817–6826.
21 Van Campen L, Amidon GL, Zografi G. Moisture sorption kinetics for water-soluble substances, III: Theoretical and experimental studies in air. *J Pharm Sci* 1983; 72: 1394–1398.
22 Defay R, Prigonine I. *Surface Tension and Adsorption*. Longmans: Green & Company; 1966: 6–8.
23 Fisher LR, Gamble RA, Middlehurst J. The Kelvin equation and the capillary condensation of water. *Nature* 1981; 290: 575–576.

24 Rozhok S, Sun P, Piner R, Lieberman M, Mirkin CA. AFM study of water meniscus formation between the AFM tip and NaCl substrate. *J Phys Chem B* 2004; 108: 7814–7819.
25 Yoshioka M, Hancock BC, Zografi G. Crystallization of indomethacin from the amorphous state below and above its glass transition temperature. *J Pharm Sci* 1994; 83: 1700–1705.
26 Dawson KJ, Kearns KL, Ediger MD, Sacchetti, MJ, Zografi G. Highly stable indomethacin glasses resist water vapor. *J Phys Chem B* 2009; 113: 2422–2427.
27 Surana R, Pyne A, Suryanarayanan R. Effect of aging on the physical properties of amorphous trehalose. *Pharm Res* 2004; 21: 867–874.
28 Taylor LS, Langkilde FW, Zografi G. Fourier transform Raman spectroscopic study of the interaction of water vapor with amorphous polymers. *J Pharm Sci* 2001; 90: 888–901.
29 Andronis V, Yoshioka M, Zografi G. Effects of sorbed water on crystallization of indomethacin from the amorphous state. *J Pharm Sci* 1997; 86: 346–351.
30 Oksanen CA, Zografi G. The relationship between the glass transition temperature and water vapor absorption by poly(vinylpyrrolidone). *Pharm Res* 1990; 7: 654–657.
31 Eichinger BE, Flory PJ. Thermodynamics of polymer solutions Part I, Natural rubber and benzene. *Trans Faraday Soc* 1968; 64: 2035–2052.
32 Orford PD, Parker R, Ring SG. Aspects of the glass transition behavior of mixtures of carbohydrates of low molecular weight. *Carbohydr Res* 1990; 196: 11–18.
33 Gunning Y, Mac Dougall A, Noel T, Parker R, Ring S. Plant polysaccharides and food quality – homogeneous, heterogeneous, and phase-separated systems. In: Roods YH, Leslie RB, Lillord PJ, eds. *Water Management in the Design and Distribution of Quality Foods (ISOPOW 7)*. Technomic Publishing Company, Lancaster, 1999: 455.
34 Hancock BC, Zografi G. The use of solution theories for predicting water vapor absorption by amorphous pharmaceutical solids: A test of the Flory-Huggins and Vrentas models. *Pharm Res* 1993; 10: 1262–1267.
35 Zhang J, Zografi G. The relationship between "BET" and "Free Volume"- derived parameters for water vapor absorption into amorphous solids. *J Pharm Sci* 2000; 89: 1063–1072.
36 Bates S, Zografi G, Engers D, Morris K, Crowley K, Newman A. Analysis of amorphous and nanocrystalline solids from their X-ray diffraction patterns. *Pharm Res* 2006; 23: 2333–2349.
37 Kilburn D, Claude J, Schweizer T, Alam A, Ubbink J. Carbohydrate polymers in amorphous states: An integrated thermodynamic and nanostructure investigation. *Biomacromolecules* 2005; 6: 864–879.
38 Dlubek G, Buchhold R, Hubner Ch, Nakladal A. Water in local free volumes of polyimides: A positron lifetime study. *Macromolecules* 1999; 32: 2348–2355.
39 Zografi G, Newman A. Interrelationships between structure and properties of amorphous solids. *J Pharm Sci* 2017; 106: 5–27.
40 Vrentas JS, Duda JL, Ling H-C. Antiplasticization and volumetric behavior in glassy polymers. *Macromolecules* 1988; 21: 1470–1475.
41 Vrentas JS, Vrentas CM. Sorption in glassy polymers. *Macromolecules* 1991; 24: 2404–2412.

16

Mechanical Properties of Pharmaceutical Materials

16.1 Introduction

The mechanical properties of pharmaceutical materials, including active pharmaceutical ingredients (APIs), excipients, and packaging materials, are determined by their response to applied mechanical forces. Such forces can include gravitational forces that act on powders during flow from a hopper in tablet manufacturing, as well as other mechanical forces applied during milling, micronization, tablet compression, coating, encapsulation, and packaging. The way a material responds to mechanical force is also critical for maintaining the quality and integrity of drug products undergoing shipping, dispensing, and patient use. Undesirable mechanical properties can cause product defects such as friability of tablets, peeling of film coatings, chipping of capsules, or damaged packaging, all of which may result in product failure.

In this chapter, several sections are dedicated to the review of the underlying physics of mechanical properties, including the introduction of important properties such as elasticity, plasticity, and viscoelasticity. Case studies are presented to show the effects of the solid form, for example, polymorphs and amorphous forms, and moisture content on various mechanical properties. A discussion of the methodologies used to study various mechanical properties will focus on techniques, such as beam bending, thermomechanical analysis (TMA), dynamic mechanical analysis, and nanoindentation.

16.2 Stress and Strain

A fundamental understanding of the physics underlying the mechanical properties of solid materials is based on the relationship between stress and strain, where stress is defined in terms of the mechanical force applied to a solid body and strain indicates the deformation, motion, or other dimensional changes in the solid under various applied mechanical forces.

There are two types of stress: axial stress and shear stress. An axial stress is directly (e.g. vertically) applied to a solid body, as either a compressive stress or a tensile stress. A compressive stress is caused by a "pushing" force, while a tensile stress arises from a "pulling force." A shear stress occurs when a force is applied to a solid body in a horizontal manner as a sliding force. The impact of these two types of stresses on the solid body is quite distinct. The axial stress causes a change in physical dimensions in the direction of the force, maintaining the overall shape, whereas shear stresses result in the modification of material shape.

Axial stress, as tensile or compressive stress, σ, is the force applied per unit area in a particular direction as described in the following equation:

$$\sigma = \frac{P}{A} \tag{16.1}$$

where P is the applied load in units of force and A is the original cross-sectional area.

Axial strain, therefore, is defined as the degree of deformation of a solid due to tensile or compressive stress (Figure 16.1). In such an analysis, the solid body is assumed to be a continuum body with homogeneous density. Axial strain is a dimensionless parameter that indicates the response of the solid to a tensile or compressive stress, as in the following equation:

$$\varepsilon = \frac{dl}{l} \tag{16.2}$$

where ε is the strain, dl is the change of length after deformation, and l is the initial length.

Solid-State Materials in Pharmaceutical Chemistry: Properties, Characterization, and Applications, Second Edition. Stephen R. Byrn, George Zografi and Xiaoming (Sean) Chen.

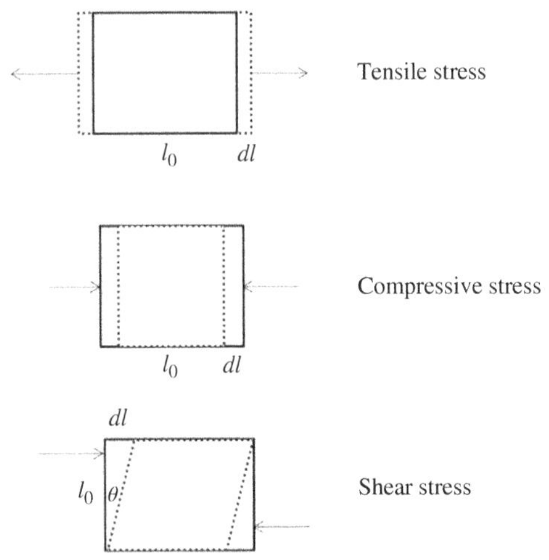

Figure 16.1 Tensile stress, compression stress, shear stress, and related strain.

Shear stress is denoted as τ and defined as the stress coplanar with a material cross section as the following equation:

$$\tau = \frac{F}{A_0} \tag{16.3}$$

where F is the shear force applied and A_0 is the cross-sectional area of material parallel to the applied shear force.

Shear strain is also a dimensional parameter described by the following equation:

$$\gamma = \frac{dl}{l_0} = \tan\theta \tag{16.4}$$

where dl is the length of deformation perpendicular to a given line, l_0 is the initial length, and θ is the angle between the sheared line and the original line. As shown in Figure 16.1, shear strain equals tan θ.

The relationship between stress and strain of a solid body can be expressed as a plot of stress versus strain, where the stress–strain curve gives a graphical description of how a material responds to mechanical force. A typical stress and strain curve for a solid body due to tensile force is given in Figure 16.2. In the initial portion of the curve under relatively low strain, the material obeys Hooke's law to a reasonable degree, so that the plot is approximately linear; such linear behavior, which is reversible, is generally defined as representing elastic deformation or elasticity. As the stress continues to increase and passes the elastic limit, the sample undergoes irreversible deformation at what is called the yield point, with the value of stress defined as the yield stress. Above the yield point, the material undergoes irreversible plastic deformation until, at some level of stress, the integrity of the material undergoes failure or fracture. This critical stress is defined as the fracture point or the failure point.

Based on the stress–strain curve, two general mechanical classes of materials can be discerned: brittle and tough. A brittle material is usually highly elastic with no, or a very short period of, plastic deformation before failure or rupture. Conversely, a tough material has a relatively long plastic deformation region before failure occurs.

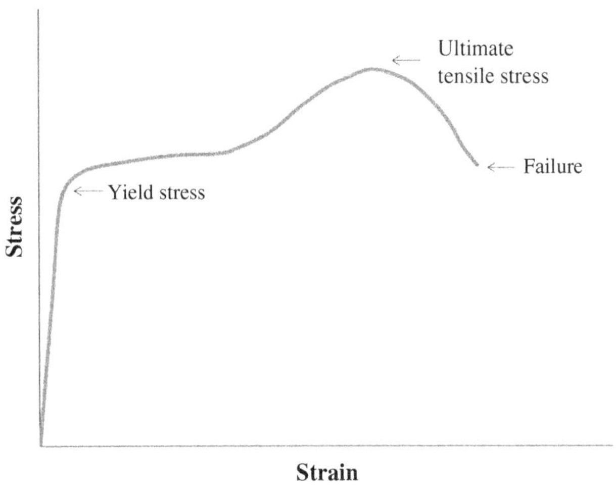

Figure 16.2 Engineering stress and engineering strain curve for a material having elastic deformation, plastic deformation, and fracture as the stress applied is increased.

16.3 Elasticity

As introduced earlier, elasticity is defined as the ability of a deformed material body to return to its original shape and size when the forces causing the deformation are removed. Most solid materials exhibit elasticity within an elastic limit, defined as the point where maximum stress can be applied to a material before the onset of permanent deformation. In a stress and strain curve, therefore, the portion of the curve before the elastic limit defines elasticity, with the slope of the linear region defined as the modulus of elasticity.

For tensile and compressive stress, the modulus of elasticity is called the Young's modulus, E, as described by the following equation:

$$E = \frac{\sigma}{\varepsilon} \tag{16.5}$$

For shear stress related, the modulus of elasticity is defined as the shear modulus, G, as in the following equation:

$$G = \frac{\tau}{\gamma} \tag{16.6}$$

The smaller the value of Young's modulus or shear modulus is, the greater the elasticity of a material. Conversely, the larger the Young's modulus and shear modulus are, the more significant is the stiffness of a material.

Table 16.1 Young's modulus of some pharmaceutical materials.

Material	Young's modulus (GPa)
Steel	220
Calcium carbonate	88
Glass	60
Calcium phosphate	48
Sorbitol	45
Lactose monohydrate	24
Spray-dried lactose	14
Theophylline	13
Acetaminophen	12
Caffeine	8.7
Aspirin	7.5
Ibuprofen	5.0
Microcrystalline cellulose	8–10
Stearic acid	3.8
Starch	3.7
High-density polyethylene	0.8
Rubber	0.002

Table 16.2 Shear modulus of some pharmaceutical materials.

Material	Shear modulus (GPa)
Steel	77
Glass	26
polyethylene	0.13
Rubber	0.0003

The Young's moduli of various pharmaceutical materials are compiled in Table 16.1 [1, 2]. As shown in the table, the Young's modulus of rubber is 0.002 GPa, and that of steel is 220 GPa. It is interesting to note that starch is more elastic than microcrystalline cellulose, with implications for their different properties under compression in pharmaceutical formulations.

The shear modulus, also called the modulus of rigidity, of some common materials related to pharmaceutical manufacturing, packaging, and storage are given in Table 16.2 [3].

The Young's modulus (E) and shear modulus (G) are related through the following equation:

$$E = 2\left(1 + \nu_{\mathrm{p}}\right)G \tag{16.7}$$

where ν_{p} is the Poisson's ratio, which describes the ratio between axial strain and transverse (shear) strain. As a material is compressed in one direction, it usually tends to expand in the other two directions perpendicular to the direction of compression. Conversely, when a material is stretched, it tends to shrink in the other transverse direction. This phenomenon is called the Poisson effect. The Poisson's ratio of microcrystalline cellulose, for example, was found to be about 0.30 [4].

16.4 Plasticity

Plasticity is the ability of a material to undergo permanent deformation under a force without breaking or rupturing.

It is an important property that, when significant, can have an important effect on pharmaceutical processes such as grinding, compaction, and extrusion. A term that provides a quantitative indication of the extent of plasticity is ductility, which is the percent maximum elongation, %EL, or maximum area reduction before disruption, %AR, as follows:

$$\%\text{EL} = \varepsilon_{\max} \times 100\% \tag{16.8}$$

where $\varepsilon_{\max}$ is the maximum strain due to plastic deformation, and

$$\%\text{AR} = \frac{(A_0 - A_f)}{A_0} \tag{16.9}$$

where A_0 is the initial area and A_f is the area right after the fracture.

At the molecular level, plastic deformation can be related to the breaking and forming of the intermolecular interactions associated with noncovalent forces, such as van der Waals interactions, hydrogen bonding, and ionic interactions.

After the removal of the stress, such plasticity results in the relocation of many atoms to a new position. For a crystalline material, plastic deformation at the microscale level is related to motions that arise at crystal dislocations, grain boundaries, or along slip planes.

Amorphous materials lack long-range order and, therefore, have higher molecular mobility and generally exhibit significant plasticity. Plastic deformation of amorphous polymeric materials typically involves chain sliding, rotation, and unkinking.

Plastic deformation of a material, such as a metal, may result in regions of strain hardening and strain softening. In strain hardening, stress increases with increasing strain, and the material displays resistance to plastic deformation, resulting in brittle fracture.

In the strain softening region, the stress drops with increasing strain due to nonhomogeneous yielding in some local areas. In those local areas, significant changes in microstructure and nanostructure can occur, including crazing. Crazing in polymers refers to a tensile cavitation process that generates voids, also called craze faces, surrounded by polymer fibrils.

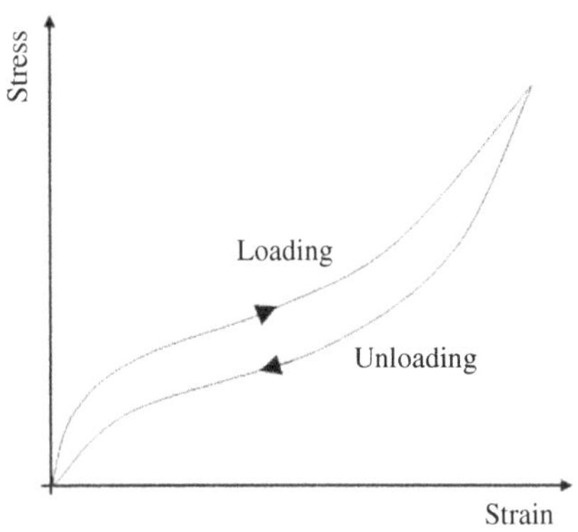

Figure 16.3 Stress–strain curves during loading and unloading for a viscoelastic material.

16.5 Viscoelasticity

Many polymers and pharmaceutical powders exhibit a combination of the mechanical properties of a solid, with an elastic modulus, and of liquids exhibiting viscous flow under stress. This property, called viscoelasticity, is a combination of an elastic component, where deformation is described by Hooke's law, and a viscous component, where an applied stress results in mechanical energy being dissipated as heat.

Viscoelastic materials display stress–strain hysteresis and time-dependent stress–strain relationships, referred to as creep and stress relaxation. Such behavior results in nonlinear and non-Hookean responses. A typical stress–strain hysteresis loop is shown in Figure 16.3, where, at the same strain value, the stress during unloading is less than that during loading. The area within the hysteresis loop due to differences in loading and unloading corresponds to the energy dissipated.

The term "creep" is used to describe phenomena that arise when the strain changes with time under constant stress. Figure 16.4 shows an example of creep that occurs in three stages. In stage I, the creep rate decreases, indicating an initial resistance to creep. The creep rate is relatively constant in stage II, where it is referred to as the stage of steady-state creep. In stage III, the creep rate increases due to necking or a reduction in the effective area caused by internal void formation. The region of elongation until fracture is called the total elongation.

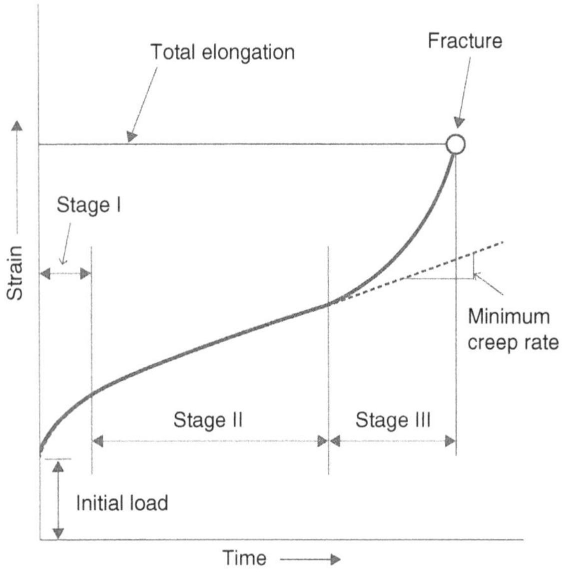

Figure 16.4 A case study of creep.

Stress relaxation commonly occurs for a viscoelastic material, where the system exhibits a decrease in tensile stress over time when the material is subjected to prolonged constant strain at a constant temperature. In such a situation, data are often presented in the form of a stress versus time plot, where the relaxation rate is the slope of the curve at any time point. Figure 16.5 shows an example of a typical stress relaxation relationship where strain is held constant, whereas the stress is decreased as a function of time.

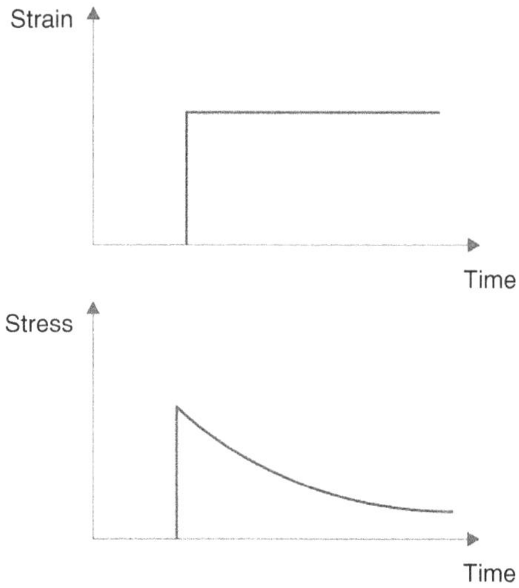

Figure 16.5 An example of stress relaxation.

The non-Hookean mechanical properties of viscoelastic materials can be modeled as a combination of springs and dashpots, as shown in Figure 16.6. The spring represents the elastic body, and the dashpot depicts the viscous behavior.

The Maxwell model is a simple model of viscoelasticity that assumes a dashpot and a spring linked in series to form a viscoelastic material. The spring behaves according to Hooke's law as in the following equation:

$$\sigma_s = E\varepsilon_s \tag{16.10}$$

where σ_s is the stress, ε_s is the strain, and E is the Young's modulus.

The dashpot models a Newtonian fluid based on the following equation:

$$\sigma_d = \eta\varepsilon_d \tag{16.11}$$

where σ_d is the stress, η is the viscosity of the liquid in the dashpot, and ε_d is the strain rate for the liquid. The ratio of the viscosity and the Young's modulus (τ) indicates the response time to stress as follows:

$$\tau = \frac{\eta}{E} \tag{16.12}$$

In the Maxwell model, the stress on each component is equal to the imposed stress, while the total strain is the sum of the strains of both components, as described in the following equations:

$$\sigma = \sigma_s + \sigma_d \tag{16.13}$$

$$\varepsilon = \varepsilon_s + \epsilon_d \tag{16.14}$$

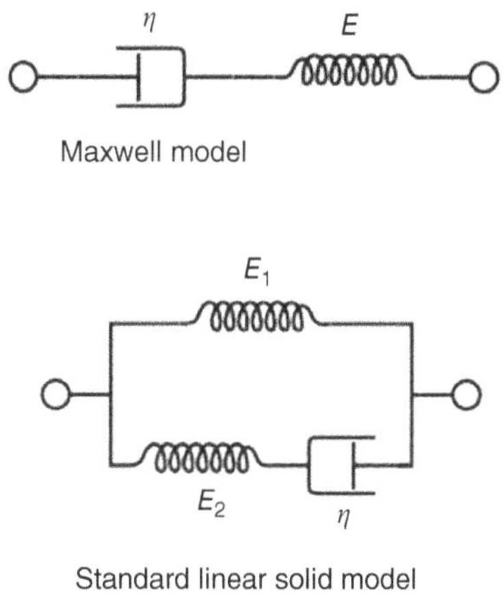

Figure 16.6 Viscoelastic models.

The Maxwell model, therefore, describes the stress relaxation as in the following equation:

$$\sigma(t) = \sigma_0 e^{-\frac{t}{\tau}} \tag{16.15}$$

where $\sigma(t)$ is the stress at time t, σ_0 is the initial stress, and τ is the relaxation time.

The Maxwell model is applicable to typical stress relaxation data but not to creep. Another model, called the standard linear solid model, is slightly more complicated in that it has a dashpot and a spring linked in one arm, as with the Maxwell model, but the Maxwell arm and the other spring are in parallel positions (Figure 16.6). In this arrangement, the Maxwell arm and the parallel spring have the same strain, and the total stress is the sum of the stress in each arm. The standard linear solid model has been found to be very successful for predicting both stress relaxation and creep.

16.6 Brittleness

The brittleness of a material is the mechanical property where fracture occurs without much plastic deformation. Brittle materials, such as glass and ceramics, tend to absorb relatively little energy prior to fracture, and, to date, no widely accepted quantitative term is available to define brittleness. Consequently, different researchers have developed different ways to measure brittleness, and, therefore, the nomenclature used in different publications seems similar, but the underlying meaning could be very different.

Table 16.3 Brittle indices for some excipients and drugs.

Material	Brittle index
Avicel PH101	0.22
Lactose β anhydrous	0.33
Ibuprofen	0.34
Sodium chloride	0.45
Aspirin	0.56
Adipic acid	0.88
Paracetamol, direct compressible	1.08
α-Lactose monohydrate	1.45
Sucrose	2.88
Paracetamol	3.65

Brostow et al. developed a quantitative way to define brittleness for elastic and viscoelastic materials [5], where the brittleness B is described by the following equation:

$$B = \frac{1}{\epsilon_b E'} \tag{16.16}$$

where ε_b is the tensile elongation at break and E' is the storage modulus. The storage modulus is related to material elasticity and can be measured using a dynamic mechanical analyzer (DMA). A higher value of B indicates greater brittleness.

Lawn and Marshall established a brittleness index and defined it as the ratio of indentation hardness to the critical stress intensity factor [6]. The hardness, H, is obtained by using an indentation hardness test, whereas the critical stress intensity factor is obtained from single-edge notched beam tests. As shown in Table 16.3, paracetamol and sucrose are much more brittle than Avicel PH101, an excipient with good compressibility. It is interesting to note that lactose monohydrate is significantly more brittle than lactose anhydrous (1.45 vs. 0.33).

Another way to evaluate brittleness for pharmaceutical materials is to calculate the brittle fracture index (BFI). The concept was developed by Hiestand et al. by applying crack theory [7], where the BFI is calculated from the tensile strength of a tablet (T_0) with a hole cut into its center, and the tensile strength of a tablet (T) without a hole made of the same material(s) using the following equation:

$$\text{BFI} = 0.5\left[\frac{T}{T_0} - 1\right] \tag{16.17}$$

The center hole acts as a weak point from which the fracture propagates during the hardness test. The tensile strength of the tablet with and without a center hole can be calculated with the following equations:

$$T_0 = \frac{2F_0}{\pi Dh} \tag{16.18}$$

where F_0 is the diametral compression load needed to cause tensile failure of the tablet with the center hole, D is the tablet diameter, and h is the tablet thickness,

$$T = \frac{2F}{\pi Dh} \tag{16.19}$$

where F is the compressional load needed to cause tensile failure for the tablet without a hole.

BFI values range from 0 to 1 and can be used as an indicator of the ability of a material to relieve stress through plastic deformation. A high BFI value indicates high brittleness, whereas a low BFI value indicates the ability of the material to relieve localized stresses. The BFI of some pharmaceutical materials at a solid fraction of 0.9 is given in Table 16.4 [8]. Regarding excipients, it can be seen that starch is much more brittle than microcrystalline cellulose, while, for the API

Table 16.4 Brittle fracture index of some pharmaceutical materials.

Material	BFI
Microcrystalline cellulose	0.03
Acetaminophen	0.03
Ibuprofen (lot A)	0.06
Lactose, spray-dried	0.12
Aspirin	0.19
Ibuprofen (lot B)	0.40
Phenacetin	0.43
Caffeine	0.47
Starch	0.70
Erythromycin	0.70
Methenamine	0.80

ibuprofen, two different lots show different BFI values. A material with the BFI value less than 0.3 is considered a non-brittle material.

16.7 Hardness

In material science, hardness is the property of a material that provides resistance stress-induced indentation, scratching, abrasion, or penetration. Hardness, as a broad and loosely defined term, is directly related to the resistance of a material to plastic deformation. The value of hardness is generally expressed on a relative scale and, therefore, is dependent on the test method. The most popular method for testing material hardness is the indentation test.

An indentation test involves slowly applying a compressive force to a surface to create an indentation into the surface of the material. The depression caused by the indentation is measured, and hardness is scaled based on the applied force and the degree of depression. Different indentation tests are distinguished from one another by the shape of the indenter, the method of calculation, and, hence, the hardness value. Widely used indentation tests include the Brinell, Vickers, and Rockwell methods.

The Brinell hardness test utilizes a hardened steel or tungsten carbide ball as the indenter. The indenter with a diameter of 1 – 10 mm is forced into the surface of a test sample (Figure 16.7), and the diameter of the indentation left in the surface is measured, after removing the test force. Brinell hardness, HB, is obtained from the test force divided by the curved surface area of the indentation, as follows:

$$\mathrm{HB} = \frac{F}{\frac{\pi}{2}D(D - \sqrt{D^2 - d^2})} \tag{16.20}$$

where HB is the Brinell hardness, D is the diameter of the ball, d is the diameter of the indentation circle, and F is the applied force. The applied force is accurately controlled and maintained for a specific dwell time, normally 10 – 30 s. The diameter of the indentation is measured by a low-amplification microscope. A Brinell hardness number is usually given with the test conditions utilized, as, for example, "50 HB 10/500/30," which means that the material has a Brinell hardness of 50, and the result is obtained from a 10 mm ball, at 500 kg force, with 30 s of dwell time.

The Vickers hardness test utilizes a square-based diamond pyramid as the indenter (Figure 16.7), where the angle between the opposite faces is 136°. The hardness number is calculated from the load and the surface area of the indentation as follows:

$$\mathrm{VHN} = \frac{2F \sin\frac{\theta}{2}}{L^2} \tag{16.21}$$

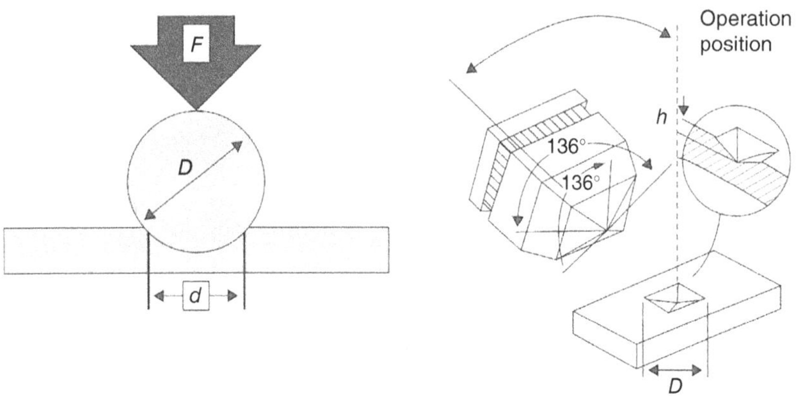

Figure 16.7 Models of Brinell hardness and Vickers hardness.

where VHN is the Vickers hardness number, F is the applied force, and L is the length of the indentation diameter. When θ is set at 136°, VHN is calculated as the following equation:

$$\mathrm{VHN} = \frac{1.854\,F}{L^2} \tag{16.22}$$

The Vickers hardness test can measure a continuous scale of hardness, from very soft materials to very hard ones. For example, the VHN of aluminum, a soft metal, is about 25, whereas the VHN of hard steel is about 1000.

The Rockwell hardness test uses the depth of the indentation instead of the relationship between the applied force and the indentation area for ranking hardness. It has the advantage of speed and convenience. During the test, a minor load is first applied to set the sample. Then, a major load is applied to the sample for a required dwell time, and the depth of the indentation is recorded. The indenter is either a 120° sphero-conical diamond or a steel ball, 1/16 to ½ inch in diameter. The type of indenter, force, and dwell time must be specified when a Rockwell hardness is reported.

Pharmaceutical materials are usually presented as powder made of particles with a size from several to hundreds of micrometers. Hardness test of pharmaceutical particles is usually carried out by nanoindentation test, which follows the same principle as the indentation test but is conducted on a relatively small scale. The nanoindentation test is discussed in a later session.

Hardness values determined from indentation tests have a different meaning from the hardness normally associated with tablets as a pharmaceutical dosage form. The hardness of a tablet is a measurement related to tensile strength instead of resistance to deformation, which is discussed in a later section.

16.8 Powder Compression

The review of mechanical properties presented in previous sections focuses on materials assumed to be a continuum body, where the solids are supposed to have a homogeneous density. This assumption is applicable to metals, minerals, and polymers. However, most pharmaceutical materials are utilized as powders, commonly made into compacts and tablets by the processes of dry granulation and tableting.

The response of a powder bed to mechanical forces is different from that of a continuum solid body and can be divided into several stages:

- Particle rearrangement
- Particle fragmentation and plastic and elastic deformation
- Particle binding and fusion
- Elastic recovery

Those stages are assumed to be sequential, but overlapping of different stages is common. A powder can be considered as a special case of a dispersed system, where the solid phase is dispersed in air as a gaseous phase. At the initial stages of compression, particles rearrange to produce a more compact degree of packing. As a result, the volume is reduced, the powder bed porosity is decreased, and the bulk density is increased. After a certain applied pressure is surpassed, the particles reach a maximum possible packing structure, and any additional particle rearrangement becomes impossible. Consequently, deformation and particle fragmentation occur through both elastic and plastic deformation. Particle fragmentation generates smaller particles, which undergo secondary particle rearrangements followed by deformation. This deformation further brings particles to a more densely packed bed and reduces porosity. The closeness of particles and the increase of surface area due to fragmentation and deformation then promote particle binding and fusion, where particles are linked through solid bridges and interlocking structures. Eventually, a powder bed becomes a cohesive mass as a solid compact or a tablet with a well-defined shape. After removing the compression pressure, the compact or the tablet may undergo some elastic recovery and volume expansion. The degree of binding of the compact or the tablet can be measured by its tensile strength. Consequently, the product of compression is a compact or tablet with a characteristic physical dimension, porosity, and tensile strength. The compression process, therefore, involves a progressive reduction of porosity and an increase of tensile strength, ranging from fragmentation to deformation, in response to the mechanical pressure. For example, microcrystalline cellulose undergoes plastic deformation when compressed, whereas, for inorganic excipients, such as calcium phosphate, particle fragmentation is the dominating event in compression. In essence, the compression process can be described as a physical process, primarily involving porosity change.

The term "porosity" is defined as the ratio of void spaces in a material relative to the total volume, as follows:

$$E = \frac{V_{\text{void}}}{V_{\text{total}}} \tag{16.23}$$

where E is the porosity, V_{void} is the volume of pores, and V_{total} is the volume of pores and particles.

Porosity can be estimated by comparing the true density and bulk density. True density, which is determined by the molecular structure of the solid, is the density of the solid material excluding the volume due to the presence of pores. For example, the true density of a crystalline material is related to its crystal packing and can be calculated from its unit cell. Bulk density, on the other hand, is the material weight divided by the total volume. Porosity (E), therefore, can be calculated from true density and bulk density as in the following equation:

$$E = 1 - \frac{\rho_{\text{bulk}}}{\rho_{\text{true}}} \tag{16.24}$$

where ρ_{bulk} is the bulk density and ρ_{true} is the true density.

Bulk density is obtained by simply dividing the mass of a powder bed by the volume. True density is usually calculated from the crystal structure or measured with a helium pycnometer.

Hiestand proposed that the ability of a formulation to exhibit good tablet bond depends on [9]:

1. viscoelastic or plastic responses that are sufficient to provide an increase in true area of contact upon compaction, and
2. the ability of the resultant intermolecular interactions to offset elastic rebound upon removal of the stress.

Tabletability, therefore, can be considered a combination of two processes: compressibility and compactibility.

16.9 Powder Compression Models and Compressibility

Compressibility is the ability of a material to deform under pressure, typically assessed through plots of porosity versus compression pressure.

The evaluation of compressibility can be assisted by using powder compression models.

The Heckel equation is the most well known and frequently used model to define the compression of pharmaceutical materials [10]. The model is established on the hypothesis that compression of powder is analogous to a first-order chemical reaction, where the pores are the "reactant" and densification of the bulk is the "product." The Heckel equation was originally developed to define plastic deformation in metal and is given as follows:

$$ln\frac{1}{E} = kP + A \tag{16.25}$$

where E is the porosity of the powder bed, P is the applied compression pressure, and A and k are constants.

Plots of $1/E$ versus P are called Heckel plots, which usually have three different regions for the compression of a powder. An initial nonlinear region corresponds to the first stage of particle rearrangement and particle fragmentation. The second region, dominated by plastic deformation, is linear and agrees with the Heckel equation, while the third region is nonlinear and related to elastic deformation.

The parameter A in the Heckel equation is related to low-pressure densification by interparticulate motion, whereas the inverse of the slope k is referred to as the Heckel parameter, which is also called the yield pressure, P_y, and is related to the plasticity of the particles under compression. A greater slope of the Heckel plot and a lower Heckel parameter indicate better plasticity, which favors compressibility and indicates the ease of densification.

The Shapiro general compression equation can be considered a refined Heckel equation [11]. Here, an exponential term is included to describe the initial nonlinear region of a Heckel plot.

$$\ln E = \ln E_0 - kP - f\sqrt{P} \quad (16.26)$$

where E is the porosity of the powder bed, E_0 is the initial porosity of the powder bed, P is the applied compression pressure, k equals the Heckel parameter, and f is an indication of the initial curvature in region I.

Instead of evaluating porosity, the Kawakita equation relates the volume reduction of a powder bed to the applied pressure as follows [12]:

$$\frac{P}{C} = \frac{1}{ab} + \frac{P}{a} \quad (16.27)$$

where C is the degree of volume reduction, P is the applied pressure, and a and b are constants.

C is defined as the following equation:

$$C = \frac{V_0 - V}{V_0} \quad (16.28)$$

The constants a and b can be derived from a plot of P/C versus P. The constant a is the maximum strain or the maximum porosity value of the powder bed, whereas the constant b is equal to the reciprocal of the pressure when the value of C reaches one-half of the maximal value.

All three equations consider changes in the physical dimensions of the sample under the compression pressure, which can then be used to evaluate compressibility.

16.10 Compactibility and Tensile Strength

Compactibility, the ability of a powder bed to convert from small particles into a coherent, well-defined compact, can be estimated by measuring the mechanical strength of the formed compact, since, once formed, it would display a certain level of tensile strength. Tensile strength is a measure of the ability of a given compact to resist forces that tend to cause it to break apart. The tensile strength is determined by three factors: intermolecular forces that attract particles, solid bridges, and mechanical interlocking. The intermolecular forces include van der Waals, hydrogen bonding, and ionic interactions. The solid bridges are formed when particles are fused together through partial melting at high pressures at the point of contact, and mechanical interlocking occurs when needle-like and irregular particles hook and twist to interlock during compaction.

There are several ways to measure tensile strength for a tablet. The most widely used is the diametrical compression test. During this test, a local tensile stress is applied in the transverse direction of a tablet. The maximum force, called the crushing force that causes failure, is utilized to calculate the tablet tensile strength. The diametrical compression test is convenient to operate, provides data rapidly, and is highly reproducible.

Tensile strength is calculated based on the crushing force and tablet dimensions of a flat face round tablet according to the following equation, derived by Fell and Newton [13]:

$$\mathrm{TS} = \frac{2F}{\pi D h} \quad (16.29)$$

where TS is the tensile strength, F is the crushing force, D is the diameter of the tablet, and h is the thickness.

The compactibility of a material or a formulation can be evaluated by studying the relationship between tensile strength and solid fraction, where solid fraction is the fraction of the solid volume compared to the total volume. Solid fraction and porosity are complementary to each other in value and, when added together, must equal 1. The relationship between tensile strength and compact solid fraction has been reported to be independent of compression speed for many common

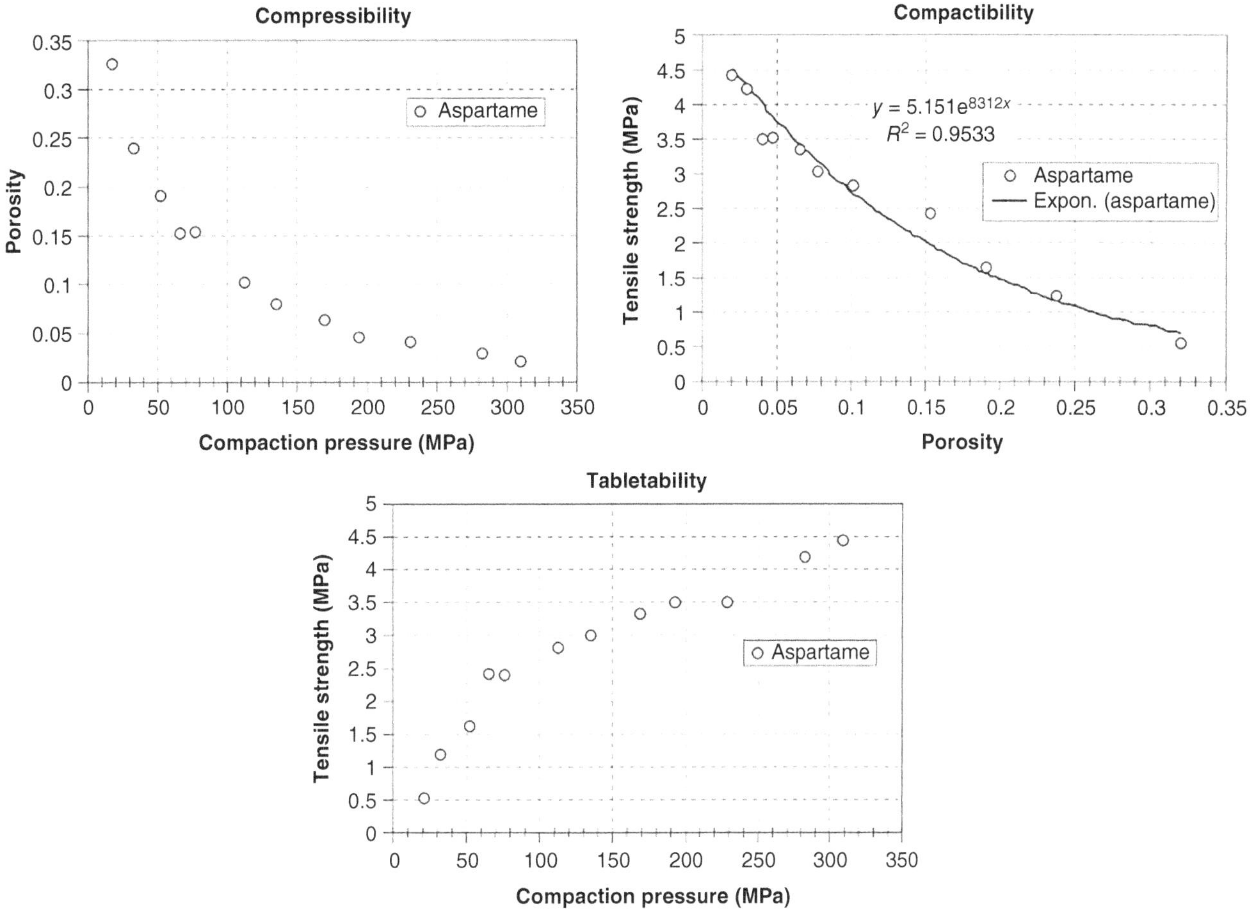

Figure 16.8 Compressibility, compactibility, and tabletability of aspartame. *Source*: [14]/with permission of Elsevier.

pharmaceutical powders. Therefore, the solid fraction of a given sample can be utilized to predict the tensile strength of tablets prepared over a broad range of tableting speeds.

Compressibility and compatibility both contribute to tabletability of a powder, wherein tabletability describes the capability of the powder to be made into a tablet with a specified strength under a certain compression pressure. It is represented by a plot of tensile strength versus compression pressure.

An example of evaluating compressibility, compatibility, and tabletability is given in Figure 16.8 [14].

16.11 Effect of Solid Form on Mechanical Properties

As has been discussed in Chapters 2, 3, and 5, pharmaceutical materials can have multiple solid forms, including amorphous, polymorphs, and hydrates, where each solid form has its characteristic internal structure and mechanical properties. The characteristics that contribute to the internal structure include molecular structure, molecular conformation, molecular interactions, and molecular mobility.

Crystalline forms of a pharmaceutical material are generally harder, more brittle, less elastic, and less plastic than amorphous materials.

Since a crystalline material has a highly ordered three-dimensional structure, a molecular network within the crystal is rigid, tightly interacting, and sometimes physically interlocked. In contrast, amorphous materials have a liquid-like structure, with much greater molecular mobility. As a result, the presence of amorphous content in a material favors plastic deformation and promotes compressibility in tablet manufacturing. Excipients generally used for direct compaction, therefore, need to have a significant amount of amorphous component to facilitate plastic deformation. For example, the widely used tablet excipient α-lactose monohydrate has relatively poor binding properties and a high degree of brittleness, whereas spray-dried lactose, α-lactose anhydrous, containing 10–15% amorphous content, usually gives excellent tensile strength [15]. Owing to its desirable mechanical properties, spray-dried lactose is the form of lactose used for direct compression tablets. Microcrystalline cellulose (MCC), also a widely used excipient for tablet compression, has about 37–50% amorphous content [16], whereas Dipac, a direct compressible excipient, contains 3% amorphous maltodextrin and 97% crystalline sucrose [17].

For various crystal forms of a material, different crystal structures and hence different crystal packings may result in distinct mechanical properties. Sun and Grant reported that the tableting properties of sulfamerazine polymorphs are related to their crystal structure, where polymorphs, Forms I and II, displayed different mechanical properties in tablets [18]. Form I particles tend to exhibit more plastic deformation during compression. Therefore, the tablets made of Form I showed greater tensile strength. The differences in tableting performance were attributed to differences in their crystal structures, where Form I has slip planes parallel to the ac plane of the unit cell and Form II does not have slip planes (Figure 16.9). It can be concluded that the slip planes in Form I favor greater plastic deformation and enhance compressibility and compactability.

Crystal polymorphism also affects the mechanical properties of polymers. Cocca et al. studied poly (lactic acid) that can have either α and α′ crystal forms to exist [19]. The (crystal form has a higher Young's modulus and a lower elongation than the α′ form, which means the α′ form exhibits better plasticity. However, the (form provides a better barrier to moisture, another factor that can influence tablet properties.

The formation of hydrates can greatly change the mechanical properties of a material, as in the case of magnesium stearate, a very popular lubricant used in tablet formulations. Magnesium stearate has four hydration states: an anhydrous form, a monohydrate, a dihydrate, and a trihydrate. It has been reported that the dihydrate is a better lubricant than the anhydrous form and the monohydrate, very likely because the water situated between oriented fatty acid layers in the crystal favors shear between these layers and then allows appropriate amounts of deformation (Figure 16.10) [20].

For amorphous solids, very different mechanical properties are associated with glassy states and rubbery states, below and above the glass transition temperature, respectively. Below the glass transition temperature, the material is hard, rigid, and brittle, like a crystalline solid. Above the glass transition temperature, the amorphous phase exhibits increased molecular mobility and becomes rubbery, soft, and flexible, which favors deformation.

For example, the shear modulus of polystyrene, an amorphous polymer, is reduced slightly as temperature is increased below the glass transition, but, above the glass transition at 107 °C, the shear modulus decreases dramatically (Figure 16.11) [21].

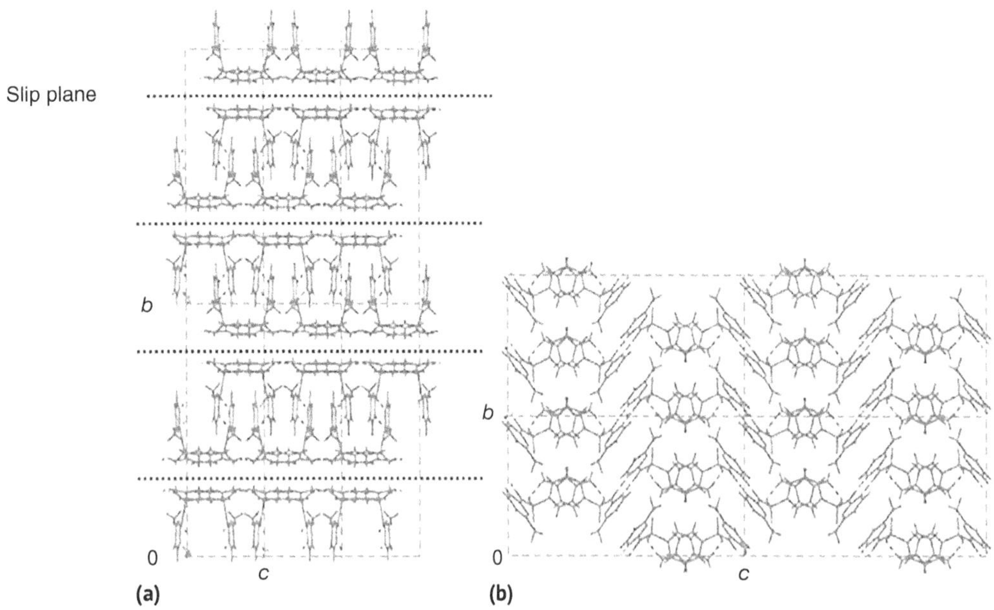

Figure 16.9 Crystal structure of sulfamerazine polymorphs (a: I; b; II) and presence of slip planes. *Source*: [18]/with permission of Springer Nature.

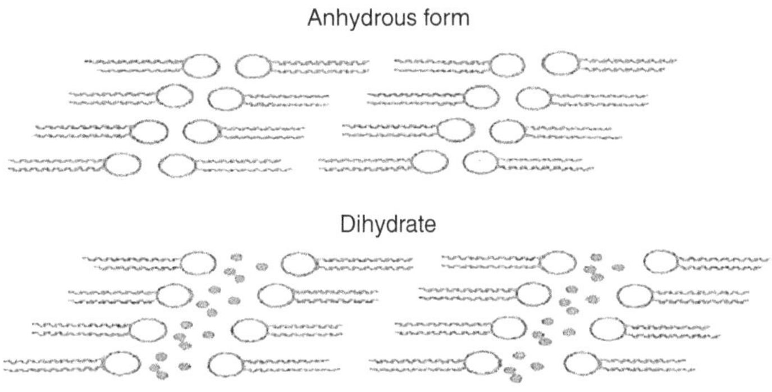

Figure 16.10 Comparison of magnesium stearate anhydrous form and dehydrate in crystal packing.

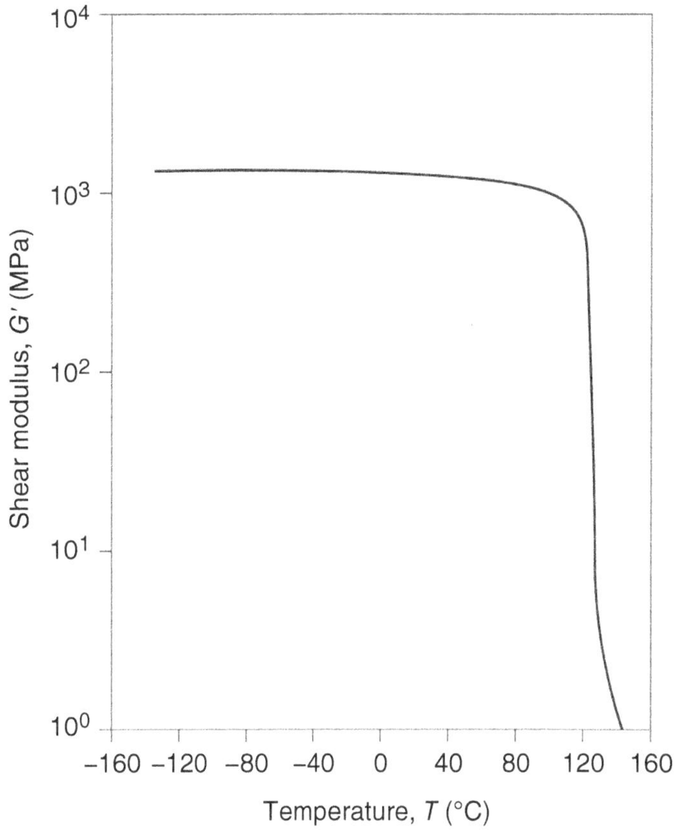

Figure 16.11 Shear modulus of polystyrene as a function of temperature. *Source*: Reproduced from [21]/with permission of Hanser.

16.12 Effect of Moisture on Mechanical Properties

The presence of moisture in a sample can affect the mechanical properties of a material in many ways, depending on the material characteristics and the manner in which water interacts with the material.

As discussed in Chapter 15, solid water interactions can take place in several different ways:

- Adsorption to the particle surface
- Absorbed into the crystal lattice to form hydrates
- Absorbed into amorphous materials to form a solid solution
- Capillary condensation into microcavities
- Deliquescence

The amount of water adsorbed to a surface is determined by surface properties, surface area, the relative humidity, and temperature. Water adsorbed into the surface to form a monolayer, or multiple layers, would be expected to modify particle interactions due to the alteration in surface energy and possible changes in surface conductivity. Tightly bound adsorbed water layers, for example, can reduce the distance between particles and increase intermolecular attractive forces. In addition, adsorbed layers of water on one particle may interact with water molecules on other particle surfaces to cause physical binding of neighboring particles. The presence of water has been shown to increase tablet tensile strength. Down and McMullen, for example, have reported that the crushing strength of sodium chloride compacts increased with increasing relative humidity before reaching a critical relative humidity of 75%, where deliquescence occurred [22].

The absorption of water into a crystal lattice is a phase transformation that results in the formation of a new solid phase, for example, a hydrate from an anhydrous form or a higher hydrate from a lower hydrate, where a new crystalline solid phase could present different mechanical properties.

Absorption of water into amorphous materials has been shown to act as a plasticizer that reduces the glass transition temperature proportionally to the amount of moisture absorbed. This plasticizer effect leads to less viscosity, reduced relaxation times, greater diffusivity, and therefore greater viscoelasticity and more plastic deformation.

Sun studied the effect of moisture on compaction properties of microcrystalline cellulose, where the compaction properties were relatively insensitive to moisture contents below 3.3% water since 3.3% water approximately corresponds to the equivalent of monolayer coverage [23]. Above 3.3% water, the plasticity of MCC was increased; however, the bonding strength was reduced. Therefore, the interplay of plasticity and bonding strength resulted in a tablet tensile strength that peaked in the range of 3.3–5.6% moisture.

Teng et al. observed that directly compressed tablets containing MCC had an increased tensile strength as the moisture content was increased [24]. It was further suggested that insufficient moisture content may cause tablet chipping or lamination owing to a relatively high yield force and elastic recovery. This occurs when less than 3% w/w of moisture is internally chemisorbed by the particles, not acting effectively as a plasticizer. It was also suggested that additionally absorbed water promotes the formation of pendular bonds on the particle surfaces, which helps to enhance plastic deformation and increase inter-particle bonding. Khan et al. were able to show that moisture is sorbed into the amorphous part of MCC and acts as a plasticizer [25].

In addition, reduced glass transition temperatures may promote recrystallization of some materials that could increase the binding of particles.

As discussed in Chapter 15, capillary condensation in microcavities or micropores can cause water to condense at pressures less than the expected saturation pressure of the vapor, and the liquid formed within the cavity can act as a bridge to link and bind particles together. Deliquescence, which leads to liquid water, could result in significant dissolution or softening of a portion of the material, thus promoting the formation of solid bridges due to a sintering effect. It is often observed that a freely flowing powder bed may cake after substantial exposure to high levels of humidity because of such behavior.

The effects of moisture on tablet tensile strength can go in either direction, in that tensile strength usually increases initially as moisture content is increased. It eventually plateaus out or decreases as excess water builds up in the sample.

Nokhodchi et al., for example, showed that moisture content significantly affects the tensile strength of tablets made with hydroxyl propyl methylcellulose (HPMC) K4M, with the tensile strength increasing from 5.6 to 8.5 Mpa at a compression force of 10 kN when the moisture content changed from 10 to 15% w/w [26].

Excess water may disturb or reduce intermolecular attraction force and compromise tablet tensile strength, as shown by Nokhodchi et al., where the crushing strength of ibuprofen tablets initially increased with increasing moisture content, reaching a maximum at 2.5% w/w and then decreasing as moisture content went from 2.5% to 10% w/w [27].

The presence of moisture also benefits the compressibility of a tablet formulation by acting as a lubricant and mitigating the surface irregular microstructure, reducing frictional forces, facilitating particle rearrangement, promoting particle slippage, and promoting particle rearrangements and plastic deformation during compression. Esezobo and Pilpel, for example, studied the effect of moisture on the compressibility of oxytetracycline formulations, where an increase in moisture content resulted in a high Heckel plot slope, indicating a greater degree of densification at low pressure and enhanced compressibility [28]. Similar trends were observed for maltodextrin by Li and Peck, where an increase in powder moisture content reduced the yield pressure and improved compressibility [29].

As mentioned earlier, residual water has a significant effect on the lubricant properties of magnesium stearate. In this regard, Mueller et al. reported that increasing the moisture content of magnesium stearate caused a decrease in die-wall friction [30]. For example, a 3.3% moisture content produces a die-wall friction of 13.2 kg/cm^2 for a lactose formulation with 0.1% magnesium stearate, while the die-wall friction was decreased to 1.5 kg/cm^2 for a batch of magnesium stearate with 6.7% moisture content.

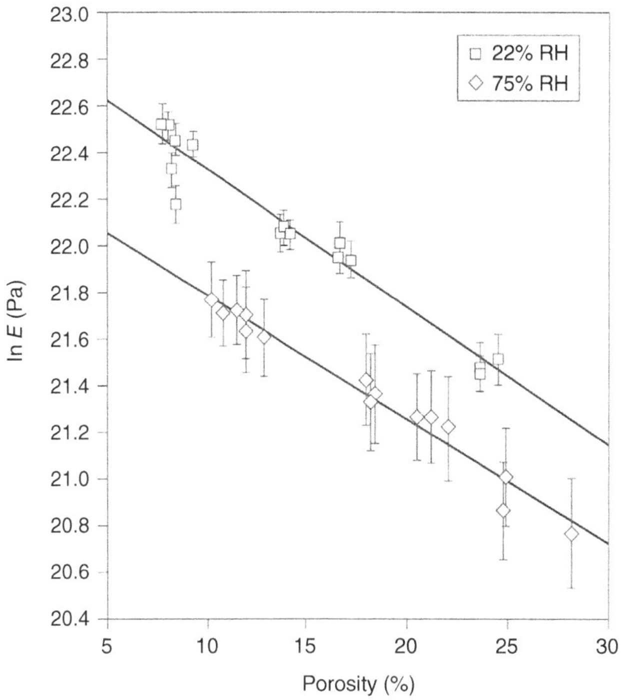

Figure 16.12 Natural log of the mean Young's modulus of the un-notched beams of MCC as a function of porosity (at 22% and 75% relative humidity). *Source*: Adapted from [33].

Wada and Matsubara reported that the lubrication properties of magnesium stearate are approximately linearly increased with increasing moisture content [31]. It was assumed that water enters the crystal lattice and weakens the interactions between oriented fatty acid layers, which favors greater shearing and, hence, lubrication.

Proper moisture content is also critical for controlling the mechanical properties of gelatin capsule shells. For example, it has been reported that a water content of less than 10% in the capsule shell causes capsules to become brittle and fracture easily [32].

The effect of water is not always beneficial, for sometimes residual water can cause strong aggregation or "concretion" of compacts or tablets during storage. Process-induced disorder, leading to partially amorphous states and greater absorption of water, can cause particle aggregation through surface crystallization, fusion, and concretion. The resulting increase in hardness (tensile strength) has been observed to result in poor dissolution.

Residual water also affects the elasticity of a material. For example, the Young's modulus of MCC compacts was shown to be decreased when humidity was increased at various porosities, as shown in Figure 16.12 [33]. The extrapolated Young's modulus at zero porosity for 22% RH and 75% RH was found to be 6.7 and 3.8 GPa, respectively.

16.13 Methods for Testing Mechanical Properties

16.13.1 Beam Bending

The beam bending test, also called a flexure test, is widely used to measure Young's modulus, where a rectangular beam or a compressed rectangular powder compact is subjected to transverse forces. The forces cause bending of the sample, which is measured by a central deflection. There are two ways to support and load the beam, including four-point bending and three-point bending (Figure 16.13). For a four-point bending, the beam is supported at two points and loaded at two points, whereas for a three-point bending, the beam is supported at two points and loaded at one point. The loading point(s) must be approximately symmetrical relative to the midpoint of the beam. The loading forces and the deflection at the midpoint are used to calculate the Young's modulus, based on equations derived by Church for four-point bending (equation [16.30]) and Bassam et al. for three-point bending (equation [16.31]) [2].

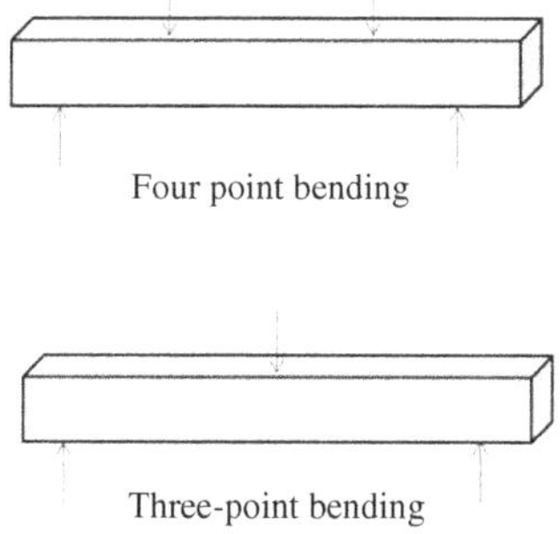

Figure 16.13 Geometries of four-point bending (top) and three-point beam bending (bottom).

$$E = \frac{F}{\xi} \frac{6a}{h^3 b}\left[\frac{l^2}{8} + \frac{al}{2} + \frac{a^2}{3}\right] \tag{16.30}$$

$$E = \frac{Fl^3}{4\xi h^3 b} \tag{16.31}$$

In equations (16.30) and (16.31), F is the sum of the loading forces, ξ is the central deflection, h and b are the height and width of the beam, l is the distance between two loading points, and a is the distance between the neighboring loading point and the support point. For both methods, the Young's modulus can be calculated from a single-point measurement or from the plot of total load versus central deflection.

Four-point beam bending tends to have more uniform bending in the central section of the beam and a less significant contribution from shear pressure than three-point beam bending. However, three-point beam bending can be performed using smaller sample sizes.

For a compacted powder, the Young's modulus is affected by porosity and usually increases as porosity decreases. Since particles are more densely packed as porosity decreases, particles are generally more resistant to deformation due to strong binding and physical interlocking. The Young's modulus at zero porosity is extrapolated from deformation measurements at various porosities.

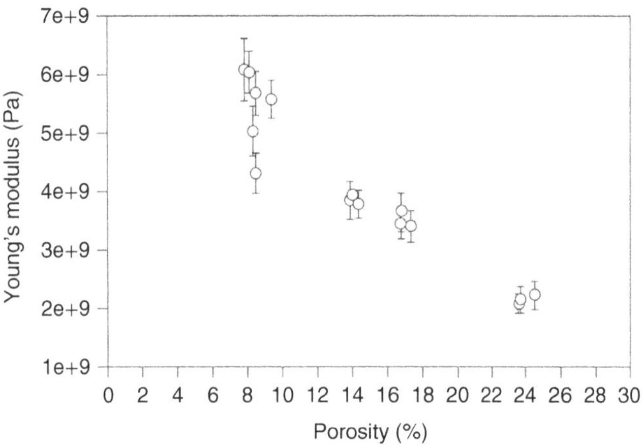

Figure 16.14 Mean Young's modulus of the un-notched beams as a function of porosity at 22% relative humidity. *Source*: [34]/with permission of Elsevier.

Bassam et al. studied 15 representative tableting excipients over a range of porosities using a four-point beam bending technique [2]. The calculated Young's modulus at zero porosity for the excipients ranged from 3.71 to 88.28 GPa, with the order of increasing stiffness being starch < microcrystalline cellulose < sugars < inorganic fillers (calcium carbonate, calcium phosphate).

Hancock et al. used a three-point beam bending technique to determine the elasticity of compacts made of microcrystalline cellulose using a thermomechanical analyzer [34]. The Young's modulus of compacts of microcrystalline cellulose with a range of porosities is given in Figure 16.14, showing that the Young's modulus increases as porosity decreases with an approximately linear relationship.

16.13.2 Thermomechanical Analyzer

TMA measures dimensional changes of a material as a function of temperature and/or time under a defined mechanical force. The material can be a solid, liquid, or semisolid. The temperature can be changed by heating or cooling, or the sample can be measured isothermally.

A simplified diagram of a TMA is given in Figure 16.15. Typically, a test material is put on a sample support in the furnace, and a test probe is made to rest upon the material to sense changes in length. The probe is linked to a sensitive position transducer, and the transducer records dimensional changes. The temperature in the furnace is measured by a thermocouple. A purging gas, such as helium, is often utilized to enhance heat flow and to mitigate oxidation during the measurement. The mechanical force is applied using static weights or a force motor. The force motor has the advantage of providing a broader range of loads with better control.

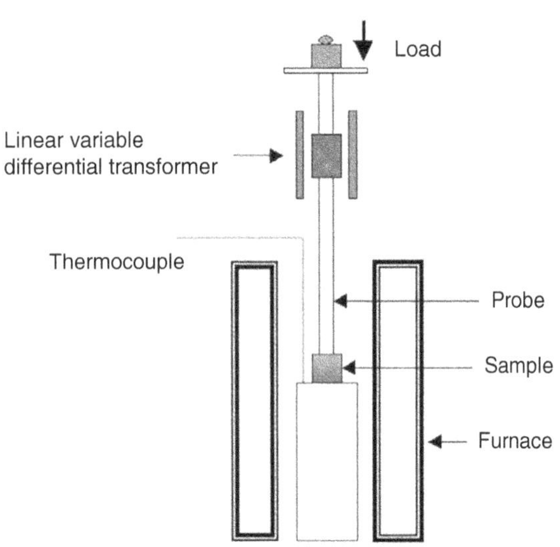

Figure 16.15 A thermal mechanical analyzer.

Temperature calibration is usually carried out using small flat pieces of pure metals, such as indium, sandwiched between thin discs of quartz. Melting of the metals causes a sharp displacement of the probe. Position calibration is carried out by heating a sample whose expansion coefficient is accurately known, such as aluminum. The aluminum samples are in the form of small blocks or cylinders with dimensions of 5–10 mm.

To measure thermal expansion, a very small amount of load is utilized to ensure proper contact of the probe with the sample while minimizing deformation. Ways to study deformation include compression, penetration, indentation, dilatometry, flexure test, and tension test (Figure 16.16). Besides studying thermal expansion, TMA has been used to investigate thermal transitions such as the glass transition and melting.

The applications of TMA are listed as follows:

- Material property characterization, such as glass transition temperature, softening temperature, and melting temperature.
- Studying the process effects on material properties and reliability of the process.
- Differentiating between different materials and or different suppliers.

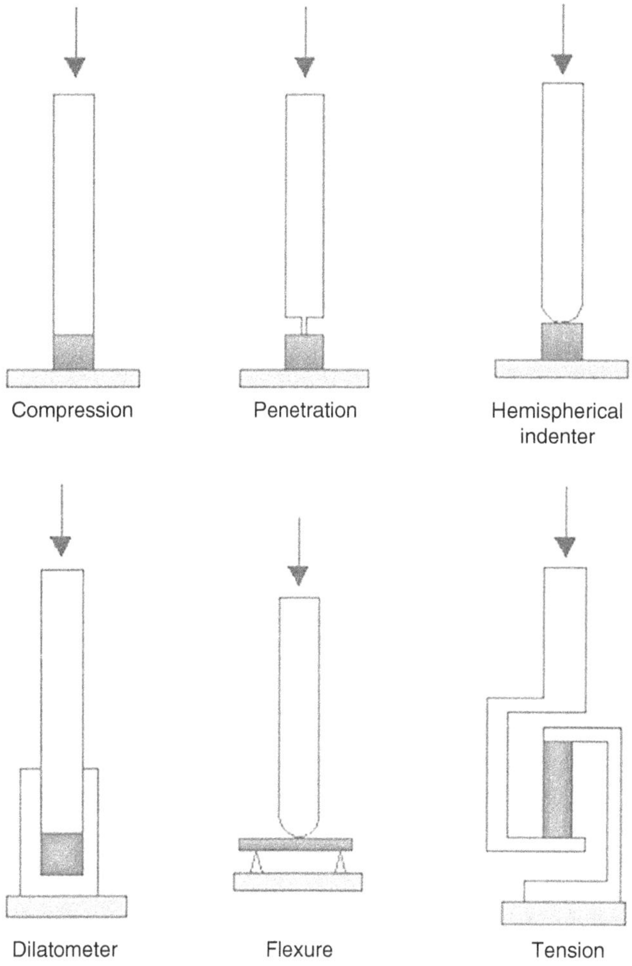

Figure 16.16 Different ways of loading for the thermal mechanical analyzer.

16.13.3 Dynamic Mechanical Analyzer

A DMA is a widely used instrument to study the viscoelastic properties of pharmaceutical materials by analyzing their response to an oscillating force. During an analysis, a sinusoidal stress is applied to a sample to generate a sinusoidal strain. The amplitude of the deformation and the lag between the stress and the strain sine waves are used to evaluate the extent of viscoelasticity.

During a DMA test, the sinusoidal deformation is applied to a sample with defined geometry, where the deformation is controlled by a force motor. The force motor generates a sinusoidal wave, and the wave is transmitted to the sample via a drive shaft. For viscoelastic materials, the strain response to the sinusoidal stress comprises two components: a storage modulus and a loss modulus. The storage modulus is an in-phase component and is the measure of the sample's elastic behavior, whereas the loss modulus is an out-of-phase component that measures viscosity. The ratio of the loss modulus to the storage modulus, which is called the loss tangent, is related to the phase lag between dynamic stress and dynamic strain.

The complex dynamic modulus, E, is made of the elastic modulus, E', and loss modulus E'' as in the following equation:

$$E = E' + iE'' \tag{16.32}$$

The loss tangent can be determined by the following equation:

$$\tan\delta = \frac{E''}{E'} \tag{16.33}$$

where δ is the phase lag.

Though the storage modulus and the Young's modulus both measure elasticity, the values of the two moduli are not necessarily the same.

Damping refers to the dissipation of energy for a viscoelastic material under cyclic load, as indicated by the tangent of the phase angle. It is the measure of how well a material can absorb energy.

Depending on the application, the choice of sample geometry can include tension, single cantilever, dual cantilever, three-point bending, simple shear, and compression.

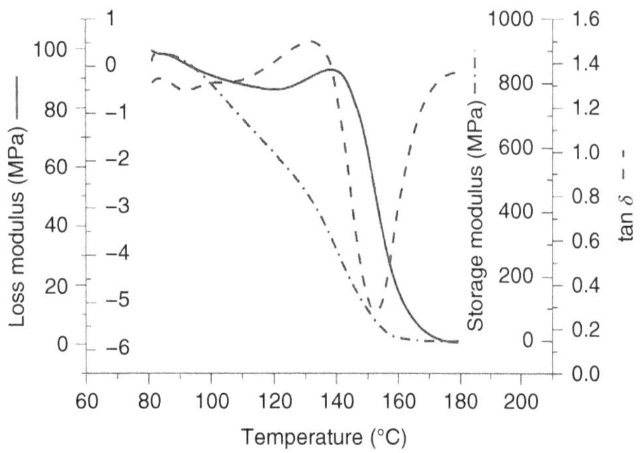

Figure 16.17 DMA results for Eudragit L polymer film to reflect the glass transition. *Source*: [35]/with permission of Elsevier.

The procedure to carry out a DMA experiment is as follows:

- Prepare the sample with a defined geometry.
- Load the sample into a DMA furnace.
- Set the starting temperature, the final temperature, and the heating rate.
- Apply the same sinusoidal stress throughout the entire temperature range.
- Record the material response to deformation.
- Determine E', E'', and tan δ.
- Detect transition temperatures based on noticeable changes in the curves obtained.

The data can be presented as either storage modulus, E', versus temperature or tan δ versus temperature.

The glass transition temperature of an amorphous material is reflected as a major change in DMA responses with temperature, as reflected in a large decrease in E' versus temperature and a peak in the tanδ plot. Figure 16.17 shows the DMA results for an Eudragit L polymer film, where the glass transition temperature is about 150 °C [35].

16.13.4 Nanoindentation

A nanoindentation test can measure samples in the micrometer and submicrometer range, allowing the measurement of mechanical properties of very small particles. During a nanoindentation test, a precisely defined tip is brought in contact with a sample, and a force with a small load is applied to measure its mechanical properties. Since the indentation is on a very small scale, it is only detectable by an atomic force microscope (AFM) or a scanning electron microscope. It is important to note that there are significant challenges to measuring the area of the indentation accurately on such a small scale. In fact, the area of indentation is calculated based on the geometry of the tip and the depth of penetration that can usually be recorded precisely, for example, by an AFM. As a result, various mechanical properties such as elastic modulus and hardness can be extracted from a load versus displacement curve from nanoindentation tests.

During a nanoindentation test, the applied force and the depth of penetration are recorded as the force increases from zero to a certain maximum and then decreases back to zero. The response of depth of penetration to the applied load is related to the mechanical properties of a material. For an elastic material, the loading and unloading curves are expected to superimpose with each other. If a material has plastic deformation, loading and unloading will exhibit different curves. As the force is removed from the indenter, the material tends to recover to a certain degree due to the relaxation of elastic strains of the material. The evaluation of the initial portion of the elastic response during unloading provides an estimation of the elastic modulus of the tested material.

The indenter tip for a nanoindentation tester has a shape of sphere, cone, four-sided pyramid (Vickers), or three-sided pyramid (Berkovich).

The hardness value, described previously, is related to the applied force and residual indentation area by the following equation:

$$H = \frac{P}{A_r} \tag{16.34}$$

where P is the applied force and A_r is the indentation area, which is derived from the depth of penetration and the geometry of the indenter. The hardness value obtained is defined by the type of indenter used, for example, Brinell hardness or Vickers diamond hardness, like a regular indenter, as discussed previously.

Loading and unloading displacement curves can be used to differentiate between materials, as illustrated in Figure 16.18.

Owing to the need for high precision in nanoscale measurements, temperature control and the control of vibrational disturbances are essential for successful measurements. It is also critical to ensure that the specimen and the instrument are in thermal equilibrium. The instrument should be insulated against vibration and acoustical noise by using a very heavy base and loading column. Samples are usually mounted on a metal holder, and the holder is then placed on a stage with precise positional control. Force is applied to the indenter shaft with an electromagnetic coil or the expansion of a piezoelectric element. Displacements are recorded using a changing capacitance or inductance signal.

Nanoindentation testing has been used to study the correlation between mechanical properties and crystal packing. For example, Ramos and Bahr studied the mechanical behavior of sucrose using nanoindentation, specifically about the (100) and (001) faces of its single crystals [36]. The elastic modulus on the (100) face was found to be 38 GPa, and that of the (001) face was 33 GPa. The measured hardness of sucrose was estimated to be approximately 1.5 GPa, and the maximum shear stress was about 1 GPa, prior to any plastic deformation.

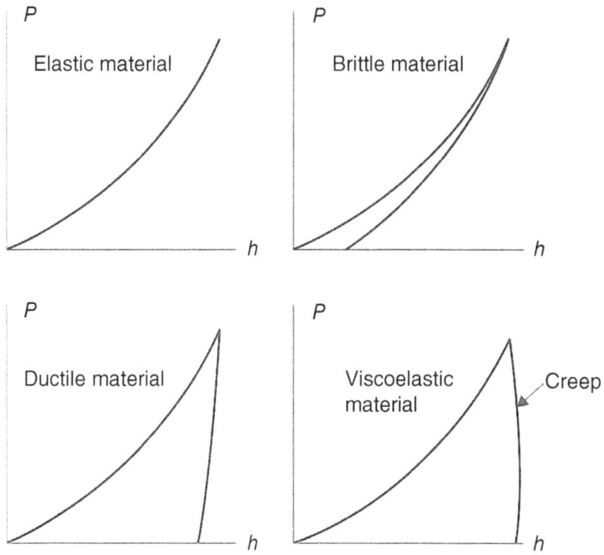

Figure 16.18 Displacement curves of material with different mechanical properties by nanoindentation test.

Liao et al. studied some process-dependent material properties of pharmaceutical solids by nanoindentation [37]. Analysis of force-displacement curves of KCl crystals obtained from quenched crystallization revealed that the hardness of these crystals was 10 times greater than that of KCl crystals obtained from solvent crystallization. It was also shown that acetaminophen crystals obtained from quenched crystallization were 25% harder than the recrystallized samples in solvents. For both materials, the elastic moduli of the samples obtained from quenched crystallization were also much greater than those observed for recrystallized samples.

Taylor et al. reported that nanoindentation of single crystals of five pharmaceutical ingredients can be used to predict their milling behavior [38].

In summary, the mechanical properties of pharmaceutical active ingredients and excipients can significantly affect the manufacturing process and product quality. Physical form and moisture content significantly influence the mechanical properties of a pharmaceutical material, including elasticity, plasticity, viscoelasticity, brittleness, hardness, compatibility, and tabletability. Consequently, the measurement of mechanical properties for various pharmaceutical solids can greatly facilitate ingredient selection, formulation development, and process development.

References

1 Schnaare RL, Block LH, Rohan LC. Rheology. In: *Remington: The Science and Practice of Pharmacy*, 21e. Baltimore: Lippincott William & Wilkins, 2005: 338–360.
2 Bassam F, York P, Rowe RC, Roberts RJ. Young's modulus of powders used as pharmaceutical excipients. *Int J Pharm* 1990; 64(1): 55–60.
3 Anonymous. Modulus of Rigidity. http://www.engineeringtoolbox.com/modulus-rigidity-d_946.html. Access 2016 Jan 12.
4 Roberts RJ, Rowe RC, York P. The Poisson's ratio of microcrystalline cellulose. *Int J Pharm* 1994; 105(2): 177–180.
5 Brostow W, Lobland HEH, Narkis M. The concept of materials brittleness and its applications. *Polym Bull* 2011; 67(8): 1697–1707.
6 Lawn BR, Marshall DB. Hardness, toughness, and brittleness: an indentation analysis. *J Am Ceram Soc* 1979; 62 (7–8): 347–350.
7 Heistand EN, Wells JE, Poet CB, Ochs JF. Physical process of tableting. *J Pharm Sci* 1977; 66(4): 510–519.
8 Hiestand EN. Rationale for the measurement of tableting indices. In: Alderborn G, Nystrom D, eds. *Pharmaceutical Powder Compaction Technology*. New York: Marcel Dekker, 1996.
9 Hiestand EN. Dispersion forces and plastic deformation in tablet bond. *J Pharm Sci* 1985; 74(7): 768–770.
10 Heckel RW. Density-pressure relationships in powder compaction. *Trans Metall Soc AIME* 1961; 221(4): 671–675.
11 Klevan I, Nordström J, Bauer-Brandl A, Alderborn G. On the physical interpretation of the initial bending of a Shapiro-Konopicky-Heckel compression profile. *Eur J Pharm Biopharm* 2009; 71(2): 395–401.
12 Kawakita K, Lüdde KH. Some considerations on powder compression equations. *Powder Technol* 1971; 4(2): 61–68.

13 Fell JT, Newton JM. Determination of tablet strength by the diametral-compression test. *J Pharm Sci* 1970; 59(5): 688–691.
14 Sun CC. A material-sparing method for simultaneous determination of true density and powder compaction properties—Aspartame as an example. *Int J Pharm* 2006; 326(1): 94–99.
15 Vromans H, Bolhuis GK, Lerk CF, Van de Biggelaar H, Bosch H. Studies on tableting properties of lactose. VII. The effect of variations in primary particle size and percentage of amorphous lactose in spray dried lactose products. *Int J Pharm* 1987; 35(1): 29–37.
16 De Figueiredo LP, Ferreira FF. The Rietveld method as a tool to quantify the amorphous amount of microcrystalline cellulose. *J Pharm Sci* 2014; 103(5): 1394–1399.
17 Ahjel SW, Lupuliasa DU. Directly compressible adjuvants-a pharmaceutical approach. *Farmacia* 2008; 56(6): 591–599.
18 Sun C, Grant DJ. Influence of crystal structure on the tableting properties of sulfamerazine polymorphs. *Pharm Res* 2001; 18(3): 274–280.
19 Cocca M, Di Lorenzo ML, Malinconico M, Frezza V. Influence of crystal polymorphism on mechanical and barrier properties of poly(l-lactic acid). *Eur Polym J* 2011; 47(5): 1073–1080.
20 Ertel KD, Carstensen JT. Chemical, physical, and lubricant properties of magnesium stearate. *J Pharm Sci* 1988; 77(7): 625–629.
21 Osswald TA. Injection molding materials. In: Osswald TA, Turng LS, Gramann P, eds. *Injection Molding Handbook*, 2e. Cincinnati: Hanser, 2008.
22 Down GR, McMullen JN. The effect of interparticulate friction and moisture on the crushing strength of sodium chloride compacts. *Powder Technol* 1985; 42(2): 169–174.
23 Sun CC. Mechanism of moisture induced variations in true density and compaction properties of microcrystalline cellulose. *Int J Pharm* 2008; 346(1): 93–101.
24 Teng CO, Alkan MH, Groves MJ. The effect of adsorbed water on compaction properties and the dissolution of quinacrine hydrochloride from compacted matrices by soy protein. *Drug Dev Ind Pharm* 1986; 12 (11–13): 2325–2336.
25 Khan F, Pilpel N, Ingham S. The effect of moisture on the density, compaction and tensile strength of microcrystalline cellulose. *Powder Technol* 1988; 54(3): 161–164.
26 Nokhodchi A, Ford JL, Rowe PH, Rubinstein MH. The influence of moisture content on the consolidation properties of hydroxypropylmethylcellulose K4M (HPMC 2208). *J Pharm Pharmacol* 1996; 48(11): 1116–1121.
27 Nokhodchi A, Rubinstein MH, Larhrib H, Guyot JC. The effect of moisture on the properties of ibuprofen tablets. *Int J Pharm* 1995; 118(2): 191–197.
28 Esezobo S, Pilpel N. Some formulation factors affecting tensile strength, disintegration and dissolution of uncoated oxytetracycline tablets. *J Pharm Pharmacol* 1976; 28(1): 8–16.
29 Li LC, Peck GE. The effect of moisture content on the compression properties of maltodextrins. *J Pharm Pharmacol* 1990; 42(4): 272–275.
30 Muller BW. Polymorphism of magnesium stearate and the influence of the crystal structure on the lubricating behavior of excipients. *Acta Pharm Suec* 1981; 18(2): 74–75
31 Wada Y, Matsubara T. Pseudopolymorphism and lubricating properties of magnesium stearate. *Powder Technol* 1994; 78(2): 109–114.
32 Chang RK, Raghavan KS, Hussain MA. A study on gelatin capsule brittleness: moisture transfer between the capsule shell and its content. *J Pharm Sci* 1998; 87(5): 556–558.
33 Rowe RC, Roberts RJ. The mechanical properties of powders. In: Ganderton D, Jones T, McGinity J, eds. *Advances in Pharmaceutical Sciences (Vol 7)*. San Diego: Academic Press, 1995.
34 Hancock BC, Clas SD, Christensen K. Micro-scale measurement of the mechanical properties of compressed pharmaceutical powders. 1: The elasticity and fracture behavior of microcrystalline cellulose. *Int J Pharm* 2000; 209(1): 27–35.
35 Fadda HM, Khanna M, Santos JC, Osman D, Gaisford S, Basit AW. The use of dynamic mechanical analysis (DMA) to evaluate plasticization of acrylic polymer films under simulated gastrointestinal conditions. *Eur J Pharm Biopharm* 2010; 76(3): 493–497.
36 Ramos KJ, Bahr DF. Mechanical behavior assessment of sucrose using nanoindentation. *J Mater Res* 2007; 22(7): 2037–2045.
37 Liao X, Scott Wiedmann T. Measurement of process-dependent material properties of pharmaceutical solids by nanoindentation. *J Pharm Sci* 2005; 94(1): 79–92.
38 Taylor LJ, Papadopoulos DG, Dunn PJ, Bentham AC, Dawson NJ, Mitchell JC, Snowden MJ. Predictive milling of pharmaceutical materials using nanoindentation of single crystals. *Org Proc Res Dev* 2004; 8(4): 674–679.

17

Solubility and Dissolution

17.1 Introduction

The solubility of drugs in various solvents, particularly water, is a critical physical property that affects their stability, bioavailability, and therapeutic activity in various drug products. For example, all drugs must dissolve into biological fluids after administration before absorption can occur across biological membranes into the systemic circulation, whereas drugs in solution generally have much greater rates of chemical reactivity and instability than when in the solid form. Additionally, a drug must generally be solubilized before it can be analyzed by various analytical techniques such as HPLC, NMR, and mass spectrometry and must be in solution during many processes, such as crystallization and spray drying. Furthermore, most parental products must be administered as solutions, particularly when using them for administration by the intravenous route. It is well recognized that the solubility of a drug is primarily determined by such factors as crystal form, amorphous content, pH, temperature, and the presence of surfactants. Of particular importance in the selection of drug candidates is often the need to enhance the solubility and dissolution of normally poorly soluble drugs to improve oral absorption and bioavailability. In this regard, solubility data in combination with permeability are utilized to categorize the potential oral bioavailability of new chemical entities and drug candidates by using the Biopharmaceutical Classification System (BCS) [1].

While solubility values are generally taken as the amount of drug in solution in equilibrium with a solid and therefore a measure of the thermodynamic potential for a drug to dissolve under a given set of conditions, the term "drug dissolution" is used to describe the rate at which a drug will dissolve under a given set of conditions. Thus, the rate of dissolution is a critical parameter to establish, for example, the relationship of dissolution to ultimate bioavailability and in vivo performance. In addition, in vitro dissolution tests play a crucial role in guiding formulation development and optimization, with dissolution data being a major part of the analytical information required for assuring the quality control (QC) of tablets and capsules as commercial products. Dissolution testing, especially in biorelevant media, in some cases, can be used as a surrogate tool to predict in vivo performance.

In this chapter, we will review the theoretical background related to drug solubility and drug dissolution and introduce various techniques for determining solubility and dissolution with an emphasis on applications for the evaluation of drug substances and drug products.

17.2 Principal Concepts Associated with Solubility

Solubility is a broad term that is often unclearly defined in many circumstances. Some of the important aspects of solubility, in this regard, include equilibrium solubility, apparent solubility, kinetic solubility, and intrinsic solubility, as shown in Table 17.1.

Equilibrium solubility represents the maximum amount of a drug that can dissolve in a solvent under a certain experimental condition, while maintaining a thermodynamically stable system. Depending on the drug concentration relative to the equilibrium solubility, a drug solution can exist in three distinct states: undersaturated, saturated, and supersaturated. For an undersaturated solution, additional solute can still dissolve, whereas, in a saturated solution, the solid and solute reach an equilibrium state. A supersaturated solution is a thermodynamically unstable system containing more solute than the equilibrium solubility, with a thermodynamic tendency to crystallize. Equilibrium solubility, generally defined

Solid-State Materials in Pharmaceutical Chemistry: Properties, Characterization, and Applications, Second Edition. Stephen R. Byrn, George Zografi and Xiaoming (Sean) Chen.

Table 17.1 Description of different solubility concepts.

Concept	Measurement	Material
Equilibrium solubility	Shake flask, until no concentration and solid form change observed, days	Crystalline
Apparent solubility	Shake flask, time course measurement, intend to reach equilibrium, hours to days	Metastable forms, amorphous, DMSO stock solution
Kinetic solubility	Mixing of stock solution with a buffer, minutes to a few hours	DMSO stock solution
Intrinsic solubility	Shake flask, titration, hours or days	Crystalline powder of salt forms or free form

as the concentration of a solute in a saturated solution at a certain temperature, is usually measured by the shake-flask method, whereby crystalline materials are added to a solvent in amounts that leave an excess of solid in equilibrium with the solution.

However, it is challenging to measure equilibrium solubility for amorphous form and other metastable forms since, once dissolved, the molecules generally undergo solution-mediated transformation to more thermodynamically stable forms with lower solubility. Therefore, the solubility of such systems is usually reported as apparent solubility.

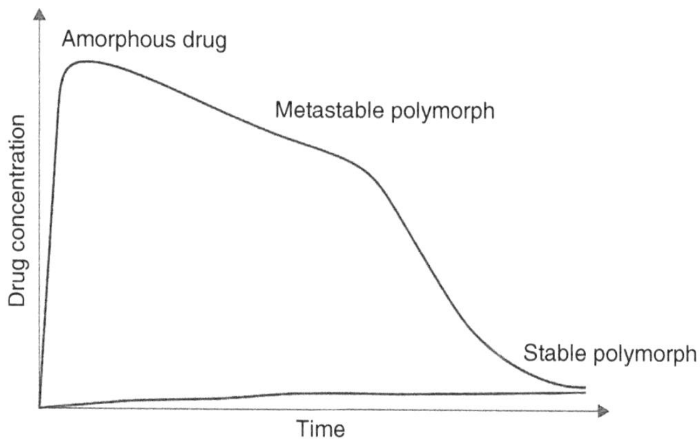

Figure 17.1 Apparent solubility measurement for an amorphous material.

Apparent solubility of a metastable form is generally measured by detecting the amount of material dissolved within a relatively short period of time, where the material commonly reaches a peak solubility when the concentration in solution tends to decrease as the more stable form crystallizes, as demonstrated in Figure 17.1.

Kinetic solubility is usually measured under a high-throughput setting, for example, in an early stage of drug development, to obtain a preliminary indication of the drug candidate's solubility required for preliminary screening and testing. It is typically obtained using a dimethyl sulfoxide (DMSO) stock solution added to a buffer solution while increasing the concentration until precipitation occurs. Kinetic solubility is generally significantly higher than the equilibrium solubility since, most often, metastable polymorphic forms or the amorphous form precipitate at first.

Intrinsic solubility is most often used to define the equilibrium solubility of the free form of an ionizable compound, such that the solubility value obtained is independent of pH, and, therefore, is the lowest solubility for an ionizable compound. The solubility value at two pH units below the pK_a for acidic compounds and above the pK_a for basic compounds is generally considered a good approximation of the intrinsic solubility. Ionizable compounds have pH- and pK_a-dependent solubility, which is related to intrinsic solubility, as described in Chapter 4.

17.3 Prediction of Aqueous Drug Solubility

As discussed earlier, drug solubility in aqueous media is most often directly related to drug dissolution, absorption, and bioavailability, with a general trend that the more water-soluble drugs are, the more likely they will be absorbed when administered orally, assuming that membrane permeability is not the rate-determining factor. Consequently, much effort has been put in to understand the relationship between molecular structure and equilibrium solubility of a drug candidate. For this purpose, many mathematical models have been developed to predict the solubility of drugs from their molecular structure as well as other fundamental physical properties such as drug lipophilicity, crystallinity, and pK_a when the drug is ionizable. Such information can be obtained experimentally if enough of the compound is available, or, alternatively, computer software can be utilized to calculate those properties based on molecular structure.

Significant contributions to the development of predictive models have come from the work of Ran and Yalkowsky [2], who have developed several empirical mathematical models to predict drug solubility in water using lipophilicity and crystallinity data. The models have a strong thermodynamic basis, but the exact models have been derived from the statistical analysis of solubility data obtained for a very large sample of organic compounds.

It is generally assumed that the solubility of an organic liquid in water is primarily determined by its lipophilicity or hydrophobicity, as can be described by using the Hansch equation [3]:

$$\log S = A \log K_{ow} + B \tag{17.1}$$

where $\log K_{ow}$ is the octanol/water partition coefficient, also designated as $\log P$, and A and B are constants related to the liquid.

Since a majority of drugs exist as crystalline solids, any prediction of drug solubility should also take into consideration the energetics associated with the crystal lattice, as reflected in the melting point. Taking this into account, Ran and Yalkowsky have developed a general solubility equation for crystalline nonelectrolytes based on both lipophilicity and lattice energy [2]:

$$\log S = 0.5 - \log P - 0.01(mp - 25) \tag{17.2}$$

where mp is the melting point of the drug in degrees Celsius. Here, therefore, solubility of the crystal is taken to be determined by overcoming the lattice energy (reflected in mp) and the release of energy due to solute–water interactions (reflected in $\log P$).

The general solubility equation is based on the van't Hoff equation and the assumption that for a solute that is completely miscible with a solvent, the ideal solubility of its crystalline form can be described by the following equation:

$$\log X_u^{ideal} = -\Delta H_m \frac{(T_m - T)}{2.303RT_mT} \tag{17.3}$$

where X_u^{ideal} is the ideal solubility, T_m is the melting point, T is the temperature, and ΔH_m is the molar heat of melting. Since the solid and the melt are in equilibrium at melting, the relationship between the entropy of melting and the heat of melting follows the equation:

$$\Delta S_m = \frac{\Delta H_m}{T_m} \tag{17.4}$$

Combination of equations (17.3) and (17.4) yields the following equation:

$$\log X_u^{ideal} = -\Delta S_m \frac{(T_m - T)}{2.303RT} \tag{17.5}$$

Since solubility measurements are usually carried out at room temperature, 298 K, and most organic compounds have ΔS_m around 56.5 J/degree·mole according to Walden's rule [4], equation (17.5) becomes:

$$\log X_u^{ideal} = -\Delta S_m \frac{(T_m - T)}{2.303RT} = -0.01(T_m - 298) = -0.01(mp - 25) \tag{17.6}$$

where mp is the melting point given in Celsius units.

As mentioned earlier, the lipophilicity of a compound can be characterized by its octanol/water partition coefficient K_{ow}, which is defined as:

$$K_{ow} = \frac{C_o}{C_w} \tag{17.7}$$

where C_o and C_w are the equilibrium molar concentrations of the solute in octanol and water, respectively, where activity coefficients are taken as one. The equation can be further approximated as

$$K_{ow} = \frac{S_o}{S_w} \tag{17.8}$$

where S_o and S_w are the equilibrium solubilities of a liquid solute in octanol and water, respectively. Since most unionized organic compounds are fairly miscible in octanol, the equilibrium solubility in octanol, S_o, can be approximately treated as:

$$\log S_o = \log \frac{\text{Molarity of Pure Octanol}}{2} = \log \frac{6.3}{2} = 0.5 \tag{17.9}$$

The solubility of the liquid solute in water, therefore, can be expressed as:

$$\log S_w = 0.5 - \log K_{ow} \tag{17.10}$$

While the solubility of the crystalline drug in water, S, can be described as

$$S = S_w \times X_u^{\text{ideal}} \tag{17.11}$$

A combination of equations (17.10) and (17.6) yields:

$$\log S = \log S_w + \log X_u^{\text{ideal}} = 0.5 - \log K_{ow} - 0.01(mp - 25) \tag{17.12}$$

Substitution of $\log K_{ow}$ with $\log P$ yields equation (17.2), the general solubility equation.

The use of such a general solubility equation involves making many simplified assumptions, and, therefore, it is best used to predict an approximate solubility. Having said this, the average absolute error for the 150 compounds studied is only 0.43 [2]. It is indeed a simple equation that only needs the input of $\log P$ and melting point, pinpointing the two key physical properties that affect aqueous solubility, lipophilicity, and the lattice energy. According to the equation, an increase in $\log P$ of one unit or the increase in melting point by 100 °C will result in a 10-fold decrease in aqueous solubility. Either factor may dominate depending on the chemical structure of the compound.

For example, it has been reported by Bergström et al. that lipophilicity is the dominating factor for compounds with very poor aqueous solubility, wherein the analysis of 15 marketed drugs with poor aqueous solubility, ranging from 2.9 nM to 1.1 μM, indicated that the solubility of these drugs was more limited by solvation than by solid-state properties [5].

On the other hand, compounds having a $\log P$ less than 4 have their solubility primarily limited by the lattice energy. Wassvik et al. reported that molecules with extended ring structures and large conjugated systems tend to be less soluble, since these structural features lead to rigidity in crystal packing and, therefore, higher lattice energy [6].

Commercially available software packages such as ACD Solubility DB (Advanced Chemistry Development, Inc.) and WSKOWWIN (The Syracuse Research Corporation) have added their own correction factors to Yalkowsky's equation to improve prediction accuracy [7].

The program, HYBOT, takes a different approach for solubility calculation by deriving estimates of solubility from the number of hydrogen bond donors (HD), hydrogen bond acceptors (HA), and polarizability of the molecule [8]:

$$\log \frac{1}{S_w} = -0.42 + 0.17 \times \text{polorizability} + 0.13 \times \text{HA} + 0.08 \times \text{HD} \tag{17.13}$$

17.4 Solubility of Pharmaceutical Solid Forms

Polymorphs have different lattice energies and, therefore, give distinctly different equilibrium solubility values, as has been discussed in Chapter 2. It is recognized that the range of solubility differences for polymorphs is generally limited. For example, Pudipeddi and Serajuddin presented such data for many polymorphic systems and showed that solubility ratios were less than two in most cases, with an average value of 1.7 (Table 17.2) [9]. Two exceptional cases reported were ritonavir and premafloxacin, which have solubility ratios of 4 and 23, respectively. The relatively small advantage in solubility in most cases, however, plus the risk for interconversion to the more stable form, limits the use of metastable crystalline forms for enhancing dissolution and oral bioavailability.

If a compound is ionizable, salt formation generally offers significant improvement in solubility and sometimes in oral bioavailability. As discussed in Chapter 4, different salt forms of a compound usually give different solubilities since their lattice energies are generally different. Table 17.3, for example, shows that the solubility of various salts compared to their free form can be increased by from 30 to 100 times [10].

Recently, significant efforts have been made to use cocrystals to overcome the low solubility of some crystalline drugs. Cocrystals are formed between a drug and a conformer via strong hydrogen bonding within the lattice to create a stoichiometric crystalline phase. The solubility of the cocrystal is governed by the solubility products of the API and guest molecule in a manner similar to salt forms. It is generally recognized that conformers with high solubility tend to more effectively increase the cocrystal solubility. For example, Good and Rodriguez-Hornedo showed that the aqueous solubility for seven carbamazepine cocrystals was 2–152 times greater than the solubility of the stable carbamazepine dihydrate form [11].

Table 17.2 Example of solubility ratio of polymorphs.

Compound	Crystal forms	Solubility ratio
Furosemide	II/I	1.0
	III/I	1.8
Gepirone	II/I	2.0
Glibenclamide	II/I	1.6
	IV/II	1.6
	IV/I	2.6
Indomethacin	α/γ	1.1
Phenylbutazone	B/A	1.1
	C/A	1.2
	D/A	1.1
	E/A	1.1
Piroxicam	I/III	1.3
	II/III	1.3
Premafloxacin	I/III	23.1
Ranitidine	I/IV	1.4
	II/IV	1.4
	III/IV	1.3
Ritonavir	I/II	4.0
Sulfathiazole	III/I	1.7
Tolbutamide	IV/II	1.5
	I/II	1.1

Source: Adapted from [9].

Table 17.3 Solubility of a basic compound and its salt in water.

Form	Solubility in water (mg/mL)
Free base	0.32
Acetate	24.20
Hydrochloride	31.70
Tartrate	25.61
Sulfate	10.81

Source: Adapted from [10].

Remenar et al. developed cocrystals of itraconazole with various 1,4-dicarboxylic acids and reported solubilities in 0.1 N HCl that were approximately the same as those of the amorphous form drug found in the commercial formulation and much higher than those found for the crystalline free base (4–20 fold) [12].

It is well known that amorphous forms and amorphous solid dispersions (ASDs) offer significant advantages in solubility enhancement because of a lack of crystallinity and hence a high-energy state. The enhancement of apparent solubility can usually be one to several magnitudes of order higher, as shown in Table 17.4, where up to a 200-fold increase was reported [13].

Despite the solubility advantage of using amorphous forms, such forms, when used alone, have limited use in drug products because of significant tendencies to crystallize under various storage conditions. The formation of miscible ASDs (see Chapter 6) with hydrophilic polymers can greatly enhance physical stability upon storage and also delay the precipitation, which tends to occur during dissolution because of the supersaturated solution that has been created upon dissolution. Such inhibition of crystallization in the solid state and in solution has led to the widespread use of ASDs as a drug delivery platform for poorly soluble drugs. Usually, the crystalline form readily crystallizes after the amorphous material is released into aqueous media. Thus, the enhanced solution concentrations attained from the amorphous form generally reach a peak value followed by a decrease in concentration until reaching the solubility of the crystal form (Figure 17.1). The lapse in time before crystallization occurs is dependent on the type of polymer used and environmental conditions such as temperature and degree of agitation. For this reason, it is difficult to obtain true equilibrium solubility for amorphous forms, so the measurement of solubility for amorphous and any high-energy form should be monitored over a short period of time, generally within 2–3 hours, as will be discussed in more detail in the following section.

17.5 Solubility Determination Using the Shake-Flask Method

Owing to the possibility of multiple solid forms and the tendency for solution-mediated transformations to occur, the physical characteristics of a solid must be understood before choosing a method for determining solubility. Generally, aqueous solubility measurements can be carried out in unbuffered or buffered media, where, for unbuffered media, the pH should be measured to account for any changes in pH due to self-buffering. A buffered medium is used to measure solubility at a specific pH controlled by the buffer, where the influence of salt formation and counterions is neglected. When the solubility of the tested material is very significant, the pH may be shifted and should be monitored.

The simplest and widely used method of measuring solubility is the shake-flask method, where a quantity of solid drug is mixed with a solvent in a container and agitated (shaken) over a given period of time with excess undissolved solid present.

The solvent can be an aqueous buffer, a water–cosolvent mixture, or a lipid. The method can be applied to many types of vessels, ranging from high-throughput miniature wells on a small scale to various types of dissolution vessels on a larger scale. Agitation is usually provided to promote mixing, and the solid is added to the liquid until the excess of solid

Table 17.4 Solubility comparison of amorphous and crystalline drugs.

Drug/excipient	Solubility (μg/mL)		Solubility ratio	Measuring condition
	Crystal	Amorphous		
Itraconazole/HPMC 1/1	0.56	5.4	9.6	0.1 N HCl at 37 °C, 1 hour
Ibuprofen/poloxamer P188 1/1	1.8	10.5	5.8	pH 6.8 water at 37 °C, 20 minutes
Pranlukast/gelatin 1/1	5	59	12	pH 7 buffer at 37 °C, 2 hours
JNJ-25894934/HPMCP/TPGS 4/27/9	< 1	44	> 44	SIF at 37 °C, 1 hour
Carbamazepine/PEG6000 1/1	16.2	18.7	1.2	0.1 N HCl at 37 °C, 90 minutes
Nifedipine/PVP 1/3	6	68	11	Water at 25 °C, 30 minutes
ER-34122/HPMC 1/1	0.03	6	200	pH 6.8 buffer at 37 °C, 1 hour
Compound 1/HPMCAS-M 1/3	5	718	144	MFD at 37 °C, 90 minutes
Compound 2/HPMCAS-M 2/1	10	120	12	MFD at 37 °C, 180 minutes
Compound 6/HPMCAS-L 1/1	9.5	86.4	9	MFD at 37 °C, 90 minutes
Tacrolimus/HPMC 1/1	2	50	25	0.1 N HCl at 37 °C, 2 hours

Source: [13]/with permission of Elsevier.

Abbreviation: HPMCAS: hydroxypropyl methylcellulose acetate succinate; PVP: polyvinyl pyrrolidone; SIF: simulated intestine fluid; MFD: model fasted duodenal.

is present to ensure saturation. After equilibrating for a certain period of time at a constant temperature, phase separation through filtration or by a centrifuge is carried out to obtain the supernatant for analysis. A prolonged mixing time, typically exceeding 1 day, is generally preferred for testing equilibrium solubility to ensure complete dissolution of the stable form or the full transformation of any metastable form that may be present. If enough solid sample exists after the experiment, it is recommended that the residual solute be collected at the end to check if there were any form changes during the solubility study. If there are needs in studying such phase conversions during the solubility measurement, more frequent sampling is required, especially at early time points, to understand the solution-mediated transformation process. This approach helps to obtain more accurate results for the apparent solubility of a metastable form and to gain information concerning the stability of the metastable form in the vehicle.

Drug concentration in solution is usually analyzed by HPLC or UV–Vis spectrophotometry. In certain circumstances, specialized analytical methods are employed when these techniques are insufficient for accurately measuring solubility. For example, a Mettler-Toledo Lasentec S400 FBRM (focused beam reflectance measurement) probe was used to measure solubilities of the various forms of axitinib in ethanol from 15 to 95 °C in a pressurized jacketed glass vessel at a test temperature sometimes above the boiling point of ethanol [14]. The use of the FBRM probe provides a continuous measurement of the dissolution process and avoids solvent evaporation or temperature fluctuations that arise during common sampling by filtration or centrifugation.

Since many factors, such as purity, polymorphism, pseudopolymorphism, residual solvents, mixing method, and temperature, can affect the measured solubility, great care in controlling these variables is essential. The pH-dependent solubility characteristics of ionizable compounds have been fully discussed in Chapter 4, but regarding the shake-flask method it is particularly important to emphasize the importance of confirming the phase identity of the excess solid by X-ray powder diffraction to check for any changes potentially occurring at different pH values.

17.6 High-throughput Screening of Solubility

High-throughput screening of solubility is essentially a miniature version of the shake-flask method using microplates. Material handling and sample analysis are usually controlled robotically. Examples of sample presentations include solutions in DMSO, slurries in heptane, and solid materials. A carrier solvent, like DMSO or heptane, needs to be removed before adding a testing solvent to initiate the solubility measurement. Solubilities are then analyzed by HPLC or a multi-wavelength UV plate reader [15, 16].

High-throughput solubility screening using microplates offers the advantage of testing numerous compounds while requiring only minimal material, making it especially valuable during the early stages of drug discovery when material availability is very limited.

This approach, for example, has been used successfully for rapidly screening the solubility of a group of BCS Class II drugs in animal gastrointestinal (GI) fluids and in simulated human GI fluids [17]. Drugs were dissolved in DMSO and dispensed into a 96-well plate. After removing DMSO by evaporation, the drugs were mixed with selected GI fluids and, after allowing time for equilibration, filtered for HPLC analysis. The solubility values reported in this screening study exhibited good agreement with literature data obtained by conventional methods. It appears that this manner of screening allows for increased throughput, reduced amount of solvent required, and minimized amount of drug consumed.

Zhou et al. introduced a high-throughput equilibrium solubility assay using a similar miniaturized shake-flask approach, with the aid of a streamlined HPLC analysis [18]. The test of 85 marketed drugs and numerous Novartis internal compounds revealed that the results from high-throughput studies correlated very well with the conventional shake-flask thermodynamic solubility data. The study also showed that the presence of a small amount of residual DMSO (0.5–5%) can result in a significant overestimation of equilibrium solubility, that is, as much as a sixfold increase.

Alsenz et al. reported a high-throughput method that eliminated the use of DMSO, where solid drugs were suspended in heptane as slurries and dispensed into microtiter plate wells [19]. Heptane was removed by evaporation before a buffer was introduced for the solubility test. Since most drugs have very little solubility in heptane, the use of heptane has advantages over the use of DMSO, where the higher solubility in DMSO, upon removal, may cause the formation of amorphous or metastable forms.

To avoid such complications of having solvents such as DMSO, compounds as solids can be dispensed directly into a microtiter plate with a minimal quantity of about 1 mg. Using this approach, Wyttenbach et al. determined the solubility of different crystal modifications of caffeine, carbamazepine, and piroxicam in 17 pharmaceutically related solvents at a volume of 100 μL, where ultraperformance liquid chromatography was utilized to assay the drug concentration, and high-throughput transmission X-ray powder diffraction was used to identify the residual solid drug on the filter plate [20].

Since manually weighing large numbers of samples is very labor intensive, the introduction of automatic powder dispensing greatly improves the time and convenience for high-throughput studies. Kane et al., for example, reported a system for dispensing submilligram doses of active pharmaceutical powders for early-stage solubility assays, where the system dispensed powder plugs with weights ranging from 10 μg to 4 mg into microplates with well openings as small as 1.5 mm [21].

17.7 Solubility Measurement of Metastable Forms

Solubility measurement of stable forms is relatively straightforward by using the shake-flask method, whereas the study of metastable forms can be compromised by solution-mediated phase transformation. As previously discussed, metastable forms include metastable polymorphs, anhydrous forms capable of forming hydrates, amorphous materials, salt forms in a pH environment that may trigger dissociation, or cocrystals in aqueous media. As also mentioned earlier, a conventional shake-flask method can be tailored to measure the solubility of metastable forms by collecting samples more frequently to monitor the time course of dissolution and the precipitation process.

Veesler and colleagues, for example, studied the solubility of irbesartan Form A, a metastable polymorph, in water by continuously monitoring the conductivity of the solution within a 3-hour time window, where conductivity gave an indirect measurement of drug concentration [22]. Figure 17.2 shows the change in concentration of irbesartan as a function of time with an initial solid concentration of 50 mg/L at 40 °C. The first stage represents the dissolution of Form A until it reaches the apparent solubility of Form A, 11 mg/L. Although the apparent solubility was stable for a period of about 1 hour, after that, the concentration of irbesartan started to decrease gradually, with the formation of Form B, the stable form. Figure 17.2 shows that Form B has a solubility of 5 mg/L in water at 40 °C.

Another approach to obtain real-time continuous measurement of solubility is to use a fiber optic detection system with a miniaturized dissolution apparatus. Apparently, this approach can offer the advantage of investigating the solubility of metastable crystal forms and amorphous forms. For example, Avdeef disclosed a miniaturized apparatus comprising six thermally controlled channels with a rapid in-situ fiber optic detection system [23], where sample quantities can be as low as 10–100 μg, while there is no need for sample preparation and dilution. Using the micro dissolution apparatus, it was possible to show that the solubility of itraconazole in water is 9 ng/mL by extrapolating solubilities obtained for itraconazole in

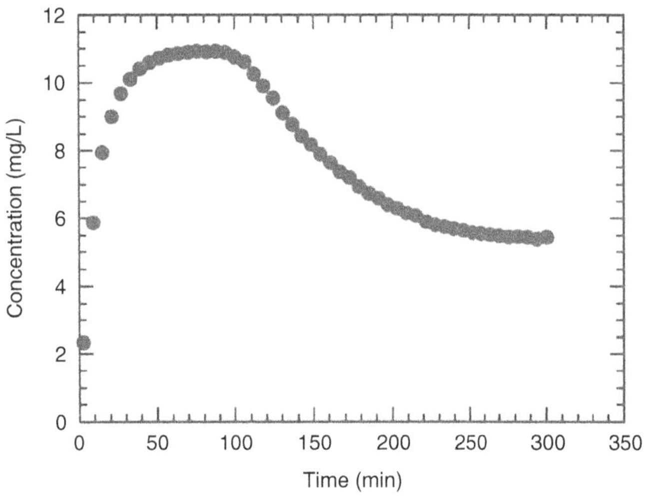

Figure 17.2 Variation of concentration versus time of irbesartan Form A at 40 °C in water by monitoring the conductivity change, indicating a solution-mediated phase transformation to the stable form B. *Source*: [22]/with permission of American Chemical Society.

solutions containing different levels of hydroxypropyl-β-cyclodextrin (HP-β-CD) to zero concentration of the complexing agent.

Another method, called the bracketing method, can be used to determine the apparent solubility of quickly converted metastable polymorphs. Here, the method involves putting small single crystals of a polymorph into a series of solutions with bracketed known concentrations. The dissolution or the growth of the single crystal, as observed under an optical microscope, provides an indication of whether the solution is undersaturated or oversaturated. By narrowing down the bracket, the solubility of the polymorph can be determined. The method was successfully utilized to measure the solubility of carbamazepine Form III in methanol and ethanol [24]. Since Form III is quickly converted to Form I in ethanol and methanol, it is difficult to obtain reliable results by using the conventional shake-flask method.

It is well known that the rate of solution-mediated transformation of a metastable form is related to its solubility relative to that of the stable form, its solubility in a particular solvent, and the solute–solvent interaction. The transformation, therefore, can be mitigated or slowed down by choosing a solvent in which the compound has a lower solubility or has strong molecular interaction with the drug, such as hydrogen bonding. For example, sulfamerazine Form I, a metastable polymorph, was shown to be stable in ethanol, 2-propanol, and water for hundreds of hours, while it was converted to Form II in acetonitrile within a few hours [25]. The study concluded that a solvent with strong hydrogen bonding propensity, while giving a low solubility can retard the solvent mediated transformation for metastable forms such as sulfamerazine Form I. The retardation of the transformation can allow a more accurate determination of the solubility difference between a polymorph pair, leading to important information concerning the thermodynamic stability of the two polymorphs.

Compared to studies with polymorphs, determining the apparent solubility of amorphous forms is much more challenging due to the much greater levels of supersaturation from amorphous forms than from metastable polymorphs. Murdande and colleagues described a method of measuring amorphous drug solubility by adding polymers in solution to inhibit crystallization [26]. Among several polymers studied, hydroxypropyl methylcellulose acetate succinate (HPMCAS) was found to be the most effective in inhibiting precipitation, and the apparent solubility data of several amorphous drugs in the presence of HPMCAS, including danazol, griseofulvin, and iopanoic acid, were found to correlate well with the theoretically calculated values. Interestingly, these maximum concentrations can be maintained for a period of 40–90 minutes.

17.8 Kinetic Solubility Measurement

Kinetic solubility measurements provide a way to measure solubility under a high-throughput setting for early discovery compounds. Typically, a compound is dissolved in DMSO as a 10 mM solution, and an aliquot of the compound solution is added to a pH 7 buffered solution, usually in a 96-well plate.

After mixing for a short period of time, the compound tends to precipitate as the solubility limit is reached, and the sample is analyzed by HPLC after filtration or tested by the measurement of turbidity. The dilution required is usually on the order of 50–100 times.

Lipinski et al. have provided a detailed description of how kinetic solubility is measured in the drug discovery stage [27].

Typically, rapid precipitation of the drug at an apparent solubility limit leads to a concentration of drug that is greatly supersaturated, and, therefore, reflects the solubility of the fastest precipitating species, which would be expected to have extremely small particle sizes. However, because of this high degree of supersaturation, eventually further precipitation will occur due to the solution-mediated transformation to more stable forms. Taking such an uncertainty into account, however, this method has proved to be a fast method and suitable for an early drug discovery setting where a vast number of compounds are being screened using automatic and high-throughput means. However, this method offers only a rough ranking order of solubility for quick assessment and generally overpredicts the true solubility value due to uncertainties in the solid form of the precipitate and the potential presence of residual DMSO. Of particular significance is the possibility of amorphous forms being produced in the procedure. For instance, Hoelke et al. reported that significant portions of the compounds that they studied precipitated out as amorphous materials after carrying out kinetic solubility measurements [28].

17.9 Solubility Determination of Drugs in Polymer Matrices

As already discussed for poorly soluble drugs, solid dispersions of the drug and the polymer provide a viable approach to enhance aqueous solubility, promote dissolution, and improve bioavailability. Furthermore, for transdermal delivery systems, a drug dissolved in a polymer adhesive matrix can form a solid solution with a high level of supersaturation that can promote drug fluxes through the skin.

There are several types of solid dispersions existing, including completely miscible molecular dispersion, dispersions with some phase separation, and nanoparticulate dispersions. Molecular dispersions are generally considered to be the most stable dispersions since a single phase of the drug and the polymer exists as a solid solution. Drugs used in molecular dispersions of pharmaceutical interest are usually in a supersaturated and thermodynamically unstable system since the drug load is usually above the equilibrium drug solubility in the polymer matrix. Consequently, drug molecules in the polymer matrix eventually tend to crystallize, resulting in a loss of solubility and, hence, a lack of dissolution enhancement. For a transdermal system, crystallization of the drug during storage could lead to a reduced rate of drug release, which is detrimental to the product quality.

The determination of drug solubility in an amorphous polymer matrix can provide some important understanding of the kinetic stability of the molecular dispersion, benefiting the formulation development of oral dosage forms and transdermal systems.

Directly measuring drug solubility in a polymer matrix is difficult because its high viscosity may impede the attainment of equilibrium, and measurement must generally be carried out at high temperatures. Many indirect methods are available for the estimation of drug solubility in polymers, including the detection of the first appearance of crystal from the supersaturated state, the use of thermal methods, the measurement of monomer solubility, and the measurement of diffusional flux. These methods are not very accurate and at best can only provide a rough estimate of the solubility.

Crystal detection in a supersaturated system is relatively straightforward, but it is extremely time-consuming. Solid solutions of the drug in polymer with different concentrations are prepared by casting a thin film after dissolving the drug and the polymer in a common solvent and observing any crystallization after a period of storage. The presence of crystals is detected using an optical microscope, by differential scanning calorimetry, or by X-ray diffractometry. Ma et al. utilized an optical microscope to detect crystallization of a norethindrone solid solution in an acrylate adhesive at various drug loadings [29]. After 1 month of storage at 23 °C, the polymer matrix containing 8% w/w of the drug or less exhibited no drug crystals. Drug crystals, however, were visible at levels of the drug above 10%. It was concluded that the solubility of this drug in the polymer system was between 8% and 10% w/w. The disadvantage of this method is the need for a long storage time for crystallization to occur.

Another approach to determine drug solubility in a polymer is based on the Flory–Huggins model [30]. In this technique, the proportions of the solid drug and the dissolved component are assessed through thermal analysis at elevated temperatures above the polymer glass transition temperature and by extrapolating the solubility data from elevated temperatures to room temperature. The solubility measurement usually involves equilibrating the drug and the polymer for a certain period to allow dissolution of the crystalline drug into the amorphous polymer matrix. The time required to reach equilibrium can be shortened by creating a supersaturated state through comilling, as reported by Mahieu and coworkers [31]. This equilibrium saturated state is considered reached when demixing of the drug and the polymer is

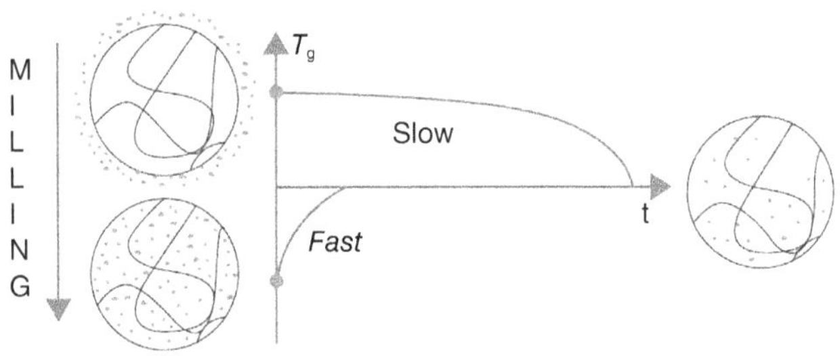

Figure 17.3 Comilling-assisted drug solubilization in polymer matrix. *Source*: [31]/with permission of American Chemical Society.

detected (Figure 17.3). Owing to the greater molecular mobility and strong plasticizing effects associated with these supersaturated conditions, the equilibrium state can be attained relatively quickly.

Using a similar comilling approach, an improved thermal analytical method has been reported to determine the solubilities of indomethacin and nifedipine in polyvinyl pyrrolidone (PVP), polyvinyl acetate (PVAc), and their copolymer (PVP/VA) [32]. In this method, a drug–polymer mixture was prepared by first cryomilling and then annealing at a different temperature using differential scanning calorimetry. The annealing was used to measure any undissolved crystalline drug and to test the upper and lower bounds of the equilibrium solution temperature to yield the activities of the drug in the polymer. The annealing temperature was chosen to be close to the glass transition temperature. The activities of indomethacin and nifedipine in the three polymers were very well fitted to the Flory–Huggins model. The study has revealed that nifedipine was less soluble than indomethacin in each polymer studied. For both drugs, the dissolving powers of the polymers have the ranking order PVP >PVP/VA/PVAc.

Bellantone et al. developed a thermodynamic model to determine the equilibrium solubility of a drug in a solid polymer at or near room temperature [33]. The model utilizes the Gibbs energy change ΔG_{ss} associated with the formation of a binary drug–polymer solid solution from the unmixed polymer and the drug crystals. The model also considers the contributions of heat capacity differences between the solid solution and the corresponding unmixed components. The Gibbs energy change and heat capacity differences are derived from thermal analysis data of the formed solid solution and each component before mixing. The study concludes that heat capacity effects need to be included for the determination of drug solubility in polymers near room temperature.

Jasti et al. reported another novel method to determine drug solubility in polymeric matrices [34]. The method utilizes the relationship between the thermodynamic activity of a drug and its steady-state flux. The steady-state flux from a reference saturated solution across a test membrane is compared to the flux from a drug-loaded polymeric dispersion. Based on this method, the solubility of lidocaine in an acrylate pressure-sensitive adhesive was $20.8 \pm 0.5\%$ (w/w).

In summary, accurate measurement of drug solubility in polymer matrix is not straightforward, whereas the methods discussed previously can potentially provide a rough estimation.

17.10 Dissolution Testing

Oral drug absorption is determined by in vivo dissolution and the intestinal membrane permeation of the drug. In vitro dissolution testing is often carried out to assess the kinetic processes by which a drug dissolves with the intention of correlating it with in vivo performance. Dissolution testing is mainly utilized to study various drug products for the purpose of assessing batch-to-batch product quality, guiding new formulation development, and monitoring product quality and performance after certain changes in formulation, process, site, and scale-up.

In a nutshell, dissolution is a time-dependent process that involves many physical changes, including disintegration to primary particles from tablets or capsules, displacement of drug molecules from the solid surface, and the diffusion of molecules across the diffusion layer adjacent to the solid surface. Figure 17.4 depicts the dissolution process of an immediate-release (IR) tablet.

The dissolution process can be described by the Nernst and Brunner equation [35]:

$$\frac{dC}{dt} = \frac{DS}{Vh}(C_s - C) \tag{17.14}$$

where D is the diffusion coefficient, h is the thickness of the diffusion layer, V is the volume of the dissolution medium, C_s is the saturated solubility, C is the concentration in solution, and S is the surface area. The Nernst and Brunner equation indicates that the dissolution rate is determined by the solid surface area, the drug's intrinsic solubility, and the drug concentration in the dissolution medium.

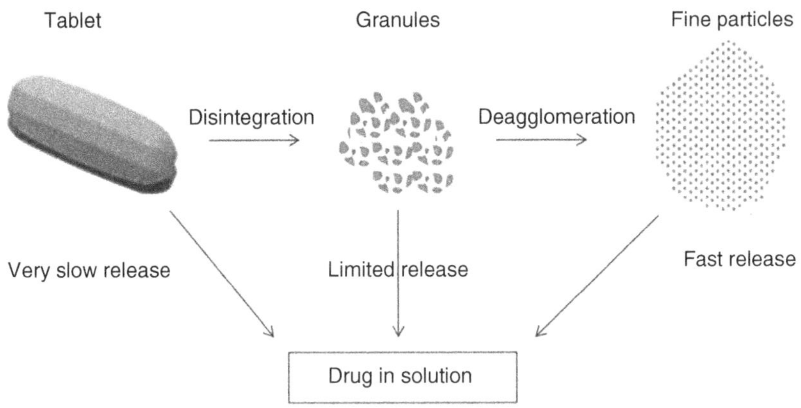

Figure 17.4 Illustration of the dissolution of an immediate-release tablet.

Table 17.5 Dissolution steps for different oral dosage forms.

Dose form	Steps in dissolution
Immediate-release tablets	Disintegration, solubilization of drug particles
Immediate-release capsules	Solubilization of capsule shell, disintegration, solubilization of drug particles
Controlled-release tablets	Diffusion-based: solubilization of drug particles, diffusion into medium/solubilization of matrix Erosion-based: erosion of matrix, solubilization of drug particles, diffusion into medium
Multiparticulates	Disintegration, solubilization of drug particles within particulates, diffusion of drug from particulates

Dissolution studies are routinely used to assess the drug release profiles of pharmaceutical products, including tablets, capsules, and transdermal patches.

The dissolution test, usually carried out by measuring the cumulative amount of drug that is released into solution as a function of time, generally involves several steps that could occur sequentially or in parallel. For example, the dissolution of an IR tablet consists of disintegration and drug solubilization, with the overall dissolution rate depending on the slower of these two steps. For a poorly soluble drug, its tablets can have a very rapid disintegration but will display limited drug dissolution due to poor solubility.

For IR products, the goal of the dissolution test in vitro is to study the first two steps, disintegration and solubilization (Table 17.5), whereas regarding controlled-release dosage forms, the method of carrying out a dissolution test is dependent on the release mechanism and targeted duration of release (Table 17.5).

For a typical dissolution test, a dosage form is added to a dissolution vessel with a fixed volume in a specific apparatus and agitated at a certain rate. The amount of drug dissolved during the dissolution test is monitored by UV spectrophotometry or HPLC and is commonly reported as a plot of percentage of the drug released versus sampling time, that is, a dissolution profile.

Dissolution methods are usually developed to fit the release profiles of the product, that is, IR or controlled release. If a proper in vitro/in vivo correlation (IVIVC) can be established, the dissolution test can then be used as a surrogate for the assessment of drug absorption in the body.

A successful dissolution study relies on appropriate method development, where the goal is to select appropriate dissolution media, media volume, type of apparatus, agitation rate, and assay method. The most critical steps are to identify a suitable medium and an appropriate dissolution apparatus.

Concerning a dissolution method used for product QC, a medium providing sufficient sink conditions is required so that the dissolution is not limited by solubility and complete release is ensured. A sink condition is generally taken as the volume of medium that is at least three times that required to form a saturated solution of drug substances. Dissolution tests under sink conditions provide dissolution data that more accurately reflects the formulation and process variables instead of drug substance solubility.

It is generally recognized that dissolution testing for oral dosage forms should be carried out using conditions that mimic the physiological environmental pH using 0.1 N HCl or a pH 4–7.4 buffer at 37°C, as outlined by pharmacopoeia and regulatory agency guidelines [36]. The use of a physiologically related dissolution medium more readily enables the interpretation of dissolution data in terms of in vivo pharmacokinetic data. The selection of a dissolution medium also requires consideration of solubility, solution stability, capability to discriminate pertinent differences between samples, reproducibility, and the environmental conditions to which the dosage form is expected to be exposed after administration. The most common dissolution media are a diluted HCl solution and phosphate buffer solution up to pH 6.8.

For poorly soluble compounds, aqueous solutions within the range of physiologic pH usually cannot provide adequate solubility, so that surfactants are often used to generate sink conditions. Commonly used surfactants include Tween 80, sodium lauryl sulfate, Cremophor, Triton X100, and cetyltrimethylammonium bromide. The type and level of surfactant must be established during dissolution method development.

For an ionizable compound, the effects of pH and surfactant concentration on compound solubility are evaluated to determine their combined effects.

Seven types of U.S. Pharmacopoeia (USP) compendia dissolution apparatuses are available for testing various dosage forms, as shown in Table 17.6.

Apparatus I, basket apparatus, and Apparatus II, paddle apparatus, have a simple design and are widely used in QC testing of capsules and tablets with IR, delayed-release, and extended-release (ER) properties, and both methods use a large volume of dissolution media (500–1000 mL). Apparatus III, reciprocating cylinder, and Apparatus IV, flow-through cell apparatus, are considered to have hydrodynamics that are closer to physiological conditions in the gastrointestinal tract. Apparatus III is most applicable for testing IR and ER dosage forms, whereas Apparatus IV is generally used for ER and poorly soluble API. Both Apparatuses III and IV offer the flexibility of changing the dissolution medium during the dissolution test, which mimics changes that take place as the product moves in vivo from the stomach to the small intestine. This is executed using a series of dissolution media of varying pH. Owing to a greater complexity, Apparatuses III and IV are less preferred for use in QC but have more utility in troubleshooting or the study of IVIVCs.

The paddle-over disk (Apparatus V) and cylinder (Apparatus VI) apparatuses are used for testing transdermal dosage form, whereas Apparatus VII, a reciprocating holder, is suitable for evaluating nondisintegrating oral-modified dosage forms.

The agitation rate is another factor affecting dissolution that must be evaluated during dissolution method development, since, for example, higher stirring rates usually lead to higher dissolution rates. For example, Wurster and Taylor reported that a power law relationship exists between the dissolution rate and the stirring rate [37]. It is generally recommended that mild agitation conditions are used for maximum discrimination between levels of dissolution to detect products with likely poor in vivo performance. The common agitation rate for the basket apparatus is 50–100 rpm, while that for the paddle method is 50–75 rpm. Katori et al. estimated that the agitation strength in human intestine matches closely with 10–30 rpm in the paddle method, which suggests that such a low agitation rate is more biorelevant [38].

Table 17.6 Type of USP dissolution apparatus and application.

USP apparatus	Description	Rotation speed (rpm)	Dosage form
I	Basket	50–120	Tablets, capsules (immediate-release, modified-release), chewable tablets
II	Paddle	25–100	Tablets, capsules (immediate-release, modified-release), chewable tablets
III	Reciprocating cylinder	6–35	Pellets, beads, multiparticulates for modified release, soft gelatin capsules, suppositories, implants
IV	Flow-through cell	NA	Soft gelatin capsules, suppositories, implants, modified-release tablets, poorly soluble drugs
V	Paddle-over disk	25–50	Transdermal dosage form
VI	Cylinder	NA	Transdermal dosage form
VII	Reciprocating holder	30	Nondisintegrating oral-modified release dosage form, transdermal dosage form

Dissolution method development involving the evaluation of the dissolution medium, apparatus, agitation rate, dissolution time, and sampling points is expected to have sufficient discriminatory power to detect drug product changes, so it can be justified as a QC method. Such product changes could involve changes in drug substance, excipients, formulation, manufacturing process, and storage conditions.

In an ideal situation, the dissolution method should be able to discriminate product changes that may affect product performance in vivo. In many situations, IVIVC cannot be established, and variations in dissolution profiles may not correlate with product performance in vivo. In evaluating the discriminatory power, therefore, changes in the drug product are evaluated against any variation in the dissolution data, so if there is a measurable difference detected for key variables, then the dissolution method can be considered discriminating. Product changes can include deliberate or unplanned changes in drug substance, such as crystal form, particle size, product composition, manufacturing processes, and storage conditions.

Depending on the dosage and release profile, dissolution is evaluated with a single point or multiple points. For immediate release, a single-point specification may be adequate in a time window of 30–60 minutes, whereas for ER dosage forms multiple time points are needed to construct a whole profile over a period of several hours, sometimes up to several days. At least three time points are required to specify the in vitro release profile. Acceptance criteria are established in the form of "Q factors" based on the in vitro dissolution data of pivotal clinical batches. Q factors are described as the minimum amount of drug dissolved in a specified time, expressed as a percentage of the labeled strength. Single Q factors or multiple Q factors can be established for a product.

The dissolution from a solid dosage form involves many steps and several parallel events, and many mathematical models and statistical tools have been developed to interpret dissolution data. The mathematical models generally work quite well if the drug release is controlled by simple phenomena such as diffusion or drug solubility.

Dissolution testing can also be utilized for the characterization of the effects of polymorphism, pseudopolymorphism, particle size, and surface area. Bonfilio et al., for example, reported a significant influence of polymorphism on the dissolution properties of glimepiride tablets using a discriminating dissolution method, as shown in Figure 17.5 [39]. The study suggested that the composition of polymorphs for a drug substance must be controlled to ensure consistent product quality.

Miniscale dissolution testing also has shown some advantages in API characterization and compound selection in the preclinical stage of drug development. This type of test is carried out using a small volume of a medium and at low agitation speeds (50 rpm), only requiring a few milligrams of material. Takano has shown that the results from such miniscale dissolution testing can be utilized to predict the fraction of a dose absorbed in humans for 12 drugs [40].

It is a general practice in such dissolution testing that samples are withdrawn at different time intervals for analysis using HPLC or UV spectrophotometry, which often leads to several disadvantages including disrupting the hydrodynamics of dissolution, a reduction of dissolution volume by removal of samples, labor intensiveness for sampling and analysis, and limited data points due to not sampling continuously.

The use of fiber optics has emerged as a promising alternative detection technique in dissolution testing. Here, a fiber optics probe is submerged into a dissolution medium, where light travels through and passes a detection gap. The absorption of light by dissolved molecules in the detection gap is recorded to monitor the dissolution process continuously. The detection of dissolved molecules in this way also offers advantages such as small displacements of volume, a simple light path, reduced labor, and improved accuracy. There are several probe configurations that are available, including a shaft probe, a rod probe, and an arch-shaped probe. A shaft probe is incorporated inside the impeller shaft. The rod probe is a solid metal

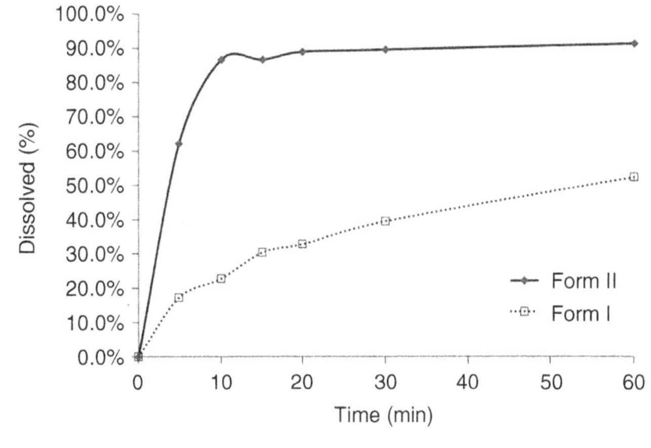

Figure 17.5 Dissolution profiles of tablets containing Forms I and II of glimepiride. *Source*: [39]/with permission of Elsevier.

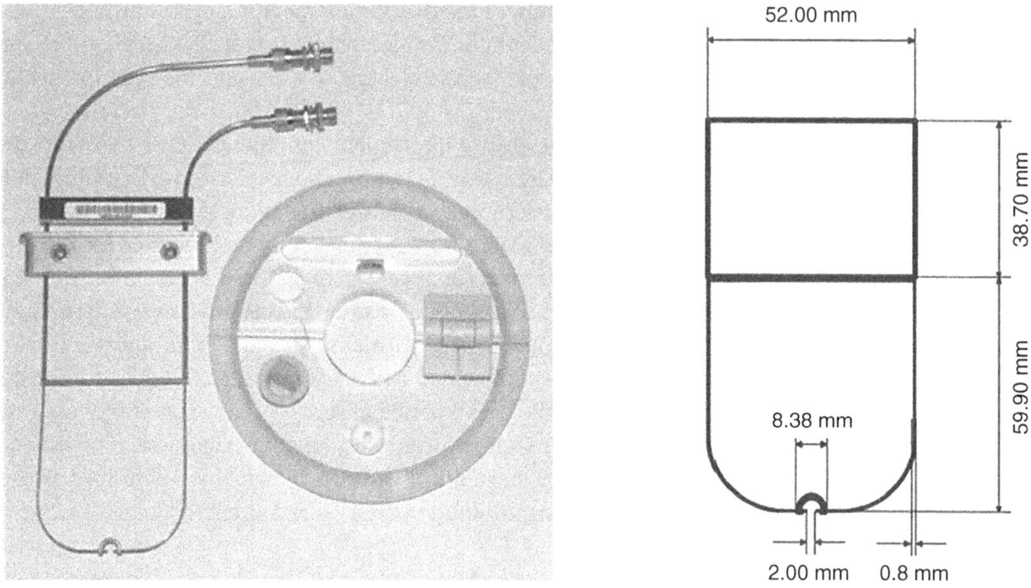

An arch-shaped fiber optics

Figure 17.6 An example of an arch-shaped probe fiber optics for dissolution studies. *Source*: Reproduced from [41]/with permission of Elsevier.

rod with a detection window near the submerged tip, whereas the arch-shaped probe has two small vertical tips pointed toward each other to form a detection window [41] (Figure 17.6).

17.11 Nonsink Dissolution Test

Dissolution testing developed according to pharmacopeial standards is carried out under a sink condition to ensure complete drug release. However, dissolution under sink conditions may lack the power to evaluate the solubility enhancement of amorphous materials or the effects of drug particle size on drug dissolution.

Nonsink dissolution testing provides a means to evaluate the maximum solution concentration achieved by an amorphous material and provides information concerning the physical stability of amorphous materials during the dissolution process. For nonsink dissolution, the amount of a solid used exceeds the amount required to reach the equilibrium solubility of the crystalline material, which can allow the amorphous material to maintain supersaturation over long time periods. A prolonged window to maintain supersaturation means an improved benefit of solubility enhancement and, consequently, enhanced in vivo performance. Measuring dissolution under nonsink conditions, therefore, can provide good estimates of dissolution rate, amorphous API apparent solubility, and the precipitation rate, all at the same time. The nonsink dissolution profiles of a fenofibrate:HPMCAS solid dispersion and fenofibrate alone, for example, were shown to be substantially different. The API alone gave a flat dissolution profile with 0.35 μg/mL dissolved after 60 minutes, whereas, in contrast, the API in the solid dispersion dissolved rapidly to reach a supersaturation state, maintained for an extensive period of time, and then gradually precipitated out as a low energy form, with the peak solution concentration of the API from the solid dispersion at 800 μg/mL. This type of profile has been described as a "spring and parachute" phenomenon, reflecting the fast dissolution due to significant solubility enhancement (spring) and a long period where supersaturation is maintained (parachute).

Nonsink dissolution studies have been applied for the investigation of the dissolution of amorphous materials in the presence of polymers by Alonzo et al. [42]. It was shown that the addition of low concentrations of polymers to the dissolution medium delayed crystallization of the amorphous felodipine, leading to the generation of supersaturated solutions, where the maximum supersaturation generated in the presence of each polymer was close to 7–9 μg/mL (Figure 17.7). For a dispersion made with povidone, the solution concentration peaked at 7.5 μg/mL and displayed a rapid decline due to precipitation. Conversely, the solutions dispersions made with HPMC and HPMCAS maintained concentrations above 7

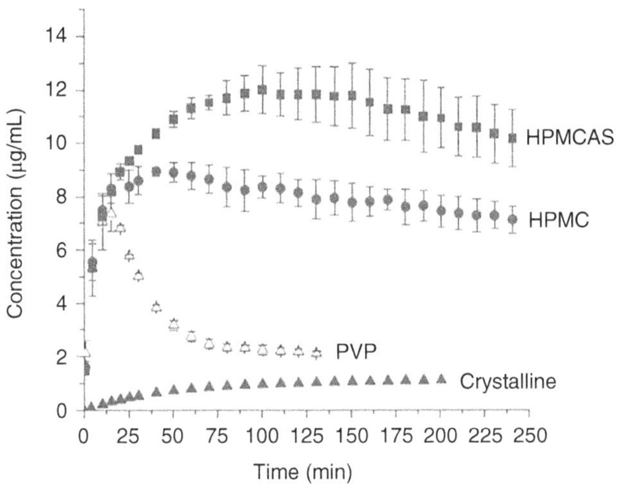

Figure 17.7 Dissolution profiles of amorphous felodipine in the presence of various polymers at 37 °C, in comparison to the crystalline material. *Source*: Adapted from [42].

and 10 μg/mL, respectively, for at least 4 hours after which a slow decline in solution concentration occurred. It was concluded from these nonsink studies that HPMC and HPMCAS act as better crystallization inhibitors than PVP.

It is important to recognize that drug release from ASDs can also be affected by several parallel or competing processes, for example, release of drug, release of carrier, drug crystallization from the solid state, and drug crystallization from the supersaturation solution. Of particular importance is the fact that the drug release rate can be controlled by the properties of both the drug and the polymer carrier and often depends on the ratio of polymer to drug. In addition, drug nanoparticles and microparticles can form upon dissolution and may play important roles in the dissolution mechanism of some dispersion formulations.

The dissolution process for ASDs under nonsink conditions can be divided into three stages. In stage I, the drug is released from the particles of the solid dispersion, and the concentration increases rapidly. Eventually, the concentration exceeds the solubility of the crystalline material, and the solution becomes supersaturated. Stage II starts when the drug concentration reaches a point where spontaneous nucleation begins. Drug dissolution and precipitation occur simultaneously in stage II, and drug concentration reaches a maximum when the rates of the two processes are equal. In stage III, drug concentration begins to decrease below the maximum concentration attained and when crystal growth occurs until the equilibrium solubility of the crystal is reached.

17.12 Biorelevant Dissolution Test

Like nonsink dissolution, biorelevant dissolution is widely utilized to link dissolution profiles of a poorly soluble drug to its biopharmaceutical performance via in vitro dissolution experiments under simulated gastrointestinal conditions. It usually involves pH shift experiments and the use of simulated GI fluids corresponding to the fasted and fed states. On the one hand, the pH shift from an acidic pH in the stomach to a neutral pH in the intestine may cause a solubility cliff, especially for compounds as weak bases, which may lead to drug precipitation. On the other hand, the presence of bile salts such as sodium taurocholate and phospholipids in the intestinal fluids may promote the drug solubility to form micelles or microemulsions or potentially attenuate drug precipitation by acting as a precipitation inhibitor. The goal of biorelevant dissolution is to simulate the GI environment and GI transition time for an oral formulation and provide a comprehensive evaluation of multiple events related to in vivo dissolution. Those events include disintegration of the dosage form in the gastric fluid, dissolution of the drug in the gastric fluid, potential precipitation due to supersaturation in the gastric fluid, potential solubility decrease when transiting into the small intestine leading to another supersaturation event, re-dissolution events of drug precipitates owing to bile salts and phospholipids present in the intestinal fluid. Biorelevant dissolution has been found very useful to rank-order solid forms or prototype formulations in terms of drug release in the GI. The drug release could become a complex event as a combined effect of drug dissolution, supersaturation, emergence of various colloidal structures, and kinetics of drug precipitation. Obviously, biorelevant dissolution studies significantly reduce the burden and the bottlenecks of fully relying on animal studies to guide formulation development.

As would be expected, simulated gastrointestinal fluids are designed to mimic in vivo fluids in terms of key characteristics such as pH, ionic strength, and presence of phospholipids and bile salts. Owing to the increased understanding of gastrointestinal physiology and dynamics of fluid composition at different stages of food intake and digestion, the proposed compositions of simulated gastrointestinal fluids keep evolving. Jantratid and Dressman reported some of the improvements made over the original Fasted-State Simulated Intestinal Fluid (FaSSIF) and Fed-State Simulated Intestinal Fluid (FeSSIF) proposed in 1998, to bring the composition and characteristics closer to those of aspirates attained from the human gastrointestinal tract, as described in Table 17.7 [43].

Table 17.7 Commonly used simulated gastrointestinal fluid.

Medium	pH	Buffer component	Salt	Bile salt	Phospholipid	Other components
Fasted-State Simulated Gastric Fluid	1.6	HCl (q.s)	34.2 mM sodium chloride	0.08 mM sodium taurocholate	0.02 mM lecithin	0.1 mg/mL pepsin
Fed-State Simulated Gastric fluid	5	17.12 mM acetic acid 29.75 mM sodium acetate	237 mM sodium chloride	NA	NA	Milk (1:1)
Fasted-State Simulated Intestine Fluid Version 1	6.5	28.66 mM sodium dihydrogen phosphate 13.8 mM sodium hydroxide	106 mM sodium chloride	3.0 mM sodium taurocholate	0.75 mM lecithin	NA
Fasted-State Simulated Intestine Fluid Version 2	6.5	19.12 mM maleic acid 34.8 mM sodium hydroxide	68.62 mM sodium chloride	3.0 mM sodium taurocholate	0.2 mM lecithin	NA
Fed-State Simulated Intestine Fluid Version 1	5.0	144 mM acetic acid 101 mM sodium hydroxide	173 mM sodium chloride	15 mM sodium taurocholate	3.75 mM lecithin	NA
Fed-State Simulated Intestine Fluid Version 2	5.8	55.02 mM maleic acid; 81.65 mM sodium hydroxide	125.5mM sodium chloride	10 mM sodium taurocholate	2.0 mM lecithin	5 mM glyceryl monooleate; 0.8 mM sodium oleate

Biorelevant dissolution usually involves one stage in a single medium or two stages of dissolution, with the first stage at a simulated gastric fluid and the second stage at a simulated intestinal fluid. It is obvious that a standard USP apparatus is applicable when samples are sufficient, whereas various small-scale systems offer advantages to test in the early development phase with minimal materials available. For example, Fagerberg and coworkers applied the microDISS Profiler (Pion Inc.) in the dissolution rate and apparent solubility measurement in biorelevant media for a series of poorly soluble drugs, with only 10–20 mL of dissolution medium being used [44]. It is interesting to note that most of the poorly soluble drugs displayed a much higher dissolution rate and apparent solubility in the FaSSIF and FeSSIF than in their respective buffers (Figure 17.8). Notably, neutral compounds or weak bases exhibited higher solubility in FeSSIF than FaSSIF, whereas a less profound difference existed for weak acid compounds.

In another example, O'Dwyer et al. outlined a small-scale dissolution system based on the MicroFlux system made by Pion Inc. in assessing the intraluminal performance of poorly soluble and ionizable compounds in the fasted state. At stage I, the donor chamber in the MicroFlux system was filled with 15 mL of hydrochloric acid solution (pH 2), and the dissolution duration was 30 minutes at 150 rpm after introducing a drug formulation. To start with stage II, 5 mL of four times concentrated FaSSIF was added into the donor chamber to elevate pH to 6.8, and the dissolution continued for 210 minutes at 150 rpm for stage II. The drug dissolution was being monitored in real time by a UV probe. It is worthwhile to note that the flux, presumably to mimic permeability, can be assessed at the same time in the acceptor chamber in the same MicroFlux system [45].

It is not surprising to note that biorelevant dissolution studies are mainly applied as a screening tool in developing ASDs, which is one of the most promising types of supersaturating enabled delivery platforms to enhance solubility and bioavailability, in terms of polymer type, drug loading, tertiary component, particle size, way to manufacture, and so on. For example, Arora et al. carried out biorelevant dissolution studies such as pH shift experiments that led to a better understanding of the in vivo release and precipitation behavior of ritonavir amorphous formulations. It is interesting to note that ritonavir precipitated readily as soon as it transited from acidic pH media to basic pH conditions and maintained its concentration close to its amorphous solubility in the tested media, for example, 38.5 μg/mL at pH 6.8 phosphate buffer, 60 μg/mL at FaSSIF V2, and 180 μg/mL at FeSSIF V2 [46]. Using a two-stage biorelevant dissolution method, Nunes

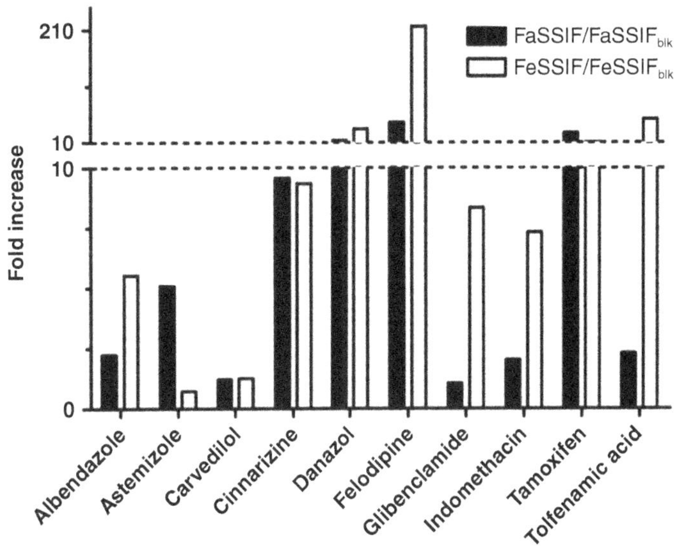

Figure 17.8 Increase in solubility obtained in FaSSIF and FeSSIF compared to the respective buffers for a series of poorly soluble drugs. *Source*: [44]/with permission of American Chemical Society.

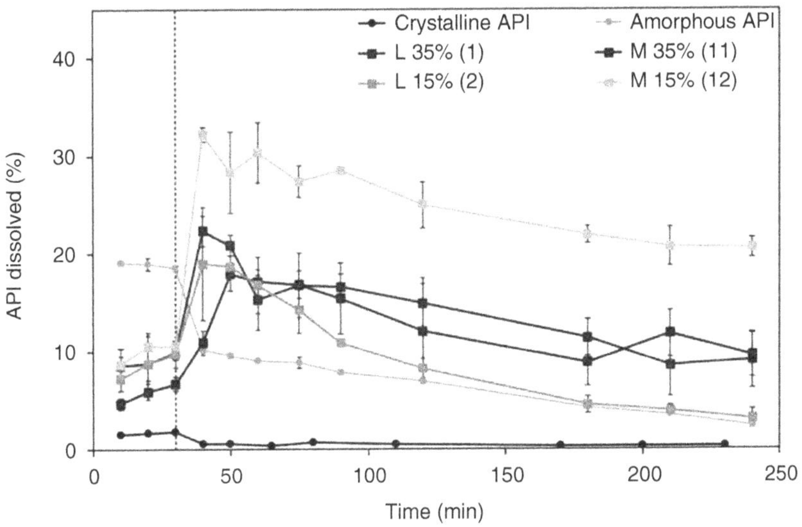

Figure 17.9 Biorelevant dissolution of itraconazole ASD in HPMCAS-L-grade and M-grade in comparison to amorphous and crystalline itraconazole API. *Source*: [47]/with permission of American Chemical Society.

and colleagues studied the release mechanisms of itraconazole ASD in HPMCAS with different polymer grades and drug loads [47]. In gastric media, interestingly, drug release followed a diffusion-controlled mechanism, and drug-rich colloids barely existed. After entering the intestinal media, the drug took a polymer dissolution-controlled release, where a rapidly dissolving polymer, not congruent with the drug, promoted the immediate release of the drug, therefore leading to a liquid–liquid phase separation and consequent formation of drug-rich colloids. The drug-rich colloids indeed enhance the apparent solubility and potentially serve as a drug reservoir to replenish the drug into the in vivo medium, hence promoting the absorption of the poorly soluble drug. As shown in Figure 17.9, biorelevant dissolution experiments indeed provided a clear ranking order of different formulations of itraconazole, with the ASD of 15% drug load in HPMCAS-M-grade performing the best and the crystalline API as the worst, not surprisingly.

Biorelevant dissolution studies for a poorly soluble drug could also provide indication of potential food effect. It is obvious that an early understanding of food effect would provide guidance on how to maximize overall drug bioavailability and how to design an appropriate clinical study. Fang et al., for example, developed a unique biorelevant dissolution method using USP Apparatus 4 to support early phase formulation development, including the intended prediction of potential food effect [48]. By comparing the dissolution results of several drugs obtained in FaSSIF and FeSSIF,

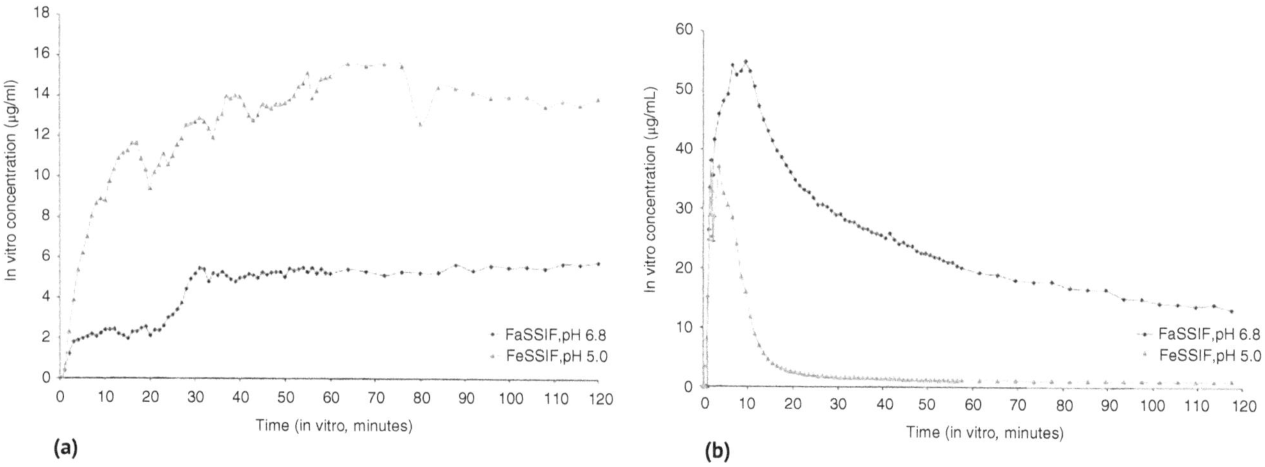

Figure 17.10 Dissolution profiles in FaSSIF and FeSSIF for danazol 200 mg capsules (a) and Compound D 25 mg tablets (b) using USP Apparatus IV at 37 °C. *Source*: [48]/with permission of American Chemical Society.

respectively, characteristic differences have been found to correlate with animal or human pharmacokinetic data. Figure 17.10 shows the dissolution profiles of danazol 200 mg capsules in FeSSIF and FaSSIF. In FeSSIF, danazol gave twice more in drug concentration and three times more area under the curve over the course of 2-hour dissolution experiments than in FaSSIF, which agrees with the report that danazol exhibited a strong positive food effect, having at least three times more oral bioavailability in the fed state than in the fasted state. Interestingly, Compound D, a BCS Class II compound as a weak acid, gave dissolution profiles with a rapid rise to reach a peak concentration and then gradually descend, likely owing to the formulation as a high-energy amorphous dispersion. It is obvious to note that more drug was dissolved and maintained in solution in FaSSIF than in FeSSIF, indicating a negative food effect that was later confirmed by a human pharmacokinetic study.

In summary, as highlighted by Lennernaš et al., biorelevant dissolution studies could be used to address several key biopharmaceutical issues during drug product development [49], including:

- optimization of formulation features (salt forms, particle sizes, amorphous formulations),
- surrogate for solubilization and precipitation in vivo,
- selection of formulation concepts for the later clinical phases,
- prediction or understanding of food effects, and
- linkage of various formulations through the development stage.

17.13 Intrinsic Dissolution Studies

Intrinsic dissolution rate measurements are carried out to characterize the dissolution of pure drug substances, using methods described in USP chapter 1087 [50].

Such studies are generally carried out by maintaining a constant surface area of the solid and under sink conditions. Two types of apparatus are recommended by the USP: a rotating-disk system and a fixed-disk system (Figure 17.11), where the rotating-disk system, also called the “Wood’s apparatus,” is used more commonly than the fixed-disk system.

For the use of Wood’s apparatus [51], a pure drug substance is compressed into a die with an opening of known diameter to produce a drug disk with a known surface area. The resulting disk is made by applying sufficient pressure to prevent any disintegration during the dissolution test. The die containing the drug disk is then mounted to the shaft of a dissolution apparatus, and the assembly is rotated at constant speed in a buffered medium held at constant temperature. Typically, intrinsic dissolution is expressed as milligrams per minute per cm^2.

Modified Wood apparatuses are supplied by many vendors, which provide a 0.5 cm^2 surface area. The apparatus consists of a punch and a die made of hardened steel. The base of the die is attached to a polished steel plate through three threaded holes to form a die cavity, where the polished steel plate ensures a smooth surface after compression. Sufficient sample is

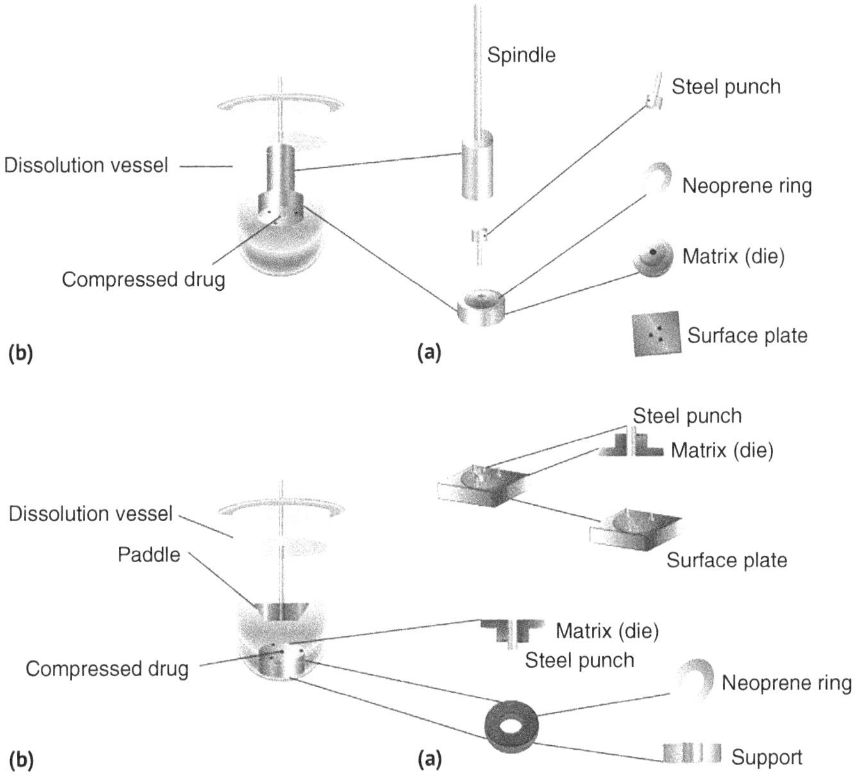

Figure 17.11 Rotating-disk system (top) and fixed-disk system (bottom): (a) unassembled apparatus and (b) configuration of the method in the dissolution equipment. *Source*: Adapted from [51].

added to the die cavity, and the apparatus is moved to a bench-top press to make a compacted pellet. The steel plate is then removed to expose the pellet for dissolution.

For a fixed-disk system, the sample preparation through the compression step is very similar to that used in the rotating-disk apparatus. A support is provided to maintain the disk in a fixed position at the bottom of the dissolution vessel with a flat surface, which ensures that the area of exposure is fixed. The agitation is provided by a paddle, as in Apparatus II.

Since the surface area is held constant, the dissolution rate should be linearly related to the solubility of the substance. Yu et al. carried out a feasibility study utilizing intrinsic dissolution rates to rank-order drugs for BCS classification. A good correlation between the intrinsic dissolution rate and BCS solubility classification was attained for 15 model compounds, with 0.1 $mg/min/cm^2$ as the solubility boundary between different BCS classes [52].

Since solubility is affected by solid-state properties such as crystallinity, the presence of the amorphous form, polymorphism, hydration, and solvation, intrinsic dissolution is an essential tool for carrying out solid-state material characterization. For example, Shah et al. discovered a metastable form of etoposide having an intrinsic dissolution rate 1.7 times greater than that of the stable polymorph. Intrinsic dissolution studies of this type also provide a significant advantage for analyzing metastable forms because solution-mediated phase transformation can be mitigated [53].

17.14 Summary

In summary, the aqueous solubility of a drug in many cases is directly related to its oral bioavailability. Solubility tests are widely utilized to differentiate the properties of drug molecules during drug discovery, mainly by using high-throughput screening and kinetic solubility techniques. Solubility directly reflects the thermodynamic stability of various solid forms of an API. Solubility studies using the shake-flask method are a major part of various toolkits that are available for characterizing solid forms. Estimation of the solubility of a drug in a polymer matrix and dissolution studies under nonsink conditions or biorelevant media are of particular importance in the development of solid dispersion formulations.

References

1 Löbenberg R, Amidon GL. Modern bioavailability, bioequivalence and biopharmaceutics classification system. New scientific approaches to international regulatory standards. *Eur J Pharm Biopharm* 2000; 50(1): 3–12.

2 Ran Y, Yalkowsky SH. Prediction of drug solubility by the general solubility equation (GSE). *J Chem Inf Comput Sci* 2001; 41(2): 354–357.

3 Hansch C, Quinlan JE, Lawrence GL. Linear free energy relationships between partition coefficients and the aqueous solubility of organic liquids. *J Org Chem* 1968; 33(1): 347–350.

4 Walden P. Ueber die Schmelzwaerme, spezifische Kohaesion und Molekulargroesse bei der Schmelztemperatur. *Z Elektrochem Angew Phys Chem* 1908; 14(43): 713–724.

5 Bergström CA, Wassvik CM, Johansson K, Hubatsch I. Poorly soluble marketed drugs display solvation limited solubility. *J Med Chem* 2007; 50(23): 5858–5862.

6 Wassvik CM, Holmén AG, Draheim R, Artursson P, Bergström CA. Molecular characteristics for solid-state limited solubility. *J Med Chem* 2008; 51(10): 3035–3039.

7 Cappelli CI, Manganelli S, Lombardo A, Gissi A, Benfenati E. Validation of quantitative structure–activity relationship models to predict water-solubility of organic compounds. *Sci Total Environ* 2013; 463: 781–789.

8 Schaper KJ, Kunz B, Raevsky OA. Analysis of water solubility data on the basis of HYBOT descriptors. *QSAR Comb Sci* 2003; 22 (9–10): 943–958.

9 Pudipeddi M, Serajuddin ATM. Trends in solubility of polymorphs. *J Pharm Sci* 2005; 94(5): 929–939.

10 Paulekuhn GS, Dressman JB, Saal C. Trends in active pharmaceutical ingredient salt selection based on analysis of the orange book database. *J Med Chem* 2007; 50(26): 6665–6672.

11 Good DJ, Rodríguez-Hornedo N. Solubility advantage of pharmaceutical cocrystals. *Cryst Growth Des* 2009; 9(5): 2252–2264.

12 Remenar JF, Morissette SL, Peterson ML, Moulton B, MacPhee JM, Guzman HR, Almarsson O. Crystal engineering of novel cocrystals of a triazole drug with 1, 4-dicarboxylic acids. *J Am Chem Soc* 2003; 125(28): 8456–8457.

13 Kawakami K. Modification of physicochemical characteristics of active pharmaceutical ingredients and application of supersaturatable dosage forms for improving bioavailability of poorly absorbed drugs. *Adv Drug Delivery Rev* 2012; 64(6): 480–495.

14 Campeta AM, Chekal BP, Abramov YA, Meenan PA, Henson MJ, Shi B, Singer RA, Horspool KR. Development of a targeted polymorph screening approach for a complex polymorphic and highly solvating API. *J Pharm Sci* 2010; 99(9): 3874–3886.

15 Glomme A, März J, Dressman JB. Comparison of a miniaturized shake-flask solubility method with automated potentiometric acid/base titrations and calculated solubilities. *J Pharm Sci* 2005; 94(1): 1–6.

16 Chen T, Shen H, Zhu C. Evaluation of a method for high throughput solubility determination using a multi-wavelength UV plate reader. *Comb Chem High Throughput Screen* 2002; 5(7): 575–581.

17 Guo J, Elzinga PA, Hageman M, Herron JN. Rapid throughput solubility screening method for BCS class II drugs in animal GI fluids and simulated human GI fluids using a 96-well format. *J Pharm Sci* 2008; 97(4): 1427–1442.

18 Zhou L, Yang L, Tilton S, Wang J. Development of a high throughput equilibrium solubility assay using miniaturized shake-flask method in early drug discovery. *J Pharm Sci* 2007; 96(11): 3052–3071.

19 Alsenz J, Meister E, Haenel E. Development of a partially automated solubility screening (PASS) assay for early drug development. *J Pharm Sci* 2007; 96(7): 1748–1762.

20 Wyttenbach N, Alsenz J, Grassmann O. Miniaturized assay for solubility and residual solid screening (SORESOS) in early drug development. *Pharm Res* 2007; 24(5): 888–898.

21 Kane NR, Broce B, Gonzales-Zugasti J, Lewis WP, LeQuesne M, Lemmo AV. A system for dispensing sub-milligram doses of active pharmaceutical powders for early stage solubility assays. *J Ass Lab Auto* 2004; 9(4): 218–227.

22 Veesler S, Lafferere L, Garcia E, Hoff C. Phase transitions in supersaturated drug solution. *Org Pro Res Dev* 2003; 7(6): 983–989.

23 Avdeef A. Solubility of sparingly-soluble ionizable drugs. *Adv Drug Delivery Rev* 2007; 59(7): 568–590.

24 O'Mahony MA, Croker DM, Rasmuson AC, Veesler S, Hodnett BK. Measuring the solubility of a quickly transforming metastable polymorph of carbamazepine. *Org Process Res Dev* 2013; 17(3): 512–518.

25 Gu CH, Young V, Grant DJ. Polymorph screening: Influence of solvents on the rate of solvent-mediated polymorphic transformation. *J Pharm Sci* 2001; 90(11): 1878–1890.

26 Murdande SB, Pikal MJ, Shanker RM, Bogner RH. Solubility advantage of amorphous pharmaceuticals: I. A thermodynamic analysis. *J Pharm Sci* 2010; 99(3):1254–1264.

27 Lipinski CA, Lombardo F, Dominy BW, Feeney PJ. Experimental and computational approaches to estimate solubility and permeability in drug discovery and development settings. *Adv Drug Delivery Rev* 2012; 64: 4–17.

28 Hoelke B, Gieringer S, Arit M, Saal C. Comparison of nephelometric, UV spectroscopic and HPLC methods for high-throughput determination of aqueous solubility in microtiter plates. *Anal Chem* 2009; 81(8): 3165–3172.

29 Ma X, Taw J, Chiang CM. Control of drug crystallization in transdermal matrix system. *Int J Pharm* 1996; 142(1): 115–119.

30 Flory PJ. *Principles of Polymer Chemistry*, 13e. Ithaca, New York: Cornell University Press; 1953.

31 Mahieu A, Willart JF, Dudognon E, Danede F, Descamps M. A new protocol to determine the solubility of drugs into polymer matrixes. *Mol Pharm* 2013; 10(2): 560–566.

32 Sun YE, Tao J, Zhang GG, Yu L. Solubilities of crystalline drugs in polymers: an improved analytical method and comparison of solubilities of indomethacin and nifedipine in PVP, PVP/VA, and PVAc. *J Pharm Sci* 2010; 99(9): 4023–4031.

33 Bellantone RA, Patel P, Sandhu H, Choi DS, Singhal D, Chokshi H, Malick AW, Shah N. A method to predict the equilibrium solubility of drugs in solid polymers near room temperature using thermal analysis. *J Pharm Sci* 2012; 101(12): 4549–4558.

34 Jasti BR, Berner B, Zhou SL, Li X. A novel method for determination of drug solubility in polymeric matrices. *J Pharm Sci* 2004; 93(8): 2135–2141.

35 Brunner E. *Reaktionsgeschwindigkeit in heterogenen Systemen*. Gottingen:Georg-Augusts-Universitat; 1903.

36 Anand OM, Lawrence XY, Conner DP, Davit BM. Dissolution testing for generic drugs: an FDA perspective. *AAPS J* 2011; 13(3): 328–335.

37 Wurster DE, Taylor PW. Dissolution kinetics of certain crystalline forms of prednisolone. *J Pharm Sci* 1965; 54(5): 670–676.

38 Katori N, Aoyagi N and Terao T. Estimation of agitation intensity in the GI tract in humans and dogs based on in vitro vivo correlation. *Pharm Res* 1995; 12(2): 237–243.

39 Bonfilio R, Pires SA, Ferreira L, de Almeida AE, Doriguetto AC, de Araújo MB, Salgado H. A discriminating dissolution method for glimepiride polymorphs. *J Pharm Sci* 2012; 101(2): 794–804.

40 Takano R, Sugano K, Higashida A, Hayashi Y, Machida M, Aso Y, Yamashita S. Oral absorption of poorly water-soluble drugs: computer simulation of fraction absorbed in humans from a miniscale dissolution test. *Pharm Res* 2006; 23(6): 1144–1156.

41 Zhang Y, Bredael G, Armenante PM. Dissolution of prednisone tablets in the presence of an arch-shaped fiber optic probe in a USP dissolution testing apparatus 2. *J Pharm Sci* 2013; 102(8): 2718–2729.

42 Alonzo DE, Gao Y, Zhou D, Mo H, Zhang GG, Taylor LS. Dissolution and precipitation behavior of amorphous solid dispersions. *J Pharm Sci* 2011; 100(8): 3316–3331.

43 Jantratid E, Dressman J. Biorelevant dissolution media simulating the proximal human gastrointestinal tract: an update. *Dissolut Technol* 2009; 16(3): 21–25.

44 Fagerberg JH, Tsinman O, Sun N, Tsinman K, Avdeef A, Bergstrom CA. Dissolution rate and apparent solubility of poorly soluble drugs in biorelevant dissolution media. *Mol Pharm* 2010; 7(5): 1419–1430.

45 O'Dwyer PJ, Box KJ, Imanidis G, Vertzoni M, Reppas C. On the usefulness of four in vitro methods in assessing the intraluminal performance of poorly soluble, ionisable compounds in the fasted state. *Eur J Pharm Sci* 2022; 168: 106034.

46 Arora S, Pansari A, Kilford P, Jamei M, Gardner I, Turner DB. Biopharmaceutic in vitro in vivo extrapolation (IVIV_E) informed physiologically-based pharmacokinetic model of ritonavir Norvir tablet absorption in humans under fasted and fed state conditions. *Mol Pharm* 2020; 17(7): 2329–2344.

47 Nunes PD, Pinto JF, Henriques J, Paiva AM. Insights into the release mechanisms of ITZ: HPMCAS amorphous solid dispersions: the role of drug-rich colloids. *Mol Pharm* 2021; 19(1): 51–66.

48 Fang JB, Robertson VK, Rawat A, Flick T, Tang ZJ, Cauchon NS, McElvain JS. Development and application of a biorelevant dissolution method using USP apparatus 4 in early phase formulation development. *Mol Pharm* 2010; 7(5): 1466–1477.

49 Lennernaš H, Lindahl A, Van Peer A, Ollier C, Flanagan T, Lionberger R, Nordmark A, Yamashita S, Yu L, Amidon GL, Fischer V. In vivo predictive dissolution (IPD) and biopharmaceutical modeling and simulation: future use of modern approaches and methodologies in a regulatory context. *Mol Pharm* 2017; 14(4): 1307–1314.

50 Issa MG, Ferraz HG. Intrinsic dissolution as a tool for evaluating drug solubility in accordance with the biopharmaceutics classification system. *Dissolut Technol* 2011; 18(3): 6–11.

51 Wood JH, Syarto JE, Letterman H. Improved holder for intrinsic dissolution rate studies. *J Pharm Sci* 1965; 54(7): 1068

52 Yu LX, Carlin AS, Amidon GL, Hussain AS. Feasibility studies of utilizing disk intrinsic dissolution rate to classify drugs. *Int J Pharm* 2004; 270(1): 221–227.

53 Shah JC, Chen JR, Chow D. Metastable polymorph of etoposide with higher dissolution rate. *Drug Dev Ind Pharm* 1999; 25(1): 63–67.

18

Physical Stability of Solids

18.1 Introduction

The development of pharmaceutical products requires that, upon administration, the drug will provide desired therapeutic potency, safety, and bioavailability in a consistent manner. This means that the ingredients of the final dosage form must not undergo any changes during manufacturing, storage, and use of the product that would alter its quality and performance. Such changes are physical instabilities if they are brought about by phase transformations involving noncovalent bond interactions and chemical instabilities if they involve chemical reactions that occur because of covalent bond changes. Such instabilities in the solid state generally arise when there is a thermodynamic net loss in free energy associated with the process and where sufficient molecular diffusional, translational, and rotational motion can occur. This chapter will deal with some aspects of solid-state physical instability, whereas the next chapter will deal with solid-state chemical instability. Physical instabilities of significant pharmaceutical interest in solids primarily arise because of a change in the physical form of the solid. Examples include melting, polymorphic phase transformations, loss of crystallinity, crystallization from amorphous solid forms, hydration and dehydration of solids, eutectic melting, and protein denaturation. Such physical instabilities, in turn, can lead to changes in particle size and morphology, manufacturability (poor powder flow, mixing, and compaction), loss of bioavailability because of changes in dissolution rate, and loss of protein potency because of denaturation and aggregation. As we shall see in the next chapter, physical instabilities, particularly the loss of crystallinity, can often lead to increased rates of solid-state chemical degradation.

In previous chapters, we have discussed several phenomena that are associated with physical instability, as defined earlier, including polymorphic transformations (Chapter 2), crystal hydrates and solvates (Chapter 3), properties of amorphous solids (Chapter 6), crystal mesophases (Chapter 7), the hydration and dehydration of solids (Chapter 15), and mechanical properties (Chapter 16). In this chapter, we will attempt to build on our understanding of these phenomena to develop a broad unifying conceptual picture of the underlying basis of most physical instabilities. Central to addressing physical instabilities is the recognition in more fundamental terms of what is meant by "stability" and "instability." The stable form of a substance is that form that exists at the lowest possible state of free energy for a given temperature, pressure, or concentration, as schematically represented in Figure 18.1 by what is known as a potential energy landscape [1]. Here, the deeper the potential energy "well" the greater the energy barrier required for any transition to occur, and hence the greater the stability. Very often, less stable "metastable" forms can exist at higher free energies than the stable form, but often with energy barriers large enough to allow the form to maintain significant apparent stability over a given time frame and under a particular set of conditions. Amorphous solid forms, for example, are metastable relative to the corresponding crystalline form, whereas higher energy polymorphic crystalline forms are also metastable. Eventually, under certain conditions, a metastable state would be expected thermodynamically to convert to the stable form, bringing about physical instability unless sufficient diffusional motion is not possible. As illustrated in Figure 18.1, molecules often can fall into a variety of unstable states with very small energy barriers and hence exhibit very limited practical stability. Such transformations would include the spontaneous transformation of high-energy glassy states to more ordered glassy states below the glass transition temperature, known as annealing or physical aging (see Chapter 6 for more details). Some high-energy polymorphic crystal forms may exhibit very small energy barriers, with the resultant difficulty of isolating and maintaining them under practical conditions.

Solid-State Materials in Pharmaceutical Chemistry: Properties, Characterization, and Applications, Second Edition. Stephen R. Byrn, George Zografi and Xiaoming (Sean) Chen.

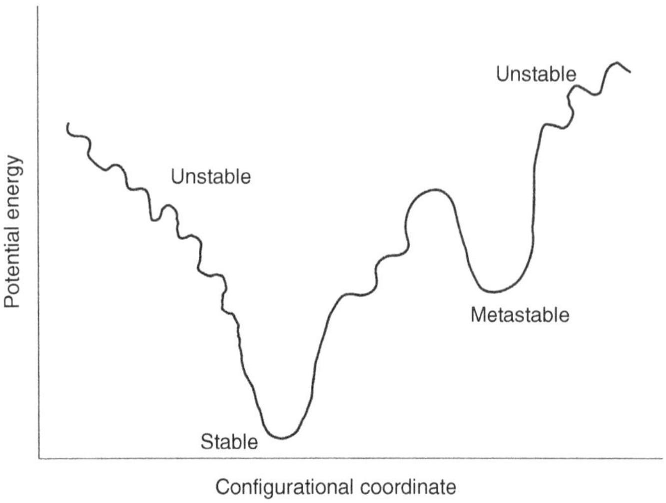

Figure 18.1 Typical potential energy landscape showing stable, metastable states, and unstable states.

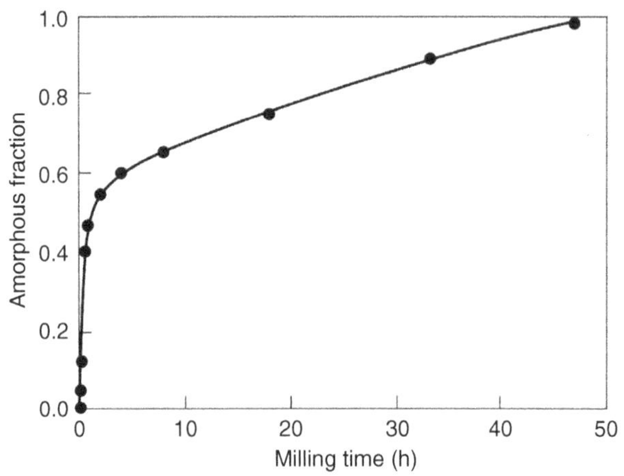

Figure 18.2 Amorphization of crystalline linaprazan as a function of milling time. *Source*: [4]/with permission of Elsevier.

18.2 Underlying Basis for Physical Instability in Pharmaceutical Systems

Whenever any substance, API or excipient, is used in a thermodynamically metastable or unstable state, there must be concern that physical instability will occur at some point. For this reason, the predominant solid form of API used in pharmaceutical products is the crystalline form and preferably the most stable form if polymorphs exist. Likewise, if anhydrous crystals of the API can easily form crystal hydrates when exposed to water in either the liquid or vapor state, one would prefer the more stable hydrate form for a particular relative humidity and temperature. If, for some reason, a higher energy state is either purposefully used or inadvertently produced during processing, we would expect that physical instability will occur unless some strategy is used to prevent the instability over the desired time of storage and use. Purposeful use of such states of any substance generally occurs when one is trying to enhance the performance of the substance in a pharmaceutical product because of the higher energy of that form. For example, less stable polymorphs or amorphous forms, because of their higher energy relative to the stable crystalline form, offer the strong potential to enhance the dissolution rate and oral bioavailability of poorly water-soluble API, particularly from the amorphous state [2, 3]. Yet, unless excipients are added and conditions are used to inhibit crystallization of the API in the solid-state during storage or during the in vitro and in vivo dissolution process, crystallization will likely occur, offsetting the advantages of using these higher energy states. Of perhaps broader importance and impact, consider the situation where a stable crystalline substance undergoes certain processes that inadvertently cause the crystal to lose part or all of its highly ordered structure, for example, milling, heating, or drying (see Chapter 6). For example, in Figure 18.2, we can observe the increase in the fraction of amorphous structure introduced into a crystalline API subjected to various ball milling time periods [4]. Such "process-induced disorder," when introduced into the crystal lattice at any level, generally leads to high-energy regions within the solid, which are now susceptible to conversion back to the original stable crystal form or to new crystal forms at various temperatures and relative humidities. It has also been shown that certain organic crystals, such as aspirin, phenacetin, and flufenamic acid, when physically mixed with high surface area inorganic excipients such as magnesium aluminum silicate or activated carbon spontaneously, lose crystallinity when stored over long periods in closed containers [5]. It was shown that this was caused by sublimation of molecules from the crystal and the subsequent condensation of the vapor to produce amorphous material on the surface of the excipient with enhanced dissolution properties.

Central to this chapter, therefore, will be a discussion about (i) the nature of crystalline disorder, with examples of how pharmaceutical processing induces such disorder in crystals; (ii) the underlying principles that influence solid-state and solvent-mediated phase transformations from disordered states that contribute to physical instability, for example, crystallization of molecules that occurs from more disordered and higher energy solids; and (iii) the examples of how disorder in crystals initiates a variety of physical instabilities of pharmaceutical interest.

18.3 Disorder in Crystals

Situations Giving Rise to Disorder in Crystalline Solids – A loss of crystallinity, leading to higher energy disordered structures within the crystal, can arise in several ways. Such disorder may occur during the crystallization process in the form of defects, usually from the solution state, or in the form of amorphous structures introduced by various solid-state processes carried out during the handling and manufacturing of API and solid dosage forms. These would include (i) any process that might give rise to local melting due to increased temperature followed by rapid cooling, (ii) local dissolution of crystalline material followed by rapid removal of solvent and precipitation, (iii) sublimation of molecules followed by rapid cooling, and (iv) mechanical stresses on the crystalline solid that cause the breakdown of regions of the lattice into disordered structures. For the purposes of discussion, we can divide the nature of such disorder into (i) the appearance of crystal defects and (ii) the creation of locally amorphous structures. It is also of interest to consider the relationship that exists between the formation of defects and the amorphous state.

Crystal Defects – It is well established that, despite the symmetrical order associated with any crystal lattice, various types of imperfections typically exist in the lattice with significant potential impact on the properties of the crystal. Such defects introduce disorder into the crystal lattice and, therefore, tend to increase the free energy of the solid with increasing defect density (number of defects per unit area of surface for surface defects and the number of defects per unit volume for bulk defects). It is generally believed that crystal defects are normally sites for the initiation of various physical phenomena, such as melting, dissolution, and vapor adsorption on crystal surfaces [6]. These imperfections generally arise during crystallization or because of additional processing of the isolated crystals, such as by milling, irradiation, or dehydration. Crystal defects are generally categorized as (i) point defects, (ii) line defects, and (iii) planar or interfacial defects, as schematically represented in Figure 18.3. Point defects occur at and around a single lattice point within the lattice structure. They can occur because of (i) the absence of an atom or molecule from its expected lattice position, (ii) the presence of the atom or molecule in an interstitial unexpected space within the lattice, and (iii) the presence of another atom or molecule (impurity) as a substitutional defect; the relatively large molecular size of most individual organic molecules and chemical intermediates in the crystallization process that can end up as impurities in the crystal often lead to rather large point defects. More extensive two- and three-dimensional line defects arise from discontinuities or dislocations within the lattice that give rise to distortions of adjacent crystal planes. Interfacial or planar defects in the form of grain boundaries are another form of dislocation characterized by a disruption of the crystallographic direction of individual crystallites. Typically, grain boundaries that occur at the interface between crystallites in a polycrystalline solid lead to disruption in the lattice and to interfacial regions of high energy relative to that expected for the lattice. For small-molecule organic materials of pharmaceutical interest, crystal lattices are particularly susceptible to imperfections by virtue of disruptions in the molecular stacking that occur within the lattice during crystallization and because of the relatively weak noncovalent intermolecular interactions that occur in such systems. The presence of such crystal defects typically gives rise to mechanical strains within the lattice, with a significant influence on the mechanical properties of the crystal. Dislocations

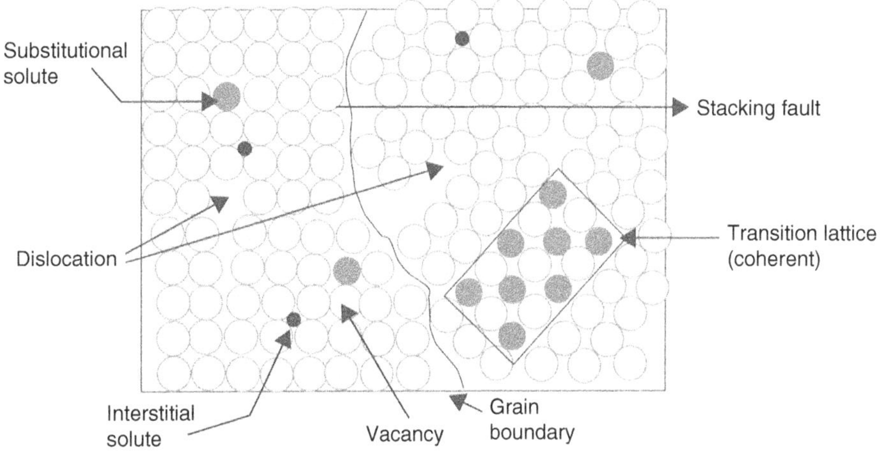

Figure 18.3 Illustration of different types of crystal defects and dislocations.

can move under stress with bond breaking and bond making, which in the case of metals leads to their well-known malleability under stress until the dislocations "pile up" and lead to brittle fracture. Organic molecules often exhibit preferential plastic shear and fracture along certain crystal planes containing dislocations when undergoing milling or compaction, with significant effects on the resulting particle properties. As another example of the effects of higher energy associated with crystal defects, it has been shown that several dislocations on the surface of well-defined potassium perchlorate crystals, determined by observing etch pits produced at defects by a mixture of sodium sulfite and sulfuric acid, correlated very well with the initial rates of dissolution of potassium perchlorate crystals into 95% ethanol [7]. It has also been shown that water vapor adsorption onto highly crystalline surfaces, such as the 001-crystal plane of NaCl, initially occurs at crystal defects, which then act as sites for the initiation of deliquescence [8]. As mentioned earlier, a common method of determining the density of defects in certain materials involves exposing the crystal to a solvent that selectively dissolves molecules at defect sites, which can then be microscopically identified and characterized. It is not always clear, however, how quantitative this technique is and to what extent important bulk defects, as opposed to surface defects, are not detected. The quantitative determination of defect density in highly crystalline materials has attracted significant attention in the use of inorganic and organic optical and electrical devices, where defects play a particularly important functional role. The variety of techniques reported with such systems includes high-resolution transmission electron microscopy, radiotracer self-diffusion, solid-state NMR, and X-ray diffraction.

Process-Induced Disorder – From the previous discussion the introduction of disorder into a crystal lattice in the form of defects can occur during the crystallization process, and that such disorder, and the higher energy contained by molecules in the defect, can lead to the preferential initiation of a variety of physical transformations, the greater the defect density the greater the expected effects on solid-state properties. It seems reasonable, therefore, to expect that as the density of defects increases due to further processing of the crystal, a point will be reached, the "critical defect density," where the "coalescence" or "nucleation" of such defects will occur and regions of molecules in the amorphous state will arise within the crystal with the properties of a glass having a well-defined glass transition temperature. From Chapter 6 we learned that there are a variety of processes that can lead to the fully formed amorphous state, including mechanical stress, rapid precipitation from solution, vapor and liquid solidification by cooling, and the dehydration of crystal hydrates and solvates. We would expect, therefore, that partially amorphous regions could be inadvertently produced in a crystal lattice by such processing, often at levels not easily detectable. For example, the introduction of partial disorder can be brought about by (i) the disruption of the crystal lattice through mechanical stress as in normal milling used to reduce particle size, as shown in Figure 18.2, roller compaction used to increase the bulk density of powders and compaction of powders to form tablets; (ii) the dehydration or desolvation of crystal hydrates or solvates, respectively, with partial collapse of the crystal lattice during drying; (iii) the disruption of the crystal structure by local increases in temperature and melting during tablet compaction followed by cooling and rapid solidification of molecules; (iv) sublimation of solids with relatively low melting temperature, followed by re-solidification to amorphous forms upon cooling; and (v) the rapid precipitation of molecules after partial dissolution that might occur during the addition of liquids, as in high shear granulation, spray drying, or the polymer film-coating process, followed by rapid drying.

From this list of possible sources of process-induced disorder given earlier, we can characterize two general categories or mechanisms. In the first category, "solid-state amorphization," one starts with the crystal and stresses it by mechanical force, or by removal of solvent molecules that are associated with the crystal lattice until amorphous structures appear. In the second general case, the processing of the crystal causes molecules to liquefy, vaporize, or dissolve (highly disordered states), and then to re-condense directly into an amorphous form because of rapid changes in local temperature or other local conditions. In the first category one would expect a transformation of crystal surface and bulk defects to the amorphous state once a critical defect density is attained, while in the second case the liquefaction, vaporization, or dissolution of molecules would likely be initiated at surface defect sites, causing a supersaturated solution to form, followed by condensation or precipitation of such molecules directly into amorphous regions. It should also be recognized that it is possible that mechanically induced disorder, which is very common in pharmaceutical systems, may be caused to some extent by local melting of the crystal during processing, in addition to direct effects of mechanical stress.

Any reading of the pharmaceutical literature related to process-induced disorder indicates a very strong emphasis on the creation of amorphous structure (solid-state amorphization) by mechanically stressing crystalline material. This is because processes such as milling, roller compaction, and tablet compaction are so central to the formulation and manufacture of solid dosage forms and because significant evidence exists for the introduction of instabilities after such processing (to be discussed more fully in the following text). A simple theoretical model describing the mechanically induced amorphization of inorganic materials in terms of the mechanical energy introduced into the crystal [9] has been applied to a series of organic molecules with the goal of relating tendencies of different molecules for amorphization to their crystal properties [10]. It is assumed that the application of mechanical stress to a crystal initially causes fracture and particle size reduction that is initiated at crystal defect sites. Further energy input creates more defects and dislocations that contain stored energy, which is then dissipated at a critical defect density to produce amorphous structures. From a thermodynamic perspective, increasing mechanical stress on a crystal and increasing defect density lead to an increase in the free energy of the crystal, ΔG_d, until at a critical defect density, ρ_c, ΔG_d becomes greater than the free energy of the amorphous state, ΔG_{am}, causing spontaneous amorphization. Using this model and a variety of well-defined single-crystal structural, mechanical, and thermodynamic parameters for different crystalline APIs, theoretical estimates were made of the critical defect density of defects associated with the onset of amorphization by milling (on the order of 10^{16}–10^{17} defects per m^2), and these were compared to estimates of their experimental susceptibility for amorphization. It was shown, for example, that indomethacin, with a strong tendency to undergo solid-state amorphization, had a predicted critical defect density, under the same conditions, that was about an order-of-magnitude lower than that of aspirin, which did not undergo amorphization under these conditions. An analysis of the contributions of temperature changes to amorphization under mechanical stress was also included.

It is of interest to determine the degree of disorder at which defects reach the critical density required for the formation of an amorphous structure and to differentiate effects due to defects and amorphous regions. In one study, using X-ray diffraction measurement with associated computational techniques, it was shown that raffinose pentahydrate, held at slightly elevated temperature for short times, easily lost two of the waters of hydration, with the formation of vacancies (defects) within the lattice and the retention of something very close to the original crystal structure of the pentahydrate [11]. It was shown further that with increased heating time and the further loss of water molecules, the defect density increased to a weight fraction of about 14% before the first appearance of amorphous structure. Interestingly, rehydration of the sample before this point easily restored the two waters of crystal hydration without significant structural change, while further heating and drying led to complete amorphization of raffinose. A key to these experiments was the ability to use X-ray diffraction to differentiate between disorder due to defects and that due to the appearance of an amorphous structure.

18.4 Quantitative Determination of Partially Amorphous Material in Crystals

Although the presence of defects in crystals plays a very important role in determining many solid-state properties, it is generally agreed that it is the inadvertent creation of partially amorphous material in solids through processing that leads to most major instabilities not expected with highly crystalline solids. The presence of amorphous material can be identified by the appearance of a glass transition temperature or by the detection of molecular mobility that is characteristic of amorphous solids (see Chapter 6 for more details). To quantitate the amount of amorphous material, present in a processed crystal, one must measure a characteristic that changes proportionally either to the loss of crystallinity or to the appearance of the amorphous state. For example, as illustrated in Figure 18.4, whereas a highly crystalline material exhibits sharp

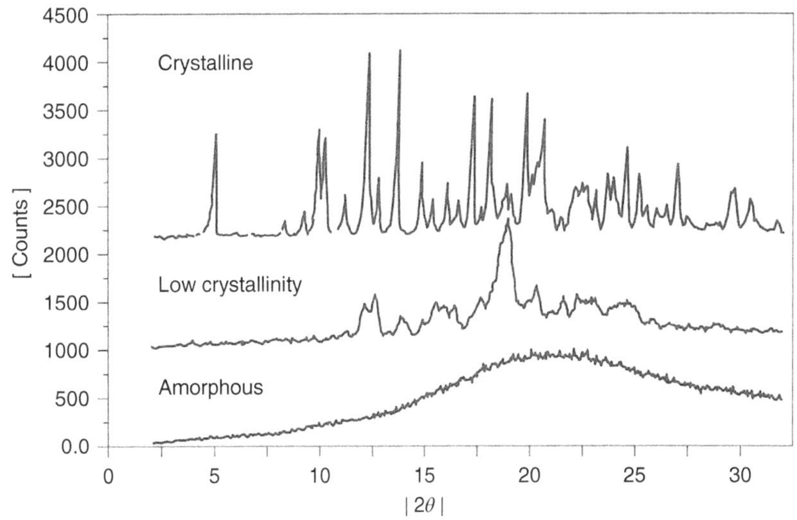

Figure 18.4 Illustration of the powder X-ray diffraction patterns for fully crystalline, partially crystalline, and amorphous forms of an organic solid.

diffraction peaks characteristic of specific planes of symmetry, and totally amorphous structures lose these diffraction peaks, producing a characteristic "halo" in the region of 2θ equal to about 20°, a partially crystalline material will show peak broadening with reduced intensity in proportion to the loss of crystallinity. As discussed in Chapter 15, water vapor under ambient conditions will be adsorbed onto highly crystalline surfaces, even very polar materials, only to the extent of a few molecular layers or in amounts at levels of less than 0.1%. On the other hand, when exposed to a completely amorphous form of this solid, it can absorb water into its bulk structure at levels of greater than 1%, depending on the polarity of the solid and the relative humidity. Thus, exposing a partially amorphous solid to various relative humidities should produce an amount of absorbed water that is proportionally intermediate to values for the fully amorphous and fully crystalline materials.

To carry out a useful, yet simple quantitative analysis of process-induced disorder, most laboratories start by selecting calibration standards of highly crystalline and purely amorphous forms of the material and measuring the particular property of interest, for example, X-ray powder diffraction or water vapor absorption. These forms are then physically mixed to various proportions from 0% to 100% amorphous or crystalline content, and the same measurements are carried out. In most cases, one can construct a calibration curve that can then be used to assess the unknown disorder in a processed sample. For example, water vapor absorption was measured at various relative humidities for various samples containing known amounts of physically mixed amorphous and crystalline sucrose (percent disorder) [12]. As can be seen in Figure 18.5, plots of water vapor absorption at each relative humidity are quite linear with different slopes depending on the relative humidity; the greater the slope, the greater the sensitivity of the measurement. Thus, by determining the water vapor absorption isotherm of a processed sample of sucrose at various relative humidities, it was possible to obtain a direct estimate of the percent crystallinity. Similar quantitative analyses of disorder in processed materials have been reported using other techniques that monitor the presence of amorphous material or the loss of crystallinity, such as X-ray diffraction, FT-IR, and FT-Raman spectroscopy, solid-state nuclear magnetic resonance spectroscopy, dynamic mechanical analysis, modulated differential scanning calorimetry, and isothermal microcalorimetry. In assessing the validity of any method used to determine the degree of disorder in a processed solid, it would be useful to use more than one technique, and especially techniques that might sense the loss of crystallinity in one case, and the increase in disorder in other cases. For example, one study looking at disorder introduced into ground samples of crystalline sucrose found good agreement between measurements made by monitoring powder X-ray diffraction changes (changes in the amount of crystallinity) and the heats of crystallization by DSC of samples heated above T_g (changes in the amount of disorder) [12]. Interestingly, water vapor sorption studies revealed estimates of much less disorder by virtue of much less water absorption than expected. It was concluded that absorbed water initially triggered surface crystallization, which then prevented full water uptake into the remaining solid, thus stressing the importance of understanding the basis of different techniques. In another study, process-induced disorder in milled samples of indomethacin and simvastatin was quantitated by three methods: differential scanning calorimetry, powder X-ray diffraction, and FT-Raman spectroscopy [13]. The results obtained with the different methods were reasonably close in some cases but not in others. The differences in quantitative values obtained appeared to be quite dependent on the methods used to obtain the amorphous calibration standard. In attempting to quantitate the degree of process-induced disorder, therefore, several important questions must be considered and addressed to confirm the validity of such analyses. First, does a physical mixture of amorphous and crystalline material used in establishing the calibration curve (two separated states) represent a partially process-induced disordered amorphous material with an unknown distribution of the disorder? For example, it is not clear whether the distribution of defects and amorphous material in the process-induced material in some cases

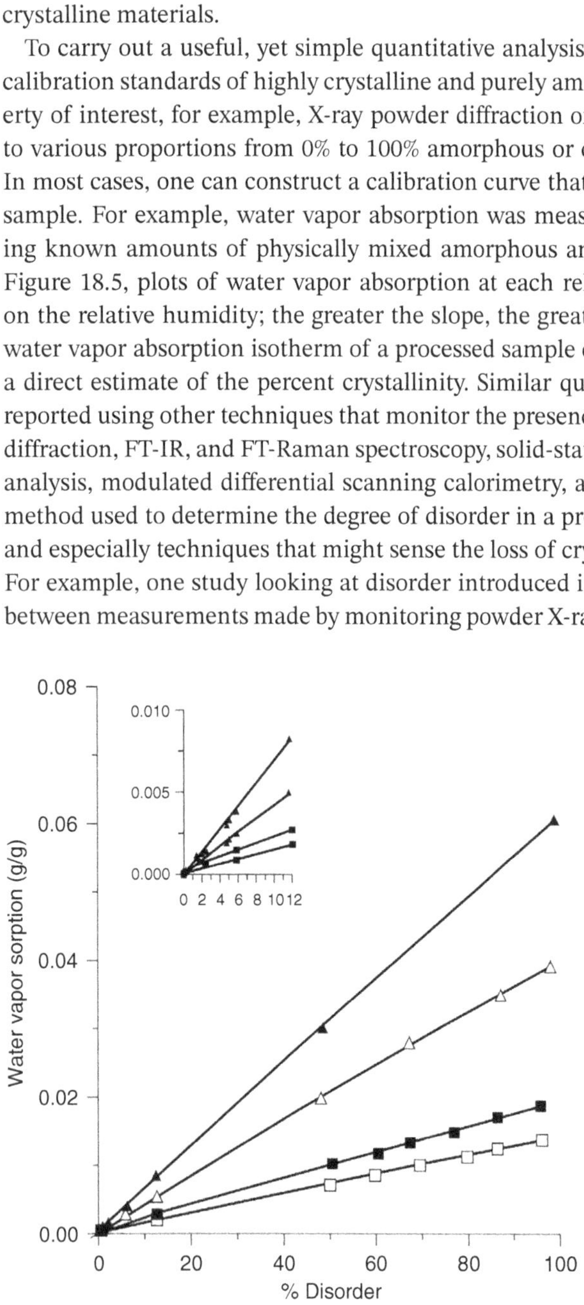

Figure 18.5 Water vapor sorption for various physical mixtures of amorphous and crystalline sucrose exposed to different relative humidities: (□) 7.6%RH; (■) 11.3%RH; (△) 21.5%RH; (▲) 32.4%RH. The inset shows 0–12% disorder. *Source*: [12]/with permission of Elsevier.

would be biased toward the surface regions as compared to the bulk of the solid. Second, in view of this, how does one choose calibration standards of crystalline and amorphous samples that truly represent the disordered and crystalline structure in the processed material? For example, if the disorder is created by milling, should the calibration material be prepared by milling, or will amorphous material prepared by rapidly cooling the melt be adequate? Third, what are the lower limits of detecting disorder for any technique, and do these techniques differentiate between defects and amorphous material at lower levels of disorder, that is, less than about 10% disorder? For example, reports using the water vapor absorption, FT-Raman, and isothermal microcalorimetry methods claim detection levels of 1–2% disorder, which is a level where defects may be the dominant form of disorder. Although such questions remain to be addressed more fully, the analytical assessment of disorder in processed solids is central to an understanding of the underlying basis for much physical instability, as will be discussed more fully in the following text.

18.5 Phase Transformation

As mentioned earlier, and as will also be discussed in more detail in the following text, most solid-state physical instabilities during storage at various temperatures and relative humidities appear to arise because of phase transformations from higher energy disordered states to lower energy, more stable states. In pharmaceutical situations, the most prevalent situations involve the process-induced formation of supercooled liquids or supersaturated solutions that then undergo spontaneous crystallization. If a higher energy form is preferred for a particular pharmaceutical use, then it is important to understand the underlying principles that might give rise to crystallization to be able to prevent it from occurring during the shelf life of the product. On the other hand, if the crystalline form is preferred and disorder is inadvertently introduced by processing, subsequent crystallization could give rise to significant instabilities, as will be discussed more fully in the following text. Minimally, we would want to anneal out any disorder before storage by promoting crystallization. Since it will be argued in this chapter that most physical transformations involving phase changes in crystals are initiated at disordered amorphous sites within a solid, it will be useful to review a few fundamental concepts associated with crystallization from the amorphous state: (i) directly from the amorphous state (supercooled liquid or glass) as a solid-solid transformation and (ii) solvent-mediated crystallization from high-energy solids via supersaturated dissolved states.

18.5.1 Solid-state Crystallization

As shown in Chapter 6, molecules in the amorphous form exhibit two states below and above the glass transition temperature, T_g, the supercooled equilibrium liquid above T_g and the unstable glassy state below T_g. Consequently, we would expect crystallization from the amorphous state, in the absence of solvent, to occur by nucleation and crystal growth mechanisms, most often as is observed with crystallization from the melt. Conceptually, therefore, it will be useful to outline some of the major features of the classical theory of homogeneous nucleation and crystal growth to better understand the key factors involved in dealing with such crystallization. Here, it is assumed that molecules in the liquid state under certain conditions must first undergo spontaneous nucleation, the formation of aggregates or nuclei consisting of a few hundred molecules, followed by the growth of macroscopic-size crystals. The major thermodynamic factor driving nucleation and crystal growth is the free energy difference per unit volume between molecules in the amorphous state and those of the crystal, ΔG_v, as depicted in Figure 18.6, which is generally negative, thus favoring crystallization. However, the formation of nuclei requires the creation of new surfaces between the nuclei and amorphous matrix, a process which is thermodynamically unfavorable; when nuclei form, there must be an increase in overall surface free energy, ΔG_s, which for a spherical nucleus of radius, r, can be described as

$$\Delta G_s = 4\pi r^2 \sigma \tag{18.1}$$

where σ is the surface free energy per unit area of surface (the surface tension in liquids) and $4\pi r^2$ is the surface area.

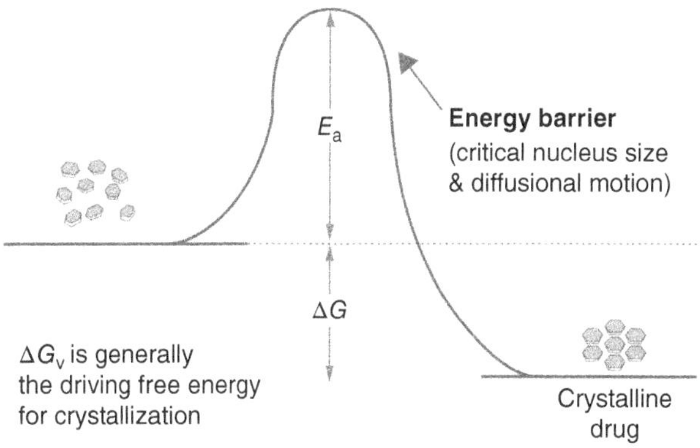

Figure 18.6 Illustration of the energetics of solid-state nucleation and crystal growth for an amorphous solid.

Thus, the overall free energy of homogeneous nucleation, ΔG, required to create a nucleus of radius, r, can be described by combining the two free energy terms, one favoring nucleation (ΔG_v) and the other opposing nucleation (ΔG_s),

$$\Delta G = \frac{4}{3}\pi r^3 \Delta G_v + \Delta G_s = \frac{4}{3}\pi r^3 \Delta G_v + 4\pi r^2 \sigma \quad (18.2)$$

where $\frac{4}{3}\pi r^3$ is the volume of a sphere. It can be shown further, as illustrated in Figure 18.6, that the net change in ΔG will be positive until a critical radius, r_c, of nuclei is reached, above which the loss of free energy due to ΔG_v overcomes ΔG_s and spontaneous nucleation occurs. One can express the overall free energy change, ΔG^*, required to reach the critical radius in terms of ΔG_v and σ as

$$\Delta G^* = \frac{16\pi\sigma^3}{3(\Delta G_v)^2} \quad (18.3)$$

The term ΔG_v in equation (18.3) can be further expressed in terms of the heat of fusion, ΔH_f, the melting temperature, T_m, and the degree of supercooling, ΔT, equal to $T_m - T$, as,

$$\Delta G^* = \frac{16\pi\sigma^3 T_m{}^2}{(3\Delta H_f^2)\ (\Delta T)^2} \quad (18.4)$$

Here, we can see that the greater the lattice energy, as expressed by the positive value of ΔH_f, and the greater the degree of supercooling, the smaller the critical radius required for nucleation will be. In equation (18.4), we can also observe the significant importance of surface effects, as reflected in the value of σ (raised to the third power) in inhibiting nucleation as opposed to the free energy difference between the amorphous and crystalline states (raised to the second power).

In addition to requiring a favorable overall loss in ΔG, at and above the critical radius r_c, molecules must have sufficient molecular mobility in the form of translational and rotational diffusive motion to undergo nucleation. Therefore, an additional important free energy barrier to nucleation arises because of the decrease in diffusional molecular mobility that occurs in amorphous solids as temperature is decreased. This diffusional motion is directly related to viscosity and is generally expressed as a relaxation time; the greater the viscosity, the greater the relaxation time and the slower the diffusional motion (see Chapter 6 for a more detailed discussion of molecular mobility). Since relaxation times at and above T_g are generally greater than 100 seconds, we would expect crystallization above T_g to occur over relatively short time scales. On the other hand, as one goes below T_g into the glassy state crystallization rates should be slowed considerably. For example, to inhibit crystallization from the amorphous state over storage periods of 2–3 years would require that diffusional motions be reduced to timescales of 2–3 years or exhibit relaxation times approaching 10^8 seconds. Expressing this kinetic energy barrier due to molecular mobility as $\Delta G'$, we can approximate the overall rate of nucleation, I, as

$$I \approx e^{-\frac{\Delta G^*}{kT}} e^{-\frac{\Delta G'}{kT}} \quad (18.5)$$

and can conclude, as shown in Figure 18.7, therefore, that, as temperature is decreased (greater supercooling), the thermodynamic barriers to nucleation will decrease, while such cooling will increase the barriers to nucleation as molecular mobility is decreased [14].

Similar analysis of thermodynamic and kinetic factors that affect the rate of crystal growth, u, once nucleation occurs, can be expressed as,

$$u = \frac{D}{\lambda\left[1 - e^{\left(-\frac{\Delta G_v}{kT}\right)}\right]} \quad (18.6)$$

where D is the diffusion constant and λ is the jump distance across the growth interface. Thus, again, we see the opposite effects of thermodynamic and kinetic energy requirements for crystallization.

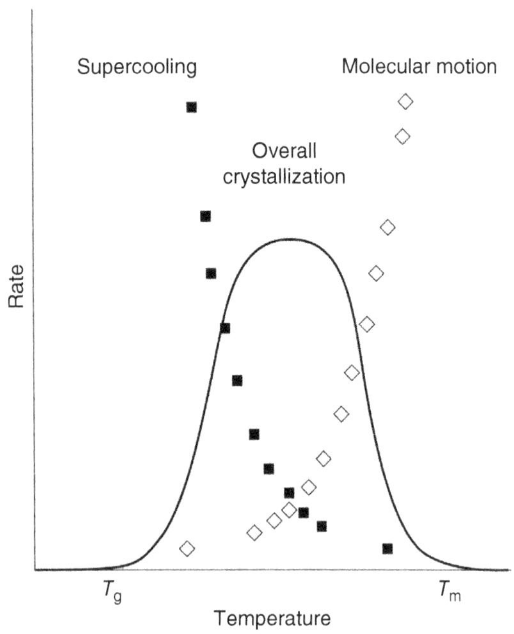

Figure 18.7 Illustration of the rates of overall crystallization of an amorphous solid as a function of temperature, and the corresponding rates of supercooling and molecular motion. *Source*: [14]/with permission of Springer Nature.

In speaking about molecular mobility in amorphous solids, it is important to recognize that in addition to the primary diffusive motions mentioned earlier, amorphous solids exhibit secondary Johari–Goldstein (JG) motions involving hindered noncooperative motion of single molecules (discussed more fully in Chapter 6). Although such motions are not directly involved in the crystal growth process, they do appear to be involved in nucleation and are required as precursors for triggering cooperative diffusive motions, maintaining significantly short relaxation times well below T_g, at temperatures much lower than those required to slow down diffusive motions. This relationship of nucleation and crystal growth to temperature and its relationship to diffusive and JG motions is illustrated in Figure 18.8, where $T_{g\beta}$ represents the temperature at which JG motions disappear [15]. It should be mentioned at this point that, although nucleation and growth mechanisms are the major mechanisms of solid-state crystallization, it has been shown that some amorphous organic molecules, held at temperatures below T_g, exhibit more direct nondiffusive crystal growth [16] and that surface nucleation and crystal growth, greatly affected by rapid surface diffusion, can also play a role, particularly when the specific surface areas are significant [17].

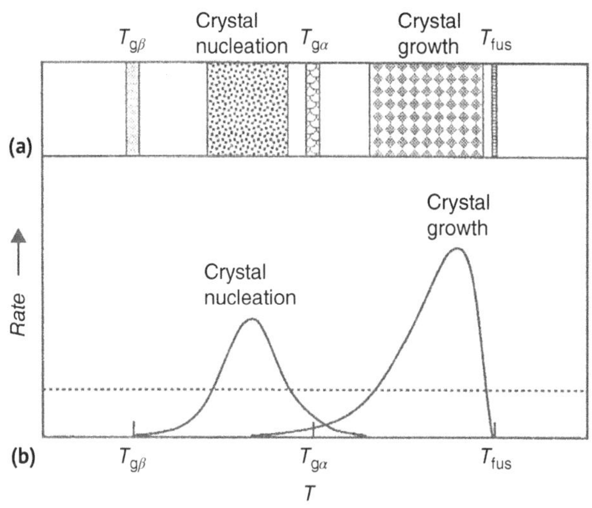

Figure 18.8 Schematic representation of the temperature dependence of the rate of crystal nucleation and growth relative to the melting temperature, T_{fus}; the glass transition temperature, $T_{g\alpha}$; and the temperature for the initiation of β-Johari–Goldstein secondary motions, $T_{g\beta}$. *Source*: [15]/with permission of Elsevier.

With these concepts in mind, we can begin to see how different factors might influence the rates of crystallization from the amorphous state. As already implied in Figure 18.8, the most important parameter is the storage temperature relative to T_g. Here we see that upon increasing the temperature nucleation is initiated at $T_{g\beta}$ equal to about T_g – 100 K but that crystal growth is not initiated until T is equal to about T_g – 50 K. This has led to the suggestion and rule of thumb that storage of any amorphous material at about 50 K below T_g should slow down diffusive motions and hence crystal growth over a period of years. Thus, for example, the storage temperature of a material having a T_g of 40 °C would have to be below 0°C for long-term stability, whereas a T_g of 75 °C would allow for long-term storage at about 25 °C. This rough analysis allows us to further observe that any factor that might change T_g would in turn change the value of T required to maintain a T_g – T of about 50 K. As seen in Chapter 15, the absorption of water vapor by amorphous solids causes a reduction in T_g and increased molecular mobility relative to that of the dry sample; a rule of thumb would be that each 1% of water content lowers T_g by roughly 10 K. Indeed, it has been shown that absorbed water in an amorphous API greatly accelerates crystallization in response to increases in molecular mobility [18]. As also discussed in Chapter 6, the addition of another amorphous solid, having a high T_g, and capable of mixing at the molecular level with an API, will produce a T_g for the mixture that is greater than that of the API alone. Thus, for example, by raising T_g from 40 to 75 °C, one would expect a significant loss of molecular mobility at a particular storage temperature, compared to the API alone. For example, based on principles discussed in Chapter 6, a polymer such as polyvinylpyrrolidone (PVP) with a T_g of 175 °C, when mixed with an API having a T_g of 40 °C, at a composition of 50% w/w PVP, would produce a molecular mixture having a T_g of roughly 100 °C, which would significantly lower the molecular mobility of the API relative to that of the API alone at 25 °C, thus inhibiting crystallization over a period of years [19]. In addition, it has been shown that interactions between the API and PVP, such as hydrogen bonding, which give rise to a miscible mixture, are able to directly inhibit nucleation of the API molecule and reduce the rate of nucleation [20]. With this general conceptual picture of nucleation and crystal growth from amorphous solids in mind, and from concepts developed in Chapters 6 and 15, we can conclude that there are two major factors that can affect crystallization from the amorphous state at a particular temperature: (i) thermodynamic factors that can directly slow nucleation as with the addition of specific nucleation inhibitors and (ii) the addition of molecules that form a molecular mixture that can either increase (water) or reduce (PVP and other polymers with high values of T_g) the molecular mobility of the system at expected storage temperatures.

18.5.2 Solvent-mediated Phase Transformations

As we will observe later, in more detail with a specific example, another way physical transformations can occur is through the mediation of a solvent [21]. In such cases, the high-energy regions in the solid dissolve in a solvent like water and form a solution that is supersaturated relative to that of the most thermodynamically stable form. Molecules in this supersaturated state then crystallize, initiating a phase transformation. Such phase transformations in pharmaceutical systems most often involve crystallization of metastable polymorphic forms, anhydrous crystals capable of forming hydrates or solvates, and amorphous solids. It is very likely that for any high-energy crystalline form defects and partially amorphous regions will be the sites for the initiation of supersaturation and phase transformation. Hence, any process-induced disorder in normally stable crystals could also be susceptible to increased dissolution and solvent-mediated instability. Crystallization from solution, in general, is a well-studied and important technological process with great importance in the pharmaceutical field for allowing the production of highly pure API crystals with morphological characteristics, such as crystal size and shape, which optimize solid therapeutic performance and stability. Given that solvent-mediated crystallization generally involves two steps, nucleation and crystal growth, we would expect that the process will be driven by the same thermodynamic factors that were discussed earlier with regard to solid-state crystallization, as reflected in the description of the rates of nucleation given in equations (18.1)–(18.6). Whereas the major thermodynamic driving force for solid-state crystallization is the degree of supercooling below the melting temperature, in solvent-mediated crystallization, it is the degree of supersaturation within the solvent, S, that is critical. Here, S is defined as

$$S = \frac{c}{c^*} \tag{18.7}$$

where c is the actual supersaturated concentration in solution and c^* is the equilibrium saturated solution concentration of the most stable form. Hence, for example, in a manner analogous to equation (18.4), we can express the critical change in free energy for solvent-mediated nucleation, ΔG^*, as

$$\Delta G^* = \frac{16\pi\sigma^3 v^2}{(3kT)^3(ln\ S)^2} \tag{18.8}$$

where v is the molar volume, k is the Boltzmann constant, and other terms are the same as defined for equation (18.4). The rate of nucleation, $\frac{dN}{dt}$, as a function of supersaturation with all other factors equal, is often simply expressed as

$$\frac{dN}{dt} = k(c - c^*)^n \tag{18.9}$$

where k is the overall rate constant and n is an empirical exponent generally found to be in the range of 3–4; it can be shown that spontaneous nucleation generally occurs over practical timescales when supersaturation, S, is greater than 3–4. However, it is possible for secondary nucleation to occur more readily in the presence of seeds, resulting in greatly enhanced rates at lower supersaturation. It should also be noted that, although the viscosity of the solvent and its effect on the diffusion of dissolved molecules can be an important factor in solvent-mediated nucleation and crystal growth, it will be less important under most circumstances when compared to the role of molecular mobility in solid-state crystallization, as discussed earlier. As also indicated earlier, the importance of solvent-mediated phase transformations in terms of solid-state physical stability arises from the fact that high-energy solid forms will often encounter liquid water during processing that will produce dissolution at concentrations leading to supersaturation. Thus, one can slurry a relatively insoluble metastable polymorphic form of an API in water during wet granulation and have enough dissolution to create sufficient supersaturation and conversion to the more stable crystal form. Likewise, contact of a relatively insoluble anhydrous crystal with water can produce levels of supersaturation, leading to the more stable crystal hydrate. In both cases mentioned, it is most likely that the transformations that take place begin with the enhanced dissolution that comes from defects and/or the presence of some process-induced amorphous content, followed by crystallization from solution. That amorphous regions in any crystalline sample can give rise to greatly enhanced supersaturation, which can then lead to instability, is illustrated in Figure 18.9, where we see an apparent solubility of amorphous indomethacin in water that is roughly five times that of the stable crystal [2]. We can also note that the maximum point reached is limited by the nucleation and crystal growth that is initiated at this supersaturated level. Such solvent-mediated phase transformation for amorphous solids is a very important factor in the use of amorphous forms of API molecules that are poorly soluble in water and hence have limited rates of dissolution and oral bioavailability. Once an API in the amorphous form encounters in vitro and in vivo dissolution media, we can expect behavior as seen in Figure 18.9 unless other formulation components are present to inhibit crystallization [22]. This is part of the basis for the use of polymers in the formation of API-polymer solid dispersions, in that the successful use of amorphous dispersions with polymers provides inhibition of both solid-state and solvent-mediated crystallization at high levels of supersaturation.

18.6 Examples of the Role of Process-induced Disorder in Solid-state Physical Instability in Pharmaceutical Systems

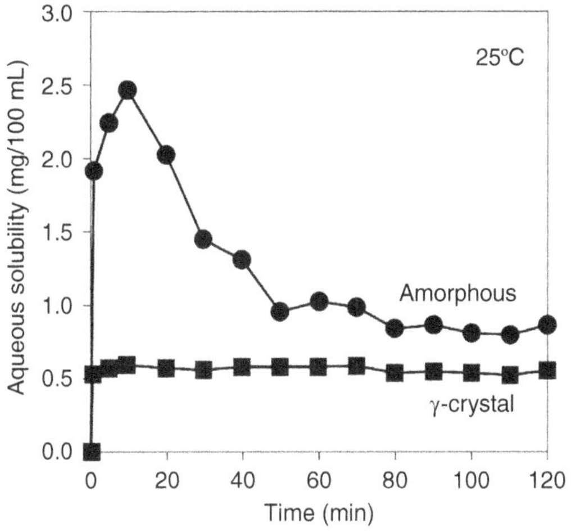

Figure 18.9 Apparent aqueous solubility of amorphous and crystalline indomethacin as a function of time. *Source*: [2]/with permission of Springer Nature.

From the discussion so far, we can conclude that crystal defects play an important role in initiating physical transformations such as phase transitions, dissolution, mechanical deformation, and vapor adsorption. Furthermore, processing of organic crystalline solids, for example, milling, heating, and dehydration, can increase defect density and produce loss of crystallinity and formation of partially amorphous material. It has also been shown that the introduction of such disorder into a crystal raises the overall free energy of the solid, increases molecular mobility, and thereby favors transformations to more thermodynamically stable states. Such process-induced disorder appears most often to be initiated at the surface of the crystal so that the level of possible instabilities caused by disorder is likely to be proportional to the specific surface area and hence the particle size. It is also generally recognized that physical instabilities are greatly affected by environmental factors, such as temperature, relative humidity, and the contact area with other formulation components such as excipients. An important consideration regarding factors influencing disorder-induced instabilities is their ability to alter the glass transition temperature, T_g, and hence the molecular mobility of amorphous regions. For example, exposing the solids to high relative humidity at constant temperature, thus increasing the amount of absorbed water, will lead to a reduction in the T_g of the system and, hence, an increase in molecular mobility at any temperature, T. This would explain the very significant acceleration of the rates of instability when disordered crystalline systems are exposed to high relative humidity. As discussed earlier, we would expect that in a partially crystalline solid essentially all the water content would be due to absorption into the amorphous regions unless crystal hydrates are formed. Generally, water content is reported based on the total weight of the solid, for example, 1% w/w. If, however, most of that water is in the amorphous regions, the actual water content should be expressed as a percent of the mass of amorphous material. For example, consider a "crystalline" solid with a water content of 1%, based on total weight, and an amorphous content of 10%. If essentially all the water is in the 10% of solid mass, the actual water content in the amorphous region would be 10% w/w, a very significant amount capable of greatly lowering T_g and increasing molecular mobility [23]. Thus, in such a situation, there has been a significant amplification of the plasticizing effects of absorbed water at relatively low overall water content with possible effects on physical instability in the form of recrystallization, polymorphic transitions, or other phase changes. In some cases, recrystallization might simply anneal out surface disorder and restore the high degree of crystallinity. In other cases, the higher molecular mobility in the amorphous regions could persist long enough to allow crystallization across the interface between particles, leading to increased particle size and to a form of particle agglomeration and "concretion" in the particle mass. A situation that often arises with milled solid materials formulated as solid oral dosage forms and held at various storage temperatures and relative humidities, for example, 40 °C/75% RH, is a spontaneous reduction of the rates of dissolution and oral bioavailability. One possible cause of such behavior would be the presence of "concretion" induced by exposure to accelerated temperature and relative humidity. That such particle size growth can be induced in milled samples upon storage has been nicely demonstrated in the work of Ward and Schultz [24] with samples of the crystalline API albuterol sulfate, ultimately intended for pulmonary delivery and hence requiring very small particle diameters. Here, as shown in Table 18.1, air jet micronization was used on a sample of albuterol sulfate having only 11% of its particles at under 5 μm and no particles at less than 2 and 1 μm, with an average diameter of 15.7 μm. Air jet micronization and storage under vacuum produced samples with 98% of the particles now under 5 μm, and 96% under 2 μm, with an average diameter of 1.5 μm. Note, however, that these samples, when stored for 24 hours at 40 °C/75% RH, lost almost all the particles having diameters less than 2 μm and a significant proportion

Table 18.1 Effects of micronization and storage on the particle size distribution of crystalline albuterol sulfate.

	Unmicronized (RT/vacuum)	Micronized (RT/vacuum)	Micronized (40 °C/75%RH-24 hours)
% under 5 μm	11	98	73
% under 2 μm	0	72	6
% under 1 μm	0	24	1
Median diameter, μm	15.7	1.5	3.9

Source: [24]/with permission of Springer Nature.

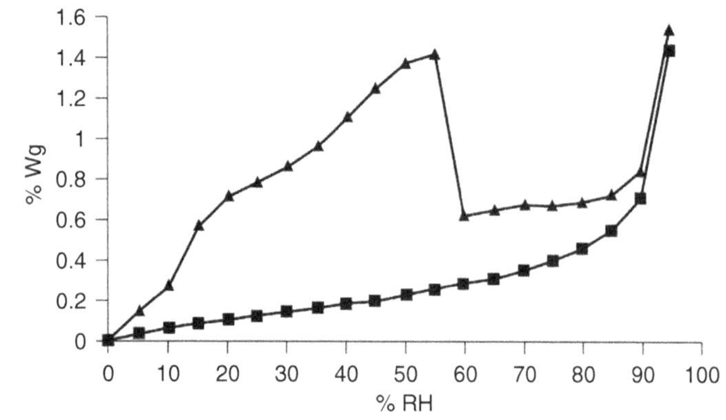

Figure 18.10 Water vapor for vacuum stored micronized albuterol sulfate, sequential scans: (▲) first scan; (■) second scan. *Source*: [24]/ with permission of Springer Nature.

under 5 μm, presumably because of the crystallization and particle growth of the smaller particles emanating from regions of disorder. Indeed, scanning electron microscopy revealed that such particle growth and agglomeration had occurred. Although powder X-ray analysis indicated no measurable loss of crystallinity upon micronization, evidence for the presence of amorphous material was obtained by DSC measurement of a glass transition temperature. Further evidence for the presence of process-induced disorder in the originally micronized system came from studies of water vapor sorption at various relative humidities. Using an automated gravimetric apparatus that allowed the measurement of water uptake at constant temperature with a controlled change of relative humidity over the range of 5–95% RH, comparisons were made of sorption by the initially processed material and then the same sample after exposure to the entire range of RH, followed by drying back to 0% RH, and then repeating the process with a second scan. As shown in Figure 18.10, the initial scan produced an isotherm at lower RH that indicated much larger amounts of water uptake than expected for a highly crystalline material, presumably because of the presence of process-induced amorphous regions. At higher RH the amount of water uptake abruptly decreased, an indication that water-induced crystallization of the amorphous regions had occurred. This interpretation is supported by the fact that a second scan of this material produced an isotherm with water uptake more characteristic of a highly crystalline material, indicating that most of the amorphous material had been eliminated by an annealing process induced by the water absorption.

Another major area where physical instability often occurs involves phase transformations that occur between anhydrous and hydrated forms of crystals. If one starts with an anhydrous crystal capable of forming hydrates at various relative humidities, or in the presence of liquid water, we can expect that the higher energy anhydrous form will be thermodynamically susceptible to conversion to a crystal hydrate form (see Chapter 3). Conversely, if a hydrate is processed by heating or drying to lose waters of hydration, a number of possible physical changes can occur: (i) the hydrate may form another lower stoichiometric hydrate; (ii) the hydrate may convert to the completely anhydrous crystal form; (iii) the removal of water will occur with no effect on the crystal structure of the hydrate and form a dehydrate; and (iv) the removal of water will cause a breakdown of the crystal structure with the formation of defects and amorphous structure, as described earlier for the dehydration of raffinose pentahydrate [11]. The formation of dehydrates and amorphous structure, and the disorder introduced, make the new solid form particularly susceptible to physical instability. One interesting study that illustrates the importance of disorder in transformations involving hydrates followed the solvent-mediated conversion of anhydrous carbamazepine to the dihydrate form after milling [25]. Here, after getting some estimates of the equilibrium solubility and dissolution rates of the two forms, suspensions of the anhydrate form were prepared, and the conversion of the anhydrous form to the hydrate during dissolution was followed. Supersaturated ratios of the anhydrous form to the dihydrate were on the order of 1.5. Comparisons were made between samples of the anhydrous form that were ground for various time periods and those unground. After considering the greater specific surface area of the ground material, it was shown that the ground material transformed to the dihydrate at a faster rate, presumably due to the mechanical activation caused by milling that introduced disorder into the anhydrous form. That these differences were due to process-induced disorder was supported by the reduction in differences between the samples after allowing the ground sample to age and crystallize, over a period of 1 month at room temperature. Powder X-ray diffraction studies indicated a measurable amount of amorphous material had been introduced by the grinding and that this was most likely the site for surface nucleation and crystal growth of the dihydrate.

18.7 Poor Wettability of Solids During Dosage Form Processing and Administration

As indicated at the beginning of this chapter, most pharmaceutical physical instabilities of solids arise during processing, storage, and use because of noncovalent phase transformations, for example, polymorphic crystal transitions, crystallization from the amorphous state (devitrification), hygroscopicity, and inadvertent disordering of crystals by milling or drying. When liquids and solids are mixed, as when powders are added to water to form solutions or suspensions, one often observes another form of physical instability, poor "wetting," leading to separation of solid and liquid, with the solid particles typically forming floating aggregates, with trapped air at the liquid–air surface. Such poor wetting also has been observed during the wet granulation process, the reconstitution of lyophilized cakes for parenteral administration, and when solid oral dosage forms, such as capsules and tablets, are administered and do not exhibit the required rapid disintegration, dissolution, and bioavailability because of a lack of wetting and penetration of liquid into the porous solid compact [26].

Wetting in most pharmaceutical situations can simply be defined as the spontaneous ability of a liquid exposed to a dry solid to displace the solid–vapor interface, causing an increase in the solid–liquid interfacial area, as schematically illustrated in Figure 18.11 for a drop of liquid placed on a flat solid surface.

Here, we can see the progressive spreading of the liquid drop quantitated in terms of a decreasing "contact angle," usually designated as θ, made by drawing a tangent outside the drop at the point of intersection of the solid–vapor, solid–liquid, and liquid–vapor interfaces. In such a convention, a contact angle of 180° represents complete nonwetting, while complete wetting is represented by a contact angle of 0°. Such angles are most often measured by placing a drop of liquid on a flat solid surface and directly measuring the angle with a goniometer [27]. As we shall see in the following text in more detail, reduced wetting is associated with the surface tension of the liquid and the lack of chemical affinity between the liquid and the solid surface. Wetting problems in pharmaceutical systems generally arise with polar liquids, such as water or aqueous solutions, which have a high surface tension, and solids exhibiting certain levels of nonpolarity and hence low tendencies to interact with water and favor spreading [28].

To mechanistically understand how polarity or nonpolarity can influence the wettability of a given solid–liquid combination, we can examine in Figure 18.12 a schematic illustration of a partially wetted solid.

First, we can observe that the process of wetting is an interfacial phenomenon controlled by the molecular structure and interactions taking place at each interface. Note that, whereas liquid molecules in the bulk phase uniformly interact with other surrounding molecules, at the liquid–vapor interface there is a distinct reduction in the number of interactions across the interface, leading to a lower surface density, which, in turn, results in a higher surface Gibbs energy and a lateral inward "surface tension," $\gamma_{lv,}$ with a lateral force per unit distance of surface which energetically opposes the expansion of the liquid–vapor interfacial area and the spreading of the drop. As mentioned earlier, the more polar the liquid, the greater the surface tension [26]. Indeed, this resistance to expansion of the liquid–vapor surface area results in a tendency for the liquid drop to assume a spherical shape (the smallest surface area for a given volume), as observed in Figure 18.11 at a contact angle of 180°. We also would expect that the molecules of the solid at the surface against air would have unbalanced forces operating on them, and, therefore, likewise, they should exhibit higher surface Gibbs energy and a lower surface density with a corresponding surface tension, γ_{sv}, not always detectable because of the rigidity and complex mechanical

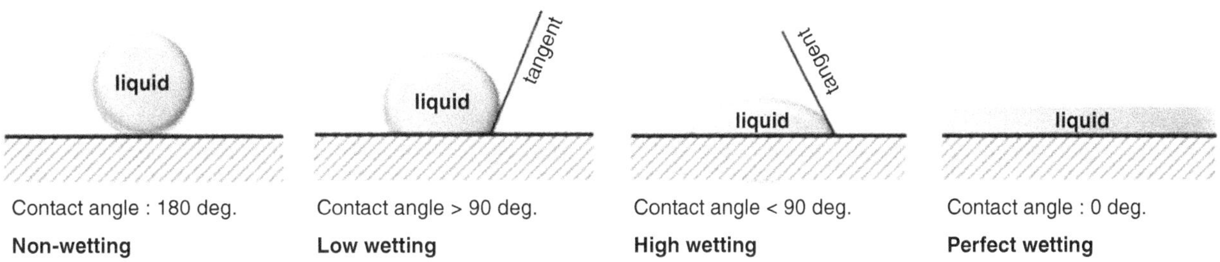

Figure 18.11 Schematic representations of the progressive wetting of a solid surface by a liquid drop.

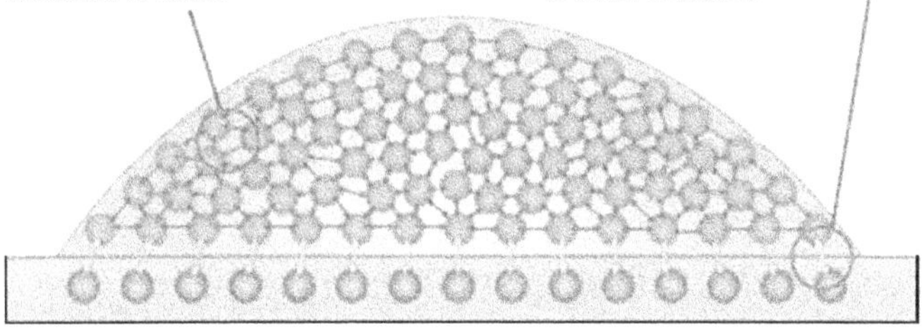

Figure 18.12 Schematic representation of a liquid drop illustrating the imbalance of cohesive bulk interactions and those at the various interfaces.

properties of most solids [29]. Since the creation of new surface area is energetically unfavorable, elimination of the solid–vapor surface by the spreading liquid contributes favorably to wetting, while spreading of the liquid over the solid surface creates a new solid–liquid interface with an interfacial tension of γ_{sl}, an energetically unfavorable process. However, in the last case, the greater the interaction between the liquid and solid molecules, for example, through hydrogen bonding, as with water and polar solids, the less will be this unfavorable contribution to wetting.

As illustrated in Figure 18.13, therefore, we can expect that the contact angle will be determined by the balance and direction of the various surface forces, represented by γ_{lv}, γ_{sv} and γ_{sl}.

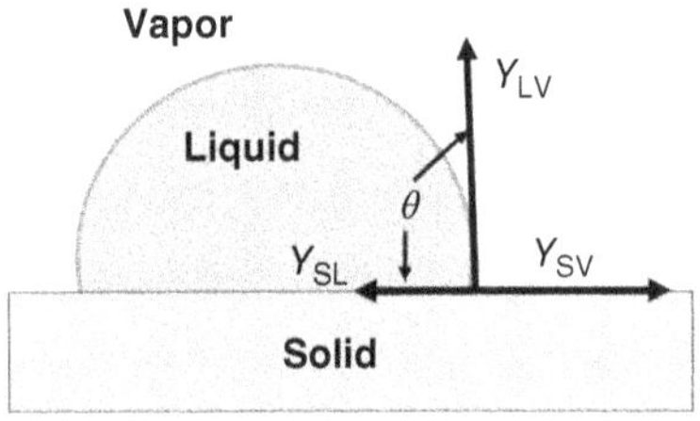

Figure 18.13 Schematic representation of the interfacial forces operating at the three-phase boundary of a liquid drop on a solid surface. *Source*: Adapted from [27].

Indeed, rigorous theoretical studies for the effects of various surface forces on the contact angle exhibited by a liquid and a molecularly flat solid surface [30], and balancing various favorable and unfavorable energetic processes, lead to the Young equation.

$$\gamma_{lv} \cos\theta = \gamma_{sv} - \gamma_{sl} \tag{18.10}$$

or

$$\cos\theta = \frac{[\gamma_{sv} - \gamma_{sl}]}{\gamma_{lv}} \tag{18.11}$$

From this equation, it is apparent that the lower γ_{lv} and γ_{sl} the lower will be the contact angle, θ (greater $\cos\theta$); that is, for the wetting of solids by polar water molecules, the more polar the solid, the lower will be γ_{sl} and, therefore, the lower θ and greater wetting. As an example, in Table 18.2, we can see some contact angles of water placed on flat surfaces of solids varying in surface polarity and hence varying in γ_{sv} and γ_{sl} [27]. Here, we can observe very high contact angles (poor wetting) for water on the hydrocarbons, paraffin, and polyethylene, as well as on the very nonpolar fatty acid tablet lubricant magnesium stearate, while the introduction of more polar oxygen groups to form poly(methyl methacrylate) increases wettability to a small extent. Interestingly, since the acetylation of the free hydroxyl group of the steroid, hydrocortisone, might be expected to decrease polarity, we would expect the observed increase in its contact angle from 43° to 57°.

The Young equation, therefore, provides a quantitative measure of the "intrinsic thermodynamic contact angle" of a liquid on an idealized molecularly homogeneous and smooth surface. However, in most pharmaceutical situations we are dealing with macroscopically rough surfaces, often with a complex porous structure produced by powder packing or compaction, and/or chemically heterogeneous surfaces containing more than one type of molecule, as well as the possibility of entrapped air between the particles or on a rough surface, as in the case of lyophilized cakes and compacted tablets. In both cases, as schematically represented in Figures 18.14 and 18.15, traditional measurement of a rough or heterogeneous surface will yield an apparent contact angle determined by the nature and extent of roughness or heterogeneity and the location of the liquid front [31, 32]. In such cases, we generally speak of an "apparent contact angle." Note in Figure 18.15 the inability of the liquid to displace the entrapped air, leading to a contact angle greater than 90°.

In the case of dispersing particles into liquids, we must also consider the wetting behavior of individual particles and of any aggregates of particles, where particle shape, size, and density play an important role [33]. To illustrate this, consider a single spherical particle floating on the liquid–vapor surface where the density of the particle is greater than that of the liquid, with an expectation that the particle will be completely wet by the liquid and sink into the bulk liquid. However, as illustrated in Figure 18.16, if the solid exhibits a certain level of hydrophobicity, and is sufficiently small, that is, less mass, the particles will float on the surface to the extent that the liquid surface tension tends to oppose gravity. It can be further shown that with many particles this lack of wetting often leads to lateral particle aggregation with entrapped air, making it more difficult to obtain complete dispersion of the powders even with agitation [26].

So far, we have mentioned particle aggregation and the entrapment of air several times. This is because it appears that in most pharmaceutical situations, poor penetration of the liquid into a solid is the dominant factor, determined by the surface tension of the liquid, the contact angle, the nature of the porous structure, and the extreme hydrophobicity of the air as schematically represented in Figure 18.17 [34].

The process by which a liquid contacts and penetrates a solid pore can best be seen by placing a cylindrical glass capillary vertically into a liquid, as schematically represented in Figure 18.18. In one case, the liquid on the left could be water, which exhibits a contact angle on glass close to 0°, forms a hemispherical meniscus with the concave side facing the air, and spontaneously rises above the level of the bulk liquid. In the case on the right, on the other hand, the liquid, which could be mercury, has a contact angle close to 180° on glass with the convex side of the meniscus facing the air, and tends to fall below the level of the bulk liquid. Hence, the water penetrates the capillary, while the mercury does not.

For such liquid movement to occur spontaneously, there must be a difference in the pressure above the liquid in the capillary and above the "flat" bulk phase, ΔP, caused by the nature of the curvature of the meniscus. It, indeed, is well established [35] that the pressure over a highly concave surface, as in the capillary on the left with contact angle less than 90° is less than that over than that over a flat surface because of unbalanced intermolecular forces on a highly curved surface, and the liquid rises until this force is matched by downward gravitational forces. For a contact angle greater than 90° and a convex meniscus, this pressure is greater than that of a flat surface and the liquid does not rise in the capillary and falls below the bulk liquid level. This pressure difference is called the capillary or Laplace pressure and, generally, can be defined in terms of the capillary radius, r, the surface tension of the liquid, γ_{lv}, and the liquid–solid contact angle, θ, as [36]:

$$\Delta P = \frac{2\gamma_{\text{lv}}\cos\theta}{r} \tag{18.12}$$

Now, let us assume that we bring a liquid in contact with a cylindrical pore positioned horizontally (absence of gravitational effects). Assuming the capillary to be uniformly cylindrical, we can obtain some measure of wettability in such a system by measuring the rate of penetration of the liquid into the capillary under a pressure gradient by assuming the application of the Poiseuille equation [37], equation (18.13), for the steady-state Newtonian volumetric flow rate, Q, as schematically represented in Figure 18.19:

$$Q = \frac{[\pi r^2 \Delta P]}{[8\eta l]} \tag{18.13}$$

Table 18.2 Contact angle of water on the "flat surface" of various solids with different levels of polarity.

Solids	Contact angle, θ (degrees)
Magnesium stearate	118
Paraffin	110
Polyethylene	90
Poly(methyl methacrylate)	78
Hydrocortisone acetate	57
Hydrocortisone	43

Source: Adapted from [27].

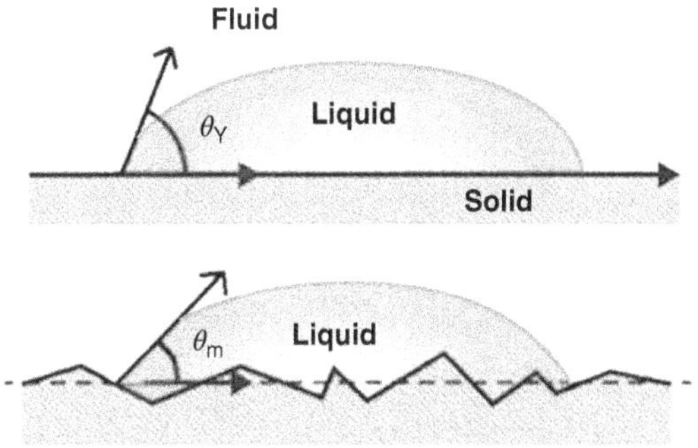

Figure 18.14 Schematic representation of the effect of rough surfaces on the apparent contact angle of a liquid on a solid surface.

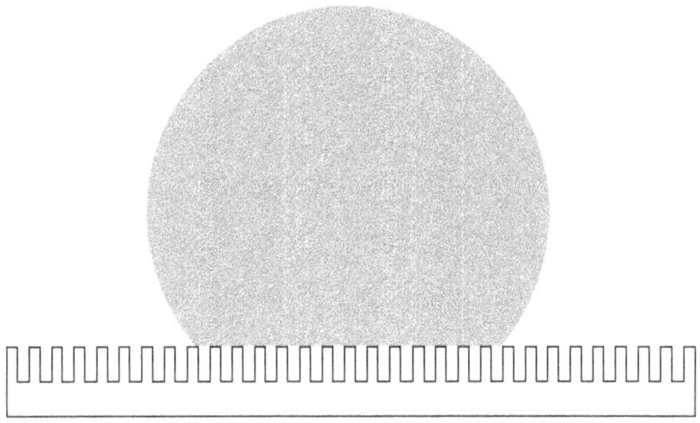

Figure 18.15 Schematic representation of the effect of entrapped air on the apparent contact angle.

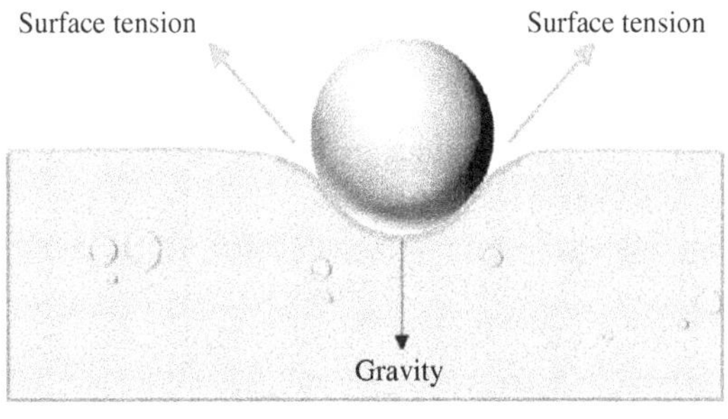

Figure 18.16 Schematic representation of the poor wetting of a solid particle due to surface forces and a finite contact angle.

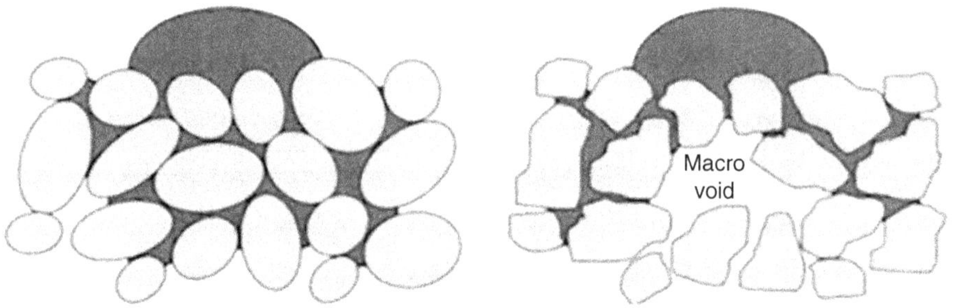

Figure 18.17 Schematic representation of the porous structure of compacted solid particles. *Source*: [34]/with permission of Elsevier.

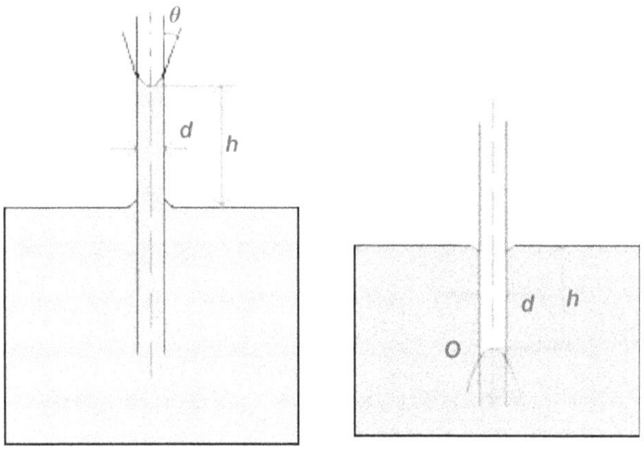

Figure 18.18 Schematic representation of capillary rise or fall in a glass cylindrical tube of a liquid with a contact angle of 0° (left) or > 90° (right).

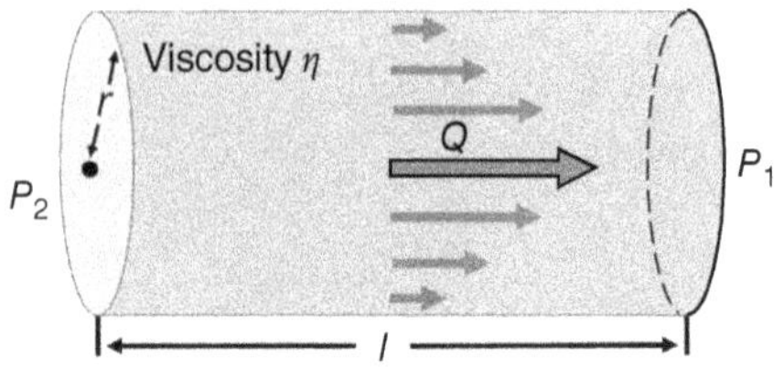

Figure 18.19 Schematic representation of laminar flow rate of a liquid in a cylindrical tube under a pressure gradient of $\Delta P = P_2 - P_1$. *Source*: Adapted from [27].

where ΔP is the pressure difference driving the process, η is the liquid viscosity, r is the capillary radius, and l is the distance traveled from the entry point at time t.

Assuming that the driving pressure, ΔP is the capillary pressure and t is time, combining equations (18.12) and (18.13) leads to the well-known Lucas–Washburn equation [38] (often just called the Washburn equation); see equation (18.14):

$$l^2 = \left[\frac{(r\gamma\cos\theta)}{(2\eta)}\right]t \tag{18.14}$$

Note that the Washburn equation predicts that for positive penetration to occur, even partially, the capillary pressure must be positive and hence the contact angle must be less than 90°, while no penetration will occur if the contact angle is 90° or greater.

Although in theory this equation should only be applicable to cylindrical pores with constant radius, it has been tested using closely packed monosized glass spheres with varying surface energy, where the porosity of randomly packed spheres is more irregular [39–41]. This was accomplished by using uncoated glass spheres and coating the beads with poly(methyl methacrylate) and poly (dimethylsiloxanic acid) to make the surface more hydrophobic and measuring the capillary pressure and rates of liquid penetration. Various pure liquids and mixtures of water and ethanol were used to vary the surface tension, contact angle, and viscosity. The direct capillary pressure measurements revealed that the penetration of packed spheres exhibited significant pressure instability, presumably due to the irregular noncylindrical geometry of the pores created by the packing of spheres [39]. However, in general, as shown in Figure 18.20, penetration rates appeared to follow the kinetics expected from the Washburn equation, despite the noncylindrical structure of the interparticle pores [41].

Using the method of Studebaker and Snow [42], which assumes that in such a case the porous system can be represented as a "bundle of cylindrical capillaries" with an average radius of r, one can measure the penetration rate of a liquid that exhibits a finite contact angle with the solid (liquid 2) and compare this at time t with the penetration for a liquid with zero contact angle (liquid 1) in the same powder bed. This then allows one to express the contact angle, θ_2, by the following equation, where it is assumed that r is the same for both systems.

$$\theta_2 = \frac{(\gamma_1 \eta_2 t_1)}{(\gamma_2 \eta_2 t_2)} \tag{18.15}$$

In Table 18.3, we compare the apparent contact angles calculated from equation (18.15) to corresponding contact angles measured directly on flat surfaces [41].

In general, the penetration into coated-siliconized and coated-PMMA beads did not occur with advancing contact angles on a flat surface of $\theta > 74–76°$, rather than at 90° as expected if the pores simply were acting as a "bundle of cylinders." However, calculated contact angles do appear to follow the expected changes with decreasing surface tension, although they were found to be consistently higher than those obtained on flat surfaces. The lack of exact agreement in contact angles would be consistent with the instabilities and unsaturation of liquid flow due to the irregular pore structure observed in the capillary pressure measurements [39]. Thus, the application of the Washburn equation to obtain apparent contact angles can be minimally useful in uncovering situations where poor wetting might occur. We can conclude at this point, therefore, that the wetting of powders and compacts is most directly influenced by the surface tension of the water and various aqueous solutions encountered, and the degree of hydrophobicity of the solid. The primary factor to consider is the situation when finite contact angles are formed due to unfavorable surface energetics. Additional complexities occur when dealing with small particles, particularly when they are closely packed and form porous structures with entrapped air.

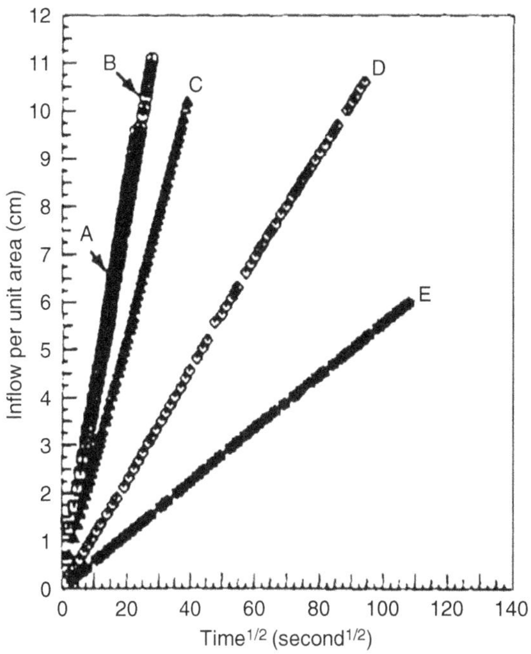

Figure 18.20 Inflow per unit area versus the square root of time for horizontal infiltration into poly(methyl methacrylate)-coated glass beads with pure liquids Key: A. (open squares) hexane; B. (open circles) octane; C. (open triangles) ethanol; D. (open diamonds) formamide; E. (closed stars) ethylene glycol. *Source*: [41]/with permission of Elsevier.

18.8 Considerations in Evaluating Solid-state Physical Stability

In this chapter, we have tried to address the important factors that give rise to phase transformations during the processing and storage of solid dosage forms that are detrimental to the performance of the product. It is suggested that such phase transformations arise primarily because regions of the processed solid have attained higher energy than that expected for the most thermodynamically stable crystalline form. Evidence has been presented to conclude that the higher energy contained within such a solid is generally associated with the presence of defects and amorphous material that are thermodynamically unstable. We differentiate between situations where the desired crystalline form is inadvertently disordered and those where the amorphous state is purposefully used to enhance product performance. Thus, in the former situation we wish to minimize disorder and maximize stability, whereas in the latter situation we wish to stabilize the disorder. As a rule, given that crystals with limited molecular motion are expected to have a high degree of stability, the presence of process-induced disorder is not often anticipated, and often is only discovered after a period of storage and late-stage testing. In particular, the signs of such instability appear to be associated with changes that are observed in particle size and morphology, in mechanical strength, particularly in compacts, and in changing dissolution patterns. A commonly observed clue for the potential of such process-induced instability in solid dosage forms is a distinct reduction in the rate of dissolution that is observed during long-term storage at various temperatures and relative humidities. Such reductions most likely arise because of the crystallization of a more stable solid form at the surface and particle agglomeration due to crystal growth across particle interfaces. In view of the long times associated with such "post-testing" of stability, special attention should be given to the possibility of predicting process-induced disorder as early as possible in the evaluation of API and the development of solid formulations. Of great importance is the need to first carefully consider the likely processing conditions that will be used to produce an appropriate form of the API and final product. Will high-energy milling be used to reduce particle size? Will the system require roller compaction? Will the system encounter solvents such as water, followed by drying, as in wet granulation and polymer film-coating processes? Is there the possibility that the presence of amorphous excipients could give rise to local molecular mixing at the particle–particle interface that would stabilize the disorder, as discussed earlier, in the context of stabilizing amorphous forms by using amorphous polymers to form amorphous dispersions? Indeed, this may be the basis for many API–excipient interactions leading to instability. It would then be important to consider the propensity of the crystalline API to undergo amorphization and to measure

Table 18.3 Apparent contact angles calculated from the Studebaker-Snow method, (θ_{app}), compared with contact angles measured for flat surfaces (θ_{flat}).

Liquid[c]	Siliconized glass		PMMA-coated glass	
	θ_{app}[a] (°)	θ_{flat}[b] (°)	θ_{app}[a] (°)	θ_{flat}[b] (°)
Octane	0	~6	0	0
Ethanol	22	22	36	0
90% ethanol[c]	57	37	36	0
80% ethanol	58	45	55	0
60% ethanol	65	54	54	0
40% ethanol	79	65	52	37
35% ethanol	80	68	–	–
30% ethanol	84	71	–	–
25% ethanol	> 90	76	–	–
20% ethanol	> 90	81	67	47
10% ethanol	> 90	99	80	56
5% ethanol	> 90	99	87	64
Water	> 90	107	90	74

Source: [41]/with permission of Elsevier.

a. Apparent contact angles, θ_{app}, labelled as being > 90°, indicates that no penetration of the bed was measured.
b. Contact angles, θ_{flat}, measured on flat surfaces.
c. All liquids appearing as % ethanol represent ethanol–water mixtures.

key characteristics of the amorphous material prepared in different ways, for example, T_g, enthalpy relaxation, water vapor absorption isotherms, apparent dissolution in water and crystallization tendencies. Early consideration should be given to the possibility that process-induced disorder might be removed by annealing at slightly elevated temperature and relative humidity without adverse effects. If physical instability is very likely under the conditions used to prepare the product, then it will be useful to consider altering the process conditions or formulation ingredients to reduce disorder. To ensure product quality, it will be helpful to have available a validated quantitative method of determining the amount of disorder, despite its limitations as discussed earlier. It should be further stated that great care must be taken when attempting to carry out studies under accelerated storage conditions, for example, 40 °C/75% RH, to predict long-term storage conditions. Whereas the measurement of rate constants for a chemical reaction in solution at elevated temperatures, and the application of the Arrhenius equation, can provide reasonable predictions at lower temperatures, solid-state physical transformations, as discussed in this chapter, involve several physical processes with widely different mechanisms and, thus, different activation energies with more complex temperature dependence. This will be discussed further in more detail in the next chapter dealing with solid-state chemical instability.

Regarding the evaluation of the potential for poor wetting issues, several approaches can be useful [26]. If one is dealing with wetting problems associated with the dispersion of powders into liquids to form solutions or suspensions, or the addition of liquid to a powder bed, as with wet granulation or reconstitution of lyophilized cakes, various methods to overcome poor wetting of addition of the liquid or the solid can be used, including spraying the liquid into the powder or by small additions of liquid or solid. In such cases, if the problem cannot be resolved by processing, it may be necessary to add some surfactants to the liquid or to add more polar excipients to the solid, including surfactants. If the problem is caused by a lack of penetration of liquid into a porous aggregate, as with tablets during disintegration and dissolution, adding polar excipients or solid surfactants would also be potentially useful. In all cases, such problems might be best handled by having a better idea of the surface energetics of the solid [27, 28] and by measuring apparent contact angles on individual API and excipients, as well as their mixtures. If water or aqueous solutions are involved, some record of the general level of hydrophobicity of potential excipients can be useful [26, 28].

References

1 Wales D. *Energy Landscapes: Applications to Clusters, Biomolecules and Glasses.* Cambridge, United Kingdom: Cambridge University Press; 2003.
2 Hancock BC, Parks M. What is the true solubility advantage for amorphous solids? *Pharm Res* 2000; 17: 397–404.
3 Newman A, Knipp G, Zografi G. Assessing the performance of amorphous dispersions. *J Pharm Sci* 2012; 101: 1355–1377.
4 Willert J-F, Durand MA, Briggner L-E, Marx A, Danede F, Descamps M. Solid-state amorphization of linaprazan by mechanical milling and evidence of polymorphism. *J Pharm Sci* 2013; 102: 2214–2220.
5 Konno T, Kinuno K. Physical and chemical changes of medicinals in mixtures with adsorbents in the solid state II: applications of reduced pressure treatment for the improvement of dissolution of flufenamic acid. *Chem Pharm Bull* 1989; 37: 2481–2484.
6 Stillinger FH, Weber TA. Packing structures and transitions in liquids and solids. *Science* 1984; 225: 983–989.
7 Burt HM, Mitchell AG. Crystal defects and dissolution. *Int J Pharm* 1981; 9: 137–152.
8 Foster MC, Ewing GE. Adsorption of water on NaCl (001) surface II: an infrared study at ambient temperatures. *J Chem Phys* 2000; 112: 6817–6826.
9 Tromans D, Meech JA. Enhanced dissolution of minerals: stored energy, amorphism, and mechanical activation. *Miner Eng* 2001; 14: 1359–1377.
10 Wildfong PLD, Hancock BC, Moore MD, Morris KR. Towards an understanding of the structurally based potential for mechanically activated disordering of small organic crystals. *J Pharm Sci* 2006; 95: 2645–2656.
11 Bates S, Kelly RC, Ivanisevic I, Schields P, Zografi G, Newman A. Assessment of defects and amorphous structure produced in raffinose pentahydrate upon dehydration. *J Pharm Sci* 2000; 89: 128–143.
12 Saleki-Gerhardt A, Ahlneck C, Zografi G. Assessment of disorder in crystalline solids. *Int J Pharm* 1994; 101: 237–247.
13 Zimper U, Aaltonen J, McGovern CM, Gordon K, Krauel-Goellner K, Rades T. Quantification of process-induced disorder in milled samples using different analytical techniques. *Pharmaceutics* 2010; 2: 30–49.
14 Saleki-Gerhardt A, Zografi G. Non-isothermal and isothermal crystallization of sucrose from the amorphous state. *Pharma Res* 1994; 11: 1166–1173.
15 Oguni M. Intra-cluster rearrangement model for -process in supercooled liquids as opposed to cooperative rearrangement of whole molecules within a cluster. *J Non-cryst Solids* 1997; 210: 171–177.
16 Shtukenburg A, Freudenthal J, Gunn E, Yu L, KahrB. Glass-crystal mode for testosterone propionate. *Cryst Growth Des* 2011; 11: 4458–4462.
17 Gunn E, Guzei IA, Yu L. Does crystal density control fast crystal growth in glasses? *Cryst Growth Des* 2011; 11: 3979–3984.
18 Andronis V, Yoshioka M, Zografi G. Effects of sorbed water on the crystallization of indomethacin from the amorphous state. *J Pharm Sci* 1997; 86: 346–351.
19 Matsumoto T, Zografi G. Physical properties of solid molecular dispersions of indomethacin with poly(ninylpyrrolidone) and poly(vinylpyrrolidone-co-vinylacetate) in relation to indomethacin crystallization. *Pharm Res* 1999; 16: 1722–1728.
20 Taylor LS, Zografi G. Spectroscopic characterization of interactions between PVP and indomethacin in amorphous molecular dispersions. *Pharm Res* 1997; 14: 1691–1698.
21 Cardew PT, Davey RJ. The kinetics of solvent-mediated phase transformations. *Proc R Soc London, Ser A* 1985; 398: 415–428.
22 Hevbare GA, Liu H, Taylor LS. Maintaining supersaturation in aqueous drug solutions of poorly-soluble drugs: implications for solubility enhancing formulations. *Cryst Growth Des* 2013; 13: 1497–1509.
23 Ahlneck C, Zografi G. The molecular basis of moisture effects on the physical and chemical stability of drugs in the solid state. *Int J Pharm* 1990; 62: 87–95.
24 Ward GH, Schultz RK. Process-induced crystallinity changes in albuterol sulfate and its effects on powder physical stability. *Pharm Res* 1995; 12: 773–779.
25 Murphy D, Rodriguez-Antron F, Langevin B, Kelly RC, Rodriguez-Hornedo N. Solution mediated phase transformation of anhydrous to dihydrate carbamazepine and the effect of lattice disorder. *Int J Pharm* 2002; 246: 121–134.
26 Newman A, Zografi G. Perspectives on the wetting of solids in pharmaceutical systems. *Pharm Res* 2023; 24: 1–20.
27 Zografi G, Tam SS. Wettability of pharmaceutical solids: estimates of solid surface polarity. *J Pharm Sci* 1976; 65: 1145–1149.
28 Williams DR. Particle engineering in pharmaceutical solids processing: surface energy considerations. *Curr Pharm Des* 2015; 21: 2677–2694.

29 Shuttleworth R. The surface tension of solids. *Proc Phys Soc Sec A* 1950; 63: 444–456.
30 Marmur A. Equilibrium contact angles: theory and measurements. *Colloids Interfaces A, Physiochem Eng Aspects* 1996; 116: 55–61.
31 Wenzel RN. Resistance of solid surfaces in wetting by water. *Trans Faraday Soc* 1936; 40: 546–551.
32 Cassie ABD, Baxter S. Wettability of porous surfaces. *Trans Faraday Soc* 1944; 40: 546–551.
33 Marmur A, Chen W, Zografi G. Characterization of particle wettability by the measurement of floatability. *J Colloid Interf Sci* 1986; 113: 114–120.
34 Hapgood KP, Lister JD, Biggs SR, Howes T. Drop penetration into porous powder beds. *J Colloid Interf Sci* 2002; 253: 353–366.
35 Li M, Callegari G, Drazer G. Effective capillary pressure, and permeability of a granular material during imbibition in a closed column. *Colloids Surf A, Physiochem Eng Asp* 2022; 648: 129280.
36 Liu H, Cao G. Effectiveness of the Young-Laplace equation at nanoscale. *Sci Rpts* 2016; 6: 23936.
37 Pfitzner J. Poiseuille and his law. *Anesthesia* 1976; 31: 273–275.
38 Washburn EW. The dynamics of capillary flow. *Phys Rev* 1921; 17: 273–283.
39 Yang YW, Zografi G, Miller EE. Capillary flow phenomena and wettability in porous media: static characteristics. *J Colloid Interf Sci* 1988; 122: 24–34.
40 Yang YW, Zografi G, Miller EE. Capillary flow phenomena and wettability in porous media: dynamic flow studies. *J Colloid Interf* 1988; 122: 35–46.
41 Yang YW, Zografi G. Use of the Washburn-Rideal equation for studying capillary flow in porous media. *J Pharm Sci* 1986; 75: 719–721.
42 Studebaker ML, Snow CW. The influence of ultimate composition upon the wettability of carbon blacks. *J Phys Chem* 1955; 59: 973–976.

19

Chemical Stability of Solids

19.1 Introduction

In the previous chapter, we examined various factors that cause the solid-state transformations of organic molecules that lead to physical instability. We would now like to consider some of the concepts and principles associated with chemical changes in the solid state that result in chemical degradation of drugs and the possible loss of therapeutic potency, production of toxic degradation products, and changes in physical appearance, for example, discoloration. The susceptibility of organic molecules to chemical change in the presence of certain reactants under various environmental conditions is a central concern for pharmaceutical scientists during the entire drug development process, from the synthesis and isolation of the API to its inclusion in formulations, which are processed to produce the final drug product [1]. Ordinarily, molecules in the solid state, particularly the crystalline state, are less susceptible to chemical degradation than when they are in solution because of the greatly reduced molecular mobility exhibited and therefore greatly reduced accessibility of reactants to reactive sites. However, such reactions do occur in solids under practical conditions of processing and storage, making solid-state stability assessment an important, albeit more complex, component of solid dosage form development [2, 3]. In this chapter we will examine (i) the types of chemical reactions that can occur in the solid state, (ii) some general principles concerning the mechanisms and kinetics of solution- and solid-state chemical reactions, (iii) processing conditions that affect chemical stability in solids by giving rise to disorder within the solid, (iv) effects of environmental conditions such as temperature and relative humidity on solid-state instability, and (v) interactions with solid excipients that directly contribute to chemical change by acting as reactants and catalysts. Emphasis will be placed on linking the chemistry of the reaction to the physical state of the system to better develop strategies that can be used to inhibit instability and to establish appropriate storage conditions and expiration dates.

19.2 Examples of Chemical Reactivity in the Solid State

It is well established that many drugs in the solid state undergo chemical degradation by mechanisms such as (i) thermal decomposition, for example, decarboxylation or dehydration; (ii) hydrolysis, for example, ester or amide hydrolysis; (iii) oxidative reactions, for example, autooxidation due to molecular oxygen, and photolytic oxidative decomposition; and (iv) rearrangement reactions such as cyclization, isomerization, and epimerization [1, 2]. Many reactions, called "topochemical reactions," occur most efficiently within the crystalline lattice and less so, if at all, in the liquid or solution states. They are generally believed to be initiated by irradiation, for example, ultraviolet radiation, or by thermal energy, and to require < 0.42 nm between reactive centers [4]. For example, the rearrangement of methyl-*p*-dimethylaminobenzoate (MSE) to trimethylammonium benzenesulfonate, as shown in Scheme 19.1, occurs most readily in the crystalline state [5].

This can be seen by the fact that the half-life for the reaction at 81 °C, below the melting temperature of 90 °C, is about 20 minutes, whereas at 95 °C in the liquid state it is about 8 hours. Melting of the crystal and the loss of crystal lattice order apparently lead to much less reactivity because the methyl transfer occurs across alternating molecular stacking of MSE within its crystal lattice and because of the 0.354 nm distance between the two loci of transfer that appear essential for reactivity, as shown in Scheme 19.2.

The fact that the half-life of the reaction at 30 °C is approximately 8 days, compared to 20 minutes at 81 °C, indicates that elevated temperatures promote the extent and rate of such reactions, presumably because of the increased energy and molecular motions of local reactive sites within the intact crystal lattice.

Solid-State Materials in Pharmaceutical Chemistry: Properties, Characterization, and Applications, Second Edition. Stephen R. Byrn, George Zografi and Xiaoming (Sean) Chen.

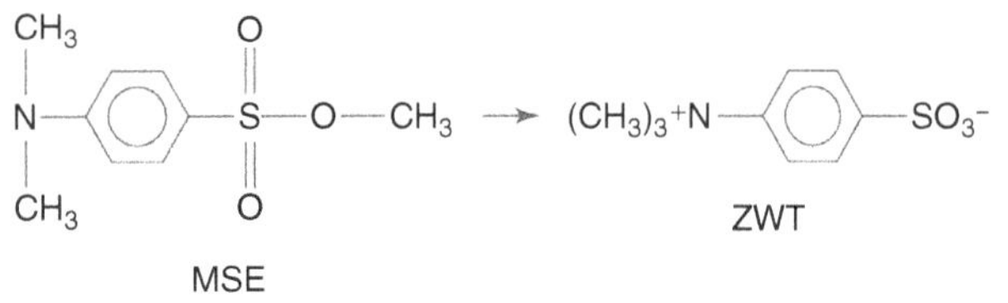

Scheme 19.1 MSE rearranges to form a zwitterionic product, trimethylammonium benzenesulfonate (ZWT).

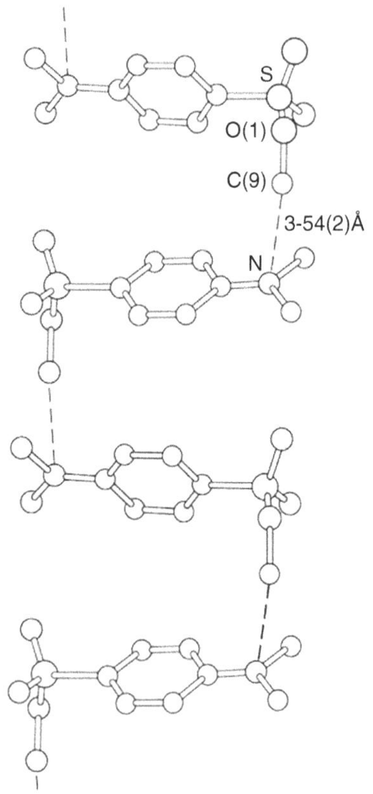

Scheme 19.2 A view of the stacking along one chain of molecules in crystals of MSE, as seen perpendicular to the (101) plane.

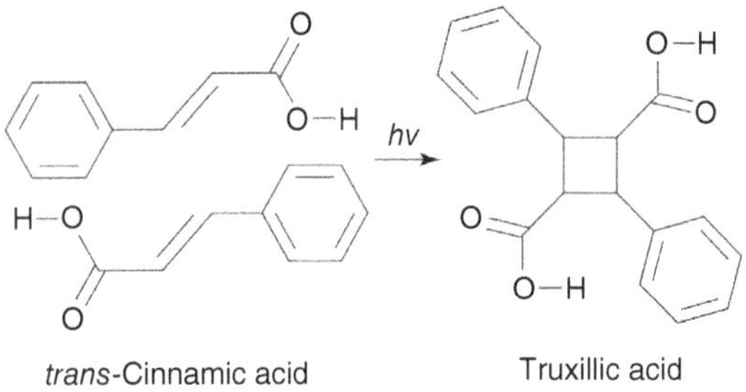

Scheme 19.3 The topochemical photocycloaddition reaction of trans-cinnamic acid.

Another interesting example of a topochemical reaction is the photochemical conversion of trans-cinnamic acid to truxillic acid, shown in Scheme 19.3, wherein electromagnetic radiation in the ultraviolet region is sufficiently energetic to trigger the covalent association of two trans-cinnamic acid molecules at the activated sites of the two double bonds [6].

Again, as shown in Scheme 19.4, we can see that the position of the pair of reactant molecules in the crystal lattice is critical for the reaction.

The two reactions shown in Schemes 19.1–19.4 are reactions that occur within the crystal lattice, only involving a single reactant molecule, and initiated by either thermal energy or electromagnetic radiation energy. It is also possible for molecules in crystalline solids to interact with molecules in the gaseous state, depending on the vapor pressure, temperature, and accessibility of the gas to the reactive center. Typically, gas molecules cannot penetrate highly crystalline solids, so such reactions are most often only observed at the surface of the solid; the greater the specific surface area, the greater the number of reactive centers. When covalent bonds are formed with the gas at the solid surface, the process is referred to as chemisorption. Chemisorption plays a particularly important role at higher temperatures in heterogeneous catalytic reactions such as the hydrogenation of unsaturated hydrocarbons in the presence of metal catalysts, as illustrated in Scheme 19.5.

Here, the hydrocarbon and hydrogen molecules are aligned at the surface by adsorption in a manner that allows dissociated chemisorbed hydrogen gas to be covalently linked across the double bond at greatly accelerated rates.

A good pharmaceutical example of the reactivity of a gas with crystalline solids is the reaction of indomethacin and ammonia, NH_3, to form the ammonium carboxylate salt, as shown in Scheme 19.6.

In this study [7], NH_3 was passed over the surface of two polymorphic forms of indomethacin, the thermodynamically most stable γ form and the less stable α form, for which crystal structures were known, and it was shown that NH_3 reacted with the α polymorph but not with the γ form. From its crystal structure it was shown that the (100) face of the α form had free carboxyl groups exposed at the surface that were able to react with NH_3, whereas the γ form, in contrast, had a crystal structure with strong hydrogen bonded carboxyl dimers and a limited number of free carboxyl groups exposed at the surface. Once the reaction at the surface of α-indomethacin was initiated, further reactivity into the bulk occurred, presumably because of the disorder introduced by the initial reaction (to be discussed more fully below).

An example of a reaction involving a gas and a solid that occurs within the crystal lattice is the photooxidation of prednisolone *tert*-butylacetate, also involving molecular oxygen, as shown in Scheme 19.7 [8].

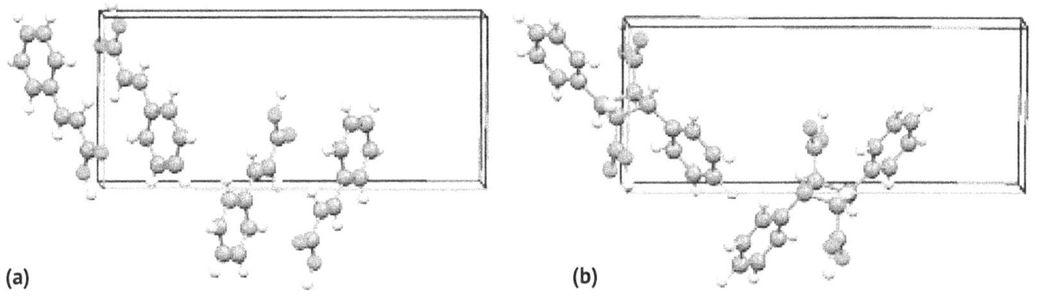

Scheme 19.4 Unit cell diagrams of cinnamic acid (a) and truxillic acid (b) for the P21/n space group.

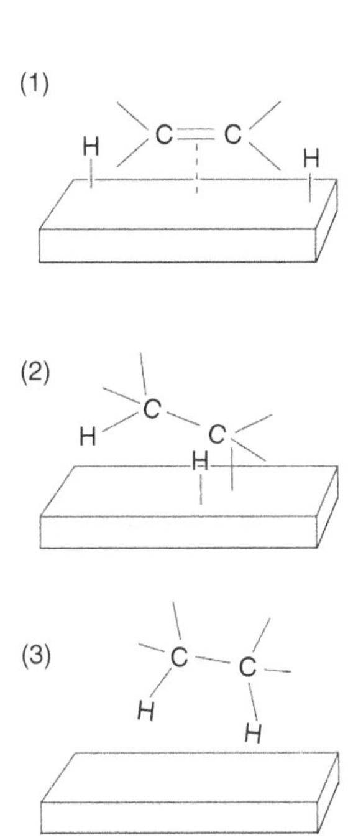

Scheme 19.5 Hydrogenation of unsaturated hydrocarbons in the presence of metal catalysts.

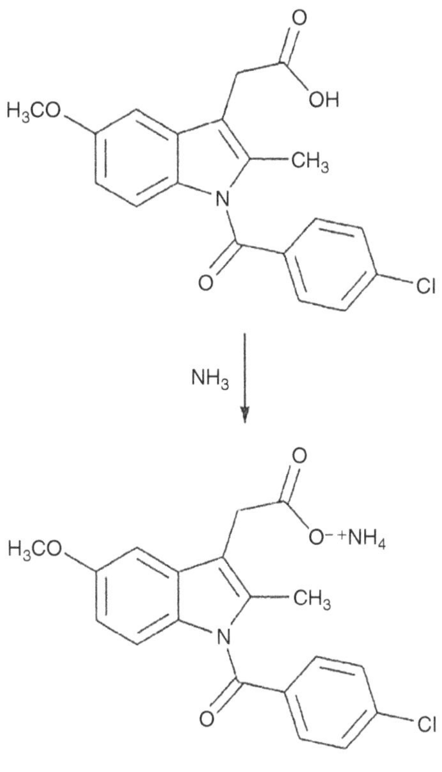

Scheme 19.6 Acid-base reaction between indomethacin and ammonia.

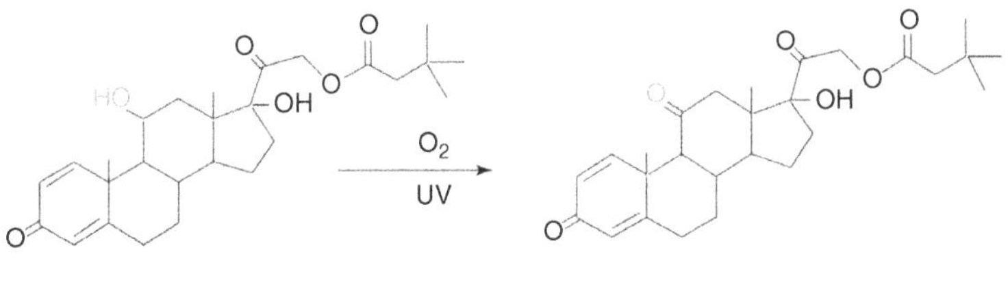

Scheme 19.7 Air oxidation of prednisolone tert-butylacetate catalyzed by UV light.

Here, it was shown that, of the five crystal forms isolated for this compound, only one form, a nonstoichiometric hydrate, underwent chemical degradation but only when the crystal was dehydrated and exposed to air and UV light. This reactive crystal form exhibited a hexagonal crystal packing arrangement containing long channels capable of forming nonstoichiometric hydrates (see Chapter 3), whereas no such channels existed for the other forms. It was suggested that reactivity in the crystalline state with molecular oxygen occurred by means of penetration into these crystal channels. The fact that such penetration of molecular oxygen to the sites of reactivity was the critical factor was shown further by the observation that only upon complete dehydration from these channels under extremely dry conditions could molecular oxygen penetrate to the site of reactivity within the hexagonal crystal. This raises the question of how the widespread reported autooxidation of many other drug molecules in the crystalline state, not involving such channels, often occurs (to be discussed).

Most reactions of pharmaceutical interest are not topochemical but, like the reaction in Scheme 19.7, can also occur in solution states by similar mechanisms. In assessing the potential for chemical reactivity of such reactions, therefore, it is possible to gain much insight into the solid-state reaction by first studying the mechanistic and kinetic details of the reaction in solution. An example of a highly studied reaction, shown in Scheme 19.8, is the hydrolysis of acetylsalicylic acid (aspirin), which readily occurs in aqueous solution and in the solid state when aspirin crystals are exposed to high relative humidity [9]. Acid–base catalysis of the degradation of aspirin, very well established mechanistically in aqueous solution, also occurs in crystalline material when acidic or basic excipients are physically mixed with aspirin.

Another example of an instability that takes place in solution and in the amorphous solid state is the cyclization of the API, quinapril hydrochloride, shown in Scheme 19.9 [10].

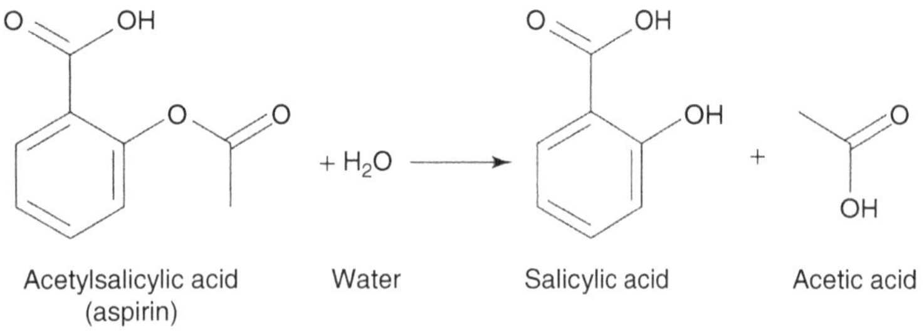

Scheme 19.8 Hydrolysis of acetylsalicylic acid (aspirin).

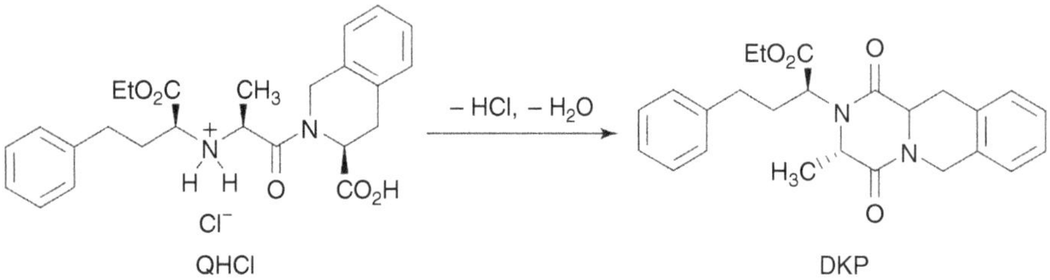

Scheme 19.9 Chemical degradation of quinapril hydrochloride by cyclization.

Here, the removal of HCl and H_2O at elevated temperatures leads to cyclization and formation of the diketopiperazine (DKP). As in the case of aspirin hydrolysis, the formation of the DKP was shown to be strongly dependent on the levels of acidity in both the solid and solution states. How other solid components of a dosage form, for example, solid acids or bases, can interact with solid API to catalyze a reaction, such as this one and that shown in Scheme 19.8, is a central question that will be addressed later in this chapter.

19.3 Some General Principles That Establish the Rate of Chemical Reactions in Solution

In general, chemical reactions that take place in solution are homogeneous, in that they occur in one phase and depend on the overall concentration of reactants in solution. Typically, therefore, the time course of the reaction in solution can very often be described by a single overall reaction rate constant determined by the analysis of potential mechanisms and mathematical modeling. Some reactions can be unimolecular, involving only the one unstable molecule undergoing, for example, thermal decomposition, rearrangement, isomerization, or cyclization. Such a reaction might be generally expressed in equation form as

$$A \rightarrow \text{products} \tag{19.1}$$

In other cases where degradation takes place because of interactions with other reactants, the reaction can be expressed as:

$$A + B \rightarrow \text{products} \tag{19.2}$$

The rate of any chemical reaction in solution, expressed as the amount of a substance undergoing chemical change per unit time, is generally proportional to the concentration of reactants and can be expressed in the form of a rate constant, k, as, for example in the case of a simple unimolecular reaction, in equation (19.1):

$$\text{Rate} = k\left(-\frac{d[A]}{dt}\right)[A] \tag{19.3}$$

where $-d[A]$ is the loss in concentration of reactant. For such a simple "first-order" reaction, it can be shown that at constant temperature the concentration of drug remaining at time, t, C_t can be expressed as,

$$C_t = C_o\exp[-kt] \tag{19.4}$$

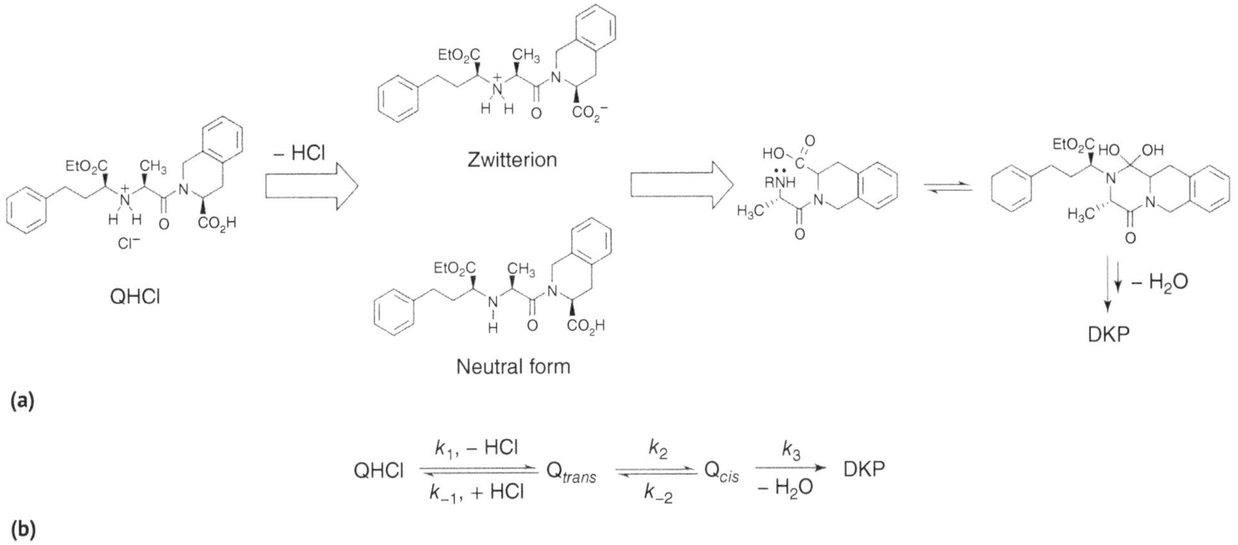

Scheme 19.10 Reaction mechanism for the cyclization of quinapril HCl.

where C_0 is the initial concentration. It is also quite common in many systems of pharmaceutical interest for the concentration of drug to be maintained constant over the time course of a reaction in solution, for example, degradation in a saturated solution in contact with excess solid, so that the system exhibits "zero-order" kinetics, where

$$C_t = C_0 - k't \tag{19.5}$$

and where k' is the zero-order rate constant. It is also possible for higher-order reactions to occur where the rate depends on the concentration of more than one component, for example, second-order reactions. Of course, it is well known that chemical reactions involving organic molecules often undergo a more complex series of reactions, generally producing intermediate states, with each intermediate reaction having a distinct rate constant under a given set of conditions. For example, the proposed mechanism for the cyclization reaction of quinapril HCl, as shown in Scheme 19.10, involves the various intermediates that precede the final cyclization product [10]. Here, it is shown that the loss of HCl, the formation of intermediate zwitterionic and neutral forms, and the cyclization with the loss of water all constitute potential rate-limiting steps. Therefore, in an evaluation of the fundamental nature of a reaction such as this one, even though the overall reaction based on loss of the original reactant may exhibit simple zero or first-order kinetics, it is useful to sort out the various reactions and to establish mechanisms and individual rate constants, as also illustrated for the solid-state reaction in the amorphous state in Scheme 19.10.

Let us now consider some of the major thermodynamic and dynamic factors that govern the extent and rates of any chemical reaction by observing a plot of energy versus reaction coordinate, as illustrated in Figure 19.1. Here we see that for the reaction between A and B molecules to occur, the molecules must have sufficient energy to reach a "transition state" and to form an activated complex $(A \cdots B)^*$ capable of then completing the transformation from reaction to products. The energy required for this process is called the activation energy, ΔE_a, the energy that must be supplied from the reactants and the surroundings to reach the transition state. In such cases, the overall reaction can be exothermic, as shown in Figure 19.1, or endothermic, indicating that the sign of the overall enthalpy of the reaction alone does not determine whether the reaction occurs. Rather, there also must be a corresponding change in entropy associated with the reaction that leads to a net loss in free energy.

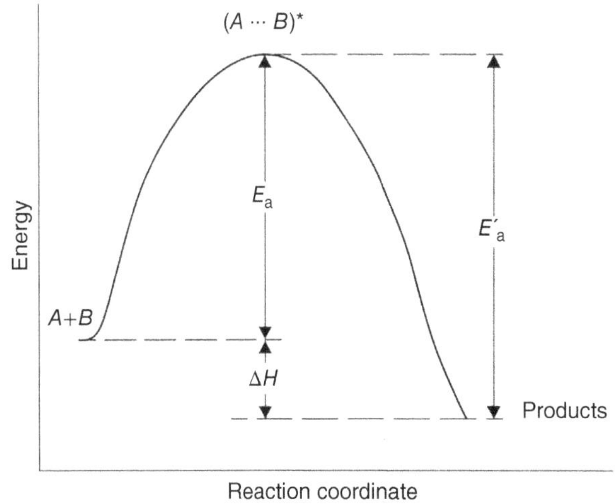

Figure 19.1 Schematic diagram of the overall energy change and energy of activation for a reaction between A and B to give products.

Of particular importance in gaining an understanding of the kinetics of any chemical reaction is the determination of the temperature dependence of the reaction rate constant, k, and the use of the Arrhenius equation:

$$k = A\ \exp\left(\frac{-\Delta E_a}{k_B T}\right) \tag{19.6}$$

where A is the preexponential frequency factor, ΔE_a is the activation energy, k_B is the Boltzmann constant, and T is temperature in Kelvin degrees. This equation has been shown to have widespread applicability to homogeneous reactions taking place in the gaseous, liquid, and solution states. In simple terms, A is the total number of collisions per second, whereas the exponential term represents the probability that a collision will lead to a reaction. The applicability of the simple Arrhenius equation to any reaction can be tested by measuring the rate constant at different temperatures and plotting the logarithm of the rate constant versus the reciprocal of temperature, where,

$$\ln k = \ln A - \frac{\Delta E_a}{k_B T} \tag{19.7}$$

and the slope of the expected linear plot leads to the value of ΔE_a and the intercept to $\ln A$. When a reaction is shown to follow simple Arrhenius kinetics, it is possible to use the equation to predict k at any other temperature and, thus, for example, to predict long-term chemical instability at room temperature from accelerated kinetic data taken at higher temperatures, assuming, of course, that the mechanism of the reaction and hence ΔE_a does not change. Although originally developed empirically, it has been given theoretical justification for solution kinetics through the "transition state theory" that includes consideration of the entropy of activation, ΔS_a^*, as well as the enthalpy of activation, ΔH_a^*, as shown in equation (19.8).

$$k = A'\ \exp\left\{\frac{\Delta S_a^*}{k_B}\right\}\exp\left\{\frac{-\Delta H_a^*}{k_B T}\right\} \tag{19.8}$$

Often, solid-state reactions appear to follow the Arrhenius equation; however, some controversy exists as to the theoretical validity of the energy of activation determined because of the relatively low mobility of the reactants and the heterogeneous distribution of energies in such systems [11]. Such contributions from the solid structure can be considered as effects on the overall entropy of the system, as in equation (19.8). When linear Arrhenius plots are obtained for solids, however, they often can be used to make empirical predictive extrapolations to other temperatures, if the mechanism does not change.

To gain some fundamental insight into the possible role of molecular mobility in solid-state chemical reactions and the meaning of the Arrhenius equation for solid systems, let us first consider situations where the liquid or solution becomes quite viscous and the reaction becomes diffusion-limited. Recall from Chapter 6 that translational and rotational diffusivity for a sphere, reflected in the value of the self-diffusion constants, D_{trans} and D_{rot}, respectively, can be expressed in terms of radius r, temperature T, and viscosity η, by the Stokes–Einstein and Debye equations, respectively, as,

$$D_{trans} = \frac{k_B T}{6\pi r\eta} \tag{19.9}$$

and,

$$D_{rot} = \frac{k_B T}{8\pi r^3\eta} \tag{19.10}$$

Thus, the lower the temperature, the larger the molecular size and the more viscous the system, the lower will be the rate of diffusion. How the structure and packing of molecules in more viscous liquids can affect chemical reactivity can be visualized through the use of free volume theory, where the interstitial or free volume, V_f, associated with a particular state, is defined as the difference between the specific volume, $V_{sp,}$ and the volume occupied by the molecule, V_o,

$$V_f = V_{sp} - V_o \tag{19.11}$$

Thus, we can conceptualize a molecule in a liquid surrounded by other molecules that form a cage around this molecule, restricting molecular motions, and where the mobility of the molecule would be expected to increase as V_f increases in relation to V_o [12]. This has been expressed quantitatively for liquids with the Doolittle equation shown as equation (19.12), where the viscosity, η, of a liquid is expressed as,

$$\eta = A\ \exp[-\frac{BV_o}{V_f}] \tag{19.12}$$

and where for low viscosity liquids, V_o/V_f, is < < < 1 and B is a constant that is near unity [13]. Conceptually, this equation indicates that for diffusive molecular motion to be initiated the value of V_f must exceed some critical free volume closely related to but greater than V_o, which we can call V^*; the greater V_f is than V^* the lower the viscosity and the greater the diffusivity. Likewise, the greater the temperature, the greater will be the free volume and diffusivity. From this discussion we now have a basis for expressing how the viscosity of a system, and the resulting reduction in molecular mobility within any form can play a critical role in affecting the rates of chemical reactivity at various temperatures by setting up an additional "physical" activational barrier, as expressed in a modified form of the simple Arrhenius equation,

$$k = A'' \exp\left\{ -\frac{V*}{V_f} - \frac{\ddot{A}E_a}{k_B T} \right\} \tag{19.13}$$

Here, we can see that if the value of V_f increasingly exceeds that of V^*, as in a liquid, the first term in equation (19.13) drops out, and the simple Arrhenius equation is applicable. However, as V_f approaches V^*, and the system becomes more viscous, the physical barrier becomes more significant, overriding the chemical activational barrier so that the corresponding reaction rate constant is greatly reduced relative to that in a solution. One example of an equation that more directly expresses equation (19.13) in terms of diffusivity is the Collins–Kimball equation [14]:

$$k_{obs} = k_{act} \frac{\alpha' D_r}{(k_{act} + \alpha' D_r)} \tag{19.14}$$

where k_{obs} is the observed rate constant, k_{act} is the rate constant obtained from the simple Arrhenius equation (equation (19.3)) for the reaction not limited by diffusion, α' is a proportionality constant, and D_r is the diffusion constant for the reactant(s). Thus, as seen by applying equation (19.9), the lower the value of the diffusion constant, that is, the greater the viscosity, the greater will be the dependence of k_{obs} on molecular mobility.

19.4 Some General Principles Governing the Rates of Solid-state Reactions

From the previous discussion and equation (19.13), it seems intuitive that a molecule located in a solid with low free volume would exhibit much less diffusivity and reactivity than one in a liquid state and, indeed, except for true solid-state reactions, as those shown in Schemes 19.1 and 19.2, chemical reactivity in solids does occur at lower rates than those in the liquid or solution states. However, many solid-state reactions of pharmaceutical interest, such as those shown in Schemes 19.8 and 19.9, do occur under normal storage conditions to an extent that is often unacceptable for pharmaceutical use, presumably because there is adequate molecular mobility on the timescale of desired shelf-life, for example, years. To examine the fundamental elements of solid-state reactivity more closely, let us first consider the proposal by Paul and Curtin that any solid-state reaction can be expected to consist of four major molecular events [15].

Step 1 requires that the reactant molecules attain a certain degree of "looseness" or molecular mobility. As we have seen in the example of the topochemical reaction shown in Schemes 19.1 and 19.2, the extent of mobility required will be very small and will depend on the reaction mechanism and the spatial aspects of the reaction within the lattice. For reactions involving more than one component, as in Scheme 19.8, molecular mobility would best approach that which occurs in a viscous solution to assure accessibility of the reactants to each other. In **Step 2**, the reactant molecules must attain a level of energy approaching the activation energy that allows the breaking of reactant covalent bonds and the making of product bonds, as discussed earlier (Figure 19.1). In **Step 3**, it is presumed that the initial amounts of product will accumulate at the site of reactivity in the form of a homogeneous solid solution in which a single-phase reaction continues to occur. Such accumulation of products at the reactive site represents a major difference with reactions in solution and, therefore, often plays a key role in determining the rate of reactivity in solids as the reaction proceeds (to be discussed). In **Step 4**, the accumulation of products may reach a point where phase separation from the solid solution creates interfaces and a heterogeneous system that can further influence the kinetics of reactivity. Consequently, therefore, as a solid-state reaction proceeds, the patterns for the kinetics of solid-state reactions would be expected to be quite complex with changing rate-determining steps and regions of apparent autocatalytic acceleration, as will be described below.

Recall that the kinetics of a reaction in solution are generally followed experimentally by measuring the change in concentration of the reactant(s) or product(s) as the reaction proceeds and plotting these changes versus time. For solid-state reactions, concentration has little meaning, and the loss of reactant(s) or gain of product(s) is generally described by the loss in mass or moles of reactant or gain in mass or moles of products. Figure 19.2 illustrates a case where the fraction of

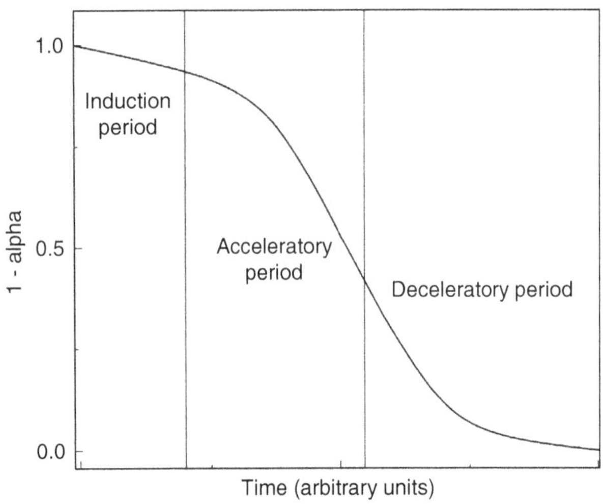

Figure 19.2 Schematic illustration of the fractional loss of reactant at various times, expressed as $1 - \alpha$, where α is the fraction of the reaction that has occurred.

reactant remaining at various times, expressed as $1 - \alpha$, where α is the fraction of the reaction that has occurred, follows the general pattern often observed with reactions from the crystalline state [16]. Initially, limited mobility and sufficient activation energy give rise to a slow and continuous reaction. If the reaction is taking place in a homogeneous phase, the rate-limiting step is usually determined by mass action, where the kinetics can be modeled as a simple zero- or first-order reaction. In some cases, in the very early period of the reaction, there may appear to be an induction period with little change, followed by an acceleration of the reaction that gives the appearance of autocatalysis. At the later stages of the reaction, as the amount of product becomes dominant, one often sees a distinct slowing down of the reaction as the amounts of product formed become dominant. An induction period followed by autocatalytic acceleration can occur for a few reasons. It is possible that the reaction is an oxidation that involves free radical chain reactions that are known to undergo autocatalysis in solution. However, solid-state reactions involving other mechanisms also exhibit such behavior, generally because of the introduction of an additional thermodynamic driving force for reactivity through phase separation of products and the appearance of an interface. In such cases, the rate-limiting process may be the formation and growth of crystal nuclei, or the rate at which the interfacial boundary of the separated product phase progresses through the solid. It is also possible for similar types of patterns to arise because the products formed are gases or liquids. For example, similar acceleration patterns can occur in solids due to the formation of a liquid product, which acts as a solvent that then facilitates the remaining reaction because of increased molecular mobility in solution. It is also possible that apparent autocatalytic behavior can occur with crystals because the formation of products into a homogeneous solid solution introduces disorder into the lattice by forming substitutional defects, thus promoting higher molecular energies and molecular mobility of reactants. An example of this last case will be discussed more fully below.

Considerable effort has been made to develop models and mathematical expressions that can be used to describe solid-state kinetic data, for example, α versus t, or $1 - \alpha$ versus t, over large time segments of any solid-state reaction, and this has been thoroughly reviewed in several publications [17, 18]. These models generally fall into four major groups where reaction rates are controlled by mass action, diffusion, nucleation of new phases, and geometric contraction of interfaces. Popular nucleation equations include the various Avrami–Erofeev equations and the Prout–Tompkins equation, whereas diffusional models can include 1-D, 2-D, and 3-D diffusivity as major determinants. Within each group, the different equations often are found to fit data over a wide portion of the reaction, however, very often more than one equation from the same group or even between different groups will be found to be applicable, making the interpretation of such fitting often no more than a curve-fitting exercise. Consequently, it is important when using various models to provide independent mechanistic information and to be clear about the practical limitations of such analyses for predicting kinetic behavior. Since in pharmaceutical studies we are primarily interested in the first 10–20% degradation of a drug to establish practical expiration dates and types and amounts of degradation products, unless practical importance can be established for interpreting later stages of the reaction, most of our attention in the remaining part of this chapter will deal with initial reaction rates that most often occur in homogeneous solid phases.

19.5 The Role of Crystal Defects in Solid-state Reactions

In Chapter 18, we presented a detailed analysis of how disorder in crystalline solids can give rise to physical instability by increasing the energy of molecules and enhancing their molecular mobility. In particular, it was shown that crystal defects that arise during crystallization or by further processing, such as milling, generally act as preferred sites for the initiation of many physical transformations, for example, melting, sublimation, adsorption, and dissolution. It follows, therefore, that enhanced solid-state chemical instability in crystalline solids would also be linked to the presence and density of crystal defects. The role of defects has attracted much attention regarding topochemical reactions as represented in Schemes 19.1–19.4. Generally, as mentioned earlier, the most critical aspect of topochemical reactions in single crystals is the proper positioning of reactive centers in the lattice at distances of less than about 0.42 nm [4]. However, molecules must contain sufficient energy and molecular motion even at this level of proximity to support reactivity. A major group of reactions where this has been shown to be important are topochemical reactions involving thermal decomposition and the nucleation and growth of a separated product phase [19]. Here, it has been shown that such reactions do not take place over the total mass of the solid but rather are initiated at localized high-energy defect sites; after the reaction occurs at a particular site, it continues to move out into the remaining part of the crystal. It has been suggested that the effects of defects will depend on the reaction mechanism and the nature of the solid, with two general classifications. In the first group, breaking of bonds during thermal decomposition occurs very locally in surface defects, and initial rates are primarily affected by the surface-to-volume ratio of the solid, reflected in crystallite size, crystal habit, and the presence of dislocations. In the second case, the reactions require charge transfer over distances longer than interatomic distances and appear to be most affected by lattice defects in the form of vacancies and impurities within the lattice. A recent study of the thermal decomposition of the crystalline explosive, HMX, octahydro-1,3,5,7-tetranitro-1,3,5,7-tetrazocine, reported molecular simulation studies that compared the rates of reaction of perfect HMX crystals as a function of temperature to those that contain various percentages of defects in the form of vacancies [20]. As shown in Figure 19.3, typical Arrhenius plots of rate constants obtained for the perfect crystal versus one with 12.5% vacancy concentration reveal greater reaction rate constants for the crystal with defects, with greater differences occurring at lower temperatures. Consequently, molecular vacancies within the crystal lattice increase reaction rate constants and lower the activation barrier. Similarly, in another recent study, it was shown that certain highly crystalline organic salts exposed to elevated relative humidity and ambient temperatures for 24 hours caused the chemical dissociation of water vapor from the atmosphere into activated oxygen species that were then chemisorbed to the solid surface [21]. Evidence was presented to show that the dissociation of water and the resulting chemisorption of the oxygen atom occurred at defect sites associated with the crystalline salt surface.

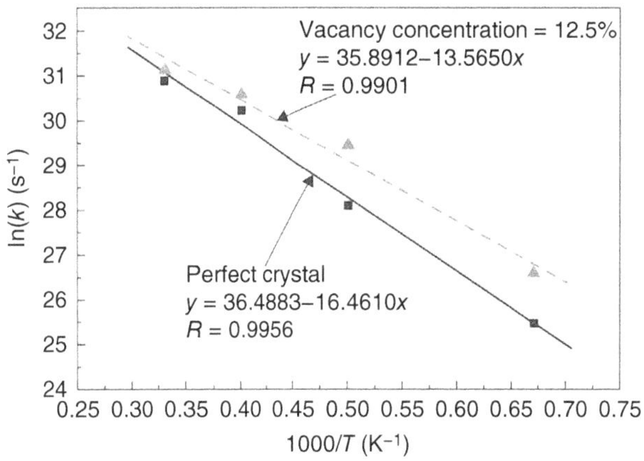

Figure 19.3 Logarithm of the rate constant versus reciprocal of temperature for the initial decomposition of HMX calculated by molecular dynamics simulations for a perfect crystal and one containing 12.5% defect vacancies. *Source:* [20]/with permission of American Chemical Society.

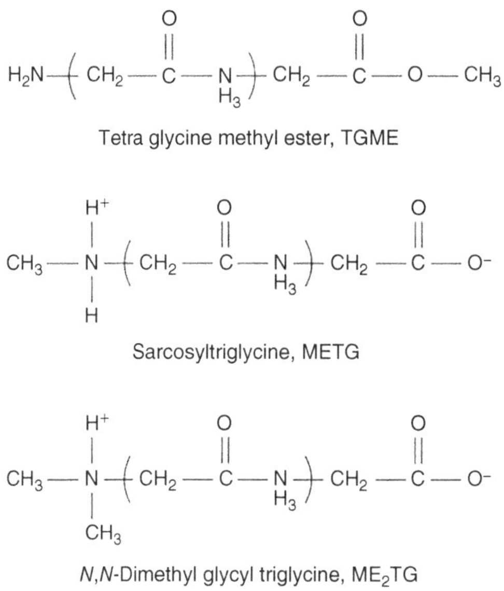

Scheme 19.11 Chemical structure of tetraglycine methyl ester (METG), sarcosyltriglycine (METG), and *N-N*-dimethyl glyceryl triglycine (ME_2TG).

The further possible importance of defects and disorder in solid-state reactions in the crystalline state can be seen in a study of the kinetics of methyl transfer in tetraglycine methyl ester (TGME) from 83 to 115°C to form two final products, sarcosyltriglycine (METG) and *N-N*-dimethyl glyceryl triglycine (ME_2TG), with the structures shown in Scheme 19.11 [22].

This reaction is characterized by the appearance of two intermediate compounds, sarcosyl triglycine methyl ester (MET GME) and tetraglycine, with structures shown in Scheme 19.12.

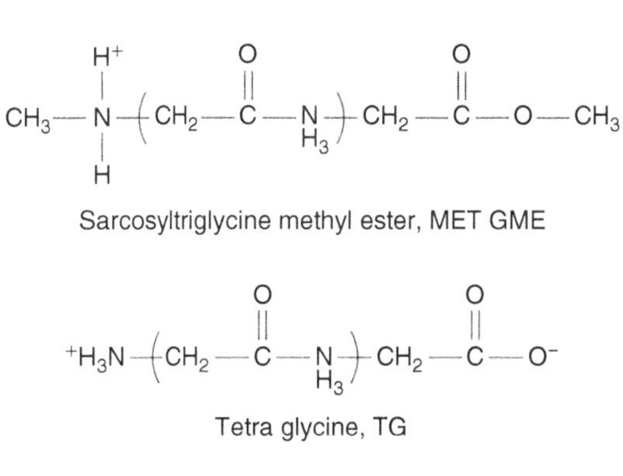

Scheme 19.12 Chemical structure of sarcosyl triglycine methyl ester (MET GME) and tetraglycine (TG).

Using HPLC analysis, it was possible to determine the change in mole fraction of intermediates and products with time, expressed as reduced time relative to the half-life of the reaction, $t/t_{0.5}$, as shown in Figure 19.4. Note that the intermediates begin to form right after time zero and then begin to disappear at $t_{0.5}$ at all temperatures.

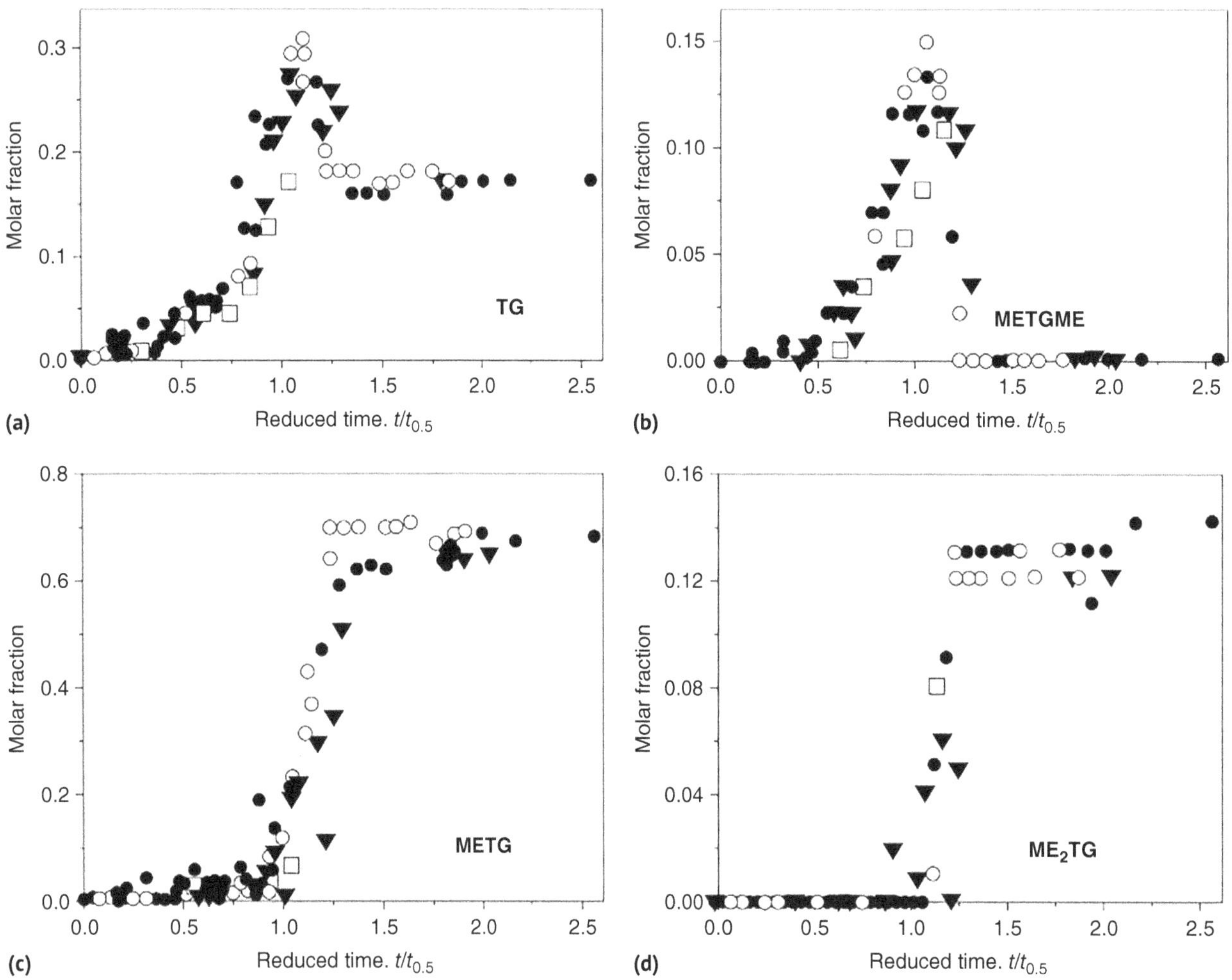

Figure 19.4 Mole fraction of the products and intermediates for the solid-state methyl transfer of TGME as a function of reduced time, $t/t_{0.5}$ at 83, 93, 100, and 115 °C. *Source:* Reproduced from [22]/with permission of American Chemical Society.

Correspondingly, the two final products begin to form at $t_{0.5}$ and then level off at distinct compositions, with METG being the major degradation product. The overall reaction in terms of the fraction of TGME remaining with time is shown in Figure 19.5. By monitoring the solid with powder X-ray analysis, it was possible to show that the first 25% of the reaction occurred in a homogeneous crystalline phase with clear evidence of an autocatalytic effect. Further analysis of kinetic data reactivity in this region provided evidence to suggest that the formation of intermediates and products in the initial stages of the reaction likely created defects and disorder that contributed to the autocatalysis. Beyond this point, X-ray analysis revealed the appearance of the crystalline product and significant additional autocatalysis, often observed with topochemical reactions in the crystalline state. Considering the reaction, therefore, to be made up of a homogeneous phase through the first 25% of the reaction and a heterophase reaction involving phase separation at later stages, overall reaction rate constants for each phase were estimated. It was assumed that a first-order process defined the kinetics in the homogeneous phase, but where the rate constant was a function as of the remaining amount of reactant (A) because of the disorder introduced, as shown in equation (19.15):

$$-\frac{dA}{dt} = \{k(A)\}A \tag{19.15}$$

It was assumed that $k(A)$ increased as the homogeneous reaction proceeded, and greater disorder was introduced. The heterogeneous part of the reaction at later times was simply modeled empirically with a widely used equation based on control by a contracting geometric front due to phase separation. Using such rate constants for the appropriate phase of the reaction, they were able to empirically account for the kinetics of this reaction in a consistent manner over a wide range of times for four temperatures, as shown in Figure 19.6.

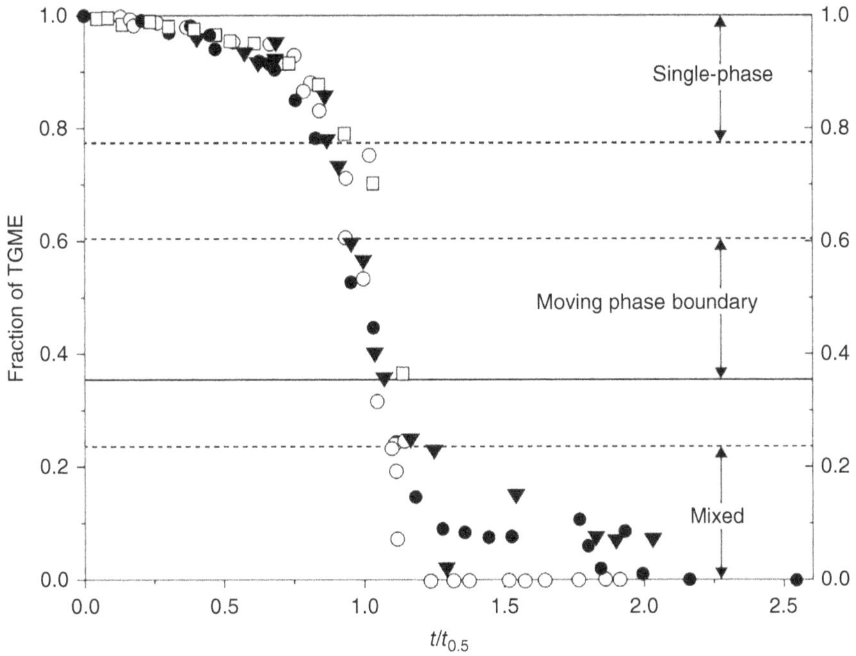

Figure 19.5 Mole fraction of tetraglycine methyl ester as a function of reduced time, $t/t_{0.5}$ at 83, 93, 100, and 115 °C. *Source:* [22]/ with permission of John Wiley & Sons.

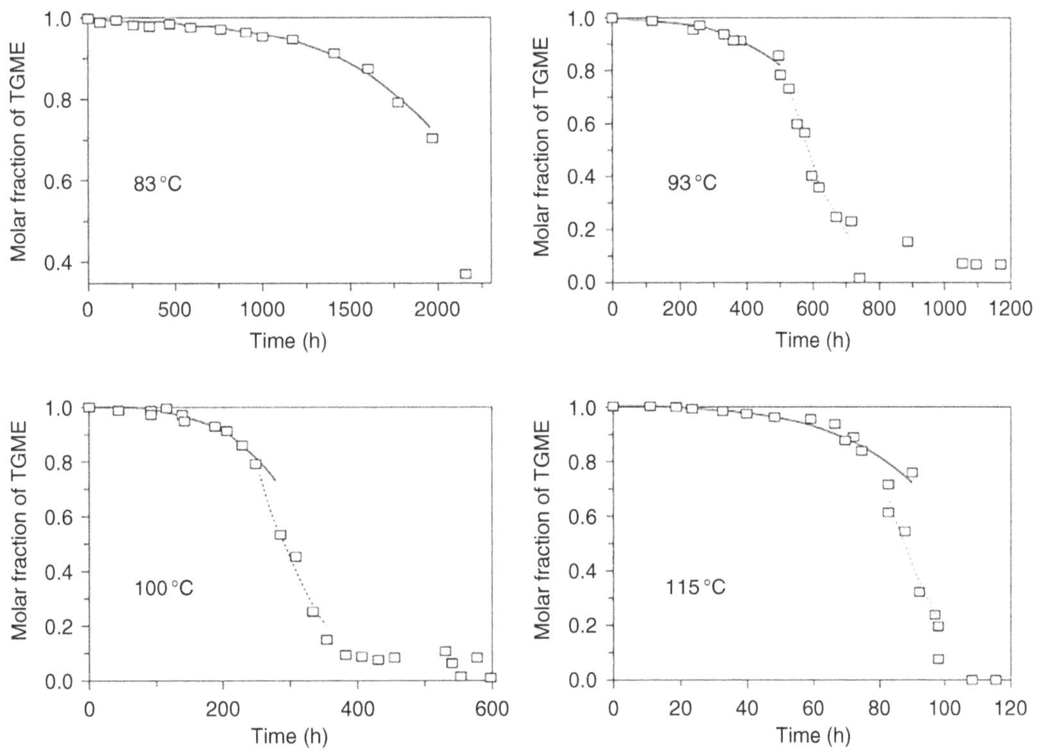

Figure 19.6 Kinetic curves for tetraglycine methyl ester reactivity at four temperatures. Symbols represent experimental points and solid lines the result of fitting data obtained in the single crystalline phase with equations based on equation (19.22). *Source:* [22]/ with permission of John Wiley & Sons.

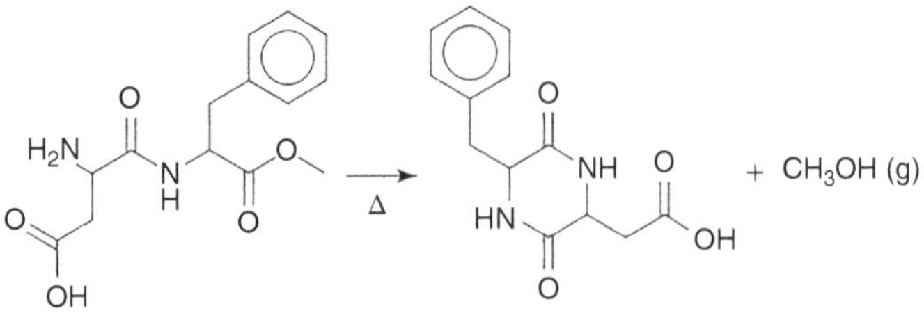

Scheme 19.13 Thermally induced aminolysis of crystalline aspartame to form diketopiperazine and methanol.

A similar example of accelerated chemical reactivity in crystals due to disorder introduced during the reaction has been reported for the thermal aminolysis of crystalline aspartame, as shown in Scheme 19.13 [23].

Here, as with the study of TGME discussed earlier, the assumption was made that a solid solution of reactants and products is produced as the reaction proceeds, leading to disorder within the crystal lattice and enhanced reactivity. In this approach, it was assumed that there were two parallel reactions taking place, one in the original crystal and the other in the disordered solid solution, produced by initial reactivity. At the initial stages, reaction 1 is dominant, whereas as the reaction proceeds reaction 2 becomes dominant. The two parallel reactions were treated kinetically, where the first reaction was defined by a single zero-order rate constant, $k^{(1)}$, and the second reaction was defined by another zero-order rate constant, $\alpha(t)k^{(2)}$, which was directly proportional to the extent of the reaction at time, t, $\alpha(t)$. Hence, the overall rate ($d\alpha/dt$) was written as

$$-\frac{d\alpha}{dt} = k^{(1)} + \alpha(t)k^{(2)} \tag{19.16}$$

Further mathematical manipulation allowed the author to successfully test the model under various conditions over a temperature range from 130 to 170°C and to show that Arrhenius behavior was being followed for both rate constants. From a pharmaceutical perspective, the key observations made in both studies were to provide strong evidence for the occurrence of continually increasing disorder, presumably due to defects, in a homogenous crystalline phase as products formed over initial phases of a reaction, with the resulting appearance of autocatalytic behavior within the crystal lattice.

19.6 Chemical Reactivity in the Amorphous Solid State

From the discussion so far, it should be clear that some level of molecular mobility plays a key role in determining the rates of all solid-state reactions. For topochemical reactions in the crystalline state, it is likely that local secondary motions associated with defects (see Chapter 6 for a discussion of secondary molecular motions) might play the critical role in view of the spatial limitations in such reactions. However, as mentioned earlier, most reactions of pharmaceutical interest in the solid state have degradative mechanisms very much like those in solution, and, therefore, diffusive motions would be expected to be the most important type of motion. So far in this discussion, we have primarily focused on the crystalline state and the important role of crystal structure and disorder in crystal defects in enhancing the rates of solid-state chemical reactivity. We would now like to explore more closely chemical reactivity in the amorphous state since its occurrence in pharmaceutical situations is widespread, most generally under the following conditions: (i) the molecule cannot be crystallized or maintained in the crystalline state, for example, quinapril HCl shown in Scheme 19.9 [10]; (ii) the amorphous form is purposefully produced to enhance pharmaceutical properties, such as API dissolution and bioavailability [24] or protein physical stability in lyophilized products [25]; and (iii) amorphous structure arises inadvertently during the processing of crystalline materials, for example, milling, drying, compaction, to produce what is generally called "process-induced disorder" [26]. Since amorphous solids represent a supercooled liquid state, completely lacking long-range order, we would expect that the formation of amorphous structure, with the accompanying increases in molecular free volume and diffusivity, would greatly increase the rates of chemical instability relative to that in the crystal, as indeed is illustrated in Figure 19.7 for the cyclization of spirapril HCl [27]. Such significantly enhanced API degradation in

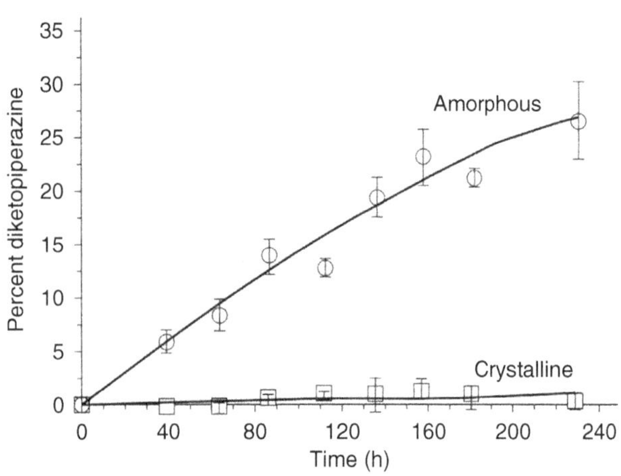

Figure 19.7 A comparison of the rate of cyclization of spirapril HCl in the crystalline and amorphous states. *Source:* Adapted from [27].

amorphous solids relative to the crystalline state has been consistently observed under a variety of conditions and, therefore, poses a major problem when a particular molecule, ordinarily susceptible to chemical degradation in solution, exists fully or partially in the amorphous state. Therefore, it is important to understand how the rates of reaction in a particular reaction are influenced by the chemical mechanisms in concert with the physical properties of the solid, for example, amorphous structure and molecular mobility.

The structure of the amorphous state, particularly the glassy state, can influence both accessibility to reactive sites and molecular mobility. An example of the importance of the molecular arrangement and spacing of molecules in crystalline solids in determining accessibility of gaseous NH_3 to react with different polymorphs of indomethacin [7] was discussed earlier (see Scheme 19.6). In the same study it was also shown that the rate of reactivity greatly increased when indomethacin was rendered completely amorphous by the quenching of the melt, presumably due to greater accessibility. From Chapter 6, we learned that such a melt-deposited glass can undergo physical ageing with time that leads to a small increase in the density of the glass, presumably due to a reduction in the amount of the less dense microstructure surrounding more dense local clusters [28]. Furthermore, it has been shown that vapor-deposited glasses generally produce highly dense glasses, significantly greater than that of melt-deposited glasses, and a marked reduction in the rate of reaction between NH_3 and the glassy indomethacin [29]. Such a reduction appears to be related to the loss of microstructure and the resulting densification, as shown in Figure 19.8.

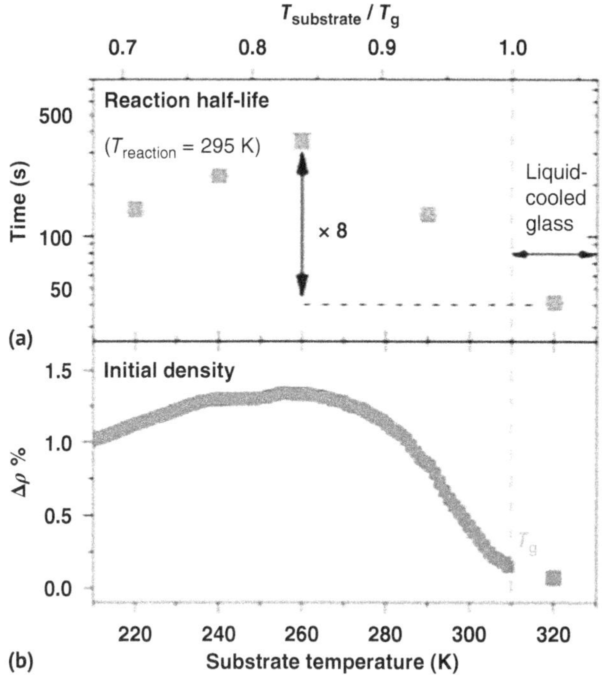

Figure 19.8 Reaction half-lives for the indomethacin glass–ammonia gas reaction (a) and indomethacin glass density (b), as a function of substrate temperature during vapor deposition. The density of the as-deposited glasses is presented relative to the LC glass. *Source:* [29]/with permission of American Chemical Society.

Recall from equation (19.14) that the effects of molecular motion on diffusion-limited chemical reactivity in liquids can be accounted for through the relevant diffusion constant D_r. Furthermore, since D_r is directly related to viscosity, η, through equations (19.9) and (19.10), and η is related to the relaxation time, τ, by equation (19.17),

$$\tau = \frac{\eta}{G_o} \tag{19.17}$$

where G_o is the limiting high frequency elastic modulus, we would expect similar correlations to exist between reactivity and τ at various temperatures and that the reaction rate would be inversely proportional to both viscosity and the relaxation time. As shown previously in Chapter 6, the temperature dependence of τ at various temperatures above and below T_g is not constant but rather depends on the temperature range considered. At temperatures greater than about 1.2 T_g in the supercooled liquid state and below T_g in the glassy state, temperature dependence generally can be described by an Arrhenius-like equation but with different dependence. At temperatures falling between T_g and 1.2 T_g, still in the supercooled liquid state, the temperature dependence of relaxation time is nonexponential in a manner that generally follows the form of equation (19.18), the Vogel, Tammen, and Fulcher (VTF) equation (see Chapter 6 for details),

$$\tau = \tau_o \exp\left\{\frac{DT_o}{T - T_o}\right\} \tag{19.18}$$

where τ_o is the relaxation time at a high temperature limit, on the order of vibrational timescales of 10^{-12} seconds, and D and T_o are constants. D has been termed the "strength parameter," a measure of the activation energy of the relaxation process, and T_o is taken to be equal to the temperature at which τ would be expected to eventually diverge to "infinity" and produce no molecular motion. Figure 19.9 illustrates the temperature dependence of τ for amorphous indomethacin, plotted as log τ versus T_g/T to normalize the temperature relative T_g (41 °C for indomethacin); the solid line represents a fit to

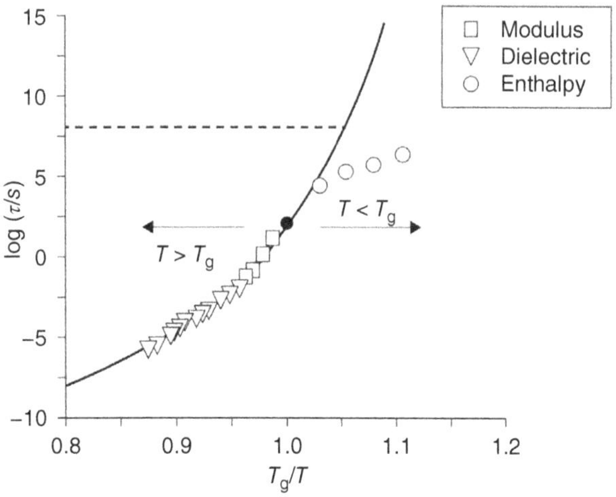

Figure 19.9 Plot of log of relaxation time for amorphous indomethacin as a function of the reduced reciprocal of temperature, T_g/T. *Source:* Adapted from [30].

the nonexponential VTF equation (see Chapter 6) [30]. Note the more than eight orders of magnitude change in τ as the temperature is reduced from 90 °C (about $1.2T_g$) to the T_g, as well as the more Arrhenius-like behavior in the glass. The horizontal dotted line represents a value for τ of 10^8 seconds or 3 years, a typically desired expiration date for solid pharmaceutical products. For example, if one extrapolates values of τ for indomethacin in the glassy state to 10^8 seconds, it appears that the storage temperature required to adequately slow down molecular mobility must be roughly 0 °C. If we assume that relaxation times on the scale of 2–3 years are required to inhibit reactivity, we can use a rule of thumb that the storage temperature should be roughly 40–50 degrees below the glass transition temperature T_g [31]. Thus, for example, to be properly stored at 25 or 40 °C with a shelf-life of a few years, the amorphous solid should have a T_g of about 75 and 90 °C, respectively. Several studies have attempted to more quantitatively correlate relaxation time for diffusive motion at a fixed temperature to the rates of chemical reactivity in an amorphous solid [10, 32–34]. In all these cases, although rough correlations are sometimes observed, it has been difficult to carry out quantitative predictions to normal storage temperatures because each chemical reaction has a unique dependence on molecular mobility that makes it difficult to generalize the extent to which the chemical activational barrier and physical barrier are interrelated. For example, some reactions may be dependent on secondary local motions, whereas others would depend on diffusive motions.

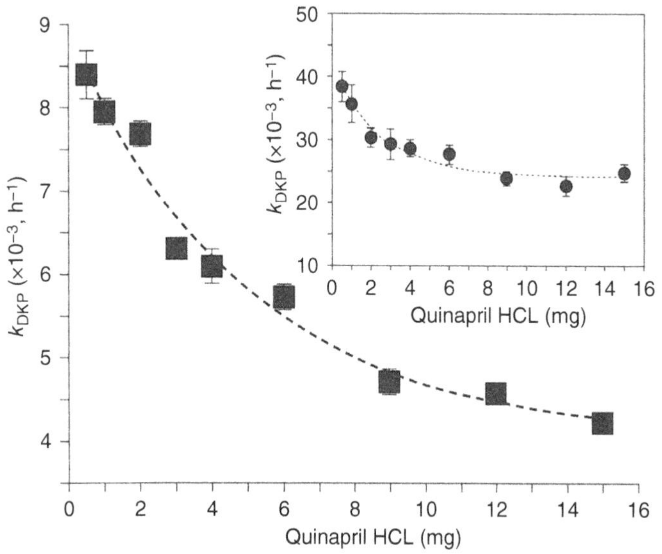

Figure 19.10 Plot of the effect of sample mass on the cyclization reaction rate constant of amorphous quinapril HCl at 80 °C. The inset shows the results of the same experiment under vacuum. *Source:* [10]/ with permission of Elsevier.

As an example of this complexity, we can consider a study of the solid-state cyclization of quinapril HCl in the dry amorphous state, as depicted in Scheme 19.10 [10]. In Scheme 19.10, we can see that in the initial stages of the reaction HCl vapor must be removed from the solid before cyclization to the DKP can proceed. It is also shown in Scheme 19.10 that this is then followed by a transformation of the more stable *trans* configuration of the intermediate product to the *cis* configuration, which allows cyclization to proceed further. Finally, in the last stages of the reaction, H_2O must be removed to complete the cyclization. In each case, a certain level of molecular mobility is required, with one of the steps likely to be rate-determining. In the case of removing HCl and H_2O in the vapor state, diffusion of the vapor through the solid sample would be critical, whereas the *trans–cis* isomerization step might only require more local intramolecular motions. To study this in more detail, kinetic studies of amorphous quinapril HCl were carried out at 80 °C, which is 10 °C below the T_g of dry quinapril HCl, where relaxation times would be expected to be on the timescale of minutes. The reaction was shown to follow simple first-order kinetics that allowed the determination of a single overall rate constant, k. Ordinarily, the first-order rate constant in the amorphous solid should not be dependent on the amount of reactant present, but, as seen in Figure 19.10, a plot of k versus the mass of amorphous quinapril HCl, the rate in a closed container greatly decreased as the sample size was increased. It was proposed that this behavior likely arises from differences in the rate of removal of HCl vapor by diffusion through the solid in proportion to sample size because the accumulation of HCl in the solid prevents further reactivity. As also seen in Figure 19.10, more rapid degradation rates occurred when the solid was exposed to a vacuum during the reaction, presumably because the removal of HCl was accelerated. Further studies were carried out with one sample weight comparing results in an open atmosphere with those obtained in one closed to the atmosphere; it was shown that the cyclization reaction in the closed system initially reacted at about the same rate as that in the open system, but at later stages it slowed down considerably, presumably due to the accumulation of HCl. Thus, it appears reasonable to suggest that the removal of HCl by diffusion through the solid mass appears to be a major rate-determining factor, further supporting the need to understand the mechanistic relationship between the physical and chemical properties of the molecule in the amorphous state. The important point here is that the specific mode of molecular mobility in any number of steps in a solid-state reaction involving the amorphous state, as well as the timescale for such motion, should be established mechanistically to describe such reaction rates more quantitatively and to make stability predictions.

19.7 Chemical Reactivity and Processed-induced Disorder

To this point, we have reviewed some important principles, with examples, regarding chemical reactivity in crystalline and amorphous solids, and have shown that disorder in crystals in the form of defects and amorphous structure is likely the major physical property of a solid that allows significant chemical degradation to occur from a pharmaceutical perspective. Although amorphous forms of an API are often purposefully used, in this section, we want to examine situations where the unexpected introduction of disorder into crystals, particularly in the form of a partially amorphous structure, leads to accelerated chemical reactivity. To set the stage for the role of process-induced disorder, as described earlier and in Chapter 18, let us begin by reviewing what happens in a crystal as any reaction is initiated and products begin to form. According to Paul and Curtin [15], the initial formation of products produces a solid solution of reactants and products, most likely at defect sites, and the disorder introduced favors greater reactivity in the crystal. That such behavior also often can lead to the amorphous structure is seen in the example of the thermal decomposition of indomethacin in its crystalline and amorphous states, as shown in Figure 19.11 [35]. These reactions were carried out at 145 °C, which is below the melting temperature of crystalline γ indomethacin (165 °C) but well above the T_g of amorphous indomethacin at 41 °C. From a strictly chemical perspective, of course, we would expect greater reactivity for both the crystalline and amorphous solids at 145 °C than would occur at room temperature, depending on the chemical activation energy for the reaction (see Figure 19.1). However, since the amorphous form is at a temperature that is roughly 100 °C above T_g, and, therefore, in the supercooled liquid state with relaxation times on the order of seconds, it is not surprising that the rate of thermal decomposition is much greater in the amorphous solid, starting immediately and then approaching a rate that might be expected to occur in the liquid state. On the other hand, the reaction in the crystal is hardly perceptible over the initial phase of the reaction, as might be expected. Note, however, that after what can be thought of as an induction period, the reaction in the crystal proceeds in a manner approaching rates comparable to that in the amorphous state. Since this reaction does not involve a free-radical chain reaction and no crystalline product forms were reported, we can conclude that the autocatalysis for the crystal likely arose because of the formation of an amorphous structure within the crystal due to the significant accumulation of reaction products. Thus, we can suggest that the initial defect sites increased in size and density as the reaction proceeded until a significant amorphous structure was produced.

It would be reasonable to extend this thinking to include the possible creation of such disorder by various manufacturing processes that can disrupt the crystal, for example, mechanical stresses or removal of organic solvent or water from crystal solvates or hydrates (see Chapter 18). It is also possible that small amounts of crystalline solids undergoing a process such as wet granulation or aqueous film coating could dissolve and form an amorphous phase upon rapid drying of the system (see Chapters 6 and 18). That milling can increase solid-state instability of crystalline powders in pharmaceutical solids was first shown by Huttenrauch in 1985 [36], where samples of crystalline ergocalciferol, a vitamin D analog, were subjected to milling for various time periods and then analyzed for oxidative decomposition with time. Considering the effects of milling on particle size reduction and hence on an increase in surface area available to oxygen, it was determined that disorder, suggested to be in the form of defects, had been introduced and was the source of increased reactivity with milling

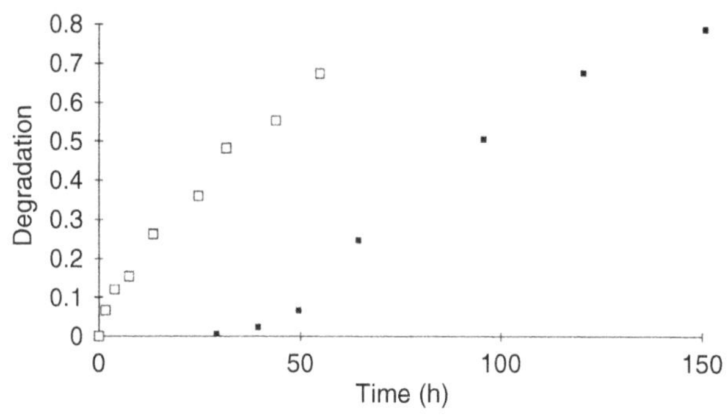

Figure 19.11 Plots of thermal degradation of indomethacin at 145 °C in the crystalline (lower plot) and amorphous (upper plot) states. *Source:* Adapted from [35].

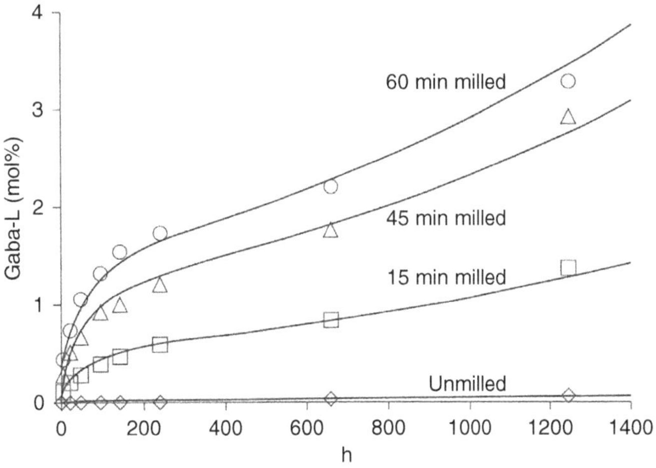

Figure 19.12 Effect of milling duration on the degradation of gabapentin at 50 °C and 11% relative humidity. *Source:* [37]/with permission of Elsevier.

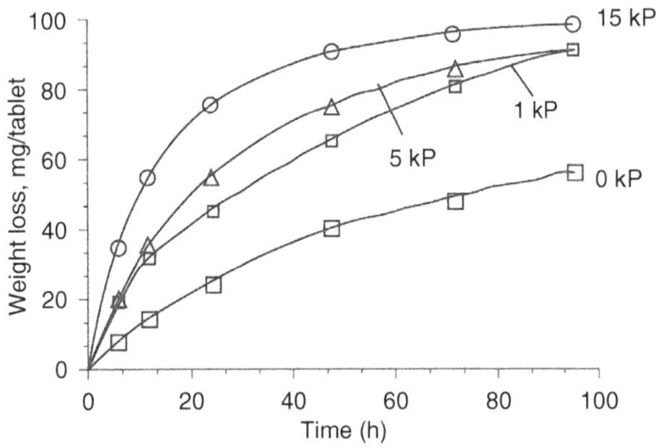

Figure 19.13 The kinetics of reaction between sodium bicarbonate and tartaric acid at 70 °C in tablets compressed at different pressures. *Source:* [38]/with permission of Elsevier.

time. In another study [37], as shown in Figure 19.12, the chemical degradation of crystalline gabapentin systematically increased as samples were milled for periods up to 60 minutes. Similarly, Figure 19.13 illustrates the rate of thermal decomposition of crystalline sodium bicarbonate in the presence of tartaric acid after the powder had been compacted into tablets under various compressional pressures [38]; the greater the pressure the greater the rate of decomposition. The significant increase in the rate and extent of reaction with pressure can be attributed to mechanical stresses that produced significant disorder within the crystals.

In another study, Shalaev et al. [39], studied the methyl transfer of TGME (see Schemes 11 and 12), for a sample of TGME, made noncrystalline by lyophilization and for crystalline samples milled for two different time periods and compared the results with those observed earlier for the unprocessed crystalline material (see Figure 19.6). As shown in Figure 19.14, the lyophilized sample underwent a much greater rate of reactivity than that for the crystalline sample, whereas the kinetics of the two milled samples seemed to first follow that of the amorphous solid and then slow down until it followed the kinetics of the crystalline sample, as shown in Figure 19.6. The solid lines associated with the milled samples were calculated by assuming a rate determined from a combination of rate constants for the purely amorphous and crystalline solids. Thus, it appears that the TGME molecules initially underwent rapid reactivity within the amorphous portion of the milled samples, followed later by the expected autocatalytic behavior in the crystal as discussed earlier.

Earlier in this chapter, when discussing the oxidative reaction in Scheme 19.7, a question was raised concerning the ability of oxygen to penetrate a crystalline solid to initiate solid-state autooxidation. In the case of the specific reaction shown in Scheme 19.7, penetration was made possible by channels created by the removal of water from the nonstoichiometric hydrate. It is certainly possible for O_2 to interact with oxidizable molecules at the crystal surface through physical adsorption and to then initiate a chain-free radical reaction that would provide a basis for the reaction to continue into the bulk phase.

As we have seen in the last two sections of this chapter, however, it is very likely that such an initiation reaction would be greatly facilitated at defect sites and, more so, if the solid contains an amorphous structure into which the O_2 can dissolve. Consider the comparison of the rate of oxidation of the crystalline and amorphous forms of the dipeptide D,L-alanine–methionine, shown in Figure 19.15 [27]. Here we see, as expected, greater reaction rates in the amorphous solid than in the crystalline form. However, despite some apparent initial induction period, the reaction in the crystal is beginning to accelerate, suggesting access of O_2 to the peptide via regions of disorder.

19.8 The Effects of Residual Water on Solid-state Chemical Reactivity

It is well recognized that, when solids susceptible to chemical reactivity are exposed to elevated relative humidity at a particular temperature, such tendencies generally will be accelerated, even when water is not one of the reactants [40–42]. As we have seen in Chapter 15, relative humidity is a measure of the concentration, or, more correctly, the activity of water vapor in the atmosphere, and the thermodynamic driving force for the interaction of water molecules with those of the solid, usually by hydrogen bonding with the polar groups of the solid. It was also shown that the amount of water taken up at a particular temperature as a function of relative humidity can be described by an equilibrium "sorption isotherm," as illustrated in Figure 15.2, with each isotherm distinctly associated with the mechanism and energetics of the water-solid interaction and the thermodynamic, dynamic, and structural properties of the solid. From a pharmaceutical perspective, the association of residual water with a solid arises in a number of practical situations: (i) the solid is exposed to an open environment containing water vapor during processing and storage; (ii) water associated with one component of a formulation in a closed container transfers via the vapor state to other drier components to maintain the appropriate overall equilibrium between relative humidity and water content for the various components of the mixture; and (iii) because of the incomplete drying of the solid during pharmaceutical processes using liquid water, such as after wet granulation, spray drying, lyophilization and aqueous film coating.

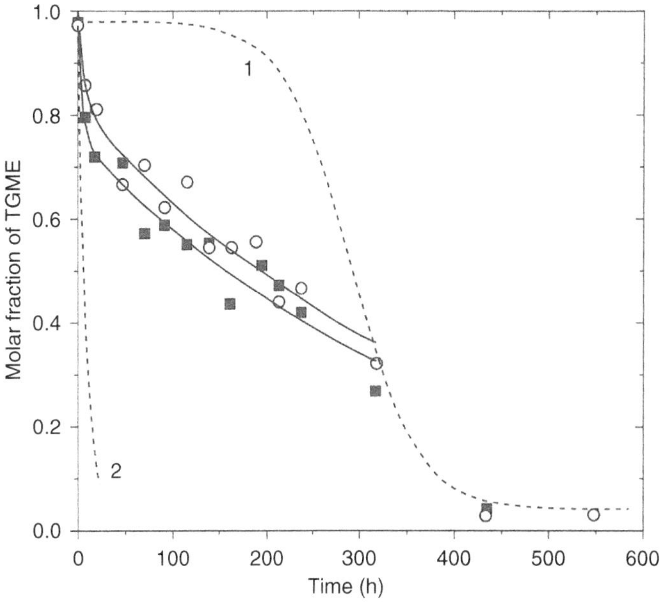

Figure 19.14 Kinetic curves for the degradation at 100 °C of (1) crystalline TGME, (2) noncrystalline lyophilized TGME, and crystalline TGME samples milled for different time periods. Points for milled samples are experimental data, and solid lines are calculated values assuming the reaction in the milled samples is taking place simultaneously in the noncrystalline and crystalline states with kinetics governed by that of curves 1 and 2. *Source*: Adapted from [40]. Reproduced with the permission of Springer.

The major mechanisms by which water is taken up by solids include (i) adsorption to the solid-vapor interface, (ii) deliquescence, and (iii) absorption into the bulk structure of the solid. Adsorption typically is the dominant mode of water uptake for highly crystalline solids, where water–solid interactions are confined to the surface of the solid. Generally, the maximum amount of water adsorbed, even at very high relative humidity, is equivalent to only three to four molecular layers or less than 0.1% w/w, an amount that generally does not appear to have any effects on the bulk properties of the solid, but significant effects on the surface properties of the solid. This leads to the general observation that highly crystalline solids are not very susceptible to chemical reactivity in the presence of adsorbed water. Since, however, most crystals contain defects at the surface, where energetics and molecular mobility are greater than in other portions of the crystal surface, adsorbed water vapor initially tends to concentrate at defects and to increase molecular mobility to an extent that might favor some reactivity due to the presence of residual water. In the case of highly water-soluble polar crystalline solids such as salts and sugars, the adsorption of water onto the crystalline solid surface similarly does not affect the bulk properties of the solid up to a certain critical relative humidity, RH_0. However, at and above this relative humidity, which is related to the water solubility of the solid and the colligative effects that occur when the solid dissolves in water, it is possible for deliquescence to occur wherein the solid dissolves in liquid water that has condensed onto the solid surface. Many drugs that exist as water-soluble salts, as well as water-soluble excipients, will have RH_0 values that fall into the range of relative humidity values encountered pharmaceutically. Once dissolved in this water, the molecules are now susceptible to solution-mediated chemical reactivity, and we can expect an abrupt increase in reactivity after the critical relative humidity is reached and exceeded (an example will be given below).

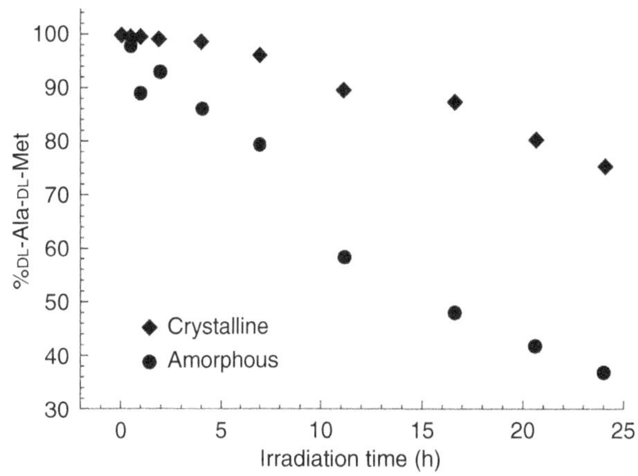

Figure 19.15 Oxidative degradation of D,L-alanine-D,L-methionine in the crystalline and amorphous states. *Source:* Adapted from [27].

Absorption of water into the bulk structure of fully or partially amorphous solids occurs by the mixing of water and solid molecules to form an amorphous solid solution. The amount of water taken up into the bulk of the solid depends on the relative humidity and the polarity of the solid. Such absorption will be offset if water-induced crystallization from the amorphous state eliminates disorder (see Chapter 18). As described in Chapters 6, 15, and 18, when molecules of miscible amorphous materials are mixed, they generally produce an amorphous solid with a single glass transition temperature, T_g, that is intermediate to those of the individual components. Since amorphous water has a T_g of 136 K, its presence within an amorphous solid can significantly reduce the T_g of the mixture, thus greatly increasing molecular mobility relative to that of the dry amorphous solid at the same temperature. An example of the effects of absorbed water on T_g is illustrated in Figure 15.16. Finally, we should also recall that water can be taken up within the crystal lattice of some anhydrous solids to form crystal hydrates where the moles of water per mole of crystal may be stoichiometric or nonstoichiometric (see Chapters 3 and 18). Generally, water, forming crystal hydrates, produces a new thermodynamic form of the solid, which would be expected to have altered physical and chemical properties, including chemical stability. In all the cases cited earlier, the extent of water–solid interaction can be reversed (desorption) by lowering the relative humidity or raising the temperature. Removal of adsorbed and absorbed water will generally reverse the effects of the original amount of water on solid-state properties, that is, reducing, for example, the adverse effects on molecular mobility or deliquescence. However, removal of water from crystal hydrates often gives rise to defects and amorphous structure within the crystal lattice that can promote molecular mobility. We also learned in Chapter 18 that water associated with crystalline solids that have undergone processing to produce defects or partially amorphous regions will concentrate in those disordered regions at amounts much greater than those predicted from the water content based on the total weight of the sample.

To characterize the possible mechanisms by which residual water can affect chemical reactivity, let us recall from earlier discussions earlier the various general steps that make up a typical solid-state reaction, as proposed by Paul and Curtin [15]: (i) sufficient molecular mobility, (ii) sufficient energy to form an activated transition state of reactant(s), (iii) the formation of a solid solution of reactants and products at initial stages, and (iv) separation of crystalline or liquid product(s) at later stages. If we assume that any reaction that can take place both in aqueous solution and in the solid state will occur more rapidly in the solution state, it follows that the appearance of liquid water through deliquescence will have a significant destabilizing effect. In such a case the liquid water can act in three possible ways, very often at the same time: (i) by dissolving the solid it increases its diffusive molecular mobility, (ii) it can be a reactant as in hydrolytic reactions, and (iii) it can affect reactivity by acting as a polar medium having dielectric properties that can support reactivity as in acid–base or free radical catalyzed reactions [38]. For one or more of these reasons, an important consideration in the selection of salts of organic molecules, used to promote aqueous solubility, is the strong potential for enhanced chemical degradation due to deliquescence. Consider in Scheme 19.14, the deliquescent hydrochloride salt of the ester, meclofenoxate (MF-HCl), with an RH_0 of about 60%, that undergoes hydrolysis to products, of which one (DMAF-HCl) is extremely deliquescent with an RH_0 of 5% [43].

It was shown in this study that the reaction at 68% RH (above the RH_0 for MF-HCl) begins almost immediately and occurs at a very rapid rate. However, at 43% RH and 50% RH (below RH_0 for MF-HCl)), there is a very long period of no reactivity, but, at some point, rapid degradation begins to occur; it occurs earlier at 50% RH than at 43% RH but proceeds at the same rate at later stages. Recognizing that one of the products of the reaction has a RH_0 of only 5%, it can be assumed that after a small amount of solid-state reactivity to occur, liquefaction of water reactivity to occur, liquefaction of water, due to the deliquescence of DMAF-HCl and dissolution of MF-HCl in this liquid water, accelerates the reaction.

We have already seen earlier and in Chapters 6, 15, and 18 that water absorbed into disordered regions of solids, that is, defects and amorphous solids, will act as a plasticizer, increasing free volume and, thus, enhancing molecular mobility relative to an anhydrous system at the same temperature. Consequently, we can expect that, in such cases, as with liquid water, absorbed water could increase reactivity without being directly involved as a reactant and that reactivity would correlate with the effects of absorbed water on T_g and relaxation time, τ. Recall the reaction depicted in Scheme 19.9, wherein quinapril HCl in the amorphous state readily undergoes cyclization to form the DKP at temperatures in the vicinity of

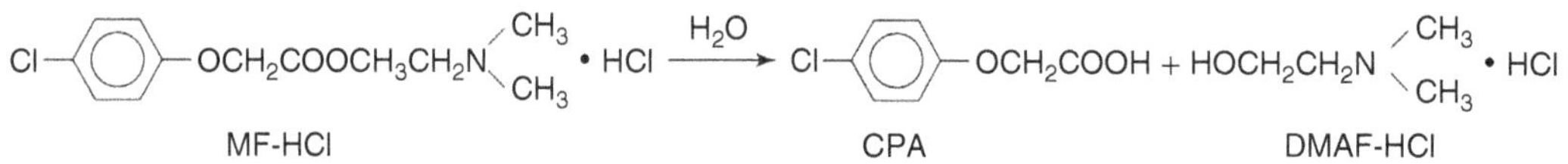

Scheme 19.14 Hydrolysis of meclofenoxate hydrochloride (MF-HCl) into p-chloro-phenoxyacetic acid (CPA) and dimethylaminoethanol hydrochloride (DMAE-HCl).

its T_g (90°C). This cyclization reaction does not involve water as a reactant, and, indeed, as seen in Scheme 19.9, water is released as a product. In Figure 19.16, rate constants for this reaction are plotted versus $T - T_g$, where T is 50°C and T_g decreases with exposure to different amounts of absorbed water to give increasing values of $T - T_g$ [27]. Note the significant acceleration of the reaction as $T - T_g$ becomes positive in the supercooled liquid state, the slower increase in reactivity at negative $T - T_g$ values in the glassy state, and the extrapolation to zero reactivity at values of $T - T_g$ roughly 40°C below 50°C. Such a value of $T - T_g$ (−40°C) is consistent with the earlier suggestion in Chapter 18 and this chapter that relaxation times at 40°–50°C below T_g increase the timescale for molecular mobility to a period of years of storage and hence should provide long-term chemical stability. In earlier discussions, it was shown that an important possible step in any solid-state reaction is the formation of a solid solution of reactants and accumulating products as the reaction proceeds, and the enhanced disorder that might contribute to an apparent autocatalytic effect. It seems reasonable to expect, therefore, that any absorption of water into these regions would also have a profound effect on the rates of reaction through plasticization.

To further illustrate the importance of disorder in solids regarding residual water-induced chemical instability, we can point to two studies of chemical reactivity where exposure to higher relative humidity caused crystallization of the amorphous material, thereby reducing molecular mobility within the solid and preventing reactivity. Recall, from Figure 19.12, that the mechanical milling of crystalline gabapentin was able to produce significant chemical instability, presumably because of the disorder introduced by the milling [37]. As shown in Figure 19.17, exposure of samples of crystalline gabapentin previously milled for 60 minutes to various relative humidities markedly increased stability, presumably because of the removal of disorder (crystallization) by the plasticizing effects of water [44]. In another example [45], it was shown that the rate of thermal degradation of sodium indomethacin in the amorphous state increased as RH was increased to a relative humidity of about 21%, but then abruptly decreased at higher RH values. It was shown independently that, indeed, crystallization of sodium indomethacin was initiated at RH values above 21%. Thus, we can conclude that a very important effect of water on any solid-state reaction will be the preferential absorption of water into disordered regions and the increase in molecular mobility produced by the plasticizing effects of the water, but before any crystallization might occur.

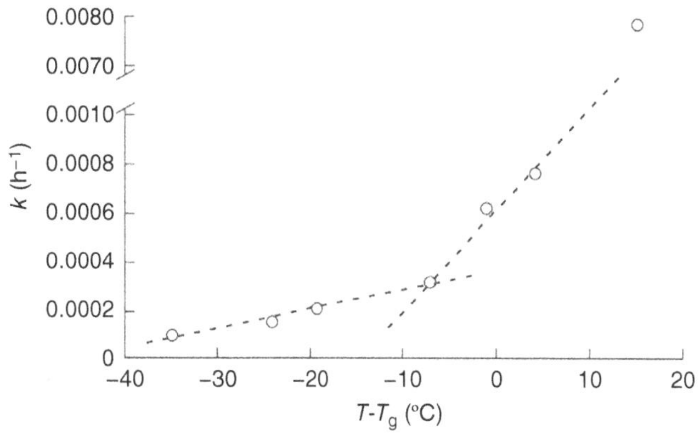

Figure 19.16 Rate constant for the cyclization of amorphous quinapril HCl at 50°C as a function of $T - T_g$, where T_g changes due to exposure to various relative humidities. *Source:* Adapted from [27].

So far, we have emphasized the role of water in promoting chemical instability by its effect on molecular mobility, either through deliquescence or by its plasticizing effects in regions of disorder introduced into a crystal through processing. It would be of interest now to consider in more detail a common situation where the crystalline form of a non-deliquescent API of limited water solubility reacts chemically with residual water to form degradation products and to explore possible relationships between the solid-state physical factors affected by the water and its role as a reactant as, for example, in hydrolytic reactions. In such situations water is consumed on a one-to-one molar basis with the API, and hence, in a closed environment with limited residual water present, the effects of water on the physical form of the solid can be complex and variable. For example, the loss of water because of reactivity might offset any effects of water on molecular mobility as a plasticizer. Another uncertainty arises when water is introduced into the solid at levels that exceed that predicted from the water adsorption isotherm, resulting in the presence of liquid water that can

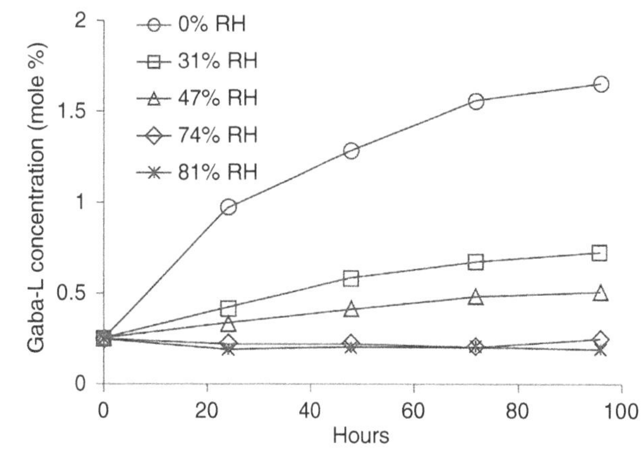

Figure 19.17 Degradation at 50°C of crystalline gabapentin milled for 60 minutes at different relative humidity. *Source:* Reproduced from [44]/with permission of Springer Nature.

give rise to some solubility of the API and, thus, to solvent-mediated chemical reactivity. To study such relationships at a more fundamental level, for example, to use mathematical modeling of kinetic data to establish mechanisms of the reaction, therefore, it is useful first to carry out experiments in an open environment of fixed relative humidity where the physical effects of residual water on the solid remain constant and then to explore the complex aspects of the system. To examine in more detail the situation where a relatively hydrophobic drug in the crystalline state undergoes solid-state hydrolysis, we can explore, as a good example, the various studies that have attempted to understand the solid-state hydrolysis of crystalline aspirin, as described in Scheme 19.8. The hydrolysis of aspirin in solution has been extensively studied over the years and generally has been shown to follow first-order kinetics with respect to aspirin concentration. Furthermore, it has been possible to characterize the reaction of aspirin mechanistically and kinetically with water under neutral pH conditions, as well as that for specific and general acid–base catalysis [46]. The first comprehensive and excellent study in the solid-state was carried out by Leeson and Mattocks [9], who followed the hydrolysis of crystalline aspirin powder in the range of 50–80 °C and 50–85% RH. Under all conditions, plotting the time course of the reaction, they obtained the typical sigmoid curve as shown in Figures 19.3 and 19.5. They developed a series of equations that assumed adsorption of water vapor by aspirin that allowed sufficient dissolution of aspirin and the initial formation of acidic products, that is, acetic and salicylic acids, to allow acid-catalyzed solution-mediated reactivity. It has also been shown that the introduction of liquid water to solid aspirin provides sufficient dissolution to promote hydrolysis [47]. The central question that could be asked is, however, whether water adsorbed at the surface in the absence of liquid water is able to sufficiently dissolve aspirin as would occur with deliquescence, given that aspirin is relatively hydrophobic and not deliquescent at ambient relative humidities. A subsequent study followed the stability of aspirin crystals at 100% RH and 25°C and showed less than 0.2% degradation for 120 days [48]. The water vapor adsorption isotherm at 25 °C, as shown in Figure 19.18, indicated less than 0.1% water adsorption at 80% RH, suggesting, as discussed previously, that a few molecular layers of water were not able to dissolve sufficient aspirin to support the reaction with water. A clue to what might be allowing small amounts of adsorbed water to support chemical reactivity comes from another study of aspirin crystals that had undergone roller compaction [49]. The authors speculated that crystal defects and dislocations introduced by roller compaction were likely the sites for the initiation of reactivity, as discussed earlier. The fact that these aspirin crystals, indeed, might contain regions of disorder, and even enough disorder to have some amorphous structure, was further suggested from the water sorption isotherm determined for the roller compacted aspirin, shown in Figure 19.18, which shows more water uptake compared to that of the unprocessed crystalline aspirin [50]. It, therefore, seems very likely that any significant chemical instability of relatively hydrophobic crystalline drugs in the presence of water vapor is likely linked with the degree of disorder present in the crystal. Although to date there has been limited success in fundamentally characterizing the kinetics of solid-state reactions in the presence of residual water in terms of chemical and physical factors, an interesting empirical equation concurrently relating the effects of temperature and relative humidity on chemical stability appears as if it may provide a way to sort this out [51]. Here, the authors determined rate constants as a function of temperature and relative humidity for the degradation of nitrazepam, which undergoes an acid–base catalyzed bimolecular reaction involving water, as shown in Scheme 19.15.

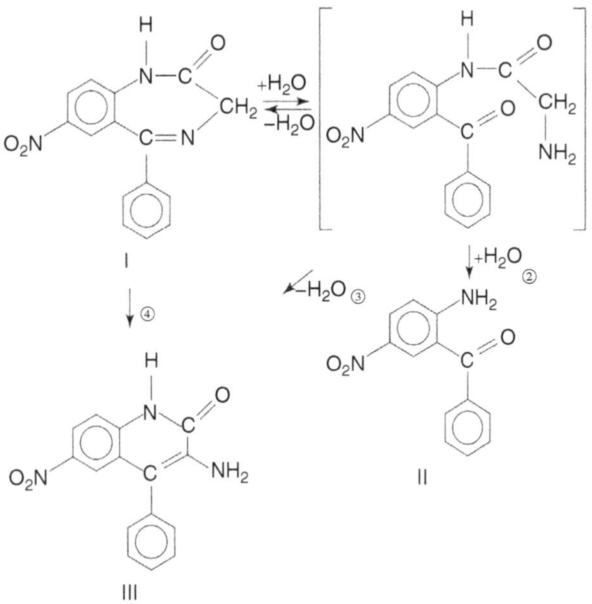

Scheme 19.15 Reaction mechanism of nitrazepam degradation.

Plotting the data as in Figure 19.19 allowed the development of correlations between the rate constant, k, temperature, T, and relative humidity, RH, by means of a three-parameter regression equation to give equation (19.19),

$$\ln k = \ln A + \left(\frac{E_a}{RT}\right) + B\%RH \tag{19.19}$$

where A and B are constants, E_a is the activation energy, and R is the gas constant. Note that the first two terms in the equation are similar to the Arrhenius equation, suggesting that this equation, similarly to equation (19.13), takes into account both the chemical and physical effects of water. Indeed, this equation has proven to be quite useful in its applicability to a range of single- and multicomponent solid systems, where the reactants are either crystalline or amorphous, and including actual manufactured products [52].

As mentioned earlier in the context of the effects of liquid water on chemical reactivity of solutions, water absorbed into disordered regions of the solid can play another important role by providing a local environment, very much like a solvent, that can favor a particular reaction. These effects might be thought of as influencing the local dielectric constant, the local acidity or basicity ("pH"), or the presence of free radicals that trigger oxidative reactions and could be central to many drug–excipient interactions that accelerate chemical reactivity, which will be discussed more fully in the following text. The significant effects of relative humidity on photolytic and autooxidation reactions appear to be very much related to the indirect effects of residual water on the local structure of disordered regions by increasing molecular mobility and producing such local "solvent" effects.

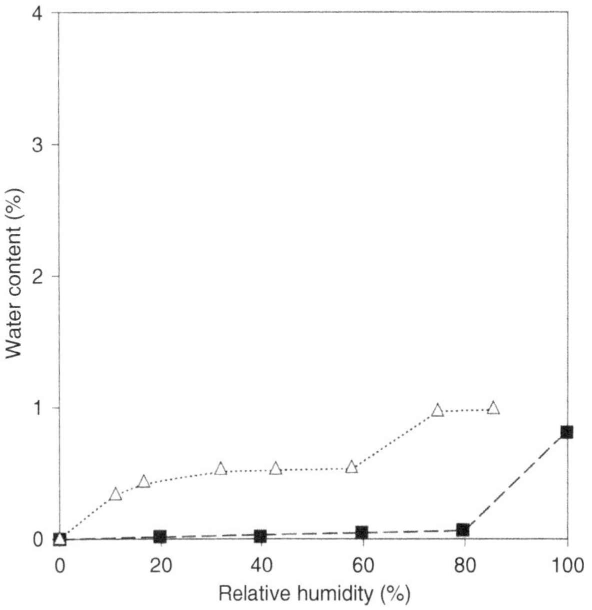

Figure 19.18 Water vapor sorption isotherms for crystalline aspirin samples at 25 °C; upper plot: roller compacted; lower plot: freshly crystallized. *Source*: Reproduced from [50]/with permission of Elsevier.

19.9 Drug–Excipient Interactions

The topic of drug–excipient interactions is a very broad one where many examples involving the effects of excipients on chemical instability of drugs have been reported [53, 54]. In this last section, we will take a broader view of such phenomena, providing some general observations and conclusions, with just a few examples, that build on what we have already learned about solid-state chemical reactivity with respect to solid-state structure and molecular mobility. Pharmaceutical excipients are materials, not having intrinsic potential as therapeutic agents, which are generally added to the formulation of drug products to enhance their manufacturability and therapeutic performance, for example, better tablet compaction properties or improved dissolution and bioavailability. They can be inorganic materials, such as salts, silica, and silicates, or organic materials, such as buffers, sugars, fatty acid derivatives, and various natural and synthetic polymers. They can act by (i) direct interaction with active ingredients, for example, acid–base or electrostatic interactions; (ii) introducing impurities that can facilitate reactivity, for example, heavy metals or peroxides; (iii) serving as catalytic surfaces for reactions, for example, acidic silica surfaces; or (iv) by simply acting as a source of residual water that can be transferred to the active ingredient. In the first three cases, excipient effects require contact between excipient and drug, so that the proportion of excipient to drug can be critical; generally, the greater the ratio of excipient to drug, the greater the excipient effects. In such cases, the surface area of contact between the drug and the excipient, and hence the relative particle size distribution of each component, is also a factor. Increasing the surface area of contact between reactants by increasing compaction pressures during tableting, for example, has been shown to facilitate reactivity. In the case, where water transfers from an excipient to the drug and facilitates degradation, the amount of water associated with the excipient and the relative affinity of the drug versus that of the excipient for such water are important [55].

An important issue regarding direct drug–excipient interactions is the extent to which process-induced disorder might be involved or even be necessary for such interactions to occur. Minimally, and based on previous discussions, for processed "crystalline solids," we would expect that the presence of defects or some amorphous structure at the surface would be required at the points of contact between drug and excipient to provide sufficient local molecular mixing and mobility. In such cases, any effects of residual water as a reactant or plasticizer, even at very low overall water content, could become significant as water preferentially concentrates in these disordered regions. That disorder can promote the role of excipients as destabilizers has been shown for many systems that have been rendered completely amorphous by lyophilization or spray drying. For example, as shown in Figure 19.20, the acid-catalyzed inversion of lyophilized sucrose to glucose and fructose in the presence of a citric acid/citrate buffer, but at water contents of less than 0.1% w/w, was shown to be significantly accelerated as the pH of the initial solution of the buffer was reduced to values from 2.40 to 1.87 [40]. Apparently, the citrate system maintains a level of acidity in the amorphous solid-state solution that is proportional to the initial solution pH.

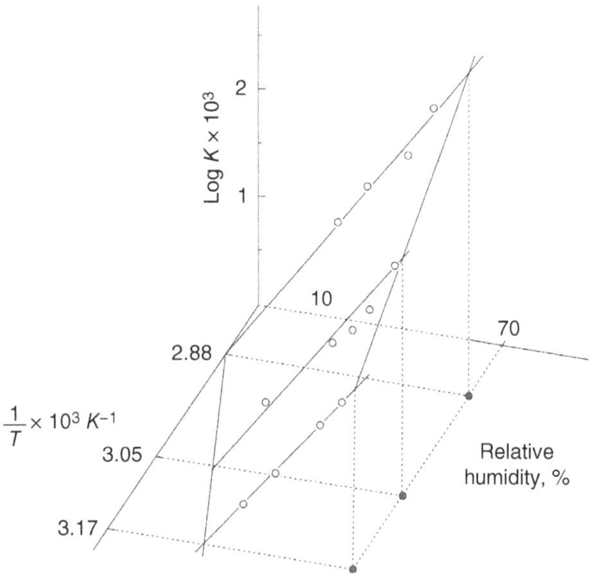

Figure 19.19 Variation of the rate constant for the thermal degradation of nitrazepam as a function of temperature and relative humidity. *Source*: Reproduced from [51]/with permission of Elsevier.

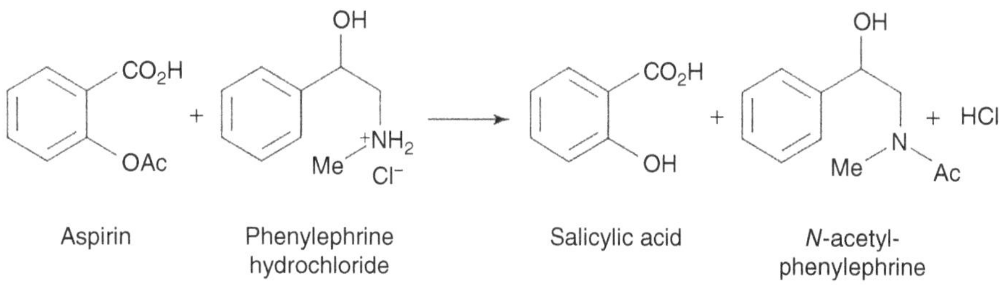

Scheme 19.16 Reaction mechanism of phenylephrine HCl with aspirin.

In many studies involving crystalline active ingredients, however, it appears that simply mixing the drug with excipients can lead to excipient effects. What is not generally reported, however, is the extent to which prior processing of the individual components and/or the amorphous nature of many excipients might have introduced sufficient disorder to promote reactivity, for example, by prior milling or drying. In this regard, we can consider an interesting study of the solid-state degradation of the drug, phenylephrine HCl, in a tablet formulation containing aspirin. As shown in Scheme 19.16, phenylephrine undergoes trans-acetylation by interacting with aspirin in the following manner [56].

Evidence of this reaction was observed at 70 °C in tablets, also containing other active ingredients and excipients, but where only phenylephrine and aspirin underwent any reactivity. Further studies were carried out with physical mixtures of phenylephrine and aspirin with the other possible active components and excipients, and it can be noted in Figure 19.21 that the rates of acetylation were significantly increased in the presence of talc (an inorganic silicate) and magnesium stearate. Certainly, in this case, there appears to be enough molecular mobility in this system for aspirin and phenylephrine to undergo the reaction and to have talc and magnesium stearate in relatively low amounts contribute to a significant increase in the rate of reaction. It should also be mentioned here that some drug–excipient interactions can act to prevent chemical degradation of the drug. For example, it has been shown that purposefully forming a miscible amorphous dispersion of antioxidants can prevent oxidation but not when the antioxidant is added as a physical mixture. Presumably, the molecular mixing of the antioxidant within the disordered drug matrix provides the necessary molecular contact for the antioxidant to preferentially undergo an oxidative free radical chain reaction and prevent oxidation of the drug. Likewise, lyophilization of protein products with buffers and disaccharides that form miscible amorphous mixtures promotes chemical and physical stability of the protein by having the buffer effective in controlling pH in the vicinity of the protein molecule and the disaccharide effective in preventing denaturation by directly interacting with the protein globular structure as the system is dried [25] (see Chapter 20). Situations that might cause the crystallization of the disaccharide, for example, generally lead to a loss of protection against denaturation and chemical degradation of the protein.

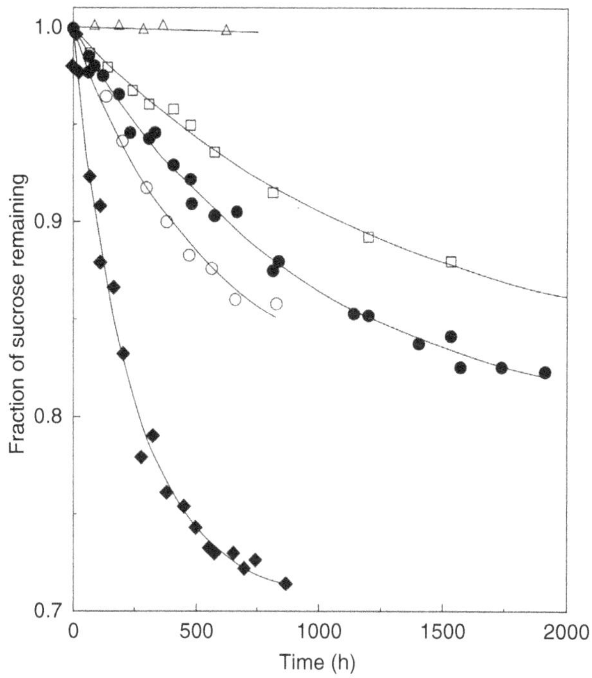

Figure 19.20 Kinetic curves for sucrose inversion at 50 °C in freeze-dried mixtures of sucrose and citric acid: (top curve) pure sucrose; (second curve from top) 10:1 sucrose: citric acid solution, pH 2.40; (third curve from top) 10:1 sucrose: citric acid, pH 2.05; (fourth curve from top) 10:1 sucrose:citric acid, pH 1.87; (bottom curve) 5:1 sucrose: citric acid, pH 1.87. *Source*: Reproduced from [40]/with permission of Springer Nature.

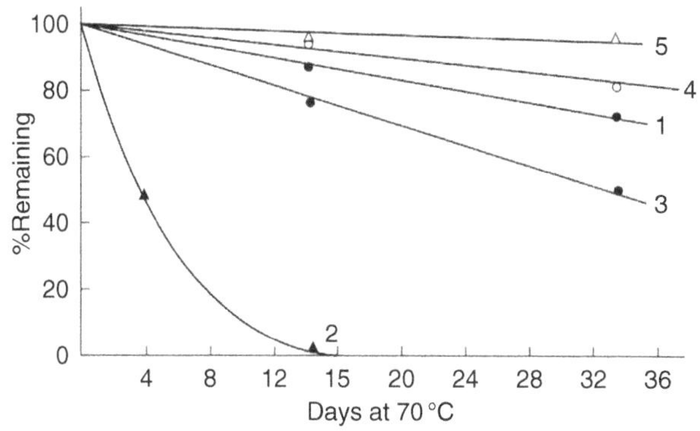

Figure 19.21 Chemical degradation of phenylephrine HCl in the presence of aspirin in powder mixtures stored at 70 °C. Key: (1) 1:20, phenylephrine HCl–aspirin; (2) 10:200:1, phenylephrine HCl–aspirin–magnesium stearate; (3) 1:20:1, phenylephrine HCl–aspirin–talc; (4) 50:150:1, phenylephrine HCl–starch–magnesium stearate; (5) 5:10:25:150:2, phenylephrine HCl–chlorpheniramine maleate–ascorbic acid–acetophenetidin–caffeine–starch. *Source*: Reproduced from [56]/with permission of Elsevier.

19.10 Summary

In this chapter, we have considered the various factors that can contribute to the initiation of solid-state chemical reactions that, in turn, can give rise to chemical drug degradation. We have seen that although every reaction under various environmental conditions and in the presence of other compounds such as excipients has a unique chemistry associated with it, some general properties associated with the form of the solid are often rate-determining. Thus, it is important to understand fully how these chemical factors are interrelated to such factors as the structural arrangement and dynamics of molecules in the solid, for example, crystal structure, degree of crystallinity or disorder, molecular mobility, and state of hydration. Such improved understanding is already beginning to provide a basis for meaningful evaluation of solid-state chemical stability including the long-term prediction of the rates of reaction from accelerated data in complex pharmaceutical formulations [2, 3].

References

1 Yoshioka S, Stella VJ. *Stability of Drugs and Dosage Forms*. New York: Kluwer Academic Publishers; 2002.
2 Byrn SR, Xu W, Newman AW. Chemical reactivity in solid-state pharmaceuticals: formulation implications. *Adv Drug Delivery Rev* 2001; 48: 115–136.

3 Waterman KC, Adami RC. Accelerated aging: Prediction of chemical stability of pharmaceuticals. *Int J Pharm* 2005; 293: 101–125.

4 MacGillivary LR, Papaefstathiou GS. *Solid-state Reactivity/Topochemistry. Encyclopedia of Supramolecular Chemistry.* New York: Marcel Dekker, Inc; 2004: 1316–1321.

5 Sukenik CN, Bonapace JA, Mandel NS, Lau PY, Wood G, Bergman RG. A kinetic and X-ray study of the solid state rearrangement of methyl p- dimethylaminobenzenesulfonate. Reaction rate enhancement due to proper orientation in a crystal. *J Am Chem Soc* 1977; 99: 851–858.

6 Nieuwendaal RC. Single crystal to single crystal topochemical photoreaction: measuring the degree of disorder in the [2+2] photodimerization of trans-cinnamic acid using single-crystal ^{13}C NMR spectroscopy. *J Phys Chem B* 2011; 115: 5785–5793.

7 Chen X, Morris KR, Griesser UJ, Byrn SR, Stowell JG. Reactivity differences of indomethacin solid forms with ammonia gas. *J Am Chem Soc* 2002; 124: 15012–15019.

8 Byrn SR, Sutton PA, Tobias B, Frye J, Main P. The crystal structure, solid-state NMR spectra, and oxygen reactivity of five crystal forms of prednisolone *tert*-butylacetate. *J Am Chem Soc* 1988; 110: 1609–1614.

9 Leeson LJ, Mattocks AM. Decomposition of aspirin in the solid state. *J Am Pharm Assoc, Sci Ed* 1958; 47: 329–333.

10 Guo Y, Byrn SR, Zografi G. Physical characteristics and chemical degradation of amorphous quinapril hydrochloride. *J Pharm Sci* 2000; 89: 128–143.

11 Galwey AK, Brown ME. Application of the Arrhenius equation to solid-state kinetics: can this be justified? *Thermochim Acta* 2002; 386: 91–98.

12 Cohen MH, Turnbull D. Molecular transport in liquids and glasses. *J Chem Phys* 1959; 31: 1164–1169.

13 Doolittle A. Studies in Newtonian flow, II, The dependence of viscosity of liquids on free space. *J Appl Phys* 1951; 22: 1471–1475.

14 Collins FC, Kimball GE. Diffusion-controlled reaction rates. *J Colloid Sci* 1949; 4: 425–437.

15 Paul IC, Curtin DY. Thermally-induced reactions in the solid state. *Acc Chem Res* 1973; 6: 217–225.

16 Shalaev EY, Zografi, G. Interrelationships between phase transformations and organic chemical reactivity in the solid state. *J Phys Org Chem* 1996; 9: 729–738.

17 Khawam A, Flanagan DR. Solid-state kinetic models: basics and mathematical fundamentals. *J Phys Chem B* 2006; 110: 17135–17328.

18 Khawam A, Flanagan DR. Basics and applications of solid-state kinetics: a pharmaceutical perspective. *J Pharm Sci* 2006; 95: 472–498.

19 Boldyrev VV. Topochemical reactions: particularities of their mechanism and outlook of their research, Part I. *J Therm Anal* 1975; 7: 685–694.

20 Zhou T-T, Huang F-L. Effects of defects on thermal decomposition of HMX *via* ReaxFF molecular dynamic simulations. *J Phys Chem B* 2011; 115: 278–287.

21 Guerrieri P, Zemlyanov D. Taylor LS. Dissociation of water on the surface of organic salts studied by X-ray photoelectron spectroscopy. *Langmuir* 2010; 26: 11998–12002.

22 Shalaev EY, Byrn SR, Zografi G. Single-phase and heterophase solid-state chemical kinetics of thermally induced methyl transfer in tetraglycine methyl ester. *Int J Chem Kinet* 1997; 29: 339–348.

23 Skwierczynski RD. Disorder, molecular mobility and solid-state kinetics: the two-environment model. *J Pharm Sci* 1999; 851234–1236.

24 Newman A, Knipp G. Assessing the performance of amorphous solid dispersions. *J Pharm Sci* 2012; 101: 1355–1377.

25 Hill JJ, Shalaev EY, Zografi G. Factors involved in the stability of native protein structure in amorphous solids in relation to levels of hydration. *J Pharm Sci* 2005; 94: 1636–1667.

26 Saleki-Gerhardt A, Ahlneck C, Zografi G. Assessment of disorder in crystalline solids. *Int J Pharm* 1994; 101: 237–247.

27 Xu W. *Investigation of Solid-state Stability of Selected Bioactive Compounds.* Ph.D. Thesis. West Lafayette, IN: Purdue University; 1997.

28 Bates S, Zografi G, Engers D, Morris K, Crowley K, Newman A. Analysis of amorphous and nanocrystalline solids from their X-ray diffraction patterns. *Pharm Res* 2006; 23: 2333–2349.

29 Qiu Y, Bieser E, Ediger MD. Dense glass packing can slow reactions with an atmospheric gas. *J Phys Chem B* 2019; 123: 10124–10130.

30 Andronis V, Zografi G. The molecular mobility of supercooled indomethacin as a function temperature and relative humidity. *Pharm Res* 1998; 15: 835–842.

31 Hancock BC, Shamblin SL, Zografi G. Molecular mobility of pharmaceutical solids below the glass transition temperature. *Pharm Res* 1995; 12: 799–806.
32 Yoshioka S, Aso Y. A quantitative assessment of the significance of molecular mobility as a determinant for the stability of lyophilized insulin formulations. *Pharm Res* 2005; 22: 1358–1364.
33 Yoshioka S, Aso Y. Correlations between molecular mobility and chemical stability during storage of amorphous pharmaceuticals. *J Pharm Sci* 2007; 96: 960–981.
34 Shamblin SL, Hancock BC, Pikal MJ. Coupling between chemical reactivity and structural relaxation in pharmaceutical glasses. *Pharm Res* 2006; 23: 2254–2268.
35 Carstensen JT, Morris T. Chemical stability of indomethacin in the solid amorphous and molten states, *J Pharm Sci* 1993; 82: 657–659.
36 Huttenrauch R. Mechanical activation of pharmaceutical systems. *Pharm Res* 1985; 2: 302–306.
37 Zong Z, Qiu J, Tnamee R, Kirsch LE. Kinetic model for solid-state degradation of gabapentin. *J Pharm Sci* 2012; 101: 2123–2133.
38 Usui F, Carstensen JT. Interactions in the solid state I: interactions of sodium bicarbonate and tartaric acid under compressed conditions. *J Pharm Sci* 1985; 74: 1293–1297.
39 Shalaev EY, Shalaeva M, Byrn, SR, Zografi G. Effects of processing on the solid-state methyl transfer of tetraglycine methyl ester. *Int J Pharm* 1997; 152: 75–88.
40 Shalaev EY, Lu Q, Shalaeva M, Zografi, G. Acid-catalyzed inversion of sucrose in the amorphous state at very low levels of residual water. *Pharm Res* 2000; 17: 366–370.
41 Shalaev EY, Zografi G. How does residual water affect the solid-state degradation of drugs in the amorphous state? *J Pharm Sci* 1996; 85: 1137–1141.
42 Ohtake S, Shalaev E. Effect of water on the chemical stability of amorphous pharmaceuticals: Small molecules. *J Pharm Sci* 2013; 102: 1139–1154.
43 Yoshioka S, Shibazaki T, Ejima A. Stability of solid dosage forms I. Hydrolysis of meclofenoxate hydrochloride in the solid state. *Chem Pharm Bull* 1982; 30: 3734–3741.
44 Zong Z, Desai SD, Kaushal AM, Barich DH, Huang H-S, Munson EJ, Suryanarayanan R, Kirsch LE. The stabilizing effect of moisture on the solid-state degradation of gabapentin. *AAPS PharmSciTech* 2011; 12: 924–931.
45 Tong P, Zografi G. Effects of water vapor absorption on the physical and chemical stability of amorphous sodium indomethacin. *AAPS PharmSciTech* 2003; 5: 1–8.
46 Edwards LJ. The hydrolysis of aspirin: A determination of the thermodynamic dissociation constant and a study of the reaction kinetics by ultra-violet spectrophotometery. *Trans Faraday Soc* 1950; 46: 723–730.
47 Carstensen JT, Li Wan Po A, The state of water in drug decomposition in the moist solid state: description and modeling. *Int J Pharm* 1992; 83: 87–94.
48 Mitrevej A, Hollenbeck RG. Influence of hydrophilic excipients on the interaction of aspirin with water. *Int J Pharm* 1983; 14: 243–250.
49 Ball MC. Solid-state hydrolysis of aspirin. *J Chem Soc, Faraday Trans* 1994; 90: 997–1001.
50 Hancock BC, Zografi G, The effects of solid-state processing on water vapor sorption by aspirin. *J Pharm Sci* 1996; 85: 246–248.
51 Genton D, Kesserling UW. Effect of temperature and relative humidity on nitrazepam stability in the solid state. *J Pharm Sci* 1977; 66: 676–680.
52 Waterman KC, Carella AJ, Gumkowski MJ, Lukuly P, MacDonald BC, Roy MC, Shamblin SL. Improved protocol and data analysis for accelerated shelf-life estimation of solid dosage forms. *Pharm Res* 2007; 24:780–790.
53 Bharate SS, Bharate SB, Bajaj AN. Interactions and incompatibilities of pharmaceutical excipients with active pharmaceutical ingredients: a comprehensive review. *J Excip Food Chem* 2010; 1: 3–26.
54 Waterman KC, Gerst P, Dai Z. A generalized relation for solid-state drug stability as a function of excipient dilution: temperature-independent behavior. *J Pharm Sci* 2012; 101: 4170–4177
55 Zografi G, Grandolfi GP, Kontny MJ, Mendenhall DW. *Int J Pharm* 1988; 42: 77–88.
56 Troup AE, Mitchner H. Degradation of phenylephrine hydrochloride in tablet formulations containing aspirin. *J Pharm Sci* 1964; 53: 375–379.

20

Solid-State Properties of Proteins

20.1 Introduction

Proteins represent a major group of macromolecules that play an important biological role as, for example, enzymes, hormones, antibodies, and receptors. They also constitute an important group of molecules that are used in the pharmaceutical field for diagnostic and therapeutic purposes, for example, insulin and growth hormone. Generally, protein drug products are administered by injection as aqueous solutions, rather than orally as solids, because of a lack of gastrointestinal stability and bioavailability. Consequently, great attention is given to the evaluation of the three-dimensional molecular structure and stability of proteins in aqueous solution. Regarding solid-state properties, there is significant interest in obtaining the crystalline forms of proteins since very detailed information about the three-dimensional structure in solution can be determined from single-crystal X-ray diffraction measurement and analysis. In the case of crystals of fully hydrated proteins (to be discussed more fully in the following text), most of the three-dimensional molecular structure in solution is retained, and, therefore, X-ray information from hydrated crystals can be directly linked to the biological activity and stability of the protein in aqueous solution. Because molecules in the crystalline state are generally more physically and chemically stable than in noncrystalline states (see Chapter 19), there has also been some interest in preparing crystalline forms of therapeutic proteins in solid products [1]. However, the preparation and use of crystalline forms on a practical scale in pharmaceutical products, except for insulin, has proven difficult and generally impractical. Rather, in situations where proteins cannot be formulated in aqueous solution because of physical or chemical instability, solid products are manufactured by freeze drying an aqueous solution, that is, lyophilization, to produce solids that can then be reconstituted as a solution just prior to administration [2]. There has also been increased interest in preparing fine particle solid forms of proteins by spray drying for use in pulmonary delivery or by the intramuscular injection of nanoparticles [3]. Most often, freeze drying and spray drying of protein solutions leads to the production of a noncrystalline solid form that can be "amorphous," and given that most molecules in the amorphous state are less stable than in the crystalline state, great attention is given to an understanding of the solid-state properties of amorphous proteins in the context of maintaining their physical and chemical stability [4]. This is particularly important in situations where the protein products are dried down to residual water levels less than that required for "full hydration," which is ordinarily required to maintain the native structure and activity of the protein [4, 5].

20.2 Solution Properties of Proteins

As with fully hydrated protein crystals, we would expect fully hydrated amorphous forms of proteins to retain most of the three-dimensional structure found in aqueous solution. Therefore, it will be useful to review some fundamental aspects of the structure and solution properties of proteins and to then see how this relates to their properties as amorphous and crystalline solids. Recall that proteins are part of a group of molecules generally known as peptides or polypeptides, which consist of two or more amino acids, where the covalent "peptide bond" between amino acids is formed as an amide group between the carboxyl and amino groups of the various amino acids, as illustrated in Figure 20.1. The length of the peptide chain can vary from a simple dipeptide to very complex proteins with molecular weights in the millions. With 20 naturally occurring amino acids available to form proteins in different combinations and permutations, as well as with different molecular weights, it is not surprising that the human body alone contains over 2 million different proteins with very

Solid-State Materials in Pharmaceutical Chemistry: Properties, Characterization, and Applications, Second Edition. Stephen R. Byrn, George Zografi and Xiaoming (Sean) Chen.

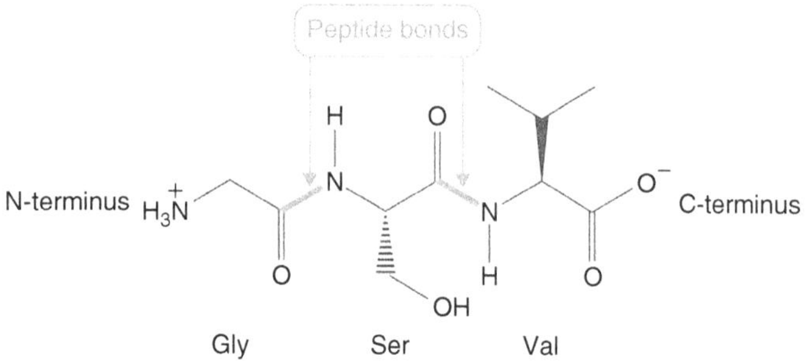

Figure 20.1 Peptide bonds between amino acid molecules.

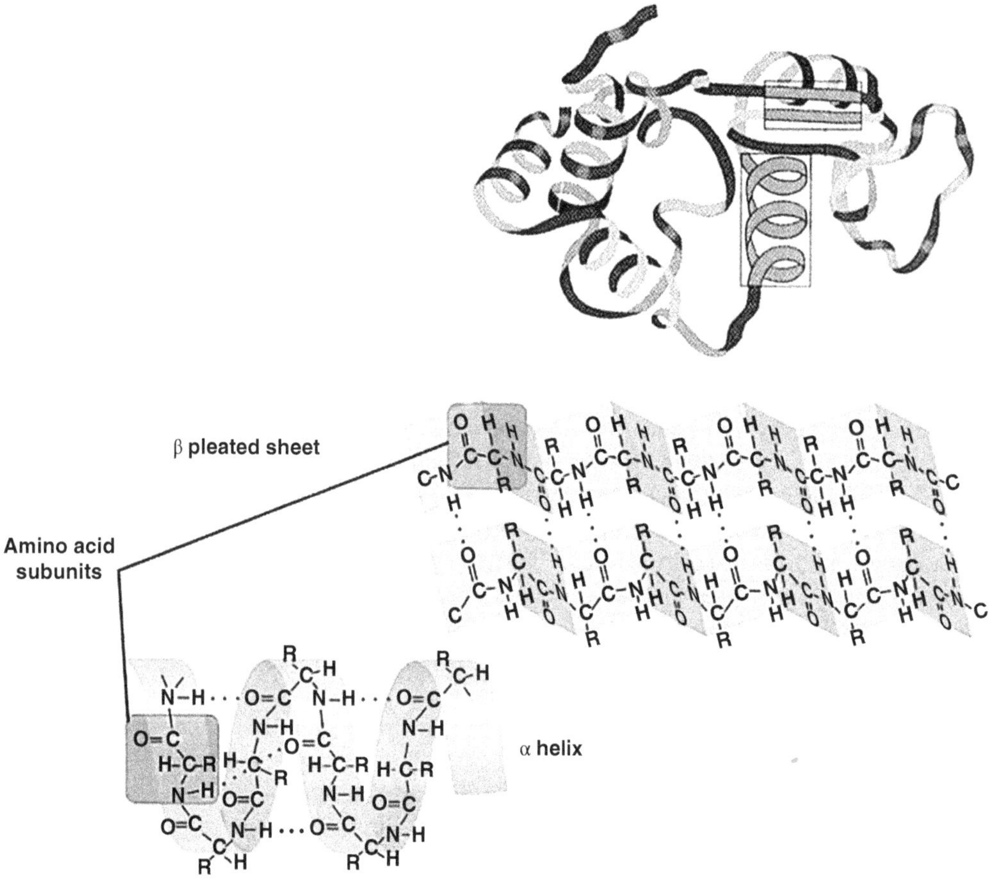

Figure 20.2 Secondary α-helix and β-pleated sheet structure of proteins.

different sequences of amino acids or "primary structures." Since various naturally occurring amino acids differ in their polarity, hydrophobicity, and net charge, proteins with different primary structures can assume very different structural features in aqueous solution and, hence, very different physical, chemical, and biological properties. Amino acids making up a particular primary structure can exhibit a "secondary structure," whereby various local segments of the protein assume certain three-dimensional forms defined by intramolecular hydrogen bonding patterns between backbone amide and carboxyl groups. The most observed secondary structures include α-helices and β-pleated sheets, as illustrated in Figure 20.2. Such structures depend on the primary sequence of any local segment, and, therefore, they play an important role in the functioning of a protein, so much so that the loss of such structure during handling, storage, and use can be a major instability [4]. Although many proteins can exist as more or less linear chains or random coils with some secondary structure, there is a strong tendency for interactions between side chains of a protein to cause the backbone chain to "fold" back on itself and to form a unique three-dimensional "tertiary structure," while retaining the local secondary structures. The shape of this tertiary structure is determined by several features of the primary structure, including hydrogen bonding,

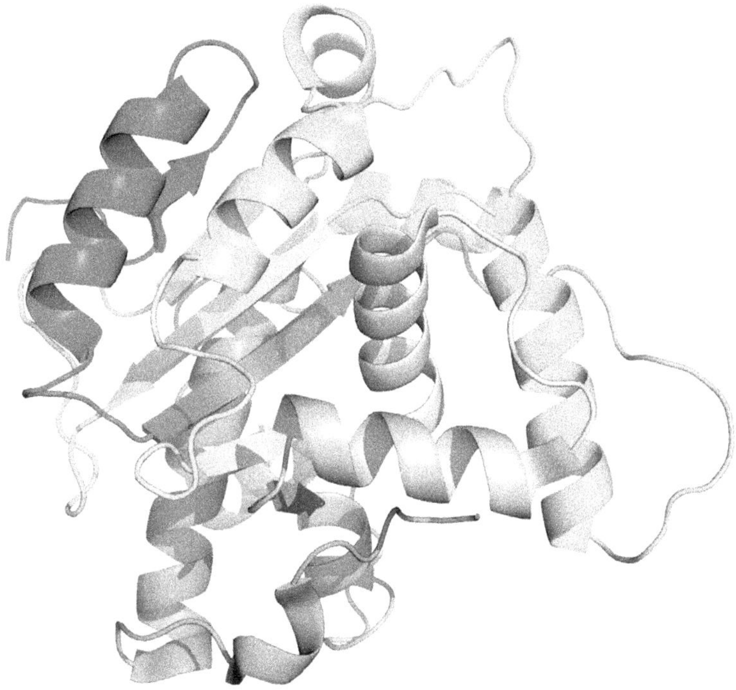

Figure 20.3 Tertiary globular structure of proteins.

salt bridges, disulfide bonds, and most importantly nonpolar hydrophobic interactions. A "globular" structure, as represented in Figure 20.3, appears to be the most common form of tertiary structure. Here, the structure is often formed and stabilized by the accumulation of hydrophobic residues within the center of the structure, with the exclusion of water, and the concentration of hydrophilic residues on the outside in contact with water. Such behavior with a single protein chain is driven by the same thermodynamic factors that cause multiple surfactant molecules to form micelles in aqueous solution. If, however, the protein is exposed to an environment devoid of water and hence more hydrophobic, such as air or other hydrophobic surfaces, there will be a strong tendency for it to "unfold" and undergo "denaturation," with resulting changes in biological activity.

The thermodynamic stability of a folded protein in its native state (N), in equilibrium with its denatured or unfolded state (D), can be expressed as an equilibrium constant, K,

$$K = \frac{[D]}{[N]} \tag{20.1}$$

where $[D]$ and $[N]$ represent the concentration of the denatured and native states, respectively. This can be further expressed in terms of the Gibbs free energy of denaturation, ΔG_μ, where

$$\Delta G_\mu = -RTlnK = \Delta H_\mu - T\Delta S_\mu \tag{20.2}$$

and where ΔH_μ and ΔS_μ represent the enthalpy and entropy changes associated with denaturation, respectively. Typically, ΔH_μ, ΔS_μ, and hence ΔG_μ can be estimated calorimetrically by measuring heat capacity changes that occur as a function of temperature. Figure 20.4 schematically presents a typical plot of ΔG_μ versus temperature, showing a parabolic curve, with a positive maximum free energy change at temperature, T_{max}, representing the minimum stability of the denatured state or the maximum stability of the native state. Typically, such free energy values for proteins are in the range of 20–60 kJ/mol, indicating that the folded state is only marginally more stable than the unfolded state at such temperatures. As also seen in Figure 20.4, the free energy of the unfolded state decreases and becomes negative at two characteristic temperatures as one raises or lowers the temperature relative to T_{max}, hence leading to denaturation. The higher temperature, T_H, is called the thermal denaturation temperature, and the lower temperature, T_C, is called the cold denaturation temperature. Consequently, there is a limited range of temperatures where the protein can exist in its native folded state. Heat denaturation occurs because heat increases the kinetic energy of the protein to break hydrogen bonds and other interactions that occur between different parts of the protein molecule, thus causing the protein to unfold. Cold denaturation is caused by the strongly temperature-dependent interaction of protein nonpolar groups with water, which, in contrast to expectations,

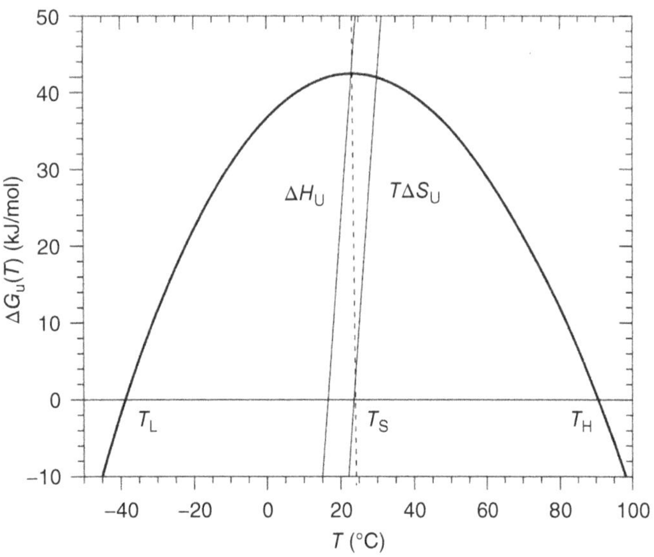

Figure 20.4 A typical protein stability free energy curve for unfolding versus temperature. *Source*: [5]/with permission of Elsevier.

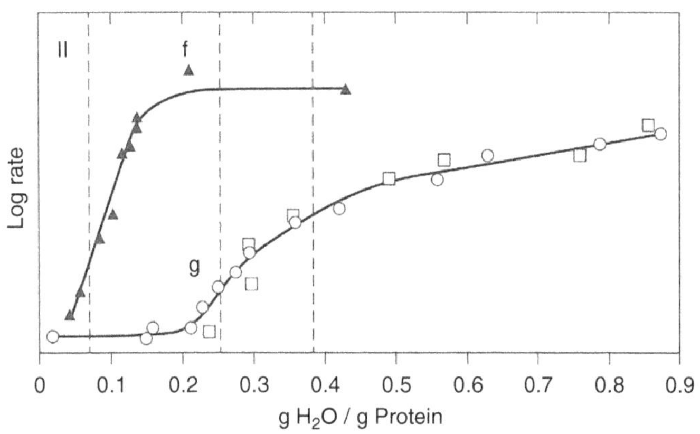

Figure 20.5 Rate of peptide hydrogen exchange (*f*) enzymatic activity/rotational relaxation of ESR probe (*g*) as a function of level of lysozyme hydration. *Source*: [7]/with permission of Elsevier.

is favorable thermodynamically; that is, the Gibbs energy of hydrophobic hydration is negative and increases in magnitude as the temperature decreases. As a result, the protein chain unfolds at a sufficiently low temperature due to the exposure of internal nonpolar groups to water. In addition, these temperatures are sensitive to solution conditions, such as pH, ionic strength, and the presence of additives such as sugars and alcohols, because of their effects on the electrostatic and hydrophobic interactions that maintain stability of the folded state. A molecular understanding of such effects is critical since tendencies for unfolding of the native protein can lead to the serious problem of protein–protein aggregation, as well as chemical instability [2, 4].

A key factor that determines the native structure (hence, biological activity and stability) of proteins in solution is the state of hydration of the protein in its folded state [6, 7]. Generally, a protein is said to be "fully hydrated" if it contains water at a concentration on the order of 40% w/w or greater. This is illustrated for the enzyme, lysozyme, in Figure 20.5, where it is shown that both enzymatic activity and rotational diffusional motion of an electron spin resonance (ESR) spectroscopy probe decrease significantly as one goes below about 40% water while more local motions involved in hydrogen exchange continue down to about 20% water. It has been shown that such changes in molecular motion and biological activity caused by dehydration are also associated with distinct changes in the secondary and tertiary structure of the protein. Such effects can be best understood by the fact that water molecules associated with proteins can exist in different states of energy. First, we can distinguish between "bound" and "unbound" water, where bound water near the protein molecule exhibits properties that are detectably different from those of bulk water, and unbound water has the properties of bulk water. One way used to distinguish these two forms of association is to observe that, whereas unbound water at levels of greater than 20–40% exhibits the normally expected freezing temperature of 273 K, bound water will exhibit distinctly higher freezing temperatures. Bound water can be further divided into water that is directly associated with the internal structure and biological activity of the protein, and water that is associated with various functional groups closer to the surface of the folded protein. Such water may be bonded to ionized groups, nonionized polar groups, and nonpolar groups with different extents of interaction energy [6, 7].

An important aspect of the solution properties of proteins that carries over to the solid state is the ability of the protein at various states of hydration to maintain a certain folded state that is stable and able to function in carrying out biological activities, for example, enzymatic activity. This suggests that the native structure of the protein must have sufficient local molecular order to retain the ability of the protein to remain folded with corresponding biological activity, while allowing sufficient local mobility of its various functional groups to carry out various biological processes. Indeed, as illustrated in Figure 20.6, and as proposed by Gregory and Lumry [8], we can think of the structure of the globular protein as containing strongly ordered and relatively immobile regions, known, respectively, as "knots" that maintain stability, and a more loosely associated "matrix" region, associated with the surface structures, that must remain sufficiently mobile to maintain biological activity. As shown in Table 20.1, globular proteins generally exhibit a variety of molecular motions that increase with temperature and eventually lead to the unfolding of the structure [9]. These range from local high-frequency harmonic motions associated with atomic vibrational fluctuations and side-chain motions to lower-frequency rigid-body anharmonic motions, as, for example, helix motions, and to larger-scale whole-molecule or segmental motions that ultimately lead to instability. Since fully hydrated amorphous and crystalline proteins tend to retain much of the structural and dynamical properties of proteins exhibited in aqueous solution, we will return to the importance of the dynamic properties of proteins after some initial discussion of some general amorphous and crystalline protein solid-state properties.

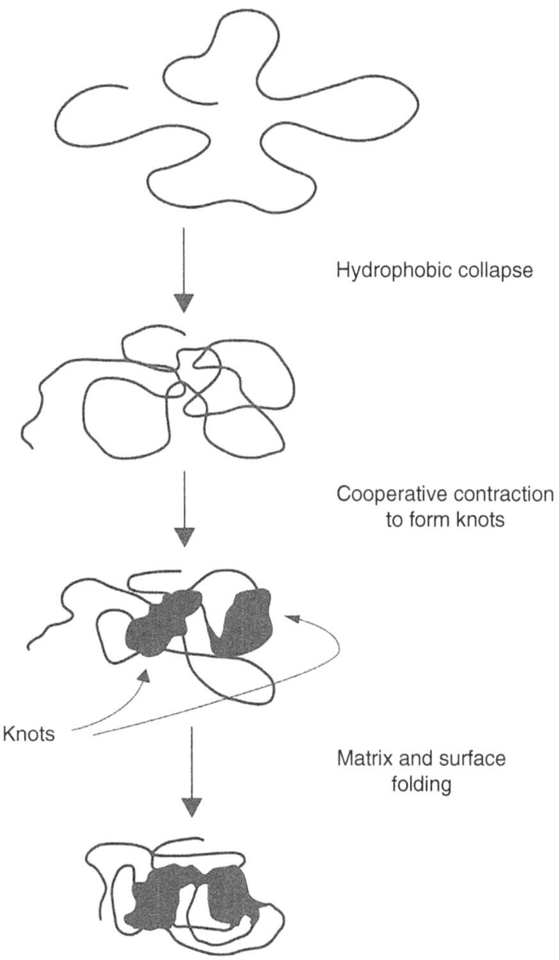

Figure 20.6 Schematic representation of the changes in protein local structure taking place during folding or unfolding. *Source*: Adapted from [8].

20.3 Amorphous Properties of Proteins

In Chapter 6, we learned that noncrystalline amorphous forms of small molecules and polymers can be produced by supercooling the liquid form below its melting temperature or by rapidly drying a solution of these molecules to form a solid. In both cases crystallization is avoided as solidification takes place and an amorphous solid that is metastable relative to the crystalline form is created. Such systems are generally characterized by a distinct temperature, the glass transition temperature, T_g, where the thermodynamic and dynamic properties of the amorphous state undergo significant change. Such properties include heat capacity and the rate of diffusive molecular rotational and translational motion. Amorphous solids below T_g are referred to as glasses, generally exhibiting viscosities greater than 10^{12} Pas and

Table 20.1 Characteristic motions in globular proteins.

Local motions (0.01 – 5 Å, 10^{-15} – 10^{-1} s)
Atomic fluctuation
Side-chain motions
Loop motions
Terminal arm motion
Rigid-body motions (1 – 10Å, 10^{-9} – 1 s)
Helix motion
Domain (hinge-bending) motions
Subunit motions
Larger-scale motions (> 5Å, 10^{-7} – 10^{4} s)
Helix-coil transition
Dissociation/association and coupled structural changes
Opening and distortional fluctuations
Folding and unfolding transition

primary α relaxation times on the order of 100 s or more (see Chapter 6). Above T_g viscosities and relaxation times rapidly decrease with increasing temperatures as the solid becomes more liquid-like. Small molecules above T_g are said to be in the "equilibrium liquid state," maintaining the equilibrium properties of the liquid, whereas polymers above T_g are said to be in a viscoelastic liquid-like "rubbery state." Glasses, formed below T_g, are in an unstable state, the structure of which depends on the manner of producing the glass and the time over which the glass is stored at a particular temperature. As discussed in more detail in Chapter 6, the value of the glass transition temperature, T_g, of a particular molecule is determined by the way in which chemical structure influences the degree of polarity, molecular size, and molecular shape. The presence of other components that can form miscible mixtures, such as water, alters the T_g in a manner dependent on the T_g of the second component and its composition (see Chapter 6). For example, as discussed in Chapter 15, water with a T_g of about 136 K significantly reduces the T_g of most organic materials with which it is miscible, the extent of change depending on the amount of water in the solid. A rough empirical rule is that each 1% of water reduces T_g by about 10 K (see Chapter 15). Typically, in pharmaceutical systems, the glass transition temperature is most conveniently measured using differential scanning calorimetry. This is done by continuously increasing the temperature from very low temperatures and observing the heat capacity change that occurs at T_g (see Figure 6.6) because of structural changes and significant increases in diffusive whole-molecule or polymer segmental molecular mobility.

As mentioned earlier, solid protein preparations are most often produced by the freeze drying or spray drying of aqueous solutions. In such cases, the solid generally forms as a noncrystalline, "amorphous" material with small percentages of residual water bound to the protein, the amount depending on the extent of drying. As shown in Figure 20.7, when such an amorphous protein sample is subjected to thermal analysis, as either a dry sample or one that is hydrated, the first scan of both samples exhibits a continuous increase in heat capacity until the heat denaturation temperature is reached [10]. In this case, there is no apparent major shift in heat capacity that could be interpreted as reflecting a glass transition, as shown in Figure 6.6. However, if the denatured protein, which loses its globular structure, is recooled to the original temperature and reheated at the same rate, a distinct glass transition temperature is now observed, as would be expected for a typical amorphous material, for example, a linear polymer or linear polypeptide. A number of studies with fully hydrated native proteins, however, have shown that very small heat capacity changes, reflecting a T_g, can be observed at around 200 K if the sample is carefully annealed [10]. Since the timescale of relaxation detected by such thermal measurement is that due to primary diffusive motions, this is a true T_g. The lack of a sharp and significant change in heat capacity for globular proteins has been attributed to the fact that a variety of local changes in molecular motion associated with the structured protein, as described in Table 20.1, occur gradually, as might be expected for a "strong" highly networked amorphous solid. On the other hand, in the denatured state, the protein acts more like a typical linear polymer, undergoing a sharp change in heat capacity at T_g, as expected for a "fragile" amorphous polymer (see Chapter 6 for a more complete discussion of fragile and strong amorphous materials). Another piece of evidence for the fact that native proteins in the amorphous state exhibit a distinct glass transition, like that seen with strong amorphous solids, can be found in studies of native proteins colyophilized with sugars, such as sucrose and trehalose, that produce miscible amorphous mixtures [11]. In such cases, it has been shown that a single well-defined T_g occurs and that at constant water content the values of T_g decrease as the composition of the sugar is increased. This plasticizing effect of the sugars indicates that the T_g of the protein under a given set of conditions must be greater than that of the sugars, and, indeed, additional studies at different water and sugar

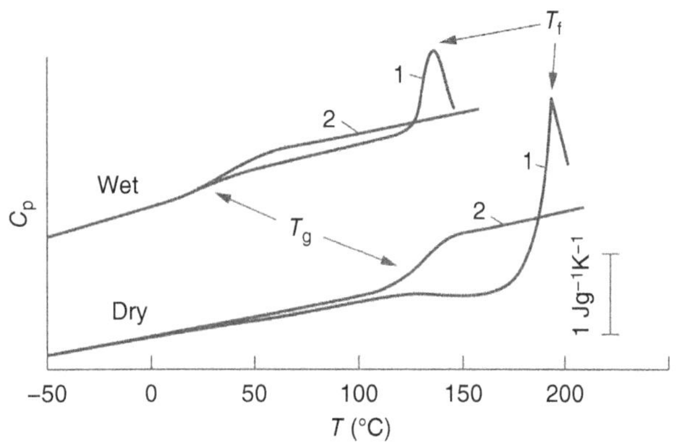

Figure 20.7 Heat capacity changes as a function of temperature for hydrated and essentially dry legumin: (1) first scan and (2) second scan. *Source*: Adapted from [10].

compositions, with extrapolation to zero sugar and zero water composition, lead to an estimation of T_g for a dry protein that is about 150 °C (423 K), a value of T_g that falls in the range of those observed for an unfolded dry denatured protein. We can conclude at this point, therefore, that amorphous proteins in the native state exhibit a glass transition temperature, as do other types of amorphous solids but because of the highly structured nature of the globular protein and the wide range of possible local motions that occur, as shown in Table 20.1, the thermal response to changes in various molecular motions is more gradual, making it often difficult to detect a distinct T_g.

20.4 Crystalline Properties of Proteins

As might be expected with most macromolecules, for example, polymers, the crystallization of proteins into well-defined crystal lattices is generally quite difficult because of the large molecular size and difficulty of producing consistent intramolecular and intermolecular order. However, the importance of using protein crystals to ascertain their tertiary structure in solution at the molecular level has provided motivation and tools that, indeed, make it possible to obtain well-defined crystals and crystal structures for many proteins [1]. Generally, crystallization of proteins occurs by producing supersaturated solutions in aqueous solution and then, with seeding and appropriate anti-solvent conditions, causing slow precipitation and crystallization. Under such conditions, as illustrated for one form of insulin in Figure 20.8, the globular structure is maintained within the individual fully hydrated protein molecules, with the molecules aligning themselves in a repeating series of unit cells [12]. Generally, the crystalline lattice that forms is held together by noncovalent interactions at a minimal number of points of contact between the various sphere-like globular structures. Because of this characteristic tendency of proteins to pack in a lattice, such crystal structures are generally disrupted at water contents less than that required for full hydration and are often quite fragile when subjected to any mechanical stress, a property very similar to that of plastic crystals, one of the types of crystal mesophases discussed in Chapter 7. Also, as discussed earlier for amorphous proteins, it has been demonstrated using calorimetry that fully hydrated crystalline proteins, such as myoglobin and lysozyme, exhibit distinct glass transition temperatures, reflecting the detection of primary diffusive motions within the lattice [13, 14]. For myoglobin, the T_g of a crystalline sample containing 48.9% w/w of water is 188 K, whereas that of a 19.0% hydrated system increases to 216 K, as would be expected with less hydration. Likewise, for lysozyme, a T_g of about 150 K observed with

Figure 20.8 Depiction of the structure of 22-n pig insulin crystals. *Source*: [12]/with permission of The Royal Society.

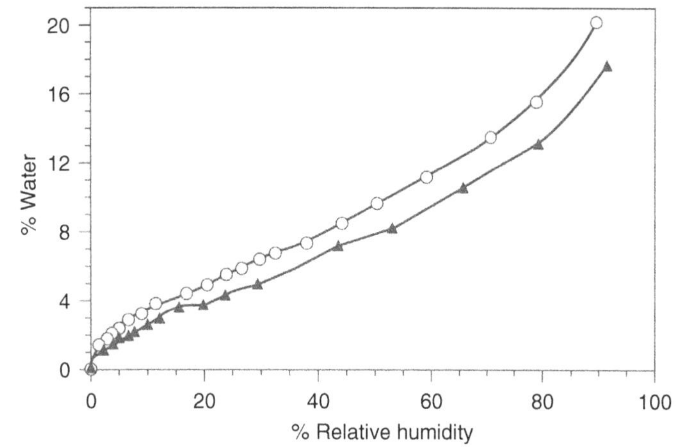

Figure 20.9 Water vapor sorption versus relative humidity for insulin: (◄) crystal; (○) amorphous. *Source*: [16]/with permission of Springer Nature.

24% w/w hydration increases to 165 and 218 K for 13.6% and 7.4% hydration. For both enzymes highly dried crystals did not exhibit a T_g below 300 K. It was also shown in these studies that the observed T_g in these crystals results from coupled diffusional motions of the water and protein molecules and not just from the motions of water alone. As a further indication of the similarity in intramolecular structure of fully hydrated protein crystals, fully hydrated amorphous proteins, and proteins in solution, fully hydrated crystalline proteins have been shown to exhibit the same variety of local internal molecular motions described in Table 20.1 [15]. The fact that a crystalline solid protein would exhibit a T_g might suggest that crystalline proteins are actually very similar to crystal mesophases, such as plastic crystals or conformationally disordered crystals, which are known to exhibit a well-defined glass transition temperature below which a glassy crystal mesophase forms (see Chapter 7). The existence of a T_g for hydrated protein crystals and, therefore, possible local structural similarities between crystalline forms and noncrystalline amorphous forms of the same protein can be further analyzed by comparing the water vapor sorption versus relative humidity profiles of crystalline and amorphous forms, as shown for crystalline and amorphous insulin in Figure 20.9 [16]. Ordinarily, water vapor adsorption for small molecules and polymers in the crystalline form is limited to a few molecular layers on the surface (we assume no crystal hydrate formation), whereas that for amorphous solids is much greater since water molecules generally can "dissolve" within the bulk phase (see Chapter 15). In contrast, we can observe in Figure 20.9 that the water vapor sorption isotherms for crystalline and amorphous insulin are very close and, therefore, that sorption by the crystal is not limited to surface adsorption.

20.5 Local Molecular Motions and the Dynamical Transitional Temperature, T_d

As discussed earlier, the local nondiffusive motions of globular proteins described in Table 20.1 appear to be important determinants of various types of biological activity. Consequently, biophysicists interested in the relationship between protein structure and biological activity have had a very strong interest in elucidating the nature and timescales of the local nondiffusive harmonic and anharmonic motions that occur with changes in temperature and at various levels of hydration, in addition to the whole-molecule diffusive motions responsible for T_g. The onset of such rapid motions can be detected by measuring mean square atomic displacements, $< \mu^2 >$, using techniques such as neutron scattering or Mossbauer spectroscopy [17, 18]. Such techniques indicate that fully hydrated proteins in the noncrystalline and crystalline states, when heated from temperatures below ~150 K to room temperature, appear to undergo a significant increase in their local dynamic motion, particularly around a characteristic temperature termed the dynamical transition, T_d. Typically, fully hydrated systems have values of T_d that occur in the range of 180 – 240 K, a temperature range similar to that observed for the glass transition temperature, T_g as discussed earlier. An indication of the importance of local motions is reflected in the fact that many important biological properties of proteins, normally correlated with the local molecular mobility of the protein, are greatly reduced, or cease to exist below T_d. Since, as with T_g, T_d appears to be highly coupled to local motions of the bound water, we would expect it to increase significantly as the degree of hydration is reduced, eventually becoming nonmeasurable for "dry" proteins, and, indeed, this is shown for amorphous ribonuclease A in Figure 20.10 [19]. Interestingly, as shown in Figure 20.11, the fully hydrated crystalline form of ribonuclease A also exhibits similar local motions in the crystalline state [15]. An indication of the importance of the solvent present in determining the extent of local motions can be seen in Figure 20.12, where values of mean square displacement $< \mu^2 >$ and apparent values of T_d of

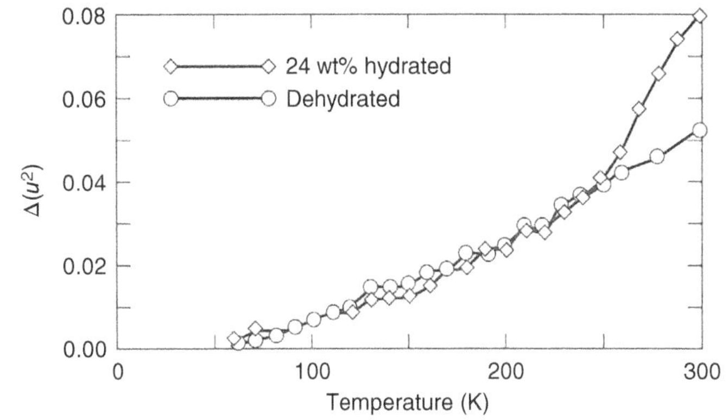

Figure 20.10 Plot of mean square displacements from dynamic neutron scattering of dry and partially hydrated amorphous ribonuclease A. *Source*: Adapted from [19].

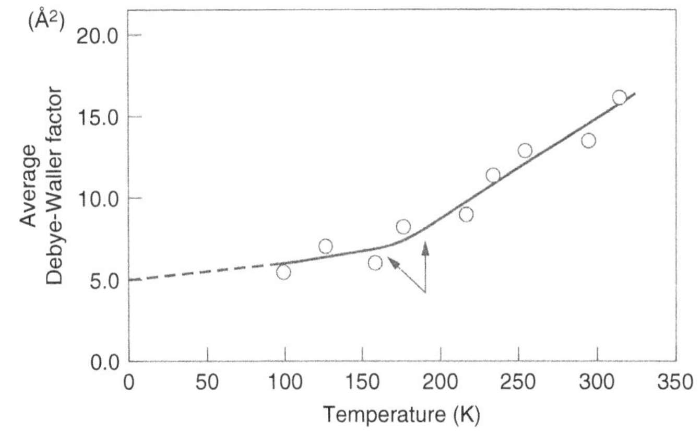

Figure 20.11 Plot of the average Debye–Waller factor for crystalline ribonuclease A as a function of temperature. *Source*: [15]/with permission of American Chemical Society.

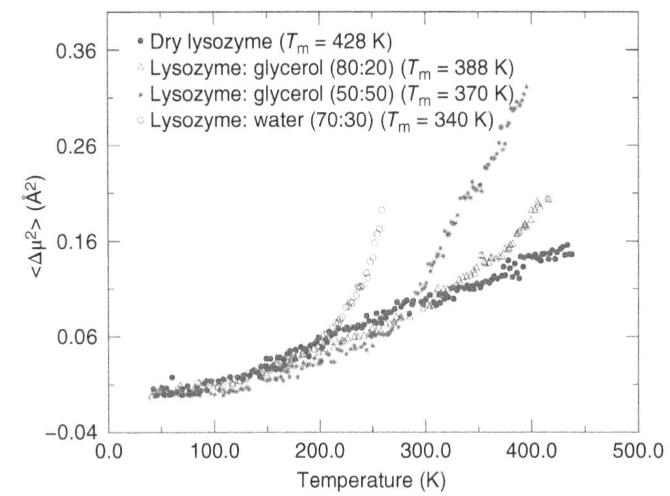

Figure 20.12 Plots of mean square displacements from dynamic neutron scattering of lysozyme as a function of temperature at different levels of solvation. *Source*: [20]/with permission of Elsevier.

amorphous lysozyme increase when water is replaced by increasing amounts of glycerol [20]. It has also been shown that T_d, normally observed in fully hydrated proteins, increases significantly as the water is replaced in the lyophilization process by the disaccharide, trehalose, as shown in Figure 20.13, for amorphous myoglobin [21].

The fact that T_d appears to reflect an important single critical temperature intrinsic to specific proteins, indicative of a sharp change from vibrational harmonic to anharmonic local motions, has been questioned in recent studies [22, 23]. First, it has been shown that although some biological functions tend to cease as the system is cooled below T_d, the onset of several biological activities, for example, enzymatic activity, in many systems does not always appear to be sharply affected by the apparent T_d. It appears that the reason for this is that the value of T_d measured by various techniques, although generally falling in the range of 180 – 240 K, is dependent on the experimental method and conditions used for making the measurement. More specifically, the value of T_d appears to depend on the timescale resolution of the instrument being utilized, for example, pico- or

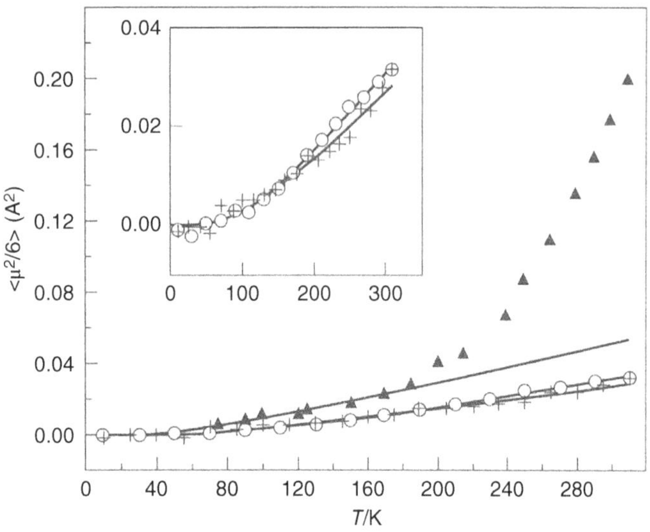

Figure 20.13 The mean square displacements from neutron scattering of myoglobin containing D20 (filled triangles), trehalose + myoglobin (open circles), and trehalose alone (+). Inset compares trehalose + myoglobin and trehalose alone. *Source*: Cordone et al. [21]. Reproduced with the permission of Elsevier.

nanoseconds, indicating that the critical nature of T_d is a result of the cutoff in timescale resolution for a particular instrument. Indeed, such instrument-dependent apparent transitions have also been observed with polymeric nonprotein systems undergoing changes in local motions. Although there is no doubt that local motions associated with the protein and any associated water molecules are important in determining protein activity, the changes in these motions with temperature occur more gradually than would be inferred from a distinct T_d. Recent studies utilizing techniques like neutron scattering and Mossbauer spectroscopy that measure dynamics on the timescale of pico- and nanoseconds have shown that, in addition to observing the instrument-dependent T_d, one can observe a significant change in local dynamics, which is instrument-independent, at the calorimetrically measured T_g of the protein under the same conditions [24]. Generally, T_g is shown to be lower than the apparent T_d values measured, although they are often so close that it is believed that T_g can be mistaken for a T_d. The fact that other glass-formers exhibit similar changes in local motions at T_g has led to the suggestion that the appearance of a T_g, which depends primarily on diffusive primary relaxations, can be observed at much faster timescales because of changes in the more rapid local β-Johari–Goldstein secondary relaxations, described in more detail in Chapter 6. Such secondary motions are known to be precursors to the onset of primary diffusive α relaxations in various amorphous solids. On the other hand, it is suggested that gradual changes in the local motions of hydrated proteins that give rise to an apparent instrument-dependent transition temperature, T_d, arise because of secondary motions associated with changes within the hydration layers coupled to the protein. That some relationship between apparent T_d values and T_g may exist is suggested further in an analysis of the data given in Figure 20.12 [20]. Here, it was shown that the measured apparent T_d for amorphous lysozyme solvated to different extents with glycerol (20:80 glycerol:protein; and 50:50 glycerol:protein) could be predicted using the Fox equation for predicting the T_g of amorphous mixtures from the T_g of each pure component and their weight fractions (see equation [6.27] in Chapter 6). If the dry protein has a T_g of 423 K, as discussed earlier, and glycerol has a T_g of 193 K, the predicted values of T_g for the mixtures are 341 K for 20: 80 glycerol:protein and 265 K for 50:50 glycerol:protein, as compared to the measured values of T_d, equal to 350 and 270 K, respectively. Whether the T_d reported in this study is actually a T_g, or that some more fundamental relationship exists between T_d and T_g, is not understood at the present time. What is important, however, is that the onset and gradual increase of local and diffusional motions as a function of temperature and degree of hydration in amorphous and crystalline proteins is in some manner related to the onset of biological activity and, presumably, the stability of such systems, as will be discussed in the following section.

20.6 Solid-state Physical and Chemical Stability of Proteins

Central to any form of instability of proteins in solution is their tendency to undergo changes in secondary and tertiary structure, that is, unfolding and denaturation, with changes in temperature and solution conditions, such as pH and ionic strength. Contact with hydrophobic surfaces, including air–water surfaces, will also facilitate the denaturation process. Such denaturation of the protein molecule causes it to be susceptible to aggregation and to exhibit a tendency to lose biological potency; it is generally believed that the formation of a "molten globule" state, intermediate in structure to the native and denatured states, is a critical step in the formation of soluble and insoluble aggregates [25, 26]. The exposure of reactive centers in protein molecules due to denaturation can also give rise to increased chemical reactivity, and hence molecular degradation, through such mechanisms as oxidation, deamination, or dimerization [4]. When such instabilities appear to occur with the formulation of proteins in aqueous solution, leading to unacceptable products, it becomes necessary to prepare solid dosage forms of the protein, which tend to be more stable during storage (see Chapters 18 and 19). In the case of products to be injected intravenously, sterile lyophilized materials intended for reconstitution at the time of administration are prepared, whereas proteins intended for pulmonary administration are formulated and processed by spray drying and used as dry powders. In both cases, an aqueous solution of the protein at the desired pH and ionic strength is dried to low levels of water content and appropriately packaged. As mentioned earlier, such processes generally produce noncrystalline or amorphous forms of the protein, and most often, as well, amorphous forms of the excipients included in the formulation. The goal of the formulator in such cases, therefore, is to process the materials in such a way as to avoid denaturation of the protein, crystallization of other amorphous formulation components during the drying process, and any further solid-state physical or chemical instability in the final package during storage under room temperature conditions. Although the technical details associated with the lyophilization and spray drying of protein solutions are beyond the scope of this chapter, it will be of interest to consider some important fundamental aspects of what happens molecularly to all formulation components, particularly proteins, during these processes that might lead to instability. This understanding can then provide a rational basis for developing practical strategies to stabilize proteins on a long-term basis.

Let us first consider some general aspects of the lyophilization process that can affect the stability of a protein [27]. During lyophilization, the temperature of the protein solution is reduced to well below the freezing point of water, typically in the range of −40 to −50 °C, causing the formation of ice. As ice continues to form, the concentration of the soluble protein and any other water-soluble components in the remaining liquid water ("the freeze-concentrate") will increase significantly. The reduction in temperature can destabilize the protein by means of cold denaturation, whereas the increase in concentration of all ingredients can change the pH and ionic strength of the remaining liquid, with possible adverse effects on the native structure and stability of the protein. There is also the possibility of adsorption of the protein to the solid-liquid interface created by ice formation, with resulting surface denaturation. Eventually, with the removal of bulk liquid water by the freezing process and the complete formation of ice, the protein in contact with the ice will become an amorphous solid; generally, in a fully hydrated amorphous state of bound water at the level of roughly 20–40%. This is then followed by the "primary drying" process, wherein this frozen sample is subjected to a vacuum that causes the ice to sublime, leaving behind a porous solid cake containing fully hydrated protein in the amorphous state and the various solid forms of other ingredients. Although these other formulation ingredients may also become amorphous, preferably forming miscible amorphous mixtures with the protein, there is the possibility of phase separation of some ingredients and even crystallization, for example, some buffer ingredients and salts. Optimally, therefore, the temperature of primary drying must be low enough to prevent structural changes in the protein as well as phase separation of other ingredients in the porous cake, that is, kept below the glass transition temperature of the freeze-concentrate, but high enough to facilitate the sublimation rate. This process is followed by "secondary drying" brought about by progressively increasing temperatures up to room temperature and until the sample retains less than about 1–2% water. The process of removing most of the bound water from an unprotected amorphous protein has been shown to disrupt the secondary and tertiary structure of the protein, often leading to denaturation and instability, particularly if the drying temperature is close to, or above the glass transition temperature at any level of hydration where significant molecular motion still exists. The spray drying of proteins to form amorphous dry powders generally involves the spraying of aqueous solutions into a heated chamber where the water in the liquid droplets rapidly evaporates, leaving behind amorphous dry powder particles. The environmental conditions faced by the formulator during spray drying are essentially the same as those that occur during secondary drying in lyophilization since dehydration at higher temperatures in the spray dryer also can lead to protein instabilities unless all local and diffusive molecular motions are kept to a minimum during the drying process.

20.7 Cryoprotection and Lyoprotection

It has been demonstrated in many studies that most proteins in aqueous solution require "cryoprotection" during lyophilization as they are being frozen, and "lyoprotection" against structural changes and instabilities as water is removed under vacuum at this low temperature. During spray drying, as water rapidly evaporates during spraying and heat-induced drying, similar lyoprotection is required. A wide range of chemically diverse substances such as disaccharides (sucrose and trehalose), amino acids (glycine and alanine), polyols (glycerol and propylene glycol), and water-soluble macromolecules, (serum albumin and polyethylene glycol) appear to act as cryoprotectants during the freezing process by stabilizing the native structure of protein dissolved in the remaining nonfrozen aqueous solution [28]. They act in this way by being preferentially excluded from contact with the surface of proteins in aqueous solution, producing a state of free energy that thermodynamically favors the existence of the native state over the denatured state [29]. The diverse chemical structure of various cryoprotectants strongly supports this proposed mechanism, which is based on solution thermodynamics. When bound water is removed from the protein by various types of drying, so that it is less than fully hydrated, and the system becomes more solid-like, a different mechanism of protection appears to be required. Here, water-soluble macromolecules, which generally do not form miscible amorphous mixtures with proteins, are generally not effective as lyoprotectants, whereas smaller polar molecules, such as sucrose and trehalose, that can form miscible mixtures are very effective. The fact that such disaccharides can form miscible mixtures by hydrogen bonding to the protein, and thereby replacing the bound water removed by drying, has given rise to the "water replacement model" for lyoprotection, wherein lyoprotectants are believed to produce a level of hydrogen bonding with the protein that allows the protein to maintain its native state during the process of dehydration [30]. However, a likely equally important contribution to the mechanism of lyoprotection by disaccharides is the ability of such small polar molecules to lower the local motions of the protein through the replacement of water with compounds that exhibit much less plasticization of these local motions, as shown in Figure 20.12, where glycerol ($T_g = 193$ K) replaces water ($T_g = 136$ K) and, in Figure 20.13, where trehalose ($T_g = 393$ K) replaces the water. That suppression of local motions does play an important role in promoting protein stability is further supported by recent research that has shown that local motions, as reflected in dynamic neutron scattering experiments, are significantly suppressed in disaccharide–protein glasses with the addition of small amounts of glycerol, with corresponding increases in protein stability [31, 32]. During processing by either lyophilization or spray drying, it is also possible that local pH conditions, which play an important role in determining secondary and tertiary protein structure, as well as chemical stability during dehydration, often require the addition of buffers that must be miscible with the protein and other components and maintain optimal pH conditions. The choice of specific buffers and amounts is critically important in maintaining product integrity [33]. We can conclude, therefore, that, although molecular motions of very dry proteins should be minimal, the process of drying during lyophilization and spray drying leads to tendencies toward denaturation unless other components are present that can interact with the protein, replace any bound water present, and produce local structural changes in the protein that lead to lower rates of local protein motions and unchanged protein structure and stability.

References

1 Shenoy B, Wang Y, Shan W, Margolin AL. Stability of crystalline proteins. *Biotechnol Bioeng* 2001; 73: 358–369.

2 Carpenter JF, Chang BS, Garzon-Rodriguez W, Randolph TW. Rational design of stable lyophilized protein formulations: Theory and practice. *Pharm Biotechnol* 2002; 13: 109–133.

3 Lee G. Spray-drying of proteins. *Pharm Biotechnol* 2002; 13: 135–158.

4 Bummer PM, Koppenol S. Chemical and physical considerations in protein and peptide stability. In: McNally EJ, ed. *Protein Formulation and Delivery*. New York: Marcel Dekker; 1999: 278.

5 Hill JJ, Shalaev EY, Zografi G. Thermodynamic and dynamic factors involved in the stability of native protein structure in amorphous solids in relation to levels of hydration. *J Pharm Sci* 2005; 94: 1636–1667.

6 Poole PL, Finney JL. Sequential hydration of dry proteins: A direct difference IR investigation of sequence homologs of lysozyme and alpha lactalbumin. *Biopolymers* 1984; 23: 1647–1666.

7 Rupley JA, Careri G. Protein hydration and function. *Adv Protein Chem* 1991; 41: 37–172.

8 Gregory RB, Lumry R. Hydrogen-exchange evidence for distinct structural classes in globular proteins. *Biopolymers* 1985; 24: 301–306.

9 Brooks CL, III, Gruebele M, Onuchic JN, Wolynes PG. Chemical physics of protein folding. *P Natl Acad Sci USA* 1998; 95: 11037–11038.

10 Sochova IV, Heat capacity and thermodynamic characteristics of denaturation and glass transition of hydrated and anhydrous proteins. *Biophys Chem* 1997; 69: 31–41.

11 Pikal MJ, Rigsbee DR, Roy ML. Solid state chemistry of proteins: I. Glass transition behavior in freeze-dried disaccharide formulations of human growth hormone (hGH). *J Pharm Sci* 2007; 96: 2765–2776.

12 Baker EN, Blundell TL, Cutfield JF, Cutfield SM, Dodson EJ, Dodson GG, Hodgkin DM, Hubbard PE, Isaaca NW, Reynolds CD. The structure of 22 n pig insulin crystals at 1.5A resolution. *Philos Trans R Soc London Ser B* 1988; 319: 369–456.

13 Miyazaki Y, Matsuo T, Suga H. Glass transition of myoglobin crystal. *Chem Phys Lett* 1993; 213: 303–308.

14 Lyazaki Y, Matsuo T, Suga H. Low-temperature heat capacity and glassy behavior of lysozyme crystal. *J Phys Chem B* 2000; 104: 8044–8052.

15 Tilton RF Jr., Dewan JC, Petsko GA. Effects of temperature on protein structure and dynamics: X-ray crystallographic studies of the protein ribonuclease-A at nine different temperatures from 98 to 320K. *Biochem* 1992; 31: 2469–2481.

16 Pikal MJ, Rigsbee DR. The stability of insulin in crystalline and amorphous solids: Observation of greater stability for the amorphous form. *Pharm Res* 1997; 14: 1379–1387.

17 Frauenfelder JM, Sligar SG, Wolynes PG. The energy landscapes and motions of proteins. *Science* 1991; 254: 1598–1603.

18 Fan J, Cooper E, Angell C. Glasses with strong calorimetric beta-glass transitions and the relation to the protein glass transition problem. *J Phys Chem* 1994; 98: 9345–9349.

19 Tsai AM, Udovic TJ, Neumann DA. The inverse relationship between protein dynamics and thermal stability. *Biophys J* 2001; 81: 2339–2343.

20 Tsai AM, Neumann DA, Bell LN. Molecular dynamics of solid-state lysozyme as affected by glycerol and water: a neutron scattering study. *Biophys J* 2000; 79: 2728–2732.

21 Cordone L, Ferrand M, Vitrano E, Zaccai G. Harmonic behavior of trehalose-coated carbon-monoxy-myoglobin at high temperature. *Biophys J* 1999; 76: 1043–1047.

22 Daniel RM, Finney JL, Smith JC. The dynamic transition in proteins may have a simple explanation. *Faraday Discuss* 2002; 122: 163–169.

23 Magazu S, Migliardo F, Benedetto A. Puzzle of protein dynamical transition. *J Phys Chem B* 2011; 115: 7736–7743.

24 Cappaccioli S, Ngai KL, Acherbak S, Paciaroni A. Evidence of coexistence of change of caged dynamics at T_g and the dynamic transition at T_d in solvated proteins. *J Phys Chem B* 2012; 116: 1745–1757.

25 Pititsyn OB. Molten globule and protein folding. *Adv Protein Chem* 1995; 47: 83–229.

26 Roberts CJ, Darrington RT, Whitley MB. Irreversible aggregation of recombinant bovine granulocyte- colony stimulating factor (bG-CSF) and implications for predicting protein shelf life. *J Pharm Sci* 2003; 95: 1095–1111.

27 Wang W. Lyophilization and development of solid protein pharmaceuticals. *Int J Pharm* 2000; 203: 1–60.

28 Park J, Nagapudi K, Vergara C, Ramachander R, Laurence JS, Krishnan S. Effect of pH and excipients on structure, dynamics and long-term stability of a model IgGI monoclonal antibody upon freeze-drying. *Pharm Res* 2013; 30: 968–984.

29 Timasheff SN. The control of protein stability and association by weak interactions with water: How do solvents affect these processes? *Annu Rev Biophys Biomol Struct* 1993; 22: 67–97.

30 Crowe JH, Crowe LM, Carpenter JF. Preserving dry biomaterials: The water replacement hypothesis, Part 2. *BioPharm* 1993; 6: 40–43.

31 Soles CL, Tsai AM, Cicerone MT. Glass dynamics and the preservation of proteins. In: Murphy RM, Tsai AM, ed. *Misbehaving Proteins*. New York: Springer, New York, 2006: 193.

32 Cicerone MT, Douglas JF. β-Relaxation governs protein stability in sugar-glass matrices. *Soft Matter* 2012; 8: 2983–2991.

33 Mutukuri TT, Ling J, Du Y, Su Y, Zhou QT. Effect of buffer salts on physical stability of lyophilized and spray-dried protein formulations containing bovine serum albumin and trehalose. *Pharm Res* 2023; 40: 1355–1371.

21

Physical and Chemical Properties of RNA Therapeutics

21.1 Introduction

Ribonucleic acid, abbreviated as RNA, exists in all living cells and represents a group of biomolecules displaying diverse structures and also serving broad functions in biological processes, including gene transcription, RNA splicing, protein synthesis, and protein regulation. For each of the biological processes, multiple types of RNA usually actively participate. Using protein synthesis as an example, ribosome RNA (rRNA) acts as a structural component of the ribosome, that is, the protein synthesis machinery in vivo, messenger RNA (mRNA) provides a sequence template, and transfer RNA (tRNA) offers the transportation of amino acids. In addition to the biological diversity, RNA molecules display chemical versatility. While the majority of RNA types work as a structure component or a carrier for genetic information, certain RNA molecules have catalyst capability to act as a nuclease in mRNA maturation through splicing to remove noncoding sequences or in removing unwanted RNA species attained endogenously or received exogenously. The diversity and versatility are owing to the extreme broad space of RNA structure aspects, considering small interference RNA (siRNA) with about 20 nucleotides, mRNA with hundreds to thousands of nucleotides, and clustered regularly interspaced short palindromic repeats (CRISPRs), a cellular gene editing machinery, as a complex of multiple types of RNA.

The unidirectional sequential feature of RNA arises from a linear structure of RNA molecules with nucleotides linked by phosphate groups to form this unique type of biopolymers, which are very hydrophilic owing to the sugar units and nucleobases, and ionizable at physiological pH due to the phosphate groups. Obviously, those unique physical properties, such as being very hydrophilic, presenting as ionic electrolytes, along with huge molecular weights, are diminishing any chance of permeability through biological membranes, which forms one of the main challenges of developing RNA molecules as therapeutic agents, on top of poor chemical stability and being susceptible to enzymatic degradation due to the ubiquitous presence of nucleases. However, those insurmountable obstacles could not slow down the steady progress in the pharmaceutical development of RNA-based therapeutics, mainly on therapeutic siRNA for some genetic diseases, through chemical modification to improve stability and using novel delivery platforms like lipid nanoparticles (LNPs) to enable RNA to enter a targeted tissue and once inside cells to exert biological activity. The pioneering work on siRNA development established the foundation for the quick development of mRNA vaccines to cope with the global COVID pandemic, which is further catalyzing the advancement of the field.

In this chapter, great attention is given to the physical and chemical properties of various RNA therapeutics, which affect functionality and stability, and also to the attributes of key delivery platforms such as mRNA LNPs, a complicated and fragile delivery system that, because of their instability, needs to be further optimized. However, little knowledge is available as to the solid-state properties of RNA molecules and LNPs, even though these properties are important, considering some RNA therapeutic agents are present in the solid state either as a lyophilized product or as a frozen material. Thus, this chapter not only provides an in-depth review of solid-state properties of these compounds but also lays out perspectives for more research to be done, which would facilitate the pharmaceutical development of RNA for many important modes of therapeutic actions, including siRNA and mRNA, and potentially CRISPR for gene therapies.

21.2 Mode of Action

Four types of RNA molecules are being widely explored as therapeutics for a broad range of diseases: antisense RNA, siRNA, mRNA, and CRISPR.

Solid-State Materials in Pharmaceutical Chemistry: Properties, Characterization, and Applications, Second Edition. Stephen R. Byrn, George Zografi and Xiaoming (Sean) Chen.

21.2.1 Antisense RNA

Antisense RNA represents a single-stranded RNA molecule comprising about 20 nucleotides that pairs with a complementary region on pre-mRNA or mRNA via Watson–Crick hydrogen binding. The pairing to form a double-stranded region offers a steric block to prevent the initiation or continuation of protein synthesis from an mRNA, leading to an arrest of pre-mRNA splicing to produce a mature mRNA or triggering the degradation of an mRNA by endogenous RNA nucleases. Consequently, downregulation of a targeted gene can occur at the post-transcriptional level. The mechanism of gene regulation by antisense RNAs exists widely in living organisms from prokaryotic to eukaryotic. For example, in eukaryotes, microRNAs act on mRNAs to regulate translation or degradation through antisense mechanisms [1].

When the molecular mechanisms of action for antisense RNA were discovered, initially, there were strong hopes in terms of producing many promising therapeutics, considering the modes of action appearing relatively simple to tackle and the binding specificity seeming to be high, along with significant genome information arising from the Human Genome program. Unfortunately, only a few antisense RNA molecules have been approved as therapeutics owing to several significant challenges, even though many research and development programs have been undertaken. In addition to the delivery challenge, one of the key obstacles is identifying a target sequence region where an antisense RNA molecule can access. Though, from the primary structure, there appear to be many possible areas, it has been realized that mRNA molecules are usually highly structured in vivo to form complex secondary and tertiary regions, in addition to many regions binding to proteins. To discover an antisense RNA drug, numerous in vitro and in vivo experiments need to be carried out to identify an effective one. For example, hundreds of antisense RNA molecules were screened via in vitro and in animal models for Hutchinson–Gilford progeria syndrome, a rare childhood premature aging disorder. The search involved finding an RNA molecule that can mitigate the pre-messenger RNA splicing defect in the LMNA gene [2].

As a meaningful example for this type of RNA therapeutics, fomivirsen, is the first antisense RNA drug approved by the FDA in 1998 to serve as an antiviral agent for the treatment of cytomegalovirus (CMV) retinitis in HIV-infected patients [3]. It consists of 21 oligonucleotides with phosphorothioate (PS) linkages to improve the stability against degradation by nucleases, with the sequence described as follows:

5′-GCGTTTGCTCTTCTTCTTGCG-3′

where G = guanosine, C = cytidine, and T = thymidine.

The antisense RNA molecule binds to the complementary sequence of mRNA encoding the major immediate early protein of human CMV, leading to the arrest of protein synthesis of several key viral proteins and subsequently inhibition of viral replication. Fomivirsen demonstrates potent and selective disruption of CMV replication in early preclinical characterization and also shows clear clinical efficacy for the treatment of CMV retinitis in patients with HIV. However, the launching of several highly effective anti-HIV therapeutics in the early 2000s led to the reduced incidence of CMV retinitis and subsequently diminished demand for fomivirsen, unfortunately ending in its market withdrawal. Despite that, fomivirsen provides a successful proof-of-concept of antisense RNA being a niche therapeutic agent and arouses the promise of this uncharted field, leading to the further development of other types of RNA therapeutics, including siRNA.

21.2.2 Small Interference RNA

Small interference RNA (siRNA) stands for a type of short, double-stranded, noncoding RNA molecule having 20–30 base pairs and a molecular weight of about 14 kDa. siRNA molecules, widely present in plant and mammalian cells, are of significance in downregulating gene expression by a mechanism known as RNA interference. RNA interference serves as a way of post-transcriptional gene silencing, a natural defense against double-stranded RNA molecules originating from either endogenous parasitic genome regions like repetitive sequence transposons in the genome or exogenous pathogenic nucleic acids from viruses.

In a natural process, a parasitic or pathogenic double-stranded RNA is cleaved by an RNase III-like enzyme called Dicer to form siRNAs [4]. Then siRNAs bind to a multiprotein complex termed RNA-induced silencing complex (RISC). In RISC, the antisense single-stranded siRNA chains are separated from the double strands, guide the binding of the complex to the target mRNA, and trigger the cleavage of the unwanted mRNA, such as the transcript of a virus genome.

siRNA therapeutics aim to take advantage of the natural RNA interference mechanism by delivering siRNA molecules into intracellular locations and downregulating the targeted mRNA. Thus, the reduction in the levels of targeted protein might lead to therapeutic effects on genetic disorders or cancers. Many siRNA therapeutics are under clinical trials, and several of them have entered the market for the treatment of genetic disorders, including patisiran, givosiran, and lumasiran [5].

For instance, patisiran, branded as Onpattro®, as a double-stranded siRNA, specifically complements a genetically conserved sequence in the 3′ untranslated region (3′ UTR) of mutant and wild-type transthyretin (TTR) mRNA [6]. The structure is given in Figure 21.1. Through RNA interference, patisiran induces the degradation of TTR mRNA, both mutant and wild-type, and causes the reduction of serum TTR protein levels and tissue TTR protein deposits, clinically to treat the polyneuropathy of hereditary TTR-mediated amyloidosis, a rare and life-threatening genetic disorder. Formulated as LNPs, patisiran is designed to deliver to hepatocytes in the liver, the main source of TTR protein being made. The pioneering work on applying LNPs to siRNA delivery leads to the successful delivery of mRNA therapeutics using the same platform.

21.2.3 Messenger RNA

mRNA, fully named as messenger RNA, is a long single-stranded ribonucleic acid that is transcribed from a respective DNA sequence and carries the coding information for protein synthesis. The molecular weight of an mRNA molecule ranges from 300 to 5000 kDa, reflecting the extreme diversity of proteins to be transcribed through the intracellular protein production machine using mRNA as the template. If mRNA molecules can be effectively delivered to the intracellular environment of a target organ, being envisioned by scientists decades ago, therapeutic proteins can be produced rapidly to exert therapeutic effects for cancer, genetic diseases, protein replacement therapies, and vaccines. Regarding the structure, a eukaryotic mRNA molecule contains a 5′ cap, a 5′ untranslated region (5′-UTR), an open-reading frame, a 3′ untranslated region (3′-UTR), and a poly A tail. Endogenous mRNA molecules are produced from DNA through a transcription process, which is highly regulated, as pre-mRNA to mature further to form mRNA. Therapeutic mRNA molecules are commonly generated by an in vitro transcription process involving multiple enzymes, which will be described in more depth in the following sections.

A notable example of mRNA therapeutics is BNT162b, an mRNA vaccine to prevent coronavirus disease 2019 (COVID-19) caused by severe acute respiratory syndrome coronavirus-2 (SARS-CoV-2) [7]. The mRNA molecule encodes a full-length SARS-CoV-2 spike glycoprotein S, which acts as an antigen to trigger an immune response and build up a human immune defense machine against the virus. In addition, it contains 5′-untranslated region (5′-UTR), 3′-untranslated region (3′-UTR), a 7-methylguanosine-derived cap (m7GpppNmN) on the 5′-end (5′ cap), and a 3′ polyadenylation region (poly(A) tail) (Figure 21.2). It should be noted that those structural elements are optimized for in vitro translational efficiency and chemical/enzymatic stability. Ultimately, the mRNA vaccine is effectively delivered to host cells after intramuscular administration via LNPs.

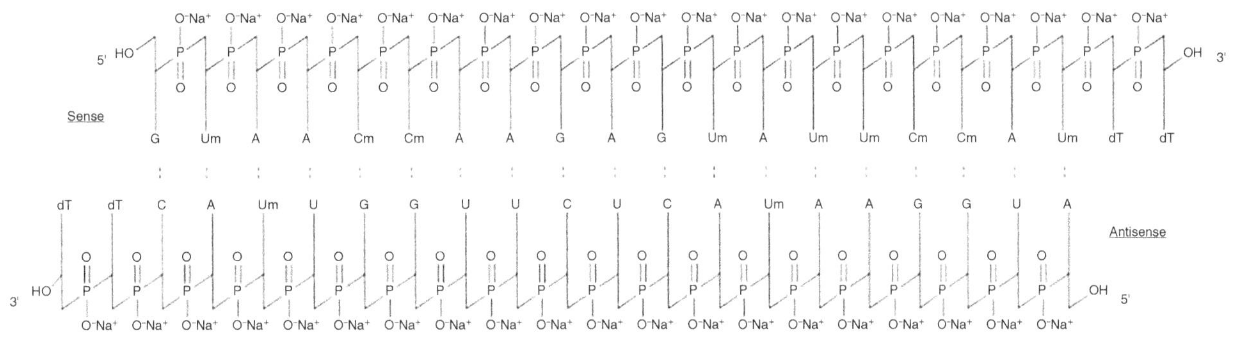

A, adenosine; C, cytidine; G, guanosine; U, uridine; Cm, 2'-*O*-methylcytidine; Um, 2'-*O*-methyluridine; dT, thymidine

Figure 21.1 Structure formula of patisiran, an RNAi therapeutic, for hereditary transthyretin amyloidosis.

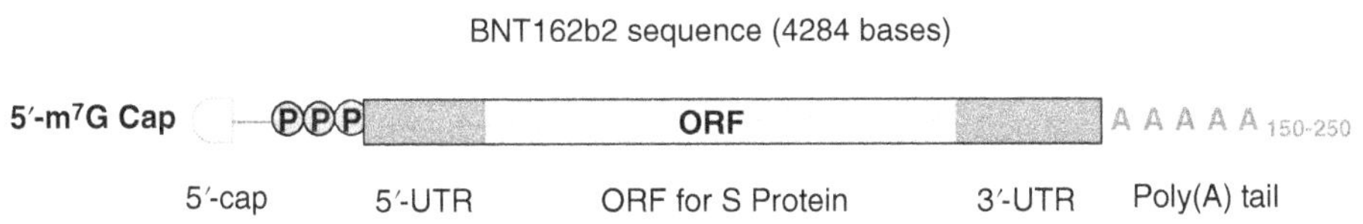

Figure 21.2 Structure components of BNT162b2, an mRNA vaccine for the prevention of COVID-19.

21.2.4 CRISPR

CRISPR, fully named as clustered regularly interspaced short palindromic repeat, along with Cas9 nuclease, is emerging as a powerful gene editing tool in molecular genetics and is being actively pursued as a novel type of RNA therapeutic [8]. CRISPR was first discovered as an antiviral mechanism for some bacterial species and archaea. When attacked by a virus, also called a phage for bacteria, the host bacteria collect small pieces of the virus's DNA and incorporate these strands into their own DNA genome to create segments known as CRISPR arrays. The way the bacteria act resembles an immune response in a mammalian system by learning about the invader and passing the information to the next self-defense event. When the virus is encountered again, the bacteria generate an RNA segment from the CRISPR array, which binds the specific regions of the virus DNA and subsequently guides Cas 9 nuclease to cut the DNA apart and destroy the replication of the virus DNA. The RNA segment is also called single-guide RNA (sgRNA). The mechanism is being explored to become a powerful gene editing tool with applications in all different species, from yeast to insects to mammalian cells. A specific gene can be removed, altered, or inserted to study biological system regulation. Since many human maladies are related to genetic disorders, CRISPR is widely conceived as a therapeutic approach for disorders including cystic fibrosis, sickle cell, and hemophilia. Considering the delivery is expected to be more challenging than mRNA and siRNA, the development of CRISPR as medicine is still in the infancy stage, whereas the success of LNPs in siRNA and mRNA delivery would be expected to benefit CRISPR gene editing to become a future goal of personalized medicine.

21.3 Building Blocks and Primary Structure

Naturally occurring RNA molecules are made of ribonucleosides, including A, T, G, U, linked by 5′-3′ phosphodiester linkages, which, by design, are susceptible to hydrolysis or enzymatic cleavage, to maintain proper levels in vivo. This instability prevents RNA molecules made of natural ribonucleosides from being developed as drugs. Chemical modifications, if allowed, offer a good solution for stability and other desirable pharmaceutical properties to enable the drug delivery, as listed in the following list:

- Improve chemical stability
- Resist degradation by a host of cellular exonucleases and endonucleases
- Enhance metabolic stability
- Mitigate immune response
- Facilitate target site accessibility

The use of phosphonothioate (PS) modification is a breakthrough achieved when developing antisense RNA therapeutics, which involves a simple sulfur atom to replace a nonbridging oxygen atom (Figure 21.3). The modification dramatically improves the stability against nuclease degradation. In addition, the PS modification improves the binding of RNAs to plasma proteins, thus delaying the excretion in the urine, which is a common problem for regular phosphodiester backbone oligonucleotides.

It is well noted that the 2′ hydroxyl group in the ribose is the main liability of RNA instability. Interestingly, the modification of the site by replacing it with F (fluoro), O-ME (O-methyl), and O-MOE (O-methoxyethyl) (Figure 21.3) can yield better chemical stability and also enzymatic stability against nucleases widely present intracellularly and extracellularly. In addition, it has been realized that the modification at that site at least maintains and potentially enhances the conformational orientation of the ribose to form an A-type backbone configuration, which favors the binding to form RNA double strands. As reported for the various 2′-modifications in siRNA molecules against Factor VII, the 2′-F modification yielded a duplex to have a melting point about 20°C above the parent duplex with naturally occurring nucleosides, indicating a much stronger binding, also showing a better gene silencing ability than the parent one [9].

The development of those modifications is indeed a breakthrough in RNA delivery to overcome the stability barrier, leading to many RNA therapeutics that can be formulated as a conventional liquid injectable stored either in the refrigerator or at ambient temperature. It is no surprise to note that the modified nucleotide, often combined with PS, is widely present in approved small oligonucleotides like antisense RNA and siRNA drugs.

As illustrated in Figure 21.1, patisiran, a siRNA drug, has modifications in 9 out of 21 nucleotides on the sense strand, including 2′-O-methylcytidine and 2′-O-methyluridine.

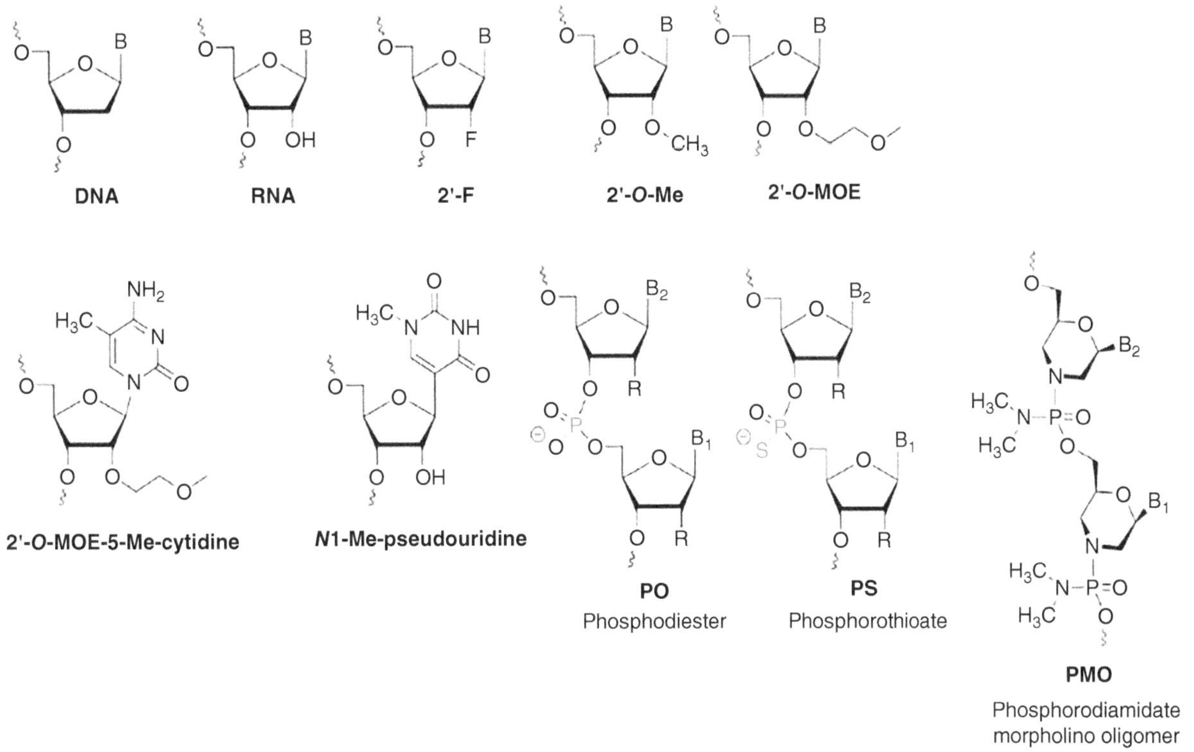

Figure 21.3 Examples of modified nucleotides and phosphonothioate modification to improve stability.

[2'-O-(2-methoxyethyl)](3'-5')(*P*-thio)(mU-mC-A-mC-mU-mU-mU-mC-A-mU-A-A-mU-G-mC-mU-G-G)

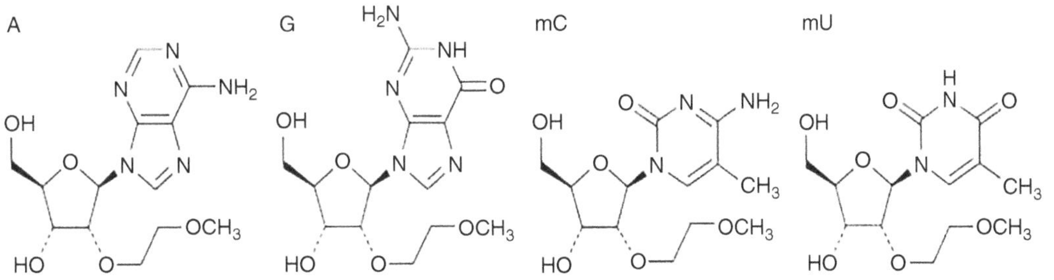

Figure 21.4 Primary structure of nusinersen, an antisense RNA, for the treatment of spinal muscular atrophy, and molecular structure of modified nucleosides including 2′-O-Me-Adenosine (A), 2′-O-Me-Guanosine (G), 2′-O-Me-5-methylcytosine (mC), and 2′-O-Me-5-methyluridine (mU).

Nusinersen, an antisense oligonucleotide approved for the treatment of spinal muscular atrophy for children and adults, represents another example, as shown in Figure 21.4, with PS modifications being used in the linkages, 2′-O-MOE being applied in every ribofuranosyl ring, and nucleotide bases being modified including methyl cytidines and uridines. The drug product presents as a sterile solution for intrathecal injection at 2.4 mg/mL, with a 30-month drug product expiration period when stored refrigerated [10], reflecting the superior chemical stability enabled by modifying the building blocks.

Nevertheless, the application of modified building blocks in mRNA therapeutics meets with some challenges. Unlike small molecular RNAs like antisense RNA and siRNA as oligonucleotides, which are made from chemical synthesis, mRNAs, having many more nucleotides, are mainly manufactured by in vitro transcription for therapeutic use. For that, the typical PS linkages cannot be used, and, in addition, commonly available nucleotide modifications such as F, O-ME, and O-MOE are not applicable for biological synthesis through various enzymes. Currently, only one modified nucleotide, 1-methyl-3′-pseudouridylyl, to replace uridine, is present in approved mRNA vaccines, mainly to lower immunogenicity and increase translation efficiency. Due to the limited chemical modification that can be used in mRNA therapeutics, it

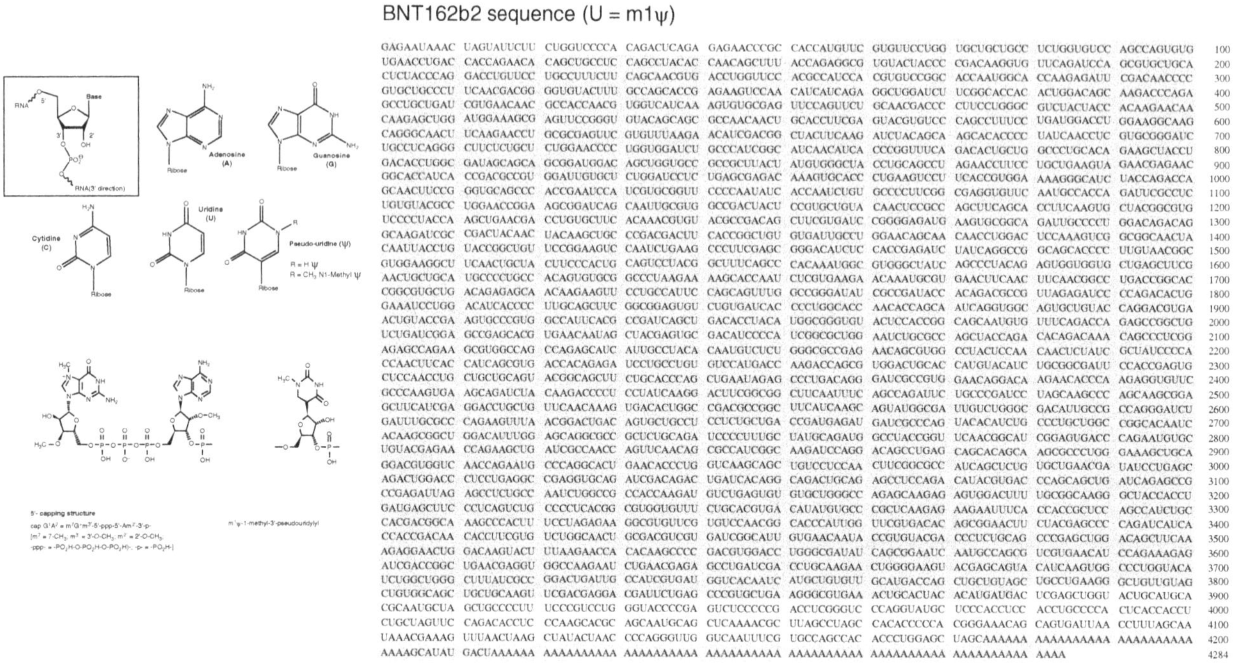

Figure 21.5 Key structure elements and primary sequence of BNT162b2, an mRNA vaccine for the prevention of COVID-19. In the sequence, A stands for adenosine, G for guanosine, C for cytosine, while U stands for N1-methyl-pseudo-uridine.

is not surprising to note that mRNA drug products are much less stable than small oligonucleotides and need to be stored either −80 or −20 °C with shelf life not more than a year, posing significant distribution challenges and economic hurdles to expand the application of this novel therapeutic strategy.

Figure 21.5 shows a typical primary structure of an mRNA therapeutic, using the COVID-19 vaccine BNT162b2 as an example [11]. The whole structure contains 4284 nucleotides, with 1-methyl-3′-pseudouridylyl in many locations, and can be divided into several elements:

- 5′-cap in position 1–2, as a modified 5′-cap1 structure, essential for translation initiation
- 5′-UTR in position 3–54, as a 5′-untranslated region derived from human alpha-globin RNA with an optimized Kozak sequence, to improve translation efficiency
- Open-reading frame, Part 1, signal peptide in position 55–102, coded for S glycoprotein signal peptide that guides translocation of the polypeptide chain into the endoplasmic reticulum as a secreted protein
- Open-reading frame, Part II, COVID-19 S protein in position 103–3879, encoding full-length SARS-COV-2 spike (S) glycoprotein containing mutations K986P and V987P
- 3′-UTR in position 3880–4174, as a 3′ untranslated region comprising sequence elements to confer RNA stability and enhanced protein expression
- Poly(A) in position 4175–4284, a 110-nucleotide poly(A) tail containing 30 adenosine units, followed by a 10-nucleotide linker and another 70 adenosine units

21.4 RNA Structure in Solution

The functional versatility of RNA arises from the dynamics of RNA secondary and tertiary structures, in response to cellular environments and interactions with DNAs and proteins. The responses, as reflected in the changes in structure characteristics, bespeak the underlying molecular interactions from hydrogen bonding, stacking, and ionic interactions determined by the structure aspects of nucleotides. Bases in the nucleotides tend to form hydrogen bonds as Watson–Crick pairs to develop helix structures. Stacking effects between neighboring bases further strengthen the helix. Moreover, ion interactions between the negatively charged phosphate backbones and metal ions like sodium and magnesium facilitate the folding to form high-order structures.

21.4.1 Hydrogen Bonding

The physical, chemical, and biological properties of an RNA molecule originate from its structure. RNA is a biological polymer made of nucleotides. Each nucleotide contains a ribose with a phosphate group and a nitrogen-containing base. The phosphate groups link nucleotides together to form a linear structure called the primary sequence, which is determined by the sequence of bases. The nucleotide bases found in RNA structure naturally are adenine (A), cytosine (C), guanine (G), and uracil (U). Those bases are strong hydrogen bond donors and acceptors, which enable the precise molecular recognition between two bases, famously named as the Watson–Crick Pair. For example, A pairs with T (in DNA) or U (in RNA) with two hydrogen bonds, while G pairs with C using three hydrogen bonds. The pairing forms the fundamental structure aspect of nucleic acids, for example, the double helix of DNA structure, secondary structure of RNA, and also plays critical roles in genetic information replication, regulation, transcription, and translation, as well as rendering the efficacies of therapeutic RNAs, for example, binding of an antisense RNA to the target sequence of an mRNA through Watson–Crick pairing to modulate the gene expression.

21.4.2 Stacking Interaction

Stacking interaction represents another important type of noncovalent interaction that affects the formation of the high-order structure of RNAs. Nucleotide bases as flat aromatic rings can form strong nonspecific face-to-face van der Waals forces between those flat bases, when orienting in a certain configuration, as a result of strong $\pi - \pi$ interactions due to delocalized π electrons. Delocalized π electrons result in a positively charged H moiety in one of the aromatic rings and a negatively charged π electronic cloud in another neighboring aromatic ring, resulting in strong attractive forces.

Consecutive stacking remains the most common type and occurs when the nucleotide bases of adjacent nucleotides in the same strand pack parallel to each other. The consecutive stacking in DNA and RNA, as described by Vicens and Kieft [12], presents an arrangement as intimate as stacked coins with no space between them for solvents, having stacked bases only ~3.4 Å apart in the helical structure. The tightly stacked bases form a helical ladder with each base pair making a small-angle turn (36° in DNA double helix B Structure) relative to a neighboring base pair, in agreement with the mechanism of $\pi - \pi$ interactions with parallel displaced or staggered stacking as one of the strong configurations. Consecutive stacking significantly contributes to the formation, stability, and rigidity of the helix structure within DNA, between DNA and RNA, and within RNA.

Other types of stacking, such as base-intercalated stacking and base-wedged stacking, contribute to the development of complex and diverse RNA three-dimensional structures [13]. Base-intercalated stacking occurs when any base from a single-stranded region of RNA inserts into two adjacent bases of another region of the same strand, resulting in the formation of secondary structures like pseudoknots. Structures such as kink-turns, hairpin loops, and junctions are driven to form by base-wedged stacking, where any base from one region wedges into two bases of another region, which are one nucleotide apart in sequence at an angle nearly perpendicular to the planes of the two bases. The two types of unique stacking mechanisms, first discovered in tRNA structures, are found to exist widely in various RNAs. It is interesting to note that the stacking motifs contribute to the diversity of RNA structures, play critical roles in the folding of RNA bulges and loops, and participate in long-range interactions with RNA molecules.

21.4.3 Ionic Interaction

As phosphate groups link nucleotides to form nucleic acids as biological polymers, the nature of phosphate group dissociations makes nucleic acids highly negatively charged molecules at physiological pH, which enables strong ionic interactions with cationic ions. It is well recognized that metal ions like Mg^{2+} and Na^{+} can have strong interactions with RNA and affect the formation of secondary and tertiary structures. For example, as revealed by several crystal structures of folded RNA entities like ribozyme domains and self-splicing group introns, individual Mg^{2+} ions are present in the crystal structures binding to multiple phosphate groups, as a key structure manifest to link adjacent helices to form close-packed three-dimensional structures [14]. The similar roles of magnesium ions are also observed in the crystal structures of other RNAs, including tRNA and the 5S ribosomal RNA domain [15]. While site-specific ions described above are essential parts of RNA structure in solution and presumably at the solid state as well, most cations are non-specifically bound and act as cationic atmospheres surrounding RNA molecules in different degrees of association. One type, as tightly bound ions, tends to cluster around the RNA surface to form a high local concentration. The other ions present as diffusively bound types and surround RNA as an ion cloud. The different types of ions act to screen the phosphate's negative charges and enable short-range interactions such as hydrogen bonding and base stacking to form and stabilize RNA high-order structures. Divalent ions, especially Mg^{2+}, are tightly bound and are found to be particularly more effective in stabilizing RNA folded structure than monovalent ions like Na^{+}, which are diffusively bound. Thus, the ionic atmosphere around the RNA delivers an electrostatic shielding to overcome the huge electrostatic repulsion of the negatively charged phosphate backbone of polynucleotides and fold into compact three-dimensional structures.

21.4.4 RNA Secondary and Tertiary Structure

The process of RNA folding can be illustrated as following scheme:

$$U \rightarrow I \rightarrow N$$

where U reflects the unfolded state, I as the intermediate, and N represents the native functional conformation.

Under the unfolded state, as commonly called the denatured state, RNA molecules present as random coils lacking any hydrogen bonds between bases. A secondary structure readily forms in the presence of a low concentration of cations to reach the intermediate state, I. Further folding into the N state relies on the increased concentration of cations and needs divalent cations like magnesium to be present.

However, due to the diversity of RNA structures and functions, the folding process is expected to take a more convoluted pathway, especially for large-molecule RNAs like mRNAs. On the one hand, there is a huge conformational landscape to navigate. On the other hand, multiple interrelated factors would drive the process in a competing fashion, including thermally driven chain fluctuation or mobility, ion-mediated electrostatic interactions, base-pairing hydrogen bonding, stacking, and other noncanonical forces.

Some fundamental understanding of RNA folding into complex three-dimensional secondary and tertiary structures derives from studies on tRNA and ribozyme molecules, which are relatively smaller than mRNA and usually have a well-defined tertiary structure, potentially allowing a crystal structure to form. Analysis of the crystal structures of those RNA molecules leads to some basic knowledge of the three-dimensional architectures and interactions of RNA molecules, including the formation of secondary structure.

Secondary structure in RNAs refers to a distinct local arrangement formed from base-paired regions and unpaired regions, including hairpins, loops, pseudoknots, bulges, and junctions (Figure 21.6).

For example, hairpins are considered the most common secondary structure and form when an RNA molecule folds back on itself to form a double-helical stem with an unpaired loop region as the cap. It should be noted that all the secondary structure regions are three-dimensional structures, though usually illustrated as two dimensional in the literature and textbooks, which may create incorrect perceptions.

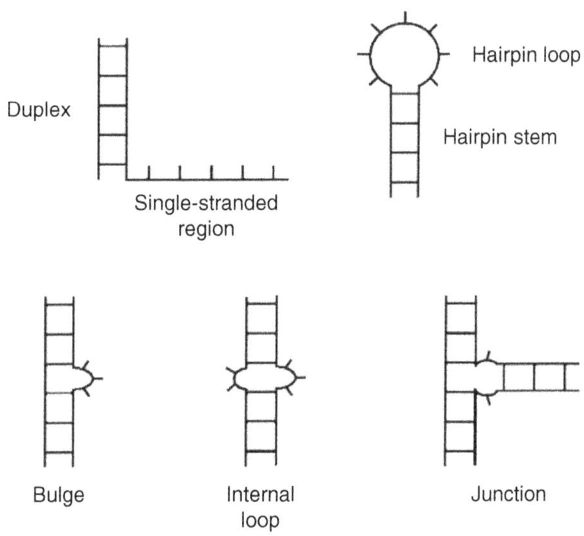

Figure 21.6 Types of secondary structure found in RNA.

The tertiary structure of an RNA molecule implies the overall three-dimensional shape that results from interactions among different regions of the RNA chain. Owing to the significant rotational freedom in the RNA backbone and many un-base-paired regions, RNA can form complex tertiary structures. Due to its relatively small size, tRNA offers a more simplified snapshot of how the tertiary structure of an RNA molecule can form, as contributed by multiple types of interactions, including base triples and base–backbone interactions. Figure 21.7 illustrates how RNA primary, secondary, and tertiary structures look, using tRNA as a case, taking a well-defined L-shaped tertiary structure.

However, RNAs with much longer chains, like most mRNAs, having a much-rugged thermodynamic landscape, favor folding into a dynamic ensemble of structures, instead of a singular three-dimensional architecture. The structures within an ensemble have comparable thermodynamic stabilities and can transition into each other reversibly, reflecting another layer of complexity in RNA three-dimensional structures.

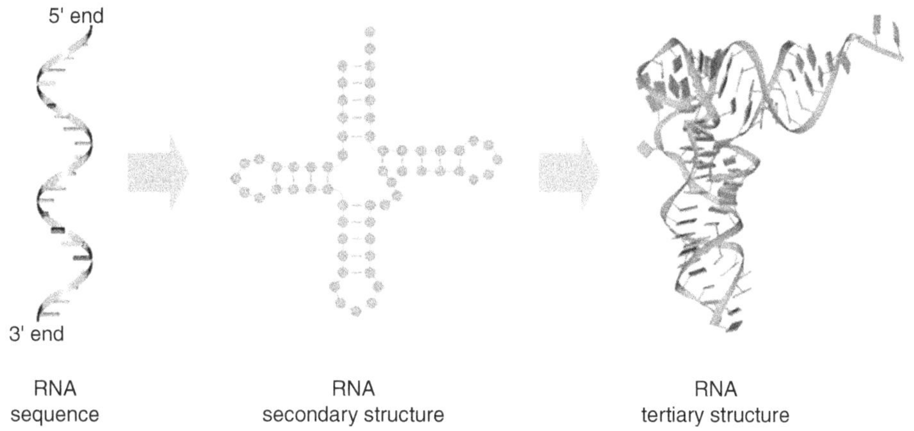

Figure 21.7 Illustration of RNA primary, secondary, and tertiary structures.

It is obvious that the formation of a high-order structure for an RNA molecule is determined fundamentally by its primary structure. It is indeed a molecular recognition process to minimize thermodynamic energy through hydrogen bonding, further enhanced by the stacking effect. Besides the primary sequence, the formation and the resulting structure characteristics of secondary and tertiary structures are strongly influenced by the environment, especially by the type of cations and the concentrations, since the RNA backbone has a strong phosphate negative charge. The neutralization by metal ions becomes one of the determining factors to dictate the folding of RNA molecules.

At a low metal ion concentration, significant un-neutralized negative charges in the backbone lead to strong coulombic repulsion and then subsequently cause the RNA to take an unfolded state with a random coil-like conformation. As the ion concentration rises to neutralize the negative charges and reduce coulombic repulsion, the RNA starts to form secondary structures like helices and loops but with minimal folding. At a high ion concentration, the coulombic repulsion is further reduced, and secondary structures are further compacted to form tertiary structures driven by attraction forces like hydrogen bonding and stacking interactions.

In addition to the concentration effect, different ions can exert different effects, promoting the formation and stabilization of tertiary structures. As reported by Tan and Chen, Mg^{2+} exhibits more effectiveness than Na^+, as exemplified by a case in which 1 M NaCl and 0.3 mM $MgCl_2$ can make approximately the same contributions to the electrostatic free energy of the folding of $tRNA^{Phe}$ and 58-nt rRNA fragment into tertiary structures [16]. Kim et al. reported that divalent ions like Mg^{2+} can give hundreds of times more efficacy in promoting and stabilizing the tertiary structures than monovalent ions such as Na^+ [17]. The phenomena can be understood in terms of counterion condensation theory [18], which assumes that multivalent cations condense more favorably around the nucleic acid than monovalent ions. Multivalent cations neutralize the phosphate negative charges more completely than monovalent ions, leading to a lower entropic cost, enabling the bridging attractive force between phosphate groups in the different secondary structure regions and resulting in a more effective folding. Other studies showed that the combination of mono- and divalent ions, often found in physiological conditions, can provide a synergy to facilitate folding of RNA [19]. As shown in Figure 21.8, sodium chloride displays a concentration-dependent acceleration of Mg^{2+} dependent folding of a ribozyme to form the active native

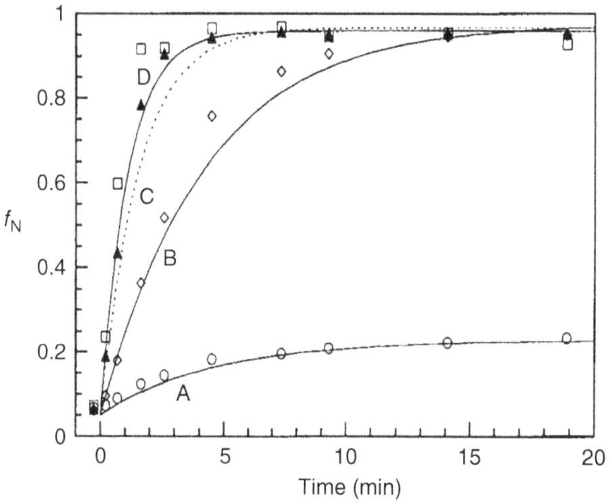

Figure 21.8 Acceleration of Mg^{2+}-dependent folding by NaCl. Folding experiments were carried out with 4 mM $MgCl_2$ and various amounts of NaCl at 30 °C. f_N stands for the fraction of native ribozyme and is fit to a first-order rate equation. (a) Open circle, no sodium chloride, rate = 0.22 $minute^{-1}$; (b) Open diamond, 0.1 M sodium chloride, rate = 0.23 $minute^{-1}$; (c) Solid triangle, 0.25 M sodium chloride, rate = 0.64 $minute^{-1}$; (d) Open square, 0.5 M sodium chloride, rate = 0.89 $minute^{-1}$. *Source*: [19]/with permission of Elsevier.

structure. The first-order rates increase from 0.22 min^{-1} at no sodium chloride to 0.89 min^{-1} at 500 mM concentration. In addition to the rate, the extent of folding completion is also dependent on the presence of sodium cations. In the absence of sodium cation, the amplitude of folding into a natural structure only reaches 20–30% [19]. The role of sodium cation can be explained by a competition between the Mg^{2+} and Na^+ for association with the RNA, which facilitates the rearrangement or refolding of the RNA into the native active structure. It has been found from a study on the ribozyme drz-spur-2 that amino acid-chelated magnesium promoted RNA folding and enhanced thermostability, at the same time, reduced magnesium-dependent degradation [20].

The degree of folding of the ensemble, as discussed earlier, is strongly influenced by the type of cationic and cationic concentration and can be impacted by the presence of molecular crowding agents like polyethylene glycol (PEG)-8000. Molecular crowding agents act through excluded volume effects to promote RNA to form a more compact conformation. Strulson et al. reported that molecular crowders such as high-molecular-weight PEG or dextran, and cosolutes like methanol can facilitate folding cooperativity of RNA under physiological ionic conditions and stabilize tertiary structures as demonstrated by the case study on $tRNA^{phe}$ [21]. As disseminated by another study on catalytic ribozyme, a molecular crowding agent, like PEG, facilitates the RNA assembly into the active folded structure with 3-to-10-fold lower concentration requirement of Mg^{2+}, resulting in better stability against hydrolysis catalyzed by Mg^{2+} [22]. It is interesting to note that some naturally occurring molecules, including amino acids, ribose, and ethylene glycol, serving as a solute, can also display a similar efficacy as PEG. The crowding-facilitating RNA assembly can be attributed to the interplay of enthalpic forces from the interactions between the RNA surface and a crowder and the entropic forces originating from volume exclusion. Molecular crowding effect on RNA folding, in terms of thermodynamics, is exemplified on an isolated GAAA tetraloop-receptor, a catalytic RNA molecule, to have up to $\Delta\Delta G^o \sim -2.5$ kcal/mol change in free energy and more than 60-fold increase in the folding equilibrium constant with an excluded volume fractions of 15% by high-molecular-weight PEG [23].

It is obvious that various molecular forces, such as hydrogen bonding, stacking, and ionic interactions, drive the formation of complex secondary structures and further promote the evolvement into highly dynamic tertiary structures. Key environmental factors, including the presence of ions and crowding agents, play a significant role in determining the kinetics and outcomes of RNA folding into a tertiary structure.

Due to its diversity and polymorphism of secondary and tertiary structures, unlike DNAs and proteins, the structure of RNAs, especially for large-molecule mRNA, is not well studied and defined, with a few unique structure features being discovered. It is interesting to note that natural mRNA molecules, as revealed by theoretical and experimental analysis, tend to have the 5′ and 3′ ends in the proximity of a few nanometers of each other, regardless of length and sequence, which has strong ramifications for how mRNA is further assembled into tertiary structures. For example, Lai et al. reported that average end-to-end distances of multiple mRNA and viral RNA, when being folded in the presence of 100 mM KCl and 1 mM $MgCl_2$, were determined by FRET to be within to 10 nm [24], compared to the theoretically predicted end-to-end of distance of unstructured RNA being linearly increased with the number of nucleotides.

21.5 RNA Stability in Solution

RNA, in general, by natural design, is inherently unstable, because its life span needs to be very well controlled for proteins, enzymes, and receptors to produce and maintain proper levels in very complex biological systems.

It is estimated that the average mRNA half-life in mammalian cells stays in the order of several hours [25], mainly decayed by intracellular ribonucleases. It is well recognized that ribonucleases are widely present and considered ubiquitous. So, aseptic techniques and decontamination approaches to get rid of ribonucleases are essential in RNA preparation, purification, and product formulation.

In addition, RNA, by its chemical structure, is inherently less stable than DNA, since its building block, made of ribose, has a highly reactive hydroxyl group on C2. The group can self-catalyze the hydrolysis of phosphodiester bonds, as shown in Figure 21.9, by acting as an effective nucleophile via an SN2-type mechanism. Water participates via providing hydroxyl or hydronium ions or by facilitating proton transfer.

Nevertheless, the hydroxyl group at the 2′ position is an essential functional group for ribosome binding, indicating that any modifications to this position would lead to ineffective mRNA translation into proteins. Thus, unlike antisense RNAs and siRNAs, the stability of therapeutic mRNAs cannot be improved by using 2′ position modified nucleotides.

The understanding of the kinetics and mechanism of RNA hydrolysis is gained through the numerous studies on dinucleotides as a simplified model by eliminating the complex effect of secondary structure in large RNAs. Both acids

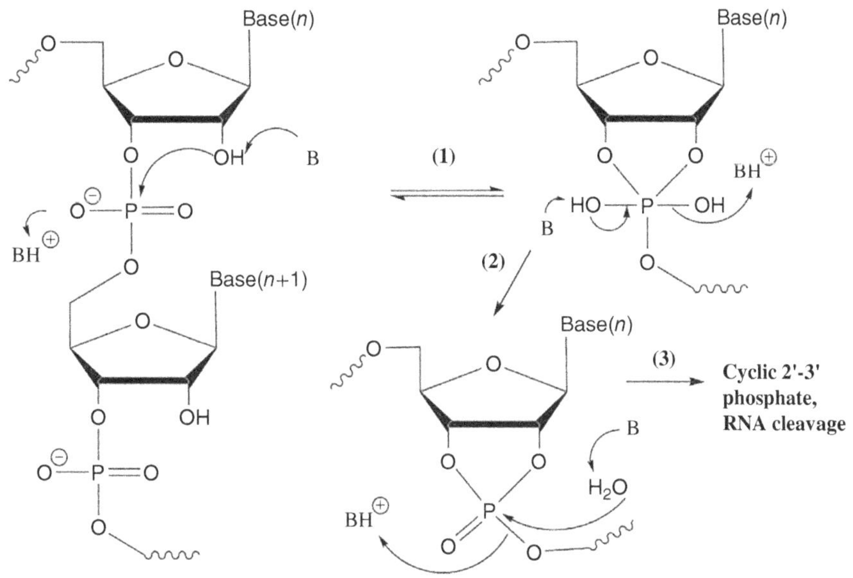

Figure 21.9 Mechanism of the degradative transesterification reaction in RNA leading to the chain breakage.

and bases can catalyze the hydrolysis of the phosphodiester bonds in RNAs and follow a first-order kinetics in bases, while the reaction order is between the first and the second order to hydronium ion in acids. This means that the cleavage of the phosphodiester has a strong pH dependence with a bell-like pH stability profile and fast reaction at acidic and basic conditions, while being more stable at close-to-neutral pH. Figure 21.10 shows a pH-rate profile of the hydrolysis of 3′,5′-UpU dinucleotide at 90 °C at different pHs, indicating the pH range with best stability against cleavage to range from pH 4 to 6 [26]. In addition to pH, temperature obviously has a substantial effect on the rate of RNA hydrolysis, giving a conventional Arrhenius-type kinetics [26].

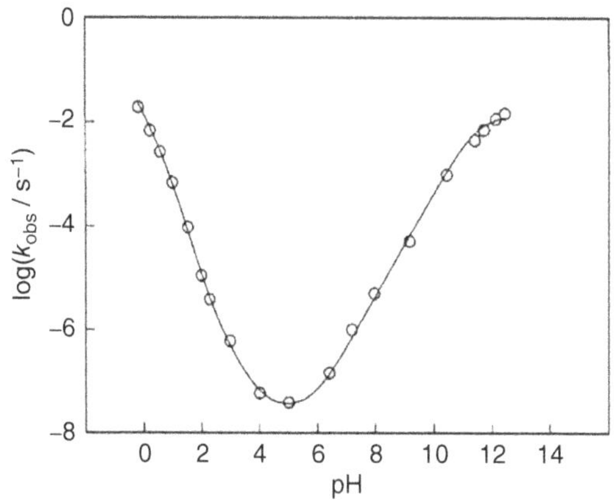

Figure 21.10 pH-rate profile for the hydrolysis of 3′,5′-UpU at 90 °C. *Source*: Adapted from [26].

As discussed in the previous section, cationic metal ions play critical roles in the formation of RNA folded structures and can also exert a profound influence on the rate of RNA hydrolysis, especially divalent ions. For example, magnesium cations can act as a Lewis acid catalyst or facilitate proton transfer to speed up the cleavage of RNA. As revealed by a study on ApG RNA dinucleotide under base-catalyzed cleavage, the presence of divalent magnesium at 5 mM promoted a threefold increase in rate constant over that observed in the absence of magnesium at pH 9.5 at room temperature. On the log scale, the rate constant proportionally increased as the concentration of magnesium ions rose [26]. Nevertheless, it has been found that Zn^{2+} is more effective than Mg^{2+} in metal ion-dependent hydrolysis of RNA phosphodiester bonds [27], and Pb^{2+} and Cu^{2+} are the most potent [28], as stronger Lewis acids. It is interesting to note that, at ambient temperature and in neutral pH, the presence of 10 μm of Pb^{2+} or Cu^{2+} resulted in the hydrolysis of single-stranded RNA [28].

Other than environmental factors such as pH, ions, and temperature, the structural characteristics of an RNA molecule play a vital role in determining its hydrolytic stability. First, as the molecule is getting bigger, more sites are available for cleavage and then less stable. Zhang et al. reported that, at alkaline pH, a 1006 nt ssRNA was degraded approximately an order of magnitude faster than another 106 nt ssRNA, aligned with the 10-fold greater number of phosphodiester bonds existing [29]. It can be envisioned that long RNA molecules like mRNA with thousands of nucleotides have much worse stability than oligonucleotides with 20–30 nucleotides.

Investigations on the non-enzymatic cleavage of small-molecule RNAs like tRNA in imidazole buffer further revealed the impact of secondary structures on the hydrolysis of RNA, using yeast $tRNA^{asp}$ as a model. It appears that the single-stranded regions are strongly preferred for the cleavage, instead of the double-stranded regions [30]. The increased resistance to hydrolysis in the double-stranded regions is due to the nucleotides being in a stacked conformation, leading to increased rigidity of the RNA structure, which prevents the conformational changes needed to allow the formation of the hydrolysis intermediate state.

The profound effect of secondary structures through Watson–Crick pairing to limit the conformational mobility of reactive groups and improve the stability is further evidenced by double-stranded RNAs showing significantly better stability against alkaline hydrolysis than single-stranded RNA [29], as well as metal-catalyzed hydrolysis by lead and copper ions [28]. Wayment-Steele et al. recognized that the increase of base-pairing would slow the rate of hydrolysis, as benchmarked by average unpaired probability and reflected by stability variations of some small RNA oligonucleotides [31].

It is worth noting that magnesium, as a key divalent cation related to RNA structure and function, has a competing effect on its stability. Magnesium catalyzes RNA cleavage but also facilitates RNA folding. Folding to form high-order structures offers protection from hydrolysis. It is proposed that there is a Goldilocks peak where magnesium concentration is optimum for RNA stability, as demonstrated in the study on yeast $tRNA^{phe}$ with a distinct Goldilocks peak near 3 mM Mg^{2+} [32]. The plot of lifetime, derived from RNA cleavage rate, versus the concentration of magnesium fits well with a three-state folding model, including unfolded, intermediate, and folded states. The Goldilocks peak correlates to the condition where the RNA gets 95% folded.

It has been noted that, in addition to hydrolysis, oxidation reactions with impurities like aldehydes, metal ions, or peroxide can occur for mRNAs in solution or solid state, leading to the backbone cleavage of the mRNA into smaller fragments and the loss of biological activities.

21.6 Solid-state Properties

Small RNA molecules like tRNA, with defined secondary structure and tertiary structure, as well as small oligonucleotides like siRNA, have the potential to form crystalline solids. Most investigations sought crystalline phases of RNAs to elucidate the three-dimensional structures and molecular interactions of RNAs but did not intend to produce crystalline materials of RNA for practical use. Thus, not much literature information is available in terms of the solid-state properties of crystalline RNAs, except for the knowledge on structural characteristics obtained from crystal structures.

Two predominant structure features revealed by various crystal structures of RNAs include significant hydration and ubiquitous presence of Mg^{2+} in crystalline phases, which are expected to exist in the solution state and at amorphous solids as well.

RNAs, if a crystal state can form, predominantly exist as a hydrate, like proteins. For example, four water molecules exist in two units of base pairs in the crystal structures of an oligonucleotide duplex r(CCCCGGGG)2 [33]. With about one water per nucleotide, water molecules are hydrogen bonded to 2′-hydroxyl groups, phosphate groups, and base atoms.

Magnesium ions, an essential part of RNA structures, in general, exist in various binding motifs in the crystalline structures. A hierarchical classification system built on published RNA crystal structures proposed by Zheng et al. yields 41 types of inner-sphere and 95 types of outer-sphere coordinating patterns [14], indicating the significance and also the complexity of the involvement of magnesium ions.

However, it is nontrivial to grow crystals even for small RNA molecules. The most challenging task for RNAs to form good-quality crystal phases, besides chemical purity, is to induce conformational homogeneity through heat-denatured RNA samples and refold over a narrow temperature gradient [34].

As would be expected, large RNA molecules like mRNA have very dynamic, high-order structures and can exist in various stable energy landscapes, which make the formation of an ordered crystal lattice extremely difficult or impossible. Thus, RNAs, as a solid in the different phases of pharmaceutical development and manufacturing, predominantly exist in the amorphous state as a polymeric salt, mainly with sodium.

For example, after purification using chromatography, oligonucleotides such as siRNAs are usually lyophilized and present as amorphous solids. mRNA COVID-19 vaccines currently present as frozen LNP suspensions at either −80 or −20 °C, presumably in the amorphous solid state complexed with lipids. However, the amorphous solid-state properties of RNAs are not well studied, and, obviously, understanding the underlying principles is essential to find means to improve the stability and advance this important type of therapeutics to many areas of unmet medical needs.

The most predominant solid-state property of RNA amorphous material is the significant hygroscopicity, reflecting its amorphous nature, plus a hydrophilic ionic polymer structure.

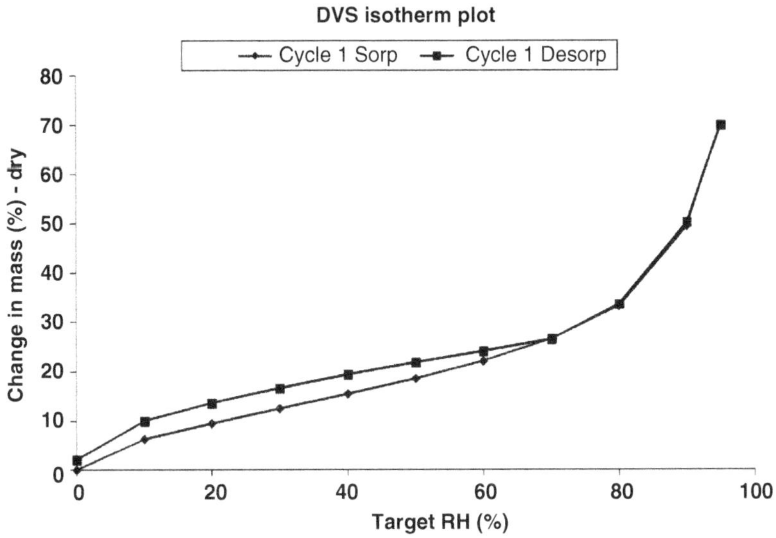

Figure 21.11 Dynamic water vapor sorption at 25 °C for a 20-mer single-stranded oligonucleotide.

RNA as an amorphous lyophilized powder often ranks very hygroscopic based on Ph Eur categorization, as demonstrated by several case studies. Murugaiah reported the water vapor sorption at 30% to 90% RH for ALN-RSV01, a 19-mer double-stranded siRNA targeted against the mRNA of the respiratory syncytial virus (RSV) nucleocapsid (N) protein as a potential anti-RSV therapeutic agent [35]. The initial water content was 4.5% w/w after drying for more than 10 hours under dry nitrogen. Exposure to 30% RH for 5 hours resulted in a weight gain of 12.3%, while weight gain reached 64.3% after equilibration at 90% RH for 12 hours. It is estimated that about 1 water molecule binds to one nucleotide initially, 3 water molecules per nucleotide at 30% RH, and 13 water per nucleotide at 90% RH. Water sorption, though very significant, appears to be reversible during desorption as shown by Appeldorff and Meldgaard in the study for a 20-mer single-stranded oligonucleotide with the dynamic vapor sorption isotherm plot given in Figure 21.11 [36]. It appears that a small hysteresis exists below 70% RH.

Since so much water can be absorbed, it can be concluded that RNA oligonucleotide can easily form a solid solution with water. Despite the amorphous nature of lyophilized oligonucleotides, very little literature is available to decode the solid-state properties, not even glass transition characteristics. There are some sporadic reports on the glass transition temperature of DNA. A study on solid calf thymus DNA with water content about 32% yielded a glass transition temperature of about 238 K from dielectric relaxation spectroscopy [37], indicating nucleic acids as amorphous solids at ambient temperature with significant water absorbed, likely existing as a supercooled liquid. Considering all the reported water content for solid RNAs is in the 20–40% range of water content, it can be concluded that RNAs, if present as amorphous solids at refrigerated conditions or ambient temperature, stay as a supercooled liquid.

It is also interesting to note that some water, about one water molecule per nucleotide, is very tightly bound and difficult to remove. As evidenced by the moisture sorption study discussed earlier shows, water cannot be removed to a level below 5%. The hydration of RNA molecules in an amorphous state is not well understood yet, and water molecules, especially the tightly bound ones, are considered part of the solid-state three-dimensional structure, as indicated by numerous crystal structures of small RNAs like tRNAs. As revealed by the crystal structures of tRNAs, water molecules are indeed an integral part of nucleic acid structures. For example, 195 water molecules are present around the two molecules of mouse $tRNA^{sec}$ in the unit cell of its crystal structure, about one water per nucleotide, considering 86 nucleotides in each tRNA [38], in agreement with the observation that one tightly bound water per nucleotide is present for dried amorphous oligonucleotide. From the structure, the 2′-hydroxyl groups of the ribose, a "hot" site affecting chemical stability and functionality, are heavily hydrated to form strong hydrogen bonding with water.

It appears that RNA likes water. Water is essential for its structure and functions but also causes its instability. The most prominent solid-state property of RNA therapeutics is its poor chemical stability, especially for mRNAs. The poor stability, obviously, is linked to the chemical structure but also related to the solid-state properties, first as an amorphous solid and second being very hygroscopic. The chemical stability of therapeutic oligonucleotides, including antisense RNAs and siRNAs, is significantly improved due to chemical modifications, including the PS linkages and chemically modified nucleotides like 2′-F or 2′-O-ME analogs, with product presentations as lyophilized products or liquids stored in a refrigerator or even at room temperature for a few examples. However, mRNAs, with many more nucleotide units and a higher molecular weight, are only practically being produced by in vitro enzymatic-based translation or the in vivo process. Considering the linear structure of mRNA with about 1000 nucleotides or often longer, any breakage of the primary sequence would lead to a non-translational, wrong, or truncated peptide. Currently, mRNA therapeutics are mainly formulated as LNPs. It has been found that the LNPs cores contain 24% water, presumably where mRNA is located, which could contribute to the poor stability of mRNA vaccines. For example, Pfizer-BioNTech BNT162b2 needs to be stored under −60 to −80 °C. To further remove water content and make mRNA into solid-state materials via lyophilization is widely explored as a means to improve the storage conditions and shelf life of mRNA. As reported by Li et al., three lyophilized mRNA LNPs formulations have water content ranging 2.8–3.1% and are stable at 2–8 °C, which demonstrates significant improvement in comparison to marketed products with storage conditions under frozen conditions like −20 or −80 °C [39].

Work in one of our laboratories (S. Byrn at Purdue University) shows that small-molecule phosphates can be incorporated into solid LNPs (50% SM-102, 38.5% cholesterol, 10% distearoylphosphatidylcholine [DSPC], and 1.5% DMG-PEF2000) and freeze-dried. The product can be loaded into capillary tubes for X-ray analysis under ambient conditions but does appear to be moderately hygroscopic. Additional studies are ongoing once larger quantities of the composition can be prepared.

21.7 Pharmaceutical Development

21.7.1 Drug Delivery Challenges and Current Platforms

There are many challenges facing the effective delivery of RNA therapeutics to the target tissues. For RNA therapeutics to act as effectively as designed, they need to get into the cytoplasm.

First, RNAs have very limited permeability through biological membranes, owing to their nature as a large and hydrophilic biopolymer usually with negative charges at physiological pH. To take into consideration, single-stranded antisense RNAs have a molecular weight from 4 to 10 k daltons, double-stranded siRNAs from 15 to 20 k daltons, and mRNAs mostly more than 500 k daltons, while small-molecule drugs usually have less than 1 k daltons in molecular weight that is regarded as a requirement to have adequate permeability.

Second, RNAs are intrinsically unstable, which is biologically advantageous to allow for rapid and tight control in response to cellular needs and environmental changes. On the one hand, RNases are ubiquitously present intracellularly and extracellularly, ready to clean up RNAs. On the other hand, RNAs are liable to hydrolysis. Therefore, chemical modifications as discussed above can result in a dramatic improvement in enzymatic stability to resist nucleases and chemical stability against hydrolysis. However, most modifications can be incorporated only through chemical synthesis, such as PS linkages, fine for antisense RNAs and siRNAs but not applicable to mRNAs that must be made through a biological process.

Notably, undesirable immune responses may be triggered as the naturally built defense system readily fights against exogenous RNAs from viruses.

Ultimately, RNAs, as a polyanion biopolymer, are highly susceptible to renal clearance.

It is no surprise to note that RNA therapeutics, predominantly up to today, exist in targeted delivery to the eye, to the spinal cord, or to the liver. The widely administered mRNA COVID-19 vaccines are considered to act at the locally delivered site after intramuscular delivery to activate the immune responses, when the local muscle cells make the spike proteins with the mRNA being delivered inside the cells.

Despite many challenges, great effort and endeavor have led to the making of two breakthroughs, including GalNAc conjugation and LNPs.

GalNAc chemical conjugation to antisense RNAs or siRNA offers a revolutionary delivery platform for hepatocyte delivery [40]. By attaching GalNAc sugars, the siRNA conjugates can be recognized by the asialoglycoprotein receptor carbohydrate recognition domain (ASGPR-CRD), which is predominantly present on the surfaces of hepatocytes but minimally on other tissue cells. The binding of GalNAc-conjugated RNAs can trigger the receptor-mediated endocytosis and enable the delivery of antisense RNAs or siRNAs into the cytoplasm. The platform potentially offers a specific delivery of RNA therapeutics for the treatment of hepatic diseases with minimal off-target risks. Several products have been approved using this creative approach, including givosiran for acute hepatic porphyria, lumasiran for primary hyperoxaluria type 1, vutrisiran for hereditary TTR-mediated amyloidosis, and inclisiran for lowering low-density lipoprotein (LDL) cholesterol [40].

Using inclisiran as an example, the triantennary GalNAc is attached to the 3′-terminus of the siRNA passenger strand to hijack the ASGPR-mediated delivery to hepatocytes, targeting the PCSK9 gene mRNA and leading to decreased concentrations of the PCSK9 protein and lowered concentrations of LDL cholesterol [41].

Besides the GalNAc platform, the LNP platform leads to the successful delivery of siRNAs, as well as mRNA therapeutics. LNPs usually comprise an ionizable lipid, a phospholipid, cholesterol, and a polyethylene glycol (PEG) lipid. The ionizable lipid is considered the most critical component for the following reasons. First, it binds to RNA in the LNPs when encapsulated at a lower pH, to offer protection from degradation by exo- and endonucleases. When at physiological pH in the biological fluids, the lipid has a neutral charge to mitigate toxicity and immunogenicity, while it acquires a positive charge in acidic endosomes and interacts with negatively charged bilayer lipids, resulting in the release of RNA into the intracellular cytosol, facilitating the therapeutic actions. To achieve the versability, it has been reported that a pK_a value of ionizable amino head of lipids between 6.2 and 6.5 gives an ideal balance between biocompatible neutral charge during circulation to maintain stability and sufficient positive charge at acidic pH to ensure high encapsulation efficiency on RNAs and eventually in vivo release from endosome [42]. Cholesterol and the phospholipid serve as structural lipids to maintain the shape and integrity of LNPs and moreover to facilitate their fusion with the cell membrane, while the PEG lipid, likely present in the surface of the LNPs, helps reduce the toxicity and immunogenicity, as well as increase the stability and circulation time in vivo.

21.7.2 How to Make RNA Therapeutics

As mentioned earlier, two distinct types of RNA therapeutics differ significantly in structure characteristics and physicochemical properties, as well as the way of manufacturing.

Antisense RNAs and siRNAs, which are RNA sometimes called oligonucleotides because of their small size, are predominantly made by chemical synthesis, on the one hand with a lower cost than biological methods and on the other hand having the opportunity to incorporate modified nucleotides and modified PS linkages. As emphasized in the previous sections, chemical modifications indeed resolve the instability problem for small RNAs.

Typically, the manufacturing process for siRNA therapeutics using chemical pathways comprises the following steps:

- Solid-phase oligonucleotide synthesis, using the phosphonamidite method, proceeding in a 3′ to 5′ direction to add a single nucleotide at a time
- Cleavage from solid support and removal of protecting groups
- Purification by preparative chromatography, either reverse phase or ion exchange chromatography
- Ultrafiltration to remove the mobile phases used in the chromatography
- Annealing of the sense strand and the antisense strand to form double-stranded siRNA
- Lyophilization

Final processing by lyophilization often yields a low-density, hygroscopic, electrostatic, and amorphous solid, typically as the sodium salt. Surely, the hygroscopicity is the most important physical attribute to control to maintain the quality of the active substance. For example, the drug substance of inclisiran, an siRNA agent to treat patients with hyperlipidemia, exists as a white to pale yellow hygroscopic powder [43]. To mitigate the moisture sorption, the substance needs to be packed in high-density polyethylene bottles with foil bags as a secondary moisture barrier and must be stored at −20 °C for long-term use. It appears that the inclisiran drug substance is very soluble in water, with at least 300 mg/mL at ambient conditions. The high solubility in water suggests no significant hurdles to formulating this type of RNA molecule into drug products. Indeed, the inclisiran drug product can be made in a relatively simple way where the active substance is dissolved in water and adjusted for pH to about 7 using sodium hydroxide or phosphoric acid. It is also interesting to note that the inclisiran drug product, a sterile aqueous solution, has a superior stability profile, in view of the granted 24-month expiration period when stored at 20 to 25 °C as prefilled syringes. The excellent stability is obviously attributed to chemical modifications in the siRNA duplex. With one exception among 44 nucleotides, the siRNA contains modifications at the 2′ ribose moiety as 2′-F or 2′-OME modification. Moreover, six of the terminal phosphodiester backbones are modified to a 3′-5′-PS linkage.

Nevertheless, siRNA-like inclisiran does degrade under stressed conditions. Forced degradation studies revealed that significant degradation was observed for inclisiran at high temperatures and under acid, base, and oxidation conditions. Typical degradations include loss of nucleotides from 3′- and 5′-termini, depurination, and desulfurization for PS linkages.

The process to produce mRNA therapeutics includes in vitro transcription of mRNA, purification, and encapsulation with lipids to create LNPs, as highlighted in the following list:

- Creation of DNA template with at least one promoter sequence and mRNA construct sequence, production by either linearizing plasmid DNA or polymerase chain reaction
- In vitro transcription, with RNA polymerases T7, T3, or SP6
- 5′ capping, to protect mRNA from 5′ to 3′ exoribonuclease, by capping enzymes

- Poly(A) tail addition, the length of poly(A) tail affects the rate of decay by 3′ to 5′ exoribonucleases, by enzymatic polyadenylation
- Purification to remove RNA polymerase, DNA templates, nucleotides, and impurities, using precipitation and chromatography
- Fabrication into LNPs

LNPs are made of an ionizable cationic lipid to complex with RNA, helper lipids including DSPC and cholesterol, and a PEG lipid. The complex nanostructured system is designed to protect the delicate RNA molecules from nucleases when delivered in vivo while also enabling intracellular delivery. To form this sophisticated system, a target RNA is combined with an ionic lipid at an acidic pH at which the positively charged ionizable lipid forms a strong charge-driven complex with the negatively charged RNA. With the presence of other lipids, through the antisolvent method and further pH adjustment above the pK_a of the ionizable lipid, RNA-containing LNPs are formed with a near-neutral surface charge, favorable for clinical use, with a sterically stabilized core-shell structure facilitated by the PEG lipid.

Even though lipid particles offer an effective way to deliver mRNA, the unique system could also contribute to the mRNA instability. For example, it was just discovered recently that the lipids in the LNPs can react with mRNA to form adducts by adding reactive lipid species to the nucleobase through covalent bonding, particularly for those ionizable amino lipids [44]. The formation of adducts results in the loss of mRNA activity since the adducts are untranslatable into proteins. The potential mechanism is due to the oxidation of the tertiary amino to form *N*-oxide and further hydrolysis of *N*-oxide to form aldehyde. To mitigate this type of instability, one way is to use temperature control by storing at −20 °C, while the other way is to control the quality of starting material, especially the ionizable lipids.

21.7.3 Characterization of mRNA LNPs

LNPs have been proven as a very effective delivery platform for short interfering RNA (siRNA), and more importantly for mRNA vaccines, pivotal to combat the COVID-19 pandemic. It is recognized that the internal structure and external architecture of nanoparticles may affect the effectiveness of drug delivery and shape pharmaceutical attributes such as particle size, physical stability, and chemical stability. It shall be noted that the characterization of mRNA LNPs relies heavily on physical test techniques, including cryogenic transmission electron microscopy (cryo-TEM), atomic force microscopy (AFM), and small-angle X-ray scattering (SAXS).

Cryo-TEM offers a powerful tool to study the particle size, morphology, and, somewhat to a certain degree, the inner structures of mRNA LNPs.

As can be seen from cryogenic electron micrographs in Figure 21.12, the LNPs, containing human erythropoietin (hEPO) mRNA formulated with DLin-MC3-DMA as the cationic lipid, are spherical, having a size from 60 to 100 nm, and contain an electron-dense core with some particles showing defects like blebs. The data clearly manifest how the change in the

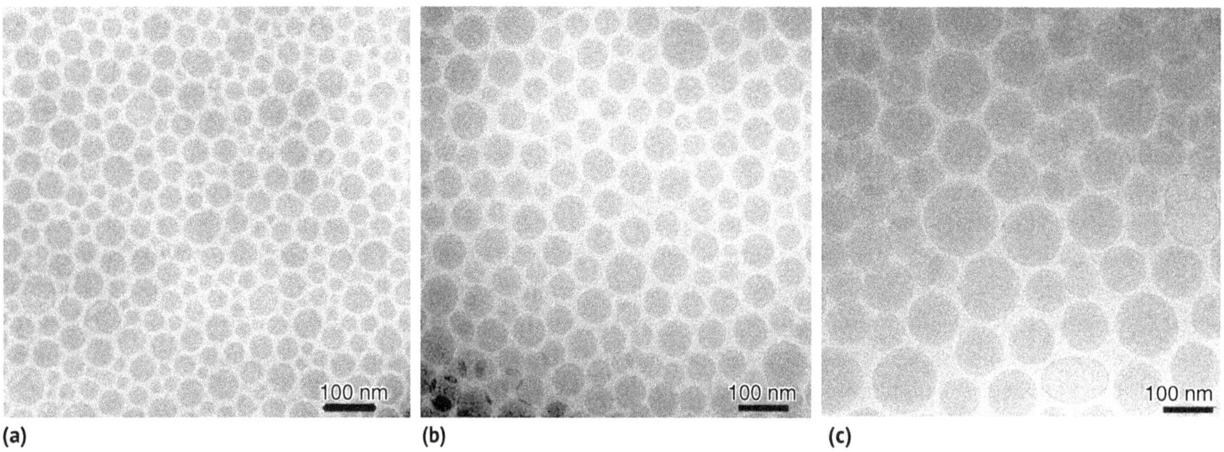

Figure 21.12 Cryo-TEM characterization of hEPO mRNA lipid nanoparticles. (a) Dlin-MC3-DMA:DSPC:Chol:DMPE-PEG2000 50:10:40:3; (b) Dlin-MC3-DMA:DSPC:Chol:DMPE-PEG2000 50:10:40:1.5; (c) Dlin-MC3-DMA:DSPC:Chol:DMPE-PEG2000 50:10:40:0.5. *Source*: Reproduced from [45]/with permission of National Academy of Sciences/CC BY 4.0.

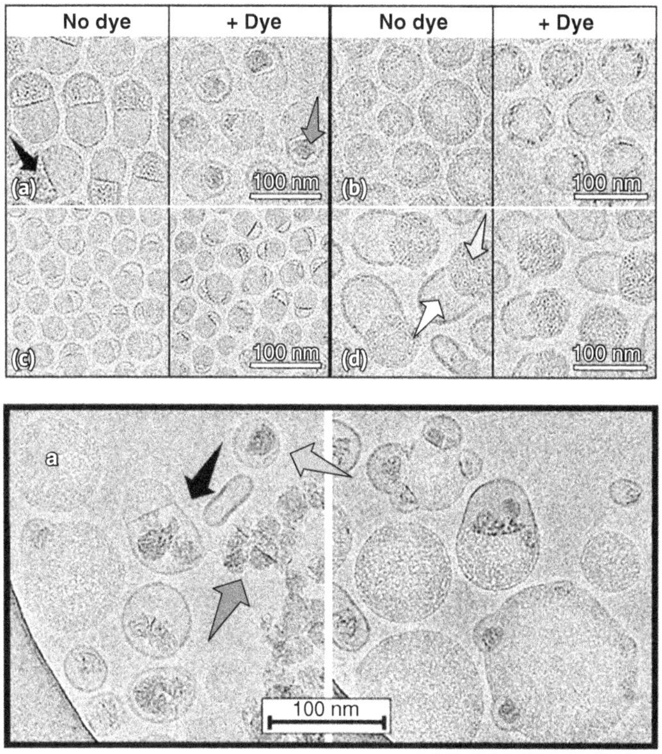

Figure 21.13 Cryo-EM images to locate encapsulated mRNA in lipid nanoparticles after different stages of dialysis (top) and after freezing-thawing (bottom). *Source:* [46]/Elsevier/CC BY 4.0.

concentration of the PEG lipid significantly affects the particle sizes and distributions of nanoparticles, which also correlates with in vivo delivery efficacy [45].

With the assistance of thionine as a dye binding specifically to mRNA, the nanostructure, specific to the location of mRNA within the nanoparticles, can be further resolved, as shown in the case study reported by Brader et al. [46]. Figure 21.13 shows that nanoparticles with bleb defects, indicating phase separation, have mRNA as a denser mottled mass in the image with dye, residing in bleb compartments, which is presumably more aqueous, while nanoparticles with spherical morphology have mRNA dispersed throughout the cores. The technique of coupling cryo-TEM with dye can be utilized to study physical changes of nanoparticles when subject to physical stress such as freeze-thawing. As illustrated in Figure 21.13, freeze-thawing of an mRNA-containing nanoparticle formulation resulted in blebs, aggregation, liberated mRNA, formation of liposomal structures, and eruption.

For example, BNT162b2, a COVID-19 mRNA vaccine branded as Comirnaty, is a preservative-free, sterile dispersion of RNA-containing LNPs in an aqueous buffer. The LNPs contain BNT162B2 mRNA, ALC-0315 as the cationic lipid, ALC-0159 as the PEG lipid, DSPC, and cholesterol in the weight ratio of 1:14.3:1.8:3.1:6.2. It appears that ALC-0315 and cholesterol are the two major components. As disclosed by the manufacturer, the RNA-containing LNPs are solid particles relatively homogeneous in submicron size, spherical in shape, and with a nearly neutral surface. Based on the schematic illustration from the manufacturer, the LNPs present as a multiple-component structure, with PEG lipid to form the external membrane, having cationic lipid to surround mRNA molecules, along with DSPC and cholesterol to support the matrix structure. However, there are debates that the illustration is too simplified and does not align with the size of mRNA molecules and how intimate interactions of the cationic lipid and mRNA are supposed to be.

A combination of SAXS and photon correlation spectroscopy analysis on mRNA-loaded Comirnaty LNP dispersions confirmed the amorphous nature of the inner core made of mRNA, cationic lipid ALC-0315, and cholesterol, while having some semisolid or mesophasic features [47]. It is hypothesized that the inner core, hydrophobic in nature, likely exists in two phases: one mRNA/ALC-0315 clusters with about 95% the lipid and 5% mRNA and matrix phase with about 70% ALC-0315, 25% cholesterol, 3–5% mRNA, which is somewhat supported by AFM and SEM analysis to be discussed in the following. The inner core, though hydrophobic in nature, is considered to have no aqueous phases but not water free. It is estimated that 15–20% of water exists in the inner core, likely binding to the mRNA molecules [47].

Similar water content, about 24% in volume, is presumably present in LNPs containing human hEPO mRNA formulated with DLin-MC3-DMA as the cationic lipid. The water likely forms cylinders along the mRNA, with the cationic lipid existing in a disordered inverse hexagonal phase to surround the cylinders [45], as revealed by combined X-ray and neutron scattering data.

SAXS data clearly show a unique structure present in the mRNA-containing LNPs with an intense peak at q~1 nm, which reflects an ordered structure with a correlation distance of ~6 nm, while the peak is absent in LNP samples without mRNA. The feature detected by SAXS can be attributed to the inverse hexagonal phase of the cationic lipids formed around the mRNA water tunnels.

There is significant interest in understanding how the fundamental structure is construed for the mRNA-containing lipid systems. So many hypotheses flow around, but no convincing models have been accepted so far. From a thorough AFM, dynamic light scattering, and cryo-TEM study, Szebeni et al. proposed mRNA lipid nanoparticles as soft, partially bilayered, covered nanoparticles with hydrogen bond-stabilized mRNA-ionizable lipid complexes [48]. Using mRNA vaccine Comirnaty for a case study, mechanical probing using AFM revealed that the nanoparticles presented as soft, elastic, and compliant structures. Cryo-TEM showed nanoparticles with a granular and solid core covered by mono- and bilayer lipids. The solid core is presumably formed by mRNA and lipid complexes. Considering roughly 10 mRNA molecules within 60–150 nm diameter nanoparticles, it is expected that mRNA molecules are highly folded with dense loops or supercoil-like structures, as hypothesized to exist as balls of yarn or threads of balls. In addition to the ionic interaction, hydrogen bonding between lipids and mRNA offers a stabilization effect on the RNA–lipid complexes [48].

21.7.4 Stability of mRNA LNPs

Indeed, LNPs offer a remarkable delivery platform to surmount numerous barriers, enabling the effective delivery of mRNA. LNPs usually comprise multiple components, including ionizable lipids, cholesterol, PEG-conjugated lipids, and phospholipids, and encapsulate mRNA to form nanosized particles. Owing to the heterogeneous nature comprising multiple phases, particle size at the nanoscale, sophisticated nanostructures, and the intrinsic instability of mRNA molecules, mRNA LNPs have an instability profile much more challenging to tackle than other biopharmaceutical products. The instability can derive from several sources:

- Degradation of mRNA
- Breakage of encapsulation
- Phase separation
- Particle growth or aggregation

mRNA is intrinsically prone to degradation mainly through inline hydrolytic cleavage. The encapsulation of mRNA with lipid formulations is indeed essential for protection from nucleases and in vivo delivery, whereas the presence of cationic lipids in the formulations may facilitate the hydrolysis of mRNA by increasing pH inside LNPs. For example, Wayment-Steele et al., using a quantitative model, predicted that a two-unit pH increase by a cationic lipid formulation could lead to an increase in cleavage rate about 100 times at refrigerated condition for a ~4000 nt mRNA encoding a SARS-CoV-2 spike glycoprotein, in comparison to the same mRNA material not formulated with lipids [31]. Tong and coworkers carried out a thermal stress study for mRNA LNP samples at various temperatures and monitored mRNA degradation, change in LNP particle sizes, and decrease in mRNA encapsulation efficiency for the stability samples up to 3 months [49]. It appears that the degradation of mRNA in the LNPs displayed strong temperature dependence, as shown in Figure 21.14. mRNA integrity remained unchanged at −20°C and decreased about 10% at 4°C after 3 months, while

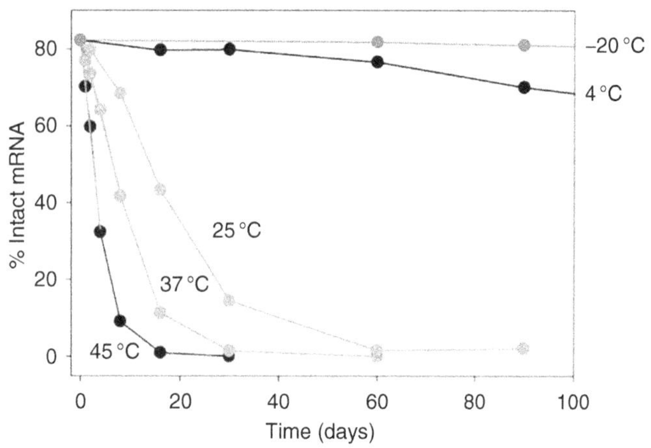

Figure 21.14 Temperature-dependent degradation of mRNA in an LNP product. *Source*: [49]/MDPI/CC BY 4.0.

the storage at 25, 37, and 45 °C resulted in a total loss of mRNA after 60, 30, and 15 days, respectively. It should be noted that the in vivo potency dropped dramatically to almost 0% when the intact mRNA decreased to below 45%. Furthermore, regardless of whether the mRNA molecule was stable at −20 °C, the potency continued to decrease to a 60% level after 3 months, which could be attributed to the undesirable particle growth [49].

Shi et al. also revealed that the long-term stability and immunogenicity of an LNP COVID-19 mRNA vaccine were determined by particle size. The mRNA LNP formulation containing 80–100 nm particles gave the best stability at 4 and −20 °C [50]. Matthessen and colleagues identified a direct correlation between LNP size increase and the loss of mRNA encapsulation when studying how mixing and shaking affect mRNA LNP drug product quality characteristics [51]. The particle size growth not only affects mRNA encapsulation but also may lead to a decrease in efficacy since larger particles with a size greater than 100 nm usually have inferior in vivo activities.

21.7.5 Lyophilization of mRNA LNPs

As would be expected, lyophilization has the potential to improve the stability of mRNA LNPs, prolong the shelf life, and attain a better storage condition other than in a freezer, by removing water from the product. It is well accepted that, in a dehydrated state, degradation like hydrolysis tends to slow down dramatically owing to less water available to react and, moreover, significantly reduced molecular mobility.

However, as mentioned in the previous section, mRNA LNPs represent a complicated delivery system. The physical integrity of the delivery system, as well as the chemical stability of mRNA, is essential for the quality and efficacy of the product. Many instability issues need to be mitigated during the freezing and drying process and through the shelf life, including phase separation, particle aggregation, and particle bursting, in addition to the degradation of mRNA. Hence, the most important goal of formulation work is to search for a proper buffer component and a sufficient lyoprotectant composition to mitigate the physical and chemical instability and develop a quality product without unacceptable loss of efficacy during the course of manufacturing and storage.

It should be noted that buffer components help maintain the pH of mRNA LNPs, which is essential for product stability. For example, a manufacturing change for the COVID-19 mRNA vaccine Comirnaty in 2021 included a reformulation to use tromethamine (Tris) buffer instead of phosphate-buffered saline, resulting in an improved vaccine product being more stable at refrigerated temperatures to allow for better flexibility for distribution. It would be expected that the benefit of Tris buffer is also applicable to lyophilized mRNA LNPs. For instance, Alejo and coworkers discovered that Tris buffer was better than phosphate-buffered saline in keeping the integrity and bioactivity of LNPs, mitigating mRNA leakage, and reducing particle aggregation for a lyophilized product containing mRNA LNPs [52]. Notably, it was found out that the use of sucrose or maltose is necessary, preferably at a 20% w/w level, to better preserve the physicochemical properties and functionality of lyophilized mRNA LNPs, as a lyoprotectant.

Lyoprotectants, widely used in lyophilized products for proteins, peptides, and monoclonal antibodies, appear to be capable of protecting mRNA LNPs from the stresses of freezing and drying and maintaining physical stability during storage, in several potential ways. First, lyoprotectants tend to form a glass matrix after lyophilization, which can offer protection for lipid membranes. Second, lyoprotectants can replace water molecules that depart during the drying process and interact intimately with lipid membranes to conserve the structure. Thus, lyoprotectant screening and level optimization are of importance to attain an acceptable lyophilized product. For example, trehalose emerged as an effective cryoprotectant capable of maintaining the stability of a lyophilized nucleic acid-loaded LNPs product up to 2 years at a concentration of 20% w/v [53]. Instead of focusing on a single lyoprotectant for formulation optimization, Li and coworkers explored the combination of lyoprotectants via experimental design and statistical modeling, and discovered a unique composition of sucrose, trehalose, and mannitol as lyoprotectants, and attained a lyophilized mRNA LNPs product with a root-shaped rigid structure with large porosity [39]. It appears that the combination led to a high eutectic point and an elevated collapse temperature, in addition to the large porosity, which allowed for a more efficient drying process with an overall process time from 8 to 18 hours, in contrast to 40 to 100 hours for a regular freezing drying process. It is interesting to note that the resulting lyophilized mRNA LNPs were stable at 2–8 °C and retained immunogenicity in vivo and in vitro during storage [39].

Lyoprotectants other than sugars are also of interest. Fan et al. did a high-throughput screening of 45 lyoprotectant and buffer conditions for an mRNA LNP product and found some unconventional formulation compositions, such as polyvinylpyrrolidone in Tris or acetate buffer, can perform as effectively as some commonly used lyoprotectants to maintain the physicochemical stability of lyophilized mRNA LNPs [54]. Hermann et al. found that a novel ultrahydrating polymer can offer exceptional efficacy in preserving mRNA LNPs during freezing and lyophilization, as a superior lyoprotective agent [55].

It is worthwhile to note that lyophilization is a lengthy process involving freezing, primary drying, and secondary drying, ranging from several days. The long process may put unwanted stress on mRNA LNPs, such as a sophisticated and relatively fragile delivery system, especially during the secondary drying at an elevated temperature. Meulewaeter et al. developed a continuous lyophilization process using spin-freezing and radiative drying to improve drying efficiency and subsequently reduce overall process time, potentially leading to a better product [56]. In the freezing step, vials were spun along their longitudinal axis to create a vertical frozen tube with a larger surface area to allow for faster heat transmission and quicker mass transfer. During the drying phase, the vials were subject to drying by an infrared source that directly heated the vials. Combination of both unique approaches resulted in a much shorter drying time [56].

In principle, lyophilization shall offer a tangible solution to the poor thermal stability of mRNA LNP products since the process can yield products with a reduced water content and presumably significantly decreased molecular mobility. The quality of the product relies not only on the stability of mRNA molecules against degradation but also on the physical integrity of the LNPs without significant changes in particle size and phase partition. It can be seen that further advancement in lyoprotectant selection and lyophilization technology can end with mRNA LNP products stable at refrigerated conditions or room temperature, with a more favorable shelf life.

21.8 Summary

RNA therapeutics represent a new class of drugs and vaccines to assert therapeutic effects through modulating gene expression. The dramatic advancement of this type of therapeutics, especially mRNA vaccines to combat the COVID-19 pandemic, is well regarded as a modern-day pharmaceutical wonder. This significant breakthrough is owing to a dramatic modification of the physical and chemical properties of RNA to reduce enzymatic liability against cleavage and improve chemical stability, preventing hydrolysis and, more importantly, the use of lipid nanoparticles to enable an effective drug delivery. In solution, RNA molecules tend to form high-order secondary and tertiary structures as stabilized by hydrogen bonding, stacking interactions, and ion interactions. The solid-state properties of RNA therapeutics, including frozen mRNA vaccines, are not well understood, except for some data showing that the type of material is highly hydrated. The presence of a significant amount of water inside mRNA LNPs could be one of the reasons to incur the thermal instability. Lyophilization to convert mRNA LNPs into a solid presentation appears to offer a potential solution to have products with more favorable storage conditions other than freezing and possibly longer shelf life.

References

1 Ambros V. The functions of animal microRNAs. *Nature* 2004; 431(7006): 350–355.

2 Puttaraju M, Jackson M, Klein S, Shilo A, Bennett CF, Gordon L, Rigo F, Misteli T. Systematic screening identifies therapeutic antisense oligonucleotides for Hutchinson–Gilford progeria syndrome. *Nat Med* 2021; 27(3): 526–535.

3 Bege M, Borbás A. Rise and fall of fomivirsen, the first approved gene silencing medicine–A historical review. *Acta Pharm Hung* 2022; 92(1): 38–44.

4 Dana H, Chalbatani GM, Mahmoodzadeh H, Karimloo R, Rezaiean O, Moradzadeh A, Mehmandoost N, Moazzen F, Mazraeh A, Marmari V, Ebrahimi M. Molecular mechanisms and biological functions of siRNA. *Int J Biomed Sci* 2017; 13(2): 48–57.

5 Zhang MM, Bahal R, Rasmussen TP, Manautou JE, Zhong XB. The growth of siRNA-based therapeutics: updated clinical studies. *Biochem Pharmacol* 2021; 189: 114432.

6 Adams D, Gonzalez-Duarte A, O'Riordan WD, Yang CC, Ueda M, Kristen AV, Tournev I, Schmidt HH, Coelho T, Berk JL, Lin KP. Patisiran, an RNAi therapeutic, for hereditary transthyretin amyloidosis. *N Engl J Med* 2018; 379(1): 11–21.

7 Teo SP. Review of COVID-19 mRNA Vaccines: BNT162b2 and mRNA-1273. *J Pharm Pract* 2022; 35(6): 947–951.

8 Chavez M, Chen X, Finn PB, Qi LS. Advances in CRISPR therapeutics. *Nat Rev Nephro* 2023; 19(1): 9–22.

9 Manoharan M, Akinc A, Pandey RK, Qin J, Hadwiger P, John M, Mills K, Charisse K, Maier MA, Nechev L, Greene EM. Unique gene-silencing and structural properties of 2′-fluoro-modified siRNAs. *Angew Chem* 2011; 123(10): 2332–2336.

10 Center for Drug Evaluation and Research. Application Number: 209531Orig1s000. Chemistry Reviews. Nusinersen Injection. https://www.accessdata.fda.gov/drugsatfda_docs/nda/2016/209531Orig1s000ChemR.pdf (accessed 24 January 2025).

11 Kim SC, Sekhon SS, Shin WR, Ahn G, Cho BK, Ahn JY, Kim YH. Modifications of mRNA vaccine structural elements for improving mRNA stability and translation efficiency. *Mol Cell Toxicol* 2022; 1: 1–8.

12 Vicens Q, Kieft JS. Thoughts on how to think (and talk) about RNA structure. *Proc Natl Acad Sci* 2022; 119(17): e2112677119.

13 Baulin E, Metelev V, Bogdanov A. Base-intercalated and base-wedged stacking elements in 3D-structure of RNA and RNA–protein complexes. *Nucleic Acids Res* 2020; 48(15): 8675–8685.

14 Zheng H, Shabalin IG, Handing KB, Bujnicki JM, Minor W. Magnesium-binding architectures in RNA crystal structures: validation, binding preferences, classification and motif detection. *Nucleic Acids Res* 2015; 43(7): 3789–3801.

15 Correll CC, Freeborn B, Moore PB, Steitz TA. Metals, motifs, and recognition in the crystal structure of a 5S rRNA domain. *Cell* 1997; 91(5): 705–712.

16 Tan ZJ, Chen SJ. Salt contribution to RNA tertiary structure folding stability. *Biophy J* 2011; 101(1): 176–187.

17 Kim HD, Nienhaus GU, Ha T, Orr JW, Williamson JR, Chu S. Mg^{2+}-dependent conformational change of RNA studied by fluorescence correlation and FRET on immobilized single molecules. *Proc Natl Acad Sci* 2002; 99(7): 4284–4289.

18 Onoa B, Tinoco Jr I. RNA folding and unfolding. *Curr Opin Struct Biol* 2004; 14(3): 374–379.

19 Heilman-Miller SL, Pan J, Thirumalai D, Woodson SA. Role of counterion condensation in folding of the Tetrahymena ribozyme II. Counterion-dependence of folding kinetics. *J Mol Biol* 2001; 309(1): 57–68.

20 Yamagami R, Bingaman JL, Frankel EA, Bevilacqua PC. Cellular conditions of weakly chelated magnesium ions strongly promote RNA stability and catalysis. *Nat Commun* 2018; 9(1): 2149.

21 Strulson CA, Boyer JA, Whitman EE, Bevilacqua PC. Molecular crowders and cosolutes promote folding cooperativity of RNA under physiological ionic conditions. *RNA*. 2014; 20(3): 331–347.

22 DasGupta S, Zhang S, Szostak JW. Molecular crowding facilitates ribozyme-catalyzed RNA assembly. *ACS Central Sci.* 2023; 9(8): 1670–1678.

23 Dupuis NF, Holmstrom ED, Nesbitt DJ. Molecular-crowding effects on single-molecule RNA folding/unfolding thermodynamics and kinetics. *Proc Natl Acad Sci* 2014; 111(23): 8464–8469.

24 Lai WJ, Kayedkhordeh M, Cornell EV, Farah E, Bellaousov S, Rietmeijer R, Salsi E, Mathews DH, Ermolenko DN. mRNAs and lncRNAs intrinsically form secondary structures with short end-to-end distances. *Nat Commun* 2018; 9(1): 4328.

25 Harpold MM, Wilson MC, Darnell Jr., JE. Chinese hamster polyadenylated messenger ribonucleic acid: Relationship to non-polyadenylated sequences and relative conservation during messenger ribonucleic acid processing. *Mol Cell Biol* 1981; 1: 188–198.

26 Li Y, Breaker RR. Kinetics of RNA degradation by specific base catalysis of transesterification involving the 2′-hydroxyl group. *J Am Chem Soc* 1999; 121(23): 5364–5372.

27 Zagórowska I, Kuusela S, Lönnberg H. Metal ion-dependent hydrolysis of RNA phosphodiester bonds within hairpin loops. A comparative kinetic study on chimeric ribo/2′-O-methylribo oligonucleotides. *Nucleic Acids Res* 1998; 26(14): 3392–3396.

28 Chatterjee A, Zhang K, Rao Y, Sharma N, Giammar DE, Parker KM. Metal-catalyzed hydrolysis of RNA in aqueous environments. *Environ Sci Technol* 2022; 56(6): 3564–3574.

29 Zhang K, Hodge J, Chatterjee A, Moon TS, Parker KM. Duplex structure of double-stranded RNA provides stability against hydrolysis relative to single-stranded RNA. *Environ Sci Technol* 2021; 55(12): 8045–8053.

30 Vlassov VV, Zuber G, Felden B, Behr JP, Giegè R. Cleavage of tRNA with imidazole and spermine imidazole constructs: a new approach for probing RNA structure. *Nucleic Acids Res* 1995; 23(16): 3161–3167.

31 Wayment-Steele HK, Kim DS, Choe CA, Nicol JJ, Wellington-Oguri R, Watkins AM, Parra Sperberg RA, Huang PS, Participants E, Das R. Theoretical basis for stabilizing messenger RNA through secondary structure design. *Nucleic Acids Res* 2021; 49(18): 10604–10617.

32 Guth-Metzler R, Mohamed AM, Cowan ET, Henning A, Ito C, Frenkel-Pinter M, Wartell RM, Glass JB, Williams LD. Goldilocks and RNA: where Mg^{2+} concentration is just right. *Nucleic Acids Res* 2023; 51(8): 3529–3539.

33 Egli M, Portmann S, Usman N. RNA hydration: a detailed look. *Biochem* 1996; 35(26): 8489–8494.

34 Ke A, Doudna JA. Crystallization of RNA and RNA–protein complexes. *Methods* 2004; 34(3): 408–414.

35 Murugaiah V. Stability indicating methods for oligonucleotide products. In Bonilla JV, Srivatsa GS, editors. *Handbook of analysis of oligonucleotides and related products*. Boca Raton: CRC Press; 2011.

36 Appeldorff L, Meldgaard M, Inventors. Roche Innovation Center Copenhagen, Applicant. Oligonucleotide formulation method. International Publication Number WO 2020/173845. 2020.

37 Neubert M, Bakule R, Nedbal J. Temperature and humidity dependence of the dielectric spectrum of Na DNA in solid state. In: *1985 5th International Symposium on Electrets (ISE 5)*, IEEE, 1985: 825–830.

38 Ganichkin OM, Anedchenko EA, Wahl MC. Crystal structure analysis reveals functional flexibility in the selenocysteine-specific tRNA from mouse. *PLoS One* 2011; 6(5): e20032.

39 Li M, Jia L, Xie Y, Ma W, Yan Z, Liu F, Deng J, Zhu A, Siwei X, Su W, Liu X. Lyophilization process optimization and molecular dynamics simulation of mRNA-LNPs for SARS-CoV-2 vaccine. *npj Vaccines* 2023; 8(1): 153.

40 Debacker AJ, Voutila J, Catley M, Blakey D, Habib N. Delivery of oligonucleotides to the liver with GalNAc: from research to registered therapeutic drug. *Mol Ther* 2020; 28(8): 1759–1771.

41 German CA, Shapiro MD. Small interfering RNA therapeutic inclisiran: a new approach to targeting PCSK9. *BioDrugs* 2020; 34(1): 1–9.

42 Yan Y, Liu XY, Lu A, Wang XY, Jiang LX, Wang JC. Non-viral vectors for RNA delivery. *J Control Release* 2022; 342: 241–279.

43 European Medicines Agency. Assessment report, Leqvio, International non-proprietary name: inclisiran. Procedure No. EMEA/H/C/005333/0000. 2020; EMA/696912/2020.

44 Packer M, Gyawali D, Yerabolu R, Schariter J, White P. A novel mechanism for the loss of mRNA activity in lipid nanoparticle delivery systems. *Nat Commun* 2021; 12(1): 6777.

45 Yanez Arteta M, Kjellman T, Bartesaghi S, Wallin S, Wu X, Kvist AJ, Dabkowska A, Székely N, Radulescu A, Bergenholtz J, Lindfors L. Successful reprogramming of cellular protein production through mRNA delivered by functionalized lipid nanoparticles. *Proc Natl Acad Sci* 2018; 115(15): E3351–3360.

46 Brader ML, Williams SJ, Banks JM, Hui WH, Zhou ZH, Jin L. Encapsulation state of messenger RNA inside lipid nanoparticles. *Biophys J* 2021; 120(14): 2766–2770.

47 Unruh T, Götz K, Vogel C, Fröhlich E, Scheurer A. SAXS/PCS study on the mesoscopic structure of lipid nanoparticle (LNP) formulations for mRNA drug delivery: Comirnaty and drug-free LNP dispersions. *ChemRxiv* 2023; https://doi.org/10.26434/chemrxiv-2023-v2t4f.

48 Szebeni J, Kiss B, Bozó T, Turjeman K, Levi-Kalisman Y, Barenholz Y, Kellermayer M. Insights into the structure of Comirnaty Covid-19 vaccine: a theory on soft, partially bilayer-covered nanoparticles with hydrogen bond-stabilized mRNA–lipid complexes. *ACS Nano* 2023; 17(14): 13147–13157.

49 Tong X, Raffaele J, Feller K, Dornadula G, Devlin J, Boyd D, Loughney JW, Shanter J, Rustandi RR. Correlating Stability-Indicating Biochemical and Biophysical Characteristics with In Vitro Cell Potency in mRNA LNP Vaccine. *Vaccines* 2024; 12(2): 169.

50 Shi R, Liu X, Wang Y, Pan M, Wang S, Shi L, Ni B. Long-term stability and immunogenicity of lipid nanoparticle COVID-19 mRNA vaccine is affected by particle size. *Hum Vacc Immunother* 2024; 20(1): 2342592.

51 Matthessen R, Van Pottelberge R, Goffin B, De Winter G. Impact of mixing and shaking on mRNA-LNP drug product quality characteristics. *Sci Rep* 2024; 14(1): 19590.

52 Alejo T, Toro-Córdova A, Fernández L, Rivero A, Stoian AM, Pérez L, Navarro V, Martínez-Oliván J, de Miguel D. Comprehensive optimization of a freeze-drying process achieving enhanced long-term stability and in vivo performance of lyophilized mRNA-LNPs. *Int J Mol Sci* 2024; 25(19): 10603.

53 Ferreira HA, Júnior SR, de Faria KK, Silva GH, Alves MT, Lobo AO, Guimarães PP. Cryoprotectant optimization for enhanced stability and transfection efficiency of pDNA-loaded ionizable lipid nanoparticles. *Int J Pharm* 2024; 665: 124696.

54 Fan Y, Rigas D, Kim LJ, Chang FP, Zang N, McKee K, Kemball CC, Yu Z, Winkler P, Su WC, Jessen P. Physicochemical and structural insights into lyophilized mRNA-LNP from lyoprotectant and buffer screenings. *J Control Release* 2024; 373: 727–37.

55 Herrmann A, Abbina S, Bathula NV, Nouri PM, Chafeeva I, Constantinescu I, Abbasi E, Abbasi U, Drayton M, Luo HD, Moon H. An ultrahydrating polymer that protects protein therapeutics and RNA-lipid nanoparticles against freezing, heat and lyophilization stress. *Adv Funct Mater* 2024; 34(42): 2406878.

56 Meulewaeter S, Nuytten G, Cheng MH, De Smedt SC, Cullis PR, De Beer T, Lentacker I, Verbeke R. Continuous freeze-drying of messenger RNA lipid nanoparticles enables storage at higher temperatures. *J Control Release* 2023; 357: 149–160.

22

Solid Form Selection of Active Pharmaceutical Ingredients

22.1 Introduction

Solid form selection of active pharmaceutical ingredients (APIs) is regarded as an essential part of drug development. The goal of form selection is to identify the most suitable solid form for toxicological studies, drug product development, and eventually commercialization. An API can exist as a free form, a salt (Chapter 4), or a cocrystal (Chapter 5), where free forms, salts, and cocrystals also can exhibit polymorphism (Chapter 2) and pseudopolymorphism (Chapter 3). It has been shown in Chapters 2–5 that each solid form has a unique physical identity and tends to display distinct pharmaceutically relevant properties, including solubility, dissolution rate, melting point, stability, hygroscopicity, crystal habit, and mechanical properties. The solid form selected for an API, therefore, affects the quality, performance, processability, and stability of drug substances and drug products.

Solid form selection can be required at various stages of development, as early as drug candidate selection and later as part of life cycle management after commercialization. It is also extremely important that proper form selection is made before a final solid form of the drug substance is produced for the initiation of toxicological studies and Phase I clinical trials. Switching solid forms during development is not desirable, but sometimes it is inevitable due to the emergence of a new stable polymorph or the need for a different solid form to enhance stability, solubility, processability, or intellectual property. Since the form change must be justified on a regulatory basis, additional "bridging" studies must be carried out with pharmacokinetic studies in animals or humans or stability evaluation.

Depending on the physical attributes of the API, solid form screening can include salt selection, cocrystal screening, polymorph screening, and hydrate/solvate screening. Sometimes, amorphous screening is deemed necessary to search for an amorphous (Chapter 6) form for various pharmaceutical uses, for example, to enhance solubility. Various characterization techniques, including X-ray powder diffraction (XRPD), thermal analysis, microscopy, and spectroscopy, are commonly utilized in combination to provide a clear identification of each solid form prepared. The physical properties of the form are further studied with regard to moisture sorption, particle size, crystallinity, solubility, and dissolution. Slurry experiments and temperature-dependent evaluation of solubility lead to the assessment of the relative stability of those forms, where physical stability and chemical stability in the solid state are quickly assessed with storage at accelerated conditions such as 40 °C/75% RH and 50–60 °C. The information obtained regarding physical properties and stability becomes the basis for the final decision to choose a suitable solid form for development.

In this chapter, we will present various screening methodologies, selection strategies, and processes by which final decisions are generally made.

22.2 Solid Form Selection

Solid form selection includes the following scope of work:

- Salt selection if the compound is ionizable
- Cocrystal screening if needed
- Screening of all possible polymorphs or pseudopolymorphs for a free form, salt form, or cocrystal
- Selecting a proper form for development based on the available solid forms

Solid-State Materials in Pharmaceutical Chemistry: Properties, Characterization, and Applications, Second Edition. Stephen R. Byrn, George Zografi and Xiaoming (Sean) Chen.

Polymorphism and pseudopolymorphism occur quite commonly with low molecular weight drugs. For example, it has been estimated that about one-third of organic compounds tend to have multiple polymorphs, whereas another one-third of those have hydrates and solvates [1]. Since different forms give different solubility, physical stability, chemical stability, processability, and bioavailability, discovering various solid forms of an API can be a valuable resource for establishing a useful design space for API development. A soluble salt form can be utilized for supporting toxicological studies, even though the salt form may not be suitable for commercial development. Chiang et al. reported that the mesylate salt of a potent pyridine-containing MK2 inhibitor was developed as part of a formulation strategy to enhance both oral bioavailability and dose escalation orally in preclinical rat studies [2]. The neutral form, having low solubility, gave limited oral bioavailability and, therefore, poor dose escalation in toxicity studies. The MK2 inhibitor is a very weak base and has a pK_a of 4, so the mesylate salt has poor physical stability and tends to disproportionate in aqueous medium and elevated humidity. Although it appears that the mesylate salt cannot be used for the commercial product, its short-term stability window in aqueous media makes it a feasible choice for toxicological dosing, providing significant enhancement in exposure, which the free form failed to achieve.

Different solid forms can be used in different formulations to match the requirement for physical stability. For example, the mometasone furoate anhydrous form is used in dry powder inhaler formulations for the treatment of asthma, while mometasone furoate monohydrate is the active ingredient for an aqueous nasal spray for the treatment of seasonal or perennial allergic rhinitis or acute rhinosinusitis. The monohydrate is more stable in aqueous media since the anhydrous form tends to convert to the monohydrate in the nasal spray formulations. Formation of mometasone furoate monohydrate was observed when a suspension containing the mometasone furoate anhydrous form was subjected to a stability test by rotating the sample for 4 weeks at room temperature and 35 °C. Use of the monohydrate removed the liability of phase transformation during storage and led to a more stable suspension product [3].

Many solid forms are patentable due to their distinct physical and chemical properties, so solid form selection offers opportunities to secure patents that help to develop line extension products.

There have been cases where the development and marketing of the same drug in generic form has been pursued by other companies using different polymorphic forms or salt forms. Ranitidine HCl, for example, can exist in two polymorphic forms: Form I (m.p. 134–140 °C) and Form II (m.p. 140–144 °C). Form I was originally discovered, but Form II was identified subsequently. Form II was then used in the originally marketed product, while generic companies attempted to launch a ranitidine product using Form I to avoid the Form II patent [4].

Solid form selection can be considered to be an informatics process since it involves the collection of significant amounts of data, data interpretation, data visualization, data mining, and finally knowledge-based decision making. The scope of the informatics required is dependent on how many forms exist or whether these forms can be discovered with reasonable means. These reasonable means include the study of conditions and solvents that are commonly utilized in processing, and exist in storage conditions. For salt forms or cocrystals, counterions or coformers should have acceptable safety profiles so as not to pose regulatory uncertainty.

Solid form selection can be considered to be a process that requires risk assessment, since indeed there is no perfect solid form that can always be identified. Stable forms, for example, are generally less soluble and may affect in vivo dissolution, while metastable forms offer greater solubility enhancement but pose challenges for physical stability. For example, the use of a metastable form of ritonavir in a semisolid formulation led to the crystallization of a stable form and caused product failure [5]. In another case, although ranitidine HCl exhibited better oral absorption than the free base and was successfully developed into a drug product, it proved to be very hygroscopic, leading to deliquescence at about 67% [6]. A cocrystal can show promising use for enhancing bioavailability, but it may be difficult to manufacture on a large scale. Risk assessment must be carried out to examine key aspects, including physical properties, control, performance, and processability (Table 22.1). The decision-making process depends on which of these aspects has a higher priority and which of the risk factors is more manageable in the development process.

The scope of solid form selection generally includes:

- design experiments for the purpose of screening and determining the amount of material required,
- screen solid forms and prepare as many forms as possible,
- identify and characterize the physical nature of each solid form,
- select two or more solid forms for scale-up and in-depth characterization,
- study chemical stability and physical properties of lead solid forms, and
- decide on a final solid form for development.

Table 22.1 Properties of solid forms.

Physical properties	Control	Performance	Processability
• Crystallinity • Hygroscopicity	• Polymorphism • Formulation of hydrates • Phase transformation • Impurities • Residual solvents • Chemical stability	• Solubility • Dissolution • Shelf life	• Mechanical properties • Micronization • Filtration • Compressibility • Compactability • Flow properties • Suspension

There is no universal strategy that is used by the pharmaceutical industry and acceptable to regulatory agencies for solid form selection. The best practice to execute solid form selection varies with companies, where the timeline and strategy for screening are tailored to the needs of each development program. If the development program is very aggressive, preliminary solid form selection can be pursued to identify a suitable solid form for the initial safety studies. It is preferable that the final solid form be selected before supplying material for Phase 2B studies, since it will be very costly to switch solid forms after Phase 2B because of the need to carry out pharmacokinetic bridging studies.

22.3 Amorphous Form Screening

The amorphous form of an API exists in a much higher energy state than that of crystalline forms, and consequently displays a significant enhancement in solubility [7]. Despite the solubility advantage, amorphous materials are not widely used as APIs due to their intrinsic chemical and physical instabilities, which cause significant challenges for maintaining drug product quality and stability during manufacture and storage. Selection of an amorphous form as API for development is considered for a compound mainly when the crystalline forms are not sufficiently water soluble, thus causing insufficient bioavailability as oral dosage forms or inadequate dissolution for reconstitution into a solution for parenteral administration. In addition, there are certain other circumstances where amorphous APIs can be useful, including the following:

- Use amorphous API in preclinical and safety studies to accelerate drug early development
- Assist in the development of amorphous solid dispersions by fully characterizing a neat amorphous API
- Serve as a 100% amorphous standard in the quantitative analysis of crystallinity
- Use as a starting material for polymorph screening

Amorphous form screening is used to search for an experimental means to prepare fully or partially amorphous materials to meet the aforementioned needs. In most cases, amorphous forms are not easily prepared due to their natural instability and, therefore, different techniques need to be explored to identify one that can yield a quality amorphous phase for a compound.

Techniques to create amorphous materials include:

- melt quenching,
- lyophilization,
- spray drying,
- dehydration/desolvation of crystal hydrates/solvates,
- high-energy milling,
- cryogenic milling, and
- acoustic levitation.

Melt quenching involves melting a crystalline drug and rapidly cooling the molten drug to avoid crystallization. The rate of cooling to "lock" a drug into the amorphous state depends on several compound properties. For example, amorphous ritonavir, ABT-229, and fenofibrate were prepared by in situ melt quenching using differential scanning calorimetry (DSC) with a cooling rate set to 20 K/min, while amorphous acetaminophen required preparation by melting in a sealed DSC pan and quenching in liquid nitrogen [8]. Crowley and Zografi reported that amorphous indomethacin was prepared by melting the γ form of indomethacin and rapidly cooling by immersion into liquid nitrogen [9]. Melt quenching is only suitable for compounds having good chemical stability around the melting temperature. For thermally labile compounds, lyophilization and spray drying are alternative approaches to prepare amorphous materials.

Lyophilization is applicable for compounds with good aqueous solubility, where an aqueous solution of a drug is frozen, and water is removed by sublimation, leading to the production of amorphous material. Zhou et al., for example, reported the use of lyophilization to prepare amorphous sucrose [8].

Spray drying involves dissolving a drug in a solvent and spraying the drug solution into a stream of hot air to rapidly remove the solvent. The rapid drying tends to promote the formation of amorphous materials by limiting the time available for crystallization.

For compounds that can form crystal hydrates or solvates, amorphous material can sometimes be prepared by desolvation or dehydration. In such cases, the removal of solvent molecules or water molecules from the lattice can lead to a collapse of the lattice, resulting in the formation of amorphous material.

High-energy milling can produce partial or fully amorphous material by destroying the crystal lattice. It has been hypothesized that milling leads to an increase in disorder that adds to the disorder inherent within the lattice up to a critical value, after which structural collapse occurs to form the disordered phase. It has been found that cryogenic milling, that is, high-energy milling under low temperature, is more effective in generating amorphous materials, as demonstrated in the case study of indomethacin where grinding in the proximity of the glass transition temperature yielded partially amorphous material whereas cryogenic milling yielded a completely amorphous phase [10].

Recently, acoustic levitation, which allows 1–3 mm diameter droplets to float freely without coming into contact with a surface, has been demonstrated to produce amorphous forms for a variety of organic molecular compounds through evaporation or vitrification by high-energy laser heating [11]. Such a containerless process eliminates extrinsic heterogeneous nucleation by surface contact that would normally occur in containers and can supercool high temperature melts several hundred degrees below their melting temperatures in a relatively rapid manner, which kinetically favors the formation of an amorphous phase.

It should be noted that amorphous materials prepared by different techniques can exhibit different qualities and different properties. According to Surana et al., for example, the method of preparation strongly influenced the pharmaceutically relevant properties of amorphous trehalose, where the resistance to crystallization can be rank ordered as trehalose prepared by dehydration < freeze-dried < spray-dried < melt-quenched [12]. Feng et al. reported that disordered materials of griseofulvin prepared by cryogenic milling are different from the material obtained from melt quenching, as evidenced by Raman spectroscopic analysis [13]. As revealed by Graeser et al., the physical stabilities of amorphous forms of simvastatin prepared by cryomilling and melt quenching were distinctly different, where the cryomilled samples recrystallized within one day at 20 °C and the quench-cooled samples showed no signs of recrystallization, even when stored at 40 °C for 3 days [14].

Further, although not the subject of this chapter, it should be noted that the previous techniques can be used to screen for amorphous drug–polymer or drug-additive dispersions. In some cases, if an amorphous solid solution of the drug and polymer can be made and cooled below the glass transition temperature, then stable amorphous formulations can be prepared.

22.4 Salt Selection

For an ionizable compound, salt selection is an essential part of drug development, where the use of salts broadens the landscape for seeking a desirable solid material for development. Salt formation offers the possibility of engineering physicochemical properties of drug substances, such as solubility, dissolution rate, crystallinity, physical stability, mechanical properties, and chemical stability.

Salt selection comprises several steps:

- Selection of counterions
- Formation of salts
- Characterization of salt forms regarding crystallinity, hygroscopicity, solubility, chemical stability, and physical stability

- Selection of a leading salt form and a backup salt form
- Polymorph screening of the selected salt forms
- Selection of a stable polymorph of the salt form for development

Counterions selected for salt formation should have acceptable regulatory status and safety profiles. It is widely accepted that two pK_a unit separation for the compound and counterion is needed to produce a stable salt form.

The formation of salts is based on carrying out a specific acid–base reaction in solution, where a solvent or a mixed solvent acts as a medium to promote the proton transfer reaction. The API and a selected counterion need to have sufficient solubility in the solvent, dependent on the required concentration of starting materials and the solubility of the salt formed. Some salts readily precipitate out from the solvents used. For example, ranitidine-free base is very soluble in organic solvents such as isopropanol, but ranitidine HCl is sparingly soluble in isopropanol and thus ranitidine HCl precipitates out readily after adding HCl to the isopropanol solution [15].

For salt forms not readily precipitated from solvents, additional steps are required to create supersaturation and to promote precipitation, which include evaporation, the addition of an antisolvent, and cooling.

To identify as many salt forms as possible for salt selection, the number of experiments can become extensive because of the many variables involved, including type of counterion, solvent system, temperature, concentration, and crystallization method. Manual screening is not feasible to handle such a huge set of experiments, so high-throughput screening has proved to be a powerful tool to execute such a labor-intensive task. But no matter what method is used, it is not possible to even come close to covering the entire landscape of possibilities. Remenar et al., for example, carried out the high-throughput salt screening of sertraline, where the screening involved over 3600 crystallization trials and yielded 18 crystalline salt forms [16]. The high-throughput approach also offered simultaneous assessment of polymorphism by incorporating many crystallization conditions for a given salt type. Among the various variables, the solvent system has proved to have a profound impact on the outcome of screening. Tarsa et al., for example, reported that the addition of 10% or more water in mixed solvents greatly increased the crystalline "hit" rate for water-insoluble drugs [17]. Conversely, in some cases salt formation of a highly water-soluble compound is facilitated in organic solvents with a minimal amount of water. The solvent composition not only determines the solubility landscape of free forms, counterions, and salt forms but also significantly influences the pK_a of the drug and counterion, which in turn affects salt formation and salt crystallization. It has been reported, for example, that the crystallization of an equal molar mixture of ciprofloxacin and succinic acid in water yielded ciprofloxacin hemisuccinate (1:1), whereas crystallization in water/ethanol (50 : 50) gave ciprofloxacin succinate (2:1) [18].

Slurrying is also suitable for salt formation due to the possibility of a solution-mediated transformation since a salt form is usually less soluble in a solvent than the API and the counterion. The crystallization of the salt form continues in the solvent at the expense of the API, and the neat-grinding and solvent-drop grinding methods have proved to be effective screening tools for the salt formation of certain APIs, as reported by Trask et al. [19].

For neat grinding, API and salt formers are mixed in a proper ratio, based on the stoichiometry, and subjected to ball milling. Solvent-drop grinding involves the addition of a small amount of organic solvent to the mixture before grinding. Trask et al. screened salts of the antibacterial drug trimethoprim and the antimalarial drug pyrimethamine with various pharmaceutically acceptable carboxylic acids after 30 minutes of neat grinding or solvent-drop grinding [19]. The neat grinding yielded six successful cases of salt formation among 16 trials. Of the six salts, two salts matched known salt forms, and the other four salts were identified as new salt forms. Solvent-drop grinding provided an even better success rate, where 14 trials yielded materials distinct from the starting materials, as evidenced by XRPD. Nine salts of the 14 cases were new salts that had not been reported before. In summary, neat grinding and solvent-drop grinding not only offer a solvent-free or solvent-sparing method for salt formation but may also yield salt forms that are not readily formed by the solvent-mediated method. Figure 22.1 provides an example of the formation of the maleate salt of pyrimethamine by the solvent-drop grinding but not by neat grinding, according to XRPD data [19].

For an API with multiple ionizable groups, the selection of a proper counterion at a particular stoichiometry can help to obtain either a monosalt or a disalt.

For a full characterization of the various salts prepared, one should use a number of techniques (Table 22.2). The scope of such characterization can include the study of crystallinity, stoichiometric relationships, hygroscopicity, solubility, chemical stability, and physical stability.

Polarized microscopy and XRPD provide a quick check for material crystallinity. No further characterization is needed for amorphous materials since crystalline salts are preferred for stability and processability. Additional assessment of

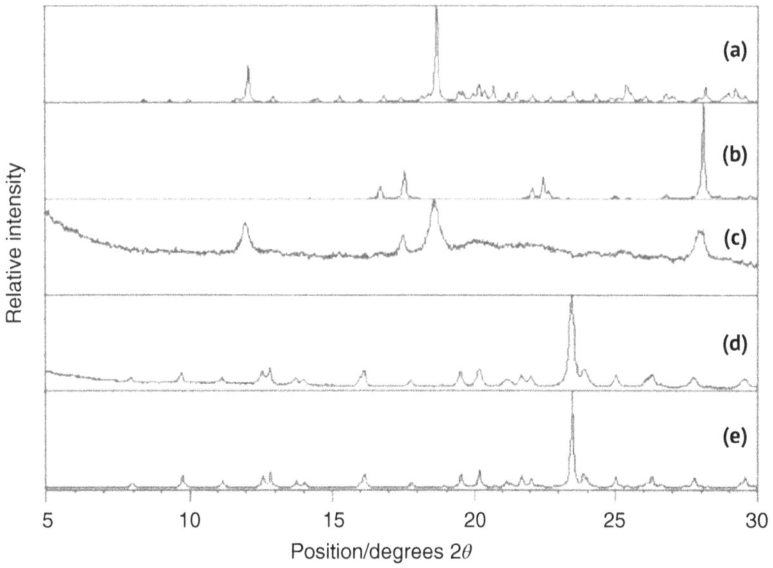

Figure 22.1 XRPD patterns of pyrimethamine samples. (a) Pyrimethamine free form, (b) maleic acid, (c) sample from neat grinding of pyrimethamine and maleic acid, (d) sample from solvent-drop grinding of pyrimethamine and maleic acid, and (e) pyrimethamine maleate. *Source*: [19]/with permission of The Royal Society of Chemistry.

Table 22.2 Physical characterization for salt selection.

Technique	Study
Microscopy	Crystallinity, shape, melting, transition
X-ray powder diffraction (XRPD)	Crystallinity, phase identification
Differential scanning calorimetry (DSC)/thermogravimetric analysis (TGA)	Melting point, heat of fusion, thermal transition, volatile components, pseudopolymorphism
Solution NMR, element analysis, ion chromatography	Stoichiometric
Infrared and Raman spectroscopy	Phase identification, interaction of API and counterion
Moisture sorption	Hygroscopicity, hydrate formation, dehydration
X-ray crystallography	Stoichiometric, crystal packing, interaction of API and counterion

crystallization should be carried out for amorphous materials to see if different solvents or crystallization conditions can yield crystalline materials.

For materials confirmed as crystalline, XRPD patterns are carefully examined to ensure that they are new crystalline phases with patterns distinct from the free form and counterions. In some cases, for example, a salt form may not readily form but, instead, the free form precipitates.

The stoichiometry of the API and counterion can be examined with several techniques, depending on what kind of counterion is used. If the counterions are organic acids or bases, solution nuclear magnetic resonance (NMR) is very useful, whereas inorganic counterions can be analyzed by ion chromatography. In addition, element analysis is an excellent way to obtain an estimation of stoichiometry.

Thermal analysis by DSC/thermogravimetric analysis (TGA) provides information concerning melting points and the presence of volatile components, as with pseudopolymorphism. Salts with low melting points, such as solvates or unstable hydrates, are usually ruled out for further study due to their high potential for instability.

Evaluation of hygroscopicity is a key component of salt selection, since hygroscopic salts, particularly deliquescent salts, can lead to serious stability issues.

The BCS classification system can be used as a guide for salt selection involving oral products, as suggested by Ku [20]. For BCS class I and III compounds, solubility generally is not a concern, where comprehensive studies on solubility and dissolution can be avoided. For BCS class II and IV compounds, however, the ranking of solubility and dissolution rate is extremely important. In addition, salt screening can be very useful for assessing potential problems with stability, processability, hygroscopicity, and polymorphism.

Many different strategies have been reported for selecting desirable salt forms for development. One such strategy, called a tier-based approach, has been reported by Morris et al. using the API, BMS180431, as a case study [21]. A total of seven salts of BMS180431 were prepared, including the sodium, calcium, zinc, magnesium, potassium, lysine, and arginine salts. The first-tier study characterized levels of hygroscopicity and led to the selection of seven salts. Additionally, three salts were dropped from consideration after the tier 2 study that assessed solubility and any crystal changes. The four remaining salts were subjected to tier 3 evaluation for stability and compatibility. As a result, the lysine salt was selected as the lead salt for development, with the arginine salt as a backup.

Another strategy used is to focus on crystallinity first, followed by the evaluation of hygroscopicity, solubility, stability, and polymorphism [22]. Regardless of the approach, finding a suitable salt for development is a complex and unpredictable process fraught with difficulty.

22.5 Cocrystal Screening

As discussed in Chapter 5, cocrystal screening offers opportunities to engineer the physicochemical properties of drug substances, especially for nonionizable compounds, where salt selection is not an option. Cocrystal screening consists of the systematic exploration of various experimental parameters that help promote cocrystal formation, including the type of coformer, the ratio of drug to coformer, solvent, temperature, the concentration of drug and coformer, and the technique used to carry out crystallization.

There is no systematic tool available to predict whether an API and a particular conformer can form a cocrystal, because, as described in Chapter 5, the formation of a cocrystal is dependent on the nature of the key molecular interactions involved, especially hydrogen bonding. Computer modeling studies that can identify potential synthons help to select likely coformers and can increase the success rate of experimental work. The study of potential synthons in this way directs the selection of coformers with complementary functional groups to those of the drug to enhance the formation of hydrogen bonding. In addition, the selection of coformers should be based on the intended application of the drug substance and the pharmaceutical acceptance of the coformer.

Owing to the significant number of experiments required because of the combination of so many parameters, high-throughput screening is generally considered to be the most efficient way to attain potential cocrystals. Multiple conformers, stoichiometry, and solvent conditions are explored in a microplate setup with minimal material consumption since only milligram quantities of material are needed at each well. Analysis of the samples in each microplate is carried out by optical microscopy, X-ray diffraction, or Raman microscopy for assessment of crystallinity and phase identification. High-throughput screening can be implemented in various ways including solvent evaporation and slurry screening.

Using the high-throughput slurry method, for example, Kojima et al. discovered novel cocrystals of indomethacin with D/L-mandelic acid, nicotinamide, lactamide, and benzamide, after screening 46 coformers [23].

After any hits are observed from high-throughput screening, additional materials (30–50 mg) are prepared by precipitation, evaporation, or slurrying for further characterization. Many cocrystals cannot readily be prepared by solvent evaporation, so manual approaches such as slurry crystallization, cogrinding, liquid-assisted grinding, and comelting are pursued as complementary approaches to obtain additional potential hits. In some cases, similar manual approaches can be used for salt discovery.

Slurry crystallization is performed simply by adding crystallization solvents to solid mixtures of a host and a guest. It ensures that both the host and the guest are at their saturated solubilities, which favors the formation of cocrystals. Gagniere et al., for example, demonstrated that the introduction of nicotinamide in dry form to a carbamazepine slurry caused a shift in the phase diagram and led to the conversion of carbamazepine crystals to cocrystals [24].

Takata et al. used the slurry crystallization technique in the cocrystal screening of two nonionizable pharmaceutical host compounds, stanolone and mestanolone, with 11 pharmaceutically acceptable guest acids [25]. The sample mixtures were prepared using lyophilization of their dimethyl sulfoxide solutions before adding a solvent to form the slurry. The screening, in this manner, discovered two new cocrystals, stanolone L-tartaric acid 1:1 cocrystal and mestanolone salicylic acid 1:1 cocrystal.

For cogrinding, stoichiometric amounts of the API and the coformer are pulverized in a ball mill for a few hours or even longer. It has been reported, for example, that acyclovir cocrystals with glutaric acid and fumaric acid, respectively, were prepared successfully via cogrinding of the drug and conformer in a molar ratio of 1:1 for 12 hours [26].

Bysouth et al. applied a modified planetary mill for the screening of pharmaceutical cocrystals and achieved success in the preparation of carbamazepine/saccharin, caffeine/oxalic acid, and caffeine/maleic acid cocrystals [27]. The grinding was performed directly in standard glass vials with 48 samples processed in parallel.

Solvent-drop grinding is to cogrind in the presence of a small amount of solvent. Lu et al. reported a thermal method for cocrystal screening, which had the advantage of using small amounts of material while allowing for rapid cocrystal screening using DSC [28].

Like salt screening, the physical characterization of cocrystals is a matrix-based approach, involving multiple techniques to provide a complete picture of crystallinity, stoichiometry, hygroscopicity, physical stability, chemical stability, solubility, dissolution, and polymorphism.

First, the materials obtained are examined by polarized light microscopy to check for crystallinity. Crystalline materials are further analyzed by XRPD to confirm crystallinity and phase identity, since XRPD is a sensitive technique to distinguish cocrystals from crystalline API and conformers. As demonstrated in Figure 22.2, carbamazepine:saccharin cocrystal and carbamazepine:nicotamide cocrystal give distinctly different XRPD patterns from the API and the corresponding conformer [29].

The stoichiometry is generally determined by solution NMR or elemental analysis, whereas thermal analysis reveals the melting point and possible pseudopolymorphism. Moisture sorption analysis evaluates the hygroscopicity.

Bruni et al. utilized a broad range of techniques to study the cocrystal properties of acyclovir:fumaric acid and acyclovir:glutaric acid, including thermal analysis, Fourier-transformed infrared (FT-IR), XRPD, and solid-state NMR [26]. Since the goal of the screening was to identify cocrystals with improved dissolution properties, intrinsic dissolution rates (IDRs) were also studied.

Usually, nonsolvated cocrystals that are not deliquescent and have a known stoichiometry and a melting temperature >125 °C are most desirable. If the purpose of cocrystal screening is to enhance solubility, aqueous solubility of the cocrystals is evaluated with an eye on their tendency to disproportionate. If the solubility of the cocrystal is much higher than the free form, solution-mediated transformation tends to occur, and, consequently, the selection of cocrystal depends on both solubility and its physical stability.

Cocrystals can show polymorphism and pseudopolymorphism. Childs and Hardcastle, for example, identified two polymorphs of a chlorzoxazone 2,4-dihydroxybenzoic acid cocrystal in their screening [30]. Karki et al. showcased two

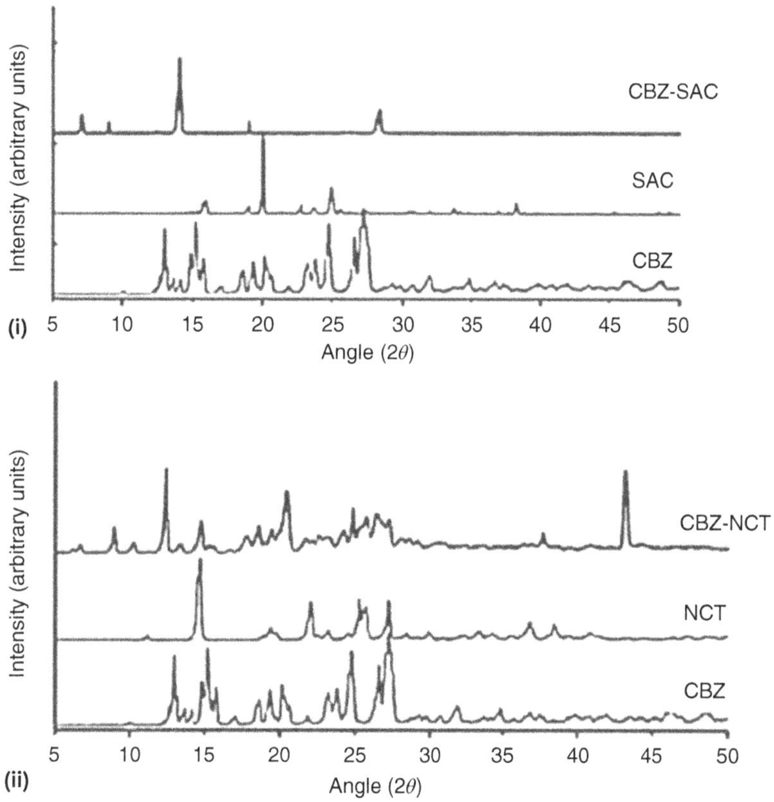

Figure 22.2 XRPD of carbamazepine:saccharin cocrystals and carbamazepine:nicotamide cocrystals, with the comparison with carbamazepine and conformer. *Source*: [29]/with permission of Elsevier.

approaches of screening hydrates of pharmaceutical cocrystals, including neat grinding using crystalline hydrates as the reactants, and liquid-assisted grinding using water as the solvent, with the latter method demonstrated to be more efficient in the studies of theophylline and caffeine cocrystals with citric acid [31].

For cocrystals with the desired properties, studies under accelerated conditions are performed to test their chemical stability regarding chemical degradation, physical stability, and possible dissociation of the cocrystal.

The decision-making process for the selection of a suitable cocrystal can be carried out using a flow chart or decision tree, as recommended by Schultheiss and Newman [32]. Key criteria are placed into different tiers so that undesired candidates can be quickly eliminated, where, therefore, one or two leading candidates can be selected for more comprehensive and resource-driven studies, such as chemical stability, polymorph screening, dissolution studies, and animal pharmacokinetic studies. The flow chart can be tailored to the goal of cocrystal screening for any API.

22.6 Polymorph Screening

The objective of polymorph screening, in general, is to discover all relevant polymorphs, solvates, and hydrates for a single chemical entity, a salt form, or a cocrystal and then select the most stable polymorphic form for further development. The polymorphic form of choice can be an anhydrous form or the most stable hydrate form. The various stages of polymorph screening include:

- full characterization of the initially discovered material,
- polymorph screening for polymorphs, solvates, and hydrates,
- polymorph characterization,
- determining physical stability in solvents and in the solid state, and
- scale-up, further characterization, and polymorph selection.

The goal of a polymorph screen is to study whether an API, as a free form, salt form, or cocrystal, can exist in multiple solid forms. The screened results provide a rationale or a knowledge base for selecting a solid form suitable for consistent manufacturing.

Owing to the lack of reliable computer-generated tools to predict how many solid forms of an API can exist, polymorph screening is still an empirical process that consists of three general steps involving form generation, form characterization, and form selection. Usually, 6–10 g of material is required to complete a comprehensive screening.

Full characterization of the starting material is essential to obtain estimates of the chemical purity and physical information, including crystal form, crystallinity, particle size, and moisture uptake. These data act as a benchmark or comparison with any additional forms that are identified. The thermal transition data obtained from DSC measurements provide information about the melting point and the possibility of recrystallization after melting or other phase transformations. Studies using thermal microscopy offer a means to evaluate possible recrystallization events upon cooling of the melt and the formation of new forms.

In addition, a solubility test in various solvents is recommended since solubility data can aid in the design of further experiments for polymorph screening. For example, the solubility data can help in the selection of solvents and antisolvents for precipitation experiments.

Solubility tests can be pursued using the shake flask method, as described in Chapter 17. A solvent addition method can also provide a quick, approximate estimation of solubility with minimal compound consumption. Aliquots of each solvent are added to an accurately dispensed API until either the materials are completely dissolved or 100 volumes of excess solvent are added.

If the goal is to identify a stable polymorph, the starting material can be used without additional sample preparation. If there is an interest in discovering metastable forms, the production of amorphous materials is generally attempted, and the amorphous materials are used as starting materials for slurrying experiments. Based on the Ostwald step rules, metastable forms tend to crystallize out first from high-energy amorphous forms [33]. Approaches to prepare an amorphous form consist of grinding, quench cooling from the melt, rapid drying, spray drying, and freeze drying. However, in some cases, such processes do not yield an amorphous form, even after extensive efforts, but instead lead to disordered materials or metastable forms.

For proper solvent selection, it is important to consider using solvents with a wide range of polarities, including solvents that are commonly used in processing the material, as well as water or water-containing mixtures. Mirmehrabi and Rohani

Table 22.3 Solvents used for the polymorph screening of formoterol fumarate.

Solvent	H-bond property	Dielectric constant at 20 °C
Water	Donor–acceptor	80
Acetonitrile	Acceptor	38
Methanol	Donor–acceptor	34
Ethanol	Donor–acceptor	24
Acetone	Acceptor	22
Isopropanol	Donor–acceptor	18
Ethylmethylketone	Acceptor	18
Pentanol	Donor–acceptor	14
Benzyl alcohol	Donor–acceptor	13
Dimethylamide	Donor–acceptor	6
Toluene	Not H-bond active	2
Cyclohexane	Not H-bond active	2

Source: [34]/with permission of Elsevier.

have provided a list of class 2 and 3 solvents from the International Conference on Harmonization (ICH) guidelines for crystallization, which are suitable for polymorph screening [15].

Jarring and colleagues selected 12 solvents with different hydrogen bonding properties and a wide span of dielectric constants for the polymorph screening of formoterol fumarate [34] (Table 22.3).

The presence of water not only can produce hydrates but also influences the nature of the crystallization of polymorphs, as demonstrated in the study of ranitidine hydrochloride [35]. Ranitidine hydrochloride has two polymorphs, Form I and Form II, differing in tautomerism. The production of pure Form I in isopropanol requires that the amount of water must be less than 0.5%, or otherwise some Form II will be produced. The presence of 2.5% water or more in isopropanol gives rise to the crystallization of fairly pure Form II.

Approaches for polymorph screening are mainly solvent based, including the use of slurrying, slow evaporation, fast evaporation, recrystallization with slow cooling, recrystallization with fast cooling, antisolvent precipitation, solvent-drop grinding, and vapor diffusion. Different approaches have distinct kinetics for crystallization, and the combined use often can promote the formation of novel forms. Sorrenti et al., for example, carried out a screening of new solid forms of the antihypertensive lisinopril by recrystallization of the commercial dihydrate form from various solvents [36]. Preparation from ethanol/water 1 : 1 resulted in an amorphous form, screening in ethanol/water 1 : 4 gave a highly crystalline dihydrate, and recrystallization from dimethyl sulfoxide yielded an anhydrous crystalline form. Interestingly, exposure of the anhydrous form to methanol vapor triggered the formation of a monohydrate form, something which had never been observed before.

The amount of material generated in a screening should be sufficient for testing by XRPD and microscopy, since both techniques are highly sensitive for the identification of crystallinity and polymorphic form. If enough materials are available, thermal analysis can generate additional information about dehydration, desolvation, melting, the amount of volatiles, and any phase transitions.

22.7 Slurrying

Slurrying is a popular technique to screen stable polymorphs and hydrates. It is also a useful technique for studying the relative stability of a polymorphic pair or an anhydrous form/hydrate pair through competitive slurry experiments, where the phase disappearing during the experiment is considered as the metastable form and vice versa. In addition, slurrying is desirable to eliminate a metastable phase in a polymorphic mixture to produce the pure stable polymorph.

Polymorph screening by slurrying is essentially based on solvent-mediated polymorphic transformation. Since a metastable form has a higher solubility than the stable polymorph, the metastable form tends to dissolve first. However, at the same time, the stable form nucleates in solution as the stable form and grows at the expense of the metastable form. Slurry experiments are carried out by mixing the sample with enough solvent to make a suspension. The suspension is constantly agitated for days or even longer, until the solid is harvested from the suspension for physical characterization. The time required for a slurrying experiment depends on the rate of the solution-mediated transformation. It has been reported, for example, that the rate is limited by the solubility of the compound in the solvent system and the nature of the solvent-solute interactions, in that higher solubility and weak solvent-solute interactions promote the transformation [37]. For the slurrying of sulfamerazine Form I, the transformation rate to the stable polymorph Form II was found to be higher in the solvent with a higher solubility, whereas a solvent with a stronger hydrogen bond propensity gave lower transformation rates.

In addition, API lots with different levels of purity should be studied to see the effect of impurities. Structurally related impurities may be purposely added to stabilize metastable forms by retarding the solution-mediated transformation. The use of pregnenolone, for example, a structurally related impurity in the crystallization of progesterone, yields crystals of metastable Form II suitable for single crystal structure analysis [38]. It has also been reported that the presence of an acetyl derivative of sulfamerazine impedes the transformation from Form I, the metastable form, to Form II, the stable form [39].

The use of a different solvent may remove the inhibitory effect of an impurity due to an increase in the solubility of the compound. Another mitigation strategy that can be used is to use less solids in the slurry so that the level of impurity in the solution can be reduced [40].

Various process-related solvents can be utilized in slurry experiments, including methanol, ethanol, isopropanol, acetonitrile, acetone, tetrahydrofuran, and toluene. Water and mixed solvents with water tend to promote the formation of hydrates in a slurry.

Cui and Yao studied various hydrate screening techniques with 10 hydrate-forming compounds and concluded that the slurry technique achieved the highest success rate (90%) [41].

In addition to solvent composition, the slurrying temperature significantly affects the outcome of polymorph screening. Campeta et al., for example, reported that axitinib, an anticancer compound under development, existed as a complex, polymorphic and highly solvated API form [42]. Polymorph screening, using common techniques, produced mostly solvates and only led to one anhydrous form, Form XXV. However, the use of high temperature slurrying (90–150 °C) changed conformational populations and enhanced the overall incidence of an anhydrous form, and led to the discovery of a stable polymorph, Form XLI.

22.8 High-Throughput Screening

High-throughput screening is widely used to screen a vast number of compounds for biological activity. Because of the greatly increased demand for comprehensive screening of polymorphs, salts, and cocrystals, it is difficult to do so using manual approaches, while automation allows the execution of numerous experiments using small amounts of API. Various solvent-based crystallization techniques can be implemented in a high-throughput setting, including evaporation, recrystallization, and slurrying. High-throughput screening allows for the variation of heating rate, cooling rate, degree of supersaturation, solvent type, and so on. Automation is also more accurate and reliable than manual work. It appears that a comprehensive polymorph screen using high-throughput procedures can allow for the identification of a landscape of forms earlier on and, therefore, provide sufficient candidates for selection. It also allows salt selection, cocrystal screening, and polymorph identification to be carried out at the same time, reducing the downstream work and helping to identify an ideal form earlier on.

For example, high-throughput crystallization studies have been reported simultaneously for salt selection and polymorph screening for sertraline, where eight crystalline salt forms and various polymorphs of these salts were identified at the same time with over 3600 crystallization trials [16].

Furthermore, the use of automation mitigates any possible cross-contamination that might occur during the transfer of solid and liquid samples.

The steps of high-throughput screening include:

- automatic crystallization experiments in a 96-well microplate or aluminum blocks holding borosilicate tubes,
- solid–liquid separation by filtration or evaporation,
- high-throughput analysis by Raman spectroscopy or transmission X-ray diffraction,
- automatic data analysis of group data by chemometrics and pattern recognition approaches, and
- data visualization and form identification.

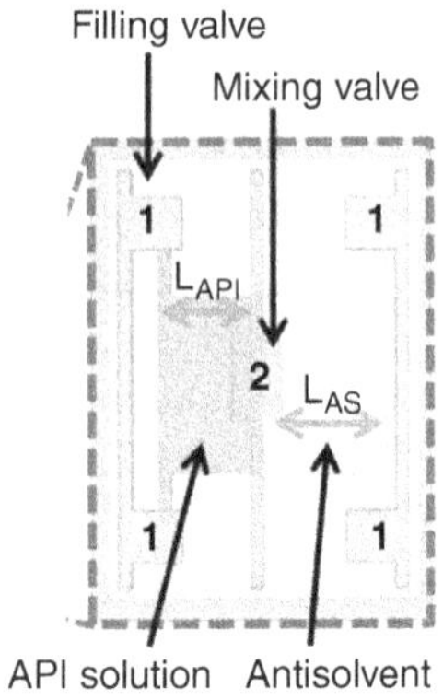

Figure 22.3 A schematic depiction of a well in a microfluidic chip. *Source*: [44]/with permission of Royal Society of Chemistry.

The traditional liquid handling system for automatic crystallization screening requires about 100 μL per well, which is much less than with manual screening, generally several milliliters. It has been reported that screening on a chip can significantly reduce the sample size even further. Goyal et al., for example, developed a microfluidic platform for an evaporation-based salt screening of pharmaceutical compounds [43]. The chip-based system allowed the mixing of API and salt former in a 24-well array with only 200 nL per well, which led to a dramatic reduction in API and solvent consumption. The well was designed for direct solid form identification using Raman spectroscopy and optical microscopy. The rate of solvent evaporation was controlled by patterned evaporation channels. The system was utilized to identify six different salts of tamoxifen and ephedrine, respectively, using on-chip Raman analysis [43].

Thorson et al. presented a similar microfluidic platform comprising 48 wells to screen for polymorphs of API through antisolvent crystallization [44]. Each well with a volume of about 37 nL contained an API solution chamber and an antisolvent chamber (Figure 22.3).

Raman spectroscopy is the leading technique to couple with high-throughput screening because it usually produces an identifiable chemical peak with different solid forms. This peak selectivity allows for a straightforward interpretation of data, data visualization, and data grouping in a high-throughput setting. Peak positions can be easily compared with pure references to provide a quick identification.

Remenar and coworkers reported a cluster classification approach of Raman spectra, where Raman spectra are processed to remove background and accentuate peaks and shoulders [16]. Peak positions, instead of peak amplitudes, are utilized to discriminate different patterns, and a hierarchical clustering algorithm is utilized for pattern recognition and spectra classification.

In addition, Raman spectroscopy can give information about the different intermolecular interactions that can take place in various solid forms.

Since the number of samples that can be studied by high-throughput techniques is enormous, a software system is needed to control the automated system, manage the workflow, coordinate the data collection and storage, provide data visualization, and pursue automatic data matching and pattern search. Unfortunately, high-throughput screening is known to miss forms because of the limited conditions used in high-throughput screening platforms. Thus, for comprehensive polymorph screens, manual screening is also required.

22.9 Crystallization in Confined Space

While high-throughput screening focuses on the improvement of efficiency, the use of small volumes in those miniature devices also limits the crystallization to a confined space. This is interesting because polymorphic compounds crystallized in a confined space often tend to display different crystallization behavior than in the bulk. This is especially true for metastable forms that, although difficult to prepare ordinarily, can be found in a confined environment.

Such differences in confined volumes are generally the result of large ratios of surface area to volume, which affect the nucleation and growth of polymorphic phases, and where heterogeneous nucleation is significantly promoted.

Crystallization in confined space can be carried out in a solvent or the melt, where confinement to the microscale includes the use of capillaries and metallic square islands. Nanoconfinement can be carried out in porous glass and polymers, which often results in the formation of polymorphs as nanocrystals.

Crystallization from a solvent in a confined space, such as capillaries, gives crystals with different characteristics than those obtained on a larger scale. First, it appears that very high supersaturation levels can be achieved since a small solvent volume (1 – 50 μL) leads to an increase in the metastable zone and a delay of nucleation. It has been reported, for example, that a 30- to 60-fold increase in supersaturation was observed for the crystallization of 5-methyl-2-[(2-nitrophenyl) amino]-3-thiophenecarbonitrile, known as ROY for its red, orange, and yellow crystals, from ethanol in glass-capillary tubes [45]. Second, the sample is present in a quiescent environment with minimal turbulence and thermal fluctuations, and, therefore, the solution will exhibit different kinetics for crystal nucleation and growth. Therefore, we can expect that high supersaturation and a quiescent environment preferentially lead to the nucleation and growth of metastable forms.

For an example, it has been reported that polymorph screening of nabumetone in 25 common organic solvents and mixtures of selected organic solvents with water in 20 mL scintillation vials only gave rise to Form I, the stable form, whereas crystallization in capillary tubes gave both Forms I and II, the metastable form (Figure 22.4). Form II was observed for approximately 20% of nearly 150 crystallizations [46].

Lee et al. reported the crystallization of glycine on confined metallic square islands, as the high-energy metastable β-glycine form, as opposed to larger islands, which promoted the crystallization of the stable form α-glycine [47].

While capillary techniques and metallic islands deliver an environment of high supersaturation so that both stable and metastable forms can grow, nanoporous host systems can actually provide a certain degree of polymorph selectivity. Figure 22.5 shows SEM images of a nanoporous glass system and a polymer monolith with cylindrical pores.

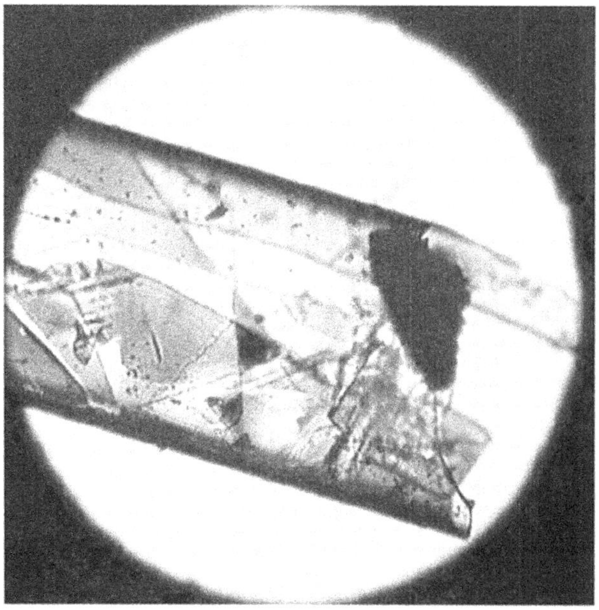

Figure 22.4 Nabumetone Form II grown in a 1.0 mm diameter capillary from water:acetone. *Source*: [46] / with permission of American Chemical Society.

Beiner et al., for example, studied the screening of acetaminophen polymorphs using a nanoporous host system made of controlled porous glasses with porosities as high as 50% with sponge-like morphology [48]. The interconnected pores had well-defined diameters of 22, 43, 60, and 103 nm. Isothermal crystallization from melt at temperatures less than 100 °C in pores with a diameter less than 60 nm yielded Form III exclusively, in the absence of a bulk reservoir. Form III is an unstable form and is considered a transient form since it is difficult to prepare and stabilize from larger volumes. For example, crystallization in pores of more than 100 nm and at temperatures held above 100°C gave rise to mixtures of Form II and Form III.

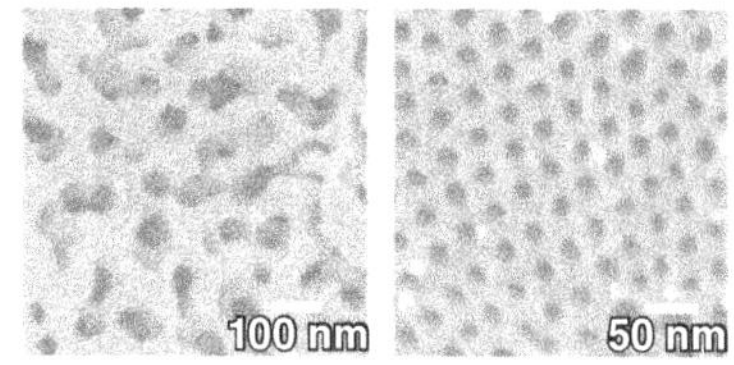

(a) (b)

Figure 22.5 SEM images of a controlled pore glass with a pore diameter of about 55 nm (left) and a platinum-coated polymer monolith with cylindrical pores about 30 nm (right).

Polymorphic selectivity under nanoscopic confinement has also been reported for anthranilic acid, ROY, and glycine. Metastable polymorph II of anthranilic acid crystallizes in nanoporous glass beads with a pore diameter of less than 23 nm, whereas polymorph III tends to form in beads with larger pore sizes and when using nonporous glass beads [49]. Although ROY has more than six polymorphs reported, only the red form (R) constantly crystallizes from pyridine solution or the melt in the pores of polycyclohexylene monoliths with a 30-nm pore size. Evaporation of glycine aqueous solution from nanoporous polystyrene-1 poly(dimethyl acrylamide) monoliths with a 22-nm pore diameter led to the exclusive formation of the metastable β-polymorph, whereas crystallization without nanoconfinement gave rise to the formation of α-polymorph, the stable form [50].

It is not well understood how nanoporous confinement affects the polymorph selectivity, but it has been proposed that the phenomenon is related to the critical nucleus sizes of different polymorphs. Apparently, polymorphs with large critical nuclei cannot fit into the nanopores, and, therefore, nucleation is inhibited [49].

22.10 Nonsolvent-based Polymorph Screening

The goal of polymorph screening is to vary crystallization conditions to control the kinetics of crystal nucleation and growth, thus separating as many polymorphs as possible. As discussed earlier, solvent properties, supersaturation levels, cooling rates, impurity levels, and agitation rate profoundly affect the generation of polymorphs. Screening without using solvents, therefore, would be expected to give rise to mechanisms and kinetics of nucleation and growth in the solid state that are significantly different from those in solvent-based screening.

Nonsolvent methods, which include thermal stresses, vapor stresses, melt crystallization, and sublimation, can be very complementary to the usual solvent-based procedures.

Vapor stress of anhydrous materials, using water vapor, for example, may lead to the formation of hydrates, whereas heat stresses starting with hydrates could help to uncover lower hydrates or anhydrous forms. The desolvation of solvates can also lead to the formation of anhydrous forms that cannot ordinarily be prepared by crystallization. For example, a polymorph screening of formoterol fumarate yielded three anhydrates, a dihydrate, a diethanolate, a disiopropanolate, and a dibensylalcholate [34]. Two of the anhydrates could not be directly obtained by slurrying and crystallization. Anhydrate A was formed by drying the dihydrate in a stream of dry nitrogen at 85 °C, whereas the drying of the diethanol or diisopropanol solvate gave rise to anhydrate C. Anhydrate B was obtained by recrystallization of anhydrate A in dry acetone and in ethylmethylketone, respectively. It was concluded that anhydrate B was the stable anhydrate form.

Melt crystallization is easily followed using thermal microscopy, where the screening only requires small amounts of API. After melting, the material is subject to isothermal or nonisothermal heating to screen conditions required to obtain polymorphs.

Crystallization from the melt often leads to the identification of metastable polymorphs that are difficult to prepare from solvent-based methods. The isolated new forms obtained from nonsolvent-based methods can then be used as seeds to prepare a desired polymorph in larger quantities by the solvent-mediated process. For example, it has been shown that screening of progesterone by crystallization from solution only yielded Forms I and Form II, whereas the metastable forms, Forms III, IV, and V, appeared only in the equilibrium melt [51]. In a second example, Form III ritonavir was discovered by melt crystallization both on the earth and in microgravity created by an orbiting crystallization system designed by Varda Space Industries and facilitated by Improved Pharma, LLC [52].

Crystallization from the melt is affected by various factors, including the free energy difference between the crystalline and amorphous states and differences in molecular mobility. The level of molecular mobility in the amorphous state is greatly affected by temperature and the glass transition temperature. It has been reported, for example, that the crystallization of amorphous indomethacin at temperatures close to or below the glass transition, 42 °C, leads to the formation of the γ-form, the stable polymorph, whereas exposure of amorphous indomethacin at temperatures above the glass transition temperature favors the formation of the α-form, the metastable polymorph [53]. It was concluded that the incubation temperature for the melt crystallization of many compounds could have a significant impact on polymorph selection [53].

Sublimation of molecules followed by cooling can lead to crystallization from the gas phase. Therefore, compounds with good thermal stability and a relatively high vapor pressure can be studied to induce crystallization of polymorphs. It has been reported, for example, that the sublimation of glycine resulted in the formation of the metastable, β-form [54].

22.11 Polymer-induced Heteronucleation

The generation of different polymorphs arises because of the possible diversity of the crystal nucleation process. Polymer-induced heteronucleation provides another dimension for empirically exploring and observing a greater variety of possible solid forms. In this method, insoluble polymers are added to the crystallization solution and utilized to promote heteronucleation, where unique surface properties of the polymer can lead to selective nucleation and growth of certain solid forms and, therefore, possibly give rise to the discovery of new polymorphs. The approach can be incorporated into traditional solution-based screening or in high-throughput screening. Because of the likely specificity of the polymer–API interaction, this technique has the potential to provide a high level of selectivity for the production of the desired form by identifying a proper solvent and polymer substrate.

It has been reported, for example, that Form III of flurbiprofen is not physically stable, and the only way to prepare it is to use melt crystallization [55]. In the presence of oxidized polyethylene, evaporation of an ethyl acetate solution gave rise to plate-like crystals of Form III, whereas a novel hydrate form was shown to grow from methanol containing poly(*n*-butyl methacrylate). However, since the hydrate was not stable in water, it transformed to Form I after being suspended in water.

Sudha and Srinivasan studied the polymer-induced selective nucleation of polymorphs of paracetamol through the rapid cooling of boiled aqueous solutions [56]. In the absence of the polymers, the pure aqueous solution resulted only in the stable monoclinic form.

Diao et al. reported selective crystallization of polymorphs induced by polymer microgels [57]. Polymer microgels generally have a tunable microstructure that offers a greater variety of possible interactions. Ethylene glycol diacrylate microgels, for example, selectively induced crystallization of Form II of carbamazepine, whereas ordinary bulk crystallization produced concomitant crystallization of Form I and Form II. In another application, the use of a microgel system led to improved polymorph selectivity of ROY polymorphs to give the R polymorph [57]. In comparison, bulk crystallization of ROY predominantly gave rise to the Y polymorph or the ON polymorph, depending on the levels of supersaturation. It was hypothesized that the microgel offered a favorable polymer–solute interaction in that the interaction promoted a particular orientation of drug molecules, thereby resulting in an entropic reduction for the formation of the nucleus associated with a certain polymorph. In other words, the polymer appears to act as a template for nucleation.

22.12 Physical Characterization

Physical characterization is an essential part of salt selection, cocrystal discovery, and polymorph screening, and, under most circumstances, sample analysis is often more challenging than sample preparation of various forms. In dealing with physical characterization, several key questions must be addressed:

- Is the sample a pure compound, or is it contaminated with organic or inorganic compounds?
- Is the sample a pure form or a mixture?
- Will the sample form change during analysis?
- Are there isomorphic forms present?
- Are differences observed due to the instrumentation used or sample presentation and not because of the sample itself?

The first important assumption to make before physical characterization is that all materials are chemically pure. Consequently, any differences observed in XRPD, DSC, or microscopy must be assumed to be due to the presence or formation of a new polymorph, hydrate, or solvate. However, contamination or inadvertent introduction of another organic or inorganic solid material during sample preparation can often complicate the sample analysis and the conclusions that are drawn. For example, the preparation of a HCl salt generally involves some acid–base neutralization, which causes some sodium chloride to be introduced into the sample. The presence or absence of a sodium chloride peak in an XRPD measurement, for example, may make the new batch look very different from previous material, and it can be mistakenly concluded to be due to a new form. The contamination of a crystalline organic compound into a drug substance may cause a minor extra peak in a DSC thermogram, which could be mistakenly considered to indicate the presence of another polymorph in minor composition. Therefore, before carrying out any characterization of physical properties, chemical purity related to organic components must be confirmed by analytical techniques such as HPLC, mass spectrometry, or NMR, whereas elemental analysis and peak matching of diffraction observed by X-ray analysis can reveal the presence of inorganic components.

Various polymorph screening techniques may lead to mixtures of various solid forms. A mixture of two known forms could give XRPD peaks and thermograms that are quite distinct from the individual components, with the possibility of appearing to be a "new" form. For a material that appears to be different in one or two characterization techniques, efforts should be made to investigate whether the material is a mixture or a single component. Hot stage microscopy can offer a quick analysis by observing if a uniform habit is present for the material and whether the material melts uniformly. Slurry experiments can also offer useful information, since, if the material changes after slurry, it is a good indication that a mixture is present or that the material is metastable. Newly developed computational tools, which allow the indexing of XRPD data, are available to indicate whether a sample is a pure single phase or a mixture [58]. Such indexing can give approximate unit cell parameters for a pure crystalline phase and thus provide evidence as to whether the material is a single crystalline phase or not.

There are many metastable forms that tend to change after sample harvesting and sample analysis, where unstable solvates or hydrates easily become desolvated or dehydrated.

Isomorphism is a phenomenon that occurs when solvates give the same XRPD patterns, because they have very similar crystal structures but are actually different forms. If isomorphic forms exist, it means that characterization of the physical form requires combining several techniques to draw a more definitive conclusion. It could be misleading if a conclusion is made based on a single technique. Since isomorphic materials give essentially the same XPRD pattern, it is important to use other tools that can discriminate such systems, including DSC/TGA, Raman, and FT-IR.

Similar characterization challenges arise when encountering nonstoichiometric channel hydrates, since in such cases the water content can easily vary with relative humidity but with minimal change in the unit cell dimensions. Consequently,

XRPD data would appear nearly identical and provide no indication of the hydration state, as shown in the case of FK041 hydrate [59], where the FK041 hydrate sample stored over silica gel for 24 hours (dehydrate) has the same XRPD pattern as the sample stored over water for 24 hours (tetrahydrate).

Since any analytical technique has its limitations, it is necessary to be able to show that the differences in any data collected are caused by the material properties and not by the method of preparing the sample. For example, XRPD responses are sensitive to the preferred orientation of certain crystal shapes, which causes peak shifts in the Bragg-Brentano optics. In DSC analysis, the choice of the amount of sample, the pan configuration, and different heating rates are all variables that may result in totally different thermograms for the same material. When different data are observed, questions should be asked as to the root causes of such data differences. The best way to come to correct conclusions concerning physical characterization, therefore, is to depend on several techniques. In other words, XRPD data should be supported by thermal analysis, microscopy, FT-IR, and Raman spectroscopy.

22.13 Thermodynamic Stability and Solid Form Selection

Determining the order in which solid forms exhibit physical stability is one of the key elements for solid form selection, since the solid form selected should maintain acceptable physical stability during product manufacturing and storage.

The order of solubility and dissolution of a solid form at a specific temperature directly reflects the order of thermodynamic stability at that temperature. For example, the polymorph screening of fusidic acid was shown to yield a metastable polymorph, Form IV, starting with the commercial drug substance (Form III) [60]. Form IV had an IDR of 0.092 mg/min/cm^2, which was two times higher than that of Form III (0.045 mg/min/cm^2). In water at 30 °C, polymorph I of sulfamerazine exhibits a solubility of 0.2269 mg/mL, whereas its polymorph II has a solubility of 0.2031 mg/mL [61]. The solubility data confirmed that the thermodynamic order of the two polymorphs, wherein polymorph II is the stable form and polymorph I is the metastable form at 30 °C. The solubility ratio of 1.117 was shown to correspond to a Gibbs free energy difference of 279 J/mol for the two forms.

In another study, water solubility data for three formoterol fumarate anhydrates and a dihydrate, shown in Table 22.4, give a clear indication of the order of stability at room temperature, where the dihydrate has the lowest solubility in water and, therefore, is the most stable form [34]. Anhydrate B is less soluble than anhydrates A and C and therefore is the most thermodynamically stable anhydrate. Anhydrate B also has a higher melting point than A and C, indicating that anhydrate B is monotropic with respect to A and C.

It is important to recognize that, although anhydrate B is the thermodynamically stable polymorph, it is metastable to the dihydrate in the presence of water at room temperature, where the relative thermodynamic stability between an anhydrate and a hydrate is dependent on water activity. Solubility experiments and competitive slurrying in solvents with different water activities are used to identify the stability window for the anhydrate and hydrate, respectively. For example, competitive slurry experiments with a mixture of formoterol fumarate forms in ethanol with 0.6–5.3% of water revealed that anhydrate B is stable at very low amounts of water content, not more than 2.7%, whereas the order of stability is reversed as more water is in the system (5.3%), with the dihydrate as the thermodynamic stable form (Table 22.5).

The thermodynamic stability order of polymorphs and pseudopolymorphs can also be probed by solution calorimetry, which measures the enthalpy of dissolution of a drug substance. The enthalpy of dissolution arises from thermal events associated with the wetting of the drug, the breakdown of the crystal lattice, and the solvation of drug molecules by solvent

Table 22.4 Melting point, heat of fusion, and water solubility of formoterol fumarate crystalline forms.

Form	Melting point (°C)	Heat of fusion (kJ/mol)	Water solubility (mM)
Anhydrate A	127	49	4.7
Anhydrate B	150	61	3.5
Anhydrate C	123	52	7.5
Dihydrate	125	126	1.5

Source: [34]/with permission of Elsevier.

Table 22.5 Competitive slurry experiments in ethanol/water for formoterol fumarate crystalline forms.

Starting solid	Seeding solids	Water amount (%)	Resulting solid
Anhydrate A	Anhydrate B + dihydrate	0.6	Anhydrate B
Anhydrate A	Anhydrate B + dihydrate	1.1	Anhydrate B
Anhydrate A	Anhydrate B + dihydrate	2.7	Anhydrate B + residual dihydrate
Anhydrate A	Anhydrate B + dihydrate	5.3	Dihydrate + residual anhydrate B

Source: [34]/with permission of Elsevier.

molecules. For polymorphs and pseudopolymorphs, where wetting effects should be similar, the enthalpy of dissolution is most likely dominated by the dissociation of molecules from the crystal lattice. Consequently, estimates of the enthalpy of dissolution primarily reflect the relative thermodynamic stability of various crystalline forms. Cosgrove et al., for example, studied the relative stability of sibenadet hydrochloride polymorphs using solution calorimetry [62]. The molar enthalpy of solution was found to be 48.5 and 40.1 kJ/mol for Polymorphs I and II, respectively, indicating that Polymorph I was the most stable of the two polymorphs at 25 °C. This conclusion was confirmed by competitive slurry experiments, where mixtures of Polymorphs I and II in a 1 : 1 ratio were slurried in ethanol, acetone, THF, water, and methanol. Analysis by DSC and XRPD revealed that either partial or complete transformation to Polymorph I had occurred for the slurried mixtures.

The relative stability of polymorphs, anhydrate/hydrate, is usually confirmed by competitive slurry experiments, where solubility differences between metastable and stable forms generally trigger solution-mediated transformation, with the growth of the stable form at the expense of the metastable form. Usually, a mixture of two polymorphs or an anhydrate/hydrate pair is subjected to slurrying in a selected solvent over a period of days, and the nature of the final form observed is confirmed by physical characterization using XRPD and DSC. The increase in the amount of one component and the decrease of the other component indicate which form is more stable under the slurring condition, temperature, and solvent.

Many commercial products contain API in hydrate forms, in that there are about 100 hydrates listed in USP monographs, whereas about 29% of the drugs listed in the European Pharmacopoeia exist as hydrate forms [41]. This generally reflects the fact that there is very little concern for safety when using hydrates in drug products. Many hydrates can form rapidly after being in contact with water or when exposed to high relative humidity, so hydrates generally are more stable than anhydrates under ambient conditions. Some compounds require water in the crystal lattice to produce a stable crystal structure, so that removal of water from such hydrates will result in the formation of amorphous or partially crystalline materials, which tend to have poor physical properties.

Azithromycin (AZI), a macrolide antibiotic, is available commercially as a dihydrate, which upon dehydration produces a very hygroscopic anhydrous form that reconverts back to the dihydrate after storage at room temperature [63]. In another example, the drug acyclovir exists in two anhydrates and two hydrates, with the hydrates having acyclovir/water ratios of 1 : 2 and 3 : 2, respectively, and the 3 : 2 hydrate is used in commercial products [64, 65].

After a particular polymorph is selected for final use, consideration must be given to conditions that can be used to produce scaled-up amounts that can be made consistently with acceptable physical purity. In addition, crystallization conditions must be optimized to produce the desirable crystal habit, crystal size, and impurity level. Controlling crystal size is particularly important since it determines what downstream processes may be needed, for example, pulverization or milling. Control of crystal habits is essential for determining the type of filtration that might be needed, since it is well recognized that needle-like crystals tend to cake during filtration. It is suggested, further, that a solid-state stability study under accelerated storage conditions be carried out to confirm that the selected form has sufficient physical and chemical stability to support an acceptable shelf life for the drug substance and drug product.

22.14 Summary

In conclusion, solid form selection, including selection of polymorphs, salts, hydrates, and cocrystals, is an important part of drug development since any solid form chosen must meet certain physical and chemical requirements for optimizing manufacture and therapeutic use. Polymorph screening is used to generate as many forms as possible, including anhydrates, hydrates, and solvates, and to select a proper form for development after a thorough evaluation and risk assessment of various forms. Slurrying is a common approach to screen for stable polymorphs, while high-throughput screening, although increasing the number of experiments required, requires extremely small amounts of material. Screening in nonsolvent systems, crystallization in confined spaces, melt-based screening, and polymer-induced nucleation expand the landscape for discovering a large number of possible solid forms. A thorough characterization of all discovered solid forms builds a knowledge base for optimal decision making.

References

1 Henck JO, Griesser UJ, Burger A. *Pharm Ind* 1997; 59: 165–169.

2 Chiang PC, South SA, Daniels S, Anderson DR, Wene SP, Albin LA, Mourey RJ, Selbo JG. Aqueous versus non-aqueous salt delivery strategies to enhance oral bioavailability of a mitogen-activated protein kinase-activated protein kinase (MK-2) inhibitor in rats. *J Pharm Sci* 2009; 98(1): 248–256.

3 Eckhart C, Etlinger T, Levine N, Yuen PH. Mometasone furoate monohydrate, process for making same and pharmaceutical compositions. 2000; 6,127,353.

4 Wu V, Rades T, Saville DJ. Stability of polymorphic forms of ranitidine hydrochloride. *Pharmazie* 2000; 55(7): 508–512.

5 Bauer J, Spanton S, Henry R, Quick J, Dziki W, Porter W, Morris J. Ritonavir: An extraordinary example of conformational polymorphism. *Pharm Res* 2001; 18(6): 859–866.

6 Ando HY, Radebaugh GW. Preformulation. In: Gennaro AR, ed. *Remington: The science and practice of pharmacy II.* Baltimore, MD: Lippincott Williams and Wilkins, 20th Ed. 2002: 700–720.

7 Hancock BC, Parks M. What is the true solubility advantage for amorphous pharmaceuticals? *Pharm Res* 2000;17(4): 397–404.

8 Zhou D, Zhang GG, Law D, Grant DJ, Schmitt EA. Physical stability of amorphous pharmaceuticals: Importance of configurational thermodynamic quantities and molecular mobility. *J Pharm Sci* 2002; 91(8): 1863–1872.

9 Crowley KJ, Zografi G. Cryogenic grinding of indomethacin polymorphs and solvates: assessment of amorphous phase formation and amorphous phase physical stability. *J Pharma Sci* 2002; 91(2): 492–507.

10 Otsuka M, Kaneniwa N. A kinetic study of the crystallization process of noncrystalline indomethacin under isothermal conditions. *Chem Pharm Bull* 1988; 36(10): 4026–4032.

11 Benmore CJ, Weber JK. Amorphization of molecular liquids of pharmaceutical drugs by acoustic levitation. *Phys Rev X* 2011; 1(1): 011004.

12 Surana R, Pyne A, Suryanarayanan R. Effect of preparation method on physical properties of amorphous trehalose. *Pharm Res* 2004; 22(7): 1167–1176.

13 Feng T, Pinal R, Carvajal MT. Process induced disorder in crystalline materials: Differentiating defective crystals from the amorphous form of griseofulvin. *J Pharm Sci* 2008; 97(8): 3207–3222.

14 Graeser KA, Strachan CJ, Patterson JE, Gordon KC, Rades T. Physicochemical properties and stability of two differently prepared amorphous forms of simvastatin. *Cryst Growth Des* 2008; 8(1): 128–135.

15 Mirmehrabi M, Rohani S. An approach to solvent screening for crystallization of polymorphic pharmaceuticals and fine chemicals. *J Pharm Sci* 2005; 94(7): 1560–1576.

16 Remenar JF, MacPhee JM, Larson BK, Tyagi VA, Ho JH, McIlroy DA, Hickey MB, Shaw PB, Almarsson Ö. Salt selection and simultaneous polymorphism assessment via high-throughput crystallization: The case of sertraline. *Org Proc Res Dev* 2003; 7(6): 990–996.

17 Tarsa PB, Towler CS, Woollam G, Berghausen J. The influence of aqueous content in small scale salt screening–improving hit rate for weakly basic, low solubility drugs. *Eur J Pharm Sci* 2010; 41(1): 23–30.

18 Paluch KJ, McCabe T, Müller-Bunz H, Corrigan OI, Healy AM, Tajber L. Formation and physicochemical properties of crystalline and amorphous salts with different stoichiometries formed between ciprofloxacin and succinic acid. *Mol Pharm* 2013; 10(10): 3640–3654.

19 Trask AV, Haynes DA, Motherwell WDS, Jones W. Screening for crystalline salts via mechanochemistry. *Chem Commun* 2006; 1: 51–53.

20 Ku MS. Use of the biopharmaceutical classification system in early drug development. *AAPS J* 2008; 10(1): 208–222.

21 Morris KR, Fakes MG, Thakur AB, Newman AW, Singh AK, Venit JJ, Spagnuolo CJ, Serajuddin AT. An integrated approach to the selection of optimal salt form for a new drug candidate. *Int J Pharm* 1994; 105(3): 209–227.

22 Newman AW, Childs SL, Cowans BA. Salt and cocrystal form selection. *Pharmaceutical Sciences Encyclopedia* 2010; 44: 1–28.

23 Kojima T, Tsutsumi S, Yamamoto K, Ikeda Y, Moriwaki T. High-throughput cocrystal slurry screening by use of in situ Raman microscopy and multi-well plate. *Int J Pharm* 2010; 399 (1): 52–59.

24 Gagniere E, Mangin D, Puel F, Valour JP, Klein JP, Monnier O. Cocrystal formation in solution: Inducing phase transition by manipulating the amount of cocrystallizing agent. *J Cryst Growth* 2011; 316(1): 118–125.

25 Takata N, Shiraki K, Takano R, Hayashi Y, Terada K. Cocrystal screening of stanolone and mestanolone using slurry crystallization. *Cryst Growth Des* 2008; 8(8): 3032–3037.

26 Bruni G, Maietta M, Maggi L, Mustarelli P, Ferrara C. Preparation and physicochemical characterization of acyclovir cocrystals with improved dissolution properties. *J Pharm Sci* 2013; 102(11): 4079–4086.

27 Bysouth SR, Bis JA, Igo D. Cocrystallization via planetary milling: Enhancing throughput of solid-state screening methods. *Int J Pharm* 2011; 411(1): 169–171.

28 Lu E, Rodríguez-Hornedob N, Suryanarayanan R. A rapid thermal method for cocrystal screening. *Cryst Eng Comm* 2008; 10(6): 665–668.

29 Tomaszewska I, Karkia S, Shur J, Price R, Fotaki N. Pharmaceutical characterisation and evaluation of cocrystals: Importance of in vitro dissolution conditions and type of conformer. *Int J Pharm* 2013; 453(2): 380–388.

30 Childs SL, Hardcastle KI. Cocrystals of chlorzoxazone with carboxylic acids. *Cryst Eng Comb* 2007; 9(5): 364–367.

31 Karki S, Friscic T, Jones W, Motherwell WDS. Screening for pharmaceutical cocrystal hydrates via neat and liquid-assisted grinding. *Mol Pharm* 2007; 4(3): 347–354.

32 Schultheiss N, Newman A. Pharmaceutical cocrystals and their physicochemical properties. *Cryst Grow Design* 2009; 9(6): 2950–2967.

33 Van Santen RA. The Ostwald step rule. *J Phys Chem* 1984; 88(24): 5768–5769.

34 Jarring K, Larsson T, Stensland B, Ymén I. Thermodynamic stability and crystal structures for polymorphs and solvates of formoterol fumarate. *J Pharm Sci* 2006; 95(5): 1144–1161.

35 Murthy K, Radatus BK, Sidhu KPS. Reissued 2002. Process for Production of an Improved Form of Form 1 Ranitidine Hydrochloride Having Improved Filtration and Drying Characteristics. US Patent 1996; 5,523,423.

36 Sorrenti M, Catenacci L, Cruickshank DL, Caira MR. Lisinopril dihydrate: single-crystal X-ray structure and physicochemical characterization of derived solid forms. *J Pharm Sci* 2013; 102(10): 3596–3603.

37 Gu CH, Young V, Grant DJ. Polymorph screening: Influence of solvents on the rate of solvent-mediated polymorphic transformation. *J Pharm Sci* 2001; 90(11): 1878–1890.

38 Lancaster RW, Karamertzanis PG, Hulme AT, Tocher DA, Lewis TC, Price SL. The polymorphism of progesterone: Stabilization of a 'disappearing' polymorph by co-crystallization. *J Pharm Sci* 2007; 96(12): 3419–3431.

39 Gu CH, Chatterjee K, Young Jr V, Grant DJW. Stabilization of a metastable polymorph of sulfamerazine by structurally related additives. *J Cryst Growth* 2002; 235(1): 471–481.

40 Gong Y, Collman BM, Mehrens SM, Lu E, Miller JM, Blackburn A, Grant DJ. Stable-form screening: Overcoming trace impurities that inhibit solution-mediated phase transformation to the stable polymorph of sulfamerazine. *J Pharm Sci* 2008; 97(6): 2230–2244.

41 Cui Y, Yao E. Evaluation of hydrate-screening methods. *J Pharm Sci* 2008; 97(7): 2730–2744.

42 Campeta AM, Chekal BP, Abramov YA, Meenan PA, Henson MJ, Shi B, Singer RA, Horspool KR. Development of a targeted polymorph screening approach for a complex polymorphic and highly solvating API. *J Pharm Sci* 2010; 99(9): 3874–3886.

43 Goyal S, Thorson MR, Schneider CL, Zhang GG, Gong Y, Kenis PJ. A microfluidic platform for evaporation-based salt screening of pharmaceutical parent compounds. *Lab Chip* 2013; 13(9): 1708–1723.

44 Thorson MR, Goyal S, Gong Y, Zhang GGZ, Kenis PJA. Microfluidic approach to polymorph screening through antisolvent crystallization. *CrystEngComm* 2012; 14(7): 2404–2412.

45 Hilden JL, Reyes CE, Kelm MJ, Tan JS, Stowell JG, Morris KR. Capillary precipitation of a highly polymorphic organic compound. *Cryst Growth Des* 2003; 3(6): 922–926.

46 Chyall LJ, Tower JM, Coates DA, Houston TL, Childs SL. Polymorph generation in capillary spaces: The preparation and structural analysis of a metastable polymorph of nabumetone. *Cryst Growth Des* 2002; 2(6): 505–510.

47 Lee AY, Lee IS, Dette SS, Boerner J, Myerson AS. Crystallization on confined engineered surfaces: A method to control crystal size and generate different polymorphs. *J Am Chem Soc* 2005; 127(43): 14982–14983.

48 Beiner M, Rengarajan GT, Pankaj S, Enke D, Steinhart M. Manipulating the crystalline state of pharmaceuticals by nanoconfinement. *Nano Letters* 2007; 7(5): 1381–1385.

49 Ha JM, Wolf JH, Hillmyer MA, Ward MD. Polymorph selectivity under nanoscopic confinement. *J Am Chem Soc* 2004; 126(11): 3382–3383.

50 Hamilton BD, Ha JM, Hillmyer MA, Ward MD. Manipulating crystal growth and polymorphism by confinement in nanoscale crystallization chambers. *Acc Chem Res* 2011; 45(3): 414–423.

51 Sarkar A, Rohani S. Investigation on polymorphic behavior of progesterone and stabilization by co-crystallization: A review. *Mini Rev Med Chem* 2014; 14(10): 853–861.

52 Parent SD, Smith PA, Purcell DK, Smith DT, Bogdanowich-Knipp SJ, Bhavsar AS, Chan LR, Croom JM, Bauser HC, McCalip A, Byrn SR. Ritonavir form III: A coincidental concurrent discovery. *Cryst Growth Des* 2022; 23(1): 320–325.

53 Andronis V, Zografi G. Crystal nucleation and growth of indomethacin polymorphs from the amorphous state. *J Non-cryst Solids* 2000; 271(3): 236–248.

54 Liu Z, Zhong L, Ying P, Feng Z, Li C. Crystallization of metastable beta glycine from gas phase via the sublimation of alpha or gamma form in vacuum. *Biophys Chem* 2008; 132(1): 18–22.

55 Grzesiak AL, Matzger AJ. New form discovery for the analgesics flurbiprofen and sulindac facilitated by polymer-induced heteronucleation. *J Pharm Sci* 2007; 96(11): 2978–2986.

56 Sudha C, Srinivasan K. Supersaturation dependent nucleation control and separation of mono, ortho and unstable polymorphs of paracetamol by swift cooling crystallization technique. *CrystEngComm* 2013; 15(10): 1914–1922.

57 Diao Y, Helgeson ME, Myerson AS, Hatton TA, Doyle PS, Trout BL. Controlled nucleation from solution using polymer microgels. *J Am Chem Soc* 2011; 133 (11): 3756–3759.

58 McClurg RB. Methods for indexing solid forms of compounds. 2013. US8576985 B2.

59 Mimura H, Gato K, Kitamura S, Kitagawa T, Kohda S. Effect of water content on the solid-state stability in two isomorphic clathrates of cephalosporin: cefazolin sodium pentahydrate (alpha Form) and FK041 hydrate. *Chem Pharm Bull* 2002; 50(6): 766–770.

60 Gilchrist SE, Letchford K, Burt HM. The solid-state characterization of fusidic acid. *Int J Pharm* 2012; 422(1): 245–253.

61 Zhang GG, Gu C, Zell MT, Burkhardt RT, Munson EJ, Grant DJ. Crystallization and transitions of sulfamerazine polymorphs. *J Pharm Sci* 2002; 91(4): 1089–1100.

62 Cosgrove SD, Steele G, Austin TK, Plumb AP, Stensland B, Ferrari E, Roberts KJ. Understanding the polymorphic behavior of sibenadet hydrochloride through detailed studies integrating structural and dynamical assessment. *J Pharm Sci* 2005; 94(11): 2403–2415.

63 Gandhi R, Pillai O, Thilagavathi R, Gopalakrishnan B, Kaul CL, Panchagnula R. Characterization of azithromycin hydrates. *Eur J Pharm Sci* 2002; 16(3): 175–184.

64 Sohn YT, Kim SH. Polymorphism and pseudopolymorphism of acyclovir. *Arch Pharm Res* 2008; 31(2): 231–234.

65 Lutker KM, Quinones R, Xu J, Ramamoorthy A, Matzger AJ. Polymorphs and hydrates of acyclovir. *J Pharm Sci* 2011; 100(3): 949–963.

23

Final Stage Drug Substance Manufacturing and Control

23.1 Introduction

Drug substance development and drug product development are key pillars of the pharmaceutical development of any new drug. The two processes are so interconnected and must be integrated well to develop a successful clinical supply, execute a sound registration study, attain a smooth regulatory approval, and ensure a robust commercial manufacture. The physical and chemical attributes of a drug substance, such as solubility, permeability, polymorph, and salt form; the quality matrix, such as potency, purity, and number of impurities; and the physical and chemical stability determine the quality, performance, safety, and the shelf life of a drug product.

A drug substance, also widely called an active pharmaceutical ingredient (API), by definition, is an effective component that is intended to exert pharmacological activity when delivered as a drug product. A drug substance is a neat and pure drug. Nowadays, drug substances have very diverse sources, including small molecules, peptides, oligonucleotides, antibodies, and mRNA vaccines. It would be expected that the ways to make those drug substances are distinctly different significantly such as chemical synthesis, fermentation, cell culture, or isolation from natural sources.

Since the majority of drug substances exist as a solid powder form, solid-state material science plays an essential role in the final stage of drug substance manufacturing, that is, crystallization, filtration, drying, and milling, to make a high-quality drug substance with

- acceptable impurity profiles and residual solvent levels,
- low risk of genotoxic and nitrosamine impurities being present,
- the desired solid form being consistently produced,
- desirable particle morphology to promote filtration, drying, or flow, and
- proper particle size and distribution to fit the targeted use in the drug product.

On the other hand, solid-state characterization tools being described in various previous chapter represent a key toolkit to guide the process development of crystallization, filtration, drying, and micronization by characterizing the form, the particle size distribution, the morphology, and the hygroscopicity and more importantly act as a key procedure to analyze the finished drug substance material to control its quality.

With a focus on small molecules drugs made by chemical synthesis, this chapter is intended to lay out key challenges in the process development of the final stage of drug substance manufacturing including purging impurities, removing residual solvent, controlling genotoxic impurities including *N*-nitrosamine, achieving chiral separation, controlling final form, and controlling particle size and shape. And it is further aimed at showcasing how to use solid form phase diagrams and solid-state form transition to yield a robust crystallization or drying process to produce drug substances with consistent quality, performance, and stability.

23.2 Crystallization and Impurity Purge

One of the most important steps of drug substance manufacturing is the purification process from a reaction product mixture, usually through crystallization to remove impurities. Impurities in a drug substance not only can compromise the quality but also would incur safety concerns if the levels are higher than an acceptable limit. Impurities from a synthetic

Solid-State Materials in Pharmaceutical Chemistry: Properties, Characterization, and Applications, Second Edition. Stephen R. Byrn, George Zografi and Xiaoming (Sean) Chen.

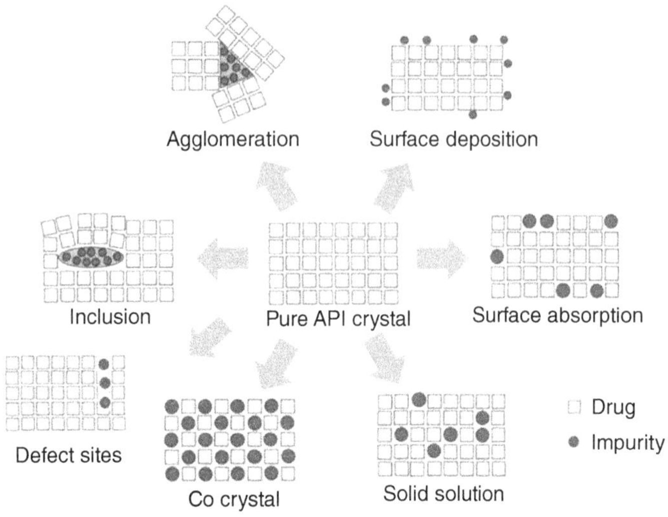

Figure 23.1 Solid-state representations of various ways of low-level impurity incorporation during crystallization. *Source*: [1]/with permission of American Chemical Society.

process can result from unreacted starting materials or intermediates, by-products from chemical synthesis, or degradation products from the drug.

It is well recognized that, to produce a small molecule drug substance, even a small number of synthesis steps can easily result in the generation of dozens of impurities, with identity and toxicological information to be investigated if the levels in the final material are above the acceptable limits suggested by regulatory agencies. The situation would become even more challenging when any new impurities appear due to a synthetic route change or during a scale-up for crystallization since they are not qualified in the initial toxicological studies. Therefore, it is of paramount importance to purge the impurities as much as possible to achieve a level as low as reasonably practical. This is usually tackled with extensive solvent screenings and empirical trials. Unfortunately, the task is very challenging since some impurities are very difficult to purge, especially when the chemical structures are like the parent drug.

In some cases, necessary process improvements to reduce and control difficult impurities rely on the understanding of how impurities are incorporated in growing drug crystals. Urwin et al. summarized the various possible solid-state representations of low-level impurity incorporation during crystallization [1], which were further expanded by Capelades and coworkers [2], as illustrated in Figure 23.1, including:

- agglomerates, which form when pockets of impurity-rich mother liquor are trapped between the intergrown drug crystal particles and become a solid after a downstream drying process,
- surface deposition, which results from residual mother liquid not completely being washed away or adsorption of impurities onto the crystal surface,
- co-precipitation, which occurs by antisolvent effects along with the drug,
- inclusion, which develops owing to the trapping of mother liquor within fast-growing crystals or crystal attrition during high-energy mixing,
- defect sites, which act as a hideout for impurities to incorporate in the crystal,
- solid solutions, which arise due to the high structural similarity to the drug and consequently solid-state partial miscibility, and
- cocrystals, which form with the drug as an impurity cocrystal.

As would be expected, the understanding of how impurities incorporate during crystallization is an essential first step to select an appropriate approach to purge impurities. For example, to mitigate surface deposition, adequate washing post filtration must be pursued to wash out the residual mother liquor or surface adsorbed impurities, whereas, to remove a co-precipitating impurity, a different crystallization solvent or a washing solvent should be explored to make the impurity more soluble. Since the formation of agglomerates is one of the common causes of impurity trapping, various anti-agglomeration strategies have been developed and reported in the literature, such as selecting a different solvent, performing crystallization with a reduced supersaturation, and changing the crystal growth dynamics.

Light microscopy (Chapter 11) offers a powerful tool to detect and identify the various impurity representations, especially for aggregates, co-precipitates, and solution entrapment. For example, as shown in Figure 23.2, Salami et al. used light microscopy and in situ process imaging to spot co-precipitates of phenylglycine impurities during crystallization of needle-like cephalexin monohydrate crystals [3].

It is interesting to note that a surface washing process through slurry experiments not only represents a meaningful way to reduce high concentrations of impurities at the crystal surface but can also reveal the presence of other mechanisms, such as inclusion or solid solution. As would be expected, the surface washing is ineffective to reject impurities for inclusion and solid solution.

Ultimately, impurity mapping through bulk dissolution experiments can further deconvolute the different mechanisms of impurity entrapment as shown in Figure 23.3. During dissolution, the concentration change of an impurity in the liquid phase can reflect the way the impurity is incorporated throughout the crystal bulk.

Generally speaking, the most challenging situation for impurity incorporation is when an impurity is inside crystals, such as the formation of a solid solution, leading to the incorporation of the impurity in the drug crystals at a non-stoichiometric level. It appears that most drug–impurity pairs do not form a fully miscible solid solution but tend to become partially miscible with a defined solid-state miscibility limit. It is interesting to note that several traditional ways of impurity purging, such as slurry aging and temperature cycling, have found some degree of success in those more intricate presentations. For example, Wang et al. reported that slurry aging of salicylic acid crystals having impurity entrapment as a solid solution led to the selective dissolution of the dirtier cores of the crystal and the recrystallization on the cleaner edges, resulting in the formation of macrotubular crystals of salicylic acid with hollow cores [4]. This type of ripening can be attributed to the slightly higher drug solubility of the dirtier portion than the cleaner region, somewhat like Ostwald ripening, but more related to the gradient of impurities instead of particle size. Notably, the dynamic dissolution–crystallization cycle can be further promoted by temperature cycling, by triggering small increases and then decreases in solubility for both the drug and the impurities.

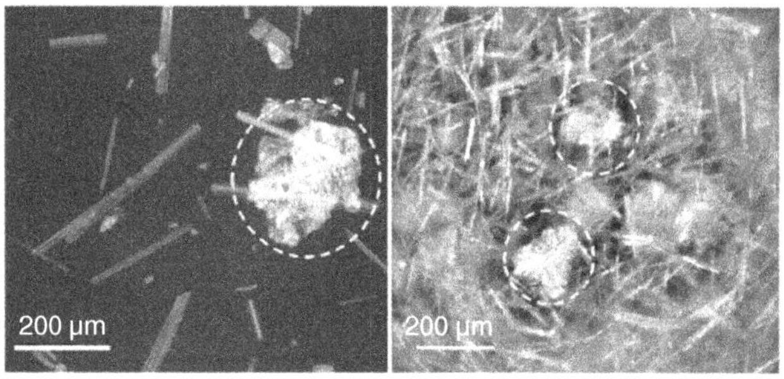

Figure 23.2 Light microscopy (left) and in situ process images (right) of needle-like cephalexin monohydrate crystals and the co-precipitates of phenylglycine impurities. *Source*: Reproduced from [3]/with permission of American Chemical Society.

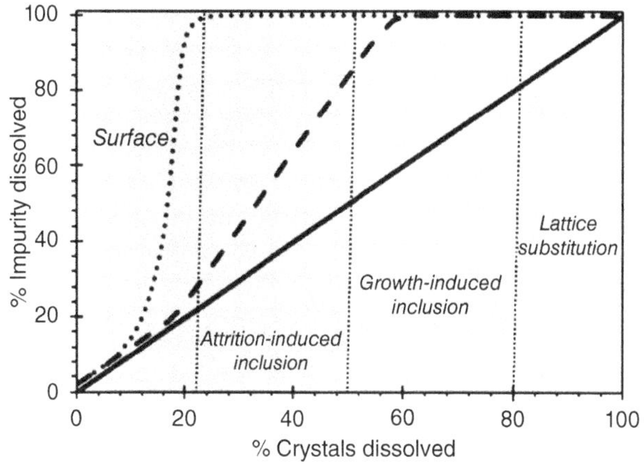

Figure 23.3 Illustrations of different impurity distribution maps to reflect various ways of an impurity being incorporated into growing crystals. *Source*: [1]/ with permission of American Chemical Society.

However, in some cases, process manipulation cannot yield the desired results, whereas solid form selection could offer another effective way to achieve purification. By changing the material into a different solid form, impurity incorporation such as solid solution formation can be dramatically changed. The different crystal forms, such as a polymorph, solvate, hydrate, salt, or cocrystal, can lead to a different interaction enthalpy with the impurity and, therefore, a dramatically different miscibility of the impurity in the crystals. For example, Tanoury and colleagues took advantage of polymorphism to effect an impurity removal for (R,R)-formoterol tartrate [5]. It appears that traditional crystallization methods yield anhydrous forms but could not purge one of the impurities below 0.2%. It is interesting to note that a slurring conversion of one of the anhydrous forms, form B, to a hydrated form, form C, resulted in a significant purification with the levels of all impurities dropping to not more than 0.1%, including the most difficult impurity to remove. Further polymorphic interconversion of the hydrate form yielded the thermodynamically most stable anhydrous form, form A.

Likewise, Black and colleagues described that a piperidene pharmaceutical intermediate can crystallize as a hydrate form with dramatically increased chemical purity over the anhydrous form, with the level of a N-linked regioisomer being purged at the acceptable level of not more than 0.3% after crystallization as the hydrate form. In contrast, 4% of the N-linked impurity persistently existed in the anhydrous form owing to the formation of a partially miscible solid solution [6]. In another example, Standley et al. discovered that the valinate of rovafovir etalafenamide, an anti-HIV drug, can serve as an intermediate salt form for purification purposes since it more readily crystallizes from the crude mother liquor and tends to yield a crystal with an excellent impurity rejection, in comparison to the free base form and the citrate salt [7]. Notably, Nordstrom and coworkers presented another interesting case of breaking the solid-state miscibility of impurities by changing the crystal form [8]. The model drug named compound B, a penultimate intermediate, tended to crystallize as an anhydrous form or a stoichiometric acetonitrile solvate. Two impurities, named IMP-1 and IMP-3, were found to be completely excluded in crystals of the acetonitrile solvate, whereas in the anhydrous form an appreciable amount of impurity was incorporated. Conversely, the sulfonic acid and sulfinic acid impurities encountered better rejection in the anhydrous form than the acetonitrile solvate. The interesting discoveries allowed for the development of a creative two-step process, with the first step to crystallize the acetonitrile solvate to get rid of IMP-1 and IMP-3, followed by a second step to convert the solvate to the anhydrous form through slurrying to clean up both the sulfonic acid and sulfinic acid impurities [8].

23.3 Removal of Residual Solvents

Like all other impurities, residual solvents in pharmaceuticals carried over from synthesis, chromatography, crystallization, spray drying, or freeze drying need to be removed to reach an acceptable level since there is no therapeutic benefit from residual solvents. Regulatory guidelines prescribe a range of specific limits of various types of organic solvents according to their safety profiles. In a nutshell:

- Class 1 solvents such as benzene, known to exhibit unacceptable toxicities, should be avoided as much as possible.
- Class 2 solvents with less severe toxicity, if in use, should be limited in final products to a safe level. For example, methanol shall be below 3000 ppm while dichloromethane shall be below 600 ppm.
- Class 3 solvents, considered less toxic, can be used, where practical, and removed to 5000 ppm or 0.5% as a common best practice. Examples of class 3 solvents include acetic acid, acetone, ethanol, ethyl acetate, dimethyl sulfoxide, and isopropanol, which are commonly used in drug substance synthesis and crystallization.

Residual solvents, as organic volatile chemicals, are usually removed through various drying processes including convection tray drying, vacuum tray drying, conical vacuum drying, or microwave-assisted drying. It appears that drying temperature, vacuum level, and agitation speed stand out as key process parameters to be fully explored during process development and optimization.

However, it is important to stress that a successful drying process not only depends on a thorough process development but also relies on a deep understanding of the physical and chemical properties of the solvent, the starting drug substance, and the endpoint, as well as the specification for the finished product regarding the levels of residual solvents, purity requirement, and polymorph control.

First, a solvent with a lower boiling point would allow for a much quicker drying. Thus, this aspect of quality should be taken into consideration during the process design of API manufacturing, especially at the final step of crystallization. In situations when a higher boiling solvent must be used, a solvent swap to a lower boiling point solvent during cake rinsing could lead to a much-favored drying condition. It is often observed that an extended drying time or an elevated drying temperature could potentially result in chemical degradation or physical transformation, which is detrimental to the quality and shelf life of the finished drug substance.

Second, a mechanistic analysis of how the solvent is bound to the API is necessary for the proper design of an overall strategy for drying. There are three distinct ways a solvent can associate with a drug substance, including

- surface binding or adsorption,
- formation of a solid solution, either totally miscible or partially miscible, and
- solvate formation with solvent incorporated into a crystal lattice

It would be expected that surface-bound solvents are relatively easy to remove, while solvents built into a crystal lattice would become much more difficult to remove since the solvent removal may need a much-needed elevated temperature to disrupt the crystal lattice, as well as solvents present in amorphous materials as a solid solution. The different ways of solvent associated with a drug substance would affect the kinetics of drying, which can be studied to create drying curves. As shown in Figure 23.4, a typical drying curve contains two regions: one constant rate period and the other falling rate period. The constant rate period exhibits a linear rate of solvent loss and is likely associated with the removal of loosely bound solvent. In contrast, the falling rate period reflects the final stage of solvent removal and is probably associated with the removal of a more tightly bound solvent molecule either in the crystal lattice of a solvate or at a solid solution with an amorphous component.

Before venturing into process development, the chemical and physical stability of the drug substance needs to be fully understood to ensure there is no chemical degradation and no unintended solid-state form conversion occurring during the drying process. The drying temperature selected needs to be safely below the point where chemical degradation may occur. For a hydrate form, the stability window in relation to temperature and relative humidity needs to be examined to ensure the drying condition will not trigger a phase transition. If needed, humidified nitrogen or air can be used to purge solvent while maintaining the physical stability of the hydrate form.

A solid-state transformation from a solvate form to a hydrate form or an anhydrate form offers a unique way to remove residual solvents. For example, Yang et al. reported a humidity-mediated transformation from the methanol trihydrate of adenosine 3′,5′-cyclic monophosphate sodium to its pentahydrate form, remarkably, resulting in the removal of methanol [9]. Notably, Pikal and colleagues discovered an even more dramatic case where the presence of water vapor resulted in a 200 times faster desolvation rate for cefamandole sodium methanolate than vacuum drying. It can be understood that the presence of water vapor triggered a rapid phase transformation from the methanolate to the hydrate as reflected by the observation of extensive crystal cracking and the subsequent formation of microcrystalline materials during the early stages of moisture exposure [10].

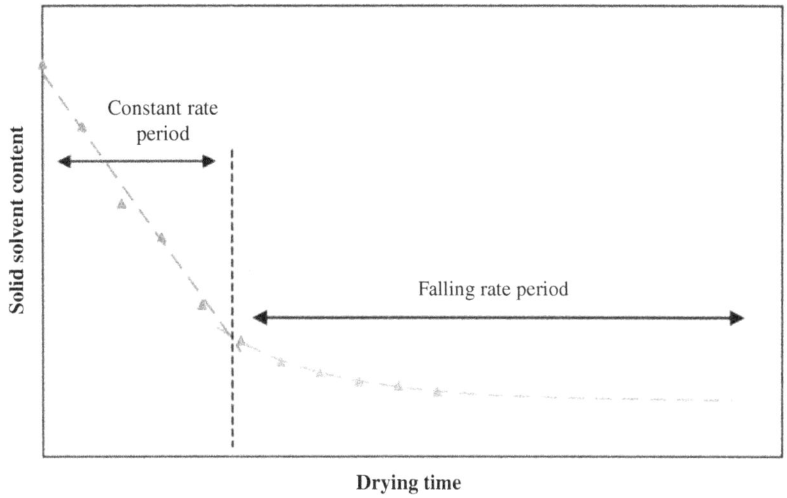

Figure 23.4 A typical drying curve of residual solvent removal.

This interesting utilization of humidified gas to facilitate the solvent removal is termed humid drying by Lamberto and Neuhaus [11], where the water vapor pressure is kept above a critical relative humidity, on the one hand, to promote the swapping of solvent molecules with water in the material to facilitate solvent removal to a desired level and, on the other hand, to maintain the hydrated form by preventing dehydration if the hydrated form needs to be preserved. For instance, humidified gas was employed during the drying of sugammadex, a modified γ-cyclodextrin, used as a reversal agent in general anesthesia, to remove methanol [11]. Sugammadex was isolated from methanol and water solvent systems as a mixed methanol solvate/water hydrate form. Direct drying under nitrogen at 40 °C was found not effective enough to remove methanol down to an acceptable level, whereas a nitrogen stream with a relative humidity of 55% at 40 °C performed well. It is obvious that the successful drying is attributed to the displacement of solvent molecules by water.

The proper control of a solid-state transformation is considered critical for the removal of residual solvents, as demonstrated in another example reported by Lamberto and coworkers [12]. The compound studied in this example was isolated as a crystalline hydrate from isopropanol and water. It is noted that a more aggressive drying condition incurred a parallel dehydration of the hydrate, which caused a solvent entrapment with 3.8 wt% isopropanol not being reduced further, even with additional drying. The entrapment is hypothesized to be caused by the formation of a poor crystalline anhydrous form, which contains a significant amount of amorphous content to retain isopropanol to form a solid solution. However, a two-stage approach resulted in a more effective way to remove isopropanol, with the first stage to dry the material under a significant level of humidity to maintain the hydrate form intact while removing isopropanol (Figure 23.5). After

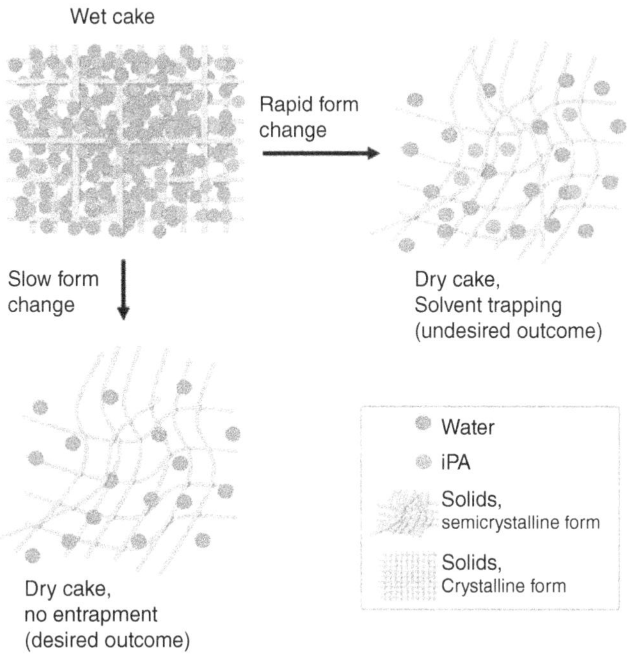

Figure 23.5 Illustrations to show phase transformation having an impact on the effectiveness of solvent removal.

solvent removal, the second stage engaged with the dehydration of the hydrate to yield the desired anhydrous form with poor crystallinity but having the isopropanol level below the acceptable limit of 0.5%.

If the solvent forms a solid solution with the drug, commonly for amorphous APIs, the design of drying temperature should take into consideration the glass transition temperature; otherwise, a high drying temperature above the glass transition may trigger particle coalescence, whereas a low temperature may cause lower drying efficiency leading to an extended drying time.

Solvent swap is an effective way to facilitate the removal of residual solvents. For instance, Lee and Kim reported that pre-treatment of a paclitaxel sample with 95% ethanol enabled the removal of methylene chloride from the initial 17 600 ppm down to less than 600 ppm through drying, an acceptable ICH limit. The further treatment of the sample with water facilitated the drying to remove ethanol to below 5000 ppm [13].

23.4 Control of Genotoxic Impurities

Pharmaceutical genotoxic impurities (GTIs), if present in drug products, may potentially increase cancer risks and, therefore, are required to be controlled to a low ppm level in drug substances and drug products, according to regulatory guidelines [14, 15]. Like the other impurities, GTIs in drug substances may increase due to the use of certain reagents and starting materials, can exist as process impurities, or be derived from degradation during manufacturing processes and storage. According to regulatory guidelines [14], potential GTIs must be controlled at a generic threshold of toxicological concerns of not more than 1.5 μg per day if the toxicological thresholds have not been qualified. For a drug product with a daily dose of 1 g, for example, the generic acceptable level of 1.5 μg per day only allows for a control level at 1.5 ppm. To tackle such a minuscule level, it would be expected that there are substantial process challenges to purify the drug substance to remove the potential GTIs. In addition, sophisticated analytical methods with a low detection limit and robust quantification are required.

As we can see, the first step to address the possibility of the presence of GTIs in a drug substance is to do a comprehensive risk assessment of synthetic pathways and purification steps to find any potential genotoxic materials being introduced into the process flow or potentially being generated during manufacturing. For example, Liu and coworkers did a risk assessment of the synthetic process for pazopanib hydrochloride, a multikinase inhibitor under development for the treatment of renal cell carcinoma, and identified five potential GTIs, including dimethyl sulfate, one of the reagents. This allowed for the development of an analytical strategy aiming to quantify dimethyl sulfate at the ppm level to support further process improvements to control the GTI level [16].

Second, it is essential to accurately quantify and unequivocally identify organic impurities present in the drug substance. If in silico analysis alerts certain compounds to have genotoxic potential, in vitro and in vivo experiments to prove with a high degree of confidence need to be designed to determine whether the compound is genotoxic or not. It is obvious that a casual designation of a compound to be potentially genotoxic with a high degree of ambiguity would cause a significant waste of resources to analyze and control it.

If a genotoxic compound does exist, thorough root cause analysis should be pursued to understand whether it is a process-related impurity or a degradation product. This would help develop strategies to eliminate it completely or prevent it from forming in the first place. As would be expected, the various approaches to purge organic impurities through crystallization mentioned in Section 23.2 are applicable if crystallization becomes an essential step to eliminate the GTI.

One of the most useful tools in risk assessment is to develop a theoretical purging factor to evaluate the risk of carryover of a potential mutagenic impurity into the final API based on the physicochemical properties of the impurity in relation to process conditions. The purging factor defines how much of a GTI is likely to be removed, as reflected by the ratio of the concentration before and after various synthetic and purification processes. Teasdale et al. identified key physicochemical properties and provided a contribution index to the purging factor for each property, as shown in Table 23.1 [17], including:

- reactivity potential to deplete GTIs through chemical reactions with reagents and solvents during workup or crystallization,
- solubility to indicate the extent of GTIs to be retained within the mother liquors during crystallization and hence to be purged from the desired intermediate or product,
- volatility to rank order the probability of GTIs to be removed during solvent evaporation, solvent distillation or drying, and
- ionizability to offer opportunities of using pH manipulation to ionize either the product or the GTI to allow for liquid/liquid separation by extraction.

Table 23.1 Purging factors and physicochemical properties of potential genotoxic impurities.

Physicochemical property	Ranking	Purging factor
Reactivity	Highly reactive	100
	Moderately reactive	10
	Low reactive/unreactive	1
Solubility	Freely soluble	10
	Moderately soluble	3
	Sparingly soluble	1
Volatility	Boiling point >20 °C below that of the reaction/process solvent	10
	Boiling point ±10 °C that of the reaction/process solvent	3
	Boiling point >20 °C above that of the reaction/process solvent	1
Ionizability	Ionization potential of GTI significantly different from that desired product	3–10
Physical process	Chromatography	10–100 depending on the extent of separation
	Other scavenger resin	Evaluated on an individual basis (3–100)

Source: [17]/with permission of American Chemical Society.

McLaughlin et al. showcased Teasdale's method in the evaluation and control of mutagenic impurities for a development compound, MK-8876 [18]. Purge factors were estimated for six impurities identified as mutagenic alerts in the synthesis of MK-8876, to compare with actual measured amounts of these impurities determined via appropriate analytical methods. The work revealed that the purge factor calculation did indeed tend to predict on the more conservative side as intended. Therefore, it can be considered a valuable tool to do risk assessment without substantial analytical investigation. Elder and coworkers applied the same method to evaluate carryover potential of GITs into the final API of pazopanib hydrochloride [19]. As to five potential GITs, the calculated purging factors align reasonably well with the measured values, even for one impurity with relatively low purging factor (<200). It is obvious that a control strategy that can be linked to the specification is necessary for such a high-risk impurity.

23.5 Control of *N*-Nitrosamines

N-nitrosamines are a group of organic impurities regarded as highly potent mutagenic and classified as probable human carcinogens. The discovery of *N*-nitrosamines in several marketed products has led to several high-profile market recalls and also resulted in substantial changes in the regulatory environment regarding the control of nitrosamines in drug substances and drug products. To perform an overall risk assessment of the potential presence of *N*-nitrosamines in drug substances, two key aspects need to be evaluated thoroughly, including nitrosamine generation ability and nitrosamine purge ability.

As illustrated in Figure 23.6, *N*-nitrosamines can form either through a nitrosating reaction or a degradation path via free radicals, with the former one considered the key risk for drug substances and with the latter one potentially occurring in drug products.

To develop a strategy to control *N*-nitrosamines, we shall first discuss how to mitigate the risk of their formation during drug substance synthesis and purification, including:

- presence of a secondary or tertiary amine,
- presence of a nitrosating agent, and
- conditions allowing for the reaction to occur.

So the first step is to examine solvents and bases used in the overall synthetic pathway, if feasible, substituting amine solvents such as dimethylacetamide, dimethylformamide, and *N*-methylpyrrolidone and any amine bases with nonamine-containing ones, to mitigate the carryovers of nitrosamine precursors present as impurities in those solvents.

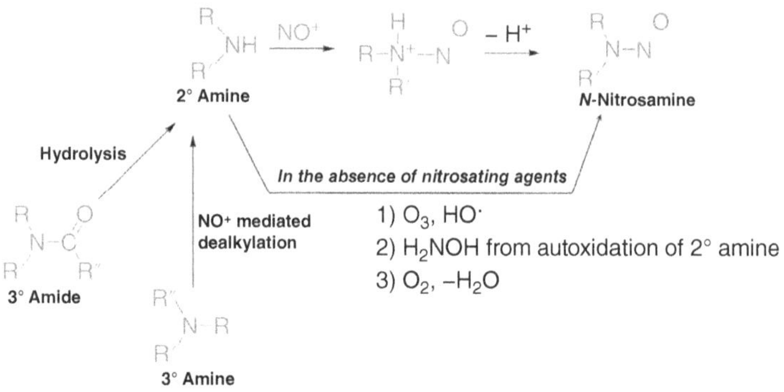

Figure 23.6 Reaction mechanism of nitrosamine formation from an amine compound with or without a nitrosating agent.

If a nitrosating step is necessary, try to redesign the synthetic pathway to avoid the presence of amine compounds or move the step earlier in the synthesis so that it will be removed by multiple steps of purification or choose a reaction condition with nitrosamine formation being effectively mitigated. Diab and coworkers found out that *N*-nitrosamines only become a significant risk to form in solution at pH less than 6 and at a higher concentration of nitrite [20]. As to the risk of generation of more complex nitrosamine drug substance-related impurities, Ashworth and coworkers proposed a unique nitrosation forced degradation study with all three potential pathways being evaluated simultaneously, including *N*-nitrosation, *C*-nitrosation, and nitrosation to non-*N*-nitrosamine [21]. If no nitrosamine-related degradants can be detected, the observation can serve as evidence of a low probability that nitrosamine products can be formed for the drug substance. Conversely, when nitrosamine-related products did exist from the forced degradation study, the information collected can provide further steps of risk assessment and offer reference samples for further evaluation, like a purging factor analysis.

It appears that the major risks of *N*-nitrosamines associated with a drug substance arise from impurity formation and subsequently carryover in the manufacturing process. As would be expected, Teasdale's method of purging factor estimation should be applicable in the risk assessment of *N*-nitrosamines present in drug substances, which aligns with the principles underlying ICH M7 option 4. Option 4 suggests a scientifically based risk assessment process to evaluate the impacts of input materials and process parameters on residual impurity purging. As described in the previous section, the purging factors are determined by chemical reactivity, solubility, volatility, ionizability, and any physical process steps capable of removing impurities. As would be expected, key physical properties such as solubility, ionizable potential, and lipophilicity have profound impacts on the extent of impurity removal by liquid–liquid extraction, crystallization, and solvent washing. For example, Ashworth et al. claimed that a number of smaller, more polar nitrosamines have significant water solubility, potentially allowing for effective purging by a simple aqueous washing [21].

23.6 Chiral Separation Through Salt and Cocrystal Formation

Besides purification steps to purge impurities and drying procedures to remove residual solvents, chiral separation is another important operation for drug substance manufacturing. Chiral separation to produce a pure enantiomer is of significance to drug substance quality. The two enantiomers of a chiral compound usually exhibit very similar physical and chemical properties; however, they often possess distinct physiological effects or different safe profiles.

Chiral separation through crystallization offers a simpler and cheaper way to purify enantiomers than chiral chromatography. The reaction of a racemic mixture with a chiral resolving agent could potentially yield two diastereomeric derivatives, as either a salt or a cocrystal. Diastereomers, unlike enantiomers, tend to have different physicochemical properties, including melting point and solubility. The solubility difference can enable differential crystallization to separate the two diastereomers. After subsequent dissociation of the salt or cocrystal, the targeted enantiomeric pure API can be produced.

For an ionizable compound, chiral resolving acid or base reagents are used to form diastereomeric salts with a subsequent preferential crystallization to yield chiral separation. For example, chiral resolution of (*R*,*S*)-ibuprofen can be attained by using (*S*)-lysine as the resolving agent in aqueous ethanol, which affords the diastereomeric (*S*)-(+)-ibuprofen/(*S*)-lysinate salt in more than 99% optical purity after two crystallizations [22].

It would be expected that the extent of the solubility difference between two diastereomeric salts impacts the efficiency of a chiral resolution process. Tan et al. established the following equation to correlate the resolution efficiency with the solubility difference [24].

$$S = 2 \times y \times de_{\text{crystal}} \tag{23.1}$$

Here, S represents the resolution efficiency, with $S = 1$ being complete resolution. y means the maximum yield as described as follows:

$$y = \frac{C_T - S_u - S_d}{C_T} \tag{23.2}$$

And $de_{crystal}$ stands for the degree of resolution as represented as follows:

$$de_{crystal} = \frac{S_u - S_d}{C_T - S_u - S_d} \tag{23.3}$$

Now, S_u and S_d are the solubilities of the desired and undesired salt, respectively. C_T is the total concentration of racemate. Based on previous equations, an effective chiral resolution is designed by screening for a resolving salt former and a solvent system to yield an appreciable difference in solubility between the undesired and the desired salt.

Since the basis of chiral resolution is a successful salt formation, the compound properties shall be assessed rationally, especially pK_a and solubility, before marching into an experimental screening investigation. As discussed in Chapter 4, a sufficient pK_a difference of two pK_a units or larger between the compound and the counter acid or base is essential for salt formation. Wang and Chen proposed a practical workflow for chiral resolution screening by diastereomeric salt formation [23], including

- assessing whether the compound is an acid or a base,
- determining the pK_a of the compound,
- selecting a list of chiral resolving agents based on pK_a difference,
- measuring the solubilities of the racemate and chiral resolving agents in a series of solvents,
- designing experiments with the combination of chiral resolving agents and solvents for diastereomeric salt formation,
- determining the ratio of two enantiomers for either the solids obtained or the supernatants,
- solid-state characterization of the resulting solids for crystallinity, and
- identifying the leading system based on solution analysis and solid-state characterization for further trials and optimization.

Following a similar workflow, Tan et al. developed a rational screening approach in a high-throughput setting, therefore, with much less material needed [24]. The approach focused on data mining on how the resolving agent, solvent composition, resolving agent stoichiometry, and racemate concentration impact the solubility difference between the undesired and desired salt and the resolution efficiency.

To tackle this important task, Borghese et al. came up with an efficient and fast screening method through the classical crystallization of diastereomeric salts with different chiral counterions at different solvent systems and the following determination of the eutectic composition of the two enantiomers in the mother liquor using chiral analysis [25]. As shown in Table 23.2, for example, the analysis of the eutectic composition from the mother liquor gives a quick ranking order of effectiveness of the chiral resolving agent for compound 2, a weak base drug. It would be expected that the wider the difference in the relative amounts of the two enantiomers, the greater the effectiveness of the potential chiral resolution, while the conditions leading to eutectic compositions close to 50:50 can be ruled out readily. It is worthwhile to point out that solvate formation can have a profound effect on the efficiency of the resolution process. For example, the crystallization of compound 2 in acetone with L-(+)-tartaric acid led to an extraordinary %S:%R ratio at 6:94 likely owing to the formation of an acetone solvate, whereas the trials with other solvents such as isopropanol, isopropanol/water, and acetone/water failed to produce chiral resolution.

Vaidya laid out several practical points that could potentially benefit screening for an optimal resolution process [26].

- Easy salt formation with the resolving agent
- Appreciable difference in solubility between the two diastereomeric salts
- High chiral purity for the resolving agent
- Good stability against racemization under the resolution process for the resolving agent
- The resolving agent being inexpensive and readily recoverable

As we shall see, the solubility difference between the undesired salt, being more soluble, and the desired salt, being less soluble, is a key parameter that affects the performance of differential crystallization. Other approaches have been reported to promote differential crystallization, including seeding and nucleation inhibition. It is easy to understand that seeding of

Table 23.2 Eutectic composition from mother liquor analysis for a weak base compound after diastereomeric salt formation.

		Eutectic composition (%S:%R)			
Entry	**Chiral acid**	**Isopropanol**	**Isopropanol/water**	**Acetone**	**Acetone/water**
1	(+)-10-Camphorsulfonic acid	27:73	24:76	27:73	25:75
2	(*R*)-2-Methyl-2-(4-nitroimidazoyl)-2-(4-methoxyphenyl)acetic acid	–	78:22	–	–
3	(*S*)-(-)-Malic acid	51:49	46:54	31:69	50:50
4	L-(-)-Dibenzoyl-L-tartaric acid monohydrate	–	60:40	–	–
5	(+)-o,o′-di-*p*-Toluyl-D--tartaric acid	–	31:69	–	30:70
6	(+)-1,2,3-Dioxophospho-rinane-5,5-dimethyl-2-hydroxy-4-phenyl-2-oxide	53:47	35:65	47:53	35:65
7	L-(+)-Tartaric acid	57:43	49:51	6:94	50:50
8	(*S*)-(+)-Mandelic acid	53:47	55:45	43:57	50:50
9	(1*R*,3*S*)-(+)-Camphoric acid	50:50	50:50	35:65	50:50
10	(1*R*,3*R*,4*R*,5*R*)-(+)-quinic acid	50:50	–	51:49	–
11	(*S*)-(+)-2-Pyrolidone-5-carboxylic acid	28:72	–	19:81	47:53
12	L-(+)-Lactic acid	24:76	68:32	28:72	28:72

Source: [25]/with permission of American Chemical Society.

the desired salt crystals promotes its differential crystallization, while the effect of small amounts of nucleation inhibitors on the resolution of racemate is more complicated to assess. The phenomenon is related to the so-called Dutch resolution, for which the simultaneous addition of several resolving agents could lead to a better outcome of chiral resolution than the use of a single agent. It is interesting to note that the potential synergic effect does not require the formation of the desired salts with each individual agent, since certain family members of the resolving agents sometimes are only slightly or not at all incorporated into the salts. It is hypothesized that those family members can act as a nucleation inhibitor to the undesired salts by widening the width of the metastable zones of supersaturation to a greater degree for the more soluble diastereomers. Dalmolen et al. discovered that the use of a single resolving agent with the addition of a small amount of additive as a nucleation inhibitor can exhibit similar functionality to the "Dutch resolution" [27].

As would be expected, diastereomeric salt formation is not applicable to nonionizable compounds or not easily ionizable drugs, including very weak acids and very weak bases, whereas cocrystal formation with a chiral coformer is considered a meaningful alternative to achieve chiral resolution. It is recognized that two mechanisms could potentially lead to chiral resolution through diastereomeric cocrystal formation and enantiospecific cocrystallization.

Diastereomeric cocrystal formation refers to the formation of a diastereomeric cocrystal pair with a chiral resolving coformer. Chiral resolution is attained by differential crystallization owing to the significant difference in solubility of the two diastereomeric cocrystals and further cocrystal dissociation to yield the enantiomeric pure drug. For example, Sańchez-Guadarrama et al. accomplished chiral resolution of praziquantel racemate (RS-PZQ) via diastereomeric cocrystal formation with L-malic acid (L-MA). It is interesting to note that R-PZQ:L-MA displays a higher melting point than S-PZQ:L-MA, 154 versus 127 °C, and is less soluble in various organic solvents including acetone, tetrahydrofuran, ethyl acetate, and chloroform [28].

Enantiospecific cocrystallization tends to occur when only one enantiomer of the target drug can cocrystallize with the resolving conformer (Figure 23.7). Since hydrogen bonding interactions, the main driving force of cocrystal formation, are strongly direction dependent, it would be expected that enantiospecific cocrystallization is more probable than the formation of diastereomeric cocrystals. Using S-mandelic acid as a conformer, Springuel and Leyssens demonstrated that the S-enantiomer of levetiracetam, an anticonvulsant used to treat epilepsy, can selectively form a cocrystal with S-mandelic acid in acetonitrile, with 70% of the S-enantiomer being separated from the racemic mixture in a single cocrystallization step [29].

Harmsen and Leyssens reported a similar method to obtain enantiopure (S)-ibuprofen through cocrystal formation. Starting from racemic ibuprofen, they discovered that levetiracetam, a chiral drug, can serve as a resolving agent and specifically form cocrystals with (S)-ibuprofen. The subsequent cocrystal dissociation further yielded pure (S)-ibuprofen API [30].

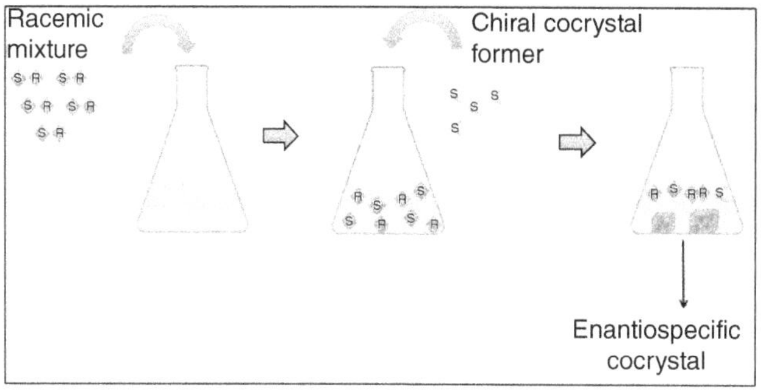

Figure 23.7 Illustration to show the formation of enantiospecific cocrystal. *Source*: [29]/with permission of American Chemical Society.

23.7 Preparation of the Final Solid Form

Besides the roles of purification, crystallization serves as a critical step to produce a crystalline API and control the final crystal form of the material for downstream drug product manufacture. The final form, also called the desired form, or the selected form, is chosen based on thermodynamic stability, solubility, dissolution, chemical stability, processability, and potential performance in the drug product, as disseminated in Chapter 2.

Just like any other process development, the journey to search for a proper crystallization condition to prepare the final form starts with the prior knowledge, such as solid-state form landscape for the drug obtained from solid form selection, the prior experience, for example, how to make the form, and design of experiments, for instance, to vary solvents, antisolvents, supersaturation level, temperature, agitation rate, and seeding. It is worthwhile to point out that the aim of the experiments is to search for a condition that the selected final form can be produced consistently with a high phase purity in an industrial crystallization setting, which is very different from the goal of solid form screening, intended to diverge conditions as wide as possible to obtain all possibly existing solid forms.

The premise to prepare the desired polymorphic form for an API in a controlled manner derives from a reasonably established phase diagram describing the stability window of the form against the other forms and a well-characterized kinetics for the form to nucleate and grow against a supersaturation profile. The phase diagram of a polymorphic system describes the solubility–temperature profiles of various polymorphs, which form the basis for delineating the supersaturation levels of drug concentrations related to each polymorph. More importantly, the diagram can provide the transition temperature of an enantiotropic system and, therefore, indicate the temperature range one form is more stable than the other. The phase diagram is important to define a process that can consistently produce the desired stable polymorph while preventing supersaturation from occurring for undesired polymorphs.

As would be expected, this approach, owing to a proper thermodynamic control, tends to be consistent and effective, as proved by a case study reported by Müller and colleagues [31]. The study illustrated a reliable process development strategy to ascertain reproducible API polymorph manufacture for a Novartis API, which has two enantiotropic polymorphs. The strategy was accomplished by employing several key practices:

- To collect reasonably accurate solubility data at different temperatures for the two polymorphs
- To calculate the transition temperature of the enantiotropic system based on the previous data
- To synthesize the API with yielded concentration at undersaturated conditions to prevent uncontrolled precipitation
- To crystallize the stable form in a controlled manner below the transition temperature based on seeding and cooling crystallization
- To control the cooling rate with a real-time monitoring of the actual supersaturation to ensure the absence of metastability as to the undesired polymorph

It is important to note that the well-designed strategy resulted in a robust process scale-up from lab to pilot and production scale, yielding acceptable polymorph control.

Interestingly, Cote and colleagues developed a similar crystallization methodology to ensure isolation of the most stable crystal form by a precise supersaturation control [32]. By tailoring the antisolvent addition rate, the drug concentration is adjusted to a high level of supersaturation relative to the desired, more stable form while below the equilibrium solubility of the metastable form, therefore, preventing the metastable form nucleation thermodynamically. This unique approach was applied successfully to a drug candidate, namely as MK-A, in terms of robust isolation of the thermodynamically most stable crystal form, Form II, while the occurrence of Form I, the metastable form, was effectively mitigated. It is important

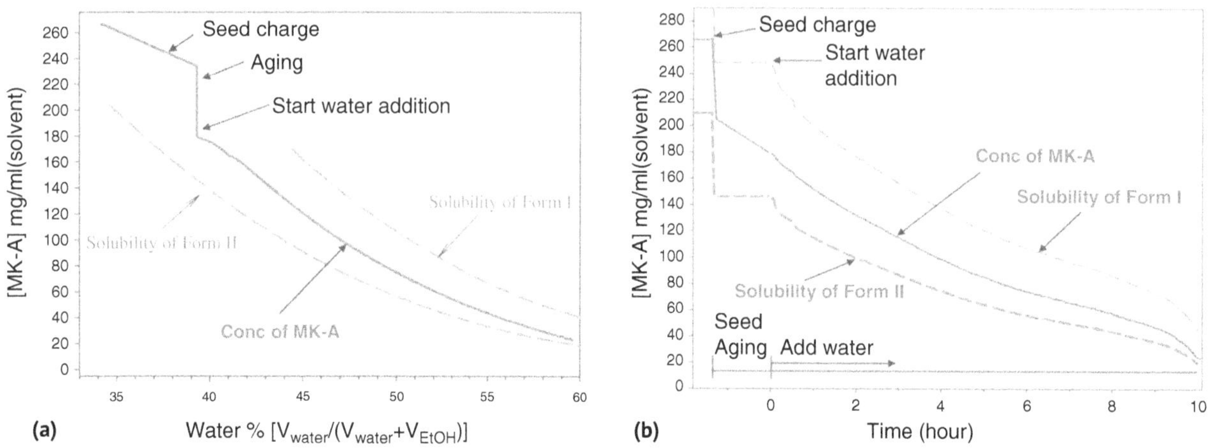

Figure 23.8 Solubility curve of MK-A polymorphs and crystallization profiles at 65 °C with a feedback-controlled addition of water as antisolvent to maintain 33% supersaturation: (a) versus % water and (b) versus time. *Source*: [32]/with permission of American Chemical Society.

to note that the precise drug concentration control was accomplished via a feedback-control loop based on concentration measurement with online Fourier-transformed infrared spectroscopy, and a reasonable production time was ensured by maximizing seed surface area using milled seed and running the crystallization at a high temperature to increase the crystal growth rate (Figure 23.8).

In terms of kinetics, supersaturation is one of the controlling factors that determines the occurrence of various polymorphs. For instance, Lu and Rohani found out that stavudine, an anti-HIV drug, likes to crystallize into Form I, the stable form, at a low supersaturation level in methanol or 2-propanol, while tends to become Form II, one of the metastable forms, at a high supersaturation level in the same solvent system [33].

Moreover, it is worthwhile to point out solvent composition could have a dramatic impact on the crystallization kinetics of a specific polymorph. For instance, Wang and coworkers discovered that the addition of a small amount of water in organic solvents could dramatically promote the crystal growth of the target polymorph of a neurokinin-1 receptor antagonist [34]. Without water in the solvents, the crystallization of the target polymorph was too slow to yield a robust and efficient production process. It is interesting to note that the increase in water content in the solvent can escalate the growth rate linearly. The finding resulted in the identification of a proper crystallization condition including the solvent composition to produce the target polymorph with approximately 100% phase purity [34]. For ranitidine hydrochloride anhydrous forms, less polar or nonpolar solvents favor the crystallization of Form I, while aqueous or more polar solvents often lead to the generation of Form II [35].

Ultimately, seeding is widely accepted as an effective means to attain polymorph control kinetically during crystallization. Sun et al. described the complexities of searching for a proper crystallization condition to prepare the desired polymorph, Form A, for an API named as compound X. Initial trials with at high supersaturation level constantly produced an undesired polymorph, Form L, with a needle-like shape and a strong tendency to agglomerate, which is associated with poor filtration and bad flow characteristics. Form A is difficult to make owing to a relatively long induction time to nucleate and a slow growth rate. It is noteworthy that the use of a high seed loading was found to effectively minimize Form L nucleation and produce Form A with a high phase purity [36].

Seeding is usually effective to promote the preferential crystallization of the selected form. For example, Batra et al. utilized seeding to achieve polymorph control to produce the stable polymorph of treprostinil, Form B, and mitigated the significant formation of Form A, a metastable form with only 0.1 kJ/mol in enthalpy difference from Form B [37].

Control of water content in crystallization solvent is considered critical to produce an anhydrous form as the final form if hydrates exist. Abhishek et al. highlighted the challenges of consistently producing the anhydrous form, Form A, for an API salt when hydrate forms did exist by distillative crystallization, even with seeding. To mitigate the risk of the appearance of hydrate form, an alternate nonaqueous antisolvent crystallization process was developed to produce Form A to supply clinical studies [38].

Likewise, if a solvate crystallizes from the same solvent system that the final form crystallizes, special attention should be applied to mitigate the solvate formation; otherwise, unexpected stability and quality issues may arise. For example,

Wibowo described a case where the formation of solvate crystals during the manufacturing of a generic API caused several production batches to have significantly small particle size and unexpectedly high residual solvent levels after drying [39].

An understanding of the relative physical stability of the solvate and the desired polymorph related to solvent composition forms the basis for designing a proper crystallization process. For instance, Desikan et al. encountered a significant challenge preparing pure Form III of compound A, a non-nucleoside reverse transcriptase inhibitor of HIV, in a *n*-propanol/water system [40]. Even with seeding, cooling crystallization in the mixed solvent system still yielded a small amount of *n*-propanol solvate while producing Form III. A unique reverse-addition process, where the drug dissolved in *n*-propanol was slowly added to the aqueous phase having seeds of Form III suspended, resulted in Form III crystals with a high phase purity. The positive outcome can be attributed to the solvent composition with a significant amount of water and a reduced amount of *n*-propanol, which makes the *n*-propanol solvate become metastable relative to Form III.

Expanding the phase diagram to include temperature as another dimension in addition to solvent composition further enhances its utility to derisk solvate formation when preparing the desired anhydrous form, as showcased in the study of an improved commercial salt formation process for gefapixant citrate (MK-7264), an investigational drug for the treatment of chronic cough [41]. Substantial process development work by Maloney and coworkers led to the selection of the methanol and isopropanol solvent mixture as the crystallization solvent to produce gefapixant Citrate Form 1, an anhydrous form and also the final form. While the desired form can be manufactured with high purity and high yield, a methanol solvate does exist and did show up in a few batches in minor quantities using an original process starting from hot methanol, followed by cooling and addition of isopropanol as an antisolvent. To mitigate the risk of solvate formation and improve the crystallization process, Maloney et al. established a phase diagram aiming to identify the conditions under which Citrate Form 1 is physically stable, via solvent-mediated slurry experiments at various solvent compositions and different temperatures, as shown in Figure 23.9. It is clear that the methanol solvate is favored in the region with a high percentage of methanol and low temperature, while Citrate Form 1 is the stable form in the space with a high percentage of isopropanol and high temperature. The phase diagram clearly reveals that the original process being carried out in methanol at 60 °C, followed by cooling and antisolvent addition, is indeed risky since the entire process takes place in a region where the methanol solvate is thermodynamically stable. As guided by the phase diagram, Maloney et al. made a simple improvement on the original process by adding the antisolvent at 60 °C prior to cooling and crystallization to ensure it within the target region, and, ultimately, developed a commercial process that maintains 2 : 1 methanol/isopropanol ratio in a co-feed crystallization process at 60 °C, followed by the addition of further isopropanol until 1 : 1 ratio before isolation at 50 °C, to have the whole process well within the target stable region for Citrate Form 1.

Sometimes, crystallization is found not to directly produce the final crystal form, while solution-mediated conversion from another intermediate form can offer an alternative solution. For example, dirithromycin tends to form solvates with nearly all of the pharmaceutically acceptable solvents, which leads to challenges in producing the pharmaceutically desirable unsolvated polymorphic form 2 by solution crystallization. Interestingly, Wirth and Stephenson discovered that slurrying dirithromycin acetone solvate in water at 50–70 °C resulted in a rapid conversion to a metastable polymorphic form 1 and a subsequent further conversion to the stable polymorphic form 2 [42].

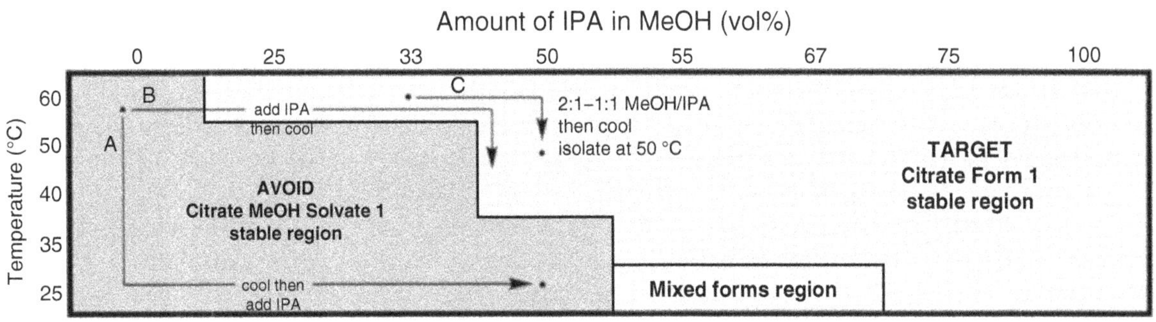

Figure 23.9 Phase diagram of Gefapixant Citrate Form 1 and Citrate methanol solvate under process-relevant conditions. Path A: original process condition; Path B: modified process; Path C: final commercial process. *Source*: [41]/with permission of American Chemical Society.

23.8 Particle Size and Shape Control During Crystallization

It is well recognized that the particle size of a drug substance has substantial influences not only on drug substance manufacturing operations such as filtration and drying but also on the performance of drug products for oral delivery and numerous downstream drug product processes including powder blending, wet granulation, roller compaction, and tablet compression. Particle size control at the crystallization stage, if it can be achieved, provides a significant advantage since further milling and micronization processes can be avoided.

The design of crystallization conditions to aim for particle size control starts from the understanding of crystallization kinetics related to secondary nucleation.

Secondary nucleation refers to the formation of new stable crystals in the presence of crystals of the same compounds as nuclei and represents a key mechanism of continuous crystallization to produce crystalline APIs. Secondary nuclei, as would be expected, can be derived from seeding or are formed during crystallization processes by attrition, shear, or surface breeding. Though the mechanism can be complicated, the kinetics is well described by the population balance equation.

It appears that crystal growth can be expressed as the following equation:

$$\frac{dx}{dt} = k_g \times S^g \tag{23.4}$$

where x is the particle size/length, t is the time, k_g stands for the crystal growth constant, S denotes the level of supersaturation, and g is the crystal growth exponential factor.

The population balance equation can describe crystallization kinetics as follows:

$$R_{\text{formation}} - R_{\text{depletion}} = \frac{\partial}{\partial t}F(x,t) + \frac{\partial}{\partial x}F(x,t)\frac{dx}{dt} \tag{23.5}$$

Here, $F(x,t)$ represents the population distribution function.

As revealed by the equation, the key to a successful crystallization to yield the desired particle size relies on a proper control of the stochastic nucleation event by initiating the process in a reproducible way such as seeding. As would be expected, the particle size distribution of an API from crystallization is determined by the:

- amount of seeding,
- supersaturation level, and
- extent of mixing.

As can be seen, seeding is a relatively straightforward means to control particle size. By removing the primary nucleation from the rate-limiting step as well as unpredictability, seeding promotes crystal growth and enables a crystallization process to be more controllable by determinants such as seed amount, seed particle size, supersaturation level, and mixing. However, control of particle size cannot be separated from control of particle shape.

One important aspect of crystal engineering is to control crystal shape. Crystal shape, along with particle size, has substantial impacts on drug substance downstream processes such as filtration and drying, while also influencing the manufacturability of drug products such as flow and compressibility. For example, needle-shaped crystals often lead to filter blockage, resulting in extra-long filtration times for a drug substance. Furthermore, an API with a needle-like shape likely would cause low bulk density and poor flow of its drug formulation and may demand a granulation step to increase density and flow.

It is well known that the crystal shape is determined by the distinct growth rates of various facets on the crystal surface. For example, needle-shaped crystals usually result from the relatively faster growth along the needle axis's direction than the surrounding side facets. As would be expected, slowing down the fast-growing facets can serve the purpose to suppress needle-like crystal formation by selecting a different solvent, varying the supersaturation level, manipulating the cooling rate of crystallization, or leveraging the impurities present in the mother liquor.

It is generally agreed that solvent selection greatly impacts the rates of crystal nucleation and growth owing to solvent-surface interactions, and potentially influences the resulting crystal shape. For example, Wang et al. reported that, to recrystallize the desired polymorph of a neurokin-1 receptor antagonist, the inclusion of a small amount water in organic solvents gave rise to a distinct crystal shape change from needles to platelets [34].

First, the degree of supersaturation plays a significant role in the relative growth rate of different crystal facets. For example, Modi et al. observed that crystallizing celecoxib at a degree of supersaturation of 190% by cooling yielded acicular crystals, whereas slow evaporation at a degree of supersaturation of 102% resulted in plate-shaped crystals [43].

Second, thermal annealing, a temperature cycling process, tends to facilitate Ostwald ripening in dissolution–crystallization cycles and could result in changes in crystal shape and particle size. To give an example, Nordstrom and coworker reported that a thermal annealing process to treat a crude batch of Compound A, a penultimate intermediate, not only finished with an effective impurity rejection but also resulted in a positive change in crystal shape from fine needles to chunks and an enlarged particle size [8], as shown in Figure 23.10 item a and b. Wilson et al. demonstrated that a temperature cycling process in combination with wet milling allowed for the successful alternation of the shape of a crystalline API to improve its processability in a downstream roll compaction operation [44]. Likewise, Eren and coworkers employed an integrated temperature cycling process with wet milling and achieved an improved aspect ratio of needle-shaped crystals for an API named as Takeda compound A [45] (Figure 23.10, items c and d).

It is obvious that controlling the particle size and shape for an API during crystallization interacts and should be integrated when designing experiments aiming to find a proper process condition to mitigate issues like particle agglomeration.

One key issue for good particle size and shape control during crystallization is to mitigate agglomeration. Particle agglomeration during crystallization may affect the impurity purging and removal of residual solvents since the mother liquid may get trapped between particles. During the early phase of crystallization, fine crystals may adhere to each other, and further growth simultaneously leads to the formation of agglomerates. In addition, cross nucleation may occur during crystallization, where a new crystal nucleates from the surface of another crystal and leading to the formation of aggregates of crystals. For example, surface nucleation and growth of dihydrate on monohydrate crystals is considered the agglomeration mechanism of azithromycin dihydrate in acetone–water mixtures [46].

Not surprisingly, increasing agitation can help reduce the formation of aggregates, as manifested in the case of paracetamol crystallization by an antisolvent method reported by Yu and colleagues [47]. During the crystallization, lower agitation speed resulted in excessive nucleation and significantly agglomerated paracetamol crystals, whereas higher agitation effectively broke up agglomerates, yielding smaller and more uniform particle sizes.

Also, additional manipulation to enhance breakage, such as ultrasound, is anticipated to help reduce agglomeration further. For example, Belca and coworkers used ultrasound to facilitate the crystallization process of an API and found that the ultrasound-assisted process resulted in significantly reduced agglomeration in comparison to the

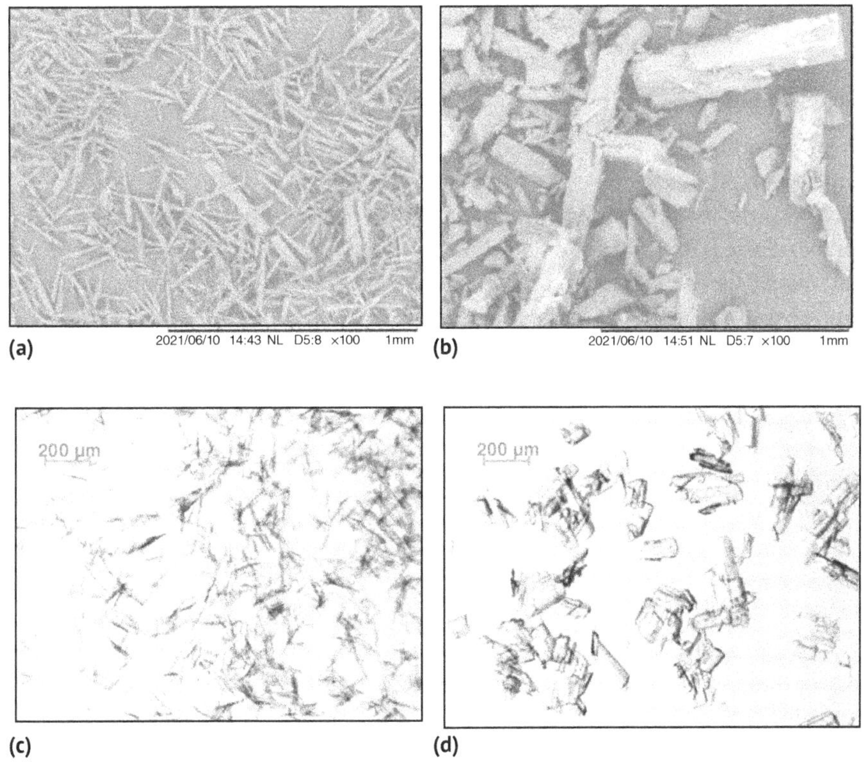

Figure 23.10 Effect of thermal cycling on particle size and shape. (1) Penultimate intermediate Compound A: (a) starting material and (b) annealed material; (2) Takeda Compound A: (c) no cycling and no milling and (d) three thermal cycling and milling. *Source*: Reproduced from [8] and [45]/with permission of American Chemical Society.

spontaneous crystallization and the seeding process [48]. Furthermore, Gielen et al. discovered that the use of ultrasound during the initial seeding stage could accelerate nucleation and inhibit agglomeration, whereas sonication as posttreatment would break agglomeration less effectively and at the same time result in significant crystal fracture and damage [49].

In addition to the benefit of preventing agglomeration, ultrasonic crystallization can potentially deliver particle size control and particle shape modification. The use of ultrasound can induce nucleation at a low supersaturation and decrease the metastable zone width, therefore leading to a shortened induction time. As exemplified in the case of phenacetin crystallization reported by Su et al. [50], power ultrasound applied in the cooling crystallization of phenacetin resulted in phenacetin crystals with a reduced particle size and a modified shape with elliptic morphology. In comparison to the silent process without ultrasound treatment to yield particles with a median size of 87.5 μm, the ultrasound implemented crystallization yielded particle sizes in a range from 15 to 45 μm median sizes when the ultrasonic intensity and duration were controlled to different levels. Likewise, Kaur Bhangu et al. described that changing ultrasonic frequency and power allowed for the manipulation of the median crystal size of paracetamol in a broad range from 10 to 50 μm [51].

Besides the ultrasonic approach, another way to generate relatively fine particles, if wanted, is to use precipitation by creating a high-level supersaturation, through either antisolvent or pH shift. Antisolvent is a solvent miscible with the solvent in which the drug is dissolved but having limited solubility for the drug. The addition of an antisolvent to a drug solution allows for a controlled way to create supersaturation in a short period of time. The pH shift method takes advantage of the pH-dependent solubility profiles of various ionizable compounds.

Another unique way to create API particles with good flowability is to generate spherical particle aggregates instead of engineering individual particle shapes from crystallization. Spherical crystallization is considered advantageous to create flowable and compressible pharmaceutical formulations without the need for milling and granulation. For example, Kawashima et al. described the direct spherical agglomeration of salicylic acid needle-like crystals after crystallization in a mixture of three partially miscible liquids, such as water, ethanol, and chloroform, with agitation. Owing to the excellent flowability, it is interesting to note that the agglomerates of salicylic acid from spherical crystallization can be made into tablets via direct compression [52]. Common techniques to create spherical particles include:

- spherical agglomeration,
- quasiemulsion solvent diffusion, and
- thermal-induced liquid–liquid phase separation.

Spherical agglomeration tends to occur when an immiscible solvent is added to a suspension to agglomerate the particles through preferential wetting and resulting capillary forces.

Quasiemulsion solvent diffusion works by antisolvent crystallization. The addition of an antisolvent to a drug solution can potentially trigger the formation of transient emulsion droplets. Further diffusion of antisolvent into the droplets tends to create a supersaturation and allows for the formation of a crust within the droplets. Often, emulsifiers or surfactants are needed to stabilize the quasiemulsion to result in spherical crystalline particles. For example, Hansen and Kleinebudde studied a quasiemulsion solvent diffusion method to prepare spherical particles for metformin hydrochloride and found out that the use of hypromellose was required in the formation of spherical agglomerates, and the further addition of emulsifiers such as polysorbate 80 and sorbitan monooleate improved the reproducibility of the process [53].

Thermal-induced liquid–liquid phase separation can lead to the formation of spherical particles from aqueous systems directly. When the temperature is raised above the eutectic temperature, liquid–liquid phase separation tends to occur, and drug-rich liquid droplets can form under proper agitation. Further quench cooling triggers the crystallization of the drug within the liquid droplets to form spherical drug particles. It appears that the particle size of the resulting products is directly related to the size of liquid droplets, which are determined by the agitation speed and the presence of surfactants. As would be expected, this method is more applicable to drugs having poor aqueous solubility with a low melting point. Sun and coworkers presented two case studies for water-insoluble drugs such as L-menthol and ibuprofen using the method [54]. As shown in Figure 23.11, the spherical crystallization of ibuprofen resulted in nearly monodisperse spherical particles. Notably, the speed of agitation can be utilized to tune the particle size, where a higher speed would lead to smaller spherical particles. It is of significance to point out that spherical crystallization can be a separate process from standard crystallization since particle aggregation leading to spheronization may result in impurity retention if applied to crude API containing impurities.

23.9 Powder Agglomeration and Caking

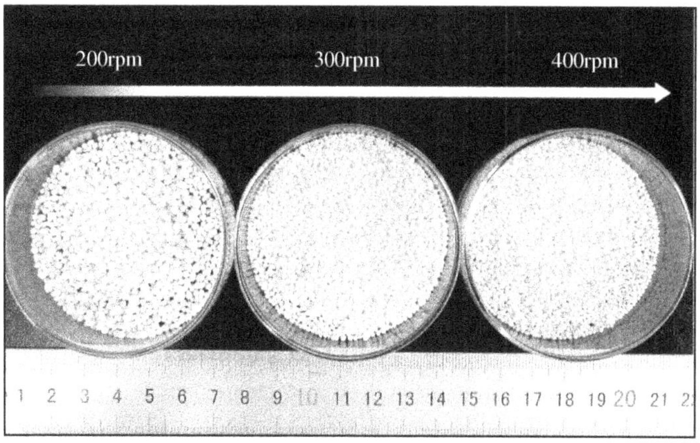

Figure 23.11 Pictures of ibuprofen spherical particles obtained from thermal-induced liquid–liquid phase separation at the agitation speeds of 200, 300, and 400 rpm. *Source*: Reproduced from [54]/with permission of American Chemical Society.

Powder agglomeration can occur during drying or sometimes from long-term storage. Unintended agglomeration can cause a significantly broad and uncontrolled particle size distribution. Agglomeration can also lead to multiple issues in material flow and transfer during a downstream drug product manufacturing process such as charging, blending, mixing, or granulation. Agglomeration can also have detrimental effects on product quality such as poor content uniformity or inconsistent drug dissolution. Figure 23.12 shows a case of powder agglomeration after drying where a substantial amount of drug was retained during sieving in comparison to a second sample with much less agglomeration [55]. Moreover, further powder agglomeration could lead to the caking of a solid bed or total solidification into a difficult, hard-to-break, coherent mass, which may cause product failure or trigger expensive reworking.

Thus, a fundamental understanding of the physical changes that trigger agglomeration in a powdered material is of importance to design a mitigation strategy to prevent or reduce the severity of powder agglomeration during API drying or long-term storage.

It is recognized that several potential mechanisms exist to trigger powder agglomeration and subsequently caking, including:

- excessive residual drug solution carried over from filtration or centrifugation allows for the formation of solid bridges upon drying,
- residual solvents retained owing to capillary forces promoting particle aggregates,
- moisture uptake incurred drug dissolution, liquid bridges, and final solid bridges formed, and
- phase transition such as crystallization of amorphous materials.

During drying, solid bridges can form to hold particles together as agglomerates when residual drug dissolved in the solvents is not completely removed by cake washing. It would be expected that the higher the solubility of the API in the wash solvent, the greater the extent of solid bridging. Thus, switching the wash to a solvent or solvent mixture with lower drug solubility represents a meaningful way to reduce particle agglomeration. For example, Birch and Marziano showed in a case study that washing the material, a model compound, with a mixture solvent, 2-methyl tetrahydrofuran/water 99/1 v/v,

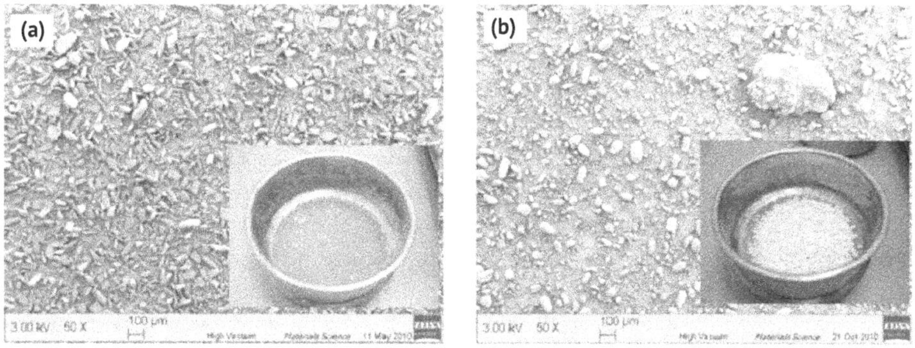

Figure 23.12 Scanning electron micrographs and pictures of sieve cuts for samples with low agglomeration (left) and more substantial agglomeration (right). *Source*: Reproduced from [55] / with permission of American Chemical Society.

resulted in significantly fewer agglomerates than washing with 2-methyl tetrahydrofuran alone, because the compound has three times less solubility in the mixed solvent than in the neat solvent [55].

Particle agglomeration of a powder during storage sometimes leads to severe caking, which likely occurs at the points of contact between particles; however, the fundamental mechanism for the formation of material bonds between particles is still not well understood. It is widely known that fluctuations in humidity and temperature and pressure exerted on particles within a powder bed by gravity or unintended mechanical forces during moving and shipping often facilitate agglomeration.

Generally speaking, water plays a significant role in promoting agglomeration since many soluble materials tend to cake if not protected from moisture, for example, table sugar or table salt. Moisture uptake can result in liquid water condensation from either deliquescence or capillary condensation, and, consequently, the formation of liquid bridges between particles. Temperature fluctuation or humidity fluctuation may trigger drug crystallization in the liquid bridges, forming solid bridges and facilitating agglomeration. The cycle can continue to proceed in many runs and, therefore, end with the formation of solid clumps or cakes. It can be expected that particle adhesion forces and physical proximity accelerate the agglomeration process. For example, needle-like or plate-like crystals tend to align with the long index direction and end with more physical contact and proximity to promote particle adhesion and potentially solid bridge formation if conditions allow. In addition, smaller particles tend to display more severe caking tendencies owing to the increased surface area and a significantly larger number of contact points between particles.

It is generally recognized that amorphous materials are more hygroscopic than crystalline materials, and water absorbed into the material can dramatically decrease the glass transition temperature, which can cause the storage temperature to be above the glass transition temperature. The material above T_g becomes a frozen liquid with a sharply decreased viscosity and significantly enhanced molecular mobility, leading to material flow which forms viscous bridges between particles to facilitate agglomeration.

Solid-state phase transformation represents another root cause for particle agglomeration. The crystallization of the new phase from either an amorphous form or another crystalline form can cause new solid bridges to form between particles. For example, Afrassiabian disclosed that the caking of anhydrous lactose can be triggered by exposure to humidity and attributed to the conversion to the monohydrate form. The degree of caking correlated well with the extent of the conversion. More than 8% of the conversion rate to lactose monohydrate resulted in the formation of hard cakes [56]. For a material with amorphous content, the sorption of water would significantly decrease the glass transition temperature, which would make the amorphous material soft and easy to flow to promote particle bonding and also to increase molecular mobility to trigger recrystallization.

23.10 Milling and Micronization

Particle control sometimes can be achieved through crystallization but not all the time, owing to the difficulties with controlling the dynamics of crystal nucleation and growth. Moreover, aggregation may occur during filtration, drying, or long-term storage. That is why milling and micronization are an essential step in drug substance manufacturing. Milling can delump, deagglomerate, or reduce particle size for drug substance powders, producing a more uniform and narrower particle size distribution.

Regarding terminology, milling refers to a group of particle size reduction techniques that use mechanical energy to break coarse particles into finer ones, while micronization describes a subset of the milling process that can break particles below 10 μm with a more aggressive energy input.

There are a variety of milling techniques available with a broad range of capabilities. Passing a powder through a sieve, a relatively simple approach, is widely used to delump and deagglomerate pharmaceutical raw materials and APIs. Conical milling is a gentle grinding technique where a rotating impeller forces the material through a conical screen surface. This method can offer sizing of pharmaceutical actives or intermediates in the hundreds of microns range and deagglomerate relatively harder particles. Hammer milling breaks particles by impact with rotating hammers and tends to produce particles from 20 to 200 μm. Pin milling works via impingement between stationary and rotating pins and can yield particles in the range of 10 to 50 μm. Ball milling grinds materials with grinding balls within a rotating container and, with significant impact and attrition, can reduce particle size to the several microns range. Jet milling, another widely used particle size reduction technique, relies on high-pressure gas to generate a high-speed aerosol flow to allow for particle–particle collision and can generate particles with sizes from 1 to 15 μm.

Selecting a proper technique depends on a clear understanding of the problem at hand. To normalize the particle size and remove agglomerates, hammer milling and pin milling can help produce particles with more uniform particle size, better flow, and denser powder bed, which are usually required for successful downstream drug product processing such as mixing, blending, granulation, tableting, and capsule filling. To enhance dissolution to ensure content uniformity of low-dose pharmaceutical products, micronization using jet milling is a common selection. One area that demands precise particle control is pulmonary delivery through drug powder inhalers. These dosage forms have a strict particle size requirement in the range of 1–5 μm. Particles of this size predominantly deposit in the large and conducting airway as desired, while particles less than 1 μm are likely to be exhaled, and particles greater than 5 μm predominantly deposit in the oropharynx.

It is worthwhile to point out that micronization, owing to its relatively high-energy input, may incur some undesired physical changes such as polymorph conversion, dehydration of hydrates, generating amorphous material, or significant buildup of static charge, leading to process failures. For example, Cote et al. noticed that, after jet milling, Form I of MK-A, a drug candidate, experienced a detectable amount of form conversion to another crystalline form, which later was identified as a more stable form and named as Form II [32].

Moreover, Lim et al. reported a milling-induced amorphization of an API having a correlation with energy inputs during jet milling. High, medium, and optimal inputs resulted in 51%, 32%, and 15% amorphous content, respectively [57]. It was recognized by Fukunaka et al. that increased particle cohesiveness incurred by micronization may cause process difficulties in milling equipment such as compaction of particles inside the mill [58].

In this regard, wet milling can offer an alternative solution. Wet milling works by passing an API suspension through a rotor-stator and can reduce particles to between 10 and 100 μm. This method can be built into the final stage of API manufacturing. One option is to telescope the suspension from a final crystallization into a wet milling chamber before filtration, to yield particles with the desired size and shape. The other option is to set an inline process along with crystallization to allow for additional control over the particle size and shape.

23.11 Impact of Process Impurities on Crystallization

As we shall see, crystallization is well regarded as an essential process for purifying synthetic intermediates and drug substances to remove/exclude organic impurities. Interestingly, the impurities present in the mother liquor, even in a very small amount, can have strong influences on crystal nucleation and growth and, therefore, affect crystallization outcomes including polymorph control, particle size, and particle shape.

Since crystallization is a molecular recognition and assembly process, the impact of impurities on crystallization can likely relate to interactions with the drug molecules in the mother liquor or crystal surfaces. An impurity can function as a nucleation inhibitor to reduce the nucleation rate, due to the additional energy needed to form nuclei. If the inhibiting effect on polymorphs is distinct, the impurity may lead to polymorph selectivity, where one polymorph is almost silenced and another polymorph proceeds freely to crystallize even though it is a metastable polymorph. Particularly, this selectivity can be attributed to the preferential adsorption of an impurity onto the surfaces of one polymorph while leaving the other free. To give an example, Liu and coworkers reported that the presence of metacetamol, a major structure-similar impurity for paracetamol, could result in the significant inhibition of the generation of paracetamol Form I, the stable form, and allowed the selective production of Form II, the metastable form. This was due to the strong adsorption of the impurity on the {011} faces (width and thickness) of Form I in molecular orientations that inhibit crystal growth [59].

Tanaka and Takiyama reported that the presence of impurities in the mother liquor of a drug candidate named Compound X resulted in the selective production of the desired crystal form, Form III, potentially through inhibiting the nucleation of the undesired crystal form, Form II. When Form II formed, likely via extended crystallization time, a significant amount of the impurities were incorporated into the crystal lattice, while, in Form III, the impurities were purged effectively. This means that the impurities have a strong interaction with Form II crystals but not with Form III crystals [60].

Moreover, instead of working as a selective inhibitor for nucleation, impurities can act as a template or seed to specifically facilitate the crystallization of one form versus another. This mechanism was proposed by Bauer et al. to be the mechanism for the well-known case of ritonavir in terms of disappearing polymorphs [61]. The sudden appearance of ritonavir Form II, which is the stable polymorph, caused the subsequent "disappearance" of ritonavir Form I in API manufacturing. Form I is the metastable form and the original final form and was the only known form during many years of development and manufacture. The appearance of Form II sounds puzzling and is putatively attributed to the incidence of a cyclic carbamate impurity, which was present in a significant amount in late batches, templating the crystallization of Form II. The carbamate degradant has a molecular conformation like the conformation of ritonavir in Form II and exhibits a lower solubility. It was hypothesized by Bauer et al. that the carbamate impurity might have acted as a seed for Form II.

Differential absorption of impurities on specific crystal facets of an API may suppress crystal growth for those facets, resulting in crystal shape changes. For example, Saleemi et al. studied the effect of metacetamol, a structurally related impurity, on the crystallization of paracetamol in isopropanol [62]. It is interesting to note that the presence of small amounts of metacetamol not only caused significant inhibition to the nucleation by prolonging the induction time and extending the metastable zone width but also introduced a continuing change in the shape of the paracetamol crystals from tabular to columnar morphology with increasing the impurity level. Likewise, Lechuga-Ballesteros and Rodriguez-Hornedo reported that the presence of hydrophobic L-amino acids at a relatively low concentration during the crystallization of L-alanine in water caused a change in crystal shape from prismatic to needle-like [63].

It is widely recognized that impurity profiles can vary significantly in the different stages of development or during scaling up due to changes in starting materials, synthetic routes, solvent compositions, and reaction conditions. The changes in type of impurities, number of impurities, and levels of impurities can incur the variability of the end product regarding polymorph, particle shape, and particle size. Sometimes, the results appear elusive, and it can be difficult to find the root cause that triggered the changes. It shall be noted that root cause analysis should focus not only on deconvoluting crystallization conditions such as supersaturation levels and solvent composition but also on studying the impurity profiles of mother liquors to see if any major structurally related impurities are present.

23.12 Summary

As discussed, crystallization is one of the most important steps during the manufacturing of APIs, to purify APIs from a reaction mixture and to yield the desired crystal form with proper polymorph control. An in-depth understanding of the thermodynamics of various solid forms associated with the API and the crystallization kinetics of those forms is required to allow for a sophisticated process development leading to the production of the desired crystal form. In addition, crystallization processes should be robust enough to purge organic impurities to acceptable levels based on safety and regulatory requirements. As disseminated in this chapter, during process development for the final stage of drug substance manufacturing, solid-state forms can offer a valuable design space. For example, strategies like the utilization of a different form could lead to a process yielding a purer API batch. A solid form phase transition could potentially allow for an alternative solution, for instance, to facilitate the removal of residual solvents. While the presence of GTIs and nitrosamines in APIs is a major regulatory concern, a risk assessment should balance both practicality and patient benefits.

References

1. Urwin SJ, Levilain G, Marziano I, Merritt JM, Houson I, Ter Horst JH. A structured approach to cope with impurities during industrial crystallization development. *Org Process Res Dev* 2020; 24(8): 1443–1456.
2. Capellades G, Bonsu JO, Myerson AS. Impurity incorporation in solution crystallization: Diagnosis, prevention, and control. *CrystEngComm* 2022; 24(11): 1989–2001.
3. Salami H, McDonald MA, Bommarius AS, Rousseau RW, Grover MA. In situ imaging combined with deep learning for crystallization process monitoring: Application to cephalexin production. *Org Process Res Dev* 2021; 25(7): 1670–1679.
4. Wang Y, Li H, Raikes M, Linehan B, Robson J, Nordstrom FL. Formation of macrotubular crystals of salicylic acid through ripening of solid solution crystals containing impurity gradients. *Cryst Grow Des* 2021; 21(7): 4100–4110.
5. Tanoury GJ, Hett R, Kessler DW, Wald SA, Senanayake CH. Taking advantage of polymorphism to effect an impurity removal: Development of a thermodynamic crystal form of (R, R)-formoterol tartrate. *Org Process Res Dev* 2002; 6(6): 855–862.
6. Black SN, Cuthbert MW, Roberts RJ, Stensland B. Increased chemical purity using a hydrate. *Cryst Grow Des* 2004; 4(3): 539–544.
7. Standley EA, Bringley DA, Calimsiz S, Ng JD, Sarma K, Shen J, Siler DA, Ambrosi A, Chang WT, Chiu A, Davy JA. Synthesis of rovafovir etalafenamide (part i): Active pharmaceutical ingredient process development, scale-up, and impurity control strategy. *Org Process Res Dev* 2021; 25(5): 1215–1236.

8 Nordstrom FL, Sirota E, Hartmanshenn C, Kwok TT, Paolello M, Li H, Abeyta V, Bramante T, Madrigal E, Behre T, Capellades G. Prevalence of impurity retention mechanisms in pharmaceutical crystallizations. *Org Process Res Dev* 2023; 27(4): 723–741.

9 Yang P, Lin C, Zhuang W, Wen Q, Zou F, Zhou J, Wu J, Ying H. Insight into a direct solid–solid transformation: A potential approach for the removal of residual solvents. *CrystEngComm*. 2016; 18(10): 1699–1704.

10 Pikal MJ, Lang JE, Shah S. Desolvation kinetics of cefamandole sodium methanolate: The effect of water vapor. *Int J Pharm* 1983; 17 (2–3): 237–262.

11 Lamberto DJ, Neuhaus J. Robust process scale-up leveraging design of experiments to map active pharmaceutical ingredient humid drying parameter space. *Org Process Res Dev* 2021; 25(2): 239–249.

12 Lamberto DJ, Diaz-Santana A, Zhou G. Form conversion and solvent entrapment during API drying. *Org Process Res Dev* 2017; 21(11): 1828–1834.

13 Lee CG, Kim JH. Improved drying method for removal of residual solvents from paclitaxel by pre-treatment with ethanol and water. *Process Biochem* 2015; 50(6): 1031–1036.

14 Guideline IH. Assessment and control of DNA reactive (mutagenic) impurities in pharmaceuticals to limit potential carcinogenic risk M7. In: *International Conference on Harmonization of Technical Requirements for Registration of Pharmaceuticals for Human use (ICH)*: Geneva (accessed 5 June 2014).

15 Galloway SM. International regulatory requirements for genotoxicity testing for pharmaceuticals used in human medicine, and their impurities and metabolites. *Environ Mol Mutagen* 2017; 58(5): 296–324.

16 Liu DQ, Chen TK, McGuire MA, Kord AS. Analytical control of genotoxic impurities in the pazopanib hydrochloride manufacturing process. *J Pharm Biomed Anal* 2009; 50(2): 144–150.

17 Teasdale A, Elder D, Chang SJ, Wang S, Thompson R, Benz N, Sanchez Flores IH. Risk assessment of genotoxic impurities in new chemical entities: Strategies to demonstrate control. *Org Process Res Dev* 2013; 17(2): 221–230.

18 McLaughlin M, Dermenjian RK, Jin Y, Klapars A, Reddy MV, Williams MJ. Evaluation and control of mutagenic impurities in a development compound: Purge factor estimates vs measured amounts. *Org Process Res Dev* 2015; 19(11): 1531–1535.

19 Elder DP, Okafo G, McGuire M. Assessment of predictivity of semiquantitative risk assessment tool: Pazopanib hydrochloride genotoxic impurities. *Org Process Res Dev* 2013; 17(8): 1036–1041.

20 Diab S, Dominey AP, Ferrini P, Hayler J, Hermitage S, Urquhart MW, Whiting MP, Wickens J. Formation of N-nitrosamines by reaction of secondary dialkylamines with trace levels of nitrite in aqueous solution: an automated experimental and kinetic modeling study using Di-n-butylamine. *Org Process Res Dev* 2023; 28(1): 293–304.

21 Ashworth IW, Curran TT, Ford JG, Tomasi S. Prediction of N-nitrosamine partition coefficients for derisking drug substance manufacturing processes. *Org Process Res Dev* 2021; 25(4): 871–883.

22 Bhattacharya A, Murphy D. Temperature selective diastereo-recognition (TSD): Enantiomeric ibuprofen via environmentally benign selective crystallization. *Org Process Res Dev* 2003; 7(5): 717–722.

23 Wang Y, Chen A. Crystallization-based separation of enantiomers. *Stereoselective synthesis of drugs and natural products* 2013; 27:1–20.

24 Tan H, Cui S, Gahm K, Luu V, Walker SD. Rational screening approach for classical chiral resolution under thermodynamic equilibrium: A case study of diphenyl-substituted N-methyl-piperazine. *Org Process Res Dev* 2011; 15(1): 53–63.

25 Borghese A, Libert V, Zhang T, Alt CA. Efficient fast screening methodology for optical resolution agents: Solvent effects are used to affect the efficiency of the resolution process. *Org Process Res Dev* 2004; 8(3): 532–534.

26 Vaidya NA. Diastereomeric crystallization—the "classical" chiral technology. *Innov. Pharm. Techol.* 2001; 82–85.

27 Dalmolen J, Tiemersma-Wegman TD, Nieuwenhuijzen JW, van der Sluis M, van Echten E, Vries TR, Kaptein B, Broxterman QB, Kellogg RM. The Dutch resolution variant of the classical resolution of racemates by formation of diastereomeric salts: Family behaviour in nucleation inhibition. *Chem Eur J* 2005; 11(19): 5619–5624.

28 Sánchez-Guadarrama O, Mendoza-Navarro F, Cedillo-Cruz A, Jung-Cook H, Arenas-García JI, Delgado-Díaz A, Herrera-Ruiz D, Morales-Rojas H, Höpfl H. Chiral resolution of RS-praziquantel via diastereomeric co-crystal pair formation with L-malic acid. *Cryst Growth Des* 2016; 16(1): 307–314.

29 Springuel G, Leyssens T. Innovative chiral resolution using enantiospecific co-crystallization in solution. *Cryst Growth Des* 2012; 12(7): 3374–3378.

30 Harmsen B, Leyssens T. Enabling enantiopurity: combining racemization and dual-drug co-crystal resolution. *Cryst Growth Des* 2018; 18(6): 3654–3660.

31 Müller M, Meier U, Wieckhusen D, Beck R, Pfeffer-Hennig S, Schneeberger R. Process development strategy to ascertain reproducible API polymorph manufacture. *Cryst Growth Des* 2006; 6(4): 946–954.

32 Cote A, Zhou G, Stanik M. A novel crystallization methodology to ensure isolation of the most stable crystal form. *Org Process Res Dev* 2009; 13(6): 1276–1283.

33 Lu J, Rohani S. Polymorphic crystallization and transformation of the anti-viral/HIV drug stavudine. *Org Process Res Dev* 2009; 13(6): 1262–1268.

34 Wang J, Loose C, Baxter J, Cai D, Wang Y, Tom J, Lepore J. Growth promotion by H_2O in organic solvent—selective isolation of a target polymorph. *J Cryst Growth* 2005; 283 (3–4): 469–478.

35 Trifkovic M, Rohani S, Mirmehrabi M. Polymorphic generation through solvent selection: ranitidine hydrochloride. *Org Process Res Dev* 2007; 11(1): 138–143.

36 Sun Z, Quon JL, Papageorgiou CD, Benyahia B, Rielly CD. Use of wet milling combined with temperature cycling to minimize crystal agglomeration in a sequential antisolvent–cooling crystallization. *Cryst Growth Des.* 2022; 22(8): 4730–4744.

37 Batra H, Penmasta R, Phares K, Staszewski J, Tuladhar SM, Walsh DA. Crystallization process development for a stable polymorph of treprostinil diethanolamine (UT-15C) by seeding. *Org Process Res Dev* 2009; 13(2): 242–249.

38 Abhishek MS, Hazra D, Steele G, Pal S. Crystallization process modifications to address polymorphic and particle size challenges in early stage development of an API salt. *Chem Eng Res Des.* 2020; 164: 400–411.

39 Wibowo C, Novoa R, Hasson M. Development of a process for crystallizing the desired crystal form of An API. In: *AIChE Annual Meeting* Philadelphia, PA 2008.

40 Desikan S, Parsons RL, Davis WP, Ward JE, Marshall WJ, Toma PH. Process development challenges to accommodate a late-appearing stable polymorph: a case study on the polymorphism and crystallization of a fast-track drug development compound. *Org Process Res Dev* 2005; 9(6): 933–942.

41 Maloney KM, Zhang SW, Mohan AE, Lee AY, Larpent P, Ren H, Humphrey GR, Desmond R, DiBenedetto M, Liu W, Lee IH. Development of a green and sustainable manufacturing process for gefapixant citrate (MK-7264). Part 6: Development of an improved commercial salt formation process. *Org Process Res Dev* 2020; 24(11): 2498–2504.

42 Wirth DD, Stephenson GA. Purification of dirithromycin. Impurity reduction and polymorph manipulation. *Org Process Res Dev* 1997; 1(1): 55–60.

43 Modi SR, Dantuluri AK, Puri V, Pawar YB, Nandekar P, Sangamwar AT, Perumalla SR, Sun CC, Bansal AK. Impact of crystal habit on biopharmaceutical performance of celecoxib. *Cryst Growth Des* 2013; 13(7): 2824–2832.

44 Wilson D, Bunker M, Milne D, Jawor-Baczynska A, Powell A, Blyth J, Streather D. Particle engineering of needle shaped crystals by wet milling and temperature cycling: Optimisation for roller compaction. *Powder Technol* 2018; 339: 641–650.

45 Eren A, Szilagyi B, Quon JL, Papageorgiou CD, Nagy ZK. Experimental investigation of an integrated crystallization and wet-milling system with temperature cycling to control the size and aspect ratio of needle-shaped pharmaceutical crystals. *Cryst Growth Des.* 2021; 21(7): 3981–3993.

46 Wu S, Shen H, Li K, Yu B, Xu S, Chen M, Gong J, Hou BH. Agglomeration mechanism of azithromycin dihydrate in acetone–water mixtures and optimization of the powder properties. *Ind Eng Chem Res* 2016; 55(17): 4905–4910.

47 Yu ZQ, Tan RB, Chow PS. Effects of operating conditions on agglomeration and habit of paracetamol crystals in anti-solvent crystallization. *J Cryst Growth* 2005; 279 (3–4): 477–488.

48 Belca LM, Ručigaj A, Teslič D, Krajnc M. The use of ultrasound in the crystallization process of an active pharmaceutical ingredient. *Ultrason Sonochem* 2019; 58: 104642.

49 Gielen B, Jordens J, Thomassen LC, Braeken L, Van Gerven T. Agglomeration control during ultrasonic crystallization of an active pharmaceutical ingredient. *Crystals.* 2017; 7(2): 40.

50 Su CS, Liao CY, Jheng WD. Particle size control and crystal habit modification of phenacetin using ultrasonic crystallization. *Chem Eng Technol* 2015; 38(1): 181–186.

51 Kaur Bhangu S, Ashokkumar M, Lee J. Ultrasound assisted crystallization of paracetamol: crystal size distribution and polymorph control. *Cryst Growth Des* 2016; 16(4): 1934–1941.

52 Kawashima Y, Okumura M, Takenaka H. Spherical crystallization: Direct spherical agglomeration of salicylic acid crystals during crystallization. *Science* 1982; 216(4550): 1127–1128.

53 Hansen J, Kleinebudde P. Improving flowability and reducing storage agglomeration of metformin hydrochloride through QESD crystallization. *Eur J Pharm Biopharm* 2021; 159: 170–176.

54 Sun M, Shichao D, Tang W, Jia L, Gong J. Design of spherical crystallization for drugs based on thermal-induced liquid–liquid phase separation: case studies of water-insoluble drugs. *Ind Eng Chem Res* 2019; 58(44): 20401–20411.

55 Birch M, Marziano I. Understanding and avoidance of agglomeration during drying processes: a case study. *Org Process Res Dev* 2013; 17(10): 1359–1366.

56 Afrassiabian Z, Saleh K. Caking of anhydrous lactose powder owing to phase transition and solid-state hydration under humid conditions: From microscopic to bulk behavior. *Powder Technol* 2020; 363: 488–499.

57 Lim RT, Hoong AY, Ng WK, Tan RB. Anomalous size evolution of partially amorphized pharmaceutical particles during post-milling storage. *Powder Technol* 2015; 286: 1–8.

58 Fukunaka T, Sawaguchi K, Golman B, Shinohara K. Effect of particle shape of active pharmaceutical ingredients prepared by fluidized-bed jet-milling on cohesiveness. *J Pharm Sci* 2005; 94(5): 1004–1012.

59 Liu Y, Gabriele B, Davey RJ, Cruz-Cabeza AJ. Concerning elusive crystal forms: The case of paracetamol. *J Am Chem Soc* 2020; 142(14): 6682–6689.

60 Tanaka K, Takiyama H. Impurities assisted with control of polymorphs during seeded crystallization. *J Chem Eng Jpn.* 2020; 53(4): 160–166.

61 Bauer J, Spanton S, Henry R, Quick J, Dziki W, Porter W, Morris J. Ritonavir: An extraordinary example of conformational polymorphism. *Pharm Res* 2001; 18: 859–866.

62 Saleemi A, Onyemelukwe II, Nagy Z. Effects of a structurally related substance on the crystallization of paracetamol. *Front Chem Sci Eng* 2013; 7: 79–87.

63 Lechuga-Ballesteros D, Rodriguez-Hornedo N. Growth and morphology of L-alanine crystals: influence of additive adsorption. *Pharm Res* 1993; 10: 1008–1014.

24

Solid-State Mixture Analysis

24.1 Introduction

Solid-state material characterization techniques play a significant role in the analysis of pharmaceutical mixtures, especially with regard to determining the composition and phase purity of various ingredients. From a quality point of view, it is ideal to have a physical form with high phase purity, so all efforts must be made to account for any impurities and eliminate them as much as possible. However, mixtures of physical forms for a compound can be encountered as the neat drug substance or in a drug product, where the mixture can be a combination of polymorphs, hydrate/anhydrate, crystalline/amorphous, or salt form/free form. The mixture can be the result of uncontrolled processing or physical transformations that occur during storage and downstream manufacturing. For example, wet granulation, using water, potentially triggers the formation of a hydrated form from an anhydrous form, whereas drying can lead to the dehydration of a hydrate, resulting in a loss of crystallinity. The interaction of a salt form with excipients in a solid dosage form sometimes triggers the disproportionation of the salt. The presence of a very small amount of a second physical form in a drug substance and a drug product does not affect the chemical purity and content, but it may lead to changes in dissolution, bioavailability, processability, and chemical stability.

Such analyses on physical forms cannot be accomplished through wet chemistry, for example, with liquid chromatography (LC) but rather require tools including X-ray powder diffraction (XRPD), Fourier-transformed infrared (FT-IR) spectroscopy, near-infrared (NIR) spectroscopy, Raman spectroscopy, and solid-state nuclear magnetic resonance (NMR) spectroscopy to analyze the composition of mixtures in a quantitative manner.

This chapter focuses on the methodologies being used for the quantitative analysis of mixtures, including sample preparation, data collection, and the establishment of calibration standards. Major attention will be directed to the determination of amorphous content, crystallinity, polymorph content, and the ratio of salt to free form, as well as the detection and identification of particulate contaminants in drug products. The last section of the chapter will be devoted to the development of process analytical technology (PAT) tools for online analysis and control during processing, which is key to the concept of "quality by design" (QbD) in pharmaceutical manufacturing (Chapter 26).

24.2 Limitations of Wet Chemistry

Wet chemistry, which includes LC, gas chromatography, UV–Vis spectrophotometry, mass spectrometry, solution NMR, and elemental analysis, is an essential part of pharmaceutical analysis in solution.

Specific applications where wet chemistry is being used include

- preformulation studies and physical/chemical profiling,
- excipient compatibility and prototype stability,
- formulation and process development,
- active pharmaceutical development (API) process development, including crystallization,
- specification development for drug substances and drug products,
- batch analysis and release testing of drug substance and drug product,
- raw material analysis and release,

Solid-State Materials in Pharmaceutical Chemistry: Properties, Characterization, and Applications, Second Edition. Stephen R. Byrn, George Zografi and Xiaoming (Sean) Chen.

- stability testing of drug substance and drug product, and
- analysis of impurities and other contaminants.

The primary purpose of such analysis is to establish the identity, purity, and potency of a drug substance or a drug product. Impurities of interest can include synthetic impurities in the drug substance after isolation, degradation products, residual solvents, water, and inorganic impurities such as heavy metals, all of which must be assessed for compliance with current good manufacturing practices (cGMP) regulatory requirements.

The major limitation of wet chemistry for use with solid drugs and products is that a sample must be dissolved before analysis, where the extraction and dissolution process often eliminates important physical information associated with the solid material. Wet chemistry, for example, is not applicable for the analysis of the following:

- Amorphous content and crystallinity
- Mixtures of polymorphs
- Mixtures of anhydrous forms and hydrates
- Mixtures of free forms and salt forms
- Crystallization from the amorphous form
- Polymorph conversion
- Hydrate formation
- Dehydration
- Acid/base reactions
- PAT

24.3 Pharmaceutical Analysis in the Solid State

The requirements of a quantitative method for solid-state analysis align with methods using wet chemistry, in that the validation of a method involves the evaluation of specificity, accuracy, precision, robustness, range, and the detection limit, which has been widely covered by various analytical textbooks and regulatory guidelines [1–3]. There are additional challenges for solid-state analysis since a material in the solid state is a heterogeneous system made up of many particles. For a multiple-component system, therefore, particle size, morphology, and bulk density can affect mixture uniformity. Moreover, the sample collected for analysis may not truly represent the entire sample, thus causing additional errors in sample preparation and analysis.

24.3.1 Sample Preparation

Samples used for the development of calibration standards should mimic "real" samples as closely as possible. The sample size, often dependent on the amount of material availability, must be adequate to meet the needs of the equipment being used for analysis. The range of sample size is typically from 100 mg to several grams, with the number of samples required being dependent on the analytical technique. For example, a calibration set for NIR analysis usually requires hundreds of samples to establish a reliable statistical model. Since solid-state materials are made of particles and mixing uniformity is essential for the accuracy of measurement, such uniformity can be enhanced in several ways. The common practice is to gently grind the materials together with a mortar and pestle and mix thoroughly. Using geometric dilution to enhance the uniformity of mixing a minor component with another major component involves multiple steps. First, a roughly equal amount of the major component is mixed with the minor component. After proper mixing, the additional major component is added to the mixed system in roughly equal amounts each time until the major component is fully utilized. Another way to ensure proper mixing of particles is to use a slurry. Bugay et al. have presented an approach, which involves slurrying the mixture in a volatile nonsolvent and collecting the mixture for analysis after filtration and drying [4]. However, it must be recognized that slurrying may trigger solution-mediated transformations, especially for high-energy forms such as amorphous solids and, therefore, should be used with caution.

24.3.2 Data Collection

The method used for data collection is specific for each instrument. Multiple measurements on any one sample help to mitigate the issue of sample uniformity, whereas spinning or rotating the sample during analysis has a similar effect.

24.3.3 Data Transformation

The purpose of data transformation is to remove or reduce extraneous variations such as equipment fluctuations in equipment performance, environmental variability, for example, temperature and relative humidity, and sample variability with regard to surface changes, density, and particle size. Those variations are unrelated to sample composition and, therefore, would lead to errors if not removed. Some techniques require minimal data transformation, including gravimetric vapor sorption and thermal analysis techniques. For spectroscopy and XRPD, however, data transformation is required in many circumstances to increase data quality for quantification, especially in situations where data analysis will be carried out through a sophisticated statistical analysis.

Common transformation approaches use baseline correction, smoothing, computing derivatives, standard normal variate (SNV), and multiplicative signal correction (MSC).

Baseline correction and smoothing are widely applied to quantitative analysis using spectroscopy and XRPD. The treatment reduces noise and eases data collection for peak height and peak area for the establishment of a calibration model.

Computing derivatives is mainly used in NIR spectroscopy. For example, it has been reported that using a first-order differential with a Savitzky-Golay filter can lead to the reduction of noise in NIR spectra and to enhanced peak resolution [5]. It is well known that the second derivative can mitigate the effects of particle size and reduce variable scattering of data, since it gives rise to a zero baseline and sharpens spectral peaks, which help to separate overlapping peaks. The use of derivative spectra is more desirable for pattern recognition methods; however, noise can be magnified by such an operation, so spectral smoothing is necessary before further data processing.

SNV transformation can mitigate the effects of particle size and density by correcting for the variance in light path for NIR and Raman spectroscopy [6].

SNV transformation is described by the following equation:

$$x_{ik,\text{SNV}} = \frac{(x_{ik} - \bar{x}_i)}{\sqrt{\frac{\sum (x_{ik} - \bar{x}_i)^2}{(k-1)}}} \tag{24.1}$$

where $k = (1, \ldots, K)$ is the wavelength, $\bar{x}_i$ is the average spectrum, x_{ik} is the spectrum at wavelength k, and $x_{ik,SNV}$ is the spectrum after SNV treatment.

MSC plays a similar role to SNV but involves a different mathematical treatment. MSC separates physical light-scattering effects from light absorbance in NIR and Raman spectra. Variations due to physical light scattering are caused by sample variations such as particle size, shape, packing, and surface characteristics. Such scattering leads to signal noise and other artifacts that complicate the accurate analysis of physical and chemical information. MSC, therefore, acts as a mathematical tool which reduces or eliminates such noise and artifacts by fitting each spectrum to the average spectrum, or "ideal" spectrum, as closely as possible [7].

MSC requires the availability of all spectra, whereas SNV treats each spectrum individually. Consequently, SNV is more properly used for process analytical related mixture analysis.

The suitability of these different methods of data transformation has to be assessed in real situations. For example, Blanco and Villar discovered that the first derivative mode, over the wavelength range of 1100–1800 nm, yielded the best results for determining the degree of crystallinity of miokamycin using NIR [8], whereas Gombás found that second-derivative spectra were more suitable for quantification of lactose crystallinity, by reducing the effects of particle size and variable scattering [9].

24.3.4 Calibration Model Development and Validation

The manner of accurate quantitative analysis in the solid state is consistent with traditional analytical methodology, which employs known samples (calibration standards) to quantify an unknown sample. The known samples should reflect the limits or ranges that will be encountered in real samples or during any process. The known samples are usually divided into two sets. One set is used to establish a calibration model, called the calibration set, whereas the other set acts as a validation set used to confirm the effectiveness of the calibration model. After the model is validated, it will be applied to the analysis of real samples or linked to process analysis.

Approaches for the establishment of suitable calibration models include univariate linear regression, multiple variable linear regression, and chemometric analysis.

Univariate linear regression relies on the application of linear regression to identify the relation between a constituent of interest, such as the concentration of a specific component, and a variable, such as absorbance at a specific wavenumber, in a simple algebraic form,

$$Y = ax + b \tag{24.2}$$

where Y is the constituent of interest, x is the response detected, a is the calibration factor or slope, and b is the intercept.

Multiple linear regression can be considered as an extension of univariate regression, necessary for the evaluation of more complicated systems. It takes into consideration the correlation of multiple variables with the constituent of interest, based on the following equation:

$$Y = a_0 + a_1x_1 + a_2x_2 + \ldots + a_nx_n \tag{24.3}$$

where x_n is the response, a_n is the regression coefficient, and a_0 is the intercept.

This type of regression generates an equation that provides the least squares of the error between the regression curve and the actual data. In general, the number of variables selected is obtained by trial and error to obtain the optimal correlation, while avoiding overfitting the data. It is applicable to cases where the components have well-resolved and characteristic peaks, for example, in the use of XRPD peaks for the quantification of polymorph composition or degree of crystallinity, where the univariate method selects a single peak while multiple linear regression considers several peaks altogether.

The disadvantages of the univariate method and multiple linear regression are the requirement that there be a linear relation between the dependent variables and the predictor variable.

More sophisticated statistical analyses are generally employed in establishing calibration models for spectroscopies such as IR, Raman, and NIR, where data are often found to have numerous, noisy, and strongly collinear spectroscopic signals. For example, chemometric analysis is useful in this regard because it takes into consideration multiple wavelengths or spectral points to build models, instead of focusing on a single characteristic peak. Chemometric data analysis, however, requires a sufficiently large data set to build models using the entire spectrum, and, therefore, the required computational work is more substantial than for linear regression. Chemometric tools available include principal component analysis (PCA), partial least squares (PLS), linear discrimination analysis, quadratic discriminant analysis, regularized discriminant analysis, artificial neural networks, conformity index, and Mahalanobis distance. PCA and PLS are the two most widely used tools in pharmaceutical quantitative analysis, whereas the conformity index and Mahalanobis distance are most applicable in process analysis using PAT tools.

PCA takes a statistical approach to identify critical variables, where the goal is to reduce the number of spectral data considered as variables. PCA identifies a small set of variables by using a statistical projection from a large quantity of spectral data. For example, Mah et al. carried out PCA of the Raman data of milled glibenclamide samples, and revealed that the peaks at 783, 800, 1090, and 1274 cm^{-1} were characteristic of amorphous glibenclamide and the peaks at 812, 824, 842, 1020, 1059, 1097, 1202, and 1249 cm^{-1} were characteristic of crystalline glibenclamide [10]. The identified principal component 1 (PC1) was shown to be positively correlated with the amorphous phase and negatively correlated with the crystalline phase. This change of PC1 values, therefore, reflected the increase of amorphous content during the process of milling (Figure 24.1). The PC1 value plateaued after 60 minutes of milling, indicating the full amorphization of glibenclamide.

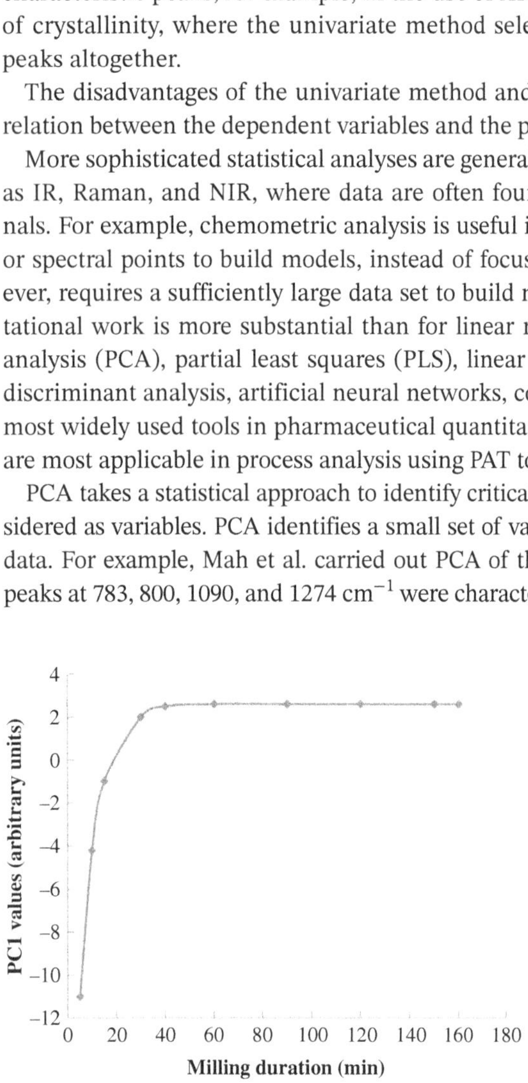

Figure 24.1 Principal component 1 (PC1) values of milled glibenclamide samples as a function of milling time. *Source*: [10]/with permission of American Chemical Society.

PLS is a widely used statistical tool to extract spectral information for the purpose of quantification, in that PLS establishes the relationship between a set of dependent variables, X, and a set of predictor variables, Y. An example could be the relationship of amorphous content of a material (Y) and some specific features of NIR data (X). PLS produces a continuum regression by incorporating regression and dimensional reduction to model the relationships between dependent variables and predictor variables. The predictor variables are also called PLS factors. A calibration data set is created from a set of samples covering the entire range of the interested dependent values for PLS analysis.

It is generally recognized that the spectral range and the number of PLS factors are critical for the establishment of an appropriate model, since the selection of a spectral range requires consideration of the location and quality of spectral information. The number of PLS factors should be assessed to avoid overfitting. In a case study of developing a NIR method for the quantitation of amorphous lactose, for example, the exclusion of the 1900–2000 nm region mitigated the effect of moisture on the PLS model since absorbed moisture is detected in that region and therefore led to a better model than that of using the whole spectra from 1100 to 2050 nm [11]. In the same case study, the PLS model based on Raman spectroscopy was adjusted when analyzing lactose samples prepared by ball milling. Significant intensity differences were observed between processed samples and calibration sets in a certain region of the spectra due to material differences in density and packing. An acceptable PLS model was obtained after narrowing down the region to 1000–1500 cm^{-1}, where processed and calibration samples give less intensity differences.

In modeling by using PLS, it is a general practice to divide the data into two sets, one to create the model and the other to test the model.

The calibration model is usually assessed through a validation data set. The predictive ability of the model is reflected by estimating the residual sum of squares. The sum is calculated from the actual response and the predicted response. Additional benchmarks include standard errors, t-distribution, and confidence intervals.

The PLS factors are evaluated through cross-validation, where the model giving a minimum standard error of cross-validation (SECV) is considered to have the optimum number of factors. SECV is generally decreased as the number of PLS factors is increased, and a certain factor SECV starts to increase. The model with the minimum SECV is considered optimal.

Just like other quantitation methods, a model built from PLS generally only fits for a certain range, so it is usually difficult to rely on one model to cover a very broad range. Savolainen et al., for example, used three models for the quantitation of amorphous lactose using NIR [11]: one for low amorphous contents (0–10%), one for 10–90% amorphous content, and one for high amorphous contents (90–100%).

Validation is carried out to confirm or test the accuracy of such an analysis, using test set validation or cross-validation approaches. For test set validation, a test set is established at the same time as the calibration set and should have a similar distribution to the calibration set. In cross-validation, one or more samples are removed from the calibration set at a time. A number of statistical error indicators are utilized to benchmark the accuracy of the analysis, including:

- correlation coefficient, R^2,
- root mean square error in calibration,
- root mean square error in prediction (RMSEP),
- standard error in calibration, and
- standard error in prediction.

RMSEC and RMSEP are utilized to quantify the error statistics.

24.4 Measurement of Amorphous Content

The measurement of amorphous content and crystallinity in a processed sample, to obtain the relative amount of either component, is an important aspect of mixture analysis. The measurement of amorphous content is generally best carried out by techniques that directly measure the amorphous portion, whereas techniques that directly measure crystallinity should be used to measure crystalline content. However, the two values are complementary to each other; therefore, the determination of either one will result in a valid prediction of the other. For example, a material having 15% amorphous content is projected to have 85% crystallinity. For crystallinity from 10% to 90%, direct measurement of either the amorphous or crystalline content is acceptable. When measuring amounts less than 10%, the relative sensitivity of each technique becomes quite different, and, therefore, measurement of amorphous content is more suitable to detect small amounts of amorphous in crystalline solids and vice versa.

Small amounts of amorphous content may not have profound effects on the therapeutic product performance but could significantly affect processability or stability. For example, since the amorphous form tends to crystallize, the amorphous regions upon crystallization may form bridges and fuse to other particles, which in turn could cause the powder in a mixing drum, for example, to cake or agglomerate. In addition, the amorphous regions can absorb significant amounts of water, further triggering chemical/physical instability.

There are several key considerations in developing a method for analyzing the amorphous content of a crystalline sample:

- Availability of amorphous and crystalline materials as calibration standards
- Assuring mixture uniformity in the calibration set
- Physical stability of the amorphous phase during sample preparation and analysis
- Selection of proper analytical techniques

The accurate measurement of amorphous content relies on the availability of amorphous and crystalline materials as standards. Ideally, those materials should have 100% purity, which is somewhat difficult to accomplish and verify in reality. There are numerous ways to make amorphous materials, including spray drying, freeze drying, and quench cooling. The prepared material should be fully characterized to ensure that it has undetectable crystalline material using XRPD and polarized microscopy. In this regard, we can generally assume that the error caused by the use of 98–99% amorphous material for calibration should be relatively small for quantifying residual amorphous content. In contrast, the phase purity of the crystalline phase has a more profound effect, so that if the goal is to measure several percentages of amorphous content, the assumption of a material with 2% amorphous content to be 100% crystalline standard can cause significant error. Reasonable effort, therefore, should be directed to ensure the crystalline standard is as fully crystalline as possible. One way to do this is to use a material prepared directly from crystallization without further processing, since any mechanical processing such as milling or micronization tends to produce amorphous content. Another way to eliminate small amounts of amorphous content in the calibration standard is to cure or anneal the sample to trigger the crystallization of amorphous material by exposing the sample to a high humidity or to organic solvent vapor. Young et al., for example, preconditioned crystalline salbutamol sulfate and lactose in containers with 84% RH at 25 °C for a period of 7 days to ensure full crystallinity before using the two materials as calibration standards [12].

Another challenging issue is the fact that amorphous materials prepared by spray drying and melt quenching are not necessarily representative of a sample that had undergone process-induced disorder. For example, the amorphous fraction or disorder generated by many processes, such as milling, tends to be concentrated most often on particle surfaces. Such differences in sample characteristics can lead to difficulty in identifying a reliable method, as demonstrated in the case of salbutamol sulfate, where Grisedale and colleagues studied the use of a spray-dried amorphous form as a calibration standard to quantify the amorphous content of milled salbutamol sulfate and found the disparity of different methods [13], as shown in Table 24.1.

Attenuated total reflectance (ATR)-FT-IR spectroscopy, XRPD, and modulated temperature differential scanning calorimetry (MTDSC) give broadly similar results but not with the type of accuracy acceptable for quality control (Table 24.1). The use of water vapor sorption method based on the spray-dried salbutamol sulfate where the sorption takes place primarily in the amorphous regions of the sample generally results in much lower figures, indicating that the sorption behavior and recrystallization behavior from a bulk amorphous material prepared by spray drying do not align with amorphous material present at the surface caused by milling. In addition, MTDSC C_p (heat capacity) and H_c (crystallization enthalpy) gave higher results, probably due to a difference in recrystallization kinetics for bulk amorphous and surface amorphous material [13].

Table 24.1 Comparison of amorphous content of salbutamol sulfate samples determined by ATR-FT-IR, XRPD, gravimetric vapor sorption (GVS), and MTDSC.

Samples	ATR-FTIR (%)			XRPD (%)	GVS (%)	MTDSC ΔC_p method (%)	MTDSC $T_{g\,mix}$ method (%)	MTDSC ΔH_c method (%)	MTDSC ΔH_c method (uncorrected) (%)
	(A)	(B)	(C)						
Crystalline	0	0	0	0	0	0	0	0	0
15 minutes milled	11	12	15	15	9	14	11	23	12
30 minutes milled	19	24	24	34	14	39	28	49	29
60 minutes milled	59	73	70	65	23	89	63	89	65
Spray-dried	100	100	100	100	100	100	100	100	100

Source: [13]/with permission of Elsevier.

As with all other types of mixture analyses, mixing uniformity is critical for the accuracy of a calibration model, especially for analysis with subsampling. Subsampling occurs when a small portion of the sample is used for analysis, such as with thermal analysis, or when the signal is collected from only a portion of the sample. For example, in Raman analysis, the laser beam is targeted at a small area of sample, whereas in NIR analysis the immersion probe only has an effective sampling area of 1–2 mm [11]. In XRPD, the X-ray beam, likewise, only covers a portion of a sample in the sample holder.

Ways to mitigate the effect of such nonuniform sampling include replicating the analysis for the same sample, spinning the sample, or rotating the sample during data collection to increase the sampling average. Savolainen et al., for example, reported that the use of a larger sample area by such means led to a decrease in the root mean square error (RMSE) of prediction from 4.3% to 2.7% when determining the amorphous content of lactose using NIR spectroscopy [11].

Content uniformity is generally of less concern for bulk techniques such as isothermal calorimetry, gravimetric gas sorption, and solid-state NMR, since these techniques detect material properties over the entire sample.

Since amorphous forms tend to transform into crystalline forms, especially, when in intimate contact with crystalline seeds, proper measures should be taken to ensure that the physical stability of the sample as a function of temperature and humidity is under control.

We can conclude that if the mixture contains more than 10% amorphous content any of the techniques discussed earlier are applicable. For convenience in terms of time and effort, it would appear that XRPD and spectroscopic techniques are most advantageous. If the level of detection required is below 10%, then detection sensitivity must be the primary focus, and in this regard isothermal calorimetry and gravimetric water vapor sorption appear to have significant advantages. It has also been reported that whole pattern fitting of powder diffraction has similar capability [14]. A few examples for the use of various techniques to detect less than 10% amorphous content are given in the following text.

XRPD is not sensitive enough to detect amorphous material since amorphous forms give a diffusive halo in the diffraction pattern. In a sample with a significant amount of crystalline phase, the presence of amorphous regions often leads to a change in baseline, which can be used to establish a calibration model. Here, a region in the diffraction pattern that lacks crystalline peaks is chosen, and the diffraction intensity is correlated to the amount of amorphous material present using linear regression. For example, in quantifying the amorphous content of the β-lactam antibiotic imipenem, Croker and McCauley made use of the intensity in the XRPD pattern at the diffraction angle of 28° 2θ, where no other crystalline peaks were present [15]. By considering the diffraction information in the amorphous halo, Chen et al. developed a whole pattern fitting approach to provide accurate analysis of relatively small amounts of amorphous content in pharmaceuticals compared to typical XRPD analysis [14]. The whole pattern fitting method allowed the use of a statistical modeling of mixture samples using the XRPD pattern of 100% amorphous and 100% crystalline materials.

Spectroscopic tools such as Raman, FT-IR, and NIR have the excellent sensitivity required to detect small amounts of amorphous content and are, therefore, superior to XRPD because the amorphous phase gives specific signals in Raman, FT-IR, and NIR spectra. The utility of these techniques is dependent on the spectroscopic difference that occurs between amorphous components and crystalline counterparts, although the signals may overlap with crystalline peaks or be broadened in some cases.

Simple linear regression or multiple linear regressions can lead to an accurate calibration model for cases where amorphous forms give distinct spectroscopic peaks from crystalline counterparts. PLS is more applicable for situations where specific peaks, required for quantitation, are difficult to detect. In such cases, PLS uses the entire spectrum or a selected region of the spectrum to develop calibration models.

Bertacche and coworkers reported that FT-IR spectra of amorphous and crystalline cyclosporine displayed very similar spectra in the range of 1125 to 1515 cm^{-1} and 3200 to 4000 cm^{-1} but with a significant difference being present from 2800 to 3000 cm^{-1} [16]. PLS modeling using this selected range was found to be more effective than the use of the whole range of spectra.

Savolainen et al. used NIR and Raman spectroscopic techniques together with PLS to determine the amorphous content of lactose and trehalose induced by milling [11]. The PLS models developed using the two techniques were found to be superior to XRPD, which has an accuracy of 10% compared to 2% for NIR and Raman spectroscopies.

Techniques such as XRPD and spectroscopies directly probe the amorphous component, whereas other techniques involve the detection of physical events that are characteristic of the amorphous form. These include heat of recrystallization, glass transition temperature, and hygroscopicity, since the magnitudes of those events correlate with the quantity of the amorphous form in a mixture.

Isothermal calorimetry detects the heat of crystallization of an amorphous component. Thermal analysis can be used to determine the glass transition temperature. The water vapor sorption technique takes advantage of the much greater amount of water absorbed by amorphous materials relative to that of the corresponding crystal (see Chapter 15).

Isothermal microcalorimetry has been proven to be one of the most sensitive techniques in detecting amorphous form with a detection limit that is potentially less than 1% [17].

The basis for the use of isothermal microcalorimetry is its ability to detect the heat of recrystallization of the amorphous component after subjecting the sample to a certain level of relative humidity under isothermal conditions, with the amount of heat released being proportional to the amount of amorphous material in the mixture. A powder sample is sealed in a glass vial with a glass tube containing a saturated salt solution to control the relative humidity, where the humidity is sufficient to trigger the recrystallization of amorphous components. The area under the power versus time curve is used to calculate the amount of amorphous content after comparing these data to those of amorphous standards. For hydrophobic compounds, moisture may not be able to trigger recrystallization, and other organic vapors may be more useful. Ahmed and colleagues, for example, reported the use of nonaqueous vapors with isothermal calorimetry to cause the crystallization of a powder having poor aqueous solubility [18]. Since isothermal calorimetry relies on the crystallization of the amorphous material for the establishment of a calibration model, obviously it is not applicable to materials that are difficult to crystallize. In such cases, simply measuring vapor sorption gravimetrically can provide a very good direct measure of amorphous content.

That such an approach is useful for determining amorphous content can be seen from the fact that, as described in Chapter 15, amorphous material absorbs water throughout the sample, whereas water is adsorbed only to the surface of crystals in relatively minute quantities. The test can be carried out in a dynamic vapor sorption system, where the weight change due to absorption is automatically recorded as the sample is exposed to a certain temperature and a certain relative humidity. A series of samples made of completely amorphous and crystalline materials are used to construct a calibration curve to cover the quantitative range of interest.

Buckton and Darcy have evaluated the use of such gravimetric studies to assess the degree of crystallinity of mixtures of amorphous lactose and crystalline lactose monohydrate, where various water sorption isotherms correlated very well with amorphous content [19]. Mixtures with 0.05%, 0.125%, 0.25%, and 0.5% amorphous lactose were subjected to multiple cycles of sorption and desorption, where it was shown that first cycle producing a sorption isotherm distinct from the remaining cycles due to weight gain of amorphous component and subsequent recrystallization that cause weight loss after reaching above 60% RH. Even the presence of 0.125% amorphous content could be differentiated from the crystalline material [19].

As mentioned earlier, organic vapors can replace water vapor in such studies when water vapor sorption is insignificant. Young et al., for example, chose *n*-octane vapor to measure the amorphous content of salbutamol sulfate and lactose [12]. A linear relationship was observed between the weight gained at a relative pressure of 0.90 and amorphous content, with detection limits being as low as 1% amorphous content [12].

Table 24.2 lists the various techniques available for the measurement of amorphous content, showing the advantages and disadvantages.

24.5 Detection of Crystallinity

Whereas in the previous section, assessing the properties of the amorphous component was used to quantitate a mixture of amorphous and crystalline components, it is also possible to directly quantitate the degree of crystallinity by measuring crystalline properties. The approach used is still based on standards of 100% amorphous and 100% crystalline material.

Sensitive techniques available to detect crystallinity include XRPD, Raman spectroscopy, and nonlinear optical imaging.

Usually, in XRPD studies, one or two diffraction peaks giving the best sensitivity, possibly the peaks with the highest intensity in the diffractogram, are chosen for the establishment of a calibration curve. Either peak height or peak area can be used for quantification.

Crowley and Zografi, for example, evaluated the diffraction peaks of γ indomethacin and found that a suitable diffraction peak at 11.8° 2θ could be detected in mixtures containing as low as 2% crystalline material [20]. A linear calibration curve was established using the peak height of this peak for quantitative analysis of amorphous content in the α and γ polymorphs of indomethacin. The peak height value was corrected for any small diffraction from the respective amorphous component.

Table 24.2 Comparison of different techniques used for amorphous content measurement.

Method	Measurement	Pro	Con	Calibration model
X-ray powder diffraction	Diffusive diffraction by amorphous	Nondestructive, suitable for routine analysis	Not sensitive and specific to amorphous, high detection limit	Simple linear regression, multiple linear regression (whole pattern fitting)
Dynamic vapor sorption	Vapor uptake by amorphous material due to the formation of a solid solution	Sensitive to amorphous content, low detection limit, small sample size	Not specific to the compound, long run time, destructive	Simple linear regression
Differential scanning calorimetry	Heat capacity changes related to amorphous form as part of the glass transition of amorphous form	Small sample size, short run time, suitable for routine analysis	Destructive, possibility of thermal changes in the sample during the scan, subsampling	Simple linear regression
Isothermal solution calorimetry	Heat of crystallization from amorphous form	Sensitive, low detection limit	Not applicable to compounds difficult to trigger crystallization, long run time, destructive	Simple linear regression
Spectroscopic: FT-IR, Raman, NIR	Spectroscopic features of amorphous form	Fast, compound specific, detection of the crystallinity, at the same time, sensitive, noninvasive, nondestructive, suitable for process analysis	Needs remarkable difference between amorphous and crystalline spectra, subsampling	Simple linear regression, multiple linear regression, partial least square
Solid-state NMR	Spectroscopic features of the amorphous form	Bulk technique, nondestructive	Long run time, not suitable for routine analysis	Simple linear regression

It has also been found that XRPD is very effective in detecting crystallinity in the presence of significant amounts of amorphous material as demonstrated by the case of sucrose, where Surana and Suryanarayanan used crystalline sucrose diffraction peaks at 12.7° and 13.1° 2θ to construct a calibration curve for crystallinity from 1% to 5% [21]. The integrated intensities of the two peaks displayed a linear relationship with the percentage of crystallinity, which gave a limit of detection of 0.9%.

Otsuka et al. developed a linear calibration curve for measuring the crystallinity of indomethacin using XRPD, based on the total intensity of three specific diffraction peaks [22]. The use of multiple peaks increased the sensitivity and reduced the impact of preferred orientation.

Liu et al. utilized seven diffraction peaks of a fluorine-containing pharmaceutical candidate to quantify compaction-induced loss of crystallinity [23]. The combined intensity of the seven peaks gave a linear relationship with the degree of crystallinity.

It has been reported that the sensitivity required to detect small percentages of crystallinity can be significantly improved by using synchrotron X-rays with a 2-D area detector, with the limit of detection being as low as 0.2% [24]. In addition, the synchrotron technique offers rapid data acquisition.

If the amorphous and crystalline phases of a material give significant spectral differences in the Raman, NIR, or IR regions, simple regression analysis or PLS can be used to quantify crystallinity. The use of the peak ratio of the absorbance spectra of respective characteristic peaks has the advantage of eliminating sampling and equipment variations. A calibration model that relates peak ratio to the component ratio, therefore, would serve as an excellent quantitative tool for crystallinity. For example, a quantitative model for measuring the crystallinity of an etravirine sample was developed by using peaks in the region of 900–850 cm^{-1}, which are markedly different for amorphous and crystalline forms [25]. The amorphous component showed a single peak at 882 cm^{-1}, whereas the peak shifted to 884 cm^{-1} in crystal Form II but remained at a similar intensity. Form II had another characteristic peak at 897 cm^{-1}, so the ratio of 897 to 884 cm^{-1} in Form II was regarded as representative of 100% crystallinity. The degree of crystallinity was then calculated by normalizing the ratio of 897 to 884 cm^{-1} in an unknown sample to the corresponding ratio in Form II. The approach eliminated the need to use standard mixtures to establish a calibration curve.

Univariate linear regression works well when crystalline and amorphous components give distinct spectral features in derivative NIR spectra, as demonstrated in the case of sucrose [26]. Figure 24.2 shows a direct relationship between the second-derivative response of the peak at 1436 nm and the degree of crystallinity of sucrose, where the response ratio of

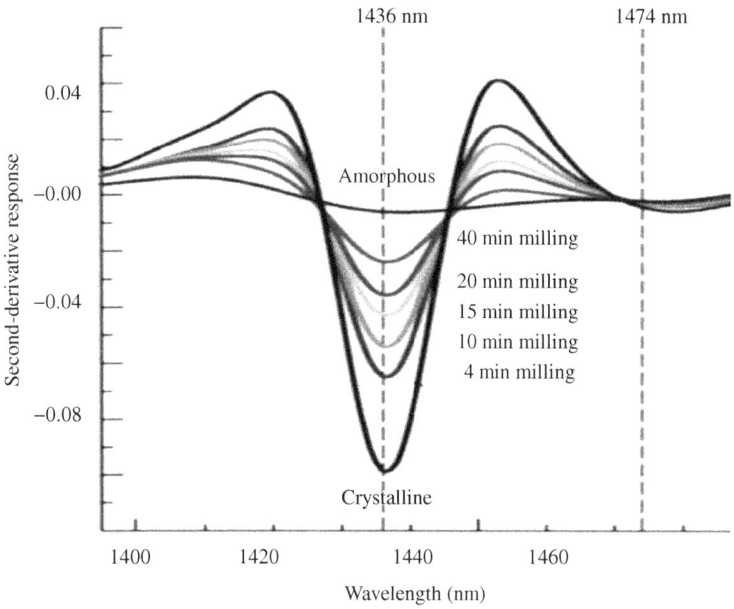

Figure 24.2 NIR spectroscopy second-derivative spectra of amorphous and crystalline sucrose, in addition to sucrose samples milled at different times. *Source*: [26]/with permission of Elsevier.

1436/1474 nm was used to construct a univariate calibration model for sucrose crystallinity. It was also shown that an increase in milling time led to decreased sucrose crystallinity.

Statistical modeling, such as PLS, has been found to be valuable in developing quantitative models for detecting crystallinity when the crystalline and amorphous forms do not have well-resolved peaks in any spectroscopic measurement.

Sinclair et al., for example, developed a calibration model for the detection of the crystalline form of ibipinabant in an amorphous drug product using Raman spectra [27]. The model, based on PLS, was shown to be capable of detecting less than 5% of a crystalline form in a tablet formulation with 1% drug load, which is equivalent to less than 0.05% of the total mass of the tablet. The model also provided an accurate assessment of the physical stability and recrystallization kinetics of the amorphous ibipinabant drug product.

Okumura and Matsuoka used PLS to construct a model for the measurement of the degree of crystallinity of indomethacin in tablets containing mannitol, where the second derivatives and MSC were used to pretreat the Raman spectra [28]. The developed quantitative model was shown to be capable of detecting the pressure-induced loss of crystallinity after tablet compression.

A new technique has been recently developed to measure the degree of crystallinity based on second harmonic generation (SHG). Kestur et al. utilized a custom-built second-order nonlinear imaging of chiral crystals (SONICC) instrument to study mixtures of crystalline naproxen and amorphous polymer, where the SHG intensity gave a dynamic linear response from 0.1% crystallinity down to 100% [29]. SONICC was demonstrated to be a sensitive technique for measuring low levels of crystallinity, with a capability similar to that of Raman spectroscopy and XRPD.

Table 24.3 compares the common techniques used for the measurement of the degree of crystallinity.

24.6 Quantification of Mixtures of Polymorphs

For a compound with multiple polymorphs, hydrates, or solvates, control of polymorph content is essential for optimizing the quality, processability, performance, and stability of drug substance and drug product. The presence of a minor component of any polymorph can be induced in the crystallization process or with further downstream processing, such as micronization, milling, and compression. It has been reported that concomitant polymorphs can be generated during crystallization, even in the production of stable polymorphs, while milling, spray drying, and freeze drying may lead to a crystal form transition to generate a minor undesired polymorph [30]. Also, the presence of water vapor or processing with water could trigger the formation of a hydrate from an anhydrate.

Table 24.3 Comparison of techniques for crystallinity measurement.

Method	Measurement	Pro	Con	Calibration model
X-ray powder diffraction	Crystalline diffraction peaks	Sensitive, widely used, quick analysis, nondestructive	Preferred orientation	Simple linear regression
Differential scanning calorimetry	Melting point, heat of fusion	Quick analysis	Potential of recrystallization of the amorphous phase, destructive	Simple linear regression
Spectroscopic: FT-IR, Raman, NIR	Spectroscopic features of the crystalline form	Fast, compound specific, process analytical capability	Time to model development, needs sophisticated statistical tools	Simple linear regression, multiple linear regression, and partial least square
Nonlinear optical imaging	Second harmonic generation signals	Fast, sensitive	Suitable for chiral crystal, complicated instrumentation setting, novel technique under development	Simple linear regression

Various solid-state characterization techniques are capable of quantifying polymorph content in a drug substance, so the technique should be selected through an evaluation of the compound properties and the intended application. Common techniques, such as XRPD, FT-IR, Raman, and NIR spectroscopies, offer good sensitivity and are not destructive, as compared to destructive techniques such as DSC. Another concern with the use of DSC for this purpose is the possible change in polymorph content during the thermal analysis.

Since XRPD is a very sensitive technique for detecting crystalline phases, it is the leading technique being used for polymorph quantification as reported in the literature.

The quantifiable level of XRPD has been reported to be in the range of 1–2.5% polymorphic content. For example, Bugay et al. applied XRPD to quantify the relative ratio of cefepime monohydrate in cefepime dihydrate, where the XRPD patterns of the monohydrate and dihydrate displayed distinct differences in the region of 12 to 15° 2θ [4]. Characteristic peaks of both forms were selected for the establishment of a univariate calibration model, and the detection limit was reported to be at 0.75% with a minimum quantifiable level of 2.5%.

FT-IR spectroscopy has also been shown to be capable of analyzing crystalline forms in a binary mixture at the level of 1–5%. Bugay et al., for example, reported a detection limit of 0.3% and a minimum quantification limit of 1% for the quantification of cefepime dihydrate in cefepime monohydrate [4]. A characteristic peak of the dihydrate at 3529 cm^{-1} was selected for establishing the calibration curve using the background-subtracted spectral integration response.

Solid-state NMR spectroscopy has a similar capability to XRPD and FT-IR spectroscopy, but its use demands a long run time, which limits its use in routine analysis.

NIR study usually involves the use of chemometric analysis. A training set is used to establish a quantitative model, and a validation set is used to demonstrate the validity of the model. The advantage of the NIR study is short analysis time and minimal sample preparation.

Siddiqui et al. have carried out a chemometric evaluation of NIR, FT-IR, and Raman spectroscopic models for the quantitation of nimodipine polymorphs. FT-IR and NIR spectroscopies provided better results than Raman spectroscopy did, as reflected in the goodness of fit, where R^2 was about 0.99 for FT-IR and NIR and 0.95 for Raman [31].

The development of methods also involves the evaluation of the suitability of calibration models, including univariate, multiple variate, and PLS analysis. The advantage of the univariate approach is that a relatively small sample set is needed, usually a series of binary mixtures with a range of polymorph content. Multiple variate and PLS models are more suitable for complex situations such as polymorphic analysis in a drug product, since they demand a relatively large sample set to establish an accurate model.

Patel and coworkers, for example, studied low levels of polymorph composition of sulfathiazole by NIR spectroscopy and identified a univariate model with sensitivity that was superior to multiple linear regression and PLS models [32]. It was discovered that NIR was a suitable technique to differentiate between sulfathiazole Form I and Form III in its NIR spectra and NIR secondary derivative spectra. The peaks at 1456 and 1970 nm in the secondary derivative clearly reflected the amount of Form III.

Jalsovsky et al. used FT-Raman spectroscopy for the quantitative analysis of polymorph content of cimetidine, where linear regression and the traditional least squares method gave satisfactory results, and PLS was found to be the best [33].

24.7 Salt and Free Form Composition

As discussed earlier, salt forms are widely used in pharmaceutical products for the enhancement of solubility, dissolution, or stability.

Of fundamental concern when using a salt in pharmaceutical systems is that the salt will undergo disproportionation when encountering an undesired microenvironmental pH. Conversely, under such conditions, a free form can be converted to a salt form unintentionally. Solid-state characterization techniques useful for quantitatively differentiating a salt from its free form include XRPD, FT-IR, FT-Raman, NIR, and solid-state NMR spectroscopies.

The analysis of salt/free form composition is very similar to that used for the analysis of polymorphs since the quantification involves detecting two different crystalline phases. The analysis typically consists of the preparation of calibration samples, the establishment of calibration models, validation, and the testing of real samples. Regarding analytical tools, spectroscopic methods, including FT-IR, Raman, and NIR, are applicable since a salt form usually exhibits significantly different spectral features due to its ionization character and different crystal packing when compared to the free form counterpart. Owing to less interference from excipients, Raman spectroscopy is the leading technique for the quantitative analysis of salt and free form composition in formulations.

Skrdla and Zhang compared FT-Raman spectroscopy, XRPD, and DSC in the quantitation of the crystalline free base form of the active pharmaceutical ingredient, MK-Q, in a binary physical mixture with the citrate salt of MK-Q and identified FT-Raman as the most sensitive technique, where FT-Raman had a detection limit less than 2%, whereas XRPD and DSC have detection limits of 6% and 7%, respectively [34]. The calibration model for the FT-Raman analysis was based on the use of the second-derivative spectra that allow for the correction of baseline drifts and signal noise. The quantitation models for the three techniques were constructed by single variable linear regression by using the unique peak or response of the individual component (Table 24.4).

Hsieh and Taylor studied the salt-to-free base conversion for benzocaine, miconazole, and sertraline, using Raman spectroscopy [35]. A unique Raman shift of the respective mesylate salt and the free base was identified and used to develop a univariate calibration model for the quantitation of the free form and salt mixture. The model used was a plot of the peak intensity ratios of the unique Raman shift for the free base and the mesylate form versus the molar ratios between the free form and the salt form, as shown in Figure 24.3, for the sertraline and sertraline mesylate [35]. The same peak ratio approach for quantification was utilized for the identification of miconazole base and its salt forms by Raman spectroscopy [36].

In the case where another ingredient is present, any interference needs to be evaluated to avoid adding complications to the calibration model. Qu and colleagues, for example, studied the process-induced phase transformation of amlodipine besylate to its free form in the presence of various excipients, where for mixtures of API with magnesium oxide, calcium silicate, and calcium carbonate, the Raman spectra within the wavenumber range of 2750–3450 cm^{-1} were selected to set up three PLC models to quantify the composition of the free form without excipient interference [37]. Since magnesium

Table 24.4 Comparison of FT-Raman, XRPD, and DSC in quantitation of the crystalline free base form in binary physical mixtures with the citrate salt of MK-Q.

Technique	Citrate salt	Free form	Model	R^2	Limit of detection for free form
FT-Raman	Second peak at 1486 cm^{-1}	Second-derivative peak at 1478 cm^{-1}	Peak ratio versus molar ratio	0.987	2%
XRPD	Diffraction peak at 18.89° 2θ	Diffraction peak at 13.10° 2θ	Peak ratio versus molar ratio	0.999	6%
DSC	Enthalpy of melting	Enthalpy of melting	Enthalpy ratio versus molar ratio	0.997	7%

Source: Adapted from [34].

stearate exhibited significant peaks in the same range, the PLS models for the mixtures with magnesium stearate were developed using spectra in the wavenumber range of 200–1000 cm^{-1}.

In a study on the disproportionation of atorvastatin calcium in the presence of acidic excipients, Christensen et al. discovered that the interference of particle size and excipient limited the choice of a Raman spectral region for quantification [38]. The Raman spectra of atorvastatin calcium and atorvastatin free form showed clear differences in the spectral regions of 600–650, 800–900, 1200–1575, and 1630–1730 cm^{-1}, however, the only region not sensitive to any interference from particle size or excipients was from 1523 to 1527 cm^{-1}.

John et al. utilized both NIR spectroscopy and XRPD to quantify the disproportionation of the HCl salt of a weak basic compound that was triggered by excipients [39]. Both techniques offered comparable sensitivity, but the NIR spectra in the region between 6300 and 6000 cm^{-1} provided a significantly greater distinction between the HCl salt and the free form. The application of the Savitzky-Golay second derivative further enhanced the spectral differences.

XRPD offers similar sensitivity to spectroscopic techniques, as shown in Figure 24.4, where the crystalline free base is easily distinguished from the HCl salt of the same compound [40]. For example, the experimental limit of detection for the free base in the salt blend was estimated to reach 5%, in formulation F1, after 4 weeks of storage at 70 °C/75%RH. One disadvantage in the use XRPD is the lack of specificity in monitoring an amorphous phase.

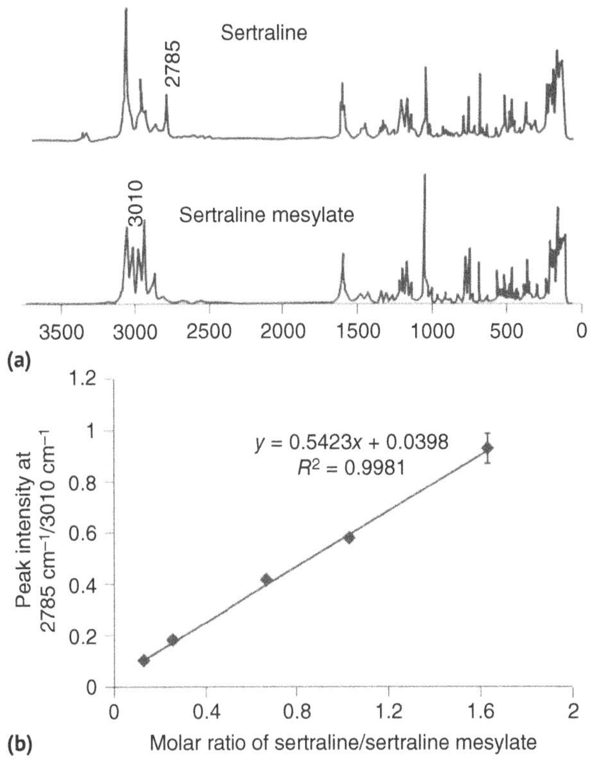

Figure 24.3 Raman spectra of sertraline and sertraline mesylate (a) and calibration curve for sertraline mesylate mixed with free base (b). *Source*: Adapted from [35].

24.8 Analysis of Particulate Contaminants in Drug Products

It is well recognized that the presence of particulate contaminants and foreign materials in a drug product may jeopardize patient safety, can compromise product quality and, however, is deemed very challenging to avoid completely. The contaminants usually are being spotted by visual examination during production or product release test and appear foreign regarding color, morphology, and particle sizes against the expected product appearance, such as black specks in a tablet surface, insoluble particulates in a drug solution, visible particulates present in a prefilled syringe, metal debris in a product detected during capsule or tablet production by a metal detector. It is not surprising to note that the occurrence of any suspected contamination event would incur concerns on product quality and, to comply with cGMP regulations, often trigger a quality investigation while the production is on hold or the product produced is in quarantine. It appears that identification of the contaminants and the source of contamination are not only time sensitive but also an essential part of risk assessment and final decision making on the batch disposition.

Obviously, particulate contaminants can be unexpectedly present in a drug product and can include metal, glass, dust, fiber, rubber, polymer, oil, mold, and degradant products. Based on the sources of the particulates, they can be divided into two categories:

- Intrinsic particulates, deriving from the manufacturing equipment, product formulation, or container system
- Extrinsic particulates, originating from the manufacturing environment and foreign to the manufacturing process and the product formulation

It is also worthwhile to mention that particulate contaminants in pharmaceutical materials and products can pose significant challenges for analysis as well. For example, most of those contaminants are insoluble in common solvents used in pharmaceutical analysis. Even dissolved, owing to the trace amount that commonly exists, the levels are usually at the

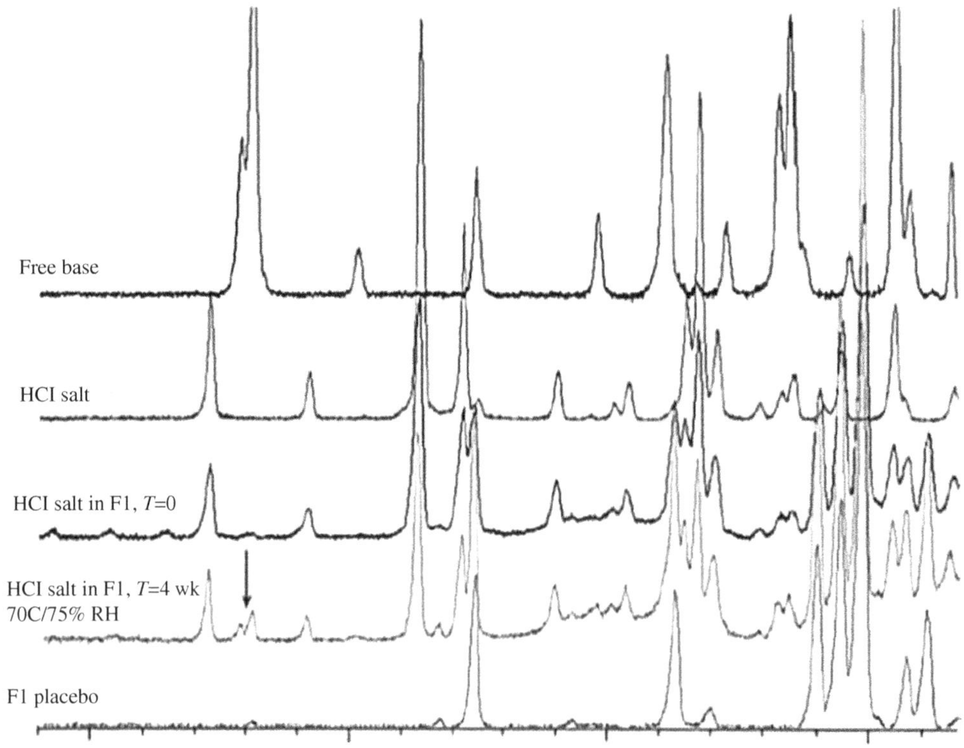

Figure 24.4 XRPD pattern of the HCl salt of a development compound in a formulation (F#1) stored at 70 °C/75% RH up to 4 weeks. The arrow points to the peaks that indicate the presence of free base caused by the disproportionation of the HCl salt. *Source*: [40]/ with permission of Springer Nature.

ppb level in sample solutions, well below the detection limit of modern chromatography techniques. Fortunately, physical characterization tools with spotting capability, such as light microscopy (Chapter 13), microscopic FT-IR (Chapter 12), and microscopic Raman (Chapter 12), often in combination with wet chemistry, can provide a meaningful path forward in those challenging forensic analysis processes.

Regarding the strategy for pharmaceutical contaminant analysis, Aldrich and coworkers developed an interesting approach [41]:

- Detection and problem definition
- Isolation and sample preparation
- Identification
- Elimination of the mechanism of particle formation

It appears that physical characterization tools like microscopy can play a significant role in detecting and isolating contaminants. For example, Aldrich and coworkers observed large visible particles present in a diluent solution used for reconstituting a lyophilized product [41]. They relied on microscopy to reveal the particles as a heterogeneous matrix and relied on FT-IR to identify one component as silicone oil. Further analysis by scanning electron microscopy and X-ray fluorescence indicated the presence of the elements copper and sulfur. Isolation of the particles leads to the analysis by mass spectrometer, revealing the presence of 2-mercaptobenzothiazole (MBT), a common vulcanizing agent, as part of the rubber stopper formulation. Thus, it was hypothesized that MBT leached from the rubber stopper reacts with the trace copper ion present in the drug formation carried over from one of the ingredients to form a highly insoluble complex.

Microscopic FT-IR and microscopic Raman can offer the advantage of in situ detection and identification of microparticle contaminants without sample isolation. For example, Cao et al. reported in situ identification of foreign particulates inside glass containers with aqueous formulated solutions by microscopic Raman, including a droplet-like particles inside a prefilled glass syringe, a fibrous particle inside a glass test tube, and a white particle inside a glass with minimal or none sample preparation needed [42].

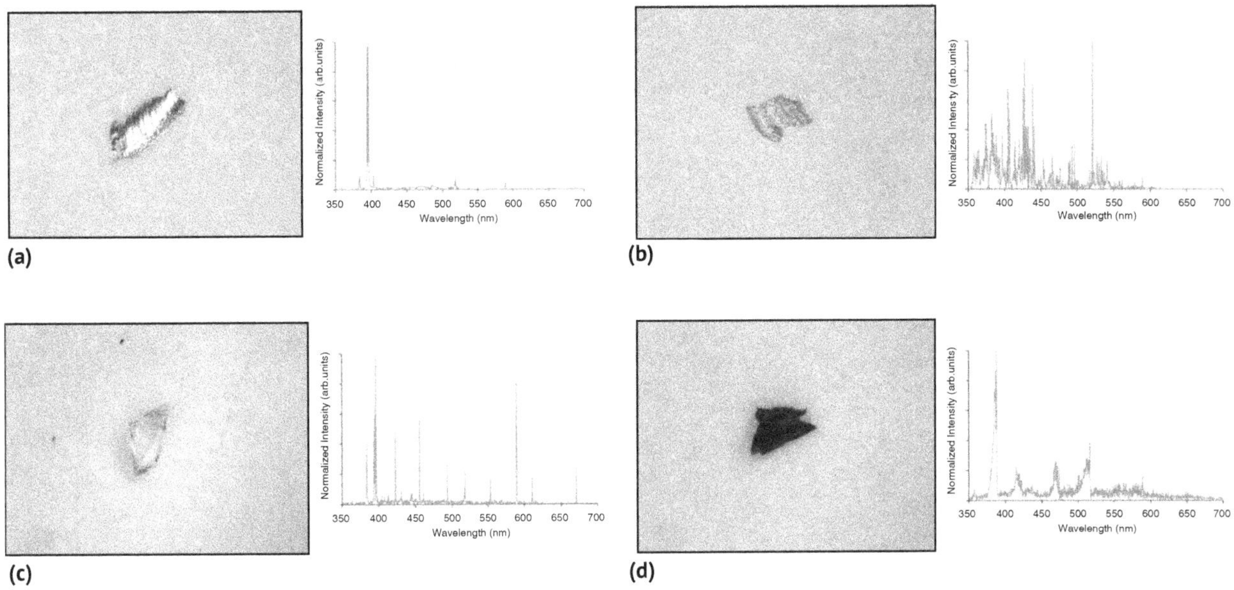

Figure 24.5 Microscopic images and LIBS spectra for various particulate contaminants. (a) Particle of aluminum alloy from a container cap; (b) particles of high alloy steel with iron, a high amount of chromium, and a low amount of manganese; (c) particles of glass; (d) particles of rubber. *Source*: [43]/Leica Microsystems.

DeRose et al. introduced a two-methods-in-one approach combining optical microscopy and laser-induced breakdown spectroscopy (LIBS) to identify particulate contaminants in pharmaceutical products [43]. As precisely directed by microscopy regarding the physical location, LIBS can provide a rapid, multielemental analysis of materials. As shown in Figure 24.5, particles made of different materials were imaged with optical microscopy and tested using LIBS. Though different metal particles appear similar in color and morphology, LIBS leads to a quick compositional analysis revealing the chemical nature of the particles.

As would be expected, this type of combination tool can find applications in identifying another common type of contaminants often encountered during tablet manufacturing: dark spots or specks in tablets. Though over-generalizing termed dark spots or specks, those unsightly blemishes or foreign-looking areas could be any color, including gray, black, and blue.

Bundenthal pinpointed some potential sources of tablet spots and specks, including [44]:

- lubricant leaked from the upper punches,
- worn or faulty dust cups or upper punch bores,
- accumulated product-lubricant mixtures from the punch barrel into the compression zone,
- powder buildup between the feeder base plates and the turret surface, subjected to continuing attrition, leading to color change,
- upstream origins from raw materials, blending steps, or granulation steps, and
- metal shavings from the tools or components of the tablet press.

Sometimes, black specks are not observed after production but could show up during stability studies as described by Carr [45]. The pharmaceutical product under investigation is a film-coated, controlled-release tablet product containing common excipients, including lactose monohydrate, hydroxypropyl methylcellulose, pregelatinized starch, sodium lauryl sulfate (SLS), colloidal silicon dioxide, and magnesium stearate with a drug load at either 0.1 or 0.4 mg per tablet. It is interesting to note that the tablets, after production, looked perfectly fine with no black specks detected. However, after 3 months, under accelerated conditions (40 °C/75% RH), black specks showed up in many tablets, as shown in Figure 24.6. Electron imaging and elemental analysis on the areas of discoloration by scanning electron microscopy (SEM) with energy-dispersive X-ray spectroscopy (EDS) revealed that the black specks are enriched with the elements sulfur and sodium. Since SLS is the sole excipient in the formulation containing sulfur and sodium, obviously, discoloration can be attributed to SLS. Further investigation disclosed that tablet discoloration was due to the dehydration reaction between the residual sulfuric acid in SLS and the carbohydrate excipients present in the formulation upon storage at the accelerated condition.

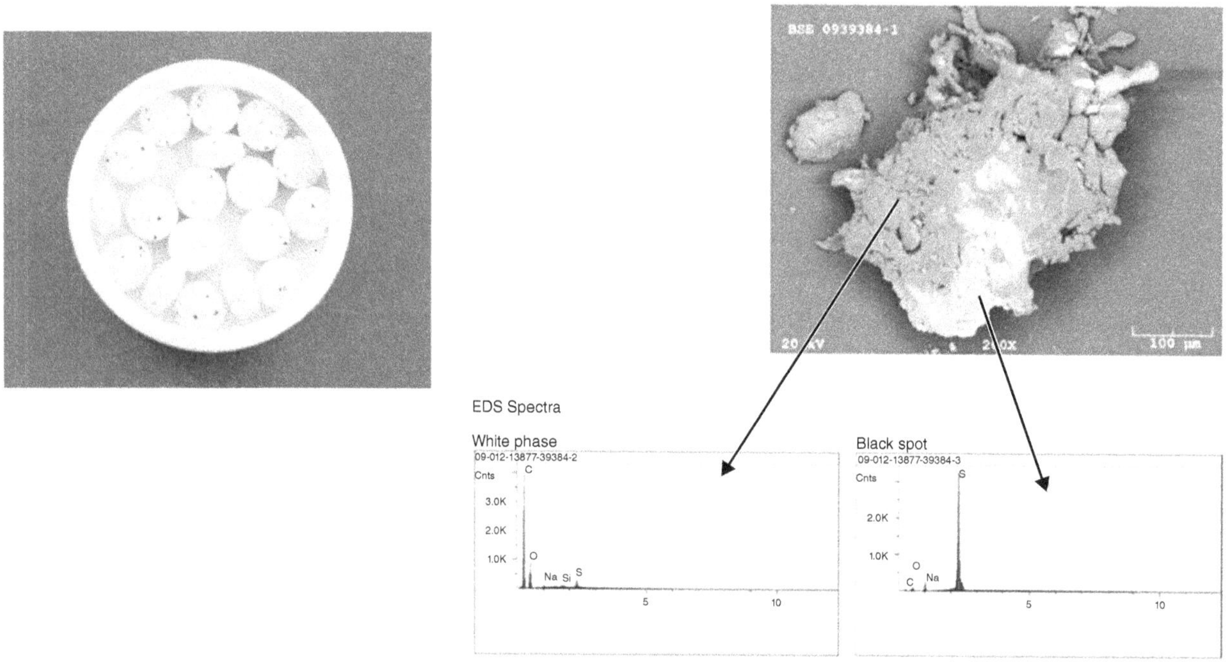

Figure 24.6 Example of tablets with black specks and SEM/EDS analysis on the specks. *Source*: [45]/MJH Life Sciences.

24.9 Process Analytical Technology

PAT is an important analytical tool for use in process-related analysis. The goal of PAT is to carry out real-time in-process measurements of critical quality and performance attributes of raw materials, intermediates, by-products, or end products during the manufacturing of drug substances and drug products.

In general, conventional manufacturing processes are monitored and controlled by offline laboratory tests that involve limited and random sample collection and analysis using wet chemistry. However, this conventional approach is not efficient and is often time consuming, in that the process is treated as a black box with very minimal information about what has occurred and how it occurred. This, in turn, poses significant challenges for process troubleshooting and improvement.

In comparison, the real-time monitoring of processes by PAT provides a better scientific understanding of a manufacturing process and potentially brings the process to a desired state to avoid batch failures. In addition, PAT acts as a key component for the assurance of using the QbD approach for the manufacturing of drug substances and drug products. The implementation of PAT, therefore, requires the establishment of a system for the design, analysis, control, and improvement of a manufacturing process.

The advantages of PAT tools over traditional in-process sampling and tests are listed as follows:

- Large sample sets to collect more reliable data
- Rapid analysis to deliver real-time information
- Reveal the progress and end points in processing
- No sample disturbance due to sampling
- No equipment operation interruption due to sampling
- Possible automatic process control

The implementation of PAT involves significant upfront work in process design and optimization, development of suitable sensor technologies and spectroscopic tools, establishment of chemometric models, validation of chemometric models, and construction of process control strategies. The development work eventually results in a much more robust monitoring and controlling of a process, based on sophisticated automation that ensures the consistent quality of end products.

Real-time process measurement can be approached in several ways, including at-line, online, and inline measurement. At-line measurement involves the withdrawal and analysis of samples very close to the process stream, whereas, for online measurements, samples are diverted from the process stream and may be sent back to the process. For inline measurements, a direct probe analyzer is placed inside the process stream so that no sample is removed and redirected.

The key component of PAT is the process analyzer that collects information related to the biological, physical, and chemical properties of the materials being processed.

The various strategies used to choose process analyzers for in-process streams include:

- determination of critical physical and chemical properties to be monitored,
- selection of an appropriate process analyzer,
- determination of the location for placement of the process analyzer in the process stream,
- development of methods for monitoring by studying various process conditions and establishing control models, and
- validation of the performance of the process analyzer and the method used for monitoring.

24.9.1 Physical and Chemical Attributes of a Process

Before the selection of a proper tool, what kind of physical and chemical attributes are critical for process success and can act as benchmarks for following the progress of the process must be well understood. Table 24.5 lists various processes and key attributes that PAT tools can offer for real-time monitoring.

As might be expected, for example, several different PAT tools are generally required to monitor the key process and quality attributes during API crystallization, including the concentration of API in the mother liquor, the particle size of crystals, the moisture content, and the polymorph content. For example, risedronate sodium (RS) is a mixed hydrate containing both lattice water and channel water [46]. Removal of the channel water has no impact on the lattice structure, but the dehydration of lattice water may lead to the collapse of the lattice. The rehydration of the dehydrated form can cause lattice expansion, so that the use of the dehydrated form in a tablet can cause tablets to crack due to this lattice expansion. Hausman et al. used online Raman spectroscopy to directly monitor the hydration state of RS during fluid bed drying of a wet granulation formulation to ensure maintenance of the hydrate [46].

24.9.2 Selection of Process Analyzers

Table 24.5 lists the tools appropriate for monitoring the physical and chemical attributes of common pharmaceutical manufacturing processes, where it can be seen that NIR and Raman spectroscopies are two of the leading techniques used to monitor various chemical components and physical properties.

Table 24.5 Chemical and physical attributes and related PAT tools for common pharmaceutical manufacturing processes.

Process	Attribute	PAT tool
Crystallization	Crystal phase	NIR, Raman
	Concentration of the drug in the mother liquid	ATR-IR
	Particle size and particle concentration	Laser-focused beam reflectance measurement (FBRM)
Drying	Moisture content	NIR
	Physical form	NIR/Raman
Wet granulation	Particle size	Laser-focused beam reflectance measurement (FBRM)
	Physical form	NIR/Raman
Blending	Blend uniformity	NIR/Raman
Tableting	Content uniformity	NIR/Raman
Coating	Uniformity	NIR/Raman
	Thickness	NIR
Freeze drying	Physical form	NIR/Raman
	Moisture content	NIR

As shown in Chapter 12, both techniques are molecular vibrational spectroscopic techniques with broad applications for the characterization of pharmaceutical materials. Both tools also have distinct capabilities for use in real-time process monitoring because of their ability to provide rapid and nondestructive measurements. Custom-made fiber-optic probes are built into process streams and linked to NIR and Raman spectrometers for PAT applications. The two techniques have similar capabilities for studying various physical and chemical attributes such as chemical content, polymorphs, hydrates, crystallinity, and content uniformity. Since the two spectroscopies rely on different selection rules, materials giving good signals in NIR spectra might show weak signals in Raman spectra, or vice versa. Selection of either technique as a PAT tool requires the evaluation of material characteristics and detection needs. NIR's sensitivity to water content and particle size enables its use in the monitoring of moisture content and particle size, whereas Raman's insensitivity to water content gives it an advantage for studying processes involving the presence of water, such as crystallization and wet granulation.

NIR spectroscopy widely serves as an excellent PAT tool for monitoring processes that involve the removal of water, a common process in the manufacturing of drug substances and drug products. Conventional monitoring of drying involves sampling from the process stream and analyzing for water content by loss on drying. A NIR probe, on the other hand, can be directly inserted into a dryer, such as a fluid bed dryer, to collect inline moisture data. Morris et al., for example, reported that real-time NIR monitoring of moisture data led to the development of an effective drying method, studying the heat and mass transfer in the system [47].

NIR spectroscopy has also been used for inline monitoring of the tablet-coating process. Traditionally, the tablet-coating process is monitored by measuring the tablet weight gain through frequent sampling from the moving tablet bed in the pan coater. Use of an NIR probe in the coater offers better specificity and accuracy, and less disruption to the coating process. As the coating proceeds and the number of coating layers increases, NIR absorbance values related to coating materials increase, while NIR absorbance values related to the core tablets decrease. Gendre et al., for example, performed inline NIR measurements inside a pan coater to monitor a coating operation in real-time, and predicted the increase in mass of coating materials and coating thickness for an ethylcellulose-based coating [48].

Laser-focused beam reflectance measurement (FBRM) acts as a key tool for inline measurement of particle size during crystallization or granulation since it has a unique ability to measure particles in suspensions and emulsions without the need for sample withdrawal and preparation. In the instrumentation of FBRM, a highly focused laser beam is directed to the external surface of a sapphire window by a rotating lens (Figure 24.7). The laser beam scans at a fixed velocity in a circular path, and backscattered light is generated when the laser hits a particle. The particle will continue to backscatter the light until the laser beam reaches the edge of the particle.

Particle size is reflected by "chord" length, defined as the line segment with endpoints lying on the external surface. The chord length is calculated by the time lag between laser emission and reflection and the scanning velocity. A single measurement can detect thousands of chords and generate a chord length distribution to reflect the change in particle number and particle size of, for example, a crystallization process. FBRM has been utilized in the characterization of the width of the metastable zone, development of a crystallization process, troubleshooting of a crystallization process, control of crystallization, and monitoring of the granulation process.

Liu et al. have investigated the crystallization with cooling of carbamazepine using in situ FBRM, where in one of the crystallization experiments, FBRM detected crystal growth and polymorph transformation [49]. Figure 24.8 shows the evolution of the chord length distribution during a slow cooling experiment of carbamazepine, where in the first 15 minutes, a steady growth in counts occurred, with chord lengths concentrated below 30 μm, which reflected the nucleation and growth of Form II. The FBRM chord length distribution remained almost constant for 15 to 21 minutes, indicating the almost completion of Form II crystallization. After 21 minutes, a decrease of the chord length distribution occurred as the chord length distribution shifted to larger particles, which was shown to be associated with the solution-mediated transformation of Form II to Form III.

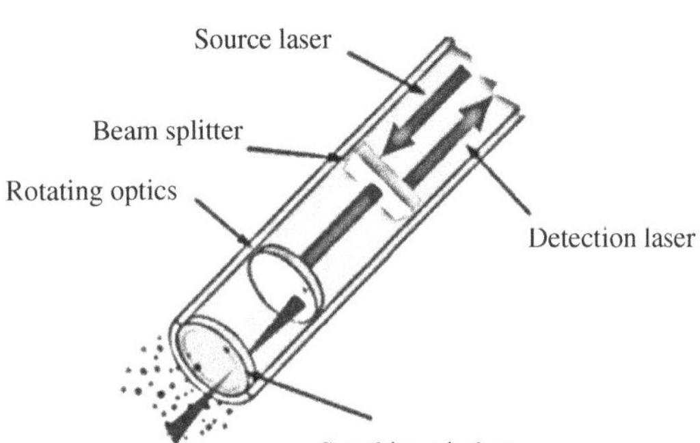

Figure 24.7 Diagram of a probe for laser-focused beam reflectance measurement (FBRM).

To monitor crystallization and solution-mediated transformation at the same time, multiple process analyzers are typically utilized. For example, Herman et al. used in situ Raman, FBRM, and ATR-FT-IR probes in a reactor to study the solvent-mediated polymorphic transformation of etiracetam Form II to Form I in methanol [50]. Raman

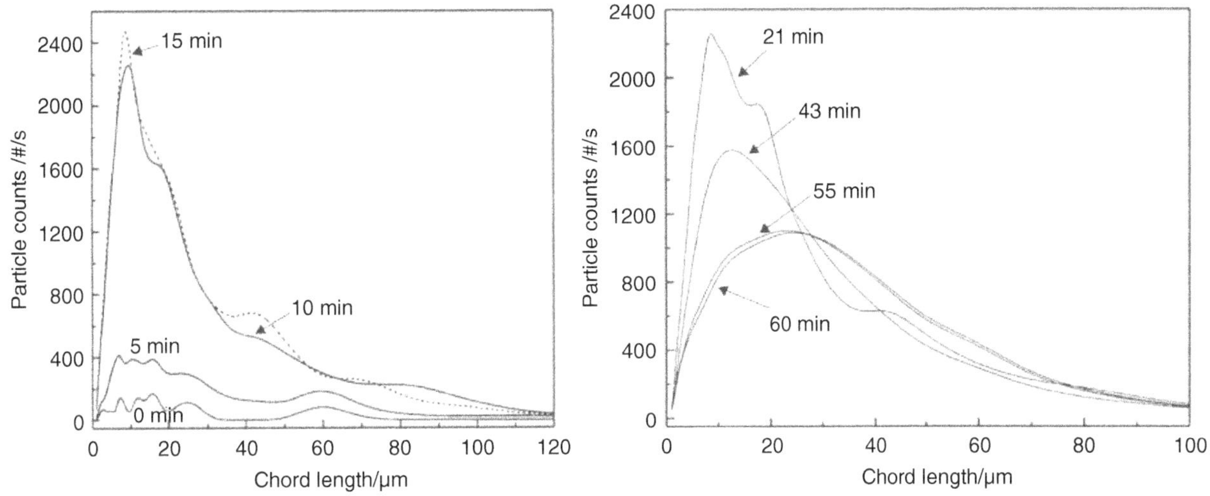

Figure 24.8 FBRM chord length distributions at various times during a crystallization experiment of carbamazepine. *Source*: [49]/with permission of American Chemical Society.

spectroscopy was able to determine the nature of the crystal form, FBRM measured the particle size distribution, and ATR-FT-IR revealed the solute concentration in solution. The combination of the three online analyzers clearly showed the three stages of the phase transformation of Form I to II: an induction phase, a growth phase, and a maturation phase. Simone et al. reported the development of a composite sensor array of Raman, NIR, ATR-UV–Vis spectroscopy, FBRM, and particle vision and measurement (PVM) in a single reactor, for use as an intelligent decision support system for crystallization and polymorphic conversion, as described in Figure 24.9 [51]. FBRM and PVM probes provided information on the particle morphology and particle size, while the Raman and NIR probes detected phase composition.

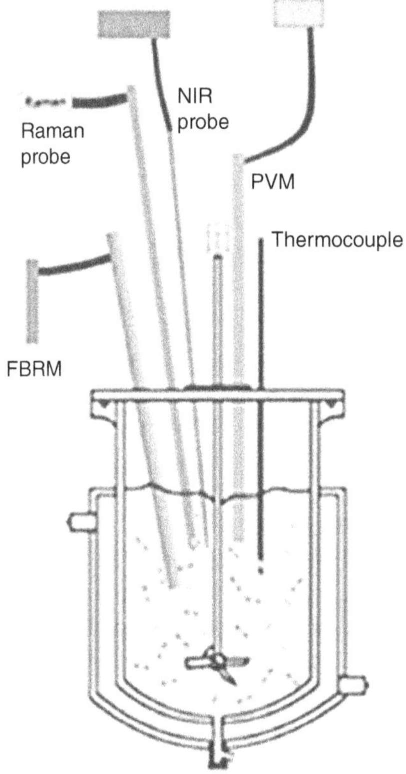

Figure 24.9 Diagram of a composite PAT array to monitor crystallization in a reactor. *Source*: [51]/with permission of American Chemical Society..

24.9.3 Location of the Process Analyzer

The location of the process analyzer is essential for a successful process analysis. An analyzer should be located in the process stream where the process is actively ongoing and also where the implementation of the analyzer does not disrupt the process and affect the integrity of the equipment.

For example, to monitor crystallization, online probes are often inserted directly into the reactor [50, 51].

For powder blending, a NIR or Raman probe is usually built into the blender, and spectroscopic signals are sent back to a linked spectrometer. In the use of Raman spectroscopy for online monitoring of low-dose blend uniformity, the fiber-optic probe head is inserted into a V-blender through the I-bar port [52]. For a direct blend formulation with 1% azimilide dihydrochloride, for example, the online Raman blend uniformity results gave an excellent correlation to the LC uniformity results of traditional thief sampling, with p-values less than 0.05.

In another application of using Raman spectroscopy for blend uniformity analysis, a glass window was built into the stainless steel mixing bowl of a planetary mixer, which allowed a fiber-optic Raman probe to collect the Raman spectra during mixing, where the inline Raman spectrometer successfully monitored the mixing of diltiazem hydrochloride pellets and paraffinic wax beads [53]. Rosas et al. reported the use of a sapphire window at the top of a bin blender to accommodate an NIR probe linked to an NIR spectrometer through an optical fiber, as depicted in Figure 24.10 [54].

A study by Shi et al. demonstrated that monitoring a blending process at multiple locations had the advantage of ensuring that a proper endpoint was reached

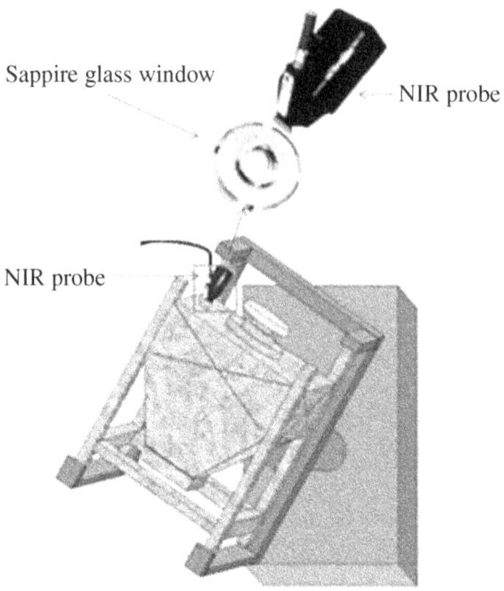

Figure 24.10 Diagram of a bin blender with an online NIR probe installed on the top. *Source*: Reproduced from [54]/with permission of Elsevier.

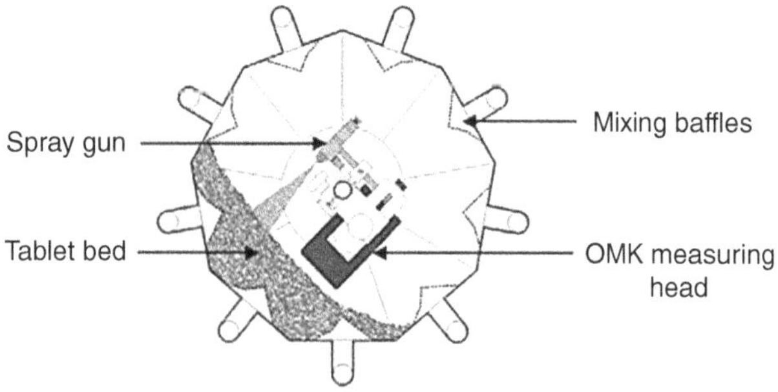

Figure 24.11 Schematic integration of the probe inside the pan coater. *Source*: [48]/with permission of Elsevier.

[55]. Two sensors, one located on the top of the bin blender and the other on the rotation axis, gave a significant difference in terms of endpoint and blending variability, for a ternary mixture of acetaminophen, lactose, and microcrystalline cellulose.

Gendre et al. developed a PAT system using NIR for inline monitoring of the film thickness during a pan coating operation [48]. The NIR probe, an OMK 500-H reflectance measuring head, was positioned at the front of the tablet bed and parallel to the direction of the spray to avoid the clogging of the detector and to maximize signal collection from the tablet bed.

24.9.4 Development of Analytical Models for Process Monitoring

Techniques like FBRM are direct methods and provide quantitative results without the use of any additional quantitative modeling. NIR and Raman, on the other hand, are spectroscopic methods, where quantitative models are established based on various statistical tools, which are dependent on the end points required for quantitation, including univariate, multivariable, or PLS approaches as described in previous sections. For example, Blanco et al. established a quantitative model to measure the content uniformity of intact pharmaceutical tablets by NIR spectroscopy [56]. The quantitation was accomplished by using a PLS calibration model with calibration samples that were prepared by varying the amount of active ingredients in the tablets.

Thirunahari et al. reported a QbD-based crystallization process for the polymorphic drug tolbutamide, using ATR-FT-IR for monitoring solute concentration and Raman spectroscopy for detecting solid form purity [57]. The calibration model of ATR-FT-IR was established using orthogonal PLS PCA to obtain a data set of 234 IR spectra collected at various tolbutamide concentrations. The FT-IR spectra in the range 1000–1650 cm^{-1} were used to construct the calibration model, as opposed to using the higher (2800–3500 cm^{-1}) and lower (650–1000 cm^{-1}) wavenumber regions due to large frequency shifts with temperature in those regions.

Gendre et al. developed PLS calibration models for an ethylcellulose-based coating system to monitor the coating process from inline NIR spectra, for both tablet-coating thickness and tablet weight gain (Figure 24.11) [48]. The best calibration model was obtained using a second-derivative transformation of inline spectra, considering both NIR regions of interest, in the 1150–1210 nm and the 1360–1440 nm NIR regions (Figure 24.12). The developed PLS model accurately predicted the coating weight gain from inline measurements and agreed with the actual mass of coating materials determined by weighing the tablets.

In many cases, statistical tools for dissimilarity analysis are utilized to assess the proximity to the endpoint instead of the quantification of the exact amount. The statistical difference between consecutive steps in the process is constantly assessed using statistical indicators such as the Mahalanobis distance, mean square difference, PLS, and Euclidean distance.

The Mahalanobis distance measures the distance between a real-time point to the point that reflects the model space, considered as the end point for a successful process. The endpoint of the model space is established by collecting spectra for the material reaching the desired property such as reaching the drying endpoint or the blend uniformity endpoint. The Mahalanobis distance is calculated from the following matrix equation:

$$\mathrm{MD}_i = \sqrt{[x_i - \bar{x}]C_x^{-1}(x_i - \bar{x})^T} \tag{24.4}$$

where $\bar{x}$ is the mean, C_x^{-1} is the inverse of the variance–covariance matrix, and T is the transpose of the matrix.

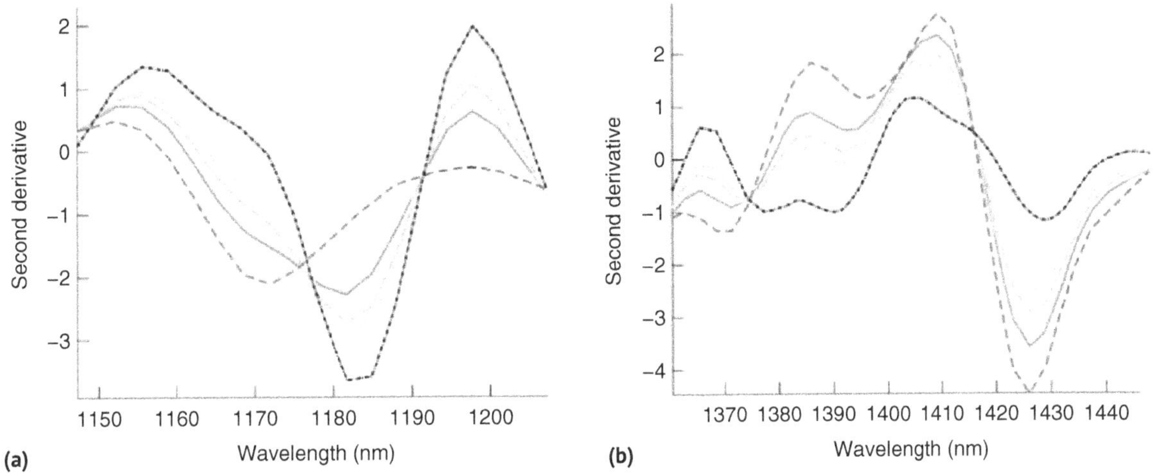

Figure 24.12 Second-derivative preprocessed NIR spectra of core tablets (dashed line), dried-coating suspension (bold dash-dotted line) and coated tablets for various theoretical mass of coating materials: 12 mg (solid line), 25 mg (dash-dotted line), 37 mg (dotted line), (a) in the 1150–1210 nm and (b) in the 1360–1440 nm NIR regions. *Source*: [48]/with permission of Elsevier.

The progress of mixing of a diltiazem hydrochloride formulation, for example, was assessed by the mean square of differences (MSDs) between two consecutive spectra, where the spectral region of diltiazem hydrochloride from 1625 to 1595 cm^{-1} was divided into sections of 5 cm^{-1} using the area of each subsection [53]. The MSD between two spectra at two consecutive times was calculated with the following equation:

$$S^2 = \frac{\sum_{i=1}^{n} (A_i^{t_k} - A_i^{t_{k-1}})^2}{n} \tag{24.5}$$

where S is the MSD, n is the number of subsections, and $A_i^{t_k}$ and $A_i^{t_{k-1}}$ are the integrated area at each subregion for the spectra measured at a given time t_k and t_{k-1}, respectively.

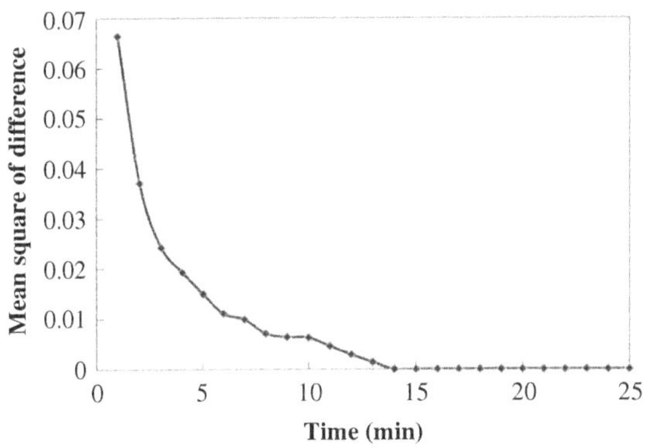

Figure 24.13 Mean square of differences between consecutive Raman spectra as a function of mixing time of a mixture consisting of diltiazem pellets and wax pellets. *Source*: Adapted from [53].

Based on the MSD profiles generated, 15 minutes of blending was shown to lead to a homogeneous mixture, as shown in Figure 24.13.

As detailed in the previous case, method development efforts also involve the selection of an appropriate spectral range for statistical analysis, where sometimes the whole spectrum works better than a selected range.

During the development of a calibration model for mirtazapine tablet, for example, Blanco and Alcalá found that selected wavelength ranges (1600–1800, 1950–2000, 2150–2200 nm) performed more poorly as a predictor than when using the whole spectra (1100–2500 nm) [59].

24.9.5 Validation

Validation of the performance of a process analyzer is a common practice for the use of PAT, where the ideal situation is to verify various characteristics, including accuracy, precision, linearity, range, repeatability, robustness, and specificity, using independent testing and statistical data analysis. The descriptions of various validation characteristics are given in regulatory guidelines [1, 2]. Bodson et al., for example, validated a NIR calibration model for the assay of diltiazem hydrochloride tablets by establishing a PLS model, where the validation results of accuracy, repeatability, intermediate precision, linearity, and limit of quantitation proved that the NIR method was suitable for its intended use [58]. Blanco and Alcalá also reported the development and validation of a content uniformity test of mirtazapine tablets by NIR spectroscopy [59]. Statistical analysis of validation sets and cross-validation offers a simple, but high level of assurance of determining the validity of a method.

Consequently, the quality of a quantitative model is often assessed in terms of the RMSE as the following equation:

$$\mathrm{RMSE} = \sqrt{\frac{\sum_{i=1}^{n} \left(y_i^{\mathrm{pred}} - y_i^{\mathrm{ref}}\right)^2}{n}} \tag{24.6}$$

where n is the number of samples of a validation set, y^{ref} is the reference value, and y^{pred} is the predicted value.

RMSE is widely used to confirm the validity of the calibration model developed by PLS.

In addition to the use of an appropriate test set of samples for validation, cross-validation with another technique offers even greater assurance of validity. For example, De Beer et al. utilized a second independent endpoint monitoring tool, NIR spectroscopy, to confirm the analysis results of using Raman spectroscopy to determine the endpoint for a powder blending process [60].

Vergote et al. used thief sampling and HPLC analysis to cross-validate the blend ending point predicted by online Raman spectroscopy [53].

24.10 Summary

In conclusion, this chapter has mainly discussed the applications of solid-state characterization techniques for the quantitative analysis of pharmaceutical mixtures. The quantitative analyses of amorphous content, degree of crystallinity, polymorph composition, salt/free form composition in drug substances and drug products are essential for quality control and assurance. Furthermore, solid-state characterization tools, including microscopy, FT-IR, and Raman, play an essential role in forensic analysis of particulate contaminants in a drug product. Ultimately, it was shown that the use of analytical techniques such as NIR and Raman spectroscopies as part of PAT offers real-time process monitoring and control.

References

1 US Department of Health and Human Services. Guidance for industry: Analytical procedures and methods validation for drugs and biologics. 2014.

2 Guideline IH. Validation of Analytical Procedures: Text and Methodology. Q2 (R1). 2005.

3 Watson DG. *Pharmaceutical Analysis: A Textbook for Pharmacy Students and Pharmaceutical Chemists*. 3rd ed. Edinburgh: Elsevier Health Sciences; 2012.

4 Bugay DE, Newman AW, Findlay WP. Quantitation of cefepime·2HCl dihydrate in cefepime·2HCl monohydrate by diffuse reflectance IR and powder X-ray diffraction techniques. *J Pharm Biomed Anal* 1996; 15(1): 49–61.

5 Savitzky A, Golay MJ. Smoothing and differentiation of data by simplified least squares procedures. *Anal Chem* 1964; 36(8): 1627–1639.

6 Barnes RJ, Dhanoa MS, Lister SJ. Standard normal variate transformation and de-trending of near-infrared diffuse reflectance spectra. *Appl Spectrosc* 1989; 43(5): 772–777.

7 Geladi P, MacDougall D, Martens H. Linearization and scatter-correction for near-infrared reflectance spectra of meat. *Appl Spectrosc* 1985; 39(3): 491–500.

8 Blanco M, Villar A. Development and validation of a method for the polymorphic analysis of pharmaceutical preparations using near infrared spectroscopy. *J Pharm Sci* 2003; 92(4): 823–830.

9 Gombás Á, Antal I, Szabó-Révész P, Marton S, Erõs I. Quantitative determination of crystallinity of alpha-lactose monohydrate by near infrared spectroscopy (NIRS). *Int J Pharm* 2003; 256(1): 25–32.

10 Mah PT, Laaksonen T, Rades T, Aaltonen J, Peltonen L, Strachan CJ. Unravelling the relationship between degree of disorder and the dissolution behavior of milled glibenclamide. *Mol Pharmaceutics* 2013; 11(1): 234–242.

11 Savolainen M, Jouppila K, Pajamo O, Christiansen L, Strachan C, Karjalainen M, Rantanen J. Determination of amorphous content in the pharmaceutical process environment. *J Pharm Pharmcol* 2007; 59(2): 161–170.

12 Young PM, Chiou H, Tee T, Traini D, Chan HK, Thielmann F, Burnett D. The use of organic vapor sorption to determine low levels of amorphous content in processed pharmaceutical powders. *Drug Dev Ind Pharm* 2007; 33(1): 91–97.

13 Grisedale LC, Jamieson MJ, Belton PS, Barker SA, Craig M, Duncan DQ. Characterization and quantification of amorphous material in milled and spray-dried salbutamol sulfate: A comparison of thermal, spectroscopic, and water vapor sorption approaches. *J Pharm Sci* 2011; 100(8): 3114–3129.

14 Chen X, Bates S, Morris KR. Quantifying amorphous content of lactose using parallel beam X-ray powder diffraction and whole pattern fitting. *J Pharm Biomed Anal* 2001; 26(1): 63–72.

15 Crocker LS, McCauley JA. Comparison of the crystallinity of imipenem samples by X-ray diffraction of amorphous material. *J Pharm Sci* 1995; 84(2): 226–227.

16 Bertacche V, Pini E, Stradi R, Statta F. Quantitative determination of amorphous cyclosporine in crystalline cyclosporine samples by Fourier transform infrared spectroscopy. *J Pharm Sci* 2006; 95(1): 159–166.

17 Gaisford S. Isothermal microcalorimetry for quantifying amorphous content in processed pharmaceuticals. *Ad Drug Del Rev* 2012; 64(5): 431–439.

18 Ahmed H, Buckton G, Rawlins DA. The use of isothermal microcalorimetry in the study of small degrees of amorphous content of a hydrophobic powder. *Int J Pharm* 1996; 130(2): 195–201.

19 Buckton G, Darcy P. The use of gravimetric studies to assess the degree of crystallinity of predominantly crystalline powders. *Int J Pharm* 1995; 123(2): 265–271.

20 Crowley KJ, Zografi G. Cryogenic grinding of indomethacin polymorphs and solvates: Assessment of amorphous phase formation and amorphous phase physical stability. *J Pharm Sci* 2002; 91(2): 492–507.

21 Surana R, Suryanarayanan R. Quantitation of crystallinity in substantially amorphous pharmaceuticals and study of crystallization kinetics by X-ray powder diffractometry. *Powder Diffr* 2000; 15(1): 2–6.

22 Otsuka M, Kato F, Matsuda Y. Comparative evaluation of the degree of indomethacin crystallinity by chemoinfometrical Fourie-Transformed near-infrared spectroscopy and conventional powder X-Ray diffractometry. *AAPS Pharmsci* 2000; 2(1): 80–87.

23 Liu J, Nagapudi K, Kiang YH, Martinez E, Jona J. Quantification of compaction-induced crystallinity reduction of a pharmaceutical solid using 19 F solid-state NMR and powder X-ray diffraction. *Drug Dev Ind Pharm* 2009; 35(8): 969–975.

24 Nunes C, Mahendrasingam A, Suryanarayanan R. Quantification of crystallinity in substantially amorphous materials by synchrotron x-ray powder diffractometry. *Pharm Res* 2005; 22(11): 1942–1953.

25 Qi S, Weuts I, De Cort S, Stokbroekx S, Leemans R, Reading M, Belton P, Craig DQ. An investigation into the crystallisation behaviour of an amorphous cryomilled pharmaceutical material above and below the glass transition temperature. *J Pharm Sci* 2010; 99(1): 196–208.

26 Luner PE, Seyer JJ. Assessment of crystallinity in processed sucrose by near-infrared spectroscopy and application to lyophiles. *J Pharm Sci* 2014; 103(9): 2884–2895.

27 Sinclair W, Leane M, Clarke G, Dennis A, Tobyn M, Timmins P. Physical stability and recrystallization kinetics of amorphous ibipinabant drug product by Fourier transform Raman spectroscopy. *J Pharm Sci* 2011; 100(11): 4687–4699.

28 Okumura T, Matsuoka M. Evaluation of the microcrystallinity of a drug substance, Indomethacin, in a pharmaceutical model tablet by chemometric FT-Raman spectroscopy. *Pharm Res* 2005; 22(8): 1350–1357.

29 Kestur US, Wanapun D, Toth SJ, Wegiel LA, Simpson GJ, Taylor LS. Nonlinear optical imaging for sensitive detection of crystals in bulk amorphous powders. *J Pharm Sci* 2012; 101(11): 4201–4213.

30 Bernstein J, Davey RJ, Henck JO. Concomitant polymorphs. *Angew Chem, Int Ed* 1999; 38(23): 3440–3461.

31 Siddiqui A, Rahman Z, Sayeed VA, Khan MA. Chemometric evaluation of near infrared, Fourier transform infrared, and Raman spectroscopic models for the prediction of nimodipine polymorphs. *J Pharm Sci* 2013; 102(11): 4024–4035.

32 Patel AD, Luner PE, Kemper MS. Low-level determination of polymorph composition in physical mixtures by near-infrared reflectance spectroscopy. *J Pharm Sci* 2001; 90(3): 360–370.

33 Jalsovszky G, Egyed O, Holly S, Hegedüs B. Investigation of the morphological composition of cimetidine by FT-Raman spectroscopy. *Appl Spectrosc* 1995; 49(8): 1142–1145.

34 Skrdla PJ, Zhang D. Disproportionation of a crystalline citrate salt of a developmental pharmaceutical compound: Characterization of the kinetics using pH monitoring and online Raman spectroscopy plus quantitation of the crystalline free base form in binary physical mixtures using FT-Raman, XRPD and DSC. *J Pharm Biomed Anal* 2014; 90: 186–191.

35 Hsieh YL, Taylor LS. Salt stability--effect of particle size, relative humidity, temperature and composition on salt to free base conversion. *Pharm Res* 2015; 32(2): 549–561.

36 Guerrieri P, Taylor LS. Role of salt and excipient properties on disproportionation in the solid-state. *Pharm Res* 2009; 26(8): 2015–2026.

37 Qu H, Savolainen M, Christensen LP, Rantanen J. Process induced phase transformations in a pharmaceutically relevant salt-free form system. *Chem Eng Sci* 2012; 77: 65–70.

38 Christensen NP, Rantanen J, Cornett C, Taylor LS. Disproportionation of the calcium salt of atorvastatin in the presence of acidic excipients. *Eur J Pharm Biopharm* 2012; 82(2): 410–416.

39 John CT, Xu W, Lupton LK, Harmon PA. Formulating weakly basic HCl salts: relative ability of common excipients to induce disproportionation and the unique deleterious effects of magnesium stearate. *Pharm Res* 2013; 30(6): 1628–1641.

40 Merritt JM, Viswanath SK, Stephenson GA. Implementing quality by design in pharmaceutical salt selection: A modeling approach to understanding disproportionation. *Pharm Res* 2013; 30(1): 203–217.

41 Aldrich DS, Borchert SJ, Abe A, Freeman JE. Pharmaceutical trace analysis. *J Res Natl Bur Stand (1977)*. 1988; 93(3): 242–245.

42 Cao X, Wen ZQ, Vance A, Torraca G. Raman microscopic applications in the biopharmaceutical industry: In situ identification of foreign particulates inside glass containers with aqueous formulated solutions. *Applied spectroscopy* 2009; 63(7): 830–834.

43 DeRose J, Scheffler K, Kartaschew K, Barbero DR. Keeping particulate contamination under control in pharmaceutical products. Leica Microsystems. Case Study. 2021; Jan 18. https://www.leica-microsystems.com/science-lab/applied/keeping-particulate-contamination-under-control-in-pharmaceutical-products/ (accessed 20 November 2024).

44 Bundenthal M. Pinpointing the source of tablet spots and specks. *PharmTech.Com*. 2014. Dec. https://www.pharmtech.com/view/pinpointing-source-tablet-spots-and-specks. (accessed 20 November 2024).

45 Carr G. Black specks in tablet stability samples. *Pharmaceutical Technology* 2015; 39(8): 6.

46 Hausman DS, Cambron RT, Sakr A. Application of on-line Raman spectroscopy for characterizing relationships between drug hydration state and tablet physical stability. *Int J Pharm* 2005; 299(1): 19–33.

47 Morris KM, Stowell JG, Byrn SR, Placette AW, Davis TD, Peck GE. Accelerated fluid bed drying using NIR monitoring and phenomenological modeling. *Drug Dev Ind Pharm* 2000; 26(9): 985–988.

48 Gendre C, Genty M, Boiret M, Julien M, Meunier L, Lecoq O, Baron M, Chaminade P, Péan JM. Development of a process analytical technology (PAT) for in-line monitoring of film thickness and mass of coating materials during a pan coating operation. *Eur J Pharm Sci* 2011; 43(4): 244–250.

49 Liu W, Wei H, Zhao J, Black S, Sun C. Investigation into the cooling crystallization and transformations of carbamazepine using in situ FBRM and PVM. *Org Process Res Dev* 2013; 17(11): 1406–1412.

50 Herman C, Haut B, Douieb S, Larcy A, Vermylen V, Leyssens T. Use of in Situ Raman, FBRM, and ATR-FTIR probes for the understanding of the solvent-mediated polymorphic transformation of II-I etiracetam in methanol. *Org Process Res Dev* 2011; 16(1): 49–56.

51 Simone E, Saleemi AN, Nagy ZK. In situ monitoring of polymorphic transformations using a composite sensor array of Raman, NIR, and ATR-UV/vis Spectroscopy, FBRM, and PVM for an intelligent decision support system. *Org Process Res Dev* 2014; 19(1): 167–177.

52 Hausman DS, Cambron RT, Sakr A. Application of Raman spectroscopy for on-line monitoring of low dose blend uniformity. *Int J Pharm* 2005; 298(1): 80–90.

53 Vergote GJ, De Beer TR, Vervaet C, Remon JP, Baeyens WR, Diericx N, Verpoort F. In-line monitoring of a pharmaceutical blending process using FT-Raman spectroscopy. *Eur J Pharm Sci* 2004; 21(4): 479–485.

54 Rosas JG, Blanco M, González JM, Alcalà M. Real-time determination of critical quality attributes using near-infrared spectroscopy: A contribution for Process Analytical Technology (PAT). *Talanta* 2012; 97: 163–170.

55 Shi Z, Cogdill RP, Short SM, Anderson CA. Process characterization of powder blending by near-infrared spectroscopy: Blend end-points and beyond. *J Pharm Biomed Anal* 2008; 47(4): 738–745.

56 Blanco M, Alcalá M, González JM, Torras E. A process analytical technology approach based on near infrared spectroscopy: Tablet hardness, content uniformity, and dissolution test measurements of intact tablets. *J Pharm Sci* 2006; 95(10): 2137–2144.

57 Thirunahari S, Chow PS, Tan RB. Quality by design (QbD)-based crystallization process development for the polymorphic drug tolbutamide. *Cryst Growth Des* 2011; 11(7): 3027–3038.

58 Bodson C, Rozet E, Ziemons E, Evrard B, Hubert P, Delattre L. Validation of manufacturing process of Diltiazem HCl tablets by NIR spectrophotometry (NIRS). *J Pharm Biomed Anal* 2007; 45(2): 356–361.

59 Blanco M, Alcalá M. Content uniformity and tablet hardness testing of intact pharmaceutical tablets by near infrared spectroscopy: A contribution to process analytical technologies. *Analytica Chimica Acta* 2006; 557(1): 353–359.

60 De Beer TR, Bodson C, Dejaegher B, Walczak B, Vercruysse P, Burggraeve A, Lemos A, Delattre L, Heyden YV, Remon JP, Vervaet C, Baeyens WR. Raman spectroscopy as a process analytical technology (PAT) tool for the in-line monitoring and understanding of a powder blending process. *J Pharm Biomed Anal* 2008; 48(3): 772–779.

25

Drug Product Development

25.1 Chemistry, Manufacture, and Control

The goal of drug product development is to design a quality drug formulation and an appropriate manufacturing process to consistently deliver the intended performance of the product with optimal quality control, which, in this regard, consists of the chemistry (C) of the various components, the manufacturing process (M), and quality control (C) at all stages. The chemistry component, the first C in CMC, includes developing a chemical and physical description of the various formulation ingredients and studying any interaction that might occur between the drug and excipients. The manufacturing component, the M in CMC, deals with the selection of processes that will optimally deliver robust and fully functional final products, whereas the second C in CMC as the control component involves developing specifications and analytical tools to measure product quality and stability. The three elements together, CMC, make up the essence of drug product development.

Strategies for drug product development vary from product to product, since composition, the choice of the manufacturing process, and analytical approaches, must all be tailored to the type of dosage form, its intended use, the route of administration, the specific components, the mode of delivery, and the stage of development. The International Conference of Harmonization guideline Q8(R2) describes the essential elements of the requirements for optimal drug product development [1]:

- Define the target product quality profile.
- Identify the potential critical quality attributes (CQAs) of the drug product.
- Determine the CQAs of the drug substance and excipients and select the type and amount of the excipient to deliver the drug product with the desired quality.
- Select an appropriate manufacturing process.
- Define a control strategy.

The target product quality profile that directs the development process is designed after comprehensive consideration of the pharmacology, pharmacokinetics, safety profile, physical-chemical properties of the various ingredients, the patient group, various stages of clinical studies, and market analysis. The target product quality profile, therefore, includes the determination of

- route of administration,
- intended use,
- dosage form,
- delivery system,
- dosage strength,
- delivery profile,
- physical appearance,
- palatability,
- bioavailability,
- shelf life,
- container/closure system.

Solid-State Materials in Pharmaceutical Chemistry: Properties, Characterization, and Applications, Second Edition. Stephen R. Byrn, George Zografi and Xiaoming (Sean) Chen

One of the primary objectives of drug product development is to identify potential CQAs of a drug product. CQAs are the physical, chemical, biological, or microbiological properties that come together to produce acceptable product quality. For example, CQAs of a tablet product typically include dose uniformity, dissolution, product purity, and stability. CQAs of a transdermal patch would include adhesive properties, drug delivery mechanisms, and chemical stability, whereas the CQAs of a lyophilized parental product must include sterility, physical stability of the lyophilized cake, reconstitution time, and chemical stability. CQAs of a dry powder inhaler (DPI) often include the emitted dose, the aerodynamic particle size distribution, and the extent of uniformity for delivery of the drug into the respiratory system.

The central theme of product development is to fabricate an API and excipients into a desired dosage form through an appropriate manufacturing process. The steps leading to a quality product include preformulation studies of the API, excipient selection, and formulation development.

Preformulation studies focus on profiling the physical and chemical properties of a drug substance, including ionization potential, lipophilicity, solubility, solution stability, solid-state stability, crystal properties, and hygroscopicity. Each excipient in a formulation is an integrated part of the delivery system, controlling drug release and delivery, aiding manufacturability, and promoting stability. For example, a hydrogel polymer such as hypromellose is a key ingredient of many controlled-release tablets as a release-modifying agent. Sodium lauryl sulfate is often included in tablet or capsule formulations to increase solubility or wettability, whereas ascorbic acid is a commonly used antioxidant to stabilize an active ingredient prone to oxidation or *N*-nitrosamine formation. Sucrose is widely used in lyophilized products to stabilize proteins and peptides during the lyophilization process. The choice of excipients is based on drug–excipient compatibility studies that provide a list of excipients that are physically and chemically compatible with the API. Formulation of a stable dosage form with a desired performance involves using a series of design of experiments, where the variables to be assessed can include drug load, type of excipients, excipient concentration, and manufacturing process. The design of such experiments leads to a number of prototype formulations, which are then evaluated regarding potency, impurities, performance, and stability. Performance assessment depends on the routes of administration and the type of dosage form. For an oral solid dosage, dissolution studies are key for evaluating performance. Lead formulations are then selected for further optimization with respect to the fine-tuning of excipient composition and various quality attributes of the API and excipients, including particle size, moisture content, and flowability. In addition to an in vitro test of potential bioavailability, such as measuring dissolution rates, in vivo studies in animals or humans are required in many cases to guide formulation selection, especially for the oral delivery of poorly soluble drugs or for the development of controlled-release oral dosage forms. Therefore, the process required for the development of the final formulation must meet not only the defined target product quality profile but also the identification of the CQAs of the API and excipients that are essential to the drug product quality and manufacturability. Assurance of those CQAs is part of a control strategy to ensure the consistent quality of the drug product and robust manufacturing.

Developing the manufacturing process is a crucial subsequent step in achieving a high-quality final drug product. The selection of the manufacturing process, unit operations, and equipment involves a multidimensional assessment of the intended use, route of administration, dosage form, properties of the active ingredient, delivery mode, batch size, and cost. For example, an API with good flow and compressibility can be directly compressed without granulation, whereas a compound sensitive to moisture must be subjected to roller compaction for granulation instead of wet granulation. In another example, for a product with more than one API, in which some of the active ingredients may be chemically or physically incompatible with each other, one might have to choose to prepare bilayer tablets.

There are two differing approaches that are generally used for process development. One is empirical and the other is more systematic. In the empirical approach, process parameters for each unit operation are selected based on prior knowledge of similar products or similar equipment. The development process, therefore, involves stepwise trial and error with one or multiple variables changing at a time until a satisfactory product is produced. Proper in-process controls are then established to avoid batch failures, and process optimization usually involves a small range of process parameters. Process validation in such cases mainly focuses on process reproducibility. The advantage of the empirical approach is its minimal resource usage. However, it lacks an understanding of the robustness of the process and the interrelationships between material attributes and process parameters to drug product CQAs. The systematic approach explores a broad range of process parameters and material attributes during process development, which often results in a significant number of development batches and a very large dataset. Data mining, using statistical tools, helps to establish the design space for process parameters and material attributes. Also, critical material quality attributes and process parameters can be established as part of a control strategy. Systematic process development leads to more robust manufacturing that can be adjusted within the design space. In addition, process analytical tools (PATs) can be part of the process control.

The systemic approach would also lead to a more robust control strategy, compared with the empirical approach. The empirical approach relies on the specification of the active ingredient, excipients, in-process control, and the specification of the final product as the primary means of control. Certificate analysis of ingredients, in-process data, and test data of the end product is assessed against the established acceptance criterion to prove the quality of the finished product. The control of the systemic approach is built on the established design space and real-time process monitoring by PATs.

In summary, developing a quality pharmaceutical product requires a knowledge-based approach and includes the design of a product, formulation development, process development, and the establishment of a control strategy.

25.2 Preformulation

Preformulation during the drug development process is primarily concerned with the physicochemical profiling of an API. Such profiling provides a general understanding of the molecular properties, material attributes, solubility, and stability of the API. The molecular properties include ionization potential, lipophilicity, and permeability, whereas the material attributes include crystallinity, polymorphism, pseudopolymorphism, hygroscopicity, and particle size. Solubility in aqueous buffers, organic solvents, and pharmaceutical cosolvents is routinely studied, along with both solution stability and solid-state stability. The information collected by preformulation provides the basic knowledge required to proceed in the development of the API, the formulation, and analytical methods. For example, the pK_a value of a drug will limit the choice of counterions for salt formation, whereas a compound with poor solubility indicates the need to develop technologies to enhance dissolution, such as solid dispersions and nanoparticles. A drug with photostability problems would require protection from light during sample analysis and in the product package.

To develop a formulation for a β-lactamase inhibitor, for example, Mortko et al. carried out comprehensive preformulation studies on the solubility, dissolution, and interconversion of four distinct crystalline phases, where the metastable hydrate B was found to be more desirable in a parental product than the stable hydrate C due to its higher solubility and consequently the reduced time required for reconstitution [2].

Kearney et al. carried out preformulation studies to aid in the development of a ready-to-use injectable solution of the antitumor agent, topotecan. The pH–solubility profile was evaluated over a pH range of 2.5–4.5 at 25°C, and the pH–stability profile was studied over a pH range of 2–4, and it was revealed that a solution pH of <2.5 was most appropriate for the formulation of a ready-to-use solution, considering both the solubility and stability requirements [3].

To formulate clotrimazole in a nanoemulsion-based system to improve its solubility and dissolution, Borhade et al. focused their preformulation work on solubility screening in lipid-based excipients and a solution stability study. Solubility screening in oils, surfactants, and cosurfactants leads to potential nanoemulsion components. In addition, surfactants and cosurfactants were evaluated for their ability to emulsify selected oily phases. Compositions of various components that produced suitable nanoemulsions were identified by constructing phase diagrams. The study of the pH–stability profile in these systems led to the selection of pH 7.6 to stabilize clotrimazole. Overall, such preformulation work resulted in the development of a stable nanoemulsion composed of Capryol 90, Solutol HS 15, and Gelucire 44/14 [4].

To resolve the instability caused by water sorption during the manufacturing and packaging of a solid dosage form containing betaine, Nunes et al. carried out an in-depth physical characterization of the anhydrous form and monohydrate form of betaine. Owing to its hygroscopicity, the significant water sorption by betaine caused particle agglomeration, transformation of the anhydrate to monohydrate, and finally deliquescence. The kinetics of water uptake of anhydrous betaine under various relative humidity conditions were studied, and a procedure was established to quantify the amount of betaine monohydrate present as a minor component with anhydrous betaine. The preformulation work helped to advance a strategy to decelerate water sorption by anhydrous betaine to slow down its conversion to betaine monohydrate [5].

In many circumstances, preformulation work is used to identify the weak points of an API, to provide a mitigation strategy for formulation development.

For example, Dong et al. carried out a preformulation study of *N*-epoxymethyl-1,8-naphthalimide, a novel synthetic antiproliferative agent, and discovered that it was susceptible to hydrolysis [6]. To minimize hydrolysis, the compound was formulated as either a lyophilized powder or a nonaqueous solution, followed by solubilization/reconstitution prior to administration.

Physical-chemical characterization of a potent 5-HT_3 receptor antagonist, CR3124, revealed that it had relatively good solubility and high permeability. However, it had poor wettability, which had to be mitigated in formulation development [7].

Fan et al. characterized a novel anti-obesity drug for oral delivery and revealed that the compound was highly hygroscopic, easily becoming deliquescent and converting to an amorphous solid, which subsequently changed to another crystal form under elevated moisture/temperature conditions [8].

During the preformulation study of MK-912, a potent α-2 adrenoceptor antagonist, Qin and Frech discovered that the rate of photodegradation of the amorphous form was 40 times faster than that of crystalline MK-912, indicating the importance of maintaining crystallinity in its product formulation [9].

One of the biggest challenges in the formulation development of biological pharmaceuticals, such as proteins and vaccines, is their tendency to undergo denaturation, which leads to the formation of biologically inactive soluble and insoluble aggregates. The main focus of preformulation studies with these macromolecules is to identify agents and conditions that can mitigate aggregation. Salnikova et al., for example, screened 30 excipients at various concentrations and in different combinations to enhance the physical stability of the vaccines *Clostridium difficile* toxoids A and B, with a high-throughput aggregate assay. Cyclodextrin, histidine, and Tween 80 were shown to be able to stabilize both toxoids [10].

Hu et al. found that preformulation characterization of a recombinant pneumolysoid protein-based pneumococcal vaccine was useful for the development of a stable formulation, where it was revealed that the protein was most stable in solution at pH 6–7 and loses its conformational integrity at temperatures above 48 °C [11]. Excipient screening assays led to the discovery that sugars such as trehalose and sucrose stabilized the pneumolysin mutant by improving the thermal transition temperature associated with denaturation and, thereby, minimizing aggregation. Research on various possible adjuvants showed that the protein antigen exhibited efficient binding to aluminum hydroxide and retained conformational stability after adsorption to the adjuvant [11].

25.3 Drug–Excipient Compatibility

The examination of drug–excipient compatibility at the early stages of formulation development provides a means of evaluating the chemical and physical stability of a drug substance in the presence of excipients at various relative concentrations, thus providing a rationale for selecting certain excipients for prototype formulations. The prototypes can then be used to obtain formulations that lead to a stable product after additional optimization. In this way, drug–excipient compatibility studies build up a scientific base to design a stable formulation for clinical and commercial development.

Excipients are an essential part of drug products. Few drug products are delivered as pure drug substances. The various roles of excipients in drug products can include the following:

- Act as bulking agents for accuracy when using low doses
- Offer desirable flow, density, compressibility, and lubrication to provide consistent production of solid unit doses such as capsules and tablets
- Promote the disintegration and dissolution of tablets and capsules
- Modulate drug release profiles
- Stabilize drug substance physically and chemically
- Promote solubilization of the drug
- Provide taste masking to improve palatability

Excipients act as general bulking agents so that an exact amount of an API can be administered over a broad dosage range. Most often, excipients are used to facilitate the stability, manufacture, and bioavailability of API in solid dosage forms. For example, microcrystalline cellulose and spray-dried lactose, because of their mechanical properties, are commonly used in tablet formulations to ensure sufficient compressibility under available processing conditions. In other cases, a disintegrant such as sodium starch glycolate is often included in immediate-release products to promote tablet disintegration and dissolution. Controlled-release agents, such as hypromellose, are widely used to produce matrix tablets that can temporally modulate drug release profiles. Some excipients are included in a formulation to promote the physical and chemical stability of drugs. Trehalose or sucrose, for example, is commonly included in lyophilized products of proteins and peptides to stabilize protein conformation and prevent denaturation. Antioxidants are often included in many products to prevent oxidation of drugs or excipients.

The goal of drug–excipient compatibility has two aspects. The first is to identify inadvertent or unintended interactions between the drug and excipients so that unacceptable excipients can be ruled out earlier in formulation development. The second is to screen for potential stabilizers that promote the physical and chemical stability of the drug substance.

The design of experiments for evaluating excipient compatibility must consider the route of administration, type of dosage form, mechanisms for delivery, and the various steps of the manufacturing process.

For solid oral dosage forms, drug–excipient compatibility studies generally focus on solid-state chemical stability and possible physical interactions. This is done by mixing a drug substance with selected excipients to form binary mixtures or, in some cases, multiple-component mixtures. The drug-to-excipient ratio is chosen to be approximately in the expected range of that to be used in the final dosage form. The mixtures are subject to an isothermal condition such as at 40 or 50 °C and withdrawn at a predefined time interval for analysis by chromatography to detect any potency loss and chemical degradation. The time frame allowed for such studies is usually several weeks to 1 month. Since water is part of some drug manufacturing processes and residual moisture is generally associated with active pharmaceutical ingredients and excipients, the effect of residual water on drug–excipient interactions is required. Storage of samples in open dishes exposed to controlled humidity conditions is the most common way to stress samples. Another practice, suggested by Serajuddin et al., is to add liquid water directly to the samples in a closed system. A drug–excipient blend is mixed with 20% water and stored at 50 °C in a closed glass vial, and the sample is analyzed after 1 and 3 weeks for chemical and physical stability [12]. One disadvantage of this approach is that liquid water may dissolve sufficient solid to cause solvent-mediated transformations that would not occur under ordinary product conditions.

Excipient-triggered chemical degradation can be divided into several categories:

- Hydrolysis promoted by hygroscopic excipients
- Oxidation triggered by residual peroxides or other oxidants in excipients
- Acid–base reactions involving excipients
- Maillard reactions between amines and reducing sugars, present as excipients

In many situations, the instability of the drug is caused by residual impurities in the excipient instead of the excipient itself. Many excipients contain residual peroxides, aldehydes, acidifying agents, alkalizing agents, and metal ions that can act to catalyze chemical reactions. It has been reported, for example, that povidone, polysorbate 80, polyethylene glycol 400, and hydroxypropyl cellulose were found to contain substantial concentrations of hydroperoxides, which led to significant lot-to-lot and manufacturer-to-manufacturer variation [13].

Wu et al. investigated the oxidative degradation of a phenylalanine-drug complex in a capsule dosage form and identified a residual sugar impurity from microcrystalline cellulose as the root cause of instability. The compound was degraded in the capsule dosage form after being stored under stressed conditions, with discoloration being observed at the same time. It was suggested that the residual sugar reacted with phenylalanine in the drug complex via the Maillard reaction, where reaction products were subsequently oxidized to form oxidative degradants [14].

Degradation caused by peroxide impurities in excipients was also encountered for raloxifene hydrochloride tablet formulations, where the degradation product was identified as an *N*-oxide derivative of the drug substance. The oxidation reaction was promoted by the peroxide impurities in povidone and crospovidone, and consequently a specification for peroxide content in povidone and crospovidone must be included as part of a control strategy to limit the formation of the degradation product [15].

Ha et al. found that a higher level of peroxides in polysorbate 80 caused a significant increase in IL-2 mutein oxidation in both the liquid and solid states, and that the presence of glutathione significantly inhibited the peroxide-induced oxidation of IL-2 mutein in a lyophilized formulation [16].

Compared with chemical incompatibility, physical interactions are more difficult to detect in the early stage of compatibility studies, but changes in appearance or poor recovery can be a helpful indicator of undesirable physical interactions between drug and excipient. If the physical interaction is due to a phase transformation, such as polymorphic conversion or a change from a crystalline to amorphous form, physical characterization techniques can be utilized.

For the development of biopharmaceutical products containing proteins and vaccines, excipient screening shall mainly focus on the identification of agents that are able to offer colloidal stability and a higher order of structural integrity. Numerous excipients can act as stabilizers in such cases, including sugars, salts, polymers, surfactants, and amino acids. Their interactions with the protein of interest should be assessed experimentally, as early as possible, to determine if there are any beneficial effects. Bhambhani et al., for example, carried out high-throughput excipient selection for a highly concentrated IgG1 monoclonal antibody solution and identified dextrose, sorbitol, and a combination of the two sugars as stabilizing excipients [17]. Cheng et al. utilized two high-throughput methods to screen for stabilizing excipients for a model IgG2 monoclonal antibody, including a kinetic study to measure aggregation by monitoring optical density, and a thermal stress test to detect structural transitions by differential scanning calorimetry (DSC) and fluorescence spectroscopy [18].

Thakkar and coworkers studied two IgG1 monoclonal antibodies and discovered that sucrose and arginine acted as stabilizers and destabilizers, respectively, for conformational stability of the two antibodies [19].

25.4 Solid Dispersions

Owing to the increased number of poorly soluble drugs coming from various high-throughput drug discovery programs, solid dispersions have been increasingly accepted as a means of increasing apparent drug solubility and dissolution and, consequently, oral absorption. Based on the phase properties of the drug embedded in a carrier matrix, where the carrier is usually a polymer matrix in the amorphous state, solid dispersions can be categorized as two general types:

- Molecular dispersions, a solid solution formed between the drug and the carrier
- Particle dispersions, drug particles as amorphous forms, nanocrystals, or microcrystals dispersed in a carrier matrix, with phase separation

An API–polymer system that exists as a molecular dispersion consists of a single phase, either in true equilibrium solution or as a supersaturated solid solution (Figure 25.1). Since the drug molecule is so intimately mixed with the carrier, usually a hydrophilic polymer, the system would be expected to exhibit better physical stability than the phase-separated amorphous particle dispersions. Although it has been shown that the nanocrystalline or microcrystalline solid dispersions of a drug also can have good physical stability, the ability to increase the apparent solubility of the API is generally inferior to that in an amorphous dispersion. The discussion that follows, therefore, will focus on amorphous solid dispersions either as molecular dispersions or as particle dispersions, where the API is in the amorphous state.

Since amorphous API in a molecular dispersion or particle dispersion is generally supersaturated with respect to the equilibrium solubility of the crystalline form, there is always a thermodynamic possibility of crystallization. The amorphous drug in a solid dispersion can be kinetically stabilized against crystallization by two mechanisms: a decrease of molecular mobility and the direct inhibition of nucleation and crystal growth. The polymer in the solid dispersion generally has a higher glass transition temperature than the drug (see Chapter 6). If the polymer is miscible with the drug and forms a single phase, the presence of polymer with a high T_g will increase the glass transition and reduce the molecular mobility at typical storage conditions. Second, molecular interactions between the polymer and the drug, especially hydrogen bond interactions, can interfere with nucleation and inhibit the crystallization of the drug. Inhibition of crystallization is important not only for maintaining solid-state physical stability during storage and handling but also for its effects on in vivo dissolution from the dispersion. Preventing or delaying crystallization during in vivo dissolution helps to maintain supersaturation and subsequently enhance oral absorption. In a molecular dispersion, a polymer interacts intimately and uniformly with drug molecules. In particle dispersions, phase separation exists between the API and polymer, precluding such direct interactions. Therefore, molecular dispersions should have a much greater inhibiting effect on crystallization than amorphous particle dispersions.

Owing to this phase separation, amorphous particle dispersions also lack the ability to reduce molecular mobility and consequently display inferior stability than molecular dispersions (Figure 25.1).

Having said this, however, although the formation of a solid solution is very much desired, it is not necessarily required to obtain beneficial effects on stability and dissolution as a drug product, in that partially miscible systems or completely

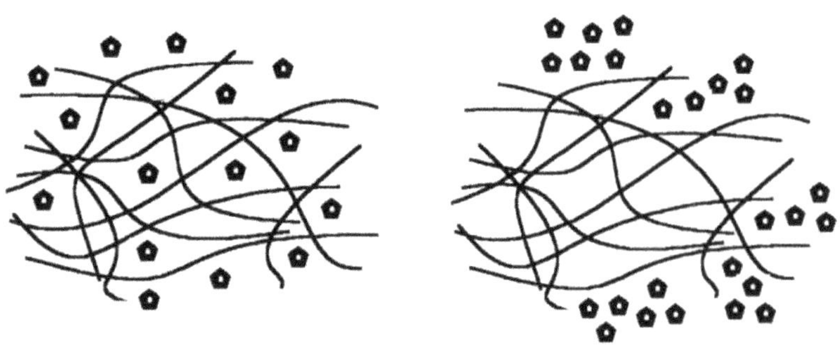

Figure 25.1 Illustrations of molecular solid dispersion and phase-separated amorphous solid dispersion.

phase-separated solid dispersions have been shown to be useful in some cases. Meng et al., for example, evaluated solid dispersions of curcumin made with povidone, Eudragit E PO, hypromellose, and PEG 8000 at a drug–polymer ratio of 70:30, where miscibility was observed only for the solid dispersion with Eudragit E PO. However, all solid dispersions showed significant dissolution enhancement, where hypromellose was more effective in maintaining the drug concentration during dissolution than the dispersion made with Eudragit E PO [20]. A solid dispersion of itraconazole with hypromellose, made by melt extrusion, is the key component of the marketed antifungal tablet Onmel, even though Six et al. reported that this solid dispersion is not a complete solid solution, probably with partial miscibility [21]. Despite this, the solid dispersion displayed a significantly increased dissolution rate compared to the physical mixture and was shown to be chemically and physically stable for a period of at least 6 months, as indicated by the absence of degradation products or recrystallization of the drug [21]. In another study, it was found that itraconazole is completely miscible with copovidone but only miscible with Eudragit E100 up to 13% drug load. The solid dispersion with Eudragit E100, however, displayed better dissolution than that of the dispersion with copovidone [22]. Hughey et al. studied a solid dispersion of a Roche Research Compound with Eudragit L100-55 and discovered that the solid dispersion contained phase-separated amorphous domains of the compound within the polymer matrix based on near-infrared spectroscopic imaging (Figure 25.2). Nonsink dissolution studies showed that the solid dispersion was three times more soluble than the crystalline drug and maintained supersaturation for at least 24 hours [23].

As we have discussed, the key quality attributes of solid dispersions are solubility enhancement and physical stability. Various in vitro studies, available to evaluate the solubility enhancement, including kinetics solubility measurement and dissolution studies under nonsink conditions, have been discussed previously (see Chapter 17). It is suggested that further animal studies must be carried out to evaluate the in vivo performance of the solid dispersion. The animal bioavailability data, for example, can provide a rank order of prototype formulations regarding performance, which can then be used to select lead formulations for further development and clinical studies.

Since the performance of a solid dispersion is achieved by maintaining the API in a high-energy state, the possibility of crystallization during processing and storage would significantly detriment the performance and affect the product quality. Thus, the physical stability of such systems must be optimized as part of formulation development, and options for packaging control should be explored as well.

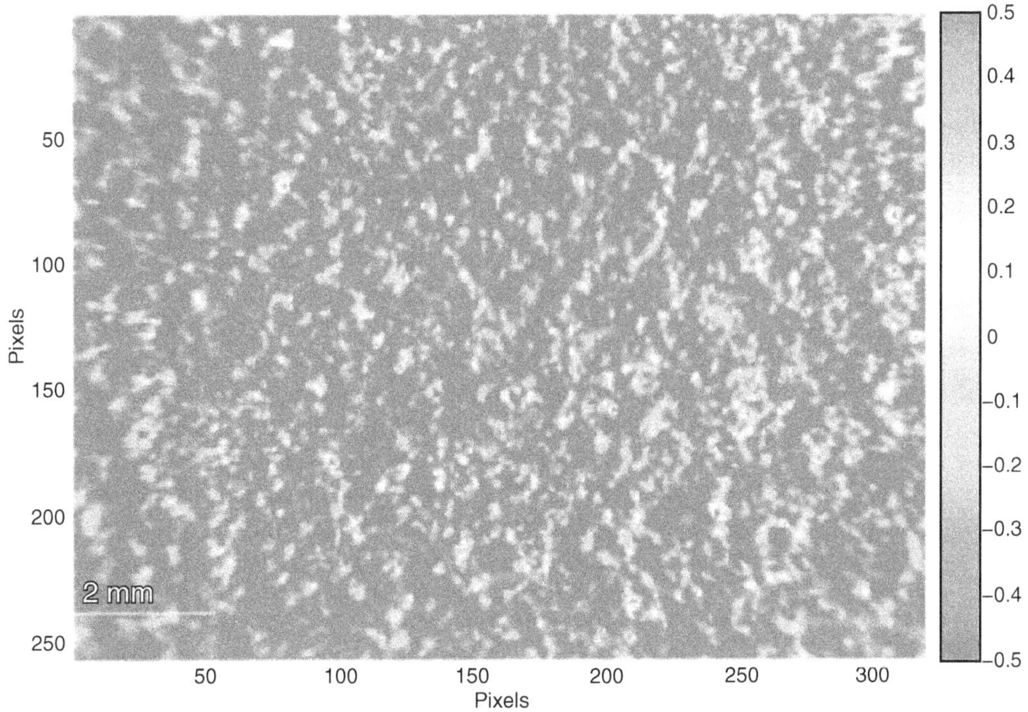

Figure 25.2 NIR image illustrating regions of localized amorphous drug (gray) and Eudragit L100-55 (dark) in a solid dispersion prepared by hot melt extrusion. *Source*: Reproduced from [23]/with permission of Springer Nature.

The selection of hydrophilic polymers must take into consideration API–polymer miscibility, the glass transition temperature, levels of hygroscopicity, hydrogen bond capabilities of the API and polymer, solubility enhancement, dissolution characteristics, and physical stability. Miscibility, as we have discussed, is a very important parameter to consider due to its expected positive impact on the consistency of quality and stability. Qualitative and quantitative methods to determine miscibility in amorphous drug–polymer systems include thermal analysis, calculation of solubility parameters, estimation of Flory–Huggins interaction parameters, and spectroscopic analysis [24]. Thermal analysis is a key technique to assess miscibility, since the formation of a solid solution for a solid dispersion generally results in a single T_g. For example, a solid dispersion of posaconazole and hypromellose acetate succinate (HPMCAS) in a 1–3 ratio gives a single glass transition of 95 °C, indicating a stable molecular dispersion [25].

The glass transition temperature of the carrier polymer is another important characteristic to consider. As indicated earlier, and discussed in Chapter 6, a polymer with a higher glass transition temperature than the API and an increase in the amount of polymer generally lead to the increase in the glass transition temperature of the solid solution, relative to that of the API alone and therefore result in reduced molecular mobility. Several common polymers used in solid dispersion formulation have relatively high T_g values. For example, copovidone has a T_g of approximately 110 °C, HPMCAS has a T_g of 120 °C, and Povidone K30 has a T_g of 165 °C. It is well recognized that the sorption of water can significantly reduce T_g (Chapters 6 and 15). Thus, although povidone has a higher T_g than copovidone and HPMCAS, its hygroscopicity makes it less favorable for solid dispersion formulations [26, 27] (Figure 25.3).

The hydrogen bonding capability of the carrier plays a crucial role in miscibility and crystallization inhibition. Both povidone and copovidone are excellent hydrogen bond acceptors, making them desirable to form molecular dispersions with drugs. Recently, enteric-coating polymers, such as HPMCAS, have been widely used for the preparation of solid dispersions because of their ability to form nanosized polymer–drug assemblies during dissolution at physiological pH values above 5. The formation of the colloidal assembly in the size range of less than 100 nm significantly inhibits the crystallization of poorly soluble drugs during dissolution and maintains significantly higher levels of supersaturation [28]. After the study of 9 compounds and 41 polymers, Curatolo concluded that HPMCAS was clearly superior to other polymers in spray-dried solid dispersions in initiating and maintaining supersaturation of a drug during dissolution [28]. The ability to maintain supersaturation was shown clearly to translate to the significant enhancement of oral absorption for one of the compounds, with six times higher C_{max} and area under the curve achieved by the solid dispersion with HPMCAS versus its crystalline counterpart.

Although binary components of drug and polymer are commonly explored for solid dispersion formulations, tertiary systems with a drug, a polymer, and another component have been reported to have advantages over binary components, where the other component can be another polymer or a surfactant. Such systems exhibit increasingly complex phase behavior, making characterization more challenging. Despite this, it has been shown that the presence of the third component can potentially enhance performance and stability or allow for an increased drug load. Prasad et al., for example, studied an amorphous tertiary solid dispersion of indomethacin, povidone, and Eudragit E100, finding that it produced greater dissolution and better physical stability than corresponding binary solid dispersions with only one of the two polymers [29]. For solid dispersions of ritonavir and lopinavir with copovidone, the inclusion of sorbitan monolaurate as the tertiary component remarkably improved the oral bioavailability in beagle dogs [30]. The ivacaftor tablet (Kalydeco™) contains a tertiary solid dispersion. The solid dispersion contains 80% active, 19.5% HPMCAS, and 0.5% sodium lauryl sulfate [31]. A solid dispersion of VX-950 prepared by spray drying contains 49.5% VX-950, 49.5% HPMCAS-HG, and 1% sodium lauryl sulfate [32]. It appears that adding a surfactant can provide some synergetic effects, reducing the polymer needed and increasing drug load. Increasing drug load is critical because poorly soluble drugs often require high doses, potentially leading to unacceptably large tablets.

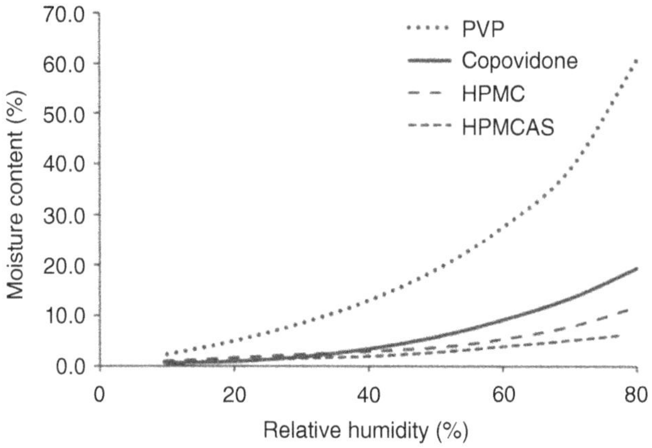

Figure 25.3 Moisture sorption of hydrophilic polymers of povidone (PVP), copovidone, hypromellose (HPMC), and hypromellose acetate succinate (HPMCAS). *Source:* [26–27].

The importance of choosing an appropriate process can be seen from reports that different processes used to produce solid dispersions from the same formulation can exhibit very different performance or stability. For example, Al-Obaidi et al. discovered that a griseofulvin solid dispersion with HPMCAS, prepared by bead milling, displayed significantly higher aqueous solubility than one prepared by spray drying. It was hypothesized that the higher surface energy of the spray-dried solid dispersion particles led to particle aggregation and very rapid recrystallization from solution [34]. It has also been reported that a posaconazole/HPMCAS solid dispersion prepared by spray drying only exhibited half of the bioavailability of the same solid dispersion made by melt extrusion [25].

To fabricate solid dispersions into solid dosage forms, such as tablets and capsules, additional excipients and processing steps are usually required, depending on the material characteristics of the solid dispersion, such as density, flowability, particle size, chemical stability, and disintegration. As mentioned earlier, solid dispersions prepared by spray drying tend to have low density and poor flow, and, consequently, a densification process such as roller compaction is often included to increase density and improve flow. Materials from melt extrusion typically are prepared as pellets or extrudates that need to be pulverized for downstream encapsulation or tableting. Fang et al., for example, reported that pellets of posaconazole freebase and HPMCAS made by melt extrusion were milled in a Fitzmill hammer mill and subsequently blended with microcrystalline cellulose, low-substituted hydroxypropyl cellulose, silicon dioxide, and magnesium stearate to produce tablets (Table 25.1) [25]. The additional excipients, acting as a filler, disintegrant, and glidant, respectively, were included for tablet manufacture to prevent particle fusion to increase compressibility and to enable proper disintegration. In another example, Kalydeco™ (Ivacaftor) tablets were prepared containing 30.5% microcrystalline cellulose, 30.4% lactose, 3% sodium croscarmellose, 0.5% sodium lauryl sulfate, and 1% magnesium stearate, in addition to a 34.1% load of solid dispersion [31]. The presence of those excipients ensured the consistent manufacture and disintegration of tablets.

25.5 Abuse-Deterrent Dosage Forms

Just as with solid dispersion, the development of abuse-deterrent dosage forms involves the application of fundamental material science and drug delivery technology, for widely abused medicines include opioids, depressants, and stimulants, in dosage forms such as capsules, tablets, transdermal patches, and transmucosal lozenges. The abuse of prescribed medicines, especially opioids, has become a major public health concern in the United States, and to mitigate such abuse issue the FDA has approved several opioids with abuse-deterrent features and issued guidance on how the evaluation and labeling of abuse-deterrent opioids should be carried out [35].

The initial development of an abuse-deterrent formulation of OxyContin® was attempted to discourage the abuse of opioids through injection and inhalation by making it more difficult to solubilize or crush the original tablets. Indeed, since such a formulation of OxyContin was launched in August 2010, the selection of OxyContin® as a primary drug of abuse decreased from 35.6% to just 12.8%, as reported by Cicero et al. [36]. Interviews of individuals who were found to abuse OxyContin®, however, revealed that 24% of the individuals found a way to overcome the tamper-resistant properties of the abuse-deterrent formulation, so improving such deterrent formulations remains a great challenge.

Most abuse-deterrent platforms developed to date are designed to make nonoral route abuse of various dosage forms more difficult or less rewarding. A common way of abusing such products is to crush or dissolve tablets, capsules, or transdermal patches for nonoral administration, including nasal insufflation, injection, and smoking.

Consequently, several drug delivery approaches have been developed to discourage such abuse, including

- physical and mechanical barriers, and
- combination with an opioid antagonist.

One way to discourage abuse is to make the dosage form very hard to crush, to grind, or to extract with water or alcohol, since this makes the tablets extremely difficult to chew, cut, smash, grind, or chip by using household means. Since the tablets cannot be converted into a powder form, snorting and smoking are also discouraged.

The hard tablets with abuse-deterrent features usually have a low porosity and typically cannot be made by conventional approaches. Melt extrusion, injection molding, and heat curing are three ways that have been reported for the manufacture of extended-release opioids with strong mechanical properties.

Melt extrusion and injection molding both involve the intimate mixing of drug with polymer in the molten state and the elimination of air during the process. It has been reported, for example, that a tablet formulation of metformin and hydroxyl propyl cellulose made into granules by melt extrusion had a porosity of 5.7%, in comparison to a porosity of

20.6% for the same tablet formulation made as granules using wet granulation [37]. Similarly, sustained-release tablets of metoprolol containing ethylcellulose and polyethylene oxide were shown to have a porosity of less than 5% after being manufactured by injection molding, since the extensive densification caused by melting led to a high tensile strength for the tablets [38].

In another study, the tablets made by melt extrusion were found to be two times harder than those produced from wet granulation and tableted under the same compression force [37]. Maddineni and colleagues found that the amount of polyethylene oxide (PEO) in a lidocaine formulation greatly increased the hardness of the extruded pellets as the concentration of PEO was increased and that pellets made of pure PEO could not be broken and could only be deformed, after reaching maximum hardness of 35 kp as limited by the equipment [39]. As disclosed in a U.S. patent application, an abuse-deterrent dosage form made by melt extrusion with PEO as the key ingredient exhibited a breaking strength of at least 500 N [40]. Bartholomaus et al. developed PEO-based tablets of tramadol and oxymorphone with improved tamper-resistant properties, where the manufacturing process included hot melt extrusion, cooling, slicing, and compression in a specially designed tablet machine, applying heat and force simultaneously [41]. The tablets effectively resisted crushing by spoons, pill crushers, or hammers, with resulting tablets not breaking until reaching the forces of about 1000 N and higher in a mechanical universal material testing system, where the tablets were flattened due to plastic deformation as shown in Figure 25.5 [42]. In another study, a heat curing process to cure tablets above the glass transition temperature of the polymer was shown to increase tablet tensile strength [43]. The curing created intimate interactions between components, so that after cooling, the polymers fused together to give plastic-like properties to the tablet with high tensile strength. After curing, the breaking strength of oxycodone extended-release tablets was increased from 90 N to about 200 N.

The ability to make tablet gel after being mixed with aqueous media can provide another barrier against abuse, and in such cases hydrophilic polymers, such as polyethylene oxide and hypromellose, are used due to their extensive ability to form a viscous gel and swell when water is added. In U.S. patent 8,309,060 B2, Bartholomaus et al. revealed that PEO in a molecular weight range of from 0.5 to 15 million g/mol could be used for this purpose to give a viscosity of 4500–17 600 cp in a 5% aqueous solution at 25°C [40]. The viscous gel is difficult to filter, making drug extraction harder and tough to pass through a needle, discouraging direct injection.

Therefore, the combination of substantial hardness and significant gel effect is the most effective approach used in several extended-release products with abuse-deterrent features, such as oxymorphone extended-release tablets (Opana), oxycodone controlled-release tablets (OxyContin®), tapentadol extended-release tablets (Nucynta), morphine extended-release tablets (Egalet-001), and oxycodone extended-release tablets (Egalet-002) [44].

Hemmingsen et al. reported the development of hydrocodone prolonged release tablets with abuse-deterrent features using injection molding technology [45]. The tablets consisted of an erodible matrix where the active ingredient was

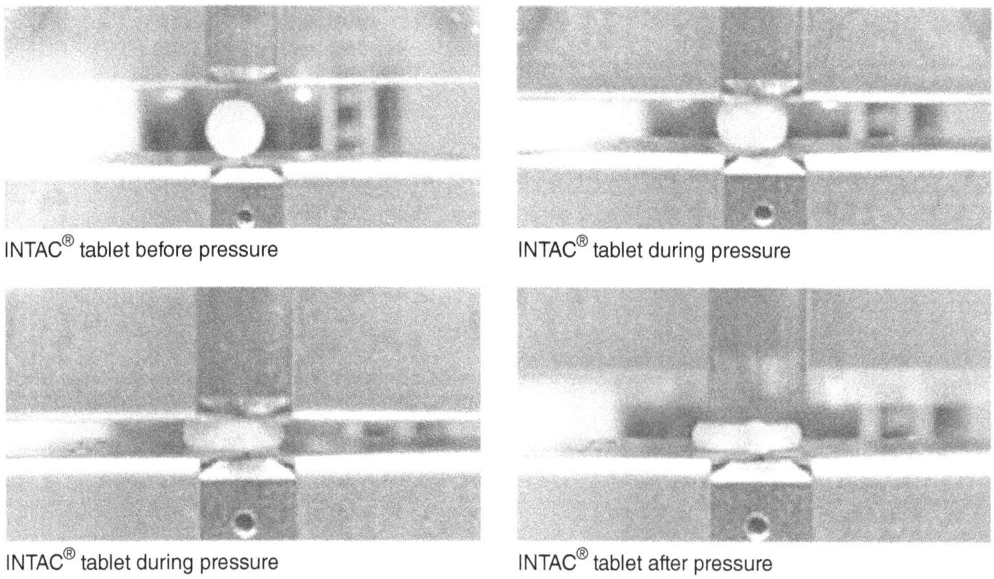

Figure 25.5 Mechanical stress testing of a tablet made by Intac® technology. *Source*: Reproduced from [42]/with permission of Drug Development & Delivery.

Figure 25.6 In vitro dissolution profiles of three 20 mg hydrocodone tablet formulations having different dimensions: (A) 6 mm, (B) 7.5 mm, and (C) 9 mm in pH 6.8 phosphate buffer. *Source*: [45]/MDPI/CC BY 4.0.

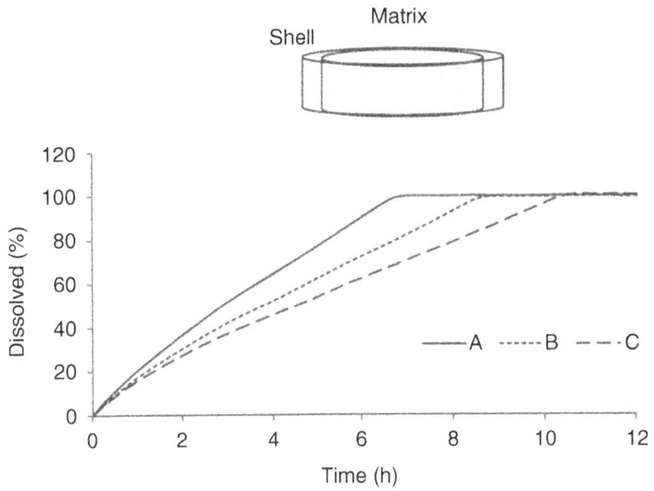

dispersed and a water-impermeable, nonerodible outer shell (Figure 25.6). The tablets have a cylindrical shape with the outside covered with the shell, except for the two ends, which are exposed to allow erosion by gastrointestinal fluids, where the dosage form was intended to provide an approximately zero-order release. Here, the erosion rate is being controlled by varying the ratio of the surface area of the exposed ends and the cylindrical length.

The effects of such geometric considerations on dissolution rate were demonstrated by a study of three 20 mg hydrocodone tablet formulations, having the same weight and composition for the matrix portion but differing in the diameter and length of the tablet (6, 7.5, 9 mm, respectively). As shown in Figure 25.6, the geometry of these tablets determines the dissolution profiles.

It has been shown that hot melt extrusion is also useful for developing immediate-release products with abuse-deterrent features. In such cases, the use of smaller dies for extrusion in combination with a pelletizer for cutting the extrudate produces small pellets or granules that are mixed with other excipients to produce immediate-release tablets. The use of a high molecular weight grade of PEO as a key ingredient led to pellets with high tensile strength and good resistance to crushing, in addition to providing gelling properties [46].

Instead of hydrophilic polymers, waxy materials are used in DETERx technology to prepare multiparticulates with abuse-deterrent properties. Here, the hydrophobic and waxy material acts as a matrix in the form of beads within which the drug is dispersed, where the beads are difficult to crush and chew, and difficult to extract with water and alcohol as well. After being melted, the beads form a viscous fatty mass that is very difficult to use for direct injection [47].

Another approach used to discourage the abuse of opioids is to develop a fixed-dose combination product of an opioid and another agent. The other agent can be an opioid antagonist, an irritant, or a dye, all designed to make abuse less rewarding. Naloxone, for example, an opioid antagonist, has been combined with buprenorphine, a partial opioid agonist-antagonist, to form sublingual tablets and films, known as Suboxone. In such a system, naloxone has virtually no oral and sublingual absorption, so it will only be effective as an opioid antagonist when injected, thus discouraging the abuse by means of extraction and injection. The same strategy has been used for a combination of extended-release morphine sulfate with naltrexone HCl, another opioid antagonist [47].

25.6 Drug-Eluting Stents

The development of drug-eluting medical devices to provide in-situ drug delivery is another pharmaceutical solid-state system where material science plays a significant role. A drug-eluting medical device is a device that is designed for localized delivery of a drug for an extended period of time. It provides a site-specific drug delivery by being placed near the target region and delivers a continuous release of drug within the therapeutic window for a defined time of treatment. Such site-specific delivery avoids the safety concern with systemic delivery and enables the delivery of drugs that are not feasible for systemic delivery. Also, due to the localized delivery, therapeutic effects can generally be achieved with a relatively safer low dose. The drawback of this type of delivery is the need for surgical intervention, thus requiring a very high standard for safety, efficacy, and precise control. The key attributes for a successful drug-eluting medical device include the following:

- Appropriate release rate related to the therapeutic need
- Physical and chemical stability of the drug and device in the body
- Reasonable dimensions consistent with the anatomy of the delivery site
- Sterilizability
- Biocompatibility

The use of drug-eluting stents (DESs) and implants is a diverse field, involving a variety of medical needs. DESs are reviewed here primarily to highlight the significance of those drug physical properties and polymer characteristics that play a role in the fabrication of those components as a unique drug delivery entity.

A DES is an implantable blood vessel scaffolding device containing a drug, which, when being implanted, releases the drug into blood vessels in a controlled manner for an extended period, usually from 90 to 180 days. Coronary stents are the most common form of interventional treatment for symptomatic coronary artery disease, where the stent is a hollow cylinder, mainly fabricated from metal meshes, inserted in a collapsed state (Figure 25.7). After insertion into the targeted spot, a mesh is stretched open by inflating an angioplasty balloon, which opens the clotted coronary artery. The inclusion of a drug in the stent reduces in-stent restenosis, which is a wound-healing process triggered by balloon-induced injury. Drugs capable of mitigating restenosis are antiproliferative and macrolide immunosuppressive agents, such as paclitaxel, sirolimus, everolimus, and zotarolimus.

Polymers used in DES need to be biocompatible and chemically stable and can maintain the physical, chemical, and mechanical integrity of the stent after sterilization. The polymers not only serve as a matrix to carry drugs but also modulate the drug release. In addition, such stents must have mechanical flexibility and integrity to withstand the significant expansion that occurs during stent implantation and the constant pulsation in the coronary vessel. The polymers are also expected to form a uniform and smooth coating in stents without cracking and flaking.

The polymers currently used in commercially available DES are proprietary and nonbiodegradable. Those polymers have been shown to be inert, durable, and biocompatible through preclinical and clinical studies as part of the development process for the associated DES. The polymer used in the Taxus DES is poly(styrene-*b*-isobutylene-*b*-styrene) triblock copolymer (SIBS). Kamath has reported that this polymer is soft, hydrophobic, and elastomeric, with elastomeric properties that are related to its structure, containing a continuous polyisobutylene phase and a discontinuous polystyrene phase [48].

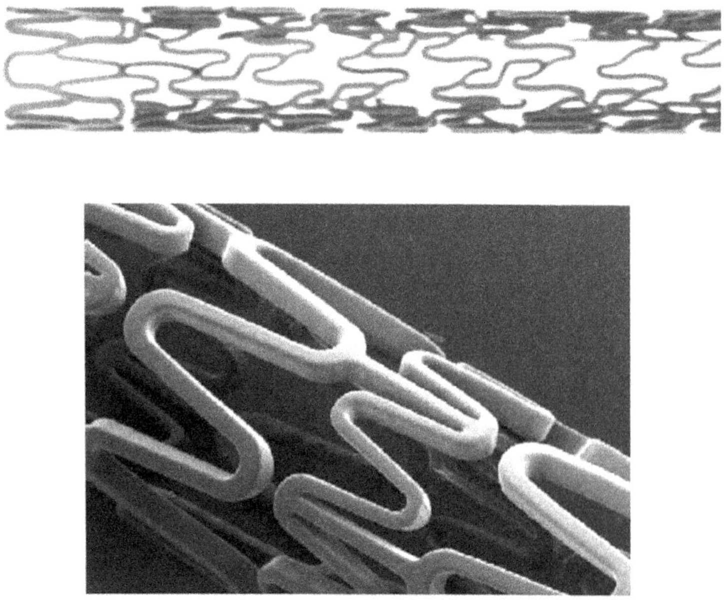

Figure 25.7 Examples of drug-eluting stents. *Source*: Reproduced from [48]/with permission of Elsevier.

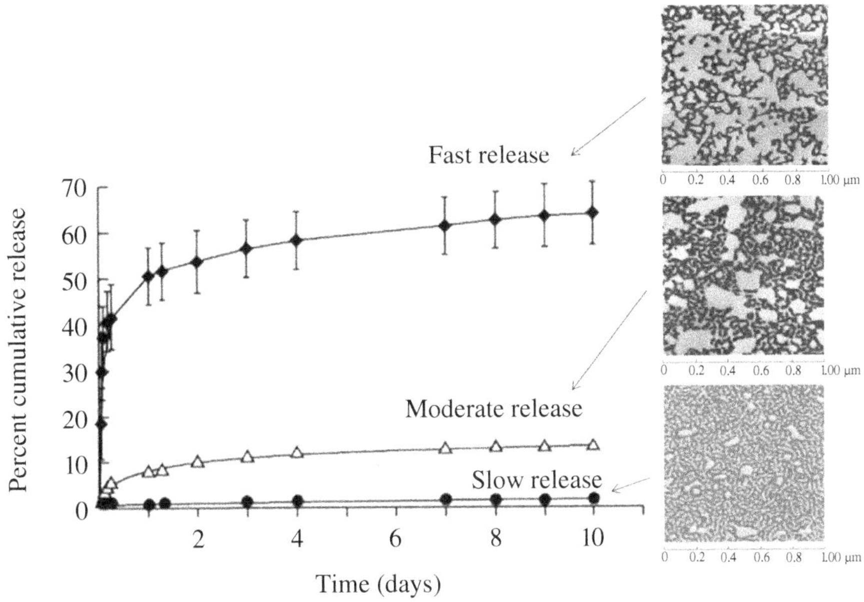

Figure 25.8 Dissolution profiles of the slow release with 8.8% w/w paclitaxel/SIBS coatings, the moderate release with 25% w/w paclitaxel/SIBS coating, and the fast release with 35% w/w paclitaxel/SIBS coatings. The atomic force microscopy images of the respective coatings are shown on the right side. *Source*: Reproduced from [48]/with permission of Elsevier.

The first goal of formulation development of DES is to study the various factors that control the dissolution of drugs from the stent and to develop lead formulations with distinct dissolution rates for clinical studies. Kamath reported, for example, that the drug to polymer ratio had a significant effect on the release rate of paclitaxel from a SIBS matrix, where the increase of drug concentration from 8.8% to 35% at a constant dose led to a significant effect on the cumulative amount of drug released (Figure 25.8) [48]. The release kinetics were described by a model developed by Higuchi for analyzing diffusion-controlled release of a drug from an inert matrix [49]. Characterization by atomic force microscopy revealed that the matrix contained phase-separated drug domains and polymer domains (Figure 25.8), where the relative size of drug domains increased as the drug load was increased. Data from XRPD proved that paclitaxel was present as an amorphous form in the coating matrix [48]. Clinical studies revealed that the formulation with the slowest release gave longer drug elution (30 days) and showed a significantly lower percent of neointimal obstruction as compared to the other two formulations with more rapid release.

Coating composition and types of polymer also affect the drug-eluting rate. For example, a zotarolimus stent coated with a phosphorylcholine polymer at a 10 μg/mm dose led to 98% drug elution in 14 days, whereas another zotarolimus stent at the same dose, but having a topcoat in addition to the matrix coating with phosphorylcholine, only delivered 90% drug after 30 days. Replacement of the phosphorylcholine polymer with C10/C19/PVP polymer (Biolinx®) yielded a much slower release stent, with 85% of the drug eluted within 60 days and continuing to elute up to 180 days [50].

The compounds used in current commercial DES have physical and chemical properties that are considered desirable for the application of restenosis prevention. Zotarolimus has extremely low water solubility and is a lipophilic compound, while the high lipophilicity of zotarolimus enables drug molecules to be sequestered in the hydrophobic core of Biolinx polymer, giving rise to a substantially slow release, which benefits its application [50]. The PVP portions of Biolinx, on the other hand, provide the hydrophilicity of the polymer, which imparts a proper balance of hydrophilicity and hydrophobicity, reported to be critical for biocompatibility and drug release [51].

The application of a drug-eluting layer into a stent is mainly carried out through spray coating, where the coating is carried out individually for each stent and not as a batch process [52]. Spray coating involves using a spray nozzle, a pump to supply coating solution, and a pressurized nitrogen source to aid the atomization, where a motor rotates and moves the stent in a coating stage while the coating solution is being sprayed. Berger, for example, reported a unique ultrasonic spray nozzle capable of producing low flow rates with a precisely shaped spray pattern and relatively small drop size, desirable for spray coating of DESs [52].

Active developments are ongoing to advance newer generation DESs with biodegradable polymers or even polymer-free options. Muramatsu et al. have provided an extensive review of the new generation DES systems currently under development [53].

25.7 Dry Powder Inhalers

Like DES, DPIs represent a complex delivery system that consists of a combination of drug and device intended for targeted delivery of a dry powder to the lung for localized or systemic therapy.

DPI systems come with many different proprietary configurations, all containing a device and a drug formulation designed to deliver precise drug doses into the lung. They can provide either premetered or device-metered dosing. For premetered dosing, preassembled doses are inserted into the device during manufacture or by the patient before use. For a device-metered DPI, multiple doses are in a reservoir, and each dose is metered through the device by the patient. The delivery of the dry powder is driven either by patient inhalation (inspiration) alone or with some external power-assistance. The formulation is usually presented as a dry blend or agglomerates and coupled with an appropriate device design to ensure dose uniformity, precise delivery, and physical integrity against storage, shipping, and in use.

Despite the diversity of DPI designs, the majority of DPIs have similar principles of operation as described in Figure 25.9 [54]. A DPI formulation commonly contains one or two micronized drug substances and an inert carrier with larger particles, with lactose as the top choice. First, a dose is added to the device and subsequently fluidized by an air flow (Figure 25.9). The airflow also generates turbulence in the device and triggers particle impaction, in the form of particle/device impaction and particle/particle impaction, causing drug particle deagglomeration, which leads to the generation of an aerosol with some drug particles separated from carrier particles. The aerosol is inhaled by the patient so that small particles are carried deeply into the lungs. The larger carrier particles are expected to collide with the oropharynx and remain there, presumably free of the smaller drug particles. It is obvious, therefore, that the material properties of the drug and the carrier can have a profound impact on DPI performance, in addition to the inspiratory airflow induced by the patient.

In this regard, the first important factor to consider must be drug particle size. For inhalation delivery, aerodynamic particle size, discussed earlier, is most relevant for evaluating the dynamic behavior of particles in the lung that ensures

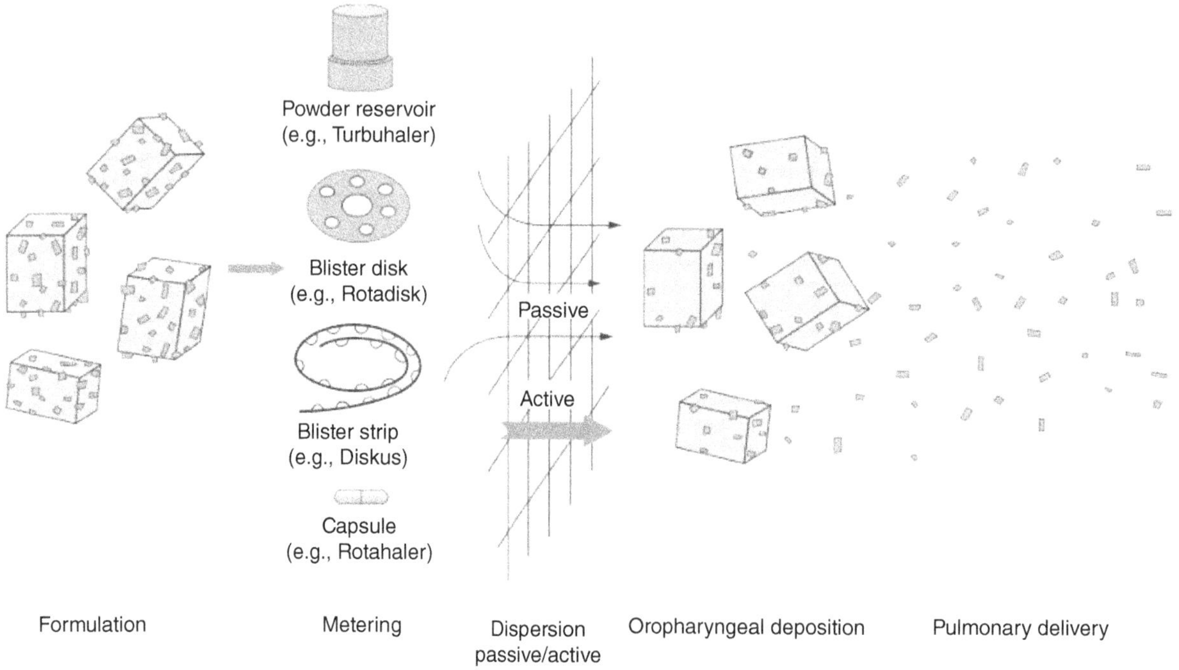

Figure 25.9 Schematic description of delivery by DPI. *Source*: [54]/with permission of Sage Publications.

proper delivery, along with factors such as gravitational settling, inertial impaction, and delivery to the lung mucosa. It is generally recognized that drug particles should be in the range of 1–5 μm for optimal delivery to the deep lung, since particles larger than 5 μm mostly impact the oropharynx and are swallowed [54]. Particles less than 1 μm, on the other hand, remain in the airstream and are exhaled during the next breathing cycle. The aerodynamic particle size is associated with the geometric particle size, though they may not be identical. The aerodynamic particle size decreases with a reduction in physical particle size or density and is also influenced by particle shape. API used in a DPI formulation is generally micronized to ensure that its aerodynamic particle size stays in 1–5 μm, thus providing good adhesion to carrier particles due to its relatively large surface area while also providing good content uniformity, since most drugs in DPI products are potent and in low doses.

The strength of the adhesion interaction between drug and carrier needs to be balanced, since weak adhesion can lead to drug being separated from the carrier during manufacture, storage, or fluidization, resulting in nonuniform dosage and variable delivery performance, and strong adhesion would cause inadequate drug/carrier separation in the airway and low deposition efficiency and underachieved efficacy.

In this regard, studying the physical interactions between drug and carrier particles is a key component of DPI product development to search for the most appropriate material characteristics, manufacturing process, and packaging control so that the particle adhesion before in-use and particle separation during inhalation delivery are optimal. Four major forces of interaction lead to the adhesion of particles, including van der Waals forces dictated by particle distance, electrostatics related to material surface characteristics, capillary forces shaped by absorbed water, and mechanical interlocking due to surface morphological irregularities.

A study of the different grades of lactose for use as a carrier for salbutamol sulfate DPI formulations led to the conclusion that the grade of lactose had a substantial effect on the performance of drug delivery. Among five different grades of lactose, anhydrous lactose provided more efficient delivery of salbutamol sulfate in comparison to the other grades of lactose. The efficient drug delivery by anhydrous lactose was related to the relatively higher concentration of fine particles [55].

A study on budesonide DPI found that the surface smoothness and particle size of different grades of lactose were interrelated, affecting the delivery performance. Anhydrous lactose, having a smooth surface and small particles, gave better delivery performance than the granulated lactose, with a rough surface and large particles [56].

Harjunen et al. evaluated four grades of lactose as a carrier for the pulmonary deposition of budesonide using a novel reservoir-based DPI, where the best performance was observed with a combination of two lactose grades, Flowlac-100 and micronized lactose [57].

Crystal engineering to improve the morphology and surface smoothness of sugar carriers has been found to benefit DPI formulation development. Zeng et al., for example, reported that the increase of the aspect ratio of lactose monohydrate crystals resulted in an increase in the respirable fraction of salbutamol sulfate from DPI formulations with lactose [58]. Similar results were observed for lactose monohydrate crystals prepared with a Carbopol 934 gel or from a water–ethanol solution, where the increase of the aspect ratio improved the deposition profiles of salbutamol sulfate and favored deep lung penetration [59]. Kaialy et al. found that recrystallized needle-shaped mannitol, where the fraction of fine particles was increased from 15% to 46%, produced a greatly enhanced performance in DPI delivery when commercial mannitol was replaced by the crystal-engineered mannitol in salbutamol sulfate DPI formulations [60]. It has also been reported that the addition of elongated fine mannitol particles as a minor tertiary component (4.7%) enhanced the performance of lactose-salbutamol sulfate DPI formulations, indicating that the needle-shaped fine mannitol particles reduced the number of physical contact points between drug and lactose particles [61].

Despite the fact that currently lactose is essentially the only excipient available for use in DPI formulations, the various grades, particle sizes, morphologies, and surface properties found with lactose offer a substantial design space for formulation development. So far, no predictive and systemic approaches or scientific tools have been developed to narrow down the choice of a certain type of lactose for DPI product formulation so that practical experiment design, prototype development, and in vitro evaluation are still required.

The physical properties of an API also affect its ability to detach from the carrier particle surface during aerosolization. Schiavone et al., for example, observed significant performance enhancement in the aerosolization and delivery of budesonide and albuterol sulfate prepared by supercritical fluid technology when compared with the respective micronized materials. The improvements were attributed to their increased surface smoothness and reduced surface energy, resulting in more balanced drug-carrier particle interactions [62].

The manufacture of DPI formulations as dry blends requires blending that leads to a uniform mixture of drug and carrier. Since most DPI products contain low doses of potent compounds, geometric mixing is preferred for obtaining a

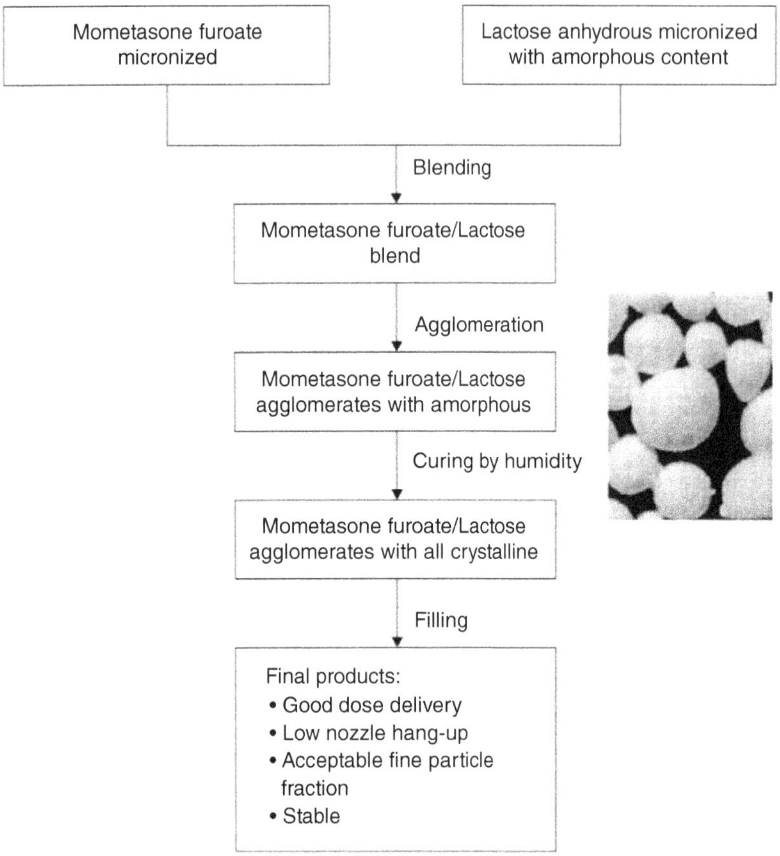

Figure 25.10 Diagram of agglomerate preparation and a picture of agglomerates. *Source*: [63]/Google LLC.

desirable content uniformity. In addition to content uniformity, as previously discussed, substantial adhesion of drug particles to carrier particles must occur to ensure the proper delivery of the drug through inhalation.

In the manufacturing process, dry powder is packed into appropriate capsules, blisters, or reservoirs, and the adhesive forces hold the API and lactose particles together. Therefore, any fluctuations in temperature and humidity, or mechanical challenges during shipping or patient in-use, can impact particle adhesion and cause desegregation. One way to mitigate the problem is to convert the dry blend into agglomerates.

Agglomerates are assemblies of drug and carrier particles with a spherical-like shape (Figure 25.10), formed from cohesive particles, by mechanical shaking or vibration. Without proper binding, such agglomerates generally will form a loose structure and tend to break into a finer powder under external mechanical force. Yang disclosed an invention that provides a method of producing agglomerates of a drug and solid carrier by producing a solid-state conversion of amorphous lactose to crystalline lactose [63]. A micronized carrier with significant amorphous content is mixed with the active ingredient (Figure 25.10), and the mixture is subjected to the agglomeration process, where the formed agglomerates are exposed to a controlled humidity to trigger crystallization of the amorphous component of the carrier. The newly formed crystals act as solid bridges to provide mechanical integrity for the agglomerates that are free flowing, sufficiently bulky, and stable under handling.

During inhalation, turbulence and collisions between agglomerates and the inhaler walls easily break these agglomerates into fine drug and excipient particles for delivery.

Recently, a thermally generated inhalation DPI system has been developed where the inhaler vaporizes the excipient-free drug to directly form an aerosol suitable for oral inhalation [64]. As described in Figure 25.11, the drug is coated as a thin film on a metallic substrate and vaporized by heating the metal for a very short period (~500 ms). The drug molecule in the vapor phase rapidly condenses and coagulates into micrometer-sized aerosol particles for inhalation. Its uniform and small particle size enables drug delivery for systemic effects in a rapid, precise, and noninvasive manner. The delivery technology in this system is similar to that used for cigarette smoking, but the drug is heated in a controlled

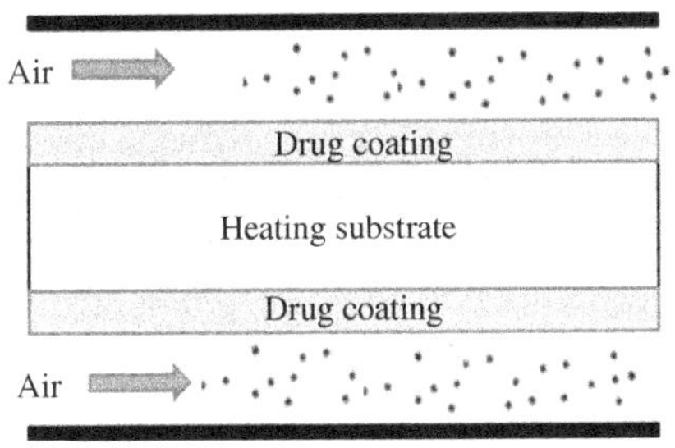

Figure 25.11 Drawing of thermally generated aerosols by a vaporization apparatus consisting of a heating substrate (metal cylinder) coated with drug in a cylindrical airway. *Source*: [64]/with permission of Elsevier.

manner to mitigate thermal degradation. A study of a number of medically significant drugs using this technology showed that aerosol particles in the size range of 1–3 μm, suitable for systemic delivery, were reliably produced without significant thermal degradation [64]. A pharmacokinetic study in dogs, for example, revealed that the delivery of rizatriptan, an antimigraine drug, through such a thermally generated aerosol, resulted in nearly instantaneous systemic drug absorption, which was much faster than with subcutaneous delivery.

Since the technique involves heating at elevated temperature, control of thermal degradation is essential for the quality and safety of the product.

One other important factor for the development of thermally generated DPI products is the solid form of the API. It appears from the literature that all studies reported involved the use of the free form as opposed to salt forms. Myers et al., for example, reported the conversion of nalbuphine hydrochloride, oxybutynin chloride, and vardenafil HCl to their respective free bases when studying the effect of film thickness on thermal aerosol generation [65]. Rabinowitz et al. reported a procedure to convert sildenafil citrate extracted from tablets to its free base, for the use in a thermally generated aerosol system [64]. Likewise, prochlorperazine base was used for the evaluation of absorption of a thermally generated aerosol, even though its maleate salt was the active ingredient of its 5 mg tablets. It appears that the free form is more suitable for thermal aerosol generation, because it tends to have lower melting points than the respective salt form, favoring vaporization. For example, prochlorperazine free base has a melting point of around 60 °C, whereas the maleate melting point is 228 °C. Also, the free form is free of counterion, which, like HCl, can generate vapors that lead to chemical degradation.

Film thickness has a significant effect on thermal degradation, where, for example, Myers et al. identified a quantitative, roughly linear relationship between coating thickness and aerosol purity, reflecting the stability during the thermal process [65]. The extent of degradation was found to be different for different compounds, depending on their thermal stability.

It is expected, therefore, that this technique should be suitable for drug molecules having low molecular weights, low melting points, and good thermal stability. A currently approved product using this technology, Adasuve, a loxapine inhalation powder, is available as a one-time use product.

25.8 Lyophilization of Biopharmaceutical Products

Biopharmaceutical products, including proteins, peptides, monoclonal antibodies, and vaccines, are an important therapeutic class of compounds used for the treatment and prevention of many diseases. Biopharmaceuticals are usually administered by the parenteral route due to their very limited oral bioavailability. However, a number of biopharmaceuticals are not suitable for formulation as solutions because they cannot maintain physical and chemical stability during storage and use. Therefore, a large proportion of all biopharmaceuticals are formulated and manufactured as lyophilized solid products to be reconstituted before administration.

The formulation development of biopharmaceutical products shall focus on the selection of excipients that can stabilize a biopharmaceutical during the freezing and drying stages of lyophilization, throughout product storage, and that also facilitate the reconstitution process.

Lyophilization is a process that removes water from a frozen sample under vacuum conditions by sublimation. It comprises four major steps: formulation of the solution and the filling of vials, freezing the sample to subzero temperatures, primary drying under vacuum, and secondary drying with gradually increasing temperatures. A sterilization step is usually included before filling the vials. Freezing leads to the formation of ice and the solidification of other ingredients to form a rigid solid "cake," where these ingredients either crystallize or form in the amorphous state. Primary drying involves the removal of most of the water as ice through sublimation with controlled vacuum and temperature conditions, thus leaving a porous cake containing noncrystalline protein, various ingredients, and residual amorphous water. The purpose of secondary drying is to remove any of the remaining water, not tightly bound to the protein and part of its structure, by drying at ambient or higher temperatures. Ideally, lyophilization leads to the production of stable porous cakes in vials, which are readily reconstituted.

The therapeutic function of biopharmaceuticals, such as proteins and antibodies, relies on maintaining the integrity of their tertiary structure, and, therefore, the formulation generally contains a buffer, a bulky agent, and various stabilizers against unfolding and aggregation of the native state of the protein. Common stabilizers include sugars, polymers, amino acids, and surfactants. The selection of specific stabilizers and their concentration depends on the chemical and physical characteristics of the biopharmaceutical.

Consequently, in the following discussion, we will focus on several key material attributes that affect the processability, stability, appearance, and performance of lyophilized products.

Since most lyophilized biopharmaceutical samples contain noncrystalline, amorphous solids, they will exhibit a glass transition temperature that will be an important determinant of the chemical and physical stability of the lyophilized cake.

The glass transition temperature of a lyophilized formulation, as with any amorphous system, is affected by its components, their concentration, and the amount of residual water. Recall from Chapters 6 and 15 that water acts as a plasticizer and lowers the glass transition temperature of amorphous systems, so that as lyophilization proceeds and water is removed, the glass transition temperature increases. Two glass transition temperatures are critical for understanding and working with the lyophilization process, including the glass transition of the frozen solution before drying, designated as T_g' and the glass transition temperature of the lyophilized cake designated as T_g. During primary drying, T_g', the glass transition temperature of the maximally freeze-concentrated amorphous phase is the main concern, since primary drying above T_g' could cause the collapse of the freeze-dried cake. T_g' is determined by the amount of unfrozen water in the cake and the glass transition temperature of the various amorphous solutes. For example, "dry" amorphous mannitol has a glass transition temperature of 13 °C [66], whereas the T_g' of a 10% frozen mannitol solution is approximately −25 to −32 °C [67]. Likewise, an IgG1/sucrose (1:1) formulation has a "dry" glass transition temperature of 93 °C and T_g' of −24.6 °C with a 50 mg/mL IgG1 frozen solution, and an IgG1/trehalose solution has a glass transition temperature of 131 °C and T_g' of −26 °C with a 50 mg/mL IgG1 frozen solution [68]. Consequently, in any formulation containing predominant amorphous components, T_g' determines the collapse temperature and the choice of the proper primary drying temperature.

The cake collapse temperature is related to the glass transition temperature for a predominantly amorphous system and is usually 1–2 °C higher. The collapse temperature (T_c) is measured using freeze-drying microscopy, where T_c is the temperature at which the cake begins to lose structure in the dried region adjacent to the ice. The transition from a glassy state to the rubbery state above T_g' results in changes in the mechanical properties of the cake that can give rise to cake collapse, so the process temperature for lyophilization should be lower than the cake collapse temperature. Such cake collapse leads to decreased surface area, higher moisture content, difficulty in reconstitution, and loss of pharmaceutical elegance [69] (Figure 25.12), whereas cake collapse during processing can cause batch failure because further drying is difficult due to the closure of porous channels that normally allow for water to leave efficiently.

Figure 25.12 A well-formed lyophilized cake and a collapsed cake. *Source:* Reproduced from [69]/with permission of Elsevier.

Secondary drying is generally carried out at temperatures lower than the glass transition temperature of the lyophilized cake, and as the moisture is desorbed, the cake's glass transition temperature increases to the final T_g of the lyophilized product.

Since sucrose and trehalose have T_g values that are much higher than those of mannitol, sucrose and trehalose are the leading choices to be used as amorphous excipients stabilizing proteins. A high glass transition temperature, as observed with sucrose and trehalose, indeed offers better biopharmaceutical stability due to the lower molecular mobility in the glassy state when stored at room temperature. For example, an IgG1 formulation with sucrose and trehalose exhibited much better stability than an IgG1 formulation with sorbitol at 25 °C, where sorbitol has a T_g of 28 °C compared to 75 and 115 °C, for sucrose and trehalose, respectively [68]. Duddu and Dal Monte highlighted the important effects of the glass transition temperature on the stability of lyophilized formulations containing a chimeric therapeutic monoclonal antibody [70]. A low-dose sucrose formulation of the antibody (sucrose/antibody 62.5/5) with a T_g of 59 °C was found to display significant aggregation, while a low-dose trehalose formulation (trehalose/antibody 62.5/5) with T_g of 80 °C was stable after storage at 60 °C. Increasing the drug load to 62.5/50 sucrose/

antibody leads to a formulation with higher T_g at 89 °C, which was still stable at 60 °C. To achieve desirable stability, therefore, it appears that the T_g of final lyophilized product should be about 30–40 °C above the storage temperature.

When possible, the use of mannitol and glycine as crystalline bulking agents increases the collapse temperature and allows primary drying to be carried out at relatively high temperatures, for example, about −10 °C. For a lyophilized formulation with excess mannitol or glycine, the collapse temperature is related to the eutectic temperature of the excipient and ice, which is about −3 to −1 °C. In comparison, the T_g' values of sucrose and trehalose formulations are usually below −30 °C. Consequently, good drying below T_g' with sucrose and trehalose takes days for complete lyophilization, while crystallized excipients like mannitol and glycine only take several hours. To take advantage of the higher drying temperatures for mannitol and glycine formulations, however, mannitol and glycine must be maximally crystallized before primary drying, through optimized freezing conditions or annealing. If a significant portion of the bulking agent remains amorphous during lyophilization, the drying temperature must be reduced, or the cake structure will be compromised due to the lowered collapse temperature. In addition, the remaining amorphous portions may affect the stability and appearance of the cake upon further storage. An annealing step can promote the crystallization of mannitol or glycine from the amorphous state during lyophilization. For example, De Beer et al. reported that mannitol crystallization readily initiated when an annealing step was introduced for a mannitol/sucrose formulation. Without such annealing, mannitol crystallization did not start until the primary drying stage [71].

Owing to the existence of multiple polymorphs for mannitol, process conditions must be controlled to promote the formation of the most desirable physical form of mannitol during lyophilization. Cao et al. reported that the freezing rate affects the composition of mannitol polymorphs, wherein rapid cooling led primarily to the formation of δ mannitol, whereas slow cooling yields a mixture of δ mannitol, β mannitol, and mannitol hemihydrate [72] (Figure 25.13).

Mannitol hemihydrate is considered an undesired form for secondary drying and storage because of the higher drying temperature required to convert it to the anhydrous form. Such high drying temperatures most likely have a detrimental effect on protein stability. In addition, the hemihydrate form is unstable during storage and tends to convert to the anhydrous form. Also, the release of water from the hydrate during storage may plasticize amorphous ingredients and cause instability. Liao et al., for example, studied the influence of process conditions on the physical state of mannitol and found that the higher annealing temperature led to δ mannitol, whereas the low annealing temperature favored the formation of mannitol hemihydrate [73].

The crystallization of mannitol or glycine is influenced by both the process conditions and the formulation composition. Johnson et al., for example, demonstrated that formulations containing lower ratios of mannitol to sucrose such as 1.5–1 and 1–1 exhibited cake collapse when primary drying was carried out at −10 °C due to the incomplete crystallization of mannitol. In comparison, formulation with a 4/1 mannitol/sucrose ratio formed a much more robust cake [74].

In general, it has been shown in many studies that sugars, such as sucrose or trehalose, are essential in protein formulations to maintain physical and chemical stability during processing and storage. Besides acting to raise T_g' and T_g, these

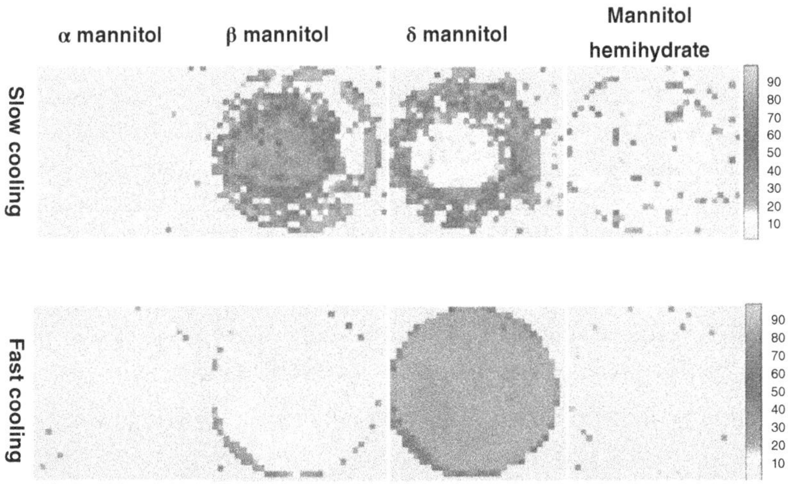

Figure 25.13 The Raman images of different crystal forms of mannitol in lyophilized mannitol samples with a slow cooling rate (1 °C/min) and a fast cooling rate (10 °C/min). *Source*: Reproduced from [72]/with permission of Springer.

disaccharides in the amorphous state appear to form hydrogen bonding with the protein, which acts as a substitute for bound water in maintaining the integrity of the native folded structure. It is generally accepted that a protein in a fully hydrated state contains about 0.3 g water per gram of protein, whereas the water content of a completely lyophilized protein sample usually has less than 1% w/w, leading to a deficiency of bound water. The hydrophilic interaction between the amorphous sugars and protein appears to replace the lost water and maintain the protein tertiary structure, thus inhibiting protein aggregation and degradation. At the same time, these sugars trap protein molecules into a miscible glassy matrix and limit the molecular mobility, slowing down diffusion-related physical and chemical changes such as protein unfolding, protein aggregation, and chemical degradation. Lueckel and coworkers, for example, discovered that the most effective protection against aggregation of interleukin-6 after lyophilization was provided by a completely amorphous formulation consisting of trehalose or sucrose either alone or in combination, with some glycine or mannitol also being present [75]. In another study, sucrose and trehalose were found to be effective stabilizers of recombinant human serum albumin (HSA) during lyophilization and storage, where sucrose and trehalose formed a completely amorphous lyophilized product, which prevented aggregation and chemical degradation [76]. Izutsu et al. reported that inositol stabilized β-galactosidase when in its amorphous form, whereas crystallization of inositol led to the loss of biological activity. Inositol crystallization can be prevented by the addition of water-soluble polymers to stabilize β-galactosidase [77]. In another example, Piedmonte et al. reported that sorbitol crystallization in a frozen Fc-fusion protein can lead to protein aggregation [78].

The water content of lyophilized products must be kept within a certain range due to its very significant plasticizing effects, since water has a T_g value of −135 °C (see Chapter 15), and a slight increase in water content can significantly decrease the glass transition temperature of the hydrated system and promote molecular mobility. Sufficiently low water contents can be achieved through proper secondary drying so that the glass transition temperature is higher than the planned storage temperature. Although for this reason most products are dried to very low water contents, for example, less than 1%, it is important to recognize, as discussed earlier, that over-drying can lead to protein denaturation and loss of activity.

For example, a humanized monoclonal antibody, lyophilized in a sucrose/histidine/polysorbate 20 formulation, produced T_g values varying from 80 °C at 1% moisture to 25 °C at 8% moisture, where the decrease of T_g at the higher moisture content led to increased rates of Asp isomerization, both above and below T_g [79]. However, samples containing intermediate moisture levels were more stable to aggregation than very dry samples and high-moisture samples. Therefore, the moisture contents chosen must be balanced for both physical and chemical stability.

Hsu et al., for example, showed that optimum residual moisture content was required to balance the physical stability and the biological stability of methionyl human growth hormone (met-hGH) and tissue-type plasminogen activator after lyophilization [80]. Samples dried to water contents below the calculated value for complete hydration by bound water exhibited opalescence upon reconstitution, suggestive of aggregation. On the other hand, samples dried to higher values of water beyond the tightly bound limit, often called a monolayer, displayed a greater loss in biological stability upon storage under stress conditions.

25.9 Oral Delivery of Peptides

Therapeutic peptides, like many biopharmaceuticals, are predominantly administered through the parenteral route owing to their very low oral bioavailability. It is obvious that oral administration usually offers better acceptance, more convenience, and better compliance for patients. Extensive efforts, therefore, have been undertaken by the pharmaceutical industry last several decades to develop platform technologies to enable oral delivery of peptides despite tremendous challenges. The challenges are reflected by the high failure rate. Even with hundreds of clinical studies in approximately 3 decades' endeavor, only two products, including Mycapssa (Octreotide) delayed-release capsule and Rybelsus (Semaglutide) tablets, have been approved by the FDA so far. These two products represent breakthroughs in the oral delivery of peptides by using permeation enhancers and leveraging API and excipient properties to overcome two major hurdles, including

- poor absorption across the gastrointestinal barrier for peptides owing to high molecular weight, significant hydrophilicity, and consequently low permeability, and
- extensive degradation by proteolytic enzymes in the gastrointestinal tract due to the peptide structure,

as further showcased in the following sections.

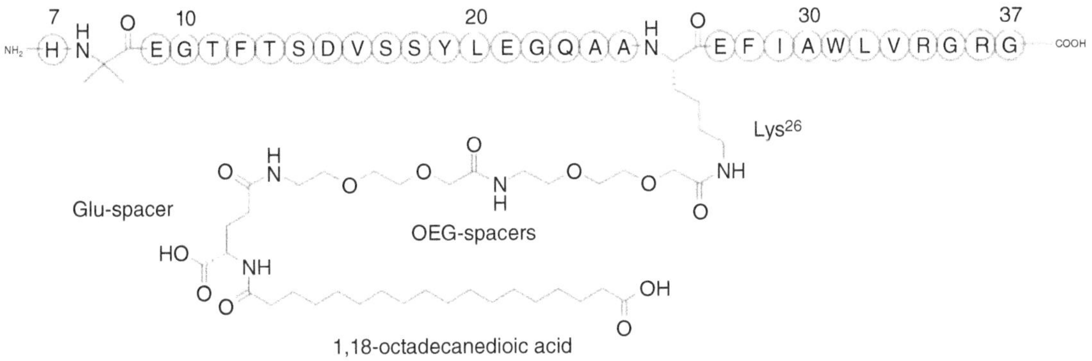

Figure 25.14 Structure diagram of semaglutide.

Table 25.2 Composition of semaglutide tablets.

Component	Function	Amount (mg/tablet)	Relative amount (%)
Semaglutide	Active ingredient	10	2.45
SNAC	Delivery agent	300	73.6
Microcrystalline cellulose	Filler	80	19.6
Povidone K90 (Kollidon 90 F)	Binder	8	1.96
Magnesium stearate	Lubricant	9.7	2.38
Total		407.7	100.0

Let's start with Rybelsus, one of the first approved peptide products for oral administration. Rybelsus, a GLP-1 analog called semaglutide combined with the permeation enhancer salcaprozate sodium (SNAC), is used alongside diet and exercise to improve blood sugar control in adults with type 2 diabetes. Semaglutide is a long-acting GLP-1 receptor agonist. The extended half-life leading to the long acting is attributed to the chemical modification of position 26 lysine by adding a C18 fatty di-acid through a hydrophilic spacer. In addition, the modification in position 8 offers stabilization against degradation by the enzyme dipeptidyl-peptidase 4 (DPP-4) (Figure 25.14). As an API, semaglutide presents as a white to almost white hygroscopic powder. Each tablet contains several common excipients for oral solid dosage form including microcrystalline cellulose, povidone, and magnesium stearate, as well as the function excipient SNAC. SNAC facilitates the absorption of semaglutide after oral administration. It is interesting to note that the SNAC-promoted absorption predominantly occurs in the stomach, as revealed by preclinical studies in dogs and clinical studies in humans [81], which, amazingly, agrees well with the formulation design to leverage the material properties of excipients.

As shown in the formulation composition for semaglutide oral tablets listed in Table 25.2, SNAC is a major component, comprising about 75% of the tablet weight [82].

First, SNAC, as a sodium salt, provides a local buffering effect and therefore protects semaglutide against enzymatic degradation. It is well recognized that pepsin is the primary digestive enzyme in the stomach and has an optimal pH range in acidic conditions from pH 2 to pH 4. Indeed, Buckley and coworkers discovered that SNAC presented in oral semaglutide tablets can increase the pH of the simulated gastric fluid (SGF) as a dissolution media from acidic to neutral within 5–15 minutes after incubating the tablets in small volumes of SGF, 1–30 mL [81].

Second, SNAC, as a permeation enhancer, not surprisingly, engages with cell membranes to promote transient enhanced absorption of semaglutide. SNAC, as an effective permeation enhancer, has a long history of development into various therapeutics, such as vitamin B12 and insulin, with a wealth of literature to refer to. The most intriguing part of the case of semaglutide is that SNAC acts on the stomach, obviously reconceptualizing the understanding of drug absorption, with the small intestine being always regarded as the primary site.

It is interesting to note that, as illustrated by a pharmacoscintigraphic imaging study in humans, semaglutide tablets containing SNAC undergo surface erosion during drug release in the stomach rather than a rapid disintegration [82]. This release profile, as would be expected, significantly promotes the localized concentration of SNAC, on the one hand, to elevate local pH to protect the drug from enzymatic degradation and, on the other hand, to promote the permeability locally with the drug also being released in the proximity. Therefore, as we can foresee, the efficient absorption of semaglutide to achieve the therapeutic level relies not only on the coformulation with SNAC but also on the performance with an erosion-type drug release.

Furthermore, it is important to note that, as disclosed by Vilhelmsen and coworkers, the dissolution profiles of semaglutide and SNAC, in combination, have substantial effects on semaglutide's oral bioavailability [82]. It was discovered that separate granulation of SNAC and semaglutide before fabricating into tablets led to faster dissolution of SNAC than the drug within the first 20 minutes, while cogranulation of SNAC and the drug before tableting caused a slower release of SNAC than the drug after the initial 5 minutes of dissolution. Surprisingly, the separate granulation tablets exhibit significantly higher bioavailability in dogs than the tablets from cogranulation, 1.0% versus 0.3%. It is obvious that achieving a 1.0% oral bioavailability for a peptide like semaglutide through a creative formulation approach is nontrivial, considering the very low oral bioavailability (<0.01%) when administered alone.

The approach to utilize erodible tablets for gastric delivery also found success in delivering LY peptide, a glucose-dependent insulinotropic polypeptide and glucagon-like-peptide-1 dual agonist, in combination with sodium caprate, another well-known permeation enhancer. The LY peptide tablets released the drug by erosion from the tablet surface with 100% release within 60 minutes at pH 6.8 and displayed a remarkable 5.8% oral bioavailability in monkeys [83].

While permeation-enhanced gastric delivery is deemed exciting, it is important to stress that permeation enhancers like SNAC and sodium caprate are equally effective in promoting the absorption of peptides in the intestine. However, as would be expected, the strategy to fabricate a dosage form would be different, as seen with Mycapssa, delayed-release octreotide capsules. Octreotide, a synthetic cyclic octapeptide analog of human somatostatin, exhibits limited intestinal absorption due to intraluminal enzymatic degradation and low intestinal permeability, as well as acid-catalyzed cleavage in the stomach. Octreotide delayed-release capsules are designed to be enteric-coated to prevent drug release and potential inactivation in the stomach. Furthermore, the formulation is an oily suspension of octreotide, including several excipients capable of transiently altering epithelial barrier integrity by opening intestinal tight junctions to promote transcellular permeation [84]. The microparticles in the suspension comprising octreotide acetate, povidone, sodium caprylate, and magnesium chloride are made by lyophilization and further dispersed in an oil blend of glycerol monocaprylate, glycerol tricaprylate, and polysorbate 80. Though the mechanisms behind the delivery platform are not disclosed, several key formulation attributes can be delineated as follows:

- Lyophilization offering intimate mixing and proximity of the drug and sodium caprylate, the permeation enhancer in the microparticles
- Dispersing in oils, protecting the hydrophilic microparticles, likely as hygroscopic amorphous dispersions, from moisture
- The oils enabling a gradual co-release of the drug and the enhancer, giving rise to a localized concentration and potentially sustained permeation enhancement

25.10 Injectable Drug–Device Combination Products

While advancements in oral delivery technology for peptides are yielding breakthrough therapeutics like semaglutide tablets, it is important to note that the majority of biopharmaceuticals must be administered as injectables. Likewise, some small-molecule drugs must be delivered as injectables, due to either poor bioavailability or the need for a specific therapeutic effect. For example, the EpiPen provides a life-saving, rapid onset of action. Traditionally, injectables are administered in hospitals or doctors' offices using vials, syringes, or infusion bags. However, the evolvement of drug–device combination products is dramatically shifting the paradigm. More and more injectable products are prescribed as pens, autoinjectors, prefilled syringes (PFSs), infusion pumps, or wearable devices, which are suitable for home use and self-administration, of significance to improve patient compliance, accessibility, and care outcomes. It was estimated that, by 2020, 1700 injectable drug–device combination products are marketed, with close to 500 products in development [85]. Notable examples include the semaglutide injection Ozempic, a prefilled single-patient use pen, and the tirzepatide injection Zepbound, prescribed as a prefilled single-dose pen.

Among various device options, PFSs are a fast-growing choice due to several advantages, including convenient administration, reduced dosing errors, less drug overfill, and a lower risk of microbial contamination. Furthermore, PFS can be integrated with autoinjectors for easier self-injection.

Because PFS is a primary packaging component with direct product contact, its material characteristics have significant impacts on the compatibility with the drug, drug product stability, and overall product performance, thereby ensuring effective drug delivery.

It is well recognized that the majority of PFS used in injectable products are made of glass. While glass is known for its chemical durability, gas impermeability, excellent resistance to moderate thermal treatments, easy inspection, and flexibility for fabrication, it also has drawbacks such as easy breakage and the need for lubricant silicone oil.

Despite being relatively inert, glass can interact with a drug product over extended periods, such as 2–3 years, potentially altering the interior glass surface and causing issues like glass delamination and leaching. Thus, the compatibility of glass syringes with a drug formulation needs to be assessed thoroughly regarding delamination of glass, detrimental surface interaction with silicone oil, or potential leachables into the drug formulations during storage and throughout the product life cycle, including shipping/handling, freeze/thaw, and temperature excursions.

Glass delamination refers to the potential for glass, under certain conditions, to shed thin, flexible fragments called glass lamellae, which could cause significant product quality issues. For example, several product recalls were associated with the issue for injectable products.

The risk of delamination is determined by glass chemistry, formulation characteristics such as pH, type of ions, and concentration of ions. It is well recognized that high salt concentration and high solution pH prompt glass delamination.

Fortunately, the pharmaceutical packaging industry is constantly improving the glass chemistry and fabrication process, aiming to reduce or eliminate the risk of glass delamination.

In glass syringes, silicone oil is commonly used as a lubricant to coat the barrel, facilitating plunger movement during injection with reasonable force. Silicone oil can leach into the product formulation and form particles. Currently, no regulatory guideline defines the acceptable level of silicone oil particles in injectable products. From a quality perspective, silicone oil particles are considered subvisible particles, for which the U.S. Pharmacopoeia sets the limits for various injectable presentations. However, both the pharmaceutical industry and regulatory agencies are concerned that silicone oil particles may induce immunogenicity or promote protein aggregation in antibody products. Two strategies are being explored to address the undesirable interactions, including formulation optimization and process improvement in siliconization. For example, the addition of surfactants to drug formulations has been reported to prevent protein adsorption to silicone oil particles and reduce aggregation. Furthermore, advanced siliconization processes such as bake-on and cross-linkable silicon have been found to limit silicone leaching into drug products.

PFS typically has a needle fixed to the syringe, protected by a needle shield, for container closure integrity and post-sterilization sterility. The syringe is therefore ready for use once the needle shield is removed for drug administration. It is important to note that the needle shield made of a synthetic elastomer is designed to have a good degree of gas permeability, which can facilitate the syringe sterilization by either steam or ethylene oxide (EO). However, this attribute may incur the permeability of water vapor or oxygen during product storage, which may trigger product instability. For example, De Bardi et al. researched the clogging of a concentrated antibody drug in staked-in needle PFSs using nondestructive neutron radiography for imaging and discovered that water vapor transmission through the needle shield could lead to the gradual evaporation of the concentrated antibody solution located in the needle lumen [86]. Consequently, the evaporation caused an exponential rise in the viscosity and resulted in the solidification of the protein solution to clog the needle when the antibody drug product in PFS was subject to storage at an accelerated storage condition (Figure 25.15). Likewise, Scheler and coworkers studied needle clogging of protein solutions in PFSs and concluded that maintaining the air pocket in the needle by mitigating any vapor diffusion promoting pressure gradient is the most effective protection from clogging [87], potentially through selecting a proper storage condition or looking for a protective secondary packaging.

It is well recognized that glass in PFS needs silicone oil for functionality, with 0.2–1 mg medical grade silicone oil being typically used per syringe by spray-on technology. However, silicone oil leached into a protein formulation or in combination with stresses like agitation or shaking can lead to protein aggregation, which could be detrimental for product quality. To mitigate the instability risk, silicone oil-free polymer syringes can serve as an alternative, such as syringes made of cyclic olefin polymers. For example, Werner et al. studied the stability of two therapeutic proteins filled in glass syringes and in silicone oil-free polymer syringes and found that high particle counts resulting from protein aggregation were observed in the glass syringes but not in polymer syringes after long-term storage [88]. Krayukhina and coworkers also discovered that silicone oil-free polymer syringes can provide more enhanced physical stability upon shipping and handling than glass syringes for therapeutic protein formulations such as abatacept (Orencia), a recombinant fusion protein, and adalimumab (Humira), a fully human recombinant immunoglobulin G1 [89]. However, it is worthwhile to point out that

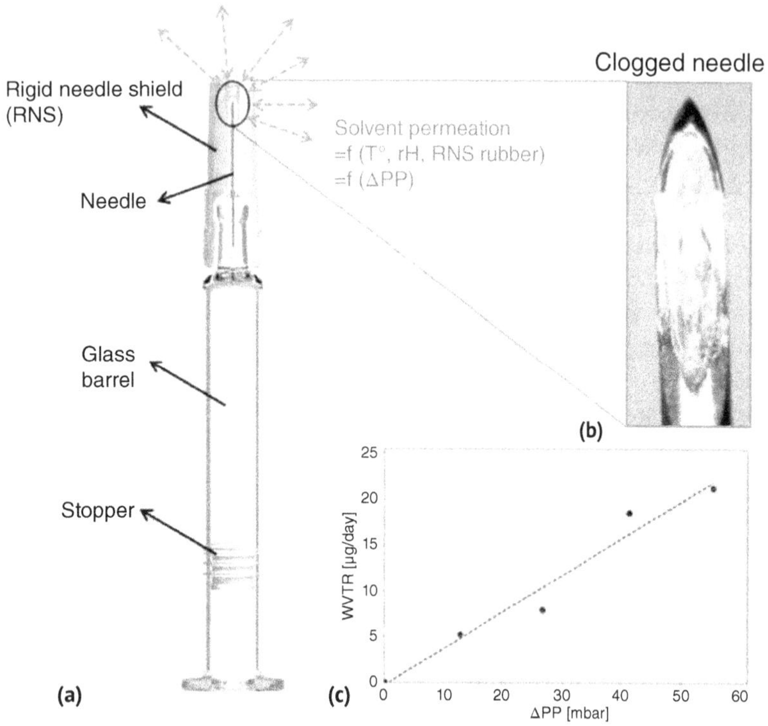

Figure 25.15 Illustration of the influence of water vapor transmission rate (WVTR) through the needle shield on clogging in staked-in needle prefilled syringes. (a) Diagram of a prefilled syringe; (b) a clogged needle; (c) correlation of partial vapor pressure difference (ΔPP) across the needle shield and WVTR. *Source*: Reproduced from [86]/with permission of Elsevier.

polymer syringes possess relatively weaker gas barrier properties in comparison to glass syringes, which might limit their usage as primary packaging components for oxygen-sensitive biopharmaceuticals [88].

Notably, Nakamura and colleagues developed a strategy for the prevention of protein oxidation by drug products in polymer syringes, through packaging control and process control [90]. It appears that the risk of oxidation-related instability arises from two sources: dissolved oxygen in the formulation and free radicals present in the polymer. It was found that the use of an oxygen absorber within the secondary packaging as a deoxygenated system helped control the oxygen level during storage, while utilizing steam sterilization mitigated the radical generation often incurred by irradiation sterilization. By using electron spin resonance, Kiminami et al. observed the formation of radicals in the syringe body after irradiation sterilization, which even persisted over the period of 12-week storage, and not surprisingly incurred instability for an EPO formulation during storage in polymer-based PFSs, including protein oxidation, aggregation, and particle formation [91]. Interestingly, steam sterilization was found as a better option for product stability than EO sterilization from a study on a model therapeutic protein, HSA, in polymer-based PFSs. Funatsu et al. reported that no chemical degradation was observed in the HAS formulation stored in steam-sterilized syringes, whereas EO adduction to cysteine and methionine in HAS reached about 45% in EO-sterilized syringes after 2 weeks of storage [92]. In summary, steam sterilization is the preferred method for polymer-based syringes when working with biopharmaceutical drugs highly sensitive to oxidation, particle formation, or aggregation. Adding a deoxygenated secondary packaging system provides further protection.

In a nutshell, developing an injectable drug–device combination product involves selecting appropriate primary packaging and optimizing the drug formulation for compatibility and shelf life.

25.11 Control of *N*-Nitrosamine in Drug Product

The discovery of *N*-nitrosamine impurities in several marketed drug products since 2018 has created a significant challenge for the pharmaceutical industry. This challenge includes the potential for market withdrawal or recall of life-saving products, stricter regulatory requirements for products awaiting approval, and the need for substantial resources for nitrosamine risk assessment across numerous drug substances, intermediates, drug products, and raw materials. For example, a

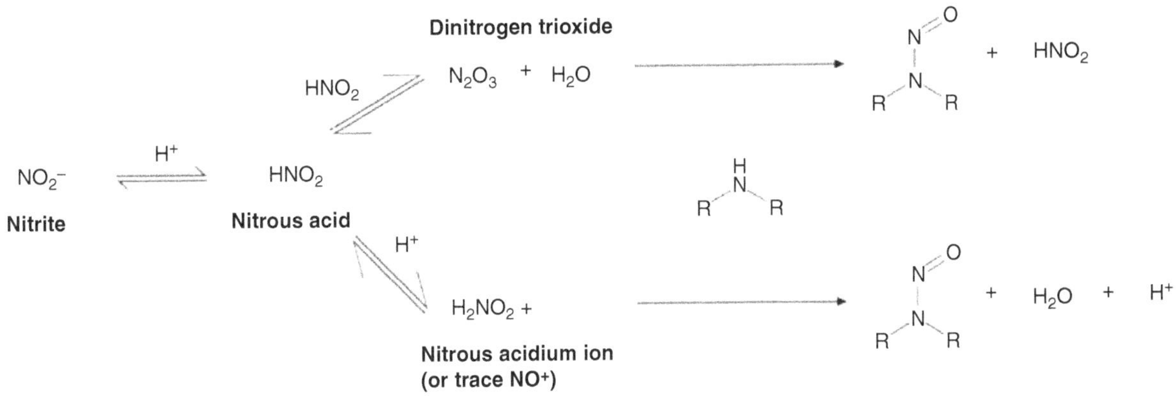

Figure 25.16 Scheme to form a *N*-nitrosamine from a secondary amine. *Source*: [94]/with permission of with permission of Elsevier.

regulatory guidance has been issued by the FDA to require market authorization holders to review all chemically synthesized commercial drug substances and drug products for the presence of *N*-nitrosamine impurities within a certain recommended timeline [93].

Certain *N*-nitrosamines, considered highly probable human carcinogens, can form by the acid-promoted reaction of secondary or tertiary amines with a nitrosating agent such as nitrite, as illustrated in Figure 25.16 [94]. It is well recognized that nitrites are common nitrosating impurities that have been found in some commonly used excipients at ppm levels, potentially leading to *N*-nitrosamine impurities generated in drug products during product manufacturing and storage. Using model amines of different chemical structures, solubilities, and p*K*a values, Moser and coworkers showed that *N*-nitrosamines can form in solid drug product formulations when nitrite is added [95]. Notably, the rate and extent of *N*-nitrosamine formation are influenced by drug solubility, nitrite levels, and microenvironmental pH. It is also interesting to point out that solid-state properties and the presence of water in the drug manufacturing process affect the rank order of *N*-nitrosamine formation: amorphous > wet granulation > direct compression > dry blends. In addition, tertiary amines are significantly less reactive with nitrite than secondary amines. The model study further revealed that the level of *N*-nitrosamine formation in solid dosage form plateaued significantly below the maximum theoretical yield based on the nitrite level. Thus, the risk assessment shall focus on whether the drug substance is a secondary or tertiary amine and the level of nitrite contamination from excipients, rather than searching for trace amine impurities.

It is worth reiterating that the presence of nitrite in excipients is considered one of the key risk factors because nitrite, under certain conditions, can react with vulnerable secondary or tertiary amine active ingredients to form *N*-nitrosamine in drug products. Boetzel and collaborators created a valuable database of nitrite levels in excipients for *N*-nitrosamine risk assessment in drug products [94]. From the database, they attained several major findings, including the following:

- Average nitrite content and batch-to-batch variability are distinct among excipient types.
- Nitrite contribution in solid dosage form is dominated by the bulking agents, leading to an average value of 1 μg/g nitrite in a typical formulation.
- Substantial differences exist in average nitrite content in batches from different excipient vendors, likely owing to differences in source materials or processing methods for excipient manufacturing.

From Figure 25.17, it is interesting to note that crospovidone, a common disintegrant, has the highest nitrite content (6.5 μg/g average) among eight widely used excipients for solid dosage formulations, whereas croscarmellose, another widely used disintegrant, has a relative lower level of nitrite (0.42 μg/g average). Therefore, choosing a proper type of excipient or selecting an excipient source with a low level of nitrite would reduce the *N*-nitrosamine presence or its formation in drug products.

Another control strategy, as proposed by Nanda et al., is to inhibit *N*-nitroamine formation through formulation approaches [96]. As illustrated by a model study, several antioxidants, including ascorbic acid, sodium ascorbate, α-tocopherol, caffeic acid, and ferulic acid, were shown to deliver greater than 80% inhibition when formulated at a 1% level [96]. Using bumetanide as a model secondary amine drug, Shakleya et al. discovered that keeping a basic microenvironmental pH in bumetanide tablets can also mitigate the nitrosating to occur and consequently result in the reduced formation of bumetanide nitrosamine, on the top to the use of an antioxidant [97].

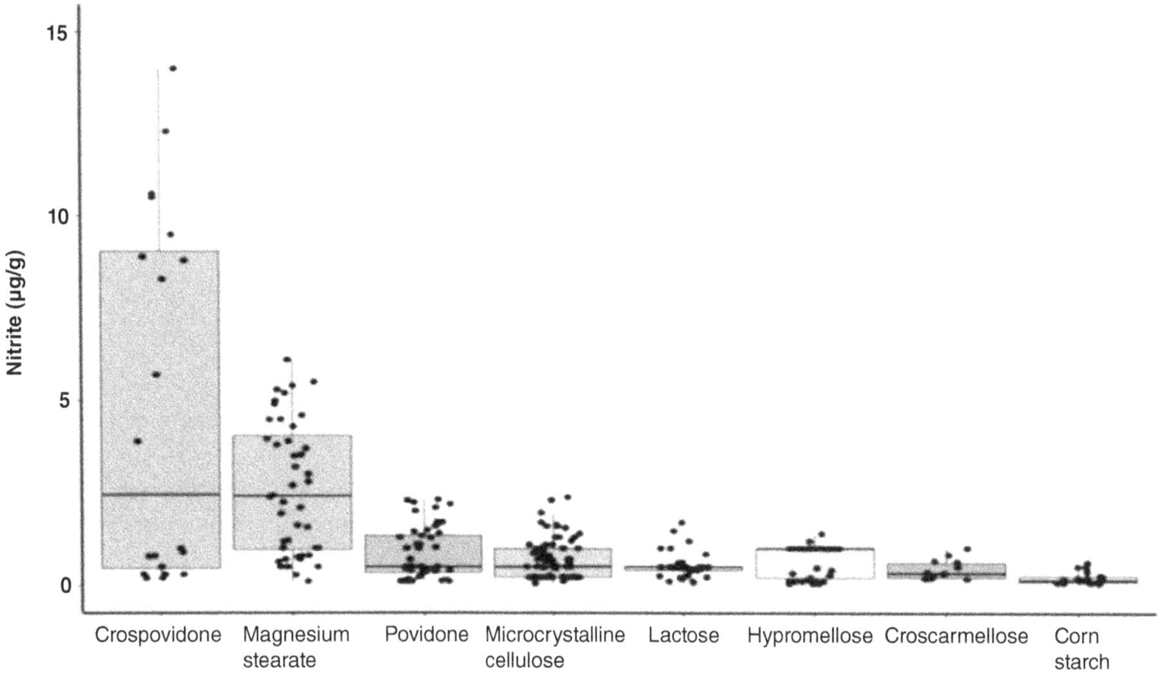

Figure 25.17 Nitrite level for selected excipients commonly used in solid dosage forms. *Source*: [94]/with permission of with permission of Elsevier.

25.12 Summary

The material properties of active pharmaceutical ingredients, excipients, and packaging components significantly affect the quality, processability, stability, and performance of pharmaceutical products. Linking material properties to product characteristics is crucial in product development to fully understand the chemistry, manufacturing, and controls involved. Preformulation work focuses on characterizing the physical and chemical properties of the active ingredient to guide the development of a product strategy that meets targeted profiles. Formulation and process development for a specific product must prioritize key product profile components and quality attributes, such as:

- solubility and physical stability for a solid dispersion,
- tensile strength and gelling for abuse-deterrent dosage forms,
- material integrity and sustained release for DESs,
- aerodynamic particle size and delivery performance of DPIs,
- physical stabilization against aggregation in lyophilized biopharmaceuticals,
- permeability enhancement for oral delivery of peptides,
- compatibility with primary packaging components for injectable drug–device combination products, and
- proper control of *N*-nitrosamine impurities in drug products in accordance with risk assessments, especially on the structure characteristics of the API and the sources of excipients.

References

1 ICH Harmonised Tripartite Guideline. *Pharmaceutical Development Q8 (R2). International Conference on Harmonisation of Technical Requirements for Registration of Pharmaceuticals for Human Use*; 2009. https://database.ich.org/sites/default/files/Q8%28R2%29%20Guideline.pdf (accessed 18 April 2025).

2 Mortko CJ, Sheth AR, Variankaval N, Li L, Farrer BT. Risk assessment and physicochemical characterization of a metastable dihydrate API phase for intravenous formulation development. *J Pharm Sci* 2010; 99(12): 4973–4981.

3 Kearney AS, Patel K, Palepu NR. Preformulation studies to aid in the development of a ready-to-use injectable solution of the antitumor agent, topotecan. *Int J Pharm* 1996; 127(2): 229–237.

4 Borhade V, Pathak S, Sharma S, Patravale V. Clotrimazole nanoemulsion for malaria chemotherapy. Part I: Preformulation studies, formulation design and physicochemical evaluation. *Int J Pharm* 2012; 431(1): 138–148.

5 Nunes C, Han J, Monkhouse DC, Suryanarayanan R. Preformulation studies to meet the challenges in the manufacture of betaine solid dosage form. *J Pharm Sci* 2004; 93(1): 38–47.

6 Dong Y, Ng WK, Surana U, Tan RB. Solubilization and preformulation of poorly water soluble and hydrolysis susceptible N-epoxymethyl-1,8-naphthalimide (ENA) compound. *Int J Pharm* 2008; 356(1): 130–136.

7 Cappelli A, Travagli V, Zanardi I, Anzini M, Giorgi G, Donati A, Aggravi M, Casolaro M, Fresta M, Paccagnini E, Makovec F. Physicochemical and biopharmaceutical characterization of endo-2-(8-methyl-8-azabicyclo[3.2.1]oct-3-yl)-2,3-dihydro-1H-benz[e]isoindol-1-one (CR3124) a novel potent 5-HT3 receptor antagonist. *J Pharm Sci* 2006; 95(12): 2706–2721.

8 Fan Y, Yang M, Wang Y, Li Y, Zhou Y, Chen X, Shan L, Wei J, Gao C. Preformulation characterization and in vivo absorption in beagle dogs of JFD, a novel anti-obesity drug for oral delivery. *Drug Dev Ind Pharm* 2015; 41(5): 801–811.

9 Qin X, Frech P. Liquid chromatography/mass spectrometry (LC/MS) identification of photooxidative degradates of crystalline and amorphous MK-912. *J Pharm Sci* 2001; 90(7): 833–844.

10 Salnikova MS, Joshi SB, Rytting JH, Warny M, Middaugh CR. Preformulation studies of Clostridium difficile toxoids A and B. *J Pharm Sci* 2008; 97(10): 4194–4207.

11 Hu L, Joshi SB, Liyanage MR, Pansalawatta M, Alderson MR, Tate A, Robertson G, Maisonneuve J, Volkin DB, Middaugh CR. Physical characterization and formulation development of a recombinant pneumolysoid protein-based pneumococcal vaccine. *J Pharm Sci* 2013; 102(2): 387–400.

12 Serajuddin A, Thakur AB, Ghoshal RN, Fakes MG, Ranadive SA, Morris KR, Varia SA. Selection of solid dosage form composition through drug-excipient compatibility testing. *J Pharm Sci* 1999; 88(7): 696–704.

13 Wasylaschuk WR, Harmon PA, Wagner G, Harman AB, Templeton AC, Xu H, Reed RA. Evaluation of hydroperoxides in common pharmaceutical excipients. *J Pharm Sci* 2007; 96(1): 106–116.

14 Wu Y, Dali M, Gupta A, Raghavan K. Understanding drug-excipient compatibility: oxidation of compound A in a solid dosage form. *Pharm Dev Technol* 2009; 14(5): 556–564.

15 Hartauer KJ, Arbuthnot GN, Baertschi SW, Johnson RA, Luke WD, Pearson NG, Rickard EC, Tingle CA, Tsang PK, Wiens RE. Influence of peroxide impurities in povidone and crospovidone on the stability of raloxifene hydrochloride in tablets: identification and control of an oxidative degradation product. *Pharm Dev Technol* 2000; 5(3): 303–310.

16 Ha E, Wang W, Wang YJ. Peroxide formation in polysorbate 80 and protein stability. *J Pharm Sci* 2002; 91(10): 2252–2264.

17 Bhambhani A, Kissmann JM, Joshi SB, Volkin DB, Kashi RS, Middaugh CR. Formulation design and high-throughput excipient selection based on structural integrity and conformational stability of dilute and highly concentrated IgG1 monoclonal antibody solutions. *J Pharm Sci* 2012; 101(3): 1120–1135.

18 Cheng W, Joshi SB, He F, Brems DN, He B, Kerwin BA, Volkin DB, Middaugh CR. Comparison of high-throughput biophysical methods to identify stabilizing excipients for a model IgG2 monoclonal antibody: conformational stability and kinetic aggregation measurements. *J Pharm Sci* 2012; 101(5): 1701–1720.

19 Thakkar SV, Joshi SB, Jones ME, Sathish HA, Bishop SM, Volkin DB, Middaugh CR. Excipients differentially influence the conformational stability and pretransition dynamics of two IgG1 monoclonal antibodies. *J Pharm Sci* 2012; 101(9): 3062–3077.

20 Meng F, Trivino A, Prasad D, Chauhan H. Investigation and correlation of drug polymer miscibility and molecular interactions by various approaches for the preparation of amorphous solid dispersions. *Eur J Pharm Sci* 2015; 71: 12–24.

21 Six K, Berghmans H, Leuner C, Dressman J, Van Werde K, Mullens J, Benoist L, Thimon M, Meublat L, Verreck G, Peeters J. Characterization of solid dispersions of itraconazole and hydroxypropyl methylcellulose prepared by melt extrusion, Part II. *Pharm Res* 2003; 20(7): 1047–1054.

22 Six K, Verreck G, Peeters J, Brewster M, Van Den Mooter GV. Increased physical stability and improved dissolution properties of itraconazole, a class II drug, by solid dispersions that combine fast- and slow-dissolving polymers. *J Pharm Sci* 2004; 93(1): 124–131.

23 Hughey JR, DiNunzio JC, Bennett RC, Brough C, Miller DA, Ma H, Williams III RO, McGinity JW. Dissolution enhancement of a drug exhibiting thermal and acidic decomposition characteristics by fusion processing: a comparative study of hot melt extrusion and KinetiSol® dispersing. *AAPS PharmSciTech* 2010; 11(2): 760–774.

24 Meng F, Dave V, Chauhan H. Qualitative and quantitative methods to determine miscibility in amorphous drug-polymer systems. *Eur J Pharm Sci* 2015; 77: 106–111.

25 Fang LY, Harris D, Krishna G, Moton AE, Prestipino RC, Steinman M, Wan J, Waskin HA, inventors; Moton Jr Allen E, Prestipino Russell C, assignee. High density compositions containing posaconazole and formulations comprising the same. United States patent application US 12/999,547. 15 April 2009.

26 Friesen DT, Shanker R, Crew M, Smithey DT, Curatolo WJ, Nightingale JA. Hydroxypropyl methylcellulose acetate succinate-based spray-dried dispersions: an overview. *Mol Pharm* 2008; 5(6): 1003–1019.

27 Bühler V. Kollidon® Polyvinylpyrrolidone excipients for the pharmaceutical industry. Ludwigshafen, Germany: BASF SE; 2008.

28 Curatolo W, Nightingale JA, Herbig SM. Utility of hydroxypropyl methylcellulose acetate succinate (HPMCAS) for initiation and maintenance of drug supersaturation in the GI milieu. *Pharm Res* 2009; 26(6): 1419–1431.

29 Prasad D, Chauhan H, Atef E. Amorphous stabilization and dissolution enhancement of amorphous ternary solid dispersions: combination of polymers showing drug-polymer interaction for synergistic effects. *J Pharm Sci* 2014; 103(11): 3511–3523.

30 Rosenberg J, Reinhold U, Liepold B, Berndl G, Breitenbach J, Alani LL, Ghosh S, inventors; Abbott Laboratories, assignee. Solid pharmaceutical dosage formulation. US patent 8377952 B2. 19 February 2013.

31 Zha J, inventor; Vertex Pharmaceuticals Incorporated, assignee. Pharmaceutical compositions for use in the treatment of cystic fibrosis. US patent 20140221424 A1. 7 August 2014.

32 Bittorf KJ, Katstra JP, Gaspar F, inventors; Vertex Pharmaceuticals Incorporated, assignee. Pharmaceutical compositions. US patent 20090247468 A1. 1 October 2009.

33 Shah N, Iyer RM, Mair HJ, Choi DS, Tian H, Diodone R, Fähnrich K, Pabst-Ravot A, Tang K, Scheubel E, Grippo JF. Improved human bioavailability of vemurafenib, a practically insoluble drug, using an amorphous polymer-stabilized solid dispersion prepared by a solvent-controlled coprecipitation process. *J Pharm Sci* 2013; 102(3): 967–981.

34 Al-Obaidi H, Lawrence MJ, Shah S, Moghul H, Al-Saden N, Bari F. Effect of drug–polymer interactions on the aqueous solubility of milled solid dispersions. *Int J Pharm* 2013; 446(1): 100–105.

35 Food and Drug Administration. Abuse-deterrent opioids—evaluation and labeling. Guidance for industry. April 2015.

36 Cicero TJ, Ellis MS. Effect of abuse-deterrent formulation of OxyContin. *New Engl J Med* 2012; 367(2): 187–189.

37 Pinto E, Yang H, Pittman A, Tewari D, Durig T. Klucel HPC in hot melt extrusion applications. http://www.ashland.com/Ashland/Static/Documents/ASI/PC_10539_Klucel_HPC_Hot-melt_Extrusion.pdf (accessed 1 February 2016).

38 Quinten T, De Beer T, Almeida A, Vlassenbroeck J, Van Hoorebeke L, Remon JP, Vervaet C. Development and evaluation of injection-molded sustained-release tablets containing ethylcellulose and polyethylene oxide. *Drug Dev Ind Pharm* 2011; 37(2): 149–159.

39 Maddineni S, Battu SK, Morott J, Soumyajit M, Repka MA. Formulation optimization of hot-melt extruded abuse deterrent pellet dosage form utilizing design of experiments. *J Pharm Pharmacol* 2014; 66(2): 309–322.

40 Bartholomaus JH, Kugelmann H, Arkenau-Marić E, inventors; Grunenthal Gmbh, assignee. Abuse-proofed dosage form. US Patent 8309060 B2. 13 November 2012.

41 Bartholomaus JH, Arkenau-Marić E, Galia E. Opioid extended-release tablets with improved tamper-resistant properties. *Expert Opin Drug Del* 2012; 9(8): 879–891.

42 Bartholomaus J, Ashworth J, Stahlberg HJ. Innovative formulation technology protecting intended drug action. *Drug Dev Deliv* 2012; 12(8): 69–75.

43 McKenna WH, Mannion RO, O'Donnell EP, Huang HH, inventors; Purdue Pharma LP, assignee. Tamper resistant dosage forms. US Patent 8894988 B2. 25 November 2014.

44 Simon K, Worthy SL, Barnes MC, Tarbell B. Abuse-deterrent formulations: transitioning the pharmaceutical market to improve public health and safety. *Ther Adv Drug Saf* 2015; 6(2): 67–79.

45 Hemmingsen PH, Haahr AM, Gunnergaard C, Cardot JM. Development of a new type of prolonged release hydrocodone formulation based on Egalet® ADPREM technology using in vivo–in vitro correlation. *Pharmaceutics* 2011; 3(1): 73–87.

46 Bartholomäus J, Schwier S, Brett M. New abuse deterrent formulation (ADF) technology for immediate-release opioids. *Drug Dev Del* 2013; 13: 76–81.

47 Moorman-Li R, Motycka CA, Inge LD, Congdon JM, Hobson S, Pokropski B. A review of abuse-deterrent opioids for chronic nonmalignant pain. *Pharm Ther* 2012; 37(7): 412.

48 Kamath KR, Barry JJ, Miller KM. The Taxus drug-eluting stent: a new paradigm in controlled drug delivery. *Adv Drug Delivery Rev* 2006; 58(3): 412–436.

49 Higuchi T. Mechanism of sustained-action medication. Theoretical analysis of rate of release of solid drugs dispersed in solid matrices. *J Pharm Sci* 1963; 52(12): 1145–1149.

50 Waseda K, Ako J, Yamasaki M, Koizumi T, Sakurai R, Hongo Y, Koo BK, Ormiston J, Worthley SG, Whitbourn RJ, Walters DL. Impact of polymer formulations on neointimal proliferation after zotarolimus-eluting stent with different polymers: insights from the RESOLUTE trial. *Circ Cardiovasc Interv* 2011; 4(3): 248–255.

51 Wilcox JN. Importance of polymer biocompatibility for drug eluting stent (DES) outcomes. *BioInterface Conference* 17–19 October 2010.

52 Berger HL. Using ultrasonic spray nozzles to coat drug-eluting stents. *Med Device Technol* 2005; 17(9): 44–49.

53 Muramatsu T, Onuma Y, Zhang YJ, Bourantas CV, Kharlamov A, Diletti R, Farooq V, Gogas BD, Garg S, García-García HM, Ozaki Y. Progress in treatment by percutaneous coronary intervention: the stent of the future. *Rev Esp Cardiol (Engl Ed)* 2013; 66(6): 483–496.

54 Telko MJ, Hickey AJ. Dry powder inhaler formulation. *Respir Care* 2005; 50(9): 1209–1227.

55 Larhrib H, Zeng XM, Martin GP, Marriott C, Pritchard J. The use of different grades of lactose as a carrier for aerosolised salbutamol sulphate. *Int J Pharm* 1999; 191(1): 1–14.

56 Donovan MJ, Smyth HD. Influence of size and surface roughness of large lactose carrier particles in dry powder inhaler formulations. *Int J Pharm* 2010; 402(1): 1–9.

57 Harjunen P, Lehto VP, Martimo K, Suihko E, Lankinen T, Paronen P, Järvinen K. Lactose modifications enhance its drug performance in the novel multiple dose Taifun DPI. *Eur J Pharm Sci* 2002; 16(4): 313–321.

58 Zeng XM, Martin GP, Marriott C, Pritchard J. The influence of carrier morphology on drug delivery by dry powder inhalers. *Int J Pharm* 2000; 200(1): 93–106.

59 Larhrib H, Martin GP, Marriott C, Prime D. The influence of carrier and drug morphology on drug delivery from dry powder formulations. *Int J Pharm* 2003; 257(1): 283–296.

60 Kaialy W, Momin MN, Ticehurst MD, Murphy J, Nokhodchi A. Engineered mannitol as an alternative carrier to enhance deep lung penetration of salbutamol sulphate from dry powder inhaler. *Colloids Surf B Biointerfaces* 2010; 79(2): 345–356.

61 Kaialy W, Nokhodchi A. Engineered mannitol ternary additives improve dispersion of lactose-salbutamol sulphate dry powder inhalations. *AAPS J* 2013; 15(3): 728–743.

62 Schiavone H, Palakodaty S, Clark A, York P, Tzannis ST. Evaluation of SCF-engineered particle-based lactose blends in passive dry powder inhalers. *Int J Pharm* 2004; 281(1): 55–66.

63 Yang TT, inventor; Schering Corporation, assignee. Preparation of powder agglomerates. US patent US 8173172 B2. 8 May 2012.

64 Rabinowitz JD, Wensley M, Lloyd P, Myers D, Shen W, Lu A, Hodges C, Hale R, Mufson D, Zaffaroni A. Fast onset medications through thermally generated aerosols. *J Pharmacol Exp Ther* 2004; 309(2): 769–775.

65 Myers DJ, Timmons RD, Lu AT, Hale RL, Solas DW, Soni P, Rabinowitz JD. The effect of film thickness on thermal aerosol generation. *Pharm Res* 2007; 24(2): 336–342.

66 Kim AI, Akers MJ, Nail SL. The physical state of mannitol after freeze-drying: Effects of mannitol concentration, freezing rate, and a noncrystallizing cosolute. *J Pharm Sci* 1998; 87(8): 931–935.

67 Pyne A, Surana R, Suryanarayanan R. Crystallization of mannitol below Tg′ during freeze-drying in binary and ternary aqueous systems. *Pharm Res* 2002; 19(6): 901–908.

68 Chang L, Shepherd D, Sun J, Ouellette D, Grant KL, Tang XC, Pikal MJ. Mechanism of protein stabilization by sugars during freeze-drying and storage: native structure preservation, specific interaction, and/or immobilization in a glassy matrix? *J Pharm Sci* 2005; 94(7): 1427–1444.

69 Wang DQ, Hey JM, Nail SL. Effect of collapse on the stability of freeze-dried recombinant factor VIII and alpha-amylase. *J Pharm Sci* 2004; 93(5): 1253–1263.

70 Duddu SP, Dal Monte PR. Effect of glass transition temperature on the stability of lyophilized formulations containing a chimeric therapeutic monoclonal antibody. *Pharm Res* 1997; 14(5): 591–595.

71 De Beer TR, Wiggenhorn M, Hawe A, Kasper JC, Almeida A, Quinten T, Friess W, Winter G, Vervaet C, Remon JP. Optimization of a pharmaceutical freeze-dried product and its process using an experimental design approach and innovative process analyzers. *Talanta* 2011; 83(5): 1623–1633.

72 Cao W, Xie Y, Krishnan S, Lin H, Ricci M. Influence of process conditions on the crystallization and transition of metastable mannitol forms in protein formulations during lyophilization. *Pharm Res* 2013; 30(1): 131–139.

73 Liao X, Krishnamurthy R, Suryanarayanan R. Influence of processing conditions on the physical state of mannitol—implications in freeze-drying. *Pharm Res* 2007; 24(2): 370–376.

74 Johnson RE, Kirchhoff CF, Gaud HT. Mannitol-sucrose mixtures—versatile formulations for protein lyophilization. *J Pharm Sci* 2002; 91(4): 914–922.

75 Lueckel B, Helk B, Bodmer D, Leuenberger H. Effects of formulation and process variables on the aggregation of freeze-dried interleukin-6 (IL-6) after lyophilization and on storage. *Pharm Dev Technol* 1998; 3(3): 337–346.

76 Han Y, Jin BS, Lee SB, Sohn Y, Joung JW, Lee JH. Effects of sugar additives on protein stability of recombinant human serum albumin during lyophilization and storage. *Arch Pharm Res* 2007; 30(9): 1124–1131.

77 Izutsu K, Yoshioka S, Kojima S. Physical stability and protein stability of freeze-dried cakes during storage at elevated temperatures. *Pharm Res* 1994; 11(7): 995–999.

78 Piedmonte DM, Summers C, McAuley A, Karamujic L, Ratnaswamy G. Sorbitol crystallization can lead to protein aggregation in frozen protein formulations. *Pharm Res* 2007; 24(1): 136–146.

79 Breen ED, Curley JG, Overcashier DE, Hsu CC, Shire SJ. Effect of moisture on the stability of a lyophilized humanized monoclonal antibody formulation. *Pharm Res* 2001; 18(9): 1345–1353.

80 Hsu CC, Ward CA, Pearlman R, Nguyen HM, Yeung DA, Curley JG. Determining the optimum residual moisture in lyophilized protein pharmaceuticals. *Dev Biol Stand* 1991; 74: 255–270.

81 Buckley ST, Bækdal TA, Vegge A, Maarbjerg SJ, Pyke C, Ahnfelt-Rønne J, Madsen KG, Schéele SG, Alanentalo T, Kirk RK, Pedersen BL. Transcellular stomach absorption of a derivatized glucagon-like peptide-1 receptor agonist. *Sci Transl Med* 2018; 10(467): eaar7047.

82 Vilhelmsen T, Eliasen H, Hansen T, inventors; Novo Nordisk AS, assignee. Compositions of GLP-1 peptides and preparation thereof. United States patent US 11,759,501. 19 September 2023.

83 Tran H, Dogra M, Huang S, Aihara E, ElSayed M, Aburub A. Development and evaluation of C10 and SNAC erodible tablets for gastric delivery of a GIP/GLP1 peptide in monkeys. *Int J Pharm* 2024; 650: 123680.

84 Brayden DJ, Maher S. Transient Permeation Enhancer®(TPE®) technology for oral delivery of octreotide: a technological evaluation. *Expert Opin Drug Deliv* 2021; 18(10): 1501–1512.

85 DeGrazio F, Paskiet D. Injectable combination product development: Facilitating risk-based assessments for efficiency and patient centric outcomes. *J Pharm Sci* 2020; 109(7): 2101–2115.

86 De Bardi M, Müller R, Grünzweig C, Mannes D, Rigollet M, Bamberg F, Jung TA, Yang K. Clogging in staked-in needle pre-filled syringes (SIN-PFS): influence of water vapor transmission through the needle shield. *Eur J Pharm Biopharm* 2018; 127: 104–111.

87 Scheler S, Knappke S, Schulz M, Zuern A. Needle clogging of protein solutions in prefilled syringes: a two-stage process with various determinants. *Eur J Pharm Biopharm* 2022; 176: 188–198.

88 Werner BP, Schöneich C, Winter G. Silicone oil-free polymer syringes for the storage of therapeutic proteins. *J Pharm Sci* 2019; 108(3): 1148–1160.

89 Krayukhina E, Tsumoto K, Uchiyama S, Fukui K. Effects of syringe material and silicone oil lubrication on the stability of pharmaceutical proteins. *J Pharm Sci* 2015; 104(2): 527–535.

90 Nakamura K, Abe Y, Kiminami H, Yamashita A, Iwasaki K, Suzuki S, Yoshino K, Dierick W, Constable K. A strategy for the prevention of protein oxidation by drug product in polymer-based syringes. *PDA J Pharma Sci Tech* 2015; 69(1): 88–95.

91 Kiminami H, Krueger AB, Abe Y, Yoshino K, Carpenter JF. Impact of sterilization method on protein aggregation and particle formation in polymer-based syringes. *J Pharm Sci* 2017; 106(4): 1001–1007.

92 Funatsu K, Kiminami H, Abe Y, Carpenter JF. Impact of ethylene oxide sterilization of polymer-based prefilled syringes on chemical degradation of a model therapeutic protein during storage. *J Pharm Sci* 2019; 108(1): 770–774.

93 US Food and Drug Administration. Guidance for industry, Control of nitrosamine impurities in human drugs. https://www.fda.gov/drugs/guidance-compliance-regulatory-information/guidances-drugs (accessed 22 November 2024).

94 Boetzel R, Schlingemann J, Hickert S, Korn C, Kocks G, Luck B, Blom G, Harrison M, Francois M, Allain L, Wu Y. A nitrite excipient database: a useful tool to support N-nitrosamine risk assessments for drug products. *J Pharm Sci* 2023; 112(6): 1615–1624.

95 Moser J, Ashworth IW, Harris L, Hillier MC, Nanda KK, Scrivens G. N-nitrosamine formation in pharmaceutical solid drug products: Experimental observations. *J Pharm Sci* 2023; 112(5): 1255–1267.

96 Nanda KK, Tignor S, Clancy J, Marota MJ, Allain LR, D'Addio SM. Inhibition of N-nitrosamine formation in drug products: a model study. *J Pharm Sci* 2021; 110(12): 3773–3775.

97 Shakleya D, Asmelash B, Alayoubi A, Abrigo N, Mohammad A, Wang J, Zhang J, Yang J, Marzan TA, Li D, Shaklah M. Bumetanide as a model NDSRI substrate: n-nitrosobumetanide impurity formation and its inhibition in bumetanide tablets. *J Pharm Sci* 2023; 112(12): 3075–3087.

26

Quality by Design

26.1 Introduction

This chapter provides an overview of the role of solid-state chemistry in quality by design (QbD). Solid-state chemistry plays a critical role in many of the steps required to achieve QbD. Put another way, an intimate knowledge of the solid-state chemistry of the drug and the formulation is needed to achieve QbD. QbD is defined by the Food and Drug Administration (FDA) and ICH Q8 as "a systematic approach to development that begins with predefined objectives and emphasizes product and process understanding and process control, based on sound science and quality risk management." QbD is achieved by building in quality from the development phase and throughout a product's life cycle. In a QbD system, the product is designed to meet patient requirements, the process is designed to meet product quality attributes, critical sources of process variability are identified and controlled, and the process is continually monitored to ensure consistent quality over time.

26.2 QbD Wheel

Figure 26.1 shows a slightly modified QbD wheel [1]. This wheel is often used to illustrate the activities involved in achieving QbD.

Product design (on the inner portion at 3 o'clock) is the first step in QbD. This step begins with the first manufactured toxicological and clinical supplies. Control is needed, as discussed below, because there must be consistency and low variability between different lots used for toxicological studies and clinical studies. In this step, the product is designed with the patient's needs in mind. For example, for an oral dosage form of a poorly soluble drug, a product that is able to dissolve in the intestinal tract must be designed. For lopinavir/ritonavir, an important protease inhibitor, a solid dosage form that was stable and could be easily handled by patients with HIV was designed. Initially, a gel-cap containing ritonavir was designed, but this dosage form had to be stored at refrigerator temperatures due to the instability of the drug. Later, a tablet that was stable at ambient temperatures was developed using amorphous lopinavir/ritonavir.

As indicated in Figure 26.1, product design involves selecting the right polymorph/solid form, the right dosage form, and the correct components and involves designing the product with the desired stability. Product design also requires an understanding of the potential sources of variability and how to control variability. Often, product design can be achieved using small-scale experiments, especially if the focus is on understanding the solid-state chemistry, including which excipients need to be inert and not interact. For lopinavir/ritonavir, an amorphous dispersion with a high glass transition temperature achieved the desired stability and avoided the need to store the product in the refrigerator.

Drs. Byrn and Henck from Purdue University and SSCI, respectively, have published a paper outlining an accelerated development strategy [2]. This paper shows that it is possible to do some optimization of the solid form even in an accelerated timeframe. Form optimization generally follows the procedure outlined in the ICH Q6A guidance on specifications. Similarly, a team from Merck published a paper in the *Journal of Medicinal Chemistry* [3]. This review presented a convincing rationale for the early pursuit of the optimum solid form. They pointed out that chemists currently focus on properties and structure and that animal studies are routinely performed quite early in discovery/development. Clearly, a product must be designed that can be administered to animals in large enough doses to determine toxicity and establish safety margins for human trials. Early optimization or partial optimization of formulation is becoming more and

Solid-State Materials in Pharmaceutical Chemistry: Properties, Characterization, and Applications, Second Edition. Stephen R. Byrn, George Zografi and Xiaoming (Sean) Chen.

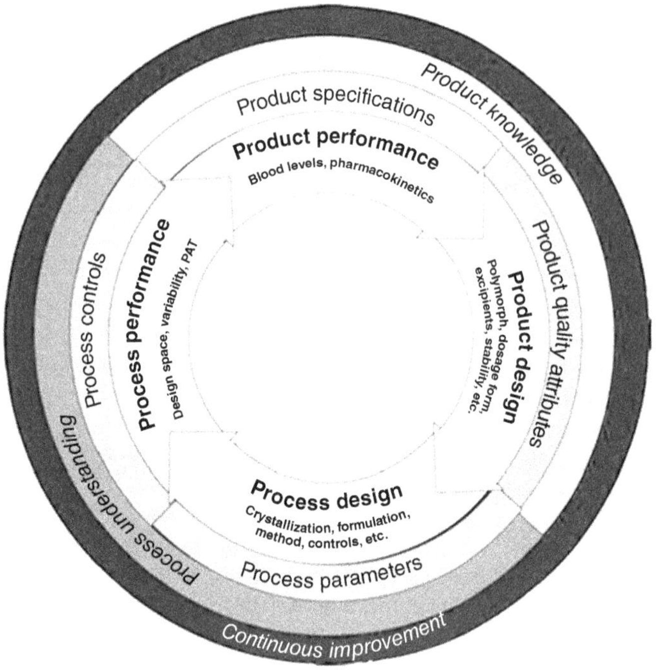

Figure 26.1 QbD wheel modified to illustrate the importance of polymorphism in the product design step.

more important as the number of poorly soluble drug candidates increases. In fact, there is now a significant amount of information that indicates that, in some cases, without formulation optimization, toxicological studies to determine dose-limiting toxicity are not possible. In such cases, the compound cannot move forward. One interesting suggestion in the Merck approach is the idea that flexibility must be maintained in the screens used for solid-state forms. These screens must be tailored to the target for the drug, since one size will not fit all. The Merck approach also suggests a continuum of formulation options for poorly soluble compounds, ranging from salts to a suspension to cosolvents/surfactants to lipids to nanoparticles to amorphous forms. In some cases, it may be expedient to go directly to amorphous formulations or salts for early trials to save time and ensure maximum solubility. These approaches are also consistent with a review of QbD by Lawrence Yu of the Office of Generic Drugs [4]. Yu indicated that for BCS class II drugs (Low Solubility/High Permeability), the formulation should be designed to overcome dissolution rate problems using particle size reduction, salts, precipitation inhibitors, metastable forms, solid dispersions, complexation, or lipid technologies.

Next is process design (inner portion at 6 o'clock). A process for making the designed product needs to be developed. This process must, at a minimum, first work with an acceptable level of variability on a small scale to prepare clinical supplies. Every effort should be made to select excipients that do not influence the variability of the product. An understanding of how the process affects critical quality attributes must be developed. Also, process controls must be initiated. The overall goal is to reduce the variability of the critical quality attributes in this step. If possible, a design space should begin to be developed. The design space is usually fully developed near commercialization, but certainly an early design space can be approached even at this stage. For example, if a melt extrusion process is being used to make an amorphous product, various parameters need to be controlled, including temperatures, extrusion speed, temperature of cooling, and milling speed.

In some cases, processes and products are linked. Docherty, Kougoulos, and Horspool reviewed and defined Pharmaceutical Material Science [5]. They suggested that this area is focused on streamlined selection of active pharmaceutical ingredients (APIs), solid forms, reduced numbers of complex formulations, rapid drug product design, and intellectual property creation. They also indicated that the particle engineering toolbox contains, in addition to conventional crystallization methods, five processes that produce unique solid materials: (i) supercritical fluid technologies; (ii) mixing intensification process technologies; (iii) segmented flow tubular reactors; (iv) droplet to particle methods; and (v) solution atomization sonocrystallization technologies.

Next, the performance of the process is monitored (inner portion at 9 o'clock). Initial monitoring may include both online/inline instruments and offline measurements. For example, a crystallization process can be monitored online/inline using Raman [6] or infrared spectroscopy, UV/Vis spectrometry, and in situ particle size analysis. Offline monitoring is usually done using powder X-ray diffraction. Blending, drying, and granulation can also be monitored using a range of techniques led by near-infrared spectroscopy (NIR). Presumably, the lopinavir/ritonavir extrusion process is monitored by following temperatures, extrusion speed, temperature of cooling, and milling speed. Additionally, it is likely that X-ray diffraction is used to make sure the product is amorphous, and it is possible that the T_g can be determined using differential scanning calorimetry (DSC). Other parameters may also be addressed.

Finally, the product performance is monitored by measuring dissolution rates and the blood levels in animals and humans (toxicokinetics and pharmacokinetics). Additionally, an in vitro in vivo correlation (IVIVC) dissolution test may be developed to provide a laboratory method to obtain a preliminary measurement of product performance. The product performance must show low variability, and the process must be validated to the extent that it is reasonably safeguarded from sources of variability.

As shown on the wheel, the second level of QbD involves identifying the product critical quality attributes (3 o'clock on the second circle). A strategy for determining product quality attributes was outlined in the Example Quality Overall Summary provided on the FDA website by the Office of Generic Drugs and is reproduced as Figure 26.2. This summary does not contain data but is meant to provide information to be used in the justification of a drug product. Figure 26.2 shows a strategy to identify critical quality attributes.

The far-right panel of Figure 26.2 shows desired product quality. The five items listed here are similar to the product design. Thus, the product is designed to have purity, content uniformity, acceptable dissolution, acceptable stability, and acceptable tablet characteristics. In this hypothetical example, tablets with these parameters will meet the desired product

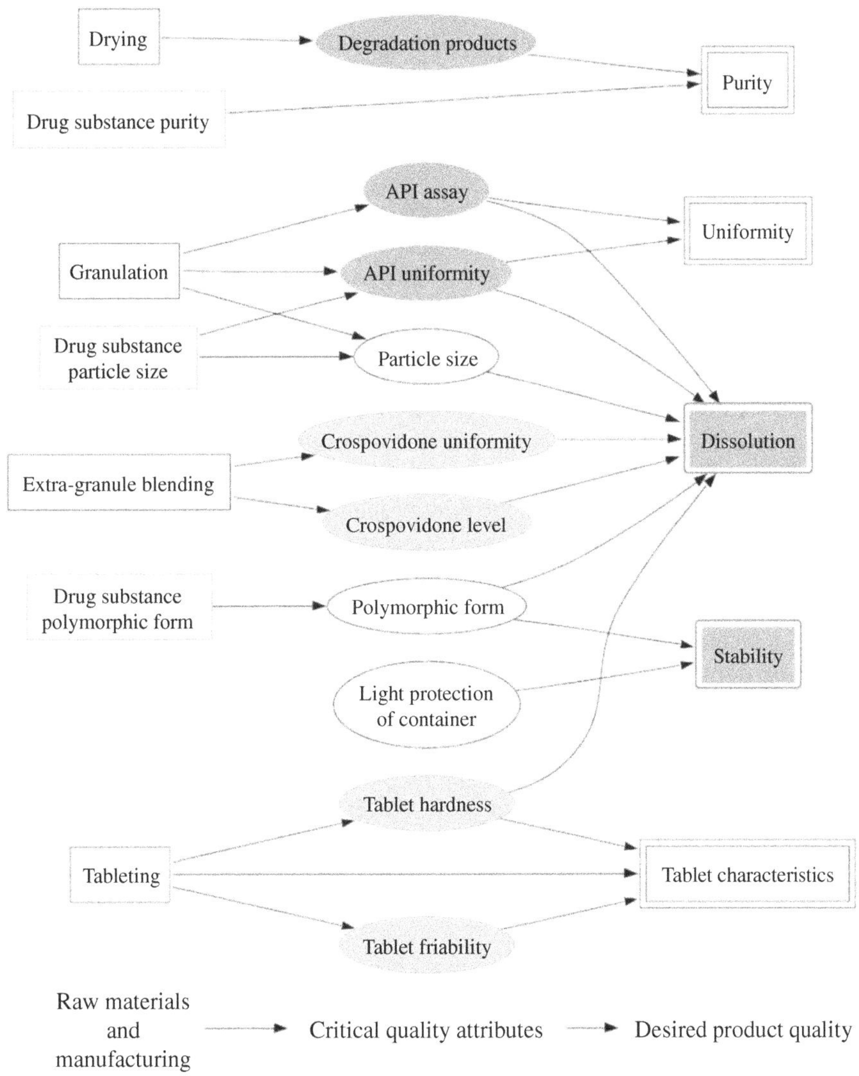

Figure 26.2 FDA strategy for identifying critical quality attributes.

quality. For example, lopinavir/ritonavir tablets must be of acceptable purity, have acceptable content uniformity, and especially have acceptable dissolution and stability. Finally, the tablets must have acceptable performance for the consumer.

The middle panel of Figure 26.2 lists the critical quality attributes that must be achieved/addressed to achieve the desired product quality. This section is based on a wet granulation process. The arrows in this diagram illustrate which attributes affect the desired product quality. This is exactly what is required by the QbD Wheel. Thus, degradation products and drug substance purity affect final purity. API assay and API uniformity in the granulation, which is in turn related to blend uniformity in the pregranulation blend, affect uniformity (content uniformity) in the drug product. Drug product dissolution is affected by API assay and API uniformity in that the amount dissolved depends on these API attributes. The particle size of the API determines the dissolution rate. Also, the uniformity and level of disintegrant determines dissolution rate. The polymorphic form is absolutely critical as it affects the dissolution rate and stability. In this particular example, the light protection provided by the container is critical to stability because the API is sensitive to light-induced degradation. Tablet hardness is related to both the dissolution rate and the characteristics of the tablet. Finally, tablet friability is related to the tablet characteristics.

The left panel lists the raw materials and manufacturing steps that affect the critical quality attributes. Drying (and possibly other manufacturing steps) affects the degradation products formed during manufacturing by the API. Granulation affects API assay, API uniformity, and particle size of the granules (not the API). Drug substance particle size affects API uniformity in that for low-dose drugs, if the API particle size is too large, then content uniformity will be more variable. Drug substance particle size also affects dissolution rate, as shown by the Noyes–Whitney equation. The uniformity of the extragranular blend affects disintegrant uniformity (crospovidone is a disintegrant). The drug substance polymorph affects dissolution and stability. Tableting affects tablet hardness and tablet friability. It is noteworthy that this FDA document does not implicate excipients other than the disintegrant (crospovidone) as affecting product quality. This is consistent with the standard formulation approach, which is to select excipients that are inert and do not affect product variability.

To summarize, as the second circle on the wheel indicates, the critical quality attributes need to be determined. Critical quality attributes have been variously defined as material attributes [4] but could include other analytical characteristics including, for example, residual solvents.

The second level of the wheel at 6 o'clock indicates that process parameters need to be determined. For the granulation process outlined in Figure 26.2, process parameters include blending time, drying, granulation time, levels of binder, levels of disintegrant, granule particle size distribution, extra granular blending time, various settings on the tablet machine like dwell time, and other parameters. Some of these parameters could be measured using process analytical technology (PAT) approaches. For example, drying and blending can be monitored by online/inline NIR.

Next, process controls need to be implemented. Process controls can be designed to reduce input variation or correct for input variation or both. Process controls address and minimize variability. Process controls also include in-process monitoring. Process controls can range from offline analysis to PAT tools equipped with feedback/feedforward controls.

All of the information mentioned previously is used to arrive at a meaningful set of product specifications. These specifications are based on the product design, the desired product quality, the critical quality attributes, and the sources of variability identified during the entire process to date. The strategy to set specifications is also outlined in ICH Q6A. This guidance provides several excellent flow charts, some of which were discussed in Chapter 1.

As shown in Figure 26.1, the third level of the QbD wheel contains two activities: product knowledge and process understanding. Product knowledge and process understanding are gained through the process validation work that is done, as well as through the analysis of the variability of the process and the causes of process variation. Product and process understanding can also be improved by purposely performing slight variations in various input parameters and materials and monitoring the effect of these changes on drug product quality and variability. Product knowledge and process understanding can also be enhanced by risk analysis of the process and the sources of variability.

The final level of QbD is continuous improvement. This is carried out under the existing change control system. If a design space has been developed (see the following text), the change control becomes much easier from a regulatory viewpoint.

26.3 Learning Before Doing

In many respects, drug development and QbD are a learning process. Pisano carried out an extensive study of the drug development process [7]. He described the development process as a learning process since knowledge of the molecule and its properties must be gathered to develop a drug product. Importantly, QbD starts with an analysis of the molecular structure of the drug. Of particular importance is knowledge of the structure, the functional groups, the partition coefficient, the molecular weight, the permeability, and the solid-state properties of the drug. The QbD approach was termed "leaning before doing" (LbD) by Pisano. One of the themes of his book is that the LbD is the best way to do drug development.

Pisano's research showed that doing it right the first time is the greatest single factor in reducing time to market. This is because of the complexity of most molecules under development and because of the stricter regulatory requirements. Pisano's studies, which focused on Big Pharma, found that glitches in about 10% of the projects delayed development by 3 months or more. For biotech companies, the rate of development problems is even larger because funding restrictions reduce the ability to learn before doing.

Specifically, LbD is approached by gaining:

- knowledge of the system using theory, algorithms, and, if possible, computer-aided simulations,
- knowledge of the underlying causal variables and their relationship to performance, especially as they relate to solid-state chemistry,
- knowledge of the future manufacturing environment and the new variables introduced by that environment, and
- knowledge of how those variables affect process performance and behavior.

Pisano showed that LbD enhances performance. Often, LbD can be done during an accelerated development regime.

Under the traditional approach, pharmaceutical development is empirical and random (Table 26.1). It focuses on taking various lots of API that have been made and optimizing them. The QbD/LbD approach uses simplified formulations and focuses on solid-state chemistry. A powder with defined solid-state chemistry is placed in a capsule if possible. Multivariate experiments are used to find the solid form with the best properties and to control the formation of that solid form in the last step of the API synthesis using crystallization/spray drying/evaporation/and so on. The manufacturing process is developed based on the initial experiments. A design space is developed based on various crystallization/precipitation experiments. The manufacturing process is then adjustable within this design space. Similarly, a quality system is developed that accommodates an adjustable design space. Next, process controls are developed based on the design space. For example, in a spray drying process, flow rates of the feed solution and the drying temperature are controlled to produce products with the proper solid form and particle size. Product specifications are set by following a simplified version of the ICH Q6A flow charts. The specifications are set to make sure that clinical supplies are equivalent to each other and well defined. Again, focus is placed on particle size and solid form. Finally, the control strategy (see ICH Q6A and Q11 draft) is established based on the knowledge of solid-state chemistry and Good Manufacturing Practice principles. A control strategy can include controls on input material attributes, controls based on the design space, in-process controls, and release testing based on specifications.

Table 26.1 Outlines of traditional and QbD/LbD approaches.

Aspects	Traditional	QbD/learning before doing (LbD)
Pharmaceutical development	Empirical, random, focus on optimization. Minimal attention to solid-state chemistry	Simplified formulations. Focus on solid-state chemistry. Multivariate experiments. Focus on control strategy and robustness
Manufacturing process	Fixed based on first runs	Based on learning from small-scale experiments and so on. Adjustable within design space, supported by robust quality systems
Process control	In-process testing based on early runs or a validated process	Simple strategies based on solid-state chemistry for clinical supplies. Later PAT was utilized and process operations were tracked and trended
Product specification	Based on characteristics of early lots. Little molecular analysis	Initial specifications based on the molecular structure of the solid form manufactured. Based on desired product performance
Control strategy	By testing and inspection	Risk-based control strategy, real-time release if possible. Some control of excipients is needed for large-scale manufacturing

26.4 Risk-based Orientation

Risk analysis and management is an important part of QbD. Risk is defined as the combination of the probability of the occurrence of harm and the severity of that harm. Areas of particular concern/interest for risk analysis include quality-related recalls, lot failures, and out-of-specification (OOS) results. The goal is to identify these problems before they happen and develop a strategy to prevent their occurrence.

The ICH guidance Q9 outlines a typical quality risk management process (see Figure 26.3). Risk management is typically done in teams, which for solid-state chemistry would include chemists, pharmacists, analytical chemists, engineers, and regulatory experts. Quality risk management will make special efforts to define the level of risk. In this figure, risk management begins with risk assessment, which includes risk identification, risk analysis, and risk evaluation. In the solid-state area, one of the major risks is the risk of changes in the solid form during manufacture or storage. Another risk is chemical degradation. Risk assessment and risk evaluation are greatly assisted by forced degradation experiments where the solid form is stressed at high temperatures and humidities to determine if there is any risk of transformation. In these studies, three questions are asked: (i) What might go wrong?; (ii) What is the probability it will go wrong?; (iii) What are the consequences? In most cases, preliminary studies of risk can provide descriptors such as high risk, medium risk, and low risk. Among the most important risks in solid-state chemistry is the risk of the appearance/crystallization of a new form out of a formulation. Additionally, there is the risk of transformation of the solid form of the API in the formulation to another solid form with significantly different properties (i.e. dissolution rate). Finally, there is a risk of degradation of the solid form (chemical instability).

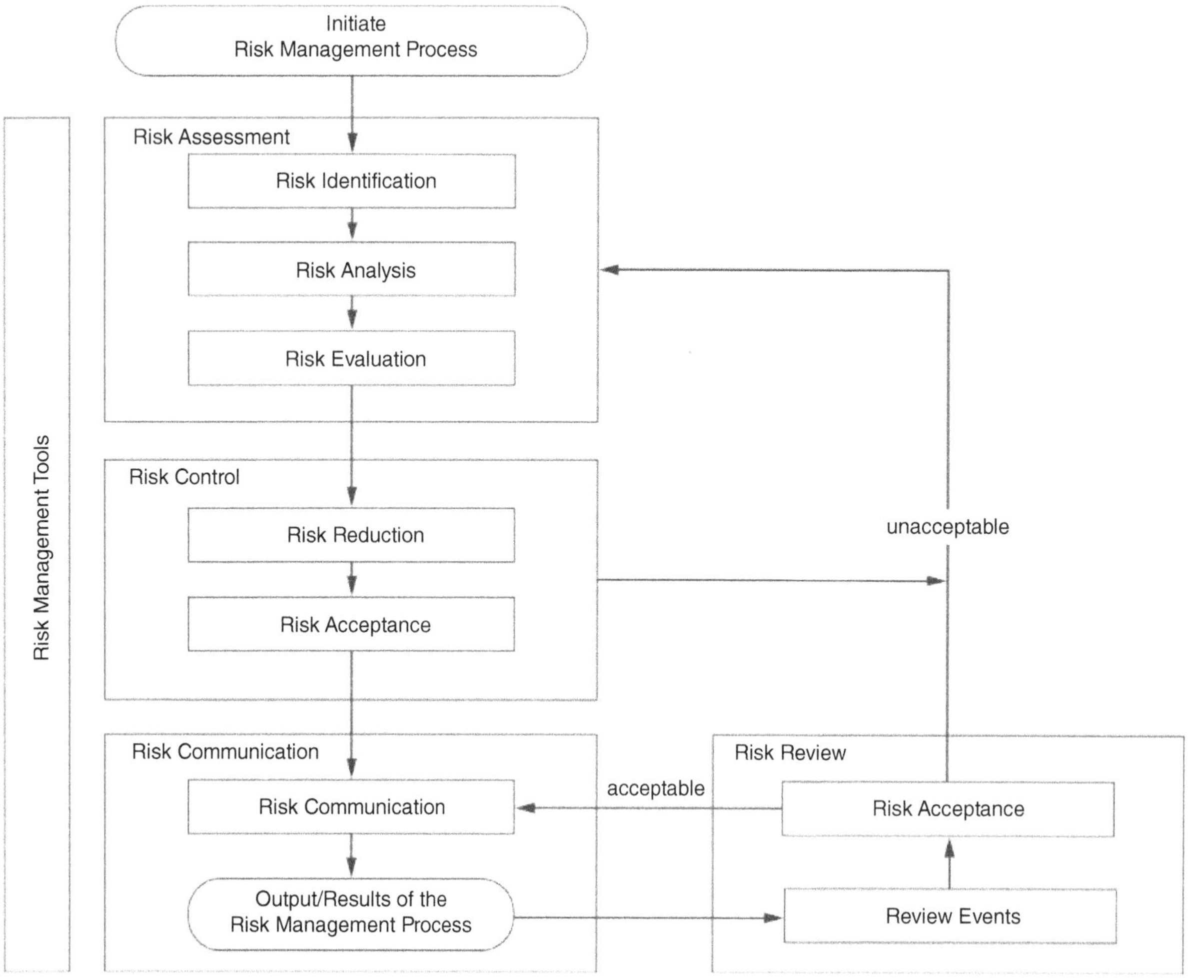

Figure 26.3 Typical quality risk management process outlined in ICH Q9.

The next step is risk control. Risk control asks the following questions: (i) Is the risk above the acceptable level?, (ii) What can be done to reduce or eliminate risks?, (iii)What is the appropriate balance among benefits, risks, and resources?, and (iv) Are new risks introduced as a result of controlling identified risks? Once these questions are answered, two approaches are taken, which are risk reduction and risk acceptance. For solids, risk reduction might involve packaging the drug product in foil–foil packages that mitigate instability associated with high humidity.

The third step is risk communication. It is important to make sure risks are communicated to the highest levels of the company. It is of utmost importance to communicate risky situations to all decision makers in the process.

The fourth step is risk review. In this step, the risk situation is reviewed, and new knowledge is input into the process. This could result in an increase or decrease in the level of risk and the acceptability of the risk.

Quality risk management needs to be integrated into the normal operations of the firm and documented. In particular, quality risk management should be incorporated into the following activities: (i) development; (ii) facilities, equipment; (iii) materials management; (iv) production; (v) laboratory control and stability; and (vi) packaging and labeling.

It is important to realize that risk management depends on the degree of understanding of the process. For well-understood processes where there is full process understanding, each process critical control point (PCCP) is known, and there is full monitoring capability; the ability to assess risk and detect deviations is high. For processes that are not understood and for which the PCCPs are not fully known, and for which only partial monitoring is available, it is much more difficult to assess risk and detect deviations. For intermediate situations, there is an intermediate ability to detect risk and deviations.

26.5 API Attributes and Process Design

During development, efforts need to be focused on the attributes of the API that may potentially impinge on the performance or manufacturing of the drug product. Once these attributes are known, the process for manufacturing can be selected. At the end of the preformulation process, a preliminary development plan can be developed. Figure 26.4 shows the general strategy used to develop a manufacturing process. As shown in this figure, the API solid form, the dose, and compatible excipients are evaluated. These components are used to carry out material characterization, which focuses on flowability/cohesion and mechanical properties. All of these factors are evaluated to ultimately determine whether the process should be direct compression, dry granulation, or wet granulation. In this process, particular emphasis is placed on the API attributes since it is possible to fill the API directly into capsules in a large percentage of cases. It is important to determine the API flowability and cohesiveness for this purpose. If flow aids or other excipients are needed, their compatibility must also be examined at this stage. Finally, it is important to define the API's critical attributes.

Morris et al. utilized X-ray diffraction to define the critical attributes of acetaminophen with respect to flow [8]. Figure 26.5 illustrates two morphologies of acetaminophen and their mass flow rates.

Analysis of the X-ray patterns of the API from these two processes showed that the relative height of the 020 peak in the diffraction pattern correlated with the mass flow rate (Figure 26.6). This is because the relative height of the 020 peak represented the morphology of the crystals, which in turn correlated with the flow rate. As discussed in the X-ray chapter, the relative height of peaks such as the 020 peak depends on a phenomenon called preferred orientation. Preferred orientation, in turn, depends on the morphology of the crystals.

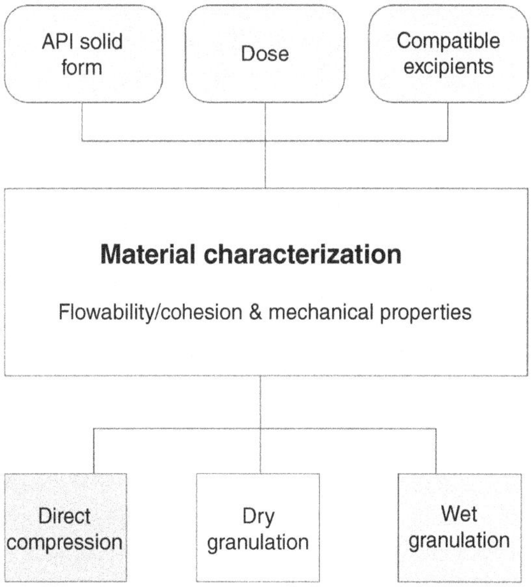

Figure 26.4 Overall QbD strategy to determine a manufacturing process.

26.6 Development and Design Space

It is important to define a preliminary design space during development. This requires developing as much knowledge as possible of the critical quality attributes early in development. The preliminary design space should focus on defining the solid-state form. Especially important is to select a solid form that meets the solubility

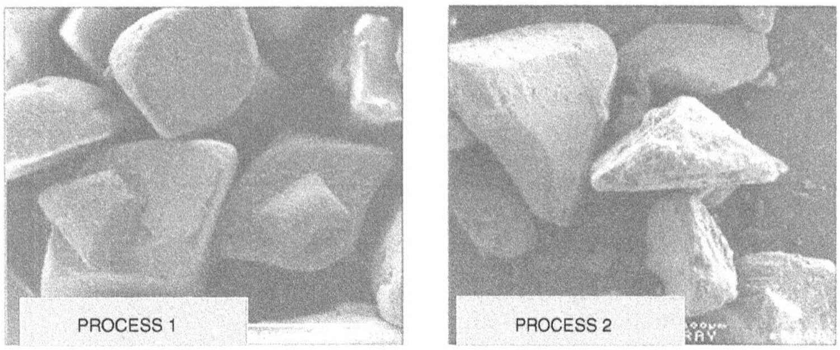

SEM of P1 and P2 APAP

Wildfong and Morris, LOL, 2003

	Tap D (g/mL)	AVG PS (um)	CHG. (CE-9/g)	Mass Flow (g/s)
P1	0.847	250.5	−5.64 (0.60)	10.20
P2	0.810	212.0	−1.56 (0.19)	7.66

Figure 26.5 SEM and mass flow properties of acetaminophen (APAP) prepared by two processes.

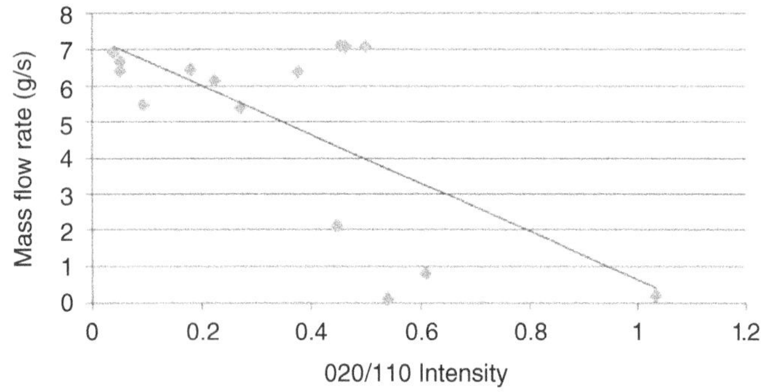

Figure 26.6 Correlation of the relative intensity of the 020 peak with mass flow rate.

and stability criteria for efficacy and safety. Recently, in the laboratory of one of the authors (Byrn SR), hand filling solid forms into capsules and measuring their dissolution rate and bioavailability in animals was carried out to make sure the form chosen has adequate bioavailability. Additionally, the process that produces the desired form needs to be developed so that clinical supplies can be made. This may involve developing an understanding of the crystallization process or other processes that produce the form. Process variables like cooling rate and concentration need to be defined to prepare the desired form. If crystallization is involved, it might be useful to determine the metastable zone width for the desired crystal form. The metastable zone width is usually shown on a concentration vs. temperature graph as the concentration area between the solubility curve and the spontaneous nucleation curve. That is, the metastable zone width is the concentration range where controlled crystallization can take place. In some cases, the process may involve crystallizing the most stable form in the last step of the synthetic process and then converting this form into an amorphous dispersion in a spray drying step.

Preparing investigational new drug (IND) clinical supplies can be as simple as filling the desired form into capsules. If a multicomponent formulation is needed, it is important to explore the compatibility of the excipients with the solid form using stress experiments. It is worthwhile to define the maximum variability of excipients with the desired solid form. Ideally, a noninteracting formulation in which the solid form does not interact with the excipients is preferred. It is also most advantageous to design a formulation where the solid form does not change during the manufacturing process. Enalapril maleate provides an interesting exception to the rule. For this formulation, a salt switch takes place during the granulation of enalapril maleate with sodium carbonate. The resulting product is an amorphous form with good solubility [9].

Once initial safety is established in an IND trial, then the formulation and the process can be refined, and a more detailed design space can be developed. In fact, it is advantageous to refine the formulation and develop a more detailed design space while the IND trial is ongoing. There is typically a time slot of 4 or more weeks where the IND safety trial takes place. During this time, the formulation can be revised, and more design space studies can be carried out. The design space should be as portable as possible since scale-up will be required for Phases II and III clinical trials.

During these initial studies about design space, NIR or Raman can be used for process monitoring. For example, Raman was used to monitor the crystallization of flufenamic acid [6]. NIR has been used extensively to monitor drying and blending.

For the early design space, it is important to establish a practical process with endpoints. For example, the crystallization process needs to have a defined endpoint (e.g. temperature). Yield is not as critical during these early stages since a process that maximizes yield may also maximize the amount of precipitated impurities. As mentioned earlier, it is useful to have a portable design space with monitoring tools like NIR or Raman spectroscopy. In some cases, in situ particle size analysis is also useful.

The development report is a critical report in drug development and QbD. The development report is outlined in ICH Q8 Pharmaceutical Development. This report outlines the scientific understanding of the formulation and provides an opportunity to establish the design space as well as specifications and manufacturing controls. The development report describes how the dosage form was selected and provides an understanding of the development of the formulation, and it also provides an opportunity to describe real-time methods that are used and how they were developed. The development report views the drug substance (API) as a component of the drug product and focuses on ICH Q6A for the procedure used to select the correct polymorph of the API. The ICH Q8 guidance defines a design space as the multidimensional combination and interaction of input variables (e.g. material attributes) and process parameters that have been demonstrated to provide assurance of quality.

The development report also describes the use of online monitoring (PAT) to control processes so that they achieve meaningful endpoints. For example, fluid bed drying monitored by NIR is a powerful method for online monitoring. Drying in a fluid bed drier typically depends on four parameters: (i) air temperature, (ii) air flow, (iii) relative humidity of the input air, and (iv) possibly load in the drier. For most applications, the batch size is held constant. The air temperature, air flow, and relative humidity are variables that can be controlled. These variables are also controllable within limits.

Figure 26.7 shows a fluid bed drier equipped with an MM55 NIR device. Using this configuration, the rate of drying of various formulations could be readily measured.

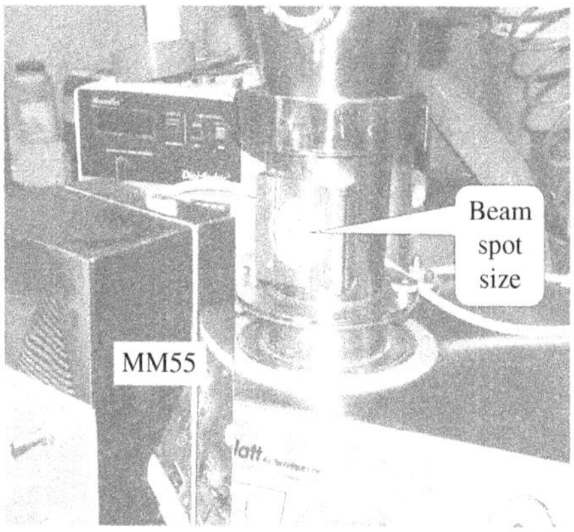

Figure 26.7 Fluid bed drier equipped with an NIR (MM55).

Figure 26.8 shows the drying curve for a standard wet granulation formulation of acetaminophen (APAP) using the equipment setup shown in Figure 26.7 [10]. It is clear from this plot that the curve fits the expected drying curve with a linear region where evaporative drying/cooling takes place, followed by an exponential region where diffusion-limited drying takes place.

Figure 26.9 shows a contour plot derived from the equations shown in Figure 26.8. The values of the constants were determined from the data and then used to generate the contour plot in Figure 26.9. This plot clearly shows that increased temperature reduces drying time. A similar plot showed that increased airflow reduced drying time.

Figure 26.10 shows a cross-section of a contour plot of drying time for a range of air velocities and temperatures. It is clear that the drying time drops from 1400 to 900 s as the temperature and air velocity increase.

Using the knowledge that NIR online sensing provides a direct way to measure moisture content, the Purdue Advanced Manufacturing team, working with the Consortium for Advanced Manufacturing (CAMP), designed a process for drying ibuprofen granulations that reduced drying time. Figure 26.11 shows the drying curves for two identical granulations, one dried with 60 °C inlet air (top tracing) and the other dried at 80 °C until evaporative drying ceased and then at 60 °C. Ibuprofen melts at 76 °C, so the ability to monitor online allowed drying at above the melting temperature of the drug is critical. This fast drying strategy was implemented at a large scale and significantly increased the throughput of the fluid bed drying step. This increased throughput allowed one firm to avoid buying additional dryers, saving several million dollars.

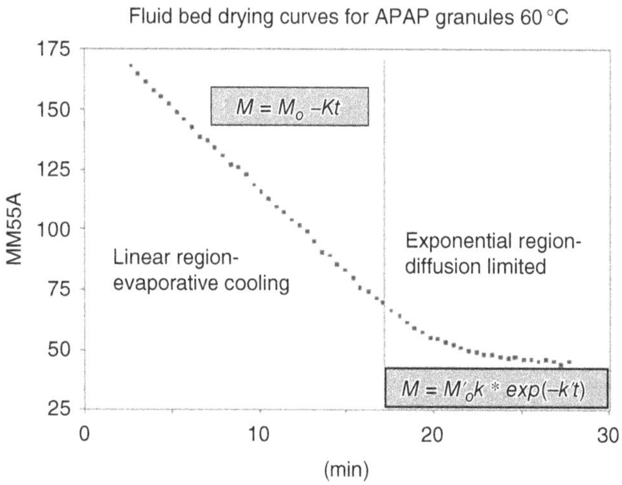

Figure 26.8 Drying curve as measured by MM55 reading versus time at 60 °C.

Importantly, several design spaces could be envisioned depending on how a firm would define the limits. One

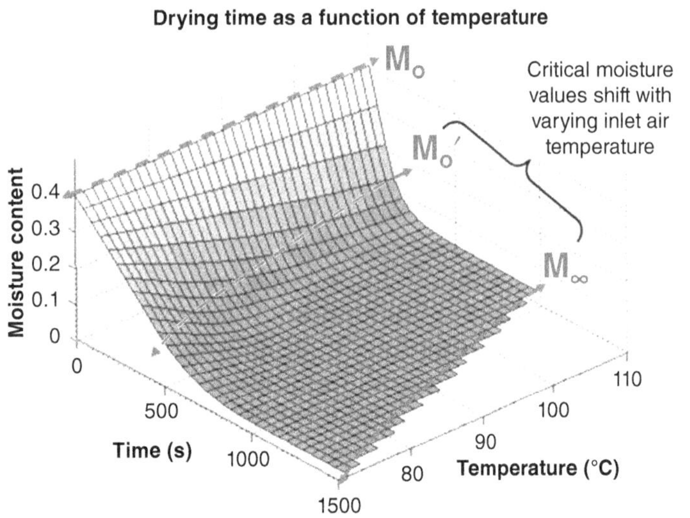

Figure 26.9 Drying time versus temperature.

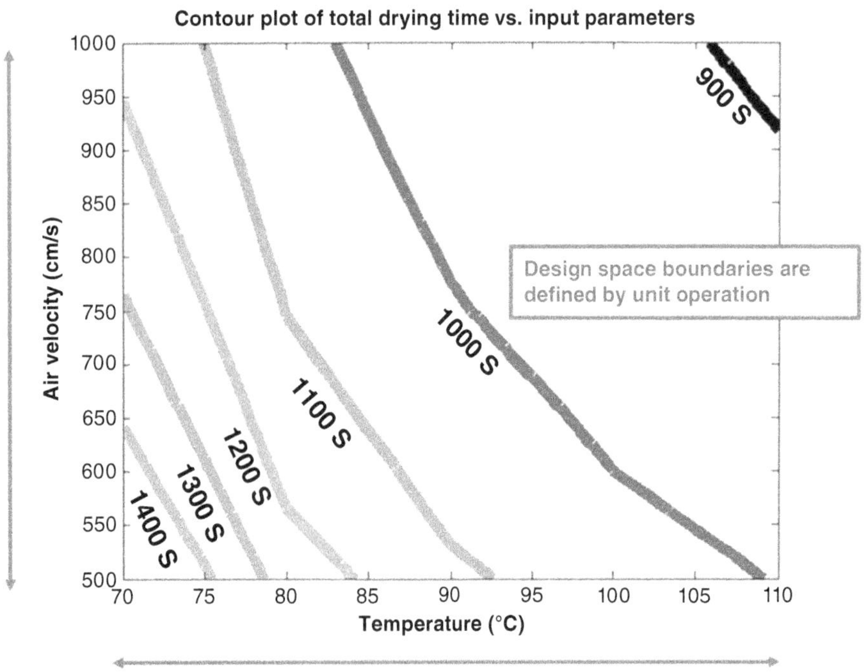

Figure 26.10 Contour plot showing drying time for various temperatures and air velocities.

possible design space is based on Figures 26.10 and 26.11. The design space for this system allows drying at any flow rate and any temperature as defined in the figures until the evaporative cooling ceases. At this point, the temperature is dropped to 60 °C, and a fixed flow rate is maintained until reaching the end point, which is indicated by an MM55 NIR reading of about 85.

26.7 Process Design – Crystallization

An important element of QbD is process design. One of the most important areas of process design is crystallization control. Crystallization control is important because the solid form determines many properties including dissolution rate, solubility, stability, and other properties such as flow. Crystallization control has been an important aspect of API manufacturing for many years, and there are many papers addressing this topic. Hu et al. carried out a study involving online monitoring of the crystallization of flufenamic acid [11]. Flufenamic acid exists in several forms, but this research focused on Forms I and III [11]. Forms I and III are enantiotropic to each other. Form I is most stable above 42 °C, and Form III is most stable below 42 °C. The Raman spectra of solid Forms I and II were analyzed along with the Raman spectra of a

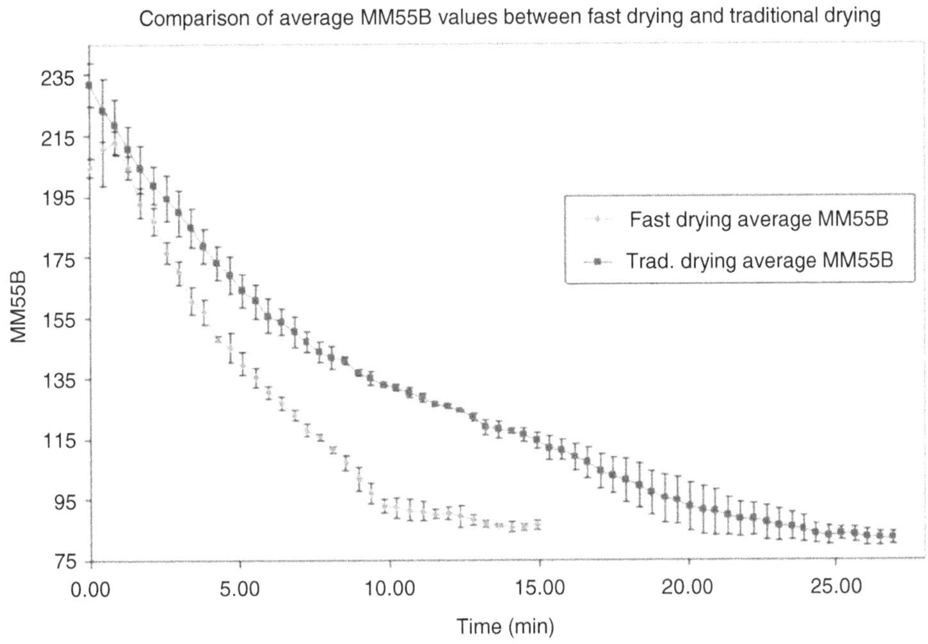

Figure 26.11 Fast drying and traditional drying of ibuprofen granulation.

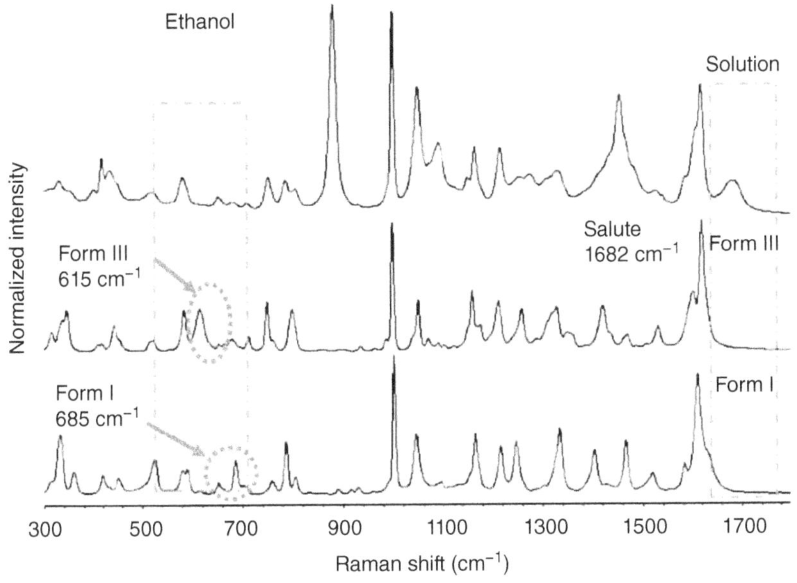

Figure 26.12 Raman spectra of solid Forms I and II and a solution of flufenamic acid. The designated wavelengths are those used to analyze each form.

solution of flufenamic acid. As can be seen from Figure 26.12, an analytical method focusing on the designated lines can be developed to analyze solutions containing Form I or III in suspension.

Figure 26.13 shows the results of monitoring the crystallization process using these absorption lines. As can be seen, first Form I crystallizes. Then, as the temperature is lowered below the transition point, the crystallized Form I converts to Form III. At the end of the process, Form III is obtained. It is interesting to note that, without the online Raman monitor, an investigator would think that this crystallization proceeded directly to Form III, the most stable form at room temperature.

26.8 Phase Transformations During Wet Granulation

One of the areas of risk in product and process design is wet granulation. In some cases, phase transformations occur during wet granulation. In such cases, the solid form produced is different from the solid form specified during API manufacture and different from the solid form listed on the components and compositions table for the formulation. Although

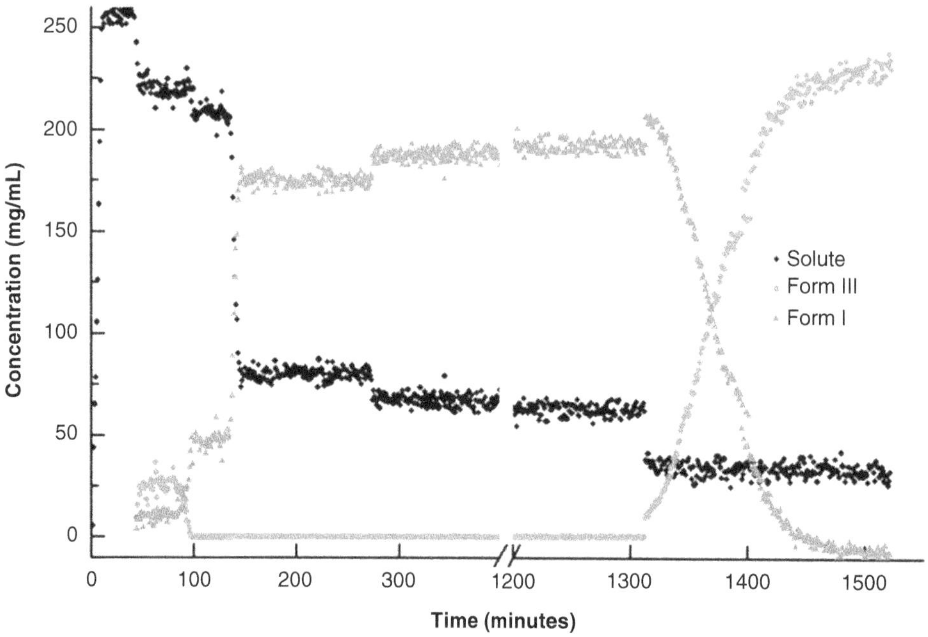

Figure 26.13 Concentration profile of flufenamic acid during a crystallization process starting with a solution at 54 °C. The solution was cooled in stages to 25 °C.

there can be advantages to transformations during wet granulation, if this is not known then the process is by definition not understood. Clearly, if a process is not understood, then this is a prescription for disaster, including OOS results, lot failures, and recalls.

Flufenamic acid is an excellent system for studying phase transformations induced by wet granulation. To carry out these studies, the group at X-ray optical systems (XOS) was involved because XOS had the capability of doing X-ray experiments on an optical bench. This allowed the installation of a "granulator" (a vessel with a beryllium window equipped with a stirrer) on the optical bench. Figure 26.14 shows the system used.

Figure 26.15 shows the X-ray data gathered using this equipment. It is clear from this data that Form I is transforming to Form III during the wet granulation mixing process. This is not surprising since Form III is the most stable form at room temperature [12].

In related studies, Abbott reported a solution-mediated phase transformation from anhydrate to amorphous for Abbott-232 during wet granulation [13].

26.9 Dissolution Tests with an IVIVC for QbD

IVIVC dissolution tests for QbD of BCS class II compounds are especially important because they can aid formulation development. If an IVIVC test can be developed early, then a formulator can rapidly develop a bioavailable formulation for clinical trials. This greatly accelerates the entire development process.

Initial work in the laboratory of one of the authors (Byrn SR) has focused on itraconazole, an antifungal agent that is BCS class 2. Dissolution tests were performed using a standard six-station dissolution bath using paddles/baskets at 50 rpm. Itraconazole concentrations were determined using standard UV analysis, using a calibration curve or by in situ UV analysis using an immersion probe. Various surfactants and pH dissolution media were used, including 0.1 N HCl with 0.1%, 1%, and 2% sodium lauryl sulfate or cetyltrimethylammonium bromide (CTAB).

The itraconazole marketed formulation, Sporanox, along with three other dispersions, was prepared at a ratio of 1 part drug: 2 parts polymer. The formulations were analyzed by X-ray diffraction and DSC to determine their structures. Dissolution testing was carried out on several drug–polymer dispersions. Figure 26.16 shows the results of studies in a USP basket apparatus using hand-filled capsules containing 100 mg of itraconazole.

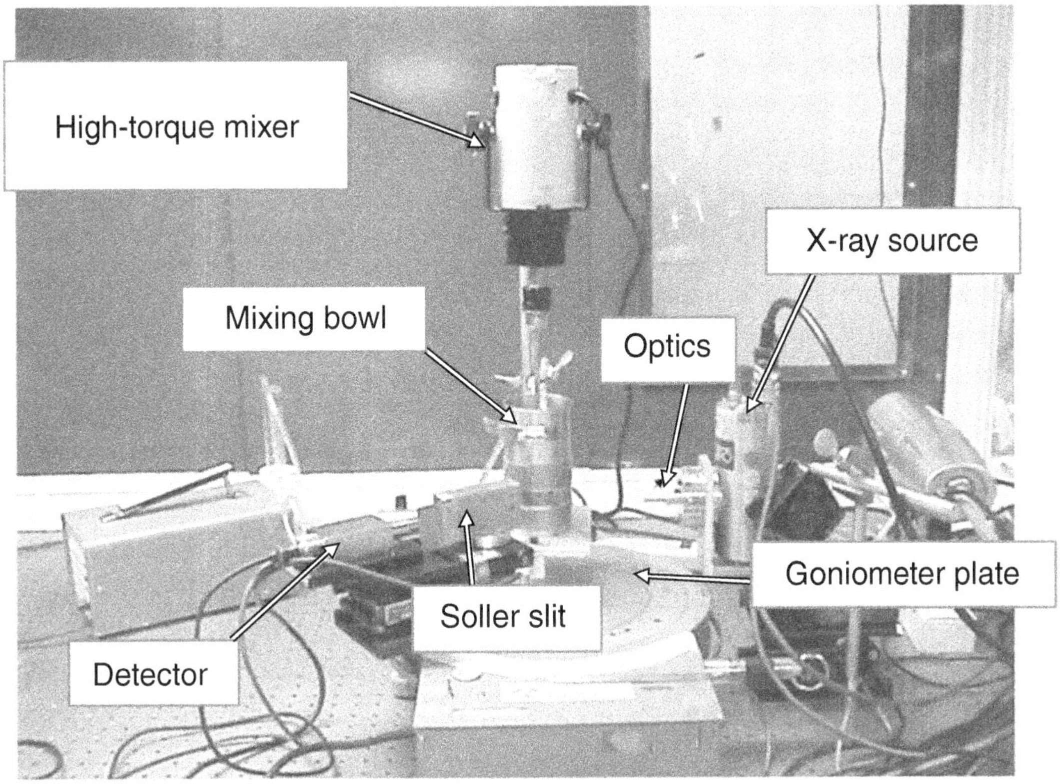

Figure 26.14 Instrument configuration used to measure X-ray diffraction during wet granulation.

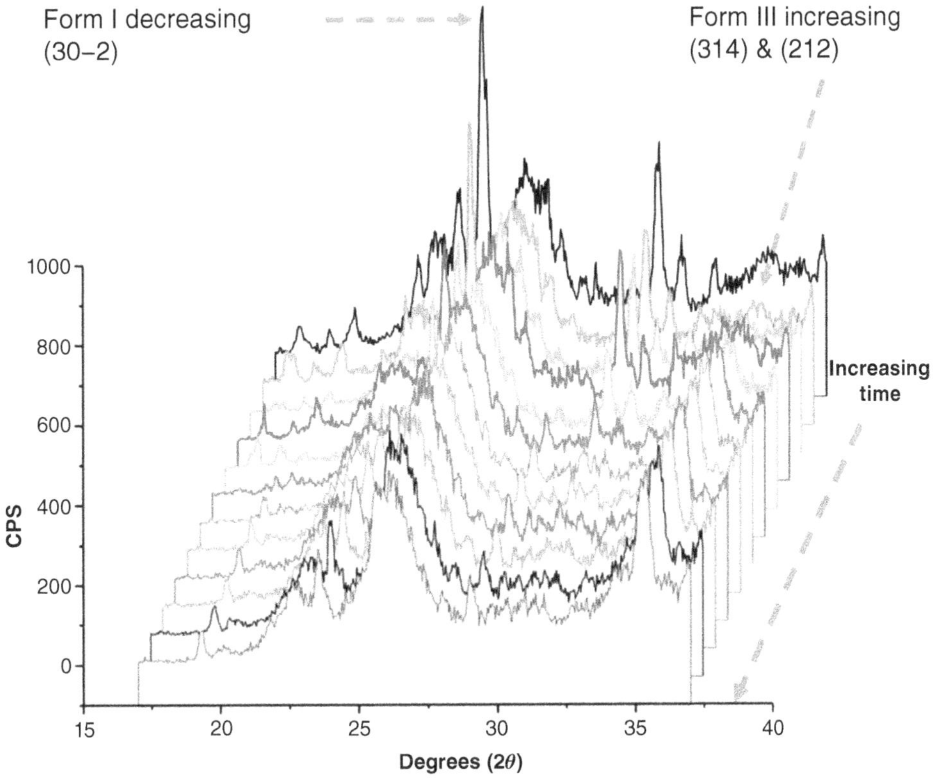

Figure 26.15 Changes in the X-ray pattern of flufenamic acid Form I with time during wet granulation.

These data support previous studies and further show that the polymer and the method of manufacture (commercial Sporanox) can significantly influence the dissolution rate. It is interesting to note that slight changes in the polymer hydroxyl propyl methyl cellulose (HPMC) versus hydroxyl propyl methyl cellulose phthalate (HPMCP), as well as changes in particle size of the dispersion, have a dramatic effect on release rate.

Six pigs weighing 20–25 pounds were used to study the pharmacokinetics of the four formulations: (i) itraconazole (crystals), (ii) itraconazole:HPMC dispersion (1:2), (iii) itraconazole:HPMCP dispersion (1:2), and (iv) Sporanox. Animals contained a surgically implanted indwelling jugular catheter. Fasted animals were dosed with 100 mg of the four formulations. Blood was collected using the "Pig-turn" system, which allows the animals to have freedom of movement during the experiment. Blood levels of itraconazole were determined using HPLC-MS. Figure 26.17 shows the blood levels. Comparison of Figures 26.16 and 26.17 shows that the blood levels in pigs correlated with the dissolution rate giving an IVIVC.

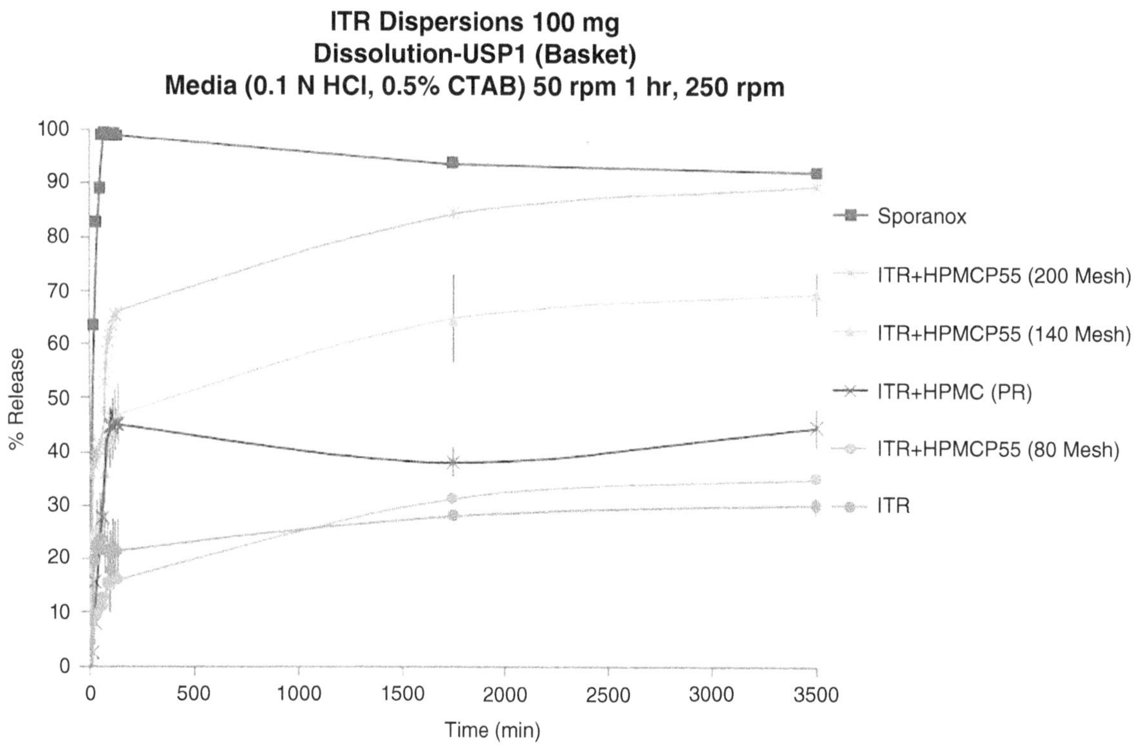

Figure 26.16 Dissolution studies of itraconazole solid dispersion in 0.1 N HCl, 0.5% CTAB using a USP basket apparatus.

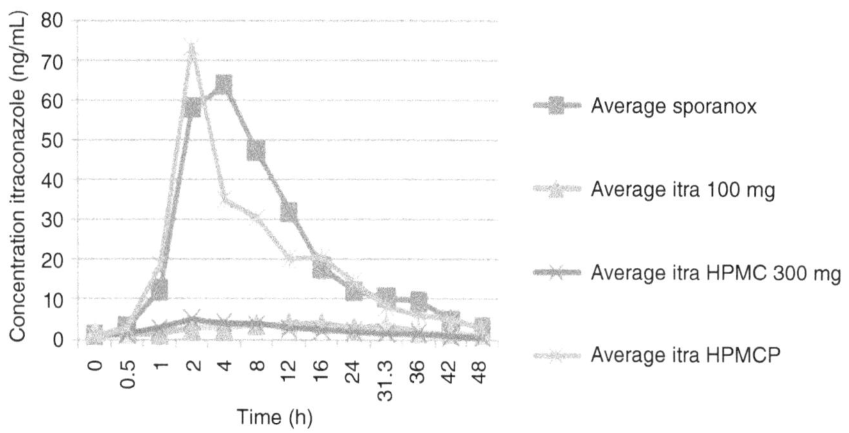

Figure 26.17 Blood levels of itraconazole in the four formulations.

26.10 Summary

Solid-state chemistry plays an important role in QbD. It is especially important in the product and process design stage. Additionally, if the solid-state chemistry is done correctly, it can accelerate drug development and reduce time to Phase II. Furthermore, understanding the solid-state chemistry of a system facilitates LbD and risk assessment. The acetaminophen case study showed that the faces of a crystal can affect flow, and the fluid bed drying case study showed how a design space can provide flexibility and reduce drying time. The crystallization case study showed how online monitoring can provide information that would not have been available from conventional studies. The wet granulation and IVIVC studies also showed other advantages of QbD studies.

References

1 Byrn SR. Predictive approaches to establishing, understanding, and communicating risk with emphasis on early development. In: Templeton AC, Byrn SR, Haskell RJ, Prisinzano TE, eds. *Discovering and Developing Molecules with Optimal Drug-Like Properties*. New York: Springer, 2015: 195–208.

2 Henck JO, Byrn SR. Designing a molecular delivery system within a preclinical timeframe. *Drug Discov Today* 2007; 12(5–6): 189–199.

3 Palucki M, Higgins JD, Kwong E, Templeton A. Strategies at the interface of drug discovery and development: early optimization of the solid-state phase and preclinical toxicology formulation for potential drug candidates. *J Med Chem* 2010; 53: 5897–5905.

4 Yu LX. Pharmaceutical quality by design: product and process development, understanding, and control. *Pharm Res* 2008; 25(4): 781–791.

5 Docherty R, Kougoulos T, Horspool K. Materials science and crystallization: the interface of drug substance and drug product. *Am Pharm Rev* 2009; 12(6): 34.

6 Hu Y, Wikström H, Byrn SR, Taylor LS. Estimation of the transition temperature for an enantiotropic polymorphic system from the transformation kinetics monitored using Raman spectroscopy. *J Pharm Biomed Anal* 2007; 45(4): 546–551.

7 Pisano GP. *The Development Factory: Unlocking the Potential of Process Innovation*. Boston: Harvard Business School Press; 1996.

8 Morris KR, Nail SL, Peck GE, Byrn SR, Griesser UJ, Stowell JG, Hwang SJ, Park K. Advances in pharmaceutical materials and processing. *Pharm Sci Tech Today* 1998; 1(6): 235–245.

9 Merslavic M, Razen J, Rotar A. Stable formulation of enalapril salt, a process for the preparation thereof and the use thereof. US patent 5350582 A. 24 November 1992.

10 Morris KR, Stowell JG, Byrn SR, Placette AW, Davis TD, Peck GE. Accelerated fluid bed drying using NIR monitoring and phenomenological modeling. *Drug Dev Ind Pharm* 2000; 26(9): 985–988.

11 Hu Y, Liang JK, Myerson AS, Taylor LS. Crystallization monitoring by Raman spectroscopy: simultaneous measurement of desupersaturation profile and polymorphic form in flufenamic acid systems. *Ind Eng Chem Res* 2005; 44(5): 1233–1240.

12 Davis TD, Morris KR, Huang H, Peck GE, Stowell JG, Eisenhauer BJ, Hilden JL, Gibson D, Byrn SR. In situ monitoring of wet granulation using online X-ray powder diffraction. *Pharm Res* 2003; 20(11): 1851–1857.

13 Wardrop J, Law D, Qiu Y, Engh K, Faitsch K, Ling L. Influence of solid phase and formulation processing on stability of Abbott-232 tablet formulations. *J Pharm Sci* 2006; 95: 2380–2393.

Index

Solid-State Materials in Pharmaceutical Chemistry: Properties, Characterization, and Applications, Second Edition. Stephen R. Byrn, George Zografi and Xiaoming (Sean) Chen.

d

n

t

Nicola Casagli • Veronica Tofani •
Kyoji Sassa • Peter T. Bobrowsky •
Kaoru Takara
Editors

Understanding and Reducing Landslide Disaster Risk

Volume 3 Monitoring and Early Warning

Springer

Editors
Nicola Casagli
Earth Science Department
University of Florence
Florence, Italy

National Institute of Oceanography
and Applied Geophysics—OGS
Trieste, Italy

Veronica Tofani
Earth Sciences Department
University of Florence
Florence, Italy

Peter T. Bobrowsky
Geological Survey of Canada
Sidney, BC, Canada

Kyoji Sassa
International Consortium on Landslides
Kyoto, Japan

Kaoru Takara
Graduate School of Advanced Integrated Studies
in Human Survivability (Shishu-kan)
Kyoto University
Kyoto, Japan

Associate Editors
Ping Lu
College of Surveying and Geo-Informatics
Tongji University, China

Kazuo Konagai
International Consortium on Landslides, Japan

Michele Calvello
University of Salerno, Italy

Katsuo Sasahara
Kochi University, Japan

Andrea Segalini
University of Parma, Italy

ISSN 2662-1894 ISSN 2662-1908 (electronic)
ICL Contribution to Landslide Disaster Risk Reduction
ISBN 978-3-030-60310-6 ISBN 978-3-030-60311-3 (eBook)
https://doi.org/10.1007/978-3-030-60311-3

Cover illustration: Landslide along the A6 highway in the province of Savona, Italy occurred in November 2019 after intense and prolonged rainfall (Civil Protection Centre, University of Florence. All rights reserved).

This Springer imprint is published by the registered company Springer Nature Switzerland AG
The registered company address is: Gewerbestrasse 11, 6330 Cham, Switzerland